Bionic Recognition and Biosensors: A Theme Issue in Honor of Professor Hong-Yuan Chen

Bionic Recognition and Biosensors: A Theme Issue in Honor of Professor Hong-Yuan Chen

Editors

Huangxian Ju
Jingjuan Xu

MDPI

Basel • Beijing • Wuhan • Barcelona • Belgrade • Novi Sad • Cluj • Manchester

Editors
Huangxian Ju
Nanjing University
Nanjing
China

Jingjuan Xu
Nanjing University
Nanjing
China

Editorial Office
MDPI AG
Grosspeteranlage 5
4052 Basel, Switzerland

This is a reprint of articles from the Special Issue published online in the open access journal *Chemosensors* (ISSN 2227-9040) (available at: https://www.mdpi.com/journal/chemosensors/special_issues/hongyuanchen).

For citation purposes, cite each article independently as indicated on the article page online and as indicated below:

Lastname, A.A.; Lastname, B.B. Article Title. *Journal Name* **Year**, *Volume Number*, Page Range.

ISBN 978-3-7258-1473-2 (Hbk)
ISBN 978-3-7258-1474-9 (PDF)
doi.org/10.3390/books978-3-7258-1474-9

Cover image courtesy of Hong-Yuan Chen

Contents

chemosensors

MDPI

Article

Droplet-Based Microfluidic Platform for High Spatiotemporal Resolved Single-Cell Signaling Profiling

Yingnan Sun [1,*], Qingqing Tian [1], Yongshu Liu [1], Kunming Xing [2], Yuyan Li [1], Yumin Liu [1] and Shusheng Zhang [1,*]

1 Shandong Province Key Laboratory of Detection Technology for Tumor Makers, School of Medicine, Linyi University, Linyi 276005, China
2 Linyi People's Hospital, Linyi 276100, China
* Correspondence: syingnan@126.com (Y.S.); shushzhang@126.com or zhangshusheng@lyu.edu.cn (S.Z.)

Abstract: A small indentation embedded in a microchannel creates a surface energy well (SEW) for a confined droplet due to surface energy release. Inspired by this, we developed a SEW-based microfluidic platform to realize high spatiotemporal-resolved signal profiling at the single-cell level applying droplet stimulus on a single chip. The method allows for controlled droplet replacement within only 3 s with almost 100% exchange efficiency, reliable single-cell patterning of adherent cells and successive treatment of adherent cells with reagent droplets. Furthermore, the PDGFR/Akt pathway served as a model system for evaluating the performance of the SEW-based method in determining the effects of ligand stimulation duration (3 s to 3 min) on receptor phosphorylation. The novel strategy offers a general platform for probing the temporal dynamics of single cells, as well for monitoring rapid chemical reactions in various applications.

Keywords: droplet microfluidics; single cells; cell signaling profiling; high spatiotemporal resolution

Citation: Sun, Y.; Tian, Q.; Liu, Y.; Xing, K.; Li, Y.; Liu, Y.; Zhang, S. Droplet-Based Microfluidic Platform for High Spatiotemporal Resolved Single-Cell Signaling Profiling. *Chemosensors* **2022**, *10*, 521. https://doi.org/10.3390/chemosensors10120521

Academic Editor: Huangxian Ju

Received: 7 November 2022
Accepted: 6 December 2022
Published: 8 December 2022

1. Introduction

Because the majority of important cellular processes involved in proliferation, differentiation and reprogramming are transient in nature, functional temporal analysis with high temporal resolution is being increasingly sought in research on signaling events involving networks of diverse biochemical reactions that lead to diverse cellular responses depending on stimulus duration [1–3]. Furthermore, because cells are highly heterogeneous and dynamic in response to stimulus perturbation, single-cell analysis would contribute significantly to advancing cancer research, since a drug targeted for one tumor cell phenotype is not necessarily effective against other phenotypes [4–8]. The current, well-established individual cellular analysis technologies involve dissociation and isolation procedures, which can be roughly classified as flow/mass cytometry [9–11], and microfluidics-based techniques [12–14]. However, a drawback of these methods is the inability to probe single-cell events while retaining the adherent cell phenotype in situ with sufficient temporal or spatial resolution required for characterizing the fast kinetics of cellular events [15–19]. Hence, robust temporal profiling of single-cell dynamics while retaining cellular spatial configuration remains challenging for the majority of available techniques.

To overcome these limitations, several technologies have been successfully developed. These bridge the divide between temporal profiling of single cells with precise spatial confinement and the phenotype of adherent cells, and achieve flow confinement within a living cell. Specifically, these technologies can be broadly classified as flow-based microfluidics comprising individual addressable cell culture chambers [20], digital microfluidics with electrode arrays for droplet manipulation [21], nanopipettes with asymmetric nanopore electrodes for current amplification [22,23], atomic force microscopy (AFM) tips with microchannels for force flow confinement [24,25] and micropipettes with electrodes for hydrodynamic flow confinement [15,26]. All of these methods allow for the perturbation and analysis of single adherent cells with high temporal resolution and controlled spatial precision, an important property that is lacking in most analytical systems. Despite the fact that the new detection strategies based on these platforms enable the acquisition of detailed information on single cells with high spatial and temporal resolution, the fabrication and assembly of extremely fine probes and auxiliary components for observation and control remain challenging because of the stringent requirements for the construction process and operational complexity. Furthermore, with the exception of microfluidics, other in-situ technologies are not well suited for applications such as extracellular excitation and imaging analysis, and their throughput is limited by the device complexity, configuration and end-point detection mode. Therefore, the ability to perform multiple manipulation procedures on living adherent cells with high spatiotemporal resolution at the single-cell level remains limited.

An alternative droplet-based microfluidics system embedded with indentations in a wide and low channel has been previously reported for sequential droplet manipulation, including trapping and replacement, based on modulating droplet confinement [27,28]. Because the indentations are beneficial for surface energy release and capable of holding droplets locally, herein they are referred to as "surface energy wells" (SEWs), in analogy with particle trapping in potential energy wells, which have also been termed Hele-Shaw cells or anchors in other studies [29,30]. This finding inspired us to explore whether the SEW structure can be utilized to achieve rapid droplet replacement over living cells for rapid biochemical reaction monitoring. To date, only one study has been reported on the application of a SEW-based device on long-term cell culturing and controlled stimulation, wherein adherent cells were confined within a gelation scaffold to resist the flush of liquid flow and maintain in situ growth while various reagents were successively introduced [31]. Although this platform is capable of monitoring cell death dynamics upon various drug exposure regimes, a low time resolution of several minutes compromised its quantification performance.

In this study, we extended the SEW-based droplet microfluidics approach to achieve high spatiotemporal resolution for in situ single-cell signal profiling on a single chip, by leveraging the SEW structure for droplet manipulation and single-cell patterning. To the best of our knowledge, this study is the first comprehensive evaluation of the temporal resolution and exchange efficiency of a chip entailing the SEW structure. Furthermore, in comparison to the above-mentioned study [31], the SEW chip was designed and optimized for achieving living adherent single-cell patterning without resorting to hydrogel scaffolds, as well as for tracing the rapid dynamics of protein phosphorylation events whose unfolding is typically in the order of seconds.

The specific achievements of this study include the following: (1) A microfluidic chip comprising dual inlets and a SEW array are controlled by varying only the flow rate of the carrier oil. (2) Droplet replacement was evaluated as rapidly as 3 s and the content exchange efficiency reached 100%, indicating that complete termination of the original stimulus can be achieved with a single droplet replacement. (3) Adherent cells were cultured on the floor of the chamber under continuous medium perfusion, and single-cell cultures were achieved by harnessing the hydrodynamic flow variations in the SEW region. (4) The well-established PDGFR/Akt pathway was chosen as a model system of fast cellular events to monitor the dynamics of receptor phosphorylation in response to a cell stimulus with a time resolution of seconds. Therefore, we propose that this method can potentially be applied as an important new strategy for probing the temporal dynamics of fast cellular events at the single-cell level in a wide range of applications.

2. Materials and Methods

2.1. Device Fabrication and Assembly

The materials for microfluidic device fabrication included an SU-8 2035 positive photoresist (Microchem, Westborough, MA, USA), Si wafers (Shuguang, Tianjin, China), and a polydimethylsiloxane elastomer kit (PDMS, Sylgard 184, Dow Corning). All tubing and fittings were acquired from Suzhou CChip. An SU-8 model with two patterns of equal depth (35 μm high) was prepared employing soft photolithography on a silicon wafer. Next, a PDMS layer was obtained by casting the PDMS precursor (base to curing agent weight ratio of 10:1) onto the prepared SU-8 mold, followed by curing for 30 min at 85 °C. Then, the cured PDMS layer was gently peeled off the mold to obtain a PDMS-negative mode structure, followed by cutting into two parts along the borders, that is, the top and bottom layers of the chip. The top layer comprised inlets, an outlet and an SEW array, and the bottom layer contained fluidic channels for aqueous media and oil flow. After creating the inlets and outlet in the top layer, the cross marks were aligned accordingly and the two layers were bonded after 90 s of plasma treatment in the high-power mode (Harrick PDC-002). In the final step, the chip was connected to the pressure pumps by tubing and fittings for medium perfusion.

2.2. Surface Modification

For the chip to allow for both stable droplet generation and adherent cell growth, two types of surface treatments were applied on separate sections of the internal microchannels. (1) For the droplet generation section: to render the internal channel surface hydrophobic, a dilute solution of 1H1H2H2H-perfluorodecyltrichlorosilane (Sigma-Aldrich) in FC40 oil (3M Fluorinert) (1% *v/v*, 10 μL) was injected via the inlet to fill the generation section without entering the downstream trapping chamber. The chip was maintained at 20 °C for approximately 20 min and then rinsed twice with pure FC40 to remove residual chemicals and deliver a hydrophobic PDMS surface. (2) For the trapping chamber: to further accelerate cell adhesion from the suspension onto the bottom of the chamber, a fibronectin solution (15 μg/mL, 350 μL) (Solarbio, Beijing, China) was perfused into the trapping chamber via the outlet without entering the upstream transporting channel [32]. The chip was incubated overnight at 4 °C to allow fibronectin to coat the plasma-treated inner surface. The coating solution was then removed from the chip via suction and the chip was washed twice with PBS via the same outlet using a syringe.

2.3. Cell Culture and Viability

The MCF-7 and 4T1 cell lines used in this study were purchased from Beyotime (Shanghai, China). All cell culture reagents were purchased from Solarbio Science & Technology (Beijing, China), including DMEM, fetal bovine serum (FBS), penicillin/streptomycin (P/S), PBS and Trypsin-EDTA. Hoechst 33342 (Thermo Fisher, Waltham, MA, USA) and fluorescein diacetate (FDA, Sigma-Aldrich, St. Louis, MO, USA) were used as the cell staining reagents The adherent cells were cultured in DMEM containing 100 U/mL P/S supplemented with 10% FBS and maintained at 37 °C in a humidified incubator containing 5% CO_2. Upon reaching 70–80% confluency, the cells were trypsinized and centrifugated, and then resuspended in culture medium for cell seeding onto the chip. Similar to the fibronectin coating operation, the cell suspension (~0.5–1 $\times$ 10^6 cells/mL) was injected into the trapping chamber via the outlet without entering the upstream channel. The cells were allowed to settle uniformly and attach slightly for 20 min under static conditions, followed by continuous perfusion of fresh culture medium to ensure nutrient and air supply for the cells in the narrow culture chamber. Importantly, the flow rate was maintained at 1 μL/min for a duration of 2 h. Eventually, the cells were well-attached with visible microvilli, appearing polygonal in shape. Hoechst 33342 (10 μg/mL) and FDA (5 mg/mL) solutions were used in combination to asses cell viability. After incubation for 10 min, the staining solution was removed by gentle washing with fresh growth media and recovered for fluorescence imaging.

2.4. Single-Cell Immunocytochemistry

The cell immunocytochemistry reagents included: (1) fixing solution: 4% paraformaldehyde (Solarbio, No.P1110); (2) blocking buffer: 10% (*v*/*v*) normal goat serum (Solarbio, No.SL038) and 0.3% (*v/v*) TritonX-100 (Sigma-Aldrich, No.T8787) diluted in PBS; (3) primary antibody for Akt (Thr308): Phospho-Akt (Thr308) Rabbit mAb (CTS, No.13038T); (4) primary antibody for Akt (Ser473): Anti-Akt1 (Phospho-Ser473) rabbit polyclonal antibody (BBI, No.D155022); (5) second antibody staining solution (second-Ab): Alexa Fluor 555-conjugated Goat anti-rabbit IgG (BBI, No.D110070); (6) ligand solution: recombinant human platelet-derived growth factor-BB (PDGF-BB) (BBI, No.C600154).

Before stimulation, cells were pre-cultured with AKT inhibitor VIII in complete medium (2 μM, Beyotime, SF2784) for 1 h. The immunocytochemistry procedure involved ligand stimulation, fixation, permeabilization, blocking, staining and washing. Specifically, ligand and fixing solutions were applied in the form of droplets, while the other liquids were injected sequentially with continuous flow. Unless otherwise stated, droplet operations and immunocytochemistry incubations were performed inside a bench incubator. The general procedure was as follows: (1) Single cells pre-inhibited with AKT inhibitor were stimulated by transported droplets of ligand solution (10 ng/mL in serum-free DMEM); stimulus duration was varied depending on the experiment. (2) Immediately after stimulation, the ligand droplets were replaced by fixing solution droplets to fix the single cells, followed by incubation for 10 min. (3) The fixing droplets were then flushed away with blocking buffer, followed by incubation for 1 h. (4) The blocked cells were stained with primary antibodies by perfusing primary antibody Thr308 solution (1:800 dilution) followed by incubation overnight at 4 °C. (5) The unbound primary antibodies were removed by injecting PBS. (6) The primary antibodies were stained by secondary IgG555 antibodies by injecting secondary antibody solution (1:200 dilution) and incubating for 2 h. Finally, the cells were rinsed twice with PBS.

2.5. Experimental Setup

The entire microfluidic device functioned within a Dolomite microfluidic system (Royston, UK), which was equipped with: (1) a high-speed digital microscope (Meros, 3200531) for capturing high-speed droplet production and replacement with a frame rate of ~4100 fps and an exposure time of 0.05 ms; (2) flow control center software (Basic, Dolomite) for accurate and simultaneous control of Dolomite modules; (3) pressure pumps (Mitos p-pump) with a wide flow rate range (from several nL/min to 1 mL/min) and a precise pressure controller (from 0 to 10 bar) connected to flow sensors to allow for highly stable pulseless liquid flow and instant responses. In addition, the bench incubator (WH-SCI-01, Suzhou Wenhao, Suzhou, China) used for cell loading and culturing was connected to the Dolomite system for continuous infusion and in situ observation. Microscopy images were acquired using an inverted fluorescence microscope (IX73, Olympus, Tokyo, Japan), and the grey values of the fluorescence images were processed and determined using ImageJ software.

3. Results and Discussion

3.1. Chip Design for Single-Cell Stimulus

The SEW-based microfluidic platform described herein was designed to perform droplet manipulation for in situ profiling of single-cell signaling with high spatiotemporal resolution by controlling the oil flow rate. Figure 1 shows a detailed schematic of the implementation principle and images of the device and droplet manipulation. The device comprised of a bottom layer with microchannels and a top layer with an SEW array (Figure 1a). For function partitioning, the device assembly included a generation section with a flow-focusing junction, delivery section with a widened channel and a trapping chamber with SEW structures. With regard to the SEW structure, the adjustment in the height confinement of droplets enables their trapping in the chamber due to surface energy release [24]. For a spherical droplet of constant volume, the surface energy is minimal and increases as the droplet is squashed into a flattened shape because of the droplet's surface energy modulation forced by the reduction of the channel's height. The flattened droplet will then be temporarily trapped by an indentation (a portion in the channel with a larger height) located downstream in the wide and thin channel. Therefore, the indentation can be named the surface energy well (SEW). Furthermore, droplet manipulation within the SEW includes replacement, parking and passing modes, depending on the hydrodynamic drag force of the carrier oil's flow rate, which is proportional to its mean velocity. Figure 1b shows the replacement mode functioning under an appropriate carrier oil flow rate v_{oil} ($v_1 < v_{oil} < v_2$): droplet A is first trapped by an SEW until droplet B appears, collides with droplet A and replaces it; the process is then repeated with droplet C replacing droplet B. On the other hand, droplets under the SEWs undergo parking ($v_{oil} < v_1$) or passing ($v_{oil} > v_2$) modes (Figure S2). It should be noted that the replacement mode enables high temporal and spatial resolution, which has not been addressed in previous studies. A single substitution of one droplet by another in the carrier oil under the SEWs is extremely rapid, allowing for the instantaneous action of distinct reagent droplets on the same SEW region. Therefore, a series of stimulus droplets can be trapped for a very short time over individual adherent cells located in the corresponding position of an SEW and sequentially replaced with termination droplets to achieve rapid immunocytochemistry at the single-cell level (Figure 1c). The SEW configuration is depicted in Figure 1c: the channel height (H) and indentation depth (h) were set to 35 μm, being considerably smaller than the channel width (W = 1.8 mm) and length (L = 3 mm). A confined droplet with a volume larger than V_1 and V_2 (Figure S3) adopts radius R, which is larger than the height of 35 μm. Furthermore, 200 μm center-to-center spacing between two SEWs allowed for the simultaneous observation of single cells. These dimensions were designed considering both droplet volume and oil velocity to avoid sacrificing cell proliferation for high spatiotemporal resolved droplet manipulation.

The complete device shown in Figure 1d was fabricated by the successive application of SU-8 monolayer photolithography, PDMS casting and plasma-assisted bonding. To further render the surface of the generation section hydrophobic and promote cell adherence in the trapping chamber, fluoroalkylsilane and fibronectin were injected and aspirated via the inlet and outlet to modify the upstream and downstream regions, respectively. During droplet operation (Figure 1b), the device operates within a Dolomite microfluidic system equipped with a high-speed digital microscope and flow control center software (Figure 1e). A bench incubator fitted to the Dolomite system (Figure 1f) provides a suitable environment for cell growth. Furthermore, the carrier oil composition must be optimized with regard to the oil and surfactant type and content to realize stable droplet formation, droplet fusion resistance, biocompatibility and to prevent the oil from acting on adherent cells [33,34]. Herein, HFE 7500 oil containing 2% surfactant was adopted. The experiment commenced with the generation of monodisperse aqueous droplets in carrier oil, followed by their delivery downstream with the oil flow, and finally, their arrival in the trapping chamber to be trapped by the SEW array (Figure 1g).

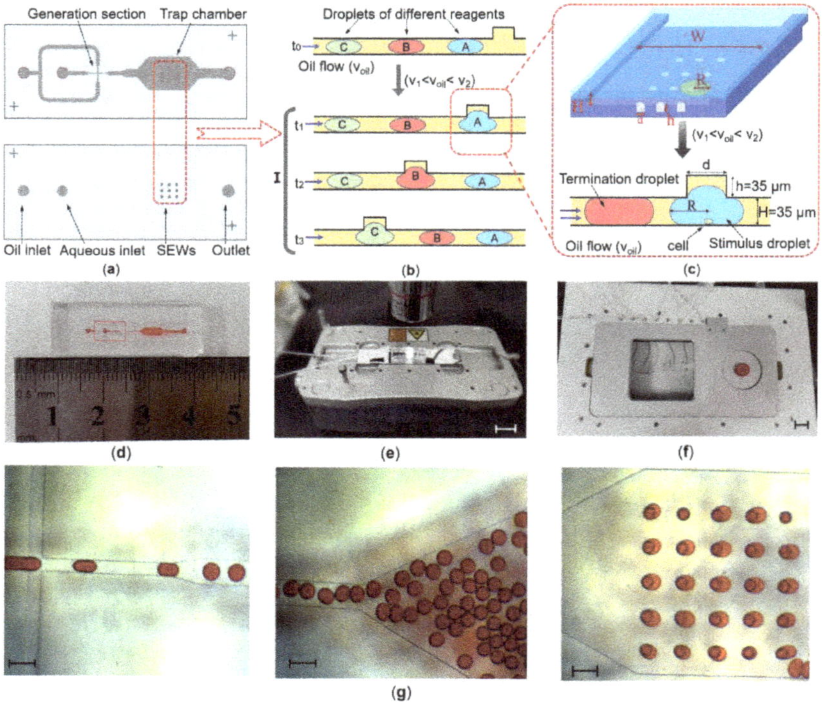

Figure 1. (**a**) Chip design: schematic top view of the microfluidic chip showing the layout of the bottom and top plasma-bonded layers, featuring dual inlets for two-phase flow. The bottom layer comprises the generation section with a flow-focusing junction and a trapping chamber. The top layer is composed of an SEW array, outlet, aqueous liquid and oil inlets for droplet generation and flow rate control. (**b**) Replacement mode: schematic side view of droplets being trapped under the SEW structure due to reduced surface energy. With an increase in the oil flow rate, the trapped droplet (reagent A) is replaced by the oncoming droplet (reagent B), and the sequence is repeated to replace droplet B with droplet C. (**c**) Side-view schematic showing adherent single cells cultured on the bottom layer under the SEWs. The cells were probed for protein phosphorylation with high spatiotemporal control: the ligand droplet (blue) becomes trapped by the SEW, thereby covering the cells for several seconds, followed by replacement by the termination droplet (red). (**d**) Image of the assembled chip filled with red ink. (**e**) Dolomite equipment comprising pressure pumps for microfluidic chip operation and a high-speed digital microscope for observation. (**f**) Bench incubator used for loading cells onto the microfluidic chip and incubation. (**g**) Sequential high-speed microscope images of red ink droplets being generated, delivered to the trapping chamber and trapped (left to right). Scale bars: 200 μm.

3.2. *Trapping and Release Performance of the Chip*

The dual-inlet configuration enables the simultaneous production of monodisperse droplets with a specific volume, and droplet manipulation through delivery velocity control, both of which are dependent on the oil flow rate. Consequently, the complexity of device operation is reduced. However, a new challenge arises: ensuring a constant volume of upstream droplets when the oil flow rate is changed to manipulate the downstream droplets. Therefore, a feasible solution is provided by synchronously increasing or decreasing the oil and aqueous flow rates within a certain range. As shown in Figure 2a, the droplet diameter is positively correlated with the liquid velocity at a specific oil velocity. Figure 2b shows generated droplets with diameters of 85, 95, 120 and 150 μm as examples, demonstrating that a roughly constant droplet volume can be achieved by simultaneously adjusting the oil and liquid velocity.

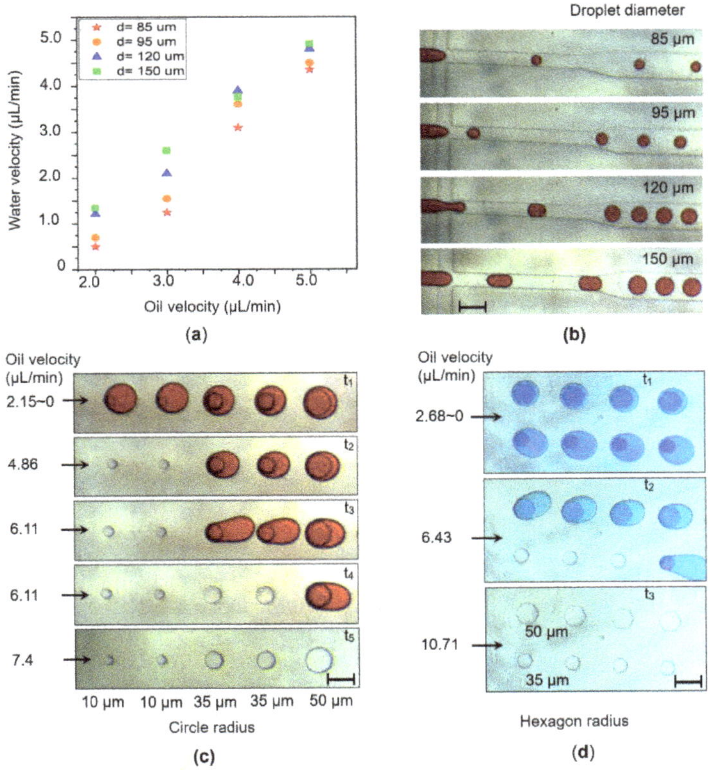

Figure 2. (**a**) The droplet diameter can be kept constant by changing the flow rates of the two phases simultaneously within a certain range, an important prerequisite for the dual-inlet design of the chip. (**b**) High-speed microscopy images of droplets with specific diameters generated by applying varying oil and liquid (red ink) flow rate combinations. Scale bars: 200 μm. (**c**) Circular SEW test: droplets of equal size are pre-trapped by the circular SEWs at a low oil flow rate (t_1). As the oil flow rate gradually increases (t_2~t_5), the trapped droplets are sequentially released, starting from the smallest SEW (10 μm radius), and finally from the largest SEW (50 μm radius). Scale bars: 100 μm. (**d**) The same test was performed using hexagonal SEWs of varying sizes. It was clear that the critical oil rate required for droplet release from the hexagonal SEWs was greater than that from the circular SEWs. Scale bars: 100 μm.

To further determine the optimal oil velocity for cell profiling, SEWs with distinct shapes were adopted to evaluate the corresponding hydrodynamic droplet capture force. Specifically, circular and hexagonal indentations of varying radii were selected as characterization parameters for analyzing droplet trapping and release. As shown in Figure 2c, the droplets were maintained at a volume of roughly 0.8 nL by manipulating the oil and aqueous flow velocities ($v_{aqueous}$ = 2 μL/min, v_{oil} = 2.15 μL/min). Initially, the droplets are instantly captured by all of the circular SEWs and remain in place due to the low oil flow rate. As the carrier oil velocity gradually increases to 7.4 μL/min, the droplets are successively released from SEWs of increasing radius. In addition, the same droplet trapping and release test was performed using hexagonal SEWs with circumradii of 35 and 50 μm. Figure 2d shows that for hexagonal SEWs, the critical release velocity and fluctuations in the release are considerably higher than for the circular holes, which hampers fast and accurate control of droplet replacement. Taking into account both the oil velocity and fluid shear, the circular SEW with a radius of 50 μm was finally chosen for profiling single-cell signaling to achieve both rapid and stable droplet replacement.

3.3. *High-Spatiotemporal Replacement Performance*

The developed SEW-based platform is capable of realizing droplet replacement with both high temporal resolution and high content replacement efficiency, which are lacking in most analysis platforms. Herein, we selected the replacement duration time and content exchange ratio as characterization parameters to demonstrate the high spatiotemporal control achieved with the SEW structure. Colored inks were used to aid in the visualization of the rapid process, captured using a high-speed microscope. Figure 3a (top row) shows rapid reagent replacement at a low oil flow rate (v_{oil} = 2.5 μL/min): the black ink droplet is pre-trapped by an SEW, and the colorless PBS droplet is transported by the carrier oil. At frame 46, the PBS droplet makes contact with the black droplet, gradually forcing it out of the SEW until it completely dislodges it at frame 50; finally, the PBS droplet remains parked on the SEW site while the black droplet flows away. Based on a frame rate of 24 fps, one droplet replacement was calculated to be complete within 0.2 s. Furthermore, rapid reagent replacement under a high oil flow rate (v_{oil} = 5.3 μL/min) is shown in Figure 3a (bottom row), applying a frame rate of 96 fps (Figure S4): the red ink droplet replaces the original black ink droplet from under an SEW site and parks there, with a temporal resolution of only 0.04 s. The replacement duration was recorded for varying oil flow rates at 10 positions and a column scatter plot was generated (Figure 3b). Furthermore, the replacement time was determined for a single SEW structure, whereby the dark blue droplet shown in Figure 3c was replaced by the light blue one, which was calculated to be 1.25 s. In all of these cases, the spatial resolution of reagents was the coverage area of a single sub-nanoliter droplet. Therefore, the spatiotemporal resolution of this platform depends on the carrier oil velocity and droplet volume, and it is sufficiently high to enable the rapid single-cell profiling of most transient cellular processes.

An important prerequisite for achieving high temporal resolution is ensuring a high reagent replacement efficiency, ideally with 100% content exchange ratio; this was evaluated based on the residual quantity of fluorescent dyes after droplet replacement. Figure 3d shows the sequential replacement of fluorescent droplets under a 2 × 2 SEW array: red fluorescent droplets (ROX, 50 μg/mL) were initially trapped by the SEWs, and then replaced with green (FN-488, 30 μg/mL), followed by blue (7-Hydroxycoumarin, 25 mg/mL) and finally non-fluorescent (PBS, 1×) droplets under an oil flow rate of 4.2 μL/min. Bright-field and fluorescence images were recorded after each droplet replacement and ImageJ software was used to split the fluorescence channels and determine their corresponding gray values. Figure 3e shows the change in fluorescence with droplet replacement. Two aspects of this histogram are worth noting: (1) At the beginning, the fluorescence image of the ROX droplets exhibited low values in both the blue and green channels. (2) Finally, the fluorescence image of the PBS droplet was obtained under UV excitation and the gray values in all three channels were zero. Hence, the integrated analysis results implied that the gray values of the two channels other than the dye fluorescence channel were neither residual nor background fluorescence, but originated from the intrinsic luminescence spectrum of each fluorescent dye. Accordingly, it can be concluded that single-drop reagent replacement displaces nearly 100% of the original solution under the SEW structure. Notably, such a complete exchange prevents cross contamination from successive stimulation exchanges, and upon replacement with a termination reagent, allows for immediate and complete termination. This feature is crucial for achieving high-time-resolution when monitoring transient biochemical processes, in addition to profiling cell signaling networks.

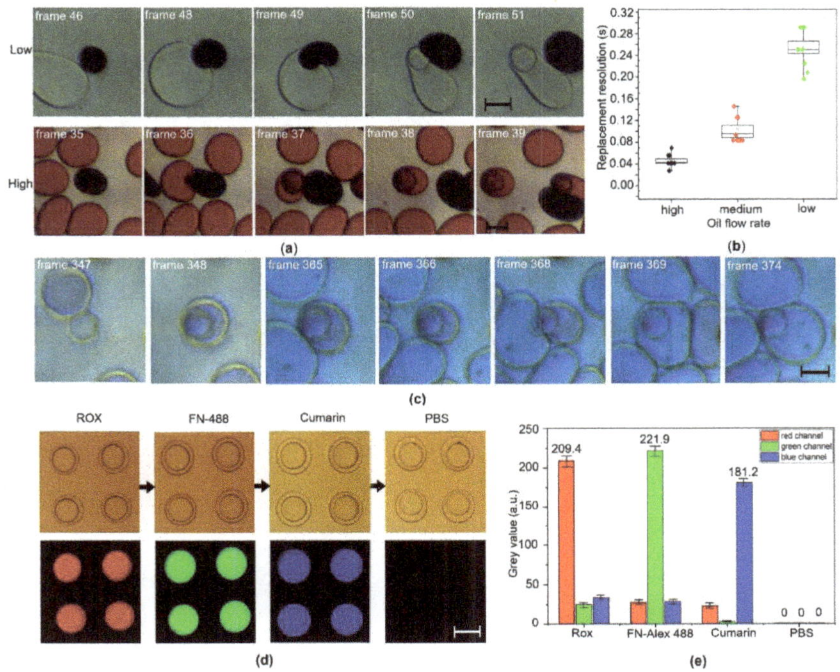

Figure 3. (**a**) Time lapse images of one SEW during the replacement of two aqueous droplets, black to colorless, under a lower oil flow rate (top), and black to red under a higher flow rate (bottom). (**b**) Column scatter plot of the replacement times at 10 positions under varying oil flow rates. (**c**) Time lapse images of droplet replacement (dark blue by light blue) for a single SEW structure. (**d**) Sequential replacement of differently colored droplets under an SEWs array (2 × 2). The top and bottom rows are the bright-field and corresponding fluorescence images, respectively. At the appropriate oil flow rate, ROX droplets are initially trapped under the SEWs, followed by successive replacement with FN-488, 7-Hydroxycoumarin and finally PBS droplets. (**e**) Statistical distribution of the mean gray values of the fluorescence images for three channels, split up and determined using ImageJ software. Scale bars: 100 μm.

3.4. Single Cell Patterning on the Chip

Another crucial step in the high spatiotemporal analysis of cells is obtaining both favorable growth of adherent cells and single-cell patterns that correspond to SEW locations. To achieve this, the cell suspension was loaded into the chamber, followed by the continuous perfusion of fresh culture medium at an extremely slow flow rate to establish a long-term culture of adherent cells in the narrow chamber (Figure S5). After culturing for a period of time, the cells sedimented uniformly and their attachment was promoted by the fibronectin coating (Figure 4a—1). Next, PBS was continuously perfused at a higher flow rate to wash the chamber lightly by removing the residual suspension, and to detach loosely attached cells outside the SEW regions (Figure 4a—2), thereby realizing single-cell patterning corresponding to the SEWs array (Figure 4a—3). The fundamental principle behind the formation of single-cell patterns entails the dissimilar vertical restrictions on transverse liquid flow, arising from the height difference in the narrow channel, which alters the flow velocity in the trapping chamber (Figure 4a). As a result, compared to the SEW region where the channel height was 70 μm, the higher velocity in the region with a height of 35 μm exerts a higher hydrodynamic drag force against the adherent cells, leading to cell detachment. In this study, cell loading and long-term cultivation were performed using a bench incubator with accessibility to medium perfusion and in situ observation. MCF-7 cells with a seeding density ranging from 0.5×10^6 to 1×10^6 cells/mL (100 μL) were loaded (Figure 4b—t_1) and allowed to adhere loosely on the bottom of the chamber under a culture medium perfusion rate of 1 μL/min (Figure 4b—t_2) to ensure appropriate cell density and cell adhesion strength on the inner surface. After culturing for 3 h, the PBS flow rate was increased to ~10–20 μL/min to obtain a single cell array (Figure 4b—t_3) that corresponded to the SEW array. Culturing was continued for an additional hour to strengthen cell adherence.

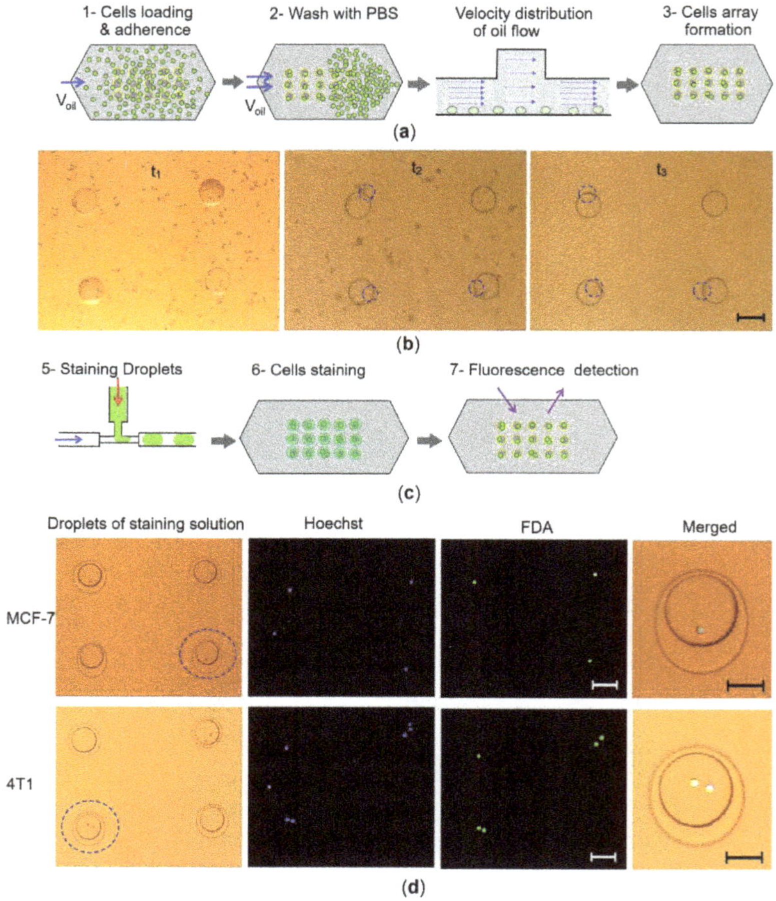

Figure 4. (**a**) Schematic diagram of cell loading, sedimentation, adherence to the bottom of the chamber and growth. Due to the oil flow velocity distribution, with an increasing flow rate of culture medium, single cells are successfully patterned according to the SEW array. (**b**) The bright-field pictures showed the adherence growth of MCF-7 cells on the fibronectin-modified surface. t_1: cell suspension was cultured in the chamber under continuous medium flow. t_2: cells uniformly adhered to the bottom under continuous medium flow at a relatively lower rate for 2 h. t_3: after the culture medium flow rate was increased, single cells appeared in an arrangement corresponding to the cells marked by the dashed blue circles in t_2. Scale bars: 100 μm. (**c**) Schematic diagram for staining single cells by loading various staining droplets; the cells can be viewed directly using a microscope. (**d**) The bright-field and fluorescence images of MCF-7 and 4T1 cell arrays were observed under a 10× objective lens, scale bar: 100 μm, while the merged images (marked by the dashed blue circles in the first column) were observed under larger magnification (20× objective), scale bars: 50 μm. These results provided early information on single cell morphology and viability. Bright droplet = staining solution, blue = cell nuclei (stained by Hoechst) and green = live cells (stained by FDA).

Moreover, to probe the analytical capabilities of the droplet-based method, droplets of cell viability reagents were generated (Figure 4c—5), subsequently trapped by the SEW structures, thereby covering the cells, and then incubated for a certain duration for cell staining (Figure 4c—6), followed by droplet release and fluorescence detection (Figure 4c—7). As shown in Figure 4d, the bright-field photos in the first column indicated that the Hoechst and FDA droplets were trapped by the SEWs on the roof of the chamber, while covering the MCF-7 cells on the bottom. The blue and green fluorescence images, observed under a 20× objective lens, indicated successful nuclear staining and single cell viability of MCF-7. The combined pictures in the right-most column observed under a 40× objective lens, indicated that the single adherent cells maintained their morphological integrity and cell viability for subsequent proliferation. Moreover, to demonstrate the versatility of the new method, 4T1 cells were analyzed in the same manner. The images in Figure 4d show similar single cell patterning under the SEWs with no significant impact on the cell viability. Notably, the successful cell staining using droplets confirmed the feasibility and validity of treating adherent cells with droplets because the SEW structure enables droplet manipulation (Figure S6).

3.5. Single Cell Immunocytochemistry on the Chip

Having established that the SEW-based microfluidics system allows for droplet replacement with high spatiotemporal resolution and adherent cell staining with reagent droplets, we chose the well-characterized PDGFR/PI3K/Akt pathway as a model system to evaluate the performance of the SEW-based method for investigating fast cell signaling events [35–37]. The diagram in Figure 5a illustrates signal transduction in the PDGFR/PI3K/Akt pathway: when the ligand (PDGF) binds to the receptor (PDGFR), the tyrosine of the latter is phosphorylated, serving as a binding site for phosphoinositide 3-kinase (PI-3K). PtdIns(3,4,5)P3 (PIP3) is then generated in response to PI-3K activation, recruiting Akt to the plasma membrane and leading to Akt phosphorylation; these processes have emerged as key regulators of downstream cellular processes [38]. In this experiment, two Akt serine phosphorylation sites involving Ser473 and Thr308 were probed both in MCF-7 and 4T1 cells to evaluate the time-dependent effects of cellular stimuli [39].

Because Akt inhibition can block Akt activation in a concentration-dependent manner [21], at the beginning of on-chip cell preparation, the culture medium was supplemented with 2 μM AKT inhibitor VIII for 3 h to temporarily block the PI3K/Akt pathways in tumor cells. Subsequently, phosphorylation of these inhibited cells was determined as described in the Section 2.5 and schematically depicted in Figure 5b. It is worth mentioning that stimulation and termination in the first two steps were performed using droplets to treat localized cells after trapping, while the subsequent steps entailed solution perfusion. First, under HFE oil (0.8 μL/min) and PDGF-BB solution (0.5 μL/min) flow, the ligand droplets were generated and transported to the chamber at a relatively low speed and then parked there separately for 0, 3 s, 5 s, 10 s and 3 min, then immediately replaced by fixation droplets formed from HFE and formalin flow at rates of 2.2 μL/min and 1.7 μL/min followed by 10 min of incubation. Subsequently, blocking solution (BSA), primary antibody for phospho-Akt (Thr308) and secondary antibody (IgG-Alexa Fluor 555) were sequentially perfused into the chamber for immunocytochemistry reagent exchange, rinsing with PBS at each interval. Finally, the influence of stimulation time on the phosphorylation level was determined based on the fluorescence of individual cells.

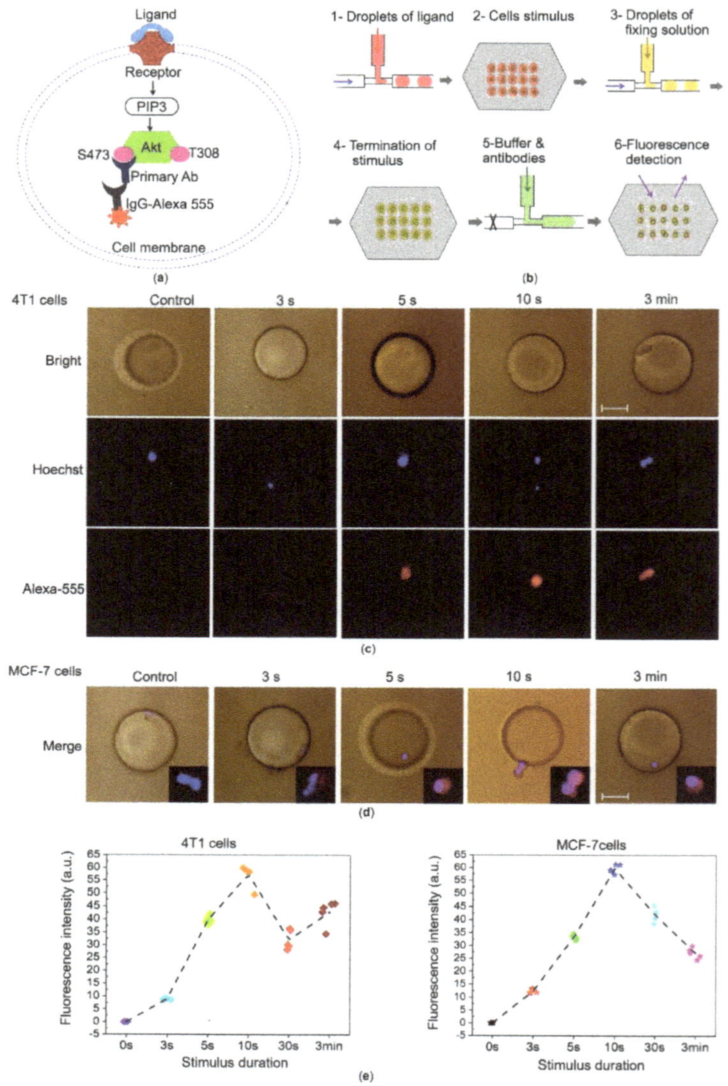

Figure 5. Intracellular protein phosphorylation in the PDGFR/PI3K/Akt pathway induced by PDGF-BB stimulation. (**a**) Schematic diagram illustrating the signal transduction pathway. (**b**) Schematic diagram depicting single-cell treatment with the ligand and sequential immunocytochemistry steps for probing protein phosphorylation. (**c**) 4T1 single cells were stimulated with PDGF-BB droplets for varying time periods, followed by replacement with fixing droplets. Representative pictures of the 4T1 cells under a SEW after the procedure: bright-field row: single cells remained intact and adherent; Hoechst row: blue fluorescence was used for cell localization and validation; Alexa-555 row: the obtained red fluorescence intensities correspond to phosphorylation levels. (**d**) MCF-7 single cells were subjected to the same immunocytochemistry steps. (**e**) The fluorescence intensities of five single cells measured at each stimulus duration; the mean values are connected with a dashed line. Scale bars: 50 μm.

As expected, the individual cells in the control group without PDGF-BB stimulus did not exhibit receptor phosphorylation, as evidenced by the Hoechst(+) and Alex-55(−)

fluorescence intensities shown in Figure 5c. Increasing the stimulation time to 3 s resulted in the phosphorylation of PDGFR being triggered on the cell membrane, resulting in low phosphorylation levels, as indicated by the weak red fluorescence in the second column of Figure 5c. With a prolonged stimulation of 5 s, the red fluorescence of individual cells was significantly enhanced, indicating a higher abundance of phosphorylated states. With prolonged stimulation to 10 s, the red fluorescence of individual cells continued to be enhanced, indicating a sustained increase in phosphorylation level. In a long-term control trial, the red fluorescence of single cells stimulated for 3 min did not change significantly compared to that at 5 s, indicating that complete phosphorylation was achieved at approximately 5 s; the response at the bulk or average level was detected using flow cytometry (Figure S7). In addition, to prove the versatility of the SEW-based platform, MCF-7 cells underwent the same protocol, with the exception of the Anti-Akt primary antibody for Ser473 recognition, and similar phosphorylation results were obtained after 3 s, 5 s, 10 s and 3 min of stimulus, as shown in Figure 5d. Figure 5e shows the fluorescence intensities of the secondary antibody for five single cells measured at each stimulus duration, and the mean values are connected with a dashed line. These results suggest that receptor phosphorylation of PDGFR may be triggered within 3 s of ligand stimulation with PDGF-BB and completed within only 10 s of stimulus time. Gaining insights into single-cell responses to transient stimuli has not been feasible using traditional methods. Specifically, the result of cell phosphorylation being completed within only 10 s of stimulus time is also found within the previous literature report [23]. However, the initiation time of receptor phosphorylation of PDGFR in a stimulated cell is as fast as 3 s, which to our knowledge is the first time that this has been observed. Moreover, to evaluate the effects of the shear stress during droplet replacement on the signaling pathway [40], Akt phosphorylation of MCF-7 cells was also observed under 'extreme' conditions, whereby culture medium droplets containing the inhibitor were trapped and replaced across single cells 240 times during 1 min. Figure S8 shows that the phosphorylation response of cells treated in this manner did not differ significantly from that of undisturbed cells, which further suggests that the SEW-based strategy did not alter the signaling pathway. We propose that the SEW-based platform may be an important new technique for probing temporal dynamics of fast cellular events at a single-cell level for a wide range of applications.

4. Conclusions

In conclusion, with the aim of realizing droplet trapping and replacement using an SEW structure, we have successfully constructed a multifunction droplet-based microfluidic platform that allows for droplet manipulation, single cell patterning and high spatiotemporal resolved profiling of signaling in a single chip. The SEWs located on the roof of the chamber not only enable droplet replacement with high spatiotemporal resolution, but also near-complete exchange of reagent droplets. Furthermore, single-cell patterning was reliably obtained on the floor of the chamber by harnessing the hydrodynamic flow differences in the SEW region. Moreover, the ability to treat cells using trapped droplets as described for the designed platform, would facilitate subsequent single-cell analysis according to the immunohistochemistry protocol. Considering such advantages, the SEW-based device was applied to analyze the phosphorylation states of PDGFR and the downstream signaling protein Akt, and evaluate the effects of time-dependent stimuli on tumor cells. The present study provides a general platform for investigating the temporal dynamics of signal profiling at the single-cell level, based on adherent cell manipulation. Furthermore, the new strategy also has significant potential for rapid chemical reaction monitoring for a wide range of applications.

Supplementary Materials: The following supporting information can be downloaded at: https://www.mdpi.com/article/10.3390/chemosensors10120521/s1, Supplementary figures including chip dimensions, passing and parking modes, critical volume, time resolution calculated using varying frame rates, cell culture in droplets, cell staining with droplets, flow cytometry results and the control test results involving continuous replacement.

Author Contributions: Conceptualization, Y.S. and S.Z.; methodology, Y.S. and Q.T.; validation, K.X., Y.L. (Yongshu Liu), Y.L. (Yuyan Li) and Y.L. (Yumin Liu); investigation, Y.S., Q.T., Y.L. (Yongshu Liu), Y.L. (Yuyan Li) and Y.L. (Yumin Liu); data curation, Y.S., K.X. and S.Z.; writing—original draft preparation, Y.S., Q.T., K.X. and Y.L. (Yongshu Liu); writing—review and editing, Q.T., K.X., Y.L. (Yongshu Liu), Y.L. (Yuyan Li) and Y.L. (Yumin Liu); supervision, Y.S. and S.Z.; funding acquisition, Y.S. and S.Z. All authors have read and agreed to the published version of the manuscript.

Funding: The authors are appreciative for the financial support from the National Natural Science Foundation of China (grant 22076074), the Natural Science Foundation of Shandong Province (2021KJ065) and the National Natural Science Foundation of China (grant 21804065, 22076073).

Institutional Review Board Statement: Not applicable.

Informed Consent Statement: Not applicable.

Data Availability Statement: Any additional data in support of the findings of this study besides those provided as Supplementary Materials are available from the corresponding author upon reasonable request.

Acknowledgments: The authors wish to thank Haonan Sun and Qian Wang from Linyi University for their technical support regarding device fabrication.

Conflicts of Interest: The authors declare no conflict of interest.

References

1. Lamanna, J.; Scott, E.Y.; Edwards, H.S.; Chamberlain, M.D.; Dryden, M.D.M.; Peng, J.; Mair, B.; Lee, A.; Chan, C.; Sklavounos, A.A.; et al. Digital microfluidic isolation of single cells for -Omics. *Nat. Commun.* **2020**, *11*, 5632–5644. [CrossRef]
2. Tay, S.; Hughey, J.J.; Lee, T.K.; Lipniacki, T.; Quake, S.R.; Covert, M.W. Single-cell NF-(kgr)B dynamics reveal digital activation and analogue information processing. *Nature* **2010**, *466*, 267–271. [CrossRef] [PubMed]
3. Kamenz, J.; Ferrell, J.E. The Temporal Ordering of Cell-Cycle Phosphorylation. *Mol. Cell* **2017**, *65*, 371–373. [CrossRef] [PubMed]
4. Aibar, S.; González-Blas, C.B.; Moerman, T.; Huynh-Thu, V.A.; Imrichova, H.; Hulselmans, G.; Rambow, F.; Marine, J.-C.; Geurts, P.; Aerts, J.; et al. SCENIC: Single-cell regulatory network inference and clustering. *Nat. Methods* **2017**, *14*, 1083–1086. [CrossRef]
5. Bedard, P.L.; Hansen, A.R.; Ratain, M.J.; Siu, L.L. Tumour heterogeneity in the clinic. *Nature* **2013**, *501*, 355–364. [CrossRef] [PubMed]
6. Peng, A.; Mao, X.; Zhong, J.; Fan, S.; Hu, Y. Single-cell multi-Omics and its prospective application in cancer biology. *Proteomics* **2020**, *20*, 1900271–1900279. [CrossRef]
7. Clark, I.C.; Delley, C.L.; Sun, C.; Thakur, R.; Stott, S.L.; Thaploo, S.; Li, Z.; Quintana, F.J.; Abate, A.R. Targeted single-cell RNA and DNA sequencing wth fluorescence activated droplet merger. *Anal. Chem.* **2020**, *92*, 14616–14623. [CrossRef]
8. Du, L.; Liu, H.; Zhou, J. Picoliter droplet array based on bioinspired microholes for in situ single-cell analysis. *Microsystems Nanoeng.* **2020**, *6*, 33–41. [CrossRef]
9. Tajik, M.; Baharfar, M.; Donald, W.A. Single-cell mass spectrometry. *Trends Biotechnol.* **2022**, *40*, 2182–2200. [CrossRef]
10. Bendall, S.C.; Nolan, G.P.; Roederer, M.; Chattopadhyay, P.K. A deep profiler's guide to cytometry. *Trends Immunol.* **2012**, *33*, 323–332. [CrossRef]
11. Krutzik, P.O.; Crane, J.M.; Clutter, M.R.; Nolan, G.P. High-content single-cell drug screening with phosphor specific flow cytometry. *Nat. Chem. Biol.* **2008**, *4*, 132–142. [CrossRef] [PubMed]
12. Junkin, M.; Kaestli, A.J.; Cheng, Z.; Jordi, C.; Albayrak, C.; Hoffmann, A.; Tay, S. High-Content Quantification of Single-Cell Immune Dynamics. *Cell Rep.* **2016**, *15*, 411–422. [CrossRef]
13. Samlali, K.; Ahmadi, F.; Quach AB, V.; Soffer, G.; Shih, S.C.C. One cell, one drop, one click: Hybrid microfluidics for mammalian single cell isolation. *Small* **2020**, *16*, 2002400–2002412. [CrossRef] [PubMed]
14. Zhang, K.; Chou, C.-K.; Xia, X.; Hung, M.-C.; Qin, L. Block-Cell-Printing for live single-cell printing. *Proc. Natl. Acad. Sci. USA* **2014**, *111*, 2948–2953. [CrossRef]
15. Brimmo, A.T.; Menachery, A.; Sukumar, P.; Qasaimeh, M.A. Noncontact Multiphysics Probe for Spatiotemporal Resolved Single-Cell Manipulation and Analyses. *Small* **2021**, *17*, 2100801. [CrossRef]
16. Tanay, A.; Regev, A. Scaling single-cell genomics from phenomenology to mechanism. *Nature* **2017**, *541*, 331–338. [CrossRef] [PubMed]
17. Mann, M.; Jensen, O.N. Proteomic analysis of post-translational modifications. *Nat. Biotechnol.* **2003**, *21*, 255–261. [CrossRef]
18. Lou, Q.; Ma, Y.; Zhao, S.-P.; Du, G.-S.; Fang, Q. A flexible and cost-effective manual droplet operation platform for miniaturized cell assays and single cell analysis. *Talanta* **2020**, *224*, 121874–121882. [CrossRef]
19. Xie, R.; Liu, Y.; Wang, S.; Shi, X.; Zhao, Z.; Liu, L.; Liu, Y.; Li, Z. Combinatorial perturbation sequencing on single cells using microwell-based droplet random pairing. *Biosens. Bioelectron.* **2023**, *220*, 114913–114921. [CrossRef]

20. Blazek, M.; Santisteban, T.S.; Zengerle, R.; Meier, M. Analysis of fast protein phosphorylation kinetics in single cells on a microfluidic chip. *Lab Chip* **2014**, *15*, 726–734. [CrossRef]
21. Ng, A.; Chamberlain, M.; Situ, H.; Lee, V.; Wheeler, A.R. Digital microfluidic immunocytochemistry in single cells. *Nat. Commun.* **2015**, *6*, 7513–7524. [CrossRef]
22. Ying, Y.-L.; Hu, Y.-X.; Gao, R.; Yu, R.-J.; Gu, Z.; Lee, L.P.; Long, Y.-T. Asymmetric Nanopore Electrode-Based Amplification for Electron Transfer Imaging in Live Cells. *J. Am. Chem. Soc.* **2018**, *140*, 5385–5392. [CrossRef] [PubMed]
23. Yu, R.; Ying, Y.; Gao, R.; Long, Y. Confined Nanopipette Sensing: From Single Molecules, Single Nanoparticles, to Single Cells. *Angew. Chem. Int. Ed.* **2019**, *58*, 3706–3714. [CrossRef] [PubMed]
24. Guillaume-Gentil, O.; Grindberg, R.V.; Kooger, R.; Dorwling-Carter, L.; Martinez, V.; Ossola, D.; Pilhofer, M.; Zambelli, T.; Vorholt, J.A. Tunable Single-Cell Extraction for Molecular Analyses. *Cell* **2016**, *166*, 506–516. [CrossRef]
25. Guillaume-Gentil, O.; Potthoff, E.; Ossola, D.; Franz, C.M.; Zambelli, T.; Vorholt, J.A. Force-controlled manipulation of single cells: From AFM to FluidFM. *Trends Biotechnol.* **2014**, *32*, 381–388. [CrossRef]
26. Sarkar, A.; Kolitz, S.; Lauffenburger, D.A.; Han, J. Microfluidic probe for single-cell analysis in adherent tissue culture. *Nat. Commun.* **2014**, *5*, 3421. [CrossRef]
27. Dangla, R.; Lee, S.; Baroud, C.N. Trapping Microfluidic Drops in Wells of Surface Energy. *Phys. Rev. Lett.* **2011**, *107*, 124501. [CrossRef]
28. Fradet, E.; McDougall, C.; Abbyad, P.; Dangla, R.; McGloin, D.; Baroud, C.N. Combining rails and anchors with laser forcing for selective manipulation within 2D droplet arrays. *Lab Chip* **2011**, *11*, 4228–4234. [CrossRef]
29. Hele-Shaw, H. Flow of water. *Nature* **1898**, *58*, 520. [CrossRef]
30. Abbyad, P.; Dangla, R.; Alexandrou, A.; Baroud, C.N. Rails and anchors: Guiding and trapping droplet microreactors in two dimensions. *Lab Chip* **2010**, *11*, 813–821. [CrossRef]
31. Sart, S.; Tomasi, R.F.-X.; Amselem, G.; Baroud, C.N. Multiscale cytometry and regulation of 3D cell cultures on a chip. *Nat. Commun.* **2017**, *8*, 469–481. [CrossRef]
32. Lin, M.; Mao, S.; Wang, J.; Xing, J.; Wang, Y.; Cai, K.; Luo, Y. Adsorption force of fibronectin controls transmission of cell traction force and subsequent stem cell fate. *Biomaterials* **2018**, *162*, 170–182. [CrossRef]
33. Herrada, M.; Ponce-Torres, A.; Rubio, M.; Eggers, J.; Montanero, J. Stability and tip streaming of a surfactant-loaded drop in an extensional flow. Influence of surface viscosity. *J. Fluid Mech.* **2022**, *934*, A26–A37. [CrossRef]
34. Sousa, A.M.; Pereira, M.J.; Matos, H.A. Oil-in-water and water-in-oil emulsions formation and demulsification. *J. Pet. Sci. Eng.* **2022**, *210*, 110041–110058. [CrossRef]
35. Basciani, S.; Mariani, S.; Spera, G.; Gnessi, L. Role of platelet-derived growth factors in physiology and medicine. *Genes Dev.* **2008**, *22*, 1276–1312.
36. Yudushkin, I. Control of Akt activity and substrate phosphorylation in cells. *IUBMB Life* **2020**, *72*, 1115–1125. [CrossRef] [PubMed]
37. Yua, C.; Suna, P.; Zhoua, Y.; Shena, B.; Zhoua, M.; Wua, L.; Kong, M. Inhibition of AKT enhances the anti-cancer effects of Artemisinin in clear cell renal cell carcinoma. *Biomed. Pharmacother.* **2019**, *118*, 109383–1092391. [CrossRef]
38. Lemmon, M.A.; Schlessinger, J. Cell Signaling by Receptor Tyrosine Kinases. *Cell* **2010**, *141*, 1117–1134. [CrossRef]
39. Manning, B.D.; Cantley, L.C. AKT/PKB Signaling: Navigating Downstream. *Cell* **2007**, *129*, 1261–1274. [CrossRef]
40. Lee, S.; Gallaire, F.; Baroud, C.N. Interface-induced recirculation within a stationary microfluidic drop. *Soft Matter* **2012**, *8*, 10750–10758. [CrossRef]

chemosensors

Article

A Sensitive Immunochromatographic Test Strip Based on Hydrophobic Quantum Dots Incorporated into Mg/Fe Nanoflowers for HCG Detection

Hao Liu [1], Qing Zhang [2], Ning Bao [3] and Shou-Nian Ding [1,*]

[1] Jiangsu Province Hi-Tech Key Laboratory for Bio-Medical Research, School of Chemistry and Chemical Engineering, Southeast University, Nanjing 211189, China
[2] Key Laboratory of Consumer Product Quality Safety Inspection and Risk Assessment for State Market Regulation, Chinese Academy of Inspection and Quarantine, Beijing 100176, China
[3] School of Public Health, Nantong University, Nantong 226019, China
* Correspondence: snding@seu.edu.cn

Abstract: As the most widely used disposable paper-based diagnostic tool in the world, immunochromatographic test strips (ICTS) have occupied more and more positions in the field of rapid diagnosis due to their ease of operation and affordability. Therefore, the development of an easily prepared, sensitive, and accurate signal reporter is of great significance for the detection of some low-abundance biomarkers in clinical diagnosis. Herein, Mg/Fe layered double hydroxide nanoflowers (MF NFs) were selected as adsorption templates and sulfhydryl-functionalized, followed by one-step loading of hydrophobic CdSe/ZnS quantum dots in the organic phase via a metal-thiol covalent bond. After coating the reporter with branched polyethyleneimine (PEI), a novel ICTS fluorescent reporter was prepared. The modification of PEI not only improved the hydrophilicity of MF@CdSe/ZnS NFs but also introduced amino functional groups on the surface of the reporter for subsequent conjugation with antibodies. X-ray photoelectron spectroscopy, UV-vis absorption, X-ray diffraction, fluorescence spectroscopy, and infrared spectroscopy were used to characterize the composition of MF@CdSe/ZnS NFs. Under the optimal experimental conditions, the detection range of MF@CdSe/ZnS@PEI-ICTS for the model analyte HCG was 0.1–500 mIU/mL, and the limit of detection (LOD) reached was 0.1 mIU/mL. The potential for practical application was validated by the detection of HCG in spiked healthy human serum, showing overall recoveries between 90.48 and 116.1% with coefficients of variation that ranged from 3.66 to 12.91%.

Keywords: layered double hydroxide nanoflowers; immunochromatography test strip; hydrophobic CdSe/ZnS QDs; human chorionic gonadotropin

Citation: Liu, H.; Zhang, Q.; Bao, N.; Ding, S.-N. A Sensitive Immunochromatographic Test Strip Based on Hydrophobic Quantum Dots Incorporated into Mg/Fe Nanoflowers for HCG Detection. *Chemosensors* **2023**, *11*, 114. https://doi.org/10.3390/chemosensors11020114

Academic Editor: Chunsheng Wu

Received: 8 November 2022
Revised: 30 December 2022
Accepted: 1 February 2023
Published: 3 February 2023

1. Introduction

POCT (point-of-care testing) is a test that can be performed on site or nearby, according to medical and individual patient needs, without the constraints of place, time, testing environment, and testing facilities. POCT currently facilitates in vitro testing of various biological indicators, reduces the volume of samples tested, greatly shortens the turnaround time of samples tested, and allows for flexible and versatile application scenarios. As a well-known disposable paper-based diagnostic tool, immunochromatography test strips (ICTS) occupy an increasingly high position in the field of rapid diagnosis, so they are widely used in the qualitative, semi-quantitative, and even quantitative detection of various proteins and small molecules [1,2]. The initial design of ICTS was first reported by Plotz and Singer in 1956 and was widely employed in urine-based pregnancy recognition [3,4]. However, due to the limited detection sensitivity of colloidal gold-based ICTS, they are usually used to qualitatively detect some high-abundance biomarkers (>10^{-12} M), which largely limits the application of ICTS in the field of low-abundance biomarker detection. In the case of the recent global COVID-19 pandemic, testing a single serological antibody or antigen can only indicate a person's past or recent exposure to SARS-CoV-2 and cannot identify the "hot spot" of infection and effectively control the disease in a timely manner. Therefore, multiple and extensive tests are necessary. Ideally, close contacts should be tested weekly or daily to be promptly isolated and to minimize the spread of the virus in the community. The main disadvantages of ICTS based on colloidal gold are low sensitivities, limited precision quantification abilities, and low detection sensitivities. As a result, some false-negative cases can be mistaken for being non-infectious, creating a significant risk of further spreading in the community. In addition to colorimetric methods, signal types used in ICTS include fluorescent reporters [5], phosphorescence reporters [6], magnetic signals [7], and electrochemical signals [8]. Among them, the fluorescent reporter provides better signal contrast and lower background interference, making it have higher sensitivity and lower detection limits for the same detection time. Therefore, the development of higher performance fluorescent labels will be of great benefit to ICTS in dealing with multiple and complex detection.

Among the fluorescence reporter molecules, colloidal quantum dots (QDs) have the advantages of wide absorption, a narrow emission band, size/shape, composition, and surface properties that can be controlled artificially and accurately, and their luminous efficiency is better than that of conjugated molecules (polymers) or inorganic phosphors [9,10]. Its applications in biomarkers, photovoltaic devices, and light-emitting diodes have been extensively and deeply studied [11–13]. Foubert et al. conducted a comparative study of colloidal gold and QDs as labels for multiplex screening assays for toxins, and the results showed that QDs-based ICTS consume fewer immunological reagents, have a lower-false negative rate (<5%), higher sensitivity, and better economic efficiency [14]. Deng et al. used QDs to detect miRNAs rapidly and sensitively, and the results showed that the detection method was also 10 times more sensitive than traditional colloidal gold-based miRNA detection test strips [15]. Although studies have pointed out that ICTS based on QDs is more sensitive than colloidal gold-based ICTS, it is still difficult to detect many low-abundance biomarkers [16,17]. Embedding QDs in large numbers into individual nanoparticles is a useful strategy for improving sensitivity [10,18,19]. At present, there are four main approaches: in situ self-assembly [20,21], in situ polymerization [22], layer-by-layer assembly [23], and porous nanoparticle-based incorporation [10,18]. Due to the presence of many alkyl stabilizers on the surface of QDs, this type of quantum dot can only be dispersed in non-polar solvents (chloroform, hexane, etc.). However, either in situ or template-based assembly processes require pre-capped ligand exchange for nanocrystals because these ligands on the surface of oil-soluble quantum dots are usually hydrophobic and lack reactive functional groups [24,25]. This process may cause irreversible physicochemical damage to QDs. Compared with the other three methods, the porous nanoparticle-based incorporation strategy has the advantages of high loading capacity, good uniformity, and reduced aggregation [26,27]. Several previous papers have demonstrated that nanoassemblies with high homogeneity and surface loading density can be prepared directly in the organic phase by ligand-driven assembly of metal-containing nanocrystals with thiol-capped colloids using metal-thiol covalent bonding [28–30]. For instance, Hu et al. fully used dendritic SiO_2 as templates to enrich high-density hydrophobic QDs for the ultrasensitive determination of C-reactive proteins [10].

Layered double hydroxides (LDHs) consist of stacked, positively charged, brucite-like octahedral metal hydroxide layers and interlayer anions and water molecules. It is a class of layered materials with positively charged metal hydroxide layers and charge-balancing anions [31]. In recent years, due to its low toxicity, unique layered structure, good biocompatibility, high anion exchange capacity, adjustable particle size, and other advantages, have been widely studied in the field of biomedicine [32,33]. LDH nanoparticles, especially spherical LDH particles with porous structures, have been proven to be excellent adsorbents due to their low cost, large specific surface area, and strong adsorption capacity [34–36]. Herein, for the first time, surface thiol-functionalized MF NFs (SH-MF) were employed as adsorption templates to load hydrophobic CdSe/ZnS QDs in an oil-phase solvent. As illustrated in Figure 1, MF NFs were used as adsorption templates, and many hydrophobic CdSe/ZnS QDs were loaded via thiol-metal covalent bonding. It is necessary to transfer the MF@CdSe/ZnS NFs from the oil phase to the water phase. Some studies have shown that coating or embedding oil-phase nanocrystals with hydrophilic polymers is a feasible method to solve the above problems [37,38]. For example, Chen et al. developed a novel ICTS system of QDs-doped polystyrene nanoparticles with satisfactory recovery and reproducibility. The test strip can simultaneously detect cytokeratin-19 fragments and carcinoembryonic antigens in human serum, with limits of detection of 0.16 and 0.35 ng/mL, respectively. Therefore, it is completely feasible to use an optimized amount of branched polyethyleneimine (PEI) to modify the fluorescent reporter in this work. Finally, a novel high-density oil-phase CdSe/ZnS QD that incorporated MF NFs fluorescent signal reporters (MF@CdSe/ZnS@PEI) was successfully prepared. These nanoflowers were further bound to antibodies and successfully applied to the ICTS for the highly sensitive detection of HCG.

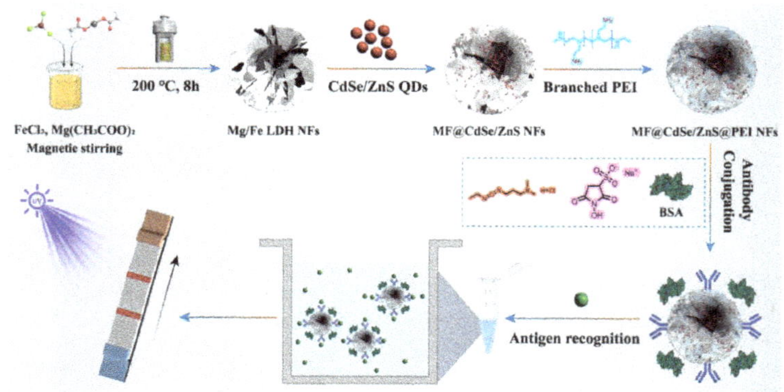

Figure 1. Schematic illustration of MF@CdSe/ZnS@PEI-ICTS for HCG detection.

2. Materials and Methods

2.1. Reagents and Characterization

Mg $(OAc)_2 \cdot 4H_2O$, $FeCl_3 \cdot 6H_2O$, ethylene glycol, Cd $(Ac)_2$, Zn $(Ac)_2$, ethanol, ammonia water, oleic acid (OA), selenium powder, sulfur powder, 1-octadecene KH_2PO_4, $Na_2HPO_4 \cdot 12H_2O$, NaCl, KCl, Tween-20, and $CHCl_3$ were purchased from Sinopharm Chemical Reagent Co., Ltd. Tri-n-octyl phosphine, γ-mercaptopropyltrimethoxysilane, EDC·HCl and Sulfo-NHS were purchased from Shanghai Aladdin Biochemical Technology Co., Ltd. (Shanghai, China). Carbohydrate antigen 125 (CA125), alpha fetoprotein (AFP), carbohydrate antigen 199 (CA199), and carcinoembryonic antigen (CEA) were purchased from Shanghai Linc-Bio Science Co., Ltd (Shanghai, China). The anti-α-HCG (Ab_1), anti-β-HCG (Ab_2), goat anti-mouse IgG antibodies, and the parts that make up the ICTS were purchased from Shanghai Joey Biotechnology Co. Ltd (city, country). All buffers in this work were prepared fresh at the time of use. A Bruker ASCENDTM 400WB spectrometer (Bruker, Newark, DE, USA) was used to characterize the ^{13}C CP/MAS NMR spectra of thiol-modified MF NFs at 400 MHz, reported in parts per million (ppm). UV-vis absorption spectra were obtained on a Shimadzu UV 2600 spectropolarimeter (Kyoto, Japan) and were used to characterize the absorbance values of hydrophobic CdSe/ZnS QDs, MF@CdSe/ZnS NFs, and MF@CdSe/ZnS@PEI NFs. Scanning electron microscope (SEM) photographs were taken under a field emission scanning electron microscope (Thermo Scientific, Waltham, MA, USA) operating at 20 kV to characterize MF NFs. Transmission electron microscopy (TEM) images were taken under a field emission high-resolution transmission electron microscope Talos F200X (Thermo Scientific, Waltham, MA, USA) for the characterization of hydrophobic CdSe/ZnS QDs, MF@CdSe/ZnS NFs, and MF@CdSe/ZnS

@PEI NFs, and elemental analysis. Fourier transform infrared (FT-IR) spectra were collected from a Nicolet 5700 (Thermo Nicolet Corporation, Waltham, MA, USA) infrared spectrometer with a characterization range of 4000–400 cm^{-1} for the characterization of Mg/Fe NFs, MF@CdSe/ZnS NFs, and MF@CdSe/ZnS@PEI NFs. Here, the surface potential characterization of hydrophobic CdSe/ZnS QDs, MF@CdSe/ZnS NFs, and MF@CdSe/ZnS@PEI NFs at 25 °C was performed using a Zetasizer Nano-ZS from Malvern Instruments (Malvern, UK). Powder X-ray diffraction (XRD) analysis of MF NFs and MF@CdSe/ZnS NFs was performed using a Rigaku Ultima IV multifunctional horizontal X-ray diffractometer (Rigaku, Tokyo, Japan). X-ray photoelectron spectroscopy (XPS, PreVac, Rogow, Poland) studies were performed using the XPS-2 system to characterize the elemental composition of MF NFs and MF@CdSe/ZnS NFs.

2.2. Preparation Procedures of the Red-Emitting Hydrophobic CdSe/ZnS QDs

The synthesis procedures for CdSe/ZnS QDs are based on previously reported literature with some minor changes [39]. An amount of 1.0 mmol Cd $(Ac)_2$ (Sinopharm Chemical, Shanghai, China) and 4.0 mmol Zn $(Ac)_2$ (Sinopharm Chemical) were placed in a four neck round bottom flask that contained 9 mL of OA (Sinopharm Chemical) and 10 mL of ODE (Sinopharm Chemical). The mixture was first heated to 180 °C to form a homogeneous solution, and after degassing at 150 °C, the mixture was heated to 300 °C. Then 0.4 mL of TOP-Se solution (1 M) was rapidly injected into the mixture solution first, and then 4 mL of TOP-S solution (1 M) was injected continuously at 30 s intervals. Finally, the temperature was set at 280 °C for subsequent growth.

2.3. Preparation Procedures of MF@CdSe/ZnS@PEI NFs [36]

An amount of 50.0 mg of MF NFs were uniformly dispersed in absolute ethanol, followed by 0.5 mL of ammonia water and 0.5 mL of γ-mercaptopropyltrimethoxysilane, and the mixture was homogeneously sonicated and vigorously stirred at room temperature for 6 hours. After the reaction, thiol-modified MF NFs (SH-MF) can be obtained by centrifugation. The SH-MF product was thoroughly washed alternately with absolute ethanol and deionized water to remove unreacted reagents and impurities, then redispersed in 50 mL of $CHCl_3$, to which excess red-emitting hydrophobic CdSe/ZnS QD dispersion was added, and the mixture was sonicated for 30 min to obtain a clear and transparent solution. The complex was washed repeatedly with chloroform to remove unattached CdSe/ZnS QDs. Afterwards, the obtained MF@CdSe/ZnS NFs were properly dried under N_2 flow and redispersed in 20 mL of $CHCl_3$, which contained 2 mg of branched PEI, under sonication for 30 min. After centrifugation and washing in deionized water, the MF@CdSe/ZnS@PEI NFs were dispersed in the PBS solution (0.01 M, pH = 7.4) and stored in a refrigerator for later use.

2.4. Preparation Procedures of Ab_2-MF@CdSe/ZnS@PEI

Anti-β-HCG was assembled to form bioconjugates with MF@CdSe/ZnS@PEI via EDC/NHS chemical cross-linking. Firstly, the freshly prepared MF@CdSe/ZnS@PEI NFs (1 mg/mL) were dispersed in PBS solution (0.01 M, pH = 7.4), and a Sulfo-NHS solution with a concentration of 5 mg/mL was added and shaken at room temperature for 4 h. After the reaction was completed, MF@CdSe/ZnS@PEI NFs were washed several times with ethanol and water to remove excess PEI and impurities and then redispersed in MES buffer solution (0.01 M, pH = 6.0) for use. Next, the carboxyl-modified MF@CdSe/ZnS@PEI NFs were dispersed into a 0.5 mL MES buffer solution (0.01 M, pH = 6.0) to prepare a concentration of 2 mg/mL, followed by adding 4 mg EDC (Energy Chemical) and 6 mg Sulfo-NHS (Energy Chemical). The above mixture was shaken and reacted at room temperature for half an hour. Afterwards, 100 μg of Anti-β-HCG (Ab_2) was added and incubated at 4 °C overnight. Finally, the Ab_2-MF@CdSe/ZnS@PEI bioconjugates were collected by centrifugation and washed several times with HEPES buffer solution (0.01 M, pH = 7.4) containing 1% BSA to remove unbound proteins. It was then redispersed in HEPES buffer solution and stored in the refrigerator.

2.5. Fabrication Procedures of the MF@CdSe/ZnS@PEI-ICTS

The pretreatment steps and preparation processes of MF@CdSe/ZnS@PEI-ICTS are basically consistent with those reported in our previous literature [40,41]. The test strip is mainly composed of four parts: a pre-treated sample pad, an NC membrane, an absorbent pad, and a black polyvinyl chloride (PVC) back card. Anti-α-HCG (Linc-Bio, Ab_1, 1.5 mg/mL) and goat anti-mouse IgG antibodies (Linc-Bio, 1 mg/mL) were applied to NC membranes by fiber pen to constitute the test (T) and control (C) lines, respectively. The prepared NC membranes were then dried at 37 °C overnight. Finally, the sample pad, NC membrane, and absorbent pad were sequentially connected and laid on a black PVC backing card, each adjacent section overlapping about 1–2 mm, and finally cut into pieces with a width of about 3.9 mm by a chopper.

2.6. Detection of HCG Standard Samples with the MF@CdSe/ZnS@PEI-ICTS

Firstly, HCG standard solutions of different concentrations (0, 0.01, 0.1, 1, 10, 50, 100, 200, and 500 mIU/mL) were prepared, and 40 μL of each concentration solution and 20 μL of Ab_2-MF@CdSe/ZnS@PEI NFs were premixed. After that, the mixture was added dropwise to the sample pad of the test strip. Each concentration was tested 3 times in parallel. After 15 min, under the excitation of a 365 nm UV lamp, qualitative results can be obtained by observing the red band on the test strip. Photographs were taken with a mobile phone camera for qualitative measurement, and the captured images were processed using ImageJ software version 1.50d to obtain the intensities of the T- and C-lines [42,43].

3. Results and Discussion

3.1. Preparation and Characterization of MF@CdSe/ZnS@PEI NFs

As shown in Figure 2a,b, the MF NFs prepared in this work consist of many nanosheets, which are interconnected to form an open structure with a size of about 260 nm. $Mg(OAc)_2 \cdot 4H_2O$ and $FeCl_3 \cdot 6H_2O$ were used as precursors, and ethylene glycol was used as a solvent. MF NFs were synthesized by the solvothermal method in an autoclave [44–46]. In addition, we applied Fourier transform-infrared spectroscopy (FT-IR), X-ray diffraction (XRD), and X-ray photoelectron spectroscopy (XPS) methods (Figure 3a,c and Figure S5) to characterize the MF NFs prepared in this work, and the results showed that they were consistent with the basic characteristics of LDH-like materials [35,36]. MF@CdSe/ZnS NFs can be obtained by sonicating oil-soluble CdSe/ZnS quantum dots (7 nm, Figure S1) of uniform size with SH-MF in chloroform solution for 30 min. As shown in Figure 2c, MF NFs as carriers can bond high-density loaded hydrophobic CdSe/ZnS QDs via Cd-S covalent bonding. After the MF@CdSe/ZnS NFs were modified with an optimized amount of branched PEI, we employed energy dispersive X-ray spectroscopy (EDS) elemental mapping to investigate its elemental composition. From Figure 2d–l, the elements contained in the MF@CdSe/ZnS@PEI NFs composite were all determined, which can prove the successful preparation of MF@CdSe/ZnS@PEI. In Figure 3a, the characteristic peaks located at 2920 cm^{-1} ($-CH_2$), 1458 cm^{-1} ($-CH_2$), and 1380 cm^{-1} ($-CH_3$) are consistent with bare CdSe/ZnS QDs (Figure S2), indicating that MF@CdSe/ZnS NFs were successfully prepared. For MF@CdSe/ZnS@PEI, the characteristic peaks of amino groups at 3448 cm^{-1} and 1640 cm^{-1} indicate that PEI exists on the surface of MF@CdSe/ZnS NFs, implying the successful introduction of amino functional groups. The UV-vis absorption data in Figure 3b show that the absorption peaks of CdSe/ZnS QDs, MF@CdSe/ZnS, and MF@CdSe/ZnS@PEI are all around 625 nm, which indicates that QDs loaded into MF NFs do not exhibit an obvious absorption peak shift, and the properties are not affected. As shown in Figure 3c, XRD measurements were applied to characterize the crystal composition of MF NFs and MF@CdSe/ZnS NFs. As indicated by the black lines in Figure 3c, MF NFs have five main peaks located at 9.26°, 22.1°, 34.1°, 38.0°, and 59.7°, respectively, and the corresponding crystal planes are (003), (006), (012), (015), and (110) [36]. The three broad peaks located at 26.1°, 43.3°, and 50.6° in the figure correspond to the (111), (220), and (311) plane diffractions, respectively. These crystal plane data are generally consistent with the cubic sphalerite structure of bulk CdSe (JCPDS file No. 88–2346) [47–49]. By observing Figure S3, the chemical shift peaks of the methylene belonging to 3-MPTMS can be determined, which are 17.30, 20.48, and 32.92 ppm, respectively, which indicates that PEI has been successfully modified to the MF@CdSe/ZnS NFs. Meanwhile, it can be seen from Figure S4 that with the loading of hydrophobic QDs, the zeta potential value changes from negative to positive; this result further proves that the loading of hydrophobic QDs changes the surface potential of MF NFs. We performed fluorescence characterization of hydrophobic CdSe/ZnS QDs, MF@CdSe/ZnS, and MF@CdSe/ZnS@PEI NFs using a spectrofluorometer. Characterization results show that the emission wavelength of CdSe/ZnS QDs is 642 nm, while the emission wavelengths of MF@CdSe/ZnS and MF@CdSe/ZnS@PEI NFs are 658 and 662 nm, respectively. Compared with CdSe/ZnS QDs, the emission wavelengths of MF@CdSe/ZnS and MF@CdSe/ZnS@PEI NFs are about 20 nm redshifted we think this may be caused by the larger overall particle size of oily quantum dots loaded by MF NFs. XPS was used to characterize the main elemental composition of MF@CdSe/ZnS NFs, and it can be seen from Figure 4a that C, O, Mg, Cd, and Zn elements were determined. High-resolution XPS spectra of C 1s, O 1s, Mg 2p, Cd 3d, and Zn 2p are recorded in Figure 4b–f. Among them, three peaks with binding energies of about 284.6 eV, 286.1 eV, and 288.0 eV can be seen in the C 1s spectrum, which correspond to the C-C coordination of the surface adventitious carbon and the C-O and C = O in the acetate anion, respectively. For O 1s, the peak at 531.4 eV in the spectrum can be attributed to the C-O and C=O bonds of oleic acid, while 532.0 eV corresponds to the bidentate carboxylate. Two peaks belonging to the Mg-OH (49.0 eV) and Mg-O (50.0 eV) bonds can be found in the spectrum of Mg 2p, respectively. Figure 4e shows that the characteristic doublets of the Cd 3d spectrum at 404.8 eV and 411.7 eV come from Cd $3d_{5/2}$ and Cd $3d_{3/2}$, respectively. Among them, the binding energy of Cd $3d_{5/2}$ is consistent with the value reported in the literature (the binding energy of bulk CdSe is 405.6 eV). Figure 4f shows the noisy signal of Zn 2p, which was deconvolved into two components. The peak at 1020.6 eV is related to Zn $2p_{3/2}$, and 1044.0 eV is related to Zn $2p_{1/2}$.

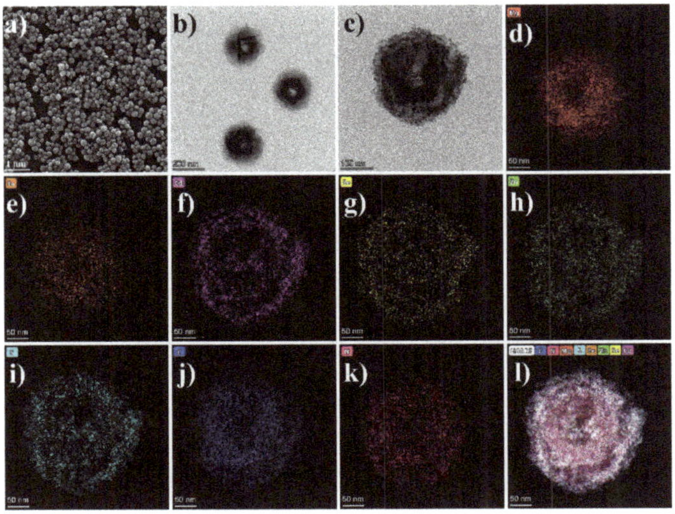

Figure 2. (**a**,**b**) Field emission scanning electron microscope (SEM) and transmission electron microscope (TEM) images of Mg/Fe LDH NFs. (**c**) TEM images of MF@CdSe/ZnS@PEI NFs. (**d–l**) EDS elemental mapping images of MF@CdSe/ZnS@PEI NFs.

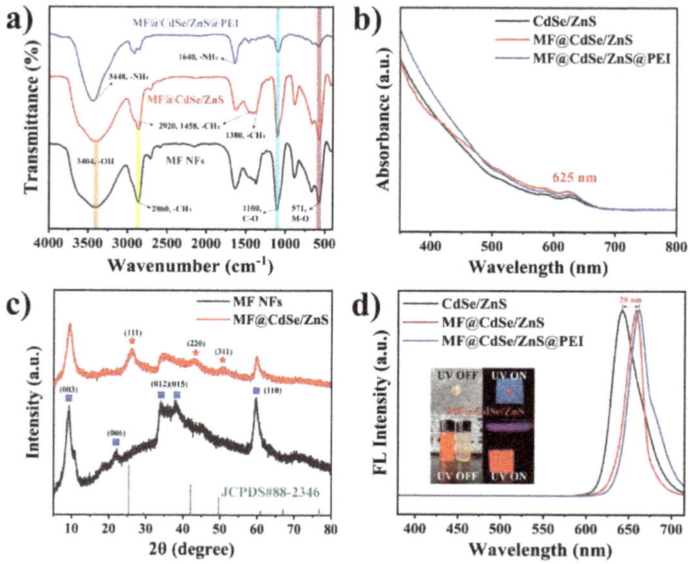

Figure 3. (**a**) Fourier transform-infrared spectra of Mg/Fe LDH NFs, MF@CdSe/ZnS NFs, and MF@CdSe/ZnS@PEI NFs. (**b**) UV-vis absorption spectra of hydrophobic CdSe/ZnS QDs, MF@CdSe/ZnS NFs, and MF@CdSe/ZnS@PEI NFs. (**c**) XRD patterns of Mg/Fe LDH NFs and MF@CdSe/ZnS NFs. (**d**) Fluorescence spectra of hydrophobic CdSe/ZnS QDs, MF@CdSe/ZnS, and MF@CdSe/ZnS@PEI NFs. (Inset is the fluorescence photo of MF@CdSe/ZnS@PEI NFs solid and dispersed in water).

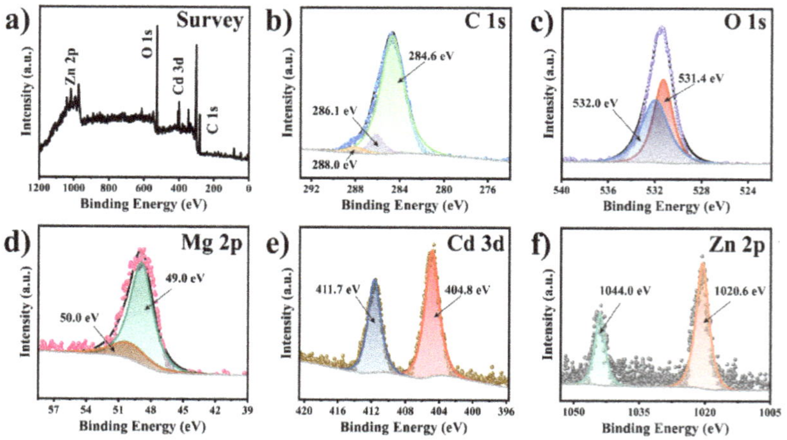

Figure 4. (**a**) XPS survey spectrum of MF@CdSe/ZnS NFs. (**b**–**f**) High-resolution XPS spectra of C 1s, O 1s, Mg 2p, Cd 3d, and Zn 2p.

3.2. *Performance of MF@CdSe/ZnS@PEI-ICTS for HCG Detection*

Before using MF@CdSe/ZnS@PEI-ICTS to detect different concentrations of HCG standards, optimizing the concentration of branched PEI coated with MF@CdSe/ZnS is critical to achieve the detection. It can be seen from Figure S6 that when the concentration exceeds 0.5 mg/mL, it will cause the MF@CdSe/ZnS NFs to exhibit mutual adhesion. The branched PEI concentration used in this work was finally determined to be 0.1 mg/mL. MF@CdSe/ZnS@PEI-ICTS is based on the independent immunoreactions of T and C probes to realize the fluorescence detection of HCG with different concentration gradients. The capture antibody Ab_1, which specifically reacts with the antigen HCG, is painted on the T-line, and goat anti-mouse immunoglobulin is painted on the C-line. The reaction on the T-line is based on a probe-coupled antibody-antigen-coated antibody double-antibody sandwich immunoreaction (i.e., Ab_2-MF@CdSe/ZnS@PEI-HCG-Ab_1), and the reaction on the C-line is based on direct immunoreactivity with probe-conjugated secondary antibody and goat anti-mouse immunoglobulin (i.e., Ab_2-MF@CdSe/ZnS@PEI-IgG). When Ab_2-MF@CdSe/ZnS@PEI was mixed with HCG samples of different concentrations and then dropped on the sample pad, the droplets were chromatographed on the NC membrane under capillary action. If there is HCG in the solution, Ab_2-MF@CdSe/ZnS@PEI-HCG will react with Ab_1 to form Ab_2-MF@CdSe/ZnS@PEI-HCG-Ab_1. When the droplet continued to flow to the C-line, Ab_2-MF@CdSe/ZnS@PEI that was not bound to Ab_1 would have a direct immune reaction with goat anti-mouse IgG to form Ab_2-MF@CdSe/ZnS@PEI-IgG. The qualitative results of MF@CdSe/ZnS@PEI-ICTS were obtained based on the visual observation of the line color on the NC membrane under a 365 nm UV lamp. After about 15 min, as shown in Figure 5a, the NC membranes of all test strips can display the C-line normally, which proves that the detection results are valid. Visual inspection of the test strips under UV light gives a limit of detection (LOD) of approximately 0.1 mIU/mL. At the same time, with the increase in the concentration of HCG standard solution, the red fluorescence color of the T-line on the NC membranes gradually deepened, which means that more Ab_2-MF@CdSe/ZnS@PEI was captured on the T-line. With the assistance of ImageJ software version 1.50d, we can obtain the grayscale signal intensities of the detection and control lines on the NC membrane. To compensate for potential intensity variations caused by acquisition conditions such as lighting and camera settings, background subtraction is required, and the ratio of T-line intensity to C-line intensity (T/C) is used in this work to quantify the signal (Figure S7). First, the RGB image taken by the phone is converted to an 8-bit grayscale image type and inverted. Then, the rectangular box tool is used to measure the average intensity of the T-line, C-line, and background areas. Subtract the average intensity of the background area from the average intensity of the T- and C-lines, take the T/C value as the ordinate, and take the HCG concentration as the abscissa, and a linear relationship between the two can be fitted (Figure 5b). The linear regression equation is $y = 0.0957 \times -0.139$ ($R^2 = 0.98$). In addition, we also verified the potential of MF@CdSe/ZnS@PEI-ICTS in clinical application by HCG standard sample addition and recovery experiments. Healthy human serum was spiked with different concentrations of HCG (10, 50, and 100 mIU/mL), and the experiment was repeated three times on test strips. We selected a batch of serum from healthy people, added different concentrations of HCG standard samples (i.e., 10, 50, and 100 mIU/mL) to them, and after incubation with Ab_2-MF@CdSe/ZnS@PEI, applied them to MF@CdSe/ZnS@PEI-ICTS to detect the recovery rate of standard addition. As shown in Table S1, the spiked recoveries were between 90.48 and 116.1%, and the coefficient of variation was below 15%, which indicated that MF@CdSe/ZnS@PEI-ICTS has the potential to be used in the analysis of clinical samples [50]. In addition, we list and compare the analytical performance of different HCG detection methods in Table S2. Compared with electrochemical and electrochemiluminescence methods, the sensitivity of MF@CdSe/ZnS@PEI-ICTS appears to be poor. However, as a POCT tool, MF@CdSe/ZnS@PEI-ICTS still has the advantages of being fast and convenient, and it has a wider linear range and a lower detection limit than similar testing methods. Finally, we selected four biomarkers (AFP, CA125, CA199, and CEA) at a concentration of 1 μg/mL to evaluate and validate the cross-reactivity of MF@CdSe/ZnS@PEI-ICTS. The results are shown in Figure S8. Except for HCG, other biomarkers could not generate corresponding red fluorescence on the T-line, which proves that MF@CdSe/ZnS@PEI-ICTS has good selectivity and low cross-reactivity.

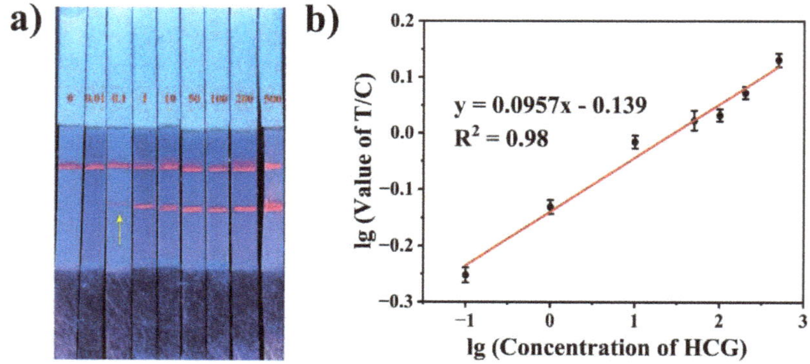

Figure 5. (**a**) Image of the MF@CdSe/ZnS@PEI-ICTS when detecting HCG standard solution (0, 0.01, 0.1, 1, 10, 50, 100, 200, and 500 mIU/mL). (**b**) Linear response of MF@CdSe/ZnS@PEI-ICTS for detection of HCG.

4. Conclusions

In conclusion, a novel fluorescent signal reporter that can be used in ICTS was prepared by implanting hydrophobic CdSe/ZnS QDs into Mg/Fe LDH nanoflowers in one step through the metal-thiol covalent bonds in the organic phase. By modifying the MF@ CdSe/ZnS with branched PEI, which not only improved its dispersion in the buffer but also introduced amino functional groups on the surface of the reporter to facilitate subsequent antibody conjugation. Under optimal experimental conditions, HCG was used as the model analyte with a linear detection range from 0.1 to 500 mIU/mL and a visual limit of detection of 0.1 mIU/mL. Compared with commercial pregnancy diagnostic strips, the addition of high-density hydrophobic CdSe/ZnS QDs greatly improved the detection sensitivity of HCG. The potential for practical application was validated by detecting HCG in spiked healthy human serum, showing overall recoveries of between 90.48 and 116.1% with coefficients of variation between 3.66 and 12.91%. Moreover, the experimental results found that MF@CdSe/ZnS@PEI-ICTS has good selectivity and little cross-reactivity with AFP, CA125, CA199, and CEA. This work preliminarily explores the possibility of Mg/Fe LDH NFs as signal carriers in ICTS, which greatly improves the practicality of ICTS.

Supplementary Materials: The following supporting information can be downloaded at: https://www.mdpi.com/article/10.3390/chemosensors11020114/s1, Figure S1: TEM images of oil-soluble, red-emitting CdSe/ZnS QDs. Figure S2: FT-IR spectra of oil-soluble, red-emitting CdSe/ZnS QDs. Figure S3: Solid-state ^{13}C CP/MAS NMR spectrum of SH-MF NFs. Figure S4: ζ-potential results of MF NFs, hydrophobic CdSe/ZnS QDs, MF@CdSe/ZnS, and MF@CdSe/ZnS@PEI NFs. Figure S5: XPS spectra of (a) C 1s, (b) Mg 2p, (c) Fe 2p, and (d) O 1s of MF NFs. Figure S6: MF@CdSe/ZnS NFs coated with different concentrations of branched PEI. Figure S7. Image processing procedures of MF@CdSe/ZnS@PEI-ICTS. Step 1: convert the photo taken by the mobile phone into an 8-bit grayscale image. Step 2: obtain the inverse image of the grayscale image by using the Invert option. Step 3: define the average intensity of the T-line, C-line, and background regions in the inverse image. Figure S8: Cross-reactivity of MF@CdSe/ZnS@PEI-ICTS for different biomarkers. Table S1: Recovery efficiency and coefficient of variation (CV) for the MF@CdSe/ZnS@PEI-ICTS for HCG (mIU/mL) to be detected in human serum samples (n = 3). Table S2. Summary of HCG detection with some different detection methods [51–56].

Author Contributions: Methodology, investigation, writing—original draft, writing—review and editing, visualization, H.L.; data curation and methodology, Q.Z.; methodology, N.B.; conceptualization, methodology, resources, funding acquisition, writing—review and editing, supervision, S.-N.D. All authors have read and agreed to the published version of the manuscript.

Funding: This work was supported by the National Key Research and Development Program of China (2017YFA0700404) and the National Natural Science Foundation of China (22174015).

Informed Consent Statement: Not applicable.

Data Availability Statement: Not applicable.

Conflicts of Interest: The authors declare no conflict of interest.

References

1. Wu, L.; Qu, X. Cancer biomarker detection: Recent achievements and challenges. *Chem. Soc. Rev.* **2015**, *44*, 2963–2997. [CrossRef] [PubMed]
2. Mahmoudi, T.; de la Guardia, M.; Baradaran, B. Lateral flow assays towards point-of-care cancer detection: A review of current progress and future trends. *TrAC Trends Anal. Chem.* **2020**, *125*, 115842. [CrossRef]
3. Plotz, C.M.; Singer, J.M. The latex fixation test: II. Results in rheumatoid arthritis. *Am. J. Med.* **1956**, *21*, 893–896. [CrossRef]
4. Brucker, M.C.; Macmullen, N.J. What's New in Pregnancy Tests. *J. Obstet. Gynecol. Neonatal Nurs.* **1985**, *14*, 353–359. [CrossRef] [PubMed]
5. Fang, C.-C.; Chou, C.-C.; Yang, Y.-Q.; Wei-Kai, T.; Wang, Y.-T.; Chan, Y.-H. Multiplexed Detection of Tumor Markers with Multicolor Polymer Dot-Based Immunochromatography Test Strip. *Anal. Chem.* **2018**, *90*, 2134–2140. [CrossRef] [PubMed]
6. He, H.; Liu, B.; Wen, S.; Liao, J.; Lin, G.; Zhou, J.; Jin, D. Quantitative Lateral Flow Strip Sensor Using Highly Doped Upconversion Nanoparticles. *Anal. Chem.* **2018**, *90*, 12356–12360. [CrossRef] [PubMed]
7. Liu, X.; Wang, K.; Cao, B.; Shen, L.; Ke, X.; Cui, D.; Zhong, C.; Li, W. Multifunctional Nano-Sunflowers with Color-Magnetic-Raman Properties for Multimodal Lateral Flow Immunoassay. *Anal. Chem.* **2021**, *93*, 3626–3634. [CrossRef] [PubMed]
8. Liu, G.; Lin, Y.-Y.; Wang, J.; Wu, H.; Wai, C.M.; Lin, Y. Disposable Electrochemical Immunosensor Diagnosis Device Based on Nanoparticle Probe and Immunochromatographic Strip. *Anal. Chem.* **2007**, *79*, 7644–7653. [CrossRef] [PubMed]
9. Jia, J.; Ao, L.; Luo, Y.; Liao, T.; Huang, L.; Zhuo, D.; Jiang, C.; Wang, J.; Hu, J. Quantum dots assembly enhanced and dual-antigen sandwich structured lateral flow immunoassay of SARS-CoV-2 antibody with simultaneously high sensitivity and specificity. *Biosens. Bioelectron.* **2022**, *198*, 113810. [CrossRef] [PubMed]
10. Huang, L.; Liao, T.; Wang, J.; Ao, L.; Su, W.; Hu, J. Brilliant Pitaya-Type Silica Colloids with Central–Radial and High-Density Quantum Dots Incorporation for Ultrasensitive Fluorescence Immunoassays. *Adv. Funct. Mater.* **2018**, *28*, 1705380. [CrossRef]
11. Wang, T.; Zhuang, J.; Lynch, J.; Chen, O.; Wang, Z.; Wang, X.; LaMontagne, D.; Wu, H.; Wang, Z.; Cao, Y.C. Self-Assembled Colloidal Superparticles from Nanorods. *Science* **2012**, *338*, 358–363. [CrossRef] [PubMed]
12. Xia, Y.; Nguyen, T.D.; Yang, M.; Lee, B.; Santos, A.; Podsiadlo, P.; Tang, Z.; Glotzer, S.C.; Kotov, N.A. Self-assembly of self-limiting monodisperse supraparticles from polydisperse nanoparticles. *Nat. Nanotechnol.* **2011**, *6*, 580–587. [CrossRef]
13. Yi, C.; Zhang, S.; Webb, K.T.; Nie, Z. Anisotropic Self-Assembly of Hairy Inorganic Nanoparticles. *Acc. Chem. Res.* **2017**, *50*, 12–21. [CrossRef]
14. Foubert, A.; Beloglazova, N.V.; De Saeger, S. Comparative study of colloidal gold and quantum dots as labels for multiplex screening tests for multi-mycotoxin detection. *Anal. Chim. Acta* **2017**, *955*, 48–57. [CrossRef]
15. Deng, H.; Liu, Q.; Wang, X.; Huang, R.; Liu, H.; Lin, Q.; Zhou, X.; Xing, D. Quantum dots-labeled strip biosensor for rapid and sensitive detection of microRNA based on target-recycled nonenzymatic amplification strategy. *Biosens. Bioelectron.* **2017**, *87*, 931–940. [CrossRef] [PubMed]
16. Gong, X.; Cai, J.; Zhang, B.; Zhao, Q.; Piao, J.; Peng, W.; Gao, W.; Zhou, D.; Zhao, M.; Chang, J. A review of fluorescent signal-based lateral flow immunochromatographic strips. *J. Mater. Chem. B* **2017**, *5*, 5079–5091. [CrossRef] [PubMed]
17. Yang, H.; Li, D.; He, R.; Guo, Q.; Wang, K.; Zhang, X.; Huang, P.; Cui, D. A Novel Quantum Dots–Based Point of Care Test for Syphilis. *Nanoscale Res. Lett.* **2010**, *5*, 875. [CrossRef] [PubMed]
18. Hu, J.; Zhang, Z.-L.; Wen, C.-Y.; Tang, M.; Wu, L.-L.; Liu, C.; Zhu, L.; Pang, D.-W. Sensitive and Quantitative Detection of C-Reaction Protein Based on Immunofluorescent Nanospheres Coupled with Lateral Flow Test Strip. *Anal. Chem.* **2016**, *88*, 6577–6584. [CrossRef]
19. Gao, F.; Lei, C.; Liu, Y.; Song, H.; Kong, Y.; Wan, J.; Yu, C. Rational Design of Dendritic Mesoporous Silica Nanoparticles' Surface Chemistry for Quantum Dot Enrichment and an Ultrasensitive Lateral Flow Immunoassay. *ACS Appl. Mater. Interfaces* **2021**, *13*, 21507–21515. [CrossRef] [PubMed]
20. Li, X.; Li, W.; Yang, Q.; Gong, X.; Guo, W.; Dong, C.; Liu, J.; Xuan, L.; Chang, J. Rapid and Quantitative Detection of Prostate Specific Antigen with a Quantum Dot Nanobeads-Based Immunochromatography Test Strip. *ACS Appl. Mater. Interfaces* **2014**, *6*, 6406–6414. [CrossRef] [PubMed]
21. Duan, H.; Huang, X.; Shao, Y.; Zheng, L.; Guo, L.; Xiong, Y. Size-Dependent Immunochromatographic Assay with Quantum Dot Nanobeads for Sensitive and Quantitative Detection of Ochratoxin A in Corn. *Anal. Chem.* **2017**, *89*, 7062–7068. [CrossRef] [PubMed]
22. Joumaa, N.; Lansalot, M.; Théretz, A.; Elaissari, A.; Sukhanova, A.; Artemyev, M.; Nabiev, I.; Cohen, J.H.M. Synthesis of Quantum Dot-Tagged Submicrometer Polystyrene Particles by Miniemulsion Polymerization. *Langmuir* **2006**, *22*, 1810–1816. [CrossRef]
23. Wang, C.; Yang, X.; Gu, B.; Liu, H.; Zhou, Z.; Shi, L.; Cheng, X.; Wang, S. Sensitive and Simultaneous Detection of SARS-CoV-2-Specific IgM/IgG Using Lateral Flow Immunoassay Based on Dual-Mode Quantum Dot Nanobeads. *Anal. Chem.* **2020**, *92*, 15542–15549. [CrossRef] [PubMed]
24. Bai, F.; Wang, D.; Huo, Z.; Chen, W.; Liu, L.; Liang, X.; Chen, C.; Wang, X.; Peng, Q.; Li, Y. A Versatile Bottom-up Assembly Approach to Colloidal Spheres from Nanocrystals. *Angew. Chem. Int. Ed.* **2007**, *46*, 6650–6653. [CrossRef]

25. Zhuang, J.; Wu, H.; Yang, Y.; Cao, Y.C. Controlling Colloidal Superparticle Growth Through Solvophobic Interactions. *Angew. Chem. Int. Ed.* **2008**, *47*, 2208–2212. [CrossRef]
26. Gao, X.; Nie, S. Doping Mesoporous Materials with Multicolor Quantum Dots. *J. Phys. Chem. B* **2003**, *107*, 11575–11578. [CrossRef]
27. Yang, F.; Skripka, A.; Tabatabaei, M.S.; Hong, S.H.; Ren, F.; Huang, Y.; Oh, J.K.; Martel, S.; Liu, X.; Vetrone, F.; et al. Magnetic Photoluminescent Nanoplatform Built from Large-Pore Mesoporous Silica. *Chem. Mater.* **2019**, *31*, 3201–3210. [CrossRef]
28. Huang, L.; Wu, Q.; Wang, J.; Foda, M.; Liu, J.; Cai, K.; Han, H. A brilliant sandwich type fluorescent nanostructure incorporating a compact quantum dot layer and versatile silica substrates. *Chem. Commun.* **2014**, *50*, 2896–2899. [CrossRef]
29. Lu, Z.; Gao, C.; Zhang, Q.; Chi, M.; Howe, J.Y.; Yin, Y. Direct Assembly of Hydrophobic Nanoparticles to Multifunctional Structures. *Nano Lett.* **2011**, *11*, 3404–3412. [CrossRef]
30. Wu, Q.; Chen, L.; Huang, L.; Wang, J.; Liu, J.; Hu, C.; Han, H. Quantum dots decorated gold nanorod as fluorescent-plasmonic dual-modal contrasts agent for cancer imaging. *Biosens. Bioelectron.* **2015**, *74*, 16–23. [CrossRef]
31. Wang, Q.; O'Hare, D. Recent Advances in the Synthesis and Application of Layered Double Hydroxide (LDH) Nanosheets. *Chem. Rev.* **2012**, *112*, 4124–4155. [CrossRef] [PubMed]
32. Song, L.; Shi, W.; Lu, C. Confinement Effect in Layered Double Hydroxide Nanoreactor: Improved Optical Sensing Selectivity. *Anal. Chem.* **2016**, *88*, 8188–8193. [CrossRef]
33. Yu, Y.; Lu, C.; Zhang, M. Gold Nanoclusters@Ru(bpy)3^{2+}-Layered Double Hydroxide Ultrathin Film as a Cathodic Electrochemiluminescence Resonance Energy Transfer Probe. *Anal. Chem.* **2015**, *87*, 8026–8032. [CrossRef] [PubMed]
34. Zong, Y.; Li, K.; Tian, R.; Lin, Y.; Lu, C. Highly dispersed layered double oxide hollow spheres with sufficient active sites for adsorption of methyl blue. *Nanoscale* **2018**, *10*, 23191–23197. [CrossRef]
35. Mubarak, M.; Islam, M.S.; Yoon, D.-Y.; Lee, J.H.; Park, H.J.; Bae, J.-S.; Lee, H.-J. Flower-like Mg/Fe-layered double oxide nanospheres with ultrahigh adsorption efficiency for anionic organic dyes. *Colloids Surf. A Physicochem. Eng. Asp.* **2021**, *618*, 126446. [CrossRef]
36. Mubarak, M.; Jeon, H.; Islam, M.S.; Yoon, C.; Bae, J.-S.; Hwang, S.-J.; Choi, W.S.; Lee, H.-J. One-pot synthesis of layered double hydroxide hollow nanospheres with ultrafast removal efficiency for heavy metal ions and organic contaminants. *Chemosphere* **2018**, *201*, 676–686. [CrossRef] [PubMed]
37. Yang, Q.; Gong, X.; Song, T.; Yang, J.; Zhu, S.; Li, Y.; Cui, Y.; Li, Y.; Zhang, B.; Chang, J. Quantum dot-based immunochromatography test strip for rapid, quantitative and sensitive detection of alpha fetoprotein. *Biosens. Bioelectron.* **2011**, *30*, 145–150. [CrossRef]
38. Chen, Z.; Liang, R.; Guo, X.; Liang, J.; Deng, Q.; Li, M.; An, T.; Liu, T.; Wu, Y. Simultaneous quantitation of cytokeratin-19 fragment and carcinoembryonic antigen in human serum via quantum dot-doped nanoparticles. *Biosens. Bioelectron.* **2017**, *91*, 60–65. [CrossRef]
39. Cho, J.; Jung, Y.K.; Lee, J.-K. Kinetic studies on the formation of various II–VI semiconductor nanocrystals and synthesis of gradient alloy quantum dots emitting in the entire visible range. *J. Mater. Chem.* **2012**, *22*, 10827–10833. [CrossRef]
40. Xu, L.-D.; Zhang, Q.; Ding, S.-N.; Xu, J.-J.; Chen, H.-Y. Ultrasensitive Detection of Severe Fever with Thrombocytopenia Syndrome Virus Based on Immunofluorescent Carbon Dots/SiO_2 Nanosphere-Based Lateral Flow Assay. *ACS Omega* **2019**, *4*, 21431–21438. [CrossRef] [PubMed]
41. Xu, L.D.; Zhu, J.; Ding, S.N. Immunoassay of SARS-CoV-2 nucleocapsid proteins using novel red emission-enhanced carbon dot-based silica spheres. *Analyst* **2021**, *146*, 5055–5060. [CrossRef]
42. Kim, M.; Kim, M.S.; Kweon, S.H.; Jeong, S.; Kang, M.H.; Kim, M.I.; Lee, J.; Doh, J. Simple and Sensitive Point-of-Care Bioassay System Based on Hierarchically Structured Enzyme-Mimetic Nanoparticles. *Adv. Healthc. Mater.* **2015**, *4*, 1311–1316. [CrossRef]
43. Kim, M.S.; Kweon, S.H.; Cho, S.; An, S.S.A.; Kim, M.I.; Doh, J.; Lee, J. Pt-Decorated Magnetic Nanozymes for Facile and Sensitive Point-of-Care Bioassay. *ACS Appl. Mater. Interfaces* **2017**, *9*, 35133–35140. [CrossRef]
44. Huang, P.; Liu, J.; Wei, F.; Zhu, Y.; Wang, X.; Cao, C.; Song, W. Size-selective adsorption of anionic dyes induced by the layer space in layered double hydroxide hollow microspheres. *Mater. Chem. Front.* **2017**, *1*, 1550–1555. [CrossRef]
45. Géraud, E.; Rafqah, S.; Sarakha, M.; Forano, C.; Prevot, V.; Leroux, F. Three Dimensionally Ordered Macroporous Layered Double Hydroxides: Preparation by Templated Impregnation/Coprecipitation and Pattern Stability upon Calcination. *Chem. Mater.* **2008**, *20*, 1116–1125. [CrossRef]
46. Wang, B.; Wu, H.; Yu, L.; Xu, R.; Lim, T.-T.; Lou, X.W. Template-free Formation of Uniform Urchin-like α-FeOOH Hollow Spheres with Superior Capability for Water Treatment. *Adv. Mater.* **2012**, *24*, 1111–1116. [CrossRef]
47. Kim, K.M.; Jeon, J.H.; Kim, Y.Y.; Lee, H.K.; Park, O.O.; Wang, D.H. Effects of ligand exchanged CdSe quantum dot interlayer for inverted organic solar cells. *Org. Electron.* **2015**, *25*, 44–49. [CrossRef]
48. Granados-Oliveros, G.; Pineros, B.S.G.; Calderon, F.G.O. CdSe/ZnS quantum dots capped with oleic acid and L-glutathione: Structural properties and application in detection of Hg^{2+}. *J. Mol. Struct.* **2022**, *1254*, 132293. [CrossRef]
49. Cingarapu, S.; Yang, Z.; Sorensen, C.M.; Klabunde, K.J. Synthesis of CdSe/ZnS and CdTe/ZnS Quantum Dots: Refined Digestive Ripening. *J. Nanomater.* **2012**, *2012*, 312087. [CrossRef]
50. Jin, Y.; Chen, Q.; Luo, S.; He, L.; Fan, R.; Zhang, S.; Yang, C.; Chen, Y. Dual near-infrared fluorescence-based lateral flow immunosensor for the detection of zearalenone and deoxynivalenol in maize. *Food Chem.* **2021**, *336*, 127718. [CrossRef]
51. Zhang, A.; Guo, W.; Ke, H.; Zhang, X.; Zhang, H.; Huang, C.; Yang, D.; Jia, N.; Cui, D. Sandwich-format ECL immunosensor based on Au star@BSA-Luminol nanocomposites for determination of human chorionic gonadotropin. *Biosens. Bioelectron.* **2018**, *101*, 219–226. [CrossRef] [PubMed]

52. Hong, G.; Zhang, D.; He, Y.; Yang, Y.; Chen, P.; Yang, H.; Zhou, Z.; Liu, Y.; Wang, Y. New photothermal immunoassay of human chorionic gonadotropin using Prussian blue nanoparticle-based photothermal conversion. *Anal. Bioanal. Chem.* **2019**, *411*, 6837–6845. [CrossRef]
53. Zhang, T.; Wang, H.-B.; Zhong, Z.-T.; Li, C.-Q.; Chen, W.; Liu, B.; Zhao, Y.-D. A smartphone-based rapid quantitative detection platform for lateral flow strip of human chorionic gonadotropin with optimized image algorithm. *Microchem. J.* **2020**, *157*, 105038. [CrossRef]
54. Wang, X.; Xue, C.-H.; Yang, D.; Jia, S.-T.; Ding, Y.-R.; Lei, L.; Gao, K.-Y.; Jia, T.-T. Modification of a nitrocellulose membrane with nanofibers for sensitivity enhancement in lateral flow test strips. *RSC Adv.* **2021**, *11*, 26493–26501. [CrossRef]
55. Danthanarayana, A.N.; Finley, E.; Vu, B.; Kourentzi, K.; Willson, R.C.; Brgoch, J. A multicolor multiplex lateral flow assay for high-sensitivity analyte detection using persistent luminescent nanophosphors. *Anal. Methods* **2020**, *12*, 272–280. [CrossRef]
56. Liang, H.; Ning, G.; Wang, L.; Li, C.; Zheng, J.; Zeng, J.; Zhao, H.; Li, C.-P. Covalent Framework Particles Modified with MnO_2 Nanosheets and Au Nanoparticles as Electrochemical Immunosensors for Human Chorionic Gonadotropin. *ACS Appl. Nano Mater.* **2021**, *4*, 4593–4601. [CrossRef]

chemosensors

Article

An Aptamer Biosensing Strategy for Label-Free Assay of Dual Acute Myocardial Infarction Biomarkers Built upon AuNPs/Ti_3C_2-MXenes

Xiaona Mi [1,2], Huiling Li [3] and Yifeng Tu [2,*]

1 College of Materials Science and Engineering, Yancheng Institute of Technology, Yancheng 224051, China
2 College of Chemistry, Chemical Engineering and Material Science, Soochow University, Suzhou 215123, China
3 The First Affiliated Hospital, Nursing College, Soochow University, Suzhou 215006, China
* Correspondence: tuyf@suda.edu.cn

Abstract: The sensitive quantification of cardiac troponin I (cTnI) and myoglobin (Myo) in blood is essential for an early emergency diagnosis of acute myocardial infarction (AMI). Attributed to AuNPs and a titanium element on the surface of the AuNPs/Ti_3C_2-MXenes hybrid, each respective aptamer strand can be immobilized on. In this work, a nanohybrid was deposited on amino-functionalized indium tin oxide (ITO) via an Au–N bond; thereafter, it could catch cTnI-specific, thiol-functionalized DNA aptamer through Au–S self-assembly or Myo-aptamer via adsorption and metal chelate interaction between phosphate groups and titanium for specific recognition. Both using $[Fe(CN)_6]^{3-/4-}$ as a signaling probe, the differential pulse voltammetric (DPV) current of the cTnI-aptasensor decreased after binding with cTnI, while the other responded to Myo via the impedimetric measurement. These developed biosensors enable the response to the femtogram/mL level cTnI or nanogram/mL level Myo. Remarkably, the proposed aptasensors exhibit high sensitivity and specificity for targets and display great potential for applications in clinic diagnosis.

Keywords: electrochemical biosensor; aptamer; AuNPs/Ti_3C_2-MXenes; cardiac troponin I; myoglobin; acute myocardial infarction

Citation: Mi, X.; Li, H.; Tu, Y. An Aptamer Biosensing Strategy for Label-Free Assay of Dual Acute Myocardial Infarction Biomarkers Built upon AuNPs/Ti_3C_2-MXenes. *Chemosensors* **2023**, *11*, 157. https://doi.org/10.3390/chemosensors11030157

Academic Editor: Manel del Valle

Received: 20 January 2023
Revised: 16 February 2023
Accepted: 21 February 2023
Published: 24 February 2023

1. Introduction

With high fatal risk, acute myocardial infarction (AMI) seriously threatens human lives as one of the most severe cardiovascular emergencies [1,2]. The early diagnosis of AMI based on the detection of cardiac biomarkers in blood is critical to guarantee the patient's survival. Clinical cardiological studies have revealed that Cardiac Troponin I (cTnI) is a gold standard biomarker due to its high cardiac specificity [3]. Compared to a normal blood level (0 to 0.01 ng/mL), it rose significantly in peripheral blood in 3–4 h after the start of an AMI attack and peaked at 12–14 h. Here, 0.01–0.04 ng/mL indicates minor myocarditis, and over 0.06 ng/mL suggests myocardial necrotic damage [4]. However, one issue is that the cTnI level is usually very low within first 2 h of the onset of symptoms and impedes the early diagnosis of AMI [5]. Recently, the multiplexed cardiac biomarker assay used in the early diagnosis of AMI has been proposed. Myoglobin (Myo) has received much attention because it is quickly released into the blood within 1 h of AMI [6]. Its concentration would drastically increase by up to 600 ng/mL within 4–5 h, while its normal range is 6–100 ng/mL [7]. Therefore, the consolidated assay of cTnI and Myo can promote responsivity in the early diagnosis of AMI [8]. In view of the nanogram-level concentrations of cTnI and Myo in normal blood, the requirement of high sensitivity in the detection of cTnI and Myo seems to be imperative.

The typical clinical practice method for their detection in blood is usually the enzyme-linked immunosorbent assay (ELISA) [1]; however, this method is expensive, laborious, and time-consuming. More methods such as surface-enhanced Raman spectroscopy [9], fluorescence [10], or electrochemistry [8,11] have been developed as alternatives in recent years, including various sensing methods, and electrochemical aptasensors have gained significant attention. The electrochemical techniques possess natural advantages originated from their uniqueness in principle. They generally have high sensitivity, and the response only occurs on the tiny electrode surface; therefore, only a very small sample and reagent consumption is needed, and thus the detection cost is extremely low. Moreover, electrochemical testing only takes a few seconds to some minutes, which is faster than other methods. Based on the high specificity of aptamer, the developed electrochemical sensor is basically free from the interference of coexisting substances, therefore various pretreating procedures can be ignored to directly determine the sample, which greatly simplifies the operation.

Ti_3C_2-MXene nanosheets have been employed to improve the sensitivity of electrochemical biosensors, using their assimilating metallic conductivity and excellent catalytic ability to amplify the electrical signal [12–16]. Chen et al. found that the immobilized tyrosinase on Ti_3C_2-MXene nanosheets could support a mediator-free electrochemical biosensor for an ultrasensitive and rapid detection of phenol [17]. Due to abundant transition metal (e.g., titanium) on the basal plane of Ti_3C_2-MXenes, it provides the opportunity to interact with ssDNA strands via chelation, making it a unique nano-bio-interface for biosensor construction [18]. Li et al. demonstrated that Ti_3C_2-MXene nanosheets can catch tetrahedral DNA nanostructures (TDNs) onto its surface through the coordination between titanium and phosphate groups of DNA for the highly sensitive detection of gliotoxin [19]. Recently, one report showed a more enhanced electrocatalytic activity of gold decorated Ti_3C_2-MXene nanosheets (AuNPs/Ti_3C_2-MXenes) than itself [20]. This hybrid is more likely to be a robust matrix for sensitive electrochemical detection.

The synthesized nanohybrid AuNPs/Ti_3C_2-MXenes in our previous work exhibits excellent biocompatibility and electrocatalytical activity. The AuNPs are generated in situ via the reduction of $HAuCl_4$ by the reducibility of activated Ti species on the surface of Ti_3C_2-MXene nanosheets [21]. The anchored AuNPs on the surface of Ti_3C_2-MXene nanosheets ensure the hybrid with improved conductivity. Moreover, the decoration of AuNPs can act as not only a spacer to prevent the pile-up of Ti_3C_2-MXene nanosheets from irreversibly restacking, but also as a connector to fix enough cTnI aptamer through the Au–S bond. Meanwhile, there is abundant titanium atoms on the basal plane of Ti_3C_2-MXene nanosheets which strongly coordinate with phosphate groups of the Myo aptamer. All these components present opportunities to construct biosensors with excellent performance. Based on this, in this work, an ultrasensitive label-free cTnI and Myo aptasensing system is developed using the leveraging of nanohybrids as the matrix for anchoring aptamer strands (Scheme 1). The nanohybrid was deposited on amino-functionalized ITO via an Au–N bond; then, the thiol-functionalized, cTnI-specific aptamer (SH-Apt_{cTnI}) was further immobilized through Au–S self-assembly, as illustrated in Scheme 1A. It is found that, after the specific binding of cTnI, the formed sensing interface showed high sensitive response toward the redox of $[Fe(CN)_6]^{3-/4-}$ as a probe. Thus, the quantitatively decreased DPV signal is observed due to the protein layer, which inhibited the diffusion of the redox probe. Again, the adsorption of ssDNA strands on Ti_3C_2-MXene nanosheets endows its bio-affinity toward a DNA aptamer [19,22] to facilitate the development of an electrochemical aptasensor using the aptamer of Myo (Apt_{Myo}), as illustrated in Scheme 1B. The impedance of the resultant sensor diminished quickly in the presence of Myo, since the strong affinity of the specifically recognized Myo with an aptamer led to the formation of a rigid structure, resulting in its release from the sensor surface. The change in impedance bears the responsibility of the aptasensor's response to Myo. This bifunctional label-free electrochemical sensing strategy used in coupling the catalytic activity and physicochemical property of AuNPs/Ti_3C_2-MXenes with two aptamers offers a new concept for the accurate measurement of dual cardiac biomarkers in clinical diagnosis.

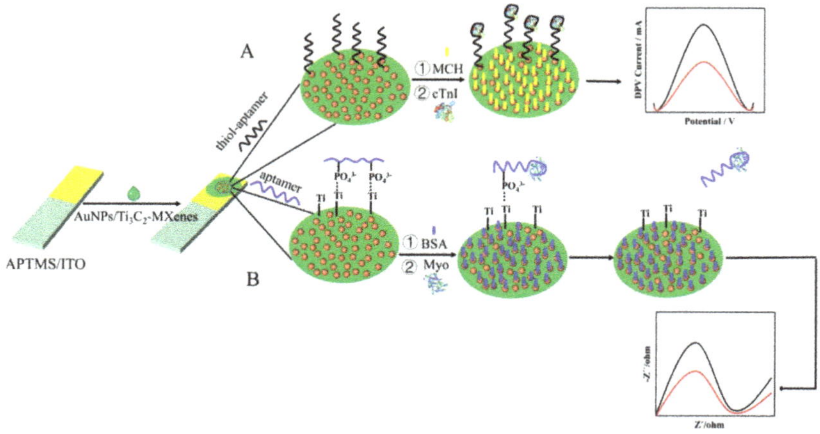

Scheme 1. Diagram of the preparation and operation process of aptasensor for dual cardiac biomarkers based on nano-hybrid functionalized substrate electrode. "A" represents the preparation process of the cTnI aptasensor using DPV current output upon scanning potential (V); "B" represents the preparation process of the Myo aptasensor with AC impedance (Z) as output signal.

2. Materials and Methods

2.1. Materials

Cardiac troponin I (cTnI) was purchased from Abcam Co., Ltd. (Shanghai, China). Cardiac myoglobin (Myo) was purchased from Nanjing Oukai Biotechnology Co., Ltd. (Nanjing, China). ELISA Kit for cTnI and Myo assays were purchased from Dongshang Co., Ltd. (Shanghai, China). Potassium ferricyanide ($K_3[Fe(CN)_6]$, 99.5%), potassium ferrocyanide ($K_4[Fe(CN)_6].3H_2O$, 99.5%), and sodium chloride (NaCl) were purchased from Macklin Biochemical Co., Ltd. (Shanghai, China). BSA was obtained from Shanghai Sangon Biotech Co., Ltd. (Shanghai, China). 6-Mercapto-1-hexanol (MCH) was purchased from Sigma–Aldrich (Shanghai, China). Chloroauric acid ($HAuCl_4{\cdot}3H_2O$) was purchased from Shanghai Sinopharm Chemical Reagent Co., Ltd. (Shanghai, China). Titanium Carbide (Ti_3AlC_2, 98%) was purchased from 11 Technology Co., Ltd. (Jilin, China). Human serum samples were collected in Second Affiliated Hospital of Soochow University (Suzhou, China) and stored under −20 °C. All other chemicals and reagents were of analytical grade. Ultrapure water (resistivity ≥18 MΩ cm^{-2}) was obtained from a Milli-Q purification system and used throughout the experiment. All oligonucleotides were synthesized by Shanghai Sangon Biotech. Co., Ltd. (Shanghai, China). The sequences of cTnI and Myo aptamers are as follows:

cTnI:

5′-SH-(CH2)$_6$-CGTGCAGTACGCCAACCTTTCTCATGCGCTGCCCCTCTTA-3′ [23].

Myo:

5′-CCCTCCTTTCCTTCGACGTAGATCTGCTGCGTTGTTCCGA-3′ [22].

0.1 M PBS solution (pH 7.0) was prepared by the mix of stock solutions of 0.1 M Na_2HPO_4 and 0.1 M NaH_2PO_4 in a specific proportion. The stock aptamer solution of cTnI or Myo was prepared using the aqua solution containing 10 mM Tris-HCl and 50 mM $MgCl_2$ (pH = 7.4). The immobilization of the aptamer was manipulated afterwards to dilute the stock solution with 0.1 M phosphate-buffered saline (PBS) (pH 7.0).

2.2. Apparatus

All electrochemical experiments were carried out using a conventional three-electrode system with an ITO-based working electrode or sensor, a Pt wire counter electrode, and a saturated calomel electrode (SCE) as a reference electrode. Electrochemical impedance spectroscopy (EIS), cyclic voltammetry (CV), and differential pulse voltammetry (DPV) were enforced on an RST-5200 Electrochemical Workstation (Risetest Instruments of Suzhou, Suzhou, China). All the electrochemical tests were carried out in 0.1 M PBS (pH 7.0) containing 5 mM $[Fe(CN)_6]^{3-/4-}$ and 0.1 M NaCl. The CV measurements were implemented in a potential range from −0.2 to 0.6 V with a canning rate of 50 mV/s. The DPV measurements were performed in a potential range from −0.3 to 0.7 V under the following conditions: a pulse amplitude of 100 mV, a pulse width of 20 ms, and a sampling width of 10 ms. EIS measurements were performed in the frequency range of 0.1–10^5 Hz. The atomic force microscopy (AFM) characterization was fulfilled using Dimension Icon Atomic Force Microscopy (Brucker, Germany). Those images were taken in tapping mode using an RTP cantilever (7 nm of tip radius, 2 N/m of elastic coefficient).

2.3. Fabrication of Aptamer/AuNPs/Ti_3C_2-MXenes Aptasensors

The synthesis of the AuNPs/Ti_3C_2-MXene hybrid, pretreatment of ITO glass, and the preparation of amino-functionalized ITO (APTMS/ITO) all can refer to our previous paper [20]. For the fabrication of the aptamer/AuNPs/Ti_3C_2-MXene biosensors, 50 µL of as-prepared AuNPs/Ti_3C_2-MXene hybrid suspension was dropped onto the surface of the APTMS/ITO electrode and left for 30 min at room temperature, which, when followed by washing with ultrapure water and dried under N_2 gas flow, gained a AuNPs/Ti_3C_2-MXenes/APTMS/ITO electrode. Subsequently, 10 µL of aptamer solution was cast onto it before waiting for 12 h at 4 °C to fix the aptamer (Apt/AuNPs/Ti_3C_2-MXenes/APTMS/ITO). Thereafter, for the cTnI aptasensor, 10 µL of 10 mM MCH (in 0.1 M PBS, pH 7.0) solution was necessary to treat the sensor for 30 min to block the unoccupied sites and prevent the entanglement of ssDNA aptamers; moreover, for the Myo aptasensor, 10 µL of 3% BSA (in 0.1 M PBS, pH 7.0) solution was placed on the resulting electrode for 30 min to block the unoccupied sites. Lastly, the obtained aptasensors were stored at 4 °C for further use.

2.4. cTnI and Myo Detection

A total of 10 µL of the sample solution was placed onto the aptasensor and incubated at 25 °C for 60 min before being washed with 0.1 M PBS (pH 7.4). Then, the sensor was imbedded into the electrochemical cell for DPV or EIS measurement.

3. Results and Discussion

3.1. Monitoring the Preparation of Aptasensors

The preparation and characterization of the AuNPs/Ti_3C_2-MXene hybrid, as well as its use to prepare the basal electrode of the aptasensors, have been presented in our previous paper [20]. Here, the interfacial character of the aptasensor in different preparing steps was explored using AFM (Figure 1A). The AFM image of a bare ITO electrode shows a grainy topology (Figure 1Aa) with an average surface roughness (R) of 5.74 nm. Thereafter, it (Figure 1Ab) revealed a smoother surface with a decreased R value of 4.66 nm when the ITO was covered using APTMS. After the AuNPs/Ti_3C_2-MXene hybrid was deposited on, the surface became much rougher (Figure 1Ac), as indicated by a significantly increased R value of 10.30 nm. The anchoring of the aptamer (as thiol–Apt_{cTnI}) induced a larger R value (Figure 1Ad, 15.10 nm), proving that lots of aptamer strands were attached onto the surfaces of the AuNPs/Ti_3C_2-MXenes through the Au–S bond [24]. As illustrated in Figure 1Ae, after treatment with MCH, the R value increased by up to 23.10 nm, indicating a much rougher morphology, which can be attributed to the fact that MCH would regulate the orientation of the ssDNA aptamer, which tended to arrange them vertically [25]. After this, to incubate the cTnI (Figure 1Af) on a sensor, a smoother morphology than the one displayed in Figure 1Ae, with a decreased R value of 18.70 nm, was observed as a result of the connection of globular and high molecular weight proteins [26]. The alteration of the surface roughness of the electrode during the preparation process indicated the successful construction of the aptasensor.

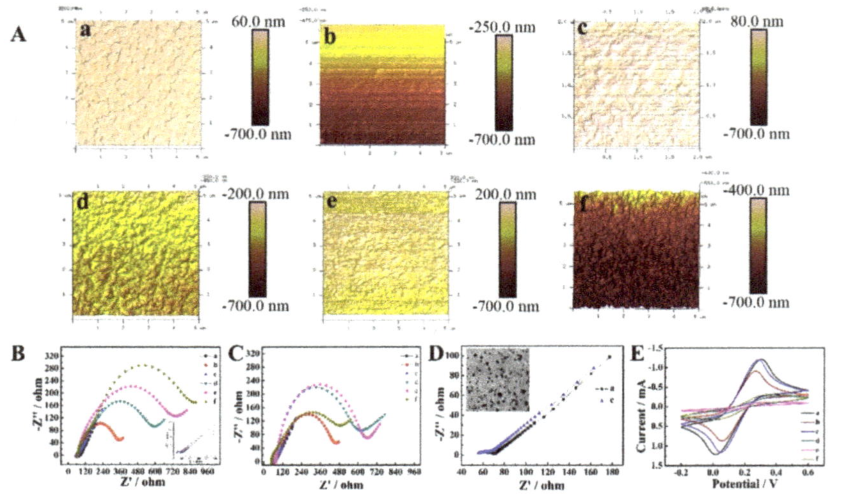

Figure 1. (**A**) Three-dimensional AFM images and (**B**) EIS curves of (**a**) bare ITO, (**b**) APTMS/ITO, (**c**) AuNPs/Ti_3C_2-MXenes/APTMS/ITO, (**d**) after to catch cTnI aptamer, (**e**) final cTnI sensor, and (**f**) responded to cTnI, the inset in (**B**) shows the enlarged EIS curves (**a**) and (**c**); (**C**,**D**) EIS curves and (**E**) CV curves of $[Fe(CN)_6]^{3-/4-}$ on (a) bare ITO, (**b**) APTMS/ITO, (**c**) AuNPs/Ti_3C_2-MXenes/APTMS/ITO, (**d**) after to catch Myo aptamer, (**e**) final Myo sensor, and (**f**) responded to Myo, the inset in (**D**) is the TEM image of AuNPs/Ti_3C_2-MXenes.

Moreover, the EIS and CV can reflect the process of sensor preparation. As shown in Figure 1B–E, the electron transfer resistance (R_{et}, it can be obtained from the equivalent circuit of the semi-circular part of EIS, which represents the impedance of the electrode surface for electron transfer) significantly increased, and the CV current obviously decreased after covering the bare ITO with hydrolyzed APTMS due to its electro-inactive feature [27] (both curves a and b). In contrast, obviously decreased R_{et} and increased CV currents (both curve c) were witnessed after the decoration of the AuNPs/Ti_3C_2-MXenes (the microscopic image of TEM is presented in Figure 1D), suggesting better conductivity and a larger specific surface area. Afterwards, the R_{et} values increased once more when the aptamer (thiol–Apt_{cTnI} or Apt_{Myo}) was fixed (curve d), or when the CV currents decreased in reverse after the successive linkage of the aptamer (as Apt_{Myo}) (curve d). When treating the cTnI sensor with MCH or the Myo sensor with BSA (both curve e), the EIS of the former increased due to the molecular orientation [25], and the latter experienced minor changes; moreover, the CV current of the latter slightly decreased due to the hampered electron transfer [28]. Finally, if the cTnI sensor had been used to detect the target, the readout signal (EIS) would further increase according to the concentration (curve f in Figure 1B); whereas, for the Myo sensor, the EIS decreased (curve f in Figure 1C), and the CV current recovered in degree (curve f in Figure 1E). These results suggest that all components of the aptasensor were successfully integrated and endowed with the desired sensing ability as expected.

3.2. Optimization of the Conditions for Sensor Preparation and Operation

The dosage of the AuNPs/Ti_3C_2-MXene hybrid significantly influences the performance of the aptasensor [7]. As shown in Figure 2A (black curve), the DPV signal is elevated along with the increasing dosage of the AuNPs/Ti_3C_2-MXene hybrid, reaching a maximum of 0.25 mg/mL, when then declined if further enlarged. This implies that overdosing the hybrid on the electrode surface would inversely impede the electron transfer, owing to more serious interface resistance [29]. Using EIS, the dosage of the AuNPs/Ti_3C_2-MXene hybrid can be optimized too, acquired a result of 0.25 mg/mL, which led to a smallest R_{et} (Figure 2A, blue curve) that was totally identical to the results obtained using DPV.

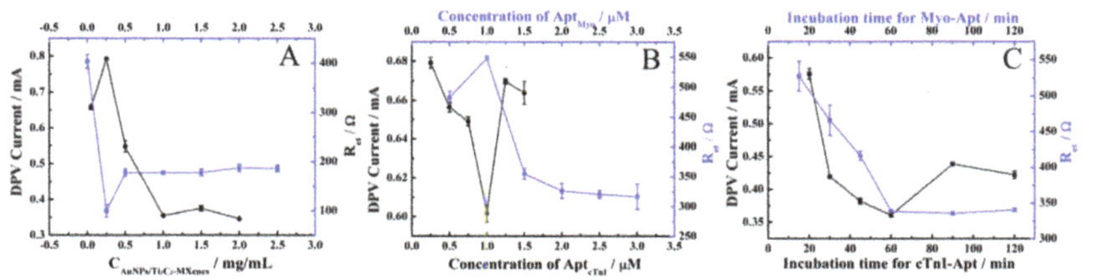

Figure 2. The optimization (**A**) and the content of AuNPs/Ti_3C_2-MXenes; (**B**) the concentration of Apt_{cTnI} or Apt_{Myo}; (**C**) the time for cTnI or Myo incubation.

SH-Apt_{cTnI} is used as a recognitive host for cTnI, and its quantitative use also must be optimized to achieve a maximal sensing outcome. As depicted in Figure 2B (black curve), with the same used volume, the minimal DPV peak current of $[Fe(CN)_6]^{3-/4-}$ was observed when the concentration of SH-Apt_{cTnI} for enabling the sensor was 1.0 μM. This means that this usage quantity is likely to ensure a binding saturation, and we speculate that an excess of SH-Apt_{cTnI} is prone to lead to its detachment, eventually being detrimental to the analytical performance of the resulting sensor [30]. Similarly, the most suitable concentration of the Apt_{Myo} solution for Myo aptasensor construction was also 1.0 μM (Figure 2B blue curve).

Meanwhile, the incubation time of the target of the corresponding aptasensor is also an important factor when aiming to achieve the best sensing performance. As shown in Figure 2C (black curve), a very significant decrease in DPV value occurs after the loading of the cTnI sample solution within an incubation time of 60 min, suggesting that the binding of cTnI on its aptamer reached a full equilibrium during this period [31]. The results indicate that the longer time negatively affected the performance of the aptasensor. Consistently, the optimal incubation time for the Myo aptasensor is also 60 min (Figure 2C blue curve).

3.3. The Analytical Performance of Prepared Aptasensors

Under the optimal conditions, the analytical performance of the proposed aptasensing strategy was validated by detecting cTnI or Myo.

Figure 3A shows that the DPV current on the prepared Apt_{cTnI} sensor gradually decreased along with elevating the cTnI concentration. This is attributed to the insulativity of the formed aptamer–cTnI complex, which obstructed the electron transfer between the electrochemical probe and the AuNPs/Ti_3C_2-MXene substrate electrode [8]. There was a good linear regression of the DPV peak current upon the logarithm of the cTnI concentration from 0.24 fg/mL to 24 ng/mL, as illustrated in Figure 3B, which perfectly covers the clinically relevant range. The linear regression equation is $I = 0.456 - 0.0177logC$ (fg/mL) ($R^2 = 0.999$), with a limit of detection (LOD) of 0.14 fg/mL (S/N = 3). Visibly, the LOD of the prepared Apt_{cTnI} sensor lays well below the normal level of cTnI and can easily identify the myocardial necrotic damage.

Table 1 lists the performance of the proposed aptasensor and offers a comparison between it and other currently available cTnI sensors. The results reveal that the developed aptasensor possesses a higher sensitivity than all other reported ones. This is mainly ascribed to the following aspects: (i) the superior electrochemical activity of the AuNPs/Ti_3C_2-MXenes greatly promotes the electron transfer and amplifies the response; (ii) the prepared AuNPs/Ti_3C_2-MXene hybrid has excellent biocompatibility, which can strengthen aptamer anchoring using the Au–S bond; (iii) SH-Apt_{cTnI} possesses a highly specific and sensitive recognition toward cTnI.

Surprisingly, after the Myo is specifically recognized by the aptasensor, the R_{et} value decreased, whereas the CV current increased, respectively (curve f in Figure 1C,E). This suggests that the electron transfer on the sensing interface was recovered. This can only because the peeled off of Apt_{Myo} from sensor surface due to the formation of the aptamer–Myo complex. As the result, the sensor can feasibly facilitate the highly sensitive detection capability of Myo because the degree of its output recovery strongly depends on the concentration of the target.

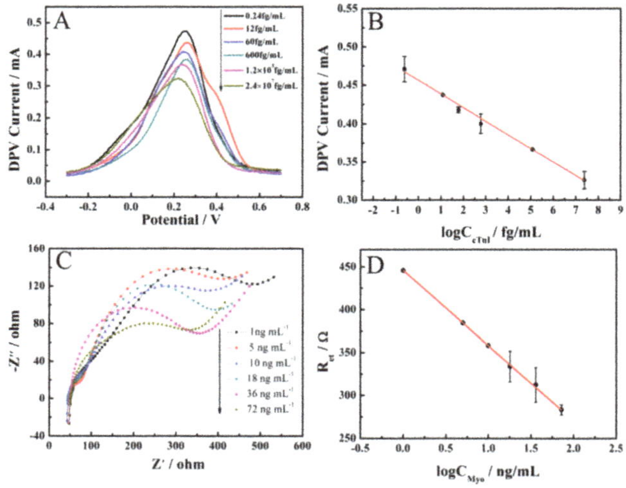

Figure 3. (**A**) The DPV responses of Apt_{cTnI} sensor for detecting cTnI. (**B**) The linear relationship between DPV current and the logarithm of cTnI concentration. (**C**) The EIS responses of Apt_{Myo} sensor for detecting Myo. (**D**) The linear relationship between Ret and the logarithm of Myo concentration.

Table 1. Performance comparison of the developed sensor with those in the literature for cTnI and Myo detection.

Electrode Materials	Target	Method	Detection Range	LOD	Refs.
HsGDY@NDs	cTnI	EIS	0.01 pg/mL–100 ng/mL	6.29 fg/mL	[8]
Fc-SiNP		SWV	0.024 ng/mL–240 ng/mL	24 pg/mL	[23]
Ti/AuNPs		DPV	0.024 ng/mL–26.4 ng/mL	4.32 pg/mL	[32]
Fc-COFNs		DPV	10 fg/mL–10 ng/mL	2.6 fg/mL	[33]
PCN-AuNPs		SWV	0.1 pg/mL–103 ng/mL	0.01 pg/mL	[2]
Au/Zr–C		i-t	0.01 pg/mL-100 ng/mL	1.24 fg/mL	[34]
Pt@Pd DNs/NH_2-HMCS		i-t	100 fg/mL–100 ng/mL	15.4 fg/mL	[35]
AuNPs/Ti_3C_2-MXenes		DPV	0.24 fg/mL–24 ng/mL	0.14 fg/mL	This work
AuNPs/BNNSs		DPV	0.1 μg/mL–100 μg/mL	34.6 ng/mL	[36]
DApt-CS conjugate		DPV	1.67 ng/mL–666.67 ng/mL	0.45 ng/mL	[37]
Poly-o-phenylenediamine	Myo	DPV	10 ng/mL–1780 ng/ml	10 ng/mL	[38]
AuNP-PEI		SWV	9.96 ng/mL–72.8 ng/mL	6.29 ng/mL	[39]
magnetic beads		GMI	1 ng/mL–10 ng/mL	0.5 ng/mL	[40]
AuNPs@rGO		DPV	1 ng/mL–1400 ng/mL	0.67 ng/mL	[41]
AuNPs/Ti_3C_2-MXenes		EIS	1 ng/mL–70 ng/mL	0.2 ng/mL	This work

The analysis of Myo was carried out under optimal conditions using the EIS test. Figure 3C shows the EIS responses of aptasensor after being incubated with differently concentrated Myo. It can be observed that the R_{et} value of the aptasensor is regressively decreased upon increasing Myo concentration. In Figure 3D, a good linear relationship (R^2 = 0.999) was obtained between the R_{et} and the logarithm of the Myo concentration ($logC_{Myo}$) in a range of 1–72 ng/mL. The linear regression equation is R_{et} = 446 − 87.6logC (ng/mL) with an LOD of 0.2 ng/mL (S/N = 3), which also lays well below the normal value of Myo in blood. This Apt_{Myo}/AuNPs/Ti_3C_2-MXene-hosted sensor exhibits superior analytical performance compared with other bioassays for Myo, as listed in Table 1. Remarkably, the low LOD of the designed aptasensor highlights the potential role of AuNPs/Ti_3C_2-MXenes in biosensing territory.

3.4. Specificity, Stability and Reproducibility of the Proposed Aptasensors

A diagnostic biosensor must possess high specificity for a target molecule against other interfering materials. Therefore, the specificity of the AuNPs/Ti_3C_2-MXene-based aptasensors for detecting their respective target were evaluated in the presence of 100-fold high concentrated inspected materials, including C-reactive protein (CRP), Albumin (Alb) and N-terminal pro-brain nitric peptide (NT-pro-BNP), as well as Myo or cTnI, and their mixture with cTnI or Myo was also examined. Except the corresponding target, the response of the coexistent biomolecules was almost zero, as illustrated in Figure 4A,D. These results indicate the high specificity of both two electrochemical aptasensors, which is apparently from the proper specific aptamer on the modified electrode.

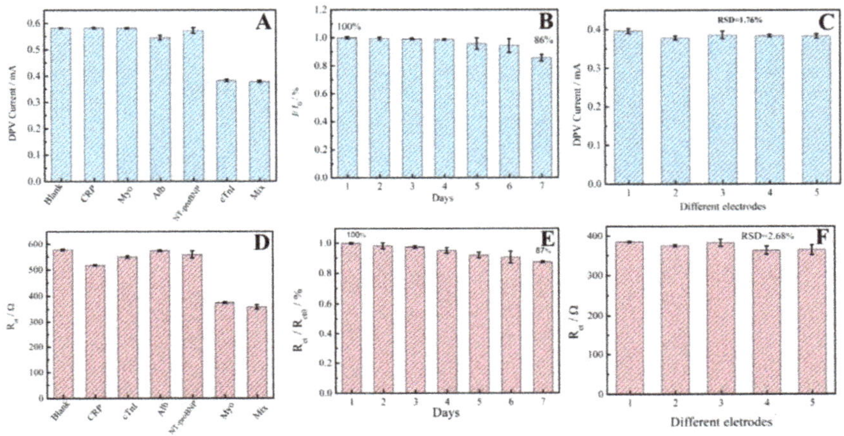

Figure 4. The (**A**) selectivity (in the presence of 60 pg /mL CRP, Myo, Alb and NT-proBNP), (**B**) stability and (**C**) reproducibility of the cTnI aptasensor for cTnI (600 fg/mL) detection, all detected values are presented as blue bars. The (**D**) selectivity (in the presence of 50 ng/mL CRP, cTnI, Alb and NT-proBNP), (**E**) stability and (**F**) reproducibility of the Myo aptasensor for Myo (5 ng/mL) detection, all detected values are presented as red bars.

The storage stability of the developed aptasensors was examined by monitoring their response toward cTnI or Myo once per day. The results show that they retained up to 86% to 87% of their initial responses after seven days of use, respectively (Figure 4B,E). The aptasensors showed good reproducibility for 1.76% or 2.68% of RSD for five parallel prepared sensors, respectively, as illustrated in Figure 4C,F. The appreciable stability and reproducibility stem from the effective immobilization efficiency of aptamers on AuNPs/Ti_3C_2-MXene-modified electrodes through interactions such as the covalent linkage of Au–S and perhaps existing chelation between the Ti of AuNPs/Ti_3C_2-MXenes and PO_4^{3-} of the DNA aptamer.

3.5. Detection of cTnI or Myo in Human Serum Sample

To explore the practicability of developed aptasensors, they were used to detect cTnI or Myo in real human serum samples, with the recovery test using a standard addition method (Table 2). Plainly, the recovery of the cTnI aptasensor is in the range of 99.0–106% (RSD in the range of 0.60–1.98%). The recovery for Myo ranges from 92.0% to 108% (RSD in the range of 1.40–5.50%). The results demonstrate that the sensors were feasibly able to determine cTnI or Myo in human serum. The accuracy of the designed aptasensors was also verified by comparing the results with enzyme-linked immunosorbent assay (ELISA). As listed in Table 2, the contents of cTnI and Myo in four human serum samples detected using the proposed sensors were in good agreement with that of the ELISA.

Table 2. Quantitative results of cTnI and Myo in human serum samples.

Sample	Found cTnI (ng/mL)	Added cTnI (ng/mL)	Total Detected cTnI (ng/mL)	Recovery (%)	ELISA (ng/mL)	RSD (%)
1	0.0667 ± 0.0028	0.01	0.0773 ± 0.0015	106	0.07	1.98
2	0.110 ± 0.007	0.01	0.120 ± 0.001	100	0.09	0.61
3	0.0121 ± 0.0010	0.01	0.0220 ± 0.0020	99.0	<0.029	0.91
4	0.0243 ± 0.0021	0.01	0.0344 ± 0.0020	101	<0.029	0.60
Sample	**Found Myo (ng/mL)**	**Added Myo (ng/mL)**	**Total Detected Myo (ng/mL)**	**Recovery (%)**	**ELISA (ng/mL)**	**RSD (%)**
1	15.5 ± 0.39	5.0	20.7 ± 1.13	104	15.7	5.50
2	24.9 ± 0.63	5.0	29.5 ± 0.75	92.0	25.0	2.54
3	69.6 ± 0.60	0.5	70.1 ± 0.98	100	70.1	1.40
4	14.1 ± 0.10	5.0	19.5 ± 1.06	108	14.1	5.43

4. Conclusions

The bifunctional and label-free electrochemical aptasensing strategy with AuNPs/Ti_3C_2-MXenes as substrate has been investigated for the sensitive quantification of the dual biomarkers of AMI disease, cTnI and Myo. The AuNPs/Ti_3C_2-MXene hybrid not only serves as a matrix to anchor the aptamer, but also provides excellent electroactivity to accelerate the electron transfer to promote the sensing ability of the aptasensors. Both using $[Fe(CN)_6]^{4-/3-}$ as redox probe, the cTnI aptasensor shows a high DPV response with a low LOD of 0.14 fg/mL and a wide linear range; meanwhile, the Myo assay is successful from the recovered EIS due to the detachment of the Myo aptamer, which acquired a low LOD of 0.2 ng/mL. Of course, the specificity of the two sensors is also excellent because of the high recognition of the target by the aptamer. Besides the high sensitivity and specificity, what is more meaningful is that the designed aptasensors are label-free ones; thus, they are simpler and quicker in response than that of the ELISA method, which was based on the labeling technique. However, so far, we can only operate on two sensors separately, due to the limitation of the function of the equipment. What we are reporting here is only a phased work, and its purpose is to verify the feasibility of these dual biomarkers' synchronous detection. As for its application in the real world, it still depends on further development and research. This will not hinder the application of this technology, as long as the equipment is upgraded. According to our study, an array composed of two sensors can feasibly solve the problem of the synchronous detection of two biomarkers in a single operation. Only needs an implemented macro instruction on the instrument to start each sensor in a time-sharing with a jumper circuit which is used to realize the gating of the sensors. Alternatively, the two sensors can be submerged totally with each other, because they adopt a different driving principle and do not interfere with each other. This research suggests that the proposed aptasensing strategy might be a promising platform for the early and accurate screening and diagnosis of AMI in clinical applications.

Author Contributions: Experiment, Data analysis, writing—original draft, X.M.; investigation, conceptualization, Writing—review and editing, H.L.; project instruction, design and methodology, experimental verification, writing—review and editing, funding acquisition, Y.T. All authors have read and agreed to the published version of the manuscript.

Funding: This work is financially supported by the National Natural Science Foundation of China (21675115, 21375091).

Institutional Review Board Statement: Not applicable.

Informed Consent Statement: Not applicable.

Data Availability Statement: Not applicable.

Conflicts of Interest: The authors declare no conflict of interest.

References

1. Ji, J.; Lu, W.; Zhu, Y.; Jin, H.; Yao, Y.; Zhang, H.; Zhao, Y. Porous Hydrogel-Encapsulated Photonic Barcodes for Multiplex Detection of Cardiovascular Biomarkers. *ACS Sens.* **2019**, *4*, 1384–1390. [CrossRef] [PubMed]
2. Khushaim, W.; Peramaiah, K.; Beduk, T.; Vijjapu, M.T.; Filho, J.I.D.O.; Huang, K.-W.; Mani, V.; Salama, K.N. Porous graphitic carbon nitrides integrated biosensor for sensitive detection of cardiac troponin I. *Biosens. Bioelectron. X* **2022**, *12*, 100234. [CrossRef]
3. Mi, X.; Li, H.; Tan, R.; Tu, Y. Dual-Modular Aptasensor for Detection of Cardiac Troponin I Based on Mesoporous Silica Films by Electrochemiluminescence/Electrochemical Impedance Spectroscopy. *Anal. Chem.* **2020**, *92*, 14640–14647. [CrossRef] [PubMed]
4. Yuan, Z.; Wang, L.; Chen, J.; Su, W.; Li, A.; Su, G.; Liu, P.; Zhou, X. Electrochemical strategies for the detection of cTnI. *Anal.* **2021**, *146*, 5474–5495. [CrossRef]
5. Maisel, A.; Mueller, C.; Neath, S.-X.; Christenson, R.H.; Morgenthaler, N.G.; McCord, J.; Nowak, R.M.; Vilke, G.; Daniels, L.B.; Hollander, J.E.; et al. Copeptin Helps in the Early Detection of Patients with Acute Myocardial Infarction. *J. Am. Coll. Cardiol.* **2013**, *62*, 150–160. [CrossRef]
6. Yoo, S.S.; Kim, S.Y.; Kim, K.S.; Hong, S.; Oh, M.J.; Nam, M.G.; Kim, W.-J.; Park, J.; Chung, C.-H.; Choe, W.-S.; et al. Controlling inter-sheet-distance in reduced graphene oxide electrodes for highly sensitive electrochemical impedimetric sensing of myoglobin. *Sens. Actuators B Chem.* **2020**, *305*, 127477. [CrossRef]
7. Zhang, J.X.; Tang, C.; Chen, D.N.; Jiang, L.Y.; Wang, A.J.; Feng, J.J. Ultrasensitive Label-Free Sandwich Immunoassay of Cardiac Biomarker Myoglobin Using Meso-SiO2@Ploydapamine@PtPd Nanocrystals and PtNi Nanodendrites for Effective Signal Amplification. *Appl. Surf. Sci.* **2023**, *608*, 155216. [CrossRef]
8. Wang, C.; Li, J.; Kang, M.; Huang, X.; Liu, Y.; Zhou, N.; Zhang, Z. Nanodiamonds and hydrogen-substituted graphdiyne heteronanostructure for the sensitive impedimetric aptasensing of myocardial infarction and cardiac troponin I. *Anal. Chim. Acta* **2021**, *1141*, 110–119. [CrossRef]
9. Zhang, D.; Huang, L.; Liu, B.; Su, E.B.; Chen, H.Y.; Gu, Z.Z.; Zhao, X.W. Quantitative Detection of Multiplex Cardiac Bi-omarkers with Encoded SERS Nanotags on a Single T Line in Lateral Flow Assay. *Sens. Actuators B* **2018**, *277*, 502–509. [CrossRef]
10. Guo, X.; Zong, L.; Jiao, Y.; Han, Y.; Zhang, X.; Xu, J.; Li, L.; Zhang, C.-W.; Liu, Z.; Ju, Q.; et al. Signal-Enhanced Detection of Multiplexed Cardiac Biomarkers by a Paper-Based Fluorogenic Immunodevice Integrated with Zinc Oxide Nanowires. *Anal. Chem.* **2019**, *91*, 9300–9307. [CrossRef]
11. Wang, B.; Shi, S.W.; Yang, X.L.; Wang, Y.; Qi, H.L.; Gao, Q.; Zhang, C.X. Separation-Free Electrogenerated Chemilumi-nescence Immunoassay Incorporating Target Assistant Proximity Hybridization and Dynamically Competitive Hybridization of a DNA Signal Probe. *Anal. Chem.* **2020**, *92*, 884–891. [CrossRef]
12. Alwarappan, S.; Nesakumar, N.; Sun, D.; Hu, T.Y.; Li, C.-Z. 2D metal carbides and nitrides (MXenes) for sensors and biosensors. *Biosens. Bioelectron.* **2022**, *205*, 113943. [CrossRef] [PubMed]
13. Lorencova, L.; Gajdosova, V.; Hroncekova, S.; Bertok, T.; Blahutova, J.; Vikartovska, A.; Parrakova, L.; Gemeiner, P.; Kasak, P.; Tkac, J. 2D MXenes as Perspective Immobilization Platforms for Design of Electrochemical Nanobiosensors. *Electroanalysis* **2019**, *31*, 1833–1844. [CrossRef]
14. Aguedo, J.; Lorencova, L.; Barath, M.; Farkas, P.; Tkac, J. Electrochemical Impedance Spectroscopy on 2D Nanomaterial MXene Modified Interfaces: Application as a Characterization and Transducing Tool. *Chemosensors* **2020**, *8*, 127–148. [CrossRef]
15. Zhang, H.X.; Zhuang, T.T.; Wang, L.; Du, L.; Xia, J.F.; Wang, Z.H. Efficient Au Nanocluster@Ti3C2 Heterostructure Lumi-nophore Combined with Cas12a for Electrochemiluminescence Detection of miRNA. *Sens. Actuators B* **2022**, *370*, 132428. [CrossRef]
16. Wei, Z.H.; Zhang, H.X.; Wang, Z.H. High-Intensity Focused Ultrasound Combined with Ti3C2–TiO2 to Enhance Electro-chemiluminescence of Luminol for the Sensitive Detection of Polynucleotide Kinase. *ACS Appl. Mater. Interfaces* **2023**, *15*, 3804–3811. [CrossRef]
17. Wu, L.; Lu, X.; Dhanjai; Wu, Z.-S.; Dong, Y.; Wang, X.; Zheng, S.; Chen, J. 2D transition metal carbide MXene as a robust biosensing platform for enzyme immobilization and ultrasensitive detection of phenol. *Biosens. Bioelectron.* **2018**, *107*, 69–75. [CrossRef]
18. Zhang, Q.; Wang, F.; Zhang, H.; Zhang, Y.; Liu, M.; Liu, Y. Universal Ti_3C_2 MXenes Based Self-Standard Ratiometric Fluorescence Resonance Energy Transfer Platform for Highly Sensitive Detection of Exosomes. *Anal. Chem.* **2018**, *90*, 12737–12744. [CrossRef]
19. Wang, H.; Li, H.; Huang, Y.; Xiong, M.H.; Wang, F.; Li, C. A Label-Free Electrochemical Biosensor for Highly Sensitive De-tection of Gliotoxin based on DNA Nanostructure/MXene Nanocomplexes. *Biosens. Bioelectron.* **2019**, *142*, 111531. [CrossRef]
20. Mi, X.N.; Li, H.; Tan, R.; Feng, B.N.; Tu, Y.F. The TDs/Aptamer cTnI Biosensors based on HCR and Au/Ti3C2-MXene Amplification for Screening Serious Patient in COVID-19 Pandemic. *Biosens. Bioelectron.* **2021**, *192*, 113482. [CrossRef]
21. Wang, L.; Zhang, H.; Zhuang, T.; Liu, J.; Sojic, N.; Wang, Z. Sensitive electrochemiluminescence biosensing of polynucleotide kinase using the versatility of two-dimensional Ti3C2TX MXene nanomaterials. *Anal. Chim. Acta* **2022**, *1191*, 339346. [CrossRef] [PubMed]
22. Wang, Q.; Liu, W.; Xing, Y.; Yang, X.; Wang, K.; Jiang, R.; Wang, P.; Zhao, Q. Screening of DNA Aptamers against Myoglobin Using a Positive and Negative Selection Units Integrated Microfluidic Chip and Its Biosensing Application. *Anal. Chem.* **2014**, *86*, 6572–6579. [CrossRef] [PubMed]
23. Jo, H.; Gu, H.; Jeon, W.; Youn, H.; Her, J.; Kim, S.-K.; Lee, J.; Shin, J.H.; Ban, C. Electrochemical Aptasensor of Cardiac Troponin I for the Early Diagnosis of Acute Myocardial Infarction. *Anal. Chem.* **2015**, *87*, 9869–9875. [CrossRef] [PubMed]

24. Lu, H.-J.; Pan, J.-B.; Wang, Y.-Z.; Ji, S.-Y.; Zhao, W.; Luo, X.-L.; Xu, J.-J.; Chen, H.-Y. Electrochemiluminescence Energy Resonance Transfer System between RuSi Nanoparticles and Hollow Au Nanocages for Nucleic Acid Detection. *Anal. Chem.* **2018**, *90*, 10434–10441. [CrossRef] [PubMed]
25. Zhao, F.; Xie, Q.; Xu, M.; Wang, S.; Zhou, J.; Liu, F. RNA aptamer based electrochemical biosensor for sensitive and selective detection of cAMP. *Biosens. Bioelectron.* **2015**, *66*, 238–243. [CrossRef]
26. Mishra, S.K.; Kumar, D.; Biradar, A.M.; Rajesh. Electrochemical Impedance Spectroscopy Characterization of Mercapto-propionic Acid Capped ZnS Nanocrystal Based Bioelectrode for the Detection of the Cardiac Biomarker—Myoglobin. *Bioelectrochemistry* **2012**, *88*, 118–126. [CrossRef]
27. Li, D.; Tan, R.; Mi, X.; Fang, C.; Tu, Y. An electrochemiluminescent biosensor for noninvasive glucose detection based on cluster-like AuAg hollowed-nanoparticles. *Microchem. J.* **2021**, *167*, 106271. [CrossRef]
28. Ge, X.-Y.; Zhang, J.-X.; Feng, Y.-G.; Wang, A.-J.; Mei, L.-P.; Feng, J.-J. Label-free electrochemical biosensor for determination of procalcitonin based on graphene-wrapped Co nanoparticles encapsulated in carbon nanobrushes coupled with AuPtCu nanodendrites. *Microchim. Acta* **2022**, *189*, 110. [CrossRef]
29. Lv, H.; Li, Y.; Zhang, X.; Li, X.; Xu, Z.; Chen, L.; Li, D.; Dong, Y. Thionin functionalized signal amplification label derived dual-mode electrochemical immunoassay for sensitive detection of cardiac troponin I. *Biosens. Bioelectron.* **2019**, *133*, 72–78. [CrossRef]
30. Rauf, S.; Mani, V.; Lahcen, A.A.; Yuvaraja, S.; Beduk, T.; Salama, K.N. Binary transition metal oxide modified laser-scribed graphene electrochemical aptasensor for the accurate and sensitive screening of acute myocardial infarction. *Electrochim. Acta* **2021**, *386*, 138489. [CrossRef]
31. Dong, L.; Yin, L.; Tian, G.; Wang, Y.; Pei, H.; Wu, Q.; Cheng, W.; Ding, S.; Xia, Q. An enzyme-free ultrasensitive electrochemical immunosensor for calprotectin detection based on PtNi nanoparticles functionalized 2D Cu-metal organic framework nanosheets. *Sens. Actuators B* **2020**, *308*, 127687. [CrossRef]
32. Lopa, N.S.; Rahman, M.; Ahmed, F.; Ryu, T.; Sutradhar, S.C.; Lei, J.; Kim, J.; Kim, D.H.; Lee, Y.H.; Kim, W. Simple, low-cost, sensitive and label-free aptasensor for the detection of cardiac troponin I based on a gold nanoparticles modified titanium foil. *Biosens. Bioelectron.* **2019**, *126*, 381–388. [CrossRef]
33. Song, Z.; Song, J.; Gao, F.; Chen, X.; Wang, Q.; Zhao, Y.; Huang, X.; Yang, C.; Wang, Q. Novel electroactive ferrocene-based covalent organic frameworks towards electrochemical label-free aptasensors for the detection of Cardiac Troponin I. *Sens. Actuators B* **2022**, *368*, 132205. [CrossRef]
34. Chen, K.; Zhao, H.; Wang, Z.; Zhou, F.; Shi, Z.; Cao, S.; Lan, M. Sandwich-type electrochemical aptasensor based on Au-modified conductive octahedral carbon architecture and snowflake-like PtCuNi for the sensitive detection of cardiac troponin I. *Biosens. Bioelectron.* **2022**, *212*, 114431. [CrossRef] [PubMed]
35. Wang, Z.; Zhao, H.; Chen, K.; Li, H.; Lan, M. Sandwich-type electrochemical aptasensor based on hollow mesoporous carbon spheres loaded with porous dendritic Pd@Pt nanoparticles as signal amplifier for ultrasensitive detection of cardiac troponin I. *Anal. Chim. Acta* **2021**, *1188*, 339202. [CrossRef]
36. Adeel, M.; Rahman, M.; Lee, J.-J. Label-free aptasensor for the detection of cardiac biomarker myoglobin based on gold nanoparticles decorated boron nitride nanosheets. *Biosens. Bioelectron.* **2019**, *126*, 143–150. [CrossRef]
37. Taghdisi, S.M.; Danesh, N.M.; Ramezani, M.; Emrani, A.S.; Abnous, K. A novel electrochemical aptasensor based on Y-shape structure of dual-aptamer-complementary strand conjugate for ultrasensitive detection of myoglobin. *Biosens. Bioelectron.* **2016**, *80*, 532–537. [CrossRef] [PubMed]
38. Shumyantseva, V.V.; Bulko, T.V.; Sigolaeva, L.V.; Kuzikov, A.V.; Archakov, A.I. Electrosynthesis and binding properties of molecularly imprinted poly-o-phenylenediamine for selective recognition and direct electrochemical detection of myoglobin. *Biosens. Bioelectron.* **2016**, *86*, 330–336. [CrossRef] [PubMed]
39. Zapp, E.; Westphal, E.; Gallardo, H.; de Souza, B.; Vieira, I.C. Liquid crystal and gold nanoparticles applied to electrochemical immunosensor for cardiac biomarker. *Biosens. Bioelectron.* **2014**, *59*, 127–133. [CrossRef]
40. Yang, Z.; Wang, H.; Dong, X.; Yan, H.; Lei, C.; Luo, Y. Giant magnetoimpedance based immunoassay for cardiac biomarker myoglobin. *Anal. Methods* **2017**, *9*, 3636–3642. [CrossRef]
41. Singh, S.; Tuteja, S.K.; Sillu, D.; Deep, A.; Suri, C.R. Gold nanoparticles-reduced graphene oxide based electrochemical immunosensor for the cardiac biomarker myoglobin. *Microchim. Acta* **2016**, *183*, 1729–1738. [CrossRef]

chemosensors

MDPI

Article

Aptamer Sensor Based on Hybrid Chain Reaction and CRISPR-Cas9 System for STX Detection

Qinguo Zhao, Guizhen Li and Xuemei Li *

Collaborative Innovation Center of Tumor Marker Detection Technology, Equipment and Diagnosis-Therapy Integration in Universities of Shandong, Shandong Province Key Laboratory of Detection Technology for Tumor Makers, School of Chemistry and Chemical Engineering, Linyi University, Linyi 276005, China
* Correspondence: xuemei_li@yeah.net

Abstract: In recent years, pollution incidents caused by red tide occur frequently, and the red tide biotoxins brought by it make the food safety problem of seafood become a difficult problem to be solved urgently, which has caused great damage to the mariculture industry. Red tide toxin is also known as "shellfish toxin". Saxitoxin (STX), is one of the strongest paralytic shellfish toxins and is also one of the most toxic marine toxins, which is extremely harmful. Aiming at the problems existing in the current research on the detection of red tide biotoxin in complex water bodies, this research developed an aptamer sensor based on hybrid chain reaction and a CRISPR-Cas9 gene editing system to detect the toxins of the clam and analyzed the feasibility of this method for the detection of the toxins of the clam. The results showed that the linear range of this method is 5.0 fM to 50 pM, and the detection limit is 1.2 fM. Meanwhile, the recovery rate of this sensor for the detection of toxins is 102.4–104.1% when applied in shellfish extract, which shows significant specificity and the reliability of this detection method.

Keywords: hybrid chain reaction; CRISPR-Cas9; saxitoxin; fluorescent detection

Citation: Zhao, Q.; Li, G.; Li, X. Aptamer Sensor Based on Hybrid Chain Reaction and CRISPR-Cas9 System for STX Detection. *Chemosensors* **2023**, *11*, 183. https://doi.org/10.3390/chemosensors11030183

Academic Editor: Alina Vasilescu

Received: 3 February 2023
Revised: 27 February 2023
Accepted: 6 March 2023
Published: 9 March 2023

1. Introduction

Since the 20th century, red tides have erupted frequently around the world. Biotoxins produced by red tide organisms have not only seriously damaged marine fishery resources and aquaculture, and worsened the marine environment, but also could be transmitted to the upper layer through the food chain, directly affecting the food safety of marine products [1]. The formation of red tide biotoxin is closely related to the red tide of toxic algae in the sea and belongs to marine natural organic matter [2]. Therefore, the early screening of marine products, water sources, etc., and the efficient and accurate detection of various marine biotoxins have become the key link in the prevention of marine biotoxin poisoning [3]. The eutrophication of water leads to the excessive reproduction of algae in the water. Shellfish filter the toxic microalgae, and after bioaccumulation and amplification, they are transformed into shellfish toxins. Therefore, red tide toxins, are also known as "shellfish toxins" [4]. According to the symptoms caused by the toxins they produce, marine biotoxins can be divided into the following categories: diarrheal shellfish poisoning (DSP), paralytic shellfish poisoning (PSP), neurotoxic shellfish poisoning (NSP), and amnesic shellfish poisoning (ASP), azaspiracid shellfish poisoning (AZP), ciguatera fish poisoning (CFP) and other toxins [5]. Paralytic shellfish poisoning (PSP) is a kind of marine toxin with the widest distribution, the highest frequency, and the greatest harm in the world that is formed by the metabolism of marine poisonous flagellates. Its toxicology is mainly to inhibit nerve conduction through the influence of the sodium channel [6]. When people ingest food containing the paralytic shellfish toxin, the toxin will be released rapidly and present toxic effects. The incubation period is only a few minutes or hours. Symptoms include limb muscle paralysis, headaches, and nausea. In severe cases, muscle paralysis, dyspnea, and even asphyxia lead to death [7].

Saxitoxin (STX) is one of the most powerful paralytic shellfish poisonings and also one of the most toxic marine toxins [8]. It can accumulate in bivalves and fiber fish through fiber filtration, thus entering the food chain. When the toxin content exceeds the safety standard, human consumption of such shellfish products often produces the risk of poisoning [9]. The oral lethal dose for humans is 5.7 mg kg^{-1}, therefore approximately 0.57 mg of saxitoxin is lethal if ingested, and the lethal dose by injection is about 0.6 mg kg^{-1}. The human inhalation toxicity of aerosolized saxitoxin is estimated to be 5 mg min m^{-3}. It is suggested that saxitoxin can enter the body via open wounds, and a lethal dose is 50 mg per person by this route [10].

At present, the conventional methods used for the detection of marine red tide toxins mainly include biological and chemical methods [11]. Biological detection can detect various algal toxins through the biological monitoring of mice. This method has obvious disadvantages, for example, it can only measure the toxicity of the toxin, but cannot determine its composition and content. Chemical detection is to know the type and toxicity of toxins through the qualitative and quantitative analysis of the toxin components in the samples [12]. Among them, enzyme-linked immunoassay is a widely used immunoassay, which is widely used in the monitoring of red tide toxins, such as PSP, DSP, NSP, and other biotoxins. However, due to the cross-reaction of antibodies, lack of standard toxins, interference of analogs, and other factors, using these methods may occur false positive results [13]. In recent years, instrumental analysis has been widely used because of its sensitivity, accuracy, and rapidity, such as HPLC chromatography, is one of the most widely used analysis methods of red tide biotoxin, with high accuracy, low detection limit, and fewer samples used [14]. UV or fluorescence detectors are usually required for HPLC detection. Due to the lack of UV chromogenic groups in many toxins and the lack of standard samples of red tide biotoxins, the development of HPLC is greatly limited [15]. Liquid chromatography–tandem mass spectrometry (LC-MS/MS) is a relatively new detection technology that combines the advantages of chromatography and mass spectrometry. This method first uses chromatography to separate compounds and then uses mass spectrometry to detect them [16]. Compared with HPLC analysis, LC-MS/MS has no complex derivation and sample purification process and can detect a wider range with a lower detection limit. However, there are some challenges in this method, such as complex pre-treatments, and so on. Capillary electrophoresis (CE) technology is actually a new type of detection technology that combines electrophoresis and chromatography [17]. The above detection methods have many shortcomings, and a new detection method needs to be developed urgently.

Biosensor is a new biological analysis and detection method that combines biotechnology and electronic technology [18]. In comparison to the commonly used chemical and biological analysis methods, this method can accurately quantify, has low sample purity requirements, saves sample pretreatment time, reduces detection cost, and effectively improves the laboratory's detection ability [19]. In addition, it also has good sensitivity and specificity, is easy to operate, achieves miniaturization, and has other advantages. Therefore, the application of biosensors to the detection of marine toxins has attracted more and more attention [20,21]. In recent years, with the rapid development of biotechnology, material science, and nanotechnology, and the emergence of technologies such as the systematic evolution of ligands by exponential enrichment (SELEX), biosensors have made great progress in both their theoretical basis and their design and processing [22]. By using this technology, one can select a specific nucleic acid aptamer (aptamer) with a high affinity for the target substance from the random single-stranded nucleic acid sequence library [23]. The nucleic acid aptamer can bind with a variety of target substances with high specificity and selectivity after being screened and enriched by SELEX and can have the same sensitivity as an antigen–antibody reaction, so it is widely used in the field of biosensors [24]. At the same time, aptamers also have some very obvious advantages, as follows: (1) low cost, short cycle, and easy access, (2) easy to modify, (3) it has good stability and good thermal and chemical properties, (4) strong specificity compounds with very similar chemical structures can be identified, (5) it has a small molecular weight, strong tissue penetration, and good biocompatibility, which are conducive to its fixation on the surface of the sensor [25,26]. Because of these advantages, more and more studies on aptamers have been carried out in recent years, including in disease diagnosis, drug research and development, biochips, and food detection [27,28]. Therefore, a nucleic acid aptamer biosensor has been rapidly developed in recent years as a new method for detecting shellfish toxins [29,30].

The hybridization chain reaction (HCR) has attracted increasing attention because it can occur under mild conditions without the use of enzymes. In a typical HCR, the initiator triggers a cascade of hybridization events between two species of DNA hairpins, leading to the formation of a nicked double helix with tens to hundreds of repeated units until the hairpins are exhausted [31]. As a powerful signal amplification tool, HCR has been used to construct a variety of simple, sensitive, and economical biosensors for detecting nucleic acids, small molecules, cells, and proteins [32]. The characteristics of HCR, such as no enzyme and constant temperature, enable the whole reaction process to be completed through simple instruments, thus reducing the experimental cost [33]. In addition, HCR has high sensitivity and specificity and specificity and has high compatibility with a variety of detection technologies. These advantages make it play an increasingly important role in the construction of biosensors [34,35].

Clustered regularly interspaced short palindromic repeats (CRISPR) regular clusters of spaced short palindromic repeats and CRISPR associated (Cas) form a system, which is a natural defense system used by bacteria to prevent bacteriophage DNA injection and plasmid transfer [36]. In 2012, the group of Doudna J A and Charpentier E demonstrated for the first time that CRISPR-Cas9 can recognize and cleave target DNA in vitro [37]. The natural CRISPR-Cas9 system is composed of three parts: SpCas9 (hereinafter referred to as Cas9), crRNA, and CRISPR RNA (tracrRNA) [38]. The crRNA is homologous with the target duplex DNA sequence and is partially complementary to the tracrRNA to form a complex protein, guiding the Cas9 protein to recognize and cut the target duplex at specific sites. The key to achieve this is the existence of a protospacer adjacent motif (PAM) sequence on the DNA chain for the binding process to proceed smoothly [36,39,40]. To simplify the process of gene editing, crRNA and tracrRNA are usually integrated into a single guide RNA (sgRNA) to identify the target duplexes [39]. The PAM sequence is located upstream of the DNA binding region and immediately adjacent to the binding region. The sequence is composed of NGG3 bases, of which the type of the first base can be freely changed, but the last two bases must be guanine. The modified CRISPR-Cas9 system has become an important tool for researchers in gene editing [41,42]. The target sites for CRISPR-Cas9 system cutting are predetermined by sgRNA sequences to achieve more precise cutting, and the cutting sequence is more flexible due to different sgRNA designs. The CRISPR-Cas9 system has been gradually developed as a new biosensor tool with its advantages of low cost, convenient operation, high efficiency, strong specificity, and other advantages. Incorporating CRISPR-Cas systems with various nucleic acid amplification strategies enables the generation of amplified detection signals, enrichment of low-abundance molecular targets, and improvements in analytical specificity and sensitivity [43]. This detection system will have great application prospects in disease diagnosis, environmental assessment, rapid food quality assessment, environmental monitoring, and other fields [44].

Biosensors for detection based on nucleic acid isothermal amplification technology and CRISPR-Cas system have been reported, such as strand displacement amplification (SDA), rolling circle amplification (RCA), and loop-mediated isothermal amplification (LAMP) [45–48]. However, a number of biosensors have been reported for the detection of nucleic acid targets. The establishment of efficient methods for non-nucleic acid target detection would further broaden the scope of this technique, but up to now, the concerning research is limited. Based on the above research, this research will aim at the problems existing in the current research on the detection of red tide biotoxins in complex water bodies, and use the isothermal amplification technology and gene editing system to prepare an aptamer sensor based on the hybridization chain reaction and CRISPR-Cas9 system to detect the toxins in the clam, so as to realize the highly sensitive detection of micro or trace red tide toxins in complex biological systems and marine environments, which is a research system of great research value.

2. Materials and Methods

2.1. Materials

All oligonucleotides were commercially synthesized by Sangon Biotech Co., Ltd. (Shanghai, China). Other sequences were listed in Table 1. Magnetic beads (MBs) were purchased from So-Fe Biomedicine (Shanghai, Beijing). Saxitoxin (STX), okadaic acid (OA), yessotoxin (YTX), and pectenotoxin (PTX) standards were purchased from National Research Council of Canada (Ottawa, Canada). Cas9 nuclease was purchased from New England Biolabs (Beijing, China) Ltd. N-(3-Dimethylaminopropyl)-N′-ethylcarbodiimide hydrochloride (EDC), and N-Hydroxysulfosuccinimide sodium salt (NHS) were purchased from Shanghai Yuanye Bio-Technology Co., Ltd. (Shanghai, China).

Table 1. DNAs in this paper.

Name	Sequence (5′-3′)
S1	5′NH2 C6-TTG AGG GTC GCA TCC CGT GGA AAC AGG TTC ATT G
S2	CAA TGA ACC TGT TTC CAC GGG ATG CGA CCC TCA A
H1	ATC CCG TGG AAA CAG GTT CAT TGA ACT CTC AAT GAA CCT GAT TCC A
H2	FAM- CAA TGA ACC TGT TTC CAC GGG A /iBHQ1dT/ TGG AAT CAG GTT CAT TGA GAG TT
sgRNA	GCA AUG AAC CUG UUU CCA GUU UUA GAG CUA GAA AUA GCA AGU UAA AAU AAG GCU AGU CCG UUA UCA ACU UGA AAA AGU GGC ACC GAG UCG GUG CUU UU

2.2. Instrumentation

F-4600 fluorescence spectrophotometer (Hitachi, Japan); PB-10 acidity (sartorius, Beijing, China); KQ218 Ultrasonic cleaner (Kunshan, China); LX-200 Mini centrifuge (Haimen Qer); VM-10 vortex oscillator (Beijing Plichi); WD-9413B gel imaging analyzer (Beijing Liuyi Biotechnology Co., Ltd.); High-speed frozen centrifuge (Zhuhai dark horse); PCR gene amplifier (Shandong Biologix Biotech Co., Ltd.); electrophoresis instrument, electrophoresis tank (Beijing Liuyi Biotechnology Co., Ltd.); nanoparticle potentiometer (Malvern, UK).

2.3. Preparation of MBs Aptasensor

Firstly, 2 μM S1 (aptamer sequence) and 2 μM S2 (aptamer complementary sequence) were mixed in PBS annealing buffer (0.1 M, pH = 7.4), placed in a PCR gene amplifier, set to 90 °C for 5 min and 65 °C for 10 min, then gradually cooled to room temperature and stored at 4 °C in a refrigerator for storage.

Next, carboxy-modified magnetic beads were sonicated for 3–5 min, taken to 80 microliters in centrifuge tubes, washed three times with 300 μL imidazole–HCl buffer solution (0.1 mM, pH = 6.8), then added 100 μL NHS (0.2 M) and EDC (0.5 M), respectively, and incubated in a 37 °C incubator for shaking slowly for 40 min to activate the MBs.

Lastly, the activated MBs were washed with PBS (0.1 M, pH = 7.4), at 11,000 rpm for 5 min, repeated 3 times to remove NHS and EDC, added 200 μL of synthesized double-stranded probes, incubated for 15 h with 37 °C oscillation, then removed the redundant double-stranded probes by magnetic separation. The MB-aptamer probe was washed 3 times with 300 μL of PBS (0.01 M, pH = 7.4), resuspended in 200 μL, and stored at 4 °C in a refrigerator for storage.

2.4. STX Detection

Firstly, 25 μL MBs solution and 25 μL of STX standard (4 nM) were taken in a centrifuge tube, gently reversed and mixed, and incubated for 1 h at 37 °C. Then, the supernatant solution containing S2 was separated from the MBs aptasensor by using magnetic separation, then a certain amount of S2 and the diluted H1 and H2 were annealed in PCR.

The procedure was set to 95 °C for 5 min and gradually cooled to room temperature for at least 1 h [47]. Then, the three components were mixed in a centrifuge tube with a total volume of 20 µL, and the HCR reaction system was constructed for 1 h at room temperature. A tube added a certain volume of Cas9 and sgRNA with a final ratio of 2:1, 1 × NEBufferTM 3.1, added DEPC water to the total volume of 20 µL. After incubation at 37 °C for 10 min, the HCR system products were added into it and then incubated for 1 h at 37 °C and 65 °C for 10 min to stop the reaction incubation. Lastly, the complex was diluted to 200 µL and the fluorescence spectrum was measured by using an F-4700 fluorescence spectrophotometer [48,49]. (Hitachi Co., Ltd., Tokyo, Japan).

3. Results

3.1. Scheme of the Developed MBs Aptasensor

As shown in Scheme 1, after the 5′ terminal amino modified aptamer chain (S1) and complementary chain (S2) synthesized a double chain probe based on the principle of base pairing, the MBs aptamer biosensor probe was constructed by reacting with the carboxyl modified magnetic bead and fixed it on the surface of the MBs. When STX existed, the aptamer S1 could be specifically recognized, and combined with it, the aptamer complementary chain S2 was released into the solution. The supernatant was taken out through magnetic separation technology, then mixed with hairpin DNA1 (H1) and hairpin DNA2 (H2) (modified fluorescent agent and quenching agent). The aptamer complementary chain S2 acted as a promoter to trigger the HCR reaction. Firstly, the H1 hairpin structure was opened, thus promoting the self-assembly between H1 and H2, and formed a long, discontinuous double-chain complex. After that, the Cas9-sgRNA complex was added. Under the guidance of sgRNA, the discontinuous double chain complex was specifically recognized, and the Cas9 protein played an active role to achieve accurate cutting, and the quenching agent was separated from the fluorescent agent to achieve the effect of fluorescence recovery. According to the mechanism of the proposed HCR, the sensitivity is predominately determined by the HCR amplification efficiency. Therefore, the rate-limiting step is the H1/H2 complementary reaction with S2, and the more thorough this reaction, the higher the response signal.

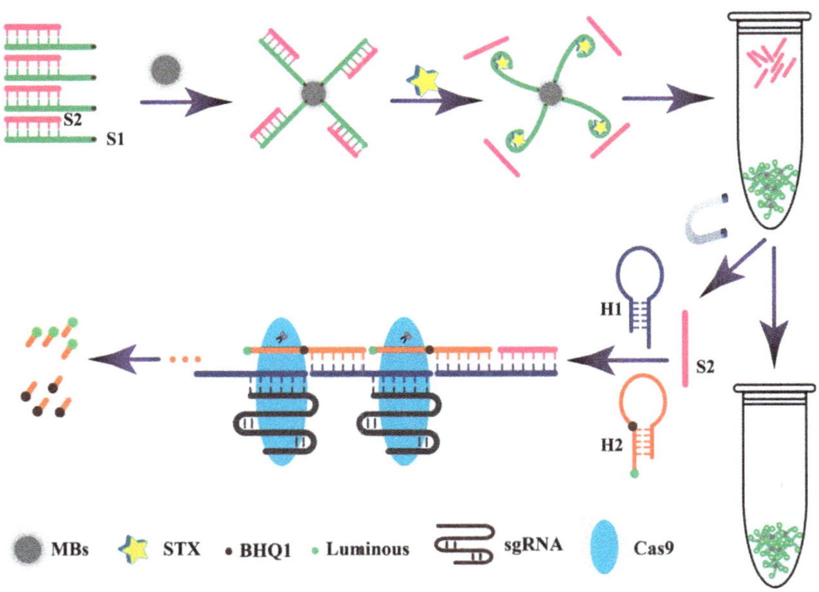

Scheme 1. Schematic illustration of the aptasensor based on hybrid chain reaction and a CRISPR-Cas9 gene editing system.

3.2. Characterization of MBs Aptasensor

In Figure 1A, the voltages of MBs and MBs + DNA were 22.4 mV and 25.1 mV, respectively. The increase in charge was mainly due to the addition of a negatively charged DNA probe to the surface of the MBs. The formed complex was loaded with the double-stranded probe. The potential values of MBs + DNA were significantly lower than that of MBs, demonstrating the successful synthesis.

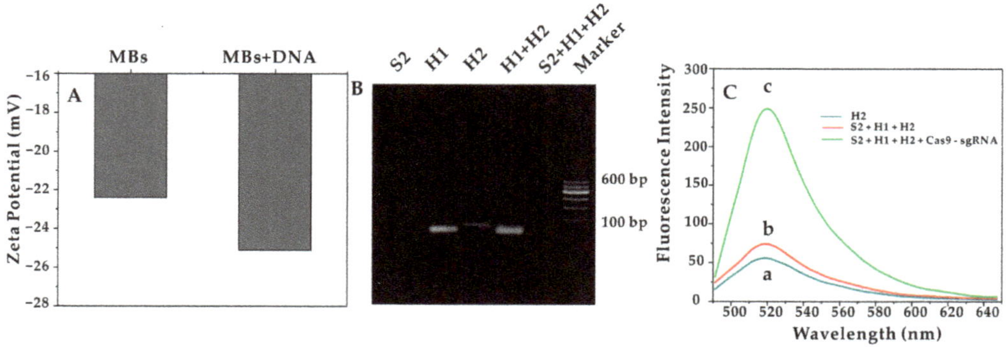

Figure 1. (**A**) Zeta potential of MBs and MBs + DNA. (**B**) Agarose gel characterization of the HCR products. (**C**) Fluorescence intensity after the introduction of CRISPR-Cas9.

3.3. Feasibility Analysis

3.3.1. Gel Electrophoresis Confirmation of the Assembly of dsDNAs by HCR

The dsDNAs formed by HCR were verified by 3% agarose gel electrophoresis with TAE buffer as the running buffer at a constant voltage of 110 v. The agarose gels were then counterstained and imaged with Redsafe. As shown in Figure 1B, from left to the right, the lanes were S2, H1, H2, H1 + H2, S2 + H1 + H2, and marker. The mixed band in lane 5 and the only band in the other lanes indicated the successful assembly of the interrupted dsDNA after the HCR reaction. Notably, the bands H1 and H2 could be clearly seen in lane 4, indicating that only the mixture of H1 and H2 would not self-hybridize without promoters.

3.3.2. The Construction of the HCR and CRISPR-Cas9 System

As mentioned above, after the successful characterization of the HCR system, the influence of fluorescence recovery after the addition of the Cas9-sgRNA complex was studied. When the STX did not exist, the aptamer (S1) on the sensor did not specifically bind to it, so the promoter (S2) could not be released into the supernatant. Because there was no promoter (S2), the HCR reaction system cannot be triggered, and the fluorescence intensity was low, as shown in Figure 1C (a). When the STX existed, the aptamer was quickly recognized and specifically combined with it, and the promoter (S2) was released into the supernatant and triggered the HCR reaction system, with a slight increase in fluorescence intensity, as shown in Figure 1C (b). The continued addition of the preprepared Cas9-sgRNA complex to the system is shown in Figure 1C (c). The fluorescence intensity increased significantly, indicating that under the guidance of the sgRNA, the activity of the Cas9 nuclease was triggered to achieve accurately cutting the dsDNAs, which had been formed after the HCR reaction. Thus, the effect of fluorescence recovery was achieved.

3.4. Optimization of Reaction Conditions

Several conditions, such as the concentration ratio of Cas9 and sgRNA, the incubation temperature, and the incubation time of the Cas9:sgRNA complex, were further optimized to improve the detection sensitivity.

Firstly, the influence of the ratios of Cas9:sgRNA from 1:4 to 4:1 on the formation of the Cas9-sgRNA complex was investigated. Studies have shown that the higher or lower proportion of the Cas9:sgRNA complex cannot achieve efficient assembly, which affected the catalytic cleavage process of double chains and thus affected the magnitude of fluorescence intensity. As shown in Figure 2A, the fluorescence intensity reaches its maximum value at 2:1. Therefore, the ratio of 2:1 was selected as the optimal condition for Cas9:sgRNA.

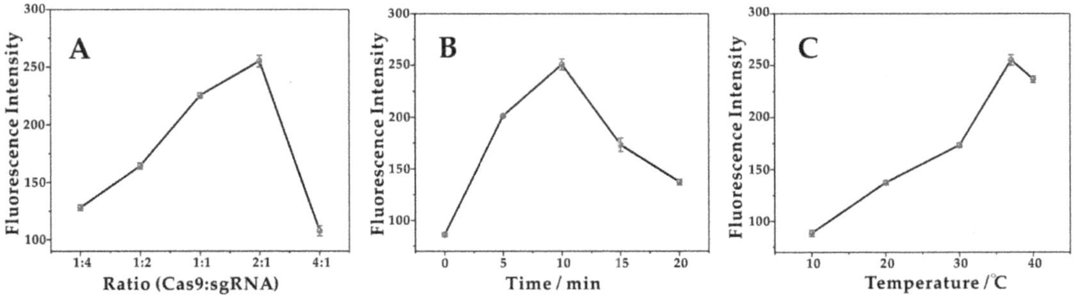

Figure 2. (**A**) Optimization of the concentration ratio of Cas9:sgRNA. The fluorescence intensity reaches a maximum at a ratio of 2:1. (**B**) Optimization of the incubation time for the formation of Cas9-sgRNA. The fluorescence intensity reaches a maximum at 10 min. (**C**) Optimization of the incubation temperature for the formation of Cas9-sgRNA. The fluorescence intensity reaches a maximum at 37 °C.

Next, the influence of the incubation time for the formation of the Cas9-sgRNA complex was investigated. The incubation of the Cas9-sgRNA complex at different times (0 min, 5 min, 10 min, 15 min, and 20 min) yielded different fluorescence intensities as shown in Figure 2B. Without a prior incubation of both Cas9 and sgRNA for a certain time, the Cas9-sgRNA complex did not achieve effective assembly, with a poor shear effect and low fluorescence intensities. Continuing to increase the incubation time, the fluorescence intensity increased significantly, reaching a maximum at 10 min, indicating that both achieved efficient assembly after 10 min of incubation and the complex formed was sufficiently stable for precise cutting. The decrease in the fluorescence intensity with increasing time is due to the decreased specificity of the formed complex for recognizing the duplex. Therefore, 10 min was chosen as the optimal incubation time.

Finally, the influence of the incubation temperature on the formation of the Cas9-sgRNA complex was investigated. At low temperatures, the Cas9 protein fails to exert full activity, and even loses activity, resulting in low fluorescence intensities. With the increase in temperature, the fluorescence intensities increased gradually and reached the peak at 37 °C, as shown in Figure 2C. Continuing to raise the temperature, the fluorescence intensity was significantly reduced, and it was considered that the high temperature will change the spatial structure of the Cas9 protein, thus affecting the activity. Therefore, 37 °C was chosen as the optimal incubation temperature.

3.5. Analysis of STX Using the Developed Aptasensor

Under optimal conditions, the sensitivity of the method for STX detection at different concentrations was evaluated. With the increasing of STX concentration, the fluorescence intensity increased, and this aptasensor demonstrated great linear responses to STX in the range of 5.0 fM to 50 pM, with the correlation coefficient (R^2 = 0.995), and the limit of detection (LOD) of 1.2 fM (Figure 3). In addition, as shown in Table 2, the LOD of the developed aptasensor in this study was lower than that of most reported detection methods for STX, and the linear detection range was relatively wider than that of the reported methods [50–53].

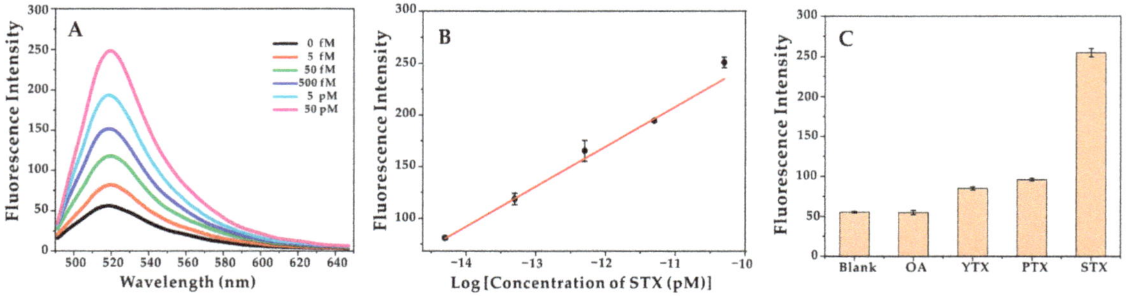

Figure 3. (**A**) Fluorescence intensity over a range of STX concentrations (5 fM, 50 fM, 500 fM, 5 pM, 50 pM). (**B**) The linear relationship between the intensity of fluorescence and the logarithm of STX concentration. (**C**) Fluorescence intensity of the aptasensor in the presence of different marine toxins, the concentration of STX is 50 pM, and others (OA, YTX, PTX) are 500 pM.

Table 2. Comparison of different aptasensor performances for STX detection.

Method	LOD	Linear Range	References
Colorimetric aptasensor	42.46 pM	78.13–2500 pM	[50]
SERS aptasensor	11.7 nM	10–200 nM	[51]
Label-free colorimetric aptasensor	0.1423 nM	0.1457–37.30 nM	[52]
Electrochemical aptasensor	0.03 nM	1.0–200 nM	[53]
Fluorescent aptasensor	1.2 fM	5 fM–50 pM	This work

3.6. Selectivity and Stability of the Aptasensor

In addition, in order to evaluate the selectivity of the aptasensor, other marine toxins, including OA, YTX, and PTX were applied to the aptasensor to obtain the fluorescence spectrum. As shown in Figure 3C, the fluorescence intensity of STX was significantly higher than other marine toxins, indicating that the developed aptasensor had high specificity and good selectivity. It should be noted that its specificity comes from (1) the recognition function of the Cas9-sgRNA complex, and (2) the S2 being separated from the aptamer to successfully initiate the HCR reaction.

To further assess the utility, accuracy, and repeatability of this aptasensor, three different concentrations of STX standard samples were added to the mussel tissue extraction to analyze the developed aptasensor. As shown in Table 3, good mussel tissue extraction recovery (102.4–104.1%) and low relative standard deviation (RSD) demonstrated the sensor's accuracy and reproducibility for actual STX samples. These results indicate that the proposed biosensor could be successfully applied to real sample tests.

Table 3. The detection of recovery rate of three different concentrations of STX in mussel tissue.

Samples	Spiked (pM)	Fluorescence Intensity	Rate of Recovery (%, Mean)	RSD (%)
Blank	0	54.6 ± 5.3	—	—
1	0.5	164.4 ± 4.1	104.1%	2.5%
2	5	201.5 ± 8.7	102.4%	4.3%
3	50	242.9 ± 3.3	103.2%	1.4%

4. Conclusions

In this study, signal amplification is achieved by introducing HCR and a CRISPR-Cas9 gene editing system for the detection of STX. Different from the previous reports that directly cut a piece of double-stranded DNA using the characteristics of the CRISPR-Cas9 system, this study developed an efficient, specific, low detection limit and high selectivity detection method by precisely cutting the long discontinuous DNA after HCR construction. With these strategies, the LOD is as low as 1.2 fM, which is lower than most reported STX detection methods. The high recovery of STX detection in mussel tissue extract indicates that the sensor can be used for actual sample detection. Therefore, this sensor has brought innovative applications to marine biotoxin detection in various fields, such as water and food quality control.

Author Contributions: Investigation, data curation, Q.Z.; writing—original draft preparation, Q.Z.; writing—review and editing, X.L. and G.L.; supervision, writing, funding acquisition and resources, X.L. All authors have read and agreed to the published version of the manuscript.

Funding: This research was funded by the National Natural Science Foundation of China (No. 2187060746) and the Taishan Scholars Program of Shandong Province (No. tsqn201812101).

Institutional Review Board Statement: Not applicable.

Informed Consent Statement: Not applicable.

Data Availability Statement: The authors confirm that the data supporting the findings of this study are available within the article.

Conflicts of Interest: The authors declare no conflict of interest.

References

1. Martin, D.F.; Martin, B.B. Red tide, red terror. Effects of red tide and related toxins. *J. Chem. Educ.* **1976**, *53*, 614. [CrossRef] [PubMed]
2. Schrope, M. Red tide rising: Algal blooms can make life miserable for coastal dwellers and wreak havoc on marine ecosystems. Mark Schrope reports on Florida's efforts to predict these red tides. *Nature* **2008**, *452*, 24–27. [CrossRef] [PubMed]
3. Patel, S.S.; Lovko, V.J.; Lockey, R.F. Red Tide: Overview and Clinical Manifestations. *J. Allergy Clin. Immunol. Pract.* **2020**, *8*, 1219–1223. [CrossRef] [PubMed]
4. Farabegoli, F.; Blanco, L.; Rodriguez, L.P.; Vieites, J.; Cabado, A. Phycotoxins in Marine Shellfish: Origin, Occurrence and Effects on Humans. *Mar. Drugs* **2018**, *16*, 188. [CrossRef] [PubMed]
5. Hinder, S.L.; Hays, G.C.; BroOKS, C.J.; Davies, A.P.; Edwards, M.; Walne, A.W.; Gravenor, M.B. Toxic marine microalgae and shellfish poisoning in the British isles: History, review of epidemiology, and future implications. *Environ. Health* **2011**, *10*, 54. [CrossRef]
6. Kellmann, R.; Mihali, T.K.; Jeon, Y.J.; Pickford, R.; Pomati, F.; Neilan, B.A. Biosynthetic intermediate analysis and functional homology reveal a saxitoxin gene cluster in cyanobacteria. *Appl. Environ. Microbiol.* **2008**, *74*, 4044–4053. [CrossRef]
7. Adams, N.G.; Robertson, A.; Grattan, L.M.; Pendleton, S.; Roberts, S.; Tracy, J.K.; Trainer, V.L. Assessment of sodium channel mutations in Makah tribal members of the U.S. Pacific Northwest as a potential mechanism of resistance to paralytic shellfish poisoning. *Harmful Algae* **2016**, *57*, 26–34. [CrossRef]
8. Thottumkara, A.P.; Parsons, W.H.; Du bois, J. Saxitoxin. *Angew. Chem. Int. Ed. Engl.* **2014**, *53*, 5760–5784. [CrossRef]
9. Walsh, J.J.; Tomas, C.R.; Steidinger, K.A.; Lenes, J.M.; Chen, F.R.; Weisberg, R.H.; Zheng, L.; Landsberg, J.H.; Vargo, G.A.; Heil, C.A. Imprudent fishing harvests and consequent trophic cascades on the West Florida shelf over the last half century: A harbinger of increased human deaths from paralytic shellfish poisoning along the southeastern United States, in response to oligotrophication? *Cont. Shelf Res.* **2011**, *31*, 891–911.
10. Zheng, W.; Liu, X.; Li, Q.; Shu, Z.; Li, Z.; Zhang, L. A simple electrochemical aptasensor for saxitoxin detection. *RSC Adv.* **2022**, *12*, 23801–23807. [CrossRef]
11. Gerssen, A.; Pol-Hofstad, I.E.; Poelman, M.; Mulder, P.P.; Van Den Top, H.J.; De Boer, J. Marine toxins: Chemistry, toxicity, occurrence and detection, with special reference to the Dutch situation. *Toxins* **2010**, *2*, 878–904. [CrossRef]
12. Humpage, A.R.; Magalhaes, V.F.; Froscio, S.M. Comparison of analytical tools and biological assays for detection of paralytic shellfish poisoning toxins. *Anal. Bioanal. Chem.* **2010**, *397*, 1655–1671. [CrossRef] [PubMed]
13. Mcnamee, S.E.; Medlin, L.K.; Kegel, J.; Mccoy, G.R.; Raine, R.; Barra, L.; Ruggiero, M.V.; Kooistra, W.H.; Montresor, M.; Hagstrom, J.; et al. Distribution, occurrence and biotoxin composition of the main shellfish toxin producing microalgae within European waters: A comparison of methods of analysis. *Harmful Algae* **2016**, *55*, 112–120. [CrossRef] [PubMed]
14. Marrouchi, R.; Dziri, F.; Belayouni, N.; Hamza, A.; Benoit, E.; Molgo, J.; Kharrat, R. Quantitative determination of gymnodimine-A by high performance liquid chromatography in contaminated clams from Tunisia coastline. *Mar. Biotechnol.* **2010**, *12*, 579–585. [CrossRef]
15. Lee, J.S.; Yanagi, T.; Kenma, R.; Yasumoto, T. Fluorometric determination of diarrhetic shellfish toxins by high-performance liquid chromatography. *Agric. Biol. Chem.* **1987**, *51*, 877–881.
16. Ciminiello, P.; Dell'aversano, C.; Iacovo, E.D.; Fattorusso, E.; Forino, M.; Tartaglione, L. LC-MS of palytoxin and its analogues: State of the art and future perspectives. *Toxicon* **2011**, *57*, 376–389. [CrossRef] [PubMed]

17. Keyon, A.S.; Guijt, R.M.; Gaspar, A.; Kazarian, A.A.; Nesterenko, P.N.; Bolch, C.J.; Breadmore, M.C. Capillary electrophoresis for the analysis of paralytic shellfish poisoning toxins in shellfish: Comparison of detection methods. *Electrophoresis* **2014**, *35*, 1496–1503. [CrossRef]
18. Naresh, V.; Lee, N. A Review on Biosensors and Recent Development of Nanostructured Materials-Enabled Biosensors. *Sensors* **2021**, *21*, 1109. [CrossRef]
19. De Oliveira, W.F.; Dos Santos Silva, P.M.; Coelho, L.C.B.B.; Dos Santos Correia, M.T. Biomarkers, biosensors and biomedicine. *Curr. Med. Chem.* **2020**, *27*, 3519–3533. [CrossRef]
20. Campas, M.; Prieto-Simon, B.; Marty, J.L. Biosensors to detect marine toxins: Assessing seafood safety. *Talanta* **2007**, *72*, 884–895. [CrossRef]
21. Reverte, L.; Solino, L.; Carnicer, O.; Diogene, J. Campas M. Alternative methods for the detection of emerging marine toxins: Biosensors, biochemical assays and cell-based assays. *Mar. Drugs* **2014**, *12*, 5719–5763. [CrossRef] [PubMed]
22. Zhuo, Z.; Yu, Y.; Wang, M.; Li, J.; Zhang, Z.; Liu, J.; Wu, X.; Lu, A.; Zhang, G.; Zhang, B. Recent Advances in SELEX Technology and Aptamer Applications in Biomedicine. *Int. J. Mol. Sci.* **2017**, *18*, 2142. [CrossRef] [PubMed]
23. Sun, H.; Zhu, X.; Lu, P.Y.; Rosato, R.R.; Tan, W.; Zu, Y. Oligonucleotide aptamers: New tools for targeted cancer therapy. *Mol. Ther. Nucleic Acids* **2014**, *3*, e182. [CrossRef]
24. Eissa, S.; Ng, A.; Siaj, M.; Tavares, A.C.; Zourob, M. Selection and identification of DNA aptamers against okadaic acid for biosensing application. *Anal. Chem.* **2013**, *85*, 11794–11801. [CrossRef]
25. Fomo, G.; Waryo, T.T.; Sunday, C.E.; Baleg, A.A.; Baker, P.G.; Iwuoha, E.I. Aptameric Recognition-Modulated Electroactivity of Poly(4-Styrenesolfonic Acid)-Doped Polyaniline Films for Single-Shot Detection of Tetrodotoxin. *Sensors* **2015**, *15*, 22547–22560. [CrossRef] [PubMed]
26. Sun, H.; Zu, Y. A highlight of recent advances in aptamer technology and its application. *Molecules* **2015**, *20*, 11959–11980. [CrossRef]
27. Barthelmebs, L.; Jonca, J.; Hayat, A.; Prieto-Simon, B.; Marty, J.L. Enzyme-Linked Aptamer Assays (ELAAs), based on a competition format for a rapid and sensitive detection of Ochratoxin A in wine. *Food Control* **2011**, *22*, 737–743. [CrossRef]
28. Sefah, K.; Phillips, J.A.; Xiong, X.; Meng, L.; Van Simaeys, D.; Chen, H.; Martin, J.; Tan, W. Nucleic acid aptamers for biosensors and bio-analytical applications. *Analyst* **2009**, *134*, 1765–1775. [CrossRef]
29. Tian, Y.; Du, L.; Zhu, P.; Chen, Y.; Chen, W.; Wu, C.; Wang, P. Recent progress in micro/nano biosensors for shellfish toxin detection. *Biosens. Bioelectron.* **2021**, *176*, 112899. [CrossRef] [PubMed]
30. Zhao, Y.; Li, L.; Yan, X.; Wang, L.; Ma, R.; Qi, X.; Wang, S.; Mao, X. Emerging roles of the aptasensors as superior bioaffinity sensors for monitoring shellfish toxins in marine food chain. *J. Hazard. Mater.* **2022**, *421*, 126690. [CrossRef]
31. Bi, S.; Yue, S.; Zhang, S. Hybridization chain reaction: A versatile molecular tool for biosensing, bioimaging, and biomedicine. *Chem. Soc. Rev.* **2017**, *46*, 4281–4298. [CrossRef]
32. Li, X.; Wang, Y.; Wang, L.; Wei, Q. A surface plasmon resonance assay coupled with a hybridization chain reaction for amplified detection of DNA and small molecules. *Chem. Commun.* **2014**, *50*, 5049–5052. [CrossRef] [PubMed]
33. Li, H.; Wang, X.; Wei, S.; Zhao, C.; Song, X.; Xu, K.; Li, J.; Pang, B.; Wang, J. Applications of hybridization chain reaction optical detection incorporating nanomaterials: A review. *Anal. Chim. Acta* **2022**, *1190*, 338930. [CrossRef]
34. Wu, J.; Lv, J.; Zheng, X.; Wu, Z.S. Hybridization chain reaction and its applications in biosensing. *Talanta* **2021**, *234*, 122637. [CrossRef]
35. Augspurger, E.E.; Rana, M.; Yigit, M.V. Chemical and Biological Sensing Using Hybridization Chain Reaction. *ACS Sens.* **2018**, *3*, 878–902. [CrossRef]
36. Gratz, S.J.; Wildonger, J.; Harrison, M.M.; O'connor-Giles, K.M. CRISPR/Cas9-mediated genome engineering and the promise of designer flies on demand. *Fly* **2013**, *7*, 249–255. [CrossRef]
37. Jinek, M.; Chylinski, K.; Fonfara, I.; Hauer, M.; Doudna, J.A.; Charpentier, E. A programmable dual-RNA–guided DNA endonuclease in adaptive bacterial immunity. *Science* **2012**, *337*, 816–821. [CrossRef]
38. Zhu, Y. Advances in CRISPR/Cas9. *Biomed. Res. Int.* **2022**, *2022*, 9978571. [CrossRef] [PubMed]
39. Hisano, Y.; Ota, S.; Kawahara, A. Genome editing using artificial site-specific nucleases in zebrafish. *Dev. Growth Differ.* **2014**, *56*, 26–33. [CrossRef] [PubMed]
40. Shademan, B.; Nourazarian, A.; Hajazimian, S.; Isazadeh, A.; Biray Avci, C.; Oskouee, M.A. CRISPR Technology in Gene-Editing-Based Detection and Treatment of SARS-CoV-2. *Front. Mol. Biosci.* **2021**, *8*, 772788. [CrossRef]
41. Hu, J.; Jiang, M.; Liu, R.; Lv, Y. Label-free CRISPR/Cas9 assay for site-specific nucleic acid detection. *Anal. Chem.* **2019**, *91*, 10870–10878. [CrossRef]
42. Liu, Y.; Li, S.; Zhang, L.; Zhao, Q.; Li, N.; Wu, Y. A sensitive and specific method for microRNA detection and in situ imaging based on a CRISPR–Cas9 modified catalytic hairpin assembly. *RSC Adv.* **2020**, *10*, 28037–28040. [CrossRef]
43. Feng, W.; Newbigging, A.M.; Tao, J.; Cao, Y.; Peng, H.; Le, C.; Wu, J.; Pang, B.; Li, J.; Tyrrell, D.L.; et al. CRISPR technology incorporating amplification strategies: Molecular assays for nucleic acids, proteins, and small molecules. *Chem. Sci.* **2021**, *12*, 4683–4698. [CrossRef]
44. Huang, Z.; Liu, S.; Pei, X.; Li, S.; He, Y.; Tong, Y.; Liu, G. Fluorescence Signal-Readout of CRISPR/Cas Biosensors for Nucleic Acid Detection. *Biosensors* **2022**, *12*, 779. [CrossRef] [PubMed]

45. Pardee, K.; Green, A.A.; Takahashi, M.K.; Braff, D.; Lambert, G.; Lee, J.W.; Ferrante, T.; Ma, D.; Donghia, N.; Fan, M. Rapid, Low-Cost Detection of Zika Virus Using Programmable Biomolecular Components. *Cell* **2016**, *165*, 1255–1266. [CrossRef] [PubMed]
46. Li, L.; Li, S.; Wu, N.; Wu, J.; Wang, G.; Zhao, G.; Wang, J. HOLMESv2 A CRISPR-Cas12b-Assisted Platform for Nucleic Acid Detection and DNA Methylation Quantitation. *ACS Synth. Biol.* **2019**, *8*, 2228–2237. [CrossRef]
47. Yin, K.; Ding, X.; Li, Z.; Zhao, H.; Cooper, K.; Liu, C. Dynamic Aqueous Multiphase Reaction System for One-Pot CRISPR-Cas12a-Based Ultrasensitive and Quantitative Molecular Diagnosis. *Anal. Chem.* **2020**, *92*, 8561–8568. [CrossRef] [PubMed]
48. Wang, R.; Zhao, X.; Chen, X.; Qiu, X.; Qing, G.; Zhang, H.; Zhang, L.; Hu, X.; He, Z.; Zhong, D.; et al. Rolling Circular Amplification (RCA)-Assisted CRISPR/Cas9 Cleavage (RACE) for Highly Specific Detection of Multiple Extracellular Vesicle MicroRNAs. *Anal. Chem.* **2020**, *92*, 2176–2185. [CrossRef]
49. Kang, Y.; Su, G.; Yu, Y.; Cao, J.; Wang, J.; Yan, B. CRISPR-Cas12a-Based Aptasensor for On-Site and Highly Sensitive Detection of Microcystin-LR in Freshwater. *Environ. Sci. Technol.* **2022**, *56*, 4101–4110. [CrossRef]
50. Zhao, Y.; Li, L.; Ma, R.; Wang, L.; Yan, X.; Qi, X.; Wang, S.; Mao, X. A competitive colorimetric aptasensor transduced by hybridization chain reaction-facilitated catalysis of AuNPs nanozyme for highly sensitive detection of saxitoxin. *Anal. Chim. Acta* **2021**, *1173*, 338710. [CrossRef] [PubMed]
51. Cheng, S.; Zheng, B.; Yao, D.; Wang, Y.; Tian, J.; Liu, L.; Liang, H.; Ding, Y. Determination of Saxitoxin by Aptamer-Based Surface-Enhanced Raman Scattering. *Anal. Lett.* **2019**, *52*, 902–918. [CrossRef]
52. Li, L.; Zhao, Y.; Yan, X.; Qi, X.; Wang, L.; Ma, R.; Wang, S.; Mao, X. Development of a terminal-fixed aptamer and a label-free colorimetric aptasensor for highly sensitive detection of saxitoxin. *Sens. Actuators B Chem.* **2021**, *344*, 130320. [CrossRef]
53. Ullah, N.; Chen, W.; Noureen, B.; Tian, Y.; Du, L.; Wu, C.; Ma, J. An Electrochemical Ti(3)C(2)T(x) Aptasensor for Sensitive and Label-Free Detection of Marine Biological Toxins. *Sensors* **2021**, *21*, 4938. [CrossRef] [PubMed]

chemosensors

MDPI

Communication

Screening the Specific Surface Area for Metal-Organic Frameworks by Cataluminescence

Zenghe Li [1], Danning Pei [1], Rui Tian [1,*] and Chao Lu [1,2,*]

[1] State Key Laboratory of Chemical Resource Engineering, Beijing University of Chemical Technology, Beijing 100029, China; lizh@mail.buct.edu.cn (Z.L.); pdanning@163.com (D.P.)
[2] Green Catalysis Center, College of Chemistry, Zhengzhou University, Zhengzhou 450001, China
* Correspondence: tianrui@mail.buct.edu.cn (R.T.); luchao@mail.buct.edu.cn (C.L.)

Abstract: Metal-organic frameworks (MOFs) are famous for their large surface area, which is responsible for the dispersed active sites and decent behaviors in gas adsorption, storage, and catalytic reactions. However, it remains a great challenge to acquire a cost-effective and accurate evaluation on the surface area for the MOFs. In this work, we have proposed cataluminescence (CTL) to evaluate the specific surface area for the MOFs, based on the adsorption–desorption and the catalytic reaction of ethanol. Aluminum-based MOFs with large-pore (*lp*), narrow-pore (*np*), and medium-pore (*mp*-130, *mp*-140, and *mp*-150 synthesized under 130, 140, and 150 °C) have been prepared. Distinguished CTL signals were acquired from ethanol in the presence of these MOFs: *lp* > *mp*-150 > *mp*-130 > *mp*-140 > *np*. Note that the CTL intensities were positively correlated with the specific surface areas of these MOFs acquired by the Brunauer–Emmett–Teller (BET) method. The distinct specific surface area of MOFs determined the capacity to accommodate and activate ethanol, leading to the varied CTL intensity signals. Therefore, the proposed CTL could be utilized for the rapid and accurate evaluation of the specific surface area for MOFs. It is believed that this CTL strategy showed great possibilities in the structural evaluation for various porous materials.

Keywords: cataluminescence; MOFs; structural evaluation; specific area

Citation: Li, Z.; Pei, D.; Tian, R.; Lu, C. Screening the Specific Surface Area for Metal-Organic Frameworks by Cataluminescence. *Chemosensors* **2023**, *11*, 292. https://doi.org/10.3390/chemosensors11050292

Academic Editor: Chung-Wei Kung

Received: 6 April 2023
Revised: 11 May 2023
Accepted: 12 May 2023
Published: 14 May 2023

1. Introduction

Metal-organic frameworks (MOFs) are a class of crystalline coordination materials that are composed of organic connectors and inorganic nodes [1,2]. MOFs have been regarded as excellent candidates for gas adsorption, storage, and the corresponding catalytic reactions [3,4]. The diverse functionalities of MOFs are highly dependent on their high porosity and large surface area [5,6]. The variations in the pore structures of MOFs, such as the known "breathing" behaviors, with the changed textural properties also manifesting as the specific surface area [7–9]. It is noteworthy that the large surface area of MOFs favored good dispersion of active centers, facilitating the adsorption, enrichment, and reaction of analytes [10,11]. The value of the surface area has been regarded as the key factor to determine the capacity of MOFs to accommodate and concentrate guest molecules for gas adsorption and the subsequent reaction [12–15]. Moreover, the large surface area of MOFs is beneficial for the synthesis reaction and functional modification, offering an infinite platform to prepare MOF-based composites for different applications [11,16,17]. Therefore, it is a topic of interest and significance to estimate the surface area of MOFs, in order to analyze the adsorption, storage, and catalytic performances for MOFs.

The Brunauer–Emmett–Teller (BET) method is known as the most commonly applied technique to evaluate the specific surface area and pore volume for a material [18–20]. The BET measurements are generally implemented through the isothermal adsorption of nitrogen at 77 K, and the adsorption capacity of the material can be evaluated by the adsorption–desorption curves [21–24]. However, N_2 diffuses slowly in the samples at the low temperature ranges, around 77 K, and thus a long equilibration time is required to achieve fully equilibrated nitrogen isotherms for data acquisition [21]. The complex experimental equipment and time-consuming process have hampered the applications of BET to evaluate the textural structure of the MOFs. Moreover, it usually requires large quantities of samples to achieve accurate statistical results, and insufficient samples could result in largely derived results [25]. These requirements mean that BET failed to provide high-throughput analyses to meet the needs of the rapid development for MOFs in recent years. Therefore, it is critical to establish an accurate and cost-effective strategy to evaluate the surface area and adsorption capacity of MOF materials.

Cataluminescence (CTL) is derived from the luminescence emissions of the species generated during a chemical reaction [26–28]. It requires a featured catalyst with a large specific surface area and abundant active sites during a chemical reaction, and the CTL emissions are dependent on the adsorption and catalytic behaviors of the catalyst [29–32]. Herein, we have employed CTL emissions to evaluate the surface area of MOFs, taking the catalytic oxidation of ethanol as a model system. The aluminum-based MOFs with the large-pore (*lp*), narrow-pore (*np*), and medium-pore phases synthesized under different temperatures (short as *mp*-130, *mp*-140, and *mp*-150) were prepared. The CTL intensities in the presence of these different MOF catalysts showed distinct, repeatable, and highly sensitive signals, according to a tendency as follows: *lp* > *mp*-150 > *mp*-130 > *mp*-140 > *np*. It should be noted that the CTL intensities of ethanol are highly correlated with the specific surface areas acquired by the BET measurements for the MOFs (Scheme 1). A positive relationship between the CTL intensity (I) and specific surface area from BET measurements (SA) could be drawn as follows: $I = -4.94 \times 10^4 \exp (4.39 \times 10^{-3} SA) + 6.88 \times 10^4$. Such a correlation could be ascribed to the distinct specific surface area of MOFs that is responsible for the adsorption, desorption, and catalytic oxidation towards ethanol during the CTL measurements. Therefore, we have proposed CTL as a fast, sensitive, and cost-effective approach to screen the specific surface area of MOF materials in the absence of sophisticated data analysis or a time-consuming process. Based on the distinct adsorption and catalytic performances of a material with varied structures, it is anticipated that the CTL could be further used to evaluate the structural variations of other materials.

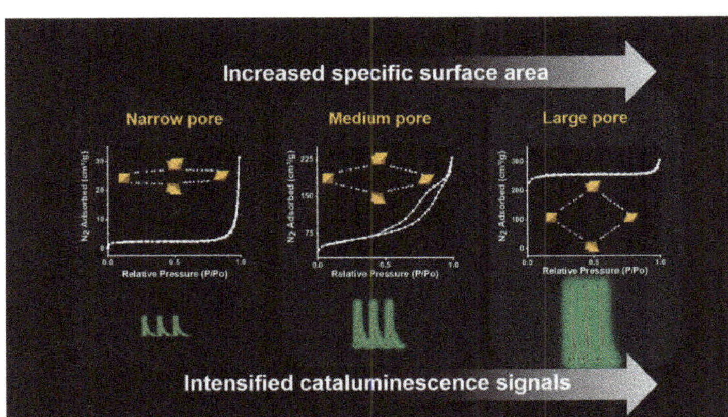

Scheme 1. Schematic representation of cataluminescence to evaluate the specific surface area for the MOFs.

2. Materials and Methods

2.1. Chemicals and Materials

All the reagents used in this work were of analytical grade and used without further purification. $Al(NO_3)_3 \cdot 9H_2O$ was provided by Beijing Chemical Works. $AlCl_3 \cdot 6H_2O$ was purchased from Xilong Science Co., Ltd. N, N-dimethylformamide (DMF) was purchased from Tianjin Fuyu Fine Chemical Co., Ltd. (Tianjin, China). 2, 2′-Bipyridine-5, 5′-dicarboxylic acid (H_2bpydc) was supplied by Shanghai Titan Scientific Co., Ltd. (Shanghai, China). Acetic acid glacial was purchased from Tianjin Damao Chemical Reagent Factory (Tianjin, China). NaOH was provided by Fuchen (Tianjin) Chemical Reagent Co., Ltd. (Tianjin, China).

2.2. Synthesis of the Narrow-Pore (np) Form of Al-Based MOF

The $Al_2(OH)_2$(bpydc)·$(H_2O)_{0.5}$ (short as *np*) was synthesized by an environmentally friendly hydrothermal method, and the unit cell parameters of np phase are a = 5.36 Å, b = 4.58 Å, c = 26.76 Å, α = 90°, β = 93.3°, and γ = 90° [7]. Specifically, $Al(NO_3)_3 \cdot 9H_2O$ (0.1 mmol, 0.375 g) and H_2bpydc (0.25 mmol, 0.061 g) were dissolved in H_2O (12.5 mL) containing 2.5 mmol of NaOH. The pH of the mixture was ~3.5. The mixture was placed in a 23 mL Teflon-lined stainless autoclave and heated at 120 °C for 24 h. After cooling to room temperature, the synthesized *np* was collected by centrifugation and dried at 60 °C.

2.3. Synthesis of the Large-Pore (lp) Form of Al-Based MOF

The macroporous Al(OH)(bpydc) (labelled as *lp*) was prepared through a hydrothermal method, and the unit cell parameters of the *lp* phase are determined as a = 23.59 Å, b = 6.91 Å, and c = 19.84 Å [33]. Briefly, the mixture of $AlCl_3 \cdot 6H_2O$ (0.151 g, 0.625 mmol), H_2bpydc (0.153 g, 0.625 mmol), DMF (10 mL), and glacial acetic acid (859 µL, 15.0 mmol) was placed in 23 mL of Teflon-lined stainless autoclave. The pH of the mixture was ~2.7. This mixture was heated at 120 °C for 24 h, followed by the centrifugation and drying procedure to acquire the MOFs with the *lp* phase. Note that the *lp* phase is the macroporous form of *np*, and the *lp* phase can be converted to *np* phase after hydrothermal treatment in water gradually [7].

2.4. Synthesis of the Medium-Pore (mp) Phase of Al-Based MOF

The medium-pore (*mp*) phases were synthesized by hydrothermal treatment of the *lp* phase under different temperatures. In total, 0.15 g of the *lp* phase was dispersed into 15 mL of water and transferred into a Teflon-lined stainless autoclave. The hydrothermal treatment was carried out at 150 °C, 140 °C, and 130 °C for 12 h, respectively. The as-prepared *mp* phases were cooled to room temperature, centrifuged, and dried to acquire *mp*-150, *mp*-140, and *mp*-130, respectively.

2.5. Apparatus and Characterization

The morphologies of the MOFs were investigated using a JEOL (JEOL, Akishima, Japan) S-4700 scanning electron microscope (SEM). The crystalline structures of the MOFs were characterized by X-ray diffraction (XRD) on Bruker (Karlsruhe, Germany) D8 ADVANCE equipped with graphite-monochromatized Cu Kα radiation (λ = 1.5406 Å).

2.6. Fourier Transform Infrared (FT-IR) Measurements

The FT-IR spectra for the samples before and after reactions were implemented on a Nicolet 6700 FTIR spectrometer (Thermo, Waltham, MA, USA) with KBr particles, and the spectra were collected in the range of 2200–400 cm^{-1}. In-situ Fourier transform infrared (FT-IR) spectra were acquired through a Nicolet 380 system (Thermo, Waltham, MA, USA) containing a controlled environment chamber with a CaF_2 window. The in-situ cell was loaded with the sample and placed in the diffuse reflectance device. The temperature was first programmed to 150 °C and maintained for 30 min under air purging. After cooling down, a background in the range of 30–150 °C was collected. Afterwards, ethanol was fed to the sample to reach an adsorption equilibrium, and the samples were purged until no changes in the infrared spectra. The IR signals were scanned through a programmed heating procedure from room temperature to 150 °C at a heating rate of 10 °C/min. The spectra were collected in the range of 4000–650 cm^{-1}, with a resolution of 8 cm^{-1} and an integration order of 64.

2.7. Gas Chromatography–Mass Spectrometry (GC–MS) Measurements

The oxidation products of ethanol by the MOF catalysts were investigated by gas chromatography–mass spectrometry (GC–MS). It was performed on a Thermo Trace 1300-ISQ GC-MS system (Thermo, Waltham, MA, USA) equipped with a TR-5MS column (length of 30 m, inner diameter of 0.25 mm, and the film thickness of 0.25 μm). The tail gas produced during the CTL reaction was collected through 10 mL of acetonitrile (liquid chromatography gradient) for about 4 h.

2.8. Brunauer–Emmett–Teller (BET) Measurements

Desorption isotherms were measured at 77 K by the Brunauer–Emmett–Teller (BET) method using an Autosorb-IQ-MP nitrogen adsorption device (Quantachrome, Boynton Beach, FL, USA).

2.9. Thermogravimetric Analysis (TGA) Measurements

Thermogravimetric analysis (TGA) of the samples was performed on a simultaneous thermal analyzer 6000 (PerkinElmer) at a heating rate of 10 °C/min in nitrogen from 30 °C to 800 °C. The data were analyzed in the range of 45–450 °C in the supporting information.

2.10. Cataluminescence (CTL) Measurements

The CTL signals were detected by a biophysical chemiluminescence (BPCL) luminescence analyzer (Institute of Biophysics, Chinese Academy of Sciences, Beijing, China), and the volatile organic gases were transported by air pump (Beijing Zhongxing Huili Co., Ltd., Beijing, China). First, 0.01 g of MOF catalyst was dispersed into 200 μL of deionized water. After ultrasonic treatment for 5 min, a uniform suspension was acquired. The suspension was evenly spread onto the surface of the ceramic rod, followed by natural air-drying. Afterwards, 50 μL of absolute ethanol was injected into the vaporization chamber, and the reactant was brought into the reaction chamber by an air pump. The flow rates of the carrier gas were set in the range of 160–400 mL·min^{-1}, and the heating voltages of the ceramic rod varied from 100 to 125 V. Finally, the CTL signals were output to the computer for data analysis, with the voltage of the BPCL analyzer set at −1000 V and an integration time of 1 s.

3. Results and Discussion

3.1. Structural and Morphological Characterization of Porous MOFs

Al-based MOFs with the different pore sizes were prepared through the hydrothermal method, involving large-pore (*lp*), narrow-pore (*np*), and the medium-pore (*mp*) phases synthesized under different temperatures. The crystallinity and structural integrity of these MOFs were characterized by XRD. The characteristic peak of the typical (002) plane in the *np* phase could be observed at 7.2° (Figure 1, black line), and this peak could be utilized to evaluate the unit cell dimensions for MOFs [7,34]. For the *lp* phase, the typical peak of the (002) plane shifted to 6.3° (Figure 1, purple line), suggesting varied cell sizes for the *lp* phase. Both the *lp* and *np* phases showed phase purity as observed from XRD patterns. The *mp* phases could be acquired by dispersing the *lp* phase in water, followed by thermal treatment at 150 °C, 140 °C, and 130 °C (labelled as *mp*-130, *mp*-140, and *mp*-150), respectively. It should be noted that two peaks appeared before 10° in the XRD patterns of these *mp* phases, located at ~6.4/7.4°, ~6.4/7.5°, and ~6.4/7.5° for *mp*-130, *mp*-140, and *mp*-150 (Table S1), respectively. These peaks suggested the gradual transformation between the *np* and *lp* phases, and that different phases coexisted in the *mp* samples. The morphological features of these MOF catalysts were observed by SEM images (Figure S1). The MOFs in the *lp* phase showed a conical shape with a smooth surface. Cracks appeared for the *mp* phases under the heating treatment, and the increased treating temperatures resulted in a severe crack from the *mp*-130 to the *mp*-150. The *np* phase showed a cracked morphology. Furthermore, TGA measurements were performed for these MOFs (Figure S2). It could be observed that all the MOFs with different phases did not show significant weight loss until the loss of the organic ligands at the temperatures of 373–386 °C (Table S2). The *lp* phase exhibited slight weight loss below 160 °C, attributing to the loss of water in the pore channels. These results were in accordance with the previous reports [7], suggesting the good thermal stability and potential applications of the MOFs in the CTL measurements.

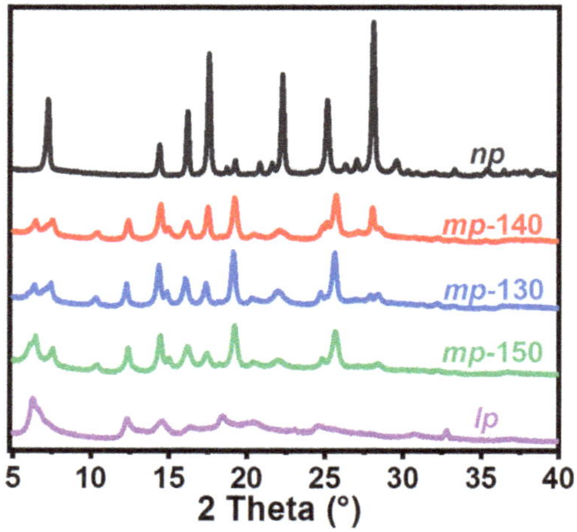

Figure 1. XRD patterns of MOF materials with *np*, *mp*-140, *mp*-130, *mp*-150, and *lp* phases.

3.2. Cataluminescence of Ethanol Catalyzed by MOFs

The synthesized MOFs were coated on the ceramic heating rod, and ethanol (50 µL) was injected into the flow path system to be catalyzed to emit luminescence. The cataluminescence (CTL) signals for the MOFs with the different phases were monitored continuously. Firstly, we have optimized the reaction temperatures and rates of carrier gas flow for the CTL system, taking *lp* phase as an example. It could be observed that the CTL intensities increased with the heating voltages, and the highest signal-to-noise (S/N) ratio can be acquired at 115 V (Figure S3a). Similarly, the flow rate was chosen at 240 mL·min^{-1} with the strongest CTL intensity and stable S/N values (Figure S3b). MOFs with the different phases were used for the CTL measurements. Good repeatability and reproducibility could be acquired from the parallel CTL experiments (Figure 2). These signals are stable and repeatable. Taking the *np* phase as an example, the average value and the standard deviation of the measurements were 23,715 and 259.72, respectively. It should be noted that the CTL signals of ethanol changed significantly upon the catalytic reaction by the MOFs with the different phases. Specifically, the CTL signal of ethanol catalyzed by the *lp* phase reached ~67,560, while the intensity for the *np* phase was only ~23,715. The MOFs in the *mp* forms showed the dependent CTL intensities on the varied treatment temperatures. Accordingly, the CTL intensity variations of these MOFs followed the sequence as follows: *lp* > *mp*-150 > *mp*-130 > *mp*-140 > *np*. These differences in the CTL measurements could be ascribed to the distinct adsorption–desorption and oxidative reaction behaviors of ethanol on the MOFs with differed structural features. Moreover, IR measurements for MOFs before and after CTL measurements were implemented (Figure S4). It could be observed that the absorption peaks at 1690 and 1310 cm^{-1} were attributed to the vibration of carboxylic acid C=O and C−O bonds. The aromatic C=C bond appeared at 1430 cm^{-1}, and the absorption peaks at 1400, 1480, and 1590 cm^{-1} can be attributed to stretching of $-CO_2$. After the CTL experiments, no obvious changes could be observed from the IR spectra, illustrating the stability and reusability of these MOFs as catalysts.

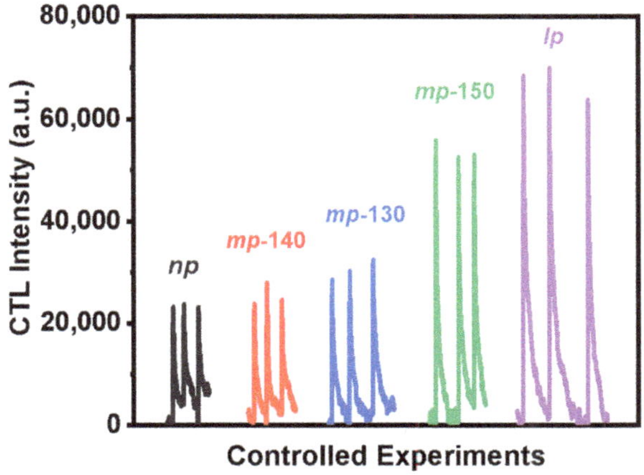

Figure 2. Comparison of CTL intensities by ethanol catalyzed by the MOFs with the different phases (air flow rate of 240 mL·min^{-1}, working voltage of 115 V, and integration time of 1 s).

In order to explore the CTL mechanism of ethanol in the presence of MOFs, we have recorded the CTL spectra of the reaction system under a series of wavelength filters. It is evident that the maximum CTL intensities could be observed around 550 nm and 620 nm for ethanol on the surfaces of *lp*, *mp*, and *np* phases (Figure 3a,b and Figure S5). These emissions could be ascribed to the characteristic emission peak of CH_3CHO^*, according to literature reports [30,35]. We have further studied the intermediates produced during the oxidative reaction of ethanol by in-situ FT-IR spectroscopy. Taking the *np* phase as an example, the absorbance bands ascribed to the −OH (at ~3300 cm^{-1}) and the corresponding C−H tensile vibration (at ~3000 cm^{-1}) decayed gradually with the increased reaction temperatures from 30 °C to 150 °C (Figure 3c). This decrease indicated the gradual consumption of ethanol during the reaction. The peaks around 1735 cm^{-1} corresponded to the C=O peak in the transition state of CH_3CHO. These peaks showed a trend increase and subsequent decrease during the reaction, suggesting that the CH_3CHO acted as an intermediate. These phenomena could also be validated by the GC–MS spectra with the peaks at 1.62/1.52 min for CH_3CHO (Figure S6). Moreover, the absorption bands at 1586 and 1513 cm^{-1} indicated the formation of CH_3COOH (Figure 3d), as an oxidized product of CH_3CHO. Finally, these intermediates were reacted to form gaseous CO_2, manifested as the increased peak at 2356 cm^{-1} (Figure 3c,d). These results demonstrated an oxidative reaction mechanism of ethanol catalyzed by MOFs (Scheme S1). Briefly, the adsorbed ethanol was firstly catalyzed to generate the intermediate CH_3CHO^*. The intermediate CH_3CHO^* was transformed to CH_3CHO, followed by the conversion to the CH_3COOH, and finally CO_2. Therefore, the reaction of ethanol in the presence of different MOFs could be revealed by the CTL signals.

3.3. Cataluminesence of Ethanol Catalyzed by MOFs

We have performed the BET measurements on the MOFs with the different phases. The *np* and *lp* phases showed the characteristics of type I adsorption isotherms of microporous substances (Figure 4a), and the pore structures were relatively regular. The N_2 adsorption curves of *mp*-150, *mp*-140, and *mp*-130 phases according to the type IV adsorption isotherms of microporous substances, manifested as the irregular pore structures [21,24]. The N_2 adsorption curve of the *mp* phase belonged to the H4 type hysteresis loop, and the adsorption branch was composed of the I-type microporous and II-type mesoporous isotherms [36,37]. This phenomenon could be ascribed to the fact that the framework transition of MOFs was a gradual process instead of an abrupt change, and multiple phases coexisted in the *mp* phases.

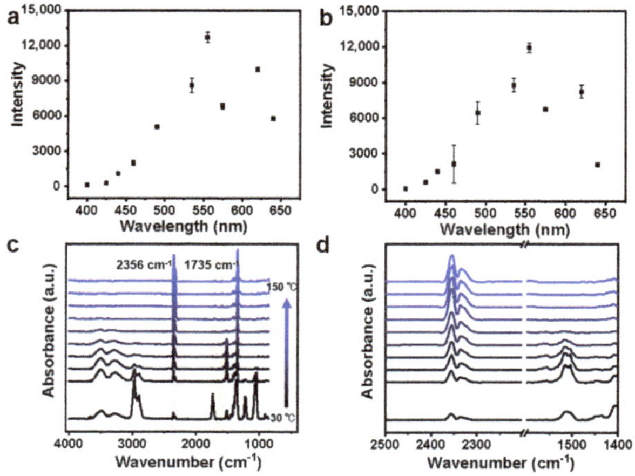

Figure 3. CTL intensities of ethanol in the presence of (**a**) *lp* and (**b**) *np* phases. In−situ FT−IR spectra in the range of (**c**) 4000−850 cm^{-1} and (**d**) 2500−1400 cm^{-1} for ethanol under increased reaction temperatures during the oxidative reaction of ethanol in the presence of MOFs with *np* phase.

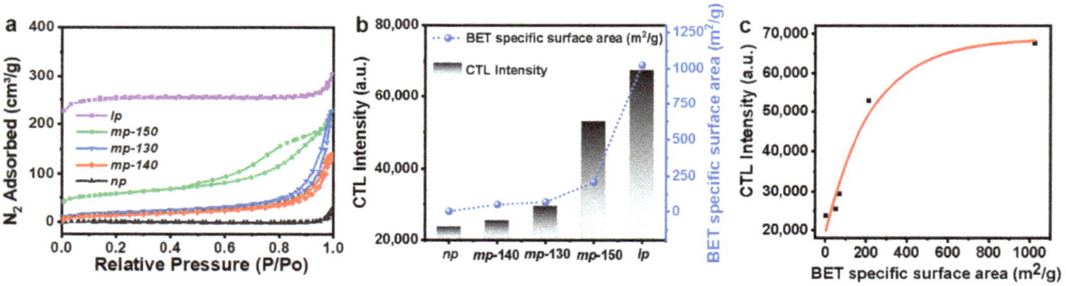

Figure 4. (**a**) N_2 adsorption and desorption curves, (**b**) variations of CTL intensities and specific surface area of different MOFs, and (**c**) correlation between the CTL intensities and specific surface area of different MOFs.

The adsorption behaviors of these MOFs were further analyzed by the calculations of pore size and specific surface area. The pore sizes of MOFs with the different phases have been estimated (Figure S7). It could be observed that the *lp* phase showed the largest pore sizes of 20.59 Å, corresponding to the highest CTL intensity of ethanol. The *mp*-150 showed a prominent pore size of 19.16 Å, corresponding to a relatively strong CTL intensity of ethanol. Moreover, although the *mp*-130, *mp*-140, and *np* phases shared similar pore sizes, the CTL signal varied. These results suggested that the pore sizes of MOFs were not the determinant factor for the CTL signals of ethanol. The specific surface areas of MOFs have been further investigated. The values of specific surface area for these MOFs increased from 2.66 m^2/g for the *np* phase to 1024.88 m^2/g for the *lp* phase, in accordance with previous reports [7,38]. The values of the specific surface area of these MOFs accorded to a tendency of *np* < *mp*-140 < *mp*-130 < *mp*-140 < *lp* (Table S3). It is noteworthy that the specific surface areas of the MOFs were positively correlated with the CTL intensities of ethanol (Figure 4b). The larger the specific surface area of the MOFs, the higher the CTL signal intensities that could be acquired for the oxidation of ethanol. For a better illustration, a quantitative relationship between the surface area and the activity of MOFs was generally needed [39]. The CTL intensity (I) is positively correlated with the specific surface area from BET measurements (SA): $I = -4.94 \times 10^4 \exp(4.39 \times 10^{-3}\, SA) + 6.88 \times 10^4$ (Figure 4c). Such a correlation could be ascribed to the distinct adsorption–desorption behaviors of the

MOFs towards ethanol. The different specific surface areas of the MOFs determined the quantities of adsorbed ethanol and thus the varied CTL intensity signals. These results demonstrated the ability of CTL measurement as an indicator to evaluate and discriminate the specific surface area of the MOFs in a cost-effective approach.

The versatility of the proposed strategy on screening the specific surface area of MOFs has been further validated by the other gas molecules, such as methanol and isopropanol. It can be observed that the CTL signals of methanol and isopropanol increased from the *np*, *mp*-150, to *lp* phases, in accordance with the signals of ethanol discussed above (Figure S8). These results demonstrated the accuracy of the proposed CTL method to evaluate the specific surface area of MOFs. Interestingly, the CTL intensity varied with the molecular structures of alcohols in the presence of a same MOF material (Figure S9). Methanol, with the smallest molecular size, showed the highest CTL intensity on these MOFs. With the increase in molecular size from methanol to isopropanol, the CTL intensities decayed accordingly. Therefore, it could be concluded that the highest CTL signal could be acquired for the smallest methanol molecules in the presence of MOFs with the largest pore structures. These results suggested the varied accommodation effect of guest molecules in the host MOFs, leading to the distinguished adsorption–desorption behaviors in the MOFs with different phases. It is anticipated that the proposed strategy could also be utilized to screen the molecules with the different sizes by MOFs.

4. Conclusions

In summary, we have proposed the CTL strategy to screen and evaluate the surface area of MOFs in an efficient and accurate approach. The CTL intensities from the reaction of ethanol varied consistently with the surface area and the adsorption capacity for the MOFs with different phases. The larger the specific surface area for MOFs, the higher the CTL intensities of ethanol that could be acquired after a catalytic oxidation. This work has successfully established an efficient platform to evaluate the structures of MOFs through the luminescence signals from a catalytic chemical reaction. It is anticipated that our approach could be utilized for the rapid screening on the structural and adsorption characteristics for the porous materials.

Supplementary Materials: The following supporting information can be downloaded at: https://www.mdpi.com/article/10.3390/chemosensors11050292/s1. Figure S1: SEM images of (a) *np*, (b) *mp*-140, (c) *mp*-130, (d) *mp*-150, and (e) *lp* phases of MOFs. Figure S2: TGA analysis for different MOF materials. Figure S3: Comparisons of the cataluminescence (CTL) intensities and the signal-to-noise (S/N) values of ethanol in the presence of the *lp* phase at (a) different temperatures (air flow rate of 240 mL·min^{-1} and integration time of 1 s) and (b) different flow rates (working voltage of 115 V and integration time of 1 s). Figure S4: FT-IR spectra for different MOF materials before and after CTL measurements. Figure S5: CTL spectrum for the oxidation of ethanol in the presence of the *mp*-150 phase. Figure S6: Gas chromatography–mass spectrometry (GC–MS) measurements for the reaction of ethanol by the (a) *np*, (b) *mp*-150, and (c) *lp* phases. Figure S7: Pore size distribution analysis for *lp*, *mp*-150, *mp*-130, *mp*-140, and *np* phases. Figure S8: CTL intensities for (a) methanol, (b) ethanol, and (c) isopropanol in the presence of MOFs with the different phases (the integration time of 0.1 s). Figure S9: CTL intensity variations for methanol, ethanol, and isopropanol in the presence of MOFs with the different phases (the integration time of 0.1 s). Scheme S1: Reaction mechanism and corresponding measurements for the CTL emissions from oxidation of ethanol in the presence of MOFs. Table S1: Peak position analysis for the MOFs with different phases from XRD patterns. Table S2: Weight loss analysis for different MOF materials from TGA measurements. Table S3: CTL intensities and specific surface areas acquired by the Brunauer–Emmett–Teller (BET) measurements for different MOF materials.

Author Contributions: Conceptualization, Z.L., R.T. and C.L.; validation, Z.L. and D.P.; investigation, D.P., R.T. and C.L.; writing—original draft preparation, Z.L., D.P., R.T. and C.L.; writing—review and editing, R.T. and C.L.; supervision, C.L.; funding acquisition, R.T. and C.L. All authors have read and agreed to the published version of the manuscript.

Funding: This work was supported by the National Natural Science Foundation of China (21974008, 21838007 and U22A20397) and Beijing Natural Science Foundation (2222018).

Institutional Review Board Statement: Not applicable.

Informed Consent Statement: Not applicable.

Data Availability Statement: Not applicable.

Conflicts of Interest: The authors declare no competing financial interest.

References

1. Kurisingal, J.F.; Li, Y.; Sagynbayeva, Y.; Chitumalla, R.K.; Park, D.-W. Porous aluminum-based DUT metal-organic frameworks for the transformation of CO_2 into cyclic carbonates: A computationally supported study. *Catal. Today* **2019**, *352*, 227–236. [CrossRef]
2. Kustov, L.M.; Isaeva, V.I.; Přech, J.; Bisht, K.K. Metal-organic frameworks as materials for applications in sensors. *Mendeleev Commun.* **2019**, *29*, 361–368. [CrossRef]
3. An, B.; Li, Z.; Song, Y.; Zhang, J.; Lin, W. Cooperative copper centres in a metal-organic framework for selective conversion of CO_2 to ethanol. *Nat. Catal.* **2019**, *2*, 709–717. [CrossRef]
4. Nath, K.; Ahmed, A.; Siegel, D.J.; Matzger, A.J. Microscale determination of binary gas adsorption isotherms in MOFs. *J. Am. Chem. Soc.* **2022**, *144*, 20939–20946. [CrossRef]
5. Rowsell, J.L.C.; Yaghi, O.M. Strategies for hydrogen storage in metal-organic frameworks. *Angew. Chem. Int. Ed.* **2004**, *44*, 4670–4679. [CrossRef] [PubMed]
6. Eddaoudi, M.; Yaghi, O.M. Systematic design of pore size and functionality in isoreticular MOFs and their application in methane storage. *Science* **2002**, *295*, 469–472. [CrossRef] [PubMed]
7. Liang, J.; Li, X.; Xi, R.; Shan, G.; Li, P.-Z.; Liu, J.; Zhao, Y.; Zou, R. A Robust aluminum metal-organic framework with temperature-induced breathing effect. *ACS Mater. Lett.* **2020**, *2*, 220–226. [CrossRef]
8. Senkovska, I.; Hoffmann, F.; FröBa, M.; Getzschmann, J.; BöHlmann, W.; Kaskel, S. New highly porous aluminium based metal-organic frameworks: Al(OH)(ndc) (ndc=2,6-naphthalene dicarboxylate) and Al(OH)(bpdc) (bpdc=4,4′-biphenyl dicarboxylate). *Microporous Mesoporous Mater.* **2009**, *122*, 93–98. [CrossRef]
9. Wang, H.; Warren, M.; Jagiello, J.; Jensen, S.; Ghose, S.K.; Tan, K.; Yu, L.; Emge, T.J.; Thonhauser, T.; Li, J. Crystallizing atomic xenon in a flexible MOF to probe and understand its temperature-dependent breathing behavior and unusual gas adsorption phenomenon. *J. Am. Chem. Soc.* **2020**, *142*, 20088–20097. [CrossRef]
10. Huang, X.; Huang, Z.; Zhang, L.; Liu, R.; Lv, Y. Highly efficient cataluminescence gas sensor for acetone vapor based on UIO-66 metal-organic frameworks as preconcentrator. *Sens. Actuators B Chem.* **2020**, *312*, 127952. [CrossRef]
11. Jiao, L.; Seow, J.Y.R.; Skinner, W.S.; Wang, Z.U.; Jiang, H.-L. Metal-organic frameworks: Structures and functional applications. *Mater. Today* **2019**, *27*, 43–68. [CrossRef]
12. Kreno, L.E.; Leong, K.; Farha, O.K.; Allendorf, M.; Duyne, R.P.V.; Hupp, J.T. Metal-organic framework materials as chemical sensors. *Chem. Rev.* **2012**, *112*, 1105–1125. [CrossRef] [PubMed]
13. Zhang, X.; Xiong, B.; Li, J.; Qian, L.; Liu, L.; Liu, Z.; Fang, P.; He, C. Dependence of dye molecules adsorption behaviors on pore characteristics of mesostructured MOFs fabricated by surfactant template. *ACS Appl. Mater. Interfaces* **2019**, *11*, 31441–31451. [CrossRef]
14. Srivastava, S.; Shet, S.P.; Priya, S.S.; Sudhakar, K.; Tahir, M. Molecular simulation of copper based metal-organic framework (Cu-MOF) for hydrogen adsorption. *Int. J. Hydrogen Energy* **2022**, *47*, 15820–15831. [CrossRef]
15. Tokalıoğlu, Ş.; Yavuz, E.; Demir, S.; Patat, Ş. Zirconium-based highly porous metal-organic framework (MOF-545) as an efficient adsorbent for vortex assisted-solid phase extraction of lead from cereal, beverage and water samples. *Food Chem.* **2017**, *237*, 707–715. [CrossRef] [PubMed]
16. He, H.; Li, R.; Yang, Z.; Chai, L.; Jin, L.; Alhassan, S.I.; Ren, L.; Wang, H.; Huang, L. Preparation of MOFs and MOFs derived materials and their catalytic application in air pollution: A review. *Catal. Today* **2021**, *375*, 10–29. [CrossRef]
17. Deng, X.; Qin, Y.; Hao, M.; Li, Z. MOF-253-supported Ru complex for photocatalytic CO_2 reduction by coupling with semidehydrogenation of 1,2,3,4-Tetrahydroisoquinoline (THIQ). *Inorg. Chem.* **2019**, *58*, 16574–16580. [CrossRef] [PubMed]
18. Gómez-Gualdrón, D.A.; Moghadam, P.Z.; Hupp, J.T.; Farha, O.K.; Snurr, R.Q. Application of consistency criteria to calculate BET areas of micro- and mesoporous metal-organic frameworks. *J. Am. Chem. Soc.* **2016**, *138*, 215–224. [CrossRef] [PubMed]
19. Datar, A.; Chung, Y.G.; Lin, L.-C. Beyond the BET analysis: The surface area prediction of nanoporous materials using a machine learning method. *J. Phys. Chem. Lett.* **2020**, *11*, 5412–5417. [CrossRef]
20. Thommes, M. Physisorption of gases, with special reference to the evaluation of surface area and pore size distribution (IUPAC Technical Report). *Pure Appl. Chem.* **2015**, *87*, 1051–1069. [CrossRef]
21. Islamoglu, T.; Idrees, K.B.; Son, F.A.; Chen, Z.; Lee, S.-J.; Li, P.; Farha, O.K. Are you using the right probe molecules for assessing the textural properties of metal-organic frameworks? *J. Mater. Chem. A* **2022**, *10*, 157–173. [CrossRef]
22. Brunauer, S.; Emmett, P.H. The use of van der waals adsorption isotherms in determining the surface arer of iron synthetic ammonia catalysts. *J. Am. Chem. Soc.* **1935**, *57*, 1754–1755. [CrossRef]

23. Walton, K.S.; Snurr, R.Q. Applicability of the BET method for determining surface areas of microporous metal-organic frameworks. *J. Am. Chem. Soc.* **2007**, *129*, 8552–8556. [CrossRef]
24. Ambroz, F.; Macdonald, T.J.; Martis, V.; Parkin, I.P. Evaluation of the BET theory for the characterization of meso and microporous MOFs. *Small Methods* **2018**, *2*, 1800173. [CrossRef]
25. Bau, S.; Witschger, O.; Gensdarmes, F.; Rastoix, O.; Thomas, D. A TEM-based method as an alternative to the BET method for measuring off-line the specific surface area of nanoaerosols. *Powder Technol.* **2010**, *200*, 190–201. [CrossRef]
26. Kockler, K.B.; Frisch, H.; Barner-Kowollik, C. Making and breaking chemical bonds by chemiluminescence. *Macromol. Rapid Commun.* **2018**, *39*, 1800516. [CrossRef] [PubMed]
27. Kwon, M.S.; Jang, G.; Bilby, D.; Lee, T.S.; Kim, J. Design principles of chemiluminescence (CL) chemodosimeter for self-signaling detection: Luminol protective approach. *RSC Adv.* **2014**, *4*, 46488–46493. [CrossRef]
28. Meng, F.; Lu, Z.; Zhang, R.; Li, G. Cataluminescence sensor for highly sensitive and selective detection of iso-butanol. *Talanta* **2019**, *194*, 910–918. [CrossRef] [PubMed]
29. Wang, S.; Yuan, Z.; Zhang, L.; Lin, Y.; Lu, C. Recent advances in cataluminescence-based optical sensing systems. *Analyst* **2017**, *142*, 1415–1428. [CrossRef] [PubMed]
30. Cheng, W.; Guan, W.; Lin, Y.; Lu, C. Rapid discrimination of adsorbed oxygen and lattice oxygen in catalysts by the cataluminescence method. *Anal. Chem.* **2022**, *94*, 1382–1389. [CrossRef]
31. Zhang, L.; Song, H.; Su, Y.; Lv, Y. Advances in nanomaterial-assisted cataluminescence and its sensing applications. *Trends Anal. Chem.* **2015**, *67*, 107–127. [CrossRef]
32. Zeng, N.; Long, Y.; Wang, J.; Sun, J.; Ouyang, N.; Na, N. An Acetone Sensor Based on Plasma-Assisted Cataluminescence and Mechanism Studies by Online Ionizations. *Anal. Chem.* **2019**, *91*, 15763–15768. [CrossRef] [PubMed]
33. Bloch, E.D.; Britt, D.; Lee, C.; Doonan, C.J.; Uribe-Romo, F.J.; Furukawa, H.; Long, J.R.; Yaghi, O.M. Metal insertion in a microporous metal-organic framework lined with 2,2′-bipyridine. *J. Am. Chem. Soc.* **2010**, *132*, 14382–14384. [CrossRef] [PubMed]
34. Yuan, S.; Huang, L.; Huang, Z.; Sun, D.; Qin, J.-S.; Feng, L.; Li, J.; Zou, X.; Cagin, T.; Zhou, H.-C. Continuous variation of lattice dimensions and pore sizes in metal-organic frameworks. *J. Am. Chem. Soc.* **2020**, *142*, 4732–4738. [CrossRef] [PubMed]
35. Ye, Q.; Gao, Q.; Zhang, X.-R.; Xu, B.-Q. Cataluminescence and catalytic reactions of ethanol oxidation over nanosized $Ce_{1-x}Zr_xO_2$ ($0 \leq x \leq 1$) catalysts. *Catal. Commun.* **2006**, *7*, 589–592. [CrossRef]
36. de Souza, M.J.B.; Silva, T.H.A.; Ribeiro, T.R.S.; da Silva, A.O.S.; Pedrosa, A.M.G. Thermal and catalytic pyrolysis of polyvinyl chloride using micro/mesoporous ZSM-35/MCM-41 catalysts. *J. Therm. Anal. Calorim.* **2020**, *140*, 167–175. [CrossRef]
37. Tanhaei, M.; Mahjoub, A.R.; Safarifard, V. Energy-efficient sonochemical approach for the preparation of nanohybrid composites from graphene oxide and metal-organic framework. *Inorg. Chem. Commun.* **2019**, *102*, 185–191. [CrossRef]
38. Yan, B.; Qin, S.-J. A facile indicator box based on Eu^{3+} functionalized MOF hybrid for the determination of 1-naphthol, a biomarker for carbaryl in urine. *Sens. Actuators B Chem.* **2018**, *259*, 125–132.
39. Visa, A.; Mracec, M.; Maranescu, B.; Maranescu, V.; Ilia, G.; Popa, A.; Mracec, M. Structure simulation into a lamellar supramolecular network and calculation of the metal ions/ligands ratio. *Chem. Cent. J.* **2012**, *6*, 91. [CrossRef]

chemosensors

MDPI

Article

Direct Electrochemical Analysis in Seawater: Evaluation of Chloride and Bromide Detection

Yuqi Chen and Richard G. Compton *

Physical & Theoretical Chemistry Laboratory, University of Oxford, Oxford OX1 3QZ, UK; yuqi.chen@sjc.ox.ac.uk
* Correspondence: richard.compton@chem.ox.ac.uk; Tel.: +44-(0)-1865-275957; Fax: +44-(0)-1865-275410

Abstract: Chloride and bromide are two of the most abundant anions found in seawater, and knowledge of their concentrations is essential for environmental monitoring. However, the analysis of chloride and bromide in seawater is challenging due to the complex nature of the seawater matrix. From an electrochemical perspective, we investigate the suitability of three types of electrode (Au, glassy carbon and Pt) for the analysis of Cl^- and/or Br^- in seawater. With the understanding of their electrochemical behaviours in artificial seawater (ASW), optimal voltammetric procedures for their detection are developed. The results show that the Au electrode is unsuitable for use as a Cl^- and/or Br^- sensor due to its dissolution and passivation in ASW. The use of glassy carbon resulted in poorly defined chloride and bromide signals. Finally, platinum was found to be a good candidate for chloride detection in artificial seawater using square wave voltammetry, and the results obtained in natural seawater via electrochemical measurement were in good agreement with those obtained via ion chromatography. Platinum electrodes are also recommended for bromide analysis.

Keywords: electrochemical analysis; electrochemical sensor; platinum electrode; glassy carbon electrode; gold electrode; chloride quantification; bromide quantification; amperometry; seawater

Citation: Chen, Y.; Compton, R.G. Direct Electrochemical Analysis in Seawater: Evaluation of Chloride and Bromide Detection. *Chemosensors* **2023**, *11*, 297. https://doi.org/10.3390/chemosensors11050297

Academic Editor: Matiar R Howlader

Received: 21 April 2023
Revised: 8 May 2023
Accepted: 16 May 2023
Published: 18 May 2023

1. Introduction

Water scarcity is a worldwide concern, while climate change exacerbates the water shortage problem [1]. As seawater consists of 96.5% of all known water on the earth and covers 71% of Earth's surface [2], desalinated seawater is a promising solution to the freshwater shortage, and has been implemented in the Near and Middle East, North Africa, North America and in Mediterranean countries [3]. Meanwhile, noting that the main chemicals in resultant concentrated seawater, such as salt (sodium chloride), magnesium, potassium and bromine, are basic raw materials for the chemical industry, extracting these highly concentrated and stable chemicals from concentrated seawater reduces energy consumption and makes comprehensive use of natural resources [4,5]. In this context, to maximise efficiency and utility, an important step is measuring the concentration of targeted ions in seawater before desalination. More generally, it is also meaningful to analyse the compounds in seawater, which reveals the local environment, geology and potential impact on humans. Chloride and bromide are the two anions addressed in this paper.

As shown in Table 1, chloride is the most concentrated anion in seawater. The concentration of chloride has been defined as 'chlorinity' [6,7], and it is so dominant that it was originally used to estimate salinity (the total salts dissolved in seawater) [8]. Salinity is an indicator of the ocean ecosystem, as the relative salt concentration within the body of a creature living in the sea versus the water is what determines the direction of osmotic pressure and the ease of water absorbed into or removed from the body [9]. Meanwhile, measuring the specific chloride level in natural seawater systems is also necessary for both environmental studies and corrosion science. First, toxic substances such as Hg (0) and silver nanoparticles (AgNPs) are more prone to oxidation in the presence of a high concentration of chloride, leading to soluble compounds [10–13]. Secondly, marine chloride can directly lead to metal corrosion [14,15], which is a major ongoing challenge in shipbuilding and maritime transport.

Bromine is an essential chemical with wide applications, such as insecticides, flame retardants, refrigerants, photosensitive materials and oilfield mining [16]. About 99% of bromine on Earth exists in the form of Br^- in seawater [17], and bromide is one of the most concentrated anions in seawater (Table 1) [18]. With seawater containing about 89 trillion tonnes of bromine [19], it is one of the main sources of bromine, the extraction of which needs to be informed by measurements of the local concentrations of bromide in seawater. Meanwhile, a key step in seawater desalination is also to remove bromide and check its concentration in the resultant water, because it is toxic and carcinogenic [20]. In addition, as the local concentrations of chloride and bromide depend on location and season [21,22], the ratio between them in normal seawater is used to monitor pollution from agricultural and industrial sources [23,24]. However, the levels of chloride are hundreds of times higher than those of bromide by weight in natural seawater [25,26], which can lead to challenges in the determination of Br^- [27,28]. Thus, it is necessary to develop methods that can measure Cl^- and/or Br^- sensitively and selectively in authentic seawater.

Table 1. Concentrations of major ions in surface seawater [18].

Ion	Molar Concentration/mM
Na^+	4.81×10^2
Mg^{2+}	5.41×10^1
Ca^{2+}	1.05×10^1
K^+	1.05×10^1
Sr^{2+}	9.20×10^{-2}
Cl^-	5.59×10^2
$SO_4{}^{2-}$	2.89×10^1
$HCO_3{}^-$	1.89
$CO_3{}^{2-}$	1.89×10^{-1}
Br^-	8.63×10^{-1}
$B(OH)_4{}^-$	8.39×10^{-2}
$B(OH)_3$	3.42×10^{-1}
F^-	7.00×10^{-2}

As reported by Guo et al. [10], there are several well-established methods to measure the chloride level in seawater. Traditionally, the argentometric method (Mohr's method) [29,30] and colourimetry using potassium chromate to indicate the endpoint of adding silver nitrate [31] are applied for neutral solutions. Meanwhile, potentiometric titration using silver nitrate [32–34] is considered to be easier and more precise than the colourimetric method, but both involve large sample volumes and toxic chemicals. Thus, instrumental rather than wet methods are currently commonly used for the measurement of chlorinity (Cl, g kg^{-1}). Notably, a salinometer based on the electrical conductivity of seawater, together with temperature, provides salinity (S) readings. To be more specific, *S* is calculated from the ratio of the conductivity of seawater to that of a KCl solution (32.4356 g/kg), with both measurements made at 15 °C and under one-atmosphere pressure [7,35–37]. It can then be converted to chlorinity using the Knudsen equation (Equation (1)) [6].

$$S\left(g\,Kg^{-1}\right) = 1.80665\,Cl\left(g\,Kg^{-1}\right) \quad (1)$$

However, there is necessarily a lack of accuracy, as it is an estimated value depending on salinity that primarily indicates the overall ion concentration rather than the chemical composition [7,35]. For bromide quantification, there are several techniques available, e.g., kinetic spectrophotometry [38–45], chromatography [46,47], capillary electrophoresis [48–52] and electrochemistry [53–56], but they have drawbacks from different perspectives, which have been discussed by Chen et al. [57]. In brief, kinetic spectrophotometry may not be ideal for natural samples, as the oxidising agent could react with other ions present [38,43] and dilution of the sample is required [38]. Although chromatography and capillary electrophoresis can result in relatively accurate quantification, they are costly and time-consuming in terms of the equipment, sample preparation (dilution) and analysis [41,58].

Electrochemistry provides the benefits of robust, low-cost and simple instrumentation with competitive selectivity and sensitivity, which makes it a good analytical method for bromide and chloride quantification. Although potentiometry has been widely applied as an electrochemical technique [53–56,59,60], the complexity of electrode fabrication [54] and the logarithmic relationship between potential and the analyte make it challenging to deliver a precise and sensitive detection. Meanwhile, for existing potentiometric measurements, selectivity, especially for bromide detection with the presence of chloride at hundreds of times the level in seawater, remains a challenge [53,54]. Indication of the misfunction of potentiometric electrodes is another essential requirement that cannot be easily met. Thus, amperometry is a technique that can avoid the drawbacks of potentiometry mentioned above [61–65], but traditionally has required analytes with a low concentration so that seawater cannot be directly used. Nevertheless, these problems have recently been solved by Guo et al. [10] and Chen et al. [57], with the first two bespoke regent-free amperometric sensors being proposed for chloride and bromide sensing, respectively, in authentic seawater.

In this paper, we assessed, compared and contrasted different electrode materials for the direct analysis of Cl^- and/or Br^- in seawater. Three electrodes, made of Au, glassy carbon and Pt, were investigated, and the full voltametric responses seen in undiluted seawater were described and analysed to understand their fundamental (electro-)chemistry using both artificial seawater (ASW) and natural seawater (NSW). The implications for the electrochemical detection of bromide and chloride were assessed, and through optimal voltametric procedures, we identified two anions of interest.

2. Experimental Section

2.1. Chemicals and Reagents

Deionised water with a resistivity of 18.2 MΩ cm at 298 K (Millipore, Millipak Express 20, Watford, UK) was utilised to prepare all synthetic solutions. All the chemicals used in the experiments were of analytical grade [66]. Artificial seawater (ASW, pH = 8.1) was prepared using the composition shown in Table 2. For the chemicals used for ASW, sodium bicarbonate ($NaHCO_3$, 99.7%) was purchased from Acros Organics, UK, whereas sodium chloride (NaCl, 99.5%), potassium chloride (KCl, 99.5%), magnesium chloride dihydrate ($MgCl_2$ $6H_2O$, 99.0%), calcium chloride dihydrate ($CaCl_2$ $2H_2O$, 99.0%), strontium chloride hexahydrate ($SrCl_2$ $6H_2O$, 99.0%), sodium sulphate (Na_2SO_4, 99.0%), potassium bromide (KBr, 99.0%), sodium fluoride (NaF, 99.0%) and boric acid (H_3BO_3, 99.5%) were purchased from Sigma-Aldrich, UK. The "blank" solution with the same ionic strength as seawater [66] was prepared using sodium nitrate ($NaNO_3$, ≥99.5%), procured from Merck KGaA, Germany.

Table 2. Chemical composition of synthetic seawater [66].

Composition	Molar Concentration (mol dm^{-3})
NaCl	4.20×10^{-1}
KCl	9.39×10^{-3}
$MgCl_2$ $6H_2O$	5.46×10^{-2}
$CaCl_2$ $2H_2O$	1.05×10^{-2}
$SrCl_2$ $6H_2O$	6.38×10^{-5}
Na_2SO_4	2.88×10^{-2}
$NaHCO_3$	2.38×10^{-3}
KBr	8.40×10^{-4}
NaF	7.15×10^{-5}
H_3BO_3	4.85×10^{-5}

2.2. Electrochemical Apparatus and Methods

Electrochemical measurements were conducted using a μAutolab II potentiostat made by Metrohm-Autolab BV in Utrecht, Netherlands. The experiments were performed using a standard three-electrode set-up at a constant temperature of 25.0 ± 0.2 °C in a Faraday cage. The reference and counter electrodes were a mercury–mercurous sulphate electrode (MSE +0.64 V vs. SHE, BASi, West Lafayette, IN, USA) and a graphite rod in all experiments. Three different working electrodes were used, namely, a gold macro-electrode (diameter of 1.60 ± 0.01 mm, geometric area of 0.02 cm^2, BASi, West Lafayette, IN, USA), a glassy carbon macro-electrode (GCE, diameter of 3.00 ± 0.01 mm, geometric area of 0.07 cm^2, BASi, West Lafayette, IN, USA) and a platinum macro-electrode (diameter of 1.66 ± 0.01 mm, geometric area of 0.02 cm^2, BASi, West Lafayette, IN, USA). Before each use, the working electrode was polished with 1.0, 0.3 and 0.05 μm alumina lapping compounds from Bucher, Germany. To remove oxygen from the solutions before electrochemical measurements, high-purity N_2 flow from BOC Gases plc, UK was used as needed. Cyclic voltammetry (CV) was conducted to investigate the electrochemical behaviour of the three working electrodes in ASW. Then, for the quantification of either bromide or chloride in ASW and/or natural seawater, square wave voltammetry (SWV) was applied.

2.3. Sample Collection and Ion Chromatography Analysis

There were two authentic seawaters used in this study, and the collection of these is described by Chen et al. [57]. Ion chromatography analysis was conducted using a Dionex (Thermo Scientific, Sunnyvale, CA, USA) ICS-5000+ SP instrument, utilising a Dionex IonPac AS23-4um analytical column for the separations. Before analysis, two seawater samples were each mixed with deionised water with a resistivity of 18.2 MΩ cm at 298 K at a dilution factor of 200.

3. Results and Discussion

In the following sections, the electrochemical behaviour of the gold electrode in ASW is first reported and interpreted using observations from pure chloride and bromide solutions reported in the literature. The inferred dissolution of gold in halide solutions is concluded to make Au a less-than-optimal electrode material. Thus, glassy carbon electrodes (GCEs) and Pt electrodes are then considered as alternatives for Cl^- and/or Br^- detection in ASW. In addition to using cyclic voltammetry to assess the merits of the two electrodes as sensors, square wave voltammetry is also attempted/implemented to enhance their sensitivity, focussing particularly on using the GCE towards Br^- and the Pt electrode towards Cl^-, leading to recommendations for the optimal electrode for and approach to each of the two ions. Finally, a bespoke method using a Pt electrode for chloride sensing is applied for quantitative analysis in two samples of natural seawater, while ion chromatography is conducted for validation of the measurements.

3.1. Cyclic Voltammetric Study of a Au Electrode in ASW

The electrochemical behaviour of a Au electrode in simple halide solutions has been widely studied since the last century [67–69], not least since gold is a noble metal, and methods of extracting, refining, electroplating, electro-etching and electropolishing it have been developed and optimised. As summarised by Nicol [68], chloride and bromide are two ligands that, when present at the millimolar level, coordinate with gold cations and decrease the potential for gold electro-solubilisation in acidic conditions [67,68]. To be specific, during anodic voltametric scanning in solutions containing 0.1 M $HClO_4^-$ with either 10 mM Cl^- or Br^-, respectively, the oxidation of Au occurred at similar potentials, ca. 1.0 V vs. SCE, with the voltametric peaks attributed to the conversion of Au into Au (I) or Au (III) [67]. The general mechanism of the electrochemical dissolution of gold in halide solutions is shown below [67,70]:

$$Au + X^- \rightleftharpoons (AuX^-)_{ad} \quad (2)$$

$$\left(\mathrm{AuX}^-\right)_{ad} + \mathrm{X}^- \rightleftharpoons \left(\mathrm{AuX}_2^-\right)_{ad} + \mathrm{e}^- \tag{3}$$

$$\left(\mathrm{AuX}_2^-\right)_{ad} \rightleftharpoons \mathrm{AuX}_2^-(dissolution) \tag{4}$$

$$\left(\mathrm{AuX}_2^-\right)_{ad} + 2\mathrm{X}^- \rightleftharpoons \left(\mathrm{AuX}_4^-\right)_{ad} + 2\mathrm{e}^- \tag{5}$$

$$\left(\mathrm{AuX}_4^-\right)_{ad} \rightleftharpoons \mathrm{AuX}_4^-(dissolution) \tag{6}$$

where X represents Cl or Br. It is worth mentioning that the voltametric peak in perchloric acid corresponding to the oxidation of gold to $AuCl_2^-$ or $AuBr_2^-$ significantly overlapped the peak for $AuCl_4^-$ or $AuBr_4^-$, respectively, and the relative amounts of Au(I) and Au(III) formed during oxidation was reported to be hard to determine [67,68,70–72]. Meanwhile, passivation due to the formation of an oxide film was also observed in both halide solutions, at sufficiently positive potentials which were dependent on the pH of the solutions [67,69,71,73,74]. From the perspective of sensing, the similarity between chloride and bromide presents a significant challenge in terms of selectivity using a gold electrode. Secondly, the unclear mechanism and corresponding species at different stages, namely the dissolution and passivation of gold, make it hard to develop a reliable understanding of the development of a sensor. Moreover, detection in seawater, which presents a much more complex matrix, may be expected to involve further difficulty and uncertainty. Nevertheless, in the following, we report the electrochemical behaviour of gold electrodes in seawater, ultimately leading to confirmation of the conclusion that Au is a poor choice as a Cl^- and/or Br^- sensor.

We next turn to experiments on the voltammetry at a gold electrode in seawater where chloride is present at the hundred milli-molar level and the bromide concentration is typically less than 1 mM. Cyclic voltammetric experiments were conducted using a Au electrode in three degassed artificial seawater (ASW) samples (Figure 1). ASW1 contained 0.571 M chloride and 0.84 mM bromide (black line); ASW2 contained 0.484 M chloride and 0.84 mM bromide (blue line); and ASW3 contained 0.571 M chloride and 1.84 mM bromide (green line). Note that for synthetic seawater, the recommended chloride level is in the range of 0.484 M–0.593M [10,75,76] and the recommended bromide level is 0.84 mM [75]. The potential sweep started at −0.45 V vs. MSE, and swept towards 0.94 V before returning to −0.45 V at a scan rate of 20 mVs^{-1}.

For ASW1, the voltammetry in Figure 1 shows a large anodic peak with a peak potential of 0.90 V vs. MSE, followed by abrupt passivation on the forward voltametric scan. The peak is seen to have a shoulder at ca 0.59 V vs. MSE. The abrupt decrease in signal is attributed to the formation of gold oxide on the electrode surface, inhibiting further electron transfer. The main peak and the shoulder are tentatively assigned to the oxidation of Au to $AuCl_4^-$ and $AuBr_4^-$, and also possibly to the hidden oxidation to $AuCl_2^-$ and $AuBr_2^-$, described by Equations (2)–(6) [67,70]. The dissolved gold was observed to be partly plated onto the surface of the counter-electrode (Figure S1, Supporting Information Section S1) and led to colouration of the solution (Figure S2, Supporting Information Section S1) after 20 repeats of the scan, confirming the dissolution of the gold electrode, which would be a serious demerit in terms of its use as a sensor. On changing the scan direction from anodic to cathodic, a small anodic Peak 2 was observed at 0.62 V vs. MSE due to partial de-passivation following a loss of material via dissolution during the reverse scan. On continuing to scan reductively, a broad peak (Peak 3) with splitting appears at ca. −0.1 V vs. MSE, and is thought to be related to the solubilisation of the gold electrode. The other two solutions shows qualitatively analogous responses, as seen in Figure 1. Comparison of the voltammograms measured using the three different solutions suggests that the large peak is chloride-related, with the size of the signal qualitatively (but not quantitatively) increasing with the chloride levels. Note that because of the potential and concentration competition between chloride-driven electro-dissolution of the gold and the formation of gold oxide, no simple linear relationship is expected, nor observed, between Peak 1 and the chloride concentration in the seawater. This further discourages the use of gold electrodes for the quantification of chloride in seawater. No clear feature for the oxidation of bromide is visible in any of the voltammograms recorded despite the difference in bromide concentrations. This likely reflects the tiny bromide concentration in comparison with the chloride levels so that any bromide signal is masked by the chloride response. The shoulder noted on Peak 1 is ascribed in the literature to the onset of partial passivation [67,68,71,73], not to bromide oxidation, but in either case, is too ill-defined to be analytically useful.

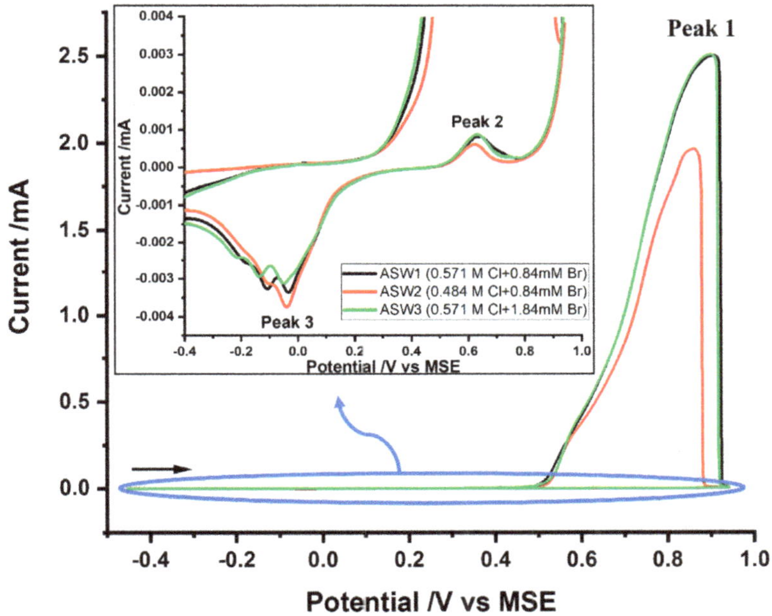

Figure 1. Cyclic voltammograms measured at a scan rate of 20 mV s^{-1} using the Au macro-electrode in degassed ASW1 which contains 0.571 M chloride and 0.84 mM bromide (black line); ASW2 contains 0.484 M chloride and 0.84 mM bromide (red line) and ASW3 contains 0.571 M chloride and 1.84 mM bromide (green line). The start potential is −0.45 V vs. MSE, and the stopping potential is 0.94 V vs. MSE. Insert: an enlarged CV focusing on the reactivation and reduction peaks.

To summarize, the dissolution and passivation of a Au electrode in artificial seawater were inferred from observations in pure chloride and bromide solutions. It is concluded that a gold electrode is unsuitable for use as a Cl^- and/or Br^- sensor.

3.2. Cyclic Voltammetry at Glassy Carbon and Platinum Electrodes in ASW

Next, cyclic voltammetry of ASW using a glassy carbon electrode (GCE) and a Pt electrode was investigated, building on an earlier report [57]. Cyclic voltammograms were first recorded with a scan rate of 100 mVs^{-1} using either the Pt electrode or the GCE in degassed ASW (see Figure 2a). For the Pt electrode, the potential sweep started at −0.45 V vs. MSE, and swept towards 1.2 V before returning to −0.45 V (red line). For the GCE, the voltammogram started at the same potential of −0.45 V vs. MSE, followed by an anodic sweep up to 2.0 V followed by a cathodic scan to −1.3 V before a further anodic sweep back to −0.45 V (blue line). The potential ranges were chosen to obtain both forward and reverse peaks of the redox reactions assigned to the analytes, chloride and bromide, as discussed as below, and to explore the anodic potential window before solvent oxidation.

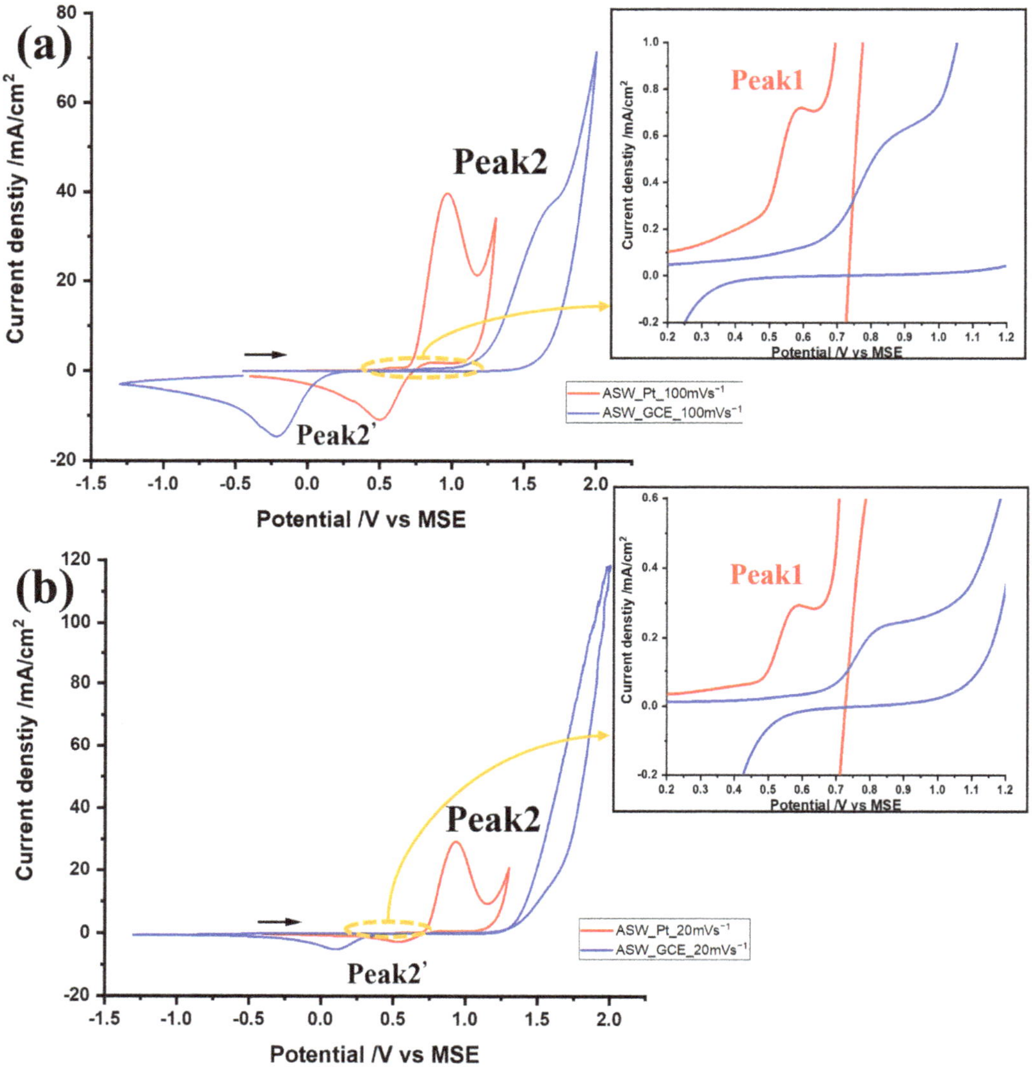

Figure 2. Cyclic voltammograms using a Pt electrode (red) and GCE (blue) in degassed ASW with different scan rates of (**a**) 100 mVs^{-1} and (**b**) 20 mVs^{-1}. Insert: expanded bromide oxidation peaks. The start potential was −0.45 V vs. MSE.

The dominant oxidation peak (Peak 2) was observed at 0.94 V vs. MSE using the Pt electrode; the corresponding feature appears as a shoulder on the solvent decomposition at ca 1.66 V vs. MSE using the GCE. As the insert shows, a first oxidation peak (Peak 1) was recorded at 0.59 V vs. MSE using the Pt electrode and 0.86 V vs. MSE using the GCE. Referring to the previous study [57], Peak 2 is attributed to the oxidation of chloride with the possible formation of hypochlorite/hypochlorous acid ($HOCl/OCl^-$), and Peak 1 is attributed to bromide oxidation with hypobromous acid ($HOBr/OBr^-$), which is among the possible products reported by Yu et al. [77]. The Peak 2′ corresponds to the reduction of the products from the chloride and bromide oxidation. GC offers a wider anodic potential window compared to platinum.

Inspection of the wave-shapes of the dominant peaks in Figure 1a shows that the redox reaction of chloride is relatively much more electrochemically reversible on Pt than on GC, as evidenced by the reduced peak-to-peak separation between the anodic and cathodic peaks, and also the steepness of the current–voltage curves. The Tafel slopes were calculated as described in Supporting Information Section S2 [78–81], with the β value for chloride oxidation being 0.64 at the Pt electrode and 0.22 at the GCE. To confirm this relative assessment of reversibility, cyclic voltammetry, with the same potential window for each electrode, was conducted at a slower scan rate of 20 mVs^{-1} (see Figure 2b). For the Pt electrode (red line), the oxidation of bromide (Peak 1) occurred at 0.59 V vs. MSE and chloride oxidation appeared at 0.94 V vs. MSE (Peak 2), followed by a reduction peak at 0.60 V vs. MSE (Peak 2′). In contrast to the voltammogram of the Pt electrode that only showed insignificant peak shifts when the slower scan rate was applied, that obtained at the GCE showed more noticeable changes (blue line), with the peak-to-peak separation decreasing with scan rate, as particularly seen in the peak potential of Peak 2′. Again, the chloride oxidation of Peak 2 is heavily merged with the solvent decomposition and is not resolved from it. These observations suggest that chloride oxidation is electrochemically irreversible in GC but nearly reversible in Pt.

From the above, it can be seen that Pt and GC electrodes show signals of similar magnitude (current density) for chloride oxidation, but the proximity of the voltametric feature to solvent decomposition prevented the identification of a clear peak at the GC electrode, especially at lower scan rates. For the Pt electrode, a plot of the peak current of chloride oxidation obtained experimentally vs. the square root of the scan rate is shown in Figure 3, where it is compared with the diffusion-controlled response calculated using the Randles–Ševčík equation (Equation (7)) for an irreversible one-electron reaction at 298K:

$$I_p = 2.99 \times 10^5 \sqrt{\beta}\left(D_{Cl-}\right)^{1/2} C_{Cl-} A v^{1/2} \tag{7}$$

where the concentration of chloride in ASW is $C_{Cl-} = 0.56 \times 10^{-3} molcm^{-3}$ [75], the diffusion coefficient of chloride (D_{Cl-}) in seawater is 1.8×10^{-5} cm^2/s [82] and the area (A) of the electrode is 0.02 cm^2. The transfer coefficient reported above ($\beta = 0.64$) was used. Figure 3 shows a gradual increase in signal with scan rate, but less than expected for a fully diffusional signal. This likely reflects the extremely high concentration of chloride, resulting in currents in the order of milliamperes, where the oxidation rate may be largely limited by the availability of adsorption sites for the likely chlorine atom intermediate, together with a contribution to the current from migration effects, since electrolysis is essentially 'unsupported'.

In the case of bromide, signals are seen at less positive potentials for both electrode materials, but these are much smaller in size relative to the chloride peak reflecting the relative concentrations of the two species. Importantly, as the GCE leads to the bromide oxidation peak being more separated from the onset of chloride oxidation compared to the Pt electrode, this gives a superior resolution for Br^- detection. Meanwhile, a Pt electrode is preferable for chloride detection, with a well-defined chloride oxidation peak and a greater degree of electrochemical reversibility. Thus, the use of a GCE for the analysis of bromide ions in ASW was explored further, whilst the merits of using a Pt electrode for the quantitative analysis of chloride were also investigated. In the following sections, the behaviour of the GCE will be discussed first, followed by that of the Pt electrode.

3.3. Analysis of Bromide at the Glassy Carbon Electrode

In this section, a GCE is briefly assessed for the oxidation of bromide ions in ASW from an analytical perspective. The study was carried out via cyclic voltammetry with different scan rates of 20, 50, 100, 200 and 400 mVs^{-1}, where the sweep of potential started at −0.45 V vs. MSE, and then, swept towards 1.1 V before returning to −0.45 V. The voltammograms are shown in Figure 4, where the two features identified above were again observed, namely, bromide oxidation and the corresponding reduction. Although these increased with increasing scan rate, the oxidation of bromide cannot be distinguished from the oxidation peak of chloride, which is also found to interfere with the oxygen evolution reaction (OER). In addition to the proximity to chloride oxidation and the OER, the presence of a pre-wave (see later) also prevented quantitative Tafel analysis and the application of the Randles–Ševčík equation. Attempts were made to increase its resolution using square wave voltammetry but, as reported Supporting Information Section S3, these were unsuccessful. Our attention therefore returned to the cyclic voltammetric data and a standard addition approach was adopted [58].

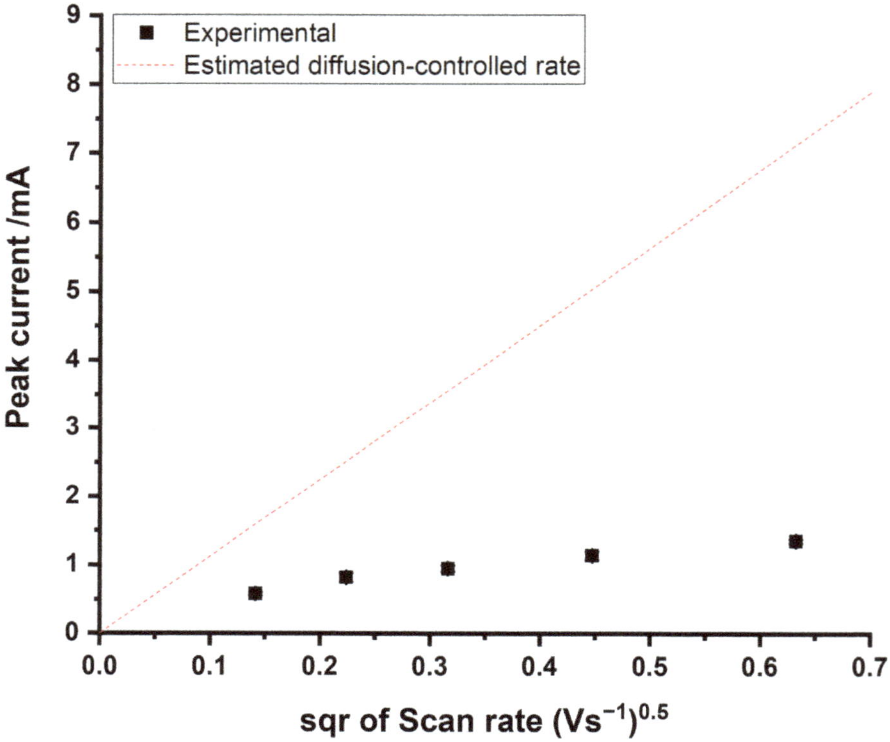

Figure 3. The plot of the peak current against the square root of the scan rate. Experimental data: black squares. Estimated diffusion-controlled rate (red dotted line).

As shown in Figure 5a, voltammograms were recorded with a scan rate of 100 mVs^{-1} representing a compromise between a larger Faradaic signal and excessive capacitive charging. The potential scan of the voltammograms started at a potential of −0.45 V vs. MSE, with a return potential of 1.1 V, before returning −0.45 V. Figure 5 shows that, as expected, Peak 1 increases with the addition of bromide, whilst there is a clear pre-wave to Peak 1 at potentials of around 0.54 V vs. MSE. Additions of 0.2–1.0 mM KBr to the ASW solutions were conducted. The measurements taken with each addition were repeated three times. To analyse the data, the oxidation peak currents of Peak 1 were recorded. Because of the pre-wave to Peak 1, a baseline correction was required for each measurement, examples of which are shown in Figure S5, Supporting Information Section S4. Figure 4b shows a plot of peak current against the concentration of bromide added to the original ASW, indicating good linearity (R^2 = 0.996). As the intercept on the x-axis corresponds with the unknown initial concentration of the analyte [58], the intercept of −0.84 ± 0.06 mM shows excellent agreement with the known bromide concentration (0.84 mM [75]) in the ASW samples.

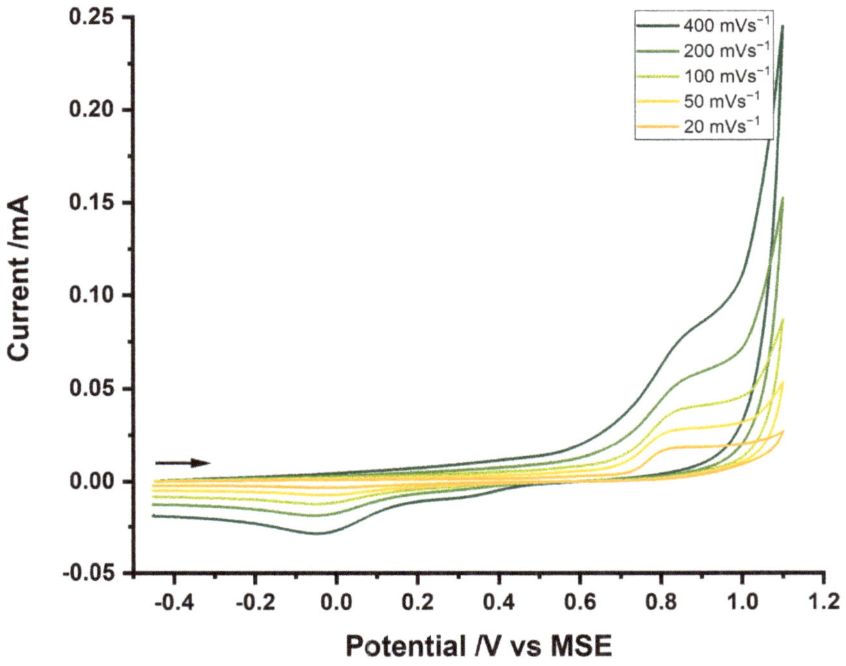

Figure 4. Cyclic voltammograms recorded at the GCE in degassed ASW at variable scan rates of 20, 50, 100, 200 and 400 mV s^{-1}.

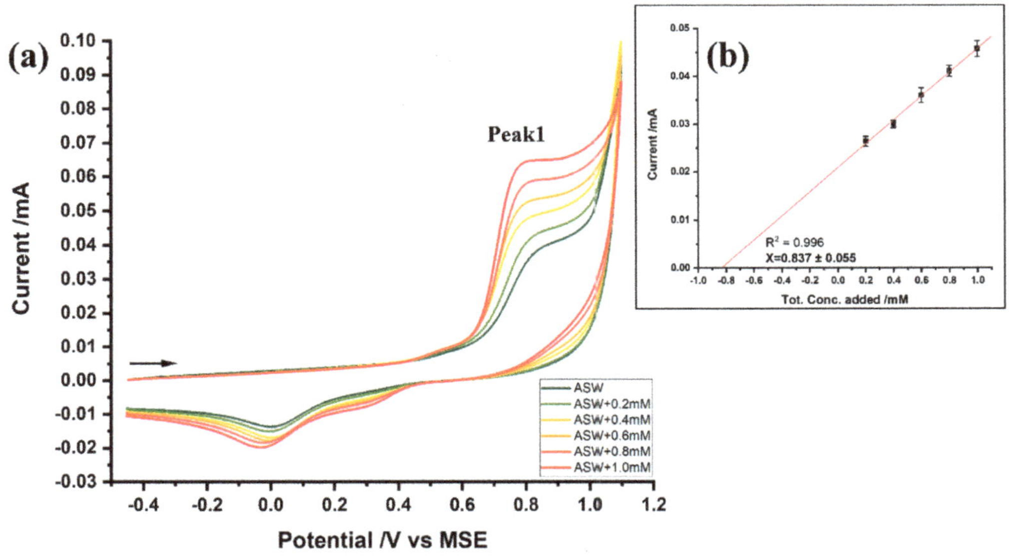

Figure 5. (**a**) Cyclic voltammograms recorded at the GCE in degassed ASW and with standard additions at a scan rate of 100 mV s^{-1}. The start potential was −0.45 V vs. MSE. (**b**) Graphical analysis of standard additions.

Although this suggests the possibility of using a GCE as a bromide sensor, the proximity of bromide oxidation to that of chloride, and of the OER to Peak 2, and the challenges of making a suitable and unambiguous baseline correction, make this method less attractive than the determination of bromide using a platinum electrode [57], where these problems are mitigated.

We next consider the analysis of chloride in seawater using Pt.

3.4. Analysis of Chloride at the Platinum Electrode

As discussed previously, for chloride detection, Pt is a more suitable electrode material compared to GC due to the distinct peak for chloride oxidation and the higher level of electrochemical reversibility. Therefore, the quantitative analysis of chloride in ASW using a Pt electrode was investigated next. This study was conducted in ASW with six levels of chloride concentration from 0.484 M to 0.593 M, while the other components followed a standard recipe (Table 2 [75,76]). Note that the detection range was expanded slightly beyond the recommended range (from 32 g kg^{-1} to 37 g kg^{-1}) to a range of 31 g kg^{-1} to 38 g kg^{-1} to obtain a potentially wider concentration range of the analyte [10]. To achieve a best-resolved chloride signal, square wave voltammetry was implemented, with the corresponding parameters optimised as shown in Supporting Information Section S5. A step potential of 1 mV, a frequency of 5 Hz and an amplitude of 50 mV were then employed. To obtain electroanalytical responses, the potential sweep started from −0.45 V and stopped at 1.25 V vs. MSE for oxidative scanning in six ASW solutions of variable chloride content in the range specified above (Figure 6).

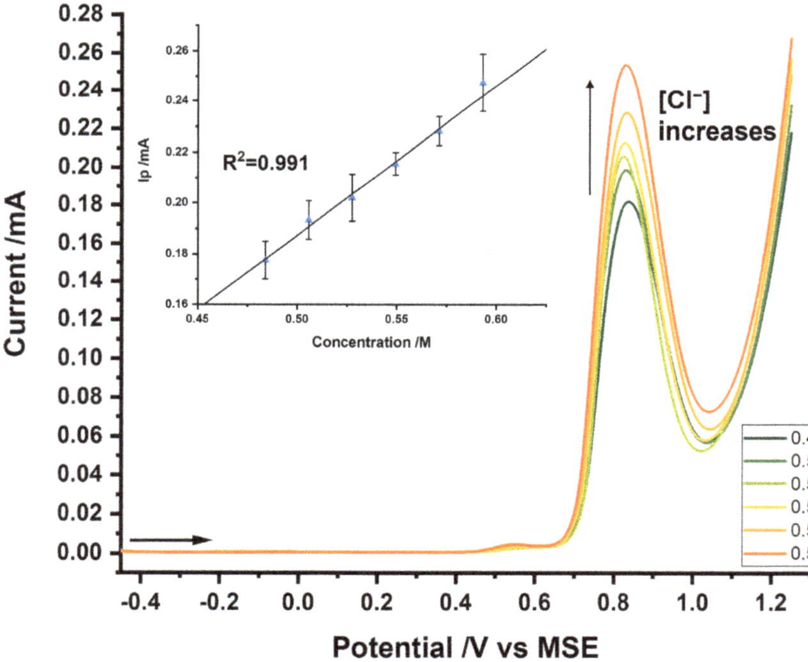

Figure 6. Square wave voltammograms (step potential: 1 mV, frequency: 5 Hz, and amplitude: 50 mV) recorded at the Pt electrode in degassed ASW with various chloride concentrations ranging from 0.484M to 0.593M. See text for details of the scan range and the start potential. The inset shows the calibration curve of the chloride oxidation peak height against chloride concentrations, with a regression coefficient R^2 of 0.991. Each data point consists of 3 repeats.

The dominant chloride oxidation peak was recorded at 0.83 V vs. MSE. To analyse the data, the oxidation peak currents of the chloride were measured. A plot of the peak current against the chloride concentration in the ASW samples is shown in the insert of Figure 6.

The peak height increased linearly in the concentration range of chloride studied (from 0.484 M to 0.593 M). The best-fit line (R^2 = 0.991) was determined as follows:

$$Ip = 0.587\left[Cl^-\right] - 0.106 \quad (8)$$

where Ip is the peak current (mA) of chloride oxidation and $[Cl^-]$ is the concentration (M) of chloride in ASW.

3.5. Real Sample Analysis

To investigate the method in authentic seawater, an analysis of two different samples was performed. Ion chromatography measurements were first made for the chloride concentration of the two samples, with a result of 0.579 ± 0.004 M for sample 1 and 0.580 ± 0.001 M for sample 2. The concentrations of chloride obtained via IC were then compared as follows, with the values measured electrochemically, and as performed in ASW via SWV with optimal parameters: a step potential of 1 mV, a frequency of 5 Hz and an amplitude of 50 mV. As shown in Figure 7, voltammograms were conducted, starting at a potential of −0.45 V and stopping at 1.25 V vs. MSE, to record the chloride oxidation signal. The oxidation peaks of chloride in the three solutions all appear at 0.83 V vs. MSE, with no additional features being observed in the real samples. With the three measurements, the average oxidation peak currents of the chloride obtained in sample 1 and sample 2 were recorded and analysed using the calibration made above in ASW (Equation (8)). The electrochemically measured chloride concentrations in sample 1 and sample 2 are 0.579 ± 0.003 and 0.581 ± 0.005, respectively. The developed electrochemical analysis resulted in good agreement with IC.

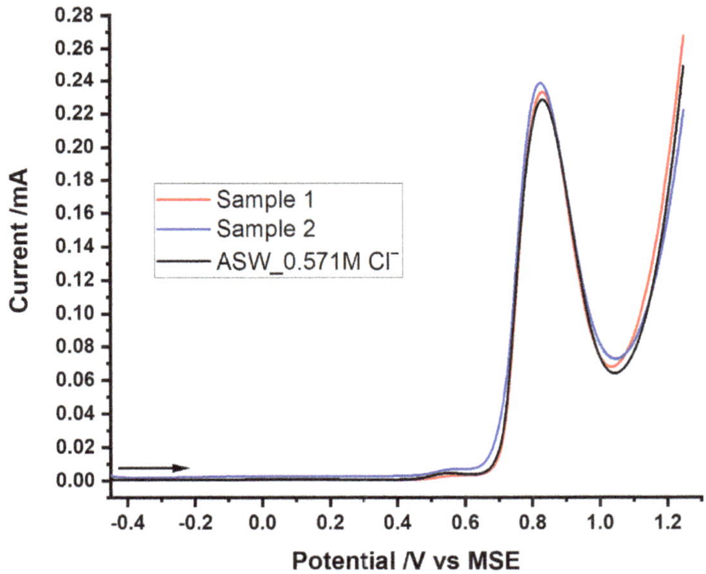

Figure 7. Square wave voltammograms (step potential: 1 mV, frequency: 5 Hz and amplitude: 50 mV) recorded at the Pt electrode in degassed sample 1 (red line), sample 2 (blue line) and ASW with a chloride concentration of 0.571 M (black line). See text for details of the scan range and the start potential.

4. Conclusions

In conclusion, three electrodes, made of Au, glassy carbon and Pt, were investigated for the analysis of Cl^- and/or Br^- in seawater. First, the dissolution and passivation of a Au electrode in ASW, which has a high chloride concentration, indicated that gold was an unsuitable electrode material. This agrees with independent studies on pure chloride and bromide solutions. For GC, while the proximity of the voltametric feature to solvent decomposition (OER) prevented the identification of a clear chloride oxidation peak, the poorly resolved bromide signal from the chloride oxidation and the OER, and the challenges of making an appropriate baseline correction, make the material less attractive than using platinum for chloride detection in both ASW and NSW using SWV. The results measured electrochemically in two real samples show a good match with those measured independently via ion chromatography. Platinum electrodes are also recommended for bromide detection using a previously developed method for seawater [57].

Supplementary Materials: The following supporting information can be downloaded at: https://www.mdpi.com/article/10.3390/chemosensors11050297/s1. References [78–81] are cited in the Supplementary Information Section S2: Calculation of Transfer Coefficients for the Chloride Oxidation from Tafel Analysis.

Author Contributions: Y.C.: validation, formal analysis, investigation, writing—original draft and visualisation; R.G.C.: writing—review and editing, supervision, project administration, funding acquisition, conceptualisation, methodology and resources. All authors have read and agreed to the published version of the manuscript.

Funding: This research received no external funding.

Institutional Review Board Statement: Not applicable.

Informed Consent Statement: Not applicable.

Data Availability Statement: The study did not report any data.

Conflicts of Interest: The authors declare that they have no known competing financial interests or personal relationships that could have appeared to influence the work reported in this paper.

References

1. Water Scarcity. Available online: https://www.unwater.org/water-facts/water-scarcity (accessed on 14 March 2023).
2. Spellman, F.R. *The Handbook of Nature*; Bernan Press: Lanham, MD, USA, 2019.
3. Bint El Hassan, H.E.S.; Guerrier, D.; Closas, A.; Demilecamps, C.; kassim, Y.; Lacirignola, C.; Lamaddalena, N.; Mateo-Sagasta, J.; Rivera, N. *Water Around the Mediterranean*; The Center for Mediterranean Integration (CMI): Marseille, France, 2018.
4. Pérez-González, A.; Urtiaga, A.M.; Ibáñez, R.; Ortiz, I. State of the art and review on the treatment technologies of water reverse osmosis concentrates. *Water Res.* **2012**, *46*, 267–283. [CrossRef] [PubMed]
5. Kim, D.H. A review of desalting process techniques and economic analysis of the recovery of salts from retentates. *Desalination* **2011**, *270*, 1–8. [CrossRef]
6. Wallace, W.J. *The Development of the Chlorinity/Salinity Concept in Oceanography*; Elsevier Science Ltd.: New York, NY, USA, 1974.
7. Millero, F.J.; Feistel, R.; Wright, D.G.; McDougall, T.J. The composition of Standard Seawater and the definition of the Reference-Composition Salinity Scale. *Deep Sea Res.* **2008**, *55*, 50–72. [CrossRef]
8. Stewart, R.H. *Introduction to Physical Oceanography*; Department of Oceanography, Texas A&M University: College Station, TX, USA, 2008.
9. Shumway, S.E. Effect of salinity fluctuation on the osmotic pressure and Na^+, Ca^{2+} and Mg^{2+} ion concentrations in the hemolymph of bivalve molluscs. *Mar. Biol.* **1977**, *41*, 153–177. [CrossRef]
10. Guo, Y.; Compton, R.G. A bespoke chloride sensor for seawater: Simple and fast with a silver electrode. *Talanta* **2021**, *232*, 122502. [CrossRef]
11. Yamamoto, M. Stimulation of elemental mercury oxidation in the presence of chloride ion in aquatic environments. *Chemosphere* **1996**, *32*, 1217–1224. [CrossRef]
12. Levard, C.; Mitra, S.; Yang, T.; Jew, A.D.; Badireddy, A.R.; Lowry, G.V.; Brown, G.E., Jr. Effect of Chloride on the Dissolution Rate of Silver Nanoparticles and Toxicity to *E. coli*. *Environ. Sci. Technol.* **2013**, *47*, 5738–5745. [CrossRef] [PubMed]
13. Batchelor-McAuley, C.; Tschulik, K.; Neumann, C.; Laborda, E.; Compton, R. Why are silver nanoparticles more toxic than bulk silver? Towards understanding the dissolution and toxicity of silver nanoparticles. *Int. J. Electrochem. Sci.* **2014**, *9*, 1132–1138.
14. Gilberg, M.R.; Seeley, N.J. The identity of compounds containing chloride ions in marine iron corrosion products: A critical review. *Stud. Conserv.* **1981**, *26*, 50–56. [CrossRef]
15. Otieno, M.; Beushausen, H.; Alexander, M. Chloride-induced corrosion of steel in cracked concrete—Part I: Experimental studies under accelerated and natural marine environments. *Cem. Concr. Res.* **2016**, *79*, 373–385. [CrossRef]
16. Downs, A.J.; Adams, C.J. *The Chemistry of Chlorine, Bromine, Iodine and Astatine: Pergamon Texts in Inorganic Chemistry, Volume 7*; Elsevier: Amsterdam, The Netherlands, 2017; Volume 7.

17. Leri, A.C.; Hakala, J.A.; Marcus, M.A.; Lanzirotti, A.; Reddy, C.M.; Myneni, S.C.B. Natural organobromine in marine sediments: New evidence of biogeochemical Br cycling. *Glob. Biogeochem. Cycles* **2010**, *24*, 1–15. [CrossRef]
18. Pilson, M.E. *An Introduction to the Chemistry of the Sea*; Cambridge University Press: Cambridge, UK, 2012.
19. Gribble, G.W. Amazing Organohalogens: Although best known as synthetic toxicants, thousands of halogen compounds are, in fact, part of our natural enviornment. *Am. Sci.* **2004**, *92*, 342–349. [CrossRef]
20. Dorji, P.; Kim, D.I.; Hong, S.; Phuntsho, S.; Shon, H.K. Pilot-scale membrane capacitive deionisation for effective bromide removal and high water recovery in seawater desalination. *Desalination* **2020**, *479*, 114309. [CrossRef]
21. Lewis, E.L. The practical salinity scale of 1978 and its antecedents. *Mar. Geod.* **1982**, *5*, 350–357. [CrossRef]
22. Lagerloef, G.S.; Swift, C.T.; Le Vine, D.M. Sea surface salinity: The next remote sensing challenge. *Oceanog.* **1995**, *8*, 44–50. [CrossRef]
23. Anfält, T.; Twengström, S. The determination of bromide in natural waters by flow injection analysis. *Anal. Chim. Acta* **1986**, *179*, 453–457. [CrossRef]
24. Vengosh, A.; Pankratov, I. Chloride/Bromide and Chloride/Fluoride Ratios of Domestic Sewage Effluents and Associated Contaminated Ground Water. *Ground Water* **1998**, *36*, 815–824. [CrossRef]
25. Morgan, J.J.; Stumm, W. *Aquatic Chemistry: An Introduction Emphasizing Chemical Equilibria in Natural Waters*; John Wiley & Sons: New York, NY, USA, 1981.
26. Salameh, E.; Tarawneh, A.; Al-Raggad, M. Origin of high bromide concentration in the water sources in Jordan and in the Dead Sea water. *Arabian J. Geosci.* **2016**, *9*, 414. [CrossRef]
27. Jain, A.; Chaurasia, A.; Sahasrabuddhey, B.; Verma, K.K. Determination of bromide in complex matrices by pre-column derivatization linked to solid-phase extraction and high-performance liquid chromatography. *J. Chromatogr. A* **1996**, *746*, 31–41. [CrossRef]
28. Borges, E.P.; Lavorante, A.F.; Reis, B.F.d. Determination of bromide ions in seawater using flow system with chemiluminescence detection. *Anal. Chim. Acta* **2005**, *528*, 115–119. [CrossRef]
29. Grasshoff, K.; Wenck, A. A modern version of the Mohr-Knudsen titration for the chlorinity of sea water. *ICES Mar. Sci. Symp.* **1972**, *34*, 522–528. [CrossRef]
30. Hong, T.-K.; Kim, M.-H.; Czae, M.-Z. Determination of chlorinity of water without the use of chromate indicator. *Int. J. Anal. Chem.* **2010**, *2010*, 602939. [CrossRef] [PubMed]
31. Reitemeier, R. Semimicroanalysis of saline soil solutions. *Ind. Eng. Chem. Anal. Ed.* **1943**, *15*, 393–402. [CrossRef]
32. Reeburgh, W.S.; Carpenter, J.H. Determination of chlorinity using A differential potentiometric. *Limnol. Oceanogr.* **1964**, *9*, 589–591. [CrossRef]
33. Clesceri, L.S.; Greenberg, A.E.; Eaton, A.D. *Standard Methods for the Examination of Water and Wastewater*, 20th ed.; APHA American Public Health Association: Washington, DC, USA, 1998.
34. Willard, H.H.; Merritt, L.L., Jr.; Dean, J.A.; Settle, F.A., Jr. *Instrumental Methods of Analysis*, 7th ed.; Wadsworth Publishing Company: Belmont, CA, USA, 1988.
35. Lewis, E. The practical salinity scale 1978 and its antecedents. *IEEE J. Ocean.* **1980**, *5*, 3–8. [CrossRef]
36. Lewis, E.; Perkin, R. The Practical Salinity Scale 1978: Conversion of existing data. *Deep Sea Res.* **1981**, *28*, 307–328. [CrossRef]
37. Unesco. *Tenth Report of the Joint Panel on Oceanographic Tables and Standards*; Unesco Division of Marine Sciences: Sidney, BC, Canada, 1980.
38. Uraisin, K.; Takayanagi, T.; Oshima, M.; Nacapricha, D.; Motomizu, S. Kinetic-spectrophotometric method for the determination of trace amounts of bromide in seawater. *Talanta* **2006**, *68*, 951–956. [CrossRef]
39. Jones, D.R. Applying the phenol red colorimetric method for bromide analysis to reducing waters. *Talanta* **1993**, *40*, 43–51. [CrossRef]
40. Anagnostopoulou, P.I.; Koupparis, M.A. Automated flow-injection phenol red method for determination of bromide and bromide salts in drugs. *Anal. Chem.* **1986**, *58*, 322–326. [CrossRef]
41. Emaus, W.J.M.; Henning, H.J. Determination of bromide in sodium chloride matrices by flow-injection analysis using blank peak elimination and kinetic discrimination. *Anal. Chim. Acta* **1993**, *272*, 245–250. [CrossRef]
42. Uraisin, K.; Nacapricha, D.; Lapanantnoppakhun, S.; Grudpan, K.; Motomizu, S. Determination of trace amounts of bromide by flow injection/stopped-flow detection technique using kinetic-spectrophotometric method. *Talanta* **2005**, *68*, 274–280. [CrossRef] [PubMed]
43. Yonehara, N.; Chaen, S.-i.; Tomiyasu, T.; Sakamoto, H. Flow injection-spectrophotometric determination of trace amounts of bromide by its catalytic effect on the 4,4′-bis(dimethylamino)diphenylmethane-chloramine T reaction. *Anal. Sci.* **1999**, *15*, 277–281. [CrossRef]
44. Shishehbori, M.R.; Elah, J.R. Kinetic spectrophotometric method For trace amounts determination of bromide in pharmaceutical samples using Janus Green-Bromate system. *Int. J. Ind. Chem.* **2011**, *2*, 27–32.
45. Sheibani, A.; Shishehbore, M.R.; Ardakani, Z.T. Kinetic spectrophotometric determination of bromide in clidinium-c drug. *Chin. Chem. Lett.* **2011**, *22*, 595–598. [CrossRef]
46. Akaiwa, H.; Kawamoto, H.; Osumi, M. Simultaneous determination of bromide and chloride in natural waters by ion-exchange chromatography and direct potentiometry with an ion-selective electrode. *Talanta* **1982**, *29*, 689–690. [CrossRef]

47. Seefeld, S.; Baltensperger, U. Determination of bromide in snow samples by ion chromatography with electrochemical detection. *Anal. Chim. Acta* **1993**, *283*, 246–250. [CrossRef]
48. Fabre, H.; Blanchin, M.D.; Bosc, N. Capillary electrophoresis for the determination of bromide, chloride and sulfate as impurities in calcium acamprosate. *Anal. Chim. Acta* **1999**, *381*, 29–37. [CrossRef]
49. Stålberg, O.; Sander, K.; Sänger-van de Griend, C. The determination of bromide in a local anaesthetic hydrochloride by capillary electrophoresis using direct UV detection. *J. Chromatogr. A* **2002**, *977*, 265–275. [CrossRef]
50. Pascali, J.P.; Trettene, M.; Bortolotti, F.; Paoli, G.d.; Gottardo, R.; Tagliaro, F. Direct analysis of bromide in human serum by capillary electrophoresis. *J. Chromatogr. B* **2006**, *839*, 2–5. [CrossRef]
51. Fukushi, K.; Watanabe, K.; Takeda, S.; Wakida, S.-I.; Yamane, M.; Higashi, K.; Hiiro, K. Determination of bromide ions in seawater by capillary zone electrophoresis using diluted artificial seawater as the buffer solution. *J. Chromatogr. A* **1998**, *802*, 211–217. [CrossRef]
52. Kocatürk, N.; Öztekin, N.; Bedia Erim, F. Direct determination of bromide ions in seawater by capillary zone electrophoresis using polyethyleneimine-coated capillaries. *Anal. Bioanal. Chem.* **2003**, *377*, 1207–1211. [CrossRef]
53. Rechnitz, G.A.; Kresx, M.R. Potentiometric measurements with choride-sensitive and bromide-sensitive membrane electrodes. *Anal. Chem.* **1966**, *38*, 1786–1788. [CrossRef]
54. Zahran, E.M.; Hua, Y.; Li, Y.; Flood, A.H.; Bachas, L.G. Triazolophanes: A new class of halide-selective ionophores for potentiometric sensors. *Anal. Chem.* **2010**, *82*, 368–375. [CrossRef] [PubMed]
55. Isildak, Ö.; Özbek, O.; Yigit, K. A bromide-selective PVC membrane potentiometric sensor. *Bulg. Chem. Commun.* **2020**, *52*, 448–452. [CrossRef]
56. Vlascici, D.; Plesu, N.; Fagadar-Cosma, G.; Lascu, A.; Petric, M.; Crisan, M.; Belean, A.; Fagadar-Cosma, E. Potentiometric sensors for iodide and bromide based on Pt(II)-Porphyrin. *Sensors* **2018**, *18*, 2297. [CrossRef]
57. Chen, Y.; Compton, R.G. A bespoke reagent-free amperometric bromide sensor for seawater. *Talanta* **2023**, *253*, 124019. [CrossRef]
58. Harris, D.C. *Quantitative Chemical Analysis*; Macmillan: London, UK, 2010.
59. Sekerka, I.; Lechner, J.F. Ion Selective Electrode for Determination of Chloride Ion in Biological Materials, Food Products, Soils, and Waste Water. *J Assoc Off Anal Chem.* **2020**, *61*, 1493–1495. [CrossRef]
60. Montemor, M.; Alves, J.; Simoes, A.; Fernandes, J.; Lourenço, Z.; Costa, A.; Appleton, A.; Ferreira, M. Multiprobe chloride sensor for in situ monitoring of reinforced concrete structures. *Cem. Concr. Compos.* **2006**, *28*, 233–236. [CrossRef]
61. Arai, K.; Kusu, F.; Noguchi, N.; Takamura, K.; Osawa, H. Selective determination of chloride and bromide ions in serum by cyclic voltammetry. *Anal. Biochem.* **1996**, *240*, 109–113. [CrossRef] [PubMed]
62. Qin, X.; Wang, H.; Miao, Z.; Wang, X.; Fang, Y.; Chen, Q.; Shao, X. Synthesis of silver nanowires and their applications in the electrochemical detection of halide. *Talanta* **2011**, *84*, 673–678. [CrossRef]
63. Bujes-Garrido, J.; Izquierdo-Bote, D.; Heras, A.; Colina, A.; Arcos-Martínez, M.J. Determination of halides using Ag nanoparticles-modified disposable electrodes. A first approach to a wearable sensor for quantification of chloride ions. *Anal. Chim. Acta* **2018**, *1012*, 42–48. [CrossRef]
64. Malongo, T.K.; Patris, S.; Macours, P.; Cotton, F.; Nsangu, J.; Kauffmann, J.-M. Highly sensitive determination of iodide by ion chromatography with amperometric detection at a silver-based carbon paste electrode. *Talanta* **2008**, *76*, 540–547. [CrossRef]
65. Cunha-Silva, H.; Julia Arcos-Martinez, M. Development of a selective chloride sensing platform using a screen-printed platinum electrode. *Talanta* **2019**, *195*, 771–777. [CrossRef] [PubMed]
66. Millero, F.J. The Physical Chemistry of Seawater. *Annu. Rev. Earth Planet. Sci.* **1974**, *2*, 101–150. [CrossRef]
67. Qi, P.H.; Hiskey, J.B. Electrochemical behavior of gold in iodide solutions. *Hydrometallurgy* **1993**, *32*, 161–179. [CrossRef]
68. Nicol, M.J. The anodic behaviour of gold. *Gold Bull.* **1980**, *13*, 46–55. [CrossRef]
69. Nicol, M.J. The anodic behaviour of gold. *Gold Bull.* **1980**, *13*, 105–111. [CrossRef]
70. Loo, B.H. In situ identification of halide complexes on gold electrode by surface-enhanced Raman spectroscopy. *J. Phys. Chem.* **1982**, *86*, 433–437. [CrossRef]
71. Gaur, J.; Schmid, G. Electrochemical behavior of gold in acidic chloride solutions. *J. Electroanal. Chem. Interfacial Electrochem.* **1970**, *24*, 279–286. [CrossRef]
72. Cadle, S.; Bruckenstein, S. A ring-disk study of the effect of trace chloride ion on the anodic behavior of gold in 0.2 M H_2SO_4. *J. Electroanal. Chem. Interfacial Electrochem.* **1973**, *48*, 325–331. [CrossRef]
73. Nicol, M.J.; Schalch, E. *Reports Nos. 1844 and 1848*; National Institute for Metallurgy: Johannesburg, South Africa, 1976.
74. Heumann, T.; Panesar, H.S. Beitrag zur Frage nach dem Auflösungsmechanismus von Gold zu Chlorkomplexen und nach seiner Passivierung. *Z. Phys. Chem.* **1965**, *229O*, 84–97. [CrossRef]
75. Morel, F.M.M.; Rueter, J.G.; Anderson, D.M.; Guillard, R.R.L. Aquil: A chemically defined phytoplankton culture medium for trace metal studies. *J. Phycol.* **1979**, *15*, 135–141. [CrossRef]
76. Price, N.M.; Harrison, G.I.; Hering, J.G.; Hudson, R.J.; Nirel, P.M.; Palenik, B.; Morel, F.M. Preparation and chemistry of the artificial algal culture medium Aquil. *Biol. Oceanogr.* **1989**, *6*, 443–461. [CrossRef]
77. Yu, J.; Yang, M.; Batchelor-McAuley, C.; Barton, S.; Rickaby, R.E.M.; Bouman, H.A.; Compton, R.G. Rapid Opto-electrochemical Differentiation of Marine Phytoplankton. *ACS Meas. Sci. Au* **2022**, *2*, 342–350. [CrossRef] [PubMed]
78. Li, D.; Lin, C.; Batchelor-McAuley, C.; Chen, L.; Compton, R.G. Tafel analysis in practice. *J. Electroanal. Chem.* **2018**, *826*, 117–124. [CrossRef]

79. Compton, R.G.; Banks, C.E. *Understanding Voltammetry*, 3rd ed.; World Scientific: Singapore, 2018.
80. Guidelli, R.; Compton, R.G.; Feliu, J.M.; Gileadi, E.; Lipkowski, J.; Schmickler, W.; Trasatti, S. Defining the transfer coefficient in electrochemistry: An assessment (IUPAC Technical Report). *Pure Appl. Chem.* **2014**, *86*, 245–258. [CrossRef]
81. Guidelli, R.; Compton, R.G.; Feliu, J.M.; Gileadi, E.; Lipkowski, J.; Schmickler, W.; Trasatti, S. Definition of the transfer coefficient in electrochemistry (IUPAC Recommendations 2014). *Pure Appl. Chem.* **2014**, *86*, 259–262. [CrossRef]
82. Poisson, A.; Papaud, A. Diffusion coefficients of major ions in seawater. *Marine Chemistry* **1983**, *13*, 265–280. [CrossRef]

chemosensors

MDPI

Article

MoS_2@Au as Label for Sensitive Sandwich-Type Immunoassay of Neuron-Specific Enolase

Yingying Wang [1], Huixin Wang [1], Yaliang Bai [1], Guanhui Zhao [2], Nuo Zhang [2], Yong Zhang [3], Yaoguang Wang [1,*] and Hong Chi [1,*]

1 Shandong Provincial Key Laboratory of Molecular Engineering, School of Chemistry and Chemical Engineering, Qilu University of Technology (Shandong Academy of Sciences), Jinan 250353, China
2 Collaborative Innovation Center for Green Chemical Manufacturing and Accurate Detection, Key Laboratory of Interfacial Reaction & Sensing Analysis in Universities of Shandong, School of Chemistry and Chemical Engineering, University of Jinan, Jinan 250022, China
3 Provincial Key Laboratory of Rural Energy Engineering in Yunnan, School of Energy and Environment Science, Yunnan Normal University, Kunming 650500, China
* Correspondence: wangyaoguang@qlu.edu.cn (Y.W.); chihong@qlu.edu.cn (H.C.); Tel.: +86-531-82767872 (Y.W.)

Abstract: Neuron-specific enolase (NSE) has gained extensive attention as a reliable target for detecting small cell carcinoma of lungs. In this paper, an electrochemical immunoassay method based on molybdenum disulfide (MoS_2) is proposed to detect NSE sensitively. By an in-situ growth method, MoS_2 and Au nanoclusters (Au NCs) were composited to form a MoS_2@Au nanozyme, and then the secondary antibodies were modified. Primary antibodies were immobilized on amino-reduced graphene oxides to capture NSE. The flower-like MoS_2 nanozyme provided abundant sites to load Au NCs and catalyze the decomposition of H_2O_2, which were beneficial to amplify an amperometric response as well as build up sensitivity. Under optimum conditions, the detection range of this strategy was 0.1 pg·mL^{-1}–10 ng·mL^{-1} and the limit of detection was 0.05 pg·mL^{-1}. This sensing strategy achieved the prospect of sensitively detecting NSE. Moreover, the prepared electrochemical immunosensor provides a theoretical basis and technical support for the detection of other disease markers.

Keywords: molybdenum disulfide; nanozyme; neuron-specific enolase; electrochemical immunosensor; detection

Citation: Wang, Y.; Wang, H.; Bai, Y.; Zhao, G.; Zhang, N.; Zhang, Y.; Wang, Y.; Chi, H. MoS_2@Au as Label for Sensitive Sandwich-Type Immunoassay of Neuron-Specific Enolase. *Chemosensors* **2023**, *11*, 349. https://doi.org/10.3390/chemosensors11060349

Academic Editor: Paolo Ugo

Received: 22 April 2023
Revised: 13 June 2023
Accepted: 17 June 2023
Published: 19 June 2023

1. Introduction

According to global cancer data compiled by the international agency for research on cancer of the World Health Organization, there were 19.29 million new cancer cases around the world in 2020. Lung cancer has a high incidence and ranks first among all cancer cases. The proportion of early-onset cancers has been on the rise since 1990. The low cure rate of cancer is mainly due to late clinical diagnosis, and it is extremely difficult for patients in the advanced stage of cancer to be completely cured. Therefore, early diagnosis and the treatment of cancer is a key factor in reducing fatality rates. Tumor markers as biomarkers are very important for the early diagnosis of cancer. Lung cancer, with a high fatality rate, is mainly divided into small cell carcinoma of the lung (SCLC) and non-small cell carcinoma the lung (NSCLC) [1]. SCLC is an invasive neuroendocrine tumor with the characteristics of rapid proliferation, swift division, and poor prognosis, resulting in a low early cure rate of SCLC patients [2–4]. Therefore, early diagnosis is expected to improve the survival rate of SCLC patients. Neuron-specific enolase (NSE) is a unique type of acidic protease produced by neurons and neuroendocrine cells, and its normal level is 4–12 ng·mL^{-1} in human serum [5–7]. The concentration of NSE in human serum can distinguish SCLC from NSCLC and be adjunctive to clinical treatment in certain circumstances [8]. Therefore, the rapid and accurate detection of NSE is of great significance for the diagnosis and therapy of lung cancer.

At present, the detection methods of NSE mainly include electrochemical analysis [9], colorimetry [10], fluorescence analysis [11], liquid chromatography-tandem mass spectrometry [12], surface-enhanced Raman scattering [13], and immunochromatographic [14]. Xu et al. [15] designed a colorimetric analysis strategy to detect NSE, employing MnO_2 as label materials. MnO_2 catalyzed the oxidation of 3, 3′, 5,5′-tetramethylbenzidine (TMB) and gained blue oxidation products. In the light of this color reaction, visual detection of NSE was realized. Gao et al. [13] introduced sandwich nanoparticles as surface enhanced Raman scattering probes, which improved the stability of bare gold nanoparticles and significantly improved the sensitivity in measuring actual samples. Xiao et al. [14] proposed an immunochromatographic analysis platform that consisted of quantum dot beads and nitrocellulose membranes to quantify NSE and CEA. The sensitivity of NSE detected by this method was up to 97%. Among them, electrochemical immunosensors attracted more attention from researchers on account of their merits, such as high specificity, fine accuracy, and satisfactory stability [16]. The mechanism of the immunosensor for detecting NSE is that the output signal can be recorded before and after the specific recognition of the target through the sensor. In recent years, the performance of sensors has been greatly improved owing to the application of functionalized nanomaterials. An immunoassay strategy was proposed based on nitrogen-doped carbon quantum dots (N-CQDs) which were prepared by hydrothermal carbonization. Reduced graphene oxide (rGO) modified by N-CQDs effectively enhanced the luminescence intensity and improved the performance of the immunosensor in measuring NSE [17]. Tang et al. designed N atom-doped graphenes (NGR) functionalized with hollow porous Pt-skin Ag–Pt alloy as a signal amplification platform [18]. The signal was amplified by the hollow Pt skins with abundant active sites and NGR with well-catalytic activity. Electrochemical immunosensors are more prominent in NSE detection because of their advantages of high sensitivity, simple operation, and simple equipment. Amani et al. [19] synthesized poly p-phenylenediamine and graphene (PPD-GR) nanocomposites for fabricating sensors. PPD-GR nanocomposites were modified in screen-printed electrodes and then PPD-GR electro-catalyzed oxidation of ascorbic acid (AA) to amplify signal amplification. With high catalytic activity, PtCu nanoparticles were utilized to catalyze the reaction of iodide and trigger cascade reactions, which could improve the performance of immunosensors. Due to the PtCu nanoprobe reducing the concentration of iodide in the solution, mercaptan substances in the electrode surface cannot be catalyzed, resulting in an increase of the SWV signal. On the grounds of the above mechanism, the immunoassay of NSE was realized. [20]. Yu et al. [21] obtained single-walled carbon nanotubes-NSE (NSE-SWNTs) by immobilizing NSE in single-walled carbon nanotubes (SWNTs) with ample antigen domains. When specific recognition occurred, free NSE and NSE on NSE-SWNTs competed with NSE-Ab_1, which resulted in a difference in the amount of anti-NSE-Ab_1 labeled by an enzyme on the electrode, leading to the change of electrochemical signals. Given the above research, the application of nanomaterials was crucial to improving the practicability and sensitivity of the sensors in the analysis and detection of targets.

As a kind of special nanomaterials, nanozymes can mimic the activity of natural enzymes and possess better application prospects than other nanomaterials in sensing analysis, antibacterial performance, and other fields [22,23]. In addition, the activity of nanozymes can be controlled by changing the size, morphology, and surface modification of nanozymes. Nanozymes are favored by researchers for their outstanding environmental adaptability, fine stability, and low cost [24,25]. Zhou et al. [26] prepared Cu/N co-doped carbon-based nanozymes (Cu/N NS) by calcination, which catalyzed the decomposition of H_2O_2 and obtained superoxide radicals that could oxidize TMB to oxTMB in blue. A multi-enzyme cascade colorimetric sensing platform was designed based on a complex consisting of Cu/NC NS, β-galactosidase, and galactose oxidase. Wang et al. [27] prepared a hollow porous carbon-frame (Au/Co@HNCF) nanozyme modified with Au/Co bimetallic nanoparticles from zeolite imidazole frame-67 (ZIF-67) etched by pyrolytic Au (III). Au/Co NPs with excellent uric acid (UA) oxidase activity catalyzed the oxidation of UA, leading to significant changes in the electrochemical signal. Remarkably, the layered and ordered porous carbon structure and interface characteristics showed high selectivity in detecting UA in complex environments. Therefore, the constructed electrochemical analysis platform based on Au/Co@HNCF realized the sensitive detection of UA. Yang et al. [28] proposed a dual-model analysis strategy based on a bismuth-based MOF (BiO-BDC-NH_2) nanozyme for the detection of trace/ultra-trace Cr^{6+}. The BiO-BDC-NH_2 nanozyme with peroxidase-like activity could effectively build up the properties of surface adsorption and electron conversion, which was beneficial to efficiently catalyzing the color reaction of a TMB-H_2O_2 system. As TMB was oxidized, the color of the solution became darker. According to this change, a colorimetric detection platform was established. In the meantime, the electrochemical signal decreased as a result of the oxidization of TMB, which accomplished a transformation from the visual colorimetric analysis platform to the electrochemical analysis platform. Bi et al. [29] designed a boron-doped graphdiyne (B-GDY) nanozyme as an effective bactericidal agent. The change of the mimic activity of B-GDY was mainly related to B doping. The change of the GDY electron cloud distribution was in favor of the rapid transfer of electrons to H_2O_2, which caused the O–O bond to break easily. The introduction of B atoms increased the number of defect sites of GDY and improved the adsorption properties of hydroxyl radicals. Therefore, B-GDY could produce reactive oxygen species (ROS) efficiently to achieve bacteriostasis. Analogously, •OH was produced in the process of glutathione decomposition catalyzed by the PdFe/GDY complex nanozyme, which effectively inhibited bacterial proliferation [30]. Biocompatibility is one of the problems that limit the further application of nanozymes.

It was reported that a molybdenum-based nanozyme is the key to solving this problem. In recent years, based on the strength of the multi-phase and polyvalence characteristics of molybdenum-based nanomaterials, researchers have gradually applied them to catalysis, energy storage, sensing, and other fields [31,32]. As the typical molybdenum-based nanomaterials, molybdenum disulfide (MoS_2) consisting of vertically stacked S–Mo–S interlayers possesses a graphene-like 2D layered structure [33]. Owing to the weak interaction forces between the interlayers, MoS_2 is easily stripped into thin layers and facilitates modification [34]. MoS_2 owns multi-enzyme activities, which can catalyze diverse reactions of different substrates [35]. Furthermore, abundant dangling bands at the edge of MoS_2 are capable of providing plentifully active sites [36,37]. Benefiting from excellent biocompatibility, low levels of toxicity, and wonderfully enzyme-like activity, MoS_2 is favored by researchers in the fields of sensors and tumor therapy [38,39]. Ma et al. [40] employed ultra-thin 2H-MoS_2/Co_3O_4 nanocomposites to detect nitroaromatic compounds. Ultra-thin 2H-MoS_2 possessed a nitroreductase-mimicking activity. After the modification of Co_3O_4 nanosheets, the oxygen vacancy of 2H-MoS_2/Co_3O_4 was increased, thus improving the nitroreductase-like activity. 2H-MoS_2/Co_3O_4 provided the basis for the fabrication of an electrochemical sensor. Zhang et al. [41] fabricated an aptamer sensor, applying MoS_2 quantum dots and Cu nanowires composites (MoS_2 QDs@Cu NWs) as the labeled materials to detect amyloid-beta oligomers (AβO). When AβO was not present, MoS_2 QDs as bio-enzymes catalyzed the reaction of produce 4-chloro-1-naphthol, producing precipitates, which significantly reduced photocurrents. Cu NWs, as an LSPR sensitizer, and MoS_2 QDs amplified the signal together. The tubular C@MoS_2 nanozyme owned a fine oxidase-like activity [42]. Because of Hg-S chelation, the C@MoS_2 nanozyme possessed an Hg^{2+}-enhanced oxidase-like activity. Based on this mechanism, C@MoS_2 was used to construct a colorimetric sensing strategy to detect Hg^{2+}. Zhang et al. [43] proposed a cascade platform based on antioxidant MoS_2 nanosheets to inhibit the production of ROS and effectively alleviate the process of hepatic fibrosis. MoS_2, which mimics the activity of multiple enzymes, could block the transfer of electrons from cytochrome c to H_2O_2 and remove excess ROS under pathological conditions. Therefore, the MoS_2 nanosheets reduced cell damage caused by ROS. MoS_2 loading in carbon nanotubes (CNTs@MoS_2 NSs) combined with hydrogels to inhibit the reproduction of bacteria and promote wound healing [44]. The application of CNTs and near-infrared light treatment significantly increased the peroxidase activity of MoS_2 NSs, resulting in the enhancement of antibacterial performance. In addition, because of the superoxidase-like activity and catalase-like activity of the nanozyme, MoS_2 catalyzed the generation of O_2, which was of benefit to effectively promote wound healing.

In this study, a MoS_2@Au complex nanozyme was successfully synthesized for the detection of NSE. The NSE secondary antibody (Ab_2) was coupled with the MoS_2@Au nanozyme to form a MoS_2@Au-Ab_2 bioconjugate by electrostatic interaction between Au and the amino group. A sandwich-type immunosensor was fabricated, in which an NSE primary antibody (Ab_1) was immobilized on the rGO-tetraethylene pentamine (rGO-TEPA) bound in the electrode. Whereafter, NSE and the MoS_2@Au-Ab_2 bioconjugate was successively modified on the electrode with antigen–antibody reaction. Ultimately, the performance of the immunosensor was measured through catalytic signals toward H_2O_2. Applications of the MoS_2@Au nanozyme amplified the current response, and rGO-TEPA provided abundant sites to anchor Ab_1. The constructed immunosensor with fine selectivity and reproducibility provided potential applications for the sensitive detection of other disease markers.

2. Materials and Methods

2.1. Apparatus and Reagents

NSE and NSE antibodies were purchased from Linc-Bio Science Co. Ltd. (Shanghai, China). Reduced graphene oxide-tetraethylene pentamine (rGO-TEPA) was purchased from XFNANO Nanotechnology Co., Ltd. (Nanjing, China). Bovine serum albumin (BSA) was purchased from Macklin Biochemical Co., Ltd. (Shanghai, China). $HAuCl_4{\cdot}4H_2O$, $NaBH_4$, $(NH_4)_6Mo_7O_{24}{\cdot}4H_2O$, glutaraldehyde (GA), polyvinylpyrrolidone K30 (PVP), and thiourea (CH_4N_2S) was purchased from Sinopharm Chemical Reagent Co., Ltd. (Shanghai, China).

Electrochemical measurements were performed on a CHI760E electrochemical workstation (Chenhua Instruments Shanghai Co., Ltd., Shanghai, China). X-ray diffraction patterns (XRD) were obtained by the X-ray Powder Diffractometer (Rigaku, Japan). Scanning electron microscopy (SEM) images and energy dispersion spectrum (EDS) were acquired using a field emission SEM (ZEISS, Oberkochen, Germany).

2.2. Synthesis of MoS_2 and MoS_2@Au Nanozyme

Firstly, the MoS_2 nanozyme was prepared through a hydrothermal process in the light of the reported literature with some modifications [45]. Firstly, 1.05 g of $(NH_4)_6Mo_7O_{24}{\cdot}4H_2O$ and 2.28 g of CH_4N_2S was dissolved in 30 mL deionized water, and then the mixture solution was transferred into a 50 mL Teflon-line autoclave. After 10 h of reaction at 180 °C, the autoclave cooled to room temperature. The final product was washed with water and ethanol successively. Additionally, it was collected after drying in a vacuum overnight.

Secondly, the MoS_2@Au nanozyme was prepared by referring to the literature [46]. A total of 2 mg of MoS_2 and 12 mL of ethanol containing 0.6 g of PVP were mixed. After sonicating for 0.5 h, $HAuCl_4{\cdot}4H_2O$ (500 μL, 0.02842 mol·L^{-1}) and $NaBH_4$ (500 μL, 0.06 mol·L^{-1}) were introduced into the above solution under magnetic agitation. Thirty minutes later, the MoS_2@Au nanozyme was washed with ethanol.

2.3. Fabrication of the Immunosensor

The preparation process of the MoS_2@Au-Ab_2 bioconjugate is shown in Scheme 1A. Ab_2 (200 μL, 10 μg·mL^{-1}) was introduced into 2 mL of a phosphate buffered solution (PBS, pH 7.4) which dispersed with 4 mg of MoS_2@Au. After incubation at 4 °C for 12 h, the MoS_2@Au-Ab_2 bioconjugate was washed at 4 °C. The obtained precipitate was redispersed in 2 mL of PBS. Moreover, the mixture was stored at 4 °C.

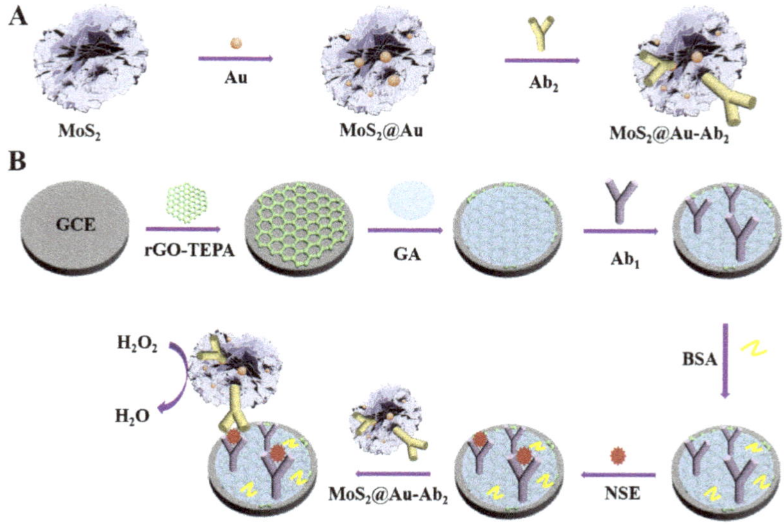

Scheme 1. Schematic diagram for the preparation of the MoS_2@Au-Ab_2 bioconjugate (**A**) and fabrication of the electrochemical immunosensor (**B**).

The fabrication process of the immunosensor is shown in Scheme 1B. First of all, aluminum oxide powder, with sizes of 1.0, 0.3, and 0.05 μm, was employed to polish glassy carbon electrodes (GCE), while rinsing the powder off GCE with ultrapure water. Subsequently, the bare GCE was modified by rGO-TEPA suspension (6 μL). A total of 40 min later, GA (6 μL, 2.5%) was dropped onto the modified GCE and kept for 1 h. Afterward, Ab_1 (6 μL, 10 μg·mL^{-1}) was introduced onto the above modified GCE, while incubating at 4 °C for 1.5 h. After that, 1% BSA was present on the surface of electrodes to stem the nonspecific binding sites. After reacting at 4 °C for 1 h, the electrodes were modified by different concentrations of NSE and stored at 4 °C for 2 h. Ultimately, the MoS_2@Au-Ab_2 bioconjugate was introduced onto the surface of the modified electrodes with incubating at 4 °C for 2 h. Whereafter, the electrodes were rinsed and then dried for measuring subsequently.

2.4. The Detection of NSE

In all of the electrochemical measurements, this work adopted a conventional three-electrode system, consisting of a modified GCE with a diameter of 4 mm, a saturated calomel electrode, and a platinum wire, which was used as the working electrode, the reference electrode, and the counter electrode, respectively. A total of a 5.0 mmol·L^{-1} of $[Fe(CN)_6]^{3-/4-}$ solution with 0.1 mol·L^{-1} KCl was employed to measure the electrochemical impedance spectroscopy (EIS). The amperometric responses were recorded by amperometric i–t curve. After stabilization of the background, 5 mmol·L^{-1} of H_2O_2 was introduced into the PBS under stirring, while recording the current response.

3. Results and Discussion

3.1. Characterization of MoS_2 and MoS_2@Au Nanozyme

The XRD result of the prepared MoS_2 is exhibited in Figure 1A. The diffraction peak at 32.6° and 58.3° corresponded to (100) and (110) crystal planes, respectively, which belonged to hexagonal MoS_2 (JCPDS 37-1492) [47–49]. In addition, the morphology of MoS_2 was characterized by SEM. Observed from Figure 1B, the MoS_2 nanoflowers were stacked by MoS_2 nanosheets. The flower-like MoS_2 also had some stacking. Flower-like MoS_2 provided abundant sites to load Au nanoclusters (Au NCs) and catalyze the decomposition of H_2O_2. The raspberry-like Au NCs are exhibited in Figure 1C. Au NCs anchored on the MoS_2 nanozyme were composed of Au nanoparticles. In addition, the diameter of Au NCs ranged between 45 nm and 100 nm, which were employed to immobilize Ab_2. As shown in Figure 1D(1–3), three elements were presented in the sample: Mo, S, and Au. Figure 1C–E clearly show that the three elements were distributed evenly in the complex nanozyme. In brief, the above analyses demonstrated that the MoS_2@Au nanozyme was synthesized successfully. Moreover, the characterization of the peroxidase-mimetic activity of MoS_2@Au and the i–t curve are shown in Figure S1.

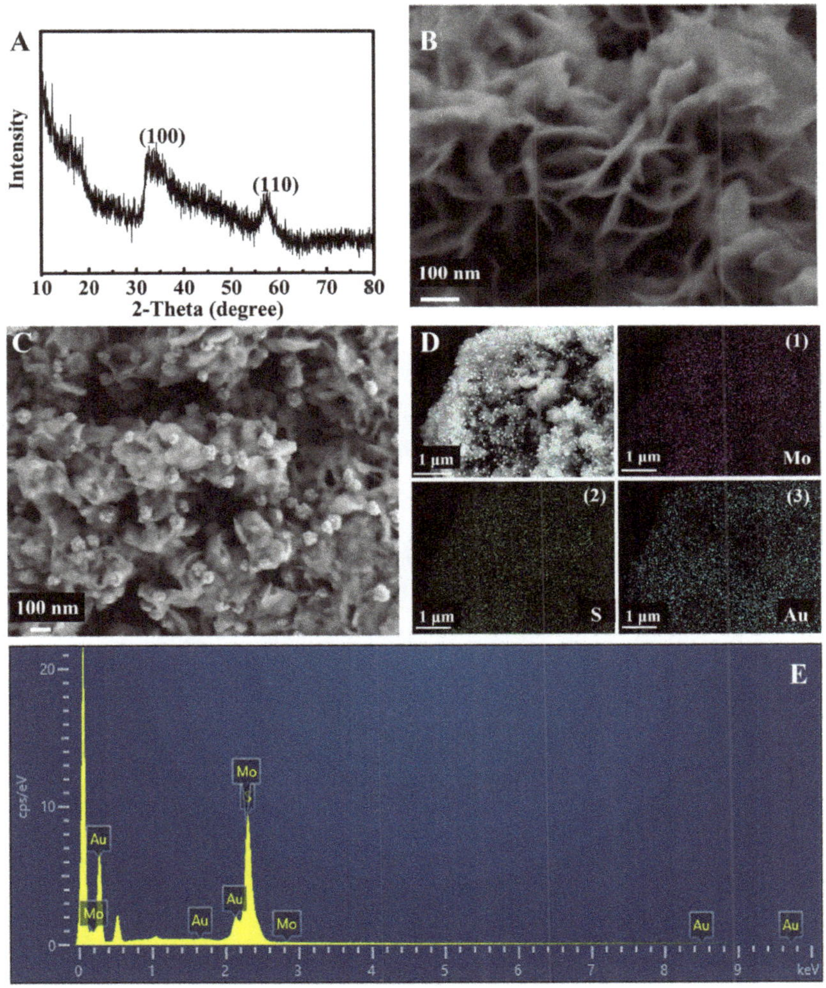

Figure 1. XRD pattern (**A**) and SEM (**B**) of the MoS_2 nanozyme; SEM (**C**) and EDS mapping images of the MoS_2@Au nanozyme (**D**), (1–3); EDS spectrum of the MoS_2@Au nanozyme (**E**).

3.2. Optimization of Experimental Conditions

In order to obtain the optimum analytical performance of the immunosensor, this work optimized the pH of the PBS solution (4.5, 5.7, 6.6, 7.4, and 8.3), the concentration of rGO-TEPA (0.5, 1.0, 1.5, 2.0, and 2.5 mg·mL^{-1}), and the incubation time of the NSE antibody and NSE (0.5, 1.0, 1.5, 2.0, 2.5, and 3.0 h). In the process of optimization, the immunosensors were fabricated with 1 ng·mL^{-1} of NSE. Under acidic conditions, MoS_2

owns a fine electrocatalytic activity; however, the NSE and NSE antibody can only maintain excellent stability under near-neutral conditions [31]. In order to obtain the optimal pH value, a PBS buffer solution in the pH range of 4.5 to 8.3 was configured for detecting NSE. The measuring result is shown in Figure 2A, and the optimum amperometric response occurred at pH 5.7. Hence, a PBS buffer solution with pH 5.7 was selected for the subsequent study.

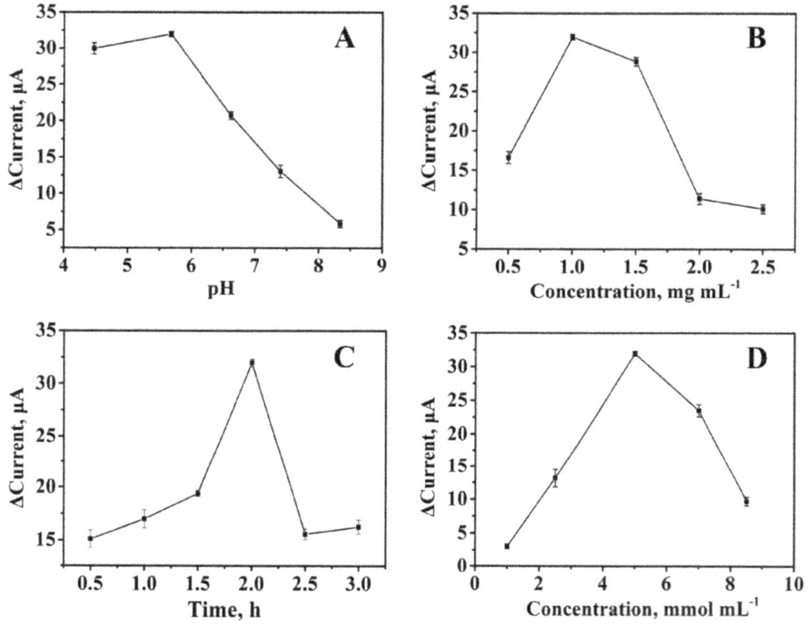

Figure 2. Effects of the pH value of PBS (**A**), the concentration of rGO—TEPA (**B**), the incubation time (**C**), and the concentration of H_2O_2 (**D**) on the current responses of the immunosensor. Error bar = SD (n = 5).

The amount of rGO-TEPA modified on the GCE is also a significant factor in current responses. Observed from Figure 2B, as the concentration range is between 0.5 mg·mL^{-1} and 2.5 mg·mL^{-1}, the current responses possessed a peak value, appearing at 1.0 mg·mL^{-1}. This may be attributed to the outstanding capability of electron transfer at 1.0 mg·mL^{-1}. Hence, 1.0 mg·mL^{-1} of rGO-TEPA was chosen for measurement throughout this research.

Moreover, the incubation time of NSE and the detector antibody also was investigated in this work. Figure 2C shows that amperometric responses gradually increased with the incubation time ranging from 0.5 h to 2.0 h. Nevertheless, amperometric responses decreased inch by inch at an incubation time ranging from 2.0 h to 3.0 h. Based on the above results, an incubation time of 2.0 h was chosen for follow-up testing. Additionally, the impact of varying concentrations of both H_2O_2 and Au NPs was investigated (Figures 2D and S2). The results have demonstrated that the optimal concentrations of H_2O_2 and Au NPs were 5 mmol·mL^{-1} and 28.42 mmol·mL^{-1}, respectively.

3.3. *Characterization of the Immunosensor*

Observed from Figure 3, EIS spectrum was composed of two parts: the semicircle portion was on behalf of the electron transfer limited process, and the linear portion represented the diffusion limited process. [50,51]. Compared with the undecorated electrode (curve a), the resistance of GCE decorated by rGO-TEPA (curve b) was barely changed, which indicates that rGO-TEPA scarcely impeded the electrode conductivity. The SEM analysis of both the bare GCE and rGO-TEPA modified GCE was utilized to confirm the successful modification of rGO-TEPA, as shown in Figure S3. After GA was modified on the electrode (curve c), the resistance increased because of the ability to block the electron transfer of GA. Subsequently, the resistance further increased with the immobilization of Ab_1 (curve d) by the reaction between amino groups which, respectively, belonged to GA and Ab_1. Afterward, the resistance was increased for introducing BSA (curve e). When the Ab_1 captured the NSE (curve f), the resistance increased significantly, proving that the protein greatly hindered the transfer of electrons. In the end, owing to the outstanding capability of MoS_2@Au, the resistance decreased with the modification of the MoS_2@Au-Ab_2 bioconjugate (curve g).

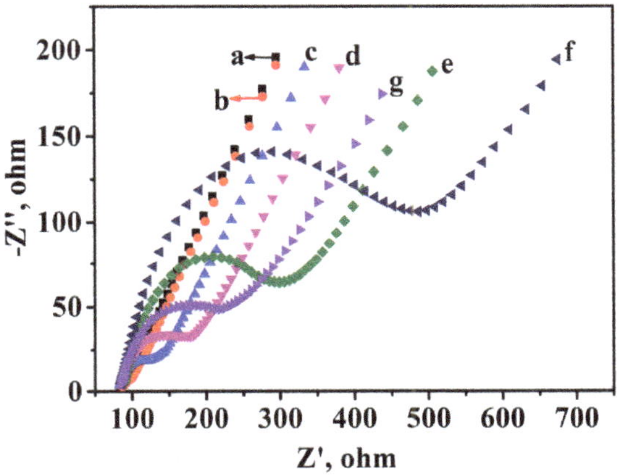

Figure 3. EIS recorded from 1 to 10^5 Hz for bare GCE (a), GCE/rGO−TEPA (b), GCE/rGO−TEPA/GA (c), GCE/rGO−TEPA/GA/Ab_1 (d), GCE/rGO−TEPA/GA/Ab_1/BSA (e), GCE/rGO−TEPA/GA/Ab_1/BSA/NSE (f), and GCE/rGO−TEPA/GA/Ab_1/BSA/NSE/MoS_2@Au−Ab_2 (g).

Under optimum conditions, a suite of electrochemical immunosensors was constructed to detect different concentrations of NSE. Electrochemical signals of the immunosensor were researched through adding the H_2O_2 into a continually stirring buffer. Observed from Figure 4A, as the NSE concentration increased, the amperometric responses increased. This could be attributed to the greater concentration of NSE and the more MoS_2@Au-Ab_2 bioconjugate modification on the electrode through antigen- and antibody-specific binding. Observed from Figure 4B, the current signal exhibited a linear relation toward the logarithmic value of NSE concentrations varying from 0.1 pg·mL^{-1} to 10 ng·mL^{-1}. I regression equation was $I = 3.07$ lg$c$ + 31.79, and the correlation coefficient was 0.998. The low limit of detection (LOD) was 0.05 pg·mL^{-1}, according to the definition of the detection limit prescribed by the International Union of Pure and Applied Chemistry (IUPAC) [52]. The calculation could be acquired from the SI. Observed from Table 1, compared with the reported methods, this proposed strategy exhibited satisfactory performances, such as a lower LOD and reasonable sensitivity. This may be mainly ascribed to the following reasons. In the beginning, a mass of Ab_1 can be immobilized due to abundant amino groups on rGO-TEPA [51]. Additionally, the MoS_2@Au nanozyme provides ample sites to bind with Ab_2. Moreover, an unexceptionable capability in decomposing H_2O_2 is conducive to improving the sensitivity of immunosensors.

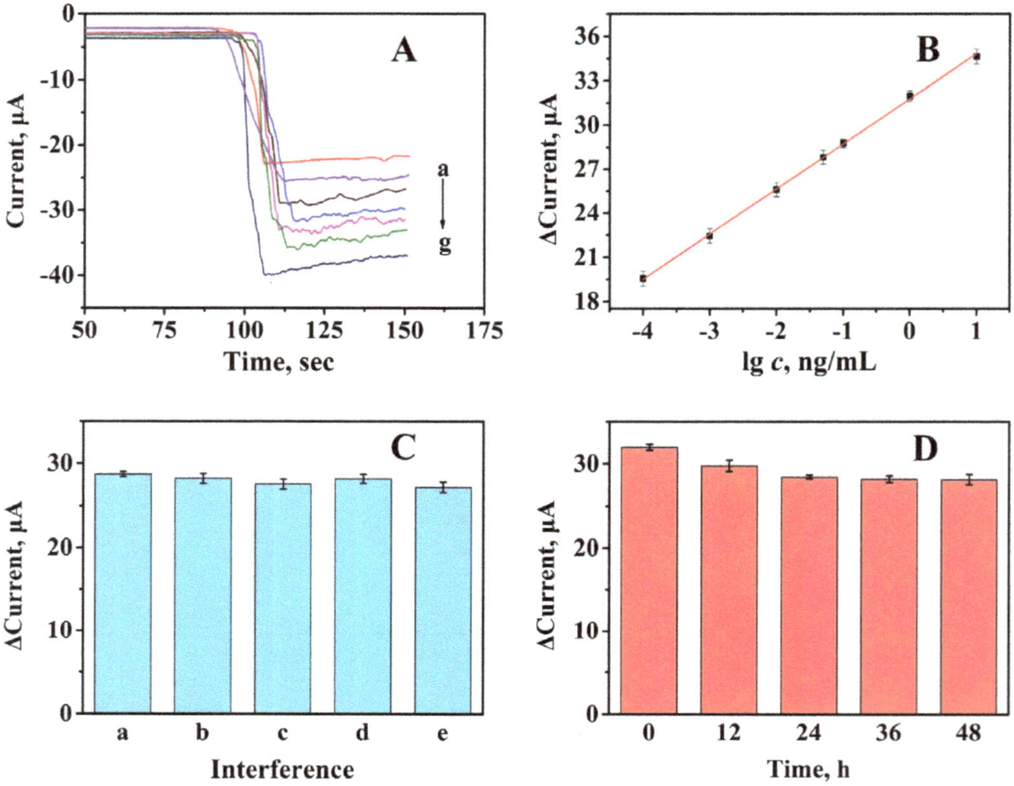

Figure 4. Current responses of the proposed immunosensor toward different concentrations of NSE (**A**): a–g: 0.0001, 0.001, 0.01, 0.05, 0.1, 1, and 10 ng·mL^{-1}; calibration curve of the immunosensor toward different concentrations of NSE (**B**). Selectivity of the immunosensor for the detection of NSE (**C**): NSE (a); NSE + CEA (b), NSE + cyfra21-1 (c), NSE + BSA (d), and NSE + CEA + cyfra21-1 + BSA (e). Stability of the immunosensor (**D**). The values of (**B–D**) are absolute values. Error bar = SD (n = 5).

Table 1. The proposed immunosensor compared with other methods in detecting NSE.

Method	Technique	Type	Sample	Linear Range ($ng \cdot mL^{-1}$)	LOD ($pg \cdot mL^{-1}$)	Selectivity	Ref.
Photoelectric immunosensor based on TiO_2/CdS	Photoelectric analysis	Label-free immunosensor	NSE	0.01–100	2.49	PSA, CEA, cTnI	[53]
Fluorescent method based on immunochromatographic test strip and quantum dot beads	Fluorescence analysis	Sandwich-type immunosensor	NSE	5–50	42.6	PSA, CA72-4, AFP	[14]
Electrochemical immunosensor based on Au-silica nanocomposite	Electrochemical analysis	Label-free immunosensor	NSE	0.1–2000	50.0	——	[54]
Photoelectrochemical immunosensor based on MoS_2 nanosheets integrated with gold nanostars	Photoelectrochemical analysis	Label-free immunosensor	NSE	0.005–1.5	3.50	AFP, CEA, PSA, LAD, GST	[55]
Photothermal immunoassay based on prussian blue nanoparticles-loaded liposomes	Photothermal analysis	Sandwich-type immunosensor	NSE	0.1–100	53.0	L-Cys, Gly, Glc, HSA, collagen, IgG	[56]
Electrochemical immunoassay based on MoS_2@Au nanozyme.	Electrochemical analysis	Sandwich-type immunosensor	NSE	0.0001–10	0.05	CEA, BSA cyfra21-1	This paper

3.4. Reproducibility, Selectivity and Stability

In the case of evaluating the reproducibility of the proposed immunoassay, five immunosensors were fabricated to measure 1 ng·mL^{-1} NSE under identical conditions. The relative standard deviation (RSD) of the current response for the above immunosensors was 1.51%, which demonstrated that the prepared immunosensor possessed satisfactory reproducibility.

The cytokeratins21-1 (cyfra21-1), carcinoembryonic antigen (CEA), and BSA were employed as interferences for the purpose of evaluating the selectivity of the proposed immunosensor. In the process of this measurement, 0.1 ng·mL^{-1} NSE with a 10 ng·mL^{-1} interference was detected, and the results are exhibited in Figure 4C. Moreover, the response to 10 ng·mL^{-1} interferences is shown in Figure S4. The current responses varied between 2% and 5%, demonstrating that the selectivity of this strategy was quite well.

The Stability of the proposed immunosensor was tested. As shown in Figure 4D, after 12 h at 4 °C, the amperometric response clearly decreased. However, the amperometric response decreased tardily as the storage time increased, and 48 h later, the amperometric response dropped to 87% of the initial amperometric response. Those above showed that the immunosensor owned certain stability in 48 h.

3.5. Detection of NSE in Serum Samples

The concentration of NSE in the human serum sample was measured through the standard addition method for evaluating the practicability of the proposed immunosensor. The obtained serum sample was diluted 20 times with a PBS solution (pH = 7.4). Observed from Table 2, the recovery was from 97.6% to 99.9% and the RSD ranged from 1.48% to 3.35%, manifesting that the proposed immunosensor was expected to monitor real samples.

Table 2. Detecting concentrations of NSE in the serum sample.

Concentration of NSE in Serum Sample (ng·mL^{-1})	The Addition Content (ng·mL^{-1})	The Detection Content (ng·mL^{-1})	RSD (%)	Recovery (%)
0.51	0.50	0.97, 0.96, 1.00, 1.02, 1.04	3.35	97.6
	2.00	2.52, 2.46, 2.53, 2.55, 2.48	1.48	99.9
	8.00	8.55, 8.35, 8.47, 8.22, 8.53	1.64	98.9

4. Conclusions

In brief, a sandwich-type immunosensor which employed the MoS_2@Au nanozyme was constructed to achieve sensitivity detection of NSE. The rGO-TEPA offered luxuriant sites for immobilizing enough Ab_1. Furthermore, the catalase activity and structural properties of MoS_2 efficiently amplified the amperometric response as well as enhanced the sensitivity of the proposed strategy. Under optimal experimental conditions, the linear detection range was 0.1 pg·mL^{-1}–10 ng·mL^{-1}, and regression equation was $I = 3.07\ \mathrm{lg}c + 31.79$, with a correlation coefficient of 0.998. This immunosensor possessed low LOD (0.05 pg·mL^{-1}), high interference resistance, and outstanding reproducibility, which is expected to be applied to the early diagnosis of NSE. Furthermore, the proposed immunosensor holds tremendous promise for determining other disease markers.

Supplementary Materials: The following supporting information can be downloaded at: https://www.mdpi.com/article/10.3390/chemosensors11060349/s1. Figure S1: (A) UV−vis absorption spectra of TMB with different substrates. (B) The i−t curve of 0.5 mg·mL^{-1} MoS_2@Au; Figure S2: Effect of the concentration of Au NPs on the current responses of immunosensor; Figure S3: The SEM of bare GCE (A) and rGO−TEPA/GCE (B−C); Figure S4: The response of immunosensor in detecting interference: (a) 10 ng·mL^{-1} CEA; (b) 10 ng·mL^{-1} cyfra21−1; (c) 10 ng·mL^{-1} BSA; (d) 10 ng·mL^{-1} CEA + 10 ng·mL^{-1} cyfra21−1 + 10 ng·mL^{-1} BSA. Calculation of detection limit.

Author Contributions: Conceptualization, Y.W. (Yaoguang Wang) and H.C.; methodology, Y.W. (Yingying Wang), H.W., Y.Z. and Y.W. (Yaoguang Wang); software, Y.W. (Yingying Wang) and H.W.; validation, Y.B., G.Z. and Y.Z.; formal analysis, Y.W. (Yingying Wang), G.Z. and N.Z.; investigation, Y.W. (Yingying Wang) and Y.B.; resources, Y.W. (Yaoguang Wang) and H.C.; data curation, Y.W. (Yingying Wang) and G.Z.; writing—original draft preparation, Y.W. (Yingying Wang); writing—review and editing, Y.W. (Yaoguang Wang) and H.C.; visualization, H.W., Y.B. and N.Z.; supervision, N.Z., Y.W. (Yaoguang Wang) and H.C.; project administration, Y.Z. and Y.W. (Yaoguang Wang); funding acquisition, Y.W. (Yaoguang Wang). All authors have read and agreed to the published version of the manuscript.

Funding: This research was funded by the National Natural Science Foundation of China (Nos. 22006080 and 22264025), the Shandong Provincial Natural Science Foundation (Nos. ZR2021YQ07, ZR2020QB094, and ZR2020QB097), the Science, Education and Industry Integration Innovation Pilot Project from Qilu University of Technology (Shandong Academy of Sciences) (2022JBZ02-04), and the Program for Scientific Research Innovation Team in Colleges and Universities of Shandong Province.

Institutional Review Board Statement: Not applicable.

Informed Consent Statement: Not applicable.

Data Availability Statement: The data are available upon request to the corresponding authors.

Conflicts of Interest: The authors declare no conflict of interest.

References

1. Blandin Knight, S.; Crosbie, P.A.; Balata, H.; Chudziak, J.; Hussell, T.; Dive, C. Progress and prospects of early detection in lung cancer. *Open Biol.* **2017**, *7*, 170070. [CrossRef] [PubMed]
2. Du, J.; Li, Y.; Wang, L.; Zhou, Y.; Shen, Y.; Xu, F.; Chen, Y. Selective application of neuroendocrine markers in the diagnosis and treatment of small cell lung cancer. *Clin. Chim. Acta* **2020**, *509*, 295–303. [CrossRef]
3. Sutherland, K.D.; Proost, N.; Brouns, I.; Adriaensen, D.; Song, J.; Berns, A. Cell of origin of small cell lung cancer: Inactivation of Trp53 and Rb1 in distinct cell types of adult mouse lung. *Cancer Cell* **2011**, *19*, 754–764. [CrossRef] [PubMed]
4. Gazdar, A.F.; Bunn, P.A.; Minna, J.D. Small-cell lung cancer: What we know, what we need to know and the path forward. *Nat. Rev. Cancer* **2017**, *17*, 725–737. [CrossRef]
5. Huang, X.; Miao, J.; Fang, J.; Xu, X.; Wei, Q.; Cao, W. Ratiometric electrochemical immunosensor based on L-cysteine grafted ferrocene for detection of neuron specific enolase. *Talanta* **2022**, *239*, 123075. [CrossRef] [PubMed]
6. Marangos, P.J.; Schmechel, D.E. Neuron specific enolase, a clinically useful marker for neurons and neuroendocrine cells. *Ann. Rev. Neurosci.* **1987**, *10*, 269–295. [CrossRef]
7. Ma, E.; Wang, P.; Yang, Q.; Yu, H.; Pei, F.; Zheng, Y.; Liu, Q.; Dong, Y.; Li, Y. Electrochemical immunosensors for sensitive detection of neuron-specific enolase based on small-size trimetallic Au@Pd^Pt nanocubes functionalized on ultrathin MnO_2 nanosheets as signal labels. *ACS Biomater. Sci. Eng.* **2020**, *6*, 1418–1427. [CrossRef]
8. Dong, H.; Liu, S.; Liu, Q.; Li, Y.; Li, Y.; Zhao, Z. A dual-signal output electrochemical immunosensor based on Au-MoS_2/MOF catalytic cycle amplification strategy for neuron-specific enolase ultrasensitive detection. *Biosens. Bioelectron.* **2022**, *195*, 113648. [CrossRef]
9. Yu, X.; Li, X.; Zhang, S.; Jia, Y.; Xu, Z.; Li, X.; Chen, Z.; Li, Y. Ultrasensitive electrochemical detection of neuron-specific enolase based on spiny core-shell Au/Cu_xO@CeO_2 nanocubes. *Bioelectrochemistry* **2021**, *138*, 107693. [CrossRef]
10. Lin, Y.; Kannan, P.; Zeng, Y.; Qiu, B.; Guo, L.; Lin, Z. Enzyme-free multicolor biosensor based on Cu^{2+}-modified carbon nitride nanosheets and gold nanobipyramids for sensitive detection of neuron specific enolase. *Sensor. Actuat. B-Chem.* **2019**, *283*, 138–145. [CrossRef]
11. Kalkal, A.; Kadian, S.; Kumar, S.; Manik, G.; Sen, P.; Kumar, S.; Packirisamy, G. Ti_3C_2-MXene decorated with nanostructured silver as a dual-energy acceptor for the fluorometric neuron specific enolase detection. *Biosens. Bioelectron.* **2022**, *195*, 113620. [CrossRef]
12. McKitterick, N.; Bicak, T.C.; Cormack, P.A.G.; Reubsaet, L.; Halvorsen, T.G. Facilitating serum determination of neuron specific enolase at clinically relevant levels by coupling on-line molecularly imprinted solid-phase extraction to LC-MS/MS. *Anal. Chim. Acta* **2020**, *1140*, 210–218. [CrossRef]
13. Gao, X.; Zheng, P.; Kasani, S.; Wu, S.; Yang, F.; Lewis, S.; Nayeem, S.; Engler-Chiurazzi, E.B.; Wigginton, J.G.; Simpkins, J.W.; et al. Paper-based surface-enhanced raman scattering lateral flow strip for detection of neuron-specific enolase in blood plasma. *Anal. Chem.* **2017**, *89*, 10104–10110. [CrossRef] [PubMed]

14. Xiao, K.; Wang, K.; Qin, W.; Hou, Y.; Lu, W.; Xu, H.; Wo, Y.; Cui, D. Use of quantum dot beads-labeled monoclonal antibody to improve the sensitivity of a quantitative and simultaneous immunochromatographic assay for neuron specific enolase and carcinoembryonic antigen. *Talanta* **2017**, *164*, 463–469. [CrossRef]
15. Xu, R.; Xu, K.; Du, Y.; Li, J.; Ren, X.; Ma, H.; Wu, D.; Li, Y.; Wei, Q. Dual direct z-scheme heterojunction with growing photoactive property for sensitive photoelectrochemical and colorimetric bioanalysis. *Anal. Chem.* **2022**, *94*, 9888–9893. [CrossRef]
16. Khatri, R.; Puri, N.K. Electrochemical biosensor utilizing dual-mode output for detection of lung cancer biomarker based on reduced graphene oxide-modified reduced-molybdenum disulfide multi-layered nanosheets. *J. Mater. Res.* **2022**, *37*, 1451–1463. [CrossRef]
17. Zheng, X.; Mo, G.; He, Y.; Qin, D.; Jiang, X.; Mo, W.; Deng, B. Electrochemiluminescent immunoassay for neuron specific enolase by using amino-modified reduced graphene oxide loaded with N-doped carbon quantum dots. *Microchim. Acta* **2019**, *186*, 817. [CrossRef]
18. Tang, C.; Wang, P.; Zhou, K.; Ren, J.; Wang, S.; Tang, F.; Li, Y.; Liu, Q.; Xue, L. Electrochemical immunosensor based on hollow porous Pt skin AgPt alloy/NGR as a dual signal amplification strategy for sensitive detection of neuron-specific enolase. *Biosens. Bioelectron.* **2022**, *197*, 113779. [CrossRef] [PubMed]
19. Amani, J.; Maleki, M.; Khoshroo, A.; Sobhani-Nasab, A.; Rahimi-Nasrabadi, M. An electrochemical immunosensor based on poly p-phenylenediamine and graphene nanocomposite for detection of neuron-specific enolase via electrochemically amplified detection. *Anal. Biochem.* **2018**, *548*, 53–59. [CrossRef] [PubMed]
20. Zhang, C.; Ma, Z. PtCu nanoprobe-initiated cascade reaction modulated iodide-responsive sensing interface for improved electrochemical immunosensor of neuron-specific enolase. *Biosens. Bioelectron.* **2019**, *143*, 111612. [CrossRef]
21. Yu, T.; Cheng, W.; Li, Q.; Luo, C.; Yan, L.; Zhang, D.; Yin, Y.; Ding, S.; Ju, H. Electrochemical immunosensor for competitive detection of neuron specific enolase using functional carbon nanotubes and gold nanoprobe. *Talanta* **2012**, *93*, 433–438. [CrossRef] [PubMed]
22. Wang, Z.; Zhang, Y.; Wang, X.; Han, L. Flow-homogeneous electrochemical sensing system based on 2D metal-organic framework nanozyme for successive microRNA assay. *Biosens. Bioelectron.* **2022**, *206*, 114120. [CrossRef]
23. Han, L.; Zhang, H.; Li, F. Bioinspired nanozymes with pH-independent and metal ions-controllable activity: Field-programmable logic conversion of sole logic gate system. *Part. Part. Syst. Char.* **2018**, *35*, 1800207. [CrossRef]
24. Gai, P.; Pu, L.; Wang, C.; Zhu, D.; Li, F. CeO_2@NC nanozyme with robust dephosphorylation ability of phosphotriester: A simple colorimetric assay for rapid and selective detection of paraoxon. *Biosens. Bioelectron.* **2023**, *220*, 114841. [CrossRef] [PubMed]
25. Zhu, D.; Zhang, M.; Wang, C.; Gai, P.; Li, F. Metal-free carbon nanozyme as nicotinamide adenine dinucleotide oxidase mimic over a broad pH range for coenzyme regeneration. *Chem. Mater.* **2022**, *34*, 11072–11080. [CrossRef]
26. Zhou, X.; Wang, M.; Chen, J.; Su, X. Cascade reaction biosensor based on Cu/N co-doped two-dimensional carbon-based nanozyme for the detection of lactose and beta-galactosidase. *Talanta* **2022**, *245*, 123451. [CrossRef]
27. Wang, K.; Wu, C.; Wang, F.; Liao, M.; Jiang, G. Bimetallic nanoparticles decorated hollow nanoporous carbon framework as nanozyme biosensor for highly sensitive electrochemical sensing of uric acid. *Biosens. Bioelectron.* **2020**, *150*, 111869. [CrossRef] [PubMed]
28. Yang, Q.Y.; Wan, C.Q.; Wang, Y.X.; Shen, X.F.; Pang, Y.H. Bismuth-based metal-organic framework peroxidase-mimic nanozyme: Preparation and mechanism for colorimetric-converted ultra-trace electrochemical sensing of chromium ion. *J. Hazard. Mater.* **2023**, *451*, 131148. [CrossRef]
29. Bi, X.; Bai, Q.; Wang, L.; Du, F.; Liu, M.; Yu, W.W.; Li, S.; Li, J.; Zhu, Z.; Sui, N.; et al. Boron doped graphdiyne: A metal-free peroxidase mimetic nanozyme for antibacterial application. *Nano Res.* **2021**, *15*, 1446–1454. [CrossRef]
30. Wang, T.; Bai, Q.; Zhu, Z.; Xiao, H.; Jiang, F.; Du, F.; Yu, W.W.; Liu, M.; Sui, N. Graphdiyne-supported palladium-iron nanosheets: A dual-functional peroxidase mimetic nanozyme for glutathione detection and antibacterial application. *Chem. Eng. J.* **2021**, *413*, 127537. [CrossRef]
31. Zu, Y.; Yao, H.; Wang, Y.; Yan, L.; Gu, Z.; Chen, C.; Gao, L.; Yin, W. The age of bioinspired molybdenum-involved nanozymes: Synthesis, catalytic mechanisms, and biomedical applications. *View* **2021**, *2*, 20200188. [CrossRef]
32. Chen, Y.; Zhang, G.; Ji, Q.; Liu, H.; Qu, J. Triggering of low-valence molybdenum in multiphasic MoS_2 for effective reactive oxygen species output in catalytic fenton-like reactions. *ACS Appl. Mater. Inter.* **2019**, *11*, 26781–26788. [CrossRef]
33. Lv, Q.; Chen, L.; Liu, H.; Zou, L. Peony-like 3D-MoS_2/graphene nanostructures with enhanced mimic peroxidase performance for colorimetric determination of dopamine. *Talanta* **2022**, *247*, 123553. [CrossRef] [PubMed]
34. Yu, Y.; Lu, L.; Yang, Q.; Zupanic, A.; Xu, Q.; Jiang, L. Using MoS_2 nanomaterials to generate or remove reactive oxygen species: A review. *ACS Appl. Nano Mater.* **2021**, *4*, 7523–7537. [CrossRef]
35. Chen, T.; Zou, H.; Wu, X.; Liu, C.; Situ, B.; Zheng, L.; Yang, G. Nanozymatic antioxidant system based on MoS_2 nanosheets. *ACS Appl. Mater. Inter.* **2018**, *10*, 12453–12462. [CrossRef]
36. Guo, X.; Wang, Y.; Wu, F.; Ni, Y.; Kokot, S. A colorimetric method of analysis for trace amounts of hydrogen peroxide with the use of the nano-properties of molybdenum disulfide. *Analyst* **2015**, *140*, 1119–1126. [CrossRef]
37. Hwang, J.H.; Islam, M.A.; Choi, H.; Ko, T.J.; Rodriguez, K.L.; Chung, H.S.; Jung, Y.; Lee, W.H. Improving electrochemical Pb^{2+} detection using a vertically aligned 2D MoS_2 nanofilm. *Anal. Chem.* **2019**, *91*, 11770–11777. [CrossRef]
38. Wang, J.; Sui, L.; Huang, J.; Miao, L.; Nie, Y.; Wang, K.; Yang, Z.; Huang, Q.; Gong, X.; Nan, Y.; et al. MoS_2-based nanocomposites for cancer diagnosis and therapy. *Bioact. Mater.* **2021**, *6*, 4209–4242. [CrossRef] [PubMed]

39. Sethulekshmi, A.S.; Saritha, A.; Joseph, K.; Aprem, A.S.; Sisupal, S.B. MoS_2 based nanomaterials: Advanced antibacterial agents for future. *J. Control. Release* **2022**, *348*, 158–185. [CrossRef]
40. Ma, Y.; Deng, M.; Wang, X.; Gao, X.; Song, H.; Zhu, Y.; Feng, L.; Zhang, Y. 2H-MoS_2/Co_3O_4 nanohybrid with type I nitroreductase-mimicking activity for the electrochemical assays of nitroaromatic compounds. *Anal. Chim. Acta* **2022**, *1221*, 340078. [CrossRef]
41. Zhang, J.; Zhang, X.; Gao, Y.; Yan, J.; Song, W. Integrating CuO/g-C_3N_4 p-n heterojunctioned photocathode with MoS_2 QDs@Cu NWs multifunctional signal amplifier for ultrasensitive detection of AβO. *Biosens. Bioelectron.* **2021**, *176*, 112945. [CrossRef]
42. Feng, L.; Zhang, L.; Gong, Y.; Du, Z.; Chen, X.; Qi, X.; Zhang, X.; Mao, G.; Wang, H. Hollow C@MoS_2 nanotubes with Hg^{2+}-triggered oxidase-like catalysis: A colorimetric method for detection of Hg^{2+} ions in wastewater. *Sens. Actuat. B-Chem.* **2022**, *361*, 131725. [CrossRef]
43. Zhang, X.; Zhang, S.; Yang, Z.; Wang, Z.; Tian, X.; Zhou, R. Self-cascade MoS_2 nanozymes for efficient intracellular antioxidation and hepatic fibrosis therapy. *Nanoscale* **2021**, *13*, 12613–12622. [CrossRef] [PubMed]
44. Li, Y.; Fu, R.; Duan, Z.; Zhu, C.; Fan, D. Adaptive hydrogels based on nanozyme with dual-enhanced triple enzyme-like activities for wound disinfection and mimicking antioxidant defense system. *Adv. Healthc. Mater.* **2022**, *11*, 2101849. [CrossRef] [PubMed]
45. Wei, F.; Cui, X.; Wang, Z.; Dong, C.; Li, J.; Han, X. Recoverable peroxidase-like Fe_3O_4@MoS_2-Ag nanozyme with enhanced antibacterial ability. *Chem. Eng. J.* **2021**, *408*, 127240. [CrossRef]
46. Chang, M.; Hou, Z.; Wang, M.; Wang, M.; Dang, P.; Liu, J.; Shu, M.; Ding, B.; Al Kheraif, A.A.; Li, C.; et al. Cu_2MoS_4/Au heterostructures with enhanced catalase-like activity and photoconversion efficiency for primary/metastatic tumors eradication by phototherapy-induced immunotherapy. *Small* **2020**, *16*, 1907146. [CrossRef] [PubMed]
47. Hoshyargar, F.; Yella, A.; Panthöfer, M.; Tremel, W. Diffusion-driven formation of MoS_2 nanotube bundles containing MoS_2 nanopods. *Chem. Mater.* **2011**, *23*, 4716–4720. [CrossRef]
48. Hu, L.; Ren, Y.; Yang, H.; Xu, Q. Fabrication of 3D hierarchical MoS_2/polyaniline and MoS_2/C architectures for lithium-ion battery applications. *ACS Appl. Mater. Inter.* **2014**, *6*, 14644–14652. [CrossRef]
49. Sheng, M.; Bin, X.; Yang, Y.; Tang, Y.; Que, W. Defect engineering-driven phase structure design of 2H@1T MoS_2 for electrochemical hydrogen evolution reaction. *Mater. Lett.* **2022**, *311*, 131624. [CrossRef]
50. Cui, M.; Wang, Y.; Wang, H.; Wu, Y.; Luo, X. A label-free electrochemical DNA biosensor for breast cancer marker BRCA1 based on self-assembled antifouling peptide monolayer. *Sens. Actuat. B-Chem.* **2017**, *244*, 742–749. [CrossRef]
51. Wang, Y.; Li, Y.; Ma, H.; Guo, A.; Du, B.; Yan, T.; Wei, Q. An ultrasensitive electrochemical immunosensor for the detection of CD146 based on TiO_2 colloidal sphere laden Au/Pd nanoparticles. *Analyst* **2015**, *140*, 3557–3564. [CrossRef] [PubMed]
52. Fassel, V. Nomenclature, symbols, units, and their usage in spectrochemical analysis—II. Data interpretation. *Spectrochim. Acta B* **1978**, *33*, 241–245.
53. Xing, Z.; Zhang, S.; Wang, H.; Ma, H.; Wu, D.; Fan, D.; Ren, X.; Wei, Q.; Ju, H. Addressable label-free photoelectric sensor array with self-calibration for detection of neuron specific enolase. *Anal. Chem.* **2022**, *94*, 6996–7003. [CrossRef] [PubMed]
54. Wang, B.; Liang, T.; Li, J.; Yu, H.; Chu, X. Fabrication of immunosensor based on au-silica nanocomposite for neuron-specific enolase detection. *Int. J. Electrochem. Sci.* **2017**, 7607–7615. [CrossRef]
55. Liu, R.; Wang, Y.; Wong, W.; Li, H.; Li, C. Photoelectrochemical immunoassay platform based on MoS_2 nanosheets integrated with gold nanostars for neuron-specific enolase assay. *Microchim. Acta* **2020**, *187*, 480. [CrossRef] [PubMed]
56. Zhi, L.; Sun, A.; Tang, D. In situ amplified photothermal immunoassay for neuron-specific enolase with enhanced sensitivity using prussian blue nanoparticle-loaded liposomes. *Analyst* **2020**, *145*, 4164–4172. [CrossRef]

chemosensors

MDPI

Article

DNAzymes-Embedded Framework Nucleic Acids (FNAzymes) for Metal Ions Imaging in Living Cells

Dan Zhu [1], Jiaxuan Huang [1], Yanting Xia [1], Shao Su [1], Xiaolei Zuo [2,3], Qian Li [2,*] and Lianhui Wang [1,*]

1 State Key Laboratory of Organic Electronics and Information Displays & Jiangsu Key Laboratory for Biosensors, Institute of Advanced Materials (IAM), Nanjing University of Posts and Telecommunications, Nanjing 210023, China; iamdzhu@njupt.edu.cn (D.Z.); iamjxhuang@163.com (J.H.); iamytxia@163.com (Y.X.); iamssu@njupt.edu.cn (S.S.)

2 School of Chemistry and Chemical Engineering, National Center for Translational Medicine, Shanghai Jiao Tong University, Shanghai 200240, China; zuoxiaolei@sjtu.edu.cn

3 Shanghai Key Laboratory for Nucleic Acids Chemistry and Nanomedicine, Institute of Molecular Medicine, Renji Hospital, School of Medicine, Shanghai Jiao Tong University, Shanghai 200127, China

* Correspondence: liqian2018@sjtu.edu.cn (Q.L.); iamlhwang@njupt.edu.cn (L.W.)

Abstract: Simultaneous and non-destructive quantitative detection of intracellular metal ions holds great promise for improving the accuracy of diagnosis and biological research. Herein, novel multicolor DNAzymes-embedded framework nucleic acids (FNAzymes) were presented, which can easily enter cells and achieve simultaneous and quantitative detection of intracellular physiologically related Cu^{2+} and Zn^{2+}. Two types of DNAzymes, specific to Cu^{2+} and Zn^{2+}, were encoded in the framework nucleic acids (FNAs) via self-assembly. With the formation of a well-ordered FNAzyme nanostructure, the fluorophore and the quencher were close to each other; therefore, the fluorescence was quenched. In the presence of Cu^{2+} and Zn^{2+}, the integrated FNAzymes would be specifically cleaved, resulting in the release of fluorophores in cells. Consequently, the fluorescence in living cells could be observed by a confocal microscope and semi-quantitatively analyzed by flow cytometry with low-nanomolar sensitivity for both metal ions. The FNAzymes have high uniformity and structural accuracy, which are beneficial for intracellular detection with excellent reproducibility. This proposed method offers new opportunities for non-destructive, semi-quantitative, multi-target detection in living cells.

Keywords: Framework Nucleic Acids (FNAs); DNAzyme; metal ions; intracellular imaging; flow cytometry

Citation: Zhu, D.; Huang, J.; Xia, Y.; Su, S.; Zuo, X.; Li, Q.; Wang, L. DNAzymes-Embedded Framework Nucleic Acids (FNAzymes) for Metal Ions Imaging in Living Cells. *Chemosensors* **2023**, *11*, 358. https://doi.org/10.3390/chemosensors11070358

Academic Editor: Chunsheng Wu

Received: 25 May 2023
Revised: 20 June 2023
Accepted: 23 June 2023
Published: 25 June 2023

1. Introduction

Physiologically related metal ions play a significant role in regulating cellular functions and life activities [1]. Among these ions, Cu^{2+} and Zn^{2+} are the second and third most abundant transition metal ions (after Fe^{3+}) inside the human body, and they participate in a variety of intra- and extracellular physiological processes, such as neural signal transmission, cellular metabolism, and respiration [2,3]. An increasing number of studies have demonstrated that the concentrations of Cu^{2+} and Zn^{2+} in vivo are correlated with many nutrition-related or enzyme-related diseases, such as genetic disorders, epilepsy, Parkinson's disease, and even cancers [4]. For instance, the relative concentration ratio of Cu^{2+}/Zn^{2+} has been considered to be an important indicator for numerous diseases, such as colorectal cancer, lung cancer, and relapsing-remitting multiple sclerosis [5]. In the cell signaling pathway, Cu^{2+} and Zn^{2+} always function complementarily to each other to regulate biological processes [6]. Therefore, quantitative detection of Zn^{2+} and Cu^{2+} simultaneously at the cellular level holds great hope for better understanding the working mechanisms of physiologically related metal ions in biological processes and has potential in biomedical applications.

The most widely used methods for conventional quantitative detection of metal elements involve inductively coupled plasma-mass spectrometry (ICP-MS) [7], inductively coupled plasma-atomic emission spectrometry (ICP-AES) [8], and atomic absorption spectrometry (AAS) [9]. These methods are commonly cumbersome and cell-destructive, and thus, they are not suitable for live-cell detection. In recent years, fluorescent sensors have turned out to be powerful tools for intracellular and super-resolution imaging [10–12]. Great efforts have been focused on the development of fluorescent probes for the imaging of intracellular metal ions such as Ca^{2+}, Cu^{2+}, Zn^{2+}, and Mg^{2+} [13–15]. Nevertheless, tedious organic synthesis steps are often required to fabricate these probes. Most small molecule probe-based approaches can only detect an individual ion due to their limited specificity towards metal ions, which restricts their applications in multiple detections. Therefore, it developing effective techniques for the non-destructive quantitative determination of multiple metal ions in living cells with high sensitivity and selectivity is in high demand.

DNAzymes are a class of functional nucleic acids that exhibit catalytic activity in the presence of specific targets [16,17]. The biocompatibility, low cytotoxicity, and ease of coding allow DNAzymes to be a powerful tool for intracellular analysis. Numerous DNAzymes specific for various metal ions (such as Na^+, Pb^{2+}, $UO_2{}^{2+}$, and K^+) have been selected for sensitive detection in vitro and in vivo [18,19]. However, due to the simple structure and the existing RNA site, DNAzymes are vulnerable to nucleases in the cell environment. To increase the stability of DNAzyme in living cells, nanocarriers have been used to assist the transfer of DNAzyme into cells, such as gold nanoparticles, two-dimensional nanomaterials, and polymers [20–22]. However, these assays also suffer from some inherent limitations. For instance, the sizes of carriers, the number of signaling molecules on carriers, and the carriers themselves could vary among different batches. Additionally, multiple-step modifications of small molecules (such as thiolated probes, fluorescent probes, or other probes) exacerbate the inhomogeneity between probes. Considering the complexity of the intracellular environment, developing probes with high uniformity and structural precision is urgent for improving the accuracy of quantification in living cells. Self-assembled framework nucleic acids (FNAs) are nano-functional materials constructed by several DNA strands through highly specific base-pairing, which have aroused widespread concern due to their ease of designing, low cytotoxicity, and good stability [23–25]. Compared to other nanocarriers, FNAs have high uniformity and structural accuracy, which is beneficial for detection with excellent repeatability [26–28]. It has been demonstrated that FNAs could penetrate cell membranes by a caveolin-dependent pathway without the assistance of transfection agents and maintain good stability in living cells, which allows FNAs to be powerful nanocarriers for intracellular applications [29].

In this work, we designed DNAzymes-embedded framework nucleic acids (FNAzymes) for the simultaneous semi-quantitative determination and imaging of intracellular Cu^{2+} and Zn^{2+}. Cu^{2+}-specific DNAzymes [30] and Zn^{2+}-specific DNAzymes [31] were employed as recognition elements, and tetrahedral FNAs were used as nanocarriers for cellular transport (Figure 1A). Benefiting from the high structural uniformity of FNAs, the incorporation of DNAzymes with FNAs might provide a platform for intracellular detection with enhanced precision. The fluorescence inside cells was observed using a laser scanning confocal microscope (LSCM). The fluorescence was analyzed using flow cytometry, which allowed for the semi-quantitative determination of intracellular metal ions (Figure 1B).

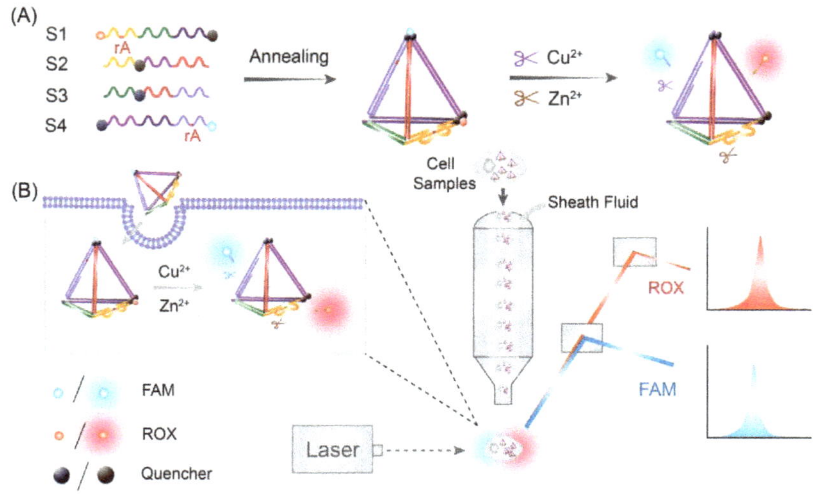

Figure 1. (**A**) Scheme of the assembly and metal ions in response to the multifunctional tetrahedral DNA nanostructure-based nanoprobe with embedded DNAzymes (FNAzymes). (**B**) The simultaneous intracellular semi-quantification of Cu^{2+} and Zn^{2+} by flow cytometer.

2. Materials and Methods

2.1. Chemicals

DNA samples (listed in Table S1) were synthesized by TaKaRa Inc. (Dalian, China). Tris (hydroxymethyl) aminomethane (Tris), NaCl, KCl, $CaCl_2$, $Cu(CH_3COO)_2{\cdot}H_2O$, $Zn(CH_3COO)_2{\cdot}2H_2O$, $FeCl_3$, $Mn(CH_3COO)_2{\cdot}4H_2O$, $MgCl_2{\cdot}6H_2O$, $Cd(NO_3)_2{\cdot}6H_2O$, $NiSO_4{\cdot}6(H_2O)$, $Co(CH_3COO)_2{\cdot}4H_2O$, and diethylpyrocarbonate (DEPC) were purchased from China National Sinopharm Chemical Reagent Co., Ltd. (Shanghai, China). Human cervical cancer cell line (HeLa) was purchased from Jiangsu Keygen Biotech Corp. Ltd. (Nanjing, China). Additionally, 3-(4,5-Dimethylthiazol-2-yl)-2,5-diphenyltetrazolium bromide (MTT) was purchased from Sigma-Aldrich (St. Louis, MO, USA). All solutions were treated with DEPC-treated ultrapure water to prevent RNA degradation by RNA enzyme. Buffer A: 20 mM Tris-HCl, 50 mM $MgCl_2$, pH 8.0. Buffer B: 50 mM Tris-HCl, 100 mM NaCl, 2 mM $MgCl_2$, pH 7.4.

2.2. Assembly of Multifunctional FNAzymes

The FNAzymes were prepared by adding the four DNA strands (strands S1–S4) in equimolar quantities together in buffer A. The solution was heated to 95 °C for 10 min, followed by rapid cooling to 4 °C. The FNAzymes were loaded onto an 8% polyacrylamide gel electrophoresis (PAGE, 8%) and imaged using a GBOX-F3-E system (Syngene UK, Massachusetts, England). The prepared FNAzymes were stored at 4 °C for further use.

2.3. AFM Imaging

The FNAzymes were diluted to 20 nM with Buffer A. Subsequently, 40 µL of APTES (0.5%) was incubated on freshly cleaved mica for a duration of 5 min. Following this, the mica surface was rinsed with Milli-Q water and dried using N_2 gas. Next, 10 µL of 20 nM FNAzymes were incubated on the treated mica for 5 min. After thorough washing, the assembly buffer was added to reach a final volume of 100 µL. The sample was then subjected to scanning using a Jscanner attached to a Multimode NanoscopeIIIa AFM (Vecco/Digital Instruments) operating in peak force mode.

2.4. Simultaneous Detection of Cu^{2+} and Zn^{2+} In Vitro

FNAzymes were firstly diluted to a concentration of 50 nM in buffer B. Various concentrations of Cu^{2+} and Zn^{2+} were then added and allowed to incubate together at 37 °C for 20 min. Fluorescence intensity of samples was recorded by RF-5301PC (Shimadzu, Tokyo, Japan). For FAM, the excitation wavelength was 494 nm, and the fluorescence spectrum was collected from 505 to 600 nm. For ROX, the excitation wavelength was 580 nm, and the fluorescence spectrum was collected from 595 to 660 nm.

2.5. Stability of FNAzymes

The FNAzymes were concentrated to 1 μM in 5 U/L of DNase I or 50% cell lysate and incubated at 37 °C. The samples collected at different time points were characterized by 8% PAGE and analyzed using ImageJ software. For control, nude Cu^{2+}-specific DNAzyme was prepared by hybridizing Enzyme-1 and Substrate-1 in a 1:1 and incubated in 1 μM concentration in 5 U/L DNase I or 50% cell lysate at the same conditions.

2.6. Cell Viability

The cytotoxicity of the FNAzymes was evaluated by an MTT assay. HeLa cells were seeded in a 96-well plate at a concentration of 1×10^5 cells per well. The plates were allowed to maintain in an air incubator containing 5% CO_2/95% air at 37 °C for 24 h and then rinsed with fresh medium. Different concentrations of FNAzymes (0, 1, 3, 5, 10, 20, 50, 80, 160, and 320 nM) were added for another 24 h incubation, followed by the addition of 0.5 mg/mL MTT (final concentration). Four hours later, the media containing MTT was discarded by washing. The DMSO (150 μL) was then introduced to dissolve the formazan crystals in each well. The absorbance at 490 nm was recorded using a microplate reader (EL × 808, BioTek, Winooski, VT, USA) for the final product.

2.7. Confocal Fluorescence Imaging

HeLa cells were incubated in a cell culture dish (MatTek, Ashland, MA, USA) with 1.0×10^5 cells per well. Certain amounts of Cu^{2+} and Zn^{2+} were added to each well for a 1 h incubation and then washed with PBS buffer three times. Then, 50 nM of FNAzymes were diluted in fresh medium and introduced for a 3 h incubation at 37 °C, followed by washing with PBS buffer three times. Images were monitored using an FV1000 LSCM (Olympus Optical Co., Ltd., Tokyo, Japan) with various laser transmitters.

2.8. Flow Cytometric Assay

HeLa cells were first treated with various concentrations of Cu^{2+} or Zn^{2+} (100, 200, 800, 3000, and 5000 nM) for 1 h, followed by incubating with 50 nM FNAzymes for another 3 h and washing with PBS buffer three times. Afterward, the HeLa cells were trypsinized and suspended in a new medium. The fluorescence intensity was analyzed using a BD FACSCanto flow cytometer (Merck Millipore, Darmstadt, Germany) in real time. The fluorescence intensities of FAM and ROX were obtained by 488 nm and 561 nm excitation, respectively.

2.9. Inductively Coupled Plasma-Mass Spectrometer (ICP-MS) Test

For the inductively coupled plasma-mass Spectrometer (ICP-MS) test, the trypsinized cells were counted with a cell counter and then transferred to a centrifuge tube and collected by centrifugation. After being dealt with aqua regia (3:1 hydrochloric acid/nitric acid, 200 μL) for 12 h, the cells were treated with ultrasound for 30 min to fully dissolve the two kinds of metal elements for the ICP-MS analysis (NexION 300D, PerkinElmer, Waltham, MA, USA). The concentrations of the two metal ions inside cells were calculated from the result of ICP-MS detection.

3. Results and Discussion

3.1. Simultaneous Detection of Cu^{2+} and Zn^{2+} In Vitro

As illustrated in Figure 1A, four strands (S1, S2, S3, and S4) containing Cu^{2+}-specific DNAzyme (marked in green on S3 and S4), Zn^{2+}-specific DNAzyme (marked in yellow on S1 and S2), fluorophores (FAM and ROX), and quenchers (Dabcyl and BHQ-2) are self-assembled into a multifunctional FNAzyme by rapid thermal annealing. The formation of a well-defined tetrahedral nanostructure brought the fluorophore and quencher together; thus, the fluorescence of FAM and ROX was efficiently quenched by Dabcyl and BHQ-2, respectively. In the presence of Cu^{2+} and Zn^{2+}, specific cleavage sites can be identified, and the site of an adenine ribonucleotide (rA) can be cleaved. Therefore, the fluorophore-labeled portions of the substrate strands (S1 and S4) were released, and the corresponding fluorescence was recovered.

The formation of FNAzymes was first determined by 8% native PAGE (Figure S1). The relatively low mobility in band 1 suggested that a yield of over 90% of the nanostructure was successfully formed with a single thermal annealing step. The monodispersed DNA structure could be observed under AFM imaging (Figure S2). The feasibility of FNAzymes in simultaneously responding to Cu^{2+} and Zn^{2+} in vitro was evaluated by fluorescence measurement (Figure 2A,B). The well-defined nanostructure of FNAzymes brought the fluorophores (FAM and ROX) quite close to the quenching molecules (DABCYL and BHQ-2), resulting in high fluorescence quenching efficiency. In this state, the fluorescence of fluorophores could not be detected (curve a colored gray in Figure 2A,B). In the presence of 400 nM Cu^{2+} and Zn^{2+}, the observed fluorescence intensity of FAM at 520 nm (excited at 494 nm, curve b in Figure 2A) and ROX at 605 nm (excited at 580 nm, curve b in Figure 2B) was significantly increased, suggesting that the cleavage of rA sites by target metal ions caused the release of fluorophore-labeled substrate strands (S4 and S1 in Figure 2) and the recovery of fluorescence. Kinetic studies in Figure 2C,D indicated that the FNAzymes respond rapidly to the target Cu^{2+} and Zn^{2+} within 20 min, suggesting that the FNAzymes were able to respond to the target Cu^{2+} and Zn^{2+} with a fast response. Considering the fact that the change in fluorescence intensity is proportional to the change in concentration at low concentrations ($\Delta F = k\Delta c$), we measured fluorescence intensities to calculate the cleavage ratio of probes. Fluorescence change (ΔF) was calculated as the fluorescence following cleavage (signal) minus the fluorescence prior to cleavage (background). To test the ratio and rate of the cleavage, FNAzymes with 50 nM free FAM or 50 nM free ROX were assembled to simulate 100% cleavage. In detail, 50 nM FNAzymes-1 were assembled by 50 nM S1, S2, S3, and S4-2 (without FAM labeling). In order to simulate 100% cleavage, an amount of 50 nM of FAM-labeled DNA (F1) was added. The fluorescence at 520 nm was recorded, and the $\Delta F_{520\ nm}$ was calculated to be 98.2, while the $\Delta F_{520\ nm}$ in the presence of 400 nM Cu^{2+} was 23.8 (20 min incubation), indicating ~24.2% cleavage ratio of Cu^{2+}-responsive DNAzyme (Figure S3). Similarly, 50 nM FNAzymes-2 were assembled by 50 nM S1-2 (without ROX labeling), S2, S3, and S4. To simulate 100% cleavage, an amount of 50 nM of ROX-labeled DNA (F2) was added. The fluorescence at 605 nm was recorded, and the $\Delta F_{605\ nm}$ was calculated to be 103.1, while the $\Delta F_{605\ nm}$ in the presence of 400 nM Zn^{2+} was 32.4 (20 min incubation), indicating ~34.4% cleavage ratio of Zn^{2+}-responsive DNAzyme. The cleavage rates within the first 20 min were calculated to be 0.61 nM/min for Cu^{2+}-responded cleavage and 0.86 nM/min for Zn^{2+}-responded cleavage.

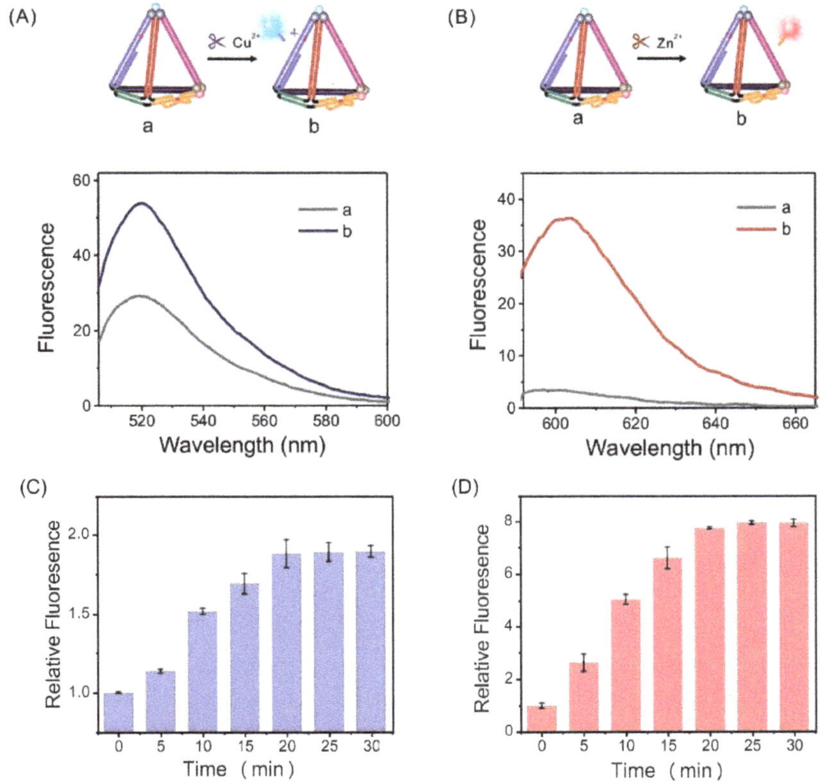

Figure 2. (**A**,**B**) Fluorescence response of the FNAzymes in the presence (a) or absence (b) of (**A**) 400 nM Cu^{2+} and (**B**) 400 nM Zn^{2+}. (**C**,**D**) Kinetics of the FNAzymes in the existence or absence of (**C**) 400 nM Cu^{2+} and (**D**) 400 nM Zn^{2+}. Cu^{2+} (FAM), $\lambda_{ex}/\lambda_{em}$: 494 nm/520 nm; Zn^{2+} (ROX), $\lambda_{ex}/\lambda_{em}$: 580 nm/605 nm.

Having demonstrated the feasibility of the simultaneous response of metal ions by FNAzymes, the detection sensitivity and selectivity in vitro were further tested. The observed FAM and ROX fluorescence increased gradually with the increase in concentrations of the target Cu^{2+} and Zn^{2+} (Figues 3 and S4), respectively, suggesting that the cleavage of DNAzyme by the corresponding target metal ions leads to fluorescence recovery. Strong linear correlations were observed between the maximum fluorescence intensities and the log C of the target metal ions ranging from 10 nM to 5000 nM with high correlation coefficients (R^2 = 0.9905 for Cu^{2+} and 0.9849 for Zn^{2+}, respectively). The detection limits of the FNAzymes for Cu^{2+} and Zn^{2+} were estimated to be 6.8 nM and 0.9 nM, respectively, indicating that the FNAzymes showed excellent detecting performance. The selectivity of FNAzymes was evaluated by detecting Cu^{2+} and Zn^{2+} in the presence of other relevant metal ions (Mn^{2+}, Li^{+}, Ni^{+}, Fe^{3+}, Cr^{2+}, and Co^{2+}) at a concentration of 1000 nM. A 3.3~11.7-fold enhancement in the fluorescence signal was observed for the target Cu^{2+} and Zn^{2+} compared to that for other ions (Figure S5). This indicated that DNAzymes could differentiate the target ions from others. However, the selectivity for the target metal ions was moderate or comparable as compared to other studies [32,33], which might be caused by the relatively low cleavage ratio of FNAzymes in the presence of metal ions and the relatively high background signal. Taken together, FNAzymes displayed good sensitivity and selectivity in vitro in signaling the Cu^{2+} and Zn^{2+} ions.

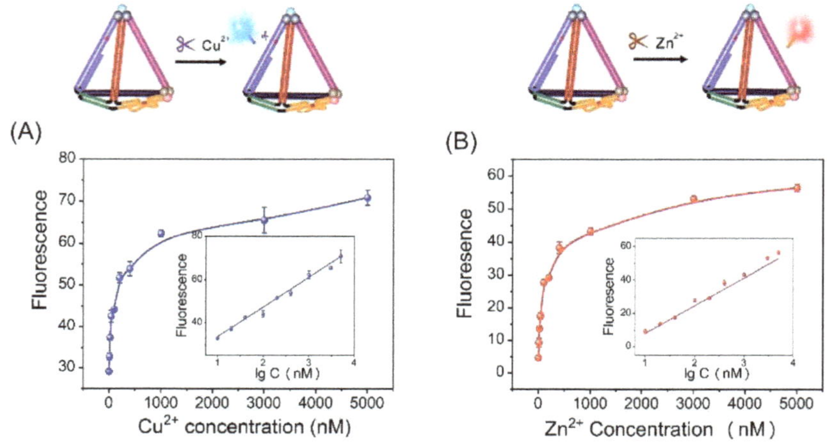

Figure 3. Plots of fluorescence peak intensities versus different concentrations of (**A**) Cu^{2+} and (**B**) Zn^{2+}. Insets are the corresponding calibration curve. Each experiment was repeated at least three times. Cu^{2+} (FAM), $\lambda_{ex}/\lambda_{em}$: 494 nm/520 nm; Zn^{2+} (ROX), $\lambda_{ex}/\lambda_{em}$: 580 nm/605 nm.

3.2. Stability and Cytotoxicity of the FNAzymes

Since the nanoprobes will be applied in a complex biological environment, the structural integrity of the FNAzymes was monitored in a DNase I solution and characterized with PAGE gel analysis. DNase I is a well-known endonuclease that efficiently digests single- or double-stranded DNA efficiently. From Figure 4A,B, a band of FNAzymes was visible for 12 h, indicating that the rigid structure was maintained against degradation by DNase I. As a control, the nude duplex Cu^{2+}-specific DNAzyme could not be visualized on PAGE gel after 8 h, indicating a much lower structural stability of the duplex structure. The enhanced stability of FNAzymes was in agreement with recently published data, which was attributed to the enzymatic resistance ability of rigid DNA nanostructures. These results further illustrated that the nude duplex DNAzyme could not work for a long time in the complex biological environment without protection due to its instability in cells [34,35]. Then, the stability of FNAzymes was further investigated in 50% cell lysate (human cervical cancer) and characterized by 8% PAGE gel analysis (Figure S6). The clear band observed after 12 h of incubation also demonstrates the good biostability of FNAzymes in cellular components. Within a period of three hours, it is possible that up to 80% of the structure would remain intact, allowing for semi-quantitative imaging in living cells. As a comparison, the nude duplex Cu^{2+}-specific DNAzyme was significantly degraded after four hours. Furthermore, the toxicity of the FNAzymes was evaluated by a standard colorimetric MTT assay. The HeLa cells were incubated with various concentrations of FNAzymes ranging from 0 to 320 nM at 37 °C for 24 h (Figure 4C). Even under the highest concentration at 320 nM, HeLa cells maintained high activity (>85%), indicating that the FNAzymes appeared to have no obvious toxicity or side effects on the living cells. With high stability in the cellular environment and low cellular cytotoxicity, the FNAzymes were applicable for intracellular detection of Cu^{2+} and Zn^{2+}.

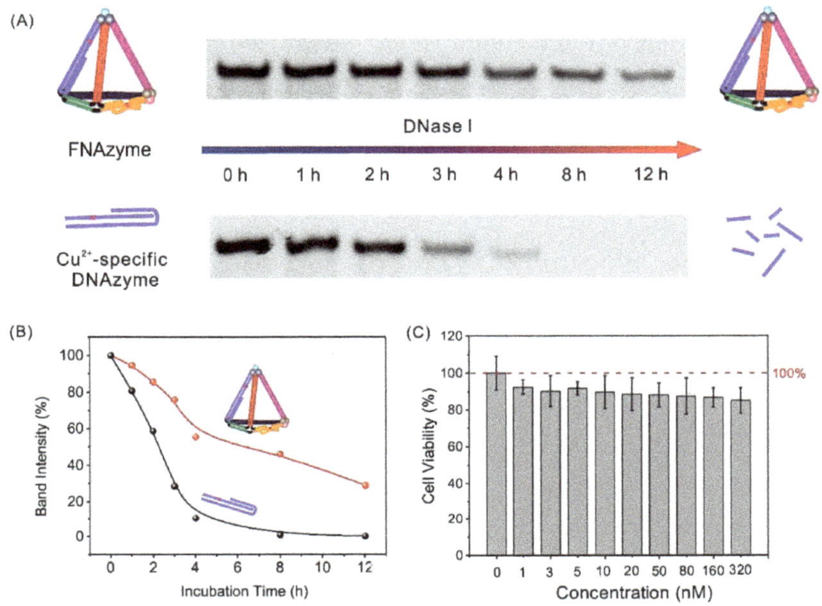

Figure 4. (**A**) Electrophoresis characterization for the stability of the FNAzymes (the upper row) and the nude Cu^{2+}-dependent DNAzyme (the lower row) treated with 5 U/L DNase I at 37 °C from 0 to 12 h. The concentrations of probes are all 1 μM. (**B**) Band intensity of FNAzymes (red dots) and the nude Cu^{2+}-dependent DNAzymes (black dots) after incubating with 5 U/L DNase I at 37 °C for different times in gel image (**A**). (**C**) Viability of HeLa cells incubated with various concentrations of FNAzymes for 24 h, respectively. Three experiments were conducted in each group, respectively.

3.3. Intracellular Imaging of Cu^{2+} and Zn^{2+} with FNAzymes

As demonstrated above, FNAzymes have the advantages of high sensitivity, high selectivity, fast detection speed, low cytotoxicity, and satisfactory stability in an intracellular environment. Hence, we aim to detect Cu^{2+} and Zn^{2+} in living cells using HeLa cells as the model. Firstly, HeLa cells without metal ion treatment were observed by confocal imaging. Under excitation of an appropriate voltage (590 V), almost no fluorescence signal was observed for cells that had not been treated with metal ions (Figure 5A). To verify that fluorescence was caused by metal ions in cells, HeLa cells were exposed to 5 μM Cu^{2+} or 5 μM Zn^{2+} for 1 h to enable the uptake of the metal ions, followed by incubation with FNAzymes for 3 h and observation through confocal imaging under the same experimental conditions. Cells incubated with Cu^{2+} ions showed bright green fluorescence (Figure 5B), while those incubated with Zn^{2+} ions exhibited red fluorescence (Figure 5C). Then, HeLa cells incubated with 5 μM Cu^{2+} and 5 μM Zn^{2+} were observed by confocal imaging. A bright green fluorescence signal for Cu^{2+} could be seen in the "FAM" channel under 480 nm excitation, and a red fluorescence signal for Zn^{2+} could be observed in the "ROX" channel under 559 nm excitation (Figure 5D). The bright-field images in Figure 5 indicated that HeLa cells were viable during the imaging process, while the overlap image confirms the simultaneous intracellular detections of Cu^{2+} and Zn^{2+}. From the inductively coupled plasma-mass spectrometer (ICP-MS) results, the intracellular concentrations of Cu^{2+} and Zn^{2+} under the experimental conditions are estimated to be 101.12 nM and 96.22 nM, respectively (assuming 2000 μm^3 as the volume of a cell) [22]. The results indicated that FNAzymes performed well in imaging Cu^{2+} and Zn^{2+} simultaneously in single living cells.

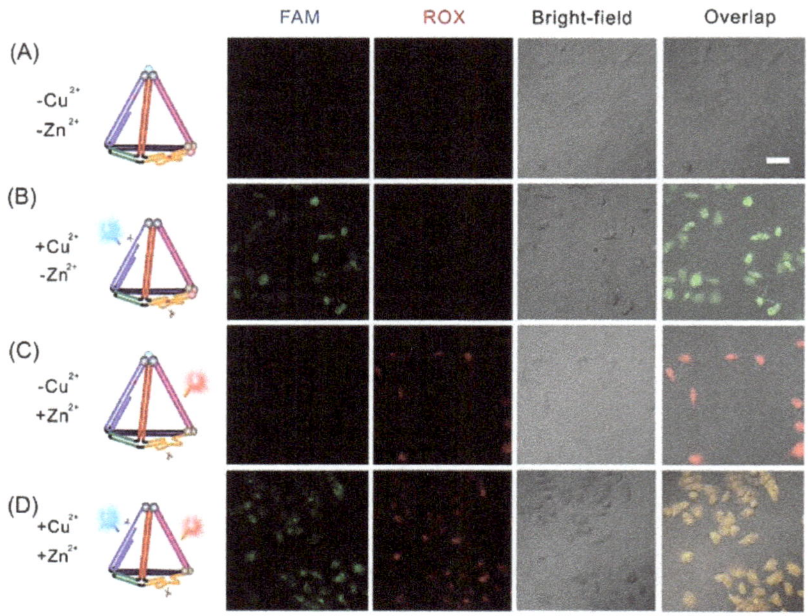

Figure 5. Fluorescence imaging of metal ions in living HeLa cells after treatment with (**A**) Cu^{2+} and Zn^{2+}; (**B**) Cu^{2+} only; (**C**) Zn^{2+} only; or (**D**) no metal ions and incubated with FNAzymes. The green channel (the first row) is FAM fluorescence (λ_{ex}: 480 nm), and the red channel (the second row) is ROX fluorescence (λ_{ex}: 559 nm). Scale bar: 50 μm.

3.4. Semi-Quantitative Detection of Cu^{2+} and Zn^{2+} in Living Cells

To semi-quantitatively detect physiologically related metal ions in cells, analytical flow cytometry was employed to analyze the intracellular fluorescence intensity in a large population of cells. Different concentrations of Cu^{2+} and Zn^{2+} ranging from 0 to 5 μM were allowed to incubate with HeLa cells for 1 h, respectively. As the concentrations of metal ions in the cells increased, the signal intensity peak in flow cytometry shifted accordingly, which agreed with the corresponding fluorescence intensities observed in the confocal images (Figure 6A,B). The results indicated that FNAzymes could effectively monitor intracellular metal ions. The fluorescence signals were semi-quantified by the fluorescence intensities of the flow cytometry and the added concentrations of Cu^{2+} and Zn^{2+} (Figure 6C,D). An approximate linear relationship between fluorescence intensities and concentrations of added metal ions was observed. The excellent reproducibility of imaging sensitivity was proposed to be attributed to the high structural uniformity of FNAzymes, where signaling molecules were attached to FNAzymes at a 1:1 ratio with high precision. These results suggested that FNAzymes could be used as an effective tool for the semi-quantitative and simultaneous determination of intracellular metal ions.

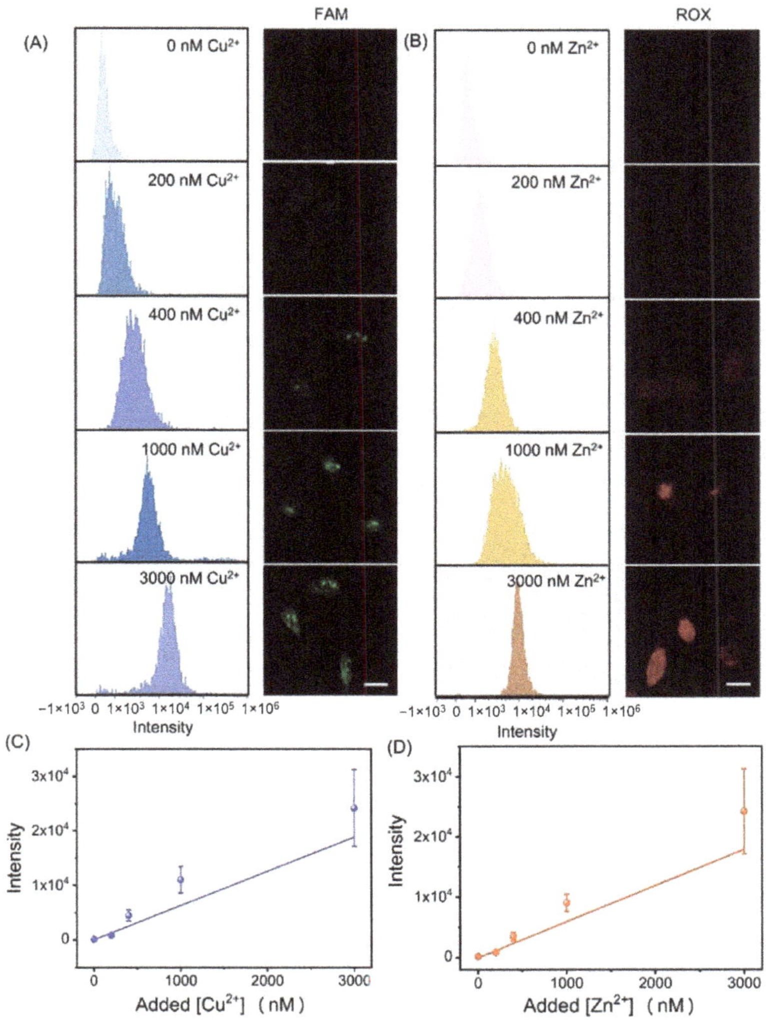

Figure 6. Semi-quantitative detection of intracellular Cu^{2+} and Zn^{2+} with FNAzymes. (**A**,**B**) Flow cytometry (left) and confocal images results (right) of HeLa cells incubated with 50 nM FNAzymes and different concentrations (0, 200, 400, 1000, and 3000 nM) of (**A**) Cu^{2+} and (**B**) Zn^{2+}. Scale bar: 25 μm. (**C**,**D**) Fluorescence intensity of the FNAzymes corresponding to various concentrations of added (**C**) Cu^{2+} and (**D**) Zn^{2+}. Error bars represented standard deviations of at least three independent experiments.

4. Conclusions

In conclusion, we have proposed elaborate FNAzymes for the non-destructive simultaneous and quantitative detection of physiologically related Cu^{2+} and Zn^{2+} inside living cells by combining the highly active DNAzyme and DNA tetrahedron with good intracellular stability. As a promising nanocarrier for entering cells, FNAzymes with favorable biocompatibility can be easily prepared in one step with a high yield of over 90%, making them suitable for non-destructive intracellular detection. The excellent high structural uniformity endows FNAzymes with excellent and highly reproducible detection properties, which improve the accuracy of quantitative detection of intracellular ions. The excellent high structural uniformity endows FNAzymes with excellent and highly reproducible detection properties, which improve the accuracy of quantitative detection of intracellular ions. Due to the amplification effect of DNAzymes, the assay displays low-nanomolar sensitivity for signaling Cu^{2+} and Zn^{2+} with a fast response time within 20 min. Enhancing the selectivity could be achieved by further designing a fluorescent probe with more efficient quenchers or fluorophores. We envision that this method could offer great opportunities for the quantitative detection of multiple targets in living cells and contribute to advancing our knowledge and understanding of the functions of Cu^{2+} and Zn^{2+} in biological research.

Supplementary Materials: The following supporting information can be downloaded at: https://www.mdpi.com/article/10.3390/chemosensors11070358/s1. Table S1: Sequences in the experiment; Figure S1: Native PAGE (8%) analysis of the self-assembled products; Figure S2: AFM image of the FNAzymes; Figure S3: The Fluorescence change of FNAzyme after metal ions-responded cleavage; Figure S4: Fluorescence spectra of the FNAzymes with various concentrations of metal ions; Figure S5: Selectivity evaluation of the FNAzymes; Figure S6: Electrophoresis characterization of the FNAzymes and DNAzymes incubated in 50% cell lysate.

Author Contributions: Conceptualization, D.Z. and L.W.; methodology, J.H. and S.S.; software, validation, and investigation, J.H. and Y.X.; resources, X.Z.; writing—original draft preparation, D.Z. and J.H.; writing—review and editing, D.Z. and Q.L.; supervision, X.Z. and L.W.; project administration, Q.L. and L.W.; funding acquisition, D.Z. and L.W. All authors have read and agreed to the published version of the manuscript.

Funding: This work was financially supported by the National Natural Science Foundation of China (62071251, 62235008, 62227803, 62288102), Natural Science Foundation of Jiangsu Province (BK20201376, BK20212012), "Belt and Road" Innovation Cooperation Project of Jiangsu (BZ2022011), Qinglan Project of Jiangsu Province of China, Young Scientific and Technological Talents Project of Jiangsu Association for Science and Technology (TJ-2021-028), and Project of State Key Laboratory of Organic Electronics and Information Displays, Nanjing University of Posts and Telecommunications (No. GZR2022010026).

Institutional Review Board Statement: Not applicable.

Informed Consent Statement: Not applicable.

Data Availability Statement: Not applicable.

Conflicts of Interest: The authors declare no conflict of interest. The funders had no role in the design of the study; in the collection, analyses, or interpretation of data; in the writing of the manuscript; or in the decision to publish the results.

References

1. Bolognin, S.; Messori, L.; Zatta, P. Metal ion physiopathology in neurodegenerative disorders. *Neuromol. Med.* **2009**, *11*, 223–238. [CrossRef] [PubMed]
2. Krezel, A.; Maret, W. The functions of metamorphic metallothioneins in zinc and copper metabolism. *Int. J. Mol. Sci.* **2017**, *18*, 1237. [CrossRef]
3. Winge, D.R.; Jensen, L.T.; Srinivasan, C. Metal-ion regulation of gene expression in yeast. *Curr. Opin. Chem. Biol.* **1998**, *2*, 216–221. [CrossRef] [PubMed]
4. Stelmashook, E.V.; Isaev, N.K.; Genrikhs, E.E.; Amelkina, G.A.; Khaspekov, L.G.; Skrebitsky, V.G.; Illarioshkin, S.N. Role of zinc and copper ions in the pathogenetic mechanisms of alzheimer's and parkinson's diseases. *Biochemistry* **2014**, *79*, 391–396. [CrossRef]
5. Krueger, W.H.H.; Gonye, G.E.; Madison, D.L.; Murray, K.E.; Kumar, M.; Spoerel, N.; Pfeiffer, S.E. Tpo1, a member of a novel protein family, is developmentally regulated in cultured oligodendrocytes. *J. Neurochem.* **1997**, *69*, 1343–1355. [CrossRef]
6. Islamoglu, Y.; Evliyaoglu, O.; Tekbas, E.; Cil, H.; Elbey, M.A.; Atilgan, Z.; Kaya, H.; Bilik, Z.; Akyuz, A.; Alan, S. The relationship between serum levels of Zn and Cu and severity of coronary atherosclerosis. *Biol. Trace Elem. Res.* **2011**, *144*, 436–444. [CrossRef] [PubMed]
7. Milne, A.; Landing, W.; Bizimis, M.; Morton, P. Determination of Mn, Fe, Co, Ni, Cu, Zn, Cd and Pb in seawater using high resolution magnetic sector inductively coupled mass spectrometry (HR-ICP-MS). *Anal. Chim. Acta* **2010**, *665*, 200–207. [CrossRef]
8. Cui, Y.; Chang, X.; Zhai, Y.; Zhu, X.; Zheng, H.; Lian, N. ICP-AES determination of trace elements after preconcentrated with p-dimethylaminobenzaldehyde-modified nanometer SiO_2 from sample solution. *Microchem. J.* **2006**, *83*, 35–41. [CrossRef]
9. Lin, P.H.; Danadurai, K.S.K.; Huang, S.D. Simultaneous determination of cobalt, nickel and copper in seawater with a multi-element electrothermal atomic absorption spectrometer and microcolumn preconcentration. *J. Anal. Atom. Spectrom.* **2001**, *16*, 409–412. [CrossRef]

10. Sheth, V.; Chen, X.; Mettenbrink, E.M.; Yang, W.; Jones, M.A.; M'Saad, O.; Thomas, A.G.; Newport, R.S.; Francek, E.; Wang, L.; et al. Quantifying intracellular nanoparticle distributions with three-dimensional super-resolution microscopy. *ACS Nano* **2023**, *17*, 8376–8392. [CrossRef]
11. Wang, S.; Deng, S.; Cai, X.; Hou, S.; Li, J.; Gao, Z.; Li, J.; Wang, L.; Fan, C. Superresolution imaging of telomeres with continuous wave stimulated emission depletion (STED) microscope. *Sci. China Chem.* **2016**, *59*, 1519–1524. [CrossRef]
12. Bao, L.; Ding, L.; Hui, J.; Ju, H. A light-up imaging protocol for neutral ph-enhanced fluorescence detection of lysosomal neuraminidase activity in living cells. *Chem. Commun.* **2016**, *52*, 12897–12900. [CrossRef]
13. Li, B.; Kou, J.; Mei, H.; Gu, X.; Wang, M.; Xie, X.; Xu, K. A hemicyanine-based "turn-on" fluorescent probe for the selective detection of Cu^{2+}ions and imaging in living cells. *Anal. Methods-UK* **2020**, *12*, 4181–4184. [CrossRef]
14. Zhang, X.; Song, Z.; Chao, Q.; Li, Q.; Kong, R.; Fan, G.; Luo, X. A DNAzyme-based normalized fluorescence strategy for direct quantification of endogenous zinc in living cells. *Chem. Commun.* **2022**. *58*, 577–580. [CrossRef]
15. Patil, M.; Keshav, K.; Kumawat, M.K.; Bothra, S.; Sahoo, S.K.; Srivastava, R.; Rajput, J.; Bendre, R.; Kuwar, A. Monoterpenoid derivative based ratiometric fluorescent chemosensor for bioimaging and intracellular detection of $Zn^{2+}$and Mg^{2+} ions. *J. Photoch. Photobio. A* **2018**, *364*, 758–763. [CrossRef]
16. Chen, Y.; Qiu, D.; Zhang, X.; Liu, Y.; Cheng, M.; Lei, J.; Mergny, J.; Ju, H.; Zhou, J. Highly sensitive biosensing applications of a magnetically immobilizable covalent G-quadruplex-hemin DNAzyme catalytic system. *Anal. Chem.* **2022**, *94*, 2212–2219. [CrossRef]
17. Guo, Y.; Chen, J.; Cheng, M.; Monchaud, D.; Zhou, J.; Ju, H. A thermophilic tetramolecular G-quadruplex/hemin DNAzyme. *Angew. Chem. Int. Ed.* **2017**, *56*, 16636–16640. [CrossRef] [PubMed]
18. McGhee, C.E.; Loh, K.Y.; Lu, Y. DNAzyme sensors for detection of metal ions in the environment and imaging them in living cells. *Curr. Opin. Biotech.* **2017**, *45*, 191–201. [CrossRef]
19. Wang, Q.; Wang, Z.; He, Y.; Xiong, B.; Li, Y.; Wang, F. Chemical and structural modification of RNA-cleaving DNAzymes for efficient biosensing and biomedical applications. *Trac-Trend Anal. Chem.* **2023**, *159*, 116910. [CrossRef]
20. Hu, L.; Fu, X.; Kong, G.; Yin, Y.; Meng, H.-M.; Ke, G.; Zhang, X.-B. DNAzyme-gold nanoparticle-based probes for biosensing and bioimaging. *J. Mater. Chem. B* **2020**, *8*, 9449–9465. [CrossRef]
21. Khan, S.; Burciu, B.; Filipe, C.D.M.; Li, Y.; Dellinger, K.; Didar, T.F. DNAzyme-based biosensors: Immobilization strategies, applications, and future prospective. *ACS Nano* **2021**, *15*, 13943–13969. [CrossRef] [PubMed]
22. Wu, P.; Hwang, K.; Lan, T.; Lu, Y. A DNAzyme-gold nanoparticle probe for uranyl ion in living cells. *J. Am. Chem. Soc.* **2013**, *135*, 5254–5257. [CrossRef] [PubMed]
23. Gerber, P.P.; Donde, M.J.; Matheson, N.J.; Taylor, A.I. XNAzymes targeting the SARS-CoV-2 genome inhibit viral infection. *Nat. Commun.* **2022**, *13*, 6716. [CrossRef]
24. Ge, Z.; Li, Q.; Fan, C. Framework nucleic acids for cell imaging and therapy. *Chem. Res. Chin. Univ.* **2020**, *36*, 1–9. [CrossRef]
25. Ge, Z.; Gu, H.; Li, Q.; Fan, C. Concept and development of framework nucleic acids. *J. Am. Chem. Soc.* **2018**, *140*, 17808–17819. [CrossRef]
26. Liu, Q.; Ge, Z.; Mao, X.; Zhou, G.; Zuo, X.; Shen, J.; Shi, J.; Li, J.; Wang, L.; Chen, X.; et al. Valency-controlled framework nucleic acid signal amplifiers. *Angew. Chem. Int. Ed.* **2018**, *57*, 7131–7135. [CrossRef]
27. Zhou, X.; Zhao, M.; Duan, X.; Guo, B.; Cheng, W.; Ding, S.; Ju, H. Collapse of DNA tetrahedron nanostructure for "off-on" fluorescence detection of DNA methyltransferase activity. *ACS Appl. Mater. Interfaces* **2017**, *9*, 40087–40093. [CrossRef] [PubMed]
28. Qing, Z.H.; Hu, J.L.; Xu, J.Y.; Zou, Z.; Lei, Y.L.; Qing, T.P.; Yang, R.H. An intramolecular catalytic hairpin assembly on a DNA tetrahedron for mRNA imaging in living cells: Improving reaction kinetics and signal stability. *Chem. Sci.* **2020**, *11*, 1985–1990. [CrossRef]
29. Liang, L.; Li, J.; Li, Q.; Huang, Q.; Shi, J.; Yan, H.; Fan, C. Single-particle tracking and modulation of cell entry pathways of a tetrahedral DNA nanostructure in live cells. Angew. *Chem. Int. Ed.* **2014**, *53*, 7745–7750. [CrossRef]
30. Liu, J.; Lu, Y. A DNAzyme catalytic beacon sensor for paramagnetic Cu^{2+} ions in aqueous solution with high sensitivity and selectivity. *J. Am. Chem. Soc.* **2007**, *129*, 9838–9839. [CrossRef]
31. Endo, M.; Takeuchi, Y.; Suzuki, Y.; Emura, T.; Hidaka, K.; Wang, F.; Willner, I.; Sugiyama, H. Single-molecule visualization of the activity of a Zn^{2+}-dependent DNAzyme. *Angew. Chem. Int. Ed.* **2015**, *54*, 10550–10554. [CrossRef] [PubMed]
32. Ning, P.; Jiang, J.; Li, L.; Wang, S.; Yu, H.; Feng, Y.; Zhu, M.; Zhang, B.; Yin, H.; Guo, Q.; et al. A mitochondria-targeted ratiometric two-photon fluorescent probe for biological zinc ions detection. *Biosens. Bioelectron.* **2016**, *77*, 921–927. [CrossRef] [PubMed]
33. Su, M.; Liu, C.; Zhang, Y.; Rong, X.; Wang, X.; Li, X.; Wang, K.; Zhu, H.; Zhu, B. Rational design of a water-soluble TICT-AIEE-active fluorescent probe for mercury ion detection. *Anal. Chim. Acta* **2022**, *1230*, 340337. [CrossRef] [PubMed]
34. Li, F.; Liu, Y.; Dong, Y.; Chu, Y.; Song, N.; Yang, D. Dynamic assembly of DNA nanostructures in living cells for mitochondrial interference. *J. Am. Chem. Soc.* **2022**, *144*, 4667–4677. [CrossRef] [PubMed]
35. Lv, M.; Wu, Z.; Yu, R.; Jiang, J. Three-dimensional DNA nanostructures for dual-color microRNA imaging in living cells via hybridization chain reaction. *Chem. Commun.* **2020**, *56*, 6668–6671. [CrossRef] [PubMed]

chemosensors

Review

DNA-Programmed Biomolecular Spatial Pattern Recognition

Yue Wang [1,2,3], Lei Ren [1,2,3], Hongzhen Peng [2], Linjie Guo [2,*] and Lihua Wang [1,2,3]

1 Division of Physical Biology, CAS Key Laboratory of Interfacial Physics and Technology, Shanghai Institute of Applied Physics, Chinese Academy of Sciences, Shanghai 201800, China
2 Institute of Materiobiology, Department of Chemistry, College of Science, Shanghai University, Shanghai 200444, China
3 University of Chinese Academy of Sciences, Beijing 100049, China
* Correspondence: guolinjie@sinap.ac.cn

Abstract: Molecular recognition based on non-covalent interactions between two or more molecules plays a crucial role in biological systems. Specific biological molecule recognition has been widely applied in biotechnology, clinical diagnosis, and treatment. The efficiency and affinity of molecular recognition are greatly determined by the spatial conformation of biomolecules. The designability of DNA nanotechnology makes possible the precise programming of the spatial conformation of biomolecules including valency and spacing, further achieving spatial pattern recognition regulation between biomolecules. This review summarizes recent achievements with DNA-based molecular spatial pattern recognition systems, the important factors affecting spatial pattern recognition, and their applications in biosensing, bioimaging, and targeted therapy. The future challenges in and development of this field are discussed and prospected. This review will provide valuable guidance for the creation of new DNA tools to enhance the efficiency and specificity of biomolecular recognition.

Keywords: biomolecular recognition; DNA technology; spatial pattern; biosensing; targeted therapy

Citation: Wang, Y.; Ren, L.; Peng, H.; Guo, L.; Wang, L. DNA-Programmed Biomolecular Spatial Pattern Recognition. *Chemosensors* **2023**, *11*, 362. https://doi.org/10.3390/chemosensors11070362

Academic Editor: Pilar López-Cornejo

Received: 19 May 2023
Revised: 15 June 2023
Accepted: 24 June 2023
Published: 27 June 2023

1. Introduction

In living organisms, molecular recognition based on the specific interaction between molecules is a crucial foundation for maintaining life, as mutual recognition and interaction between biomolecules are essential for processes such as cell signaling, metabolism, cell proliferation, and differentiation [1]. For example, cell-surface receptor molecules can interact with signaling molecules, triggering intracellular signal transduction to regulate cellular physiology and metabolism [2,3]. Enzymes can recognize their substrate molecules, catalyzing the conversion of the substrate [4,5]. In the field of biotechnology, specific molecular recognition also has a wide range of applications in the diagnostic and biosensing fields. For example, the recognition of specific proteins on the surface of tumor cells can facilitate early diagnosis and targeted treatment of tumors [6,7]. Biosensors leverage molecular recognition principles to identify particular molecules, harnessing the specificity of enzymes to detect their corresponding substrates [8,9].

In biomolecular recognition, non-covalent interactions such as hydrogen bonds, ionic interactions, and van der Waals forces between molecules in the binding site are usually involved. These non-covalent interactions usually occur based on the spatial conformation of biomolecules [10]. For example, proteins on the cell surface cluster into a specific pattern and interact with ligands through specific structural domains, triggering downstream biological processes such as immune response and cell adhesion. Therefore, the conformation of biomolecules plays an important role in biomolecular recognition. Recently, by linking biomolecules to organic/inorganic or biological scaffolds, researchers have constructed ex vivo molecular recognition systems. Various spatial conformations were constructed by adjusting the valence and spacing of biomolecules on the scaffold [11]. Efficient biomolecular spatial pattern recognition and relevant applications, including targeted diagnosis, can ultimately be achieved.

In addition to polymers [12], nanoparticles [13], and dendritic polymers [14], nucleic acids have emerged as novel scaffold materials for biomolecule coupling. With the development of DNA chemical synthesis technology, various chemical groups, including fluorescent dyes, thiols, and biotin, can be modified on DNA chains. Directional coupling and the precise spatial arrangement of various biomolecules can be achieved. In addition, DNA molecules have excellent biocompatibility and biological stability. In particular, DNA technology enables a class of one-dimensional to three-dimensional nanostructures formed by the precise Watson–Crick base pairing of DNA [15]. DNA nanostructures exhibit unparalleled monodispersity and atomic-level precision when compared to other nanoparticles [16]. Excellent chemical addressability, precise assembly, and highly ordered conformation make DNA nanostructure the perfect template for a variety of nanoscale molecules, such as small molecules, proteins, liposomes, and nanoparticles [17,18]. Thus, biomolecular spatial pattern recognition can be easily programmed by DNA [19]. This review provides a comprehensive overview of DNA-based molecular spatial pattern recognition systems and DNA biomolecule conjugation chemistry. We emphasize the crucial factors influencing spatial pattern recognition and explore their applications (Scheme 1). Furthermore, we examine the challenges encountered in the advancement of DNA-programmed biomolecular spatial pattern recognition and propose future directions for this field.

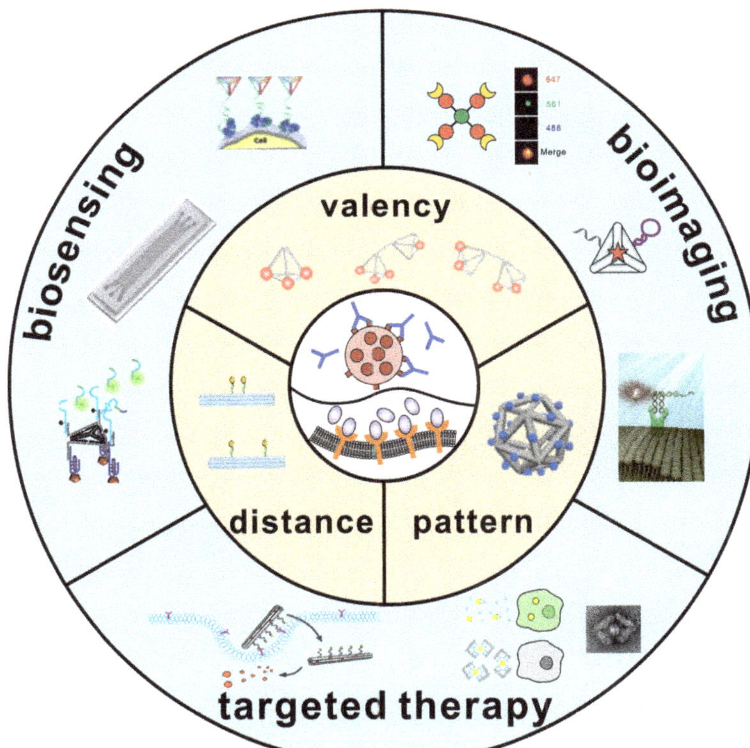

Scheme 1. Schematic illustration of DNA-programmed biomolecular spatial pattern recognition.

2. Biomolecular Systems Based on Spatial Pattern Recognition

Specific recognition between antibodies and antigens is a common molecular recognition in biological systems and is typically used to recognize and clear foreign substances or abnormal cells in living organisms (Figure 1a). Antibodies, which are proteins generated by the immune system, possess the ability to identify and attach themselves to particular antigen molecules, creating a complex interaction with the antigen. The recognition and binding between antibodies and antigens are highly specific, with only antigens that precisely match the antibody being capable of recognition and binding [20]. In addition, antigen molecules on the surface of tumor cells or viruses are often multivalent and arranged into specific graphic patterns. For example, the number and arrangement of spikes on the surface of enveloped viruses differ among various viral species, leading to distinct characteristics in terms of infection mechanisms and behaviors. Therefore, multivalent antibodies are often designed to match the spatial conformation of antigens to enhance the affinity of antigens and antibodies through molecular spatial pattern recognition in antibody-based virus neutralization and tumor therapy [21]. For example, the conformation of complementary-determining regions (CDRs) of natural antibodies can be reconstructed on gold nanoparticles to achieve a high binding affinity with target antigens [22].

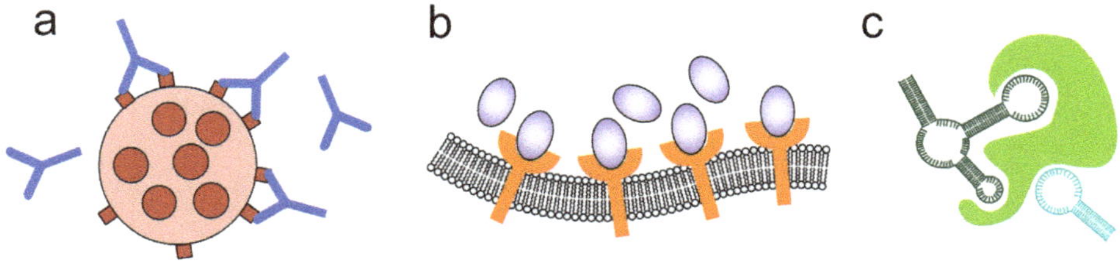

Figure 1. Biomolecular spatial pattern recognition based on antibody–antigen (**a**), receptor–ligand (**b**), and aptamer–target (**c**).

Receptor–ligand recognition is another ubiquitous molecular recognition in vivo (Figure 1b). Receptors are proteins on the cell membrane that can bind to specific ligands, triggering biological reactions inside the cell. Ligand molecules often include metal ions, nucleic acids, proteins, etc. [23]. The recognition between receptor and ligand molecules is highly selective. Ligands with specific chemical and structural features bind to specific receptors. This selectivity ensures the specificity and effectiveness of biological reactions [24]. Since the receptor molecules often cluster on the cell membrane, spatial pattern recognition between receptor and ligand has been applied in the studies of receptor-mediated biological processes. Palma's group proposed a biomimetic nanoscale array fabrication strategy based on triangular origami. By assembling integrin-specific binding ligands and epidermal growth factor (EGF) in a multivalent form in patterned nanoscale arrays, they studied receptor–ligand recognition in the spread of cancer cells with nanoscale spatial resolution and single-molecule control. They demonstrated that the synergistic effect of integrin and EGF in the spread of melanoma cells is related to the number and ratio of the two ligands [25]. Dutta et al. designed a six-helix DNA origami tension probe by customizing different numbers of ligands on DNA origami. They used single-molecule force spectroscopy to measure the tension signal of platelets and found that the total tension signal of the platelets increases with the number of modified ligands on the probe [26].

Nucleic acid aptamers, characterized by unique secondary and tertiary structures, are single-strand nucleic acid oligomers that possess the capability to selectively bind to specific target molecules (Figure 1c). These aptamers are commonly referred to as chemical antibodies and offer several advantages over traditional antibodies due to their high specificity [27]. First, the molecular weights of nucleic acid aptamers are relatively small and their structures are simple. Therefore, they penetrate cell membranes more easily and can bind to target molecules within the cell. Second, aptamers are generated from an in vitro process known as the systematic evolution of ligands by exponential enrichment (SELEX), which is more economical and can be prepared on a large scale. Last, nucleic acid aptamers can maintain their structural stability over a wide range of temperatures and pH. Their resistance to degradation by proteases makes them well-suited for utilization in intricate biological systems [28]. In addition, aptamers can be easily modified with different functional groups. Therefore, nucleic acid aptamers have been increasingly used in the research and application of biomolecular recognition [29]. Significantly, drawing inspiration from natural multivalent interactions, researchers have ventured into exploring multivalent aptamers that exhibit enhanced binding affinity, heightened specificity, and extended circulation duration compared to monomeric aptamers [30]. The rational conformational design of multivalent aptamers is key to their target recognition. For example, human thrombin aptamers with defined distances and orientations were established to study the effect of the aptamer geometry on binding properties. It has been proven that optimization of the spatial pattern of the aptamers can improve the affinity of binding efficiently [31].

3. Approaches for the Construction of DNA-Based Multivalent Biomolecules

When establishing a DNA-based biomolecular recognition system, the conjugation of DNA and biomolecules is the first and most important step. There are several highly specific and efficient methods for modifying biomolecules on DNA, including biotin–avidin interactions, and click chemistry reactions. The approaches are mainly divided into non-covalent conjugation and covalent conjugation based on the mechanism of the reaction (Table 1) [32,33].

3.1. Non-Covalent Conjugation

The non-covalent conjugation between DNA and biomolecules is realized through the interaction between the connecting molecules on DNA and ligands such as biotin–avidin, Ni^{2+}–NTA–Histag, etc.

The non-covalent interaction between biotin and avidin is a highly specific and high-affinity binding interaction, commonly employed in the conjugation of DNA with biomolecules. With the maturation of solid-phase DNA synthesis technology, biotin-modified oligonucleotides have been commercialized. The common strategy is to conjugate biotin-modified DNA with biomolecules modified with tetrameric streptavidin protein under mild conditions. Using this strategy, different ligands have been ligated to DNA nanostructures [34–36]. However, this method has some drawbacks. For example, the large size of tetrameric streptavidin may cause steric hindrance, which will affect the efficiency of ligation. The structure of the tetramer may lead to uncertain stoichiometry between DNA and biomolecules. These problems may be solved by using monomeric avidin, but the binding affinity would be weakened accordingly.

Ni^{2+}–NTA–Histag non-covalent interaction is another commonly used linking strategy. In the presence of Ni^{2+} ions, molecules labeled with multiple histidine residues (Histags) are bound to NTA-labeled DNA to construct DNA–biomolecule complexes [37]. This interaction is reversible in the presence of strong chelators such as ethylenediaminetetraacetic acid (EDTA). As the protein labeled with multiple histidine residues can be easily fused with the target protein through protein engineering, this method is more commonly used for the coupling of protein and DNA [38].

In addition, non-covalent interactions between proteins are also used to construct DNA–protein complexes. Protein A/G is a protein ligand that binds specifically to the Fc region of antibodies. The binding between protein A/G and the Fc region of antibodies is achieved through a combination of electrostatic interactions, hydrophobic interactions, and hydrogen bonding. These forces enable the formation of a stable complex between protein A/G and the Fc region of antibodies. A ProA/G-dRep fusion protein was developed for the conjugation of DNA and antibody with precise stoichiometry. In addition, protein-binding cyclic peptides can also bind to certain proteins through non-covalent interactions. Gothelf's group selected the FC-III, a 13-amino-acid cyclic peptide binder of the human immunoglobulin G (Ig G) Fc domain, to direct DNA protein conjugation. Quantitative conversion of DNA is achieved at low stoichiometries and the reaction can be performed in complex biological matrixes, such as cell lysates [39]. Although the non-covalent interaction between proteins can precisely control the ratio of DNA to protein in conjugation, it is not universally applicable to other biomolecules because such interactions occur mostly in antibodies. Some DNA binding proteins that bind to the DNA of specific sequences non-covalently, such as Zinc-finger protein and transcription activator-like effector nucleases (TALEN), also have potential in DNA biomolecule conjugation [40,41].

3.2. Covalent Conjugation

Compared with non-covalent conjugation, covalent conjugation provides a much stronger binding between DNA and biomolecules, further improving the stability of DNA–biomolecule complexes and promoting their applications in physiological environments.

The most commonly used strategy of covalent conjugation is to use the native functional group on the biomolecules. For example, covalent reactions occur between the amine group in lysine and different chemical groups modified on DNA, such as carboxyl. The thiol group is also an active group for covalent reactions. In addition to direct reactions such as acid–base condensation reactions, some heterobifunctional crosslinking agents, such as succinimide 3 (-2-pyridyl dithionyl) -propionate (SPDP) and sulfonyl succinimide 4-(n-maleimide methyl) cyclohexane-1-carboxylate (Sulfo-SMCC) were also introduced in the conjugation of two different chemical groups [42,43]. The use of a native chemical group of biomolecules for conjugation is simple but lacks selectivity. When more than one functional group used for ligation exists in biomolecules, it is difficult to control the valency and direction of the coupled DNA. Therefore, click chemistry with high selectivity, fast reaction rates, and few side reactions has been more and more applied in the conjugation of DNA and biomolecules [44,45]. Cu(I)-catalyzed azide–alkyne cycloaddition (CuAAC), the first widely accepted click reaction, was first proposed by Sharpless in 2001 [46]. Alkyne-labeled DNA strands have been conjugated with the genetically incorporated azide group on glycoproteins or enzymes to form DNA–biomolecule complexes. However, the requirement for copper ions as a catalyst limits its application in vivo. Later, a copper-free click reaction was proposed by Bertozzi and widely used for biomolecule conjugation [47]. One of the most commonly used reactions is strain-promoted alkyne–azide cycloaddition (SPAAC), which conjugates azide and dibenzocyclooctyne (DBCO). Knappe and colleagues developed a versatile DNA origami functionalization platform based on SPAAC, realizing the in situ conjugation of carbohydrates, small molecules, peptides, polymers, and proteins to DNA scaffolds [48].

Another method for covalent linkage is to introduce a tag protein on the biomolecules, which provides a self-catalytic reaction site for reaction with modified DNA. The commonly used tag proteins are O6-alkylguanine-DNA-alkyltransferase (SNAP-tag) [49] and haloalkane dehalogenase (Halo-tag) [50]. SNAP-tag transfers the benzyl group on O6-benzyl-guanine-modified DNA to cysteine, while phenylalanine in Halo-tag undergoes displacement reaction with chlorine atoms on DNA, forming a covalent linkage between the ligand and DNA. Since these two reactions are orthogonal, multiple biomolecules can be simultaneously modified on DNA using different tags [51].

Table 1. Summary of approaches for the construction of DNA-based multivalent biomolecules.

	Type	Modification	Reference
Non-covalent	Biotin–avidin interaction	Biotinylated protein and avidin-modified DNA	[34]
	Ni^{2+}–NTA–Histag interaction	Protein bearing histidine clusters and NTA-modified DNA	[38]
	Protein–protein interaction	Protein A/G or protein binding peptide-modified DNA	[39]
	Protein–DNA interaction	Protein bearing Zinc-finger protein/TALEN	[41]
Covalent	Heterobifunctional crosslinking	Amine and thiol modification	[43]
	Click chemistry	Azide and alkyne modification	[48]
	Tag-protein-mediated conjugation	Protein fused with SNAP-/Halo-tag	[51]

4. Factors Affecting Spatial Pattern Recognition

After conjugating DNA with biomolecules, the spatial conformation of these biomolecules can be arranged precisely according to design. The interaction between receptor and ligand molecules with different spatial conformations can be studied with the help of many techniques including X-ray crystallography, nuclear magnetic resonance, and surface plasmon resonance. So far, researchers have investigated several factors that influence spatial pattern molecular recognition by leveraging the programmability of DNA. By optimizing these factors, they have successfully achieved efficient molecular recognition.

4.1. Valency

Spatial pattern recognition often involves multivalent interactions between multiple biomolecules. In nature, multivalent interactions are characterized by the binding of multiple ligands on one biological entity with multiple receptors on another, exhibiting characteristics that are not present in monovalent interactions [52,53]. In contrast to weak monovalent binding, multivalent interactions substantially augment the molecular-level binding between receptors and ligands. For example, the multivalent interaction between a virus and its host cell allows the virus to stably adhere to the cell surface, achieving efficient invasion [54]. Immunoglobulin M (IgM) is the first defense for the body against foreign pathogens, typically binding to antigens in the form of pentamers, and subsequently activating the complement response [55]. Hence, through chemical synthesis that mimics endogenous multivalent biomolecular arrays, researchers can investigate the pivotal role of multivalent interactions in molecular spatial pattern recognition. Moreover, it becomes possible to construct novel multivalent biomolecular patterns that can alter the efficiency of molecular recognition through mechanisms beyond the capabilities of natural substances [56].

Low-valence multivalent biomolecules can be constructed by the site-specific conjugation of molecules on single- or double-strand DNA, which is also called DNA functionalization or modification. These multivalent biomolecules, such as proteins, small molecules, and nucleic acids, can achieve the selective recognition and quantitative measurement of the target molecules. In addition to single- and double-strand DNA, DNA nanostructures are more widely used in the construction of multivalent biomolecules. DNA nanostructures are self-assembled from multiple DNA strands with high precision and predictability, with one-dimensional to three-dimensional sizes and geometries [15,16]. For example, the DNA tetrahedron is a three-dimensional framework DNA nanostructure, usually formed by the hybridization of four DNA single strands [57]. Small molecules, peptides, nucleic acids, and other biomolecules can be ligated to the four DNA single strands through covalent or non-covalent connections, forming multivalent biomolecules with different valencies (primarily 1–4 valence) [58–60]. Li et al. connected 1–4 unmethylated CpG oligodeoxynucleotide sequences to the vertices of a DNA tetrahedron. The multivalent CpG specifically recognizes Toll-like receptor 9 (TLR9) on the surface of macrophages and successfully activates downstream immune regulatory functions [61] (Figure 2a). Other DNA nanostructures composed of a few single-strand DNA materials such as DNA G-quadruplex and DNA tile, can also be used as scaffolds for low-valence multivalent biomolecule construction. Liu et al. designed trivalent and tetravalent nucleic acid aptamers based on the J1 connection and 4×4 DNA tile structures, both of which showed strong affinity when binding to target cells. Based on this, two types of nucleic acid aptamers were linked to the end of a DNA tile dimer to construct an octavalent double-specific nucleic acid aptamer, which mediated the connection between the two types of cells [62].

Sometimes, it is necessary to construct tens or even hundreds of multivalent biomolecules for molecular recognition, where DNA origami serves as a good candidate for a multivalent scaffold. DNA origami is a type of bottom-up synthesized nanostructure ranging in size from tens of nanometers to sub-micrometers. A typical DNA origami structure contains about 200 staple strands with unique sequences and positions. By conjugating biological molecules to these staple strands, multivalent biomolecules with hundreds of valencies can, in principle, be constructed [63]. Song's group precisely arranged 10 to 90 SARS-CoV-2 RBDs (receptor binding domains) on a ~74 nm DNA soccer-ball origami structure and studied the multivalent molecular recognition between RBD and ACE2 (angiotensin-converting enzyme 2) on the host cell. It was found that both the affinity and the rate of the DNA-based multivalent RBDs binding to the host cell increase with the RBD number [64] (Figure 2c). Furthermore, certain DNA nanostructures, such as DNA tetrahedra, can serve as fundamental units for constructing higher-order nanostructures. By attaching biomolecules to each DNA nanostructure monomer, multivalent molecular recognition can be achieved, thereby expanding the capabilities and applications of these DNA-based architectures. Yang's group arranged sub-10 nm DNA tetrahedrons on a microfluidic chip and ligated nucleic acid aptamers (SYL3C) to them. The multivalent nucleic acid aptamer formed on the interface increased the affinity with human colon cancer cells (SW480) by about four times compared to free nucleic acid aptamers [65]. DNA nanostructures self-assembled via the hybridization chain reaction (HCR) or rolling circle amplification (RCA) can also serve as multivalent nanoscaffolds. Tan's group designed and constructed several nanostructures with multivalent aptamers such as "nanocentipede" and "nanoflower", which specifically bind to target cells with high affinity through multivalent recognition [66,67] (Figure 2b). Nonetheless, molecular recognition based on these structures can only be studied qualitatively due to the unpredictable nature of the reactions.

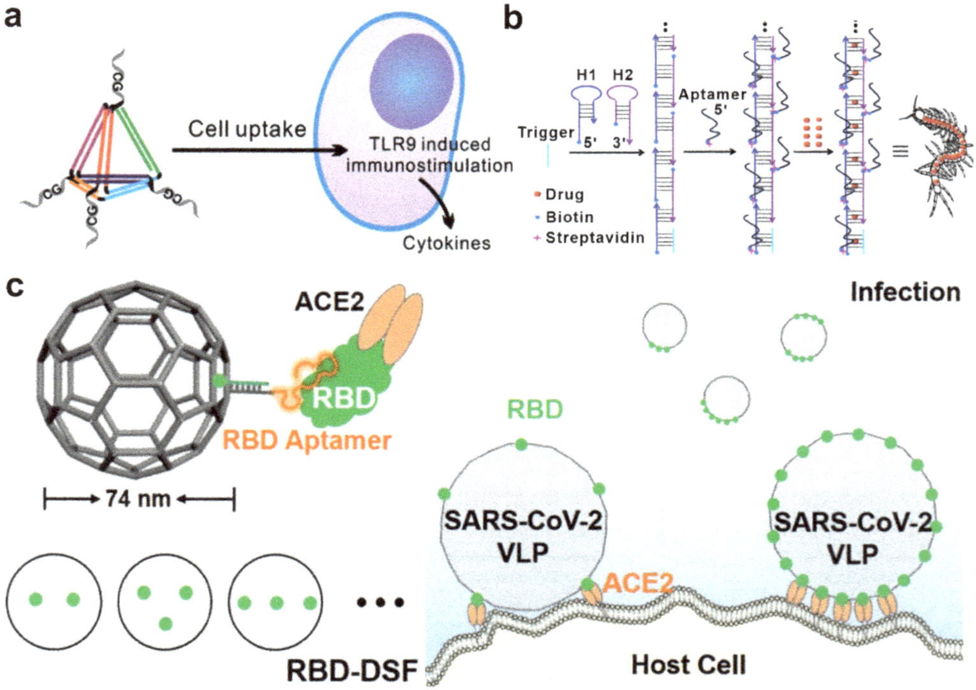

Figure 2. Biomolecular spatial recognition based on valency regulation. (**a**) Schematic showing of the assembly of a CpG-bearing DNA tetrahedron and its immunostimulatory effect. Reprinted with permission from [61], copyright 2011, American Chemical Society. (**b**) The nanocentipede consists of multivalent aptamers as a function of targeting moieties to target cells. Reprinted with permission from [66], copyright 2016, American Chemical Society. (**c**) Schematic illustration for the assembly of RBDs on a DSF and the interaction with the host cell. Reprinted with permission from [64], copyright 2022, American Chemical Society.

In general, a higher number of valence states in biomolecules provides an advantage in molecular recognition due to the multivalent synergistic effect. Nevertheless, several studies have indicated the presence of a valence threshold in specific molecular recognition systems, wherein the affinity no longer improves beyond a certain number of valence states, despite an increase in their quantity. Consequently, it becomes desirable, from both economic and practical perspectives, to determine the precise threshold of valence states for optimal molecular recognition. In the DNA-based multivalent biomolecular systems mentioned above, the structures formed through HCR and RCA result in a substantial amplification of valence states due to the repetitive units of the structure. However, the lack of control in their reactions hinders the assurance of precise and uniform valence states. Conversely, multivalent biomolecules based on DNA tetrahedra and DNA origami provide accurate manipulation of valence states, offering greater advantages in research on biomolecular recognition that relies on spatial conformation.

4.2. Distance

The distances between molecules play a crucial role in molecular recognition, particularly in spatial pattern recognition, as they provide valuable information about specific interaction patterns. Biomolecules, in particular, can exhibit even distribution on the cell membrane or localize in microdomains due to the fluidity of the phospholipid bilayer. Consequently, the distances between biomolecules can vary depending on the cellular state [68]. Understanding and analyzing these distances is essential for deciphering molecular recognition processes and spatial patterns [69]. DNA nanostructures enable precise localization of biomolecules. Among them, DNA origami with strong designability, span scales from nanometers to hundreds of nanometers, perfectly matching the scale range of molecular arrangements on the cell surface. In addition, the remarkable spatial addressability of DNA origami allows for the precise anchoring of biomolecules at the nanometer scale. These capabilities enable researchers to investigate biomolecular networks with controlled distances, facilitating in-depth studies in distance-dependent molecular recognition.

In 2008, Yan's group first utilized the spatial addressing ability of self-assembled DNA nanoscaffolds to construct bivalent aptamers with different distance intervals for recognizing thrombin protein and visualized this interaction at the single-molecule level. The bivalent aptamers with the highest thrombin binding activity were found to have a distance of 5.8 nm between the two aptamers [70] (Figure 3a). Similarly, Shaw et al. modified DNA origami with the Eph receptor tyrosine kinase ligand ephrin-A5, and constructed a series of "nano-rulers" with different ligand spacing to study the role of distances between ephrin-A5 in Eph receptor recognition and receptor-mediated signal transduction [71]. Furthermore, various biomolecules such as caspase-9 monomers and cell-binding ligand RGD have been strategically positioned on diverse DNA nanostructures. These investigations aim to elucidate the impact of spacing on the molecular recognition capacity and the effectiveness of downstream signal transduction [72,73].

An antibody molecule typically consists of two antigen-binding fragments (Fab), with each Fab region capable of binding to an antigen. Consequently, a single antibody molecule can simultaneously bind to two identical or different antigens. This bivalent binding mechanism plays a vital role in enhancing the antibody's affinity and specificity, thereby increasing its effectiveness in immune responses. As a result, the distance between antigens becomes crucial in immune reactions. By precisely controlling the spacing of antigen molecules, it is possible to optimize antigen–antibody recognition and interactions, enhance the efficacy and selectivity of immune responses, and provide guidance and principles for the design and optimization of molecular vaccines. For example, Shaw et al. used DNA origami to precisely control the distance between antigens and characterized the binding of antibodies with identical antigen-binding domains. They found that the antibodies bound to two antigens at distances of 3–17 nm, and the binding affinity varied with the distance between the antigens, with the highest affinity observed for antigens at a distance of approximately 16 nm [74]. Fan's group constructed artificial antigen epitopes of

3–20 nm by anchoring antigens onto triangular DNA origami and used high-speed atomic force microscopy to image the antigen–antibody interaction at the single-molecule level, providing dynamic evidence for the antigen-binding process of IgG from monovalent to bivalent [75] (Figure 3b). In addition to antibody binding, antigens can also activate B-cell receptor (BCR) signaling pathways to activate antigen-specific B cells. To further explore the effect of antigen spacing on IgM–BCR activation, Bathe's group conjugated the clinical vaccine immunogen eOD-GT8 (an HIV-1 glycoprotein-120 external domain) to the surface of icosahedral and six-helix bundle DNA origami at different numbers and distances. It was shown that the activation of B cells was maximized with an antigen spacing of 25–30 nm. This work provides optimization principles for the design of molecular vaccines based on B-cell immune responses [76] (Figure 3c).

Overall, the influence of spacing on biomolecular recognition varies according to the specific recognition system. It is necessary to investigate the optimal distance within specific types of recognition biomolecules and environments to attain the highest efficiency in molecular recognition, which can ultimately be applied in biomedical fields such as disease diagnosis, vaccine design, and other related areas.

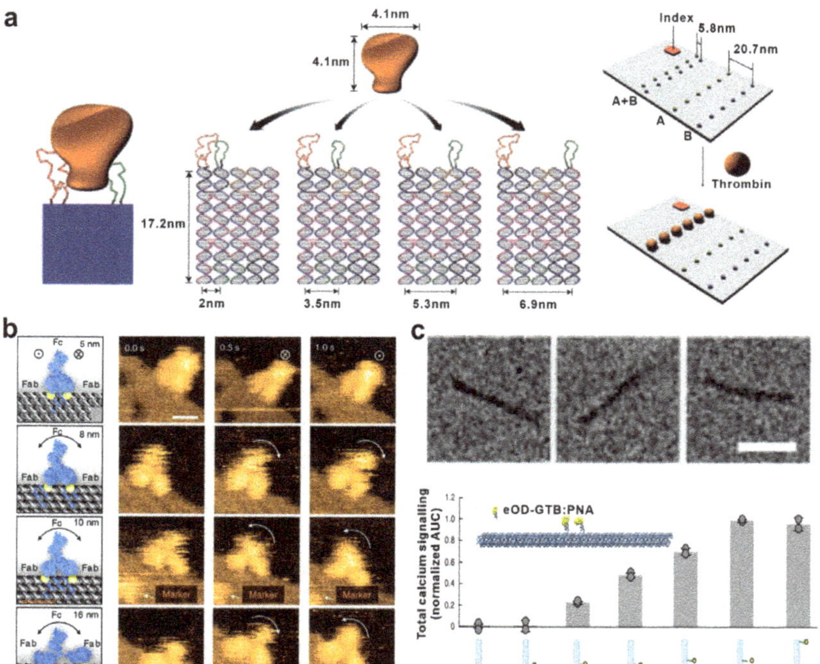

Figure 3. Biomolecular spatial recognition based on distance regulation. (**a**) Schematic showing a rigid DNA tile (blue) spatially separating two aptamers (red and green) at a controlled distance for bivalent binding. Reprinted with permission from [70], copyright 2008, Springer Nature. (**b**) Snapshot HS-AFM images of single IgGs captured by DNA origami epitopes with designed distances of 5, 8, 10, and 16 nm, respectively. Reprinted with permission from [75], copyright 2020, Springer Nature. (**c**) TEM images of six-helix bundle DNA origami and the total calcium signaling in B cells stimulated with DNA-eOD-GT8 dimers with inter-antigen distances between 7 nm and 80 nm. Reprinted with permission from [76], copyright 2020, Springer Nature.

4.3. Pattern Arrangement

Molecular recognition in the biological system is based on specific interactions between ligands and receptors, which often occur on the cell membrane and exhibit topological cluster characteristics. As an illustration, surface antigens found on viruses and bacteria often exhibit a specific spatial topology [77]. Moreover, the response of T-cell receptors (TCRs) to antigens is intricately linked to the precise positioning of the antigen in three-dimensional space [78]. By orienting multiple biomolecules into a certain geometric pattern, multivalent ligands with various topological structures can be formed and specifically recognize target molecules. DNA nanostructures, ranging from one-dimensional to three-dimensional architectures, offer unparalleled advantages over other materials when it comes to the precise spatial arrangement of molecules. Through DNA structure design and modification site engineering, biomolecular patterns of diverse shapes can be constructed. Moreover, by precisely controlling the spatial arrangement of multivalent ligands on DNA, the affinity with target molecules can be significantly enhanced.

The algebraic topology of biomolecules can be arranged on a DNA framework to match the receptor patterns on the cell surface. Fan's group topologically rearranged 1–3 aptamers on a DNA tetrahedral framework including point, line, and surface configurations, targeting overexpressed epithelial cell adhesion molecule (EpCAM) on tumor cell membranes. The multivalent ligands distributed in a surface pattern showed the strongest ability to recruit receptor aggregation on the cell membrane, with a 19-fold increase in affinity compared to free aptamers [79] (Figure 4a). By utilizing DNA tetrahedral dimers, researchers were able to broaden the topological configurations of the aptamers. This expansion resulted in altered molecular recognition properties and binding strength by inducing receptor aggregation on the cell membrane [80]. Shen and colleagues built a suite of nuclear pore complex (NPC) mimics by programmably arranging multiple nucleoporin proteins on DNA origami. They designed two octagonal-shaped DNA origami channels to mimic the NPC's eightfold rotational symmetry. Different numbers of nucleoporin proteins formed different topologies in the structure and showed different binding affinity to the capsid of human immunodeficiency virus 1 (HIV-1), which provides a mechanistic insight for elucidating how viruses enter the nucleus [81] (Figure 4b).

Some biomolecules combine into unique shapes and configurations for the specific recognition of target molecules. A noteworthy example is the diverse shapes of surface antigens found on viruses. These epitope configurations serve as specific markers for the viruses and play a vital role in enabling the immune system to recognize and mount an attack against the viral pathogens. The precise design of DNA nanostructures can be used to achieve a highly matched spatial conformation and achieve precise regulation of molecular recognition. A star-shaped DNA nanostructure carrying 10 envelope protein domain III (ED3) aptamers was developed to precisely match the trivalent and pentavalent ED3 epitope on the dengue virus surface, which can be used for virus detection and inhibition [82] (Figure 4c). The aptamers against the SARS-CoV-2 S protein were assembled into a trimeric complex on a DNA nanocage to match the pattern of the S protein trimer in the spatial arrangement, resulting in significantly higher affinity to the S protein trimer [83] (Figure 4d).

Precisely modulating the spatial arrangement of multivalent biomolecules to match the patterns of the target can maximize the strength of multivalent interactions and further enhance the affinity between biomolecules. However, the precise spatial arrangement of biomolecules on DNA nanostructures to achieve spatial pattern recognition can only be accomplished when the exact distribution of target molecules is known. For target molecules with unknown or imprecise distribution, a relatively flexible DNA scaffold may be constructed to enable dynamic and adaptive matching of biomolecules, thereby enhancing the affinity of biomolecular recognition.

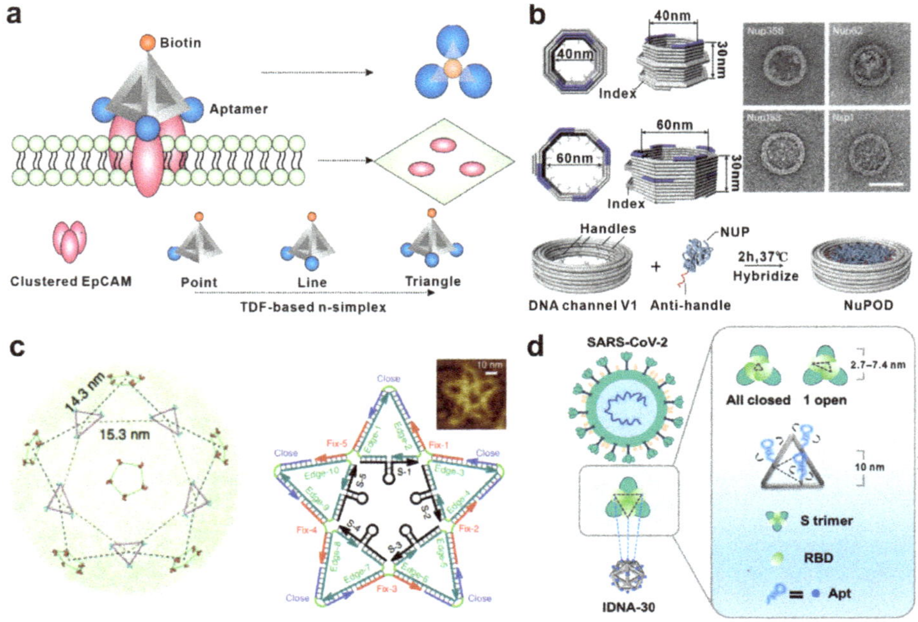

Figure 4. Biomolecular spatial recognition based on pattern arrangement. (**a**) Scheme of the DNA-tetrahedra-based n-simplexes and recruitment-binding-induced EpCAM cluster on the cell membrane. Reprinted with permission from [79], copyright 2019, American Chemical Society. (**b**) Design of DNA origami channels as mimics of the NPC scaffold and the attachment of nup-DNA conjugates to a DNA origami channel by DNA hybridization. Reprinted with permission from [81], copyright 2023, Springer Nature. (**c**) Distribution of DENV ED3 clusters and the structure of the star-shaped DNA nanostructure carrying 10 envelope protein domain III (ED3) aptamers. Reprinted with permission from [82], copyright 2019, Springer Nature. (**d**) Spatially matched IDNA-30 against SARS-CoV-2. Reprinted with permission from [83], copyright 2022, American Chemical Society.

5. Application of DNA-Based Spatial Pattern Recognition

DNA scaffolds have been utilized to precisely attach diverse biomolecules, enabling spatially controlled molecular recognition characterized by specificity and strong affinity. Since molecular recognition is crucial in many biomedical fields, such as immunology and drug development, emerging studies have focused on developing DNA-programmed multivalent biomolecules for biomedical applications, such as biosensing, bioimaging, and targeted therapy.

5.1. Biosensing

Based on the interactions between biosensors and the target substance, biosensing technology finds extensive application in the detection and analysis of target molecules within biological systems. Efforts have been made to improve the efficiency and sensitivity of the biosensor, such as developing new construction materials, optimizing the biorecognition element, and introducing a signal amplification strategy [84–87]. Among them, DNA-based biomolecular recognition can be conducive to highly sensitive biosensors in detecting cancer biomarkers [88]. The overexpression of distinct membrane proteins in tumor cells, which can be specifically recognized by corresponding ligands, plays a vital role in the early diagnosis of cancer. As an illustration, in liquid biopsy, various membrane proteins, such as EpCAM, which are overexpressed on the surface of circulating tumor cells (CTCs), serve as novel biomarkers, offering a broader understanding of the primary tumor tissue for precision medicine. Yang's group combined microfluidic chips and DNA-tetrahedron-conjugated EpCAM aptamers to ensure highly ordered and perpendicular ligand orientation at the nanoscale, avoiding the non-specific binding and crowding effects of traditional microfluidic interfaces (Figure 5a). The chip they developed improved the capture efficiency of circulating tumor cells (CTCs) by nearly 60% compared to chips with monovalent ligand modification. Additionally, 83% cell release can be achieved by DNase I treatment, with a cell survival rate of 91% [65]. In addition, the spatial topology of the EpCAM aptamer was optimized with DNA tetrahedrons, and a capture efficiency of up to 97% for MCF-7 cells injected into whole blood was achieved, which was higher than the antibody capture method (~50%). CTCs from the whole blood samples of cancer patients were successfully captured, with 3–10 CTCs per milliliter for non-metastatic patients and 38–44 CTCs per milliliter for metastatic patients [79] (Figure 5b). In addition to DNA tetrahedra, DNA-origami-based multivalent aptamers were also developed, which match the spatial distribution of target protein clusters and detect CTCs with high affinity [89]. In comparison to DNA tetrahedra, DNA origami employed in CTC detection offers more binding sites. This enables the conjugation of recognition molecules with higher valence and different types. As a result, the affinity for binding with CTC is significantly enhanced. The binding equilibrium constant K_d is reduced from 7 nM (for DNA tetrahedra) to 260 pM (for DNA origami). However, the smaller size and greater flexibility of DNA tetrahedra make them more amenable to be modified at the interface. By combining with a microfluidic system, the thermodynamics of molecular recognition at the modified interface were greatly improved. The strategy can be used in a wide variety of areas, including non-invasive testing, and the interfacial regulation of cellular behavior.

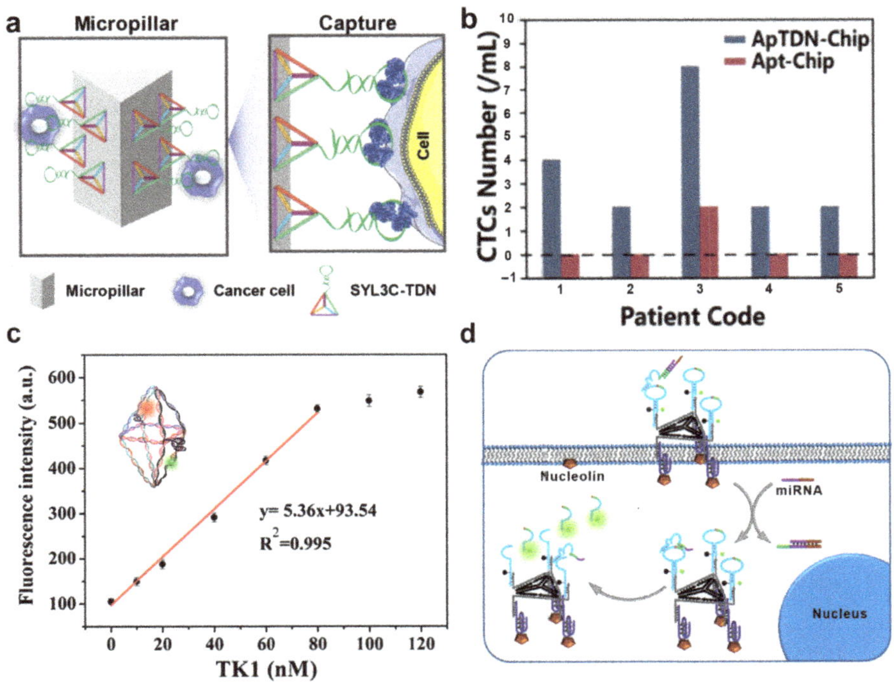

Figure 5. Application of DNA-based spatial pattern recognition in biosensing. (**a**) Engineered multivalent aptamers on the microfluidic interface for CTCs capture. (**b**) CTCs capture performance of the chip with multivalent aptamers for clinical samples. Reprinted with permission from [65], copyright 2020, John Wiley and Sons. (**c**) TK1 mRNA detection curve of a DNA-octahedron-based

fluorescence nanoprobe. Reprinted with permission from [90], copyright 2018, American Chemical Society. (**d**) DNAzyme walking and amplification imaging of miRNAs within living cells. Reprinted with permission from [91], copyright 2019, American Chemical Society.In addition to biosensors that directly detect molecules on the cell surface, high-affinity molecular recognition also mediates the entry of biosensors into cells for the efficient detection of intracellular biomolecules. After binding to specific molecules on the cell surface, DNA-based multivalent biomolecules can enter cells via endocytosis and detect target molecules inside cells. Lu et al. constructed a DNA octahedron with a divalent AS1411 aptamer, which can simultaneously detect and image two tumor-related mRNAs in living cells, distinguishing target cancer cells from normal cells. The fluorescence signal of the octahedron with the divalent AS1411 aptamer was stronger than that of the octahedron without aptamers, indicating that the DNA-based biosensor can be more effectively taken up by cancer cells through modification with the AS1411 aptamer [90] (**c**). Xue et al. integrated a DNAzyme walker into a triangular DNA scaffold functionalized with multivalent aptamers for highly sensitive detection of cell miRNA. The movement of the DNAzyme walker was activated by the target miRNA and produced signal amplification [91] (**d**).

Despite its polyanionic nature, DNA can cross the negatively charged membrane to enter living cells by assembling into specific nanostructures. It was reported that DNA nanostructures approach the membrane primarily with their corners to minimize electrostatic repulsion and their binding affinity and cellular internalization frequency depended on the corner angle of the structures [92,93]. Therefore, three-dimensional DNA nanostructures with corner structures, such as tetrahedra and octahedra, offer greater advantages for intracellular sensing applications than two-dimensional planar structures such as triangles and squares.

5.2. Bioimaging

Molecular recognition has a wide range of applications in biological imaging, including cell imaging, tumor imaging, and molecular localization. Fluorescent probes can be used to detect specific receptors on the surface of tumor cells, assisting in tumor diagnosis and treatment. For example, EpCAM aptamers with different numbers and directions were assembled on a DNA framework to construct a series of probes for the imaging of tumor cells. The directional-yet-flexible probe exhibited adaptability to the receptor distribution on cell surfaces and high affinity against target tumor cells and has the potential of becoming an excellent imaging probe for EpCAM-positive tumors [94] (Figure 6a). Multicolor probes were constructed by arranging different fluorescent molecules and aptamers on DNA tetrahedra. Molecular recognition facilitated the achievement of multiplexed cell imaging and classification of various tumor cells by harnessing the programmability of fractal DNA frameworks [95]. In terms of tumor imaging in vivo, Ding's research group has employed rational design to create DNA-based probes that feature precisely organized tumor-targeting components along with imaging molecules such as fluorescent groups and magnetic resonance contrast agents. Through this approach, they have successfully achieved specific and non-invasive tumor imaging in mouse models using DNA nanostructures [96,97] (Figure 6b).

The biology and chemistry of cells can also be studied by probes based on biomolecular pattern recognition. For example, introducing imaging molecules into specific cellular structures can help in understanding the distribution and dynamic changes of molecules within cells. Xu's group developed DNA-programmed plasma rulers for imaging the dimerization of the RTK on the cell membrane. The binding of aptamers to cell-surface receptors was regulated by DNA, further mediating the dimerization and dissociation of the receptors. Real-time imaging of RTK dimerization and dissociation processes was realized by detecting the scattered signal caused by the plasma coupling effect of gold nanoparticles. This method provides a more persistent and higher resolution tool for studying protein oligomerization, transport, and dynamics processes, which is of great significance for the in-depth interpretation and precise regulation of cellular signal transduction [98] (Figure 6c).

Overall, in biological imaging systems, DNA origami has an advantage in multimodal imaging due to its numerous modification sites, by simultaneous modifications such as fluorescence and radiolabeling. Conversely, DNA double strands and DNA tetrahedra are more suitable for dynamic imaging due to their structural flexibility.

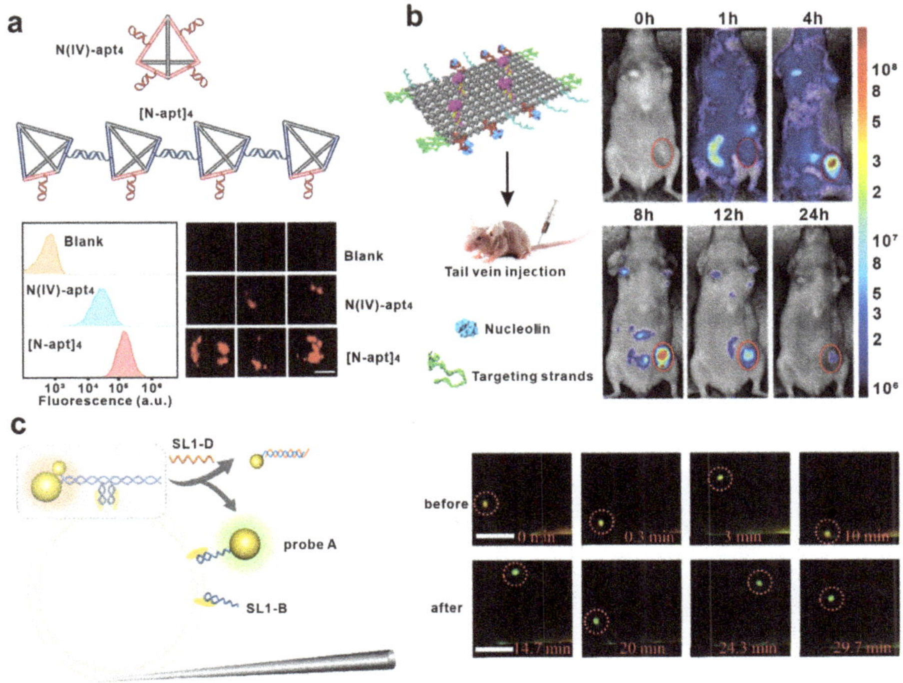

Figure 6. Application of DNA-based spatial pattern recognition in bioimaging. (**a**) MCF-7 imaging with tetravalent aptamers and linear aptamer oligomer. Reprinted with permission from [94], copyright 2022, John Wiley and Sons. (**b**) Imaging of a mouse bearing a human breast tumor before and after intravenous injection of a Cy5.5-labeled nanorobot. Reprinted with permission from [96], copyright 2018, Springer Nature. (**c**) Schematic illustration of DNA-programmed single-molecule manipulation on giant plasma membrane vesicle and representative images of the vesicle over time before and after the DNA strand displacement. Reprinted with permission from [98], copyright 2023, American Chemical Society.

5.3. *Targeted Therapy*

One of the main challenges in the treatment of diseases, especially tumors, is the lack of effective and precise targeted delivery systems. Molecular recognition can be used to identify target molecules that are related to the occurrence of diseases. Consequently, there has been extensive research on utilizing modified DNA as drug carriers for targeted therapy, aiming at specific molecules.

Commonly used chemotherapy drugs such as doxorubicin (DOX) can intercalate into DNA base pairs. Nucleotide drugs such as siRNA and antisense oligodeoxynucleotide (ASO) can be hybridized with DNA by Watson–Crick complementary base pairing [99]. After being conjugated with ligands for recognition, DNA specifically delivers these drugs into target cells, without toxic side effects on normal cells. Lin's group ligated the nucleic acid aptamer AS1411 to DNA tetrahedra (t-FNAs) to target the overexpressed nucleolin protein on tumor cells. Compared to tetrahedra without AS1411, AS1411–tFNA accumulated in the nuclei of MCF-7 cells. Through the additional loading of 5-fluorouracil (5-FU) onto t-FNAs, the successful induction of apoptosis in MCF-7 cells was achieved [100].

In addition to AS1411, other nucleic acid aptamers such as Gint4.T and GMT8 (which specifically recognize platelet-derived growth factor receptor beta and human glioblastoma cells U87MG, respectively) were also ligated to t-FNAs, and the loaded chemotherapeutic drug paclitaxel was target-released in tumor cells [101]. Ge and colleagues reported a DNA origami nanostructure modified with targeting ligand DUPA as an antibody–drug conjugate (ADC) analog for the targeted therapy of prostate cancer. The platform efficiently delivers DOX to prostate-specific membrane antigen (PSMA)-positive cells, and the therapeutic effect is dependent on the number of ligands connected [102] (Figure 7a). Compared to other organic and inorganic nanomaterials, DNA nanostructures demonstrate significant advantages in the targeted delivery of drugs because they can provide large payload capacity and improve cellular internalization with high specific recognition. However, the expression of surface biomarkers on cancer cells is heterogeneous, which causes the inaccurate identification of cancer cells by a single biomarker. To address this issue, Ju's research group developed a dual-receptor-mediated DNA nanocarrier for siRNA delivery. DNA is self-assembled into a double-lock structure with aptamers Sgc8c and Sgc4f that bind to two specific receptors on the cell surface. The delivery system is active only when both of the aptamers were recognized by cells, thus improving the accuracy of cell identification and greatly reducing the off-target toxicity [103].

Functional molecules can be simultaneously modified on DNA frameworks, providing a powerful platform for multimodal cancer therapy. Zhang et al. combined a DNA tetrahedron modified with a nucleic acid aptamer and a metal–organic framework (MOF) nanoparticle to design a hybrid nanocarrier for loading DOX. The chemotherapy drug was released in cancer cells rich in adenosine triphosphate (ATP) or vascular endothelial growth factor (VEGF). At the same time, a photosensitizer, Zn (II) protoporphyrin IX (Zn (II)-PPIX), was introduced to produce high-intensity fluorescence efficiently and selectively in tumor cells, inducing the production of reactive oxygen species (ROS) and conducting photodynamic therapy on malignant tumor cells [104]. A nanomachine for DNA logic gate operation across cell membranes was designed, realizing precise photodynamic therapy for solid tumors in vivo. This DNA nanomachine consists of upconversion nanoparticle cores (UCNPs), DNA assemblies, and sgc8 aptamer. The overexpressed PTK-7 protein on the cancer cell membrane and the high-expressed miRNA-21 inside the cancer cell serve as two signals trigging the production of intracellular ROS [105] (Figure 7b). A DNA nanorobot was constructed based on a rectangular DNA origami structure for delivering and releasing thrombin in tumors. Thrombin molecules were placed inside the lumen of the tubular DNA nanorobot and then closed by a predesigned anchor chain containing the nucleic acid aptamer AS1411. The nucleic acid aptamer on the outer surface can guide and trigger the release of thrombin, activating clotting at the tumor site and ultimately resulting in tumor necrosis and growth inhibition [96]. The DNA-modified biomolecules not only act as guides for targeted delivery but also function as triggers for drug release or ROS production, thereby significantly enhancing the spatiotemporal control of targeted therapy.

Tumor immunotherapy, as an emerging method for cancer therapy, activates the immune system for better recognition and attacking of cancer cells, inhibiting the growth and metastasis of tumors. Tumor-specific antigens specifically expressed by tumor cells can be recognized by the immune system as foreign and trigger an immune response to eliminate these tumor cells. For example, T cells acquire adaptive immunity through the binding of T-cell receptor (TCR) to the major histocompatibility complex (pMHC) on the surface of antigen-presenting cells (APCs). By designing a series of DNA nanoscale junctions (DNJs) based on DNA tetrahedra with different sizes, the intermembrane distance at the interface of APC and T cells was precisely controlled. It was demonstrated that the axial distance of the immunological synapse plays an important role in T-cell recognition and activation, providing a basis for T-cell immunology research [106]. Recently, researchers constructed a set of self-adjuvant carriers known as framework nucleic acids (FNAs), which possess regulated rigidity and size. These FNAs were utilized to investigate the impact of epitope spacing on the efficacy of peptide vaccines. When epitopes were assembled on FNAs of appropriate size, the recognition between the epitopes and B-cell receptors (BCRs) as well as the immunogenicity, could be efficiently enhanced [107] (Figure 7c). The recognition and activation of T cells and B cells with antigen epitopes can be regulated and optimized through the design of DNA nanostructures. However, current research focuses primarily on regulating distances and there has been a lack of investigations into patterned arrangement recognition for immune cell activation.

In addition to the targeted treatment of tumors, molecular recognition also plays an important role in targeted antiviral therapy. By interfering with the specific molecular interactions between viruses and host cells, the infection of host cells by viruses could be prevented. SARS-CoV-2 infection was inhibited by anchoring neutralizing aptamers on a DNA scaffold in a pattern that matches the spatial configuration of the viral S protein [83]. Dietz's group conjugated antivirus antibodies to a DNA icosahedral shell in a modular fashion. Viruses were recognized and trapped in the shell, realizing the neutralization of hepatitis B virus and adeno-associated virus [108] (Figure 7d). The virus-trapping strategy mediated by DNA provides a novel approach to antiviral therapy. The use of DNA-based agents potentially circumvents neutralization, phagocytosis, and degradation by pathways of the innate and adaptive immune system targeting protein structures. However, nucleic-acid-specific reactions, such as the activation of pattern-recognition receptors recognizing DNA, may occur in vivo. Thus, assessing the potential adverse effects in organisms remains to be an important challenge in application.

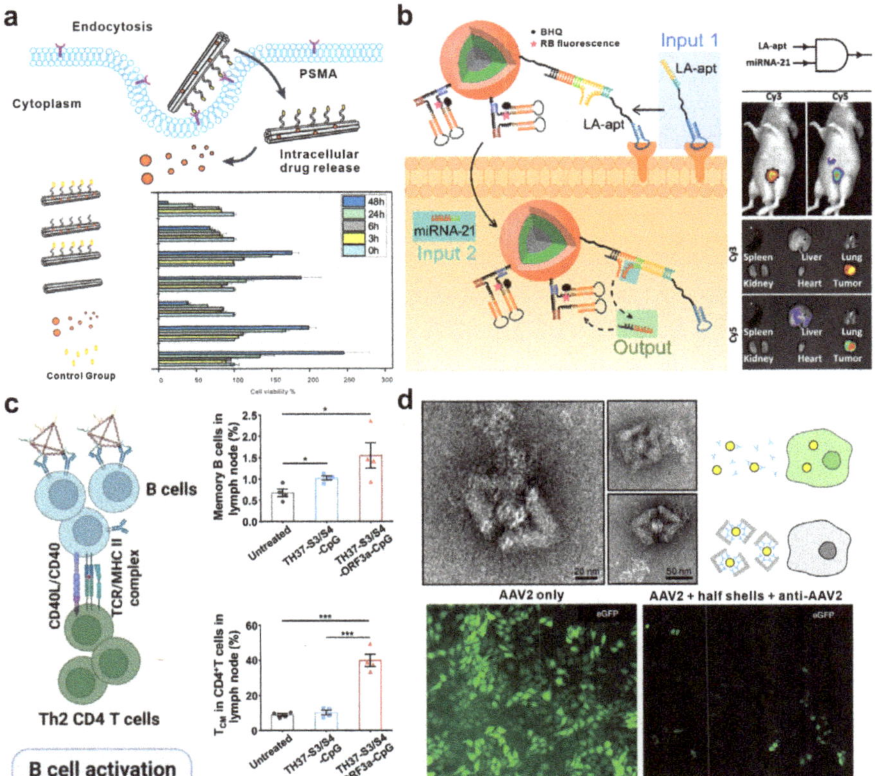

Figure 7. Application of DNA-based spatial pattern recognition in targeted therapy. (**a**) Schematic illustration of the ligand-modified six-helical-bundle DNA origami as drug carriers and its working principle. Reprinted with permission from [102], copyright 2020, John Wiley and Sons. (**b**) Structural

diagram of the transmembrane DNA computation and in vivo fluorescence imaging of mice based on different inputs. Reprinted with permission from [105], copyright 2021, American Chemical Society. (**c**) Immune responses induced by FNA-constructed COVID-19 epitope vaccines. The significance of difference was evaluated by p value; * $p < 0.05$, *** $p < 0.001$. Reprinted with permission from [107], copyright 2023, John Wiley and Sons. (**d**) Negative stain TEM images demonstrating the capture of AAV2 virus particles within antibody-modified DNA origami half shells and fluorescent microscopy images showing the anti-infective effect. Reprinted with permission from [108], copyright 2021, Springer Nature.

6. Conclusions and Perspectives

Advancements in DNA-programmed biomolecular spatial pattern recognition have led to the synthesis of multivalent biomolecules that demonstrate enhanced affinity to receptors, both in vivo and in vitro. These biomolecules exhibit exceptional performance in various applications, including biological detection and targeted therapy. However, DNA-based biomolecular recognition still faces many challenges in future applications: (1) Due to the small differences in the expression levels of some biomolecules in different cells, it is necessary to further optimize the spatial arrangement of biomolecules on DNA scaffolds to achieve precise control of recognition affinity and improve the sensitivity of molecular detection. (2) Presently, DNA-based multivalent ligands are capable of precisely targeting receptors with predetermined conformations on the cell surface. Nevertheless, the cell membrane exhibits mobility, and the distribution of receptors on its surface undergoes dynamic changes. Consequently, the development of adaptive multivalent biomolecules that leverage the adjustability of DNA to accurately match molecules on the cell membrane surface remains an ongoing challenge. (3) In the specific detection of membrane proteins on the cell surface, the cellular uptake of DNA scaffolds may affect the detection sensitivity. To address this issue, spatial pattern recognition can be designed on interfaces such as microfluidic chips and microbeads to eliminate the influence of cellular uptake. (4) Matrices are complex in the biological system and the biomolecular recognition in vivo can be affected by heterogeneity, steric hindrance, and so on. The factors that influence molecular spatial pattern recognition have not been fully elucidated. Computational simulations, such as the simulation of dissipative particle dynamics, may help the construction of a biomolecular recognition model.

Despite these challenges, the studies of DNA-programmed biomolecular spatial pattern recognition have provided compelling evidence for the power of this approach to reveal important physiological processes in cells and promoted the development of biosensing and drug delivery. We believe that with further developments in DNA nanotechnology, synthetic biology, microfluidics, and information technology, more accurate and adaptive control over biomolecular recognition will be realized, providing a powerful tool for the clinical diagnosis and treatment of diseases.

Author Contributions: Conceptualization, L.W. and L.G.; writing-original draft preparation, Y.W. and L.R.; writing—review and editing, L.G.; visualization, H.P.; supervision, L.W.; funding acquisition, L.W. All authors have read and agreed to the published version of the manuscript.

Funding: This research was funded by the National Key R&D Program of China 2021YFA1200900, the National Natural Science Foundation of China 22022410, 82050005, 12105352.

Institutional Review Board Statement: Not applicable.

Informed Consent Statement: Not applicable.

Data Availability Statement: There was no new data created or analyzed in this study. Data sharing does not apply to this article.

Conflicts of Interest: The authors declare no conflict of interest.

References

1. Persch, E.; Dumele, O.; Diederich, F. Molecular recognition in chemical and biological systems. *Angew. Chem. Int. Ed.* **2015**, *54*, 3290–3327. [CrossRef] [PubMed]
2. Akira, S.; Uematsu, S.; Takeuchi, O. Pathogen recognition and innate immunity. *Cell* **2006**, *124*, 783–801. [CrossRef]
3. Ruoslahti, E.; Pierschbacher, M.D. New Perspectives in Cell Adhesion: RGD and Integrins. *Science* **1987**, *238*, 491–497. [CrossRef]
4. Ciferri, A. Critical issues in molecular recognition: The enzyme–substrate association. *Soft Matter.* **2021**, *17*, 8585–8589. [CrossRef] [PubMed]
5. Chen, K.; Zhao, Y. Molecular recognition of enzymes and modulation of enzymatic activity by nanoparticle conformational sensors. *Chem. Commun.* **2022**, *58*, 1732–1735. [CrossRef]
6. Wu, X.; Chen, J.; Wu, M.; Zhao, J.X. Aptamers: Active targeting ligands for cancer diagnosis and therapy. *Theranostics* **2015**, *5*, 322–344. [CrossRef] [PubMed]
7. Weigelt, B.; Peterse, J.L.; van't Veer, L.J. Breast cancer metastasis: Markers and models. *Nat. Rev. Cancer* **2005**, *5*, 591–602. [CrossRef]
8. Tong, P.; Zhang, L.; Xu, J.; Chen, H. Simply amplified electrochemical aptasensor of ochratoxin A based on exonuclease-catalyzed target recycling. *Biosens. Bioelectron.* **2011**, *29*, 97–101. [CrossRef]
9. Lin, J.; Ju, H. Electrochemical and chemiluminescent immunosensors for tumor markers. *Biosens. Bioelectron.* **2005**, *20*, 1461–1470. [CrossRef]
10. Jiao, Y.; Qiu, Y.; Zhang, L.; Liu, W.G.; Mao, H.; Chen, H.; Feng, Y.; Cai, K.; Shen, D.; Song, B.; et al. Electron-catalysed molecular recognition. *Nature* **2022**, *603*, 265–270. [CrossRef]
11. Zhao, M.; Wu, T.; Xiao, X.; Liu, Y.; Su, X. New advances in molecular recognition based on biomolecular scaffolds. *Anal. Bioanal. Chem.* **2013**, *405*, 5679–5685. [CrossRef] [PubMed]
12. Wu, Y.; Sefah, K.; Liu, H.; Wang, R.; Tan, W. DNA aptamer-micelle as an efficient detection/delivery vehicle toward cancer cells. *Proc. Natl. Acad. Sci. USA* **2010**, *107*, 5–10. [CrossRef]
13. Sheng, W.; Chen, T.; Tan, W.; Fan, Z.H. Multivalent DNA Nanospheres for enhanced capture of cancer cells in microluidic devices. *ACS Nano* **2013**, *7*, 7067–7076. [CrossRef] [PubMed]
14. Geng, Z.; Wang, L.; Liu, K.; Liu, J.; Tan, W. Enhancing anti-PD-1 Immunotherapy by Nanomicelles Self-Assembled from Multivalent Aptamer Drug Conjugates. *Angew. Chem. Int. Ed.* **2021**, *60*, 15459–15465. [CrossRef]
15. Seeman, N.; Sleiman, H. DNA nanotechnology. *Nat. Rev. Mater.* **2018**, *3*, 17068. [CrossRef]
16. Ge, Z.; Gu, H.; Li, Q.; Fan, C. Concept and Development of Framework Nucleic Acids. *J. Am. Chem. Soc.* **2018**, *140*, 17808–17819. [CrossRef]
17. Ouyang, X.; Wang, M.; Guo, L.; Cui, C.; Liu, T.; Ren, Y.; Zhao, Y.; Ge, Z.; Guo, X.; Xie, G.; et al. DNA Nanoribbon-Templated Self-Assembly of Ultrasmall Fluorescent Copper Nanoclusters with Enhanced Luminescence. *Angew. Chem. Int. Ed.* **2020**, *59*, 11836–11844. [CrossRef] [PubMed]
18. Liu, N.; Liedl, T. DNA-Assembled Advanced Plasmonic Architectures. *Chem. Rev.* **2018**, *118*, 3032–3053. [CrossRef]
19. Chen, S.; Xu, Z.; Yang, W.; Lin, X.; Li, J.; Li, J.; Yang, H. Logic-Gate-Actuated DNA-Controlled Receptor Assembly for the Programmable Modulation of Cellular Signal Transduction. *Angew. Chem. Int. Ed.* **2019**, *58*, 18186–18190. [CrossRef]
20. Holliger, P.; Hudson, P.J. Engineered antibody fragments and the rise of single domains. *Nat. Biotechnol.* **2005**, *23*, 1126–1136. [CrossRef]
21. Chiu, G.N.; Edwards, L.A.; Kapanen, A.I.; Malinen, M.M.; Dragowska, W.H.; Warburton, C.; Chikh, G.G.; Fang, K.Y.; Tan, S.; Sy, J.; et al. Modulation of cancer cell survival pathways using multivalent liposomal therapeutic antibody constructs. *Mol. Cancer Ther.* **2007**, *6*, 844–855. [CrossRef]
22. Yan, G.; Wang, K.; Shao, Z.; Luo, L.; Song, Z.; Chen, J.; Jin, R.; Deng, X.; Wang, H.; Cao, Z.; et al. Artificial antibody created by conformational reconstruction of the complementary-determining region on gold nanoparticles. *Proc. Natl. Acad. Sci. USA* **2018**, *115*, E34–E43. [CrossRef] [PubMed]
23. Kiessling, L.L.; Gestwicki, J.E.; Strong, L.E. Synthetic multivalent ligands in the exploration of cell-surface interactions. *Curr. Opin. Chem. Biol.* **2000**, *4*, 696–703. [CrossRef] [PubMed]
24. Kawai, T.; Akira, S. The role of pattern-recognition receptors in innate immunity: Update on Toll-like receptors. *Nat. Immunol.* **2010**, *11*, 373–384. [CrossRef] [PubMed]
25. Huang, D.; Patel, K.; Perez-Garrido, S.; Marshall, J.F.; Palma, M. DNA Origami Nanoarrays for Multivalent Investigations of Cancer Cell Spreading with Nanoscale Spatial Resolution and Single-Molecule Control. *ACS Nano* **2019**, *13*, 728–736. [CrossRef]
26. Dutta, P.K.; Zhang, Y.; Blanchard, A.T.; Ge, C.; Rushdi, M.; Weiss, K.; Zhu, C.; Ke, Y.; Salaita, K. Programmable Multivalent DNA-Origami Tension Probes for Reporting Cellular Traction Forces. *Nano Lett.* **2018**, *18*, 4803–4811. [CrossRef]
27. Tan, W.; Donovan, M.J.; Jiang, J. Aptamers from cell-based selection for bioanalytical applications. *Chem. Rev.* **2013**, *113*, 2842–2862. [CrossRef]
28. Iliuk, A.B.; Hu, L.; Tao, W.A. Aptamer in bioanalytical applications. *Anal. Chem.* **2011**, *83*, 4440–4452. [CrossRef] [PubMed]
29. Chen, S.; Zhang, L.; Yuan, Q.; Tan, J. Current Advances in Aptamer-based Biomolecular Recognition and Biological Process Regulation. *Chem. Res. Chin. Univ.* **2022**, *38*, 847–855. [CrossRef]
30. Lin, M.; Zhang, J.; Wan, H.; Yan, C.; Xia, F. Rationally Designed Multivalent Aptamers Targeting Cell Surface for Biomedical Applications. *ACS Appl. Mater. Interfaces* **2021**, *13*, 9369–9389. [CrossRef]

31. Aghebat Rafat, A.; Sagredo, S.; Thalhammer, M.; Simmel, F.C. Barcoded DNA origami structures for multiplexed optimization and enrichment of DNA-based protein-binding cavities. *Nat. Chem.* **2020**, *12*, 852–859. [CrossRef] [PubMed]
32. Kong, G.; Xiong, M.; Liu, L.; Hu, L.; Meng, H.M.; Ke, G.; Zhang, X.B.; Tan, W. DNA origami-based protein networks: From basic construction to emerging applications. *Chem. Soc. Rev.* **2021**, *50*, 1846–1873. [CrossRef] [PubMed]
33. Zhao, D.; Kong, Y.; Zhao, S.; Xing, H. Engineering Functional DNA-Protein Conjugates for Biosensing, Biomedical, and Nanoassembly Applications. *Top. Curr. Chem.* **2020**, *378*, 41. [CrossRef]
34. Mallik, L.; Dhakal, S.; Nichols, J.; Mahoney, J.; Dosey, A.; Jiang, S.; Sunahara, R.; Skiniotis, G.; Walter, N. Electron Microscopic Visualization of protein assemblies on flatted DNA origami. *ACS Nano* **2015**, *9*, 7133–7141. [CrossRef]
35. Kuzuya, A.; Kimura, M.; Numajiri, K.; Koshi, N.; Ohnishi, T.; Okada, F.; Komiyama, M. Precisely Programmed and Robust 2D Streptavidin Nanoarrays by Using Periodical Nanometer-Scale Wells Embedded in DNA Origami Assembly. *Chembiochem* **2009**, *10*, 1811–1815. [CrossRef] [PubMed]
36. Numajiri, K.; Kimura, M.; Kuzuya, A.; Komiyama, M. Stepwise and reversible nanopatterning of proteins on a DNA origami scaffold. *Chem. Commun.* **2010**, *46*, 5127–5129. [CrossRef]
37. Shen, W.; Zhong, H.; Neff, D.; Norton, M. NTA Directed Protein Nanopatterning on DNA Origami Nanoconstructs. *J. Am. Chem. Soc.* **2009**, *131*, 6660–6661. [CrossRef]
38. Ouyang, X.; De Stefano, M.; Krissanaprasit, A.; Bank Kodal, A.L.; Bech Rosen, C.; Liu, T.; Helmig, S.; Fan, C.; Gothelf, K.V. Docking of Antibodies into the Cavities of DNA Origami Structures. *Angew. Chem. Int. Ed.* **2017**, *56*, 14423–14427. [CrossRef]
39. Nielsen, T.B.; Thomsen, R.P.; Mortensen, M.R.; Kjems, J.; Nielsen, P.F.; Nielsen, T.E.; Kodal AL, B.; Clo, E.; Gothelf, K.V. Peptide-Directed DNA-Templated Protein Labelling for The Assembly of a Pseudo-IgM. *Angew. Chem. Int. Ed.* **2019**, *58*, 9068–9072. [CrossRef]
40. Nakata, E.; Liew, F.F.; Uwatoko, C.; Kiyonaka, S.; Mori, Y.; Katsuda, Y.; Endo, M.; Sugiyama, H.; Morii, T. Zinc-finger proteins for site-specific protein positioning on DNA-origami structures. *Angew. Chem. Int. Ed.* **2012**, *51*, 2421–2424. [CrossRef]
41. Praetorius, F.; Dietz, H. Self-assembly of genetically encoded DNA-protein hybrid nanoscale shapes. *Science* **2017**, *355*, eaam5488. [CrossRef] [PubMed]
42. Fisher, P.D.E.; Shen, Q.; Akpinar, B.; Davis, L.K.; Chung, K.K.H.; Baddeley, D.; Saric, A.; Melia, T.J.; Hoogenboom, B.W.; Lin, C.; et al. A Programmable DNA Origami Platform for Organizing Intrinsically Disordered Nucleoporins within Nanopore Confinement. *ACS Nano* **2018**, *12*, 1508–1518. [CrossRef] [PubMed]
43. Fu, J.; Liu, M.; Liu, Y.; Woodbury, N.W.; Yan, H. Interenzyme substrate diffusion for an enzyme cascade organized on spatially addressable DNA nanostructures. *J. Am. Chem. Soc.* **2012**, *134*, 5516–5519. [CrossRef]
44. El-Sagheer, A.H.; Brown, T. Click chemistry with DNA. *Chem. Soc. Rev.* **2010**, *39*, 1388–1405. [CrossRef]
45. Tang, W.; Becker, M.L. "Click" reactions: A versatile toolbox for the synthesis of peptide-conjugates. *Chem. Soc. Rev.* **2014**, *43*, 7013–7039. [CrossRef]
46. Kolb, H.C.; Finn, M.G.; Sharpless, K.B. Click Chemistry: Diverse Chemical Function from a Few Good Reactions. *Angew. Chem. Int. Ed.* **2001**, *40*, 2004–2021. [CrossRef]
47. Laughlin, S.T.; Bertozzi, C.R. Metabolic labeling of glycans with azido sugars and subsequent glycan-profiling and visualization via Staudinger ligation. *Nat. Protoc.* **2007**, *2*, 2930–2944. [CrossRef]
48. Knappe, G.A.; Wamhoff, E.C.; Read, B.J.; Irvine, D.J.; Bathe, M. In Situ Covalent Functionalization of DNA Origami Virus-like Particles. *ACS Nano* **2021**, *15*, 14316–14322. [CrossRef] [PubMed]
49. Nakata, E.; Dinh, H.; Ngo, T.A.; Saimura, M.; Morii, T. A modular zinc finger adaptor accelerates the covalent linkage of proteins at specific locations on DNA nanoscaffolds. *Chem. Commun.* **2015**, *51*, 1016–1019. [CrossRef]
50. Koßmann, K.J.; Ziegler, C.; Angelin, A.; Meyer, R.; Skoupi, M.; Rabe, K.S.; Niemeyer, C.M. A Rationally Designed Connector for Assembly of Protein-Functionalized DNA Nanostructures. *Chembiochem* **2016**, *17*, 1102–1106. [CrossRef]
51. Saccà, B.; Meyer, R.; Erkelenz, M.; Kiko, K.; Arndt, A.; Schroeder, H.; Rabe, K.S.; Niemeyer, C.M. Orthogonal Protein Decoration of DNA Origami. *Angew. Chem. Int. Ed.* **2010**, *49*, 9378–9383. [CrossRef] [PubMed]
52. Fasting, C.; Schalley, C.A.; Weber, M.; Seitz, O.; Hecht, S.; Koksch, B.; Dernedde, J.; Graf, C.; Knapp, E.-W.; Haag, R. Multivalency as a Chemical Organization and Action Principle. *Angew. Chem. Int. Ed.* **2012**, *51*, 10472–10498. [CrossRef] [PubMed]
53. Mammen, M.; Choi, S.-K.; Whitesides, G.M. Polyvalent Interactions in Biological Systems: Implications for Design and Use of Multivalent Ligands and Inhibitors. *Angew. Chem. Int. Ed.* **1998**, *37*, 2754–2794. [CrossRef]
54. Muller, M.; Lauster, D.; Wildenauer, H.H.K.; Herrmann, A.; Block, S. Mobility-Based Quantification of Multivalent Virus-Receptor Interactions: New Insights Into Influenza A Virus Binding Mode. *Nano Lett.* **2019**, *19*, 1875–1882. [CrossRef]
55. Czajkowskya, D.M.; Shaoa, Z. The human IgM pentamer is a mushroom-shaped molecule with a flexural bias. *Proc. Natl. Acad. Sci. USA* **2009**, *106*, 14960–14965. [CrossRef]
56. Hernandez-Lopez, R.A.; Yu, W.; Cabral, K.A.; Creasey, O.A.; Pazmino, M.D.P.L.; Tonai, Y.; De Guzman, A.; Mäkelä, A.; Saksela, K.; Gartner, Z.J.; et al. T cell circuits that sense antigen density with an ultrasensitive threshold. *Science* **2021**, *371*, 1166–1171. [CrossRef]
57. Goodman, R.P.; Berry, R.M.; Turberfield, A.J. The single-step synthesis of a DNA tetrahedron. *Chem. Commun.* **2004**, *12*, 1372–1373. [CrossRef] [PubMed]

58. Lee, H.; Lytton-Jean, A.K.; Chen, Y.; Love, K.T.; Park, A.I.; Karagiannis, E.D.; Sehgal, A.; Querbes, W.; Zurenko, C.S.; Jayaraman, M.; et al. Molecularly self-assembled nucleic acid nanoparticles for targeted in vivo siRNA delivery. *Nat. Nanotechnol.* **2012**, *7*, 389–393. [CrossRef]
59. Tian, T.; Xiao, D.; Zhang, T.; Li, Y.; Shi, S.; Zhong, W.; Gong, P.; Liu, Z.; Li, Q.; Lin, Y. A Framework Nucleic Acid Based Robotic Nanobee for Active Targeting Therapy. *Adv. Funct. Mater.* **2021**, *31*, 2007342. [CrossRef]
60. Zhang, T.; Tian, T.; Lin, Y. Functionalizing Framework Nucleic-Acid-Based Nanostructures for Biomedical Application. *Adv. Mater.* **2022**, *34*, e2107820. [CrossRef]
61. Li, J.; Pei, H.; Zhu, B.; Liang, L.; Wei, M.; He, Y.; Chen, N.; Li, D.; Huang, Q.; Fan, C. Self-Assembled Multivalent DNA Nanostructures for Noninvasive Intracellular Delivery of Immunostimulatory CpG Oligonucleotides. *ACS Nano* **2011**, *5*, 8783–8789. [CrossRef]
62. Liu, X.; Yan, H.; Liu, Y.; Chang, Y. Targeted cell-cell interactions by DNA nanoscaffold-templated multivalent bispecific aptamers. *Small* **2011**, *7*, 1673–1682. [CrossRef] [PubMed]
63. Dey, S.; Fan, C.; Gothelf, K.V.; Li, J.; Lin, C.; Liu, L.; Liu, N.; Nijenhuis, M.A.D.; Saccà, B.; Simmel, F.C.; et al. DNA origami. *Nat. Rev. Methods Prim.* **2021**, *1*, 13. [CrossRef]
64. Zhang, J.; Xu, Y.; Chen, M.; Huang, Y.; Song, T.; Yang, C.; Yang, Y.; Song, Y. Elucidating the Effect of Nanoscale Receptor-Binding Domain Organization on SARS-CoV-2 Infection and Immunity Activation with DNA Origami. *J. Am. Chem. Soc.* **2022**, *144*, 21295–21303. [CrossRef] [PubMed]
65. Zhang, J.; Lin, B.; Wu, L.; Huang, M.; Li, X.; Zhang, H.; Song, J.; Wang, W.; Zhao, G.; Song, Y.; et al. DNA Nanolithography Enables a Highly Ordered Recognition Interface in a Microfluidic Chip for the Efficient Capture and Release of Circulating Tumor Cells. *Angew. Chem. Int. Ed.* **2020**, *59*, 14115–14119. [CrossRef] [PubMed]
66. Li, W.; Yang, X.; He, L.; Wang, K.; Wang, Q.; Huang, J.; Liu, J.; Wu, B.; Xu, C. Self-Assembled DNA Nanocentipede as Multivalent Drug Carrier for Targeted Delivery. *ACS Appl. Mater. Interfaces* **2016**, *8*, 25733–25740. [CrossRef]
67. Zhu, G.; Hu, R.; Zhao, Z.; Chen, Z.; Zhang, X.; Tan, W. Noncanonical self-assembly of multifunctional DNA nanoflowers for biomedical applications. *J. Am. Chem. Soc.* **2013**, *135*, 16438–16445. [CrossRef]
68. Simons, K.; Toomre, D. Lipid rafts and signal transduction. *Nat. Rev.* **2000**, *1*, 31–41. [CrossRef]
69. de Castro, M.A.G.; Wildhagen, H.; Sograte-Idrissi, S.; Hitzing, C.; Binder, M.; Trepel, M.; Engels, N.; Opazo, F. Differential organization of tonic and chronic B cell antigen receptors in the plasma membrane. *Nat. Commun.* **2019**, *10*, 820. [CrossRef]
70. Rinker, S.; Ke, Y.; Liu, Y.; Chhabra, R.; Yan, H. Self-assembled DNA nanostructures for distance-dependent multivalent ligand-protein binding. *Nat. Nanotechnol.* **2008**, *3*, 418–422. [CrossRef]
71. Shaw, A.; Lundin, V.; Petrova, E.; Fördős, F.; Benson, E.; Al-Amin, R.A.; Herland, A.; Blokzijl, A.; Högberg, B.; Teixeira, A. Spatial control of membrane receptor function using ligand nanocalipers. *Nat. Methods* **2014**, *11*, 841–846. [CrossRef] [PubMed]
72. Rosier, B.; Markvoort, A.J.; Gumi Audenis, B.; Roodhuizen, J.A.L.; Den Hamer, A.; Brunsveld, L.; De Greef, T.F.A. Proximity-induced caspase-9 activation on a DNA origami-based synthetic apoptosome. *Nat. Catal.* **2020**, *3*, 295–306. [CrossRef] [PubMed]
73. Zhang, K.; Deng, R.; Sun, Y.; Zhang, L.; Li, J. Reversible control of cell membrane receptor function using DNA nano-spring multivalent ligands. *Chem. Sci.* **2017**, *8*, 7098–7105. [CrossRef] [PubMed]
74. Shaw, A.; Hoffecker, I.T.; Smyrlaki, I.; Rosa, J.; Grevys, A.; Bratlie, D.; Sandlie, I.; Michaelsen, T.E.; Andersen, J.T.; Högberg, B. Binding to nanopatterned antigens is dominated by the spatial tolerance of antibodies. *Nat. Nanotechnol.* **2019**, *14*, 184–190. [CrossRef]
75. Zhang, P.; Liu, X.; Liu, P.; Wang, F.; Ariyama, H.; Ando, T.; Lin, J.; Wang, L.; Hu, J.; Li, B.; et al. Capturing transient antibody conformations with DNA origami epitopes. *Nat. Commun.* **2020**, *11*, 3114. [CrossRef]
76. Veneziano, R.; Moyer, T.J.; Stone, M.B.; Wamhoff, E.-C.; Read, B.J.; Mukherjee, S.; Shepherd, T.R.; Das, J.; Schief, W.R.; Irvine, D.J.; et al. Role of nanoscale antigen organization on B-cell activation probed using DNA origami. *Nat. Nanotechnol.* **2020**, *15*, 716–723. [CrossRef]
77. Harrison, S.C. Looking Inside Adenovirus. *Science* **2010**, *329*, 1026–1027. [CrossRef]
78. Joglekar, A.V.; Li, G. T cell antigen discovery. *Nat. Methods* **2021**, *18*, 873–880. [CrossRef]
79. Li, M.; Ding, H.; Lin, M.; Yin, F.; Song, L.; Mao, X.; Li, F.; Ge, Z.; Wang, L.; Zuo, X.; et al. DNA Framework-Programmed Cell Capture via Topology-Engineered Receptor-Ligand Interactions. *J. Am. Chem. Soc.* **2019**, *141*, 18910–18915. [CrossRef]
80. Yin, F.; Li, M.; Mao, X.; Li, F.; Xiang, X.; Li, Q.; Wang, L.; Zuo, X.; Fan, C.; Zhu, Y. DNA Framework-Based Topological Cell Sorters. *Angew. Chem. Int. Ed.* **2020**, *59*, 10406–10410. [CrossRef]
81. Shen, Q.; Feng, Q.; Wu, C.; Xiong, Q.; Tian, T.; Yuan, S.; Shi, J.; Bedwell, G.J.; Yang, R.; Aiken, C.; et al. Modeling HIV-1 nuclear entry with nucleoporin-gated DNA-origami channels. *Nat. Struct. Mol. Biol.* **2023**, *30*, 425–435. [CrossRef] [PubMed]
82. Kwon, P.S.; Ren, S.; Kwon, S.J.; Kizer, M.E.; Kuo, L.; Xie, M.; Zhu, D.; Zhou, F.; Zhang, F.; Kim, D.; et al. Designer DNA architecture offers precise and multivalent spatial pattern-recognition for viral sensing and inhibition. *Nat. Chem.* **2020**, *12*, 26–35. [CrossRef] [PubMed]
83. Zhang, J.; Xu, Y.; Huang, Y.; Sun, M.; Liu, S.; Wan, S.; Chen, H.; Yang, C.; Yang, Y.; Song, Y. Spatially Patterned Neutralizing Icosahedral DNA Nanocage for Efficient SARS-CoV-2 Blocking. *J. Am. Chem. Soc.* **2022**, *144*, 13146–13153. [CrossRef] [PubMed]
84. Dai, Z.; Liu, S.; Ju, H.; Chen, H. Direct electron transfer and enzymatic activity of hemoglobin in a hexagonal mesoporous silica matrix. *Biosens. Bioelectron.* **2004**, *19*, 861–867. [CrossRef]

85. Liu, S.; Dai, Z.; Chen, H.; Ju, H. Immobilization of hemoglobin on zirconium dioxide nanoparticles for preparation of a novel hydrogen peroxide biosensor. *Biosens. Bioelectron.* **2004**, *19*, 963–969. [CrossRef]
86. Dong, H.; Gao, W.; Yan, F.; Ji, H.; Ju, H. Fluorescence resonance energy transfer between quantum dots and graphene oxide for sensing biomolecules. *Anal. Chem.* **2010**, *82*, 5511–5517. [CrossRef]
87. Lei, J.; Ju, H. Signal amplification using functional nanomaterials for biosensing. *Chem. Soc. Rev.* **2012**, *41*, 2122–2134. [CrossRef]
88. Feng, Q.M.; Guo, Y.H.; Xu, J.; Chen, H. A surface-confined DNA assembly amplification strategy on DNA nanostructural scaffold for electrochemiluminescence biosensing. *Biosens. Bioelectron.* **2018**, *100*, 571–576. [CrossRef]
89. Mao, M.; Lin, Z.; Chen, L.; Zou, Z.; Zhang, J.; Dou, Q.; Wu, J.; Chen, J.; Wu, M.; Niu, L.; et al. Modular DNA-Origami-Based Nanoarrays Enhance Cell Binding Affinity through the "Lock-and-Key" Interaction. *J. Am. Chem. Soc.* **2023**, *145*, 5447–5455. [CrossRef]
90. Zhong, L.; Cai, S.; Huang, Y.; Yin, L.; Yang, Y.; Lu, C.; Yang, H. DNA Octahedron-Based Fluorescence Nanoprobe for Dual Tumor-Related mRNAs Detection and Imaging. *Anal. Chem.* **2018**, *90*, 12059–12066. [CrossRef]
91. Xue, C.; Zhang, S.; Li, C.; Yu, X.; Ouyang, C.; Lu, Y.; Wu, Z.S. Y-Shaped Backbone-Rigidified Triangular DNA Scaffold-Directed Stepwise Movement of a DNAzyme Walker for Sensitive MicroRNA Imaging within Living Cells. *Anal. Chem.* **2019**, *91*, 15678–15685. [CrossRef] [PubMed]
92. Ding, H.; Li, J.; Chen, N.; Hu, X.; Yang, X.; Guo, L.; Li, Q.; Zuo, X.; Wang, L.; Ma, Y.; et al. DNA nanostructure-programmed like-charge attraction at the cell-membrane interface. *ACS Cent. Sci.* **2018**, *4*, 1344–1351. [CrossRef] [PubMed]
93. Peng, X.; Fang, S.; Ji, B.; Li, M.; Song, J.; Qiu, L.; Tan, W. DNA nanostructure-programmed cell entry via corner angle-mediated molecular interaction with membrane receptors. *Nano Lett.* **2021**, *21*, 6946–6951. [CrossRef]
94. Guo, L.; Zhang, Y.; Wang, Y.; Xie, M.; Dai, J.; Qu, Z.; Zhou, M.; Cao, S.; Shi, J.; Wang, L.; et al. Directing Multivalent Aptamer-Receptor Binding on the Cell Surface with Programmable Atom-Like Nanoparticles. *Angew. Chem. Int. Ed.* **2022**, *61*, e202117168.
95. Xie, M.; Guo, L.; Xing, S.; Cao, S.; Zhao, Z.; Liang, K.; Li, J.; Luo, S.; Zhang, Y.; Wang, L. Cell imaging with multi-color DNA framework probes. *Chem. Commun.* **2021**, *57*, 11318–11321. [CrossRef]
96. Li, S.; Jiang, Q.; Liu, S.; Zhang, Y.; Tian, Y.; Song, C.; Wang, J.; Zou, Y.; Anderson, G.J.; Han, J.Y.; et al. A DNA nanorobot functions as a cancer therapeutic in response to a molecular trigger in vivo. *Nat. Biotechnol.* **2018**, *36*, 258–264. [CrossRef]
97. Han, L.; Wang, Y.; Tang, W.; Liu, J.; Ding, B. Bioimaging Based on Nucleic Acid Nanostructures. *Chem. Res. Chin. Univ.* **2021**, *37*, 823–828. [CrossRef]
98. Wang, J.; Song, J.; Zhang, X.; Wang, S.M.; Kang, B.; Li, X.L.; Chen, H.; Xu, J. DNA-Programed Plasmon Rulers Decrypt Single-Receptor Dimerization on Cell Membrane. *J. Am. Chem. Soc.* **2023**, *145*, 1273–1284. [CrossRef]
99. Xie, S.; Sun, W.; Fu, T.; Liu, X.; Chen, P.; Qiu, L.; Qu, F.; Tan, W. Aptamer-Based Targeted Delivery of Functional Nucleic Acids. *J. Am. Chem. Soc.* **2023**, *145*, 7677–7691. [CrossRef]
100. Li, Q.; Zhao, D.; Shao, X.; Lin, S.; Xie, X.; Liu, M.; Ma, W.; Shi, S.; Lin, Y. Aptamer-Modified Tetrahedral DNA Nanostructure for Tumor-Targeted Drug Delivery. *ACS Appl. Mater. Interfaces* **2017**, *9*, 36695–36701. [CrossRef]
101. Shi, S.; Fu, W.; Lin, S.; Tian, T.; Li, S.; Shao, X.; Zhang, Y.; Zhang, T.; Tang, Z.; Zhou, Y.; et al. Targeted and effective glioblastoma therapy via aptamer-modified tetrahedral framework nucleic acid-paclitaxel nanoconjugates that can pass the blood brain barrier. *Nanomedicine* **2019**, *21*, 102061. [CrossRef] [PubMed]
102. Ge, Z.; Guo, L.; Wu, G.; Li, J.; Sun, Y.; Hou, Y.; Shi, J.; Song, S.; Wang, L.; Fan, C.; et al. DNA Origami-Enabled Engineering of Ligand-Drug Conjugates for Targeted Drug Delivery. *Small* **2020**, *16*, e1904857. [CrossRef] [PubMed]
103. Ren, K.; Liu, Y.; Wu, J.; Zhang, Y.; Zhu, J.; Yang, M.; Ju, H. A DNA dual lock-and-key strategy for cell-subtype-specific siRNA delivery. *Nat. Commun.* **2016**, *7*, 13580. [CrossRef] [PubMed]
104. Zhang, P.; Fischer, A.; Ouyang, Y.; Wang, J.; Sohn, Y.S.; Nechushtai, R.; Pikarsky, E.; Fan, C.; Willner, I. Aptamer-modified DNA tetrahedra-gated metal–organic framework nanoparticle carriers for enhanced chemotherapy or photodynamic therapy. *Chem. Sci.* **2021**, *12*, 14473–14483. [CrossRef] [PubMed]
105. Zhang, Y.; Chen, W.; Fang, Y.; Zhang, X.; Liu, Y.; Ju, H. Activating a DNA Nanomachine via Computation across Cancer Cell Membranes for Precise Therapy of Solid Tumors. *J. Am. Chem. Soc.* **2021**, *143*, 15233–15242. [CrossRef]
106. Du, Y.; Lyu, Y.; Lin, J.; Ma, C.; Zhang, Q.; Zhang, Y.; Qiu, L.; Tan, W. Membrane-anchored DNA nanojunctions enable closer antigen-presenting cell-T-cell contact in elevated T-cell receptor triggering. *Nat. Nanotechnol.* **2023**. [CrossRef]
107. Shen, F.; Xiong, Z.; Wu, Y.; Peng, R.; Wang, Y.; Sun, L.; Fan, C.; Liu, Z. Precise Epitope Organization with Self-adjuvant Framework Nucleic Acid for Efficient COVID-19 Peptide Vaccine Construction. *Angew. Chem. Int. Ed.* **2023**, *62*, e202301147. [CrossRef]
108. Sigl, C.; Willner, E.M.; Engelen, W.; Kretzmann, J.A.; Sachenbacher, K.; Liedl, A.; Kolbe, F.; Wilsch, F.; Aghvami, S.A.; Protzer, U.; et al. Programmable icosahedral shell system for virus trapping. *Nat. Mater.* **2021**, *20*, 1281–1289. [CrossRef]

chemosensors

Article

A Paper-Based Multicolor Colorimetric Aptasensor for the Visual Determination of Multiple Sulfonamides Based on Aptamer-Functionalized Magnetic Beads and NADH–Ascorbic Acid-Mediated Gold Nanobipyramids

Meiling Ping [1], Wenchao Lv [1], Chen Yang [1], Qian Chen [1], Zongwen Wang [2,*] and Fengfu Fu [1,*]

1 Key Laboratory for Analytical Science of Food Safety and Biology of MOE, College of Chemistry, Fuzhou University, Fuzhou 350108, China
2 State Key Laboratory of Ecological Pest Control for Fujian and Taiwan Crops, College of Plant Protection, Fujian Agriculture and Forestry University, Fuzhou 350002, China
* Correspondence: 000q813035@fafu.edu.cn (Z.W.); fengfu@fzu.edu.cn (F.F.)

Abstract: It is crucial that simple and high-throughput methods for determining multiple, or groups of, sulfonamides (SAs) be developed since they are widely used in animal husbandry and aquaculture. We developed a paper-based multicolor colorimetric aptasensor to detect 3 SAs: sulfaquinoxaline (SQ), sulfamethoxypyridazine (SMP) and sulfamethoxydiazine (SMD). Using a broad-specificity aptamer as a bioreceptor, we reduced the growth of nicotinamide adenine dinucleotide I (NADH)–ascorbic acid (AA)-mediated gold nanobipyramids (AuNBPs) to generate a multicolor signal. We also used a paper-based analytical device (PAD) system to deposit AuNBPs for a sensitive color signal read out. The aptasensor can detect more color changes corresponding to the concentrations of SQ, SMP and SMD and has higher sensitivity, better specificity and stability. It can also be used to determine SQ, SMP and SDM individually, or collectively, or any two together with a visual detection limit of 0.3–1.0 μM, a spectrometry quantification limit (LOQ) of 0.3–0.5 μM and a spectrometry detection limits (LOD) of 0.09–0.15 μM. The aptasensor was successfully used to determine SQ, SMP and SDM in fish muscle with a recovery of 89–94% and a RSD $n = 5$) < 8%, making it a promising method for the rapid screening of total SQ SMP and SDM residue in seafood.

Keywords: aptasensor; antibiotics residue seafood; sulfonamides; visual detection

Citation: Ping, M.; Lv, W.; Yang, C.; Chen, Q.; Wang, Z.; Fu, F. A Paper-Based Multicolor Colorimetric Aptasensor for the Visual Determination of Multiple Sulfonamides Based on Aptamer-Functionalized Magnetic Beads and NADH–Ascorbic Acid-Mediated Gold Nanobipyramids. *Chemosensors* **2023**, *11*, 386. https://doi.org/10.3390/chemosensors11070386

Academic Editors: António M. Peres and Danila Moscone

Received: 10 June 2023
Revised: 7 July 2023
Accepted: 8 July 2023
Published: 9 July 2023

1. Introduction

Sulfonamides (SAs), one of the oldest groups of synthetic antibiotics, offers low-cost and broad-spectrum activity against bacterial infection [1,2], and up to now they have been widely used as feed additives to treat infectious diseases in animal husbandry and aquaculture [3–5]. However, the inappropriate or excessive use of SAs inevitably results in excess residue in animal-derived food since animals cannot completely metabolize or degrade SAs [6–8]. Such excess residue poses a danger to human health [9,10], and thus a maximum SA residue limit (MRL) of 100 μg/Kg was established by the European Union to ensure consumption safety [11]. At present, the groups contain 19 commonly used SAs, and different SAs are usually used in rotation to avoid generating drug resistance. (Their full names, abbreviations and chemical structures are summarized in Figure S1 in Supporting Information, SI). Thus, it is essential to develop simple and high-throughput assays to determine multiple SAs, preferably groups of SAs, to achieve the rapid screening of residues in animal-derived food.

High-performance liquid chromatography–mass spectrometry (HPLC–MS) simultaneously determines groups of SAs with a high sensitivity, which makes it a main technique for determining SA residues [12–15]. However, HPLC–MS requires time-consuming pretreatment of samples and a costly apparatus, which do not meet the requirements of rapid screening. An immunoassay, including enzyme-linked immunosorbent assay (ELISA), is easy to use and does not require a costly apparatus. Thus, some immunoassays, including a few high-throughput ELISAs that use broad-specificity anti-SA antibodies as bioreceptors, have been developed for the rapid screening of multiple SA residues [16–21]. However, immunoassays have obvious weaknesses such as poor thermal stability, higher cost and risk of false positives. Compared to antibodies, aptamers offer advantages such as a similar or higher affinity to target molecules, higher stability, lower cost, and ease of modification. So far, some aptamer-based methods have been developed for the rapid determination of SAs [22–32]. Unfortunately, most previous aptamer-based methods determined only one kind of SA such as sulfadimethoxine (SDM) or sulfamethazine (SMZ) [22–32], not multiple or groups of SAs, which means they did meet the requirements for the rapid screening of SA residues in food. Recently, we isolated a broad-specificity SA aptamer and further established a colorimetric method for the visual determination of multiple SAs by using the isolated aptamer as bioreceptor together with dye displacement [33]. However, the high-throughput colorimetric method has poorer sensitivity because of the dye displacement.

Gold nanobipyramids (AuNBPs) generate strong light absorption and vivid color changes in response to the tunable aspect ratio due to localized surface plasmon resonance (LSPR) [34,35]. Therefore, colorimetric assays based on AuNBPs can exhibit vivid multiple color changes, which can be distinguished more easily by naked eye observation, giving them higher visual sensitivity and accuracy [36,37]. In particular, AuNBPs with reduced nicotinamide adenine dinucleotide I (NADH)–ascorbic acid (AA)-mediated growth have more colorful and clearer color changes corresponding to AA concentrations in the lower range [38], which provides a promising color signal for developing sensitive colorimetric methods. Encouraged by the above results, we developed a paper-based multicolor colorimetric aptasensor based on a broad-specificity SA aptamer, the NADH–AA-mediated AuNBP growth system with a paper-based analytical device (PAD), to provide a sensitive, high-throughput assay for the rapid screening and semi-quantitative determination of multiple SA residues in animal-derived food by naked eye observation.

2. Materials and Methods

2.1. Chemicals and Apparatus

The 19 commonly used SAs—sulfaquinoxaline (SQ), sulfamethoxypyridazine (SMP), sulfamethoxydiazine (SMD), sulfachloropyridazine (SCP), sulfapyridine (SPD), sulfamethazine (SMZ), sulfadimethoxine (SDM), sulfisomidine (SIM), sulfamonomethoxine (SMM), sulfamerazine (SMR), sulfathiazole (ST), sulfabenzamide (SB), sulfadoxine (SDX), sulfadiazine (SD), sulfamethizole (SMT), sulfisoxazole (SIZ), sulfaguanidine (SG), sulfapyrazole (SPA) and sulfacetamide (SAA)—and 6 other antibiotics—cefotaxime (CTX), tetracycline (TC), enrofloxacin (ENR), kanamycin (KANA), ampicillin (AMP) and oxacillin (OXA)—were purchased from Aladdin Biochemical Technology Co., Ltd. (Shanghai, China). All oligonucleotides (see Table S1 in SI), which were purified with HPLC, were purchased from Sangon Biotech Co., Ltd. (Shanghai, China). The ELISA Kit for detection of SQ and SMD was obtained from Suzhou Kuaikang Biotechnology Co., Ltd. (Suzhou, China).

The buffers used in the experiment are as follow: TES buffer was a mixture of 10 mM Tris, 1.0 mM EDTA and 2.0 M NaCl (pH 9.5); TEST buffer is the TES buffer contained 0.01% Tween-20 (pH 9.5); binding buffer was a mixture of 10 mM Tris-HCl, 3.0 M NaCl, 10 mM $MgCl_2$ and 0.05% Tween-20 (pH 7.35–7.45); PBST buffer was a mixture of 2.0 mM KH_2PO_4, 10 mM $Na_2HPO_4 \cdot 12H_2O$, 137 mM NaCl, 3.0 mM KCl and 0.01% Tween-20 (pH 7.4); PBST–BSA buffer was a PBST buffer containing 0.02% bovine serum albumin (BSA); substrate buffer was a mixture of 0.6 M Tris-HCl buffer and 0.005% of BSA (pH 9.5). Other reagents, materials and apparatuses used in the experiment are summarized in the SI.

2.2. Fabrication of the Paper-Based Analytical Device (PAD) and Preparation of Gold Seeds

The PAD used in the experiment was fabricated according to the previous method with a minor modification [39]. Briefly, a highly absorbent cotton layer and then a nylon membrane were laid flat on an acrylic plate, and then a processed 96-hole acrylic plate was placed on the top layer (see Figure S2 in SI).

The solution for the gold seeds was prepared using the previous method [40]. Briefly, 0.25 mL of 25 mM freshly prepared $NaBH_4$ cold solution was fleetly added to a 10 mL solution, of 50 mM hexadecyl trimethyl ammonium chloride (CTAC), 5 mM citric acid and 0.25 mM $HAuCl_4$, under vigorous stirring for 2 min. Subsequently, the solution was incubated at 80 °C for 90 min to form gold seeds and then stored at 5 °C in the dark and used within 60 days.

2.3. Detection of Multiple SAs with PAD-Based Multicolor Colorimetric Aptasensor

First, 1.5 μM of broad-specificity aptamer sequence (aptamer) and 1.5 μM of link DNA 1 (cDNA-1) modified by a biotin on its 3′ end were mixed at equal volume in a TES buffer, incubated for 10 min at 95 °C and naturally cooled to room temperature to form the aptamer/cDNA-1 conjugate. Meanwhile, 3.0 μL of streptavidin-modified magnetic beads (MBs, ~1 μm, 10 g/L) were taken, put into a vial, washed 3 times with 30 μL of TEST buffer and re-dispersed in 30 μL of TES buffer. Then, 30 μL of aptamer/cDNA-1 conjugate was added, and the whole was incubated for 30 min at room temperature and vigorously agitated to immobilize the aptamer/cDNA-1 conjugate on the MB surfaces (aptamer/cDNA-1/MBs). The aptamer/cDNA-1/MBs were washed 2 times with 60 μL of TEST buffer and then re-dispersed in 30 μL of binding buffer. The aptamer/cDNA-1/MBs dispersion was stored at 5 °C for use within one month.

In another vial, 3.0 μL of the same MBs were added and washed 3 times with 30 μL of TEST buffer and then re-dispersed in 30 μL of TES buffer. Subsequently, 30 μL of 2.0 μM cDNA-1 solution (dissolved in TES buffer) was added, and the whole was incubated for 30 min under room temperature and vigorous agitation to immobilize cDNA-1 on the MB surfaces (cDNA-1/MBs). The cDNA-1/MBs were then washed 2 times with 60 μL of TEST buffer and then re-dispersed in 30 μL of binding buffer. The cDNA-1/MBs dispersion was stored at 5 °C for use within one month.

To determine the target SAs (SQ, SMP and SMD), 30 μL of the above aptamer/cDNA-1/MBs dispersion and 20 μL of target SAs standard or sample solution were mixed in a vial for 45 min to perform the specific binding of the SAs and aptamer, which will release SAs/aptamer complex into the solution from the aptamer/cDNA-1/MBs. The solution was separated from the MBs by a magnet and collected into a vial. Subsequently, 30 μL of the cDNA-1/MBs dispersion, 50 μL of link DNA 2 (cDNA-2) modified with biotin on its 5′ end, and 15 μL of binding buffer were added, and the whole solution was incubated for 1 h under room temperature and vigorous agitation to obtain the conjugate of cDNA-1/MBs, aptamer and cDNA-2 (cDNA-1/MBs/aptamer/cDNA-2), which was separated and collected with a magnet. After being washed two times with 60 μL binding buffer, the cDNA-1/MBs/aptamer/cDNA-2 was re-dispersed in 45 μL of PBST buffer. Next, 15 μL of streptavidin-modified alkaline phosphatase (streptavidin-ALP) solution was added, and the whole solution was agitated for 30 min to bind streptavidin-ALP to cDNA-1/MBs/aptamer/cDNA-2 (cDNA-1/MBs/aptamer/cDNA-2/ALP), which was washed 3 times with PBST-BSA buffer. Then 100 μL of 6.0 mM L-ascorbic acid 2-phosphate trisodium salt (AAP) dissolved in substrate buffer was added, and the whole solution was incubated for 20 min at 37 °C to perform the ALP enzymolysis of AAP to obtain AA. Finally, 5 μL of 660 mM HCl was added to stop the enzymolysis, and the AAP solution was separated and collected with a magnet.

In 86.4 μL of AAP enzymolysis solution, 86 μL of AuNBP growth solution (190 mM cetyltrimethylammonium bromide, 730 μM $HAuCl_4$, 200 μM $AgNO_3$, 40 mM HCl, and 2.0 mM NADH) and 0.6 μL of the abovementioned gold seed solution were added. The mixture was incubated for 15 min at 50 °C to obtain AuNBPs. After that, 120 μL of the

AuNBPs solution was dropwise added to the PAD with 5 min standing to deposit AuNBPs on a nylon membrane. After washing the membrane with pure water, the color of the deposited AuNBPs (in a wet state) was immediately observed by the naked eye and recorded by a camera. Simultaneously, the AuNBP-deposited nylon membrane was positioned in a 96-well plate frame, and the extinction spectra of the AuNBPs were measured by a microplate reader in the range of 400–1000 nm. The concentration of target SAs was calculated according to the AuNBPs color or the maximum localized surface plasmon resonance absorption wavelength (λ_{LSPR}).

2.4. Detection of Fish Muscle Sample

The SAs in the fish muscle samples were extracted using the Industry Standards for Entry-Exit Inspection and Quarantine of China (SN/T 5140-2019). Briefly, about 0.1 g of dried fish muscle, vacuum freeze-dried at −46 °C, was extracted with 5 mL of ethyl acetate for 10 min under ultrasonic agitation. Then, the extract was separated and collected via centrifugation, and the residue was extracted once again in the same manner. The two extracts were combined, and the total was evaporated to near dryness using a pressured nitrogen blowing concentrator. After dissolving the residue with 2 mL of binding buffer, 5 mL of hexane was added, and the whole extract was fully vortexed for 2 min to remove lipids by discarding the hexane solution. Finally, the remaining solution was filtered through a 0.22 μm filter and then directly used for the colorimetric detection of SAs according to the procedure in Section 2.3. The samples that spiked with SAs in advance were also analyzed by the same procedure to investigate the recovery. The concentration of each SQ, SMP and SMD in the sample was detected by using the corresponding SAs as calibration. The sum of SQ, SMP and SMD or the sum of any two was screened or semi-quantitatively detected using a mixture of SQ and SMD (1:1) as calibration.

3. Results and Discussion

3.1. The Principle of the PAD-Based Multicolor Colorimetric Aptasensor

The detailed strategy of the PAD-based multicolor colorimetric aptasensor is shown in Scheme 1. A competitive aptamer-binding system was employed to recognize the target SAs (SQ, SMP and SMD), a sensitive NADH-AA-mediated system was employed to regulate AuNBP growth to generate vivid multicolor signals, and a PAD system was used to deposit AuNBPs so that the color signals could be sensitively and stably read out. In the competitive aptamer-binding system, Xu et al. [33] reported a broad-specificity SA aptamer that was used as a bioreceptor and had binding affinity to SQ, SMP, SMD, SCP and SPD (higher to SQ, SMP and SMD and lower to SCP and SPD). A link probe 1 (cDNA-1) modified with biotin on the 3′ end, was designed to hybridize with part of the aptamer to form an aptamer/cDNA-1 conjugate. This was immobilized on the MB surfaces (aptamer/cDNA-1/MBs) via a specific interaction between the biotin and streptavidin to achieve convenient separation/collection. Upon the addition of target SAs (SQ, SMP and SMD), the aptamer specifically bound with the target SAs to form an SAs/aptamer complex and thus separated from the aptamer/cDNA-1/MBs. By removing the excess aptamer/cDNA-1/MBs with a magnet, the released SAs/aptamer in solution hybridized with the cDNA-1-modified MBs (MBs/cDNA-1) and link probe 2 (cDNA-2), modified with biotin on the 5′ end, to form an MB/cDNA-1/aptamer/cDNA-2 conjugate. It captured streptavidin-ALP via the biotin on the 5′ end of cDNA-2 to form an MB/cDNA-1/aptamer/cDNA-2/ALP conjugate, which specifically hydrolyzes AAP to generate AA. The presence of more target SAs resulted in more MBs/cDNA-1/aptamer/cDNA-2-ALP, thus generating more AA.

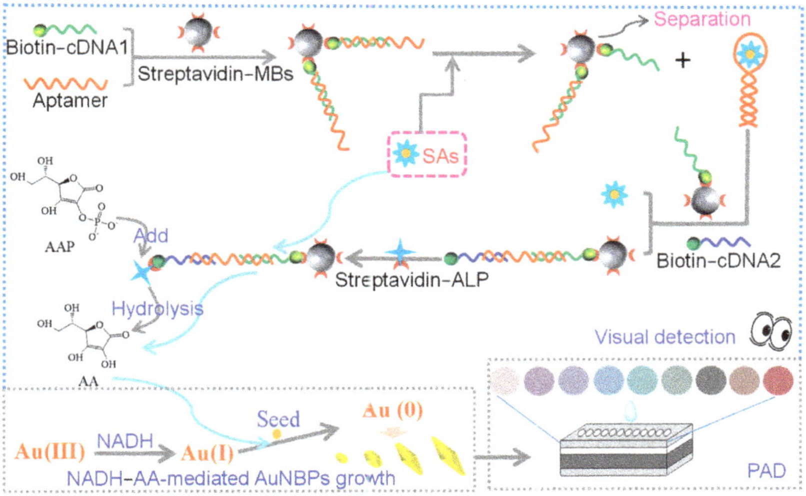

Scheme 1. Schematic illustration of the principle for the PAD-based multicolor aptasensor.

In the NADH-AA-mediated AuNBP growth system, the AA generated in a competitive aptamer-binding system sensitively promoted AuNBP growth to generate AA-corresponding colorful multicolor changes [38]. The PAD system was used to deposit AuNBPs on a nylon membrane to concentrate and stabilize its colors to achieve a sensitive read out of AuNBP multicolor changes. Thus, the combination of the above competitive aptamer-binding system, NADH–AA-mediated AuNBP growth system and PAD system provided high throughput and a sensitive and stable multicolor colorimetric sensing platform to determine each SQ, SMP and SMD, or the sum of the 3 SAs or the sum of any two.

3.2. The Feasibility of the Experimental Strategy

In this study, the cDNA-1 and cDNA-2 were designed to hybridize from the 5′ and 3′ ends aptamer, respectively, to form the competitive aptamer-binding system. To confirm that the hybridization of the cDNA-1, cDNA-2 and aptamer was successful, gel electrophoresis was used to characterize each hybrid product. The results in Figure S3 (see SI), show that, compared with cDNA-1 and aptamer separately, their hybrid products showed a band on the site that had shorter migration distances (about 80 pb), indicating that the hybridization was successful. Meanwhile, after the addition of a target SA (SQ), the hybrid products of cDNA-1 and aptamer also showed a weak band on the near site of aptamer, indicating that SQ could bind specifically with an aptamer to produce SA/aptamer. Similarly, compared with cDNA-2 and aptamer separately, their hybrid products also showed a band on the site with shorter migration distances of about 80 pb. The hybrid products of cDNA-1, cDNA-2 and aptamer showed a band that had the shortest migration distances (about 100 pb), indicating the successful hybridization among cDNA-1, cDNA-2 and aptamer. These experimental results demonstrated the feasibility of a competitive aptamer-binding system.

The feasibility of the NADH-AA-mediated AuNBP growth system was confirmed by regulating AuNBP growth with different AA concentrations and investigating the morphologies of AuNBPs with transmission electron microscopy (TEM). The TEM images in Figure 1 show that as the AA concentration increased from 38.5 to 227 μM, the shape of the AuNBPs changed step by step from spheroid to rice-like oval, to truncated bipyramid, to standard bipyramid. Correspondingly, the AuNBP solution exhibited vivid color changes from pale pink to blue-green to brownish red. Meanwhile, the λ_{LSPR} of the AuNBP solution gradually shifted from 504 to 796 nm, and the λ_{LSPR} showed a good linear relationship with AA concentrations in the range of 56.0–90.5 μM (see Figure 2). All the above results demonstrated that the NADH-AA–mediated system can mediate AuNBPs growth sensitively to generate many more color changes, which could make naked-eye observation more accurate and sensitive.

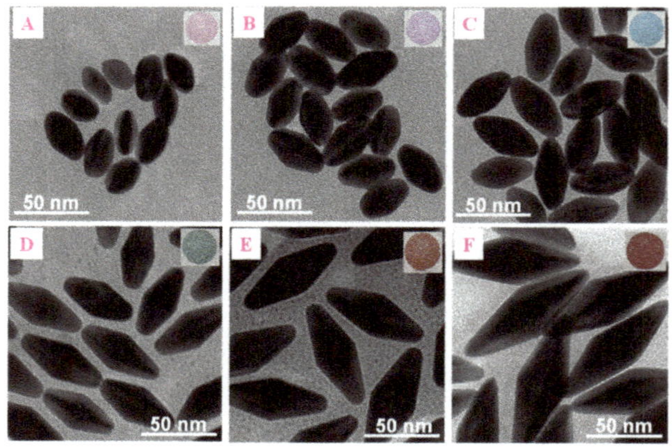

Figure 1. TEM images of AuNBPs obtained from a NADH–AA-mediated growth system at different AA concentrations. (**A**) 38.5, (**B**) 56.0, (**C**) 67.0, (**D**) 82.5, (**E**) 101, and (**F**) 227 μM. The inset is the color of the corresponding AuNBPs deposited on the PAD.

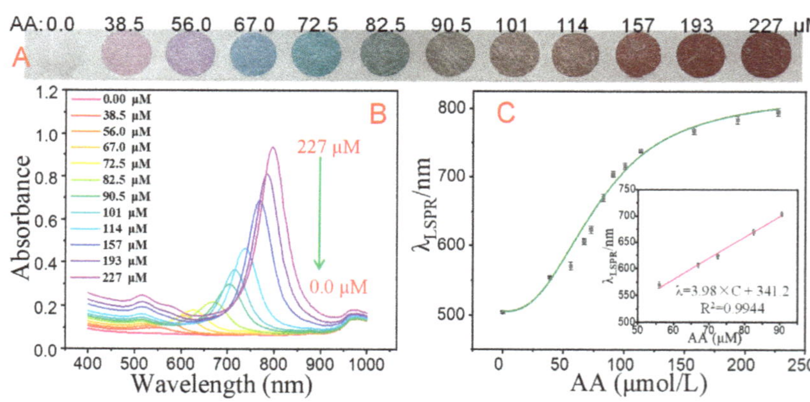

Figure 2. Photographs (**A**) and the extinction spectra of AuNBPs (**B**) obtained from the NADH–AA-mediated growth system at different AA concentrations (0.0, 38.5, 56.0, 67.0, 72.5, 82.5, 90.5, 101, 114, 157, 193 and 227 μM) and the variation of the λ_{LSPR} of AuNBPs versus AA concentrations (**C**). The insert in (**C**) is the linear relationship between the λ_{LSPR} of the AuNBPs and AA concentrations in range of 56.0–90.5 μM.

3.3. *Optimization of PAD-Based Multicolor Colorimetric Aptasensor*

In the NADH–AA-mediated AuNBPs growth system, acidity lowered the reduction ability of AA, which affected the growth rate of the AuNBPs, which on turn affected the λ_{LSPR} of the AuNBPs, i.e., the sensitivity of the multicolor colorimetric aptasensor. In this study, HCl was used to control acidity, and the concentration was optimized in the range of 440–880 mM. The experimental results (Figure S4 in SI) showed that the λ_{LSPR} of the AuNBPs increased when the HCl concentration increased from 440 to 660 mM, whereas it decreased when the concentration was higher than 660 mM. Thus, the acidity of the NADH–AA-mediated AuNBP growth system was controlled by adding HCl in a final concentration of 660 mM.

The streptavidin–ALP concentration in the competitive aptamer-binding system directly affected the ALP amount loaded onto the MBs/cDNA-1/aptamer/cDNA-2/ALP, which was used to hydrolyze AAP to generate AA and affecting the sensitivity and linear range of the aptasensor. The streptavidin–ALP concentration was optimized in the range of 21–27 μg/L. The experimental results (Figure S5 in SI) show the λ_{LSPR} shift of the generated AuNBPs:

$$\Delta\lambda_{LSPR} = \lambda_{LSPR(+)} - \lambda_{LSPR(-)},$$

where $\lambda_{LSPR\,(+)}$ is the λ_{LSPR} of the generated AuNBPs in the presence of target SAs, and $\lambda_{LSPR\,(-)}$ is the λ_{LSPR} in the absence of target SAs). The λ_{LSPR} shift had the largest value when the streptavidin–ALP concentration was 25.3 μg/L, which was selected as the optimal concentration.

3.4. Specificity of the Multicolor Colorimetric Aptasensor

As we mentioned above, the aptamer used in this study had a higher binding affinity to SQ, SMP and SMD, and a relatively lower binding affinity to SCP and SPD [33]. To evaluate the specificity of the multicolor colorimetric aptasensor to the targeted SAs (SQ, SMP and SMD), it was used to test a 1.0 μM concentration of the 19 commonly used SAs and a 100 μM concentration of the 6 types of antibiotics mentioned in Section 2.1. As Figure 3 shows, only SQ, SMP and SMD induced an obvious λ_{LSPR} shift ($\Delta\lambda_{LSPR}$) of the AuNBPs, and exhibited a blue or blue-green color. The other 16 SAs (SCP, SPD, SMZ, SDM, SIM, SMM, SMR, ST, SB, SDX, SD, SMT, SIZ, SG, SPA and SAA) exhibited the same color (pale pink) as that of the control AuNBPs with no λ_{LSPR} shift ($\Delta\lambda_{LSPR} \approx 0$). In particular, the 6 other antibiotics also did not induce an obvious λ_{LSPR} shift, and they exhibited the same color (pale pink) as that of the control AuNBPs even though the antibiotic concentrations were much higher than those of the 3 target SAs (SQ, SMP and SMD). The above results demonstrated that the proposed multicolor colorimetric aptasensor had good selectivity for SQ, SMP and SMD, and thus could be used to detect them singly, or the sum of all three or the sum of any two with no interference of other antibiotics.

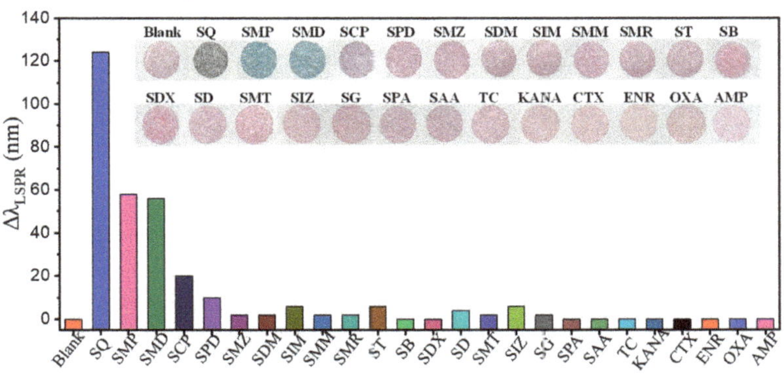

Figure 3. The photographs and $\Delta\lambda_{LSPR}$ of multicolor colorimetric aptasensor for detecting 19 commonly used different SAs and 6 other types of antibiotics. The SA concentration was 1.0 μM, and the that of the 6 other antibiotics was 100 μM.

3.5. Analytical Performance of the Multicolor Colorimetric Aptasensor

To investigate the analytical performance of the multicolor colorimetric aptasensor, different concentrations of SQ, SMP and SMD were determined. As the photographs and extinction spectra in Figure 4 show, when the SQ concentration increased from 0.0 to 2.5 μM, the colors of the AuNBPs changed step by step from pale pink to bluish violet, to green-blue, to green, to atrovirens, to lime-green, to brownish red to deep brown, and were accompanied by an AuNBP λ_{LSPR} of shift from 557 to 716 nm. When the concentration of SQ was 0.3 μM, the color change int the AuNBPs could be clearly confirmed with the naked eye, i.e., the visual detection limit (LOD) of the aptasensor was 0.3 μM for SQ. The good linear relationship between the λ_{LSPR} of the AuNBPs and SQ concentrations was observed in the range of 0.3–1.0 μM; thus, the spectrometry quantification limit (LOQ) was 0.3 μM (the lowest concentration in the linear range) and the spectrometry LOD was 0.09 μM (LOQ/3.3) for SQ (Table 1). As the concentrations of SMP and SMD increased, the AuNBPs exhibited similar stepped color changes and were accompanied by a shift in the AuNBP λ_{LSPR} from about 550 to 700 nm (see Figures S6 and S7 in SI). The visual detection limit of the aptasensor was 0.5 μM for SMP and 1.0 μM for SMD. The spectrometry LOQ was 0.3 μM for SMP and 0.5 μM for SMD, and the spectrometry LOD was calculated to be 0.09 μM for SMP and 0.15 μM for SMD (Table 1). The stability of the multicolor aptasensor was investigated by repeatedly detecting 0.5 μM of SQ and SMP, and 2.0 μM of SMD 5 times. The relative standard deviation (RSD, n = 5) was calculated to be 5% for SQ, and 6% for SMP and SMD.

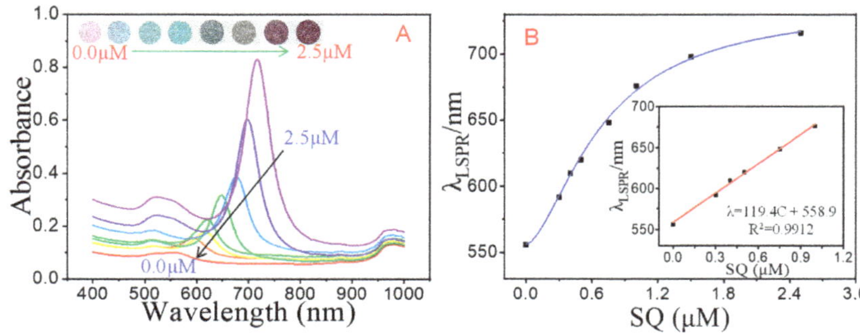

Figure 4. The photographs and UV-vis extinction spectra (**A**) of the multicolor colorimetric aptasensor for detecting different concentrations of SQ (0.0, 0.3, 0.4, 0.5, 0.75, 1.0, 1.5 and 2.5 μM), and the variation of the λ_{LSPR} of AuNBPs versus SQ concentrations (**B**). The insert in (**B**) is the linear relationship between the λ_{LSPR} of AuNBPs and SQ concentrations in the range of 0.0–1.0 μM.

Table 1. Analytical performances of the proposed aptasensor for detecting SQ, SMP and SMD.

SAs	Linear Equation	Coefficient (R^2)	Linear Range	Visual LOD	Spectrometry	
					LOQ	LOD
SQ	$\lambda = 119.4 \times C + 558.9$	0.9912	0.3–1.0 μM	0.3 μM	0.3 μM	0.09 μM
SMP	$\lambda = 56.4 \times C + 550.4$	0.9941	0.3–1.0 μM	0.5 μM	0.3 μM	0.09 μM
SMD	$\lambda = 20.4 \times C + 550.1$	0.9951	0.5–5.0 μM	1.0 μM	0.5 μM	0.15 μM

As mentioned above, some aptamer-based colorimetric methods have been reported for the rapid determination of SAs [22,32,33]. However, previous aptamer-based colorimetric methods suffered from one or more following disadvantages: the ability to determine only one kind of SA or color change and had lower sensitivity. Compared to previous colorimetric methods, the multicolor colorimetric aptasensor developed in this study has obvious analytical advantages such as the ability of determining multiple SAs and color changes with a higher sensitivity (see Table S2 in SI).

3.6. Determination of SQ, SMP and SMD in Fish Muscle Samples

The applicability and reliability of the proposed PAD-based multicolor colorimetric aptasensor was confirmed by determining SQ, SMP and SMD in fish muscle samples (*Perca fluviatilis*) collected from coastal waters of Fujian in China. The samples, which previously spiked with one or two or all three of SQ, SMP and SMD, were also determined in the same manner to investigate recovery. The results from our method were compared with those obtained by the commercial ELISA kit. As Table 2 shows, SQ, SMP and SMD were determined by naked eye observation or UV-visible spectrometry with a recovery of 89–94% and an RSD < 8% (n = 5). The results from the proposed colorimetric aptasensor were consistent with those from the commercial ELISA kit. All the above results indicated that our PAD-based multicolor colorimetric aptasensor is reliable and can be used for the rapid screening or semi-quantitative determination of SQ, SMP and SMD individually, the sum of all three or as the sum of any two.

Table 2. Analytical results of SQ, SMP and SMD in dried fish muscle samples.

Added SAs	[1] Added Con. (μg/g)	[2] SAs Detected				
		Our Multicolor Aptasensor				[5] ELISA Kit Con. (μg/g)
		Color	UV-Vis Spectrometry			
			Con. (μg/g)	RSD (n = 5)	Rec.	
None	0.0		-	-	-	-
SQ	3.6		3.4	5%	94%	3.7
SMP	5.7		5.2	6%	91%	-
SMD	5.6		5.0	8%	89%	5.2
[3] SQ + SMD	5.8	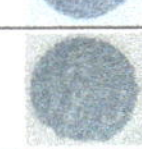	5.2	7%	90%	5.4
[4] Mixture	4.6		4.1	6%	90%	3.3

[1] The concentration of SAs in dried fish muscle sample; [2] The concentration of SAs obtained using our method and commercial ELISA kit in a dried fish muscle sample; [3] The mixture of SQ and SMD in equal proportion; [4] Mixture of SQ, SMP and SMD in equal proportion. [5] The ELISA kit could only be used to determine SQ and SMD, not SMP.

4. Conclusions

In summary, we developed a PAD-based multicolor colorimetric aptasensor to detect 3 different SAs (SQ, SMP and SMD) by using a broad-specificity SA aptamer, functionalized magnetic beads as a bioreceptor, an NADH–AA-mediated AuNBP growth system to generate color signals, and a PAD system that deposited AuNBPs more sensitively and stably so that color signals could be read out. The colorimetric aptasensor had more color changes corresponding to the concentrations of SQ, SMP and SMD; higher sensitivity; better specificity and stability; and was used to determine SQ, SMP and SMD with a visual detection limit of 0.3, 0.5 and 1.0 μM, respectively. The detection limits of UV-visible spectrometry were calculated to be 0.09, 0.09 and 0.15 μM for SQ, SMP, and SMD, respectively. By using the established aptasensor, we successfully determined SQ, SMP, SMD, the sum of SQ and SMD, and the sum of all 3 SAs in the fish muscle samples with a recovery of 89–94% and an RSD (n = 5) < 8%. This simple method provides multiple color changes and higher sensitivity, which make it a promising method for the rapid and instrument-free screening of the total residue of SQ, SMP and SMD in seafood.

Supplementary Materials: The following supporting information can be downloaded at: https://www.mdpi.com/article/10.3390/chemosensors11070386/s1: Main chemicals and apparatus, supplementary tables including detailed DNA sequences (Table S1) and comparison with previous colorimetric aptasensors (Table S2), supplementary figures including chemical structures of the 19 commonly used sulfonamides and 6 other antibiotics (Figure S1), PAD structure (Figure S2), gel electrophoresis analysis (Figure S3), optimization parameters (Figures S4 and S5), and photographs and UV-vis extinction spectra for detecting SMP and SMD (Figures S6 and S7).

Author Contributions: Conceptualization, F.F. and Z.W.; methodology, M.P. and W.L.; validation, M.P., C.Y. and Q.C.; investigation, M.P., W.L, C.Y. and Q.C.; resources, F.F.; data curation, M.P.; writing—original draft preparation, F.F. and Z.W.; writing—review and editing, F.F.; supervision, Z.W. and F.F.; project administration, F.F.; funding acquisition, F.F. All authors have read and agreed to the published version of the manuscript.

Funding: The authors gratefully acknowledge National Natural Science Foundation of China (21976029, 22276032) and Fujian Provincial Department of Science and Technology (2023Y0005) for financial support.

Institutional Review Board Statement: Not applicable.

Informed Consent Statement: Not applicable.

Data Availability Statement: Any additional data in support of the findings of this study besides those provided as Supplementary Materials are available from the corresponding author upon reasonable request.

Conflicts of Interest: The authors declare no conflict of interest.

References

1. Białk-Bielińska, A.; Stolte, S.; Arning, J.; Uebers, U.; Böschen, A.; Stepnowski, P.; Matzke, M. Ecotoxicity evaluation of selected sulfonamides. *Chemosphere* **2011**, *85*, 928–933. [CrossRef]
2. Apaydin, S.; TöröK, M. Sulfonamide derivatives as multi-target agents for complex disease. *Bioorganic Med. Chem. Lett.* **2019**, *29*, 2042–2050. [CrossRef]
3. Yu, L.; Song, C.; Zhang, C.; Fan, L.; Qiu, L.; Wu, W.; Meng, S.; Hu, G.; Chen, J. Occurrence of sulfonamides in fish in the lower reaches of Yangtze River, China and estimated daily intake for understanding human dietary exposure. *Aquaculture* **2018**, *495*, 538–544. [CrossRef]
4. Tačić, A.; Nikolić, V.; Nikolić, L.; Savić, I. Antimicrobial sulfonamide drugs. *Adv. Technol.* **2017**, *6*, 58–71. [CrossRef]
5. Sarmah, A.K.; Meyer, M.T.; Boxall, A.B. A global perspective on the use, sales, exposure pathways, occurrence, fate and effects of veterinary antibiotics (VAs) in the environment. *Chemosphere* **2006**, *65*, 725–759. [CrossRef]
6. Hamscher, G.; Bachour, G. Veterinary drugs in the environment: Current knowledge and challenges for the future. *J. Agric. Food Chem.* **2018**, *66*, 751–752. [CrossRef] [PubMed]
7. Suo, D.C.; Wang, P.L.; Xiao, Z.M.; Zhang, S.; Zhuang, H.T.; Li, Y.; Su, X.O. Multi-residue determination of 27 sulfonamides in poultry feathers and its application to a sulfamethazine pharmacokinetics study on laying hen feathers and sulfonamide residue monitoring on poultry feathers. *J. Agric. Food Chem.* **2019**, *67*, 11236–11243. [CrossRef]
8. Kim, Y.; Jung, J.; Kim, M.; Park, J.; Boxall, A.B.A.; Choi, K. Prioritizing veterinary pharmaceuticals for aquatic environment in Korea. *Environ. Toxicol. Pharmacol.* **2008**, *26*, 167–176. [CrossRef]
9. Baran, W.; Adamek, E.; Ziemiańska, J.; Sobczak, A. Effects of the presence of sulfonamides in the environment and their influence on human health. *J. Hazard. Mater.* **2011**, *196*, 1–15. [CrossRef] [PubMed]
10. Lin, H.; Zhang, J.; Chen, H.; Wang, J.; Sun, W.; Zhang, X.; Yang, Y.; Wang, Q.; Ma, J. Effect of temperature on sulfonamide antibiotics degradation, and on antibiotic resistance determinants and hosts in animal manures. *Sci. Total Environ.* **2017**, *607*, 725–732. [CrossRef] [PubMed]
11. Zhang, H.Y.; Duan, Z.J.; Wang, L.; Zhang, Y.; Wang, S. Hapten synthesis and development of polyclonal antibody-based multi-sulfonamide immunoassays. *J. Agric. Food Chem.* **2006**, *54*, 4499–4505. [CrossRef]
12. Hu, S.P.; Zhao, M.; Xi, Y.Y.; Mao, Q.Q.; Zhou, X.D.; Chen, D.W.; Yan, P.C. Non-targeted screening and determination of sulfonamides: A dispersive micro solid-phase extraction approach to the analysis of milk and honey samples using liquid chromatography high-resolution mass spectrometry. *J. Agric. Food Chem.* **2017**, *65*, 1984–1991. [CrossRef]
13. Li, T.; Wang, C.; Xu, Z.A.; Chakraborty, A. A coupled method of on-line solid phase extraction with the UHPLC-MS/MS for detection of sulfonamides antibiotics residues in aquaculture. *Chemosphere* **2020**, *254*, 126765. [CrossRef]
14. Saxena, S.K.; Rangasamy, R.; Krishnan, A.A.; Singh, D.P.; Uke, S.P.; Malekadi, P.K.; Sengar, A.S.; Mohamed, D.P.; Gupta, A. Simultaneous determination of multi-residue and multi-class antibiotics in aquaculture shrimps by UPLC-MS/MS. *Food Chem.* **2018**, *260*, 336–343. [CrossRef]

15. Zhao, X.L.; Wang, J.Y.; Wang, J.P.; Wang, S. Development of water-compatible molecularly imprinted solid-phase extraction coupled with high performance liquid chromatography-tandem mass spectrometry for the detection of six sulfonamides in animal-derived foods. *J. Chromatogr. A* **2018**, *1574*, 9–17. [CrossRef] [PubMed]
16. Zhang, Z.; Yang, M.Y.; Wu, X.Y.; Dong, S.B.; Zhu, N.F.; Gyimah, E.; Wang, K.; Li, Y. A competitive immunosensor for ultrasensitive detection of sulphonamides from environmental waters using silver nanoparticles decorated single-walled carbon nanohorns as labels. *Chemosphere* **2019**, *225*, 282–287. [CrossRef]
17. Li, Z.B.; Cui, P.L.; Liu, J.; Liu, J.X.; Wang, J.P. Production of generic monoclonal antibody and development of chemiluminescence immunoassay for determination of 32 sulfonamides in chicken muscle. *Food Chem.* **2020**, *311*, 125966. [CrossRef]
18. Wang, Z.X.; Xing, K.Y.; Ding, N.S.; Wang, S.H.; Zhang, G.G.; Lai, W.H. Lateral flow immunoassay based on dual spectral-overlapped fluorescence quenching of polydopamine nanospheres for sensitive detection of sulfamethazine. *J. Hazard. Mater.* **2022**, *423*, 127204. [CrossRef] [PubMed]
19. Galvidis, I.A.; Wang, Z.H.; Nuriev, R.I.; Burkin, M.A. Broadening the detection spectrum of small analytes using a two-antibody-designed hybrid immunoassay. *Anal. Chem.* **2018**, *90*, 4901–4908. [CrossRef] [PubMed]
20. Chen, Y.N.; Guo, L.L.; Liu, L.Q.; Song, S.S.; Kuang, H.; Xu, C.L. Ultrasensitive immuno-chromatographic strip for fast screening of 27 sulfonamides in honey and pork liver samples based on a monoclonal antibody. *J. Agric. Food Chem.* **2017**, *65*, 8248–8255. [CrossRef]
21. Jiang, W.X.; Wang, Z.H.; Beier, R.C.; Jiang, H.Y.; Wu, Y.N.; Shen, J.Z. Simultaneous determination of 13 fluoroquinolone and 22 sulfonamide residues in milk by a dual-colorimetric enzyme-linked immunosorbent assay. *Anal. Chem.* **2013**, *85*, 1995–1999. [CrossRef] [PubMed]
22. Chen, X.-X.; Lin, Z.-Z.; Hong, C.-Y.; Yao, Q.-H.; Huang, Z.-Y. A dichromatic label-free aptasensor for sulfadimethoxine detection in fish and water based on AuNPs color and fluorescent dyeing of double-stranded DNA with SYBR Green I. *Food Chem.* **2020**, *309*, 125712. [CrossRef]
23. Okoth, O.K.; Yan, K.; Liu, Y.; Zhang, J.D. Graphene-doped Bi_2S_3 nanorods as visible-light photoelectrochemical aptasensing platform for sulfadimethoxine detection. *Biosens. Bioelectron.* **2016**, *86*, 636–642. [CrossRef]
24. You, H.; Bai, L.J.; Yuan, Y.H.; Zhou, J.; Bai, Y.; Mu, Z.D. An amperometric aptasensor for ultrasensitive detection of sulfadimethoxine based on exonuclease-assisted target recycling and new signal tracer for amplification. *Biosens. Bioelectron.* **2018**, *117*, 706–712. [CrossRef]
25. Liu, Q.; Shi, T.Y.; Cheng, Y.; Wen, Z.R.; Ding, C.F.; Li, Y.Q.; Wang, K. Amplified photocurrent signal for fabricating photoelectrochemical sulfadimethoxine aptasensor based on carbon nitride photosensitization with visible/near-infrared light responsive zinc phthalocyanine. *J. Hazard. Mater.* **2021**, *406*, 124749. [CrossRef] [PubMed]
26. Mohammad-Razdari, A.; Ghasemi-Varnamkhasti, M.; Izadi, Z.; Rostami, S.; Ensafi, A.A.; Siadat, M.; Losson, E. Detection of sulfadimethoxine in meat samples using a novel electrochemical biosensor as a rapid analysis method. *J. Food Compos. Anal.* **2019**, *82*, 103252. [CrossRef]
27. Yang, Z.H.; Ding, X.F.; Guo, Q.; Wang, Y.; Lu, Z.W.; Ou, H.C.; Luo, Z.F.; Lou, X.H. Second generation of signaling-probe displacement electrochemical aptasensor for detection of picomolar ampicillin and sulfadimethoxine. *Sens. Actuators B-Chem.* **2017**, *253*, 1129–1136. [CrossRef]
28. Yang, L.; Ni, H.J.; Li, C.L.; Zhang, X.Y.; Wen, K.; Ke, Y.B.; Yang, H.J.; Shi, W.M.; Zhang, S.X.; Shen, J.Z.; et al. Development of a highly specific chemiluminescence aptasensor for sulfamethazine detection in milk based on in vitro selected aptamers. *Sens. Actuators B-Chem.* **2019**, *281*, 801–811. [CrossRef]
29. Chen, X.-X.; Lin, Z.-Z.; Yao, Q.-H.; Huang, Z.-Y. A practical aptaprobe for sulfadimethoxine residue detection in water and fish based on the fluorescence quenching of CdTe QDs by poly (dial-lyldimethylammonium chloride). *J. Food Compos. Anal.* **2020**, *91*, 103526. [CrossRef]
30. Wang, Y.R.; Yan, X.-L.; Kou, Q.-M.; Sun, Q.; Wang, Y.-X.; Wu, P.; Yang, L.-L.; Tang, J.-M.; Le, T. An ultrasensitive label-free fluorescent aptasensor platform for detection of sulfamethazine. *Int. J. Nanomed.* **2021**, *16*, 2751–2758. [CrossRef]
31. Song, K.-M.; Jeong, E.; Jeon, W.; Jo, H.; Ban, C. A coordination polymer nanobelt (CPNB)-based aptasensor for sulfadimethoxine. *Biosens. Bioelectron.* **2012**, *33*, 113–119. [CrossRef]
32. Chen, A.L.; Jiang, X.L.; Zhang, W.W.; Chen, G.; Zhao, Y.; Tunio, T.M.; Liu, J.C.; Lv, Z.Z.; Li, C.; Yang, S.M. High sensitive rapid visual detection of sulfadimethoxine by label-free aptasensor. *Biosens. Bioelectron.* **2013**, *42*, 419–425. [CrossRef] [PubMed]
33. Xu, R.-Y.; Yang, C.; Huang, L.; Lv, W.-C.; Yang, W.-J.; Wu, Y.-N.; Fu, F.-F. Broad-specificity aptamer of sulfonamides: Isolation and its application in simultaneous detection of multiple sulfonamides in fish sample. *J. Agric. Food Chem.* **2022**, *70*, 11804–11812. [CrossRef] [PubMed]
34. Kirschner, M.S.; Lethies, C.M.; Lin, X.-M.; Schatz, G.C.; Chen, L.-X.; Schaller, R.D. Size-dependent coherent-phonon plasmon modulation and deformation characterization in gold bipyramids and nanojavelins. *ACS Photonics* **2016**, *3*, 758–763. [CrossRef]
35. Lombardi, A.; Loumaigne, M.; Crut, A.I.; Maioli, P.; Del Fatti, N.; Vallée, F.; Spuch-Calvar, M.; Burgin, J.; Majimel, J.R.; Tréguer-Delapierre, M. Surface plasmon resonance properties of single elongated nano-objects: Gold nanobipyramids and nanorods. *Langmuir* **2012**, *28*, 9027–9033. [CrossRef] [PubMed]
36. Wang, H.-Q.; Rao, H.-H.; Luo, M.-Y.; Xue, X.; Xue, Z.-H.; Lu, X.-Q. Noble metal nanoparticles growth-based colorimetric strategies: From monocolorimetric to multicolorimetric sensors. *Coord. Chem. Rev.* **2019**, *398*, 113003. [CrossRef]

37. Zhang, Z.-Y.; Wang, H.; Chen, Z.-P.; Wang, X.-Y.; Choo, J.; Chen, L.-X. Plasmonic colorimetric sensors based on etching and growth of noble metal nanoparticles: Strategies and applications. *Biosens. Bioelectron.* **2018**, *114*, 52–65. [CrossRef]
38. Wang, Z.-W.; Chen, Q.; Zhong, Y.-Y.; Yu, X.-H.; Wu, Y.-N.; Fu, F.-F. A multicolor immunosensor for sensitive visual detection of breast cancer biomarker based on sensitive NADH-ascorbic-acid-mediated growth of gold nanobipyramids. *Anal. Chem.* **2020**, *92*, 1534–1540. [CrossRef]
39. Ye, X.-Y.; Zhang, F.; Yang, L.; Yang, W.-J.; Zhang, L.-Y.; Wang, Z.-W. Paper-based multicolor sensor for on-site quantitative detection of 2, 4-dichlorophenoxyacetic acid based on alkaline phosphatase-mediated gold nanobipyramids growth and colorimeter-assisted method for quantifying color. *Talanta* **2022**, *245*, 123489. [CrossRef]
40. Wang, Z.-W.; Yang, L.; Ye, X.-Y.; Huang, C.-L.; Yang, W.-J.; Zhang, L.-Y.; Wu, Z.-J.; Fu, F.-F. Multicolor visual screening of total dithiocarbamate pesticides in foods based on sulfydryl-mediated growth of gold nanobipyramids. *Anal. Chim. Acta* **2020**, *1139*, 59–67. [CrossRef]

chemosensors

MDPI

Review

Current Status and Prospect of Diabetes Diagnosis and Treatment Based on Biosensing Technology

Xinyi Xu, Xuemei Wang and Hui Jiang *

State Key Laboratory of Digital Medical Engineering, School of Biological Science and Medical Engineering, Southeast University, Nanjing 210096, China; xuewang@seu.edu.cn (X.W.)
* Correspondence: sungi@seu.edu.cn

Abstract: Diabetes mellitus has increasingly become a threat to health all over the world. This review focuses on the promoting effect of biosensing technology on the diagnosis and treatment of diabetes mellitus. Types of diabetes and their corresponding pathogeneses are first introduced, followed by the diabetes prevalence and research progress at home and abroad. To emphasize the importance of diabetes diagnosis and treatment, we secondly summarize the breakthrough technology in this field based on biosensing technology at the present stage. In terms of diagnosis, diversified ways of blood glucose detection and multiple combinations of diabetes biomarkers are discussed, while a variety of insulin administration routes and non-drug treatment means are presented in the aspect of treatment. In conclusion, the prospect of the future development of diabetes diagnosis and treatment is put forward at the end of the review.

Keywords: diabetes mellitus; bioelectronics; electrochemical sensor

Citation: Xu, X.; Wang, X.; Jiang, H. Current Status and Prospect of Diabetes Diagnosis and Treatment Based on Biosensing Technology. *Chemosensors* **2023**, *11*, 391. https://doi.org/10.3390/chemosensors11070391

Academic Editor: Constantin Apetrei

Received: 9 June 2023
Revised: 5 July 2023
Accepted: 11 July 2023
Published: 13 July 2023

1. A General Overview of Diabetes Mellitus

1.1. Types and Pathogenesis of Diabetes Mellitus

Diabetes Mellitus (DM) is a chronic metabolic disease, manifesting as metabolic disorders of glucose, protein, lipid, and other substances within the body and persistent hyperglycemia externally. It is caused by the lack of insulin secretion and/or insulin resistance, which results from the combined effects of various factors.

Most DMs can be classified as type 1 diabetes mellitus (T1DM) and type 2 diabetes mellitus (T2DM). In addition, there are monogenic diabetes mellitus, secondary diabetes mellitus, gestational diabetes mellitus, and unresolved diabetes mellitus as a more refined distinction. The classification of T1DM and T2DM is based on classic characteristics of early understanding of DM, including the age of onset, degree of loss of β-cell function, degree of insulin resistance, presence of diabetes-related autoantibodies, and insulin therapy required for survival, etc. [1]. The World Health Organization has updated the classification standard of DM by issuing guidelines regularly since 1965. However, with the increasing prevalence of obesity in the young generation, the phenotypes of T1DM and T2DM are less obvious, on the basis of which more types of diabetes subtypes are produced [2]. The combined influence of multiple genes and factors has become a typical feature of contemporary DMs, which makes it difficult for people to understand the whole spectrum of diabetes phenotypes. Based on the classical diabetes model, only the known pathogenesis of classic T1DM and T2DM is elaborated here.

T1DM is a kind of organ-specific autoimmune disease, the pathogenesis of which is that the necrosis of pancreatic β cells in the autoimmune environment leads to the decrease in insulin secretion in the body, resulting in metabolic disorders. At present, T1DM accounts for about 5–10% of the total DM cases [3]. As early as 1974, the association between T1DM and genetic factors was discovered by scientists, showing a pattern of multi-gene inheritance. Further analyses conducted recently, locations of important, relevant alleles have been identified, and the mechanism of many cytokines directly related to T1DM has been successfully explored, such as interleukin-1 (IL-1), interferon-γ (IFN-γ), tumor necrosis factor (TNF), etc. [4]. In addition to cytokines, the formation of T1DM participates more in the process of human immunity; therefore, polypeptides and proteins result in the typical biomarkers, such as C-peptide, urinary haptoglobin [5] and zinc transporter 8 autoantibody (ZnT8A) [6].

T2DM accounts for about 90% of the total DM cases. The course of T2DM is characterized by the co-existence of β-cell secretion dysfunction and insulin resistance. The initial factors of the onset of T2DM are still controversial; however, it can be concluded that the former is a necessary condition for the onset, while the latter is throughout the course of T2DM. Both glucotoxicity and lipid toxicity can cause damage to pancreatic β cells; therefore, long-term acquired high-carbohydrate and high-fat diet can promote the onset of T2DM [7]. Further, related pathogenic genes (CDKAL1, CDKN2A, CDKN2B, etc.) have also been found. Due to the high proportion of T2DM in diabetics and the complexity of its causes, the biomarkers related to T2DM are diverse and involve various parts of physiological processes. For example, glycation markers (such as albumin and cystatin C), oxidation markers (such as apolipoprotein CI), and truncation markers (such as chemokine CCL5) all have considerable reference values in the detection of T2DM.

1.2. Development Status of Diabetes Mellitus Worldwide

According to the 10th edition of the Global Diabetes Map [8] released by the International Diabetes Federation (IDF) in 2021, about 537 million adults worldwide were living with DM. The prevalence rate reached 10.5%, indicating the fact that 1 in 10 people was living with the disease (Figure 1a). The age range of diabetics is wide, covering people aged 20–79 years. Prevalence increases with age, and a similar trend is expected to continue till 2045. Further, the report estimates that for every 20 percent increase in the world population, the number of diabetics increases by about 46 percent, which is expected to rise to 643 million by 2030 and rise to 783 million by 2045. Excluding acute deaths due to COVID-19, approximately 6.7 million people died from DM or its complications in 2021, representing 12.2% of all deaths. Moreover, one-third (32.6%) of diabetes deaths occurred in people younger than 60 years of age.

In such a global situation, the number of diabetics in China increased from 90 million to 140 million in the past 10 years, ranking first in the total number of diabetics. The number of undiagnosed diabetics and the number of diabetes-related deaths (Figure 1b) also showed a rapid growth trend which simultaneously took the first place in the world. Both the treatment conditions for diagnosed diabetes patients and the screening measures for potential diabetes patients bring great pressure on the existing medical system [9]. Therefore, China is faced with a severe situation of DM prevention and control.

The data display that the number of research on DM is gradually increasing in the world (Figure 1c). In recent 5 years, the number of articles published worldwide has been on the rise, with the United States taking the leading position. Although the number of articles published in Asian countries has been on the rise [10], it reveals the lack of influence of articles published and the necessity to improve the article quality. In addition, comparing the research emphases worldwide, it can be seen that, with diabetes complications as the central axis, researchers abroad focus on effective correlation indicators such as blood glucose detection and all-cause mortality, and more attention is paid to female patients. In China, the research is more based on DM core mechanics, including blood glucose treatment and intervention, self-health management, disease management, health education, etc., showing no obvious group research preference. Similar research directions exhibit in DM information intervention and the emphasis of intelligent health management guided by disease [11].

(a)

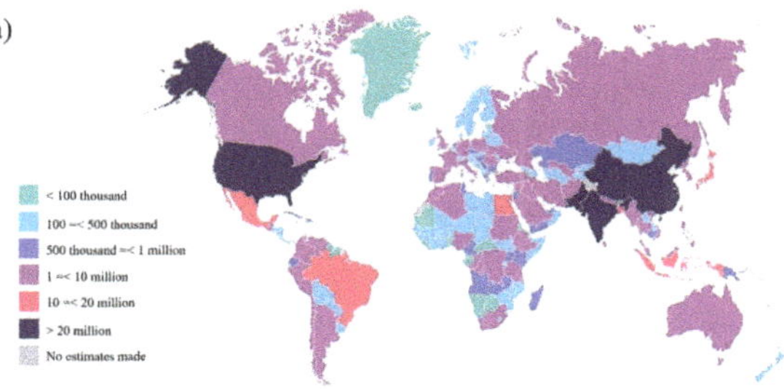

(b)

2021		
Rank	Country or territory	Number of people with diabetes (millions)
1	China	140.9
2	India	74.2
3	Pakistan	33.0
4	United States of America	32.2
5	Indonesia	19.5
6	Brazil	15.7
7	Mexico	14.1
8	Bangladesh	13.1
9	Japan	11.0
10	Egypt	10.9

(c)

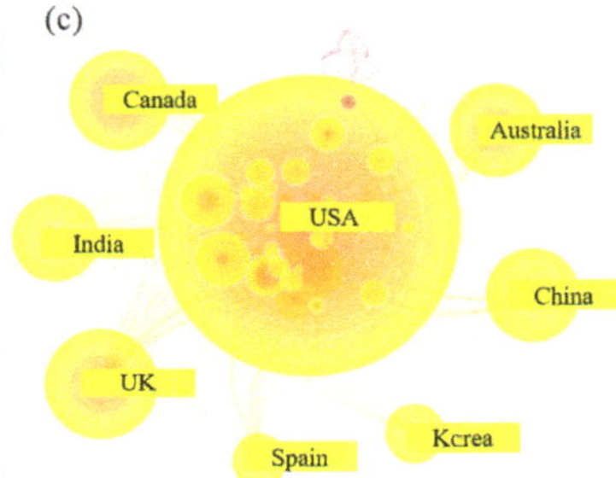

Figure 1. Status of diabetes mellitus worldwide. (**a**) The estimated total number and global distribution of diabetics in 2021 [8]. (**b**) Top 10 countries or territories with the highest number of diabetics in 2021 [8]. (**c**) Visualization analysis of the output of DM information literature in China3National Knowledge Network (CNKI) database and Web of Science Core Collection database by country or region [11]. All figures are licensed under an open-access Creative Commons CC BY license.

1.3. Importance of Diabetes Mellitus Diagnosis and Treatment

Complications developed in the late stage of DM will cause great harm to the human body. Affected by excessive blood glucose environment, large, medium, small, and micro blood vessels in the whole body could be injured, leading to various cardiovascular and cerebrovascular diseases. The respiratory system and urinary system are also common areas that suffer from complications, including tuberculosis and kidney failure. In addition, DM has a great impact on patients' eyes and feet. The complications in eyes often reflect in diabetic fundus diseases, cataracts, and retinopathy, while diabetic foot shows infection, diabetic foot ulcer (DFU), and gangrene as typical symptoms [12], which is the most well-known complication in the public view [13]. Diabetic foot is one of the most serious complications of DM, as well as the most common reason for hospitalization of diabetics and one of the leading causes of amputation worldwide. For female diabetics, various risks during pregnancy will be increased to some extent, followed by the increasing probability of birth defects in the next generation [14].

According to the recovery situation of COVID-19 in the past two years, COVID-19 patients with underlying diseases often face more severe symptoms or deterioration of the existing underlying diseases [15,16], most commonly seen in patients with cardiovascular and diabetes diseases. This discovery indicates that DM undoubtedly interferes and hinders the treatment and intervention of COVID-19 patients, and the complicated factors that doctors need to consider in the treatment process have increased. The facts above confirm the high proportion of deaths from DM and its complications mentioned in Section 1.2. Therefore, reasonable prevention, diagnosis, and treatment of DM are significant for national health. What is noteworthy is that some data speculate that COVID-19 survivors may be at an increased risk of developing new-onset DM, identified as a kind of post-acute sequelae of COVID-19 (PASC) [17].

In order to deal with the current situation of diabetics with multiple subtypes and phenotypes mentioned in Section 1.1, researchers have constantly emphasized the importance of DM reclassification. With the support of the "palette model" theory [18], reclassification was carried out based on topology-based analysis [19] and Bayesian non-negative matrix factorization clustering method [20] to achieve personalized treatment at the level of precise diabetology from pathogenesis [21] as much as possible (Figure 2). On the one hand, effective prevention of the formation of DM through genetic testing and lifestyle management can improve the quality of life of the people, fitting right in the concept of "preventive treatment of disease", which is expected to achieve if further research and development of early diagnosis technology of DM make great progress. On the other hand, for patients already suffering from DM, small portable wearable devices helping in diabetes diagnosis, multi-target and multi-channel diabetes treatment will effectively extend lives and improve patients' motivation facing the disease.

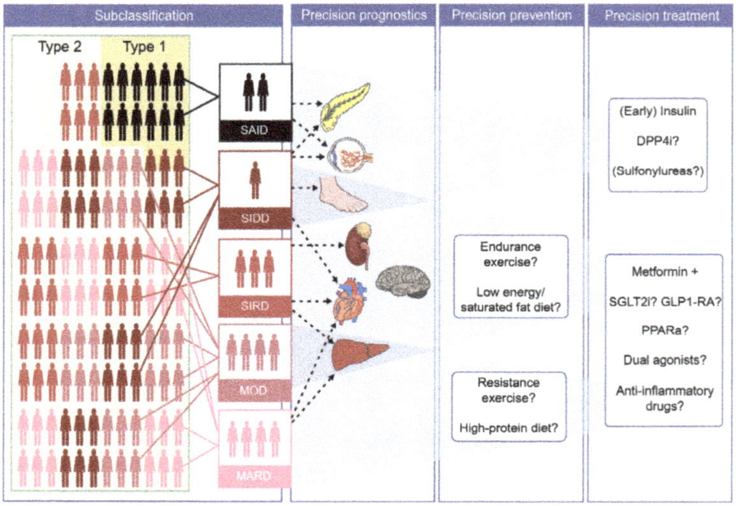

Figure 2. Future implications of precision diabetology based on novel diabetes subgroups [21]. SAID: Severe autoimmune diabetes. SIDD: Severe insulin-deficient diabetes. SIRD: Severe insulin-resistant diabetes. MOD: Mild obesity-related diabetes. MARD: Mild age-related diabetes. All figures are licensed under an open-access Creative Commons CC BY license.

2. Diagnosis of Diabetes Based on Electrochemical Biosensing Technology

According to the distribution and total number of DM patients, the demand in the global market for diabetes diagnostic equipment increased quickly, while the demand for cost-effective, convenient, highly portable, and responsive medical equipment has also increased year by year in the past decade. Both the rapid development of biochemical technology and the increasing demand for a higher standard of living have contributed to this scene.

The diagnosis of DM is inseparable from blood glucose detection, its original feature. The earliest detection for blood glucose was developed in the early 1900's by Canadian surgeon Frederick Banting and his assistant Charles Best to measure glucose in blood and urine. Clark and Lyons proposed the initial concept of a glucose sensor in 1962 [22] on the basis of it, which explored the approach for people to study glucose detection technology. After that, electrochemical technologies were introduced into the invention of sensors, and electrochemical sensing components such as enzymes and electrodes [23] were added to the detection chain to optimize the whole process. It was not until 1971 that Clemens in the United States developed the first regular glucose monitor [24] (the Ames Reflectance meter), which was a milestone in DM detection technology. Although the device was expensive and cumbersome from the point of now, it undoubtedly brought together the best technologies and efforts of numerous scientists at the time. During the evolution of the structure and mechanism of glucose meters, the successful application of screen-printing technology [25], invented in 2004, made the mass production of electrochemical glucose sensors possible and brought great efficiency improvement in the preparation of glucose sensing strips for electrochemical devices. With the development of electrochemical technology, Abbott launched the first electrochemical glucose monitor named ExacTech [26]. The clunky meters have evolved into lighter and more sensitive combinations of strips and reagents.

Nowadays, electrochemical biosensors have been widely used in glucose detection due to their high sensitivity and specificity, accounting for about 85% of the global biosensor market [27]. They are iterating into more sophisticated instruments that can be adapted to a wide range of needs.

2.1. Diverse Blood Glucose Detection

DM, as an endocrine disease, is characterized by hyperglycemia, indicating that its most direct biomarker is blood glucose. During the development history of the glucose meter mentioned, the glucose analyzer (known as Model 23A YSI analyzer) proposed by Yellow Springs Instrument Company in the United States for the amperometric detection of hydrogen peroxide was still used as a clinical diagnostic standard until the 21st century [28]. Although the model of glucose analysis has been changing, its basic chemical reaction mechanism has not changed. Among them, the redox chemical chains based on glucose oxidase (GOx) and glucose dehydrogenase (GDHs) are the most classic, which are still the basis of many current glucose biosensors [29] (Figure 3a). These typical methods are highly selective to glucose targets, and their disadvantages of oxygen dependence and electroactivity interference have also been further overcome by the breakthrough of the electrochemical biosensing technology. So far, three generations of electrochemical glucose analyzers have been developed, including enzymatic and non-enzymatic electrochemical glucose sensing methods for applications in different fields according to the different electron transfer mechanisms (Figure 3b). Enzymatic sensors are constructed by the combination of the immobilized enzyme and electrodes, in which the enzyme is acted as the sensitive element of the biosensor to realize the concentration monitoring of biomolecules. The first-generation glucose-enzyme biosensors use oxygen molecules as electron acceptors. The catalytic reaction of the enzyme will reduce oxygen in the solution and generate hydrogen peroxide, which can measure the concentration of glucose indirectly through the detection of the amount of oxygen reduction or the amount of hydrogen peroxide generated. The second-generation biosensors make use of electron transport medium instead of oxygen as the electron acceptor, overcoming the disadvantages of the first-generation biosensors which are limited by oxygen. The electron transport medium refers to small volume of soluble redox active molecules, such as ferrocene derivatives, ferrocyanide, conductive organic salts and quinones, which are able to carry out a rapid reversible redox reaction to increase the rate of enzymatic reaction. However, this kind of medium can be easily diffused from the enzyme layer into the substrate solution, contributing to the instability of the whole biosensor. Compared with previous two generations, the third-generation biosensors avoid the need of oxygen molecules or electron transfer mediators as electron acceptors, but fix the enzyme directly on the modified electrode. This change of environmental conditions of catalytic reactions makes the active site of the enzyme close to the electrode, which can directly carry out the electron transfer, improving the sensitivity and selectivity of the glucose-enzyme biosensor. The materials used to fix enzymes are often organic conductive composite films, metal or non-metallic nanoparticles. In addition to the enzymatic catalysis, various technologies were applied to develop the non-enzymatic methods [30] to detect glucose and corresponding mechanism has been found in the years of research. Chen and his co-workers have demonstrated the electrochemical detection of glucose using nonenzyme mode, such as plasmon-accelerated electrochemical reaction [31], or on Cu_2O/Au nano-composite modified surface [32]. Further, the nanopipette confinement effect is used to investigate the reaction in volume-confined glucose oxidase-based detection [33].

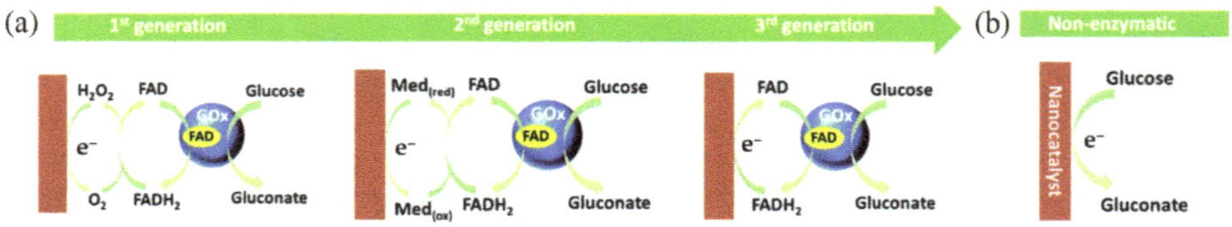

Figure 3. Schematic diagrams of principle of (**a**) enzymatic and (**b**) non-enzymatic electrochemical glucose sensors [29]. Reprint permission was requested from [29].

Blood is the most available body fluid for measuring glucose; therefore, most electrochemical glucose analyzers rely on detecting glucose directly in the blood. Although the electrochemical glucose analyzer has made a breakthrough in functionality, achieving great success in the commercialization process, it still has limitations. On the one hand, it is inconvenient that such methods have to pick the finger for sampling. On the other hand, some drugs, metabolites and other blood components may interfere with the glucose sensor [34]. Researchers began to explore wearable non-invasive technology, considering the need for health and cleaning. These non-invasive methods mainly concern the physical and chemical properties of internal and external biological fluids. A typical case of internal fluids is the detection of metabolites in interstitial fluid (ISF). In daily life, the self-regulation of physical and chemical components in ISF contributes to the homeostasis of glucose, Na^{+}, and K^{+} in the blood, which provides a new approach for people to measure blood glucose. The emergence of flexible electronic devices, like microneedle patches [35], inspired researchers who tried to use them to detect blood glucose through ISF. For example, a kind of biosensor was designed to measure blood glucose which can change the osmotic pressure of ISF by changing the electric field to induce glucose in blood vessels to seep into the skin's surface [36]. It completed the whole process by means of an electrochemical dual channel generated by a paper battery under the skin, using blood glucose seepage and glucose reverse ion electroosmosis, which showed a high correlation with the clinically measured blood glucose level (>0.9). This kind of flexible electronic device removes interference from the external environment to some extent and is one of the main directions of non-invasive real-time glucose detection in the future.

In addition to the study of internal biological fluids, the measurement methods of glucose in external biological fluids (tears, sweat, saliva, etc.) have also been developed. These fluids are used as samples of glucose, whose specific component concentration is determined by its association with the amount of glucose in the body (Figure 4); thus, glucose concentration can be calculated through the information from these fluids. Saliva contains biomarkers of clinical significance and is easy to obtain, which is of great importance for monitoring physiological and metabolic status. Laura et al. [37] proposed an integrated pacifier sensor (Figure 4b) for non-invasive chemical detection of newborn saliva, whose results achieved a relative standard deviation of about 9%. Integrated disposable PB electrodes by screen printing, which is modified with chitosan and GOx, were used for in vitro analysis of glucose in saliva. It not only breaks through the traditional thought of integrated physical parameters sensors but also avoids piercing the skin to obtain tissue fluids, commonly used in neonatal care units, making the newborn in a safer environment. The sweat can present a timely concentration tracking in various parameters with the change of blood glucose concentration, including sweat glucose concentration and pH value, although its concentration is much smaller than the glucose concentration in the blood [38]. In a typical contribution, Lee et al. [39] developed a wearable sensor for detecting glucose in sweat. When the glucose concentration of sweat was fed back, the statistics of pH value, temperature, and humidity of sweat can assist in correcting the result of it in real-time, forming a multi-mode accurate glucose concentration detection (Figure 4d). This self-checking working system allowed it to measure blood glucose concentration with low sweat volume (4 µL) in non-extreme environments, which showed almost no significant deviation compared with direct blood detection. Further, glucose and lactate contained in tears can also reflect blood glucose concentration to a certain extent [40], and contact lenses equipped with integrated lactate sensors have been invented. In terms of the newly proposed contact lens that detected glucose in tears [41], the researchers realized the detection accuracy of a wide range of concentrations, from the extremely low concentration of 0.05 mM to the high concentration of 0.9 mM, which had a good reproducibility of the results with a standard deviation of only about 4.5%.

Although the composition of biological fluids can indicate the blood glucose concentration, considering the uncertainties of human activities (such as uncontrollable sweat production and pollution risk of the unpredictable mix of new and old sweat in collection pipelines [42]) and the effects of environmental changes (such as changes in ambient temperature and humidity), biological fluids will inevitably be affected by the surrounding parameters, which may not accurately reflect the data. Therefore, how to tackle these extra problems becomes a further research topic. It is one of the development directions in the future to integrate multiple detection technologies of biological fluid components for multi-dimensional detection and calibration, which appears as multi-channel and multi-substrate biosensors. Correspondingly, with the increase in biological fluid technologies for collection and detection, researchers also need to overcome the barriers in computer science of the stable algorithm [43] to predict the level of blood metabolites through various kinds of data.

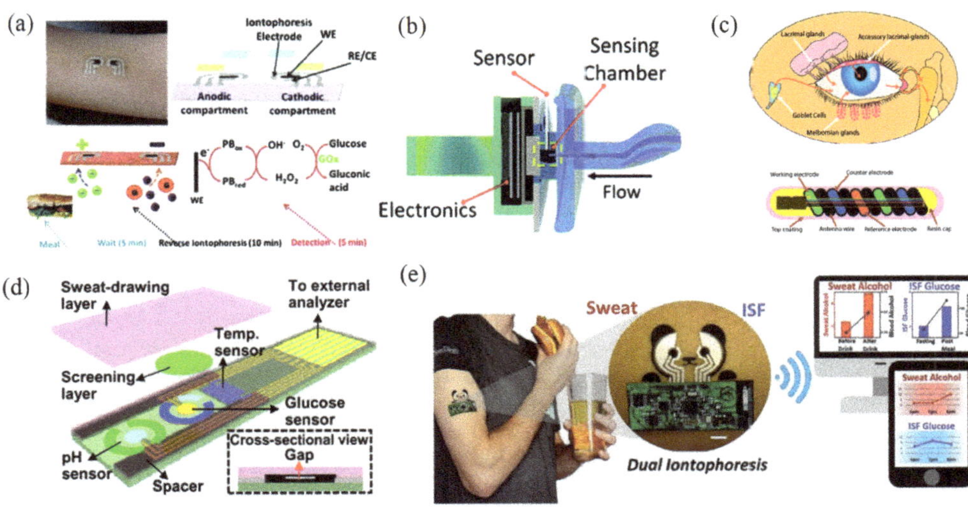

Figure 4. Some wearable non-invasive glucose sensor. (**a**) Tattoo glucose sensor based on ISF [45]. (**b**) Wireless glucose pacifier sensor [37]. (**c**) Electrochemical tear glucose sensor [46]. (**d**) Sensor arrays for sweat glucose monitoring [39]. (**e**) Wearable biosensor device for wireless real-time detection of the ISF glucose and sweat alcohol [44]. All figures are licensed under an open-access Creative Commons CC BY license.

How to collect multiple types of external fluids simultaneously has great difficulty in technology. For example, sweat and ISF can share the same sampling sites on the skin surface; however, the methods used to collect them are different. When the method of sweat stimulation is used, the perspiration production area cannot be controlled, which may result in the mix of the two collection fluids and interference with the signals. As shown in Kim's work [44], positive and reverse iontophoresis (IP) was proposed to deal with the problem. The two external fluids were obtained in parallel through the reverse IP of ISF in the cathode chamber and the IP of transdermal pilocarpine delivery in the anode chamber. Under the effect of electric repulsion, the anode delivered the pilocarpine with a positive charge and caused the delivery area to produce sweat, which was located in the pilocarpine-loaded gel in the anode compartment. At the same time, because of the negative charge of the human skin, a convective ionic electroosmotic flow to the cathode resulted in the migration of positive ions and neutral molecules (such as glucose), excluding the interference of proteins, other biomolecules, and some electroactive substances. Therefore, this system allowed selective, simultaneous, and localized sampling to be performed on different electrodes, avoiding mixing between samples (Figure 4e). The cathode side of the system is a biosensor based on glucose oxidase (GOx), and the anode side is a biosensor based on an alcohol oxidation reaction. The alcohol level (anode) and glucose level (cathode) of the blood can be effectively detected, which can be used for the blood glucose diagnosis of DM and the monitoring of the diabetic glucose curve after drinking, by completing the following reactions (1) and (2), respectively on the biosensor. Thus, it may provide a facile strategy to design a non-invasive epidermal sampling and analysis platform for effective evaluation of physiological status.

$$\text{Glucose} + \text{oxygen} \stackrel{\text{GOx}}{\rightarrow} \text{hydrogen peroxide} + \text{gluconic acid} \tag{1}$$

$$\text{Alcohol} + \text{oxygen} \stackrel{\text{AOx}}{\rightarrow} \text{hydrogen peroxide} + \text{acetaldehyde} \tag{2}$$

2.2. Diverse Selections of Biomarkers for Diabetes

The diversity of the synergistic redox reactions makes it possible to detect various biomarkers. On the one hand, there are many other biomarkers useful to indicate the occurrence of DM, just like blood glucose can do, because the endocrine disorder is a complex process. In traditional detections, biomarkers such as urine glucose, glycosylated hemoglobin, serum insulin, and insulin antibody have been used to indicate DM [47]. Studies on metabolomics markers start from the direction of lipid and glucose metabolism and have been reported that amino acids and their derivatives are significantly associated with DM [48,49], such as glycine and glutamine, the content of which is negatively correlated with the disease. In order to overcome the shortcoming that a single biomarker cannot accurately indicate diseases, Borges et al. proposed an integrative method on the basis of principal component analysis (PCA). Through the evaluation of the reliability of different biomarkers for the diagnosis in patients, different combinations of diabetes biomarkers were associated, and specific biomarker combinations, such as glycation markers and oxidation markers, were obtained. The detection of a single biomarker was expanded into the detection of a combination of biomarkers, which greatly improved the accuracy of DM detection [50].

The study of molecular markers has also been developed since most types of DM have been linked to genes. These markers are mainly microRNAs (miRNAs). For example, miRNA-375 will be highly expressed during the onset of DM, while miRNA-23a will be significantly reduced in the plasma of patients with T2DM. Recent studies have confirmed that circRNA is another reliable molecular marker to indicate the occurrence of DM and its related complications [51], which have been shown to be related to insulin regulation and disease development in patients with T2DM [52]. Furthermore, these kinds of markers suggest the occurrence of complications such as diabetic retinopathy and diabetic nephropathy, more specifically [53], which provide a more effective way to predict complications for the long-term treatment of DM patients and a new guiding basis for the comprehensive early diagnosis of DM.

On the other hand, the evolution of electrochemical techniques would bring about the way of biomarker detection updates. With researchers experimenting with new substrates and electrochemical reaction chains, the quest for biosensors has never stopped. Newfound reaction chains continuously improve the electrochemical detection scheme of various biomarkers and simplify the reaction steps, which further increases the accuracy of results and improves detection efficiency. For example, Esokkiya et al. reported an acetaldehyde (AcH) ultra-trace detection sensor based on electrochemically activated copper nitropru-

side (CuNPr) in 2022 [54]. This method is expected to overcome the difficulty and high cost of traditional breathalyzers with fuel cells for the accurate measurement of acetaldehyde. Therefore, based on technological development, people will have more feasible sections about biomarker detection objects and methods for DM in the future.

With the development of non-invasive detection, researchers found that reasonable capture and analysis of volatile organic compounds (VOCs) in human breath can obtain associated blood indicators, combined with the biochemical indicators of DM in body fluids and current breath detection methods. The ease of access to breath samples is one of the huge advantages of expiratory biochemical sensors. Acetone, acetaldehyde, carbon monoxide, alkane, methyl nitrate, and other compounds in respiration have been proven to have abnormal concentrations in patients with DM [55]. According to the method of PCA, researchers further classified VOCs into different clusters (Table 1) to indicate the occurrence of DM accurately [56]. On account of non-invasive breath testing, some researchers have tried to develop a self-powered wearable breath biosensor based on the electrochemical determination of acetone for non-invasive blood glucose monitoring. Su et al. [57] proposed an electrochemical respiratory biochemical sensor based on a triboelectric nanogenerator (TENG) (Figure 5). The periodic vertical contact and separation of the PTFE film and nylon film acted as a self-powered device in the upper chamber. Patients' breath entered the lower reaction chamber, and then the acetone in the airflow reacted with the sensitive film at the bottom of the chamber to produce electron exchange which changed the dielectric constant of the gas so that the voltage generated by the upper self-powered chamber, which would conduct to the copper electrode in the lower chamber, occurred different degrees of attenuation. From the analysis of the final load voltage displayed by the device, the acetone concentration in the breath and the breath state of the patients can be detected simultaneously. In addition to the sensors of a single biochemical indicator, Wang et al. [58] prepared a self-powered respiratory biosensor based on the transverse sliding mode of a TENG for the multi-channel detection of gas molecules synchronously. The sensor set up the sensing unit and the self-powered unit in a piece of the electronic sticker, which could complete the whole sensing process through the displacement current caused by the breakdown discharge. In a further study, researchers successfully adjusted the amplitude and spectrum of the signal, making this self-powered wireless sensing e-sticker (SWISE) realize the multi-point sensing behavior of gas sensing.

Table 1. VOC clusters for diabetes diagnosis *.

Biomarker Clusters	Healthy/T1DM/T2DM Subjects	Method Used	Research Outcome
Acetone, methyl nitrate, ethanol, and ethylbenzene	17 healthy, 8 T1DM subjects	Gas Chromatograpy	Mean Correlation Coefficients All = 0.883 Healthy Subjects = 0.836 T1DM Subjects = 0.950
2-pentyl nitrate, propane, methanol, and acetone	17 healthy, 8 T1DM subjects	Gas Chromatograpy	Mean Correlation Coefficients All = 0.869 Healthy Subjects = 0.829 T1DM Subjects = 0.990
Acetone, ethanol, and propane	130 healthy, 70 subjects with diabetes	Analog Semiconductor Sensors	Mean Correlation Coefficients All = 0.25 Healthy Subjects = 0.97 Subjects with diabetes = 0.35
Isopropanol, 2,3,4-trimethylhexane, 2,6,8-trimethyldecane, tridecane, and undecane	39 healthy, 48 T2DM subjects	Gas Chromatograpy—Mass Spectrometry	Sensitivity = 97.9% Specificity = 100%

* Data adapted from [56].

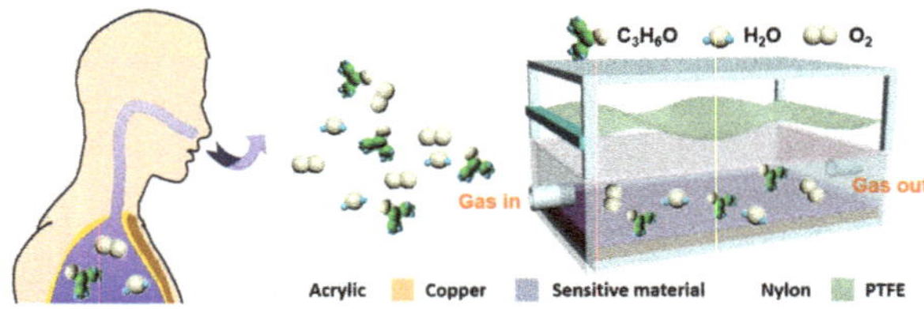

Figure 5. Schematic diagram of self-powered wearable acetone breath biosensor [57]. Reprint permission was requested from [57].

3. Novel Treatment of Diabetes Based on Electrochemical Biosensing Technology

Over the years, a variety of technologies have been applied in the treatment of DM, aiming to inhibit the progression of DM based on blood glucose control and insulin regulation. Hypoglycemic drugs have the longest history. Metformin was first reported in 1922 [59] and is currently the most classic hypoglycemic drug ingredient. From the 1920s to the 1960s, people gradually discovered the hypoglycemic ability of metformin in animals [60] and humans [61]. It is surprising that the process of hypoglycemia is harmless to the body, both in animals and humans. With the development of drug extraction technology, researchers discovered and prepared more components with hypoglycemic effects, such as the genus Aloe [62] and tetrahydrocarbazole derivatives [63]. At the same time, the discovery of insulin marked a major breakthrough in medicine and treatment for DM. The deepening knowledge of the human body and physiology made people aware of insulin [64] and its close association with DM. Therefore, in addition to controlling DM by reducing blood glucose, researchers also shifted their attention to insulin levels, including the improvement in insulin sensitivity [65], external insulin delivery, and the development of insulin analogs [66]. Combination therapies with hypoglycemic drugs and insulin control were also being perfected. The variety of means of controlling the progression of DM led to an increase in the life expectancy and the frequency of diabetes complications of most people with DM. At present, the prevention and cure of diabetic complications have become an important issue in improving the quality of life of DM patients. Researchers are still struggling at the forefront of DM treatment, seeking the well-being of patients, while patients' demand for DM treatment has changed from blood glucose reduction and regulation simply to diversified requirements such as portability, autonomy, and safety.

3.1. *Insulin Administration Regimen*

Oral hypoglycemic drugs and insulin administration are traditional methods of DM treatment; however, both of them lack intelligent regulation of blood glucose levels, often leading to poor blood glucose control and the risk of hypoglycemia. If the oral drug administration schedule is fixed, it cannot be adjusted in time in the long-term development of DM, which will result in problems such as a potential mismatch between the severity of the disease and the schedule. Insulin administration is mostly limited to subcutaneous and intravenous injections. Frequent injections may cause pain, and the injection site is prone to inflammation, increasing the risk of infection [67].

The study of the microneedle patch greatly reduces the risk of infection during insulin administration, and the environmental stimulus-responsive microneedle patch is more efficient in regulating blood glucose levels. Glucose-responsive compounds with different mechanisms (such as glucose oxidase (GOx), glucose-binding protein (GBP), etc.) are used to construct biochemical cascade reactions with glucose in vivo so as to effectively stimulate polymer dissociation in microneedles to release insulin and other glucose-regulating agents when blood glucose concentration rises. At the same time, when the content is released by the change of glucose, the internal environmental feedback can regulate the amount of the content released. For example, the pH-responsive micellar complex developed by Zhang [68] has a cascade reaction behavior triggered by changes in the pH of the environment. In the condition of hyperglycemia, the oxidation of the polymer and the reduction in the negative charge of insulin occur successively, which leads to the dissociation of the complex on the microneedle and the subsequent release of insulin. This two-step cascade of insulin release avoids the failure release of drugs in single acid environments and increases the safety of drug delivery (Figure 6a). Similarly, oxidation-responsive polymer microneedle patches were developed. Chen et al. [69] reported an intelligent insulin delivery device based on dual mineralized particles containing exendin-4 (Ex4) and glucose oxidase (GOx). While sensitive to blood glucose concentration, the patch device can achieve the transformation of glucose signal to H signal when blood glucose concentration rises, releasing drugs specifically. After the blood glucose concentration gradually returns to normal level, the release rate of drugs slows down, forming a long-term closed-loop regulation. Some studies have proposed that glucose-responsive copolymers, which can be self-assembled into nanocarriers, can decompose and release insulin in a high-glucose environment in the body [70]. The researcher used polymer vesicles, a kind of carrier with a strong hollow structure and a large loading capacity of hydrophilic drugs, to improve the efficiency of insulin use and its own biocompatibility to a certain extent. The carrier was self-assembled from an amphiphilic triblock copolymer, poly(ethylene glycol)-b-poly(3-acrylamidophenylboronic acid)-b-poly(4-(4,4,5,5-tetramethyl-1,3,2-dioxaborolan-2-yl)benzyl acrylate) (PEG-b-PPBA-b-PPBEM). The polymer vesicles have both functional groups of phenylboronic acid (PBA) sensitive to glucose and phenylboronic acid pinacol ester (PBEM) sensitive to H_2O_2, which have a dual cascade response to glucose after being coated with insulin and GOx. For the selection of patch microneedles, researchers prepared soluble microneedles based on poly(vinylpyrrolidone) (PVP) and (poly(vinyl alcohol)) (PVA), which can be inserted into the skin cuticle to deliver internal reagents to the skin barrier not only noninvasively but also safely dissolve in the skin tissue after the use. It had a practical significance that insulin can release from vesicles in an incomplete state under normal glucose concentration, while it is able to release a large amount of insulin promptly under high glucose concentration according to the results of animal experiments, which better fulfilled the goal of rapid response and long-term painless drug administration (Figure 6b). As described in Section 2.1, with the improvement in glucose-related reaction chains, adjustable means of insulin administration are gradually developed.

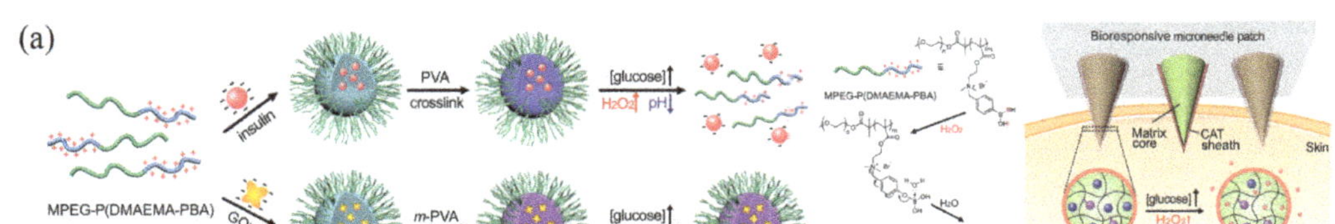

Figure 6. Schematic diagrams of glucose-responsive insulin delivery systems. (**a**) A nanoarray transdermal microneedle patch with cascade reaction of H2O2 and pH [68]. Reprint permission was requested from [68]. (**b**) Synthesis route of PEG-b-PPBA-b-PPBEM and its transdermal drug delivery approach [70]. All figures are licensed under an open-access Creative Commons CC BY license.

Insulin administration sources include external and internal routes. The external routes, including oral medication, subcutaneous injection, and microneedle administration, are discussed previously. Internally, the artificial pancreas (closed-loop glucose controller) is created to deliver insulin as an implantable pump. The drugs in closed-loop control show better blood glucose control compared to traditional insulin therapy (open-loop control). In general, the closed-loop system includes an insulin pump, a continuous glucose monitor, and a control algorithm [71] designed to measure and regulate blood glucose concentration in vivo in real time. PID algorithm [72], model predictive control (MPC) [73], controller based on pancreatic physiology (BiAP) [74], and other algorithms iteratively improve the drug delivery scheme of an artificial pancreas. Through the integration of bioelectronics, information engineering, and other disciplines, human identity as an important controller of the external open loop is liberated, which is expected to achieve long-term operation by self-help drug delivery depending on the changes of the organism's environment. At present, many varieties of closed-loop control systems have completed commercial applications. Medtronic MiniMed 670 G (Minimed Medtronic, Northridge, California) [75] is the first commercial hybrid closed-loop system. Although T1DM is treated by artificial pancreas far more than T2DM and the algorithm is more mature, the plasticity of an artificial pancreas is stronger than other open-loop controls when facing both T1DM and T2DM simultaneously.

3.2. Bioelectronic Medicine Based on Nervous System Regulation

Despite the widespread success of the drug approach in treating DM, many problems remain, such as side effects, drug resistance, and patient compliance. The increase in treatment resistance in many chronic diseases will further limit the use of drug therapies. Therefore, people begin to explore alternative treatment strategies for chronic metabolic diseases.

Recent studies in the field of glucose homeostasis show that the nervous system is greatly involved in the control of dietary intake (such as appetite) as well as the regulation of hormone pancreatic secretion and insulin sensitivity [76], providing brand new thoughts toward the treatment of DM. In the past, this type of electrostimulation therapy focused on treating inflammatory diseases by interfering with the neuroimmune reflex and has successfully improved the treatment of rheumatoid arthritis and colitis in clinical trials.

The success of electrical nerve stimulation has given researchers confidence that the use of electrical stimulation in conjunction with nervous system therapy is expected to be the non-drug treatment option. People try to abandon the traditional thinking of drug therapy and look for ways to regulate the repair from the inside out: (1) Whether they can record the electrical activity of peripheral nerves to extract metabolic information; (2) Whether blood glucose fluctuations can be improved by regulating electrical activity. This therapeutic approach, which alleviates metabolic dysfunction by stimulating relevant nerves and silences or activating nerves to control organ and tissue function, is called bioelectronic medicine [77].

Studies of the nervous system are gradually unraveling the question of how neural circuits regulate blood glucose in the human body. On the one hand, pancreatic β-cells have been found to be electrochemically active cells that are able to depolarize and release insulin in response to glucose activation. Since adjacent pancreatic β-cells are connected through the interstice, depolarization will spread throughout the network and islets. On the other hand, the activity of pancreatic β-cell is also regulated by direct innervation. Acetylcholine (ACh) released from nerve endings of efferent fibers in the vagus nerve increases insulin release in response to glucose stimulation [78]. However, the complexity of the nervous system also has resulted in its characteristics of multi-pathway, multi-target, and multi-synergistic glucose regulation circuits. In addition to the neural pathways found around the pancreatic β-cells, the central nervous system also plays a role in glucose homeostasis and numerous peripheral nerves, which directly control the endocrine function of the pancreas.

To date, researchers have only a sketchy understanding of the physiological mechanisms behind this neurometabolic interaction, while multicellular biological circuits are expected to be properly designed to replace and restore the lost function, which diseased tissues operate to complete the body's autonomous response in metabolic diseases. In terms of DM, activation or inhibition of neural activity in specific groups of neurons can be used to achieve glucose homeostasis. Over the past few years, new treatment strategies for T2DM have been developed based on electrical stimulation of peripheral nerve fibers. These studies focus on the effects of neural stimulation on food intake and insulin sensitivity recovery [79]. Joana et al. used kilohertz frequency alternating current (KHFAC) modulation to electrically stimulate the carotid sinus nerve (CSN). The glucose tolerance and insulin sensitivity of mice were improved to a certain extent under the appropriate KHFAC pulse frequency, which was designed in the form of rectangular pulses, with several clinical experiments showing good responses. Therefore, bioelectronic medicine based on nervous system regulation is expected to provide more entry points for DM treatment.

4. Summary and Prospect

The correlation between electrochemical biosensing technology and existing diagnosis and treatment technology for DM indicates that electrochemistry plays an important role in such processes, including reaction principle at the biochemical level in diagnosis and treatment technology, signal transduction at the electronic level, or the construction of dynamic models. It connects various technologies together to construct biological signal transduction pathways with different functions and application scenes, making contributions to the improvement in the efficiency, portability, timeliness, and safety of DM diagnosis and treatment (Figure 7). This is of great significance at a time when the global diabetes situation is becoming increasingly serious.

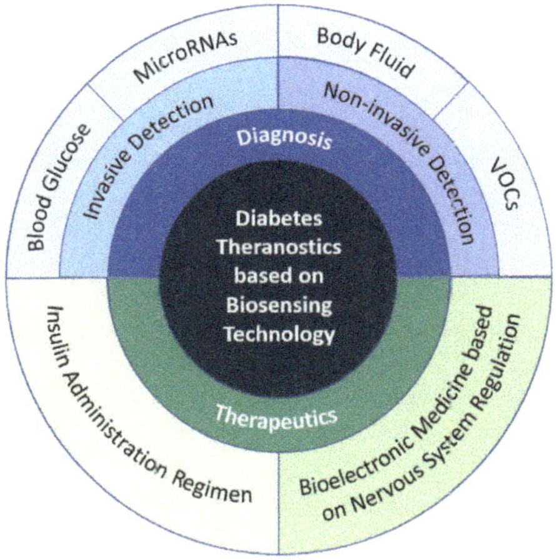

Figure 7. Diabetes theranostics based on biosensing technology.

4.1. Diagnosis of Diabetes

For diagnosis of diabetes, firstly, the diversified detection methods of blood glucose and the relevant principles of mainstream blood glucose measurements (enzymatic and non-enzymatic glucose electrochemical sensors) are introduced, including direct blood collection and indirect analyses of external body fluids (tears, saliva, sweat, tissue fluid, etc.). Followed by the proposal of unreliability and non-uniqueness of single indicators, some biochemical indicators (metabolomics markers, molecular markers, etc.) that have been proved to be associated with diabetes are listed on the basis of the principal ponent analysis. Further, breath measurement methods are mentioned in detail as an application of VOCs.

The development of glucose detection technology based on electrochemistry has made great progress in the perspective of diabetes diagnosis. Researchers have expanded more diversified diagnostic ideas based on traditional glucose-related electrochemical reactions, including but not limited to the joint application of new nanomaterial reaction chains and enzyme electrodes, the integration of non-invasive/minimally invasive electrochemical sensor platforms and wearable bioelectronics and self-powered biosensors. Although there are some problems with the development of advanced material technology and computational models, such as accuracy, reliability, and environmental impact in the realization process, it is expected to develop the sensor system with the more accurate functions required. In addition, as the limitations of single biomarker detection are gradually realized, multi-biomarker analysis technology has come to the stage. Recently, the convergence of technologies has become a trend that advance the integration of different surface chemical compositions and detection principles in a single device. In virtue of methods for simplification of the complexity in high-dimensional data, such as principal component analysis, the detection techniques of different combinations of biochemical indexes become more and more mature, preventing conflicting indicators from influencing results. Future diabetes-related sensing strategies tend to focus on the improvement in cost performance, accuracy, precision, selectivity, and stability, providing more comfortable and safe conditions for patients. Advances with the concept of miniaturization and portability, better software, and hardware facilities can be provided in booming electrochemical biosensing technology.

4.2. Treatment of Diabetes

Considering the demand for the treatment of diabetes, the advantages and disadvantages of different drug delivery methods are introduced in Section 3, drawing forth the typical cases of current controlled drug delivery methods (microneedle patch, artificial pancreas, etc.) and their general principles. For non-drug treatment methods, we have summarized the theoretical studies and empirical studies on the use of the nervous system to regulate insulin secretion in DM.

The hygiene and regulation of dosage is very important in terms of the treatment of diabetes. In view of the shortcomings of traditional oral drugs and subcutaneous injection, glucose and its Gox-related reactions show the possibility of controlled insulin release, which starts to build a long-term independent and sustained-release insulin delivery channel. Researchers have paid more attention to microneedle patches recently due to their low risk of infection and portable characteristics, promoting the progress of modified polymer micelles/vesicles in controlled slow-release insulin delivery.

Generally, the emergence of resistance to the treatment of chronic metabolic diseases prompts the urgent demand for new non-drug alternative therapies. Drawing inspiration from the successful application of spinal cord electrical stimulation to treat chronic pain, researchers turn their attention to electrotherapy based on continuous in-depth research on human metabolic neural pathways. However, on account of the high complexity of the human nervous system, the relevant neural pathways of blood glucose regulation have not been sorted out until now. While neurotherapy is promising, there are still a lot of research depths and hurdles to overcome. It is worth mentioning that the combination of existing neural pathways and the closed-loop control of the artificial pancreas has become a new direction, which is expected to simultaneously instruct the artificial pancreas to acquire information and repair the body damage to reduce the amount of insulin administration to a certain extent.

Funding: This research was funded by the National Natural Science Foundation of China (21974019, 82027806).

Institutional Review Board Statement: Not applicable.

Informed Consent Statement: Not applicable.

Data Availability Statement: No research data are available.

Conflicts of Interest: The authors declare no conflict of interest.

References

1. Leslie, R.D.; Palmer, J.; Schloot, N.C.; Lernmark, A. Diabetes at the crossroads: Relevance of disease classification to pathophysiology and treatment. *Diabetologia* **2016**, *59*, 13–20. [CrossRef] [PubMed]
2. Adler, A.; Bennett, P.; Chair, S.C.; Gregg, E.; Narayan, K.M.V.; Schmidt, M.I.; Sobngwi, E.; Tajima, N.; Tandon, N.; Unwin, N.; et al. Reprint of: Classification of diabetes mellitus. *Diabetes Res. Clin. Pract.* **2021**, 108972. [CrossRef] [PubMed]
3. Burahmah, J.; Zheng, D.; Leslie, R.D. Adult-onset type 1 diabetes: A changing perspective. *Eur. J. Intern. Med.* **2022**, *104*, 7–12. [CrossRef] [PubMed]
4. Zhang, L.W.; Ruan, M.H.; Liu, J.L.; He, C.H.; Yu, J.R. Progress on research and development in diabetes mellitus. *Yi Chuan = Hered.* **2022**, *44*, 824–839.
5. Bezen, D.; Vurgun, E.; Dursun, H. Evaluation of urinary haptoglobin level as a biomarker of diabetic nephropathy in children with type 1 diabetes. *Pediatr. Nephrol.* **2023**. [CrossRef]
6. Bhatty, A.; Baig, S.; Fawwad, A.; Rubab, Z.E.; Shahid, M.A.; Waris, N. Association of zinc transporter-8 autoantibody (ZnT8A) with type 1 diabetes mellitus. *Cureus* **2020**, *12*, e7263. [CrossRef]
7. Liang, W. Type 2 Diabetes Pathogenesis and Treatment. *Smart Healthc.* **2021**, *7*, 43–45.
8. International Diabetes Federation. *IDF Diabetes Atlas*, 10th ed.; International Diabetes Federation: Brussels, Belgium, 2021.
9. Wang, L.; Gao, P.; Zhang, M.; Huang, Z.; Zhang, D.; Deng, Q.; Li, Y.; Zhao, Z.; Qin, X.; Jin, D.; et al. Prevalence and ethnic pattern of diabetes and prediabetes in China in 2013. *JAMA* **2017**, *317*, 2515–2523. [CrossRef]
10. Ramachandran, A.; Ma, R.C.W.; Snehalatha, C. Diabetes in asia. *Lancet* **2010**, *375*, 408–418. [CrossRef]
11. Sun, X.R.; Yong, Q.M.; Peng, T.; Gao, W.J.; Yang, L. A visual analysis of the research progress in information technology-enabled diabetes management in China and abroad. *Chin. J. Clin. Healthc.* **2022**, *25*, 662–666.
12. Kalish, J.; Hamdan, A. Management of diabetic foot problems. *J. Vasc. Surg.* **2010**, *51*, 476–486. [CrossRef]
13. Matijević, T.; Talapko, J.; Meštrović, T.; Matijević, M.; Erić, S.; Erić, I.; Škrlec, I. Understanding the multifaceted etiopathogenesis of foot complications in individuals with diabetes. *World J. Clin. Cases* **2023**, *11*, 1669–1683. [CrossRef] [PubMed]
14. Walsh, J.W.; Hoffstad, O.J.; Sullivan, M.O.; Margolis, D.J. Association of diabetic foot ulcer and death in a population-based cohort from the United Kingdom. *Diabet. Med.* **2016**, *33*, 1493–1498. [CrossRef] [PubMed]
15. Mansueto, G.; Niola, M.; Napoli, C. Can COVID 2019 induce a specific cardiovascular damage or it exacerbates pre-existing cardiovascular diseases? *Pathol.-Res. Pract.* **2020**, *216*, 153086. [CrossRef]
16. Guan, W.; Ni, Z.; Hu, Y.; Liang, W.-H.; Ou, C.-Q.; He, J.-X.; Liu, L.; Shan, H.; Lei, C.-L.; Hui, D.S.C.; et al. Clinical characteristics of coronavirus disease 2019 in China. *N. Engl. J. Med.* **2020**, *382*, 1708–1720. [CrossRef] [PubMed]
17. Harding, J.L.; Oviedo, S.A.; Ali, M.K.; Ofotokun, I.; Gander, J.C.; Patel, S.A.; Magliano, D.J.; Patzer, R.E. The bidirectional association between diabetes and long-COVID-19–A systematic review. *Diabetes Res. Clin. Pract.* **2022**, *195*, 110202. [CrossRef] [PubMed]
18. McCarthy, M.I. Painting a new picture of personalized medicine for diabetes. *Diabetologia* **2017**, *60*, 793–799. [CrossRef]
19. Li, L.; Cheng, W.Y.; Glicksberg, B.S.; Gottesman, O.; Tamler, R.; Chen, R.; Bottinger, E.P.; Dudley, J.T. Identification of type 2 diabetes subgroups through topological analysis of patient similarity. *Sci. Transl. Med.* **2015**, *7*, 311ra174. [CrossRef]
20. Udler, M.S.; Kim, J.; von Grotthuss, M.; Bonàs-Guarch, S.; Cole, J.B.; Chiou, J.; Boehnke, M.; Laakso, M.; Atzmon, G.; Glaser, B.; et al. Type 2 diabetes genetic loci informed by multi-trait associations point to disease mechanisms and subtypes: A soft clustering analysis. *PLoS Med.* **2018**, *15*, e1002654. [CrossRef]
21. Herder, C.; Roden, M. A novel diabetes typology: Towards precision diabetology from pathogenesis to treatment. *Diabetologia* **2022**, *65*, 1770–1781. [CrossRef]
22. Clark, L.C., Jr.; Lyons, C. Electrode systems for continuous monitoring in cardiovascular surgery. *Ann. N. Y. Acad. Sci.* **1962**, *102*, 29–45. [CrossRef] [PubMed]
23. Updike, S.J.; Hicks, G.P. The enzyme electrode. *Nature* **1967**, *214*, 986–988. [CrossRef] [PubMed]
24. Newman, J.D.; Turner, A.P.F. Home blood glucose biosensors: A commercial perspective. *Biosens. Bioelectron.* **2005**, *20*, 2435–2453. [CrossRef] [PubMed]
25. Zhang, X.E. Screen-printing methods for biosensor production. In *Biosensors: A Practical Approach*, 2nd ed.; Cooper, J.M., Cass, A.E.G., Eds.; Oxford University Press: New York, NY, USA, 2004; pp. 41–58.
26. Kyvik, K.O.; Traulsen, J.; Reinholdt, B.; Frøland, A. The ExacTech blood glucose testing system. *Diabetes Res. Clin. Pract.* **1990**, *10*, 85–90. [CrossRef] [PubMed]
27. Shoaib, A.; Darraj, A.; Khan, M.E.; Azmi, L.; Alalwan, A.; Alamri, O.; Tabish, M.; Khan, A.U. A Nanotechnology-Based Approach to Biosensor Application in Current Diabetes Management Practices. *Nanomaterials* **2023**, *13*, 867. [CrossRef] [PubMed]
28. Vashist, S.K.; Zheng, D.; Al-Rubeaan, K.; Luong, J.H.T.; Sheu, F.S. Technology behind commercial devices for blood glucose monitoring in diabetes management: A review. *Anal. Chim. Acta* **2011**, *703*, 124–136. [CrossRef]
29. Teymourian, H.; Barfidokht, A.; Wang, J. Electrochemical glucose sensors in diabetes management: An updated review (2010–2020). *Chem. Soc. Rev.* **2020**, *49*, 7671–7709. [CrossRef]
30. Scandurra, A.; Censabella, M.; Boscarino, S.; Condorelli, G.G.; Grimaldi, M.G.; Ruffino, F. Fabrication of Cu(II)oxide-hydroxide nanostructures onto graphene paper by laser and thermal processes for sensitive nano-electrochemical sensing of glucose. *Nanotechnology* **2021**, *33*, 045501. [CrossRef]

31. Wang, C.; Nie, X.G.; Shi, Y.; Zhou, Y.; Xu, J.J.; Xia, X.H.; Chen, H.Y. Direct plasmon-accelerated electrochemical reaction on gold nanoparticles. *ACS Nano* **2017**, *11*, 5897–5905. [CrossRef]
32. Zhao, Y.; Zhao, W.; Chen, H.Y.; Xu, J.J. Dark-field microscopic real-time monitoring the growth of Au on Cu_2O nanocubes for ultra-sensitive glucose detection. *Anal. Chim. Acta* **2021**, *1162*, 338503. [CrossRef]
33. Wang, Y.; Pan, R.; Jiang, D.; Jiang, D.C.; Chen, H.Y. Nanopipettes for the electrochemical study of enhanced enzymatic activity in a femtoliter space. *Anal. Chem.* **2021**, *93*, 14521–14526. [CrossRef] [PubMed]
34. Tang, Z.; Du, X.; Louie, R.F.; Kost, G.J. Effects of drugs on glucose measurements with handheld glucose meters and a portable glucose analyzer. *Am. J. Clin. Pathol.* **2000**, *113*, 75–86. [CrossRef] [PubMed]
35. Zhu, D.D.; Tan, Y.R.; Zheng, L.W.; Lao, J.Z.; Liu, J.Y.; Yu, J.; Chen, P. Microneedle-Coupled Epidermal Sensors for In-Situ-Multiplexed Ion Detection in Interstitial Fluids. *ACS Appl. Mater. Interfaces* **2023**, *15*, 14146–14154. [CrossRef] [PubMed]
36. Chen, Y.; Lu, S.; Zhang, S.; Li, Y.; Qu, Z.; Chen, Y.; Lu, B.W.; Wang, X.Y.; Feng, X. Skin-like biosensor system via electrochemical channels for noninvasive blood glucose monitoring. *Sci. Adv.* **2017**, *3*, e1701629. [CrossRef]
37. Garcia-Carmona, L.; Martin, A.; Sempionatto, J.R.; Moreto, J.R.; Gonzalez, M.C.; Wang, J.; Escarpa, A. Pacifier biosensor: Toward noninvasive saliva biomarker monitoring. *Anal. Chem.* **2019**, *91*, 13883–13891. [CrossRef]
38. Gao, W.; Emaminejad, S.; Nyein, H.Y.Y.; Challa, S.; Chen, K.; Peck, A.; Fahad, H.M.; Ota, H.; Shiraki, H.; Kiriya, D.; et al. Fully integrated wearable sensor arrays for multiplexed in situ perspiration analysis. *Nature* **2016**, *529*, 509–514. [CrossRef]
39. Lee, H.; Song, C.; Hong, Y.S.; Kim, M.S.; Cho, H.R.; Kang, T.; Shin, K.; Choi, S.H.; Hyeon, T.; Kim, D.H. Wearable/disposable sweat-based glucose monitoring device with multistage transdermal drug delivery module. *Sci. Adv.* **2017**, *3*, e1601314. [CrossRef]
40. Gao, W.; Brooks, G.A.; Klonoff, D.C. Wearable physiological systems and technologies for metabolic monitoring. *J. Appl. Physiol.* **2018**, *124*, 548–556. [CrossRef]
41. Li, Z.; Yun, J.; Li, X.; Kim, M.; Li, J.; Lee, D.H.; Wu, A.Y.; Lee, S.W. Power-Free Contact Lens for Glucose Sensing. *Adv. Funct. Mater.* **2023**, 2304647. [CrossRef]
42. Son, J.; Bae, G.Y.; Lee, S.; Lee, G.; Kim, S.W.; Kim, D.; Chung, S.; Cho, K. Cactus-Spine-Inspired Sweat-Collecting Patch for Fast and Continuous Monitoring of Sweat. *Adv. Mater.* **2021**, *33*, 2102740. [CrossRef]
43. Anwar, F.; Ejaz, M.Y.; Mosavi, A. A comparative analysis on diagnosis of diabetes mellitus using different approaches–A survey. *Inform. Med. Unlocked* **2020**, *21*, 100482. [CrossRef]
44. Kim, J.; Sempionatto, J.R.; Imani, S.; Hartel, M.C.; Barfidokht, A.; Tang, G.D.; Campbell, A.S.; Mercier, P.P.; Wang, J. Simultaneous monitoring of sweat and interstitial fluid using a single wearable biosensor platform. *Adv. Sci.* **2018**, *5*, 1800880. [CrossRef] [PubMed]
45. Bandodkar, A.J.; Jia, W.; Yardımcı, C.; Wang, X.; Ramirez, J.; Wang, J. Tattoo-based noninvasive glucose monitoring: A proof-of-concept study. *Anal. Chem.* **2015**, *87*, 394–398. [CrossRef] [PubMed]
46. Kownacka, A.E.; Vegelyte, D.; Joosse, M.; Anton, N.; Toebes, B.J.; Lauko, J.; Buzzacchera, I.; Lipinska, K.; Wilson, D.A.; Geelhoed-Duijvestijn, N.; et al. Clinical evidence for use of a noninvasive biosensor for tear glucose as an alternative to painful finger-prick for diabetes management utilizing a biopolymer coating. *Biomacromolecules* **2018**, *19*, 4504–4511. [CrossRef] [PubMed]
47. Ogurtsova, K.; da Rocha Fernandes, J.D.; Huang, Y.; Linnenkamp, U.; Guariguata, L.; Cho, N.H.; Cavan, D.; Shaw, J.E.; Makaroff, L.E. IDF Diabetes Atlas: Global estimates for the prevalence of diabetes for 2015 and 2040. *Diabetes Res. Clin. Pract.* **2017**, *128*, 40–50. [CrossRef]
48. Ashoori, M.R.; Rahmati-Yamchi, M.; Ostadrahimi, A.; Aval, S.F.; Zarghami, N. MicroRNAs and adipocytokines: Promising biomarkers for pharmacological targets in diabetes mellitus and its complications. *Biomed. Pharmacother.* **2017**, *93*, 1326–1336. [CrossRef]
49. Lees, T.; Nassif, N.; Simpson, A.; Shad-Kaneez, F.; Martiniello-Wilks, R.; Lin, Y.G.; Jones, A.; Qu, X.Q.; Lal, S. Recent advances in molecular biomarkers for diabetes mellitus: A systematic review. *Biomarkers* **2017**, *22*, 604–613. [CrossRef]
50. Cao, Y.M. Advances in biomarkers of type 2 diabetes mellitus. *J. Med. Theory Pract.* **2021**, *34*, 2.
51. Dehghanbanadaki, H.; Asili, P.; Haji Ghadery, A.; Mirahmad, M.; Dehnavi, A.Z.; Parsaei, A.; Baradaran, H.R.; Azami, M.; Jose Justo da Silva, G.; Parvan, R.; et al. Diagnostic accuracy of circular RNA for diabetes Mellitus: A systematic review and diagnostic Meta-analysis. *BMC Med. Genom.* **2023**, *16*, 48. [CrossRef]
52. Zhang, Z.; Yang, T.; Xiao, J. Circular RNAs: Promising biomarkers for human diseases. *EBioMedicine* **2018**, *34*, 267–274. [CrossRef]
53. Fan, W.; Pang, H.; Xie, Z.; Huang, G.; Zhou, Z.G. Circular RNAs in diabetes mellitus and its complications. *Front. Endocrinol.* **2022**, *13*, 885650. [CrossRef] [PubMed]
54. Esokkiya, A.; Murugasenapathi, N.K.; Kumar, S.; Sudalaimani, S.; Santhosh, B.; Tamilarasan, P.; Sivakumar, C.; Giribabu, K. Electrochemically activated copper nitroprusside as a catalyst for sensing of carcinogenic acetaldehyde in red wine. *Sens. Actuators B Chem.* **2022**, *363*, 131798. [CrossRef]
55. Yan, K.; Zhang, D.; Wu, D.; Wei, H.; Lu, G.M. Design of a breath analysis system for diabetes screening and blood glucose level prediction. *IEEE Trans. Biomed. Eng.* **2014**, *61*, 2787–2795. [CrossRef] [PubMed]
56. Dixit, K.; Fardindoost, S.; Ravishankara, A.; Tasnim, N.; Hoorfar, M. Exhaled breath analysis for diabetes diagnosis and monitoring: Relevance, challenges and possibilities. *Biosensors* **2021**, *11*, 476. [CrossRef]
57. Su, Y.; Yang, T.; Zhao, X.; Cai, Z. A wireless energy transmission enabled wearable active acetone biosensor for non-invasive prediabetes diagnosis. *Nano Energy* **2020**, *74*, 104941. [CrossRef]

58. Wang, H.; Wang, J.; Yao, K.; Fu, J.; Xia, X.; Zhang, R.; Li, J.; Xu, G.; Wang, L.; Yang, J.; et al. A paradigm shift fully self-powered long-distance wireless sensing solution enabled by discharge-induced displacement current. *Sci. Adv.* **2021**, *7*, eabi6751. [CrossRef]
59. Werner, E.A.; Bell, J. CCXIV.—The preparation of methylguanidine, and of ββ-dimethylguanidine by the interaction of dicyanodiamide, and methylammonium and dimethylammonium chlorides respectively. *J. Chem. Soc. Trans.* **1922**, *121*, 1790–1794. [CrossRef]
60. Slotta, K.H.; Tschesche, R. Über Biguanide, II.: Die blutzucker-senkende Wirkung der Biguanide. *Berichte Der Dtsch. Chem. Ges. (A B Ser.)* **1929**, *62*, 1398–1405. [CrossRef]
61. Bailey, C.J.; Day, C. Metformin: Its botanical background. *Pract. Diabetes Int.* **2004**, *21*, 115–117. [CrossRef]
62. Kazeem, M.I.; Bankole, H.A.; Fatai, A.A.; Saibu, G.M.; Wusu, A.D. Genus Aloe as sources of antidiabetic, antihyperglycemic and hypoglycemic agents: A review. *S. Afr. J. Bot.* **2022**, *147*, 1070–1077. [CrossRef]
63. Wang, L.L.; Du, Y.; Li, S.M.; Cheng, F.; Zhang, N.-N.; Chen, R.; Cui, X.; Yang, S.-G.; Fan, L.-L.; Wang, J.-T.; et al. Design, synthesis and evaluation of tetrahydrocarbazole derivatives as potential hypoglycemic agents. *Bioorganic Chem.* **2021**, *115*, 105172. [CrossRef]
64. Rydén, L.; Lindsten, J. The history of the Nobel prize for the discovery of insulin. *Diabetes Res. Clin. Pract.* **2021**, *175*, 108819. [CrossRef]
65. Ye, J. Commentary: Inhibitors of mitochondrial respiratory chain in the treatment of type 2 diabetes. *Acta Pharm. Sin. B* **2023**, *13*, 2807–2808. [CrossRef]
66. Sebastian, S.A.; Co, E.L.; Mehendale, M.; Hameed, M. Insulin analogs in the treatment of type II diabetes and future perspectives. *Dis. -A-Mon.* **2023**, *69*, 101417. [CrossRef]
67. Zheng, J.; Wu, X.H. Research progress of microneedle patch as insulin delivery device. *Chin. J. Diabetes* **2022**, *30*, 3.
68. Zhang, Y.; Wang, J.; Yu, J.; Wen, D.; Kahkoska, A.R.; Lu, Y.; Zhang, X.D.; Buse, J.B.; Gu, Z. Bioresponsive microneedles with a sheath structure for H_2O_2 and pH cascade-triggered insulin delivery. *Small* **2018**, *14*, 1704181. [CrossRef]
69. Chen, W.; Tian, R.; Xu, C.; Yung, B.C.; Wang, G.; Liu, Y.; Ni, Q.; Zhang, F.; Zhou, Z.; Wang, J.; et al. Microneedle-array patches loaded with dual mineralized protein/peptide particles for type 2 diabetes therapy. *Nat. Commun.* **2017**, *8*, 1777. [CrossRef]
70. Tong, Z.; Zhou, J.; Zhong, J.; Tang, Q.; Lei, Z.; Luo, H.; Ma, P.; Liu, X. Glucose-and H_2O_2-responsive polymeric vesicles integrated with microneedle patches for glucose-sensitive transcutaneous delivery of insulin in diabetic rats. *ACS Appl. Mater. Interfaces* **2018**, *10*, 20014–20024. [CrossRef]
71. Nwokolo, M.; Hovorka, R. The artificial pancreas and type 1 diabetes. *J. Clin. Endocrinol. Metab.* **2023**, *108*, 1614–1623. [CrossRef]
72. Mahmud, F.; Isse, N.H.; Daud, N.A.M.; Morsin, M. Evaluation of PD/PID controller for insulin control on blood glucose regulation in a Type-I diabetes. In *AIP Conference Proceedings*; AIP Publisher: College Park, MD, USA, 2017; Volume 1788, p. 030072.
73. Magni, L.; Raimondo, D.M.; Dalla Man, C.; De Nicolao, G.; Kovatchev, B.; Cobelli, C. Model predictive control of glucose concentration in subjects with type 1 diabetes: An in silico trial. *IFAC Proc. Vol.* **2008**, *41*, 4246–4251. [CrossRef]
74. Herrero, P.; Georgiou, P.; Oliver, N.; Reddy, M.; Johnston, D.; Toumazou, C. A composite model of glucagon-glucose dynamics for in silico testing of bihormonal glucose controllers. *J. Diabetes Sci. Technol.* **2013**, *7*, 941–951. [CrossRef]
75. Bergenstal, R.M.; Garg, S.; Weinzimer, S.A.; Buckingham, B.A.; Bode, B.W.; Tamborlane, W.V.; Kaufman, F.R. Safety of a hybrid closed-loop insulin delivery system in patients with type 1 diabetes. *JAMA* **2016**, *316*, 1407–1408. [CrossRef]
76. Güemes, A.; Herrero, P.; Georgiou, P. A novel glucose controller using insulin sensitivity modulation for management of type 1 diabetes. In Proceedings of the 2019 IEEE International Symposium on Circuits and Systems (ISCAS), Hokkaido, Japan, 26–29 May 2019; pp. 1–5.
77. Güemes Gonzalez, A.; Etienne-Cummings, R.; Georgiou, P. Closed-loop bioelectronic medicine for diabetes management. *Bioelectron. Med.* **2020**, *6*, 1–7. [CrossRef]
78. Seicol, B.J.; Bejarano, S.; Behnke, N.; Guo, L. Neuromodulation of metabolic functions: From pharmaceuticals to bioelectronics to biocircuits. *J. Biol. Eng.* **2019**, *13*, 11. [CrossRef]
79. Sacramento, J.F.; Chew, D.J.; Melo, B.F.; Donegá, M.; Dopson, W.; Guarino, M.P.; Robinson, A.; Prieto-Lloret, J.; Patel, S.; Holinski, B.J.; et al. Bioelectronic modulation of carotid sinus nerve activity in the rat: A potential therapeutic approach for type 2 diabetes. *Diabetologia* **2018**, *61*, 700–710. [CrossRef]

chemosensors

MDPI

Article

Electrochemical Detection of Tumor Cell-Derived Exosomes Based on Cyclic Enzyme Scission and Hybridization Chain Reaction Dual-Signal Amplification

Die Sun, Qunqun Guo, Hui Zhang * and Chenxin Cai

Jiangsu Key Laboratory of New Power Batteries, Jiangsu Collaborative Innovation Center of Biomedical Functional Materials, School of Chemistry and Materials Science, Nanjing Normal University, Nanjing 210023, China

* Correspondence: zhangh@njnu.edu.cn

Abstract: Tumor cell-derived exosomes are considered a potential source of cancer biomarkers. Here, we developed an electrochemical sensing platform for the rapid and simple detection of exosomes, using the CCRF-CEM exosome as a model. The platform utilizes cyclic nicking enzyme cleavage and a hybridization chain reaction (HCR) for dual-signal amplification. A hairpin aptamer probe (HAP) containing an aptamer was designed for the assay. The specific binding between the aptamer and PTK7, present on the exosome surface, causes a conformational change in the HAP. This facilitates hybridization between the HAP and the linker DNA, which subsequently triggers cyclic cleavage of the nicking endonuclease towards the linker DNA. Therefore, exosome detection is transformed into DNA detection. By combining this approach with HCR signal amplification, we achieved high-sensitivity electrochemical detection of CCRF-CEM exosomes, down to 1.1×10^4 particles/mL. Importantly, this assay effectively detected tumor exosomes in complex biological fluids, demonstrating the potential for clinical diagnosis.

Keywords: exosomes; electrochemistry; cyclic enzyme scission; hybridization chain reaction

Citation: Sun, D.; Guo, Q.; Zhang, H.; Cai, C. Electrochemical Detection of Tumor Cell-Derived Exosomes Based on Cyclic Enzyme Scission and Hybridization Chain Reaction Dual-Signal Amplification. *Chemosensors* **2023**, *11*, 415. https://doi.org/10.3390/chemosensors11070415

Academic Editor: Győző G. Láng

Received: 24 June 2023
Revised: 20 July 2023
Accepted: 21 July 2023
Published: 23 July 2023

1. Introduction

Exosomes are small extracellular vesicles, ranging in size from 30 to 150 nm. They are released by various cell types and are present in most body fluids including breast milk, tears, blood, etc. [1,2]. Similar to cells, exosomes have a phospholipid bilayer structure containing diverse proteins, nucleic acids (DNA and RNA), and lipids [3,4]. In body fluids, exosomes act as carriers, transporting functional information to neighboring or distant cells and facilitating cell-to-cell communication without direct contact. This communication mode is crucial for the exchange of materials between cells [5,6].

Tumor cell-derived exosomes represent a potential source of cancer biomarkers and can be utilized for tumor screening by detecting relevant exosomal information [7]. Additionally, the exosome surface contains various membrane proteins that can serve as markers for tumor generation and reproduction [8,9]. Consequently, developing an efficient and sensitive detection platform for analyzing tumor cell-derived exosomes and their surface proteins is highly valuable for applications in areas such as basic research, clinical diagnosis, and molecular therapy [10–12].

Currently, many methods have been developed for detecting exosomes, including nanoparticle tracking analysis (NTA) [13], surface-enhanced Raman spectroscopy (SERS) [14–16], fluorescence [17–19], and electrochemical methods [20–23]. Among these techniques, electrochemical methods have gained considerable interest due to their simplicity, portability, affordability, wide measurement range, and high sensitivity [20–24]. For instance, Wang et al. [20] immobilized DNA tetrahedrons containing aptamers on the working electrode surface to capture exosomes derived from hepatoma cells, using the

$[Fe(CN)_6]^{3-/4-}$ redox couple for signal generation. Zhou et al. [21] immobilized thiolated anti-EpCAM aptamers on the working electrode to capture exosomes and used electrochemical signal probes consisting of Ag nanoparticles modified with anti-EpCAM aptamers and Cu nanoparticles modified with anti-PSMA aptamers for exosome detection. They also compared the expression levels of EpCAM and PSMA on the exosomes.

To enhance sensitivity, various signal amplification strategies have been used, including enzymatic amplification [25,26], DNA hybridization chain reaction (HCR) [27,28], DNA primer exchange reaction (PER) [29], hemin/G-quadruplex [30] and nanoparticles [23]. However, most of these methods involve the direct capture of exosomes by aptamers or antibodies on the electrode surface. The captured exosomes on the electrode surface can hinder electron transfer, which is unfavorable to electrochemical detection. To address this issue, several indirect detection methods have been developed. For example, our group [22] converted exosomes detection into DNA detection by employing aptamer recognition of exosomes. In this approach, the aptamer and partially complementary mDNA were modified on magnetic beads. The exosomes could specifically bind to the aptamer, resulting in the release of triple the amount of mDNA to amplify the signal. The signal was further amplified by adding exonuclease III for auxiliary circulation. This sensor offers a promising platform for highly sensitive detection of exosomes derived from human prostate tumor cells, with a detection limit of 7×10^4 particles/mL. Zhao et al. [31] have developed a signal amplified electrochemical method for exosome detection. The electrochemical signal was amplified by a three-dimensional DNA walker and exonuclease III-assisted signal amplification.

Herein, we report an electrochemical strategy for detecting tumor cell-derived exosomes using cyclic nicking enzyme cleavage and a hybridization chain reaction (HCR) for dual-signal amplification. We used CCRF-CEM exosomes as a model, together with tyrosine kinase-like 7 (PTK7), abundant on CCRF-CEM exosome membranes [32], as the target protein. A hairpin aptamer probe (HAP) was designed to contain the aptamer sequence sgc8 and a sequence complementary to the linker DNA, enabling conformational changes triggered by the target protein. In the presence of CCRF-CEM exosomes, sgc8 specifically bound PTK7 on the exosome membrane, causing HAP conformational changes that expose the hybridization region. The linker DNA hybridized to this HAP region, initiating nicking endonuclease-assisted cycling cleavage. Thus, small exosome quantities generated many cleaved DNA fragments (S1). Consequently, exosome detection became DNA detection. Combined with HCR amplification, the highly sensitive electrochemical detection of CCRF-CEM exosomes was achieved. Compared to previous work [22], this indirect detection strategy produces a signal DNA S1 based on HAP conformational changes, eliminating magnetic bead modification, separation and washing as well as making the method more convenient and concise. Additionally, this work has a lower detection limit (1.1×10^4 particles/mL) and wider linear range (8.0×10^4~3.2×10^8 particles/mL).

2. Experimental Section

2.1. Materials and Reagents

Nicking endonuclease Nb.BbvCI and 10 × NEBuffer were obtained from New England Biolabs (Ipswich, MA, USA). Tris (2-carboxyethyl) phosphine hydrochloride (TCEP) was purchased from Aladdin (Shanghai, China). 6-Mercapto-1-hexanol (MCH) was obtained from Sigma-Aldrich (St. Louis, MO, USA). The plasma samples from healthy mice were obtained from Jiangsu KeyGen Biotech. Co., Ltd. (Nanjing, China). RPMI-1640 medium, DMEM medium and the HPLC-purified synthetic oligonucleotides were obtained from the Sangon Biotechnology Co., Ltd. (Shanghai, China). The sequences of these oligonucleotides are listed in Table 1. All other chemicals used were of analytical grade.

A 0.1 M phosphate buffer solution (PBS) at pH 7.4, supplemented with 0.1 M NaCl, was used as the immobilization buffer. A 10 mM PBS solution at pH 7.4 containing 0.1 M NaCl was used for cleaning the gold electrode. A 10 mM Tris-HCl buffer at pH 7.4, supplemented with 5 μM hexaammineruthenium(III) chloride (RuHex), was used for electrochemical detection. All aqueous solutions were prepared using 18 MΩ·cm deionized water from a Millipore water purification system.

Table 1. The sequences of the DNA used in this work *.

Name	Sequence (5′-3′)
HAP	**ATC TAA CTG CTG CGC CGC CGG GAA AAT ACT GTA CGG TTA** **GA**A AAA AAA AAA AAT *CCT CAG* CAG TTA
Linker DNA	TCA GCA GGG AGG AAG ACA ATA TTA ACT *GC*$_{\Delta}$*T GAG* GAT AAA CG
Capture DNA	CGC AGT TAA TAT TGT C
Hairpin H1	TTC CTC CCT GCT GAC ACA GAT CAG CAG GG
Hairpin H2	TCA GCA GGG AGG AAC CCT GCT GAT CTG TG

* The sequence in bold is the sequence of aptamer sgc8. The sequence in *italic* is the recognition sequence of nicking endonuclease Nb.BbvCI. "$_{\Delta}$" is the nicking site.

2.2. Apparatus

Exosome collection was performed by ultracentrifugation using an Optima XPN (Beckman Coulter, Indianapolis, IN, USA). Transmission electron microscopy (TEM) was performed using a JEOL-2010 microscope operating at 120 kV. Exosome was negative stained with 3% uranyl acetate solution. Nanoparticle tracking analysis (NTA) was done with a ZetaView from Particle Metrix, Germany. Polyacrylamide gel electrophoresis (PAGE) used a JY600C electrophoresis apparatus and gel imaging with a Tanon-3500 system (Shanghai, China). Differential pulse voltammetry (DPV) was carried out on a CHI 760C electrochemical system (Chenhua, Shanghai, China) using a three-electrode setup with a Pt wire counter electrode, saturated calomel reference electrode (SCE), and modified gold working electrode.

2.3. Cell Culture and Exosome Extraction

All cells were incubated under standard cell culture conditions (5% CO_2 at 37 °C). CCRF-CEM (T-lymphoblast, human acute lymphoblastic leukemia) and Ramos (B-lymphocyte, human Burkitt's lymphoma) cells were cultured in an RPMI-1640 medium, while HeLa (human cervix carcinoma) cells were cultured in a DMEM medium. Both culture media were supplemented with 10% fetal bovine serum (FBS) and 100 IU/mL penicillin-streptomycin. Exosomes from different tumor cells were extracted by ultracentrifugation [33,34] as follows: After 48 h of culture in a serum-free medium, tumor cells were collected and centrifuged at 300× *g* for 10 min to remove dead cells. The resulting supernatant was centrifuged at 16,500× *g* for 20 min and filtered through a 0.22 μm filter. To precipitate the exosomes, the supernatant was ultracentrifugated at 110,000× *g* for 2 h and then washed with PBS. Finally, the collected exosomes were resuspended in PBS and stored at −80 °C until use. The plasma sample was subjected to ultracentrifugation at 110,000× *g* for 2 h, and the supernatant was collected as exosome-free plasma. All centrifugation steps were performed at 4 °C.

2.4. Detection Protocol

Before assembly, the 2 mm diameter gold electrode underwent pretreatment following a previously published method [35]. Briefly, the gold electrode was firstly polished to a mirror surface with 0.3 and 0.05 μm alumina slurry on a polishing cloth, followed by ultrasonic cleaning in ethanol and deionized water, respectively. The gold electrode was then immersed in a freshly prepared piranha solution for 15 min. After being washed thoroughly with deionized water, the electrode was electrochemically cleaned in a 0.5 M H_2SO_4 with a potential scan from −0.2 V to 1.7 V until a remarkable cyclic voltammetry was obtained. Finally, the electrode was soaked in ethanol for 0.5 h and then rinsed with a copious amount of deionized water and dried with nitrogen.

The HAP and linker DNA were heated to 95 °C for 5 min and cooled to room temperature to facilitate formation of a stable hairpin structure. Next, 10 μL of 0.5 μM thiolated probe DNA was dropped on the pre-treated gold electrode for assembly for 13 h at 4 °C. The electrode was then thoroughly rinsed with 10 mM PBS buffer. Subsequently, it was blocked with 1 mM MCH for 1 h at room temperature to eliminate nonspecific adsorption on the gold electrode surface. After thorough washing with 10 mM PBS, the gold electrode was stored at 4 °C for subsequent use.

The HAP (10 μM) and linker DNA (10 μM) were mixed with varying concentrations of exosomes at room temperature for 1 h to form aptamer–exosome complexes. Subsequently, 10 U of Nb.BbvCI and 10 × NEBuffer were added and incubated for 1 h. The gold electrode modified with probe DNA and MCH was then immersed in this solution and incubated at 37 °C for 1 h. The electrode was then thoroughly washed with 10 mM PBS and dried with nitrogen gas. Next, the electrode was immersed in 50 μL of freshly prepared reaction buffer containing H1 (0.5 μM) and H2 (0.5 μM) and incubated at room temperature for 2 h. After thorough washing with 10 mM PBS buffer, the electrode was immersed in Tris-HCl buffer containing RuHex for electrochemical measurements. Differential pulse voltammetry (DPV) measurements were performed by scanning the potential from −0.50 to +0.10 V, with a pulse amplitude of 50 mV and pulse width of 10 ms.

3. Results and Discussion

3.1. Mechanism of the Assay

In this study, the CCRF-CEM tumor cell-derived exosome was chosen as a model for the development of an electrochemical aptamer sensor. The detection mechanism is depicted in Scheme 1. The HAP used in this assay consists of two parts: the aptamer sgc8 (purple fragment) that specifically binds to the PTK7 protein (abundant on the CCRF-CEM exosome surface), and the hybridization region (light blue fragment) for signal amplification.

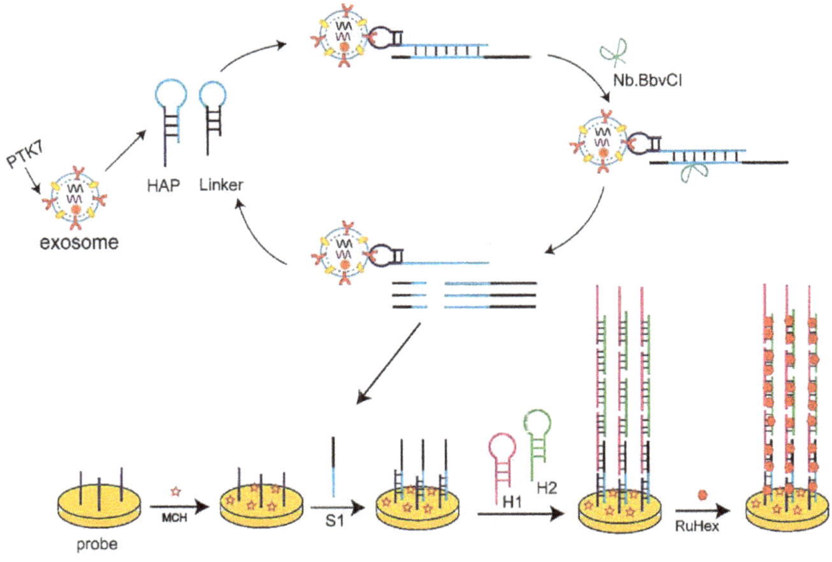

Scheme 1. Schematic of exosomes assay.

If no CCRF-CEM exosome is present, the sgc8 sequence of the HAP remains enclosed within the stem–loop structure, maintaining a stable configuration. Upon incubation with the CCRF-CEM exosomes, the sgc8 sequence of the HAP binds specifically to the PTK7 protein on the exosome membrane, leading to the conformational changes in the HAP and the release of the hybridization region. The released hybridization region opens the stem–loop structure of the linker DNA and hybridizes with it, forming a double-stranded substrate for the nicking enzyme Nb.BbvCI. Nb.BbvCI cleaves the hybridized DNA fragment [36,37], resulting in the release of the cleaved DNA fragment (S1) and hybridization region into the solution. The free hybridization region can then hybridize with more linker DNA to initiate the cyclic cleavage of Nb.BbvCI toward the linker DNA, releasing numerous S1 sequences into the solution.

The cleaved DNA fragment S1 acts as the initiator for the HCR. It hybridizes with the probe DNA on the working electrode, and in the presence of the auxiliary hairpin DNA H1 and H2, the HCR process occurs, resulting in the formation of a long DNA double helix. RuHex, an electroactive substance, can be adsorbed onto the negatively charged DNA double helix structure through electrostatic interactions, generating high electrochemical signals [22,24]. The introduction of Nb.BbvCI-assisted cycling scission and HCR dual signal amplification significantly enhances the detection sensitivity of the assay.

3.2. Characterization of the Exosomes

TEM analysis was conducted to characterize the size and morphology of the isolated CCRF-CEM exosomes. As depicted in Figure 1A, the TEM image reveals that the exosomes exhibit a characteristic cup-shaped morphology with an evident lipid bilayer membrane. The diameter of the exosomes ranges from 30 to 150 nm, which is consistent with reference values reported in the literature [31,34]. This confirms the successful collection of exosomes.

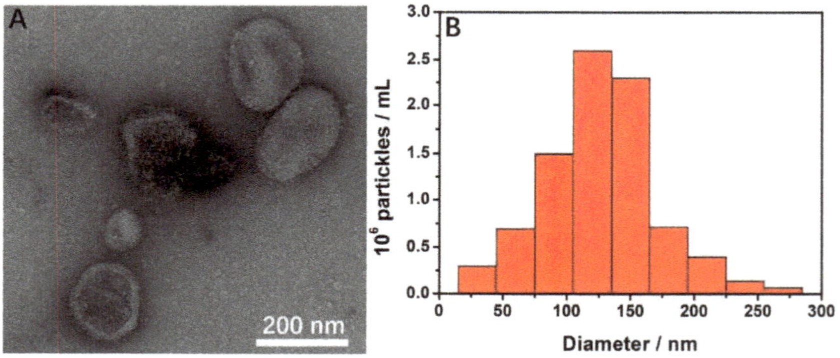

Figure 1. (**A**) TEM image and (**B**) particle size distribution of the extracted CCRF-CEM exosomes.

Additionally, NTA was employed to obtain the particle size distribution of the exosomes, as shown in Figure 1B. The y-axis represents the concentration of diluted exosomes, while the original concentration of exosomes was determined to be 1.6×10^9 particles/mL. The NTA results further support the successful isolation of exosomes.

3.3. Feasibility of the Assay

To evaluate the feasibility and effectiveness of the developed assay, the PAGE experiment was conducted. The results are presented in Figure 2.

In Figure 2A, it can be observed that in the absence of CCRF-CEM exosomes (lane 3), the HAP and linker DNA coexist without being cut by Nb.BbvCI, indicating that no hybridization occurs between the HAP and linker DNA without the presence of the target exosome. However, when CCRF-CEM exosomes were added and treated with Nb.BbvCI, the band corresponding to the linker DNA almost disappeared, and a new band appeared (lane 4). This observation suggests the target exosomes successfully trigger conformational changes in the HAP, leading to the formation of the exosome–aptamer complex and release of the hybridization region. Consequently, the stem of the linker DNA is opened by the hybridization region, allowing it to hybridize with the exosome–aptamer complex.

Nb.BbvCI then cleaves the hybridized DNA fragment, resulting in the release of S1 (new band in lane 4).

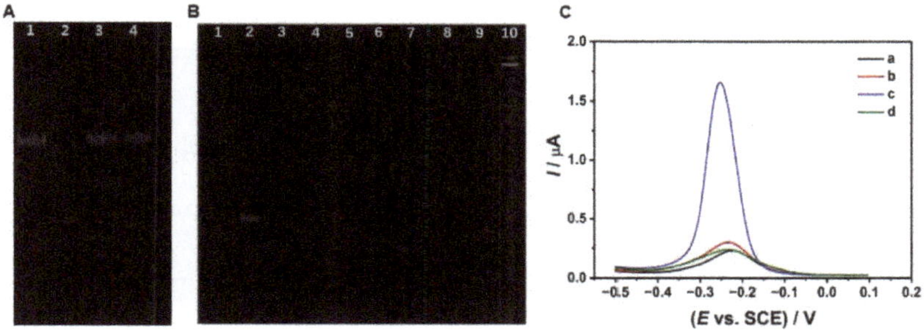

Figure 2. Gel electrophoresis of different samples: (**A**) lane 1, HAP; lane 2, linker DNA; lane 3, HAP + linker DNA + Nb.BbvCI; lane 4, HAP + linker DNA + Nb.BbvCI + CCRF-CEM exosomes; (**B**) Lane 1, probe DNA; lane 2, linker DNA; lane 3, probe DNA + linker DNA; Lane 4, S1; Lane 5, S1 + probe DNA; Lane 6, H1; Lane 7, H2; Lane 8, S1 + probe DNA + H1 + H2; Lane 9, HAP + linker DNA + Nb.BbvCI + CCRF-CEM exosomes + H1 + H2; Lane 10, DNA marker. (**C**) DPV response of gold electrode in different processes: (a) probe DNA/MCH-modified gold electrode; (b) incubation with CCRF-CEM exosomes, HAP, linker DNA and Nb.BbvCI; (c) further incubation with H1 and H2; (d) the probe DNA/MCH-modified gold electrode was treated with HAP, linker DNA and Nb.BbvCI, and then treated with H1 and H2.

Figure 2B provides further evidence of the assay's viability. Lane 3 is a mixture of the probe DNA and linker DNA. It clearly shows that the probe DNA does not hybridize with the linker DNA. However, in the presence of S1, hybridization occurs between the probe DNA and S1 (lane 5). Subsequently, S1 can further hybridize with H1 and H2, resulting in the generation of multiple new bands (lanes 8 and 9). This outcome confirms that the presence of the target exosomes successfully triggers the HCR.

Overall, the results from the PAGE experiment provide strong evidence supporting the feasibility of the proposed strategy for the detection of tumor cell-derived exosomes.

To further validate the feasibility of the developed strategy, DPV curves were recorded. As shown in Figure 2C, the gold electrode modified with the probe DNA and MCH exhibited a small cathodic peak (curve a). The small cathodic peak can be attributed to the electrochemical reduction of adsorbed RuHex to the probe DNA through electrostatic interactions [22,24]. After incubating the working electrode in the solution containing the target exosomes, HAP, linker DNA, and Nb.BbvCI, a slight increase in the peak current was observed (curve b). This increase is due to the successful triggering of HAP conformational changes by the target exosomes. The HAP then hybridizes with the linker DNA to generate the nicking site for Nb.BbvCI. Consequently, Nb.BbvCI cleaves the linker DNA, releasing S1 [36,37]. The released S1 further hybridizes with the probe DNA on the working electrode, and more RuHex molecules are absorbed, leading to an increase in the peak current.

Upon introduction of H1 and H2, a large cathodic peak is observed in curve c. This increase in the peak current can be attributed to the successful occurrence of the HCR triggered by the presence of S1. The HCR could result in the formation of a long DNA double-helix structure, facilitating the adsorption of many RuHex molecules onto the DNA through electrostatic interactions. This leads to a significantly amplified electrochemical signal. In curve d, when the target exosome is absent, the peak current remains relatively constant. This observation indicates that the HCR does not occur in the absence of target exosomes.

The DPV curves demonstrate the feasibility of the prepared electrochemical aptamer sensor for the detection of specific exosomes. The sensor utilizes the aptamer's specificity for the membrane protein of the exosomes, enabling the specific recognition and detection of the target exosomes.

Overall, the results from the PAGE and DPV experiments provide strong evidence supporting the feasibility of the proposed strategy for detecting tumor cell-derived exosomes.

3.4. Optimization of Experimental Conditions

In order to achieve excellent performance, the concentration and incubation time of Nb.BbvCI, as well as the concentrations of the probe DNA, H1, and H2, together with the HCR reaction time, were evaluated. Figure 3A illustrates the variation in the current with different concentrations of Nb.BbvCI. It can be observed that the current gradually increases with the increase in Nb.BbvCI concentration, reaching a relatively stable plateau at 10 U. Therefore, the optimal Nb.BbvCI concentration for the assay was determined to be 10 U.

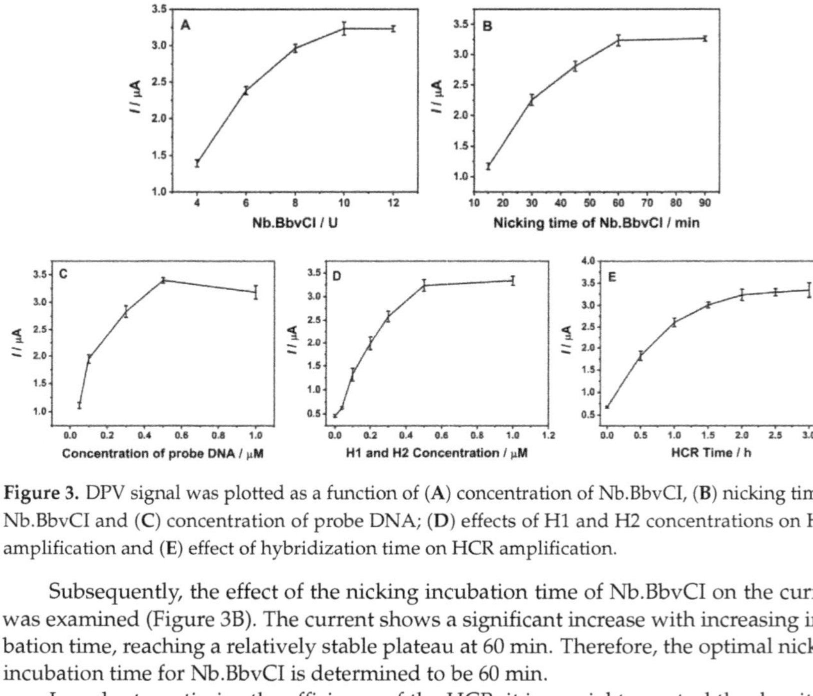

Figure 3. DPV signal was plotted as a function of (**A**) concentration of Nb.BbvCI, (**B**) nicking time of Nb.BbvCI and (**C**) concentration of probe DNA; (**D**) effects of H1 and H2 concentrations on HCR amplification and (**E**) effect of hybridization time on HCR amplification.

Subsequently, the effect of the nicking incubation time of Nb.BbvCI on the current was examined (Figure 3B). The current shows a significant increase with increasing incubation time, reaching a relatively stable plateau at 60 min. Therefore, the optimal nicking incubation time for Nb.BbvCI is determined to be 60 min.

In order to optimize the efficiency of the HCR, it is crucial to control the density of the probe DNA on the gold electrode surface. This density affects steric hindrance and, consequently, the efficiency of the hybridization process [38]. To determine the optimal probe DNA concentration, different concentrations were tested (Figure 3C). It can be observed that the peak current gradually increases with the increase in the probe DNA concentration, until reaching a maximum at 0.5 μM. However, when the concentration exceeds 0.5 μM, the peak current starts to decrease gradually. This decrease at higher probe DNA concentrations can be attributed to the limited surface area available for hybridization on the modified electrode, leading to a reduced HCR efficiency. Based on these results, the optimal probe DNA concentration is determined to be 0.5 μM.

Furthermore, the concentrations of H1 and H2, as well as the hybridization time, play a crucial role in the electrochemical signal generated by RuHex. Figure 3D demonstrates that increasing the H1 and H2 concentrations led to a significant increase in the current signal, reaching a maximum at 0.5 μM. Further concentration increases did not significantly affect the current. Figure 3E shows that the current signal gradually increases with longer hybridization time and eventually reaches a steady plateau after 2 h. Consequently, the optimal conditions for the HCR reaction were determined to be 0.5 μM for both H1 and H2 concentrations and a 2-h hybridization time.

3.5. CCRF-CEM Exosomes Detection

Under optimized conditions, various concentrations of CCRF-CEM exosomes were examined using the designed assay. As depicted in Figure 4A, the DPV signal displayed a gradual increase as the exosome concentration increased from 8.0×10^4 to 3.2×10^8 particles/mL. Figure 4B demonstrates that the DPV peak current changes (ΔI) exhibited a linear correlation with the logarithm of the exosome concentration, yielding a correlation coefficient (R^2) of 0.982. The linear regression equation was $\Delta I = -2.804 + 0.731\text{lg}c$, where ΔI represents the difference in the current signal between the initial and final states, and c represents the exosome concentration. The assay achieved a detection limit of approximately 1.1×10^4 particles/mL (3σ rule), indicating its sensitivity to CCRF-CEM exosome detection. Additionally, the assay outperformed some previous methods in terms of a broader linear range and lower detection limit, as summarized in Table 2. These findings highlight the enhanced sensitivity and improved dynamic range of the assay developed for exosome detection.

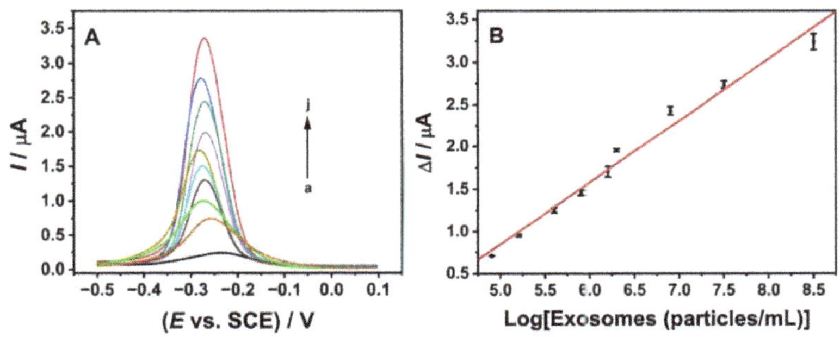

Figure 4. (**A**) DPV response of different concentrations of exosomes, the concentrations of exosomes from a to j are 0, 8.0×10^4, 1.6×10^5, 4.0×10^5, 8.0×10^5, 1.6×10^6, 2.0×10^6, 8.0×10^6, 3.2×10^7, 3.2×10^8 particles/mL; (**B**) Linear relationship between DPV signal changes and the logarithm of exosomes concentrations.

Table 2. Comparison of sensitivity and linear range of exosome detection methods.

Detection Method	Linear Range (Particles/mL)	LOD (Particles/mL)	Ref
electrochemical method	1.0×10^6~1.2×10^8	7.0×10^4	[22]
electrochemical method	1×10^6~1×10^{11}	1.58×10^5	[23]
electrochemical method	2.47×10^8~1.23×10^9 2.47×10^8~2.47×10^9	9.3×10^7 7.1×10^8	[39]
colorimetric method	8.3×10^5~5.3×10^7	3.94×10^5	[40]
fluorescence method	1×10^7~5×10^8	3.12×10^6	[17]
electrogenerated chemiluminescence	1×10^5~1×10^8	3×10^4	[41]
surface plasmon resonance	1.00×10^5~1×10^7	1.00×10^5	[42]
electrochemical method	8.0×10^4~3.2×10^8	1.1×10^4	This work

3.6. Selective and Feasible Evaluation of the Proposed Strategy

To evaluate the selectivity of the developed assay, we conducted a specificity assessment using exosomes obtained from Ramos and HeLa tumor cells, as well as a PBS buffer solution as a control. Figure 5A illustrates that the PTK7 protein content in CCRF-CEM tumor cell-derived exosomes is much higher compared to exosomes from Ramos and HeLa tumor cells. Moreover, the DPV signal from the PBS solution, which lacks the PTK7 protein, was almost negligible, consistent with findings reported in the literature [32,43]. These results provide evidence of the assay's excellent selectivity, specifically targeting the

PTK7 protein in CCRF-CEM tumor cell-derived exosomes, and are potentially useful for clinical diagnosis.

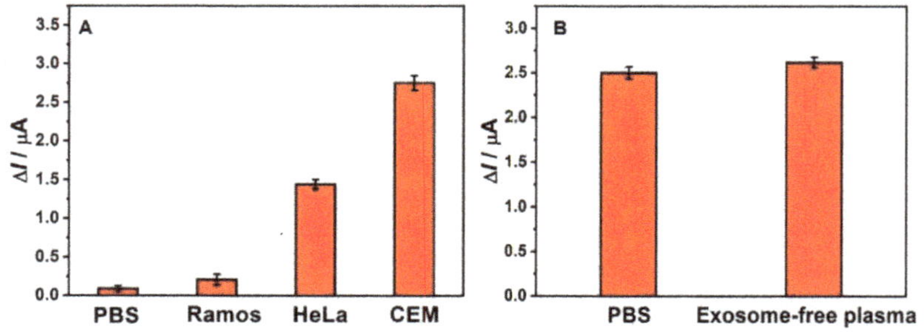

Figure 5. (**A**) DPV signal response of PTK7 content in different exosomes; (**B**) Analysis of the impact of complex environment on exosomes.

Furthermore, the reproducibility of the assay was evaluated by detecting CCRF-CEM exosomes at a concentration of 3.2×10^7 particles/mL using six parallel electrodes. The relative standard deviation (RSD) was about 2.41%. The results demonstrate the excellent reproducibility of the assay. This low RSD value highlights the consistency and reliability of the developed method for exosome detection.

To assess the feasibility of the sensing platform in real samples, we conducted detection and analysis using exosome-depleted plasma. In Figure 5B, we compare the DPV signals obtained by adding the same number of exosomes to the PBS buffer solution and exosome-depleted plasma under the same conditions. Remarkably, the current signal values detected in both environments exhibit negligible differences. This observation suggests the excellent feasibility and robust performance of the assay in complex biological matrices such as plasma, demonstrating its potential for exosome analysis in real samples.

To further validate the reliability of the assay in practical applications, we conducted tests by adding known concentrations of CCRF-CEM exosomes to exosome-depleted plasma. The detected final concentrations were 3.06×10^7, 2.42×10^7, 1.35×10^7, and 5.90×10^6 particles/mL, respectively. The results presented in Table 3 demonstrate the robustness of the assay in complex environments. The calculated recovery rates ranged from 95.63% to 103.5%, indicating the accuracy of the method for quantifying exosome concentrations. Additionally, the RSD, with values below 3.48% ($n = 3$), underscores the excellent reproducibility of the assay in complex biological matrices. These findings further support the reliability and practical applicability of the assay in real-world scenarios.

Table 3. Exosomes were analyzed in exosome-removed plasma.

Number	Added (Particles/mL)	Count (Particles/mL)	Recovery (%)	RSD (%)
1	3.20×10^7	3.06×10^7	95.63	1.38
2	2.50×10^7	2.42×10^7	96.8	2.85
3	1.30×10^7	1.35×10^7	103.8	3.48
4	6.00×10^6	5.90×10^6	98.3	2.34

4. Conclusions

In this study, an electrochemical sensing platform for the detection of CCRF-CEM exosomes was successfully developed. The assay utilizes a dual signal amplification strategy by a combination of a cyclic nicking enzyme and a HCR, eliminating the need for exosome capture on the electrode surface. The method offers several advantages, including a wider linear range, high selectivity, and good reproducibility. Notably, the limit of detection was as low as 1.1×10^4 particles/mL. Importantly, the method demonstrated applicability in complex plasma conditions, highlighting its potential for exosome-based research and diagnostic applications, especially in early clinical diagnosis of oncological diseases. This innovative approach holds promise for advancing exosome analysis and enhancing our understanding of tumor-related biomarkers.

Author Contributions: Experiment, data analysis, writing—original draft, D.S. and Q.G.; design and methodology, project instruction, funding acquisition, H.Z.; resources, writing—review and editing, H.Z. and C.C. All authors have read and agreed to the published version of the manuscript.

Funding: This research was funded by the National Natural Science Foundation of China (No. 22274077).

Institutional Review Board Statement: Not applicable.

Informed Consent Statement: Not applicable.

Data Availability Statement: The data are available upon request to the corresponding authors.

Conflicts of Interest: The authors declare no conflict of interest.

References

1. Raposo, G.; Stoorvogel, W. Extracellular vesicles: Exosomes, microvesicles, and friends. *J. Cell Biol.* **2013**, *200*, 373–383. [CrossRef] [PubMed]
2. Alptekin, A.; Parvin, M.; Chowdhury, H.I.; Rashid, M.H.; Arbab, A.S. Engineered exosomes for studies in tumor immunology. *Immunol. Rev.* **2022**, *312*, 76–102. [CrossRef] [PubMed]
3. Kalluri, R.; LeBleu, V.S. The biology, function, and biomedical applications of exosomes. *Science* **2020**, *367*, eaau6977. [CrossRef] [PubMed]
4. Pegtel, D.M.; Gould, S.J. Exosomes. *Annu. Rev. Biochem.* **2019**, *88*, 487–514. [CrossRef]
5. Stefanska, K.; Jozkowiak, M.; Angelova Volponi, A.; Shibli, J.A.; Golkar-Narenji, A.; Antosik, P.; Bukowska, D.; Piotrowska-Kempisty, H.; Mozdziak, P.; Dziegiel, P.; et al. The Role of Exosomes in Human Carcinogenesis and Cancer Therapy-Recent Findings from Molecular and Clinical Research. *Cells* **2023**, *12*, 356. [CrossRef]
6. Kalluri, R. The biology and function of exosomes in cancer. *J. Clin. Investig.* **2016**, *126*, 1208–1215. [CrossRef]
7. Guo, Y.; Ji, X.; Liu, J.; Fan, D.; Zhou, Q.; Chen, C.; Wang, W.; Wang, G.; Wang, H.; Yuan, W.; et al. Effects of exosomes on pre-metastatic niche formation in tumors. *Mol. Cancer* **2019**, *18*, 39. [CrossRef]
8. Zhang, J.; Zhu, Y.; Shi, J.; Zhang, K.; Zhang, Z.; Zhang, H. Sensitive Signal Amplifying a Diagnostic Biochip Based on a Biomimetic Periodic Nanostructure for Detecting Cancer Exosomes. *ACS Appl. Mater. Interfaces* **2020**, *12*, 33473–33482. [CrossRef]
9. Li, A.; Zhang, T.; Zheng, M.; Liu, Y.; Chen, Z. Exosomal proteins as potential markers of tumor diagnosis. *J. Hematol. Oncol.* **2017**, *10*, 175. [CrossRef]
10. Xu, B.; Chen, Y.; Peng, M.; Zheng, J.H.; Zuo, C. Exploring the potential of exosomes in diagnosis and drug delivery for pancreatic ductal adenocarcinoma. *Int. J. Cancer* **2023**, *152*, 110–122. [CrossRef]
11. Sun, Y.; Liu, J. Potential of cancer cell-derived exosomes in clinical application: A review of recent research advances. *Clin. Ther.* **2014**, *36*, 863–872. [CrossRef]
12. Hsu, M.T.; Wang, Y.K.; Tseng, Y.J. Exosomal Proteins and Lipids as Potential Biomarkers for Lung Cancer Diagnosis, Prognosis, and Treatment. *Cancers* **2022**, *14*, 732. [CrossRef] [PubMed]
13. Sharma, V.; Nikolajeff, F.; Kumar, S. Employing nanoparticle tracking analysis of salivary neuronal exosomes for early detection of neurodegenerative diseases. *Transl. Neurodegener.* **2023**, *12*, 7. [CrossRef] [PubMed]
14. Li, J.; Li, Y.; Li, P.; Zhang, Y.; Du, L.; Wang, Y.; Zhang, C.; Wang, C. Exosome detection via surface-enhanced Raman spectroscopy for cancer diagnosis. *Acta Biomater.* **2022**, *144*, 1–14. [CrossRef] [PubMed]
15. Park, J.; Hwang, M.; Choi, B.; Jeong, H.; Jung, J.-h.; Kim, H.K.; Hong, S.; Park, J.-h.; Choi, Y. Exosome Classification by Pattern Analysis of Surface-Enhanced Raman Spectroscopy Data for Lung Cancer Diagnosis. *Anal. Chem.* **2017**, *89*, 6695–6701. [CrossRef] [PubMed]
16. Shin, H.; Oh, S.; Hong, S.; Kang, M.; Kang, D.; Ji, Y.G.; Choi, B.H.; Kang, K.W.; Jeong, H.; Park, Y.; et al. Early-Stage Lung Cancer Diagnosis by Deep Learning-Based Spectroscopic Analysis of Circulating Exosomes. *ACS Nano* **2020**, *14*, 5435–5444. [CrossRef]
17. Xia, Y.; Chen, T.; Chen, G.; Weng, Y.; Zeng, L.; Liao, Y.; Chen, W.; Lan, J.; Zhang, J.; Chen, J. A nature-inspired colorimetric and fluorescent dual-modal biosensor for exosomes detection. *Talanta* **2020**, *214*, 120851. [CrossRef]
18. Zhou, J.; Lin, Q.; Huang, Z.; Xiong, H.; Yang, B.; Chen, H.; Kong, J. Aptamer-Initiated Catalytic Hairpin Assembly Fluorescence Assay for Universal, Sensitive Exosome Detection. *Anal. Chem.* **2022**, *94*, 5723–5728. [CrossRef]
19. Huang, R.; He, L.; Li, S.; Liu, H.; Jin, L.; Chen, Z.; Zhao, Y.; Li, Z.; Deng, Y.; He, N. A simple fluorescence aptasensor for gastric cancer exosome detection based on branched rolling circle amplification. *Nanoscale* **2020**, *12*, 2445–2451. [CrossRef]
20. Wang, S.; Zhang, L.; Wan, S.; Cansiz, S.; Cui, C.; Liu, Y.; Cai, R.; Hong, C.; Teng, I.T.; Shi, M.; et al. Aptasensor with Expanded Nucleotide Using DNA Nanotetrahedra for Electrochemical Detection of Cancerous Exosomes. *ACS Nano* **2017**, *11*, 3943–3949. [CrossRef] [PubMed]

21. Zhou, Y.G.; Mohamadi, R.M.; Poudineh, M.; Kermanshah, L.; Ahmed, S.; Safaei, T.S.; Stojcic, J.; Nam, R.K.; Sargent, E.H.; Kelley, S.O. Interrogating Circulating Microsomes and Exosomes Using Metal Nanoparticles. *Small* **2016**, *12*, 727–732. [CrossRef] [PubMed]
22. Dong, H.; Chen, H.; Jiang, J.; Zhang, H.; Cai, C.; Shen, Q. Highly Sensitive Electrochemical Detection of Tumor Exosomes Based on Aptamer Recognition-Induced Multi-DNA Release and Cyclic Enzymatic Amplification. *Anal. Chem.* **2018**, *90*, 4507–4513. [CrossRef] [PubMed]
23. Cheng, W.; Duan, C.; Chen, Y.; Li, D.; Hou, Z.; Yao, Y.; Jiao, J.; Xiang, Y. Highly Sensitive Aptasensor for Detecting Cancerous Exosomes Based on Clover-like Gold Nanoclusters. *Anal. Chem.* **2023**, *95*, 3606–3612. [CrossRef]
24. Bagheri Hashkavayi, A.; Cha, B.S.; Lee, E.S.; Kim, S.; Park, K.S. Advances in Exosome Analysis Methods with an Emphasis on Electrochemistry. *Anal. Chem.* **2020**, *92*, 12733–12740. [CrossRef]
25. Jeong, S.; Park, J.; Pathania, D.; Castro, C.M.; Weissleder, R.; Lee, H. Integrated Magneto-Electrochemical Sensor for Exosome Analysis. *ACS Nano* **2016**, *10*, 1802–1809. [CrossRef] [PubMed]
26. Ziaei, P.; Berkman, C.E.; Norton, M.G. Review: Isolation and Detection of Tumor-Derived Extracellular Vesicles. *ACS Appl. Nano Mater.* **2018**, *1*, 2004–2020. [CrossRef]
27. An, Y.; Jin, T.; Zhu, Y.; Zhang, F.; He, P. An ultrasensitive electrochemical aptasensor for the determination of tumor exosomes based on click chemistry. *Biosens. Bioelectron.* **2019**, *142*, 111503. [CrossRef]
28. Liu, X.; Gao, X.; Yang, L.; Zhao, Y.; Li, F. Metal-Organic Framework-Functionalized Paper-Based Electrochemical Biosensor for Ultrasensitive Exosome Assay. *Anal. Chem.* **2021**, *93*, 11792–11799. [CrossRef]
29. Huang, M.; Xiang, Y.; Chen, Y.; Lu, H.; Zhang, H.; Liu, F.; Qin, X.; Qin, X.; Li, X.; Yang, F. Bottom-Up Signal Boosting with Fractal Nanostructuring and Primer Exchange Reaction for Ultrasensitive Detection of Cancerous Exosomes. *ACS Sens.* **2023**, *8*, 1308–1317. [CrossRef]
30. Huang, R.; He, L.; Xia, Y.; Xu, H.; Liu, C.; Xie, H.; Wang, S.; Peng, L.; Liu, Y.; Liu, Y.; et al. A Sensitive Aptasensor Based on a Hemin/G-Quadruplex-Assisted Signal Amplification Strategy for Electrochemical Detection of Gastric Cancer Exosomes. *Small* **2019**, *15*, e1900735. [CrossRef]
31. Zhao, L.; Sun, R.; He, P.; Zhang, X. Ultrasensitive Detection of Exosomes by Target-Triggered Three-Dimensional DNA Walking Machine and Exonuclease III-Assisted Electrochemical Ratiometric Biosensing. *Anal. Chem.* **2019**, *91*, 14773–14779. [CrossRef] [PubMed]
32. Jiang, Y.; Shi, M.; Liu, Y.; Wan, S.; Cui, C.; Zhang, L.; Tan, W. Aptamer/AuNP Biosensor for Colorimetric Profiling of Exosomal Proteins. *Angew. Chem. Int. Ed.* **2017**, *56*, 11916–11920. [CrossRef] [PubMed]
33. Skog, J.; Wurdinger, T.; van Rijn, S.; Meijer, D.H.; Gainche, L.; Sena-Esteves, M.; Curry, W.T., Jr.; Carter, B.S.; Krichevsky, A.M.; Breakefield, X.O. Glioblastoma microvesicles transport RNA and proteins that promote tumour growth and provide diagnostic biomarkers. *Nat. Cell Biol.* **2008**, *10*, 1470–1476. [CrossRef] [PubMed]
34. Tosar, J.P.; Gambaro, F.; Sanguinetti, J.; Bonilla, B.; Witwer, K.W.; Cayota, A. Assessment of small RNA sorting into different extracellular fractions revealed by high-throughput sequencing of breast cell lines. *Nucleic Acids Res.* **2015**, *43*, 5601–5616. [CrossRef]
35. Zhang, J.; Song, S.; Wang, L.; Pan, D.; Fan, C. A gold nanoparticle-based chronocoulometric DNA sensor for amplified detection of DNA. *Nat. Protoc.* **2007**, *2*, 2888–2895. [CrossRef]
36. Zhang, X.; Xiao, K.; Cheng, L.; Chen, H.; Liu, B.; Zhang, S.; Kong, J. Visual and highly sensitive detection of cancer cells by a colorimetric aptasensor based on cell-triggered cyclic enzymatic signal amplification. *Anal. Chem.* **2014**, *86*, 5567–5572. [CrossRef]
37. Li, J.; Fu, H.E.; Wu, L.J.; Zheng, A.X.; Chen, G.N.; Yang, H.H. General colorimetric detection of proteins and small molecules based on cyclic enzymatic signal amplification and hairpin aptamer probe. *Anal. Chem.* **2012**, *84*, 5309–5315. [CrossRef]
38. Liu, Q.; Xie, X.L.; Mao, C.J.; Chen, J.S.; Niu, H.L.; Song, J.M. Electrochemiluminescent biosensor with DNA link for selective detection of human IgG based on steric hindrance. *Talanta* **2019**, *194*, 745–751. [CrossRef]
39. Kasetsirikul, S.; Tran, K.T.; Clack, K.; Soda, N.; Shiddiky, M.J.A.; Nguyen, N.T. Low-cost electrochemical paper-based device for exosome detection. *Analyst* **2022**, *147*, 3732–3740. [CrossRef]
40. Zhou, Y.; Xu, H.; Wang, H.; Ye, B.C. Detection of breast cancer-derived exosomes using the horseradish peroxidase-mimicking DNAzyme as an aptasensor. *Analyst* **2019**, *145*, 107–114. [CrossRef]
41. Zhang, H.; Wang, Z.; Wang, F.; Zhang, Y.; Wang, H.; Liu, Y. In Situ Formation of Gold Nanoparticles Decorated Ti_3C_2 MXenes Nanoprobe for Highly Sensitive Electrogenerated Chemiluminescence Detection of Exosomes and Their Surface Proteins. *Anal. Chem.* **2020**, *92*, 5546–5553. [CrossRef] [PubMed]
42. Chen, W.; Li, J.; Wei, X.; Fan, Y.; Qian, H.; Li, S.; Xiang, Y.; Ding, S. Surface plasmon resonance biosensor using hydrogel-AuNP supramolecular spheres for determination of prostate cancer-derived exosomes. *Mikrochim. Acta* **2020**, *187*, 590. [CrossRef] [PubMed]
43. Jin, D.; Yang, F.; Zhang, Y.; Liu, L.; Zhou, Y.; Wang, F.; Zhang, G.J. ExoAPP: Exosome-Oriented, Aptamer Nanoprobe-Enabled Surface Proteins Profiling and Detection. *Anal. Chem.* **2018**, *90*, 14402–14411. [CrossRef] [PubMed]

chemosensors

Perspective

Mass Spectrometry-Based Biosensing and Biopsy Technology

Fengjian Chu [1], Wei Wei [2], Nazifi Sani Shuaibu [1], Hongru Feng [2], Xiaozhi Wang [1] and Yuanjiang Pan [2,*]

1 College of Information Science and Electronic Engineering, Zhejiang University, Hangzhou 310027, China; 12231056@zju.edu.cn (F.C.); zuhudu@zju.edu.cn (N.S.S.); xw224@zju.edu.cn (X.W.)
2 Department of Chemistry, Zhejiang University, Hangzhou 310027, China; wei.wei@zju.edu.cn (W.W.); fenghongru@zju.edu.cn (H.F.)
* Correspondence: panyuanjiang@zju.edu.cn

Abstract: Sensitive and accurate detection of biomolecules by multiplexed methods is important for disease diagnosis, drug research, and biochemical analysis. Mass spectrometry has the advantages of high sensitivity, high throughput, and high resolution, making it ideal for biomolecular sensing. As a result of the development of atmospheric pressure mass spectrometry, researchers have been able to use a variety of means to identify target biomolecules and recognize the converted signals by mass spectrometry. In this review, three main approaches and tools are summarized for mass spectrometry sensing and biopsy techniques, including array biosensing, probe/pen-based mass spectrometry, and other biosensor–mass spectrometry coupling techniques. Portability and practicality of relevant mass spectrometry sensing methods are reviewed, together with possible future directions to promote the advancement of mass spectrometry for target identification of biomolecules and rapid detection of real biological samples.

Keywords: biomolecules detection; mass spectrometry; biosensors; biometrics; clinical mass spectrometry; ambient ionization mass spectrometry

Citation: Chu, F.; Wei, W.; Shuaibu, N.S.; Feng, H.; Wang, X.; Pan, Y. Mass Spectrometry-Based Biosensing and Biopsy Technology. *Chemosensors* **2023**, *11*, 419. https://doi.org/10.3390/chemosensors11080419

Academic Editor: Fabio Gosetti

Received: 28 June 2023
Revised: 19 July 2023
Accepted: 20 July 2023
Published: 26 July 2023

1. Introduction

A biosensor is an analytical device consisting of a sensing element and a signal transducer. In a general biosensing process, a biological signal is recognized and amplified by a sensing element, converted into a readable state by a signal transducer, and further processed into a digital signal. The role of the sensing element is to recognize target substances, mainly including antibodies, enzymes, nucleic acids, cells, and other biological substances, and also some synthetic substances similar to biological substances, such as aptamers, peptides, and MIPs (molecularly imprinted polymers) [1]. The function of a signal transducer is to convert the interaction between a sensing element and a target molecule into an identifiable signal [2]. For example, enzymes catalyze chemical reactions with specific substances and transform them into electrical signals; biological antibodies capture specific antigens and convert them into optical signals through labeled fluorescence [3].

Mass spectrometry is a method to analyze target compounds based on their mass-to-charge ratio. Mass spectrometry separates and detects the composition of substances by the mass difference of the atoms, molecules, or molecular fragments of the substance through the principle that charged particles are able to deflect in an electromagnetic field [4]. It typically consists of four parts: injection system, ion source, mass analyzer, and detector. With the advantages of high sensitivity, high resolution, and wide analytical range, mass spectrometry can be used with specificity for chemical analysis in food, drugs, cellular components, blood, and other fields [5]. For a classical biosensor system, mass spectrometry is an excellent signal transducer. In addition, researchers have modified the ion source or injection method of a mass spectrometry system to give it the ability to specifically detect certain biomolecules or to enhance the signal response to certain biomolecules; this process is used as a sensing element to form a mass spectrometry-based biochemical sensing system [6]. Especially after the first ambient ionization technique (Desorption Electrospray Ionization, DESI) reported by Cooks in 2004 [7], ambient ionization mass spectrometry (AMS) techniques have undergone rapid development, allowing the adaptation of multiple types of mass spectrometry interfaces in the open air [8,9], greatly enhancing the application of mass spectrometry in the field of biochemical sensing.

Depending on the device structure and ionization mechanism, mass spectrometry-based biosensing technologies can be broadly divided into 1. biochip and mass tag-based mass spectrometry sensing technology, 2. probe (pen) based mass spectrometry, and 3. integrating other biosensing technologies for mass spectrometry biosensors. Biochip-based mass spectrometry sensing techniques enable targeted screening by coupling biomolecules on a substrate, often in combination with matrix-assisted laser desorption ionization mass spectrometry (MALDI-MS) [10], which has been an effective tool for studying biomolecules since its introduction in 1987. Probe-based mass spectrometry biosensing techniques are more in situ, in vivo, and clinical. Researchers collect biological samples through probes or pens of different materials and structures, then conduct direct mass spectrometric detection online [11]. In addition, the successful integration of biospecific interaction analysis based on other biosensors (surface plasmon resonance (SPR), microfluidics) and mass spectrometry produces a powerful technique that couples the benefits of sensitive affinity capture with the ability to characterize interacting molecules [12].

It is worth noting that the concept of mass spectrometry biosensing was proposed and generalized by Ju's group in 2021 [6], and this review expands and summarizes the concept from different perspectives. The concept of mass spectrometry biosensors tends to be less frequently mentioned, and this review classifies key technologies that improve the mass spectrometric response of biochemical molecules or make them measurable by mass spectrometry, as mass spectrometry-based biosensors. This review critically reviews mass spectrometry-based biosensing technologies divided into three main sections. The first part has an emphasis on explanations of biochip and mass tag-based mass spectrometry sensing technology, including applications of microarray mass spectrometry and nanomaterial mass tags, such as MALDI for biomolecular arrays. The second part covers various probes, probe pen techniques for in situ biomass sensing, and the latest clinical references. The third part covers some integrated techniques of biosensing and mass spectrometry.

2. Biochip and Mass Tag-Based Mass Spectrometry Sensing Technology

Laser desorption-based mass spectrometry techniques include matrix-assisted laser desorption ionization (MALDI), surface-assisted laser desorption ionization (SALDI), and ambient ionization mass spectrometry (AMS) techniques using UV or IR pulsed laser ablation. All these techniques facilitate the desorption of the sample through a laser, and the biosensing process is accomplished by transferring energy or electrons to the target molecule through a specific matrix (small molecule, polymer, nanomaterial, etc.) to achieve an increase in the signal intensity of the target biomolecule (Figure 1). Research on the use of MALDI to measure biomolecules dates back to the 1990s [13] and was awarded the Nobel Prize in 2002. Over the next 30 years, a number of variants of the technique were developed, and they are used in various biochemical analyses [14–16]. With the development of laser desorption-based mass spectrometry, researchers have found that the use of microarray biochips can effectively target and detect biomolecules, including DNA, RNA, peptides, sugars, proteins, etc. [6].

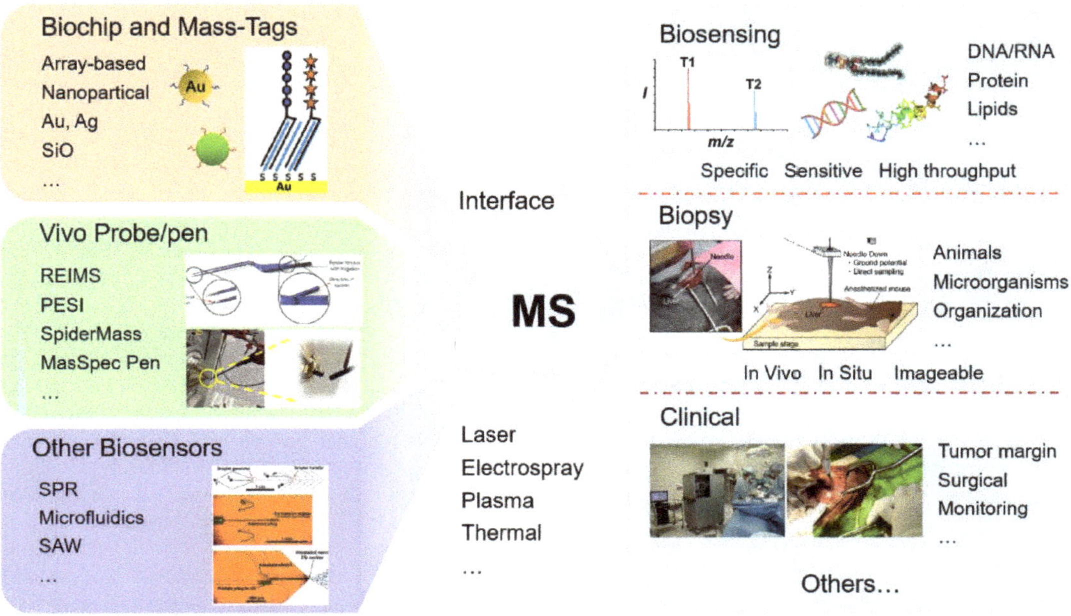

Figure 1. Schematic representation of different sensing technologies (biochip [17], in vivo probe [18,19], microfluidic [20], etc.) combined with mass spectrometry for different detection scenarios (biopsy [21], clinical [18], etc.). Copyright with permission from Wiley, Royal Society of Chemistry, American Chemical Society.

In 2002, Mrksich's group discovered that when MALDI-MS combined with self-assembled monolayers (SAMs) that are engineered to give specific interactions with biomolecules, it is well suited for characterizing biological activities [10]. They first formed SAM (self-assembled monolayers)-Au chips using oligoethylene glycol groups of alkanethiols and peptides, proteins, or carbohydrates to achieve recognition of specific biomolecules, named SAMDI-MS. Then they used SAMDI to provide ligands that interact with target proteins and enzymes for enzyme activity studies and applied this method to screen a chemical library against protease activity of anthrax lethal factor [13,22]. Becker et al. used MS to detect Ras-protein-receptor interactions on protein-oligonucleotide affixes attached to in silico sheets by DNA-directed immobilization in 2005 [23]. In 2007, Mrksich et al. expressed a membrane scaffold protein (MSP) with a hexahistidine (his_6) tag at its N terminus and prepared nanodiscs containing rhodopsin protein and the lipid 1-palmitoyl-2-oleoyl-glycero-3-phosphocholine (POPC) based on the self-assembly of lipid molecules within the membrane scaffold protein, producing a circular patch of a soluble lipid bilayer that can immobilize transmembrane proteins for the screening and characterization of transmembrane proteins [24]. Yeo et al. used gold particles carrying small molecules as reporters for target proteins based on the oligoethylene glycol SAM target protein microarray and analyzed by LDI-TOF-MS. As the number of small molecules far exceeded the number of cooperating target proteins, the biosignal was amplified, enabling ultrahigh sensitivity detection in the attomolar range [25]. In 2009, Min et al. used the MALDI laser for selective desorption on the SAMs chip surface to create patterns of cell adhesion ligands on SAMs with simple control over the ligand density [26].

After 2010, with the development of experimental instrumentation, more research groups are involved in the field. Beloqui et al. immobilized lipid-labeled oligosaccharide fragments onto MALDI sample plates by hydrophobic interactions compatible with the solution-like enzyme activity on the chip, allowing easy sample cleanup and subsequent enzymatic interaction analysis by MALDI-TOF [12]. In subsequent work, they chose commercially available ITO-coated glass sheets. Silylation with 3-aminopropyl-triethoxysilane (APTES), followed by coupling with NHS-activated stearic acid, formed a hydrophobic support layer that immobilized lipid-labeled biomolecules through hydrophobic interactions, achieving highly sensitive detection of peptides, sugars, and other biomolecules [27]. In the same year, Kuo et al. analyzed deacetylase activity in cell lysates using peptide arrays and SAMDI-MS [28]. Li et al. devised a method for detecting protein kinase A (PKA) phosphorylated cysteine peptides using the nanostructure initiator mass spectrometry (NIMS) technique [29]. Hong et al. used anti-Bcr on gold nanoparticles (AuNPs) and anti-Abl on biochips to capture the Bcr/Abl chimeric protein and quantified them in cells by LDI-TOF-MS [30]. Both et al. investigated the application of peptide microarrays in sugar donor promiscuity of pp-a-GanT2 using the SAM chip tandem IM-MS technique [31]. In 2015, Ju et al. proposed a peptide-encoded microplate for MALDI-TOF-MS analysis of protease activity with a low detection limit of 2.3 nM as well as good selectivity [32]. Lorey et al. proposed a new method for the analysis of antibody arrays using laser desorption/ionization mass spectrometry (LDI-MS) with "mass-labeled" specific small reporter molecules to detect immunocapture proteins in human plasma, with detection limits much lower than clinical methods [33]. Hu et al. proposed a MALDI-MS patterning strategy for the convenient visualization of multiple enzyme activities by caspase-activity patterned chip (Casp-PC) with ITO surface peptide arrays [34], and on this basis, they proposed a quantitative method for a variety of enzyme proteins [35]. Xu et al. established an array-based electrospray accelerated chip spray ionization device where gold nano-ions and target proteins form an immune sandwich on an indium tin oxide glass chip for the identification of membrane proteins in blood [36], and the throughput of the method was improved in subsequent studies and applied to screening for cancer markers [37]. Mrksich et al. used cysteine-terminated peptides to covalently capture metabolites bound to CoA and immobilize them on self-assembled monolayer arrays. Thus, the captured metabolites were rapidly separated from the complex mixture and directly quantified by SAMDI-MS [38–40]. Gunnarsson et al. performed multiplex DNA detection using random arrays to capture the binding of DNA-modified liposomes to surface-immobilized probe DNA, forming sequences encoding unique target DNA sequences, and analyzed them with SIMS-TOF [41]. Li et al. used peptide arrays of ITO slides for the study of thrombin activity and screening of potential inhibitors [42].

In addition to microarray biochip mass spectrometry (Figure 2), the development of nanomaterials has contributed to the refinement of mass spectrometry sensing techniques These techniques currently utilize nanomaterials directly or small molecules as mass tags and are categorized as nanomaterial-based mass tags (MT) mass spectrometry. Several nanomaterials, including AuNPs, AgNPs, PtNPs [43–45], quantum dots [46], and metal nanoclusters, can be effectively bound to specific biological moieties. This capability enables the targeted labeling of biomolecules, allowing in situ multiplexed mass spectrometry analysis of proteins [47], glycans [48], and other biomolecules [49] within biological systems [50–52]. It should be noted that Min's team has recently provided an exhaustive review on the topic of MT-encoded MS [53], thus this review will not delve into the field in greater depth.

Overall, compared to traditional biomolecular analysis processes, biochip and mass tag-based mass spectrometry sensing technology are characterized by high specificity analysis, and highly sensitive detection of targeted biomolecules can be achieved through mass tag amplification. However, there are limitations to this technology. For instance, the preparation and pre-processing of bioarray chips are complex and time-consuming. Antibodies utilized for specific recognition are susceptible to inactivation and damage during the experimental process, resulting in a lack of robustness and making them challenging to recover for multiple analyses. Therefore, shortening the pre-processing steps and enhancing the reliability and stability of sensing labels may represent the new direction for development.

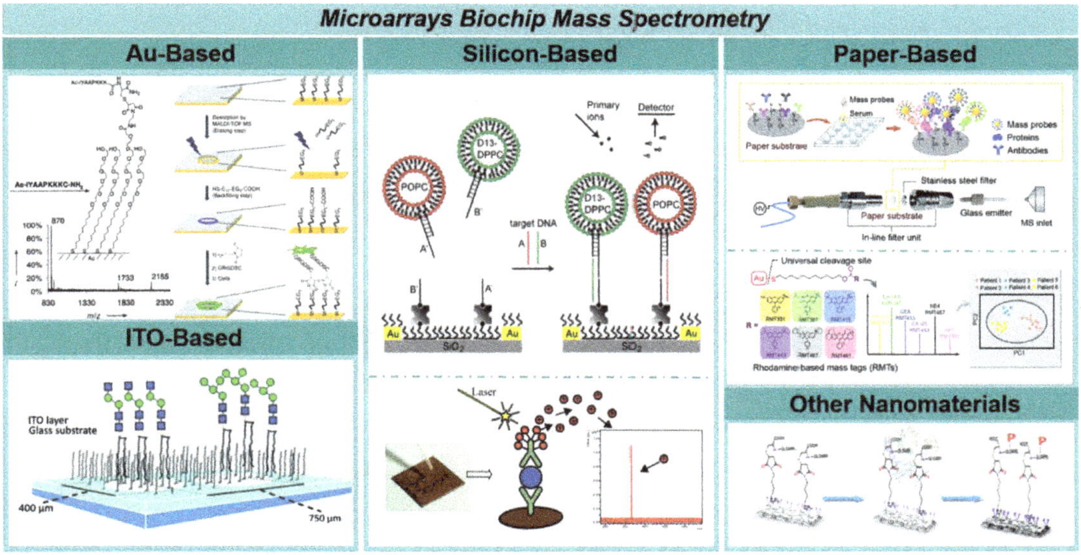

Figure 2. Biochips on different substrates, including Au-based [10,26], ITO-based [27], silicon-based [33,41], paper-based [37], and other nanomaterials-based [29]. Copyright with permission from Wiley, Elsevier, Royal Society of Chemistry, American Chemical Society.

3. Probe/Pen-Based Mass Spectrometry Sensing Technology

With the development of atmospheric pressure mass spectrometry, several in vivo mass spectrometry techniques are available for the detection of specific chemical information contained in different tissues. Such techniques often require a tip or a smaller area to keep the organism with low or no destruction. It is challenging to extract enough chemical information from biological tissues at a low loss to access mass spectrometry for accurate identification with high efficiency and throughput. We classify such online mass spectrometry techniques into biosensing and provide a categorical overview of the following two aspects: in vivo probe mass spectrometry (e.g., probe electrospray ionization, PESI) and pen-based mass spectrometry modalities (e.g., rapid evaporative ionization mass spectrometry (REIMS), MasSpec Pen, and SpiderMass).

3.1. In Vivo Probe Mass Spectrometry

Since the introduction of probe electrospray ionization technology in 2007 [54], researchers have found that the use of probes can cause less damage to the organism while extracting the target compound. PESI is widely used for in vivo non-destructive biological mass spectrometry [55,56]. Chen et al. applied PESI directly to various biological samples such as urine, mouse brain, mouse liver, and fruit, demonstrating that PESI is a practical non-invasive biomolecular detection technique [57]. Yoshimura et al. performed a real-time analysis of in vivo mice using PESI, revealing differences in hepatocyte lipid composition between normal and steatotic mice, with no significant postoperative damage in the in vivo mice [17]. Hsu et al. took advantage of the ultra-fine size of the probe tip to directly extract and ionize a mixture of metabolites from live microbial colonies grown in Petri dishes without any sample pretreatment [58]. Gong et al. used a 1 μm tungsten probe inserted directly into live cells to enrich for metabolites and used PESI for elution and ionization, resulting in the detection of single-cell metabolites [59]. Deng et al. designed a surface-coated probe nanoelectrospray ionization mass spectrometry (SCP-nanoESI-MS) based on SPEM for the analysis of target compounds in individual small organisms and fish; probe tips are at the micron level and exhibit good linearity in the analysis of real samples [60–62]. Zaitsu et al. applied PESI to the analysis of intact endogenous metabolites in the liver and brain of living mice and achieved the detection of multiple metabolites, including organic acids, sugars, and amino acids in 2015 [63]. Zaitsu et al. constructed a high-throughput metabolic mass spectrometry platform by PESI, screened 72 metabolites in mouse liver and brain, and built data processing software; in subsequent work, they used the platform to analyze extracellular neurotransmitters in mouse brain with excellent linearity and precision [64,65]. The in vivo online detection capability of PESI technology has been well established and is of great value in the study of real-time metabolomics, but the problems of low sampling efficiency of PESI probes, difficulties in the analysis of large molecules, and lack of specific identification limit the wider application of the technology.

In addition to the classical PESI technique, numerous new material-based, solid-phase microextraction (SPME)-based, and plasma-based in vivo probe mass spectrometry techniques have been proposed in the last 5 years. Ngernsutivorakul et al. combined a sampling probe with a microfluidic chip to achieve a 1000-fold increase in resolution over ordinary microdialysis probes, enabling real-time chemical monitoring in vivo [66]. Lendor et al. performed chemical biopsies of the brain by synthesizing SPME probes with functionalized hydrophilic layers, allowing quantitative analysis of multiple neurotransmitters [67]. The method is also combined with Paternò–Büchi (PB) reactions for in vivo, in situ, and microscale analysis of lipid species and C=C location isomers in complex biological tissues [68]. Lu et al. coupled a metal microprobe to a dielectric barrier discharge ionization (DBDI) with a limit of detection as low as 8 pg/mL, and then they achieved the monitoring of drug residues in different organs of live fish using this method [69]. Bogusiewicz et al. first used SPME probe MS technology for biopsy sampling of the human brain, followed by metabolic and lipidomic analysis, demonstrating higher concentrations and diversity of metabolites in the white matter [70]. Mendes et al. performed direct analysis of fruit and mouse brains using inexpensive and environmentally friendly pencil graphite rods as probes for mass spectrometry biosensing [15]. Cheng et al. coupled SPME with nanoESI-MS, achieved by surface-coated acupuncture needles, for the in vivo detection of small molecules, proteins, and peptides in plants [71].

3.2. Pen/knife-Like Mass Spectrometry Sensing Technology

Real clinical testing has placed higher demands on key components of mass spectrometry sensing, and a number of new AMS technologies have been developed to meet scenarios such as tumor margin detection during surgery. One of the more widely used is rapid evaporative ionization mass spectrometry (REIMS), which is often integrated into the clinic as a smart knife. This device cauterizes biological tissue (e.g., tumor margins) by heating the tip and continuously collects a plume for chemical information acquisition [72]. Balog et al. used this technique to analyze various tissue samples from over 300 patients, reflecting the lipidomic profile among different histological tumor types and between primary and metastatic tumors [14]. Golf et al. constructed the REIMS imaging platform for differentiating healthy/cancerous tissues and different bacterial/Fungi strains and built a spectral library [73–75]. Manoli et al. combined REIMS technology with an ultrasonic scalpel for real-time monitoring of lipid profiles during laparoscopic operations [76]. Overall, REIMS focuses on lipids with limitations in the detection of biochemical molecules such as sugars and peptides. In addition, due to the destructive nature of cauterized tissue and the relatively low spatial resolution, a high level of professionalism is required of the operator.

Based on the limitations of REIMS, a series of new clinical biosensing mass spectrometry techniques have been proposed. Fatou et al. proposed a new instrument named SpiderMass for real-time in vivo mass spectrometry detection with a miniature probe designed based on infrared laser ablation that is much less damaging than REIMS, enabling minimally invasive in vivo online analysis [77]. Fatou et al. performed a real-time in vivo pharmacokinetic study by SpiderMass, revealing the potential for DMPK (drug metabolism and pharmacokinetics) and ADME (absorption, distribution, metabolism, and excretion) analysis with the technique [78]. Subsequently, SpiderMass and the variants have been used in a variety of scenarios such as biofluidics, tumor margins, tissue biopsy, and skin cancer [79–82]. Ogrinc et al. combined a SpiderMass probe with a high-precision robotic arm to enable mass spectrometry imaging on arbitrary sample surfaces, paving the way for surgical applications of excised edges [83]. Although the SpiderMass technology reduces damage to biological tissue compared to REIMS, laser desorption still damages the sample to some extent.

Therefore, a gentler in situ sampling technique called MasSpec Pen has been proposed, which uses discrete droplets to collect chemical information from the tissue surface, and since only droplets are used as a medium, the method is completely non-destructive to the tissue. Using this technique, Zhang et al. analyzed 253 tissue samples from human cancer patients and a variety of metabolites, lipids, and proteins were identified as potential cancer biomarkers [84–86]. The MasSpec Pen was also transferred to the operating room, performing tumor margin prediction during 18 pancreatic surgeries with an accuracy of 93.8% [87].

The studies referenced above have shown that a simple modification of the ion source section can result in a mass spectrometry-based in vivo biosensing system for clinical, cancer detection, tissue biopsy, and various other scenarios (Figure 3). Notwithstanding the numerous inherent advantages associated with these techniques, they are not without limitations. Firstly, their utilization necessitates a certain proficiency level in manipulation skills. Secondly, most of these techniques are limited by the mode of ionization, where the ionization capacity depends on the proton affinity potential, so only polar molecules can be recognized.

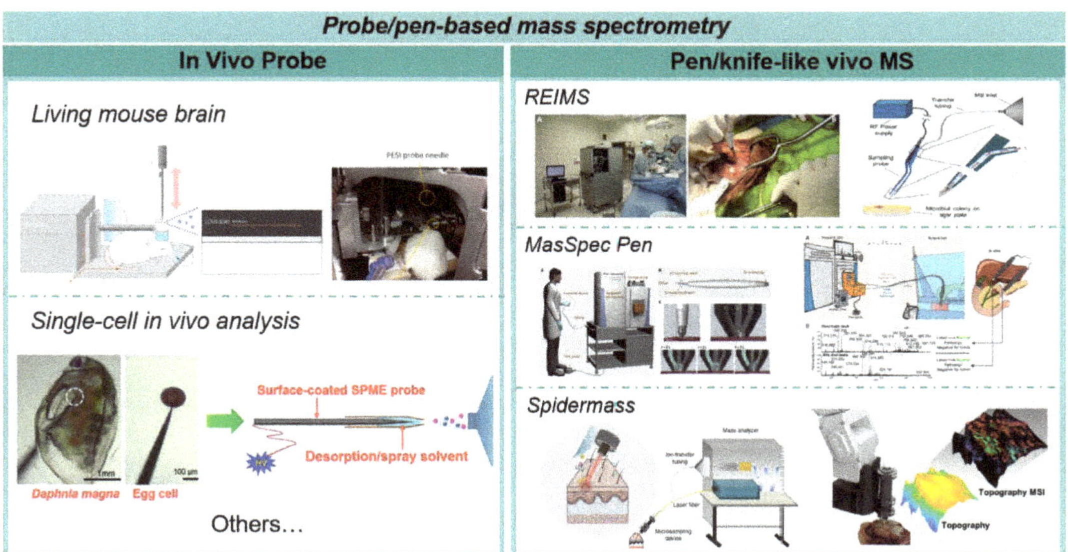

Figure 3. Highlights of in vivo analysis using MS biosensor techniques. Probe mass spectrometry, mainly PESI, for biopsies of various organisms [59,60,66] is shown on the left panel. The right panel shows pen/knife-like in vivo MS technology including REIMS [14], MasSpec Pen [84], and SpiderMass [80,83]. Copyright with permission from Wiley, Elsevier, American Chemical Society.

4. Integration of Mass Spectrometry with Other Biosensors

In recent years, some classical biosensing systems have been coupled with mass spectrometry to achieve complementary performance and functionality. The flexible modification of the interface between mass spectrometry sampling and ion source has given rise to many new coupling methods, with the main difference being whether direct coupling, indirect coupling, online, offline, etc. This chapter focuses on the integration methods between various biosensors with mass spectrometry and their applications.

One of the more widely integrated sensors with mass spectrometry is the surface plasmon resonance (SPR), and MS can provide molecular information that perfectly complements the SPR sensors [12,88]. A common offline coupling approach is to elute biomolecules on the SPR sensing chip into the mass spectrometry test. For example, Hamaloglu et al. used SPR and MALDI-MS together for the detection of Fab-anti-HSA (human serum albumin) on MUA (Mercaptoundecanoic acid) molecules array platforms [89]. Yang et al. used the SPR technique to screen TNF (Tumor Necrosis Factor) from angelicae pubescentis radix extracts, followed by quantitation and evaluation via UPLC-MS/MS [90]. Castells et al. used SPR and an approach that combines limited proteolysis mass spectrometry to analyze in detail glycan–protein interactions [91]. Compared to such offline techniques, the online coupling interface between mass spectrometry and SPR can improve detection speed and avoid sample denaturation, generating greater interest among researchers. Marchesini et al. online coupled SPR-based inhibition biosensor immunoassay (iBIA) with nano-liquid-chromatography electrospray ionization time-of-flight mass spectrometry (nano-LC ESI TOF MS) for effective screening of small molecules in organisms [92]. Zhang et al. proposed an interface for online coupled SPR with direct analysis in real time (DART) MS, and in subsequent work, a direct online coupling technology between dielectric barrier discharge (DBD)-MS and SPR was proposed for the study of various interaction of biomolecules [93,94]. Mihoc et al. used proteolytic epitope extraction mass spectrometry combined with SPR biosensor analysis to determine the molecular epitope structures and affinity of equine heme-myoglobin and apo-myoglobin to a monoclonal antibody [95]. Joshi et al. developed a method to simplify the coupling of SPR and MS by direct biochip spraying, which can selectively capture target small molecules on the SPR surface and nebulize them directly into the mass spectrometry under high voltage [96]. The combination of SPR and mass spectrometry provides protein specific binding recognition ability and highly sensitive, precise qualitative analysis ability, providing an ideal analytical tool for studying the interactions of biomolecules.

In addition to SPR, various other types of biosensors also exhibit different complementary advantages in coupling with mass spectrometry (Figure 4). For example, microfluidic chips are widely used in conjunction with nano ESI to provide excellent performance in reaction monitoring, sensitivity enhancement, biomolecule extraction, etc. [15,97–100]. Surface acoustic wave (SAW) devices are used for sample delivery and assisted ionization to enable a real-time, high-throughput quantitative and qualitative analysis of heavy metals and various biomolecules in human serum [101,102].

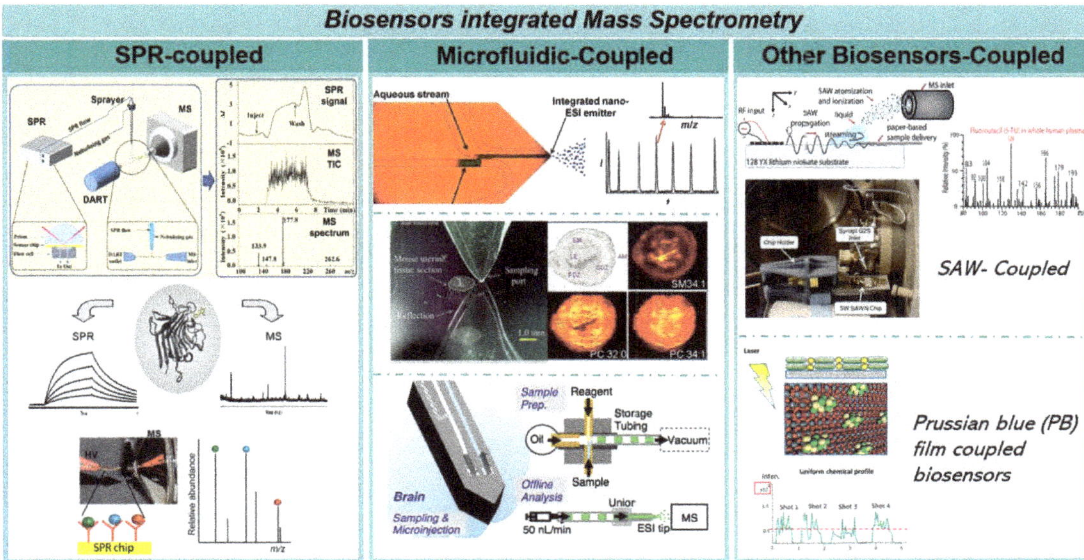

Figure 4. Highlights of different sensor technologies coupled with mass spectrometry including online and offline integration of SPR technology [91,93,96], combination of microfluidic sensing chips [16,66,100], and integration of other biosensors [101–104]. Copyright with permission from Wiley, Elsevier, Royal Society of Chemistry, American Chemical Society.

5. Conclusions

This review details several types of research in the direction of mass spectrometry-based biosensing and summarizes advanced sensing techniques in combination with mass spectrometry, including biochips, mass labels, in vivo probes, clinical biopsies, etc. Given the advantages of mass spectrometry in molecular recognition, each type of technology can exploit its specificity, such as mass-label type technology with flexible target recognition capability and in vivo probe/pen technology for clinical biopsy capability, while the coupling with other sensors can produce functional complementarity. Although mass spectrometry has the above-mentioned advantages, it still has the problems of large instrument size and poor portability, and it cannot completely replace electrochemical and optical sensors. The 21st century is the era of cross-disciplinary development, and combining various analytical methods to form a richer, more sensitive, and faster biosensing system may become the new development direction. In addition, new mass spectrometry sensing systems with greater sensitivity, better robustness, and higher throughput will be introduced and applied in various fields of chemical analysis.

Author Contributions: F.C., literature search, graph creation, data analysis, and manuscript writing. W.W. and N.S.S., literature search and manuscript writing. H.F. and X.W., language correction, and data analysis. Y.P., article organization and correction. All authors have read and agreed to the published version of the manuscript.

Funding: This work was supported by the Science and Technology Program of Zhejiang Province (LGC21B050008).

Institutional Review Board Statement: Not applicable.

Informed Consent Statement: Not applicable.

Data Availability Statement: No new data were generated for this manuscript.

Conflicts of Interest: The authors declare no conflict of interest.

References

1. Kulkarni, M.B.; Ayachit, N.H.; Aminabhavi, T.M. Biosensors and Microfluidic Biosensors: From Fabrication to Application. *Biosensors* **2022**, *12*, 543. [CrossRef]
2. Naresh, V.; Nohyun, L. A Review on Biosensors and Recent Development of Nanostructured Materials-Enabled Biosensors. *Sensors* **2021**, *21*, 1109. [CrossRef]
3. You, M.; Li, Z.; Feng, S.; Gao, B.; Yao, C.; Hu, J. Ultrafast Photonic PCR Based on Photothermal Nanomaterials. *Trends Biotechnol.* **2019**, *38*, 637–649. [CrossRef]
4. Yang, Z.; Ren, Z.; Cheng, Y.; Sun, W.; Xi, Z.; Jia, W.; Li, G.; Wang, Y.; Guo, M.; Li, D. Review and Prospect on Portable Mass Spectrometer for Recent Applications. *Vacuum* **2022**, *199*, 110889. [CrossRef]
5. Feider, C.L.; Krieger, A.; Dehoog, R.J.; Eberlin, L.S. Ambient Ionization Mass Spectrometry: Recent Developments and Applications. *Anal. Chem.* **2019**, *91*, 4266–4290. [CrossRef] [PubMed]
6. Hu, J.; Liu, F.; Chen, Y.; Shangguan, G.; Ju, H. Mass Spectrometric Biosensing: A Powerful Approach for Multiplexed Analysis of Clinical Biomolecules. *ACS Sens.* **2021**, *6*, 3517–3535. [CrossRef]
7. Wiseman, J.M.; Gologan, B.; Cooks, R.G. Mass Spectrometry Sampling Under Ambient Conditions with Desorption Electrospray Ionization. *Science* **2004**, *306*, 471–474.
8. Shi, L.; Habib, A.; Bi, L.; Hong, H.; Begum, R. Ambient Ionization Mass Spectrometry: Application and Prospective. *Crit. Rev. Anal. Chem.* **2022**, 1–50. [CrossRef]
9. Chu, F.; Zhao, G.; Li, W.; Wei, W.; Chen, W.; Ma, Z.; Gao, Z.; Shuaibu, N.S.; Luo, J.; Yu, B.; et al. Catalyst-Free Oxidation Reactions in a Microwave Plasma Torch-Based Ion/Molecular Reactor: An Approach for Predicting the Atmospheric Oxidation of Pollutants. *Anal. Chem.* **2022**, *95*, 2004–2010. [CrossRef]
10. Su, J.; Mrksich, M. Using Mass Spectrometry to Characterize Self- Assembled Monolayers Presenting Peptides, Proteins, and Carbohydrates. *Angew. Chem.* **2002**, *114*, 4909–4912. [CrossRef]
11. Kou, X.; Chen, G.; Huang, S.; Ye, Y.; Ouyang, G.; Gan, J.; Zhu, F. In Vivo Sampling: A Promising Technique for Detecting and Pro Fi Ling Endogenous Substances in Living Systems. *J. Agric. Food Chem.* **2019**, *67*, 2120–2126. [CrossRef] [PubMed]
12. Xue, J.; Liu, H. Surface Plasmon Resonance Coupled to Mass Spectrometry in Bioanalysis. *Compr. Anal. Chem.* **2021**, *95*, 89–106. [CrossRef]
13. Wang, R.; Chait, B.T.; Kent, S.B.H. Protein Ladder Sequencing: Towards Automation. *Tech. Protein Chem.* **1994**, *5*, 19–26. [CrossRef]
14. Tong, Y.; Guo, C.; Liu, Z.; Shi, K.; Zhang, H.; Liu, Y.; Wu, G.; Feng, H.; Pan, Y. In Situ Localization of Tris(2,3-Dibromopropyl) Isocyanurate in Mouse Organs by MALDI-IMS with Auxiliary Matrix Strategy. *Talanta* **2021**, *235*, 122723. [CrossRef]
15. Wang, H.; Gao, Y.; He, Q.; Liao, J.; Zhou, S.; Liu, Y.; Guo, C.; Li, X.; Zhao, X.; Pan, Y. 2-Hydrazinoterephthalic Acid as a Novel Negative-Ion Matrix-Assisted Laser Desorption/Ionization Matrix for Qualitative and Quantitative Matrix-Assisted Laser Desorption/Ionization-Mass Spectrometry Analysis of N-Glycans in Peach Allergy Research. *J. Agric. Food Chem.* **2023**, *71*, 952–962. [CrossRef]
16. Tong, Y.; Liu, Z.Z.; Lu, J.F.; Zhang, H.Y.; Shi, K.Q.; Chen, G.R.; Liu, Y.Q.; Feng, H.R.; Pan, Y.J. Detection and Quantification of Water-Soluble Inorganic Chlorine, Bromine and Iodine by MALDI-MS. *J. Anal. Test.* **2022**, *6*, 419–423. [CrossRef]
17. Beloqui, A.; Sanchez-Ruiz, A.; Martin-Lomas, M.; Reichardt, N.C. A Surface-Based Mass Spectrometry Method for Screening Glycosidase Specificity in Environmental Samples. *Chem. Commun.* **2012**, *48*, 1701–1703. [CrossRef]
18. Balog, J.; Sasi-Szabó, L.; Kinross, J.; Lewis, M.R.; Muirhead, L.J.; Veselkov, K.; Mirnezami, R.; Dezso, B.; Damjanovich, L.; Darzi, A.; et al. Intraoperative Tissue Identification Using Rapid Evaporative Ionization Mass Spectrometry. *Sci. Transl. Med.* **2013**, *5*, 194ra93. [CrossRef]
19. Mendes, T.P.P.; Lobón, G.S.; Lima, L.A.S.; Guerra, N.K.M.; Carvalho, G.A.; Freitas, E.M.M.; Pinto, M.C.X.; Pereira, I.; Vaz, B.G. Mass Spectrometry-Based Biosensing Using Pencil Graphite Rods. *Microchem. J.* **2021**, *164*, 106077. [CrossRef]
20. Kelly, R.T.; Page, J.S.; Marginean, I.; Tang, K.; Smith, R.D. Dilution-Free Analysis from Picoliter Droplets by Nano-Electrospray Ionization Mass Spectrometry. *Angew. Chem.-Int. Ed.* **2009**, *48*, 6832–6835. [CrossRef]
21. Yoshimura, K.; Chen, L.C.; Yu, Z.; Hiraoka, K.; Takeda, S. Real-Time Analysis of Living Animals by Electrospray Ionization Mass Spectrometry. *Anal. Biochem.* **2011**, *417*, 195–201. [CrossRef] [PubMed]
22. Min, D.-H.; Su, J.; Mrksich, M. Profiling Kinase Activities by Using a Peptide Chip and Mass Spectrometry. *Angew. Chem.* **2004**, *116*, 6099–6103. [CrossRef]
23. Becker, C.F.W.; Wacker, R.; Bouschen, W.; Seidel, R.; Kolaric, B.; Lang, P.; Schroeder, H.; Müller, O.; Niemeyer, C.M.; Spengler, B.; et al. Direkter Nachweis von Protein-Protein-Wechselwirkungen Durch Massenspektrometrie an Protein-DNA-Mikroarrays. *Angew. Chem.-Int. Ed.* **2005**, *117*, 7808–7812. [CrossRef]
24. Marin, V.L.; Bayburt, T.H.; Sligar, S.G.; Mrksich, M. Functional Assays of Membrane-Bound Proteins with SAMDI-TOF Mass Spectrometry. *Angew. Chem.-Int. Ed.* **2007**, *119*, 8952–8954. [CrossRef]
25. Lee, J.R.; Lee, J.; Kim, S.K.; Kim, K.P.; Park, H.S.; Yeo, W.-S. Mass Spectrometry Signal Amplification Method for Attomolar Detection of Antigens Using Small-Molecule-Tagged Gold Microparticles. *Angew. Chem.-Int. Ed.* **2008**, *120*, 9660–9663. [CrossRef]
26. Kim, Y.-K.; Ryoo, S.-R.; Kwack, S.-J.; Min, D.-H. Mass Spectrometry Assisted Lithography for the Patterning of Cell Adhesion Ligands on Self-Assembled Monolayers. *Angew. Chem.-Int. Ed.* **2009**, *121*, 3559–3563. [CrossRef]
27. Beloqui, A.; Calvo, J.; Serna, S.; Yan, S.; Wilson, I.B.H.; Martin-Lomas, M.; Reichardt, N.C. Analysis of Microarrays by MALDI-TOF MS. *Angew. Chem.-Int. Ed.* **2013**, *52*, 7477–7481. [CrossRef]

28. Kuo, H.Y.; Deluca, T.A.; Miller, W.M.; Mrksich, M. Profiling Deacetylase Activities in Cell Lysates with Peptide Arrays and SAMDI Mass Spectrometry. *Anal. Chem.* **2013**, *85*, 10635–10642. [CrossRef]
29. Li, J.; Lipson, R.H. Assays Using a NIMS Chip: Loosely Bound but Highly Selective. *Anal. Chem.* **2013**, *85*, 6860–6865. [CrossRef]
30. Hong, S.H.; Kim, J., II; Kang, H.; Yoon, S.; Kim, D.E.; Jung, W.; Yeo, W.S. Detection and Quantification of the Bcr/Abl Chimeric Protein on Biochips Using LDI-TOF MS. *Chem. Commun.* **2014**, *50*, 4831–4834. [CrossRef]
31. Both, P.; Green, A.P.; Gray, C.J.; Šardzík, R.; Voglmeir, J.; Fontana, C.; Austeri, M.; Rejzek, M.; Richardson, D.; Field, R.A.; et al. Discrimination of Epimeric Glycans and Glycopeptides Using IM-MS and Its Potential for Carbohydrate Sequencing. *Nat. Chem.* **2014**, *6*, 65–74. [CrossRef] [PubMed]
32. Hu, J.; Liu, F.; Ju, H. Peptide Code-on-a-Microplate for Protease Activity Analysis via MALDI-TOF Mass Spectrometric Quantitation. *Anal. Chem.* **2015**, *87*, 4409–4414. [CrossRef] [PubMed]
33. Lorey, M.; Adler, B.; Yan, H.; Soliymani, R.; Ekström, S.; Yli-Kauhaluoma, J.; Laurell, T.; Baumann, M. Mass-Tag Enhanced Immuno-Laser Desorption/Ionization Mass Spectrometry for Sensitive Detection of Intact Protein Antigens. *Anal. Chem.* **2015**, *87*, 5255–5262. [CrossRef]
34. Hu, J.; Liu, F.; Ju, H. MALDI-MS Patterning of Caspase Activities and Its Application in the Assessment of Drug Resistance. *Angew. Chem.-Int. Ed.* **2016**, *55*, 6667–6670. [CrossRef]
35. Feng, N.; Hu, J.; Ma, Q.; Ju, H. Mass Spectrometric Biosensing: Quantitation of Multiplex Enzymes Using Single Mass Probe and Fluorous Affinity Chip. *Biosens. Bioelectron.* **2020**, *157*, 112159. [CrossRef] [PubMed]
36. Xu, S.; Ma, W.; Bai, Y.; Liu, H. Ultrasensitive Ambient Mass Spectrometry Immunoassays: Multiplexed Detection of Proteins in Serum and on Cell Surfaces. *J. Am. Chem. Soc.* **2019**, *141*, 72–75. [CrossRef]
37. Xu, S.; Liu, M.; Feng, J.; Yan, G.; Bai, Y.; Liu, H. One-Step Hexaplex Immunoassays by on-Line Paper Substrate-Based Electrospray Ionization Mass Spectrometry for Combined Cancer Biomarker Screening. *Chem. Sci.* **2021**, *12*, 4916–4924. [CrossRef]
38. O'Kane, P.T.; Dudley, Q.M.; McMillan, A.K.; Jewett, M.C.; Mrksich, M. High-Throughput Mapping of CoA Metabolites by SAMDI-MS to Optimize the Cell-Free Biosynthesis of HMG-CoA. *Sci. Adv.* **2019**, *5*, eaaw9180. [CrossRef]
39. Techner, J.M.; Kightlinger, W.; Lin, L.; Hershewe, J.; Ramesh, A.; Delisa, M.P.; Jewett, M.C.; Mrksich, M. High-Throughput Synthesis and Analysis of Intact Glycoproteins Using SAMDI-MS. *Anal. Chem.* **2020**, *92*, 1963–1971. [CrossRef]
40. Ma, Q.; Chen, Y.; Feng, N.; Yan, F.; Ju, H. A MALDI-MS Sensing Chip Prepared by Non-Covalent Assembly for Quantitation of Acid Phosphatase. *Sci. China Chem.* **2021**, *64*, 151–156. [CrossRef]
41. Gunnarsson, A.; Sjövall, P.; Höök, F. Liposome-Based Chemical Barcodes for Single Molecule DNA Detection Using Imaging Mass Spectrometry. *Nano Lett.* **2010**, *10*, 732–737. [CrossRef]
42. Tang, W.; Gordon, A.; Wang, H.; Li, P.; Chen, J.; Li, B. Development of MALDI MS Peptide Array for Thrombin Inhibitor Screening. *Talanta* **2021**, *226*, 122129. [CrossRef] [PubMed]
43. Han, G.; Xing, Z.; Dong, Y.; Zhang, S.; Zhang, X. One-Step Homogeneous DNA Assay with Single-Nanoparticle Detection. *Angew. Chem.-Int. Ed.* **2011**, *50*, 3462–3465. [CrossRef] [PubMed]
44. Zhang, Y.; Chen, B.; He, M.; Yang, B.; Zhang, J.; Hu, B. Immunomagnetic Separation Combined with Inductively Coupled Plasma Mass Spectrometry for the Detection of Tumor Cells Using Gold Nanoparticle Labeling. *Anal. Chem.* **2014**, *86*, 8082–8089. [CrossRef]
45. Zhang, S.; Han, G.; Xing, Z.; Zhang, S.; Zhang, X. Multiplex DNA Assay Based on Nanoparticle Probes by Single Particle Inductively Coupled Plasma Mass Spectrometry. *Anal. Chem.* **2014**, *86*, 3541–3547. [CrossRef]
46. Yang, B.; Chen, B.; He, M.; Yin, X.; Xu, C.; Hu, B. Aptamer-Based Dual-Functional Probe for Rapid and Specific Counting and Imaging of MCF-7 Cells. *Anal. Chem.* **2018**, *90*, 2355–2361. [CrossRef] [PubMed]
47. Verkhoturov, D.S.; Crulhas, B.P.; Eller, M.J.; Han, Y.D.; Verkhoturov, S.V.; Bisrat, Y.; Revzin, A.; Schweikert, E.A. Nanoprojectile Secondary Ion Mass Spectrometry for Analysis of Extracellular Vesicles. *Anal. Chem.* **2021**, *93*, 7481–7490. [CrossRef] [PubMed]
48. Li, P.; Pang, J.; Xu, S.; He, H.; Ma, Y.; Liu, Z. A Glycoform-Resolved Dual-Modal Ratiometric Immunoassay Improves the Diagnostic Precision for Hepatocellular Carcinoma. *Angew. Chem.-Int. Ed.* **2022**, *61*, e202113528. [CrossRef]
49. Xu, S.; Liu, M.; Bai, Y.; Liu, H. Multi-Dimensional Organic Mass Cytometry: Simultaneous Analysis of Proteins and Metabolites on Single Cells. *Angew. Chem.-Int. Ed.* **2021**, *60*, 1806–1812. [CrossRef] [PubMed]
50. Bodenmiller, B.; Zunder, E.R.; Finck, R.; Chen, T.J.; Savig, E.S.; Bruggner, R.V.; Simonds, E.F.; Bendall, S.C.; Sachs, K.; Krutzik, P.O.; et al. Multiplexed Mass Cytometry Profiling of Cellular States Perturbed by Small-Molecule Regulators. *Nat. Biotechnol.* **2012**, *30*, 858–867. [CrossRef]
51. Newell, E.W.; Davis, M.M. Beyond Model Antigens: High-Dimensional Methods for the Analysis of Antigen-Specific T Cells. *Nat. Biotechnol.* **2014**, *32*, 149–157. [CrossRef] [PubMed]
52. Ma, W.; Xu, S.; Nie, H.; Hu, B.; Bai, Y.; Liu, H. Bifunctional Cleavable Probes for in Situ Multiplexed Glycan Detection and Imaging Using Mass Spectrometry. *Chem. Sci.* **2019**, *10*, 2320–2325. [CrossRef] [PubMed]
53. Yin, H.; Chu, Y.; Wang, W.; Zhang, Z.; Meng, Z.; Min, Q. Mass Tag-Encoded Nanointerfaces for Multiplexed Mass Spectrometric Analysis and Imaging of Biomolecules. *Nanoscale* **2023**, *15*, 2529–2540. [CrossRef]
54. Hiraoka, K.; Nishidate, K.; Mori, K.; Asakawa, D.; Suzuki, S. Development of Probe Electrospray Using a Solid Needle. *Rapid Commun. Mass Spectrom.* **2007**, *21*, 3139–3144. [CrossRef]
55. Yu, Z.; Chen, L.C.; Mandal, M.K.; Nonami, H.; Erra-Balsells, R.; Hiraoka, K. Online Electrospray Ionization Mass Spectrometric Monitoring of Protease-Catalyzed Reactions in Real Time. *J. Am. Soc. Mass Spectrom.* **2012**, *23*, 728–735. [CrossRef]

56. Mandal, M.K.; Yoshimura, K.; Chen, L.C.; Yu, Z.; Nakazawa, T.; Katoh, R.; Fujii, H.; Takeda, S.; Nonami, H.; Hiraoka, K. Application of Probe Electrospray Ionization Mass Spectrometry (PESI-MS) to Clinical Diagnosis: Solvent Effect on Lipid Analysis. *J. Am. Soc. Mass Spectrom.* **2012**, *23*, 2043–2047. [CrossRef]
57. Chen, L.C.; Nishidate, K.; Saito, Y.; Mori, K.; Asakawa, D.; Takeda, S.; Kubota, T.; Terada, N.; Hashimoto, Y.; Hori, H.; et al. Application of Probe Electrospray to Direct Ambient Analysis of Biological Samples. *Rapid Commun. Mass Spectrom.* **2008**, *22*, 2366–2374. [CrossRef]
58. Hsu, C.C.; Elnaggar, M.S.; Peng, Y.; Fang, J.; Sanchez, L.M.; Mascuch, S.J.; Møller, K.A.; Alazzeh, E.K.; Pikula, J.; Quinn, R.A.; et al. Real-Time Metabolomics on Living Microorganisms Using Ambient Electrospray Ionization Flow-Probe. *Anal. Chem.* **2013**, *85*, 7014–7018. [CrossRef]
59. Gong, X.; Zhao, Y.; Cai, S.; Fu, S.; Yang, C.; Zhang, S.; Zhang, X. Single Cell Analysis with Probe ESI-Mass Spectrometry: Detection of Metabolites at Cellular and Subcellular Levels. *Anal. Chem.* **2014**, *86*, 3809–3816. [CrossRef]
60. Deng, J.; Yang, Y.; Xu, M.; Wang, X.; Lin, L.; Yao, Z.P.; Luan, T. Surface-Coated Probe Nanoelectrospray Ionization Mass Spectrometry for Analysis of Target Compounds in Individual Small Organisms. *Anal. Chem.* **2015**, *87*, 9923–9930. [CrossRef]
61. Deng, J.; Li, W.; Yang, Q.; Liu, Y.; Fang, L.; Guo, Y.; Guo, P.; Lin, L.; Yang, Y.; Luan, T. Biocompatible Surface-Coated Probe for in Vivo, in Situ, and Microscale Lipidomics of Small Biological Organisms and Cells Using Mass Spectrometry. *Anal. Chem.* **2018**, *90*, 6936–6944. [CrossRef] [PubMed]
62. Xiao, X.; Chen, C.; Deng, J.; Wu, J.; He, K.; Xiang, Z.; Yang, Y. Analysis of Trace Malachite Green, Crystal Violet, and Their Metabolites in Zebrafish by Surface-Coated Probe Nanoelectrospray Ionization Mass Spectrometry. *Talanta* **2020**, *217*, 121064. [CrossRef]
63. Zaitsu, K.; Hayashi, Y.; Murata, T.; Yokota, K.; Ohara, T.; Kusano, M.; Tsuchihashi, H.; Ishikawa, T.; Ishii, A.; Ogata, K.; et al. In Vivo Real-Time Monitoring System Using Probe Electrospray Ionization/Tandem Mass Spectrometry for Metabolites in Mouse Brain. *Anal. Chem.* **2018**, *90*, 4695–4701. [CrossRef] [PubMed]
64. Zaitsu, K.; Eguchi, S.; Ohara, T.; Kondo, K.; Ishii, A.; Tsuchihashi, H.; Kawamata, T.; Iguchi, A. PiTMaP: A New Analytical Platform for High-Throughput Direct Metabolome Analysis by Probe Electrospray Ionization/Tandem Mass Spectrometry Using an R Software-Based Data Pipeline. *Anal. Chem.* **2020**, *92*, 8514–8522. [CrossRef] [PubMed]
65. Liu, S.; Xu, Q.; Li, Y.; Xu, W.; Zhai, Y. Coupling Handheld Liquid Microjunction-Surface Sampling Probe (HLMJ-SSP) to the Miniature Mass Spectrometer for Automated and in-Situ Surface Analysis. *Talanta* **2022**, *242*, 123090. [CrossRef]
66. Ngernsutivorakul, T.; Steyer, D.J.; Valenta, A.C.; Kennedy, R.T. In Vivo Chemical Monitoring at High Spatiotemporal Resolution Using Microfabricated Sampling Probes and Droplet-Based Microfluidics Coupled to Mass Spectrometry. *Anal. Chem.* **2018**, *90*, 10943–10950. [CrossRef]
67. Lendor, S.; Hassani, S.A.; Boyaci, E.; Singh, V.; Womelsdorf, T.; Pawliszyn, J. Solid Phase Microextraction-Based Miniaturized Probe and Protocol for Extraction of Neurotransmitters from Brains in Vivo. *Anal. Chem.* **2019**, *91*, 4896–4905. [CrossRef]
68. Deng, J.; Yang, Y.; Liu, Y.; Fang, L.; Lin, L.; Luan, T. Coupling Paternò-Büchi Reaction with Surface-Coated Probe Nanoelectrospray Ionization Mass Spectrometry for in Vivo and Microscale Profiling of Lipid C=C Location Isomers in Complex Biological Tissues. *Anal. Chem.* **2019**, *91*, 4592–4599. [CrossRef]
69. Lu, Q.; Lin, R.; Du, C.; Meng, Y.; Yang, M.; Zenobi, R.; Hang, W. Metal Probe Microextraction Coupled to Dielectric Barrier Discharge Ionization-Mass Spectrometry for Detecting Drug Residues in Organisms. *Anal. Chem.* **2020**, *92*, 5921–5928. [CrossRef]
70. Bogusiewicz, J.; Burlikowska, K.; Łuczykowski, K.; Jaroch, K.; Birski, M.; Furtak, J.; Harat, M.; Pawliszyn, J.; Bojko, B. New Chemical Biopsy Tool for Spatially Resolved Profiling of Human Brain Tissue in Vivo. *Sci. Rep.* **2021**, *11*, 19522. [CrossRef]
71. Cheng, H.; Zhao, X.; Zhang, L.; Ma, M.; Ma, X. Surface-Coated Acupuncture Needles as Solid-Phase Microextraction Probes for In Vivo Analysis of Bioactive Molecules in Living Plants by Mass Spectrometry. *Metabolites* **2023**, *13*, 220. [CrossRef] [PubMed]
72. Schäfer, K.C.; Dénes, J.; Albrecht, K.; Szaniszló, T.; Balogh, J.; Skoumal, R.; Katona, M.; Tóth, M.; Balogh, L.; Takáts, Z. In Vivo, in Situ Tissue Analysis Using Rapid Evaporative Ionization Mass Spectrometry. *Angew. Chem.-Int. Ed.* **2009**, *48*, 8240–8242. [CrossRef]
73. Golf, O.; Strittmatter, N.; Karancsi, T.; Pringle, S.D.; Speller, A.V.M.; Mroz, A.; Kinross, J.M.; Abbassi-Ghadi, N.; Jones, E.A.; Takats, Z. Rapid Evaporative Ionization Mass Spectrometry Imaging Platform for Direct Mapping from Bulk Tissue and Bacterial Growth Media. *Anal. Chem.* **2015**, *87*, 2527–2534. [CrossRef]
74. Bolt, F.; Cameron, S.J.S.; Karancsi, T.; Simon, D.; Schaffer, R.; Rickards, T.; Hardiman, K.; Burke, A.; Bodai, Z.; Perdones-Montero, A.; et al. Automated High-Throughput Identification and Characterization of Clinically Important Bacteria and Fungi Using Rapid Evaporative Ionization Mass Spectrometry. *Anal. Chem.* **2016**, *88*, 9419–9426. [CrossRef]
75. Strittmatter, N.; Rebec, M.; Jones, E.A.; Golf, O.; Abdolrasouli, A.; Balog, J.; Behrends, V.; Veselkov, K.A.; Takats, Z. Characterization and Identification of Clinically Relevant Microorganisms Using Rapid Evaporative Ionization Mass Spectrometry. *Anal. Chem.* **2014**, *86*, 6555–6562. [CrossRef]
76. Manoli, E.; Mason, S.; Ford, L.; Adebesin, A.; Bodai, Z.; Darzi, A.; Kinross, J.; Takats, Z. Validation of Ultrasonic Harmonic Scalpel for Real-Time Tissue Identification Using Rapid Evaporative Ionization Mass Spectrometry. *Anal. Chem.* **2021**, *93*, 5906–5916. [CrossRef]
77. Fatou, B.; Saudemont, P.; Leblanc, E.; Vinatier, D.; Mesdag, V.; Wisztorski, M.; Focsa, C.; Salzet, M.; Ziskind, M.; Fournier, I. In Vivo Real-Time Mass Spectrometry for Guided Surgery Application. *Sci. Rep.* **2016**, *6*, 25919. [CrossRef] [PubMed]

78. Fatou, B.; Saudemont, P.; Duhamel, M.; Ziskind, M.; Focsa, C.; Salzet, M.; Fournier, I. Real Time and in Vivo Pharmaceutical and Environmental Studies with SpiderMass Instrument. *J. Biotechnol.* **2018**, *281*, 61–66. [CrossRef]
79. Saudemont, P.; Quanico, J.; Robin, Y.M.; Baud, A.; Balog, J.; Fatou, B.; Tierny, D.; Pascal, Q.; Minier, K.; Pottier, M.; et al. Real-Time Molecular Diagnosis of Tumors Using Water-Assisted Laser Desorption/Ionization Mass Spectrometry Technology. *Cancer Cell* **2018**, *34*, 840–851.e4. [CrossRef]
80. Ogrinc, N.; Saudemont, P.; Balog, J.; Robin, Y.M.; Gimeno, J.P.; Pascal, Q.; Tierny, D.; Takats, Z.; Salzet, M.; Fournier, I. Water-Assisted Laser Desorption/Ionization Mass Spectrometry for Minimally Invasive in Vivo and Real-Time Surface Analysis Using SpiderMass. *Nat. Protoc.* **2019**, *14*, 3162–3182. [CrossRef]
81. Plekhova, V.; Van Meulebroek, L.; De Graeve, M.; Perdones-Montero, A.; De Spiegeleer, M.; De Paepe, E.; Van de Walle, E.; Takats, Z.; Cameron, S.J.S.; Vanhaecke, L. Rapid ex Vivo Molecular Fingerprinting of Biofluids Using Laser-Assisted Rapid Evaporative Ionization Mass Spectrometry. *Nat. Protoc.* **2021**, *16*, 4327–4354. [CrossRef] [PubMed]
82. Katz, L.; Woolman, M.; Kiyota, T.; Pires, L.; Zaidi, M.; Hofer, S.O.P.; Leong, W.; Wouters, B.G.; Ghazarian, D.; Chan, A.W.; et al. Picosecond Infrared Laser Mass Spectrometry Identifies a Metabolite Array for 10 s Diagnosis of Select Skin Cancer Types: A Proof-of-Concept Feasibility Study. *Anal. Chem.* **2022**, *94*, 16821–16830. [CrossRef]
83. Ogrinc, N.; Kruszewski, A.; Chaillou, P.; Saudemont, P.; Lagadec, C.; Salzet, M.; Duriez, C.; Fournier, I. Robot-Assisted SpiderMass ForIn VivoReal-Time Topography Mass Spectrometry Imaging. *Anal. Chem.* **2021**, *93*, 14383–14391. [CrossRef]
84. Zhang, J.; Rector, J.; Lin, J.Q.; Young, J.H.; Sans, M.; Katta, N.; Giese, N.; Yu, W.; Nagi, C.; Suliburk, J.; et al. Nondestructive Tissue Analysis for ex Vivo and in Vivo Cancer Diagnosis Using a Handheld Mass Spectrometry System. *Sci. Transl. Med.* **2017**, *9*, eaan3968. [CrossRef]
85. Sans, M.; Zhang, J.; Lin, J.Q.; Feider, C.L.; Giese, N.; Breen, M.T.; Sebastian, K.; Liu, J.; Sood, A.K.; Eberlin, L.S. Performance of the MasSpec Pen for Rapid Diagnosis of Ovarian Cancer. *Clin. Chem.* **2019**, *65*, 674–683. [CrossRef] [PubMed]
86. Povilaitis, S.C.; Chakraborty, A.; Kirkpatrick, L.M.; Downey, R.D.; Hauger, S.B.; Eberlin, L.S. Identifying Clinically Relevant Bacteria Directly from Culture and Clinical Samples with a Handheld Mass Spectrometry Probe. *Clin. Chem.* **2022**, *68*, 1459–1470. [CrossRef]
87. King, M.E.; Zhang, J.; Lin, J.Q.; Garza, K.Y.; DeHoog, R.J.; Feider, C.L.; Bensussan, A.; Sans, M.; Krieger, A.; Badal, S.; et al. Rapid Diagnosis and Tumor Margin Assessment during Pancreatic Cancer Surgery with the MasSpec Pen Technology. *Proc. Natl. Acad. Sci. USA* **2021**, *118*, e2104411118. [CrossRef] [PubMed]
88. Wang, K.; Xu, B.F.; Lei, C.Y.; Nie, Z. Advances in the Integration of Nucleic Acid Nanotechnology into CRISPR-Cas System. *J. Anal. Test.* **2021**, *5*, 130–141. [CrossRef]
89. Hamaloğlu, K.Ö.; Çelikbıçak, Ö.; Salih, B.; Pişkin, E. Performances of Protein Array Platforms Prepared by Soft Lithography and Self-Assemblying Monolayers-Approach by Using SPR, Ellipsometry and MALDI-MS. *J. Mol. Struct.* **2019**, *1198*, 126856. [CrossRef]
90. Yang, L.; Hou, A.; Wang, S.; Zhang, J.; Man, W.; Guo, X.; Yang, B.; Wang, Q.; Jiang, H.; Kuang, H. Screening and Quantification of TNF-α Ligand from Angelicae Pubescentis Radix by Biosensor and UPLC-MS/MS. *Anal. Biochem.* **2020**, *596*, 113643. [CrossRef]
91. Jiménez-Castells, C.; Defaus, S.; Moise, A.; Przbylski, M.; Andreu, D.; Gutiérrez-Gallego, R. Surface-Based and Mass Spectrometric Approaches to Deciphering Sugar-Protein Interactions in a Galactose-Specific Agglutinin. *Anal. Chem.* **2012**, *84*, 6515–6520. [CrossRef] [PubMed]
92. Marchesini, G.R.; Buijs, J.; Haasnoot, W.; Hooijerink, D.; Jansson, O.; Nielen, M.W.F. Nanoscale Affinity Chip Interface for Coupling Inhibition SPR Immunosensor Screening with Nano-LC TOF MS. *Anal. Chem.* **2008**, *80*, 1159–1168. [CrossRef] [PubMed]
93. Zhang, Y.; Li, X.; Nie, H.; Yang, L.; Li, Z.; Bai, Y.; Niu, L.; Song, D.; Liu, H. Interface for Online Coupling of Surface Plasmon Resonance to Direct Analysis in Real Time Mass Spectrometry. *Anal. Chem.* **2015**, *87*, 6505–6509. [CrossRef]
94. Zhang, Y.; Xu, S.; Wen, L.; Bai, Y.; Niu, L.; Song, D.; Liu, H. A Dielectric Barrier Discharge Ionization Based Interface for Online Coupling Surface Plasmon Resonance with Mass Spectrometry. *Analyst* **2016**, *141*, 3343–3348. [CrossRef]
95. Mihoc, D.; Lupu, L.M.; Wiegand, P.; Kleinekofort, W.; Müller, O.; Völklein, F.; Glocker, M.O.; Barka, F.; Barka, G.; Przybylski, M. Antibody Epitope and Affinity Determination of the Myocardial Infarction Marker Myoglobin by SPR-Biosensor Mass Spectrometry. *J. Am. Soc. Mass Spectrom.* **2021**, *32*, 106–113. [CrossRef]
96. Joshi, S.; Zuilhof, H.; Van Beek, T.A.; Nielen, M.W.F. Biochip Spray: Simplified Coupling of Surface Plasmon Resonance Biosensing and Mass Spectrometry. *Anal. Chem.* **2017**, *89*, 1427–1432. [CrossRef] [PubMed]
97. Piendl, S.K.; Schönfelder, T.; Polack, M.; Weigelt, L.; van der Zwaag, T.; Teutenberg, T.; Beckert, E.; Belder, D. Integration of Segmented Microflow Chemistry and Online HPLC/MS Analysis on a Microfluidic Chip System Enabling Enantioselective Analyses at the Nanoliter Scale. *Lab Chip* **2021**, *21*, 2614–2624. [CrossRef] [PubMed]
98. Das, A.; Weise, C.; Polack, M.; Urban, R.D.; Krafft, B.; Hasan, S.; Westphal, H.; Warias, R.; Schmidt, S.; Gulder, T.; et al. On-the-Fly Mass Spectrometry in Digital Microfluidics Enabled by a Microspray Hole: Toward Multidimensional Reaction Monitoring in Automated Synthesis Platforms. *J. Am. Chem. Soc.* **2022**, *144*, 10353–10360. [CrossRef]
99. Takagi, Y.; Kazoe, Y.; Morikawa, K.; Kitamori, T. Femtoliter-Droplet Mass Spectrometry Interface Utilizing Nanofluidics for Ultrasmall and High-Sensitivity Analysis. *Anal. Chem.* **2022**, *94*, 10074–10081. [CrossRef]
100. Li, X.; Yin, R.; Hu, H.; Li, Y.; Sun, X.; Dey, S.K.; Laskin, J. An Integrated Microfluidic Probe for Mass Spectrometry Imaging of Biological Samples**. *Angew. Chem.-Int. Ed.* **2020**, *59*, 22388–22391. [CrossRef]

101. Ho, J.; Tan, M.K.; Go, D.B.; Yeo, L.Y.; Friend, J.R.; Chang, H.C. Paper-Based Microfluidic Surface Acoustic Wave Sample Delivery and Ionization Source for Rapid and Sensitive Ambient Mass Spectrometry. *Anal. Chem.* **2011**, *83*, 3260–3266. [CrossRef] [PubMed]
102. Yoon, S.H.; Liang, T.; Schneider, T.; Oyler, B.L.; Chandler, C.E.; Ernst, R.K.; Yen, G.S.; Huang, Y.; Nilsson, E.; Goodlett, D.R. Rapid Lipid a Structure Determination via Surface Acoustic Wave Nebulization and Hierarchical Tandem Mass Spectrometry Algorithm. *Rapid Commun. Mass Spectrom.* **2016**, *30*, 2555–2560. [CrossRef]
103. Silina, Y.E.; Morgan, B. LDI-MS Scanner: Laser Desorption Ionization Mass Spectrometry-Based Biosensor Standardization. *Talanta* **2021**, *223*, 121688. [CrossRef] [PubMed]
104. Ahmad, F.; Siddiqui, M.A.; Babalola, O.O.; Wu, H.F. Biofunctionalization of Nanoparticle Assisted Mass Spectrometry as Biosensors for Rapid Detection of Plant Associated Bacteria. *Biosens. Bioelectron.* **2012**, *35*, 235–242. [CrossRef] [PubMed]

chemosensors

MDPI

Review

Framework-Enhanced Electrochemiluminescence in Biosensing

Haomin Fu, Zhiyuan Xu, Hanlin Hou, Rengan Luo, Huangxian Ju and Jianping Lei *

State Key Laboratory of Analytical Chemistry for Life Science, School of Chemistry and Chemical Engineering, Nanjing University, Nanjing 210023, China; hxju@nju.edu.cn (H.J.)
* Correspondence: jpl@nju.edu.cn

Abstract: Electrochemiluminescence (ECL) has attracted increasing attention owing to its intrinsic advantages of high sensitivity, good stability, and low background. Considering the fact that framework nanocrystals such as metal–organic frameworks and covalent organic frameworks have accurate molecular structures, a series of framework-based ECL platforms are developed for decoding emission fundamentals. The integration of fluorescent ligands into frameworks significantly improves the ECL properties due to the arrangement of molecules and intramolecular electron transfer. Moreover, the various framework topologies can be easily functionalized with the recognition elements to trace the targets for signal readout. These ECL enhancement strategies lead to a series of sensitive analytical methods for protein biomarkers, DNA, small biomolecules, and cells. In this review, we summarize recent advances in various functions of frameworks during the ECL process, and constructions of framework-based ECL platforms for biosensing. The framework-based ECL nanoemitters and enhancement mechanisms show both theoretical innovation and potential applications in designing ECL biosensing systems. Perspectives are also discussed, which may give a guideline for researchers in the fields of ECL biosensing and reticular materials.

Keywords: electrochemiluminescence; frameworks; biosensing; nanoemitters

Citation: Fu, H.; Xu, Z.; Hou, H.; Luo, R.; Ju, H.; Lei, J. Framework-Enhanced Electrochemiluminescence in Biosensing. *Chemosensors* **2023**, *11*, 422. https://doi.org/10.3390/chemosensors11080422

Academic Editor: Christos Kokkinos

Received: 28 June 2023
Revised: 23 July 2023
Accepted: 25 July 2023
Published: 28 July 2023

1. Introduction

Electrochemiluminescence (ECL) is a classic and powerful analytical technique involving a redox process at electrodes where excited states are electrochemically generated and emit light [1,2]. Benefitting from its unique light-free luminescence mechanism, ECL has many advantages for an analysis, such as a high sensitivity, good stability, and low background [3]. Nowadays, ECL is widely applied in the areas of environmental monitoring [4], cell sensing [5], imaging [6], food [7], and water safety [8]. ECL-driven tumor photodynamic therapy (PDT) was proposed through the effective energy transfer from ECL emission to photosensitizer chlorin e6 [9]. With a high spatiotemporal controllability, stable luminescence, and high photon flux of ECL, ECL microscopy may be more fascinating than fluorescence [10], bioluminescence [11], and surface-enhanced Raman scattering [12]. Over several decades of research, the variety of ECL emitters has substantially increased, and they can be broadly classified into an inorganic system (such as Ir or Ru complexes), an organic system (such as luminols), and semiconductor nanomaterials [13]. Recently, several novel nanomaterials have been used as luminophores in ECL, such as Au nanomaterials [14], quantum dots [15], and frameworks [16]. The combinations of ECL techniques and these new materials broaden the scope of ECL applications.

Frameworks containing metal–organic frameworks (MOFs), covalent–organic frameworks (COFs), and hydrogen-bonded organic frameworks (HOFs) have been developed rapidly since the 21st century [17,18]. Owing to their flexible, synthetically controllable, and adjustable structure, frameworks have been utilized in various areas such as energy storage, sewage treatment, gas separation, catalysis, and biosensing [19]. Although there are some deficiencies for electrochemical reactions in frameworks, such as an intrinsic poor electroconductivity and low mass permeability [20], frameworks have been gradually regarded as one of the most promising nanomaterials in ECL assays. Different from other ECL luminophores like metal complexes and quantum dots, predesignable structures of frameworks make them more suitable and efficient in ECL processes. For example, Yin's group designed an ECL-active MOF by using a ruthenium complex as a ligand and more intense ECL emission was observed with the aid of grapheme oxide [21]. By combining a predesigned structure with post-modification, frameworks provide various strategies to regulate their ECL signals for adapting the requirement. Furthermore, the structures of frameworks are utilized dexterously to create novel ECL enhancement mechanisms for developing sensitive and stable analytical methods. Overall, the variation of an ECL signal highly depends on the optoelectronic properties of frameworks, which are adjustable with designation or post-modification [22].

In order to design self-luminescent reticular nanoemitters, three types of methods have been developed. First, frameworks can be synthesized with ECL-active luminophore ligands, such as porphyrin [23], pyrene [24], and aggregation-induced emission luminogen [25], for constructing emitters with an improved ECL efficiency. Second, the doping of transition metal elements is a promising way to deal with an intrinsic low conductivity of MOFs in ECL reactions. Classic Ru complexes [26] and novel lanthanide ions [27] are already applied in biosensing with this method. Third, non-emitting monomers can be endowed with intense ECL emission through a rational design. Typically, by utilizing well-designed donor/acceptor units, a high ECL efficiency of COFs could be realized [28,29].

Therefore, integrating the frameworks and ECL methods is of great significance to construct high-performance biosensing platforms. In recent years, the roles of frameworks in ECL processes develop rapidly with multifunctions in biosensing. Initially, frameworks were used as carriers of classic luminophores or catalysts to accelerate the ECL reactions, but they later became ECL emitters for biosensing platform establishment, achieving a successful analysis of proteins, nucleic acids, small molecules, and cells.

In this review, the various functions of frameworks in ECL emission are first analyzed to show the rapid development in this area (Figure 1). Then, framework-enhanced ECL biosensing applications in recent years are introduced and analyzed. Finally, perspectives and potential issues are proposed, which may guide the great development of framework-based ECL biosensing systems.

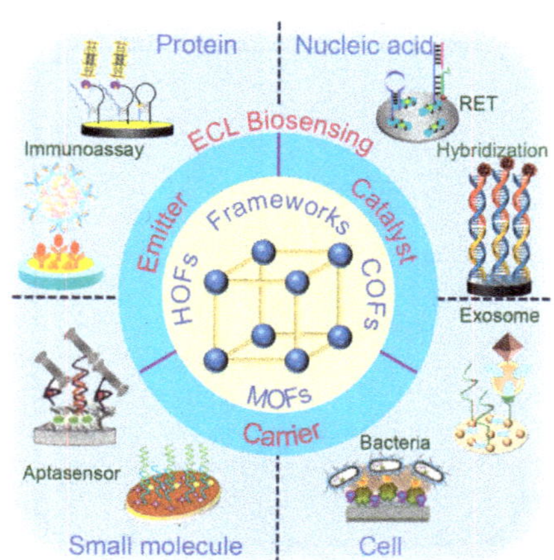

Figure 1. Schematic illustration of roles of frameworks in ECL processes and their ECL applications in biosensing.

2. The Roles of Frameworks in ECL Processes

Although the combination of frameworks with ECL demonstrates great potentials in signal readout, low conductivities and inert electrochemical properties of frameworks may result in a poor sensing performance. Therefore, frameworks were often used as a carrier of efficient ECL emitters like quantum dots (QDs) or Ru complexes, or a catalyst to strengthen ECL emission during the initial period. After the report of the electroactive MOF [16], frameworks gradually began to be used as emitters in ECL processes and various ECL-active frameworks were extensively applied in biosensing with several emerging signal amplification strategies.

2.1. The Carriers of ECL Luminophores

The outstanding features of frameworks, such as an adjustable reticular structure, large surface area, tunable pore sizes, and functionalized sites, make them competent to be used as a carrier. In early reports, classic ECL luminophores (Ru complexes, luminol, QDs, et al.) were integrated with frameworks using encapsulation or post-modifications.

The encapsulation of ECL luminophores into frameworks is a widely applied strategy to make frameworks better in an ECL performance. Through introducing guest materials, host frameworks receive an improved ECL efficiency while largely maintaining their own original properties. Therefore, encapsulation gives a flexible way to prepare frameworks with promising ECL activity. In view of the porous structure of frameworks, Qin et al. prepared $Ru(bpy)_3^{2+}$-functionalized MOF thin films using the self-assembly approach (Figure 2a). Plenty of $Ru(bpy)_3^{2+}$ molecules in Ru-MOF films showed an intense ECL emission and excellent behavior in the detection of the human-heart-type fatty-acid-binding protein [30]. Also, classic luminol-based frameworks are conducted through this method. Tang et al. synthesized porous Zn-based MOF, which loaded a large amount of luminol by encapsulating into its pores. The resulting Zn-MOF@luminol as the signal probe achieved a strong ECL signal for detecting concanavalin A [31]. Furthermore, luminophores with large sizes, such as QDs or g-C_3N_4, can be encapsulated in frameworks with high surface areas. As shown in Figure 2b, Fe-MIL-88B-NH_2@ZnSe was successfully prepared via the one-pot method. By using Fe-MIL-88B-NH_2 as an efficient coreaction accelerator, the biosensor realized the sensitive detection of a squamous cell carcinoma antigen in human serum [32]. Qin et al. designed a triethanolamine-functionalized MOF on graphene oxide nanosheets to accomplish creating a rapid label-free ECL immunosensor for the detection of human copeptin [33].

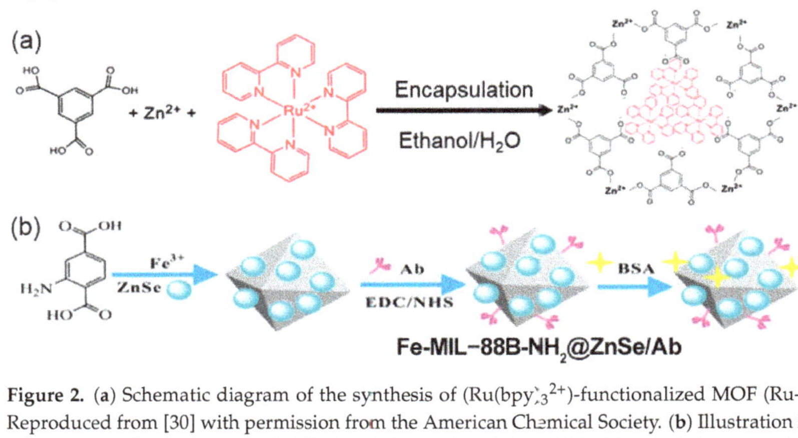

Figure 2. (**a**) Schematic diagram of the synthesis of ($Ru(bpy)_3^{2+}$)-functionalized MOF (Ru-MOF). Reproduced from [30] with permission from the American Chemical Society. (**b**) Illustration for the construction of Fe-MIL-88B-NH_2@ZnSe/Ab. Reproduced from [32] with permission from Elsevier.

On the basis of porosity and a large surface area, frameworks are considered to be suitable for post-modification with functional materials to obtain specific properties [34], which can be conducted through covalent or noncovalent bonding. For example, Wang et al. combined zeolitic imidazolate frameworks and luminol-capped Ag nanoparticles to form a luminol-AgNPs@ZIF-67 system via electrostatic interaction, which had ~115-fold-enhanced ECL compared to the luminol system [35]. In addition, QDs were merged onto MIL-53 through noncovalent adsorption and the resulting MIL-53@QDs platform demonstrated a large ECL intensity enhanced by the surface plasmon resonance process between AuNPs and CdS QDs for kanamycin and neomycin biosensing [36]. Furthermore, Liu's group developed a nanoreactor based on $Ru(bpy)_3{}^{2+}$-doped nanoporous zeolite nanoparticles (Ru@zeolite) [37], in which frameworks not only served as a carrier of Ru complexes through post-modification but were also spatially confined for efficient collision reactions in in situ ECL reactions.

2.2. The Catalyst in ECL Processes

By integrating catalytically active components, frameworks have been utilized as electrocatalysts, such as in an oxygen reduction reaction and CO_2 reduction, for a long time [38,39]. A more intense ECL emission will be observed when decisive elementary reactions are accelerated during the ECL process. For instance, Zn tetrakis(carboxyphenyl)-porphyrin (TCPP) linkers in MOF-525 acted as ECL active centers to facilitate the conversion from dissolved oxygen to singlet oxygen for enhanced ECL (Figure 3a). Based on MOF-525-Zn as signal amplifying probes, an ultrasensitive ECL sensor was proposed for the detection of protein kinase A activity with a linear range from 0.01 to 20 U mL^{-1} and detection limit of 0.005 U mL^{-1} [40]. Furthermore, the inorganic Zr–O clusters of MOF-525 simultaneously served as the recognition sites of phosphate groups for a specific bioanalysis.

On the other hand, MOFs were utilized as a coreactant accelerator to enhance the ECL of CdTe QDs through accelerating the generation of the sulfate radical anion ($SO_4{}^{\bullet -}$), which is critical in producing excited states of QDs, further realizing an ultrasensitive bioanalysis of the cardiac troponin-I antigen [41]. Similarly, 2D Fe-Zr metal–organic layers were applied for the construction of an ECL immunosensor by utilizing their peroxidase-like activity, which could effectively enhance the ECL signal of luminol through H_2O_2 catalysis [42]. Additionally, Song et al. designed a signal-amplified ECL sensor chip via the synergistic catalysis of Au–Pd bimetallic nanocrystals and mixed-valence Ce-based MOFs for the fast reduction of dissolved O_2 (Figure 3b). By integrating a three-electrode detection system into the self-assembled microfluidic chip, the developed sensor showed a high sensitivity for procalcitonin detection with the automation and portability of the detection process [43]. In a word, by introducing active catalytic sites or utilizing intrinsic properties, frameworks have nanozyme-like functions for ECL catalytic enhancement.

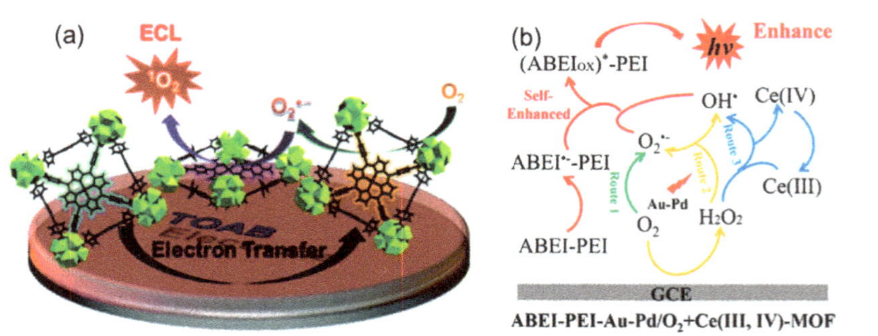

Figure 3. (**a**) Schematic illustration for the ECL catalysis mechanism of singlet oxygen based on MOF-525. Reproduced from [40] with permission from the Royal Society of Chemistry. (**b**) Supposed ECL catalysis mechanism for Au−Pd/O_2 + Ce(III, IV)-MOF system. "*" represents the excited state. Reproduced from [43] with permission from the American Chemical Society.

2.3. ECL Nanoemitters

Considering the structures of framework units, introducing ECL luminophores as linkers is thought to be a proper approach to establish framework-based ECL emitters. Due to intrinsic structural features, framework-based emitters are considered to be promising material for ECL biosensing based on the combined advantages of framework emitters and

ECL techniques [44]. Because of efficient energy migration [45], Ru-complex-based linkers have been applied for designing ECL-active frameworks since 2010. Ru(II) bipyridine ($Ru(bpy)_3{}^{2+}$) derivatives as ligands can be synthesized into frameworks using coordination with metal ions or clusters [46]. For example, functionalized Ru-based MOF nanosheets comprising carboxyl-rich tris(4,4′-dicarboxylic acid-2,2′-bipyridyl) Ru(II) and Zn^{2+} nodes exhibited a good water solubility and excellent ECL performance (Figure 4a). By employing Ru-MOF as an ECL probe, a "signal-on" ECL immunosensor was designed for the selective detection of cardiac troponin I in the range from 1 fg/mL to 10 ng/mL [47]. However, Ru complexes are costly when adjusting their structures and large in steric size, which inevitably restrict their application in direct framework synthesis. In fact, Ru complexes are more often modified onto frameworks through a post-synthesized route, which makes frameworks work like a carrier rather than a nanoemitter [48,49]. Meanwhile, other ECL-active organic ligands, such as a porphyrin derivative, perylene-3,4,9,10-tetracarboxylate, and 9,10-anthracene dibenzoate (DPA), were utilized in constructing MOF emitters for proprotein convertase subtilisin/kexin type 9, microRNAs, and MCU1 detection, respectively [50–52].

Inspired by aggregation-induced emission (AIE) luminophores, which show a stronger photoluminescence in the aggregated state than that of the isolated one [53,54], frameworks constructed with AIE molecules become attractive in ECL sensing. Typically, tetraphenylethylene (TPE)-based AIEgens are mostly reported in recent research thanks to designable molecular structures. For instance, a fiber-like MOF, synthesized with the coordination of Zn^{2+} and 1,1,2,2-tetrakis(4-(pyridin-4-yl)phenyl)ethane (TPPE), showed a more intense ECL emission than its ligand TPPE in the presence of 1,4-diazabicyclo[2.2.2]octane (DABCO) (Figure 4b,c). More significantly, different from the constant ECL intensity using a tri-n-propylamine (TPrA) coreactant, DABCO exhibited a time-dependent ECL intensity due to the intrareticular electron transfer through coordination interaction between DABCO and Zn^{2+} [55]. In another work, Wei's group synthesized a dumbbell-plate-shaped MOF consisting of 1,1,2,2-tetra(4-carboxylbiphenyl)ethylene and Zr(IV) cations, which was utilized as an ECL tag for neuron-specific enolase detection with a sandwich-type immunoreaction [56]. In addition, a two-dimensional AIEgen-based MOF was also fabricated into an efficient ECL biosensing platform [57], which restricted the intramolecular free rotation and vibration of these ligands and then reduced the non-radiative transition. The combination of AIE ligands and frameworks paved a potential way for better ECL sensors; that is, the large surface area and porous properties of MOFs make ECL reactions more effective while the AIE molecular motion is restricted by the rigid MOF structure, which is theoretically beneficial to AIE emission [58]. Similar to the term aggregation-induced emission, the strong ECL emission based on the restricted AIE molecules within MOFs has been named 'aggregation-induced ECL' (AIECL) [59], which is also successfully used in COFs [60] and polymers [61].

Identically, luminophores can be introduced into frameworks by serving as ion nodes. Due to a good photoluminescent emission and successful applications in biosensing [62], self-luminescent lanthanide MOFs (Ln-MOFs) are considered promising luminophores in ECL reactions. Dai's group synthesized La^{3+}-BTC MOFs as an ECL emitter and highly active reactor simultaneously to construct a gene sensor. With the assistance of crystal violet, a good performance toward a p53 gene analysis was obtained through the co-quenching effect mechanism [63]. Furthermore, Eu-based Ln-MOFs were prepared with 5-boronoisophthalic acid and Eu (III) ions. The ECL emission mechanism was identified to be that 5-bop was excited with ultraviolet photons to generate a triplet state, which then triggered Eu (III) ions for red emission. The Eu-MOFs showed a great sensitivity in an ECL immunoassay for Cytokeratin 21-1 detection [64].

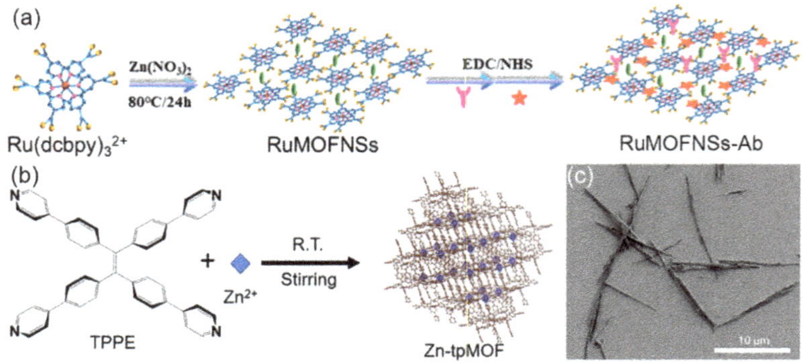

Figure 4. (**a**) Schematic illustration for the fabrication process of a Ru-MOF nanosheet. Reproduced from [47] with permission from the American Chemical Society. (**b**) The synthesis of and (**c**) SEM image of Zn-tpMOF. Reproduced from [55] with permission from Elsevier.

In order to obtain a better biosensing performance, a higher ECL efficiency is urgently needed. Conventional coreactant ECL is convenient in operation but inefficient in electron transfer due to the intermolecular route. Thanks to a shortened pathway of mass transport and electron transfer, the intramolecular electron transfer process is recognized as a promising solution [65]. Inspired by this theory, a mixed-ligand MOF (m-MOF) was designed for proof of concept by integrating it with two ligands, one as a luminophore and the other as a coreactant, on one metal node for self-enhanced ECL [66]. As shown in Figure 5a,b, the resulting m-MOF had a highly ordered crystalline unit proved by comparing the experimental PXRD pattern and theoretical simulation. Then, the m-MOF exhibited greatly enhanced ECL compared to its ligand and Zn-DPA MOF, indicating a high efficiency of the intrareticular charge transfer process (Figure 5c). Finally, the proposed stepwise ECL mechanism of the m-MOF was given as a result of local excitation in the DPA unit, which was identified through a density functional theory calculation (Figure 5d). Overall, the mixed-ligand approach successfully shortens the pathway of charge transfer, providing a new idea in ECL platform designs.

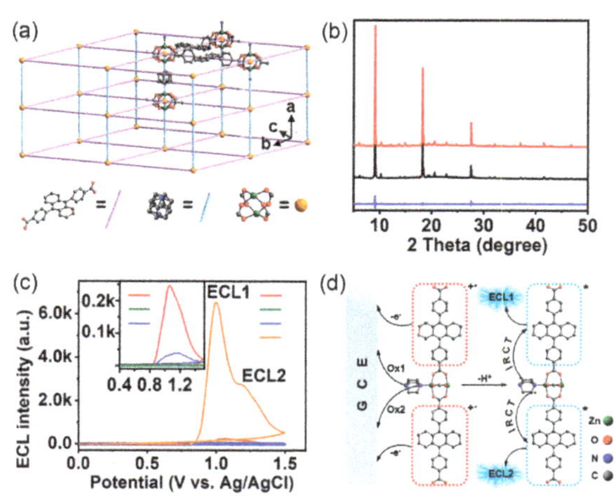

Figure 5. (**a**) Three-dimensional structure of m-MOF. (**b**) PXRD pattern of m-MOF (black), simulated result (red), and their difference (blue). (**c**) ECL curves of DPA-modified GCEs (red) in presence of DABCO, and of DPA-(green), s-MOF-(blue), and m-MOF (orange)-modified GCEs in 0.1 M PBS. (**d**) Stepwise ECL mechanism of m-MOF. "*" represents the excited state. Reproduced from [66] with permission from American Chemical Society.

As a novel member of frameworks, COFs gradually became fascinating in ECL applications. Firstly, Li et al. gave general advice on how to design COFs with highly efficient ECL [67]. Meanwhile, Lei's group provided a detailed mechanism on the enhanced ECL of COFs [68]. Based on donor–acceptor (D-A) units, a luminescent *t*-COF was synthesized as an ECL emitter by integrating triazine and triphenylamine as donor and acceptor units in the reticular skeleton, respectively (Figure 6a). Revealed with a PXRD analysis, the *t*-COF showed a crystalline structure with diffraction peaks at 2θ = 4.4, 7.7, 8.9, 11.8, and 22.5°, which were assigned to the 100, 110, 200, 210, and 001 facets, respectively (Figure 6b). Compared to the other two COFs, *t*-COF had a magnificent ECL performance in TPrA/PBS (Figure 6c), indicating the importance of a D-A structure in *t*-COF during an ECL reaction. The simulated charge density difference between the first excited state and ground state of COF demonstrated an electron density loss on the triazine units and an electron density gain on the triphenylamine units, confirming the charge transfer between triphenylamine and triazine units (Figure 6d). Furthermore, the efficient charge transfer could be identified with the movement of HOS/LUS to the Fermi level when holes/electrons were doped (Figure 6e). Finally, the competitive oxidation mechanism involved the triazine unit gaining electrons from TPrA$^{\bullet}$ while the triphenylamine unit was oxidized by oxidative TPrA$^{+\bullet}$ (Figure 6f, left) or the electrode (Figure 6f, right), leading to dual ECL emissions.

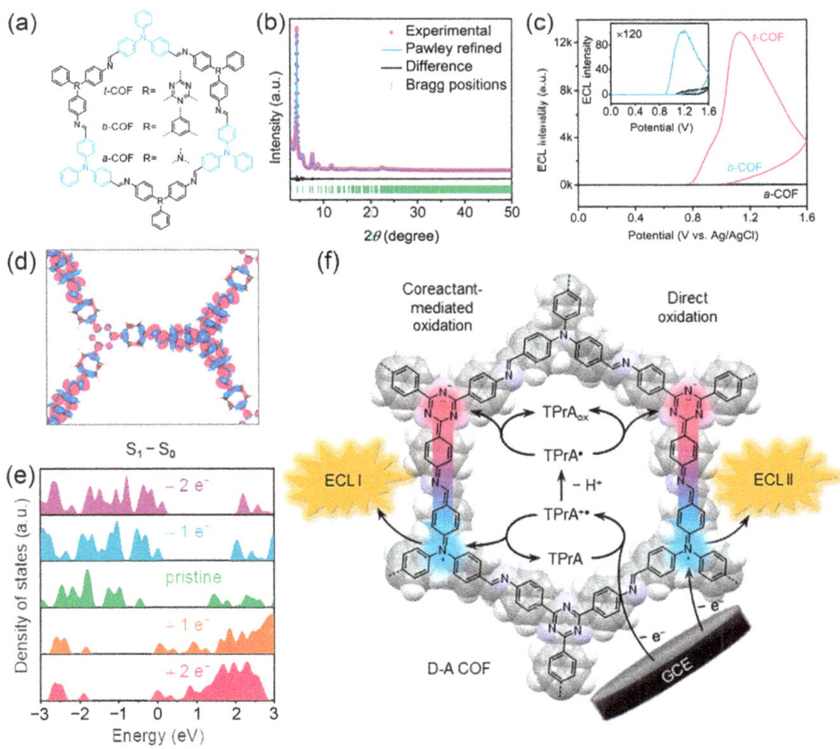

Figure 6. (**a**) Structure of TFPA-based COFs. (**b**) PXRD patterns of *t*-COF and their difference. (**c**) ECL curves of three COF-modified GCEs in the presence of 20 mM of TPrA. (**d**) The difference in charge density between the first excited state and ground state of *t*-COF. (**e**) Density of states of *t*-COF doped with different electron/hole numbers. (**f**) Competitive oxidation mechanism via intrareticular charge transfer. Reproduced from [68] with permission from the Nature Publishing Group.

HOFs comprised solely of pure organic or metal–organic units connected by intermolecular H-bonds were also found to have ECL enhancement properties compared to their monomers. Zhang et al. synthesized a triazinyl-based HOF through N···H hydrogen bond self-assembly aggregation. The resulting HOF showed a highly enhanced ECL efficiency (21.3%) relative to the $Ru(bpy)_3{}^{2+}$ standard, and was applied for ultrasensitive kanamycin biosensing [69]. Benefiting from the densely stacked structure, Lei's group proposed the HOF-based ECL enhancement mechanism via the intrareticular electron coupling (IREC) pathway [70]. Utilizing multiple H-bonds and π-interactions, HOF-101 with 1,3,6,8-tetra(4-carboxylphenyl)pyrene as a ligand was synthesized (Figure 7a). Compared with 1,3,6,8-tetracarboxypyrene-based HOF-100 and a bare electrode, HOF-101 modified GCE showed significantly enhanced ECL in the presence of TPrA due to the IREC effect (Figure 7b). Through model simulation, the charge density difference between S_1 and S_0 of HOF-101 was illustrated (Figure 7c), showing a mutual electron density depletion and accumulation of vertical stacking units. This IREC pathway in HOF-101 achieves ECL enhancement by accelerating electron transfer between anion radicals and cation radicals (Figure 7d).

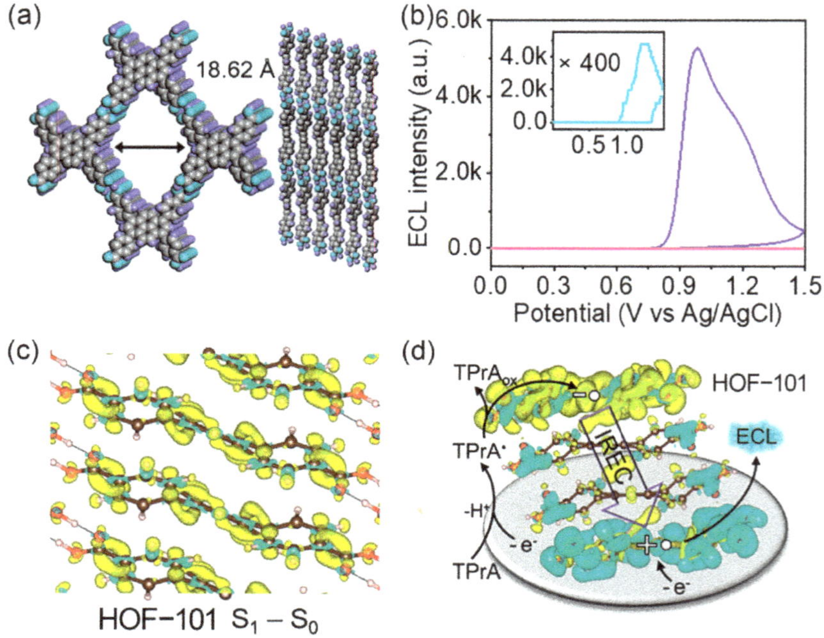

Figure 7. (**a**) Pore structure and stacking of HOF-101. (**b**) ECL curves of HOF-101 (purple), HOF-100 (blue), and a bare electrode (pink) in the presence of 20 mM of TPrA. (**c**) The difference of charge density distribution between S_1 and S_0 of HOF-101. (**d**) IREC-driven ECL mechanism of HOF-101 with a neutral charge density difference. Reproduced from [70] with permission from the Royal Society of Chemistry.

3. Framework-Enhanced ECL for Biosensing

The sensitive, specific, and reliable detection of tumor markers is vital for the early diagnosis of cancer, which brings hope to human patients for cancer prevention. Based on unique physical properties, chemical compositions, and functional methods, framework-enhanced ECL may provide an ultrasensitive and comprehensive assay for monitoring these markers. By combining with biological tools, framework-based biosensors can distinguish various biomarkers such as proteins, nucleic acid, cells, and small molecules in a clinical analysis.

3.1. Proteins

Proteins are typical biomacromolecules that are generally analyzed through immunoassays. Once a protein is clinically certified as a disease-related biomarker for a diagnosis, it will receive much attention in ultrasensitive detection. For example, to improve the survival of patients, cancer markers are of great significance in the guidance of an early tumor diagnosis and introducing appropriate targeted therapies [71]. Integrating with highly specific immunoreactions, ECL immunoassays are powerful tools for protein detections. Alpha-fetoprotein (AFP) is a well-known biomarker for the diagnosis of a liver malignant tumor [72]. Zhao et al. synthesized bimetallic NiZn MOF nanosheets to amplify cathodic luminol ECL through the synergistic effect of the bimetallic catalyst in AFP immunodetection [73]. Li et al. designed a signal-off ECL biosensor for AFP detection by utilizing the MnO_2 nanosheet/polydopamine dual-quenching effect towards a $Ru(bpy)_3^{2+}$-functionalized MOF [74].

Cytokeratin 19 fragment 21–1 is recognized as an essential biomarker of non-small-cell lung cancer with a high specificity. Wei's group constructed a "signal-on" ECL immunosensor for this biomarker detection by using a copper-doped terbium MOF as a luminescent tag, which exhibited a strong ECL emission with $K_2S_2O_8$ as a coreactant through electrocatalyzing the reduction of $S_2O_8^{2-}$ [75]. In the same group, a biocompatible tris(4,4′-dicarboxylic acid-2,2′-bipyridyl)ruthenium(II) [$Ru(dcbpy)_3^{2+}$]-functionalized γ-cyclodextrin MOF not only served as a carrier to immobilize the detection antibody via a Pd-N bond but also facilitated the electron transfer rate to amplify the ECL signal [76], providing the ultrasensitive method for an early diagnosis of lung cancer.

Through potential-resolved ECL, a reticular biosensor could detect multiple protein biomarkers in a single run. Zhang et al. developed a MOF-based ECL tag with both anodic and cathodic emission [77]. A useful strategy with the isolated anodic and cathodic coreactants was applied to improve the analytical performance of this potential-resolved ECL sensor, leading to a successful analysis of a carcinoembryonic antigen (CEA) and neuron-specific enolase (NSE) simultaneously.

According to the different roles of frameworks in ECL processes, various signal transductions can be realized in an analysis of the same targets. For example, a hollow hierarchical MOF was employed as a carrier to graft Ru complexes as a signal amplification with the catalytic hairpin assembly strategy [78], showing an excellent selectivity and high sensitivity for thrombin determination. By tuning the reaction time, a series of porphyrin Zr-MOFs (PCN-222) with different specific surface areas, pore sizes, structures, and surface charge states were synthesized (Figure 8a), which served as an ECL emitter, coreactant promoter, and connection in the ECL immunoassay [79]. Furthermore, Xiao's group designed a COF-based ECL biosensor with conductivity- and pre-reduction-enhanced ECL, which overcame the intrinsic poor conductivity of COF [80]. With the aid of the signal amplification of the aptamer/protein-proximity-binding-induced 3D bipedal DNA walker, the constructed ECL sensor realized the supersensitive detection of thrombin (Figure 8b).

In addition, some proteins can be detected by utilizing their bioactive properties. For example, telomerase can extend the length of specific DNA, indicating its possible role as a signal switch. By monitoring bioactivity, telomerase was already analyzed with several well-designed ECL methods [81,82]. In Lei's group, an ECL telomerase biosensor was proposed with a BODIPY-based MOF nanoemitter composed of pyridine-substituted BODIPY, a terephthalic acid ligand, and Zn nodes (Figure 9a) [83]. The BODIPY-based MOF showed the P6/m trigonal crystal system, reducing the over-aggregation of BODIPY for enhanced optical signals (Figure 9b). After an elaborative design, the BODIPY-based MOF ECL sensors reached a good sensitivity under different telomerase concentrations (Figure 9c). The mechanism of this sensor was that the DNA hairpin opened when telomerase appeared, allowing the MOF to approach the electrode surface for ECL signal generation (Figure 9d). Integrating with unique immunoreactions, framework-based ECL biosensors become powerful for protein detection (Table 1).

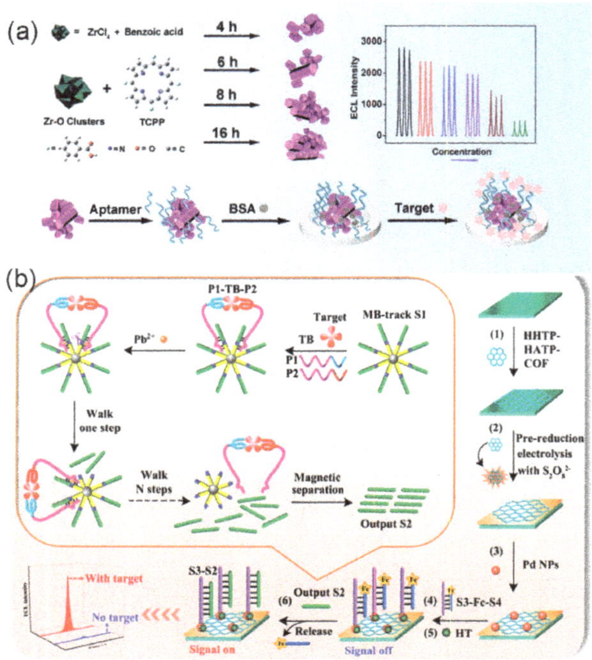

Figure 8. (**a**) Stepwise construction of PCN-222-based thrombin biosensor. Reproduced from [79] with permission from American Chemical Society. (**b**) Preparation procedure and DNA walker amplification principle of COF-based ECL sensor. Reproduced from [80] with permission from American Chemical Society.

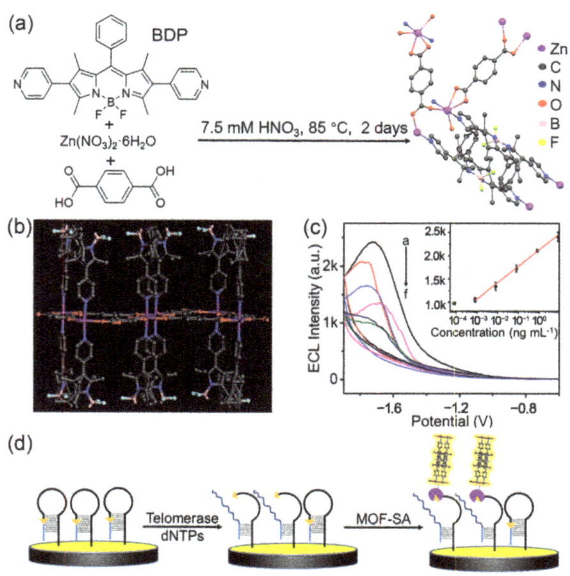

Figure 9. (**a**) Synthesis of the BODIPY-based MOF. (**b**) The spatial structure of MOFs. (**c**) ECL response under different telomerase concentrations. (**d**) Schematic diagram of the stepwise telomerase recognition. Reproduced from [83] with permission from the Royal Society of Chemistry.

Table 1. A summary of framework-enhanced ECL for detection of proteins.

Targets	Frameworks	Linear Range	LOD	Ref.
AFP	NiZn MOF	0.00005 to 100 ng/mL	0.98 fg/mL	[73]
AFP	$Ru(bpy)_3^{2+}$@TMU-3	0.01 pg/mL to 5 ng/mL	10.7 fg/mL	[74]
AFP	Magnetic MOF@CdSnS	1 fg/mL to 100 ng/mL	0.2 fg/mL	[84]
CYFRA21-1	Pd-ZIF-67	0.01 to 100 ng/mL	2.6 pg/mL	[75]
CYFRA21-1	Ru@ γ-CD-MOF	0.1 pg/mL to 50 ng/mL	0.048 pg/mL	[76]
Thrombin	Ru-UiO-66-NH_2	100 fM–100 nM	31.6 fM	[78]
Thrombin	PCN-222	50 fg/mL to 100 pg/mL	2.48 fg/mL	[79]
Thrombin	Conductive COF	100 aM to 1 nM	62.1 aM	[80]
Telomerase	BODIPY MOF	8.0×10^{-4} to 8.0 ng/mL	0.43 pg/mL	[83]
PSA	Ru-MOF	5 pg/mL to 5 µg/mL	1.78 pg/mL	[85]
PSA	MOF/Au/DNAzyme	0.5 to 500 ng/mL	0.358 ng/mL	[86]
CEA	N,B-doped Eu MOF	0.1 pg/mL to 1 µg/mL	0.06 pg/mL	[87]
NSE	J-aggregated MOF	10 pg/mL to 50 ng/mL	7.4 pg/mL	[88]
Peptide	Cu:Tb-MOF	1.0 pg/mL to 50 ng/mL	0.68 pg/mL	[89]
ALP	π-conjugated COF	0.01 to 100 U/L	7.6×10^{-3} U/L	[90]
D-dimer	RuZn MOFs	0.001~200 ng/mL	0.20 pg/mL	[91]

3.2. Nucleic Acids

In the analysis of nucleic acids, signal amplification techniques such as a catalytic hairpin assembly (CHA) [92], rolling circle amplification [93], and hybridization chain reaction [94] have been widely used for a long time. The Crisper/Cas12a technique is also utilized for an enhanced ECL signal in DNA biosensing [95]. Combined with these powerful tools, a series of framework-based ECL genosensors are being developed rapidly for ultrasensitive nucleic acid detection.

As noncoding RNAs, microRNAs (miRNAs) regulate the expression of messenger RNA by binding to complementary sequences. Once alterations in miRNA expression happen, messenger RNA expression is disrupted, which leads to potential oncogenic changes [96]. Therefore, it is crucial to construct reliable and sensitive biosensors for miRNA detection. With the structural development of frameworks, framework-based ECL genosensors for a miRNA analysis were extensively investigated. For instance, Wang et al. synthesized a Zn MOF as a self-enhanced ECL emitter with dual ligands of DPA and *N,N*-diethylethylenediamine for miRNA-21 detection [97]. DPA is a typical luminophore in ECL while DEAEA could be used as both a coreactant and a morphologic regulator, which leads to a strong and stable ECL emission with the efficient intramolecular electron transfer process. Based on CHA and ECL resonance energy transfer, this sensor realized 'signal-off'-mode signal amplification in the presence of miRNA-21. Similarly, Xue et al. developed a microRNA-141 ECL bioassay by using a dual-ligand MOF, which simultaneously contained a luminophore TPE derivative and a coreactant ligand (1,4-diazabicyclo[2.2.2]octane) in the structural unit [98]. Using a DNA triangular prism as a signal switch to detect microRNA-141, this ECL biosensor achieved a low detection limit at the level of 22.9 aM. Furthermore, a dual-wavelength multifunctional ECL biosensor was established for the rapid simultaneous detection of dual targets miRNA-141 and miRNA-155 [99]. As shown in Figure 10a, a Zr MOFs@PEI@AuAg nanocomposite exhibited intense and stable dual-wavelength ECL emissions. Since ECL emissions of the nanocomposite at two wavelengths of 535 nm and 644 nm were both quenched by resonance energy transfer, this sensor achieved a good linear relation for the miRNA analysis at two different wavelengths (Figure 10b). The experiment of ECL stability showed a low signal change, indicating a good accuracy and convincing stability in the simultaneous detection of miRNAs (Figure 10c). In addition, the classic-DNA-walker-based signal amplification strategy is also used for a MOF sheet-based ECL sensor in the detection of oral cancer overexpressed 1 gene [100].

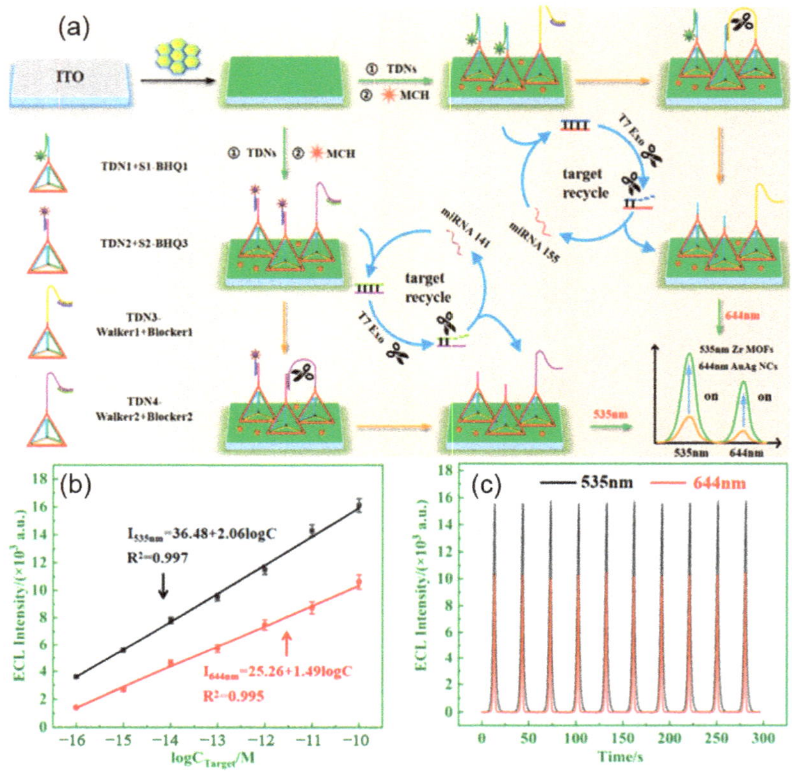

Figure 10. (**a**) Schematic illustration of ECL biosensing platform based on Zr MOFs@PEI@AuAg nanocomposite for simultaneous detection of dual microRNAs. (**b**) Plot of the ECL intensity as a function of the logarithm of target miRNAs' concentration at 535 nm and 644 nm. (**c**) ECL signal stability. Reproduced from [99] with permission from Elsevier.

To overcome an intrinsic low conductivity in MOFs, a conductive NiCo bimetal–organic framework nanorod was successfully applied in miRNA-141 detection, broadening the horizon of conductive MOFs in ECL sensing applications [101]. Furthermore, with the long-range orderly arrangement and effective intramolecular charge transfer, a pyrene-based sp^2 COF was synthesized as an efficient ECL emitter via the polycondensation of tetrakis(4-formylphenyl)pyrene and 2,2′-(1,4-phenylene)-diacetonitrile. Because of topologically linking pyrene luminophores and aggregation-induced emissive luminogens, the luminescent COF showed a strong and stable ECL emission [102], leading to a highly sensitive microRNA-21 biosensor.

As a great threat to health, viruses also receive much attention in ultrasensitive detection. For instance, the Zika virus, a member of the Flaviviridae virus family, is suspected to be associated with severe congenital malformations [103]. Mao's group quantified the Zika virus based on Zr-based metal–organic gel and Fe-MIL-88 MOFs as an electrode matrix and nanotag, respectively [104]. The double quenching effect originated from Fe-MIL-88 MOFs as both an ECL acceptor and metal active centers to consume the coreactant, resulting in a distinct turn-off signal in the presence of the virus. On the other hand, Shan's group designed a 2D MOF with an excellent ECL performance by combining the photosensitizer ZnTCPP and electroactive $[Co_2(\text{-}CO_2)_4]$ secondary building units for a Sars-CoV-2 gene analysis [105]. The ECL sensor achieved a rapid nonamplified detection of the RdRp gene of SARS-CoV-2 with an extremely low limit of detection (30 aM). Furthermore, Wu et al. designed an ECL biosensor using PCN-224/ZnO/polyacrylamide as signal tag for an accurate analysis of the HPV-16 virus [106]. With the aid of multiple target-cycling amplification technologies and HCR reactions, this method achieved a rapid and effective "signal-off" detection of the target with the detection limit of 0.13 fM.

Overall, by integrating appropriate frameworks with well-designed DNA sequences, these above methods show a great performance in a nucleic acid analysis, which expands the application of frameworks in biosensing.

3.3. Small Molecules

Compared to traditional analytical methods like chromatography and enzyme catalysis, framework-based ECL methods are more sensitive and convenient for small molecule detection. For the determination of small molecules, utilizing specific recognition between an aptamer and target is the most common strategy. For instance, the transduction of aptamer configurations alters the distance between a signal promoter and ECL luminophores, resulting in a signal change by introducing target molecules (Figure 11a). A plasmon-enhanced ECL aptasensor displayed highly sensitive detection for lincomycin [107]. Based on a suitable aptasensor, a wide range of molecules can be efficiently detected, such as kanamycin [108], sulfadimethoxine [109], and isocarbophos [110]. Apart from the aptamer, a competition-type ECL immunosensor using Pt NPs@MOFs for the quantitative detection of trenbolone was successfully constructed, demonstrating the simplicity of framework-based ECL systems [111].

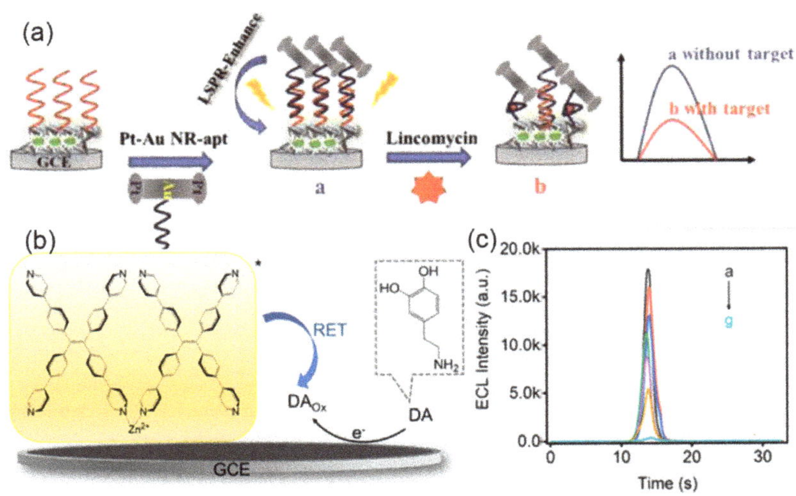

Figure 11. (**a**) Stepwise illustration of Eu MOF-based aptasensor for lincomycin detection. Reproduced from [107] with permission from American Chemical Society. (**b**) Quenching mechanism of dopamine ECL detection. (**c**) Variation of ECL intensity at different concentrations of dopamine: 0, 0.01, 0.1, 1, 10, 100, and 1000 μM from (a) to (g). Reproduced from [55] with permission from Elsevier.

Based on the quenching effect between MOF radicals and oxidized dopamine (Figure 11b), dopamine can be analyzed without the aid of an aptamer [55]. This hindrance to ECL was highly relevant to the dopamine concentration, and then was applied to construct an ECL method for the highly sensitive detection of dopamine in serum samples (Figure 11c). Similarly, uric acid [112], rutin [113], and deoxynivalenol [114] can also be directly measured with framework-based ECL sensors. In addition, a MOF/COF-mixed emitter with dual-color ECL was prepared [115]. Based on a π–π interaction between targets and a MOF/COF, diclazepam can not only be absorbed but also selectively quench ECL, achieving sensitive detection.

Furthermore, metal ions with potential harm to human health are generally analyzed with inductively coupled plasma mass spectrometry, ion chromatography, and atomic absorption spectroscopy, requiring expensive instruments and staff costs. The inhibition effect of metal ions towards ECL makes them detectable through well-designed framework-based ECL sensors [116].

3.4. Cellular Analysis

ECL-based cellular analyses [117], such as circulating tumor cells (CTCs) [118] and the cell matrix [119], have been developed for several decades. With the combination of ECL biotechnology, framework-based ECL sensors for a cell-related analysis gradually emerge. In a typical manner, Liu's group realized single-molecule movement visualization at the cellular membrane through capturing photoluminescence signals of the designed $Ru(bpy)_3^{2+}$-embedded MOF complex (RuMOF) [120]. With the aid of the nanoconfinement effect within frameworks, RuMOFs had a splendid ECL intensity at the single-molecule level, which was conducive to visualize the distribution of RuMOF-labeled-membrane PTK7 proteins at low-expressing cells, demonstrating a great potential of framework-based ECL systems in cellular monitoring.

Bacteria may cause great harm to health while existing in the human circulatory system, indicating the importance of sensitive detection. Utilizing steric hindrance on electron transfer, Vibrio parahaemolyticus [121] and Escherichia coli [122] can be successfully analyzed with ECL sensors based on Ru-MOF and NH_2-MIL-53(Al) signal reporters, respectively. In addition, an exosome as a subcellular structure is also accurately detected using well-designed ECL sensors with a different signal transduction. For example, Cui's group constructed a label-free $HepG_2$-derived exosome ECL sensor based on the selectivity of the CD63 peptide in recognizing CD63 proteins on the exosome surface and strong coordination interactions between the Zr^{4+} of Zn-TCPP/UiO-66-NH_2 and the phosphate head of exosomes (Figure 12). The ECL biosensor exhibited a good sensitivity with a detection range from 1.00×10^4 to 3.16×10^6 particles/μL, which is better than most of the existing label-free methods for detecting exosomes [123], showing the great prospects of framework-based ECL in sensitive bioassays.

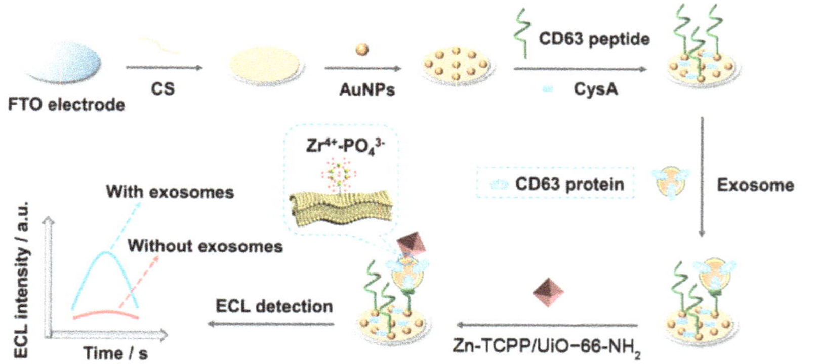

Figure 12. ECL biosensor construction and detecting $HepG_2$-derived exosomes. Reproduced from [123] with permission from American Chemical Society.

4. Conclusions and Perspectives

Frameworks are a kind of widely used material in an ECL analysis owing to the flexible structure, long-range ordered units, and controllable modification with some recognition elements. According to the functions in the ECL process, frameworks have been exploited as a carrier of luminophores, a catalyst of ECL reactions, and crystalline emitters. Based on the innovation of ECL mechanisms, reticular biosensors gain a more efficient ECL for signal amplification, which improves the sensitivity of biosensing. Different from MOFs, metal-free COFs and HOFs have been rapidly developed and constructed lots of biocompatible analytical methods. To date, framework-based ECL biosensors have been successfully utilized in the detection of proteins, nucleic acids, small molecules, and cells by integrating specific functional materials, facilitating the further development of ECL techniques.

Actually, the research on framework-based ECL systems is much more mature than several years ago but challenges still exist. To further expand their application and improve their biosensing performance, some aspects should be deliberated in future works. (1) It is crucial to explore near-infrared ECL-active frameworks for developing in vivo biosensing and bioimaging because of a good penetrability and low scattering in NIR [124]. (2) The 3D-sp^2-carbon-conjugated COFs may accelerate charge transfer through the largely conjugated electron structure, resulting in a high ECL efficiency. (3) In order to develop a single-molecule ECL imaging technique [125], frameworks at the nanoscale with an intense ECL emission are promising crystalline nanoemitters. (4) The framework characteristics of self-luminescence such as the orthogonal luminescence lifetime [126] can be endowed through the introduction of lanthanide elements, which may be used for designing ECL sensors. (5) Multivariate MOFs containing multiple metals have a greater selectivity in catalysis for the acceleration of charge transfer, which improves the efficiency of ECL reactions, leading to a strong ECL emission [127]. (6) New types of methods for reticular nanoemitter construction should be discovered, which may simplify synthesis. (7) The conductivity of framework-based emitters should be improved by integrating the redox-active ligands in the frameworks [128]. In a word, the key of ECL techniques relies heavily on the improvement of emitters, suggesting that ECL-active frameworks with a good stability, easy accessibility, and high ECL efficiency are urgently required in future research.

Author Contributions: Conceptualization, H.F. and J.L.; writing—original draft preparation, writing—review and editing, H.F., Z.X., H.H., R.L., H.J. and J.L.; visualization, H.F. and J.L.; funding acquisition, H.J. and J.L. All authors have read and agreed to the published version of the manuscript.

Funding: This research was supported by the National Natural Science Foundation of China (22274071, 22234005, 21890741), the Natural Science Foundation of Jiangsu Province (BZ2021010), the Fundamental Research Funds for the Central Universities (020514380297), and the State Key Laboratory of Analytical Chemistry for Life Science (5431ZZXM2204).

Institutional Review Board Statement: Not applicable.

Informed Consent Statement: Not applicable.

Data Availability Statement: Not applicable.

Conflicts of Interest: The authors declare no conflict of interest.

References

1. Miao, W. Electrogenerated chemiluminescence and its biorelated applications. *Chem. Rev.* **2008**, *108*, 2506–2553. [CrossRef] [PubMed]
2. Richter, M.M. Electrochemiluminescence (ECL). *Chem. Rev.* **2004**, *104*, 3003–3036. [CrossRef] [PubMed]
3. Wu, P.; Hou, X.; Xu, J.; Chen, H. Electrochemically generated versus photoexcited luminescence from semiconductor nanomaterials: Bridging the valley between two worlds. *Chem. Rev.* **2014**, *114*, 11027–11059. [CrossRef] [PubMed]
4. Chen, S.; Lei, Y.; Xu, J.; Yang, Y.; Dong, Y.; Li, Y.; Yi, H.; Liao, Y.; Chen, L.; Xiao, Y. Simple, rapid, and visual electrochemiluminescence sensor for on-site catechol analysis. *RSC Adv.* **2022**, *12*, 17330–17336. [CrossRef] [PubMed]
5. Wang, N.; Gao, H.; Li, Y.; Li, G.; Chen, W.; Jin, Z.; Lei, J.; Wei, Q.; Ju, H. Dual intramolecular electron transfer for in situ coreactant embedded electrochemiluminescence microimaging of membrane protein. *Angew. Chem. Int. Ed.* **2021**, *60*, 197–201. [CrossRef]
6. Dong, J.; Lu, Y.; Xu, Y.; Chen, F.; Yang, J.; Chen, Y.; Feng, J. Direct imaging of single-molecule electrochemical reactions in solution. *Nature* **2021**, *596*, 244–249. [CrossRef]
7. You, F.; Wei, Z.; Yuan, R.; Qian, J.; Long, L.; Wang, K. Sensitive and stable detection of deoxynivalenol based on electrochemiluminescence aptasensor enhanced by 0D/2D homojunction effect in food analysis. *Food Chem.* **2023**, *403*, 134397. [CrossRef]
8. Wang, C.; Pei, Y.; Liu, P.; Li, Y.; Wang, Z. Donor-acceptor structure-dependent electrochemiluminescence sensor for accurate uranium detection in drinking water. *ACS Sustain. Chem. Eng.* **2022**, *10*, 14665–14670. [CrossRef]
9. Chen, M.-M.; Xu, C.-H.; Zhao, W.; Chen, H.-Y.; Xu, J.-J. Single cell imaging of electrochemiluminescence-driven photodynamic therapy. *Angew. Chem. Int. Ed.* **2022**, *61*, e202117401.
10. Shen, Y.; Tian, Q.; Sun, Y.; Xu, J.-J.; Ye, D.; Chen, H.-Y. ATP-activatable photosensitizer enables dual fluorescence imaging and targeted photodynamic therapy of tumor. *Anal. Chem.* **2017**, *89*, 13610–13617. [CrossRef]

11. Yang, Y.; Hou, W.; Liu, S.; Sun, K.; Li, M.; Wu, C. Biodegradable polymer nanoparticles for photodynamic therapy by bioluminescence resonance energy transfer. *Biomacromolecules* **2018**, *19*, 201–208. [CrossRef] [PubMed]
12. Yin, B.; Ho, W.K.H.; Xia, X.; Chan, C.K.W.; Zhang, Q.; Ng, Y.M.; Lam, C.Y.K.; Cheung, J.C.W.; Wang, J.; Yang, M.; et al. A multilayered mesoporous gold nanoarchitecture for ultraeffective near-infrared light-controlled chemo photothermal therapy for cancer guided by SERS imaging. *Small* **2023**, *19*, 2206762. [CrossRef] [PubMed]
13. Wei, X.; Zhu, M.; Yan, H.; Lu, C.; Xu, J. Recent advances in aggregation-induced electrochemiluminescence. *Chem. Eur. J.* **2019**, *25*, 12671–12683. [CrossRef]
14. Wang, T.; Wang, D.; Padelford, J.W.; Jiang, J.; Wang, G. Near-infrared electrogenerated chemiluminescence from aqueous soluble lipoic acid Au nanoclusters. *J. Am. Chem. Soc.* **2016**, *138*, 6380–6383. [CrossRef]
15. Stewart, A.J.; Brown, K.; Dennany, L. Cathodic quantum dot facilitated electrochemiluminescent detection in blood. *Anal. Chem.* **2018**, *90*, 12944–12950. [CrossRef] [PubMed]
16. Jin, Z.; Zhu, X.; Wang, N.; Li, Y.; Ju, H.; Lei, J. Electroactive metal-organic frameworks as emitters for self-enhanced electrochemiluminescence in aqueous medium. *Angew. Chem. Int. Ed.* **2020**, *59*, 10446–10450. [CrossRef]
17. Furukawa, H.; Cordova, K.E.; O'Keeffe, M.; Yaghi, O.M. The chemistry and applications of metal-organic frameworks. *Science* **2013**, *341*, 1230444. [CrossRef]
18. Geng, K.; He, T.; Liu, R.; Dalapati, S.; Tan, K.T.; Li, Z.; Tao, S.; Gong, Y.; Jiang, Q.; Jiang, D. Covalent organic frameworks: Design, synthesis, and functions. *Chem. Rev.* **2020**, *120*, 8814–8933.
19. Cai, G.; Yan, P.; Zhang, L.; Zhou, H.; Jiang, H. Metal-organic framework-based hierarchically porous materials: Synthesis and applications. *Chem. Rev.* **2021**, *121*, 12278–12326. [CrossRef] [PubMed]
20. Lyu, S.; Guo, C.; Wang, J.; Li, Z.; Yang, B.; Lei, C.; Wang, L.; Xiao, J.; Zhang, T.; Hou, Y. Exceptional catalytic activity of oxygen evolution reaction via two-dimensional graphene multilayer confined metal-organic frameworks. *Nat. Commun.* **2022**, *13*, 6171. [CrossRef] [PubMed]
21. Xu, Y.; Yin, X.B.; He, X.W.; Zhang, Y.K. Electrochemistry and electrochemiluminescence from a redox-active metal-organic framework. *Biosens. Bioelectron.* **2015**, *68*, 197–203. [CrossRef]
22. Liao, X.; Fu, H.; Yan, T.; Lei, J. Electroactive metal-organic framework composites: Design and biosensing application. *Biosens. Bioelectron.* **2019**, *146*, 111743. [CrossRef] [PubMed]
23. Han, Z.; Wang, J.; Du, P.; Chen, J.; Huo, S.; Guo, H.; Lu, X. Highly facile strategy for detecting D_2O in H_2O by porphyrin-based luminescent probes. *Anal. Chem.* **2022**, *94*, 8426–8432. [CrossRef] [PubMed]
24. Zhang, Y.-J.; Yang, Y.; Wang, J.-M.; Liang, W.-B.; Yuan, R.; Xiao, D.-R. Electrochemiluminescence enhanced by isolating ACQphores in pyrene-based porous organic polymer: A novel ECL emitter for the construction of biosensing platform. *Anal. Chim. Acta* **2022**, *1206*, 339648. [CrossRef] [PubMed]
25. Zhang, B.; Kong, Y.; Liu, H.; Chen, B.; Zhao, B.; Luo, Y.; Chen, L.; Zhang, Y.; Han, D.; Zhao, Z.; et al. Aggregation-induced delayed fluorescence luminogens: The innovation of purely organic emitters for aqueous electrochemiluminescence. *Chem. Sci.* **2021**, *12*, 13283–13291. [CrossRef]
26. Bahari, D.; Babamiri, B.; Moradi, K.; Salimi, A.; Hallaj, R. Graphdiyne nanosheet as a novel sensing platform for self-enhanced electrochemiluminescence of MOF enriched ruthenium (II) in the presence of dual co-reactants for detection of tumor marker. *Biosens. Bioelectron.* **2022**, *195*, 113657. [CrossRef]
27. Dai, W.; Wang, X.; Chen, G.; Wang, X.; Hu, C.; Zhen, S.; Huang, C.; Li, Y. Facile synthesis of 2D Europium-metal organic frameworks nanosheets for highly efficient electrochemiluminescence in DNA detection. *Chem. Eng. J.* **2023**, *465*, 143037. [CrossRef]
28. Li, Y.-J.; Cui, W.-R.; Jiang, Q.-Q.; Liang, R.-P.; Li, X.-J.; Wu, Q.; Luo, Q.-X.; Liu, J.; Qiu, J.-D. Arousing electrochemiluminescence out of non-electroluminescent monomers within covalent organic frameworks. *ACS Appl. Mater. Interfaces* **2021**, *13*, 47921–47931. [CrossRef]
29. Cui, W.-R.; Li, Y.-J.; Jiang, Q.-Q.; Wu, Q.; Liang, R.-P.; Luo, Q.-X.; Zhang, L.; Liu, J.; Qiu, J.-D. Tunable covalent organic framework electrochemiluminescence from nonelectroluminescent monomers. *Cell Rep. Phys. Sci.* **2022**, *3*, 100630. [CrossRef]
30. Qin, X.; Zhao, X.; Wang, M.; Dong, Y.; Liu, J.; Zhu, Z.; Li, M.; Yang, D.; Shao, Y. Fabrication of tris(bipyridine)ruthenium(II)-functionalized metal-organic framework thin films by electrochemically assisted self-assembly technique for electrochemiluminescent immunoassay. *Anal. Chem.* **2018**, *90*, 11622–11628. [CrossRef] [PubMed]
31. Tang, T.; Yan, F.; Wang, L.; Zhao, C.; Nie, F.; Yang, G. A sandwich electrochemiluminescent assay for determination of concanavalin A with triple signal amplification based on MoS_2NF@MWCNTs modified electrode and Zn-MOF encapsulated luminol. *Microchim. Acta* **2020**, *187*, 523. [CrossRef] [PubMed]
32. Mo, G.; Qin, D.; Jiang, X.; Zheng, X.; Mo, W.; Deng, B. A sensitive electrochemiluminescence biosensor based on metal-organic framework and imprinted polymer for squamous cell carcinoma antigen detection. *Sens. Actuators B* **2020**, *310*, 127852. [CrossRef]
33. Qin, X.; Dong, Y.; Wang, M.; Zhu, Z.; Li, M.; Yang, Y.; Shao, Y. In situ growing triethanolamine-functionalized metal-organic frameworks on two-dimensional carbon nanosheets for electrochemiluminescent immunoassay. *ACS Sens.* **2019**, *4*, 2351–2357. [CrossRef] [PubMed]
34. Micheroni, D.; Lan, G.; Lin, W. Efficient electrocatalytic proton reduction with carbon nanotube-supported metal-organic frameworks. *J. Am. Chem. Soc.* **2018**, *140*, 15591–15595. [CrossRef]
35. Wang, S.; Zhao, Y.; Wang, M.; Li, H.; Saqib, M.; Ge, C.; Zhang, X.; Jin, Y. Enhancing luminol electrochemiluminescence by combined use of cobalt-based metal organic frameworks and silver nanoparticles and its application in ultrasensitive detection of cardiac troponin I. *Anal. Chem.* **2019**, *91*, 3048–3054. [CrossRef]

36. Feng, D.; Tan, X.; Wu, Y.; Ai, C.; Luo, Y.; Chen, Q.; Han, H. Electrochemiluminecence nanogears aptasensor based on MIL-53(Fe)@CdS for multiplexed detection of kanamycin and neomycin. *Biosens. Bioelectron* **2019**, *129*, 100–106. [CrossRef]
37. Lu, Y.; Huang, X.; Wang, S.; Li, B.; Liu, B. Nanoconfinement-enhanced electrochemiluminescence for in situ imaging of single biomolecules. *ACS Nano* **2023**, *17*, 3809–3817. [CrossRef]
38. Diercks, C.S.; Liu, Y.; Cordova, K.E.; Yaghi, O.M. The role of reticular chemistry in the design of CO_2 reduction catalysts. *Nat. Mater.* **2018**, *17*, 301–307. [CrossRef]
39. Xue, X.; Gao, H.; Liu, J.; Yang, M.; Feng, J.; Liu, Z.; Lin, J.; Kasemchainan, J.; Wang, L.; Jia, Q., et al. Electrostatic potential-derived charge: A universal OER performance descriptor for MOFs. *Chem. Sci.* **2022**, *13*, 13160–13171. [CrossRef]
40. Zhao, G.; Cai, C.; Cosnier, S.; Zeng, H.; Zhang, X.; Shan, D. Zirconium-metalloporphyrin frameworks as a three-in-one platform possessing oxygen nanocage, electron media, and bonding site for electrochemiluminescence protein kinase activity assay. *Nanoscale* **2016**, *8*, 11649–11657.
41. Yang, X.; Yu, Y.; Peng, L.; Lei, Y.; Chai, Y.; Yuan, R.; Zhuo, Y. Strong electrochemiluminescence from MOF accelerator enriched quantum dots for enhanced sensing of trace cTnI. *Anal. Chem.* **2018**, *90*, 3995–4002. [CrossRef] [PubMed]
42. Zhang, W.; Han, D.; Wu, Z.; Yang, K.; Sun, S.; Wen, J. Metal-organic layers-catalyzed amplification of electrochemiluminescence signal and its application for immunosensor construction. *Sens. Actuators B-Chem.* **2023**, *376*, 133004. [CrossRef]
43. Song, X.; Yu, S.; Zhao, L.; Guo, Y.; Ren, X.; Ma, H.; Wang, S.; Luo, C.; Li, Y.; Wei, Q. Efficient ABEI-dissolved O_2-Ce(III, IV)-MOF ternary electrochemiluminescent system combined with self-assembled microfluidic chips for bioanalysis. *Anal. Chem.* **2022**, *94*, 9363–9371. [CrossRef] [PubMed]
44. Zhou, J.; Li, Y.; Wang, W.; Tan, X.; Lu, Z.; Han, H. Metal-organic frameworks-based sensitive electrochemiluminescence biosensing. *Biosens. Bioelectron.* **2020**, *164*, 112332. [CrossRef] [PubMed]
45. Kent, C.A.; Mehl, B.P.; Ma, L.; Papanikolas, J.M.; Meyer, T.J.; Lin, W. Energy transfer dynamics in metal-organic frameworks. *J. Am. Chem. Soc.* **2010**, *132*, 12767–12769. [CrossRef] [PubMed]
46. Hou, C.; Li, T.; Cao, S.; Chen, Y.; Fu, W. Incorporation of a $[Ru(dcbpy)(bpy)_2]^{2+}$ photosensitizer and a $Pt(dcbpy)Cl_2$ catalyst into metal-organic frameworks for photocatalytic hydrogen evolution from aqueous solution. *J. Mater. Chem. A* **2015**, *3*, 10386–10394. [CrossRef]
47. Yan, M.; Ye, J.; Zhu, Q.; Zhu, L.; Huang, J.; Yang, X. Ultrasensitive immunosensor for cardiac troponin I detection based on the electrochemiluminescence of 2D Ru-MOF nanosheets. *Anal. Chem.* **2019**, *91*, 10156–10163. [CrossRef]
48. Hu, G.; Xiong, C.; Liang, W.; Zeng, X.; Xu, H.; Yang, Y.; Yao, L.Y.; Yuan, R.; Xiao, D. Highly stable mesoporous luminescence-functionalized MOF with excellent electrochemiluminescence property for ultrasensitive immunosensor construction. *ACS Appl. Mater. Interfaces* **2018**, *10*, 15913–15919. [CrossRef]
49. Yang, Y.; Hu, G.; Liang, W.; Yao, L.; Huang, W.; Yuan, R.; Xiao, D. A highly sensitive self-enhanced aptasensor based on a stable ultrathin 2D metal-organic layer with outstanding electrochemiluminescence property. *Nanoscale* **2019**, *11*, 10056–10063. [CrossRef]
50. Zhou, Y.; He, J.; Zhang, C.; Li, J.; Fu, X.; Mao, W.; Li, W.; Yu, C. Novel Ce(III)-metal organic framework with a luminescent property to fabricate an electrochemiluminescence immunosensor. *ACS Appl. Mater. Interfaces* **2020**, *12*, 338–346. [CrossRef]
51. Wang, J.; Yao, L.; Huang, W.; Yang, Y.; Liang, W.; Yuan, R.; Xiao, D. Overcoming aggregation-induced quenching by metal-organic framework for electrochemiluminescence (ECL) enhancement: ZnPTC as a new ECL emitter for ultrasensitive MicroRNAs detection. *ACS Appl. Mater. Interfaces* **2021**, *13*, 44079–44085. [CrossRef]
52. Yao, L.; Yang, F.; Hu, G.; Yang, Y.; Huang, W.; Liang, W.; Yuan, R.; Xiao, D. Restriction of intramolecular motions (RIM) by metal-organic frameworks for electrochemiluminescence enhancement: 2D Zr12-adb nanoplate as a novel ECL tag for the construction of biosensing platform. *Biosens. Bioelectron.* **2020**, *155*, 112099. [CrossRef]
53. Han, Z.; Zhang, Y.; Wu, Y.; Li, Z.; Bai, L.; Huo, S.; Lu, X. Substituent-induced aggregated state electrochemiluminescence of tetraphenylethene derivatives. *Anal. Chem.* **2019**, *91*, 8676–8682. [CrossRef]
54. Liu, J.; Zhang, J.; Tan, Z.; Zhuo, Y.; Chai, Y.; Yuan, R. Near-infrared aggregation-induced enhanced electrochemiluminescence from tetraphenylethylene nanocrystals: A new generation of ECL emitters. *Chem. Sci.* **2019**, *10*, 4497–4501. [CrossRef]
55. Fu, H.; Xu, Z.; Liu, T.; Lei, J. In situ coordination interactions between metal-organic framework nanoemitters and coreactants for enhanced electrochemiluminescence in biosensing. *Biosens. Bioelectron.* **2023**, *222*, 114920. [CrossRef] [PubMed]
56. Li, J.; Jia, H.; Ren, X.; Li, Y.; Liu, L.; Feng, R.; Ma, H.; Wei, Q. Dumbbell plate-shaped AIEgen-based luminescent MOF with high quantum yield as self-enhanced ECL tags: Mechanism insights and biosensing application. *Small* **2022**, *18*, 2106567. [CrossRef]
57. Yang, Y.; Hu, G.; Liang, W.; Yao, L.; Huang, W.; Zhang, Y.; Zhang, J.; Wang, J.; Yuan, R.; Xiao, D. An AIEgen-based 2D ultrathin metal-organic layer as an electrochemiluminescence platform for ultrasensitive biosensing of carcinoembryonic antigen. *Nanoscale* **2020**, *12*, 5932–5941. [CrossRef]
58. Kwok, R.T.K.; Leung, C.W.T.; Lam, J.W.Y.; Tang, B.Z. Biosensing by luminogens with aggregation-induced emission characteristics. *Chem. Soc. Rev.* **2015**, *44*, 4228–4238. [CrossRef]
59. Carrara, S.; Aliprandi, A.; Hogan, C.F.; De Cola, L. Aggregation-induced electrochemiluminescence of platinum(II) complexes. *J. Am. Chem. Soc.* **2017**, *139*, 14605–14610. [CrossRef]
60. Song, L.; Gao, W.; Han, Q.; Huang, Y.; Cui, L.; Zhang, C.-Y. Construction of an aggregation-induced electrochemiluminescent sensor based on an aminal-linked covalent organic framework for sensitive detection of glutathione in human serum. *Chem. Commun.* **2022**, *58*, 10524–10527. [CrossRef]

61. Wang, Z.; Xu, M.; Zhang, N.; Pan, J.-B.; Wu, X.; Liu, P.; Xu, J.-J.; Hua, D. An ultra-highly sensitive and selective self-enhanced AIECL sensor for public security early warning in a nuclear emergency via a co-reactive group poisoning mechanism. *J. Mater. Chem. A* **2021**, *9*, 12584–12592. [CrossRef]
62. Wiwasuku, T.; Chuaephon, A.; Habarakada, U.; Boonmak, J.; Puangmali, T.; Kielar, F.; Harding, D.J.; Youngme, S. A water-stable lanthanide-based MOF as a highly sensitive sensor for the selective detection of paraquat in agricultural products. *ACS Sustain. Chem. Eng.* **2022**, *10*, 2761–2771. [CrossRef]
63. Gao, H.; Wei, X.; Li, M.; Wang, L.; Wei, T.; Dai, Z. Co-quenching effect between lanthanum metal-organic frameworks luminophore and crystal violet for enhanced electrochemiluminescence gene detection. *Small* **2021**, *17*, 2103424. [CrossRef] [PubMed]
64. Wang, Y.; Zhao, G.; Chi, H.; Yang, S.; Niu, Q.; Wu, D.; Cao, W.; Li, T.; Ma, H.; Wei, Q. Self-luminescent lanthanide metal-organic frameworks as signal probes in electrochemiluminescence immunoassay. *J. Am. Chem. Soc.* **2021**, *143*, 504–512. [CrossRef] [PubMed]
65. Carrara, S.; Arcudi, F.; Prato, M.; De Cola, L. Amine-rich nitrogen-doped carbon nanodots as a platform for self-enhancing electrochemiluminescence. *Angew. Chem. Int. Ed.* **2017**, *56*, 4757–4761. [CrossRef]
66. Zhu, D.; Zhang, Y.; Bao, S.; Wang, N.; Yu, S.; Luo, R.; Ma, J.; Ju, H.; Lei, J. Dual intrareticular oxidation of mixed-ligand metal-organic frameworks for stepwise electrochemiluminescence. *J. Am. Chem. Soc.* **2021**, *143*, 3049–3053. [CrossRef]
67. Li, Y.; Cui, W.; Jiang, Q.; Wu, Q.; Liang, R.; Luo, Q.; Qiu, J. A general design approach toward covalent organic frameworks for highly efficient electrochemiluminescence. *Nat. Commun.* **2021**, *12*, 4735. [CrossRef]
68. Luo, R.; Lv, H.; Liao, Q.; Wang, N.; Yang, J.; Li, Y.; Xi, K.; Wu, X.; Ju, H.; Lei, J. Intrareticular charge transfer regulated electrochemiluminescence of donor-acceptor covalent organic frameworks. *Nat. Commun.* **2021**, *12*, 6808. [CrossRef]
69. Zhang, N.; Wang, X.; Xiong, Z.; Huang, L.; Jin, Y.; Wang, A.; Yuan, P.; He, Y.; Feng, J. Hydrogen bond organic frameworks as a novel electrochemiluminescence luminophore: Simple synthesis and ultrasensitive biosensing. *Anal. Chem.* **2021**, *93*, 17110–17118. [CrossRef] [PubMed]
70. Hou, H.; Wang, Y.; Wang, Y.; Luo, R.; Zhu, D.; Zhou, J.; Wu, X.; Ju, H.; Lei, J. Intrareticular electron coupling pathway driven electrochemiluminescence in hydrogen-bonded organic frameworks. *J. Mater. Chem. C* **2022**, *10*, 14488–14495. [CrossRef]
71. Ye, F.; Zhao, Y.; El-Sayed, R.; Muhammed, M.; Hassan, M. Advances in nanotechnology for cancer biomarkers. *Nano Today* **2018**, *18*, 103–123. [CrossRef]
72. Mohammadinejad, A.; Oskuee, R.K.; Eivazzadeh-Keihan, R.; Rezayi, M.; Baradaran, B.; Maleki, A.; Hashemzaei, M.; Mokhtarzadeh, A.; de la Guardia, M. Development of biosensors for detection of alpha-fetoprotein: As a major biomarker for hepatocellular carcinoma. *TrAC Trends Anal. Chem.* **2020**, *130*, 115961. [CrossRef]
73. Zhao, C.; Ma, C.; Zhang, F.; Li, W.; Hong, C.; Qi, Y. Two-dimensional metal-organic framework nanosheets: An efficient two-electron oxygen reduction reaction electrocatalyst for boosting cathodic luminol electrochemiluminescence. *Chem. Eng. J.* **2023**, *466*, 143156. [CrossRef]
74. Li, J.; Lai, W.; Ma, C.; Zhao, C.; Li, P.; Jiang, M.; Wang, M.; Chen, S.; Hong, C. MnO_2 nanosheet/polydopamine double-quenching $Ru(bpy)_3^{2+}$@TMU-3 electrochemiluminescence for ultrasensitive immunosensing of alpha-fetoprotein. *ACS Appl. Nano Mater.* **2022**, *5*, 14697–14705. [CrossRef]
75. Zhou, L.; Yang, L.; Wang, C.; Jia, H.; Xue, J.; Wei, Q.; Ju, H. Copper doped terbium metal organic framework as emitter for sensitive electrochemiluminescence detection of CYFRA 21-1. *Talanta* **2022**, *238*, 123047. [CrossRef] [PubMed]
76. Li, X.; Ren, X.; Yang, L.; Wang, W.; Fan, D.; Kuang, X.; Sun, X.; Wei, Q.; Ju, H. $Ru(dcbpy)_3^{2+}$-functionalized γ-cyclodextrin metal-organic frameworks as efficient electrochemiluminescence tags for the detection of CYFRA21-1 in human serum. *Sens. Actuators B-Chem.* **2023**, *378*, 133152. [CrossRef]
77. Zhang, P.; Shen, Q.; Wang, J.; Yu, M.; Kang, Q.; Zhang, W.; Zou, G. Intrareticular charge transfer triggered self-electrochemiluminescence of zirconium-based metal-organic framework nanoparticles for potential-resolved multiplex immunoassays with isolated coreactants. *Anal. Chem.* **2023**, *95*, 10096–10104. [CrossRef]
78. Huang, W.; Hu, G.; Liang, W.; Wang, J.; Lu, M.; Yuan, R.; Xiao, D. Ruthenium(II) complex-grafted hollow hierarchical metal-organic frameworks with superior electrochemiluminescence performance for sensitive assay of thrombin. *Anal. Chem.* **2021**, *93*, 6239–6245. [CrossRef]
79. Li, P.; Luo, L.; Chen, D.; Sun, Y.; Zhang, Y.; Liu, M.; Yao, S. Regulation of the structure of zirconium-based porphyrinic metal-organic framework as highly electrochemiluminescence sensing platform for thrombin. *Anal. Chem.* **2022**, *94*, 5707–5714. [CrossRef]
80. Zhang, J.; Yao, L.; Yang, Y.; Liang, W.; Yuan, R.; Xiao, D. Conductive covalent organic frameworks with conductivity- and pre-reduction-enhanced electrochemiluminescence for ultrasensitive biosensor construction. *Anal. Chem.* **2022**, *94*, 3685–3692. [CrossRef]
81. Xiong, C.; Liang, W.; Zheng, Y.; Zhuo, Y.; Chai, Y.; Yuan, R. Ultrasensitive assay for telomerase activity via self-enhanced electrochemiluminescent ruthenium complex doped metal-organic frameworks with high emission efficiency. *Anal. Chem.* **2017**, *89*, 3222–3227. [CrossRef]
82. Zhang, H.; Li, B.; Sun, Z.; Zhou, H.; Zhang, S. Integration of intracellular telomerase monitoring by electrochemiluminescence technology and targeted cancer therapy by reactive oxygen species. *Chem. Sci.* **2017**, *8*, 8025–8029. [CrossRef] [PubMed]
83. Xu, Z.; Wu, F.; Zhu, D.; Fu, H.; Shen, Z.; Lei, J. BODIPY-based metal-organic frameworks as efficient electrochemiluminescence emitters for telomerase detection. *Chem. Commun.* **2022**, *58*, 11515–11518. [CrossRef] [PubMed]
84. Ding, Y.; Zhang, X.; Peng, J.; Zheng, D.; Zhang, X.; Song, Y.; Cheng, Y.; Gao, W. Ultra-sensitive electrochemiluminescence platform based on magnetic metal-organic framework for the highly efficient enrichment. *Sens. Actuators B-Chem.* **2020**, *324*, 128700. [CrossRef]

85. Zheng, L.; Guo, Q.; Yang, C.; Wang, J.; Xu, X.; Nie, G. Electrochemiluminescence and photoelectrochemistry dual-signal immunosensor based on $Ru(bpy)_3^{2+}$-functionalized MOF for prostate-specific antigen sensitive detection. *Sens. Actuators B* **2023**, *379*, 133269. [CrossRef]
86. Shao, K.; Wang, B.; Nie, A.; Ye, S.; Ma, J.; Li, Z.; Lv, Z.; Han, H. Target-triggered signal-on ratiometric electrochemiluminescence sensing of PSA based on MOF/Au/G-quadruplex. *Biosens. Bioelectron.* **2018**, *118*, 160–166. [CrossRef]
87. Li, J.; Yang, H.; Cai, R.; Tang, W. Ultrahighly sensitive sandwich-type electrochemical immunosensor for selective detection of tumor biomarkers. *ACS Appl. Mater. Interfaces* **2022**, *14*, 44222–44227. [CrossRef]
88. Fang, J.; Dai, L.; Feng, R.; Wu, D.; Ren, X.; Cao, W.; Ma, H.; Wei, Q. High-performance electrochemiluminescence of a coordination driven J-aggregate K-PTC MOF regulated by metal-phenolic nanoparticles for biomarker analysis. *Anal. Chem.* **2023**, *95*, 1287–1293. [CrossRef]
89. Wang, C.; Li, Z.; Ju, H. Copper-doped terbium luminescent metal organic framework as an emitter and a co-reaction promoter for amplified electrochemiluminescence immunoassay. *Anal. Chem.* **2021**, *93*, 14878–14884. [CrossRef]
90. Cui, L.; Zhu, C.; Hu, J.; Meng, X.; Jiang, M.; Gao, W.; Wang, X.; Zhang, C. Construction of a dual-mode biosensor for electrochemiluminescent and electrochemical sensing of alkaline phosphatase. *Sens. Actuators B* **2023**, *374*, 132779. [CrossRef]
91. Zhao, J.; Du, Y.; Zhang, N.; Li, C.; Ma, H.; Wu, D.; Cao, W.; Wang, Y.; Wei, Q. Dual-quenching mechanisms in electrochemiluminescence immunoassay based on zinc-based MOFs of ruthenium hybrid for D-dimer detection. *Anal. Chim. Acta* **2023**, *1253*, 341076. [CrossRef] [PubMed]
92. Liu, Q.; Huang, Y.; Li, Z.; Li, L.; Zhao, Y.; Li, M. An enzymatically gated catalytic hairpin assembly delivered by lipid nanoparticles for the tumor-specific activation of signal amplification in miRNA imaging. *Angew. Chem. Int. Ed.* **2022**, *61*, e202214230.
93. Yao, Q.; Wang, Y.; Wang, J.; Chen, S.; Liu, H.; Jiang, Z.; Zhang, X.; Liu, S.; Yuan, Q.; Zhou, X. An ultrasensitive diagnostic biochip based on biomimetic periodic nanostructure-assisted rolling circle amplification. *ACS Nano* **2018**, *12*, 6777–6783. [CrossRef]
94. Wei, J.; Wang, H.; Wu, Q.; Gong, X.; Ma, K.; Liu, X.; Wang, F. A Smart, Autocatalytic, DNAzyme biocircuit for in vivo, amplified, microRNA imaging. *Angew. Chem. Int. Ed.* **2020**, *59*, 5965–5971. [CrossRef]
95. Liu, Y.; Wang, F.; Ge, S.; Zhang, L.; Zhang, Z.; Liu, Y.; Zhang, Y.; Ge, S.; Yu, J. Programmable T-junction structure-assisted CRISPR/Cas12a electrochemiluminescence biosensor for detection of Sa-16S rDNA. *ACS Appl. Mater. Interfaces* **2023**, *15*, 617–625. [CrossRef]
96. Hill, M.; Tran, N. MicroRNAs regulating microRNAs in cancer. *Trends Cancer* **2018**, *4*, 465–468. [CrossRef]
97. Wang, X.; Wang, X.; Hu, C.; Guo, W.; Wu, X.; Chen, G.; Dai, W.; Zhen, S.; Huang, C.; Li, Y. Controlled synthesis of zinc-metal organic framework microflower with high efficiency electrochemiluminescence for miR-21 detection. *Biosens. Bioelectron.* **2022**, *213*, 114443. [CrossRef]
98. Xue, Y.; Liao, N.; Li, Y.; Liang, W.; Yang, X.; Zhong, X.; Zhuo, Y. Ordered heterogeneity in dual-ligand MOF to enable high electrochemiluminescence efficiency for bioassay with DNA triangular prism as signal switch. *Biosens. Bioelectron.* **2022**, *217*, 114713. [CrossRef]
99. Yin, T.; Wu, D.; Du, H.; Jie, G. Dual-wavelength electrochemiluminescence biosensor based on a multifunctional Zr MOFs@PEI@AuAg nanocomposite with intramolecular self-enhancing effect for simultaneous detection of dual microRNAs. *Biosens. Bioelectron.* **2022**, *217*, 114699. [CrossRef]
100. Wei, Y.; Chen, J.; Liu, X.; Miao, C.; Jin, B. ORAOV 1 detection made with metal organic frameworks based on $Ti_3C_2T_x$ MXene. *ACS Appl. Mater. Interfaces* **2022**, *14*, 23726–23733. [CrossRef]
101. Yang, Y.; Zhang, J.; Liang, W.; Zhang, J.; Xu, X.; Zhang, X.; Yuan, R.; Xiao, D. Conductive NiCo bimetal-organic framework nanorods with conductivity-enhanced electrochemiluminescence for constructing biosensing platform. *Sens. Actuators B* **2020**, *362*, 131802. [CrossRef]
102. Zhang, J.; Yang, Y.; Liang, W.; Yao, L.; Yuan, R.; Xiao, D. Highly stable covalent organic framework nanosheets as a new generation of electrochemiluminescence emitters for ultrasensitive microRNA detection. *Anal. Chem.* **2021**, *93*, 3258–3265. [CrossRef]
103. Musso, D.; Gubler, D.J. Zika Virus. *Clin. Microbiol. Rev.* **2016**, *29*, 487–524. [CrossRef]
104. Zhang, Y.; Liu, W.; Chen, J.; Niu, H.; Mao, C.; Jin, B. Metal-organic gel and metal-organic framework based switchable electrochemiluminescence RNA sensing platform for Zika virus. *Sens. Actuators B-Chem.* **2020**, *321*, 128456. [CrossRef]
105. Li, Y.; Li, J.; Zhu, D.; Wang, J.; Shu, G.; Li, J.; Zhang, S.; Zhang, X.; Cosnier, S.; Zeng, H.; et al. 2D Zn-porphyrin-based Co(II)-MOF with 2-methylimidazole sitting axially on the paddle-wheel units: An efficient electrochemiluminescence bioassay for SARS-CoV-2. *Adv. Funct. Mater.* **2022**, *32*, 2209743. [CrossRef]
106. Wu, D.; Dong, W.; Yin, T.; Jie, G.; Zhou, H. PCN-224/nano-zinc oxide nanocomposite-based electrochemiluminescence biosensor for HPV-16 detection by multiple cycling amplification and hybridization chain reaction. *Sens. Actuators B-Chem.* **2022**, *372*, 132659. [CrossRef]
107. Li, J.; Luo, M.; Jin, C.; Zhang, P.; Yang, H.; Cai, R.; Tan, W. Plasmon-enhanced electrochemiluminescence of PTP-decorated Eu MOF-based Pt-tpped Au bimetallic nanorods for the lincomycin assay. *ACS Appl. Mater. Interfaces* **2022**, *14*, 383–389. [CrossRef]
108. Li, J.; Luo, M.; Yang, H.; Ma, C.; Cai, R.; Tan, W. Novel dual-signal electrochemiluminescence aptasensor involving the resonance energy transform system for kanamycin detection. *Anal. Chem.* **2022**, *94*, 6410–6416. [CrossRef]
109. Wang, J.; Xu, X.; Zheng, L.; Guo, Q.; Nie, G. A signal "on-off-on"-type electrochemiluminescence aptamer sensor for detection of sulfadimethoxine based on Ru@Zn-oxalate MOF composites. *Microchim. Acta* **2023**, *190*, 131. [CrossRef]
110. Shen, K.; Zhang, J.; Shen, L.; Xiong, Z.; Zhu, H.; Wang, A.; Yuan, P.; Feng, J. Hydrogen bond organic frameworks as radical reactors for enhancement in ECL efficiency and their ultrasensitive biosensing. *Anal. Chem.* **2023**, *95*, 4735–4743. [CrossRef]

111. Wang, B.; Zhao, L.; Li, Y.; Liu, X.; Fan, D.; Wu, D.; Wei, Q. Porphyrin-based metal-organic frameworks enhanced electrochemiluminescence (ECL) by overcoming aggregation-caused quenching: A new ECL emitter for the detection of trenbolone. *Anal. Chim. Acta* **2023**, *1276*, 341616. [CrossRef]
112. Tao, X.; Pan, C.; Yang, X.; Yuan, R.; Zhuo, Y. CDs assembled metal-organic framework: Exogenous coreactant-free biosensing platform with pore confinement-enhanced electrochemiluminescence. *Chin. Chem. Lett.* **2022**, *33*, 4803–4807. [CrossRef]
113. Nie, Y.; Tao, X.; Zhang, H.; Chai, Y.; Yuan, R. Self-assembly of gold nanoclusters into a metal-organic framework with efficient electrochemiluminescence and their application for sensitive detection of rutin. *Anal. Chem.* **2021**, *93*, 3445–3451. [CrossRef] [PubMed]
114. Zheng, H.; Yi, H.; Dai, H.; Fang, D.; Hong, Z.; Lin, D.; Zheng, X.; Lin, Y. Fluoro-coumarin silicon phthalocyanine sensitized integrated electrochemiluminescence bioprobe constructed on TiO_2 MOFs for the sensing of deoxynivalenol. *Sens. Actuators B-Chem.* **2018**, *269*, 27–35. [CrossRef]
115. An, X.; Jiang, D.; Cao, Q.; Xu, F.; Shiigi, H.; Wang, W.; Chen, Z. Highly efficient dual-color luminophores for sensitive and selective detection of diclazepam based on MOF/COF bi-mesoporous composites. *ACS Sens.* **2023**. [CrossRef]
116. Shan, X.; Pan, T.; Pan, Y.; Wang, W.; Chen, X.; Shan, X.; Chen, Z. Highly sensitive and selective detection of Pb (II) by $NH_2SiO_2/Ru(bpy)_3^{2+}$ UiO66 based solid-state ECL sensor. *Electroanalysis* **2020**, *32*, 462–469. [CrossRef]
117. He, Q.-N.; Ma, Z.-Y.; Yang, Y.-X.; Xu, C.-H.; Zhao, W. Recent advances in electrochemiluminescence-based single-cell analysis. *Chemosensors* **2023**, *11*, 281. [CrossRef]
118. Pan, D.; Fang, Z.; Yang, E.; Ning, Z.; Zhou, Q.; Chen, K.; Zheng, Y.; Zhang, Y.; Shen, Y. Facile preparation of WO_3-x dots with remarkably low toxicity and uncompromised activity as co-reactants for clinical diagnosis by electrochemiluminescence. *Angew. Chem. Int. Ed.* **2020**, *59*, 16747–16754. [CrossRef]
119. Ding, H.; Zhou, P.; Fu, W.; Ding, L.; Guo, W.; Su, B. Spatially selective imaging of cell-matrix and cell-cell junctions by electrochemiluminescence. *Angew. Chem. Int. Ed.* **2021**, *60*, 11769–11773. [CrossRef]
120. Li, B.; Huang, X.; Lu, Y.; Fan, Z.; Li, B.; Jiang, D.; Sojic, N.; Liu, B. High electrochemiluminescence from $Ru(bpy)_3^{2+}$ embedded metal-organic frameworks to visualize single molecule movement at the cellular membrane. *Adv. Sci.* **2022**, *9*, 2204715. [CrossRef]
121. Wei, W.; Lin, H.; Shao, H.; Hao, T.; Wang, S.; Hu, Y.; Guo, Z.; Su, X. Faraday cage-type aptasensor for dual-mode detection of Vibrio parahaemolyticus. *Microchim. Acta* **2020**, *187*, 529. [CrossRef] [PubMed]
122. Sun, Y.; Chen, Y.; Duan, Y.; Ma, F. Electrogenerated chemiluminescence biosensor based on functionalized two-dimensional metal-organic frameworks for bacterial detection and antimicrobial susceptibility assays. *ACS Appl. Mater. Interfaces* **2021**, *13*, 38923–38930. [CrossRef]
123. Wang, Y.; Shu, J.; Lyu, A.; Wang, M.; Hu, C.; Cui, H. Zn^{2+}-modified nonmetal porphyrin-based metal-organic frameworks with improved electrochemiluminescence for nanoscale exosome detection. *ACS Appl. Nano Mater.* **2023**, *6*, 4214–4223. [CrossRef]
124. Liang, T.; Guo, Z.; He, Y.; Wang, Y.; Li, C.; Li, Z.; Liu, Z. Cyanine-doped lanthanide metal–organic frameworks for near-infrared II bioimaging. *Adv. Sci.* **2022**, *9*, 2104561. [CrossRef]
125. Liu, Y.; Zhang, H.; Li, B.; Liu, J.; Jiang, D.; Liu, B.; Sojic, N. Single biomolecule imaging by electrochemiluminescence. *J. Am. Chem. Soc.* **2021**, *143*, 17910–17914. [CrossRef] [PubMed]
126. Deneff, J.I.; Rohwer, L.E.S.; Butler, K.S.; Kaehr, B.; Vogel, D.J.; Luk, T.S.; Reyes, R.A.; Cruz-Cabrera, A.A.; Martin, J.E.; Sava Gallis, D.F. Orthogonal luminescence lifetime encoding by intermetallic energy transfer in heterometallic rare-earth MOFs. *Nat. Commun.* **2023**, *14*, 981. [CrossRef]
127. Ji, Z.; Li, T.; Yaghi, O.M. Sequencing of metals in multivariate metal-organic frameworks. *Science* **2020**, *369*, 674–680. [CrossRef]
128. Qin, X.; Zhan, Z.; Ding, Z. Progress in electrochemiluminescence biosensors based on organic framework emitters. *Curr. Opin. Electrochem.* **2023**, *39*, 101283. [CrossRef]

chemosensors

Review

Electrochemical Biosensing Methods for Detecting Epigenetic Modifications

Ziyue Huang [1,2], Yanzhi Dou [3], Jing Su [4], Tie Li [3] and Shiping Song [1,5,*]

1 Division of Physical Biology, Shanghai Institute of Applied Physics, Chinese Academy of Sciences, Shanghai 201800, China; huangziyue@sinap.ac.cn
2 University of Chinese Academy of Sciences, Beijing 100049, China
3 Shanghai Institute of Microsystem and Information Technology, Chinse Academy of Sciences, Shanghai 200050, China; douyanzhi@mail.sim.ac.cn (Y.D.); tli@mail.sim.ac.cn (T.L.)
4 State Key Laboratory of Oncogenes and Related Genes, Institute for Personalized Medicine, School of Biomedical Engineering, Shanghai Jiao Tong University, Shanghai 200030, China; sujing@sjtu.edu.cn
5 Institute of Materiobiology, College of Science, Shanghai University, Shanghai 200444, China
* Correspondence: spsong@shu.edu.cn; Tel.: +86-133-8612-8515

Abstract: Epigenetic modifications are closely related to diseases and physiological health, mainly including DNA methylation, RNA methylation, histone acetylation, and noncoding RNA. Recently, a large amount of research has been conducted on the detection of epigenetic modifications. Electrochemical biosensors, with their low cost, high sensitivity, high compatibility, and simple operation, have been widely used in the detection of epigenetic biomarkers. This review discusses the detection of epigenetic biomarkers using different electrochemical sensing methods. Here we discuss various aspects, including free labels, signal labeling, signal amplification, nano-based electrodes, and the combined use of other methods. By summarizing the existing electrochemical detection methods for epigenetic modifications, this review also proposes future development trends and challenges for electrochemical biosensors in this field.

Keywords: epigenetic modifications; electrochemical biosensing; signal amplification; nanostructure modified electrodes

Citation: Huang, Z.; Dou, Y.; Su, J.; Li, T.; Song, S. Electrochemical Biosensing Methods for Detecting Epigenetic Modifications. *Chemosensors* **2023**, *11*, 424. https://doi.org/10.3390/chemosensors11080424

Academic Editor: Boris Lakard

Received: 27 June 2023
Revised: 19 July 2023
Accepted: 29 July 2023
Published: 1 August 2023

1. Introduction

Epigenetics is a discipline that studies the inheritance of genetic information through DNA methylation or chromatin conformation changes without altering the DNA sequence [1]. Epigenetic phenomena include DNA methylation, genomic imprinting, maternal effects, gene silencing, RNA editing, noncoding RNA, and more [2,3]. Currently, there is extensive research on the modifications of DNA, RNA, and histones [4–6]. The modification of DNA and RNA mainly includes DNA methylation, DNA hydroxymethylation, and RNA methylation [4,7–9]. Research on the modification of RNA and DNA mainly focuses on the modification of nucleic acid bases and sugars [10]. Research on histone modifications is also extensive, mainly focusing on histone phosphorylation and acetylation [5,11–13]. In addition to directly studying the changes in epigenetic genetic information, enzymes related to the modification of RNA, DNA, and histones are also being studied.

Abnormal DNA and RNA methylation and histone acetylation can lead to various diseases [10,14–16]. Abnormal methylation of RNA and DNA and histone acetylation are closely related to the occurrence of various cancers and can also cause metabolic and neurological diseases [17–19]. Due to the close relationship between epigenetic modifications and various diseases, epigenetic modifications has been widely studied. By detecting epigenetic biomarkers, researchers can gain a deeper understanding of the mechanisms and progression of diseases, providing more accurate and precise methods for early diagnosis and treatment. For example, detecting DNA methylation levels can improve the sensitivity and specificity of tumor detection, providing more reliable methods for early diagnosis [20]. In addition, for some difficult-to-diagnose diseases such as autism and schizophrenia, the detection of epigenetic biomarkers can provide new diagnostic and treatment approaches [21,22]. Therefore, the detection of epigenetic biomarkers is of great significance for the prevention, diagnosis, and treatment of diseases. The conventional detection methods for epigenetic substances are already mature, and commonly used methods include radioanalysis [23], chromatography [24–26], immunological analysis [27], and single-molecule sequencing [28–30]. Radioactive methods require labeling of the detection substance, which may produce harmful radiation to humans and the environment. Chromatography and sequencing analysis have the disadvantage of being time-consuming and tedious. These detection methods also require expensive instruments and are not suitable for on-site testing in complex environments.

Electrochemical biosensors do not require complex equipment, and only an electrochemical workstation, electrodes, and a computer are needed to complete the detection of target analytes [4,31]. Compared with complex detection methods such as fluorescence [32,33], surface plasmon resonance [34,35], and surface-enhanced Raman spectroscopy [36,37], electrochemical biosensing detection methods have the advantages of simple operation, low equipment cost, and high sensitivity [38]. They have been developed to detect various forms of epigenetic modifications in recent years [39–41].

The principle of electrochemical biosensors for epigenetic modification detection is to use electrodes as conversion elements and immobilization carriers, and to immobilize bio-sensitive substances such as antibodies and capture probes, or biomolecules themselves as sensitive elements on the electrodes, and to convert the signals of target molecules and their reactions into electrical signals such as current, capacitance and conductivity through specific recognition between biomolecules, so as to detect epigenetic modifications qualitatively or quantitatively. PCR methods for detecting epigenetic modifications can only be suitable for DNA or RNA, but not suitable for epigenetic-related enzymes and histone modifications. Thus, electrochemical biosensors have a wider range of applications. While PCR methods are sensitive but require complex instrumentation and equipment, electrochemical biosensors are easier to implement than PCR for in situ detection because they are sensitive, simple, affordable, easy to miniaturize, multiplex, multiplex and timely medical compatible, and can achieve the advantage of field detection in complex environments. Obviously, electrochemical biosensors are a powerful tool for quantitative analysis of various biomarkers. In particular, the integration of electrochemical sensors into portable devices makes point-of-care testing (POCT) possible, which may provide new approaches to medical diagnostics, especially in low-resource settings.

This review provides an overview of electrochemical detection methods for epigenetic modifications. It introduces the label-free method of using electrochemical detection for epigenetic modifications, the method of using signal probes based on label modification, the electrochemical detection method based on signal amplification, and the method of using nanostructure-modified electrodes. We will review the principles, characteristics, and applications of these electrochemical biosensors and discuss challenges and future development directions in this field.

2. Label-Free Methods

The direct detection method without labeling utilizes electrochemical impedance spectroscopy (EIS) detection in electrochemistry to directly read electrical signals without any modification or amplification. The direct label-free detection methods mainly include EIS and the method using electrochemically active substances as indicators. EIS is mainly based on the principle that the signal is generated by the increase of electrochemical impedance after the analyte binds to the capture material on the electrode, which is then used to quantify the analyte. The use of electrochemically active substances as indicators is based on the fact that these substances can bind with the analyte due to electrostatic forces.

The more the analyte presents, the more electrochemically active substances will bind, and the larger the electrical signal generated, which can then reflect the content of the analyte.

DNA methylation is the most common type of epigenetic modification. There have been many reported methods for detecting DNA methylation. Most DNA methylation detection methods rely on the principle of base complementary pairing. Sheppard's Group developed a biological platform that takes advantage of the stability of double-stranded targets (Figure 1) [42]. A denaturation step is added prior to detection to take advantage of the sensitivity and selectivity of the single-stranded DNA (ss-DNA) target and probe hybridization for the detection of the ss-DNA target. The authors used conductive polymer materials to modify the electrode and covalently attached probes for bio-recognition on the modified electrode surface. Due to steric hindrances presented by methyl groups, that methylation can affect the hybridization rate. EIS was used to detect the signal generated by the potassium ferricyanide reduction reaction to study the hybridization kinetics of double-stranded DNA to detect DNA methylation. This method detects target methylation through the kinetic changes in methylation DNA hybridization, providing a design concept for future methylation biosensors.

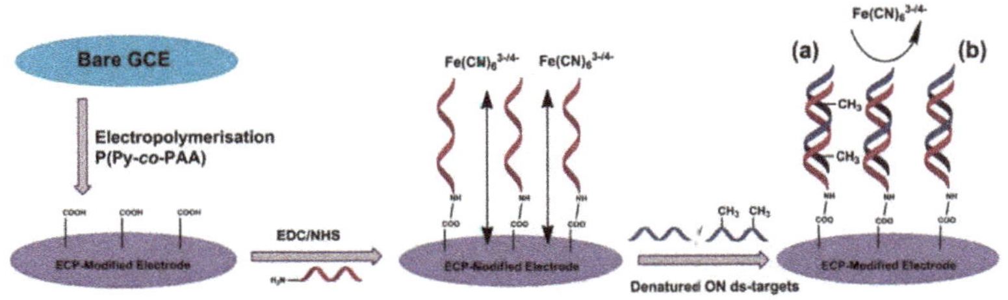

Figure 1. Schematic representation of DNA sensor based on ECP, including steps of electrocopolymerization, probe attachment, and hybridization with (a) methylated and (b) unmethylated denatured ds-ON target [42]. Copyright 2015, Elsevier.

Currently, in contrast to the extensive research on DNA methylation, there are relatively fewer reports on RNA modification detection methods because the content of abnormal RNA in total RNA is particularly low and difficult to detect. Since the content of RNA is relatively lower than that of DNA, more sensitive detection methods are needed to detect RNA. N6-methyladenosine (m6A) is a common RNA methylation modification in epigenetics. There have been a few reports of using electrochemical biosensors combined with competitive degradation to quantify such kinds of RNA methylation.

Xie's group developed a sensitive and label-free electrochemical immunosensor for m6A-RNA detection, utilizing the advantages of the sensitivity and selectivity of electrochemical biosensing technology (Figure 2) [43]. The key to achieving sensitivity and specificity in the developed method is the use of the specific interaction between antibody (Ab) and antigen (Ag). The authors used recombinant proteins tagged with histidine to modify the gold electrode surface. Histidine binds to the gold electrode surface, allowing the recombinant protein to be oriented. The specificity of the Ab crystal region and recombinant protein binding was then used to expose the Ab binding site, thereby improving the binding efficiency of the Ag-Ab, enhancing the detection signal, and enabling the detection of low-abundance m6A-RNA. The anti-m6A-Ab used in this method can bind to both m6A-RNA and m6A-DNA. M6A-DNA serves as a signaling molecule and participates in the reaction together with m6A-RNA. After binding to the Ab, RNase A is used to hydrolyze the bound m6A-RNA. The amount of m6A-RNA is quantified by the decrease in the EIS signal. The linear range of detection can be improved by using a method of competition reaction with m6A-DNA and m6A-RNA, followed by degradation. Finally, the EIS signal of the detection electrode is detected, and the decrease of signal intensity is proportional to the abundance of m6A-RNA, while the intensity of the signal is inversely proportional to the amount of m6A-RNA in the sample. This biosensor has the advantages of simplicity and sensitivity, with a wide detection range and a sensitivity of up to 0.016 nM. This method for detecting RNA methylation is not affected by chain length or base sequence and has a certain degree of universality.

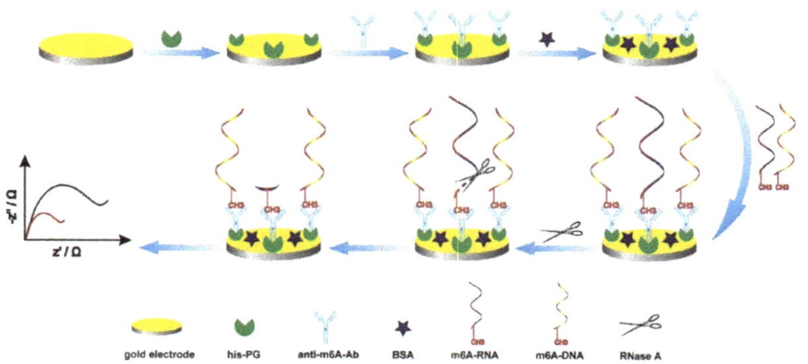

Figure 2. The procedures for fabricating the immunosensor [43]. Copyright 2018, Elsevier.

The above two methods use the principle of base complementary pairing to detect DNA methylation or RNA methylation. The methods can only detect methylated nucleic acids as epigenetic markers, and the methods can only detect single biomarkers. Electrochemical sensing methods can not only detect nucleic acids in epigenetics but also use Ag-Ab interactions to detect proteins. They can not only detect single biomarkers but also perform multiple detections of two different types of biomarkers. Due to their cell-type specificity, robustness, and ability to be released into body fluids, DNA methylation and histone acetylation can serve as biomarkers for cancer diagnosis in vitro.

Sheppard's group used graphene screen-printed electrodes to detect DNA and chromatin, because graphene screen-printed electrodes have the advantages of low cost, high signal-to-noise ratio, and no need for surface preparation [44]. The surface of the screen-printed electrode was coated with polyaniline, a conductive polymer, to avoid the influence of defects on the graphene surface while also allowing the Ab to be surface-functionalized. The anti-5-methylcytosine Ab, which specifically binds to DNA methylation, and the anti-acetylated histone H3 Ab, which specifically binds to histone acetylation, were then directly coupled to the polyaniline-modified graphene screen-printed electrode surface. This label-free method uses EIS to detect DNA and chromatin. The authors used the developed biosensor to detect endometrial cancer cell and breast cancer cell systems and found that there is no difference in total DNA methylation, but there is a difference in histone acetylation. Unlike other methods that are time-consuming or require expensive hardware, electrochemical biosensors have the advantages of simplicity, sensitivity, and portability and can be used as important tools for DNA and histone detection. The biosensor developed by the authors can simultaneously detect two epigenetic markers. It can also be used to observe the therapeutic effect of epigenetic drugs by detecting cells in vitro. This method demonstrates the potential of using the same sample for multiple epigenetic detections.

In addition to using EIS to directly detect the binding of analytes to the electrode surface, there are also some detection methods that use electrochemically active substances as indicators. Ai's group used methylene blue as an indicator to detect DNA methylation and methyltransferase activity [45]. Methylene blue can be inserted into the double-stranded DNA, providing a reliable electrochemical signal. Methylated DNA can be selectively cleaved, reducing the amount of methylene blue and producing a decrease in the electrochemical signal. Li's group chose the electroactive complex $[Ru(NH_3)_6]^{3+}$ as a signal converter [46]. The electroactive complex can bind to double-stranded DNA through electrostatic forces. Similar to methylene blue as an indicator, it can also detect methylated DNA and has a signal amplification effect, with the ability to sensitively detect DNA methylation. The activity of DNA adenine methylation methyltransferase was detected by signal closure. Nie's group prepared a coenzyme A silver ion-coordinated polymer with high electrocatalytic activity as a signal probe for high-sensitivity detection of coenzyme A and histone transferase activity [47].

Direct and label-free epigenetic detection methods have the advantages of being simple and easy to operate. The methods rely on the specificity of Ag-Ab binding or DNA base complementary pairing. As the sensitivity of all methods that use EIS to detect biomarkers is often not enough, conductive material modification or interface orientation methods have been used to improve the Ab coverage and binding efficiency on the electrode interface to some degree. Direct detection of DNA methylation distinguishes between methylation patterns based on the kinetics of hybridization, while RNA methylation is quantified based on changes in electrochemical impedance caused by enzyme degradation after competition. DNA methylation and histone acetylation are detected through Ag-Ab binding. The detection methods are similar and the quantification principle is based on the signal change of the potassium ferricyanide reduction reaction. However, although direct detection methods are simple, their sensitivity is not enough. Even though the sensitivity has been improved by interface modification, it is still not sufficient for the detection of some low-abundance epigenetic modifications.

3. Methods with Labeled Signal Probes

Due to the low sensitivity of direct detection methods, they are not suitable for the detection of all epigenetic markers, especially in the detection of low-abundance DNA and RNA modifications. Even with the use of electroactive indicators, it is still challenging to meet the high sensitivity requirements for detection. Therefore, there is a need to develop more sensitive detection methods to improve the performance of electrochemical biosensors. Common methods to enhance sensitivity involve the use of signal probe labeling. The methods of signal probes for electrochemical biosensing detection of epigenetic markers mainly include chemically modifying specific sites, using protein interactions to bind signals, and employing signal probe methods.

Chen's group took advantage of the high sensitivity of electrochemistry and used a signal labeling strategy to detect low-abundance 5-formyluracil (5fU), which is closely related to the function of DNA (Figure 3) [48]. The most critical step of this method is to signal-modify the target site of 5fU-DNA. The authors first used the azide derivative of (2-phenylimidazole) acetonitrile and the aldehyde group of 5fU to form a covalent bonding through a chemical reaction. Then, DBCO-PEG4-biotin was connected to the target DNA through a copper-free click chemistry reaction. Next, T4 polynucleotide kinase was used to catalyze the target DNA to generate a sulfhydryl group at the 5′ end. The target DNA containing 5fU can be assembled on the electrode surface by the interaction of Au-S bond. Finally, specific recognition between biotin and streptavidin was used to label horseradish peroxidase onto the surface of the above electrode. Horseradish peroxidase catalyzed the oxidation and reduction of hydroquinone to generate an electric current signal. Differential pulse voltammetry (DPV) was used to detect the current and enable high-sensitivity detection of 5fU. The specificity of detection comes from the specific recognition of azide to 5fU. Introducing a biotin label at the position of 5fU and then combining it with streptavidin-horseradish peroxidase can significantly improve the detection sensitivity. Using T4 polynucleotide kinase to directly connect the target DNA onto the electrode through a covalent bond is more direct and does not require a capture probe. This method can avoid interference from 5-formylcytosine and apyrimidinic sites in the detection of 5fU. At the same time, this direct bonding method also avoids the sequence matching problem caused by the use of capture probes in traditional connection methods. This method has a good linear range and a low detection limit.

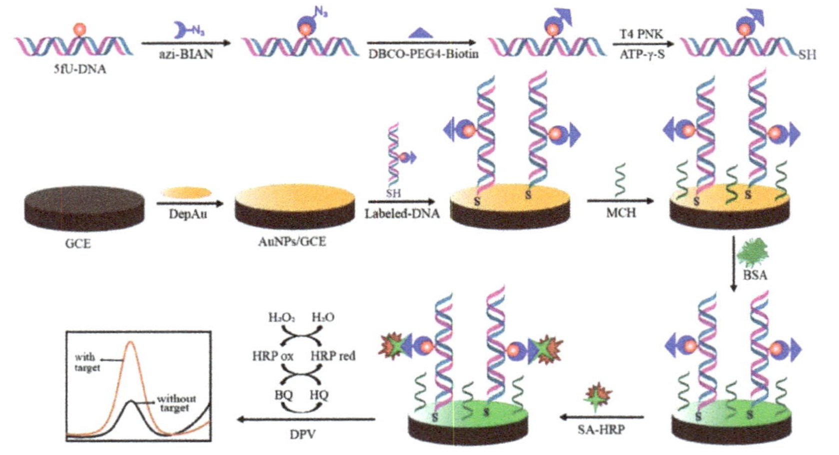

Figure 3. Schematic illustration of the electrochemical detection of 5fU [48]. Copyright 2021, American Chemical Society.

The chemical modification of signal probes requires the use of toxic chemical reagents and involves multiple steps of manipulating the target DNA during the labeling process. The labeling process may cause some loss of targets, which can introduce bias in the upstream analysis and affect the accuracy of the experimental results. Although the method uses chemical substances to recognize specific sites and achieve specific labeling, the drawbacks of the method are also evident. The complex processing steps limit the practicality of this method. Compared with the method of modifying DNA using chemical reagents, the specific binding of biological proteins is safer and more reliable. Biological proteins can selectively bind to specific groups without the need for complex processing steps, and they do not have toxic effects on the operator, making them a safer and more reliable modification labeling method.

Ai's group developed a biosensor to detect cytosine methylation of CpG dinucleotides and the activity of methyltransferase (MTase) in DNA, using a methyl-binding domain (MBD) protein that can specifically bind to CpG dinucleotides and Coomassie brilliant blue G250 (CBB-G250) as the signal label (Figure 4) [49]. This sample method can also be applied to screen for MTase inhibitors. The process of this method is to fix the capture DNA probe on the electrode using the Au-S bond firstly and then hybridize the target DNA with the capture DNA probe. Treatment of the hybridized DNA with M.SssI-MTase in the presence of the methyl donor S-adenosylmethionine can methylate CpG dinucleotide specific sites. The methylated CpG region can be specifically recognized and bound by the MBD protein. CBB-G250 can bind to the MBD protein through intermolecular forces. CBB-G250 is a common electroactive molecule that can provide redox signals to methylated DNA. When the CCGG symmetrical sequence of the hybrid molecule is specifically recognized and cut by the Hpall restriction endonuclease, it cannot be methylated by MTase. The amperometric current method is used to detect the redox signal of CBB-G250, and the obtained current signal can reflect the level of DNA methylation and the activity of MTase. The authors developed a simple, portable, and sensitive biosensor that can be used to detect DNA methylation and MTase activity, and it can also screen for methylation inhibitors.

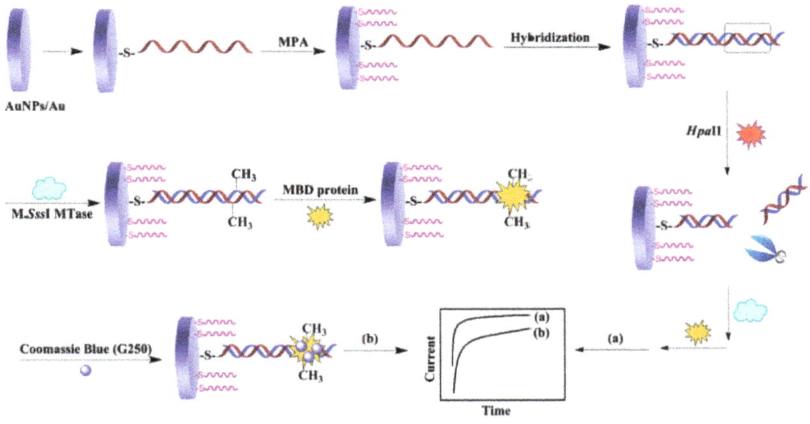

Figure 4. Schematic representation of the developed method for detection of DNA methylation and assay of M. SssI MTase activity [49]. Copyright 2013, Elsevier.

In addition to detecting DNA methylation, the above methods can also be used to detect DNA MTase. The detection of enzymes related to epigenetic nucleic acid markers is also essential for studying epigenetics. Detecting DNA MTase is different from directly detecting nucleic acid markers, as it requires a reactive subject. Using the specific binding ability of biological proteins and the non-specific binding method of specific groups and Coomassie blue does not directly combine with the label. The binding between the label and the electrode is not specific, which may result in insufficient accuracy in actual sample detection. Developing a method that can directly and specifically bind to the electrode can improve accuracy and sensitivity. The use of unique hairpin probes to generate capture probes and then combine them with signal probes can increase the specificity of the recognition process. It is also important to detect MTase for epigenetic research.

Yan's group developed a sensitive and simple method for detecting MTase activity using hairpin DNA probes (Figure 5) [50]. The key design of this method is the elaborate design of the hairpin DNA probe. The 5′ end of this hairpin DNA probe is modified with a sulfhydryl group, which can directly generate an Au-S covalent bond with the gold electrode, fixing the probe on the electrode surface. The hairpin DNA probe also has a methylation recognition site. After treatment with MTase or restriction endonucleases such as Dam MTase and Dpn I that can recognize methylated sites, the hairpin DNA probe is cleaved. The remaining DNA fragments after cleavage are still left on the electrode as a capture probe that can hybridize with a signal probe. The signal DNA probe is modified with methylene blue, which can undergo redox reactions on the electrode surface. Unlike conventional detection methods where the signal is directly labeled on the target molecule, this method cleverly designs the capture probe to release and bind with the signal probe. When the hairpin DNA probe cannot be cleaved due to methylation, the capture probe cannot be released and cannot bind with the signal probe, avoiding false-positive results. This method is simple to prepare and easy to operate, and it has good selectivity and high sensitivity. The detection limit for Dam MTase using this method is 0.07 U/mL. This method can also be applied to the screening of inhibitors and the discovery of anticancer drugs. Yuan's group evaluated the activity of DNA transferase using a commercial blood glucose meter [51]. This method used biotin-avidin-peroxidase as a label to achieve sucrose catalytic hydrolysis and used a blood glucose meter to convert the signal for detection.

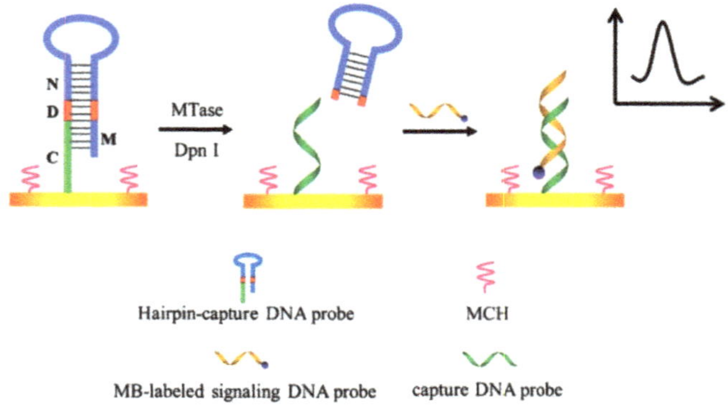

Figure 5. Schematic representation of the electrochemical sensor for the assay of MTase activity [50]. Copyright 2012, Elsevier.

Most of the label-based electrochemical detection methods for epigenetics use substances with electrochemical activity, such as CBB-G250, methylene blue, or enzymes that catalyze reactions such as the biotin-avidin-peroxidase system. The changes in current in the electrochemical or enzyme-catalyzed redox reactions are used for detection. This method has higher sensitivity and specificity compared to direct detection methods. It can detect not only epigenetic biomarkers but also enzymes that cause abnormal changes, making the range of substances detected wider and the sensitivity higher. However, the labeling method is cumbersome and may have multiple synthesis steps. The storage time of the labeled target is also greatly limited.

4. Methods Based on Signal Amplification

Signal labeling methods can improve detection sensitivity, enabling quantitative detection of low-abundance 5-methyluracil and expanding the range of electrochemical detection. However, the signal labeling operation itself involves multiple steps, increasing the complexity of actual detection operations. Additionally, the limited storage time of the target may restrict the practical application of the method. Considering that the signal labeling method has limitations in improving sensitivity, it still lacks detection capabilities for ultra-low-abundance epigenetic biomarkers. Developing a method for ultra-sensitive detection of ultra-low-abundance markers is crucial. Signal amplification methods can effectively solve this problem. Signal amplification methods mainly include using a dual signal amplification strategy combining PCR and CRISPR/Cas12 systems, redox signal amplification combined with enzyme-catalyzed amplification, and a multi-step circuit amplification design. Detection methods based on signal amplification can significantly improve detection sensitivity.

Liao's group used the difference in the thermodynamic stability of hybridization between a xeno nucleic acid (XNA) probe with m6A-RNA and A-RNA to develop a specific reverse transcription polymerase chain reaction for m6A-RNA (Figure 6) [52]. Combined with the CRISPR/Cas12a signal amplification strategy, m6A modification can be detected with high sensitivity. In the detection process, RNA is extracted from cells to obtain RNA containing m6A-RNA and A-RNA without methylation modification. XNA probes are used to hybridize with RNA. Non-methylated RNA and XNA are more stable after hybridization, and the strand displacement reactions (SDR) with the reverse transcription primer occur slowly. Due to the hybrid of m6A-RNA and XNA being unstable, the m6A-RNA can preferentially undergo SDR with the reverse transcription primer. The reverse transcription of non-methylated RNA is blocked by the XNA probe, directly magnifying the minute differences between m6A-RNA and A-RNA. The single-stranded DNA obtained from the reverse transcription of m6A-RNA is PCR amplified to generate double-stranded DNA, which is positively correlated with the m6A fraction. The authors utilized the differences in thermodynamic stability between m6A-RNA and RNA hybridization with XNA and the difficulty of SDR to amplify the small differences in the first step after PCR amplification. Then, the CRISPR/Cas12a system is used to amplify the signal. The CRISPR-derived RNA (crRNA) in the designed CRISPR can specifically target the m6A RT-PCR amplification fragment. The CRISPR/Cas12a system is activated by the specific double-stranded DNA generated in the reverse transcription polymerase chain reaction and the realized signal amplification output. A methylene blue-modified DNA probe is immobilized on the gold electrode surface, and the remaining bare area of the Au electrode is blocked by MCH before adding the activated Cas12 reaction system. After the DNA probe is cut away from the electrode surface, the methylene blue molecules on the electrode are released. The abundance of m6A-RNA is quantified by detecting the decrease in the square wave voltammetry signal. In this study, the authors used a dual signal amplification system. After amplifying the signal with RT-PCR, the CRISPR/Cas12a system was used to further amplify the signal. The m6A-RNA can be detected with ultra-high sensitivity and can sensitively detect 1% m6A sites.

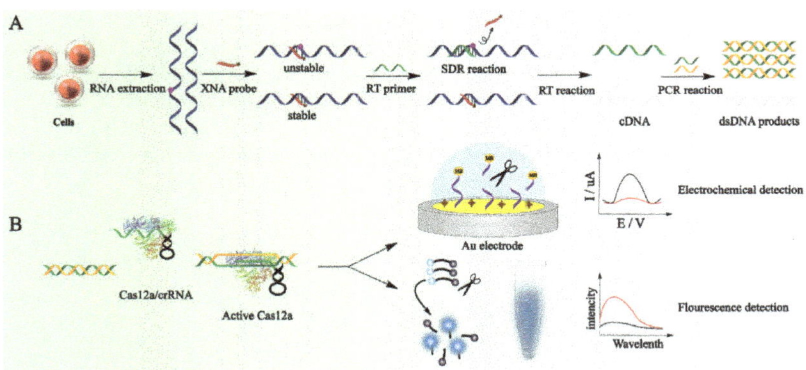

Figure 6. Schematic illustration of the MsRT-PCR and CRISPR/Cas12a integrated detection system. (**A**) The principle of MsRT-PCR. (**B**) The procedure of dsDNA products activated CRISPR/cas12a system for signal amplification [52]. Copyright 2023, Elsevier.

The method of using RT-PCR and the CRISPR/Cas12a system amplification can achieve ultra-sensitive detection of extremely low-abundance m6A-RNA. However, the principle of realizing specific detection using this signal amplification method is based on the difference in complementary pairing with exogenous nucleic acid fragments. Nucleic acid-based complementary pairing detection has strong sequence specificity and can only detect nucleic acids with specific sequences. The CRISPR/Cas12a system also requires a specific sequence design to complete the dual signal amplification. Through the dual specificity recognition of exogenous nucleic acid probes and crRNA, as well as the combined signal amplification of PCR and CRISPR/Cas12a, the specificity and sensitivity can be significantly improved. However, this method can only detect m6A-RNA with specific sequences, which also has certain limitations.

Compared with the dual signal amplification method based on nucleic acid complementary pairing, the following signal amplification method based on magnetic separation does not require a specific sequence design or signal labeling and can achieve signal amplification for the detection target. Zhang's group developed an unlabeled electrochemical magnetic biosensor for the quantitative detection of 5-hydroxymethylcytosine (5-hmC) DNA, which is closely related to cancer and is an important epigenetic biomarker for tumorigenesis (Figure 7) [53]. The core of this method is coupling with the terminal deoxynucleotidyl transferase (TDT) enzyme-catalyzed amplification and Ru(III) redox cycling, a dual signal amplification system, to significantly improve the sensitivity of the detection. In the detection process, 5-hmC is first modified and enriched. The 5-hmC specific site in the non-paired double-stranded DNA is modified by a biotinyl-cysteine derivative. After purification using a Micro-Bio-Spin P6 column, biotin-coupled 5-hmC double-stranded DNA is obtained. Because biotin has good specificity with streptavidin, the biotin-coupled double-stranded DNA is bound to magnetic beads coated with streptavidin, which links the double-stranded DNA and magnetic beads to achieve the enrichment effect. The magnetic beads linked with double-stranded DNA are elongated by terminal deoxynucleotidyl transferase treatment. $Ru(NH_3)_6{}^{3+}$ is positively charged, which can be electrostatically attracted to the negatively charged phosphate backbone of DNA. They can be combined through electrostatic interaction, allowing $Ru(NH_3)_6{}^{3+}$ to reach the vicinity of the DNA main chain. The magnetic beads linked with DNA double-strands can bind to the screen-printed electrode surface through magnetic force without immobilization. $Fe(CN)_6{}^{3-}$ is negatively charged and repels the DNA main chain, making it difficult to contact. $Ru(NH_3)_6{}^{3+}$ directly participates in the electrochemical redox process. With Ru(III) being reduced to Ru(II), Ru(II) can react with $Fe(CN)_6{}^{3-}$, causing Ru(II) to be oxidized to Ru(III) again. The regenerated Ru(III) can continue to bind with DNA to participate in the redox cycle. The redox process of Ru can further amplify the electrical signal. This electrochemical biosensing detection method has good specificity and can significantly distinguish 5-hmC from 5-methylcytosine. This method does not require specific templates or a special sequence design and is not limited to DNA with specific base sequences. Amplification is performed through molecular interactions without signal labeling. The dual signal amplification system enables a detection limit as low as 9.06 fM.

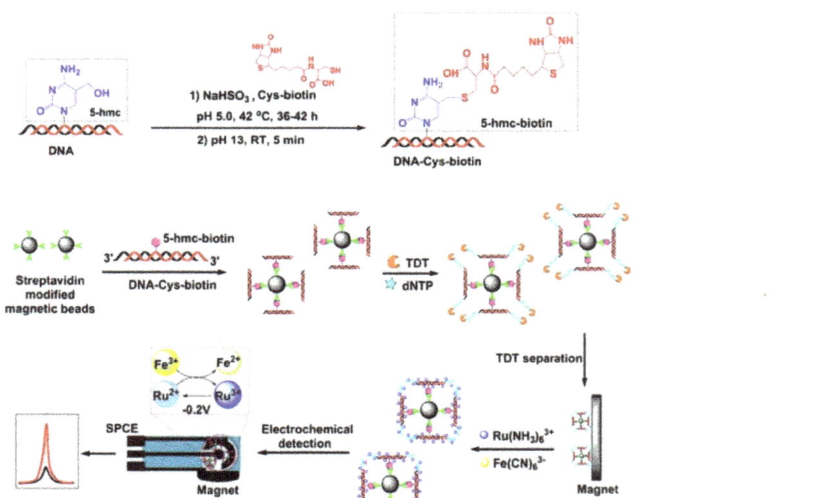

Figure 7. Schematic illustration of a label-free and immobilization-free electrochemical magneto-biosensor for 5-hmC assay [53]. Copyright 2019, American Chemical Society.

Although the electrochemical magnetic biosensor method does not require signal labeling and is not limited to specific base sequences, the detection process involves column separation technology and magnetic enrichment technology, making the steps relatively complex. The above method uses terminal deoxynucleotidyl transferase treatment, which is an enzyme-dependent and magnetic-dependent complex detection technology. Although it greatly improves detection sensitivity, the implementation of magnetic separation technology also requires specific modifications and column separation treatment for specific sites. The more complex and cumbersome the operation step, the more potential biases may be introduced. Chemical modification of specific sites requires the use of chemical reagents and long reaction times. In contrast to the above-mentioned magnetic separation technology, the use of multi-step circuit amplification technology achieves an enzyme-free and treatment-free process. Using the multi-step circuit cycling to achieve signal amplification does not require special chemical treatment and long reaction times, making the detection process as simple as possible while improving sensitivity. In addition to direct detection and signal labeling methods, signal amplification methods have also been used for sensitive detection of DNA methylation.

Based on a multi-step DNA amplification circuit design, Wang's group developed a sensitive proportional electrochemical biosensor for detecting methylated DNA (Figure 8) [54]. The multi-step circuit amplification design, which is a non-enzymatic amplification design, can use the product upstream as the initiator of the downstream circuit, effectively achieving signal amplification. The core of the non-enzymatic electrochemical biosensor designed by the authors is the design of multi-step circuit amplification, mainly including three cycles. Introduction of methylated target DNA in the first cycle triggers a Mg^{2+}-dependent DNA enzyme cycle, causing the hairpin structure to break and generate HP1*. The HP1* generated in the first cycle can act as an initiator for the second cycle, starting the CHA-1 cycle. QHP2/HP3 produced by the CHA-1 cycle can initiate the third cycle CHA-3, ultimately generating the four-way junction QHP4/QHP5. The multi-step amplification circuit is triggered by methylated target DNA and generates a DNA four-way junction after three cycles. The gold electrode surface is fixed with a capture probe, which binds to the signal probe of methylene blue. The DNA four-way junction can bind to the capture probe, undergo strand displacement reaction, replace the original signal probe of methylene blue, and connect the DNA four-way junction to the electrode interface. The DNA four-way junction has sufficient sites to introduce doxorubicin molecules. By detecting the ratio of doxorubicin and methylene blue signals using electrochemistry, the methylated target DNA can be sensitively detected. The ratiometric method is a way to detect biomarkers based on the ratio of two independent signals, rather than using a specific signal output value. Using the ratio of two signals to detect can significantly improve the accuracy and precision of detection. The oxidation-reduction of doxorubicin and the signal probe of methylene blue can be used as a pair of signal reporters. Although the ratiometric method produces stable and repeatable signals, amplification methods are used to increase the limited sensitivity. The multi-step circuit amplification method exhibited high sensitivity for target analysis with a detection limit of 4 aM.

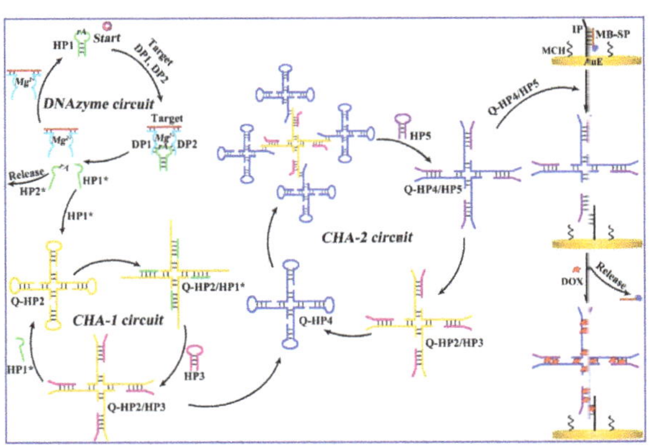

Figure 8. Schematic illustration for the ratiometric electrochemical detection of methylated target DNA based on multistep DNA amplification circuits [54]. Copyright 2022, Elsevier.

Ai's group developed an electrochemical detection method for DNA hydroxymethylation based on alkaline phosphatase-catalyzed signal amplification [55]. This method involves specific modification of the group with glycosylation and bridging with 1,4-phenylene diboronic acid to capture alkaline phosphatase. Signal amplification is achieved through enzyme-catalyzed generation of p-nitrophenol. This method can detect 5-hmC with high selectivity and sensitivity. Xie's group developed a method combining m6A-sensitive DNA enzyme and three-way junction-mediated isothermal exponential CRISPR amplification to identify methylated RNA by using deoxyribonuclease for specific cleavage of non-methylated RNA and three-way junction. Isothermal CRISPR amplification can amplify the signal to achieve high-sensitivity detection of low-abundance m6A-RNA [56]. Zhao's group combined sodium bisulfite conversion and PCR amplification and developed an electrochemical biosensor with methylation specificity using PCR amplification and enzyme-catalyzed signal amplification to quantitatively detect methylation of tumor suppressor gene promoter [57]. Our group developed a method combining glycosylation modification and enzyme-catalyzed signal amplification to detect 5-hmc DNA [58]. The developed electrochemical biosensor has high specificity and is multiplexed, capable of detecting multiple samples simultaneously. Li's group used the peroxidase property of G-quadruplex-Cu(II) metalloenzyme to catalyze hydrogen peroxide for signal amplification to electrochemically detect histone acetyltransferase activity [59].

Direct detection and labeled detection can detect epigenetic markers with high abundance, but there are still many biomarkers with extremely low abundance that are difficult to detect by simple signal labeling. The signal generated by a single label is not sufficient to meet the detection requirements. Signal amplification methods provide a good solution for this problem. Signal amplification methods can further amplify the signal when combined with other methods. For example, in m6A-RNA detection, PCR and the CRISPR/Cas12a system can be combined, or an amplification circuit and ratio signal can be combined, or magnetic bead enrichment and Ru oxidation-reduction signal amplification can be combined. Signal amplification technology is no longer just using signal labeling but also combining multiple technologies to achieve electrochemical detection methods with improved sensitivity. Compared with conventional detection methods, signal amplification technology is more complex, but it provides more accurate and precise measurements, and has greater specificity, which can detect specific targets at specific sites. More importantly, signal amplification technology greatly improves detection sensitivity.

By amplifying the signal, the sensitivity and detection limit of the sensor can be improved, enabling the detection of low-concentration target substances. In addition to using signal amplification methods, interface modification methods can also be used to enhance the stability, selectivity, and biocompatibility of biosensors, to improve signal transmission efficiency, and to reduce background noise, thereby improving the sensor's detection performance. Electrochemical biosensor interface modification and signal amplification play different roles. Signal amplification is mainly used to enhance the detection signal of the target analyte. Interface modification mainly focuses on improving the sensor surface to enhance sensor performance. They each have their advantages, but in some aspects, interface modification may be more advantageous than signal amplification. Interface modification can improve the stability of the sensor, allowing it to maintain performance over long periods of use or under different conditions, while signal amplification may introduce additional instability factors, such as enzyme inactivation or aggregation of nanomaterials. Interface modification can improve the selectivity of the sensor by modifying specific recognition elements, enabling the sensor to better distinguish between target analytes and interfering substances. In some cases, interface modification can achieve simpler operation and fewer steps, thereby improving the ease of the practical use of biosensors.

5. Methods Based on Nanostructured Modified Electrodes

As designing various signaling methods is necessary to improve the sensitivity of epigenetic analysis, developing high-performance electrodes is extremely important for electrochemical biosensing methods, providing both high sensitivity and high signal-to-noise ratio. The development of nanostructured modified electrodes becomes a more acceptable choice because nano-biomaterials and interfaces show unique performance for biosensing applications. The method of modifying nanostructures on the electrode surface is simpler and requires less sample processing. It can improve the sensitivity without complicated processing, making the detection method more concise in operation while meeting the sensitivity requirements. Nanostructure modification methods can significantly improve the signal-to-noise ratio and enhance the interference resistance of biosensors. Using nanostructure-modified electrodes can compensate for the interface defects of electrochemical biosensors, making the interface more regular, facilitating the binding of target molecules, and reducing background noise. The core of electrochemical biosensing lies in the binding of capture probes and analytes, and modifying the interface with nanomaterials is more conducive to capturing target substances. Currently, commonly used nanostructure-modified electrode methods include modifying the electrode surface with alloy nanostructures, directly generating nanostructures on the substrate surface and combining them with metal nanoparticles, modifying the electrode surface with composite materials, modifying the electrode surface with gold nanoparticles, and modifying the electrode surface with DNA tetrahedra.

Recent studies have shown that noncoding RNAs play an increasingly important role in the regulation of epigenetics. Noncoding RNAs are functional RNA molecules that cannot be translated into proteins, and some common regulatory noncoding RNAs include small interfering RNAs, miRNAs, piRNAs and long-stranded noncoding RNAs [60]. MiRNAs are short endogenous noncoding RNAs that control epigenetic remodeling. Shiddiky's group designed a gold-loaded nanoporous superparamagnetic iron oxide nanocubes as a nanomaterial to improve the performance of the electrochemical biosensor to detect miRNA (Figure 9) [61]. This nanomaterial is highly porous and has an exposed gold surface, which binds to DNA or RNA more efficiently through affinity interactions, thereby increasing the binding efficiency with ribonucleotides. This nanocube has catalytic activity towards Ru(III) and is an electroactive material that can amplify detection signals through catalytic redox processes. The iron oxide material is paramagnetic, which enables magnetic separation for sample enrichment and separation. In the specific detection process, RNA is first extracted from the sample, and the target RNA is extracted using the specific interaction between the streptavidin-modified magnetic beads and the biotin-modified capture probe. The extracted miRNA is then magnetically separated and released. The released miRNA is adsorbed onto the gold-loaded iron oxide nanocube. Since RNA molecules contain a negatively charged phosphate backbone, they can bind with positively charged Ru(III) through electrostatic interaction, resulting in a large number of Ru(III) molecules on the RNA molecules. The Ru(III) redox process is coupled with the $[Fe(CN)_6]^{3-}/^{4-}$ redox system to further amplify the signal. The quantitative information of miRNA can be obtained by detecting the Ru(III) bound to the electrode surface using chronoamperometry. This method solves the problem of amplifying miRNA that cannot be achieved by other methods and can achieve highly specific and sensitive detection even at low abundance and in the presence of similar RNA interference. This method uses nanomaterials to modify the electrode interface and combines multiple electrochemical signal amplification systems to achieve ultra-sensitive detection of miRNA with a detection limit of 100 aM.

The method of modifying electrodes with iron oxide nanocubes loaded with gold can detect miRNA without amplification. MiRNA binds to iron oxide nanocubes loaded with gold through affinity interactions. The nanoparticles themselves do not have specific binding capabilities and do not have interference resistance to other RNAs. In the detection process, the target miRNA needs to be processed first by using magnetic separation and enrichment, and then the specific capture probe is used to extract and release the target miRNA.

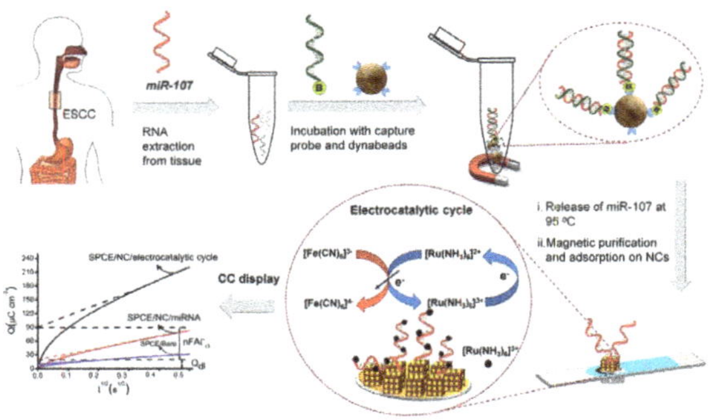

Figure 9. Schematic illustration of electrocatalytic detection of miR-107 using gold-loaded nanoporous superparamagnetic iron oxide nanocubes (Au-NPFe_2O_3NC) [61]. Copyright 2018, Elsevier.

Lim's group develops an electrochemical biosensor for the detection of 5-hmc using a combination of zinc oxide nanorods (ZnO NRs) and gold nanoparticles (Au NPs) (Figure 10) [62]. Compared to iron oxide nanocubes loaded with gold, the generated ZnO NRs loaded with Au NPs have a certain ability to distinguish target molecules. The difference in adsorption properties can effectively distinguish between target molecules and interfering molecules without additional extraction and enrichment steps. The direct generation of ZnO NRs also makes the process between nanoparticles and electrodes interference-free, resulting in a more stable electron transfer process.

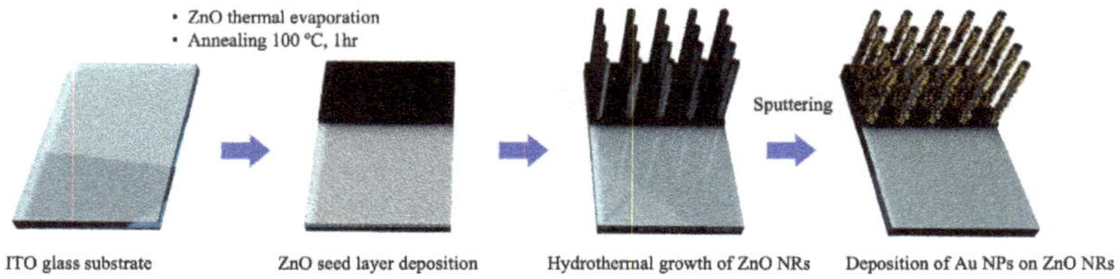

Figure 10. Schematic of the fabrication process of the Au NPs/ZnO NRs nanostructure-based sensor for 5hmC detection [62]. Copyright 2021, Wiley.

The core of the biosensor is the construction of nanostructures to detect 5-hmC, which is closely related to cancer epigenetics. AuNPs serve as excellent platforms to detect a variety of biomolecules because of their excellent biocompatibility, conductivity, and diverse surface functionalization options. The affinity of 5-hmC and 5-mC for Au is different, and their different adsorption and affinity abilities can be used for label-free electrochemical detection and quantification. ZnO NRs have good biocompatibility and are widely used in biochemical-related detection. ZnO NRs can be directly grown and synthesized on the electrode and can directly contact the electrode, avoiding interference from other substances and providing a stable connection for electron transfer, thereby improving sensing performance. They also have a large surface area that can be used for modification and connection. First, they sputtered ultrathin ZnO quantum dots as a seed layer sputtered onto indium tin oxide glass substrate. Then, they fabricate the hybrid structure by the vertical hydrothermal growth of ZnO NRs on the zinc oxide film. Finally, they modify gold nanoparticles onto the surface of the ZnO NRs resulting in the direct generation of nanomaterials on the electrode surface. Because the adsorption of 5-hmC on the gold surface is lower than that of 5-mC, a larger current signal is generated during current detection. The biosensor developed by the authors can also distinguish different levels of 5-hmC DNA very well. This method avoids selective modification of the sample and instead constructs nanostructures on the electrode surface and modifies the surface of the biosensor in a controllable manner by directly growing nanostructures on the substrate.

At the same time, the surface of the ZnO NRs can also be modified with gold nanoparticles, which can reflect the good performance and structure of the surface nanostructures. Instead of loading another capture probe on the surface, the distinction between 5-hmC and interference is based on the adsorption properties of the nanostructures.

The two methods mentioned above utilize the affinity interaction between miRNA and iron oxide nanocubes loaded with gold, as well as the adsorption property of 5-hmc on ZnO NRs modified with AuNPs, to successfully bind target molecules to the electrode surface. Both of these binding methods do not have high specificity, so pre-extraction and enrichment are required, or they can only distinguish between one interfering molecule and the target molecule. Compared to using the inherent affinity or adsorption of nanoparticles to directly bind to the analyte, nanoparticles can also provide a stable structural foundation for the electrode surface. The use of composite materials, such as Graphene oxide-Fe_3O_4-β-cyclodextrin(GO-Fe_3O_4-β-CD), can provide a stable cavity nanointerface for the electrode surface, increasing the surface area of the electrode and providing stable binding sites for specific capture antibodies. Subsequently, the specific binding of antigens and antibodies is used to detect DNA methylation of unknown sequences in the target. The specific interaction method that connects target molecules and electrodes significantly improves the specificity of detection.

Chen's group developed an electrochemical immunosensor that can use antibodies to recognize DNA bases to detect DNA methylation of unknown sequences (Figure 11) [63]. They prepared a GO-Fe_3O_4-β-CD nanocomposite material, which combines the advantages of GO-Fe_3O_4, which has good biocompatibility, a large specific surface area, and good dispersion, with β-CD, which has a hydrophobic cavity that can form stable host-guest complexes. The authors dropped the synthesized GO-Fe_3O_4-β-CD composite nanomaterial on the glassy carbon electrode surface, forming a stable nanointerface with a large number of cavities, which increased the surface area of electrode and provided many antibody binding sites. Then, they immobilized the antibody for 5-mC on the formed nanointerface. Through the specific interaction between the antigen and antibody, the target gene can be captured. After methylation DNA binds to the antibody, the $Ru(NH_3)_6^{3+}/Fe(CN)_6^{3-}$ redox amplification system is used for current amplification, significantly increasing the sensitivity of detection. The current detected by the electrochemical biosensor is related to the amount of methylated DNA. The method of modifying the interface using nanomaterials can achieve a detection limit of 0.0825 pm. Using capture antibodies instead of specific sequence DNA as capture probes can detect DNA methylation of unknown sequences, but there is an issue of poor detection specificity. The combination of electrochemical detection methods and nanotechnology can achieve high sensitivity.

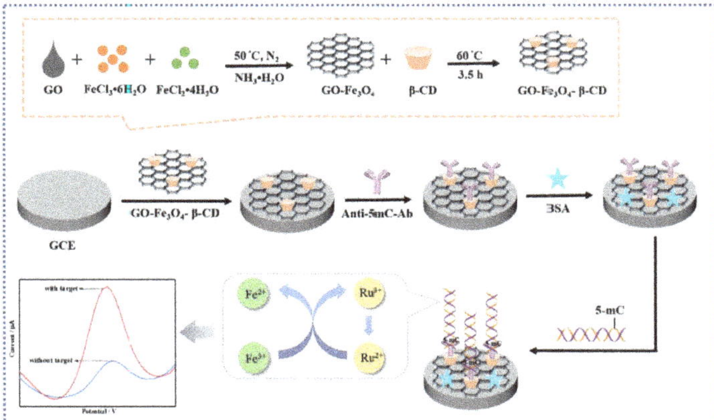

Figure 11. Schematic representation of the electrochemical immunosensor for detection of MGMT gene methylation [63]. Copyright 2022, Elsevier.

Our group used the ordered orientation, controllable spacing, good biocompatibility, and high stability of framework nucleic acids to develop electrochemical biosensors for the detection of DNA methylation (Figure 12) [64,65]. By using the nanostructure of DNA framework nucleic acids, the probe and surface can be separated from each other, making it easier to bind with the analyte in the solution, achieving high sensitivity detection of DNA methylation. We developed the method that first extracts and amplifies the circulating methylated DNA in trace plasma, with single-copy sensitivity and ultrahigh specificity. Subsequently, biotin-labeled methylation-specific primers are used for asymmetric methylation-specific PCR amplification of methylated DNA, generating a large number of biotin-labeled single-stranded amplicons. Self-assembled DNA tetrahedra modified on the gold electrode immobilize the DNA nanostructured probes on the electrode surface to capture the amplicons. Finally, horseradish peroxidase-avidin and biotin-avidin bind together, and horseradish peroxidase catalyzes the redox process to produce an electrochemical signal. The use of DNA tetrahedra-modified electrodes provides stable support for the capture probe, achieving orientational order and distance control, increasing target polarizability, and reducing non-specific adsorption of byproducts, resulting in efficient and specific hybridization with a significantly reduced signal-to-noise ratio. We used the developed method to determine the DNA methylation of the p16INK4a gene promoter in trace plasma samples from lung cancer patients, and the results demonstrated good consistency with clinical diagnosis.

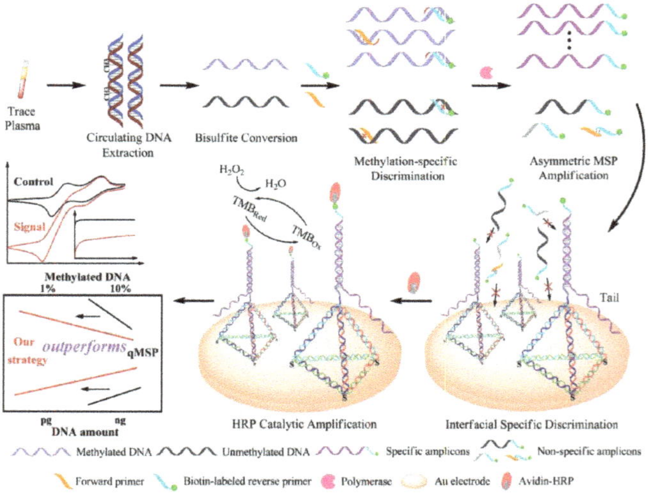

Figure 12. Schematic representation of the sequential discrimination–amplification strategy [64]. Copyright 2017, Royal Society of Chemistry.

Huang's group developed a highly sensitive electrochemical gene sensor for detecting cancer long noncoding RNA using a composite interface of graphene-like tungsten disulfide/dendritic gold nanostructures [66]. The composite nanointerface ensures the biological recognition ability of the target and capture probes. Zhang's group developed an electrochemical immunobiosensor for m6A-RNA detection using a glassy carbon electrode modified with gold nanoparticles [67]. Ai's group used a gold electrode modified with gold nanoparticles as the substrate to develop an electrochemical immunosensor for m6A-RNA detection [68]. The developed electrochemical biosensors have high sensitivity and specificity for target detection. Raouafi's group conducted simple and sensitive detection of miRNA-21 on the surface of electrode structures modified with gold nanoparticles [69].

6. Other Electrochemical Biosensing Methods

Electrochemical biosensing technology can also be combined with other techniques to detect epigenetic biomarkers. When there are many interferences in the sample, preprocessing or amplification of the sample can be chosen before electrochemical detection. Matson's group has reported using high-performance liquid chromatography (HPLC) to separate trace amounts of 7-methyl guanine and then using electrochemical detection to assess DNA methylation levels in Huntington's disease [70]. This method takes full advantage of the powerful separation capability of HPLC and the high sensitivity of electrochemical sensors to quantify trace base changes. Lu's group used electrochemical and linker-PCR technology coupling to detect the methylation of the 5′-CpG islands of the human p16Ink4a gene [71]. The PCR was used to synthesize and purify target DNA fragments, which were then detected using DNA biosensors for hybridization. This electrochemical biosensor has the advantages of fast speed, high sensitivity, and low cost. Toumazou's group applied semiconductor technology to the electrochemical detection of cancer DNA methylation, using Ion-Sensitive Field-Effect Transistors in real-time DNA methylation detection for the first time through experimental verification [72]. The introduction of semiconductor technology in disease detection can provide new solutions for disease detection.

Table 1 summarizes the methods for detecting epigenetic modifications by electrochemical biosensors mentioned above.

Table 1. Recent studies on electrochemical methods for detecting epigenetic modifications.

Epigenetic Mark	Technique	Linear Range	LOD	Ref
dsM25T DNA	EIS	-	-	[42]
m6A-RNA	EIS	0.05–200 nM	0.016 nM	[43]
5-mC DNA and acH3	EIS	DNA (4.8 pg/μL–75 ng/μL) Chromtin (8.6 pg/mL–134 ng/μL)	-	[44]
5fU-DNA	DPV	0.1–10 nM	0.075 nM	[48]
MDNA and MTase	EIS	0.1–40 U/mL	0.04 U/mL	[49]
MTase	ACV	0.1–1.0 U/mL	0.07 U/mL	[50]
m6A-RNA	SWV	0.01–0.2	0.01	[52]
5-hmC DNA	DPV	0.01–1000 pM	9.06 fM	[53]
MDNA	SWV	10 aM–to 20 pM	4 aM	[54]
miR-107	CC	100 aM–1.0 nM	100 aM	[61]
5-hmC DNA	CV	-	-	[62]
5-mC DNA	DPV	-	0.0825 pM	[63]
MDNA	CA	3–150 pg	-	[64]

ACV: Alternating current voltammetry. SWV: square wave voltammetry. CC: chroncoulmetric. CV: Cyclic Voltammetry. CA: Chronoamperometry.

7. Conclusions and Prospect

Since the association between epigenetic modifications and cancers came into the spotlight in 1964, epigenetic modifications have been the subject of investigation. In this field, the great advantages of electrochemical biosensors in terms of high sensitivity, easily operated instrumentation, and loss cost have converted them in promising tools. In this review, we have classified and summarized the methods of using electrochemical biosensors for detecting epigenetic modifications and explained the applicable scope and characteristics of different detection methods, pointing out the advantages and disadvantages of these methods.

Simple, direct, and label-free electrochemical methods can solve some epigenetic problems, but due to their insufficient detection sensitivity and accuracy, they are not suitable for detecting low amounts of RNA. The use of signal labeling, which can modify the biomarkers, improves the electrochemical biosensing sensitivity of detection. However, the labeling process is complex and can introduce bias, which can affect the detection results. Moreover, the labeling signal also has certain limitations and cannot detect markers with extremely low sensitivity. Electrochemical detection methods based on signal amplification strategies can detect low-abundance RNA, and some methods can achieve sensitivity of aM. However, signal amplification strategies require high specificity and need to be designed ingeniously, and they can only detect specific epigenetic biomarkers. The method of modifying the electrode surface with nanostructures can solve the problem of insufficient sensitivity. Nanoparticles have superior conductivity and high surface area, which can provide more binding sites to improve detection sensitivity.

It is now clear that electrochemical biosensing methods can improve the sensitivity of epigenetic analysis through signal label modification, signal amplification, or electrode surface modification with nanomaterials. However, most of the methods described in the literature were only used for proof-of-concept validation and had not yet been applied to real-life samples. Real biological samples are more complex, with excessive interference, making accurate analysis challenging. Current electrochemical biosensing methods for epigenetics mainly focus on research related to DNA methylation, with fewer reports on other epigenetic modifications and a limited research scope. In the future, it is necessary to expand the detection of other epigenetic modifications. The abundance of epigenetic biomarkers in the body is very low, and there are few detection methods that can achieve high sensitivity at low abundance. To meet the detection requirements, signal amplification and nanomaterial modification remain the main research directions for the electrochemical detection of epigenetic biomarkers. For the interference problem in the detection of real-life samples, it is necessary to consider using simple and reliable specific recognition methods while also achieving signal amplification. This will be a major focus of future research. Current detection methods are mainly focused on single epigenetic molecules, with fewer reports on the combined detection of multiple related epigenetic biomarkers. It is hoped that in the future, the high-throughput advantage of electrochemical biosensors can be developed to enable rapid and multiplex detection of epigenetic biomarkers.

With the development of technology, people have higher requirements for high sensitivity and high specificity while hoping that the detection methods can meet the needs of portability, simplicity, and ease of operation. The development of nanotechnology and electrochemical technology will promote the development of electrochemical detection of epigenetics. Ultimately, the development of these detection methods needs to solve practical problems, and the future clinical application of electrochemical detection of epigenetics can be rapidly developed.

Author Contributions: Writing—original draft, Z.H.; writing—review and editing, S.S.; figures, Y.D.; literature retrieval, Z.H., Y.D. and J.S.; funding acquisition, S.S.; supervision, S.S. and T.L. All authors have read and agreed to the published version of the manuscript.

Funding: This research was supported by the National Science and Technology Foundation (21974147; 32000921).

Institutional Review Board Statement: The study did not require ethical approval.

Informed Consent Statement: This study did not involve humans.

Data Availability Statement: Data are available upon request.

Conflicts of Interest: The authors declare no conflict of interest.

References

1. Waterland, R.A.; Michels, K.B. Epigenetic epidemiology of the developmental origins hypothesis. *Annu. Rev. Nutr.* **2007**, *27*, 363–388. [CrossRef]
2. Kim, J.K.; Samaranayake, M.; Pradhan, S. Epigenetic mechanisms in mammals. *Cell Mol. Life Sci.* **2009**, *66*, 596–612. [CrossRef]
3. Wu, Y.L.; Lin, Z.J.; Li, C.C.; Lin, X.; Shan, S.K.; Guo, B.; Zheng, M.H.; Li, F.; Yuan, L.Q.; Li, Z.H. Epigenetic regulation in metabolic diseases: Mechanisms and advances in clinical study. *Signal Transduct. Target. Ther.* **2023**, *8*, 98. [CrossRef] [PubMed]
4. Campuzano, S.; Pedrero, M.; Yánez-Sedeño, P.; Pingarrón, J.M. Advances in Electrochemical (Bio)Sensing Targeting Epigenetic Modifications of Nucleic Acids. *Electroanalysis* **2019**, *31*, 1816–1832. [CrossRef]
5. Chen, Y.; Zhou, Y.; Yin, H. Recent advances in biosensor for histone acetyltransferase detection. *Biosens. Bioelectron.* **2021**, *175*, 112880. [CrossRef]

6. Hulshoff, M.S.; Xu, X.; Krenning, G.; Zeisberg, E.M. Epigenetic Regulation of Endothelial-to-Mesenchymal Transition in Chronic Heart Disease. *Arterioscler. Thromb. Vasc. Biol.* **2018**, *38*, 1986–1996. [CrossRef]
7. Chen, X.; Sun, Y.Z.; Liu, H.; Zhang, L.; Li, J.Q.; Meng, J. RNA methylation and diseases: Experimental results, databases, Web servers and computational models. *Brief. Bioinform.* **2019**, *20*, 896–917. [CrossRef] [PubMed]
8. Xie, S.; Chen, W.; Chen, K.; Chang, Y.; Yang, F.; Lin, A.; Shu, Q.; Zhou, T.; Yan, X. Emerging roles of RNA methylation in gastrointestinal cancers. *Cancer Cell Int.* **2020**, *20*, 585. [CrossRef]
9. Zhu, H.; Zhu, H.; Tian, M.; Wang, D.; He, J.; Xu, T. DNA Methylation and Hydroxymethylation in Cervical Cancer: Diagnosis, Prognosis and Treatment. *Front. Genet.* **2020**, *11*, 347. [CrossRef]
10. Chen, Y.; Hong, T.; Wang, S.; Mo, J.; Tian, T.; Zhou, X. Epigenetic modification of nucleic acids: From basic studies to medical applications. *Chem. Soc. Rev.* **2017**, *46*, 2844–2872. [CrossRef]
11. Bhaumik, S.R.; Smith, E.; Shilatifard, A. Covalent modifications of histones during development and disease pathogenesis. *Nat. Struct. Mol. Biol.* **2007**, *14*, 1008–1016. [CrossRef] [PubMed]
12. Bird, A. Perceptions of epigenetics. *Nature* **2007**, *447*, 396–398. [CrossRef] [PubMed]
13. Shi, Y.J.; Matson, C.; Lan, F.; Iwase, S.; Baba, T.; Shi, Y. Regulation of LSD1 histone demethylase activity by its associated factors. *Mol. Cell* **2005**, *19*, 857–864. [CrossRef]
14. Jones, P.A.; Baylin, S.B. The fundamental role of epigenetic events in cancer. *Nat. Rev. Genet.* **2002**, *3*, 415–428. [CrossRef] [PubMed]
15. Lu, X.; Wang, L.; Yu, C.; Yu, D.; Yu, G. Histone Acetylation Modifiers in the Pathogenesis of Alzheimer's Disease. *Front. Cell. Neurosci.* **2015**, *9*, 226. [CrossRef]
16. Yang, M.; Zhang, Y.; Ren, J. Acetylation in cardiovascular diseases: Molecular mechanisms and clinical implications. *Biochim. Biophys. Acta Mol. Basis Dis.* **2020**, *1866*, 165836. [CrossRef]
17. Chen, D.; Wu, Y.; Tilley, R.D.; Gooding, J.J. Rapid and ultrasensitive electrochemical detection of DNA methylation for ovarian cancer diagnosis. *Biosens. Bioelectron.* **2022**, *206*, 114126. [CrossRef]
18. Campuzano, S.; Barderas, R.; Pedrero, M.; Yanez-Sedeno, P.; Pingarron, J.M. Electrochemical biosensing to move forward in cancer epigenetics and metastasis: A review. *Anal. Chim. Acta* **2020**, *1109*, 169–190. [CrossRef]
19. Chen, C.; Liu, J. Histone acetylation modifications: A potential targets for the diagnosis and treatment of papillary thyroid cancer. *Front. Oncol.* **2022**, *12*, 1053618. [CrossRef]
20. Campuzano, S.; Pingarrón, J.M. Electrochemical Sensing of Cancer-related Global and Locus-specific DNA Methylation Events. *Electroanalysis* **2018**, *30*, 1201–1216. [CrossRef]
21. Diwadkar, V.A.; Bustamante, A.; Rai, H.; Uddin, M. Epigenetics, stress, and their potential impact on brain network function: A focus on the schizophrenia diatheses. *Front. Psychiatry* **2014**, *5*, 71. [CrossRef]
22. Eshraghi, A.A.; Liu, G.; Kay, S.S.; Eshraghi, R.S.; Mittal, J.; Moshiree, B.; Mittal, R. Epigenetics and Autism Spectrum Disorder: Is There a Correlation? *Front. Cell. Neurosci.* **2018**, *12*, 78. [CrossRef]
23. Horiuchi, K.Y.; Eason, M.M.; Ferry, J.J.; Planck, J.L.; Walsh, C.P.; Smith, R.F.; Howitz, K.T.; Ma, H. Assay development for histone methyltransferases. *Assay. Drug Dev. Technol.* **2013**, *11*, 227–236. [CrossRef]
24. Peng, S.-Y.; Zhang, J.; Tian, M.-P.; Wang, Z.-L.; Shen, H.-Q. Determination of Global DNA Methylation in Biological Samples by Liquid Chromatography-Tandem Mass Spectrometry. *Chin. J. Anal. Chem.* **2012**, *40*, 1201–1206. [CrossRef]
25. Rozhon, W.; Baubec, T.; Mayerhofer, J.; Mittelsten Scheid, O.; Jonak, C. Rapid quantification of global DNA methylation by isocratic cation exchange high-performance liquid chromatography. *Anal. Biochem.* **2008**, *375*, 354–360. [CrossRef] [PubMed]
26. Ye, X.; Zhang, L.; Chen, B.; Li, J.; Yang, Q.; Huang, Q.; Zhang, J.; Gao, Y.; Li, Z.; Cai, C. A quantitative method for detecting DNA methylation over targeted genomic regions using isotope dilution liquid chromatography tandem mass spectrometry. *Talanta* **2017**, *169*, 136–140. [CrossRef]
27. Ghadiali, J.E.; Lowe, S.B.; Stevens, M.M. Quantum-dot-based FRET detection of histone acetyltransferase activity. *Angew. Chem. Int. Ed. Engl.* **2011**, *50*, 3417–3420. [CrossRef]
28. Choy, L.Y.L.; Peng, W.; Jiang, P.; Cheng, S.H.; Yu, S.C.Y.; Shang, H.; Olivia Tse, O.Y.; Wong, J.; Wong, V.W.S.; Wong, G.L.H.; et al. Single-Molecule Sequencing Enables Long Cell-Free DNA Detection and Direct Methylation Analysis for Cancer Patients. *Clin. Chem.* **2022**, *68*, 1151–1163. [CrossRef] [PubMed]
29. Han, Z.W.; Ma, F.; Zhang, C.Y. Methylation-sensitive transcription-enhanced single-molecule biosensing of DNA methylation in cancer cells and tissues. *Anal. Chim. Acta* **2023**, *1251*, 340996. [CrossRef]
30. Shipony, Z.; Marinov, G.K.; Swaffer, M.P.; Sinnott-Armstrong, N.A.; Skotheim, J.M.; Kundaje, A.; Greenleaf, W.J. Long-range single-molecule mapping of chromatin accessibility in eukaryotes. *Nat. Methods* **2020**, *17*, 319–327. [CrossRef] [PubMed]
31. Bartosik, M.; Hrstka, R. Bioelectrochemistry of nucleic acids for early cancer diagnostics—Analysis of DNA methylation and detection of microRNAs. *Rev. Anal. Chem.* **2017**, *36*, 20160022. [CrossRef]
32. Wang, G.L.; Luo, H.Q.; Li, N.B. Gold nanorods-based FRET assay for ultrasensitive detection of DNA methylation and DNA methyltransferase activity. *Analyst* **2014**, *139*, 4572–4577. [CrossRef]
33. Wei, W.; Gao, C.; Xiong, Y.; Zhang, Y.; Liu, S.; Pu, Y. A fluorescence method for detection of DNA and DNA methylation based on graphene oxide and restriction endonuclease HpaII. *Talanta* **2015**, *131*, 342–347. [CrossRef]

34. Pan, S.; Xu, J.; Shu, Y.; Wang, F.; Xia, W.; Ding, Q.; Xu, T.; Zhao, C.; Zhang, M.; Huang, P.; et al. Double recognition of oligonucleotide and protein in the detection of DNA methylation with surface plasmon resonance biosensors. *Biosens. Bioelectron.* **2010**, *26*, 850–853. [CrossRef]
35. Li, X.; Song, T.; Guo, X. DNA methylation detection with end-to-end nanorod assembly-enhanced surface plasmon resonance. *Analyst* **2015**, *140*, 6230–6233. [CrossRef] [PubMed]
36. Ishwar, D.; Venkatakrishnan, K.; Tan, B.; Haldavnekar, R. DNA Methylation Signatures of Tumor-Associated Natural Killer Cells with Self-Functionalized Nanosensor Enable Colorectal Cancer Diagnosis. *Nano Lett.* **2023**, *23*, 4142–4151. [CrossRef]
37. Zhang, Y.; Zhan, D.S.; Xu, X.Y.; Zhang, Z.; Hafez, M.E.; He, Y.; Li, Y.; Li, D.W. Label-free detection of DNA methylation by surface-enhanced Raman spectroscopy using zirconium-modified silver nanoparticles. *Talanta* **2023**, *253*, 123941. [CrossRef] [PubMed]
38. Zhang, Q.; Wu, Y.; Xu, Q.; Ma, F.; Zhang, C.Y. Recent advances in biosensors for in vitro detection and in vivo imaging of DNA methylation. *Biosens. Bioelectron.* **2021**, *171*, 112712. [CrossRef] [PubMed]
39. Li, C.C.; Wang, Z.Y.; Wang, L.J.; Zhang, C.Y. Biosensors for epigenetic biomarkers detection: A review. *Biosens. Bioelectron.* **2019**, *144*, 111695. [CrossRef]
40. Krejcova, L.; Richtera, L.; Hynek, D.; Labuda, J.; Adam, V. Current trends in electrochemical sensing and biosensing of DNA methylation. *Biosens. Bioelectron.* **2017**, *97*, 384–399. [CrossRef]
41. Hossain, T.; Mahmudunnabi, G.; Masud, M.K.; Islam, M.N.; Ooi, L.; Konstantinov, K.; Hossain, M.S.A.; Martinac, B.; Alici, G.; Nguyen, N.T.; et al. Electrochemical biosensing strategies for DNA methylation analysis. *Biosens. Bioelectron.* **2017**, *94*, 63–73. [CrossRef]
42. Zhu, B.; Booth, M.A.; Shepherd, P.; Sheppard, A.; Travas-Sejdic, J. Distinguishing cytosine methylation using electrochemical, label-free detection of DNA hybridization and ds-targets. *Biosens. Bioelectron.* **2015**, *64*, 74–80. [CrossRef]
43. Dai, T.; Pu, Q.; Guo, Y.; Zuo, C.; Bai, S.; Yang, Y.; Yin, D.; Li, Y.; Sheng, S.; Tao, Y.; et al. Analogous modified DNA probe and immune competition method-based electrochemical biosensor for RNA modification. *Biosens. Bioelectron.* **2018**, *114*, 72–77. [CrossRef]
44. Teixeira, S.R.; Abreu, C.M.; Parkes, L.; Davies, J.; Yao, S.; Sawhney, M.A.; Margarit, L.; Gonzalez, D.; Pinto, I.M.; Francis, L.W.; et al. Direct monitoring of breast and endometrial cancer cell epigenetic response to DNA methyltransferase and histone deacetylase inhibitors. *Biosens. Bioelectron.* **2019**, *141*, 111386. [CrossRef] [PubMed]
45. Xu, Z.; Wang, M.; Zhou, T.; Yin, H.; Ai, S. Electrochemical biosensing method for the detection of DNA methylation and assay of the methyltransferase activity. *Sens. Actuators B Chem.* **2013**, *178*, 412–417. [CrossRef]
46. Wang, G.L.; Zhou, L.Y.; Luo, H.Q.; Li, N.B. Electrochemical strategy for sensing DNA methylation and DNA methyltransferase activity. *Anal. Chim. Acta* **2013**, *768*, 76–81. [CrossRef] [PubMed]
47. Hu, Y.; Chen, S.; Han, Y.; Chen, H.; Wang, Q.; Nie, Z.; Huang, Y.; Yao, S. Unique electrocatalytic activity of a nucleic acid-mimicking coordination polymer for the sensitive detection of coenzyme A and histone acetyltransferase activity. *Chem. Commun.* **2015**, *51*, 17611–17614. [CrossRef] [PubMed]
48. Tang, J.; Zou, G.; Chen, C.; Ren, J.; Wang, F.; Chen, Z. Highly Selective Electrochemical Detection of 5-Formyluracil Relying on (2-Benzimidazolyl) Acetonitrile Labeling. *Anal. Chem.* **2021**, *93*, 16439–16446. [CrossRef] [PubMed]
49. Yin, H.; Zhou, Y.; Xu, Z.; Chen, L.; Zhang, D.; Ai, S. An electrochemical assay for DNA methylation, methyltransferase activity and inhibitor screening based on methyl binding domain protein. *Biosens. Bioelectron.* **2013**, *41*, 492–497. [CrossRef]
50. Su, J.; He, X.; Wang, Y.; Wang, K.; Chen, Z.; Yan, G. A sensitive signal-on assay for MTase activity based on methylation-responsive hairpin-capture DNA probe. *Biosens. Bioelectron.* **2012**, *36*, 123–128. [CrossRef]
51. Chen, Y.; Yi, H.; Xiang, Y.; Yuan, R. Commercial glucometer as signal transducer for simple evaluation of DNA methyltransferase activity and inhibitors screening. *Anal. Chim. Acta* **2018**, *1001*, 18–23. [CrossRef]
52. Pu, Q.; Ye, Y.; Hu, J.; Xie, C.; Zhou, X.; Yu, H.; Liao, F.; Jiang, S.; Jiang, L.; Xie, G.; et al. XNA probe and CRISPR/Cas12a-powered flexible fluorescent and electrochemical dual-mode biosensor for sensitive detection of m6A site-specific RNA modification. *Talanta* **2023**, *252*, 123754. [CrossRef]
53. Cui, L.; Hu, J.; Wang, M.; Li, C.C.; Zhang, C.Y. Label-Free and Immobilization-Free Electrochemical Magnetobiosensor for Sensitive Detection of 5-Hydroxymethylcytosine in Genomic DNA. *Anal. Chem.* **2019**, *91*, 1232–1236. [CrossRef]
54. Zhou, H.; Shen, H.; Dou, B.; Feng, Q.; Han, X.; Wang, P. Construction of a sensitive ratiometric electrochemical sensing platform for DNA methylation detection based on the design of multistep DNA amplification circuits. *Sens. Actuators B Chem.* **2022**, *370*, 132491. [CrossRef]
55. Zhou, Y.; Yang, Z.; Li, X.; Wang, Y.; Yin, H.; Ai, S. Electrochemical biosensor for detection of DNA hydroxymethylation based on glycosylation and alkaline phosphatase catalytic signal amplification. *Electrochim. Acta* **2015**, *174*, 647–652. [CrossRef]
56. Yu, H.; Pu, Q.; Weng, Z.; Zhou, X.; Li, J.; Yang, Y.; Luo, W.; Guo, Y.; Chen, H.; Wang, D.; et al. DNAzyme based three-way junction assay for antibody-free detection of locus-specific N(6)-methyladenosine modifications. *Biosens. Bioelectron.* **2021**, *194*, 113625. [CrossRef] [PubMed]
57. Liu, S.; Zhang, X.; Zhao, K. Methylation-specific electrochemical biosensing strategy for highly sensitive and quantitative analysis of promoter methylation of tumor-suppressor gene in real sample. *J. Electroanal. Chem.* **2016**, *773*, 63–68. [CrossRef]
58. Chen, S.; Dou, Y.; Zhao, Z.; Li, F.; Su, J.; Fan, C.; Song, S. High-Sensitivity and High-Efficiency Detection of DNA Hydroxymethylation in Genomic DNA by Multiplexing Electrochemical Biosensing. *Anal. Chem.* **2016**, *88*, 3476–3480. [CrossRef]

59. Cheng, W.; Ma, J.; Zhang, Y.; Xu, C.; Zhang, Z.; Hu, L.; Li, J. Bio-inspired construction of a semi-artificial enzyme complex for detecting histone acetyltransferases activity. *Analyst* **2020**, *145*, 613–618. [CrossRef]
60. Islam, M.N.; Masud, M.K.; Nguyen, N.T.; Gopalan, V.; Alamri, H.R.; Alothman, Z.A.; Hossain, M.S.A.; Yamauchi, Y.; Lamd, A.K.; Shiddiky, M.J.A. Gold-loaded nanoporous ferric oxide nanocubes for electrocatalytic detection of microRNA at attomolar level. *Biosens. Bioelectron.* **2018**, *101*, 275–281. [CrossRef] [PubMed]
61. Wu, G.; Jose, P.A.; Zeng, C. Noncoding RNAs in the Regulatory Network of Hypertension. *Hypertension* **2018**, *72*, 1047–1059. [CrossRef] [PubMed]
62. Sideeq Bhat, K.; Kim, H.; Alam, A.; Ko, M.; An, J.; Lim, S. Rapid and Label-Free Detection of 5-Hydroxymethylcytosine in Genomic DNA Using an Au/ZnO Nanorods Hybrid Nanostructure-Based Electrochemical Sensor. *Adv. Healthc. Mater.* **2021**, *10*, e2101193. [CrossRef] [PubMed]
63. Yang, H.; Ren, J.; Zhao, M.; Chen, C.; Wang, F.; Chen, Z. Novel electrochemical immunosensor for O(6)-methylguanine-DNA methyltransferase gene methylation based on graphene oxide-magnetic nanoparticles-beta-cyclodextrin nanocomposite. *Bioelectrochemistry* **2022**, *146*, 108111. [CrossRef] [PubMed]
64. Wang, X.; Chen, F.; Zhang, D.; Zhao, Y.; Wei, J.; Wang, L.; Song, S.; Fan, C.; Zhao, Y. Single copy-sensitive electrochemical assay for circulating methylated DNA in clinical samples with ultrahigh specificity based on a sequential discrimination-amplification strategy. *Chem. Sci.* **2017**, *8*, 4764–4770. [CrossRef]
65. Chen, S.; Su, J.; Zhao, Z.; Shao, Y.; Dou, Y.; Li, F.; Deng, W.; Shi, J.; Li, Q.; Zuo, X.; et al. DNA Framework-Supported Electrochemical Analysis of DNA Methylation for Prostate Cancers. *Nano Lett.* **2020**, *20*, 7028–7035. [CrossRef]
66. Li, X.; Peng, G.; Cui, F.; Qiu, Q.; Chen, X.; Huang, H. Double determination of long noncoding RNAs from lung cancer via multi-amplified electrochemical genosensor at sub-femtomole level. *Biosens. Bioelectron.* **2018**, *113*, 116–123. [CrossRef]
67. Yin, H.; Zhou, Y.; Yang, Z.; Guo, Y.; Wang, X.; Ai, S.; Zhang, X. Electrochemical immunosensor for N6-methyladenosine RNA modification detection. *Sens. Actuators B Chem.* **2015**, *221*, 1–6. [CrossRef]
68. Li, Z.; Li, B.; Yin, H.; Zhang, Q.; Wang, H.; Fan, H.; Ai, S. Electrochemical immunosensor based on hairpin DNA probe for specific detection of N6-methyladenosine RNA. *J. Electroanal. Chem.* **2017**, *804*, 192–198. [CrossRef]
69. Zouari, M.; Campuzano, S.; Pingarron, J.M.; Raouafi, N. Amperometric Biosensing of miRNA-21 in Serum and Cancer Cells at Nanostructured Platforms Using Anti-DNA-RNA Hybrid Antibodies. *ACS Omega* **2018**, *3*, 8923–8931. [CrossRef]
70. Thomas, B.; Matson, S.; Chopra, V.; Sun, L.; Sharma, S.; Hersch, S.; Rosas, H.D.; Scherzer, C.; Ferrante, R.; Matson, W. A novel method for detecting 7-methyl guanine reveals aberrant methylation levels in Huntington disease. *Anal. Biochem.* **2013**, *436*, 112–120. [CrossRef]
71. Hou, P.; Ji, M.; Ge, C.; Shen, J.; Li, S.; He, N.; Lu, Z. Detection of methylation of human p16(Ink4a) gene 5′-CpG islands by electrochemical method coupled with linker-PCR. *Nucleic Acids Res.* **2003**, *31*, e92. [CrossRef] [PubMed]
72. Kalofonou, M.; Toumazou, C. Semiconductor technology for early detection of DNA methylation for cancer: From concept to practice. *Sens. Actuators B Chem.* **2013**, *178*, 572–580. [CrossRef]

chemosensors

Communication

Determination of Extra- and Intra-Cellular pH Using Characteristic Absorption of Water by Near-Infrared Spectroscopy

Jiani Li [1,2], Fanfan Liang [1,2], Li Han [1,2], Xiaoxuan Yu [1], Dingbin Liu [1], Wensheng Cai [1,2] and Xueguang Shao [1,2,*]

1 Research Center for Analytical Sciences, Tianjin Key Laboratory of Biosensing and Molecular Recognition, State Key Laboratory of Medicinal Chemical Biology, College of Chemistry, Nankai University, Tianjin 300071, China; jnli@mail.nankai.edu.cn (J.L.); 2120220973@mail.nankai.edu.cn (F.L.); 1120180241@mail.nankai.edu.cn (L.H.); 2120210817@mail.nankai.edu.cn (X.Y.); liudb@nankai.edu.cn (D.L.); wscai@nankai.edu.cn (W.C.)

2 Haihe Laboratory of Sustainable Chemical Transformations, Tianjin 300192, China

* Correspondence: xshao@nankai.edu.cn; Tel.: +86-022-23503430

Abstract: Accurate determination of extra-cellular pH (pHe) and intra-cellular pH (pHi) is important to cancer diagnosis and treatment because tumor cells exhibit a lower pHe and a slightly higher pHi than normal cells. In this work, the characteristic absorption of water in the near-infrared (NIR) region was utilized for the determination of pHe and pHi. Dulbecco's modified eagle medium (DMEM) and bis (2-ethylhexyl) succinate sodium sulfonate reverse micelles (RM) were employed to simulate the extra- and intra-cellular fluids, respectively. Continuous wavelet transform (CWT) was used to enhance the resolution of the spectra. Quantitative models for pHe and pHi were established using partial least squares (PLS) regression, producing relative errors of validation samples in a range of −0.74–2.07% and −1.40–0.83%, respectively. Variable selection was performed, and the correspondence between the selected wavenumbers and water structures was obtained. Therefore, water with different hydrogen bonds may serve as a good probe to sense pH within biological systems.

Keywords: near-infrared spectroscopy; water structure; extra-cellular pH; intra-cellular pH; partial least squares regression

Citation: Li, J.; Liang, F.; Han, L.; Yu, X.; Liu, D.; Cai, W.; Shao, X. Determination of Extra- and Intra-Cellular pH Using Characteristic Absorption of Water by Near-Infrared Spectroscopy. *Chemosensors* **2023**, *11*, 425. https://doi.org/10.3390/chemosensors11080425

Academic Editor: Matjaž Finšgar

Received: 23 June 2023
Revised: 19 July 2023
Accepted: 29 July 2023
Published: 1 August 2023

1. Introduction

Cells are the basic structure and functional unit of organisms, and pH is one of the most significant physiological parameters that governs cellular activities and behaviors [1,2]. Even a slight change in pH can impact the activity of biological enzymes, which affects life processes such as ion transport, nucleic acid formation, and the release of metabolites [3]. Compared to normal cells, tumor cells exhibit a lower extra-cellular pH (pHe) and a slightly higher intra-cellular pH (pHi) [4,5]. Consequently, monitoring the pHe and pHi not only provides valuable insights into cellular behavior but also plays an important role in guiding the clinical diagnosis, treatment, and evaluation of the effectiveness of anticancer drugs [6,7].

Water is the primary component of living cells. Water molecules form a complex and flexible hydrogen bond network [8], which is highly sensitive to external factors such as pH, temperature, and solute concentration [9]. Near-infrared (NIR) spectroscopy has emerged as a powerful tool to investigate water structures [10–12] and has been well studied in experimental [13–15] and theoretical calculations [16–18]. Six absorption peaks were obtained from derivative spectra, corresponding to the rotational vibrational peak of water molecules (S_r) and the characteristic peaks of water molecules with zero to four hydrogen bonds (S_0–S_4) [19]. Gaussian fitting of NIR spectra was performed using a knowledge-based genetic algorithm, and the spectral features of nine water structures were obtained [20]. In addition, water can be applied as a probe to sense the structural variations of solutes [21–23]. Water structures in embryos with different bioactivities were investigated [24]. An increase in weakly hydrogen-bonded water molecules at lower embryonic bioactivity was found, which coincides with the changes in protein secondary structure from α-helix to β-sheet. Tsenkova et al. [25] explored the mechanism of prion protein (PrP) fibrosis induced by Mn and Cu ions through the spectral changes of water, and the PrP-Cu complex was found to prevent Cu from interacting with water, leading to an increase in protein stability, whereas PrP-Mn does not [25]. Water structures during the aggregation of the R2/wt fragment of the tau protein were investigated. It was revealed that the dissociation of hydrated water near the NH group on R2/wt facilitates the formation of β-sheet between the amide groups [26]. The nucleation of insulin fibers [27] and the diagnosis of soybean mosaic disease [28] were also studied by analyzing changes in the NIR spectra of water. In addition, water can also be used as a probe for the quantitative determinations of pH, temperature, and concentration [29–31]. Spectral changes of water were captured from temperature-dependent NIR spectra using multilevel simultaneous component analysis (MSCA). The quantitative spectra–temperature relationship (QSTR) and the quantitative spectra–concentration relationship (QSCR) were established [32]. Furthermore, MSCA was also utilized to quantitatively detect the isoelectric point, concentration, and temperature by extracting the spectral changes of proline solution [33].

In this work, the characteristic absorption of water was utilized for the determination of pHe and pHi by NIR spectra. Dulbecco's modified eagle medium (DMEM) and bis (2-ethylhexyl) succinate sodium sulfonate reverse micelles (RM) were used to simulate extra- and intra-cellular fluids, respectively. The quantitative models for pHi and pHe were established using partial least squares (PLS) regression, and pH-dependent wavenumbers were selected to study the relationship between pH and the structures of water.

2. Materials and Methods

2.1. Reagents and Sample Preparation

DMEM (D5523) was produced from Sigma Aldrich. Fetal bovine serum (FBS) and penicillin-streptomycin solution (PS) were produced from Invitrogen Gbico. $NaH_2PO_4 \cdot 2H_2O$ (99.9% purity) and $Na_2HPO_4 \cdot 12H_2O$ (99% purity) were obtained from Macklin and Aladdin, respectively. Bis (2-ethylhexyl) succinate sodium sulfonate (AOT, 99% purity) from Klamar and isooctane (IO, >99.5% purity) from Concord were used. Water was purified using the Advantage A10TM ultra-pure water instrument (Millipore, Molsheim, France).

Three groups of samples were prepared in this study. The first group consists of 38 DMEM samples with different pH values. Firstly, 50 g of DMEM powder was dissolved in ultra-pure water, followed by the addition of 10 mL PS and 50 mL FBS to prepare a DMEM solution with a total volume of 1000 mL. Secondly, phosphate buffer (PB) solutions with different pH values were obtained by mixing 0.1 $mol \cdot L^{-1}$ Na_2HPO_4 and NaH_2PO_4 in different volume ratios. Finally, a total of 38 DMEM samples with different pH values were prepared by mixing PB with the DMEM solution in the volume ratio of V[PB]:V[DMEM] = 1:4.

The second group consists of 34 RM samples. Firstly, 0.2 $mol \cdot L^{-1}$ AOT/IO solution was prepared by dissolving AOT in IO. Subsequently, a total of 34 RM samples encapsulating PB solutions with different pH values were prepared by mixing PB with AOT/IO solution in the molar ratio of n[PB]:n[AOT] = 30.

The third group consists of three DMEM samples that have been used for cell cultivation. Three kinds of cells, namely 5637, N_2A, and HUEVC, were cultured using the prepared DMEM solution. After 24 h of incubation, three DMEM samples were collected separately by centrifuging the cells.

2.2. Spectral Measurement

All NIR spectra were measured using a Vertex 70 spectrometer (Bruker Optics Inc., Ettlingen, Germany) equipped with a tungsten-halogen light source and InGaAs detector. The measurements were conducted using a quartz cuvette with an optical path of 0.5 mm. The measured spectra ranged from 12,000 to 4000 cm^{-1} with a resolution of 4 cm^{-1}. The background spectrum was collected using an empty cuvette. To improve the signal-to-noise ratio, the number of scans was set to 64. The temperature control of the instrument was achieved using a temperature controller (Bruker Spectral Instruments, Ettlingen, Germany) equipped with a 2216e temperature module, ensuring a constant temperature of 30 °C throughout the measurement.

2.3. pH Measurement

The pH values of the samples were measured using a PHS-3E pH meter (Shanghai Leici Scientific Instrument Co., Ltd., Shanghai, China) equipped with an E-301-QC pH triple composite electrode. Before measuring pH of the samples, a calibration process was conducted using standard solutions of pH 4.00, pH 6.86, and pH 9.18.

The pH of the first and third group samples was directly measured using the pH meter, considering it as pHe. For the second group, the pH of PB solution was measured before being encapsulated within AOT RM, considering it as pHi. The variation in pHi before and after encapsulation is small enough to be neglected [34,35]. The measurement was performed three times for each sample, and the average value was taken as the final result.

The pH values of the first and second groups are from 8.03 to 5.78 with a mean and standard deviation of 6.89, 0.66, 6.90, and 0.67, respectively. The pH values of the three DMEM samples are 6.49, 6.32, and 6.22, respectively.

2.4. Calculation Method

Continuous wavelet transform (CWT) was utilized to enhance the spectral resolution and remove the background in the spectra [36]. A Symmlet filter with a vanishing moment of 4 (Sym4) was adopted, and the scale parameter was set to 40 for enhancing the smooth effect. CWT with Sym4 filter is an approximate equivalence of the fourth derivative [12,37].

PLS regression was used to establish the quantitative models. The first and second groups were used separately to develop models for pHe and pHi, respectively. The third group served as an external validation for the pHe model. The performances of models were evaluated by correlation coefficient (Rcv) and root mean squared error of cross validation (RMSECV). The prediction accuracy was evaluated by correlation coefficient (R) and root mean squared error of prediction set (RMSEP) [12,38,39]. The calibration and validation sets were divided by using the even distribution of the pH values. The ratio of the calibration set to the validation set was around 5:1.

Monte Carlo uninformative variable elimination (MC-UVE), randomization test (RT), and competitive adaptive reweighted sampling (CARS) were employed for the selection of variables, aiming to reduce the interference of uninformative variables and select the pH-dependent variables [40–42].

3. Results and Discussion

3.1. Spectral Analysis

Figure 1 shows the NIR spectra of water, DMEM, AOT/IO, and RM. In the spectra of water and DMEM, two main absorption peaks can be observed at 6900 and 5200 cm^{-1}, corresponding to the first overtone of OH stretching and the combination of OH bending and stretching vibrations, respectively [19]. The peak around 6900 cm^{-1} is an overlapping absorption arising from OH groups in water molecules with different hydrogen bonds, which is sensitive to changes in concentration, temperature, and pH. Therefore, the subsequent analysis was mainly focused on the peak at 6900 cm^{-1}. In the spectra of AOT/IO and RM, the water content is significantly reduced, which leads to weaker signals at 6900 and 5200 cm^{-1}. The peaks around 5800 and 4312 cm^{-1} contain the vibrational absorption of CH groups in AOT and IO [43]. It is difficult to perform further analysis due to the broadness of the peaks. Resolution enhancement is needed.

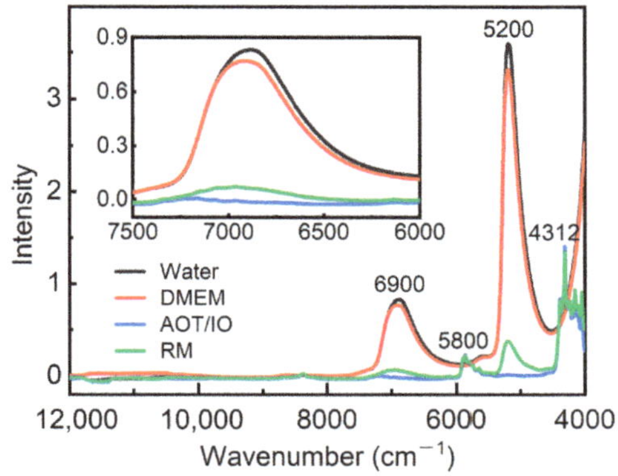

Figure 1. NIR spectra of water, DMEM, AOT/IO, and RM. The inset shows the enlarged spectra in the range of 7500–6000 cm^{-1}.

CWT was used to enhance the resolution of spectra and remove the background [36,37]. Figure 2 presents the transformed spectra of water, DMEM, and water inside RM. The insert exhibits the enlarged spectra in the range of 7500–6500 cm^{-1}, which mainly corresponds to the characteristic absorption of water. In the spectra of water and DMEM, three peaks were observed at 7110, 6976, and 6840 cm^{-1}, which can be assigned to the absorption of OH in non-bonded (NHB), weakly hydrogen-bonded (WHB), and strongly hydrogen-bonded (SHB) water molecules, respectively [44,45]. The intensity of the peak at 7110 cm^{-1} in the spectrum of DMEM slightly exceeds that of water, while the intensity of the peak at 6976 cm^{-1} is slightly lower than water, which may be attributed to the hydrophobic interaction of organic compounds in DMEM [46]. In the RM system, the weak absorption of water is covered by the strong C-H absorption. To extract the absorption of water in RM, the transformed spectrum of AOT/IO was subtracted from the spectrum of RM. The blue curve in Figure 2 represents the amplified difference spectra, which mainly reflect the absorption of water inside RM. The peaks at 7110, 6976, and 6840 cm^{-1} remain and correspond to the absorption of NHB, WHB, and SHB water inside RM, respectively.

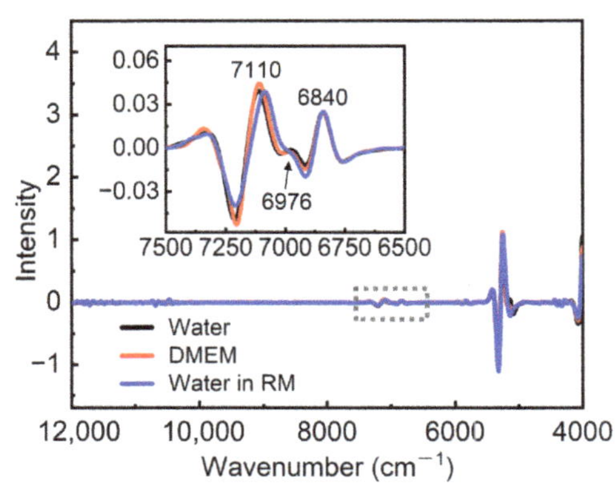

Figure 2. Transformed spectra of water, DMEM, and water inside RM. The inset shows the enlarged spectra in the range of 7500–6500 cm^{-1}.

3.2. Quantitative Models and Validation

The quantitative relationships between the spectra of DMEM and pHe, as well as the spectra of water in RM and pHi, were established using PLS regression. Different spectral ranges were used to develop optimal PLS models, including the original spectra in the full range and in the range of 7500–6500 cm^{-1} and CWT spectra in the full range and in the range of 7500–6500 cm^{-1}. Variable selection was performed by MC-UVE, RT, and CARS. Table 1 summarizes the PLS models and corresponding prediction results. Generally, a better model should have a higher value of R/Rcv and a lower value of RMSEP/RMSECV [12]. As shown in Table 1, for both pHe and pHi prediction, the CWT-MC-UVE, CWT-RT, and CWT-CARS models exhibit better performance compared to other models. In this study, the CWT-CARS models were adopted because of the fewest variables and good predictive performance.

Table 1. Parameters for the calibration and validation performance of PLS models.

Methods	Spectral Range (cm^{-1})	*n*Var [1]	*n*LV [2]	Rcv	RMSECV	R	RMSEP
pHe							
None	Full spectra	4148	7	0.9775	0.14	0.9787	0.15
None	7500–6500	520	7	0.9816	0.12	0.9903	0.10
CWT	Full spectra	4148	7	0.9743	0.15	0.9731	0.17
CWT	7500–6500	520	7	0.9853	0.11	0.9848	0.11
CWT-MC-UVE	7500–6500	95	4	0.9853	0.11	0.9910	0.09
CWT-RT	7500–6500	101	4	0.9861	0.11	0.9906	0.09
CWT-CARS	7500–6500	28	4	0.9890	0.08	0.9854	0.11
pHi							
None	Full spectra	4148	4	0.9046	0.26	0.9240	0.24
None	7500–6500	520	4	0.9623	0.17	0.9611	0.15
CWT	Full spectra	4148	4	0.9628	0.17	0.9626	0.17
CWT	7500–6500	520	5	0.9639	0.16	0.9630	0.17
CWT-MC-UVE	7500–6500	119	5	0.9787	0.13	0.9795	0.12
CWT-RT	7500–6500	127	4	0.9802	0.12	0.9828	0.12
CWT-CARS	7500–6500	21	4	0.9834	0.11	0.9807	0.12

[1] Number of spectral variables; [2] number of latent variables.

To evaluate the prediction performance of the pHe and pHi models, internal validation was performed using 16 and 14 samples, respectively. The relationships between the reference and predicted pH are shown in Figure 3. It can be seen that both quantitative models exhibit a strong linear correlation between the reference and predicted values, despite the slight deviation between the fitted line and the diagonal line. The values of R and RMSEP for the pHe model are 0.9854 and 0.11, respectively. The values of R and RMSEP for the pHi model are 0.9807 and 0.12, respectively. For a more detailed analysis of the prediction results, the reference pH, predicted pH, absolute error, and relative error for each validation sample were listed in Table 2. For the pHe model, the absolute errors of the samples range from −0.05 to 0.16, with corresponding relative errors varying from −0.74% to 2.07%. For the samples repeated three times, the maximum value of the relative standard deviation (RSD) is 1.70%. For the pHi model, the absolute errors of the samples range from −0.09 to 0.06, with relative errors varying from −1.40% to 0.83%. For the samples repeated three times, the maximum value of RSD was 1.47%. The values of relative errors and RSD indicate that both models have good prediction performance and excellent repeatability.

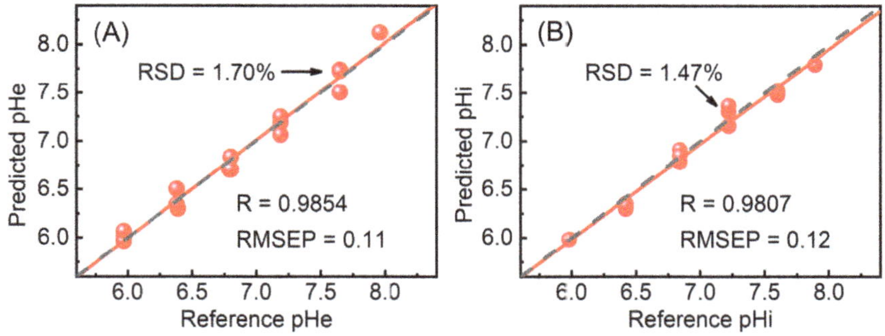

Figure 3. Relationships between the reference and predicted values of the pHe (**A**) and pHi (**B**) models. The red line represents the linear fit of all points, and the gray dashed line represents the diagonal. The maximum value of RSD is labeled in the figure.

Table 2. Predicted results for the validation samples *.

Sample Number	Reference	Prediction	Absolute Error	Relative Error (%)	RSD (%) $n = 3$
pHe					
1	7.96	8.12	0.16	2.07	-
2	7.65	7.66	0.01	0.13	1.70
3	7.19	7.17	−0.02	−0.28	1.36
4	6.80	6.75	−0.05	−0.74	1.07
5	6.38	6.45	0.07	1.10	1.66
6	5.97	6.02	0.05	0.84	0.84
7	6.49	6.36	−0.13	−2.00	-
8	6.32	6.26	−0.06	−0.95	-
9	6.22	6.41	0.19	3.05	-
pHi					
1	7.89	7.80	−0.09	−1.18	-
2	7.60	7.51	−0.09	−1.18	0.35%
3	7.22	7.28	0.06	0.83	1.47%
4	6.84	6.85	0.01	0.15	1.45%
5	6.42	6.33	−0.09	−1.40	0.42%
6	5.98	5.99	0.01	0.17	-

* For the pHe model, samples 1–6 are used for internal validation, and samples 7–9 are used for external validation; repetition (3 times) was conducted for samples 2–6 and 2–5, respectively, for pHi and pHe models.

To assess the practicality of the pHe model, external validation was conducted using three DMEM samples that had been used for cell cultivation, and the results were listed in Table 2. The absolute errors of the three samples are −0.13, −0.06, and 0.19. The relative errors are −2.00%, −0.95%, and 3.05%. The result indicates the good practicality of the pHe model. Due to the difficulty in obtaining actual intra-cellular fluid samples, no external validation for the pHi model was performed.

3.3. pH Dependence of Selected Wavenumbers

To better understand the selected variables (wavenumbers) by MC-UVE, RT, and CARS, the attribution of these wavenumbers was performed. Figure 4A shows the selected wavenumbers for the pHe model. Among the common variables selected by the three methods, 6671, 6828, 6916, and 7110 cm^{-1} are consistent with the peak positions of water molecules with three hydrogen bonds (S_3), two hydrogen bonds (S_2), one hydrogen bond (S_1), and no hydrogen bond (S_0), respectively [19,47–49]. Figure 4B depicts the selected wavenumbers for the pHi model: 6661, 6719/6823, 6904, and 7095 cm^{-1} belong to the common variables selected by the three methods and correspond to the peak positions of the S_3, S_2, and S_1 water molecules, respectively [19,47–49]. In our previous studies,

A*m*D*n* was used to describe water structures [20], where A and D represent the proton acceptor and proton donor, respectively, while *m* and *n* indicate the number of hydrogen bonds (*m*, *n* = 0, 1, or 2). Therefore, nine water structures can be defined, namely A0D0, A0D1, A1D0, A0D2, A1D1, A2D0, A1D2, A2D1, and A2D2. The possible A*m*D*n* water structures corresponding to the selected wavenumbers were summarized in Table 3. The correspondence between the water structure and the selected wavenumbers demonstrates that water molecules with different hydrogen bonds may serve as a probe to sense pH within the biological system.

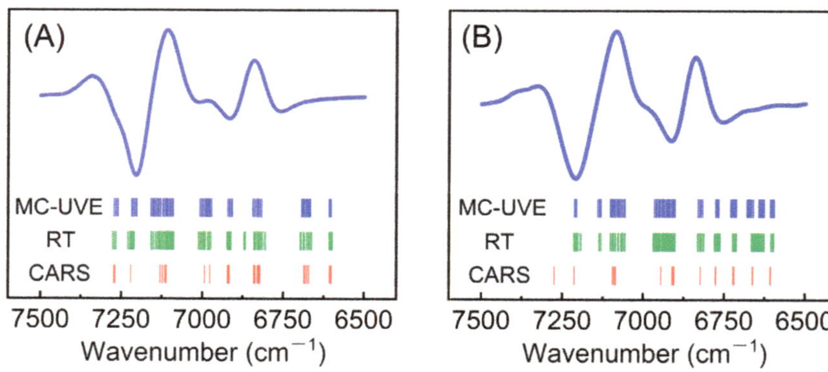

Figure 4. Selected wavenumbers for the pHe (**A**) and pHi (**B**) models. The blue curve represents the average spectrum of samples, and the short line represents the selected wavenumber.

Table 3. Possible water structures corresponding to the selected wavenumbers.

Wavenumber (cm^{-1})		Water Structure
pHe	**pHi**	
6671	6661	S_3 (A2D1)
-	6719	S_2 (A2D0)
6828	6823	S_2 (A1D1)
6916	6904	S_1 (A0D1)
7110	7095	S_0 (A0D0)

In order to investigate the pH dependence of wavenumbers listed in Table 3, the variations in peak intensities with pH were plotted in Figure 5, and a linear fit was performed. Figure 5A–D represent the variations in intensities at 6671, 6828, 6916, and 7110 cm^{-1} with pHe, and R-squared (R^2) values are 0.8278, 0.8730, 0.9438, and 0.9458, respectively. Figure 5E–I depict the variations in intensities at 6661, 6719, 6823, 6904, and 7095 cm^{-1} with pHi, and R^2 values are 0.8139, 0.8983, 0.8005, 0.8513, and 0.9464, respectively. All R^2 values exceed 0.80, indicating a good linear relationship between intensities and pH. It is demonstrated that the selected wavenumbers do have a good pH dependence. Obviously, the intensities at 7110 and 7095 cm^{-1} show a strong linear relationship with pH, with R^2 close to 0.95, suggesting that the S_0 water molecules may exhibit a greater accuracy and sensitivity than S_3–S_1 water molecules. However, due to the complexity of the spectra and overlapping of the peaks, the spectral intensities can only show an approximate dependency on pH, and the precise relationship between water structure and the respective pHe or pHi cannot be obtained. Additionally, considering the disparity between pHe and pHi systems, it is reasonable that the variation trends of peak intensities with pH for the same structure are different.

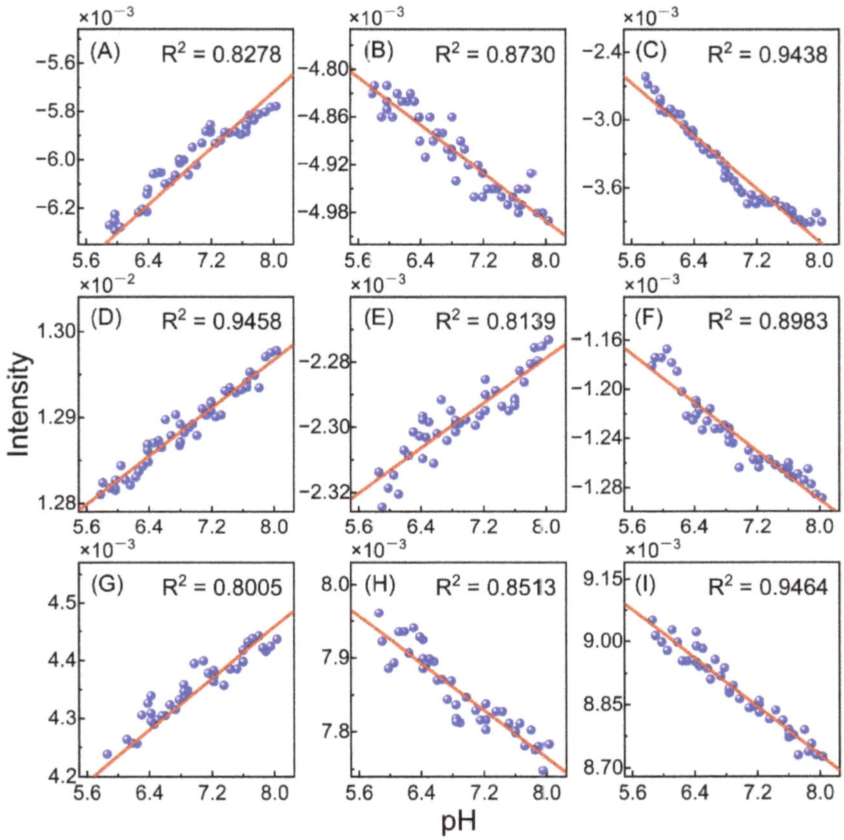

Figure 5. Relationships between pH and intensities of the peaks at 6671 (**A**), 6828 (**B**), 6916 (**C**), and 7110 cm^{-1} (**D**) for the pHe model, and the peaks at 6661 (**E**), 6719 (**F**), 6823 (**G**), 6904 (**H**), and 7095 cm^{-1} (**I**) for the pHi model. The red line represents the linear fit of all points.

4. Conclusions

The characteristic absorption of water in the NIR region was applied to detect pHe and pHi. DMEM and RM were used to simulate the extra- and intra-cellular fluids with different pH values, respectively. With the help of CWT and CARS, optimal quantitative models for pHe and pHi were built separately and exhibited good prediction performance and excellent repeatability. The correspondence between the selected wavenumbers and S_0–S_3 water structures can be found, and the good pH dependence of the selected wavenumbers was demonstrated. The findings indicate that water with different hydrogen bonds may be used as a probe to sense pH within biological systems.

Author Contributions: Conceptualization, X.S. and D.L.; methodology, J.L., F.L. and L.H.; software, J.L.; validation, X.S. and W.C.; formal analysis, J.L., F.L. and X.Y.; investigation, J.L., F.L. and L.H.; resources, X.S. and D.L.; data curation, J.L., F.L. and X.Y.; writing—original draft preparation, J.L.; writing—review and editing, X.S.; visualization, J.L.; supervision, X.S., W.C. and D.L.; project administration, X.S., W.C. and D.L.; funding acquisition, X.S. and W.C. All authors have read and agreed to the published version of the manuscript.

Funding: This research was funded by the National Natural Science Foundation of China (No. 22174075) and the Haihe Laboratory of Sustainable Chemical Transformations.

Institutional Review Board Statement: Not applicable.

Informed Consent Statement: Not applicable.

Data Availability Statement: The data presented in this study are available on request from the corresponding author.

Conflicts of Interest: The authors declare no conflict of interest.

References

1. Cialla, M.D.; Zheng, X.S.; Weber, K.; Popp, J. Recent progress in surface-enhanced Raman spectroscopy for biological and biomedical applications: From cells to clinics. *Chem. Soc. Rev.* **2017**, *46*, 3945–3961. [CrossRef]
2. Kurkdjian, A.; Guern, J. Intracellular pH: Measurement and importance in cell activity. *Annu. Rev. Plant Physiol. Plant Mol. Biol.* **1989**, *40*, 271–303. [CrossRef]
3. Sindhu, R.; Binod, P.; Pandey, A. Biological pretreatment of lignocellulosic biomass—An overview. *Bioresour. Technol.* **2016**, *199*, 76–82. [CrossRef]
4. Persi, E.; Duran, F.M.; Damaghi, M.; Roush, W.R.; Aloy, P.; Cleveland, J.L.; Gillies, R.J.; Ruppin, E. Systems analysis of intracellular pH vulnerabilities for cancer therapy. *Nat. Commun.* **2018**, *9*, 2997. [CrossRef]
5. Stubbs, M.; McSheehy, P.M.; Griffiths, J.R.; Bashford, C.L. Causes and consequences of tumour acidity and implications for treatment. *Mol. Med. Today* **2000**, *6*, 15–19. [CrossRef]
6. Suffrian, K.; Schulz, K.G.; Gutowska, M.A.; Riebesell, U.; Bleich, M. Cellular pH measurements in Emiliania huxleyi reveal pronounced membrane proton permeability. *New Phytol.* **2011**, *190*, 595–608. [CrossRef]
7. Shirmanova, M.V.; Druzhkova, I.N.; Lukina, M.M.; Dudenkova, V.V.; Ignatova, N.I.; Snopova, L.B.; Shcheslavskiy, V.I.; Belousov, V.V.; Zagaynova, E.V. Chemotherapy with cisplatin: Insights into intracellular pH and metabolic landscape of cancer cells in vitro and in vivo. *Sci. Rep.* **2017**, *7*, 8911. [CrossRef]
8. Ball, P. Water—An enduring mystery. *Nature* **2008**, *452*, 291–292. [CrossRef]
9. Ball, P. Water as an active constituent in cell biology. *Chem. Rev.* **2008**, *108*, 74–108. [CrossRef]
10. Czarnik-Matusewicz, B.; Pilorz, S. Study of the temperature-dependent near-infrared spectra of water by two-dimensional correlation spectroscopy and principal components analysis. *Vib. Spectrosc.* **2006**, *40*, 235–245. [CrossRef]
11. Dong, Q.; Guo, X.P.; Li, L.; Yu, C.; Nie, L.; Tian, W.L.; Zhang, H.; Huang, S.; Zang, H.C. Understanding hyaluronic acid induced variation of water structure by near-infrared spectroscopy. *Sci. Rep.* **2020**, *10*, 1387. [CrossRef]
12. Su, C.L.; Han, L.; An, H.L.; Cai, W.S.; Shao, X.G. Structures of water on the surface of anatase TiO_2 studied by diffuse reflectance near-infrared spectroscopy. *Spectrochim. Acta Part A Mol. Biomol. Spectrosc.* **2023**, *296*, 122674. [CrossRef]
13. Han, G.; Yu, X.Y.; Xia, D.D.; Liu, R.; Liu, J.; Xu, K.X. Preliminary clinical validation of a differential correction method for improving measurement accuracy in noninvasive measurement of blood glucose using near-infrared spectroscopy. *Appl. Spectrosc.* **2017**, *71*, 2177–2186. [CrossRef]
14. Eum, C.; Jang, D.; Kim, J.; Choi, S.; Cha, K.; Chung, H. Improving the accuracy of spectroscopic identification of geographical origins of agricultural samples through cooperative combination of near infrared and laser-induced breakdown spectroscopy. *Spectrochim. Acta Part B At. Spectrosc.* **2018**, *149*, 281–287. [CrossRef]
15. Porep, J.U.; Kammerer, D.R.; Carle, R. On-line application of near infrared (NIR) spectroscopy in food production. *Trends Food Sci. Technol.* **2015**, *46*, 211–230. [CrossRef]
16. Bec, K.B.; Huck, C.W. Breakthrough potential in near-infrared spectroscopy: Spectra simulation. A review of recent developments. *Front. Chem.* **2019**, *7*, 48. [CrossRef]
17. Bec, K.B.; Grabska, J.; Huck, C.W. Current and future research directions in computer-aided near-infrared spectroscopy: A perspective. *Spectrochim. Acta Part A Mol. Biomol. Spectrosc.* **2021**, *254*, 119625. [CrossRef]
18. Bec, K.B.; Futami, Y.; Wojcik, M.J.; Nakajima, T.; Ozaki, Y. Spectroscopic and computational study of acetic acid and its cyclic dimer in the near-infrared region. *J. Phys. Chem. A* **2016**, *120*, 6170–6183. [CrossRef]
19. Maeda, H.; Ozaki, Y.; Tanaka, M.; Hayashi, N.; Kojima, T. Near infrared spectroscopy and chemometrics studies of temperature-dependent spectral variations of water: Relationship between spectral changes and hydrogen bonds. *J. Near Infrared Spectrosc.* **1995**, *3*, 191–201. [CrossRef]
20. Tan, J.H.; Sun, Y.; Ma, L.; Feng, H.Y.; Guo, Y.C.; Cai, W.S.; Shao, X.G. Knowledge-based genetic algorithm for resolving the near-infrared spectrum and understanding the water structures in aqueous solution. *Chemom. Intell. Lab. Syst.* **2020**, *206*, 104150. [CrossRef]
21. Wang, Y.; Murayama, K.; Myojo, Y.; Tsenkova, R.; Hayashi, N.; Ozaki, Y. Two-dimensional Fourier transform near-infrared spectroscopy study of heat denaturation of ovalbumin in aqueous solutions. *J. Phys. Chem. B* **1998**, *102*, 6655–6662. [CrossRef]
22. Goto, N.; Bazar, G.; Kovacs, Z.; Kunisada, M.; Morita, H.; Kizaki, S.; Sugiyama, H.; Tsenkova, R.; Nishigori, C. Detection of UV-induced cyclobutane pyrimidine dimers by near-infrared spectroscopy and aquaphotomics. *Sci. Rep.* **2015**, *5*, 11808. [CrossRef]
23. Gao, L.L.; Zhong, L.; Zhang, J.; Zhang, M.Q.; Zeng, Y.Z.; Li, L.; Zang, H.C. Water as a probe to understand the traditional Chinese medicine extraction process with near infrared spectroscopy: A case of Danshen (*Salvia miltiorrhiza Bge*) extraction process. *Spectrochim. Acta Part A Mol. Biomol. Spectrosc.* **2021**, *244*, 118854. [CrossRef]

24. Ishigaki, M.; Yasui, Y.; Kajita, M.; Ozaki, Y. Assessment of embryonic bioactivity through changes in the water structure using near-infrared spectroscopy and imaging. *Anal. Chem.* **2020**, *92*, 8133–8141. [CrossRef]
25. Tsenkova, R.N.; Iordanova, I.K.; Toyoda, K.; Brown, D.R. Prion protein fate governed by metal binding. *Biochem. Biophys. Res. Commun.* **2004**, *325*, 1005–1012. [CrossRef]
26. Sun, Y.; Ma, L.; Cai, W.S.; Shao, X.G. Interaction between tau and water during the induced aggregation revealed by near-infrared spectroscopy. *Spectrochim. Acta Part A Mol. Biomol. Spectrosc.* **2020**, *230*, 118046. [CrossRef]
27. Chatani, E.; Tsuchisaka, Y.; Masuda, Y.; Tsenkova, R. Water molecular system dynamics associated with amyloidogenic nucleation as revealed by real time near infrared spectroscopy and aquaphotomics. *PLoS ONE* **2014**, *9*, e101997. [CrossRef]
28. Jinendra, B.; Tamaki, K.; Kuroki, S.; Vassileva, M.; Yoshida, S.; Tsenkova, R. Near infrared spectroscopy and aquaphotomics: Novel approach for rapid in vivo diagnosis of virus infected soybean. *Biochem. Biophys. Res. Commun.* **2010**, *397*, 685–690. [CrossRef]
29. Alam, M.K.; Rohrscheib, M.R.; Franke, J.E.; Niemczyk, T.M.; Maynard, J.D.; Robinson, M.R. Measurement of pH in whole blood by near-infrared spectroscopy. *Appl. Spectrosc.* **1999**, *53*, 316–324. [CrossRef]
30. Amerov, A.K.; Chen, J.; Small, G.W.; Arnold, M.A. Scattering and absorption effects in the determination of glucose in whole blood by near-infrared spectroscopy. *Anal. Chem.* **2005**, *77*, 4587–4594. [CrossRef]
31. Yadav, J.; Rani, A.; Singh, V.; Murari, B.M. Prospects and limitations of non-invasive blood glucose monitoring using near-infrared spectroscopy. *Biomed. Signal Process. Control* **2015**, *18*, 214–227. [CrossRef]
32. Cui, X.Y.; Liu, X.W.; Yu, X.M.; Cai, W.S.; Shao, X.G. Water can be a probe for sensing glucose in aqueous solutions by temperature dependent near infrared spectra. *Anal. Chim. Acta* **2017**, *957*, 47–54. [CrossRef]
33. Han, L.; Cui, X.Y.; Cai, W.S.; Shao, X.G. Three-level simultaneous component analysis for analyzing the near-infrared spectra of aqueous solutions under multiple perturbations. *Talanta* **2020**, *217*, 121036. [CrossRef]
34. Smith, R.E.; Luisi, P.L. Micellar solubilization of bio-polymers in hydrocarbon solvents III. Empirical definition of an acidity scale in reverse micelles. *Helv. Chim. Acta* **1980**, *63*, 2302–2311. [CrossRef]
35. Fujii, H.; Kawai, T.; Nishikawa, H. Determination of pH in reversed micelles. *Bull. Chem. Soc. Jpn.* **1979**, *52*, 2051–2055. [CrossRef]
36. Shao, X.G.; Leung, A.K.M.; Chau, F.T. Wavelet: A new trend in chemistry. *Acc. Chem. Res.* **2003**, *36*, 276–283. [CrossRef]
37. Shao, X.G.; Ma, C.X. A general approach to derivative calculation using wavelet transform. *Chemom. Intell. Lab. Syst.* **2003**, *69*, 157–165. [CrossRef]
38. Zhu, Z.Y.; Chen, S.B.; Wu, X.Y.; Xing, C.R.; Yuan, J. Determination of soybean routine quality parameters using near-infrared spectroscopy. *Food Sci. Nutr.* **2018**, *6*, 1109–1118. [CrossRef]
39. Zhang, M.; Zhao, C.; Shao, Q.J.; Yang, Z.D.; Zhang, X.F.; Xu, X.F.; Hassan, M. Determination of water content in corn stover silage using near-infrared spectroscopy. *Int. J. Agric. Biol. Eng.* **2019**, *12*, 143–148. [CrossRef]
40. Tang, G.; Huang, Y.; Tian, K.D.; Song, X.Z.; Yan, H.; Hu, J.; Xiong, Y.M.; Min, S.G. A new spectral variable selection pattern using competitive adaptive reweighted sampling combined with successive projections algorithm. *Analyst* **2014**, *139*, 4894–4902. [CrossRef]
41. Zhang, X.; Li, W.; Yin, B.; Chen, W.Z.; Kelly, D.P.; Wang, X.X.; Zheng, K.Y.; Du, Y.P. Improvement of near infrared spectroscopic (NIRS) analysis of caffeine in roasted Arabica coffee by variable selection method of stability competitive adaptive reweighted sampling (SCARS). *Spectrochim. Acta Part A Mol. Biomol. Spectrosc.* **2013**, *114*, 350–356. [CrossRef]
42. Zhang, M.J.; Zhang, S.Z.; Iqbal, J. Key wavelengths selection from near infrared spectra using Monte Carlo sampling–recursive partial least squares. *Chemom. Intell. Lab. Syst.* **2013**, *128*, 17–24. [CrossRef]
43. Qi, L.H.; Cai, W.S.; Shao, X.G. Effect of temperature on near infrared spectra of n-alkanes. *Acta Chim Sin.* **2016**, *74*, 172–178. [CrossRef]
44. Czarnecki, M.A.; Morisawa, Y.; Futami, Y.; Ozaki, Y. Advances in molecular structure and interaction studies using near-infrared spectroscopy. *Chem. Rev.* **2015**, *115*, 9707–9744. [CrossRef]
45. Gowen, A.A.; Amigo, J.M.; Tsenkova, R. Characterisation of hydrogen bond perturbations in aqueous systems using aquaphotomics and multivariate curve resolution-alternating least squares. *Anal. Chim. Acta* **2013**, *759*, 8–20. [CrossRef]
46. Takeuchi, M.; Kajimoto, S.; Nakabayashi, T. Experimental evaluation of the density of water in a cell by Raman microscopy. *J. Phys. Chem. Lett.* **2017**, *8*, 5241–5245. [CrossRef]
47. Tsenkova, R. Aquaphotomics: Dynamic spectroscopy of aqueous and biological systems describes peculiarities of water. *J. Near Infrared Spectrosc.* **2009**, *17*, 303–314. [CrossRef]
48. Muncan, J.; Anantawittayanon, S.; Furuta, T.; Kaneko, T.; Tsenkova, R. Aquaphotomics monitoring of strawberry fruit during cold storage-A comparison of two cooling systems. *Front. Nutr.* **2022**, *9*, 1058173. [CrossRef]
49. Tsenkova, R.; Muncan, J.; Pollner, B.; Kovacs, Z. Essentials of aquaphotomics and its chemometrics approaches. *Front. Chem.* **2018**, *6*, 363. [CrossRef]

chemosensors

Article

Fluorophores-Assisted Excitation Emission Matrix Fluorescence Method for the Origin Traceability of Lily

Huan Fang, Hailong Wu *, Tong Wang *, Yao Chen and Ruqin Yu

State Key Laboratory of Chemo/Biosensing and Chemometrics, College of Chemistry and Chemical Engineering, Hunan University, Changsha 410082, China; hfang@hnu.edu.cn (H.F.); chenyao717@hnu.edu.cn (Y.C.); rqyu@hnu.edu.cn (R.Y.)

* Correspondence: hlwu@hnu.edu.cn (H.W.); wangtong@hnu.edu.cn (T.W.)

Abstract: In this work, a fluorophores-assisted excitation/emission matrix (EEM) fluorescence method was proposed to trace the origin of lily in the Chinese market. There are few active components in lilies that have fluorescent signals, and too few characteristic variables may lead to unsatisfactory accuracy in the subsequent classification. Therefore, three fluorophores, 2-Aminoethyl diphenylborinate (DPBA), o-Phthalaldehyde (OPA) and Rhodamine B (RB), were used to enrich the information of the fluorescent fingerprint of lily, which can improve the classification accuracy. The lily samples were characterized by using EEM fluorescence coupled with the alternating trilinear decomposition (ATLD) algorithm, which was able to extract information of various fluorophores in lily samples. Two chemical pattern recognition methods, principal component analysis-linear discriminant analysis (PCA-LDA) and partial least squares-discrimination analysis (PLS-DA), were used to model and trace the origin of different lilies. When the fluorophores were added, the accuracy of the test set and prediction set obtained by the classification model increased from 71.4% to 92.9% and 66.7% to 100%, respectively. The proposed method combined fluorophores-assisted EEM fluorescence with multi-way chemometric methods to extract comprehensive information on the samples, which provided a potential method for the origin traceability of traditional Chinese medicine.

Keywords: lily; fluorophores-assisted excitation-emission matrix fluorescence; alternating trilinear decomposition; chemometrics; fluorophores

Citation: Fang, H.; Wu, H.; Wang, T.; Chen, Y.; Yu, R. Fluorophores-Assisted Excitation Emission Matrix Fluorescence Method for the Origin Traceability of Lily. *Chemosensors* **2023**, *11*, 426. https://doi.org/10.3390/chemosensors11080426

Academic Editor: Cynthia M. Dupureur

Received: 29 June 2023
Revised: 25 July 2023
Accepted: 30 July 2023
Published: 1 August 2023

1. Introduction

Lily, the dry fleshy scales of *Lilium lamcifolium* Thunb, *Lilium broumii* F.E. Brown *var. viridulum* Bake or *Lilium pumilum* DC., is a traditional Chinese medicine used for both medicine and food [1]. In recent years, the lily has received more and more attention in the fields of clinical medicine and health food. Its main active ingredients are steroidal saponins, phenolic acids, polysaccharides and so on [2–4]. It has anti-inflammatory, anti-tumor, anti-depression and anti-bacterial biological activity [5–9], among others. Therefore, the lily is widely used in clinical Chinese medicine formula and Chinese patent medicine. In China, lilies are grown in a wide range of areas, mainly in Hebei, Shanxi, Henan, Shaanxi, Hubei, Hunan, Jiangxi, Anhui and Zhejiang. The quality of the lily is affected by the growth environment, such as soil quality and climate, and there are large regional differences [10]. Qin et al. [11] have proved that the chemical components of lily grow in different regions and therefore result in there being huge differences between the lilies studied and collected from different origins. There are huge differences in the prices of lilies from different production areas, which has led to some unscrupulous traders making profits by pretending to have high quality lilies. However, in Hunan province and the surrounding area, lilies from different production areas are often sold together. The Longshan lily, as a geographical protection product, is usually counterfeited by lilies taken from other neighboring regions. Therefore, the traceability of lilies in a certain production area has significance, but their similarities will pose greater challenges to classification.

Traditional identification always relies on sensory indicators, such as appearance, color and smell, to realize the identification and analysis of origin and quality. As a result, a sensory evaluation that is based on manual experience cannot achieve fast and accurate detection. In addition to the traditional manual identification methods, a large number of strategies combining modern analytical instruments and chemometric methods have also been reported for the origin traceability and quality evaluation of the lily, such as high-performance capillary electrophoresis (HPCE) [12], near-infrared spectroscopy (NIR) [13], ultra-high performance liquid chromatography coupled with hybrid quadrupole time-of-flight mass spectrometry (UHPLC-QTOF-MS) [14], liquid chromatography-mass spectrometry-mass spectrometry (LC-MS-MS) [15], gas chromatography(GC) [16,17], etc.

Compared to other modern instruments, fluorescence has the advantages of a fast analysis speed, simple operation, low cost and high sensitivity. At the same time, the EEM fluorescence data generated by it contain more information than traditional fluorescence spectra, and it can be decomposed by multi-way analysis algorithms to obtain detailed qualitative and quantitative spectral information of analytes in complex systems [18]. The combination of EEM spectroscopy and chemometrics can give the component information of the samples and improve classification accuracy. However, the fluorescent composition of lilies has few fluorescent components, which presents a difficulty in fluorescent classification methods.

In this work, the ATLD algorithm was used to decompose the EEM fluorescence data of pure lilies. Then, PLS-DA and PCA-LDA were chosen to build classification models based on the proposed features, respectively. Due to the few extracted features, the built models may not trace the origin of lilies accurately. Thus, three fluorophores, 2-aminoethyl diphenylborinate (DPBA), o-Phthalaldehyde (OPA) and Rhodamine B (RB), were added to react with the lily samples to increase the qualitative information of the lily samples, so as to improve the classification accuracy of origin traceability. ATLD was used to decompose the obtained EEM fluorescent data of the lilies added with fluorophores. Then, PLS-DA and PCA-LDA were used to build classification models based on the extracted features, and the built models were also used to trace the origin of other lily samples. Finally, the origin traceability results obtained by the pure lilies and the lilies added with three fluorophores were compared to choose the best strategy. The specific workflow is shown in Figure 1.

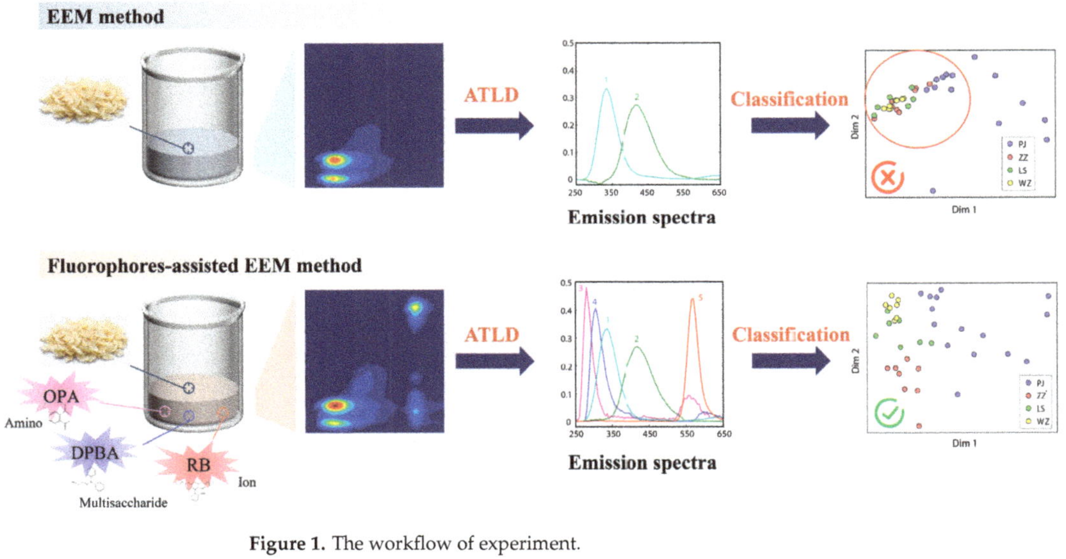

Figure 1. The workflow of experiment.

2. Materials and Methods

2.1. Sample Collection and Preparation

Sixty lily samples from four different origins were collected, and the detailed information is summarized in Table 1. Firstly, the surface soil of the fresh lily bulb samples was washed with running water, then the lily scales were split into pieces, placed in 100 °C water for 5 min, removed and the water drained, before being dried with hot air at 60 °C, crushed, and then passed through an 80-mesh sieve to obtain lily powder.

Table 1. Information of lily samples.

Place of Origin	Variety	Sample Size
Pingjiang, Hunan province (PJ) [a]	Long Ya lily (annual)	24
Zhuzhou, Hunan Province. (ZZ)	Unknown	13
Longshan, Hunan Province. (LS)	Long Yu lily	12
Wanzai, Jiangxi Province. (WZ)	long Ya lily	11

[a] Abbreviations of places of origin are presented in the parentheses. Pingjiang: PJ, Zhuzhou: ZZ, Longshan: LS, Wanzai: WZ.

Among the total of 60 lily samples, 9 samples were collected additionally as the prediction set to prove the practicality of the classification models. The remaining 51 samples were split into two parts by a random sampling method: 37 samples were selected as the training set to build models and optimize model parameters; and 14 samples were selected as the test set to test the classification performance of the established models.

An extract was prepared for each lily sample using the following method: 50 mg of the lily powder sample was placed in 2 mL of methanol aqueous solution ($CH_3OH:H_2O$ = 6:4), ultrasonicated for 30 min, and then centrifuged at 4000 r/min for 10 min. Subsequently, 200 μL of the supernatant was placed into a 10 mL volumetric flask, which was then diluted to 10 mL with methanol aqueous solution ($CH_3OH:H_2O$ = 6:4). For the lily supernatant with fluorophores, appropriate amounts of fluorophores were also added.

2.2. Chemicals and Reagents

L-Lysine, DL-Homocysteine, Chlorogenic acid, Kaempferol, 2-Aminoethyl diphenylborinate (DPBA), o-phthalaldehyde (OPA) and rhodamine B (RB) were purchased from Sigma-Aldrich. The methanol was chromatographic grade and purchased from OCEANPAK Company (Germany). Nitrogen standard solutions of Cu and Ca were purchased from the National Nonferrous Metals and Electronic Materials Analysis and Testing Center. Ultrapure water was produced by a Milli-Q instrument.

Firstly, 2.3 mg of DPBA, 5.8 mg of OPA and 4.6 mg of RB were weighed in 10 mL volumetric flasks and diluted with ultrapure water, respectively. After that, 40 μL of DPBA, 15 μL of OPA and 3 μL of RB solutions were added into each lily supernatant and diluted to 10 mL.

2.3. Instrument, Software and Their Parameters

The fluorescence detector used in this experiment is the Hitachi F-7000 fluorescence spectrophotometer. All samples were assayed using a 1.00 cm quartz cell. The instrument parameters are as follows: a test voltage of 700 V, scanning speed of 30,000 nm per minute and step size of 5 nm to obtain fluorescence signals in the excitation wavelength region of 200–600 nm and emission wavelength region of 250–650 nm; and the excitation and emission slit width is 5 nm.

All experimental data were processed in a MATLAB environment. PCA-LDA [19], PLS-DA [20,21], ATLD and other codes of this study were written by our group based on the original literature.. Data and other homemade codes pertaining to this study are available from the corresponding authors upon reasonable request.

2.4. Theory

2.4.1. Trilinear Component Model

The size of the EEM fluorescence data array generated by each sample is $I \times J$, where I is the number of excitation wavelengths and J is the number of emission wavelengths. The stacking of K samples can obtain a three-way data matrix $\underline{\mathbf{X}}(I \times J \times K)$, according to the theory of the trilinear component model, and the three-way array $\underline{\mathbf{X}}$ $(I \times J \times K)$ can be expressed as:

$$x_{ijk} = \sum_{n=1}^{N} a_{in} b_{jn} c_{kn} + e_{ijk} \; (i = 1,2,\ldots,I; j = 1,2,\ldots,J; k = 1,2,\ldots,K) \tag{1}$$

where i, j and k represent the excitation wavelength point, the emission wavelength point and the sample number, respectively; x_{ijk} is the element in $\underline{\mathbf{X}}$, which represents the fluorescence intensity under the corresponding i, j and k; N is the number of components in the system; and a_{in}, b_{jn} and c_{kn} are the elements of three profile matrices (**A**, **B** and **C**), which record the excitation spectrum profile information, emission spectrum profile information and relative concentration information of each component, respectively.

2.4.2. Alternating Quadrilinear Decomposition

Wu et al. firstly proposed the Alternating Trilinear Decomposition (ATLD) algorithm [22]. The ATLD algorithm updates three profile matrices (**A**, **B** and **C**) by alternately minimizing the following three objective functions using the Moore–Penrose generalized inverse based on singular value decomposition:

$$\sigma(\mathbf{A}) = \sum_{i=1}^{I} \left\| \mathbf{X}_{i..} - \mathbf{B} diag(a_{(i)}^{\mathrm{T}}) \mathbf{C}^{\mathrm{T}} \right\|_{\mathrm{F}}^{2} \tag{2}$$

$$\sigma(\mathbf{B}) = \sum_{j=1}^{J} \left\| \mathbf{X}_{.j.} - \mathbf{C} diag(b_{(j)}^{\mathrm{T}}) \mathbf{A}^{\mathrm{T}} \right\|_{\mathrm{F}}^{2} \tag{3}$$

$$\sigma(\mathbf{C}) = \sum_{k=1}^{K} \left\| \mathbf{X}_{..k} - \mathbf{A} diag(c_{(k)}^{\mathrm{T}}) \mathbf{B}^{\mathrm{T}} \right\|_{\mathrm{F}}^{2} \tag{4}$$

The reason why ATLD is widely used is that it converges quickly and can still give accurate analytical results when the component numbers are overestimated. More detailed information and applications can be found in related references [22,23] and will not be repeated here. In this work, ATLD is used for the decomposition of the three-way EEM data array, thereby providing the qualitative and relative concentration information of all fluorophores with physical/chemical significance. Matrix **A** corresponds to the normalized excitation spectrum, matrix **B** corresponds to the normalized emission spectrum and matrix **C** corresponds to the relative concentration.

Partial least squares-discrimination analysis (PLS-DA) is a classical supervised classification method. When the number of variables is large and the number of samples is small, it is often a preferred option. Ten-fold cross-validation is often used to determine its model complexity, i.e., the number of latent variables. Its main modeling and prediction steps are as follows: (1) modeling process: set the Y-block as a dummy matrix with one-hot encoding, in which 1 means the sample belongs to the designated class and 0 means the opposite. Then, the PLS model for classification purposes can be built. (2) Prediction process: the test data matrix is used as the input, and then the attribution of the sample can be determined according to the return values based on the principle of maximum value.

PCA is an important method used for data dimensionality reduction and feature extraction. It applies the principle of variance maximization to project high-dimensional data into a low-dimensional data space to achieve the purpose of dimensionality reduction. LDA achieves sample classification by minimizing the intra-class variance and maximizing the inter-class variance after projection. When the above two methods are combined, then supervised classification-based principal components can be achieved.

To further evaluate the overall performance of the proposed classification models, the correct classification rate (CCR) was described as follows:

$$CCR = \frac{\sum_{g=1}^{G} n_{gg}}{N} \tag{5}$$

where each element n_{gg} represents the number of samples belonging to class g and assigned to class g, G represents the number of classes and N represents the total number of samples.

For each class, Ture positive (TP) means positive samples were predicted to be positive by the model, Ture negative (TN) means negative samples were predicted to be negative by the model, False positive (FP) means negative samples were predicted to be positive by the model and False negative (FP) means positive samples were predicted to be negative by the model.

3. Results and Discussion

3.1. Experiment Principle

In order to improve the selectivity of the classification model for lilies from various origins, it is necessary to increase the fluorescence spectral information required for classification. Lilies contain many chemical active ingredients, such as steroidal saponins, phenolic acids, alkaloids, polysaccharides, amino acids and so on. Some active components have fluorescent signals while others have no fluorescent signals, and their EEM fluorescence spectroscopy is shown in Figure 2a. It can be seen from the figure that the endogenous fluorescent components in lilies are indeed relatively small, and it is difficult to provide sufficient information. Therefore, three fluorophores, DPBA, OPA and RB, were chosen to solve this problem. Their EEM fluorescence spectrum is shown in Figure 2b–d. DPBA is often used as a receptor to recognize sugar molecules. It can combine with some sugar molecules in lilies to form five-membered or six-membered rings, which leads to a shift in fluorescence peaks and a decrease in fluorescence intensity [24]. OPA is a common amino acid derivatization reagent which can react with the amino group on the amino acid in lilies to form a condensation product with strong fluorescence [25]. RB is a common fluorescent probe with a strong fluorescent signal. When it reacts with some anions and cations in the solution, the fluorescence intensity will change [26]. Many components in lilies do not have endogenous fluorescence, but they will interact with RB, such as L-Lysine, DL-Homocysteine, Cu^{2+}, Ca^{2+}, Chlorogenic acid and Kaempferol. Their effect on the fluorescence intensity of RB was also tested and the detailed information is shown in the Supplementary Material, Figure S1 and Table S1. On the basis of the fluorescence spectrum of pure lily, three fluorophores were combined to form a sensor array to obtain a fluorescent fingerprint with richer classification information of the lily, as shown in Figure 2e. Moreover, the peak intensities of pure lily (a), DPBA (b), OPA (c), RB (d) and fluorophores-lily (e) at different excitation/emission wavelengths were also listed in Table 2. It can be seen from the table that when three fluorescent probes reacted with lily, its peak intensities were not the linear sum of the four substances' peak intensities, and some of the peak intensities (e.g., 225/330 nm and 260/565 nm) were also reduced. Meanwhile, the spectral changes of adding fluorophores and reactions were also analyzed, and the results are shown in the Supplementary Material, Figure S2. Those data results showed that three fluorophores interacted with substances in the lily, thus changing and enriching the fluorescence spectral signal of the lily.

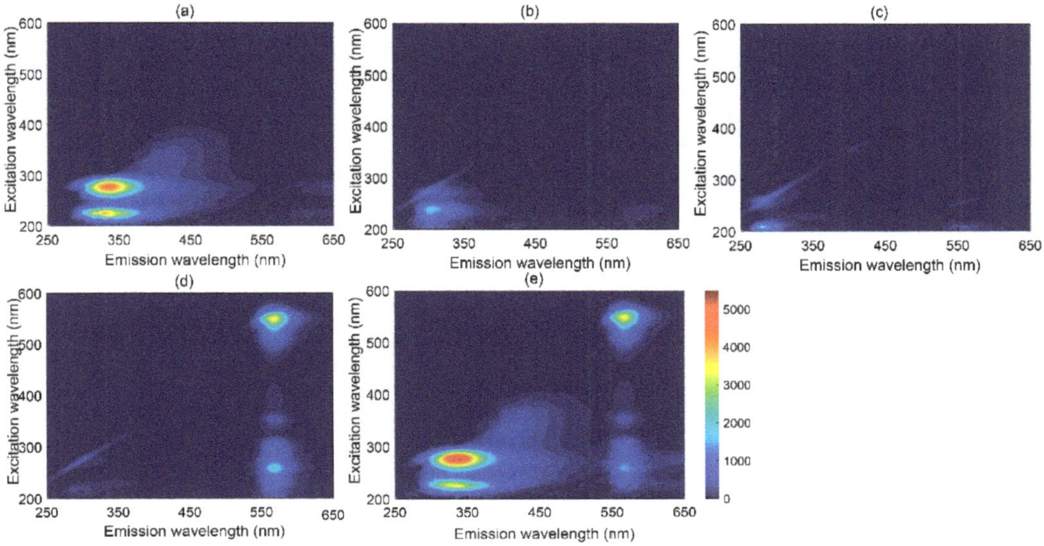

Figure 2. EEM fluorescence after removing scattering: (**a**) pure lily; (**b**) DPBA; (**c**) OPA; (**d**) RB; (**e**) fluorophores-lily.

Table 2. The peak intensities of pure lily, DPBA, OPA and RB at different excitation/emission wavelengths.

Subgraph in Figure 2	280/330 nm	225/330 nm	335/425 nm	260/565 nm	550/565 nm
a	4604	4503	364.2	113.1	3.055
b	311.9	572.4	33.72	79.44	3.850
c	99.97	187.8	56.85	157.2	3.202
d	97.41	155.1	37.99	1970	3475
e	5450	3810	564.6	2079	4192

3.2. Component Analysis and Origin Traceability of Lily

It can be seen from Figure 2a that the fluorescence signals of components in the lily overlapped significantly, and the direct modeling and classification based on the EEM fluorescence cannot accomplish the characterization and feature extraction of the fluorescent components in lily at the same time. In addition, Rayleigh scattering and Raman scattering in the spectrum will affect the trilinear structure of the EEM data. Firstly, the effect of Raman scattering was eliminated by subtracting the average value of three blank sample data, and then the interpolation method [27] was used to eliminate Rayleigh scattering. The whole process is shown in the Supplementary Material, Figure S3. Finally, ATLD was used to decompose the three-way fluorescence data array generated by the lily samples. The size of the three-way data array generated by 60 lily samples is 81 × 81 × 60. The Core Consistency Diagnostic (CORCONDIA) was used to estimate the number of components in the lily samples before ATLD decomposition. After calculation, the number of components in the lily is two, and the components here represent one or more real fluorescent molecules, which have similar structures and fluorescent properties. In addition, non-negative constraints are applied to all data in the calculation process to ensure reasonable results.

After ATLD decomposition, (a) the normalized excitation spectrum, (b) normalized emission spectrum and (c) relative concentration diagram of each component in the lily were shown in Figure 3. Component 1 presents its fluorescence peak at the excitation/emission wavelength of 335 (290)/425 nm. At the same time, the excitation spectrum has many shoulder peaks, which may be estimated to be ferulic acid and caffeic acid [28,29]. Component

2 presents its fluorescence peak at the excitation/emission wavelength of 280 (225)/330 nm, and the peak is relatively broad, which may be estimated to be catechin and epicatechin [30]. It can be seen that the decomposition of excitation and emission spectra by ATLD is useful for the characterization of active components in the lily. In addition, the relative concentration of each component reflects the concentration difference of this component among different lilies, which is an important basis for completing the origin traceability of lilies.

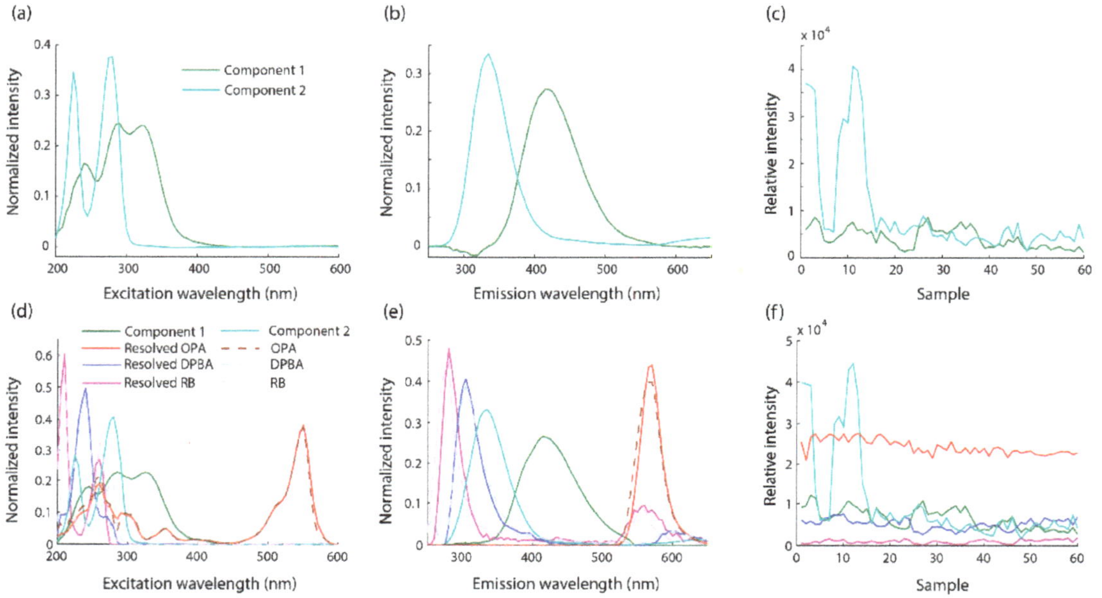

Figure 3. The resolved (**a**) normalized excitation profiles of pure lily; (**b**) normalized emission profiles of pure lily; (**c**) relative concentration profiles of pure lily; (**d**) normalized excitation profiles of fluorophores-lily; (**e**) normalized emission profiles of fluorophores-lily; (**f**) relative concentration profiles of fluorophores-lily by ATLD algorithm.

The lily EEM fluorescence data were obtained based on the pure lilies' extract. After ATLD decomposition, the relative concentration information of two components (ATLD scores) was extracted for subsequent lily classification. Based on this, two classic chemometric classification algorithms, PLS-DA and PCA-LDA, were used to build classification models and realize the origin traceability of different lilies. Firstly, the parameters of the two models are optimized. The number of latent variables (*LVs*) for PLS-DA and the number of principal components (*PCs*) for PCA-LDA were optimized according to the best correct classification rates (CCRs) obtained by 10-fold cross-validation. The optimization process of *LVs* for PLS-DA and *PCs* for PCA-LDA has been shown in the Supplementary Material, Figure S4. When *LVs* = 1 and *PCs* = 1, PLS-DA and PCA-LDA models are the best, respectively. After that, the CCR of cross-validation for PLS-DA was 59.5%, and that of PCA-LDA was 48.7%. In addition, these models were used to validate the categories of training set samples and prediction set samples, and the results are summarized in Table 3. According to the results, the two models built by EEM for lily origin traceability could not meet expectations, which may be due to the fact that there are few fluorescent components in the lily. Figure 4a is the scores plot of PLS-DA. The corresponding scatter diagram is drawn by using the scores of the first two latent variables (*LVs*) of PLS-DA. From the figure, all kinds of lilies are mixed together and cannot be distinguished, which also proves that the model built by EEM cannot complete the classification task.

Table 3. CCR (%) of cross-validation and each dataset, parameters (*LVs* for PLS-DA, *PCs* for PCA-LDA) of each classification model.

Method	PLS-DA					PCA-LDA				
	LVs	CV	Training	Test	Pre	*PCs*	CV	Training	Test	Pre
EEM	1	59.5	67.6	71.4	66.7	1	48.7	56.8	50.0	66.7
F-EEM	4	94.6	97.3	92.9	100.0	4	89.2	94.6	92.9	88.9

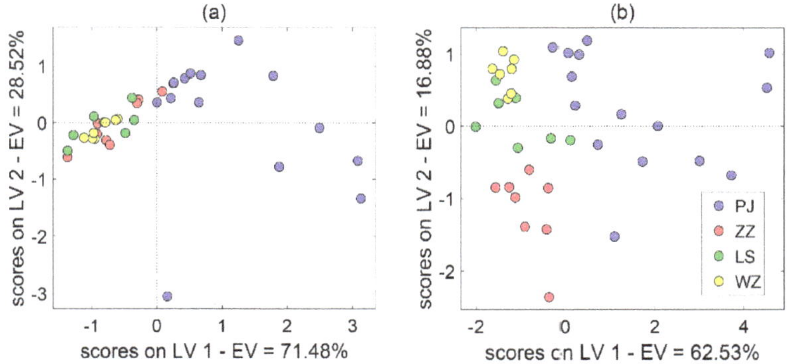

Figure 4. The plots of the scores based on the first two latent variables (*LVs*) of PLS-DA in EEM (**a**) and F-EEM (**b**).

In order to solve the problem of insufficient fluorescence information in EEM, three fluorophores that can react with fluorescent substances in lilies are used. Firstly, a certain concentration of the three fluorophores was mixed with the pure lily sample solution, and then the EEM fluorescence data were obtained by the F-7000 fluorescence instrument, and the obtained EEM fluorescence spectrum was shown in the Figure 2e. Likewise, the Core Consistency Diagnostic (CORCONDIA) was used to estimate the number of components in the lily samples before the ATLD decomposition of the data. After calculation, when three fluorophores react with the lily, the number of components increased to five, which is just equal to three fluorophores plus two components in the pure lily. In addition, non-negativity constraints are also placed on all data during calculations to ensure reasonable results. Then, the ATLD algorithm is used to decompose the three-way fluorescence data, and the detailed decomposition results are shown in Figure 3d–f. In order to qualitatively analyze the decomposed components, they were compared with the fluorescence spectra of pure DPBA, pure OPA, pure RB and pure lily. The results show that when DPBA, OPA and RB reacted with lilies, the generated three-way fluorescence data array can be decomposed by ATLD to obtain five components, which correspond to the two components obtained from the pure lily samples and three fluorophores. Their spectra are almost in perfect agreement. At the same time, after the reaction of fluorophores with pure lily samples, the concentration information for some components without fluorescent response is also displayed in Figure 3f.

In F-EEM, DPBA-OPA-RB-Lily EEM fluorescence was obtained based on the lilies' extract added with three fluorophores. After ATLD decomposition, the relative concentration information of all five components (ATLD scores) for the subsequent lily classification was extracted. Immediately after, PLS-DA and PCA-LDA were used to build classification models based on the obtained ATLD scores. Firstly, the parameters of the two models were optimized. The number of latent variables (*LVs*) for PLS-DA and the number of principal components (*PCs*) for PCA-LDA were optimized according to the best correct classification rate (CCR) obtained by 10-fold cross-validation. When *LVs* = 4 and *PCs* = 4, PLS-DA and PCA-LDA can achieve the best results, respectively. The cross-validation CCR of PLS-DA is 94.6%, and the cross-validation CCR of PCA-LDA is 89.2%. In addition, these models were also used to validate the classes of training set samples and test set samples, and the results are summarized in Table 3. The CCRs of the training set and test set obtained by the PLS-DA model are 94.6% and 97.3%, respectively. The CCRs of the training set and test set obtained by the PCA-LDA model are 89.2% and 94.6%, respectively. This shows that the proposed methods can realize the origin traceability of different types of lilies based on the fluorophores-assisted EEM fluorescence, and the obtained results are satisfactory. In order to externally validate the model, an additional 9 lily samples that were not involved in the modeling were collected as the prediction set, and the obtained CCRs were also satisfactory; the CCR of PLS-DA was 100.0% and the CCR of PCA-LDA was 88.9%. This also shows that the adaptability of the above models is satisfactory. In contrast, the model built by PLS-DA can trace the origin of lilies more accurately. In order to further understand the classification performance of the model built by PLS-DA, its detailed classification performance parameters, including the sensitivity and specificity of cross-validation and the training set and test set, are summarized in Table 4, and Table 5 lists the confusion matrix of the train set and test set. This is helpful for us to explore the main categories of misclassification. Figure 4b provides the scores plot of PLS-DA. The sample data can be visible based on the scores of the first two LVs of PLS-DA, which shows the distribution of different samples in a two-dimensional space. Figure 4 demonstrates that there is almost no overlap between different types of lilies, and lilies of the same type are basically clustered together, which also means that the classification model has good classification performance.

Table 4. Sensitivity and specificity of cross-validation and each dataset obtained by PLS-DA based on F-EEM.

Model Validation	CCR (%)	Sensitivity [a]				Specificity [b]			
		PJ	ZZ	LS	WZ	PJ	ZZ	LS	WZ
Cross-validation	94.6	93.3	100.0	85.7	100.0	100.0	96.6	100.0	96.7
Training set	97.3	100.0	100.0	85.7	100.0	100.0	100.0	100.0	96.7
Test set	92.9	100.0	100.0	66.7	100.0	100.0	100.0	100.0	90.9

[a] The sensitivity is equal to TP/(TP + FN), TP means true positives and FN means false negatives. [b] The specificity is equal to TN/(TN + FP), TN means true negatives and FP means false positives.

Table 5. Confusion matrix of train set and test set obtained by PLS-DA based on F-EEM.

		Train Set				TEST SET			
		Predicted				Predicted			
		PJ	ZZ	LS	WZ	PJ	ZZ	LS	WZ
Actual	PJ	15	0	0	0	5	0	0	0
	ZZ	0	8	0	0	0	3	0	0
	LS	0	0	6	1	0	0	2	1
	WZ	0	0	0	7	0	0	0	3

4. Conclusions

In this work, the combination of fluorophores-assisted EEM fluorescence and multi-way chemometrics was successfully applied to the characterization and origin traceability of lilies from different origins. The EEM fingerprint of pure lily contains less fluorescent information, and after ATLD decomposition, only two components with clear chemical meaning can be obtained. Too little classification information made the established classification model unable to complete the accurate origin traceability of the lily, and the CCRs were all lower than 71.4%. In order to solve this problem, three fluorophores which can react with lilies were chosen to enrich its EEM fingerprint. After decomposition by ATLD, the number of chemically meaningful components obtained increased to five. Compared to the classification models based on the EEM fluorescence data of pure lily extract, the classification accuracy of the classification model established by PLS-DA based on fluorophores-assisted EEM fluorescence data was greatly improved. The classification accuracies of cross-validation and the training set, test set and prediction set are all above 92.9%. The experimental results prove that when the studied sample contains endogenous fluorescence components, the qualitative information can be enriched by EEM scanning and adding appropriate fluorophores, thereby improving the accuracy of the built classification model. This F-EEM is expected to be more widely used in the rapid classification and authenticity identification of weakly fluorescent or even non-fluorescent systems.

Supplementary Materials: The following supporting information can be downloaded at: https://www.mdpi.com/article/10.3390/chemosensors11080426/s1, Figure S1: EEM fluorescence after removing scattering: (a) RB-L-Lysine; (b) RB-DL-Homocysteine; (c) RB-Cu^{2+}; (d) RB-Ca^{2+}; (e) RB-Chlorogenic acid; (f) RB-Kaempferol; (g) RB; Figure S2: EEM fluorescence after removing scattering: (a) DPBA; (b) OPA; (c) RB; (d) DPBA-lily; (e) OPA-lily; (f) RB-lily; (g) pure lily; Figure S3: The elimination of non-bilinear factor (Rayleigh scattering) for the EEM fluorescence data of fluorophores-lily: (a) raw data with visible Rayleigh scattering; (b) gapped data with scattering being removed in the regions; (c) repaired data fitted by an interpolation method; Figure S4: The number of latent variables (*LVs*) for PLS-DA and the number of principal components (*PCs*) for PCA-LDA were optimized according to the best correct classification rates (CCRs) obtained by 10-fold cross-validation: (a) Fluorophore-lily-PLS-DA; (b) Fluorophore-lily -PCA-LDA; (c) Lily-PLS-DA; (d) Lily-PCA-LDA; Table S1: The intensities of pure RB and RB added with different chemical components at different excitation/emission wavelength.

Author Contributions: H.F.: Investigation, writing—original draft preparation, writing—review and editing. H.W.: conceptualization, methodology, funding acquisition. T.W.: Methodology, validation, funding acquisition. Y.C.: Investigation, conceptualization, validation. R.Y.: Methodology, supervision. All authors have read and agreed to the published version of the manuscript.

Funding: This research was funded by the National Natural Science Foundation of China under Grant (No. 22174036, No. 21775039 and No. 22204049).

Institutional Review Board Statement: Not applicable.

Informed Consent Statement: Not applicable.

Data Availability Statement: Not applicable.

Conflicts of Interest: The authors declare no conflict of interest.

References

1. Pharmacopoeia, C. Commission. In *Pharmacopoeia of the People's Republic of China*; China Medical Science Press: Beijing, China, 2010.
2. Chau, C.F.; Wu, S.H. The development of regulations of Chinese herbal medicines for both medicinal and food uses. *Trends Food Sci. Technol.* **2006**, *17*, 313–323. [CrossRef]
3. Hou, Y.; Jiang, J.G. Origin and concept of medicine food homology and its application in modern functional foods. *Food Funct.* **2013**, *4*, 1727–1741. [CrossRef]
4. Wang, P.; Li, J.; Attia, F.A.K.; Kang, W.; Wei, J.; Liu, Z.; Li, C. A critical review on chemical constituents and pharmacological effects of Lilium. *Food Sci. Hum. Wellness* **2019**, *8*, 330–336. [CrossRef]
5. Mimaki, Y.; Nakamura, O.; Sashida, Y.; Satomi, Y.; Nishino, A.; Nishino, H. Steroidal saponins from the bulbs of *Lilium longiflorum* and their antitumour-promoter activity. *Phytochemistry* **1994**, *37*, 227–232. [CrossRef] [PubMed]
6. Nakamura, O.; Mimaki, Y.; Nishino, H.; Sashida, Y. Steroidal saponins from the bulbs of *Lilium speciosum* x L. nobilissimum 'Star Gazer'and their antitumour-promoter activity. *Phytochemistry* **1994**, *36*, 463–467. [CrossRef]
7. Zhou, J.; An, R.; Huang, X. Genus Lilium: A review on traditional uses, phytochemistry and pharmacology. *J. Ethnopharmacol.* **2021**, *270*, 113852. [CrossRef]
8. Wang, H.Q.; Liu, H.T.; Wang, L.; Min, L.; Chen, B.; Li, H. Uncovering the active components, prospective targets, and molecular mechanism of Baihe Zhimu decoction for treating depression using network pharmacology-based analysis. *J. Ethnopharmacol.* **2021**, *281*, 114586. [CrossRef]
9. Hui, H.; Li, X.; Jin, H.; Yang, X.; Xin, A.; Zhao, R.; Qin, B. Structural characterization, antioxidant and antibacterial activities of two heteropolysaccharides purified from the bulbs of Lilium davidii var. unicolor Cotton. *Int. J. Biol. Macromol.* **2019**, *133*, 306–315. [CrossRef]
10. Rong, L.; Lei, J.; Wang, C. Collection and evaluation of the genus Lilium resources in Northeast China. *Genet. Resour. Crop Evol.* **2011**, *58*, 115–123. [CrossRef]

11. Qin, K.; Fang, Q.; Cai, H.; Chen, Z.; Yang, G.; Cai, B. Analysis of chemical components in Baihe Zhimu Tang and its single herbs by high performance liquid chromatography-electrospray ionization-mass spectrometry. *Chin. J. Anal. Chem.* **2009**, *37*, 1759–1764.
12. Restaino, O.F.; De Rosa, M.; Schiraldi, C. High-performance capillary electrophoresis to determine intact keratan sulfate and hyaluronic acid in animal origin chondroitin sulfate samples and food supplements. *Electrophoresis* **2020**, *41*, 1740–1748.
13. Long, W.J.; Hu, Z.; Wei, L.; Chen, H.; Liu, T.; Wang, S.; Guan, Y.; Yang, X.; Yang, J.; Fu, H.Y. Accurate identification of the geographical origins of lily using near-infrared spectroscopy combined with carbon dot-tetramethoxyporphyrin nanocomposite and chemometrics. *Spectrochim. Acta A* **2022**, *271*, 120932. [CrossRef]
14. Long, W.J.; Wang, S.; Hai, C.; Chen, H.; Gu, H.W.; Yin, X.L.; Yang, J.; Fu, H.Y. UHPLC-QTOF-MS-based untargeted metabolomics revealing the differential chemical constituents and its application on the geographical origins traceability of lily bulbs. *J. Food Compos. Anal.* **2023**, *118*, 105194. [CrossRef]
15. Bian, Y.; Zheng, R.; Bayer, F.P.; Wong, C.; Chang, Y.C.; Meng, C.; Zolg, D.P.; Reinecke, M.; Zecha, J.; Wiechmann, S. Robust, reproducible and quantitative analysis of thousands of proteomes by micro-flow LC–MS/MS. *Nat. Commun.* **2020**, *11*, 157. [CrossRef] [PubMed]
16. Ferreira, D.C.; Hernandes, K.C.; Nicolli, K.P.; Souza-Silva, É.A.; Manfroi, V.; Zini, C.A.; Welke, J.E. Development of a method for determination of target toxic carbonyl compounds in must and wine using HS-SPME-GC/MS-SIM after preliminary GC× GC/TOFMS analyses. *Food Anal. Method.* **2019**, *12*, 108–120. [CrossRef]
17. Zhang, S.; Wang, H.; Zhu, M.J. A sensitive GC/MS detection method for analyzing microbial metabolites short chain fatty acids in fecal and serum samples. *Talanta* **2019**, *196*, 249–254. [CrossRef]
18. Fang, H.; Wu, H.L.; Wang, T.; Long, W.J.; Chen, A.Q.; Ding, Y.J.; Yu, R.Q. Excitation-emission matrix fluorescence spectroscopy coupled with multi-way chemometric techniques for characterization and classification of Chinese lager beers. *Food Chem.* **2021**, *342*, 128235.
19. Lenhardt, L.; Bro, R.; Zeković, I.; Dramićanin, T.; Dramićanin, M.D. Fluorescence spectroscopy coupled with PARAFAC and PLS DA for characterization and classification of honey. *Food Chem.* **2015**, *175*, 284–291. [CrossRef] [PubMed]
20. Oh, S.K.; Yoo, S.H.; Pedrycz, W. Design of face recognition algorithm using PCA-LDA combined for hybrid data pre-processing and polynomial-based RBF neural networks: Design and its application. *Expert Syst. Appl.* **2013**, *40*, 1451–1466. [CrossRef]
21. Fisher, R.A. The use of multiple measurements in taxonomic problems. *Ann. Hum. Genet.* **1936**, *7*, 179–188. [CrossRef]
22. Wu, H.L.; Shibukawa, M.; Oguma, K. An alternating trilinear decomposition algorithm with application to calibration of HPLC–DAD for simultaneous determination of overlapped chlorinated aromatic hydrocarbons. *J. Chemom.* **1998**, *12*, 1–26. [CrossRef]
23. Qing, X.D.; Zhou, H.B.; Zhang, X.H.; Wu, H.L.; Chen, C.Y.; Xu, S.W.; Li, S.S. Alternating trilinear decomposition of highly overlapped chromatograms for simultaneously targeted quantification of 15 PAHs in samples of pollution source. *Microchem. J.* **2019**, *146*, 742–752. [CrossRef]
24. Arimori, S.; Bosch, L.I.; Ward, C.J.; James, T.D. A D-glucose selective fluorescent internal charge transfer (ICT) sensor. *Tetrahedron Lett.* **2002**, *43*, 911–913. [CrossRef]
25. Wagner, B.D.; McManus, G.J. Enhancement of the fluorescence and stability of o-phthalaldehyde-derived isoindoles of amino acids using hydroxypropyl-β-cyclodextrin. *Anal. Biochem.* **2003**, *317*, 233–239.
26. Elcoroaristizabal, S.; Callejón, R.M.; Amigo, J.M.; Ocaña-González, J.A.; Morales, M.L.; Ubeda, C. Fluorescence excitation–emission matrix spectroscopy as a tool for determining quality of sparkling wines. *Food Chem.* **2016**, *206*, 284–290. [CrossRef] [PubMed]
27. Bahram, M.; Bro, R.; Stedmon, C.; Afkhami, A. Handling of Rayleigh and Raman scatter for PARAFAC modeling of fluorescence data using interpolation. *J. Chemom.* **2006**, *20*, 99–105. [CrossRef]
28. del Ácido Protocatéquico, E.A.; y el Ácido Caféico, Á.F. Antioxidant effects of protocatechuic acid, ferulic acid, and caffeic acid in human neutrophils using a fluorescent substance. *Int. J. Morphol.* **2010**, *28*, 911–920.
29. Wang, L.; Wu, H.L.; Yin, X.L.; Hu, Y.; Gu, H.W.; Yu, R.Q. Simultaneous determination of umbelliferone and scopoletin in Tibetan medicine Saussurea laniceps and traditional Chinese medicine Radix angelicae pubescentis using excitation-emission matrix fluorescence coupled with second-order calibration method. *Spectrochim. Acta A* **2017**, *170*, 104–110.
30. Senanayake, S.N. Green tea extract: Chemistry, antioxidant properties and food applications—A review. *J. Funct. Foods* **2013**, *5*, 1529–1541. [CrossRef]

chemosensors

Review

A Review on Non-Noble Metal Substrates for Surface-Enhanced Raman Scattering Detection

Ying Chen, Yuling Hu * and Gongke Li

School of Chemistry, Sun Yat-sen University, Guangzhou 510006, China; cheny2268@mail2.sysu.edu.cn (Y.C.); cesgkl@mail.sysu.edu.cn (G.L.)
* Correspondence: ceshyl@mail.sysu.edu.cn

Abstract: Surface-enhanced Raman scattering (SERS), a powerful spectroscopic technique owing to its abundant vibrational fingerprints, has been widely employed for the assay of analytes. It is generally considered that one of the critical factors determining the SERS performance is the property of the substrate materials. Apart from noble metal substrates, non-noble metal nanostructured materials, as emerging new substrates, have been extensively studied for SERS research by virtue of their superior biocompatibility, good chemical stability, outstanding selectivity, and unique physicochemical properties such as adjustable band structure and carrier concentration. Herein, in this review, we summarized the research on the analytical application of non-noble metal SERS substrates from three aspects. Firstly, we started with an introduction to the possible enhancement mechanism of non-noble metal substrates. Then, as a guideline for substrates design, several main types of materials, including carbon nanomaterials, transition metal dichalcogenides (TMDs), metal oxides, metal-organic frameworks (MOFs), transition metal carbides and nitrides (MXenes), and conjugated polymers were discussed. Finally, we especially emphasized their analytical application, such as the detection of pollutants and biomarkers. Moreover, the challenges and attractive research prospects of non-noble metal SERS substrates in practical application were proposed. This work may arouse more awareness of the practical application of the non-noble metal material-based SERS substrates, especially for bioanalysis.

Keywords: surface-enhanced Raman scattering; detection; non-noble metal nanomaterials substrate

Citation: Chen, Y.; Hu, Y.; Li, G. A Review on Non-Noble Metal Substrates for Surface-Enhanced Raman Scattering Detection. *Chemosensors* **2023**, *11*, 427. https://doi.org/10.3390/chemosensors11080427

Academic Editor: Barbara Palys

Received: 28 June 2023
Revised: 21 July 2023
Accepted: 27 July 2023
Published: 1 August 2023

1. Introduction

Surface-enhanced Raman scattering (SERS) has become an ideal approach for the real-time detection of various samples due to the advantages of unique fingerprint identification, no sample pretreatment, rapid detection, and low reagent consumption [1–4]. In fact, of the various factors that affect SERS performance, the design and fabrication of high-performance SERS substrates are the most critical to promoting the development of SERS technology. Traditionally, noble metal nanomaterials of different shapes and sizes have always been employed as SERS substrates, which can induce prominent localized surface plasmon resonance (LSPR) under electromagnetic radiation, generating huge electromagnetic field nearby, thereby drastically enhancing Raman signals of target molecules to achieve ultra-high SERS sensitivity [5,6]. As a promising complement to noble metal substrates, the introduction of non-noble metal nanostructures has injected new vitality into the traditional SERS and is of great significance for expanding the application of SERS.

So far, various non-noble metal nanomaterials with the advantages of high uniformity, good chemical stability, and significant biocompatibility have been widely studied as SERS substrates [7,8]. Compared to noble metal SERS substrates, they can not only achieve efficient electromagnetic enhancement (EM) with a low damping rate and reduced Ohmic loss [9–11] but also could adjust the charge transfer to achieve enhanced Raman scattering [12]. Moreover, recent studies have also shown that chemical enhancement

(CM) usually coexisted with EM in non-noble metal substrates, which indicated that the substrates effectively combined with the two mechanisms were expected to achieve highly sensitive SERS detection [13–15]. As a unique characteristic of non-noble metal nanomaterials, it can hinder the combination of electrons and holes by effectively transferring free electrons. Thus, the fluorescence background of probe molecules was destroyed, further improving the resolution of SERS detection [16,17]. Moreover, the high selectivity owing to the CM effect of non-noble metal SERS substrates can distinguish specific molecules from complex environments [18,19]. Based on this, non-noble metal nanomaterials, as the promising candidate nanomaterials for SERS substrates, show a huge application prospect in SERS detection. However, most reviews regarding non-noble metal SERS substrates are focused on the summary of the material engineering, such as the category of substrates and approaches to boost SERS performance, including morphology design, size adjustment, defect engineering, crystallinity, and phase structure [10,12]. In fact, non-noble metal SERS substrates also have practical application potential in many fields, but this aspect is often ignored.

In this review, we concentrate on the latest application progress of SERS active non-noble metal nanomaterials, as presented in Figure 1. Firstly, the enhancement mechanism of non-noble metal materials was discussed briefly. Subsequently, several main types of non-noble metal materials were introduced. Finally, as an important part of the review, various promising applications of non-noble metal SERS substrates for the detection of pollutants and biomarkers were summarized. Moreover, we proposed current challenges and future perspectives of non-noble SERS substrates in practical detection with a view to maximizing their great potential.

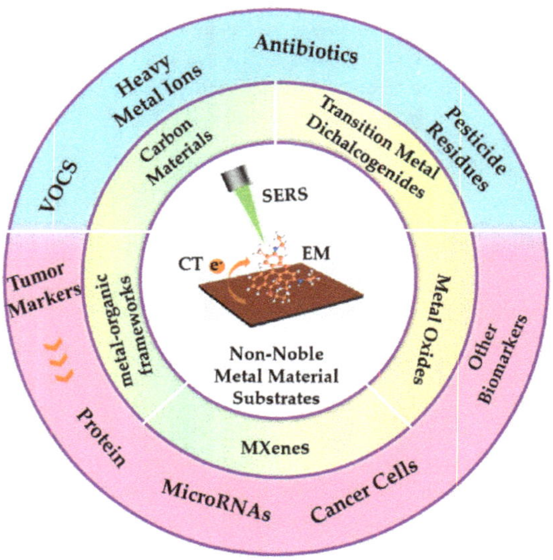

Figure 1. Overview of non-noble metal nanomaterials employed in SERS detection.

2. Raman Enhancement Mechanism of Non-Noble Metal SERS Substrates

Due to the wide variation in material preparation and properties, the mechanism of non-noble metal SERS substrates is relatively complex. According to the existing research results, the two key mechanisms that are widely accepted for describing the SERS enhancement effect include EM and CM [20,21].

EM is generally believed to result from the enhancement of local electromagnetic field generation by the collective oscillations of plasma nanoparticles or rough surfaces with surface roughness ranging from tens to hundreds of nanometers due to surface plasmon resonance (SPR) [9,22,23]. For most non-noble metal nanomaterials, the surface plasma frequency (ω_{sp}) rises in proportion to the electron density [24,25]. However, the electron density is presumed to be small in the conduction band of them. Thus, it is often stated that ω_{sp} lies in the infrared. Fortunately, the bandgap and carrier concentration of most non-precious metals is adjustable, resulting in an LSPR effect in the visible region. Meanwhile, Mie scattering can enhance the electric field in non-noble metal micro- or nanostructures at certain diameters. When the molecules adhered to the surface of non-noble metal materials satisfy the excitation conditions of Mie resonance, the generated electric near-field significantly enhances the coupling between the incident wave and the molecules, thereby enhancing the Raman signal of the absorbed molecules [24–26]. Compared to EM, CM mainly relies on the amplification of the Raman polarization tensor by chemical adsorption of molecules on the surface of non-noble metal nanomaterials and plays a more critical role in the Raman enhancement of non-noble metal SERS substrates. For the molecular-non-noble metal nanomaterial system, the enhancement of CM mainly comes from charge transfer (CT), which depends on the energy levels between a molecule and the substrate, which results in the enhanced selectivity of the substrates towards the molecules [27,28]. When the CT occurred between the molecule and the substrate material, the molecular polarizability and electron density distribution changed, resulting in the SERS effect [29].

In order to further improve the SERS performance of non-noble metal substrates, a "coupled resonance" effect was proposed. It requires that the selection of the substrates-molecule system and relevant experimental parameters must conform to the SERS rules of producing synergistic resonance effects, including molecular resonance, substrates-molecular CT resonance, and exciton resonance [30]. It is worth mentioning that exciton resonance is also a unique electronic property of most non-noble metal materials, and the SERS performance is also improved when the frequency of the incident light is close to the exciton resonance frequency.

3. Classification of Non-Noble Metal SERS Substrates

As promising complements to the current most popular plasmonic metal SERS substrates, non-noble metal materials have attracted more and more researchers to engage in the SERS field due to their many appealing attributes. Firstly, we classified and summarized non-noble metal SERS substrates, including carbon nanomaterials, transition metal dichalcogenides (TMDs), metal oxides, metal-organic frameworks (MOFs), transition metal carbides and nitrides (MXenes), and conjugated polymers.

3.1. Carbon Materials-Based SERS Substrates

Carbon materials have been widely employed as the SERS active substrates due to their excellent surface uniformity and enhancement efficiency [31]. Graphene, as a typical carbon material, is prone to coupling with biomolecules and aromatic chemicals due to the presence of a large delocalized π bond and shows obvious merits as a SERS substrate [32]. Ling et al. [33] first reported the SERS phenomenon on graphene surfaces using phthalocyanine (Pc), rhodamine 6G (R6G), protoporphyrin IX (PPP), and crystal violet (CV) as probe molecules in 2010 (Figure 2). Researchers have fabricated graphene with higher SERS activity by doping or forming heterojunctions. For example, Feng et al. [34] prepared nitrogen-doped graphene as a SERS substrate and confirmed that this substrate had a lower SERS detection limit than graphene. Ghopry et al. [35] reported a graphene heterostructure SERS substrate with high sensitivity. The detection range of R6G obtained on the substrate is 5×10^{-11}~5×10^{-12} M. In addition to graphene, its oxidized derivatives, such as graphene oxide (GO), can also be used as SERS substrates. Singh et al. [36] used GO and RGO sheets with large areas as SERS substrates for the detection of rhodamine B. The result exhibited a high EF of 10^4 and a low detection limit of 10 nM.

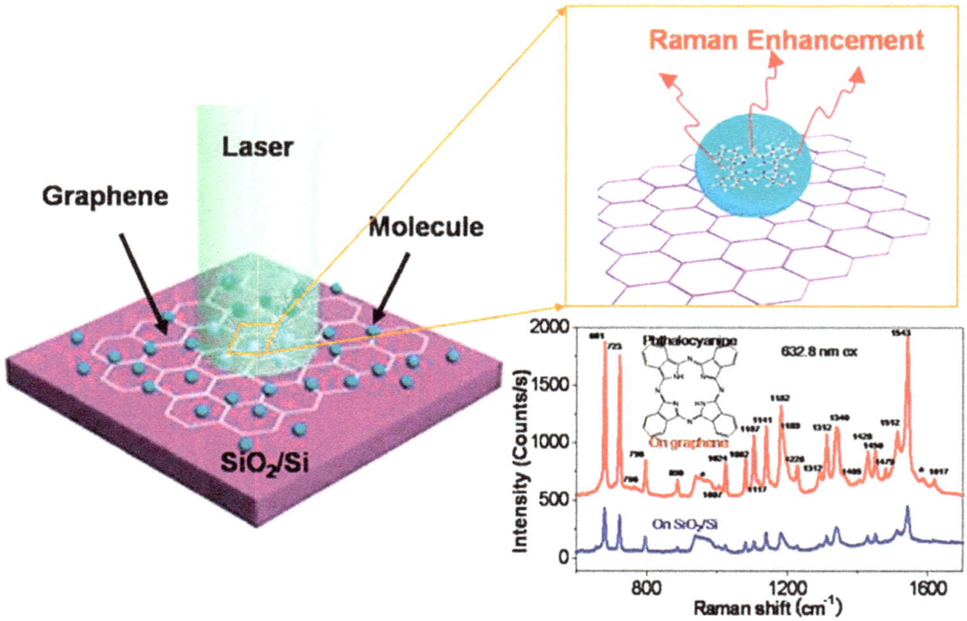

Figure 2. Schematic illustration of the molecule on graphene and a SiO_2/Si substrate, and the comparison of Raman signals of phthalocyanine deposited on graphene (red line) and on the SiO_2/Si substrate (blue line) at 632.8 nm excitation. Reprinted with permission from [33]. Copyright 2010 American Chemical Society.

3.2. TMDs-Based SERS Substrates

TMDs can be represented by a general formula MX_2, in which M is a transition metal atom (group III B-VIII), and X is a chalcogen atom (S, Se, Te). It has tunable optical and electronic properties, resulting in many excellent photoelectric properties, including SERS [37,38]. Molybdenum disulfide (MoS_2) is the most typical material in TMDs. It was also the first TMDs with SERS activity found by Ling et al. [39]. However, its enhancement effect was low. Therefore, researchers have proposed different strategies to improve the SERS performance of MoS_2. Considering the layer-dependent effect of two-dimensional (2D) MoS_2, Lee et al. [40] demonstrated that the monolayer MoS_2 produced the most obvious Raman enhancement than 3 L and bulk MoS_2. Moreover, Li et al. [41] achieved efficient CT from MoS_2 to probe molecules by adjusting the interlayer distance of MoS_2 and obtained an EF as high as 5.31×10^5 (Figure 3). Majee et al. [42] prepared interconnected and vertically oriented few-layer MoS_2 nanosheets (NTNs) as an ultrasensitive SERS substrate by adjusting the growth arrangement mode of MoS_2. In addition to the above methods, other methods, including defect/active site induction [43], chemical doping [44], and heterostructure design [45], are also useful in improving the SERS performance of MoS_2. The study of MoS_2 SERS activity has opened the way to study the SERS enhancement of TMDs. Subsequently, a series of TMDs with complex band structures and rich electronic states have been explored, such as SnS_2, $MoTe_2$, $NbTe_2$, and so on [46–48].

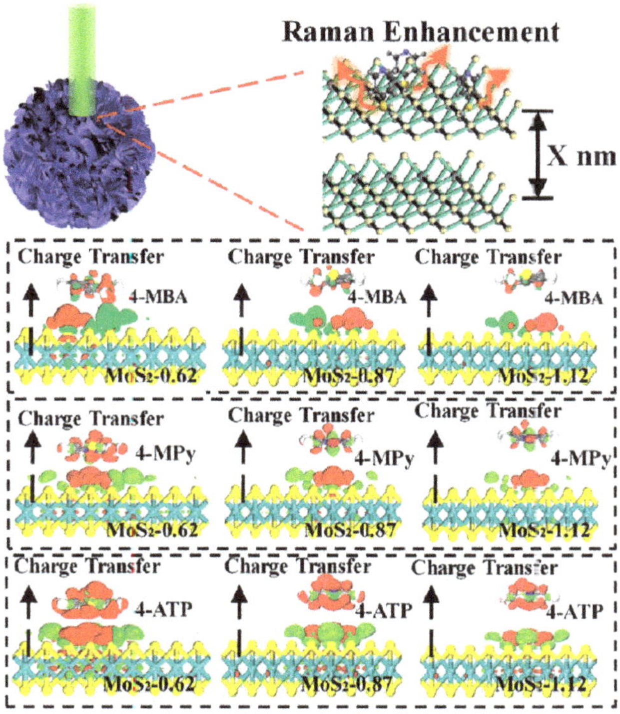

Figure 3. Scheme of the Raman enhancement and difference of total electron density distributions between MoS_2-0.62 (MoS_2-0.87 or MoS_2-1.12) and 4-MBA/4-MPy/4-ATP. The red (green) distribution stands for electron accumulation (depletion). Reprinted with permission from [41]. Copyright 2014 American Chemical Society.

3.3. Metal Oxide-Based Substrates

The promising properties such as physio-chemical stability, high uniformity, high refractive index, and tunable band gap of metal oxide make them excellent SERS substrates [49]. As early as 1982, Yamada et al. observed the SERS effect on the surface of NiO and TiO_2 [50,51]. With the continuous gushing of metal oxides, it was proved that many metal oxide materials exhibit SERS activity, including ZnO, Fe_3O_4, Cu_2O, MoO_2, WO_3, VO_2, Nb_2O_5, Ta_2O_5, Ti_3O_5, etc. [10]. Among them, MoO_2 is considered to be an ideal material for SERS substrate because it has a large number of free electrons, which can lead to the formation of the LSPR effect [52]. Wu et al. [53] synthesized ultrathin MoO_2 nanosheets (NTNs) as a SERS substrate; they demonstrated that the substrate has superior signal uniformity in the whole area with a limit of detectable concentration down to 4×10^{-8} M and EF up to 2.1×10^5. Moreover, Miao et al. [54] revealed enhanced SERS signals of probe molecules on defect-rich VO_2 NTNs. Moreover, Li et al. [55] obtained novel quasi-metallic γ-Ti_3O_5 with a strong LSPR effect by reducing TiO_2 microspheres and using it as a SERS substrate. The results showed that the reduced γ-Ti_3O_5 showed a 10,000-times increase in SERS sensitivity compared to TiO_2 microspheres, and the lowest detectable limit is up to 10^{-10} M (Figure 4). In fact, in addition to fabricating novel metal oxide substrates with high SERS activity, it is also an effective strategy to explore their enhancement mechanism to improve their sensitivity. The crystallinity of metal oxides can affect SERS performance by adjusting the CT process. Compared with crystalline metal oxides, amorphous structures are more conducive to the electron transfer process [10,56]. This could be ascribed to the fact that amorphous structures can reduce electron confinement, improve light scattering (such as ZnO nanocages), or enhance molecular adsorption (such as MoO_{3-x} quantum dots). In addition, the SERS substrates of sub-stoichiometric metal oxides have rich defect electronic states associated with surface oxygen vacancies, which can synergistically enhance excitons and CT resonance, providing new and efficient charge escape, thus improving SERS sensitivity [57,58].

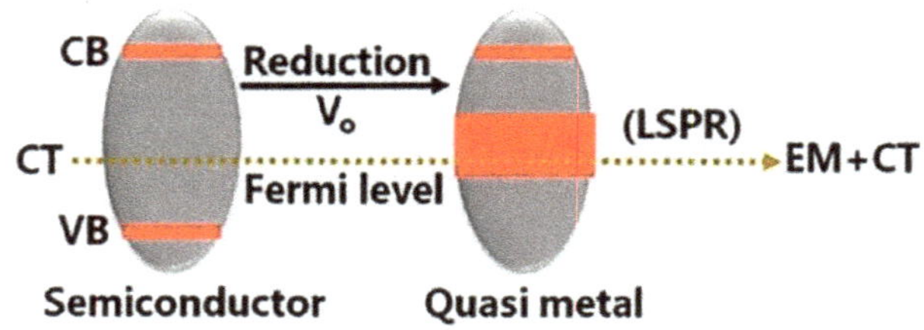

Figure 4. By introducing oxygen vacancies, TiO_2 semiconductors with wide-band gaps are transformed into quasi-metals γ-Ti_3O_5 with rich free electrons. Reprinted with permission from [55]. Copyright 2019 American Chemical Society.

3.4. MOFs-Based Substrates

MOFs are a new class of hybrid organic-inorganic supramolecular materials that have a large surface area, porous properties, chemical stability, and uniform and tunable nanostructured cavities, which strongly favor potential use as a substrate [59–61]. Although the SERS phenomenon based on noble metals and MOF hybrids has been discussed [62–64], MOF materials are generally regarded as a template to load tightly ordered and well-dispersed SERS-active NPs, and it is believed that the SERS enhancement resulted from the contributions of noble metal particles. Until Yu et al. [65] reported the SERS enhancement phenomenon of methyl orange molecule on the surface of MOFs in 2013 (MIL-100 and MIL-101), thereafter, with the continuous exploration of MOF materials, the exploration of their SERS activity also received attention. Sun et al. [66] achieved the transformation of MOF materials from non-SERS active substrates to SERS active substrates by careful modulation of metal centers, organic ligands, and framework topologies and obtained an EF of up to 10^6 (Figure 5). Moreover, Fu et al. [67] reported MIL-100 (Fe) as a SERS substrate with high SERS activity.

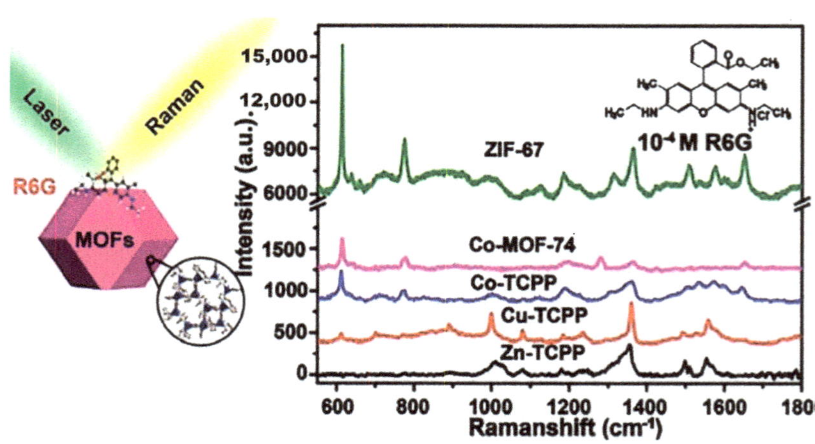

Figure 5. Schematic illustration of the molecule on MOFs substrate and the comparison of Raman signals of R6G deposited on Co-MOF-74 and on the TCPP-based MOF compounds with different metal ions. Reprinted with permission from [66]. Copyright 2018 American Chemical Society.

3.5. MXene-Based Substrates

MXenes, composed of transition metal carbides/nitrides, is a kind of semi-metal material with high electric conductivity and has the ability to generate plasmon resonance in the visible or near-infrared range. These unparalleled properties render them promising candidates for SERS substrates [32,68,69]. Sarycheva et al. [69] first demonstrated that 2D $Ti_3C_2T_x$ MXene as a SERS substrate exhibited good SERS activity with an EF of up to 10^6. An important potential advantage of MXenes is the ability to deposit materials on different substrates, including flexible materials. Soundiraraju et al. [70] prepared different SERS substrates based on paper, silicon, and glass by loading Ti_2NT_x onto them. The results showed that Ti_2NT_x had the highest enhancement efficiency up to 10^{12} on paper-based substrates (Figure 6). In addition, Yang et al. [71] introduced ultraflexible characteristics of $Ti_3C_2T_x$ MXene into SERS microfluidic chips, which significantly improved the detection sensitivity of SERS microfluidic sensors. Although the development of MXene-based substrates is still in the first stages, according to the current reported results, MXene materials will become promising novel SERS substrates.

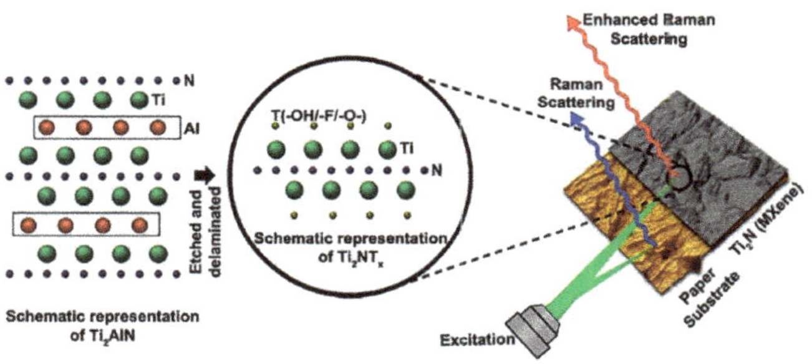

Figure 6. Illustration of the preparation of Ti_2NT_x and the detection of analysis. Reprinted with permission from [70]. Copyright 2017 American Chemical Society.

3.6. Conjugated Polymers-Based Substrates

Conjugated polymers are organic macromolecules that have attractive optical properties due to the delocalized π electron cloud on their main chain [27,72]. It is considered a promising new type of non-noble metal SERS substrate. Among them, covalent organic frameworks (COFs), as a newly emerging material of conjugated polymer, have been widely concerned by researchers in the SERS field due to their high specific surface area, excellent thermal stability, and abundant surface functional groups [73,74]. In particular, small molecule screening can be realized based on the synergistic effect of pore and surface functional groups of COFs materials. It was helpful to improve the selectivity of SERS detection. Nevertheless, the Raman enhancement ability of COFs substrates is much lower than inorganic and noble metal SERS substrates. At present, in order to gain a high analytical sensitivity, metal-COFs composites were introduced into SERS-based assays.

4. Non-Noble Metal Materials for Analytical Application

Non-noble metal nanomaterials have recently been widely used as SERS substrates in various fields due to their particular advantages. For example, they have abundant surface sites that are different from noble metal materials, which provide a large number of coordination sites for the modification and functionalization of other molecules. Meanwhile, the combination of the two can further improve the stability of the modified molecules on the non-noble metal materials surface. In addition, non-noble metal materials can also be used as flexible substrates, which makes it possible to in situ detect pesticide residues on the surface of fruits and vegetables. The good biocompatibility of non-noble metal substrates especially makes them produce great breakthroughs in bioimaging, cancer diagnosis, and treatment [75–77]. It should be mentioned that the unique catalytic properties of non-noble metal materials also enable the development of renewable materials. Herein, in this section, we will give a detailed introduction to the recent application of non-noble metal materials as SERS substrates in different fields.

4.1. Detection of Pollutants

The increase in pollutants such as volatile organic compounds (VOCs), heavy metal ions, and antibiotics every year has created an enormous burden threatening ecological cycles and public health. Thus, the detection of them was given high priority. SERS, as a universal analytical technique, has been considered a label-free and ultra-sensitive detection tool for the detection of a wide range of adsorbent molecules [78–80]. As SERS substrates with important and promising at present, non-noble metal nanomaterials have the advantages of being environmentally friendly, non-energy intensive, nontoxic, and cost-friendly, and have wide application prospects in pollutant detection. Some of these non-noble metal SERS substrates have a large surface area, which can achieve high enrichment of heavy metal ions and gases, thus improving their detection sensitivity. In addition, the inter-band-gap states electrons of non-noble metal substrates with band gaps can promote the reduction of the redox potential of the ions couple, resulting in the change of valence states of heavy metal ions so as to achieve the selective detection of them.

4.1.1. Detection of VOCs

Yang et al. [71] prepared SERS substrate by adhering ultraflexible $Ti_3C_2T_x$ MXene onto the surface of three-dimensional (3D) honeycomb arrays and introduced them into microfluidic chips (Figure 7a). $Ti_3C_2T_x$ MXene has a universal high adsorption efficiency of various gases, and its super flexible well retained the sophisticated nanomicrostructure that generated the vortex fields to extend the molecule residence time and kept the analytes inside the SERS-active hot spots, leading to the increased sensitivity. Thus, three typical VOCs with different functional groups, including 2, 4-dinitrotoluene (DNT), and benzaldehyde and indole, can be achieved by the SERS-vortexene chip with high repeatability. The LODs for DNT, benzaldehyde, and indole were about 10, 10, and 50 ppb, respectively.

4.1.2. Detection of Heavy Metal Ions

Lead, arsenic, mercury, and other heavy metals are not only the most toxic contaminants but they are considered human carcinogens [81,82]. In order to effectively monitor the above-mentioned heavy metal pollutants, a great deal of work based on SERS detection has been reported. For example, Zhang et al. [83] prepared ZnO submicron flowers (ZnO SFs) as a substrate for sensitive detection of Pb^{2+}. The ZnO SFs substrate exhibited a narrower energy band, which promoted the coupling resonance of the molecular-substrate system, thus improving the CT efficiency and achieving a significant SERS enhancement effect. Furthermore, by utilizing the free-legged DNA walker amplification strategy, an ultrasensitive biosensor was constructed to monitor Pb^{2+}. The experimental data demonstrated that this biosensor could efficiently determine Pb^{2+} with a low detection of 3.55 pM. Similarly, Parveen et al. [84] used ZnO nanoparticles (NPs) functionalized single-wall carbon nanotubes (SWCNTs) as a nanocomposite SERS sensor for the detection of Pb^{2+} ions in an aqueous medium. The prepared sensor exhibited high selectivity and low limit of detection for Pb^{2+} ions (0.225 nM). In addition, a dye-sensitized non-noble metal colloid system can also show synergistic effects in the detection of metal ions. Ji et al. [85] reported an example of using alizarin red S (ARS) sensitized colloidal TiO_2 NPs as a SERS substrate for the detection of Cr (VI) in water. In the presence of Cr (VI), the ARS molecule stably adsorbed on the surface of colloidal TiO_2 NPs can achieve self-degradation, leading to the decrease of SERS signal intensity. The detection of Cr (VI) can be achieved by monitoring the SERS intensity of the ARS probe molecule. This ARS-TiO_2 complex could be further applied to the detection of Cr (VI) in environmental samples.

4.1.3. Detection of Antibiotics

Antibiotic chemicals are not only an important source of water pollution, but also the existence of excessive antibiotics can cause the problem of antibiotic resistance, which poses a serious threat to human and animal health [86,87]. Therefore, ensuring the rational use of antibiotics and effective monitoring has always been the focus of research. Recent studies have shown that antibiotics can be effectively detected by non-noble metal SERS substrates. For example, Wang et al. [48] prepared 2D niobium ditelluride ($NbTe_2$) nanosheets (NTNs) for the detection of ciprofloxacin (CIP) and enrofloxacin (ENR) antibiotics molecules (Figure 7b). Based on the fluorescence quenching effect of $NbTe_2$ NTNs and CT between the analytes, the ultrasensitive detection of CIP and ENR was achieved. The LODs were 35.1 and 35.9 ppb, respectively. In addition, Singh et al. [88] synthesized 3D MoS_2 nanoflowers (NFs) with tunable surface area (5–20 m^2/g) as a SERS substrate to realize ultrafast SERS detection of various antibiotics. In particular, MoS_2 NFs have a narrow bandgap and high specific surface area, which significantly enhances light-harvesting and adsorption capability, leading to the improvement of their photodegradation behavior. Therefore, the photocatalytic degradation experiment-based SERS monitoring was realized. The results showed that even a very low amount (0.025 mg) of MoS_2 NFs (20 m^2/g) can tremendously decompose oxytetracycline hydrochloride (OTC-HCl) molecules within 60 min under visible light.

4.1.4. Detection of Pesticide Residues

The excessive use of pesticides has gained increasing attention in recent years owing to their enormous menace to human health. SERS is a useful ultrasensitivity detection tool, which is of great significance for the monitoring of pesticide residue levels. As excellent SERS substrates for the detection of pesticide residues, non-noble metal nanomaterials not only exhibited trace level SERS response but also could be used for in situ detection of pesticide residues on the surface of fruits and vegetables. For example, Zhang et al. [89] prepared a plasmonic MoO_2 nanosphere as a highly sensitive SERS substrate for the detection of various pesticides (Figure 7c). In order to verify the feasibility of this substrate in practical application, clenbuterol hydrochloride (CH), methyl parathion (MP), and 2, 4-dichlorophenoxyacetic acid (2, 4-D) were chosen as pesticide models. Experimental data indicated that this substrate could achieve the quantification of the above pesticide models with a LOD of 10^{-7} M. Another balsam pear-shaped CuO with dense nanoparticle protuberance was reported by Liang et al. [90]. In their work, the paraquat solution was taken as an example. Attributed to the CuO substrate's unique shape and rough surface, more "hot spots" were generated on the particle surface, which significantly enhanced the SERS effect. This substrate could detect paraquat with a detection limit of 275 $\mu g\ L^{-1}$. Moreover, Quan et al. [91] successfully prepared MoS_2/TiO_2 substrate (MTi_{20}) assembled from MoS_2 nanoflowers (NFs) and TiO_2 NPs, which exhibited ultrahigh SERS response to α-endosulfan, the LOD of α-endosulfan can reach 10^{-8} M. In addition, Gokulakrishnan et al. [92] fabricated a flexible rGO SERS active substrate for in situ detection of thiabendazole (TBZ). Owning to the large surface area, the flexible rGO SERS substrate could detect TBZ with a LOD of 1 nM. Furthermore, they further demonstrated that the flexible SERS substrates could also be used for detecting pesticides on target surfaces with complex geometries like fruits, leaves, and vegetables.

Table 1 summarizes the SERS detection results of non-noble metal substrates for various pollutants above-mentioned, such as VOCs, heavy metal ions, antibiotics, and pesticide residues.

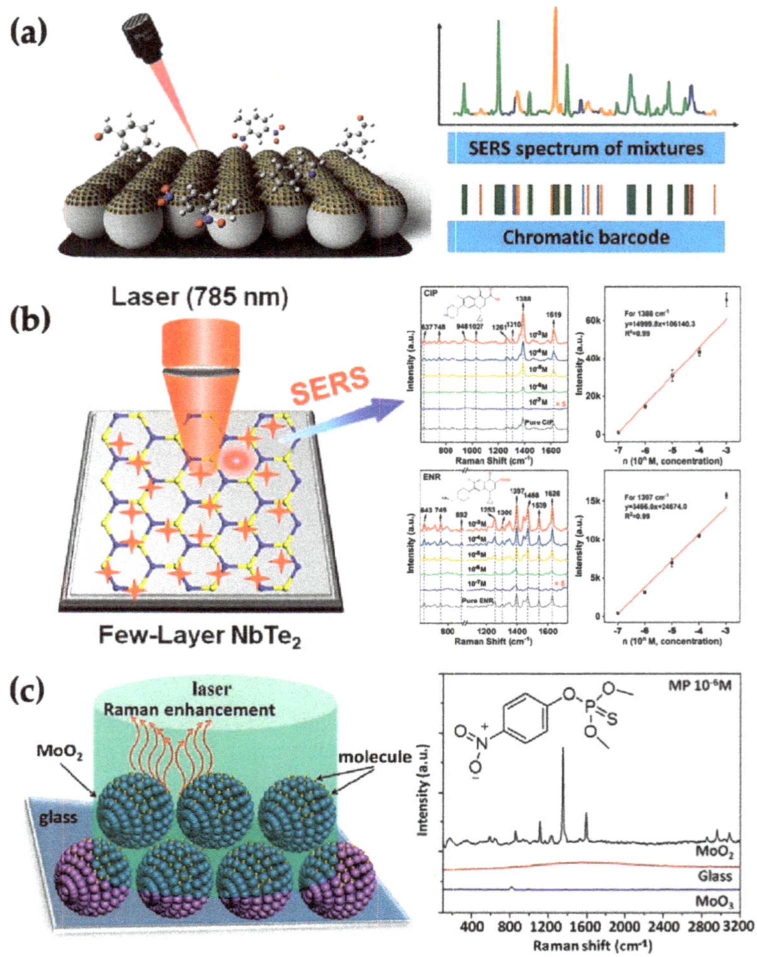

Figure 7. (**a**) Schematic diagram of multiplex VOCs mixture detection. (**b**) Schematic illustration of the ultrathin $NbTe_2$-based SERS experiment and SERS measurement of ciprofloxacin and enrofloxacin antibiotics molecules adsorbed on $NbTe_2$ NTNs substrates. (**c**) Schematic diagram of the SERS experiment and SERS spectra of methyl parathion on the MoO_2 substrates. (**a**) Reprinted with permission from [71]. Copyright 2021 American Chemical Society. (**b**) Reprinted with permission from [48]. Copyright 2020 American Chemical Society. (**c**) Reprinted with permission from [89]. Copyright 2017 American Chemical Society.

Table 1. A summary of non-noble meta-based SERS substrates for pollutant detection is mentioned in this review.

SERS Substrate	Enhancement Mechanism	Target	Linear Range	LOD	Ref
Honeycomb-like 3D $Ti_3C_2T_x$	Synergistic effect of strong adsorption capacity of substrate and EM	Benzaldehyde 2, 4-dinitrotoluene indole	/	10 ppb 10 ppb 50 ppb	[71]
ZnO SFs	Synergistic effect of CT and Mie resonance enhancement	Pb^{2+}	10 pM–100 μM	3.55 pM	[83]

Table 1. *Cont.*

SERS Substrate	Enhancement Mechanism	Target	Linear Range	LOD	Ref
ZnO@SWCNTs	Synergistic effect of "hot spot" generated by intertube and interparticle coupling, as well as CT enhancement	Pb^{2+}	0.01–100 μM	0.225 nM	[84]
ARS-TiO_2 complexes	CT enhancement	Cr (VI)	0.6–10 mM	0.6 mM	[85]
$NbTe_2$ NTNs	Synergistic effect of CT enhancement and fluorescence quenching effect of substrate	Ciprofloxacin enrofloxacin	351 ppm–35.1 ppb	35.1 ppb 35.9 ppb	[48]
3D-MoS_2 NFs	Synergistic effect of effective enrichment capability of substrate and CT enhancement	Oxytetracycline hydrochloride	/	/	[88]
MoO_2 nanospheres	LSPR effect in the visible and near-infrared (NIR) range	Clenbuterol hydrochloride methyl parathion 2,4-dichlorophenoxyacetic acid	10^{-7}–10^{-4} M	10^{-7} M	[89]
Balsam pear-shaped CuO	Synergistic effect of EM and CT enhancement	Paraquat	10^{-7}–10^{-3} M	10^{-7} M	[90]
MTi_{20}	Interfacial CT enhancement	α-endosulfan	10^{-8}–10^{-4} M	10^{-8} M	[91]
Flexible rGO	Synergistic effect of effective enrichment of substrate and CT enhancement	Thiabendazole	10^{-9}–10^{-3} M	10^{-9} M	[92]

4.2. Detection of Biomarkers

Currently, cancer remains one of the most serious health problems in the world and the leading cause of death [93,94]. Therefore, the realization of cancer screening is particularly important, which can help find early cancers and improve survival. Among the many detection techniques, SERS is a trace detection technique with a unique fingerprint spectrum, and its application can even be extended to a single molecule detection level. As a good candidate for studying and simulating biological reactions, non-noble metal SERS substrates can not only serve as a potential interface for immobilizing biomolecules but also maintain their biological activity. In this way, tumor markers, including protein, nucleic acid, and circulating tumor cells (CTCs), can be monitored by non-noble metals SERS substrates to achieve early screening of cancers.

4.2.1. Detection of Protein Tumor Markers

As an example of carbohydrate antigen 19-9 (CA19-9), which has been a widely used clinical protein marker of cancer in human serum, Jiang et al. [16] designed a SERS-based immunoassay mediated by MoS_2 for the sensitive and specific monitoring of carbohydrate antigen CA19-9. In their work, the corresponding antibodies absorbed on the MoS_2 NTNs and MoS_2 NFs modified by the probe molecule R6G, respectively. Thus, a sandwich immunostructure was constructed by the specific recognition of anti-CA19-9 and target CA19-9. The concentration of target CA19-9 was indirectly reflected by the intensity variation of the SERS signal of R6G. The sandwich immunostructure could detect target CA19-9 with a LOD of 3.43×10^{-4} IU·mL^{-1}. Furthermore, it could also be used for target CA19-9 detection in several clinical patient serum samples, which had higher sensitivity compared with the conventional chemiluminescent immunoassay (CLIA) strategy. Using a similar principle of specific recognition, Liu et al. [95] designed methylene blue (MB) modified aptamer attached to the tungsten trioxide (WO_3) film for the SERS detection of glioblastoma biomarkers vascular endothelial growth factor (VEGF) in human serum samples. In their work, MB-modified aptamer serves as a SERS signal source. After the aptamer specifically linked to VEGF, it folded into a conformationally restricted stem-loop structure, forcing the MB molecule to contact the substrate, which leads to a significant enhancement of SERS signals, thus achieving sensitive detection of VEGF. The proposed aptasensor could detect VEGF with high selectivity, while the detection limit was down to 8.7 $pg{\cdot}mL^{-1}$.

4.2.2. Detection of MicroRNAs

MicroRNAs (miRNA) is a noncoding single-stranded RNA molecule with approximately 18–25 nucleotides and has been verified as promising biomarkers for cancer diagnosis [96,97]. For example, Jiang et al. [98] developed a 3D WO_3 hollow microsphere as a SERS-active substrate for miRNA 155 detection (Figure 8a). The 3D WO_3 hollow microsphere with a small band gap and rich surface defects promoted an increase in CT, resulting in a large SERS enhancement. Then, by employing a catalytic hairpin assembly (CHA) strategy, a SERS biosensor based on a 3D WO_3 hollow microsphere was constructed for the sensitive detection of miRNA 155 with a LOD of 0.18 fM. In order to verify the feasibility of this biosensor, miRNA 155 from HeLa (cervical cancer cells) and MDA-MB-231 (human breast cancer cells) were chosen as models. Experimental data proved that this biosensor could achieve the detection of miRNA 155 in actual samples. Similar to the above miRNA 155 detection, Liu et al. [99] developed a multifunctional SERS platform composed of hexagonal boron nitride nanosheets (h-BNNS), HG DNA oligonucleotide, and copper(II) phthalocyanine (CuPc) (CuPc@HG@BN) for the detection of miR-21. Benefiting from the circle amplification of miRNA and the high SERS effect of hBNNS, the detection of miR-21 in live cells can be realized as low as 0.7 fM.

4.2.3. Detection of Cancer Cells

Circulating tumor cells (CTCs) are shed by the primary tumor, representing the characteristics of a certain tumor and playing a central role in tumor dissemination and metastases. Therefore, their detection and analysis can be very valuable in the early diagnosis of cancer [100,101]. Recently, Xu et al. [102] reported a TiO_2-based nontoxicity SERS bioprobe for CTCs detection. They used reduced bovine serum protein (rBSA) and folic acid (FA) to functionalize TiO_2 NPs and used 4-mercaptobenzoic acid (4-MBA) adsorbed on the surface of TiO_2 NPs as a signal molecule. Benefiting from the high targeting ability of FA towards cancer cells overexpress FR, the proposed SERS bioprobe can be effectively identified different kinds of cancer cells in rabbit blood, which has a LOD of 1 cell/mL. Due to the high efficiency of interfacial photo-induced charge transfer (PICT) and strong vibration coupling effect, the amorphous nanomaterial-molecular system has significant SERS activity. Inspired by this, Lin et al. [56] designed black TiO_2 NPs (B-TiO_2 NPs) with crystal-amorphous core–shell structures for the precision diagnosis and treatment of cancer. The feasibility of applying the high-sensitivity SERS bioprobe based on B-TiO_2 NPs modified with Alizarin red (AR) signal molecule, polydopamine (PD) layer, and an antibody (AB) was verified by the detection of MCF-7 drug-resistant (MCF-7/ADR) breast cancer cells (Figure 8b). Based on of FA specific recognition and ultrahigh SERS activity of B-TiO_2 NPs, a novel strategy that combined microfilter CTC isolation and SERS bioprobe detection for in situ isolating and directly detecting CTCs from peripheral blood at single-cell resolution was proposed by Xu et al. [103] The SERS bioprobe was composed by Raman reporting molecule AR and FA functionalized B-TiO_2 NPs. Benefiting by rapid separation of microfilters and the high sensitivity of SERS spectroscopy, the SERS bioprobe of B-TiO_2-AR-PEG-FA could distinguish FR-positive CTCs from peripheral blood cells efficiently within 1.5 h, and the LOD of CTCs in rabbit blood can reach to 2 cells/mL. Furthermore, this strategy has also been successfully detected in blood samples from cancer patients. In addition, Feng et al. [104] fabricated ternary heterostructure Fe_3O_4@GO@TiO_2 (MGT) NC and used it to develop a robust SERS probe for the sensitive and selective detection of triple-negative breast cancer cells (TNBCs). In their work, Abs can adsorb on the surface of MGT, and the cancer cells can be specifically captured by MGT-ABs. The introduction of CuPc molecules can reflect the SERS signals in the detection system. In this way, TNBCs models such as HCC38, MDA-MB-231, and MCF-7 detection can be monitored by the MGT-Abs-CuPc nanoprobe, and the LOD of HCC38 cell was calculated to be ~3 cells. Different from the above studies on TiO_2-based CTC SERS detection, Haldavnekar et al. [105] introduced a quantum-size 3D ZnO SERS probe for the first time in the study of labeling-free in vitro diagnosis of cancer. Due to the significant reduction of the probe size, the surface defects, such as oxygen vacancies and stacking faults, resulted in a unique exponential increase in SERS. The in vitro sensing of the 3D ZnO SERS probes was demonstrated by successfully distinguishing between cancer and non-cancer cells based on the ratio of peak intensities of lipids and proteins (I_{1445}/I_{1654}).

4.2.4. Detection of Other Biomarkers

SERS is very promising for identifying disease markers due to its unique fingerprint spectrum, high sensitivity, and large dynamic range detection [106]. To pursue more excellent biocompatibility with biological samples and spectral stability as well as reproducibility, non-noble metal substrates have been gradually applied to detect various disease markers. Huang et al. [107] implemented a signal enhancement strategy featuring graphene as a substrate for the label-free detection of key blood constituent hemoglobin (Hb) and albumin (Alb) (Figure 8c). This is the first time demonstrated that the Raman signal of biomolecules was enhanced when contacted with graphene. Similarly, Dharmalingam et al. [108] proposed the concept of atomic defect enhancement for quantum probe Raman scattering (DERS) at the molecular level detection. They significantly improved the sensitivity of DERS probes by introducing high concentrations of atomic scale defects into non-plasma quantum size probes (TiO_{2-x}). Thus, the obtained DERS probes could be used for the label-free Raman detection of biomarkers such as ATP and EGFR peptides with low Raman cross-sections. Moreover, graphene-like nanomaterials such as MoS_2 have been used to detect biomolecules. Typically, Su et al. [109] successfully synthesized Ni-doped MoS_2 NFs (Ni-MoS_2 NFs) for bilirubin detection in serum. The experimental analysis results showed that the Ni-MoS_2 NFs SERS substrate could detect bilirubin in serum samples with a detection limit as low as 10^{-7} M. In addition to the detection of biomarkers in the blood samples, Fu et al. [67] prepared an excellent SERS active substrate MIL-100 (Fe). Because of MIL-100 (Fe) unique "array sensing" characteristics, it has been successfully used for the sensitive detection of gas biomarkers of lung cancer, such as 4-ethylbenzaldehyde, acetone, and isopropanol.

Severe acute respiratory syndrome-coronavirus-2 (SARS-CoV-2) is a highly infectious virus serious that poses a serious threat to human health [110,111]. Therefore, achieving rapid, ultrasensitive, and highly reliable detection of SARS-CoV-2 in patients' bodies is of great significance for controlling the epidemic. Peng et al. [46] designed a spherical SnS_2 with the "nano-canyon" morphology as a SERS substrate for the detection of SARS-CoV-2 S protein and RNA to recognize and diagnose SARS-CoV-2. Benefiting from the unique nano-canyon structure, the capillary effect generated on the surface of SnS_2 microspheres can significantly enrich the detection molecules. At the same time, the synergistic effect of CM can achieve the EF of MB up to 3.0×10^8 and can identify various physical forms of SARS-CoV-2 with high sensitivity. This result exhibited that the EF and LOD achieved here even much better than most reported noble metal-based SERS substrates (Table 2). In addition to the SnS_2 microspheres mentioned above, Peng et al. [15] also reported semi-metal material Nb_2C and Ta_2C MXenes with remarkable SERS enhancement as SERS substrates for SARS-CoV-2 spike protein detection. As compared with Nb_2C, the Ta_2C substrate had a better selective SERS-enhanced effect on molecules. Therefore, Ta_2C was chosen as a model substrate for the detection of SARS-CoV-2 S protein. The detection limit of SARS-CoV-2 S protein was as low as 5×10^{-9} M. Similarly, Fraser et al. [47] prepared few-layered $MoTe_2$ films as a SERS substrate for β-sitosterol detection in a complex cell culture media. Among others, Li et al. [112] reported 2D hafnium ditelluride ($HfTe_2$) as a new label-free SERS substrate to detect trace uric acid, the important biomarker for gout disease, which demonstrated a reliable LOD of 10^{-4} M. In addition, Sun et al. [113] proposed to use the photo-excited D-MIL-125-NH_2 material, which could obtain stable photo-induced oxygen vacancy (PIVo) defect as a SERS substrate for the detection of disease markers. In order to evaluate the SERS activity of the D-MIL-125-NH_2 substrate, dopamine was chosen as a probe molecule. It was reported that the detection of dopamine has been linked to the diagnosis of various diseases, such as Parkinson's disease [114,115]. The experimental data demonstrated that this SERS substrate could efficiently determine dopamine with a detection limit of 6.6×10^{-7} M, which were two orders of magnitude lower than the lowest concentrations of dopamine detected on photo-excited TiO_2 particles.

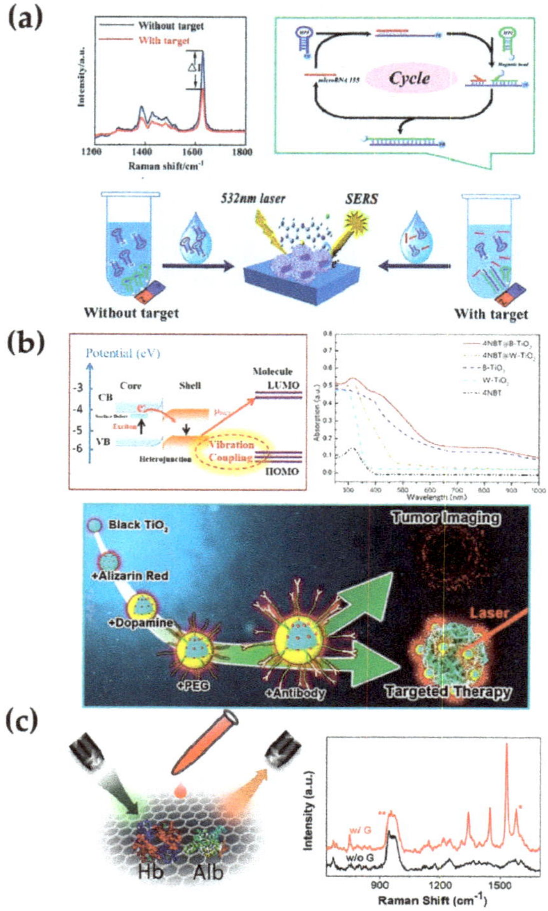

Figure 8. (**a**) Schematic diagram of the SERS Biosensor for the detection of miRNA 155. (**b**) Schematic diagram of the design process for B-TiO_2 bioprobe and their application. (**c**) Illustration of the GERS measurement and the Raman spectra of hemoglobin in contact with and in absence of graphene measured under 633 nm excitation laser. The graphene G bands at around 1582 cm^{-1} are labeled with "*", and the second-order modes of Si substrate are labeled with "**". (**a**) Reprinted with permission from [98]. Copyright 2022 American Chemical Society. (**b**) Reprinted with permission from [56]. Copyright 2019 American Chemical Society. (**c**) Reprinted with permission from [107]. Copyright 2018 American Chemical Society.

Table 2. SERS-active substrates used for the detection of SARS-CoV-2S protein and reported SERS performance.

SERS Substrates	Target	LODs	Reported SERS Performance		Ref
			Molecules	EF	
SnS_2 microspheres	SARS-CoV-2S protein	10^{-14} M	MB	3.0×10^8	[46]
Ta_2C		5×10^{-9} M	CV	1.4×10^6	[15]
Bumpy core–shell Au NPs		7.1×10^{-16} M	4-NBT	2.1×10^8 to 2.2×10^9	[116]
$Au/MgF_2/Au$		3.7×10^{-12} M	MB	2.0×10^5	[117]
Au NPLs/PDMS		2.6×10^{-10} M	R6G	6.4×10^7	[118]
Au-TiO_2		10^{-10} M	R6G	/	[119]
Au–Ag hollow nanoshells		7.5×10^{-15} M	4-MBA	/	[120]

Table 3 summarizes the SERS detection results of non-noble metal substrates for the above-mentioned biomarkers.

Table 3. A summary of non-noble meta-based SERS substrates for biomarkers detection mentioned in this review.

SERS Substrate	Enhancement Mechanism	Target	Linear Range	LOD	Ref
MoS_2 NFs	Synergistic effect of effective enrichment of substrate and laser-induced CT resonances	CA-199	5×10^{-4}–1×10^2 $IU{\cdot}mL^{-1}$	3.43×10^{-4} $IU{\cdot}mL^{-1}$	[16]
WO_3 film	CT enhancement	VEGF	10–250 pg/mL	8.7 pg/mL.	[95]
3D WO_3 hollow microsphere	CT enhancement	miRNA 155	1 fM–100 pM	0.18 fM	[98]
CuPc@HG@BN	Interface dipole interaction	miR-21	1.6 fM–2.8 pM	0.7 fM	[99]
TiO_2-AR -rBSA-FA	Synergistic effect of strong vibration coupling resonance and PICT enhancement	SY5Y H226KYSE-150 HeLa	/	1 cell/mL	[102]
B-TiO_2-AR-PD-PEG-AB	Synergistic effect of interfacial PICT and PICT resonance as well as strong vibronic coupling in amorphous shell-molecule system	MCF-7MCF-7 drug-resistant (MCF-7/ADR) breast cancer cells	/	/	[103]
B-TiO_2-AR -PEG-FA	Synergistic effect of interfacial PICT and PICT resonance as well as strong vibronic coupling in amorphous shell-molecule system	MCF-7(folate receptor (FR) positive), A549 Raw264.7 (FR negative)	1–100 cells/mL (MCF-7)	2 cells/mL (MCF-7)	[56]
MGT-ABs-CuPc	Synergistic effect of enrichment from a porous TiO_2 shell and CT enhancement	HCC38 MDA-MB-231 MCF-7	5×10^2–5×10^5 cells 5×10^2–10^6 cells 5×10^3–5×10^4 cells	3 cells / /	[104]
3D ZnO quantum	Synergistic effect of SPR effect and CT enhancement	MDAMB231 HeLa NIH3T3	/	/	[105]
Graphene	Synergistic effect of CT enhancement and π-π interaction between probe and graphene	Hemoglobin albumin	/	/	[107]
Quantum-size $TiO_{2\text{-}x}$	Synergistic effect of effective enrichment of substrate and laser-induced CT resonances	ATP EGFR	/	/	[108]

Table 3. *Cont.*

SERS Substrate	Enhancement Mechanism	Target	Linear Range	LOD	Ref
Ni-MoS_2 NFs	Synergistic effect of effective enrichment of substrate and CT enhancement as well we interface dipole–dipole interaction	Bilirubin	10^{-7}–10^{-3} M	10^{-7} M	[109]
SnS_2 microspheres	Synergistic effect of the molecular enrichment caused by capillary effect and CT enhancement	SARS-CoV-2 S protein, SARS-CoV S protein, SARS-CoV-2 RNA	10^{-14}–10^{-6} M 10^4–10^7 copies/mL 10^4–10^6 copies/mL	10^{-14} M 10^7 copies/mL 10^6 copies/mL	[46]
Ta_2C	Synergistic effect of PICT resonance enhancement and EM	SARS-CoV-2 S protein	/	5×10^{-9} M	[15]
1T′-$MoTe_2$	CT enhancement	β-sitosterol	10^{-9}–10^{-4} M	10^{-9} M	[47]
Few-layered $HfTe_2$	CT enhancement	Uric acid	100 μM–1 mM	100 μM	[117]
D-MIL-125-NH_2	Photo-induced oxygen vacancy detection in D-MIL-125-NH_2 results in modified energy bands to boost both the inter-valence CT within MOFs and interfacial CT transitions	Dopamine	6.6×10^{-4}–6.6×10^{-7} M	6.6×10^{-7} M	[118]

5. Conclusions and Prospects

Non-noble metal nanomaterials have become the most promising SERS substrates due to their unique physicochemical properties. Although significantly fewer studies have been focused on their practical application compared with noble metals-based SERS substrates, they have shown their own advantages in addressing the challenges related to conventional SERS substrates and expanding the scope of SERS detection. In this review, we briefly summarized the enhancement mechanism of non-noble metal SERS substrates. Several main types of materials were also briefly reviewed. The latest application progress of non-noble metal materials based on SERS technology for the detection of pollutants and biomarkers in recent years is highlighted and discussed in detail. Although many innovative works have been completed, non-noble metal SERS substrates are still facing many limitations and require perpetual efforts in SERS detection. The following three aspects should be considered for the design of non-noble metal SERS substrates.

(1) The lower detection sensitivity of most non-noble metal SERS substrates compared to noble metal SERS substrates (Au and Ag) is the main limiting factor for their applications. Therefore, the development of non-noble metal SERS substrates with high sensitivity can be made possible through the ability to rationally tune the multiparametric combination of resonance conditions for the target;

(2) The enhancement mechanism of non-noble metal SERS substrates is still needed to fully understand. Although more and more non-noble metal nanomaterials have been proposed, their enhancement mechanisms are different due to their various nanostructure and physicochemical properties. So far, the understanding of enhancement mechanisms is far from enough. In order to accurately control the SERS enhancement of non-noble metal nanomaterials, it is important to systematically research the enhancement mechanism of various non-noble metal nanomaterials;

(3) In complex environments, background interference caused various characteristic peaks in the SERS spectrum, resulting in difficult-to-distinguish characteristic peaks of analytes. In addition to the specific recognition strategies of binding antigens and antibodies as well as aptamers to improve the selectivity of detection, various functionalized SERS tags that have aroused widespread interest in SERS anti-interference detection may alleviate this dilemma.

Although there is still a certain gap between the detection of non-noble metal SERS substrates in the lab and practical application, the development of novel SERS-active non-noble metal nanomaterials are meaningful to enrich the development of high-performance SERS substrates and to expand the practical application range of SERS. Especially given the development of detection strips, microfluidic chips, and kits based on non-noble metal nanomaterials combined with portable Raman may be a focus of future research. In addition, non-noble metal SERS active materials with good biocompatibility, low biotoxicity, and good spectral stability have also been proven to have imaging functions. This not only provides a new strategy for the development of a SERS platform for tumor diagnosis but also expands the application of SERS platforms based on non-noble metal nanomaterials in the accurate diagnosis and treatment of cancer. We believe and expect that with the development of related research, non-noble metal materials will eventually become the next generation of multifunctional SERS substrates for overcoming various challenges in practical applications.

Funding: This research was funded by the National Natural Science Foundation of China (No. 22074162), the Guangdong Provincial Natural Science Foundation of China (No. 2022A1515011171), and the State Key Program of National Natural Science of China (No. 22134007), respectively.

Conflicts of Interest: The authors declare no conflict of interest.

References

1. Ji, W.; Zhao, B.; Ozaki, Y. Semiconductor Materials in Analytical Applications of Surface-Enhanced Raman Scattering. *J. Raman Spectrosc.* **2016**, *47*, 51–58. [CrossRef]
2. Pillai, I.C.L.; Li, S.; Romay, M.; Lam, L.; Lu, Y.; Huang, J.; Dillard, N.; Zemanova, M.; Rubbi, L.; Wang, Y. Cardiac Fibroblasts Adopt Osteogenic Fates and Can Be Targeted to Attenuate Pathological Heart Calcification. *Cell Stem Cell* **2017**, *20*, 218–232. [CrossRef] [PubMed]
3. Qian, X.M.; Nie, S.M. Single-Molecule and Single-Nanoparticle SERS: From Fundamental Mechanisms to Biomedical Applications. *Chem. Soc. Rev.* **2008**, *37*, 912–920. [CrossRef]
4. Xia, M. A Review on Applications of Two-Dimensional Materials in Surface-Enhanced Raman Spectroscopy. *Int. J. Spectrosc.* **2018**, *2018*, 4861472. [CrossRef]
5. Park, M.; Hwang, C.S.H.; Jeong, K.H. Nanoplasmonic Alloy of Au/Ag Nanocomposites on Paper Substrate for Biosensing Applications. *ACS Appl. Mater. Interfaces* **2018**, *10*, 290–295. [CrossRef] [PubMed]
6. Chang, R.; Wang, T.; Liu, Q.; Tang, J.; Wu, D. Ag Nanoparticles@Agar Gel as a 3D Flexible and Stable SERS Substrate with Ultrahigh Sensitivity. *Langmuir* **2022**, *38*, 13822–13832. [CrossRef]
7. Yang, B.; Jin, S.; Guo, S.; Park, Y.; Chen, L.; Zhao, B.; Jung, Y.M. Recent Development of SERS Technology: Semiconductor-Based Study. *ACS Omega* **2019**, *4*, 20101–20108. [CrossRef]
8. Lan, L.L.; Gao, Y.M.; Fan, X.C.; Li, M.Z.; Hao, Q.; Qiu, T. The Origin of Ultrasensitive SERS Sensing Beyond Plasmonics. *Front. Phys.* **2021**, *16*, 43300. [CrossRef]
9. Yin, Z.; Xu, K.; Jiang, S.; Luo, D.; Chen, R.; Xu, C.; Shum, P.; Liu, Y.J. Recent Progress on Two-Dimensional Layered Materials for Surface Enhanced Raman Spectroscopy and Their Applications. *Mater. Today Phys.* **2021**, *18*, 100378. [CrossRef]
10. Wang, X.J.; Zhang, E.J.; Shi, H.M.; Tao, Y.F.; Ren, X.D. Semiconductor-Based Surface Enhanced Raman Scattering (SERS): From Active Materials to Performance Improvement. *Analyst* **2022**, *147*, 1257–1272. [CrossRef]
11. Liu, Y.; Ma, H.; Han, X.X.; Zhao, B. Metal-Semiconductor Heterostructures for Surface-Enhanced Raman Scattering: Synergistic Contribution of Plasmons and Charge Transfer. *Mater. Horiz.* **2021**, *8*, 370–382. [CrossRef] [PubMed]
12. Alessandri, I.; Lombardi, J.R. Enhanced Raman Scattering with Dielectrics. *Chem. Rev.* **2016**, *116*, 14921–14981. [CrossRef] [PubMed]
13. Zhou, X.; Zhao, X.; Gu, S.; Gao, K.; Xie, F.; Wang, X.; Tang, Z. A Novel Sensitive ACNTs-MoO_2 SERS Substrate Boosted by Synergistic Enhancement Effect. *Phys. Chem. Chem. Phys.* **2021**, *23*, 20645–20653. [CrossRef]
14. Ye, Y.; Yi, W.; Liu, W.; Zhou, Y.; Bai, H.; Li, J.; Xi, G. Remarkable Surface-Enhanced Raman Scattering of Highly Crystalline Monolayer Ti_3C_2 Nanosheets. *Sci. China Mater.* **2020**, *63*, 794–805. [CrossRef]
15. Peng, Y.; Lin, C.; Long, L.; Masaki, T.; Tang, M.; Yang, L.; Liu, J.; Huang, Z.; Li, Z.; Luo, X. Charge-Transfer Resonance and Electromagnetic Enhancement Synergistically Enabling MXenes with Excellent SERS Sensitivity for SARS-CoV-2 S Protein Detection. *Nano-Micro Lett.* **2021**, *13*, 52. [CrossRef] [PubMed]

16. Jiang, J.; Liu, H.; Li, X.; Chen, Y.; Gu, C.; Wei, G.; Zhou, J.; Jiang, T. Nonmetallic SERS-Based Immunosensor by Integrating MoS_2 Nanoflower and Nanosheet Towards the Direct Serum Detection of Carbohydrate Antigen 19-9. *Biosens. Bioelectron.* **2021**, *193*, 113481. [CrossRef] [PubMed]
17. Zhang, Y.; Zheng, B.; Zhu, C.; Zhang, X.; Tan, C.; Li, H.; Chen, B.; Yang, J.; Chen, J.; Huang, Y. Single-Layer Ttransition Metal Dichalcogenide Nanosheet-Based Nanosensors for Rapid, Sensitive, and Multiplexed Detection of DNA. *Adv. Mater.* **2015**, *27*, 935–939. [CrossRef]
18. Wang, X.; She, G.; Xu, H.; Mu, L.; Shi, W. The Surface-Enhanced Raman Scattering from ZnO Nanorod Arrays and its Application for Chemosensors. *Sens. Actuator B Chem.* **2014**, *193*, 745–751. [CrossRef]
19. Yu, J.; Lin, J.; Chen, M.; Meng, X.; Qiu, L.; Wu, J.; Xi, G.; Wang, X. Amorphous $Ni(OH)_2$ Nanocages as Efficient SERS Substrates for Selective Recognition in Mixtures. *Colloid Surf. A* **2021**, *631*, 127652. [CrossRef]
20. Zhang, J.; Xing, T.; Zhang, M.; Zhou, Y. Facile Preparation of $Cu_{2-x}S$ Supernanoparticles with an Unambiguous SERS Enhancement Mechanism. *Chem. Eng. J.* **2022**, *434*, 134457. [CrossRef]
21. Karthick Kannan, P.; Shankar, P.; Blackman, C.; Chung, C.H. Recent Advances in 2D Inorganic Nanomaterials for SERS Sensing. *Adv. Mater.* **2019**, *31*, e1803432. [CrossRef]
22. Liu, J.; He, H.; Xiao, D.; Yin, S.; Ji, W.; Jiang, S.; Luo, D.; Wang, B.; Liu, Y. Recent Advances of Plasmonic Nanoparticles and their Applications. *Materials* **2018**, *11*, 1833. [CrossRef] [PubMed]
23. Guerrini, L.; Graham, D. Molecularly-Mediated Assemblies of Plasmonic Nanoparticles for Surface-Enhanced Raman Spectroscopy Applications. *Chem. Soc. Rev.* **2012**, *41*, 7085–7107. [CrossRef] [PubMed]
24. Lombardi, J.R.; Birke, R.L. Theory of Surface-Enhanced Raman Scattering in Semiconductors. *J. Phys. Chem. C* **2014**, *118*, 11120–11130. [CrossRef]
25. Han, X.X.; Ji, W.; Zhao, B.; Ozaki, Y. Semiconductor-Enhanced Raman Scattering: Active Nanomaterials and Applications. *Nanoscale* **2017**, *9*, 4847–4861. [CrossRef]
26. Ji, W.; Li, L.; Song, W.; Wang, X.; Zhao, B.; Ozaki, Y. Enhanced Raman Scattering by ZnO Superstructures: Synergistic Effect of Charge Transfer and Mie Resonances. *Angew. Chem. Int. Ed.* **2019**, *58*, 14452–14456. [CrossRef] [PubMed]
27. Demirel, G.; Usta, H.; Yilmaz, M.; Celik, M.; Alidagi, H.A.; Buyukserin, F. Surface-Enhanced Raman spectroscopy (SERS): An Adventure from Plasmonic Metals to Organic Semiconductors as SERS Platforms. *J. Mater. Chem. C* **2018**, *6*, 5314–5335. [CrossRef]
28. Zheng, X.; Wu, X.; Zhang, L.; Kang, J.; Zhou, M.; Zhong, Y.; Zhang, J.; Wang, L. High Spin Fe^{3+}-Related Bonding Strength and Electron Transfer for Sensitive and Stable SERS Detection. *Chem. Sci.* **2022**, *13*, 12560–12566. [CrossRef]
29. Wang, X.; Shi, W.; She, G.; Mu, L. Using Si and Ge Nanostructures as Substrates for Surface-Enhanced Raman Scattering Based on Photoinduced Charge Transfer Mechanism. *J. Am. Chem. Soc.* **2011**, *133*, 16518–16523. [CrossRef]
30. Yang, L.; Peng, Y.; Yang, Y.; Liu, J.; Huang, H.; Yu, B.; Zhao, J.; Lu, Y.; Huang, Z.; Li, Z. A Novel Ultra-Sensitive Semiconductor SERS Substrate Boosted by the Coupled Resonance Effect. *Adv. Sci.* **2019**, *6*, 1900310. [CrossRef]
31. Liang, X.; Li, N.; Zhang, R.; Yin, P.; Zhang, C.; Yang, N.; Liang, K.; Kong, B. Carbon-Based SERS Biosensor: From Substrate Design to Sensing and Bioapplication. *NPG Asia Mater.* **2021**, *13*, 8. [CrossRef]
32. Chen, M.; Liu, D.; Du, X.; Lo, K.H.; Wang, S.; Zhou, B.; Pan, H. 2D Materials: Excellent Substrates for Surface-Enhanced Raman Scattering (SERS) in Chemical Sensing and Biosensing. *TrAC-Trends Anal. Chem.* **2020**, *130*, 115983. [CrossRef]
33. Ling, X.; Xie, L.; Fang, Y.; Xu, H.; Zhang, H.; Kong, J.; Dresselhaus, M.S.; Zhang, J.; Liu, Z. Can Graphene Be Used as a Substrate for Raman Enhancement? *Nano Lett.* **2010**, *10*, 553–561. [CrossRef] [PubMed]
34. Feng, S.; Santos, M.C.D.; Carvalho, B.R.; Lv, R.; Li, Q.; Fujisawa, K.; Elías, A.L.; Lei, Y.; Perea-López, N.; Endo, M.; et al. Ultrasensitive Molecular Sensor Using N-Doped Graphene through Enhanced Raman Scattering. *Sci. Adv.* **2016**, *2*, e1600322. [CrossRef]
35. Ghopry, S.A.; Alamri, M.A.; Goul, R.; Sakidja, R.; Wu, J.Z. Extraordinary Sensitivity of Surface-Enhanced Raman Spectroscopy of Molecules on MoS_2 (WS_2) Nanodomes/Graphene Van Der Waals Heterostructure Substrates. *Adv. Opt. Mater.* **2019**, *7*, 1801249. [CrossRef]
36. Singh, N.S.; Mayanglambam, F.; Nemade, H.B.; Giri, P.K. Facile Synthetic Route to Exfoliate High Quality and Super-Large Lateral Size Graphene-Based Sheets and Their Applications in SERS and CO_2 Gas Sensing. *RSC Adv.* **2021**, *11*, 9488–9504. [CrossRef]
37. Khan, K.; Tareen, A.K.; Aslam, M.; Wang, R.; Zhang, Y.; Mahmood, A.; Ouyang, Z.; Zhang, H.; Guo, Z. Recent Developments in Emerging Two-Dimensional Materials and Their Applications. *J. Mater. Chem. C* **2020**, *8*, 387–440. [CrossRef]
38. Zhou, J.; Yang, T.; Chen, J.; Wang, C.; Zhang, H.; Shao, Y. Two-Dimensional Nanomaterial-Based Plasmonic Sensing Applications: Advances and Challenges. *Coord. Chem. Rev.* **2020**, *410*, 213218. [CrossRef]
39. Ling, X.; Fang, W.; Lee, Y.H.; Araujo, P.T.; Zhang, X.; Rodriguez-Nieva, J.F.; Lin, Y.; Zhang, J.; Kong, J.; Dresselhaus, M.S. Raman Enhancement Effect on Two-Dimensional Layered Materials: Graphene, h-BN and MoS_2. *Nano Lett.* **2014**, *14*, 3033–3040. [CrossRef]
40. Lee, Y.; Kim, H.; Lee, J.; Yu, S.H.; Hwang, E.; Lee, C.; Ahn, J.H.; Cho, J.H. Enhanced Raman Scattering of Rhodamine 6G Films on Two-Dimensional Transition Metal Dichalcogenides Correlated to Photoinduced Charge Transfer. *Chem. Mater.* **2015**, *28*, 180–187. [CrossRef]
41. Li, X.; Guo, S.; Su, J.; Ren, X.; Fang, Z. Efficient Raman Enhancement in Molybdenum Disulfide by Tuning the Interlayer Spacing. *ACS Appl. Mater. Interfaces* **2020**, *12*, 28474–28483. [CrossRef] [PubMed]

42. Majee, B.P.; Srivastava, V.; Mishra, A.K. Surface-Enhanced Raman Scattering Detection Based on an Interconnected Network of Vertically Oriented Semiconducting Few-Layer MoS_2 Nanosheets. *ACS Appl. Nano Mater.* **2020**, *3*, 4851–4858. [CrossRef]
43. Zuo, P.; Jiang, L.; Li, X.; Ran, P.; Li, B.; Song, A.; Tian, M.; Ma, T.; Guo, B.; Qu, L. Enhancing Charge Transfer with Foreign Molecules through Femtosecond Laser Induced MoS_2 Defect Sites for Photoluminescence Control and SERS Enhancement. *Nanoscale* **2019**, *11*, 485–494. [CrossRef] [PubMed]
44. Su, R.; Yang, S.; Han, D.; Hu, M.; Liu, Y.; Yang, J.; Gao, M. Ni and O Co-Modified MoS_2 as Universal SERS Substrate for the Detection of Different Kinds of Substances. *J. Colloid Interface Sci.* **2023**, *635*, 1–11. [CrossRef]
45. Li, M.; Fan, X.; Gao, Y.; Qiu, T. $W_{18}O_{49}$/Monolayer MoS_2 Heterojunction-Enhanced Raman Scattering. *J. Phys. Chem. Lett.* **2019**, *10*, 4038–4044. [CrossRef]
46. Peng, Y.; Lin, C.; Li, Y.; Gao, Y.; Wang, J.; He, J.; Huang, Z.; Liu, J.; Luo, X.; Yang, Y. Identifying Infectiousness of SARS-CoV-2 by Ultra-Sensitive SnS_2 SERS Biosensors with Capillary Effect. *Matter* **2022**, *5*, 694–709. [CrossRef]
47. Fraser, J.P.; Postnikov, P.; Miliutina, E.; Kolska, Z.; Valiev, R.; Svorcik, V.; Lyutakov, O.; Ganin, A.Y.; Guselnikova, O. Application of a 2D Molybdenum Telluride in SERS Detection of Biorelevant Molecules. *ACS Appl. Mater. Interfaces* **2020**, *12*, 47774–47783. [CrossRef]
48. Wang, K.; Guo, Z.; Li, Y.; Guo, Y.; Liu, H.; Zhang, W.; Zou, Z.; Zhang, Y.; Liu, Z. Few-Layer $NbTe_2$ Nanosheets as Substrates for Surface-Enhanced Raman Scattering Analysis. *ACS Appl. Nano Mater.* **2020**, *3*, 11363–11371. [CrossRef]
49. Samriti; Rajput, V.; Gupta, R.K.; Prakash, J. Engineering Metal Oxide Semiconductor Nanostructures for Enhanced Charge Transfer: Fundamentals and Emerging SERS Applications. *J. Mater. Chem. C* **2022**, *10*, 73–95. [CrossRef]
50. Yamada, H.; Yamamoto, Y.; Tan, N. Surface-Enhanced Raman Scattering (SERS) of Adsorbed Molecules on Smooth Surfaces of Metals and a Metal Oxide. *Chem. Phys. Lett.* **1982**, *86*, 397–400. [CrossRef]
51. Yamada, H.; Yamamoto, Y. Surface Enhanced Raman Scattering (SERS) of Chemisorbed Species on Various Kinds of Metals and Semiconductors. *Surf. Sci.* **1983**, *134*, 71–90. [CrossRef]
52. Chen, J.; Sun, K.; Zhang, Y.; Wu, D.; Jin, Z.; Xie, F.; Zhao, X.; Wang, X. Plasmonic MoO_2 Nanospheres Assembled on Graphene Oxide for Highly Sensitive SERS Detection of Organic Pollutants. *Anal. Bioanal Chem.* **2019**, *411*, 2781–2791. [CrossRef] [PubMed]
53. Wu, H.; Zhou, X.; Li, J.; Li, X.; Li, B.; Fei, W.; Zhou, J.; Yin, J.; Guo, W. Ultrathin Molybdenum Dioxide Nanosheets as Uniform and Reusable Surface-Enhanced Raman Spectroscopy Substrates with High Sensitivity. *Small* **2018**, *14*, e1802276. [CrossRef]
54. Miao, P.; Wu, J.; Du, Y.; Sun, Y.; Xu, P. Phase Transition Induced Raman Enhancement on Vanadium Dioxide (VO_2) nanosheets. *J. Mater. Chem. C* **2018**, *6*, 10855–10860. [CrossRef]
55. Li, Y.; Bai, H.; Zhai, J.; Yi, W.; Li, J.; Yang, H.; Xi, G. Alternative to Noble Metal Substrates: Metallic and Plasmonic Ti_3O_5 Hierarchical Microspheres for Surface Enhanced Raman Spectroscopy. *Anal. Chem.* **2019**, *91*, 4496–4503. [CrossRef]
56. Lin, J.; Ren, W.; Li, A.; Yao, C.; Chen, T.; Ma, X.; Wang, X.; Wu, A. Crystal-Amorphous Core-Shell Structure Synergistically Enabling TiO_2 Nanoparticles' Remarkable SERS Sensitivity for Cancer Cell Imaging. *ACS Appl. Mater. Interfaces* **2020**, *12*, 4204–4211. [CrossRef]
57. Zheng, X.; Zhong, H.; Wang, Z.; Li, J.; Hu, Y.; Li, H.; Jia, J.; Zhang, S.; Ren, F. Fabrication of Stable Substoichiometric WO_x Films with High SERS Sensitivity by Thermal Treatment. *Vacuum* **2022**, *198*, 110884. [CrossRef]
58. Zhong, X.; Sun, Y.; Chen, X.; Zhuang, G.; Li, X.; Wang, J.G. Mo Doping Induced More Active Sites in Urchin-Like $W_{18}O_{49}$ Nanostructure with Remarkably Enhanced Performance for Hydrogen Evolution Reaction. *Adv. Funct. Mater.* **2016**, *26*, 5778–5786. [CrossRef]
59. Xie, X.; Gao, N.; Huang, Y.; Fang, Y. SERS Monitored Kinetic Process of Gaseous Thiophenol Compound in Plasmonic MOF Nanoparticles. *ACS Appl. Mater. Interfaces* **2022**, *14*, 51468–51475. [CrossRef]
60. Zhang, Y.; Xue, C.; Xu, Y.; Cui, S.; Ganeev, A.A.; Kistenev, Y.V.; Gubal, A.; Chuchina, V.; Jin, H.; Cui, D. Metal-Organic Frameworks Based Surface-Enhanced Raman Spectroscopy Technique for Ultra-Sensitive Biomedical Trace Detection. *Nano Res.* **2023**, *16*, 2968–2979. [CrossRef]
61. Meng, J.; Liu, X.; Niu, C.; Pang, Q.; Li, J.; Liu, F.; Liu, Z.; Mai, L. Advances in Metal-Organic Framework Coatings: Versatile Synthesis and Broad Applications. *Chem. Soc. Rev.* **2020**, *49*, 3142–3186. [CrossRef] [PubMed]
62. Sugikawa, K.; Nagata, S.; Furukawa, Y.; Kokado, K.; Sada, K. Stable and Functional Gold Nanorod Composites with a Metal-Organic Framework Crystalline Shell. *Chem. Mater.* **2013**, *25*, 2565–2570. [CrossRef]
63. He, L.; Liu, Y.; Liu, J.; Xiong, Y.; Zheng, J.; Liu, Y.; Tang, Z. Core-Shell Noble-Metal@Metal-Organic-Framework Nanoparticles with Highly Selective Sensing Property. *Angew. Chem. Int. Ed.* **2013**, *52*, 3741–3745. [CrossRef]
64. Hu, Y.; Liao, J.; Wang, D.; Li, G. Fabrication of Gold Nanoparticle-Embedded Metal-Organic Framework for Highly Sensitive Surface-Enhanced Raman Scattering Detection. *Anal. Chem.* **2014**, *86*, 3955–3963. [CrossRef] [PubMed]
65. Yu, T.H.; Ho, C.H.; Wu, C.Y.; Chien, C.H.; Lin, C.H.; Lee, S. Metal-Organic Frameworks: A Novel SERS Substrate. *J. Raman Spectrosc.* **2013**, *44*, 1506–1511. [CrossRef]
66. Sun, H.; Cong, S.; Zheng, Z.; Wang, Z.; Chen, Z.; Zhao, Z. Metal-Organic Frameworks as Surface Enhanced Raman Scattering Substrates with High Tailorability. *J. Am. Chem. Soc.* **2019**, *141*, 870–878. [CrossRef]
67. Fu, J.H.; Zhong, Z.; Xie, D.; Guo, Y.J.; Kong, D.X.; Zhao, Z.X.; Zhao, Z.X.; Li, M. SERS-Active MIL-100(Fe) Sensory Array for Ultrasensitive and Multiplex Detection of VOCs. *Angew. Chem. Int. Ed.* **2020**, *59*, 20489–20498. [CrossRef]

68. Zheng, T.; Zhou, Y.; Feng, E.; Tian, Y. Surface-enhanced Raman Scattering on 2D Nanomaterials: Recent Developments and Applications. *Chin. J. Chem.* **2021**, *39*, 745–756. [CrossRef]
69. Sarycheva, A.; Makaryan, T.; Maleski, K.; Satheeshkumar, E.; Melikyan, A.; Minassian, H.; Yoshimura, M.; Gogotsi, Y. Two-Dimensional Titanium Carbide (MXene) as Surface-Enhanced Raman Scattering Substrate. *J. Phys. Chem. C* **2017**, *121*, 19983–19988. [CrossRef]
70. Soundiraraju, B.; George, B.K. Two-Dimensional Titanium Nitride (Ti_2N) MXene: Synthesis, Characterization, and Potential Application as Surface-Enhanced Raman Scattering Substrate. *ACS Nano* **2017**, *11*, 8892–8900. [CrossRef]
71. Yang, K.; Zhu, K.; Wang, Y.; Qian, Z.; Zhang, Y.; Yang, Z.; Wang, Z.; Wu, L.; Zong, S.; Cui, Y. $Ti_3C_2T_x$ MXene-Loaded 3D Substrate toward On-Chip Multi-Gas Sensing with Surface-Enhanced Raman Spectroscopy (SERS) Barcode Readout. *ACS Nano* **2021**, *15*, 12996–13006. [CrossRef] [PubMed]
72. Malik, A.H.; Habib, F.; Qazi, M.J.; Ganayee, M.A.; Ahmad, Z.; Yatoo, M.A. A Short Review Article on Conjugated Polymers. *J. Polym. Res.* **2023**, *30*, 115. [CrossRef]
73. Yang, Z.; Ma, C.; Gu, J.; Wu, Y.; Zhu, C.; Li, L.; Gao, H.; Yin, W.; Wang, Z.; Zhang, Y. SERS Detection of Benzoic Acid in Milk by Using Ag-COF SERS Substrate. *Spectrochim. Acta A Mol. Biomol. Spectrosc.* **2022**, *267*, 120534. [CrossRef] [PubMed]
74. Liu, Q.; Zhang, R.; Yu, B.; Liang, A.; Jiang, Z. A Highly Sensitive Gold Nanosol SERS Aptamer Assay for Glyphosate with a New COF Nanocatalytic Reaction of Glycol-Au (III). *Sens. Actuator B Chem.* **2021**, *344*, 130288. [CrossRef]
75. Chen, J.; Liu, C.; Hu, D.; Wang, F.; Wu, H.; Gong, X.; Liu, X.; Song, L.; Sheng, Z.; Zheng, H. Single-Layer MoS_2 Nanosheets with Amplified Photoacoustic Effect for Highly Sensitive Photoacoustic Imaging of Orthotopic Brain Tumors. *Adv. Funct. Mater.* **2016**, *26*, 8715–8725. [CrossRef]
76. Lin, C.; Liang, S.; Peng, Y.; Long, L.; Li, Y.; Huang, Z.; Long, N.V.; Luo, X.; Liu, J.; Li, Z. Visualized SERS Imaging of Single Molecule by Ag/Black Phosphorus Nanosheets. *Nano-Micro Let.* **2022**, *14*, 75. [CrossRef]
77. Liu, Z.; Chen, H.; Jia, Y.; Zhang, W.; Zhao, H.; Fan, W.; Zhang, W.; Zhong, H.; Ni, Y.; Guo, Z. A Two-Dimensional Fingerprint Nanoprobe Based on Black Phosphorus for Bio-SERS Analysis and Chemo-Photothermal Therapy. *Nanoscale* **2018**, *10*, 18795–18804. [CrossRef]
78. Barhoumi, A.; Halas, N.J. Label-Free Detection of DNA Hybridization Using Surface Enhanced Raman Spectroscopy. *J. Am. Chem. Soc.* **2010**, *132*, 12792–12793. [CrossRef]
79. Wu, D.; Chen, J.; Ruan, Y.; Sun, K.; Zhang, K.; Xie, W.; Xie, F.; Zhao, X.; Wang, X. A Novel Sensitive and Stable Surface Enhanced Raman Scattering Substrate Based on a MoS_2 Quantum Dot/Reduced Graphene Oxide Hybrid System. *J. Mater. Chem. C* **2018**, *6*, 12547–12554. [CrossRef]
80. Park, J.; Thomasson, J.A.; Gale, C.C.; Sword, G.A.; Lee, K.M.; Herrman, T.J.; Suh, C.P. Adsorbent-SERS Technique for Determination of Plant VOCs from Live Cotton Plants and Dried Teas. *ACS Omega* **2020**, *5*, 2779–2790. [CrossRef]
81. Maharjan, S.; Yun, Y.J.; Okello, V.A.; Wiederrecht, G.P.; Gosztola, D.J.; Ayitou, A.J.L. Photometric Sensing of Heavy Metal Ions Using a Naphthoquinodimethyl-Bis-Thioamide Dye: Selectivity & Photophysics of the Metal Organic Complexes. *J. Phtochem. Photobiol. A* **2022**, *424*, 113648.
82. Kamal, S.; Yang, T.C. Silver Enriched Silver Phosphate Microcubes as an Efficient Recyclable SERS Substrate for the Detection of Heavy Metal ions. *J. Colloid Interface Sci.* **2022**, *605*, 173–181. [CrossRef]
83. Zhang, H.; Huang, S.; Yang, X.; Yuan, R.; Chai, Y. A SERS Biosensor Cconstructed by Calcined ZnO Substrate with High-Efficiency Charge Transfer for Sensitive Detection of Pb^{2+}. *Sens. Actuator B Chem.* **2021**, *343*, 130142. [CrossRef]
84. Parveen, S.; Saifi, S.; Akram, S.; Husain, M.; Zulfequar, M. ZnO Nanoparticles Functionalized SWCNTs as Highly Sensitive SERS Substrate for Heavy Metal Ions Detection. *Mater. Sci. Semicond. Process.* **2022**, *149*, 106852. [CrossRef]
85. Ji, W.; Wang, Y.; Tanabe, I.; Han, X.; Zhao, B.; Ozaki, Y. Semiconductor-Driven "Turn-Off" Surface-Enhanced Raman Scattering Spectroscopy: Application in Selective Determination of Chromium (VI) in Water. *Chem. Sci.* **2015**, *6*, 342–348. [CrossRef] [PubMed]
86. Xie, Z.; Feng, Y.; Wang, F.; Chen, D.; Zhang, Q.; Zeng, Y.; Lv, W.; Liu, G. Construction of Carbon Dots Modified $MoO_3/g\text{-}C_3N_4$ Z-Scheme Photocatalyst with Enhanced Visible-Light Photocatalytic Activity for the Degradation of Tetracycline. *Appl. Catal. B Environ.* **2018**, *229*, 96–104. [CrossRef]
87. Qu, L.L.; Liu, Y.Y.; Liu, M.K.; Yang, G.H.; Li, D.W.; Li, H.T. Highly Reproducible Ag NPs/CNT-Intercalated GO Membranes for Enrichment and SERS Detection of Antibiotics. *ACS Appl. Mater. Interfaces* **2016**, *8*, 28180–28186. [CrossRef]
88. Singh, J.; Rishikesh; Kumar, S.; Soni, R.K. Synthesis of 3D-MoS_2 Nanoflowers with Tunable Surface Area for the Application in Photocatalysis and SERS Based Sensing. *J. Alloys Compd.* **2020**, *849*, 156502. [CrossRef]
89. Zhang, Q.; Li, X.; Yi, W.; Li, W.; Bai, H.; Liu, J.; Xi, G. Plasmonic MoO_2 Nanospheres as a Highly Sensitive and Stable Non-Noble Metal Substrate for Multicomponent Surface-Enhanced Raman Analysis. *Anal. Chem.* **2017**, *89*, 11765–11771. [CrossRef]
90. Liang, P.; Cao, Y.; Dong, Q.; Wang, D.; Zhang, D.; Jin, S.; Yu, Z.; Ye, J.; Zou, M. A Balsam Pear-Shaped CuO SERS Substrate with Highly Chemical Enhancement for Pesticide Residue Detection. *Mikrochim. Acta* **2020**, *187*, 335. [CrossRef]

91. Quan, Y.; Su, R.; Yang, S.; Chen, L.; Wei, M.; Liu, H.; Yang, J.; Gao, M.; Li, B. In-Situ Surface-Enhanced Raman Scattering Based on MTi_{20} Nanoflowers: Monitoring and Degradation of Contaminants. *J. Hazard. Mater.* **2021**, *412*, 125209. [CrossRef] [PubMed]
92. Gokulakrishnan; Alex, K.V.; Sekhar, K.C.; Koppole, K. Highly Sensitive, Cost-Effective, and Flexible SERS Substrate Based on Green Synthesized GO/rGO for Pesticide Detection. *ChemistrySelect* **2022**, *7*, e202200348.
93. Guo, C.; Sun, J.; Dong, J.; Cai, W.; Zhao, X.; Song, B.; Zhang, R. A Natural Anthocyanin-Based Multifunctional Theranostic Agent for Dual-Modal Imaging and Photothermal Anti-Tumor Therapy. *J. Mater. Chem. B* **2021**, *9*, 7447–7460. [CrossRef] [PubMed]
94. Shao, H.; Lin, H.; Guo, Z.; Lu, J.; Jia, Y.; Ye, M.; Su, F.; Niu, L.; Kang, W.; Wang, S. A Multiple Signal Amplification Sandwich-Type SERS Biosensor for Femtomolar Detection of MiRNA. *Biosens. Bioelectron.* **2019**, *143*, 111616 [CrossRef]
95. Liu, X.; Zhou, Y.; Zheng, T.; Tian, Y. Surface-Enhanced Raman Scattering Technology Based on WO_3 Film for Detection of VEGF. *Chem. Res. Chin. Univ.* **2021**, *37*, 900–905. [CrossRef]
96. Atlas, G. Editorial: The 16th Annual Nucleic Acids Research Web Sserver Issue 2018. *Nucleic Acids Res.* **2018**, *46*, W1–W4.
97. Brancati, G.; Grosshans, H. An Interplay of MiRNA Abundance and Target Site Architecture Determines MiRNA Activity and Specificity. *Nucleic Acids Res.* **2018**, *46*, 3259–3269. [CrossRef]
98. Jiang, L.; Hu, Y.; Zhang, H.; Luo, X.; Yuan, R.; Yang, X. Charge-Transfer Resonance and Surface Defect-Dominated WO_3 Hollow Microspheres as SERS Substrates for the MiRNA 155 Assay. *Anal. Chem.* **2022**, *94*, 6967–6975. [CrossRef]
99. Liu, J.; Zheng, T.; Tian, Y. Functionalized h-BN Nanosheets as a Theranostic Platform for SERS Real-Time Monitoring of MicroRNA and Photodynamic Therapy. *Angew. Chem. Int. Ed.* **2019**, *58*, 7757–7761. [CrossRef]
100. Yu, L.; Ng, S.R.; Xu, Y.; Dong, H.; Wang, Y.J.; Li, C.M. Advances of Lab-on-a-Chip in Isolation, Detection and Post-Processing of Circulating Tumour Cells. *Lab Chip* **2013**, *13*, 3163–3182. [CrossRef] [PubMed]
101. Feng, Z.; Wu, J.; Lu, Y.; Chan, Y.T.; Zhang, C.; Wang, D.; Luo, D.; Huang, Y.; Feng, Y.; Wang, N. Circulating Tumor Cells in the Early Detection of Human Cancers. *Int. J. Biol. Sci.* **2022**, *18*, 3251–3265. [CrossRef] [PubMed]
102. Xu, Y.; Lin, J.; Wu, X.; Xu, X.; Zhang, D.; Xie, Y.; Pan, T.; He, Y.; Wu, A.; Shao, G. A TiO_2-Based Bioprobe Enabling Excellent SERS Activity in the Detection of Diverse Circulating Tumor Cells. *J. Mater. Chem. B* **2022**, *10*, 3808–3816. [CrossRef] [PubMed]
103. Xu, X.; Lin, J.; Guo, Y.; Wu, X.; Xu, Y.; Zhang, D.; Zhang, X.; Yujiao, X.; Wang, J.; Yao, C. TiO_2-Based Surface-Enhanced Raman Scattering Bio-Probe for Efficient Circulating Tumor Cell Detection on Microfilter. *Biosens. Bioelectron.* **2022**, *210*, 114305. [CrossRef] [PubMed]
104. Feng, E.; Zheng, T.T.; He, X.X.; Chen, J.Q.; Tian, Y. A Novel Ternary Heterostructure with Dramatic SERS Activity for Evaluation of PD-L1 Expression at the Single-Cell Level. *Sci. Adv.* **2018**, *4*, eaau3494. [CrossRef]
105. Haldavnekar, R.; Venkatakrishnan, K.; Tan, B. Non Plasmonic Semiconductor Quantum SERS Probe as a Pathway for in Vitro Cancer Detection. *Nat. Commun.* **2018**, *9*, 3065. [CrossRef]
106. Wang, Z.; Zong, S.; Wu, L.; Zhu, D.; Cui, Y. SERS-Activated Platforms for Immunoassay: Probes, Encoding Methods, and Applications. *Chem. Rev.* **2017**, *117*, 7910–7963. [CrossRef]
107. Huang, S.; Pandey, R.; Barman, I.; Kong, J.; Dresselhaus, M. Raman Enhancement of Blood Constituent Proteins Using Graphene. *ACS Photonics* **2018**, *5*, 2978–2982. [CrossRef]
108. Dharmalingam, P.; Venkatakrishnan, K.; Tan, B. An Atomic-Defect Enhanced Raman scattering (DERS) Quantum Probe for Molecular Level Detection-Breaking the SERS Barrier. *Appl. Mater. Today* **2019**, *16*, 28–41. [CrossRef]
109. Su, R.; Quan, Y.; Yang, S.; Hu, M.; Yang, J.; Gao, M. Destroying the Symmetric Structure to Promote Phase Transition: Improving the SERS Performance and Catalytic Activity of MoS_2 Nanoflowers. *J. Alloys Compd.* **2021**, *886*, 161268. [CrossRef]
110. Tang, Z.; Kong, N.; Zhang, X.; Liu, Y.; Hu, P.; Mou, S.; Liljestrom, P.; Shi, J.; Tan, W.; Kim, J.S A Materials-Science Perspective on Tackling COVID-19. *Nat. Rev. Mater.* **2020**, *5*, 847–860. [CrossRef]
111. Jiang, Y.; Hu, M.; Liu, A.A.; Lin, Y.; Liu, L.; Yu, B.; Zhou, X.; Pang, D.W. Detection of SARS-CoV-2 by CRISPR/Cas12a-Enhanced Colorimetry. *ACS Sens.* **2021**, *6*, 1086–1093. [CrossRef]
112. Li, Y.; Chen, H.; Guo, Y.; Wang, K.; Zhang, Y.; Lan, P.; Guo, J.; Zhang, W.; Zhong, H.; Guo, Z. Lamellar Hafnium Ditelluride as an Ultrasensitive Surface-Enhanced Raman Scattering Platform for Label-Free Detection of Uric Acid. *Photonics Res.* **2021**, *9*, 1039–1047. [CrossRef]
113. Sun, H.; Song, G.; Gong, W.; Lu, W.; Cong, S.; Zhao, Z. Stabilizing Photo-Induced Vacancy Defects in MOF Matrix for High-Performance SERS Detection. *Nano Res.* **2022**, *15*, 5347–5354. [CrossRef]
114. Ferna'ndez, R.D.; Ruth, T.J.; Sossi, V.; Schulzer, M.; Calne, D.B.; Stoessl, A.J. Expectation and Dopamine Release: Mechanism of the Placebo Effect in Parkinson's Disease. *Science* **2001**, *293*, 1164–1166. [CrossRef] [PubMed]
115. Zhang, A.; Neumeyer, J.L.; Baldessarini, R.J. Recent Progress in Development of Dopamine Receptor Subtype-Selective Agents: Potential Therapeutics for Neurological and Psychiatric Disorders. *Chem. Rev.* **2007**, *107*, 274–302. [CrossRef] [PubMed]
116. Shim, J.E.; Kim, Y.J.; Choe, J.H.; Lee, T.G.; You, E.A. Single-Nanoparticle-Based Digital SERS Sensing Platform for the Accurate Quantitative Detection of SARS-CoV-2. *ACS Appl. Mater. Interfaces* **2022**, *14*, 38459–38470. [CrossRef]
117. Wu, P.; Luo, X.; Xu, Y.; Zhu, J.; Jia, W.; Fang, N.; Cai, C.; Zhu, J.J. Long-Range SERS Detection of the SARS-CoV-2 Antigen on a Well-Ordered Gold Hexagonal Nanoplate Film. *Anal. Chem.* **2022**, *94*, 17541–17550. [CrossRef]
118. Yue, W.; Xia, Z.; Zeng, Z.; Chen, Z.; Qiao, L.; Li, P.; He, Y.; Luo, X. In Situ Surface-Enhanced Raman Scattering Detection of a SARS-CoV-2 Biomarker Using Flexible and Transparent Polydimethylsiloxane Films with Embedded Au Nanoplates. *ACS Appl. Nano Mater.* **2022**, *5*, 12897–12906. [CrossRef]

119. Hwang, C.S.H.; Lee, S.; Lee, S.; Kim, H.; Kang, T.; Lee, D.; Jeong, K.H. Highly Adsorptive Au-TiO_2 Nanocomposites for the SERS Face Mask Allow the Machine-Learning-Based Quantitative Assay of SARS-CoV-2 in Artificial Breath Aerosols. *ACS Appl. Mater. Interfaces* **2022**, *14*, 54550–54557. [CrossRef]
120. Zhao, T.; Liang, P.; Ren, J.; Zhu, J.; Yang, X.; Bian, H.; Li, J.; Cui, X.; Fu, C.; Xing, J. Gold-Silver Alloy Hollow Nanoshells-Based Lateral Flow Immunoassay for Colorimetric, Photothermal, and SERS Tri-Mode Detection of SARS-CoV-2 Neutralizing Antibody. *Anal. Chim. Acta* **2023**, *1255*, 341102. [CrossRef] [PubMed]

chemosensors

MDPI

Review

MOF-Based Materials for Glucose Detection

Yiling Zhang †, Qian Lin †, Yiteng Song, Jiaqi Huang, Miaomiao Chen, Runqi Ouyang, Si-Yang Liu * and Zong Dai

Guangdong Provincial Key Laboratory of Sensing Technology and Biomedical Instrument, School of Biomedical Engineering, Shenzhen Campus of Sun Yat-Sen University, Sun Yat-Sen University, Shenzhen 518107, China; zhangyling27@mail2.sysu.edu.cn (Y.Z.); linq59@mail2.sysu.edu.cn (Q.L.); songyt8@mail2.sysu.edu.cn (Y.S.); huangjq59@mail2.sysu.edu.cn (J.H.); chenmm35@mail2.sysu.edu.cn (M.C.); ouyrq3@mail2.sysu.edu.cn (R.O.); daizong@mail.sysu.edu.cn (Z.D.)

* Correspondence: liusiyang@mail.sysu.edu.cn

† These authors contributed equally to this work.

Abstract: Metal–organic frameworks (MOFs), constructed by coordination between metal-containing nodes and organic linkers, are widely used in various fields due to the advantages of tunable pores, diverse functional sites, stable structure, and multi-functionality. It should be noted that MOF-based materials play a major role in glucose detection, serving as a signal transducer or functional substrate for embedding nanoparticles/enzymes. Diabetes is one of the most common and fast-growing diseases worldwide, whose main clinical manifestation is high blood sugar levels. Therefore, accurate, sensitive, and point-of-care glucose detection is necessary. This review orderly introduces general synthetic strategies of MOF-based materials (pristine MOF, nanoparticles, or enzymes-modified MOF and MOF-derived materials) and detection methods (electrochemical and optical methods) for glucose detection. Then, the review refers to the novel MOF-based glucose detection devices (flexible wearable devices and microfluidic chips), which enable non-invasive continuous glucose monitoring or low-cost microscale detection. On the basis of describing the development of glucose sensors based on MOF materials in the past five years, the review presents merits, demerits, and possible improvements of various detection methods.

Keywords: metal–organic frameworks (MOFs); glucose detection; synthesis; electrochemical sensor; optical sensor; wearable sensor

Citation: Zhang, Y.; Lin, Q.; Song, Y.; Huang, J.; Chen, M.; Ouyang, R.; Liu, S.-Y.; Dai, Z. MOF-Based Materials for Glucose Detection. *Chemosensors* **2023**, *11*, 429. https://doi.org/10.3390/chemosensors11080429

Academic Editor: Chung-Wei Kung

Received: 22 June 2023
Revised: 24 July 2023
Accepted: 31 July 2023
Published: 2 August 2023

1. Introduction

Glucose is an indispensable small molecule for human life, which not only provides energy for the human body but also controls cell activities by regulating the intracellular glucose level. For example, one of the important methods to treat cancer is blocking intracellular glucose uptake, which can effectively inhibit the activity of cancer cells [1,2]. Meanwhile, glucose is also an important intermediate of metabolism. When the blood glucose level deviates from the normal range, serious damage to the human body may be caused, including tissue damage, stroke, renal, heart attack, etc. [3]. Moreover, diabetes, a common metabolic disease, is mainly characterized by high blood glucose levels. The blood glucose disorder also occurs in pancreatic exocrine (pancreatitis, cystic fibrosis, etc.) and endocrine diseases (Cushing syndrome, acromegaly, etc.) [4]. Nowadays, the self-monitoring of blood glucose levels is the most efficient way for the management of diabetes and other related disease [5]. Except for the above-mentioned applications, glucose detection is also applied in artificial taste sensors and quality control of food and drinks [6].

The glucose sensors on the market quantify the glucose by measuring the biochemical reaction products generated from glucose with the catalysis of enzymes, including glucose oxidase-peroxidase (GOD-POD), glucose dehydrogenase (GDH), and hexokinase (HK) [7]. The GOD-POD method is usually utilized in portable glucose meters (finger-prick blood testing) and self-monitoring blood glucose meters (interstitial fluid detection), which are based on the product (H_2O_2) and the electron transfer during the glucose oxidation reaction [8]. The HK method is highly accurate and precise but needs large and expensive equipment, which is commonly used in hospital and laboratory research. In the HK method, the phosphorylation product of glucose catalyzed by hexokinase can react with nicotinamide adenine dinucleotide phosphate ($NADP^+$) under the cascade catalysis of glucose phosphate dehydrogenase (G6PDH) and generate a chemiluminescence signal [9]. Although these methods are widely reported and even commercialized, their sensitivity, stability, and non-invasion detection are still challenged. Consequently, an accurate, selective, stable, and non-invasive glucose detection method is highly demanded. With the advancements in material science, emerging functional materials with superior glucose detection performance have been increasingly reported.

Due to the tunable pores, diverse functional site, high specific surface area, and stable structure, metal–organic frameworks (MOFs), constructed by connecting metal-containing nodes with organic linkers through coordination bonds, have received extensive attention in various detections, especially in glucose detection (Figure 1) [10–14]. MOF-based materials in glucose sensors can be divided into three types, including pristine MOF, modified MOF, and MOF-derived material. Pristine MOF with inherent redox properties that catalyze the reaction of glucose is mostly utilized in non-enzymatic glucose detection. The desired detection properties can be obtained from appropriate ligands and metal ions, and it is also related to the degree of conjugation and coordination of MOF. The modification of MOF with nanoparticles and enzymes can also endow MOF-based materials with catalytic properties and improved detection ability. In addition, after the thermal or solvent treatment, the MOF-derived materials can achieve novel pore structure and detection properties. The detection mechanism of MOF-based glucose sensors can be divided into electrochemical and optical methods based on the types of output signals. The electrochemical detection is based on the redox reaction between glucose with MOF or the modified materials. The optical methods utilize the optical signal produced by the interaction with the glucose and the modified fluorescent probe, enzymes, or MOFs. The preparation strategy (pristine MOF, nanoparticles in/on MOF, enzymes in/on MOF, and MOF-derived materials) and detection mechanism (electrochemical and optical detection) of MOF-based sensors are also inseparable; therefore, this review will introduce MOF-based materials for glucose detection mainly from these two aspects: the preparation strategies and the detection methods.

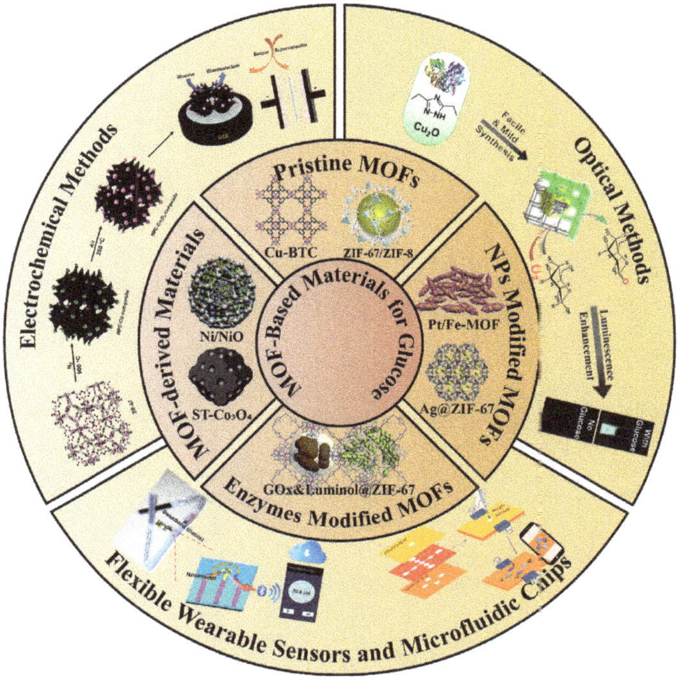

Figure 1. Different types of MOF-based materials for glucose detection. Pristine MOFs: Cu-BTC (benzene-1,3,5-tricarboxylate) (reproduced with permission from [15], copyright © 2013, Royal Society of Chemistry), ZIF-67/8 (reproduced with permission from [16], copyright © 2022, Elsevier); NPs modified MOFs: AgNPs@ZIF-67 (reproduced with permission from [17], copyright © 2018, Elsevier), Pt/Fe-MOF (reproduced with permission from [18], copyright © 2021, Springer Nature); Enzymes modified MOFs: GOx&Luminol@ZIF-67 (reproduced with permission from [19], copyright © 2023, Elsevier); MOF-derived materials: Ni/NiO (reproduced with permission from [20], copyright © 2020, Springer Nature), ST-Co_3O_4 (reproduced with permission from [21], copyright © 2021, John Wiley and Sons). The typical works of electrochemical methods (reproduced with permission from [22], copyright © 2018, Elsevier) and optical methods (reproduced with permission from [23], copyright © 2019, John Wiley and Sons) for glucose detection with MOF-based materials. Furthermore, the flexible wearable sensors (reproduced with permission from [24], copyright © 2022, Elsevier) and microfluidic chips (reproduced with permission from [25], copyright © 2022, Elsevier) for MOF-based glucose detection.

2. Synthesis Strategy of MOF-Based Materials

2.1. Pristine MOF

The pristine MOF with a large specific surface area and structural diversity is commonly utilized in nonenzymatic electrochemical detection, which avoids restricting detection and storage requirements [26]. In electrochemical detection, the valence of metal ions in pristine MOF is changed, resulting in a redox reaction. The MOF-based electrodes not only have excellent electrocatalytic performance for glucose but also have remarkable selectivity and stability. The pristine MOF used in glucose detection can be divided into single metallic MOF, like Fe [26,27], Co [28], Cu [29,30], and Ni-MOF [31,32], and bimetallic or multivalent metallic MOF. Due to the different ligand coordination abilities, valence states, and ionic radii of the two metals, the substitution of the second metal to the host metal happens, and the crystal structure, morphology, and catalytic properties are changed. On the one hand, the addition of the second metal could improve the catalytic ability and bring other unique properties. On the other hand, the extensive substitution of the second metal may lead to crystal distortion and structural collapse. The synergism and competition between the two metals improve the bimetallic MOF electrodes and increase the number of structure defects, which endow the electrodes with more binding sites and stronger electrocatalytic performance [33]. Ma et al. synthesized a novel stylophora coral-like furan-based Ni/Co-MOF (Ni/Co-FAMOF) through the simple solvothermal method [34]. Ni/Co-FAMOF, combined with bimetal Ni/Co and furan dicarboxylic acid ligand, has a high specific capacitance, capacity retention rate, and good electrochemical glucose detection performance. The excellent performance of the Ni/Co-FAMOF electrode derived from (1) ligand with rigid furan ring skeleton, which has high molecular stacking properties; (2) bimetal-containing nodes, which combine the high electrocatalytic activity of Ni and oxygen adsorption ability of Co; and (3) unique stylophora coral-like morphology, which greatly increases the active sites on the surface.

2.1.1. Hydrothermal/Solvothermal Synthesis

Hydrothermal/solvothermal synthesis is one of the most commonly employed methods to synthesize pristine MOF for glucose detection. Hydrothermal/solvothermal synthesis typically reacts the metal salts and organic ligands in a closed system, such as a Teflon-lined steel autoclave, with water or organic solvent at a relatively high temperature (100–1000 °C) and pressure (1 MPa–1 GPa). The medium provides a favorable reaction environment to promote the generation of desired products, which inhibits the formation of undesirable products. Therefore, hydrothermal/solvothermal synthesis has the advantages of high purity of products, efficiency, simplicity, versatility, and large-scale preparation [35].

The mechanisms of crystal growth by hydrothermal/solvothermal growth are dissolve crystallization and in situ crystallization (Table 1). Dissolve crystallization: the dissolved precursor mixture reaches a subcritical or supercritical state under high temperature and pressure and converts to the final product for the different solubility between product and precursor [36]. Ramaprabhu et al. synthesized copper-terephthalate (CuBDC) MOF by solvothermal method with different time durations [37]. When the reaction time exceeded 6 h, Cu-BDC started agglomerating, and the morphology changed from flat, rod-shaped into a cuboidal solid block-shape with stacked layers, which completely changed after 48 h. The changes in the morphology greatly increased the specific surface area and active sites, which further improved the detection performance in glucose detection.

In situ crystallization: in the presence of precursors (such as graphene, carbon cloth, metal nanoparticles, etc.), MOF crystals are promoted to in situ grown at active sites formed by the dehydration of precursors [38]. As shown in Figure 2A, Yang et al. used the in situ growth method to prepare NiCo-BTC nanosheets on the surface of the precursor carbon cloth (CC) with oxygen-containing functional groups [39]. The application of CC not only improved the distribution of NiCo-BTC and avoided aggregation but also increased the active site to oxidize glucose. The electrochemical performance for glucose detection was further improved, attributing to the positive synergistic effect of bimetal. Wang et al. successfully in situ grew bimetallic CuCo-MOF on nickel foam (NF) through a facile one-pot hydrothermal treatment [40]. With the positive synergistic effect of Co and Cu, CuCo-MOF obtained enhanced sensitivity of electrodes from Co and broadened linear range from Cu. The microflower-like morphological structure further improved electrochemical glucose detection properties.

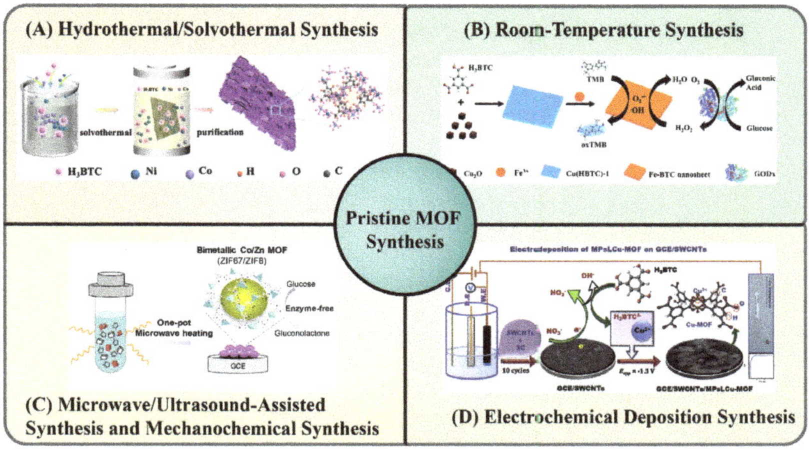

Figure 2. Synthesis methods of the pristine MOF. (**A**) Solvothermal synthesis of NiCo-BTC/CC (reproduced with permission from [39], copyright © 2022, American Chemical Society). (**B**) Room-temperature synthesis of 2D Fe-BTC (reproduced with permission from [41], copyright © 2013, Royal Society of Chemistry). (**C**) Microwave-assisted synthesis of ZIF-67/8 (reproduced with permission from [16], copyright © 2022, Elsevier). (**D**) Electrochemical deposition synthesis of Cu-BTC on SWCNTs/GCE (reproduced with permission from [42], copyright © 2020, Elsevier).

Table 1. MOF synthesized by hydrothermal/solvothermal synthesis for glucose detection.

Sample	Solvent	Method	Morphology	Ref
Ni/Co-FAMOF	H_2O	Dissolve crystallization	Stylophora coral-like	[34]
Co-MOF	ChCl	Dissolve crystallization	Nanoparticles	[43]
Co-BTC	DMF, Ethanol	Dissolve crystallization	Cuboid	[44]
NiCu-MOF	DMF	Dissolve crystallization	Nanosheets	[45]
NiCo-BTC/CC	Ethanol	In situ crystallization	Nanosheets	[39]
Co/Cu-MOF/NF	H_2O	In situ crystallization	Microflowers	[40]
Ni/Co(HHTP)MOF/CC	H_2O	In situ crystallization	Thick rods	[46]
CuCo-MOF/CFP	DMF	In situ crystallization	Book-like	[47]

BTC: benzene-1,3,5-tricarboxylate. DMF: N, N-Dimethylformamide. HHTP: triphenylene-2,3,6,7,10,11-hexaol. CFP: carbon fiber paper.

In addition to solvothermal/hydrothermal synthesis, ionothermal synthesis was also applied to synthesize MOFs for glucose detection. Compared with solvothermal/hydrothermal synthesis, which uses water or an organic solution as a solvent, ionothermal synthesis uses ionic liquids as reactants, which could also act as structure-guiding agents and charge-balancing agents. The ionic liquids, which have unique ionic composition, low freezing point, strong solubility, and good electrical conductivity, incubate the ionothermal synthesis with high safety and green environmental protection. Jiao et al. utilized the deep eutectic solvent choline chloride (ChCl) as a solvent to synthesize Co-MOF, $[Ch]_2[Co_3(BDC)_3Cl_2]$ [43]. The detection performance of the synthesized $[Ch]_2[Co_3(BDC)_3Cl_2]$ electrode towards glucose is excellent, with a rapid response.

2.1.2. Room-Temperature Synthesis

The room-temperature synthesis strategy, which reacts under room temperature and mild conditions, is simple and facile. Moreover, room-temperature synthesis is suitable for industrial large-scale production to eliminate unstable factors, including the heating and pressurization process [48]. Wang et al. prepared an electrode with Co-MOF on CC, which was fabricated by the room-temperature liquid-phase deposition strategy [49]. The room-temperature liquid-phase deposition strategy avoided post-coating procedures with polymeric binders that inevitably reduced the active surface area and the rate of charge transfer. Furthermore, the ultrathin two-dimensional (2D) Co-MOF nanosheets with well-aligned open-shelled MOF nanoarrays on CC promoted the penetration of electrolytes and the diffusion of small molecules and improved the oxidation efficiency of glucose, which endowed the electrode with excellent catalytic performance. As shown in Figure 2B, Huang et al. synthesized Cu(HBTC) using the room-temperature synthesis strategy, and Fe^{3+} ions were added to replace the Cu^{2+}, obtaining Fe-BTC with peroxidase activity. The 2D nanosheets of Fe-BTC offered higher diffusion efficiency and more active sites than three-dimensional (3D) bulk crystals, which improved the glucose detection performance [41]. Additionally, NiCo-MOF [50,51], Cu-MOF [52–54], ZIF-67 [55], and other MOFs were also synthesized using the room-temperature synthesis strategy to detect glucose.

2.1.3. Microwave/Ultrasound-Assisted Synthesis and Mechanochemical Synthesis

With the assistance of high-energy ultrasonic and electromagnetic waves to increase the reaction energy, respectively, microwave/ultrasound-assisted synthesis has the advantages of high efficiency, rapid reaction, uniform particle morphology, and high phase purity. In microwave-assisted synthesis, the interaction between the electromagnetic field of the microwave and electrons, the precursors which receive the energy from the microwave, are heated uniformly and rapidly to accelerate nucleation and crystal growth [56]. The ultrasound-assisted synthesis is based on the effect of the acoustic cavitation; the bubbles are continuously generated, grown, and collapsed in a hot spot, where the temperature reaches 5000 °C, the pressure is 1000 atm, and the heating/cooling rate is above 1010 K/s, inducing local heating and pressurization. The high temperature and pressure promote the growth of the surrounding crystal nucleus [57]. Compared with traditional hydrothermal/solvothermal synthesis, microwave/ultrasound-assisted synthesis greatly reduces the reaction time, even just in a few minutes. As shown in Figure 2C, Ni et al. used a one-pot, rapid microwave-assisted method to synthesize bimetallic ZIF-67/8 that only needs 20 min [16]. The same method was also used to synthesize Ni-MOF by Sargazi et al. [58]. The synthesized Ni-MOF electrode with a high surface area of 1381 m^2/g had high sensitivity and accuracy in glucose detection.

Except for microwave/ultrasound-assisted synthesis, mechanochemical synthesis is also used to synthesize MOF-based materials in glucose detection. After applying appropriate mechanical force, the fluidity at the molecular level of the reactants is increased, which results in the breaking of intramolecular bonds and accelerating the chemical reaction [59]. Moreover, mechanochemical synthesis requires no external heat, which is simple, green, and rapid. Lee et al. introduced an innovative rapid agitation-induced synthesis route to synthesize Ni-MOF at 10,000 rpm, which only needs 5 min [60]. The synthesized 2D ultrathin Ni-MOF nanosheet has a good electrocatalytic activity for glucose detection.

2.1.4. Electrochemical Deposition Synthesis

The electrochemical deposition method, which deposits MOF directly on the conductive substrates, effectively eliminates the decrease of charge transfer efficiency caused by the use of adhesives, such as Nafion. This strategy has the advantages of mild reaction and easy treatment, which also can effectively eliminate the influence of countering ions [61,62]. Meanwhile, changing the electrolyte and current density can effectively control the shape, size, and distribution of MOF. As shown in Figure 2D, Tominaga et al. electrodeposited three-dimensional (3D) nucleated microparticles like Cu-BTC directly on glassy carbon electrode (GCE) with single-walled carbon nanotubes (SWNTs) [42]. Hosseini et al. formed a crystalline rectangular bar-shape Co-BTC on rGO/GCE by depositing nano-flake $Co(OH)_2$ intermediates on rGO/GCE and rapidly conversing $Co(OH)_2$ to Co-BTC [63]. Compared with the one-step electrochemical deposition method, the multi-step electrochemical deposition method, intermediate deposition, and conversion can more effectively control the growth and distribution of MOF and improve the adhesion of MOF to the electrode.

2.2. Nanoparticles-Modified MOF

Compared with pristine MOF, MOF modified with other materials, including nanoparticles (NPs) and enzymes, can achieve diverse functions and properties for glucose detection (Table 2). Most pristine MOF was applied in electrochemical detection, and the choice of metal in MOF is limited to redox metals, such as Fe, Co, Cu, and Ni. Excitingly, the modified MOF broadened the choice of metals, ligands, and detection methods [64]. Because of the pore osmotic adsorption of porous MOF materials and the interaction between functional groups on MOF and NPs (van der Waals force, hydrogen bonding, electrostatic adsorption, etc.), nanoparticles can be stably and uniformly dispersed in the pores of MOF and prevent the aggregation [65]. Hereafter, the synthesis of two types of NP/MOF composites for glucose detection was reviewed (Figure 3).

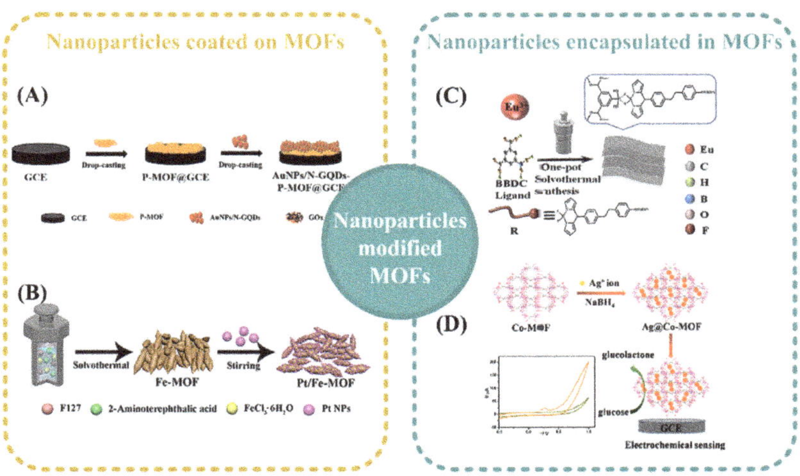

Figure 3. The synthesis of NPs-modified MOFs. NPs coated on MOFs: (**A**) AuNPs/N-GQDs coated on P-MOF as a substrate (reproduced with permission from [66], copyright © 2022, Elsevier); (**B**) PtNPs coated on Fe-MOF as a catalyst (reproduced with permission from [18], copyright © 2021, Springer Nature). NPs encapsulated in MOFs: (**C**) BODIPY encapsulated in Eu-MOF by one-pot synthesis (reproduced with permission from [67], copyright © 2022, Elsevier); (**D**) AgNPs encapsulated in Co-MOF by in situ growth (reproduced with permission from [68], copyright © 2019, American Chemical Society).

2.2.1. NPs Coated on MOFs

NPs, including noble metal nanoparticles, metal oxides, quantum dots, etc., were usually used to modify MOFs to improve detection performance [69]. According to the role played by MOFs, NPs-coated MOFs can be divided into two categories: (1) MOFs were applied as substrates to immobilize NPs with catalytic properties. (2) MOFs were used as catalytic materials, and the coated nanoparticles interacted with MOFs to improve the catalytic performance.

MOFs as Substrates

The NPs can be absorbed in the pores or on the surface of MOFs, which prevents the aggregation and reduction of the catalyst activity. Meanwhile, MOFs also can promote contact between the active site and the substrate, which is very suitable to be used as the substrate of NPs. As shown in Figure 3A, Wu et al. used the PEI functionalized MOF as the substrate to physically absorb AuNPs/N-GQDs on the surface [66]. PEI-functionalized Fe-MOF provides a large specific surface area for AuNPs/N-GQDs and facilitates the contact between uniformly dispersed active sites and the electrode surface. The enzyme-linked reaction of AuNPs/N-GQDs with peroxidase mimics (AuNPs) and nitrogen-doped graphene quantum dots (N-GQDs) realized the glucose detection.

MOFs as Catalyst

Besides simply being a substrate, MOFs also possess diverse properties, especially catalytic activities, including peroxidase activity, electrocatalytic activity, etc. Thus, MOFs can also be used in glucose detection and the detection performance as catalysts to further improve the detection performance. Lu et al. utilized the coordination between PtNPs and Fe-MOF to stabilize PtNPs on the surface of Fe-MOF (Figure 3B) [18]. The electrons on PtNPs can induct to transfer to Fe, which accelerates the redox cycle of Fe^{3+}/Fe^{2+}, which significantly improves the efficiency of the peroxidase reaction of Fe-MOF. There are many NPs on MOFs synthesized by similar methods, such as N-Co-MOF@PDA-AgNPs [70], α-CD-rGO/Ni-MOF [71], RhB-CDs@MOF-808 [72], etc.

2.2.2. Nanoparticles Encapsulated in MOFs

Except for coating NPs on the surface of MOFs, encapsulating NPs in MOFs also can be used for glucose detection, including in situ encapsulation and in situ growth [65].

As for the in situ encapsulation method, the NPs are encapsulated in MOFs by adding the pre-synthesized NPs into the mother solutions of MOF, and the MOF crystalizes around the NPs, which act as crystal nuclei [64]. For instance, Bagheri et al. synthesized CeO_2@NH_2-MIL-88B by a room-temperature one-pot strategy to form a colorimetric detection system for glucose detection [25]. The CeO_2 enhanced the peroxidase activity of NH_2-MIL-88B, exhibiting faster response speed and stronger colorimetric signal compared to horseradish peroxidase (HRP).

The in situ growth method requires two or more steps for preparation. First, the precursors of NPs, such as metal salts and polymer monomers, are encapsulated or immersed into the pores of MOF. Next, NPs are formed in situ by polymerization or redox reaction [69], generating relatively small NPs with homogeneous distribution and improved stability. Lu et al. in situ polymerized polypyrrole (PPy) in Co-Ni(Fe)-MOF nanosheets to fabricate the Co-Ni(Fe)-MOF/PPy electrode [73]. The Fe^{3+} catalyzed the in situ polymerization of PPy, avoiding the addition of additional oxidants and improving the conductivity and electrocatalytic performance of composite electrodes. Wang et al. used the in situ growth method to embed highly conductive and biocompatible AgNPs into Co-MOF (Figure 3D) [68]. The encapsulation of AgNPs enhanced the conductivity and electrocatalytic activity of the Ag@Co-MOF electrode.

Table 2. MOFs modified with NPs for glucose detection.

Sample	NPs	Function	Method	Ref
P-MOF	AuNPs/N-GODs	Substrate	Physical adsorption	[66]
UiO-66-NH_2	PPG@Ru	Substrate	EDC/NHS	[72]
Fe-MOF	PtNPs	Improve performance	Coordinate bonds	[18]
N-Co-MOF@PDA	AgNPs	Improve performance	Surface growth	[70]
Ni-MOF	αCD-rGO	Improve performance	Surface electrodeposition	[71]
Eu-MOF	BODIPY	Improve performance	In situ encapsulation	[67]
NH_2-MIL-88B	CeO_2	Improve performance	In situ encapsulation	[25]
ZIF-67	Ag@TiO_2	Improve performance	In situ encapsulation	[74]
CoNi(Fe)-MOF	PPy	Improve performance	In situ polymerization	[73]
ZIF-67	AgNPs	Improve performance	In situ generation	[17]
Cu-TCPP(Fe).	AuNPs	Improve performance	In situ generation	[75]
Co-MOF	CuNPs	Improve performance	In situ generation	[76]

PPG: poly (N-phenylglycine). EDC/NHS: cross-linker. PDA: polydopamine. αCD-rGO: α-cyclodextrin functionalized reduced graphene oxide. BODIPY: boron-dipyrromethene. TCPP: Fe(III) tetra(4-carboxyphenyl)porphyrin.

2.3. *Enzymes-Modified MOFs*

Compared with nanozymes, nature enzymes have the advantages of high catalytic efficiency and selectivity. However, natural enzymes are easily denaturalized by high temperatures, acid or alkali environments, and organic solvents, which limits their practical application. As shown in Figure 4, MOFs with a high specific surface area and stability are ideal substrates for the immobilization of enzymes, and they can not only protect the structure of enzymes from adverse conditions but also improve the selectivity of enzymes to the substrate through pore confinement [77,78]. The Immobilization methods of enzymes on MOFs are summarized in Table 3. Moreover, the fluorescence, magnetism, catalysis, and selective adsorption properties of MOFs can be combined with enzymes to achieve abundant expansive functions.

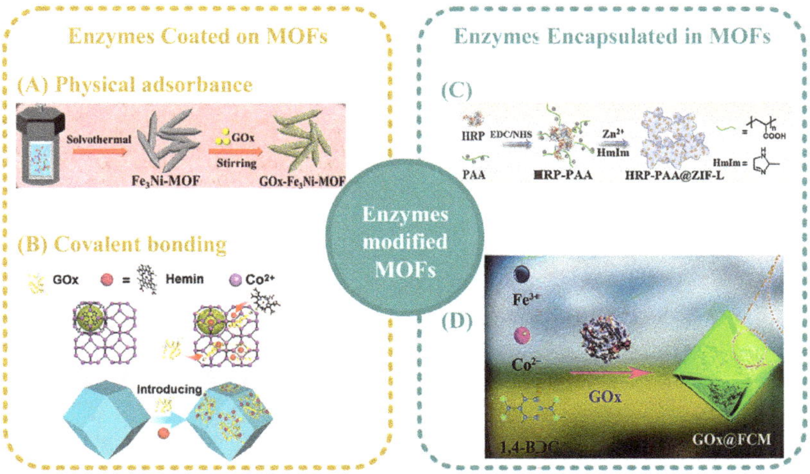

Figure 4. The synthesis of enzymes-modified MOFs. Enzymes coated on MOFs: (**A**) GOx-Fe_3Ni-MOF by physical adsorption (reproduced with permission from [79], copyright © 2022, American Chemical Society); (**B**) GOx&Hemin-ZIF-67 by the coordination between the Co^{2+} and the carbonyl group of the proteins (reproduced with permission from [24], copyright © 2022, Elsevier). Enzymes encapsulated in MOFs: (**C**) HRP-PAA@ZIF-L (reproduced with permission from [80], copyright © 2022, Elsevier); (**D**) GOx@FCM (reproduced with permission from [81], copyright © 2012, Royal Society of Chemistry).

2.3.1. Enzymes Coated on MOFs

Similar to the modification of NPs on the surface of MOFs, enzymes can also be modified on MOFs by physical adsorption and covalent bonding [82]. Wang et al. prepared bimetallic Fe_3Ni-MOFs by solvothermal synthesis, and glucose oxidase (GOx) was coated on the surface of MOF by simple physical adsorption (Figure 4A) [79]. Fe_3Ni-MOF can not only protect GOx and improve its stability but also has excellent peroxidase activity, benefiting from the bimetal sites. Zhang et al. also coated hierarchically porous HPPCN-222(Fe) with GOx by physical adsorption [83].

In addition, MOF with abundant functional groups on the surface and pores can also bond with biomolecules through covalent bonding and hydrogen bonding. Kim et al. made use of the pore adsorption of Co-MOF and hydrogen bonding, which formed between the carboxyl functional groups of the enzyme and the imine group of the ligand, 2-methylimidazole (2-MIM), to load GOx on Co-MOF successfully [84].

2.3.2. Enzymes Encapsulated in MOFs

In situ encapsulation, by which the enzyme encapsulation and MOF crystallization occur at the same time, requires mild synthesis conditions to maintain the activity of the enzymes [85]. Meanwhile, the selection of MOFs with hydrophilic skeletons and the creation of a suitable enzymatic microenvironment are more beneficial to improve the activity and loading efficiency of the enzyme during in situ encapsulation. Yang et al. enhanced the hydrophilicity of the enzyme microenvironment in ZIF-L by grafting short-chain polyacrylic acid (PAA) onto GOx (Figure 4C) [80]. PAA not only promoted the diffusion of substances by forming mesopores in ZIF-L through competitive coordination but also reduced the influence of interfacial interaction on enzyme conformation.

MOFs with peroxidase activity can also be utilized to construct enzyme-linked reaction systems with GOx, which can reduce the types of encapsulated enzymes and simplify the synthesis. Yu et al. in situ encapsulated GOx in FeCo-MOF with peroxidase activity. The layered structure of MOF, treated with tannic acid, not only protected GOx but also reduced its hindrance to material diffusion (Figure 4D) [81].

Table 3. MOFs modified with enzymes for glucose detection.

Sample	Enzymes	Function	Method	Ref
Fe_3Ni-MOF	GOx	Protection, Nanozyme	Physical adsorption	[79]
HPPCN-222	GOx	Protection, Nanozyme	Physical adsorption	[83]
ZIF-8	GOx, BHb	Protection	MIP	[86]
UiO-66-NH_2	GOx	Protection	Glutaraldehyde fixation	[72]
NiCu-MOF	GOx	Protection, Nanozyme	Glutaraldehyde fixation	[87]
Co-MOF	GOx	Protection, Nanozyme	Hydrogen bond	[84]
HP-MIL-88B-BA	GOx	Protection, Nanozyme	Specific identification	[88]
Fe-MOF	GOx	Protection, Nanozyme	EDC/NHS	[89]
Co-TCPP(Fe)	GOx	Protection, Nanozyme	EDC/NHS	[90]
dZIF-8	GOx, HRP	Protection	In situ encapsulation	[91]
ZIF-L	PAA-GOx	Protection	In situ encapsulation	[80]
Fe/Co-MOF	GOx	Protection, Nanozyme	In situ encapsulation	[81]
ZIF-67	GOx, HRP	Protection, Nanozyme	In situ encapsulation	[19]

BHb: bovine hemoglobin. MIP: molecular imprinting polymers. HP: hierarchically porous. BA: boronic acid.

2.4. *MOF-Derived Materials*

In order to increase the catalytic rate of the catalyst, MOF-derived materials with large pore structures are synthesized. MOF-derived materials with different functions can be obtained through different treatments (thermal treatment, solvent treatment, etc.), such as metal and metal ox-ide/hydroxide (pyrolysis), metal sulfide (sulfurization) and metal oxide/carbon composites (carbonization), and so on. The recent related studies were listed out in the Table 4. Due to their outstanding catalytic properties, metal and metallic oxide are widely utilized in glucose detection [92,93]. Nowadays, the traditional synthesis method, including hydrothermal synthesis and electrodeposition, are very sophisticated, but the properties of the synthesized materials still remain to be improved [94]. The metallic materials derived from MOFs can not only retain the morphological and structural characteristics of MOFs precursors but also have the advantages of large specific surface area, abundant active sites, and high porosity [95]. Moreover, MOF-derived materials can obtain additional properties from doped non-metallic elements and effectively avoid the aggregation of particles [96]. In addition, MOF-derived materials have higher stability and response ability than the original neat MOFs by creating atomically active sites, exposing more oxygen vacancy, and increasing the mobility of molecules at the solid/liquid interface, resulting in enhanced glucose detection performance.

2.4.1. Thermal Treatment

MOFs contain both inorganic metal-containing nodes and organic ligands; therefore, they possess not only metallic elements but also various non-metallic elements, such as

C, H, N, O, P, S, etc. In order to obtain the desired MOF-derived materials, the undesired element components should be removed by post-processing. One of the most commonly used methods is thermal treatment at a specific temperature and atmosphere [97]. As shown in Figure 5A, Qin et al. synthesized CuO/C core-shell nanoparticles by calcinating Cu-MOF as sacrificial templates at 400 °C in air [98]. The interaction between the carbon shell and CuO core increased the charge transfer efficiency and the oxygen vacancy content, which endowed the catalytic oxidation capacity of CuO/C for glucose. Fransaer et al. formed Co_3O_4-NiO/C composites with a "yolk-albumen-shell" structure (YASNiCo@C) by carbonization of bimetallic CoNi-MOF (Figure 5B) [99]. Furthermore, MOF-derived materials synthesized by pyrolysis also include CuO/NiO-C (Cu/Ni-MOF) [100], CuO (CuBTC) [101], and $NiFe_2O_4$ (NiFe-MOF) [102], etc.

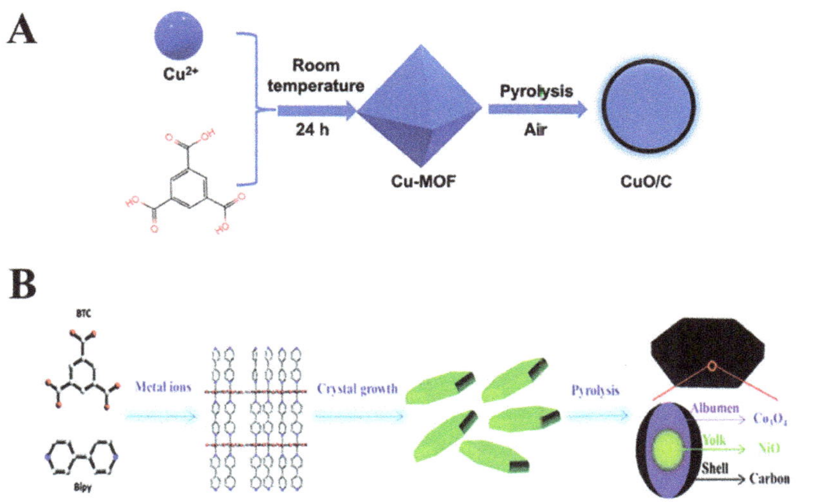

Figure 5. (**A**) The synthesis strategy of CuO/C (reproduced with permission from [98], copyright © 2022, Springer Nature). (**B**) The synthesis strategy of YASNiCo@C (reproduced with permission from [99], copyright © 2020, Royal Society of Chemistry).

2.4.2. Solvent Treatment

In addition to pyrolysis, MOF-derived materials can also be obtained by solvent treatments, i.e., reacting MOFs with specific etching solutions and obtaining doped metal compounds with special morphology and structure. Jiang et al. prepared nanorod-like porous nickel phosphate NiPO derived from spherical Ni-MOF [103]. The addition of nickel ion controlled the etching degree of Ni-MOF, and the construction of nickel phosphate conversed to flower-like, rod-like, and ribbon-like. Liu et al. formed hydrophilic hierarchically-porous nanoflowers (HHNs) by etching hydrophobic ZIF-8 with an organic weak acid, gallic acid (GA) [104]. The synergistic effect of GA with a conjugate rigid plane and free protons combined with nitrogen atoms of ZIF-8 resulted in partial damage of ZIF-8. Therefore, the HNNs kept a 3D hierarchically porous structure consisting of a 2D sheet-like structure and exposed more active sites. With the incorporation of CuNPs into the pores of HHNs, the Cu@HHNs-based electrode obtained superior catalytic activities for glucose electrocatalytic oxidation in glucose detection.

Table 4. MOF-derived materials for glucose detection.

Derivative	MOFs	Method	Ref
CuO/NiO-C	Cu/Ni-MOF	Carbonization (430 °C, N_2)	[100]
Co/MnO@HC	MnCo-MOF-74	Carbonization (900 °C, N_2)	[105]
NiO/Co_3O_4/C	NiCo-MOF	Carbonization (500 °C, Ar_2)	[106]
Ni/NCNs	Ni-MOF	Pyrolysis (500 °C, N_2)	[107]
Fe_3O_4	Fe-BDC	Pyrolysis (500 °C, N_2)	[108]
NiO	MOF-74 (Ni)	Pyrolysis (400 °C, N_2)	[109]
$ZnCo_2O_4$	ZnCo-MOF	Pyrolysis (400 °C, air)	[110]
Ni_3S_2@NCNT	Ni-MOF	Carbonization (700 °C, H_2/Ar_2), Sulfurization (Solvothermal)	[111]
E-CuO	Cu-MOF	Etch, Pyrolysis (550 °C, air)	[112]
HHN	ZIF-8	Etch (gallic acid)	[104]
Co-CuS	Cu-Co MOF	Sulfurization (Solvothermal)	[113]
Ni-HHTP	Ni-MOF	Etch (Solvothermal)	[114]

HC: hierarchical carbon. NCNs: nanoporous carbon nanorods. BDC: benzene-1,4-dicarboxylic acid. NCNT: N-doped carbon nanotube. HHTP: 2,3,6,7,10,11-hexahydroxytriphenylene.

3. Detection Mechanisms for MOF-Based Glucose Sensors

The previous section introduced four kinds of MOF-based materials for glucose detection. Different types of materials have different functions, and the response signals to glucose are also diverse. Depending on the type of response signal, detection methods can be divided into electrochemical and optical methods. The electrochemical methods include chronoamperometry (CA), linear sweep voltammetry (LSV), cyclic voltammetry (CV), differential pulse voltammetry (DPV), amperometry (AMP), etc. Optical methods include colorimetry, fluorescence (FL), chemiluminescence (CL), surface-enhanced Raman scattering (SERS), etc. In this section, the recent progress on MOF-based electrochemical and optical glucose sensors was reviewed and discussed.

3.1. *Electrochemical Methods*

Electrochemical detection is a redox-based method where an electric potential is applied between the working and reference electrodes, driving an electrochemical reaction on the surface of the working electrode [115]. This method has gained significant popularity in analytical science due to its high sensitivity, low cost, and ease of operation, making it a powerful tool for various applications [116]. In the following sections, we will discuss recent research advances in electrochemical glucose sensors based on different electrochemical detection methods.

3.1.1. Chronoamperometry

The CA method is a commonly used electrochemical method in which the electric potential of the working electrode is stepped [117]. One of the applications of CA is controlled-potential chronoamperometry. During this process, a constant potential tested from CV measurement is applied to the working electrode, and the current is monitored over time, accompanied by the oxidation or reduction of electrochemically active substances in a solution [118]. Huang et al. successfully electrodeposited MOF (Cu) onto a single-walled carbon nanotube (swnt)-modified gold wire electrode (gwe) to form a composite electrode named swnt-MOF(Cu)@gwe, which has good electrical conductivity (Figure 6A) [119]. Using the chronoamperometric technique to determine glucose, the linear range of modified electrodes was from 1 μM to 3 mM and the limit of detection (LOD) was down to 0.16 μM (S/N = 3).

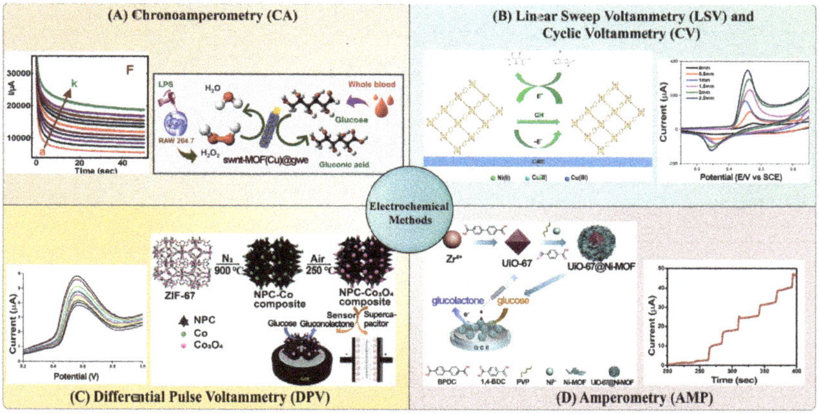

Figure 6. (**A**) Mechanism of glucose oxidation at swnt-MOF(Cu)@gwe and corresponding CA curves for glucose detection (reproduced with permission from [119], copyright © 2023, Elsevier). (**B**) Mechanism of glucose electrocatalytic oxidation at Ni@Cu-MOF nanocomposite and corresponding CV curves of glucose detection (reproduced with permission from [120], copyright © 2020, Elsevier). (**C**) Fabrication of NPC-Co_3O_4 composite and corresponding DPV curves for glucose detection (reproduced with permission from [22], copyright © 2018, Elsevier). (**D**) Synthesis of core-shell UiO-67@Ni-MOF composites and corresponding AMP curves for glucose detection (reproduced with permission from [121], copyright © 2020, Elsevier).

Although noble metals and their alloys offer a high catalytic activity for glucose oxidation, their high cost and limited availability hinder their widespread use. In contrast, transition metal materials (e.g., Mn, Fe, Co, Ni, and Cu) are more affordable and abundantly available on Earth [122]. In recent years, bimetallic transition materials have received increasing attention due to their unique properties, including versatility, efficient catalytic activity, selectivity, and stability, which often exceed monometallic materials in various applications [123]. Muthurasu et al. used bimetallic MOF to detect glucose. Due to the synergistic effects between metals, bimetallic materials often exhibit enhanced catalytic activity compared to their monometallic counterparts [124]. Finally, they obtained a bimetallic nitrogen-doped carbon nanotube (NCNT) MOF CoCu nanostructure through high-temperature calcination. The CA study demonstrates that the bimetallic NCNT MOF CoCu nanostructure substantially increases the catalytic activity towards glucose. This sensor provides a linear range from 0.05 to 2.5 mM, a sensitivity of 1027 $\mu A \cdot mM^{-1} \cdot cm^{-2}$, and a LOD of 0.15 μM ($S/N$ = 3). The same group also constructed a bimetallic Cu@Ni organic framework electrode, which also used the CA method for sensitive and selective detection of glucose [123]. CA provides a better signal-to-noise ratio in comparison to other amperometric techniques [125], but it shows a relatively slow measurement speed that requires 10 s or longer [126].

3.1.2. Linear Sweep Voltammetry and Cyclic Voltammetry

When the potential changes with time, the analyte in a solution will generate a current due to an oxidation or reduction reaction on the electrode surface at the characteristic potential. If a scanning potential changes linearly, it is called LSV; if a scanning potential changes linearly and reverses at a certain time, it is called CV.

Ozacar et al. synthesized a MOF-based composite named GOx-reduced graphene oxide (rGO)/Pt NPs@Zn-MOF-74, where rGO/Pt NPs were deposited on Zn-MOF-74 through π–π interactions, and GOx was immobilized on rGO/Pt NPs@Zn-MOF-74 via hydrogen bonds [127]. The porous structure of GOx-rGO/Pt NPs@Zn-MOF-74 allowed the substrate to be easily accessible to GOx, improving the electrochemical response. LSV curves of the electrodes indicated that the modified electrode lowered the detection potential for H_2O_2 compared to the bare glassy carbon electrode, which oxidized H_2O_2 only above 0.8 V. Kuang et al. grew the GOx@Cu-MOF packaging structure on the surface of a 3D porous conducting copper foam (CF) electrode by a one-step electrochemically assisted biomimetic mineralization method [128]. In the GOx@Cu-MOF/CF electrode, the embedded GOx catalyzes the conversion of glucose into gluconate and H_2O_2, and the H_2O_2 is immediately electrocatalytically reduced by Cu-MOF, resulting in an increase in the reduction current that is measured by CV. A clear linear relationship can be established between the reduction current and the glucose concentration in the concentration range of 0–6 mM. Zhao et al. successfully prepared a nickel-modified Cu-based MOF nanocomposite electrode (Ni@Cu-MOF) as an electrochemical detector for glucose (Figure 6B), with a sensitivity of 1703.33 μA mM^{-1} cm^{-2}, a linear range between 5 and 2500 μM, and a LOD of 1.67 μM (S/N = 3) [120]. However, CV is not an ideal technique for quantitative analysis because the limits of detection and quantification are higher than those of other types of voltammetry [129].

3.1.3. Differential Pulse Voltammetry

DPV is a technique involving the application of an amplitude potential pulse to a linear ramp potential [130]. The current is measured before the potential pulse and at the end of the pulse, and the difference between the two currents is recorded as the response [117]. Generally, pulsed techniques (e.g., DPV) are more responsive than linear scanning methods because the capacitive current is minimized, and CV is more commonly used for exploratory analysis [130].

Han et al. developed the nanoporous carbon and cobalt oxide (NPC-Co_3O_4) composite through the heat treatment of a MOF precursor, ZIF-67 (Figure 6C) [22]. DPV analysis of NPC-Co_3O_4/GCE with different glucose concentrations was performed at a scan rate of 0.1 V s^{-1} and a given potential range of 0.2–1.0 V in 0.1 M KOH. The differential pulse voltammetry response of NPC-Co_3O_4/GCE has a linear relationship with the glucose concentration. The sensor enables glucose detection in the concentration range from 5×10^{-12} to 2.05×10^{-10} M and a LOD of 2 pM (S/N = 3). It is worth noting that the sensitivity reaches up to 0.14 μA pM^{-1} cm^{-2}.

Wang et al. grew Cu-trimesic acid (Cu-BTC) MOFs on 3D-macroporous carbon (KSCs) electrodeposited Au NPs on the integrated electrode and then fixed GOx by a Au-S bond. GOD/AuNPs/Cu-BTC MOFs/3D-KSCs were obtained to construct a proportional electrochemical glucose biosensor [131]. Cu-BTC MOF can catalyze glucose oxidation and convert Cu(II) to Cu(I), which enhances the reduction peak of Cu(I) when glucose is added. The DPV response of the O_2 reduction peak current also decreases with the addition of glucose. In order to obtain good selectivity and reproducibility, $j_{O2}/j_{\text{Cu-BTC}}$ was used as the response signal. The sensor enables glucose detection in the concentration range from 44.9 μM to 19 mM and a LOD of 14.77 μM (S/N = 3). Ramaprabhu et al. fabricated a copper terephthalate (CuBDC) MOF-modified GCE to construct a non-enzymatic glucose biosensor with a sensitivity of 37.09 μA μM^{-1} cm^{-2} and a LOD of 0.077 μM (S/N = 3) [37]. Compared with the CV technique, the DPV technique displays lower signal background and better sensitivity, resulting in higher resolution at lower concentrations of an analyte [84].

3.1.4. Amperometry

AMP has become a commonly used detection method. In 1956, Clark developed the first bio-amperometric sensor for measuring dissolved O_2 in the blood consumed in an enzymatic reaction catalyzed by GOx [132]. AMP involves applying a constant reduction or oxidation potential on the working electrode and then measuring the resulting steady-state current [133].

ZIF-67-derived materials are currently used as efficient catalysts for sensors such as glucose, 4-nitrophenol (4-NP) [134], acetaminophen [2], SO_2 [135], and so on. Ruan et al. synthesized a novel N-doped carbon dodecahedron embedded with Co nanoparticles (Co@NCD) by pyrolyzing ZIF-67 in a reductive atmosphere [136]. ZIF-67-derived electrocatalysts obtained from high-temperature carbonization have improved conductivity, large surface area, and active non-precious Co and N-doped carbon heterojunctions. By using Co@NCD/GCE to detect glucose, two linear ranges were obtained from current–time curves; one was 0.0002–1.0 mM with a sensitivity of 125 μA mM^{-1} cm^{-2} and the other was 1.0–12.0 mM with a sensitivity of 23 μA mM^{-1} cm^{-2}. The LOD was calculated to be 0.11 μM (S/N = 3). Aside from that, Yin et al. prepared Au@NiCo LDH via etching ZIF-67 and reducing $HAuCl_4$ at a high temperature to construct glucose-sensing electrodes with a sensitivity of 864.7 μA mM^{-1} cm^{-2} and a LOD of 0.028 μM (S/N = 3) [137].

The use of core-shell MOF@MOF materials offers several advantages by combining the chemical, physical, and structural properties of both MOFs, leading to unexpected synergistic effects [138]. Wang et al. first synthesized the core-shell UiO-67@Ni-MOF material [121]. The composite material was synthesized by an internal extended growth method under polyvinylpyrrolidone (PVP) regulation. The pre-prepared UiO-67 was used as the growth core of shell Ni-MOF (Figure 6D). In this study, UiO-67 was selected for its expansive specific surface area and good conductivity, which facilitated efficient electron transfer within the UiO-67@Ni-MOF composite. On the other hand, Ni-MOF exhibited excellent electrochemical activity for glucose oxidation, rendering it an ideal electrocatalytic material. The findings demonstrated that the UiO-67@Ni-MOF composites displayed significantly enhanced electrocatalytic activity for glucose oxidation compared to the individual UiO-67 and Ni-MOF structures. Ni^{2+} from Ni-MOF is oxidized to Ni^{3+} in the −0.43 V alkaline electrolyte. Ni^{3+} then oxidizes glucose molecules at a voltage of 0.5 V and produces glucolactone. AMP curves show that the sensor has a fast response (<5 s), wide linear range (5 μM–3.9 mM), and low LOD (0.98 μM, S/N = 3). It has good reproducibility (RSD = 1.1%), repeatability (RSD = 1.9%), and long-term stability. It should be noted that the sensitivity and selectivity of the amperometry method are influenced by many factors, such as electrode material, electrolyte solution, temperature, oxygen, etc., and therefore need to be optimized and controlled according to the actual situation.

In addition to being classified by the electrochemical detection method, common electrochemical glucose sensors can also be divided into two types: enzymatic and non-enzymatic sensors (Tables 5 and 6). The glucose biosensor based on enzyme promotion is based on the specific reaction between glucose and active enzymes, which causes the change of an electrical signal so as to realize the detection of glucose. The electrical signals in enzymatic sensors can be generated through various processes, including O_2 consumption, electrooxidation of H_2O_2, electroreduction of H_2O_2, and oxidation of H_2O_2 by peroxidase [139]. However, enzymatic sensors may be affected by several factors, such as pH, temperature, and the presence of detergents [140]. Enzyme-free electrochemical sensors, in contrast to active enzyme-based sensors, utilize substances with glucose catalytic activity directly at the electrode to oxidize glucose. MOF-based materials have been shown to exhibit high electrocatalytic activity [141]. The presence of OH^- ions on the electrode surface increases the local pH, creating an alkaline microenvironment that enhances glucose oxidation [37]. This process involves the catalytic oxidation of glucose to gluconolactone, followed by further hydrolysis to gluconic acid. The porous structures and large active surface area of MOF-based materials significantly enhance their analytical performance [137]. Moreover, unique morphologies of MOF-based materials expose more active sites, thereby increasing the contact area with glucose and facilitating faster charge transfer. Additionally, incorporating mixed-valence metal ions or organic ligands into the MOF structure can further enhance the conductivity, catalytic activity, and, therefore, the detection performance of MOF-based materials [142].

Table 5. MOF-based non-enzymatic electrochemical glucose sensors.

Electrode Material	Electrochemical Method	Electrolyte	Linear Range (μM)	LOD (μM)	Sensitivity (μA mM^{-1} cm^{-2})	Ref
Ni-MOF	CA	0.1 M NaOH	10–2000	1.16	3.03	[143]
$Ni_3(HITP)_2$ MOF	CV	0.1 M KOH	0–10,000	–	–	[144]
Ni-MOFN	AMP	0.1 M KOH	25–3150	0.6	402.3	[60]
r-NiPO	AMP	0.1 M NaOH	1–3	1	3169	[103]
$NiO/Co_3O_4/C$	AMP	0.1 M NaOH	0.2–10,000	0.045	2820	[106]
Ni/Co-FAMOF	AMP	1 M KOH	6–1004	2	366	[34]
Ni/Co(HHTP) MOF/CC	AMP	0.1 M NaOH	0.3–2312	0.1	3250	[46]
CC@MOF-74(NiO)@NiCo LDH	AMP	1 M KOH	10–1100, 1500–9000	0.278	1699	[109]
NiCu-MOF-6	AMP	0.1 M NaOH	20–4930	15	1832	[45]
Ni@Cu-MOF	CV	0.1 M NaOH	5–2500	1.67	1703.33	[120]
NiCoBP-Br	AMP	0.1 M NaOH	0.5–6065.5	0.0665	1755.51	[47]
Ni-Co MOF/Ag/rGO/PU	AMP	0.1 M NaOH	10–660	3.28	425.9	[140]
YASNiCo@C	AMP	0.1 M NaOH	5–5000	0.75	1964	[99]
bimetallic Cu@Ni organic framework	CA	0.1 M NaOH	0–5000	0.4	496	[123]
UiO-67@Ni-MOF	AMP	0.1 M NaOH	5–3900	0.98	–	[121]
Cu-MOF	DPV	0.01 M NaOH	0.06–5000	0.01	89	[145]
CuBDC12E	DPV	0.02 M PBS (pH 7.4)	0–2000	0.077	37,090	[37]
Cu-MOF/CF	AMP	0.1 M NaOH	1–950	0.076	30,030	[146]
CuO nanorod	CA	0.1 M NaOH	up to 1250	1	1523.5	[147]
swnt-MOF(Cu)@gwe	CA	0.1 M NaOH	1–3000	0.16	–	[119]
CuO/C	AMP	0.1 M NaOH	5–25,325	1	244.71	[98]
$Cu_2(NDC)_2$/PDHP	AMP	electrolyte-simulated sweat	5–1775	2	1690	[148]
CuO polyhedrons/CC	AMP	0.1 M NaOH	0.5–800	0.46	13,575	[15]
Cu@Co-MOF	AMP	0.01 M NaOH	5–400	1.6	282.89	[76]
bimetallic NCNT MOF CoCu nanostructure	CA	0.1 M NaOH	50–2500	0.15	1027	[124]
E-NiCo-BTC MOF	AMP	0.1 M NaOH	1–1780	0.187	1789	[149]
NPC-Co_3O_4	DPV	0.1 M KOH	5×10^{-6}–2.05×10^{-4}	2×10^{-6}	0.14 μA pM^{-1} cm^{-2}	[22]
Co-MOF	CA	0.01 M NaOH	5–900	1.6	169	[150]
Co@NCD	AMP	0.1 M NaOH	0.2–12,000	0.11	125,23	[136]
Au@NiCo LDH	AMP	1.0 M NaOH	5–12,000	0.028	864.7	[137]
Co/MnO@HC	AMP	0.1 M NaOH	50–900, 1900–6900	1.31	233.8	[105]

HITP: (2,3,6,7,10,11-hexaiminotriphenylene)$_2$. MOFNs: MOF nanosheets. r-NiPO: nanorod-like nickel phosphate. LDH: layered double hydroxides. BP: 4-bromopyridine. NDC: naphthalenedicarboxylic. PDHP: pencil drawing hydrophobic paper. E-NiCo-BTC: nickel–cobalt–benzene tricarboxylic acid.

Table 6. MOF-based enzymatic electrochemical glucose sensors.

Electrode Material	Electrochemical Method	Mechanism	pH	Linear Range (μM)	LOD (μM)	Sensitivity (μA mM^{-1} cm^{-2})	Ref
PDA-GOx-HKUST-1-MWCNTs/Pt/Au	AMP	electrooxidation of H_2O_2	7.0	5–7050	0.12	178	[151]
GOD-GA-Ni/Cu-MOFs-FET	AMP	electrooxidation of H_2O_2	7.4	1–100	0.51	26.05	[87]
rGO/Pt NPs@Zn-MOF-74	LSV	electrooxidation of H_2O_2	7.4	6–6000	1.8	64.51	[127]
GO_X-AuNPs/N-GQDs-P-MOF@GCE	AMP	electroreduction of H_2O_2	4.0	2–10, 20–3000	0.7	1512.4	[66]

Table 6. *Cont.*

Electrode Material	Electrochemical Method	Mechanism	pH	Linear Range (μM)	LOD (μM)	Sensitivity (μA mM^{-1} cm^{-2})	Ref
GOx/Hemin@NC-ZIF	AMP	oxidation of H_2O_2 by peroxidase	7.2	0–20,000	10	–	[24]
GOx@Cu-MOF/CF	CV	electroreduction of H_2O_2	7.4	0–6000	–	–	[128]
GOD/AuNPs/Cu-BTC/3D-KSCs	DPV	O_2 depletion monitoring	7.0	44.9–4000, 4000–19,000	14.77	–	[131]

MWCNTs: multi-walled carbon nanotubes. GA: glutaraldehyde. FET: field-effect transistor.

3.2. Optical Methods

Optical methods for sensing glucose have been extensively studied and developed, such as colorimetry, FL, CL, SERS, etc. The optical glucose sensors based on MOF-based materials provide a visualized and cost-effective way to measure glucose concentrations compared to electrochemical glucose sensors that need relatively expensive and complicated instruments, as well as being prone to interferences [152]. However, the long-term stability and reusability of MOF-based optical glucose sensors are the major challenges for future applications. Hereafter, some typical and important optical glucose sensors are introduced and classified into colorimetry, FL, CL, and SERS-based methods (Table 7).

Table 7. MOF-based optical glucose sensors.

Materials	Detection Method	Linear Range (μM)	LOD (μM)	Ref
dZIF-8 BH	colorimetry	50–4000	–	[91]
G&L@ZIF@Paper	colorimetry	200–2000	120	[19]
GOx/Hemin@NC-ZIF	colorimetry	1000–20,000	10	[24]
Aga/GOD@Cu-hemin MOF/TMB	colorimetry	30–800	10	[49]
GOx@FCM-TA	colorimetry	5–750	0.94	[81]
Fe_3Ni-MOF	colorimetry	2–1000	1	[79]
5R@Eu-MOF	FL	0–6	0.00692	[67]
Ni-MOF	FL	8–30	4	[153]
Pt/Fe-MOF	colorimetry	3900–6400	2.3	[18]
MOF@GOx@BHb-MIPs	colorimetry	0.5–20	0.4	[86]
GOx@MOF-545(Fe)	colorimetry	0.5–20	0.28	[154]
GOx@MAF-2	CL	20–200, 500–30,000	1.4	[23]
Cu(bpy)$_2$(OTf)$_2$ nanosheets	FL	10–1000	0.41	[155]
In-aip nanosheets	FL	0–160	0.87	[156]
boric-acid Eu-MOF	FL	0.1–4	0.0643	[157]
Co-TCPP(Fe)@Luminol@GOD	CL	0.177–30.53	0.0592	[90]
Co-MOF	CL	0.04–8	0.012	[158]
AuNPs/Cu-TCPP(Fe)	SERS	160–8000	3.9	[75]
MBs@MIL-100(Fe)@Ag	SERS	20–1000	15.95	[159]
Ag NPs/UiO-66-NH_2	FL	1–200	0.5	[160]
MOF-235/β-CD	CL	0.01–3	0.01	[23]
ficin@MOF	colorimetry	1–140	0.12	[161]
GOx@Zr-PCN-222 (Fe)	colorimetry	0–5000	250	[162]
CeO_2@NH_2-MIL-88B(Fe)	colorimetry	200–15,000	80	[25]

Aga: agarose hydrogels. FCM: Fe/Co-MOF. R: BODIPY. bpy: 4,4-bipyridine. OTf: trifluoromethanesulfonate. aip: aminoisophthalic.

3.2.1. Colorimetry

Colorimetry is a fast and effective method for analyzing colored solutions or any colored substances. It is usually measured using a spectrophotometer. The need for point-of-care tests (POCT) has led to the invention of paper-based colorimetric methods, smartphone-based analyses, etc., in the last two decades [163]. The advantages of colorimetry are the convenience of use, high sensitivity, and satisfactory repeatability.

Wang et al. focused on the development of bimetal–organic frameworks (Fe_xNi_y-MOF) as catalysts with peroxidase-like activity (Figure 7A) [79]. By introducing nickel (Ni) into the framework, they achieved enhanced redox capacity and accelerated electron transfer between TMB and H_2O_2. The improved conversion efficiency between Fe^{3+} and Fe^{2+} ions, coupled with the promotion of ·OH generation, led to a significant increase in peroxidase-like activity. Based on the excellent activity of Fe_3Ni-MOF, one-step colorimetric detection of glucose is achieved by immobilizing GOx on Fe_3Ni-MOF through physical adsorption. The linear range of the glucose biosensor is 2–1000 μM, and the detection limit is 1 μM (S/N = 3).

Moving on to Chen et al.'s work, they presented a biohybrid hydrogel system encapsulating GOx and HRP in a morphology-adjusted defective ZIF biohybrid hydrogel (dZIF-8 BH) [91]. This unique system utilized the biocatalytic cascade involving the conversion of glucose into gluconic acid and H_2O_2 by GOx. The generated H_2O_2 further oxidized ABTS (single molecule 2,2′-azino-azinobis(3-ethylbenzothiazoline-6-sulfonic acid)) into ABTS·+ through the peroxidase-like activity of HRP (Figure 7B). The encapsulation within the hydrogel not only provided a stable and confined microenvironment for enzymatic reactions but also facilitated the accumulation of catalytic products, resulting in a stronger color signal for glucose detection. Moreover, the integration of a smartphone with the biosensor opens up the possibility of point-of-care diagnostics and remote monitoring of glucose levels. The dZIF-8 BH portable biosensor provides a glucose detection range from 0.05 to 4 mM. Further, Song et al. developed an integrated agarose-based hydrogel film (Aga/GOD@Cu-hemin MOF/TMB) with a linear range from 30 μM to 0.8 mM and a LOD of 0.01 mM (S/N = 3) [49]. Liu et al. encapsulated GOx and luminol in ZIF-67 to form the GOx&luminol@ZIF-67@Paper (G&L@ZIF@Paper) chip (Figure 7C) [19]. The G&L@ZIF@Paper could achieve highly sensitive and specific measurements for glucose with a linear range from 0.2 to 2 mM and a limit of detection of 0.12 mM (S/N = 3). The combination of one-pot synthesis and one-pot detection greatly improved the convenience of detection and the possibility of subsequent industrialization. Besides these advantages, the unevenness of color, susceptibility to interference, and relatively low sensitivity of the colorimetric method still need to be improved in order to achieve sensitive and accurate quantification detection of glucose.

3.2.2. Fluorescence and Chemiluminescence

Fluorescence (FL), a form of photoluminescence (PL), widely exists in gas, liquid, and solid chemical systems. This method involves the absorption and re-emission of photons by materials, where electrons absorb photons, transition to higher energy levels, and return to the ground state while emitting photons [164]. The FL method has emerged as a powerful tool in qualitative and quantitative analysis due to its rapid detection, reproducibility, high sensitivity, and selectivity.

Among FL sensors, single-emission fluorescence probes encounter challenges related to probe concentration changes, sample light scattering, excitation light fluctuations, and emission collection efficiency [165]. To address these limitations, researchers have turned to dual-emission ratio fluorescence probes, offering higher resolution and improved visualization. For example, Yin et al. developed a ratiometric fluorescence probe using Eu-MOF hollow spheres, exhibiting dual emissions at 370 nm and 623 nm, respectively (Figure 8A) [157]. This probe showed sensitive responses to glucose concentration changes through interactions with the boric acid group and H_2O_2, enabling glucose detection within the range from 0.1 μM to 4 μM with a LOD of 0.0643 μM (S/N = 3). In addition, an increase (on) of one signal (354 nm) and a decrease (off) of the other (624 nm) caused the solution color to change from bright red to blue. Similarly, Ke et al. crafted a ratiometric fluorescence sensor by grafting an R photosensitizer on Eu-MOFs, enabling the detection of F⁻, H_2O_2, and glucose in water solutions and living cells [67].

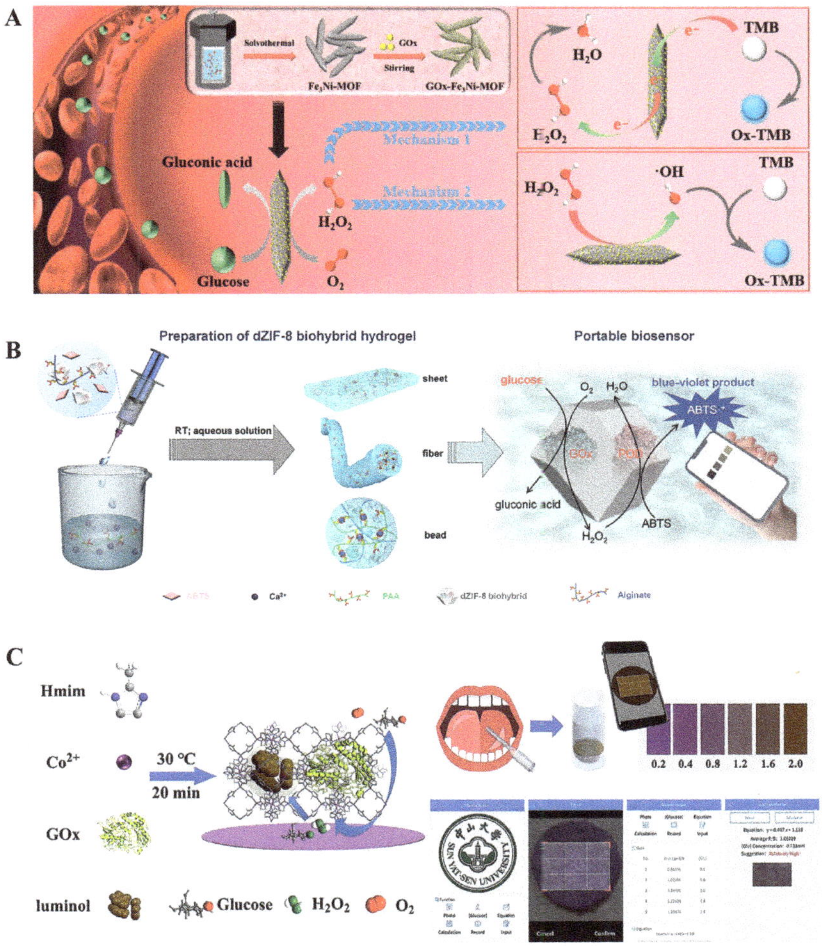

Figure 7. (**A**) Schematic diagram of the synthesis of Fe_3Ni-MOF/GOx and the reaction mechanism for the cascade oxidation of glucose (reproduced with permission from [79], copyright © 2022, American Chemical Society). (**B**) Schematic representation of the preparation of dZIF-8 BH and the colorimetric sensing mechanism based on the biocatalytic cascade of dZIFs BH (reproduced with permission from [91], copyright © 2022, American Chemical Society). (**C**) Schematic illustration of the principle of glucose detection by G&L@ZIF@Paper and the step-by-step flow for detecting glucose in saliva (reproduced with permission from [19], copyright © 2023, Elsevier).

Despite the potential of MOF-based FL sensors, their limited water stability has hindered widespread use. To overcome this challenge, Wang et al. utilized UiO-66-NH_2, known for its excellent stability in water and organic solvents, to design a dual-emission ratiometric fluorescence probe based on the inner filter effect (IFE) for glucose and cholesterol detection (Figure 8B) [160]. In the presence of GOx, glucose is converted to H_2O_2, which etches Ag NPs into silver ions (Ag^+), leading to fluorescent 2,3-diaminophenazine (DAP) generation from o-Phenylenediamine (OPD). The emission spectrum of Ag NPs/UiO-66-NH_2 and the excitation spectrum of DAP produce an effective overlap. The fluorescence intensity ratio ($F_{555\ nm}/F_{425\ nm}$) increases with the target concentration, and the color of the solution changes from blue to yellow-green, enabling higher resolution color variance and better visualization. Furthermore, the reaction reagent is integrated into filter paper, creating a user-friendly test paper that reads RGB values using a smartphone-based portable platform.

Dai et al. introduced a novel concept by combining a responsive MOF with a MOF-enzyme composite, creating a multifunctional "all-in-one" particle that incorporates both catalytic and luminescence capabilities [23]. They successfully encapsulated GOx molecules within the O_2-sensitive, luminescent Cu^I triazolate MOF (MAF-2), named GOx@MAF-2 (Figure 8C). This innovative composite offers a noble metal-free solution and provides a single particle with diverse functionalities. The increasing concentration of glucose accelerated the consumption of dissolved oxygen, resulting in the enhancive luminescence intensity of GOx@MAF-2 suspension. The luminescence intensity of the suspension shows good linearity with the logarithmic concentration of glucose in the ranges of 20–200 and 500–30,000 μM, and the detection limit is approximately 1.4 μM ($S/N = 3$).

The difference between CL and FL is that CL is a luminescence phenomenon in which the outer electrons of a molecule absorb the energy of a chemical reaction and are in an excited state, and return to the ground state by a radiative leap [166]. CL reaction involves two key steps, namely chemical excitation and luminescence. The CL reaction between luminol and H_2O_2 is generally slow and inefficient, but the reaction rate can be greatly enhanced when there are certain catalysts. CL analysis possesses the advantage of high sensitivity, fast responsibility, easy operation, inexpensive instrumentation, low background signal, and so on [158].

In 2018, Huang et al. successfully prepared β-Cyclodextrin functionalization of the metal–organic framework MOF-235 [23]. Due to the synergistic interaction between β-CD and MOF-235, the MOF-235/β-CD hybrid exhibited high catalytic activity on the luminol-H_2O_2 system. Compared to the luminol-H_2O_2 system, the CL response of the system using MOFs was enhanced more than 30 times, resulting in much lower detection limits for H_2O_2 and glucose (5 nM and 10 nM, respectively) (Figure 8D). Following in 2019, Zhang et al. synthesized a 2D-MOF nanosheet with peroxidase activity and sequentially labeled luminol and GOx on the 2D-MOF nanosheet to obtain a simple CL-functionalized glucose sensor (Co-TCPP(Fe)@Luminol@GOD) (Figure 8E) [90]. When GOx oxidizes glucose to produce gluconic acid and H_2O_2, 2D-MOF can catalyze the decomposition of H_2O_2, further oxidize luminol to generate a strong CL response, and finally realize the rapid detection of glucose with a LOD of 10.667 μg/L (~0.0592 μM).

All in all, future research is likely to focus on enhancing the water stability, dispersibility in solvents, and fluorescence performance of MOF-based FL and CL sensors under real-time environments (pressure, temperature, and mechanical stress) through improved synthetic strategies.

3.2.3. Surface-Enhanced Raman Scattering

SERS is a sensitive analytical technique that can significantly enhance the Raman signal of molecules adsorbed on the surface of precious metal nanomaterials by either electromagnetic enhancement (EM) or chemical enhancement (CM) mechanisms [167]. By introducing MOFs, there are two main factors that improve the stability of the SERS substrates: firstly, the large specific surface area of MOFs provides an abundance of sites to capture target analytes; secondly, the MOFs shell protects the NPs from oxidation and corrosion [168]. Thus, the research on the synthesis of SERS substrates based on MOFs has become more and more active in recent times.

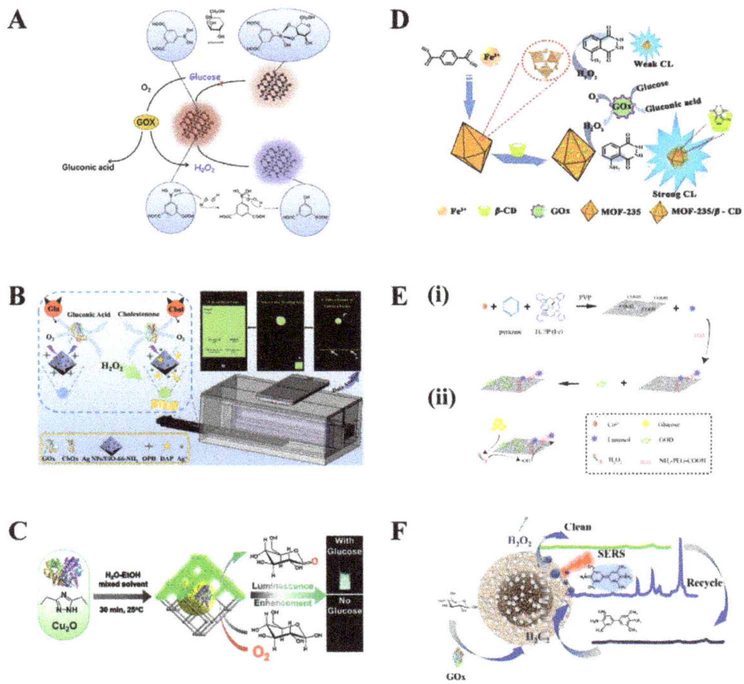

Figure 8. (**A**) Reaction mechanism of FL detection of glucose based on an Eu-MOF (reproduced with permission from [157], copyright © 2019, Elsevier). (**B**) The smartphone-based Ag NPs/UiO-66-NH_2 and OPD composite film for glucose detection (reproduced with permission from [160], copyright © 2021, American Chemical Society). (**C**) Synthesis of the GOx@MAF-2 composite and its application in glucose detection (reproduced with permission from [23], copyright © 2019, John Wiley and Sons). (**D**) CL enhancement mechanism of the luminol-H_2O_2 system by MOF-235/β-CD composite (reproduced with permission from [23], copyright © 2018, Elsevier). (**E**) Principle of the novel CL sensor for one-step ultrasensitive glucose detection (i) preparation of Co-TCPP(Fe)@luminol@GOD; (ii) process of one-step detection for glucose (reproduced with permission from [90], copyright © 2019, American Chemical Society). (**F**) Fabrication and detection principle of MMA (reproduced with permission from [159], copyright © 2022, Springer Nature).

Yang et al. modified AuNPs on the 2D metalloporphyrinic MOF (Cu-tetra(4-carboxyphenyl) porphyrin chloride(Fe(III)), Cu-TCPP(Fe)) in situ, named Au NPs/Cu-TCPP(Fe) [75]. Au NPs converted glucose in saliva to H_2O_2, and then H_2O_2 can be catalyzed by Cu-TCPP(Fe) nanosheets to oxidize leucomalachite green (LMG) into the malachite green (MG), which is a Raman-active molecule. Subsequently, the detection of MG could be performed using SERS. SERS activity is also enhanced in the presence of Au NPs. The linear range of glucose detection is 0.16–8 mM, and the LOD is 3.9 μM ($S/N = 3$) without interference from fructose, lactose, and maltose in saliva.

Cui et al. developed a highly sensitive SERS probe named MBs@MIL-100(Fe)@Ag(MMA) (Figure 8F) [159]. This probe demonstrated excellent SERS activity and peroxidase-like catalytic activity, making it ideal for glucose detection. The probe utilized the catalytic cascade reaction between MMA and GOx using TMB molecules enriched by MMA as Raman beacons, resulting in remarkable sensitivity in detecting glucose. Based on the enrichment and size screening capabilities of MIL-100(Fe), MMA has excellent anti-interference properties, allowing it to specifically enrich indicator molecules even in non-ferrous beverages or complex biological fluids (saliva) while excluding the influence of other impurities (dyes, proteins, etc.). In addition, the self-cleaning properties of MMA permitted the recycling of the probe for detection, which reduced the cost of the solution and provided a new versatile strategy for the accurate detection of glucose in complex samples such as biofluids and food and beverages.

The preparation process of MOF-based SERS substrates is simpler than the conventional SERS substrate preparation process. However, the SERS-enhanced mechanism of MOF-based substrates is still unclear. Further research is needed regarding practicality, portability, and commercialization.

4. Novel MOF-Based Glucose Detection Devices

4.1. Flexible Wearable Devices for Glucose Detection

Compared to traditional glucose sensors, flexible wearable devices have emerged as highly promising tools for in situ biomarker analysis in body fluids, such as sweat and interstitial fluids, providing a means to monitor blood glucose levels [169]. Sweat glucose has shown a qualitative correlation with blood glucose levels [170], and the integration of flexible electrodes with printed circuit boards (PCBs) has enabled the development of complete wearable devices. When coupled with mobile devices using wireless communication technology, these wearables allow for real-time data collection, transmission, and analysis [171]. It has been reported that there are about 450 million cases of diabetes worldwide, and the number may reach 700 million by 2045 [172]. Compared to traditional glucose testers, flexible wearable glucose-sensing devices offer the advantages of portability, comfort, and real-time monitoring, making them highly desirable for the growing number of diabetes cases worldwide [163].

Hu et al. developed a highly stretchable wearable electrochemical sensor called NCGP fiber for tracking glucose levels in sweat [140]. The sensor was based on a Ni-Co metal–organic framework/Ag/reduced graphene oxide/polyurethane (Ni-Co MOF/Ag/rGO/PU) composite, which was prepared using an improved wet spinning technique. The Ni-Co MOF nanosheets were coated onto the surface of reduced graphene oxide/polyurethane fiber (rGO/PU), along with conductive Ag glue (Figure 9A). This NCGP fiber sensor exhibited excellent electrocatalytic performance for glucose detection, with a linear range from 10 μM to 0.66 mM, a sensitivity of 425.9 μA mM^{-1} cm^{-2}, and a low limit of detection (LOD) of 3.28 μM (S/N = 3). Similarly, Liu et al. successfully manufactured a MOF film using a "coffee ring"-inspired approach, creating a hierarchical and oriented pore structure for improved electron transport efficiency (Figure 9B) [148]. The resulting MOF film-based sweat sensor demonstrated a linear range from 5 μM to 1775 μM, a sensitivity of 1.69 mA mM^{-1} cm^{-2}, and a LOD of 2 μM (S/N = 3). These innovative approaches offer promising solutions for non-invasive glucose monitoring, showcasing the potential of MOF-based materials in wearable sensing devices.

Wu et al. developed a multi-enzyme system by immobilizing highly loaded enzymes in a nanocage-based zeolite imidazole framework (NC-ZIF) using a dual restriction strategy to obtain GOx/Hemin@NC-ZIF [24]. The enzyme protection has two lines of defense, the outer shell of NC-ZIF and the inner nanocage of NC-ZIF. In addition to preventing enzyme leakage, the commodious internal space refrains from disrupting enzyme primitive conformation. The resulting GOx/Hemin@NC-ZIF multi-enzyme system exhibits 8.3- and 16-fold higher catalytic cascade activity than free enzymes in a solution in colorimetric and electrochemical sensors for glucose detection, and it also demonstrates long-term stability, excellent selectivity, and reusability (Figure 9C). At an applied potential of 0.6 V, the system showed a good linear range of 50–600 μM and a LOD of 2 μM (S/N = 3). Continuous glucose monitoring of sweat was successfully achieved by integrating an enzyme@NC-ZIF-based sensor and PCB into the sweatband and connecting it to a smartphone via Bluetooth.

In addition to detecting glucose in sweat, a MOF-based mouthguard sensor can also be developed to detect glucose in saliva. Moreover, MOFs and their carbon derivatives possess excellent electron-capture capacities and multifunctional structures, making them ideal as active electrode materials for triboelectric nanogenerators (TENG) [173]. By incorporating MOF-based TENG into the design of flexible wearable devices, the potential of self-powered, environmentally friendly, and powerful wearable glucose-monitoring devices becomes promising [174]. Another avenue for improvement lies in refining the wearable device from the software side. By combining flexible wearable devices with deep learning algorithms, meaningful information can be extracted from the collected data to predict subsequent blood glucose levels with greater accuracy and efficiency [175,176].

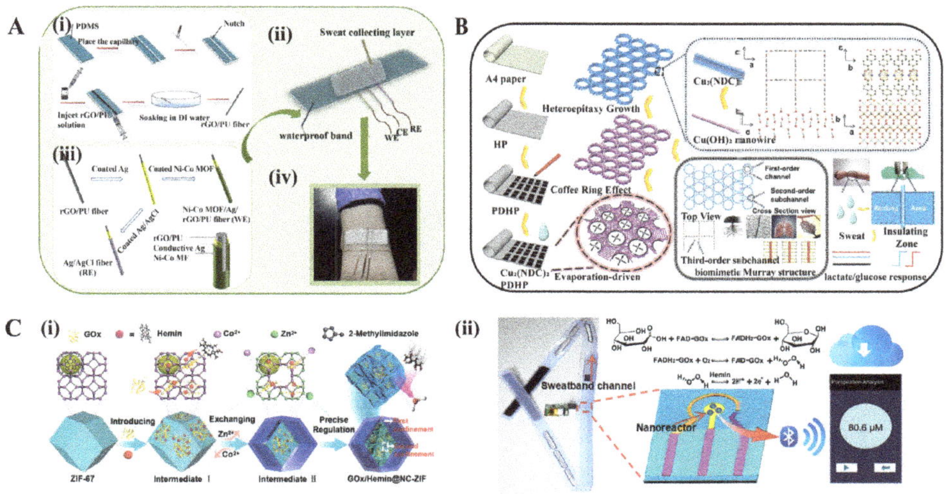

Figure 9. (**A**) Schematic illustration of (i) production of rGO/PU fiber, (ii) manufacturing process of the NCGP working electrode and Ag/AgCl fiber reference electrode, (iii) the NCGP glucose sensor integrated into the elastic fabric, and (iv) the physical image of the device attached to the volunteer's arm (reproduced with permission from [140], copyright © 2021, American Chemical Society). (**B**) The fabrication process of biomimetic Murray $Cu_2(NDC)_2$/PDHP (reproduced with permission from [148], copyright © 2018, John Wiley and Sons). (**C**) (i) Schematic illustration of the synthetic route of GOx/Hemin@NC-ZIF; (ii) Left: A photograph of a sweatband integrated with the portable prototype glucose sensor. Middle: Scheme of the GOx/Hemin@NC-ZIF catalytic cascade reaction based on the electrochemical biosensor and the all-integrated sensor fabricated on a polyimide (PI) sheet. Right: A photograph of the smartphone with an app for the perspiration analysis (reproduced with permission from [24], copyright © 2022, Elsevier).

4.2. Microfluidic Chips for Glucose Detection

The miniaturized total analytical system (μTAS), sometimes called lab-on-chip (LOC), integrates the operations of chemistry and/or biology experiments (e.g., sample handling, separation, reaction, testing, etc.) on a small chip for fast and accurate analyses of a handful of samples [177]. The first microfluidic paper-based analytical device (μPAD) was proposed by Whitesides et al. in 2007, and since then, μPADs have become one of the most popular POCT platforms [178]. A μPAD is typically made of paper with a liquid channel and a reaction zone that allows a sample to be injected into the paper and moved to the reaction zone for analysis through capillary force and absorption by the paper fibers. Moreover, it is cost-effective and easy to read [179]. In the POCT process, the analyte is sensed by the functionalized paper, and then the color signal is reported [163]. Gomez et al. utilized Fe-centered porphyrinic Zr-PCN-222(Fe) MOF for encapsulating GOx to make well-based and lateral-flow assay (LFA)-based μPADs [162]. Employing the peroxidase-like activity of Zr-PCN-222(Fe), MOF can achieve colorimetric detection of glucose on these purpose-made μPADs (Figure 10A,B). The well-based μPAD, where reaction, mixing, and analysis are all conducted in a single "well", prevents sample loss and has a lower detection limit compared to the LFA-based μPAD. However, well-based and LFA-based µPADs take too long to detect glucose, 3.5 and 2 h, respectively, and there is subjective uncertainty in the assessment of color by the naked eye. As traditional spectrometers struggle to identify paper colors, many people tend to use mobile apps. Al Lawati et al. designed a disposable 3D µPAD that can use a smartphone to record the color of a paper chip after sample injection within 3–40 min (Figure 10C) [25]. The CeO_2 NPs@NH_2-MIL-88B(Fe)/TMB/specific enzyme complex was loaded in each detection zone. Simultaneous quantification of glucose, fructose, sucrose, and maltose in food and biological samples was successfully achieved by using a distribution layer to disperse samples from the sample injection layer to the four detection zones of the detection layer.

Thin, soft, skin-compatible microfluidic systems have rocketed to popularity in recent years, which are composed of inlet/outlet ports, microfluidic channels, microreservoirs, and colorimetric sensors to collect, capture, store, and analyze sweat [180]. Apart from the paper base, combining skin-interfaced microfluidic systems with MOF-based colorimetric devices may also be a good solution for glucose detection.

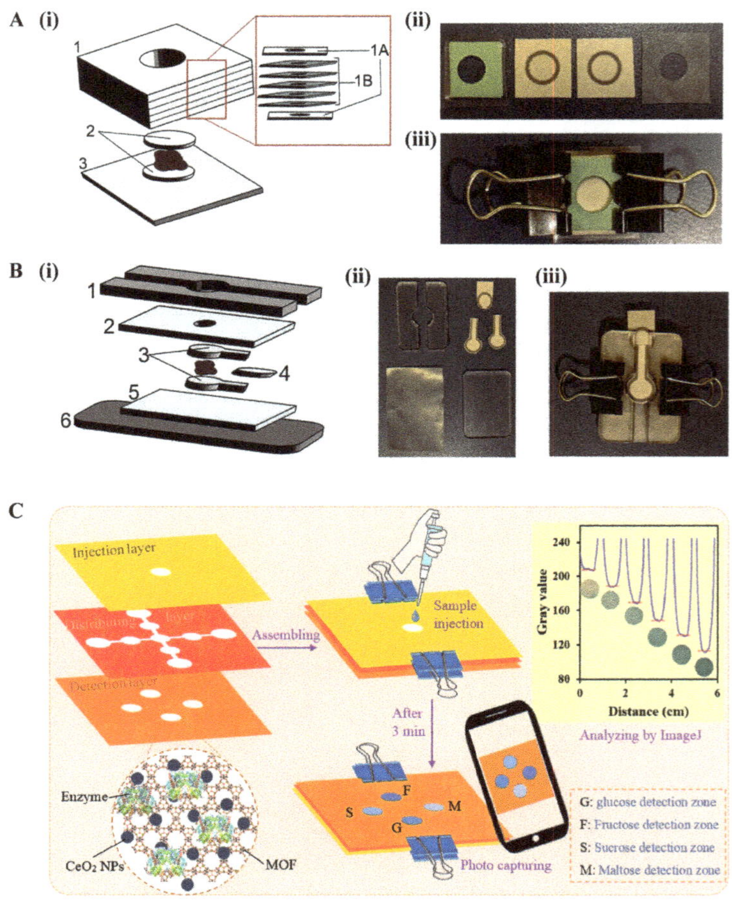

Figure 10. (**A**) (i) Schematic of the well-based µPAD; (ii) well-based µPAD layers; and (iii) assembled chip. (**B**) (i) Schematic of LFA-based µPAD; (ii) layers; and (iii) assembled chip (reproduced with permission from [162], copyright © 2019, Elsevier). (**C**) Schematic design for the simultaneous determination of different sugars using a CeO_2@NH_2-MIL-88B(Fe)-modified paper-based analytical device (reproduced with permission from [25], copyright © 2022, Elsevier).

5. Conclusions

In this review, we summarized the various preparation strategies for four types of MOF-based materials utilized in glucose detection: pristine MOFs, nano-particles in/on MOFs, enzymes in/on MOFs, and MOF-derived materials. We have also explored the two primary mechanisms for MOF-based glucose sensors: electrochemical and optical methods. Notably, this review highlights the significant advancements in flexible wearable sensors and microfluidic chips for glucose sensing. The exceptional properties of MOF materials, such as their large specific surface area, tunable pores, abundant active sites, and diverse capabilities (including catalysis and fluorescence), play crucial roles in the development of glucose sensors. For instance, for electrochemical-based glucose sensors, the large specific surface area of MOFs aids in the immobilization of GOx or promotes the conductivity of modified electrodes. On the other hand, optical-based glucose sensors benefit from the enzyme-like activities or fluorescence properties of MOFs, enabling a more visualized and convenient glucose-detection process. As we move forward, the potential of MOF-based electrochemical flexible wearable devices, such as sweatbands and mouthguards, along with optical skin-like biosensors and skin-compatible microfluidic systems, appears highly promising for the future of glucose and other biological analyte detection.

Author Contributions: Conceptualization, Y.Z., Q.L., S.-Y.L. and Z.D.; writing—original draft preparation, Y.Z. and Q.L.; writing—review and editing, Y.S., S.-Y.L. and Z.D.; visualization, J.H., M.C. and R.O.; funding acquisition, S.-Y.L. and Z.D. All authors have read and agreed to the published version of the manuscript.

Funding: This research was funded by the National Natural Science Foundations of China (22274169), the Scientific Technology Project of Shenzhen City (JCYJ20210324120601004), the Guangdong Natural Science Foundation (2022A1515011318), and the Guangdong Science and Technology Plan Project Grant (2020B1212060077).

Data Availability Statement: No new data were created or analyzed in this study. Data sharing is not applicable to this article.

Conflicts of Interest: The authors declare no conflict of interest.

References

1. Pliszka, M.; Szab lewski, L. Glucose transporters as a target for anticancer therapy. *Cancers* **2021**, *13*, 4184. [CrossRef]
2. Wang, K.; Wu, C.; Wang, F.; Jing, N.; Jiang, G. Co/Co_3O_4 Nanoparticles Coupled with Hollow Nanoporous Carbon Polyhedrons for the Enhanced Electrochemical Sensing of Acetaminophen. *ACS Sustain. Chem. Eng.* **2019**, *7*, 18582–18592. [CrossRef]
3. American Diabetes Association. 10. Cardiovascular disease and risk management: Standards of medical care in diabetes—2021. *Diabetes Care* **2020**, *44*, S125–S150.
4. Resmini, E.; Minuto, F.; Colao, A.; Ferone, D. Secondary diabetes associated with principal endocrinopathies: The impact of new treatment modalities. *Acta Diabetol.* **2009**, *46*, 85–95. [CrossRef] [PubMed]
5. American Diabetes Association. 6. Glycemic targets: Standards of medical care in diabetes-2021. *Diabetes Care* **2021**, *44*, S73–S84. [CrossRef] [PubMed]
6. Do, H.H.; Cho, J.H.; Han, S.M.; Ahn, S.H.; Kim, S.Y. Metal-organic-framework- and mxene-based taste sensors and glucose detection. *Sensors* **2021**, *21*, 7423. [CrossRef] [PubMed]
7. Mohamad Nor, N.; Ridhuan, N.S.; Abdul Razak, K. Progress of enzymatic and non-enzymatic electrochemical glucose biosensor based on nanomaterial-modified electrode. *Biosensors* **2022**, *12*, 1136. [CrossRef]
8. van Enter, B.J.; von Hauff, E. Challenges and perspectives in continuous glucose monitoring. *Chem. Commun.* **2018**, *54*, 5032–5045. [CrossRef]
9. Galant, A.L.; Kaufman, R.C.; Wilson, J.D. Glucose: Detection and analysis. *Food Chem.* **2015**, *188*, 149–160. [CrossRef]
10. Gao, Q.; Bai, Q.; Zheng, C.; Sun, N.; Liu, J.; Chen, W.; Hu, F.; Lu, T. Application of metal-organic framework in diagnosis and treatment of diabetes. *Biomolecules* **2022**, *12*, 1240. [CrossRef]
11. Adeel, M.; Asif, K.; Rahman, M.M.; Daniele, S.; Canzonieri, V.; Rizzolio, F. Glucose detection devices and methods based on metal–organic frameworks and related materials. *Adv. Funct. Mater.* **2021**, *31*, 2106023. [CrossRef]
12. Tian, L.; Huang, Z.; Lu, X.; Wang, T.; Cheng, W.; Yang, H.; Huang, T.; Li, T.; Li, Z. Plasmon-mediated oxidase-like activity on Ag@ZnS heterostructured hollow nanowires for rapid visual detection of nitrite. *Inorg. Chem.* **2023**, *62*, 1659–1666. [CrossRef] [PubMed]
13. Shang, H.; Zhang, X.; Ding, M.; Zhang, A.; Wang, C. A smartphone-assisted colorimetric and photothermal probe for glutathione detection based on enhanced oxidase-mimic CoFeCe three-atom nanozyme in food. *Food Chem.* **2023**, *423*, 136296. [CrossRef]

14. Wang, N.; Zhang, L.; Li, Z.; Zhou, C.; Lv, Y.; Su, X. A sensing platform for on-site detection of glutathione S-transferase using oxidized Pi@Ce-doped Zr-based metal-organic frameworks(MOFs). *Talanta* **2023**, *259*, 124537. [CrossRef]
15. Cheng, S.; Gao, X.; DelaCruz, S.; Chen, C.; Tang, Z.; Shi, T.; Carraro, C.; Maboudian, R. In situ formation of metal-organic framework derived CuO polyhedrons on carbon cloth for highly sensitive non-enzymatic glucose sensing. *J. Mater. Chem. B* **2019**, *7*, 4990–4996. [CrossRef]
16. Li, X.; Dong, H.; Fan, Q.; Chen, K.; Sun, D.; Hu, T.; Ni, Z. One-pot, rapid microwave-assisted synthesis of bimetallic metal-organic framework for efficient enzyme-free glucose detection. *Microchem. J.* **2022**, *179*, 107468. [CrossRef]
17. Meng, W.; Wen, Y.; Dai, L.; He, Z.; Wang, L. A novel electrochemical sensor for glucose detection based on Ag@ZIF-67 nanocomposite. *Sens. Actuators B Chem.* **2018**, *260*, 852–860.
18. Li, J.; Zhao, J.; Li, S.; Chen, Y.; Lv, W.; Zhang, J.; Zhang, L.; Zhang, Z.; Lu, X. Synergistic effect enhances the peroxidase-like activity in platinum nanoparticle-supported metal-organic framework hybrid nanozymes for ultrasensitive detection of glucose. *Nano Res.* **2021**, *14*, 4689–4695. [CrossRef]
19. Lin, Q.; Huang, J.; Zhang, Y.; Chen, M.; Xu, Y.; Zou, X.; Liu, S.-Y.; Dai, Z. A smartphone-assisted "all-in-one" paper chip for one-pot noninvasive detection of salivary glucose level. *Chem. Eng. J.* **2023**, *468*, 143608. [CrossRef]
20. Ma, X.; Tang, K.-L.; Yang, M.; Shi, W.; Zhao, W. Metal-organic framework-derived yolk-shell hollow Ni/NiO@C microspheres for bifunctional non-enzymatic glucose and hydrogen peroxide biosensors. *J. Mater. Sci.* **2021**, *56*, 442–456. [CrossRef]
21. Sun, Q.; Ding, J.; Chen, D.; Han, C.; Jiang, M.; Li, T.-T.; Hu, Y.; Qian, J.; Huang, S. Silica-Templated Metal Organic Framework-Derived Hierarchically Porous Cobalt Oxide in Nitrogen-Doped Carbon Nanomaterials for Electrochemical Glucose Sensing. *ChemElectroChem* **2021**, *8*, 812–818. [CrossRef]
22. Haldorai, Y.; Choe, S.R.; Huh, Y.S.; Han, Y.-K. Metal-organic framework derived nanoporous carbon/Co_3O_4 composite electrode as a sensing platform for the determination of glucose and high-performance supercapacitor. *Carbon* **2018**, *127*, 366–373. [CrossRef]
23. Xu, Y.; Liu, S.-Y.; Liu, J.; Zhang, L.; Chen, D.; Chen, J.; Ma, Y.; Zhang, J.-P.; Dai, Z.; Zou, X. In Situ Enzyme Immobilization with Oxygen-Sensitive Luminescent Metal-Organic Frameworks to Realize "All-in-One" Multifunctions. *Chem. Eur. J.* **2019**, *25*, 5463–5471. [CrossRef] [PubMed]
24. Wang, Q.; Chen, M.; Xiong, C.; Zhu, X.; Chen, C.; Zhou, F.; Dong, Y.; Wang, Y.; Xu, J.; Li, Y.; et al. Dual confinement of high-loading enzymes within metal-organic frameworks for glucose sensor with enhanced cascade biocatalysis. *Biosens. Bioelectron.* **2022**, *196*, 113695. [CrossRef] [PubMed]
25. Hassanzadeh, J.; Al Lawati, H.A.J.; Bagheri, N. On paper synthesis of multifunctional CeO_2 nanoparticles@Fe-MOF composite as a multi-enzyme cascade platform for multiplex colorimetric detection of glucose, fructose, sucrose, and maltose. *Biosens. Bioelectron.* **2022**, *207*, 114184. [CrossRef]
26. Mao, X.; Lu, Y.; Zhang, X.; Huang, Y. beta-Cyclodextrin functionalization of metal-organic framework MOF-235 with excellent chemiluminescence activity for sensitive glucose biosensing. *Talanta* **2018**, *188*, 161–167. [CrossRef]
27. Chen, C.; Xiong, D.; Gu, M.; Lu, C.; Yi, F.-Y.; Ma, X. MOF-Derived Bimetallic CoFe-PBA Composites as Highly Selective and Sensitive Electrochemical Sensors for Hydrogen Peroxide and Nonenzymatic Glucose in Human Serum. *ACS Appl. Mater. Interfaces* **2020**, *12*, 35365–35374. [CrossRef]
28. Wen, Y.; Meng, W.; Li, C.; Dai, L.; He, Z.; Wang, L.; Li, M.; Zhu, J. Enhanced glucose sensing based on a novel composite Co(II)-MOF/Acb modified electrode. *Dalton Trans.* **2018**, *47*, 3872–3879. [CrossRef] [PubMed]
29. Ling, W.; Liew, G.; Li, Y.; Hao, Y.; Pan, H.; Wang, H.; Ning, B.; Xu, H.; Huang, X. Materials and Techniques for Implantable Nutrient Sensing Using Flexible Sensors Integrated with Metal-Organic Frameworks. *Adv. Mater.* **2018**, *30*, e1800917. [CrossRef]
30. Bagheri, N.; Dastborhan, M.; Khataee, A.; Hassanzadeh, J.; Kobya, M. Synthesis of g-C_3N_4@CuMOFs nanocomposite with superior peroxidase mimetic activity for the fluorometric measurement of glucose. *Spectrochim. Acta Part A* **2019**, *213*, 28–36. [CrossRef]
31. Zhang, X.; Xu, Y.; Ye, B. An efficient electrochemical glucose sensor based on porous nickel-based metal organic framework/carbon nanotubes composite (Ni-MOF/CNTs). *J. Alloys Compd.* **2018**, *767*, 651–656. [CrossRef]
32. Li, W.; Lv, S.; Wang, Y.; Zhang, L.; Cui, X. Nanoporous gold induced vertically standing 2D NiCo bimetal-organic framework nanosheets for non-enzymatic glucose biosensing. *Sens. Actuators B Chem.* **2019**, *281*, 652–658. [CrossRef]
33. Rezki, M.; Septiani, N.L.W.; Iqbal, M.; Adhika, D.R.; Wenten, I.G.; Yuliarto, B. Review—Recent advance in multi-metallic metal organic frameworks (MM-MOFs) and their derivatives for electrochemical biosensor application. *J. Electrochem. Soc.* **2022**, *169*, 017504. [CrossRef]
34. Song, S.; Ma, X.; Li, W.; Zhang, B.; Shao, B.; Chang, X.; Liu, X. Novel stylophora coral-like furan-based Ni/Co bimetallic metal organic framework for high-performance capacitive storage and non-enzymatic glucose electrochemical sensing. *J. Alloys Compd.* **2023**, *931*, 167413. [CrossRef]
35. Stock, N.; Biswas, S. Synthesis of metal-organic frameworks (MOFs): Routes to various mof topologies, morphologies, and composites. *Chem. Rev.* **2012**, *112*, 933–969. [CrossRef]
36. Demazeau, G. Solvothermal processes: Definition, key factors governing the involved chemical reactions and new trends. *Z. Für Naturforschung B* **2010**, *65*, 999–1006. [CrossRef]
37. Ghosh, A.; Fathima, T.K.S.; Ramaprabhu, S. Effect of Coordinated Solvent Molecules in Cu-MOF on Enzyme Free Sensing of Glucose and Lactate in Physiological pH. *J. Electrochem. Soc.* **2022**, *169*, 057524. [CrossRef]

38. Demazeau, G.; Largeteau, A. Hydrothermal/solvothermal crystal growth: An old but adaptable process. *Z. Anorg. Allg. Chem.* **2015**, *641*, 159–163. [CrossRef]
39. Zha, X.; Yang, W.; Shi, L.; Li, Y.; Zeng, Q.; Xu, J.; Yang, Y. Morphology control strategy of bimetallic mof nanosheets for upgrading the sensitivity of noninvasive glucose detection. *ACS Appl. Mater. Interfaces* **2022**, *14*, 37843–37852. [CrossRef]
40. Du, Q.; Liao, Y.; Shi, N.; Sun, S.; Liao, X.; Yin, G.; Huang, Z.; Pu, X.; Wang, J. Facile synthesis of bimetallic metal-organic frameworks on nickel foam for a high performance non-enzymatic glucose sensor. *J. Electroanal. Chem.* **2022**, *904*, 115887. [CrossRef]
41. Yuan, A.; Lu, Y.; Zhang, X.; Chen, Q.; Huang, Y. Two-dimensional iron MOF nanosheet as a highly efficient nanozyme for glucose biosensing. *J. Mater. Chem. B* **2020**, *8*, 9295–9303. [CrossRef]
42. Arul, P.; Gowthaman, N.S.K.; John, S.A.; Tominaga, M. Tunable electrochemical synthesis of 3D nucleated microparticles like Cu-BTC MOF-carbon nanotubes composite: Enzyme free ultrasensitive determination of glucose in a complex biological fluid. *Electrochim. Acta* **2020**, *354*, 136673. [CrossRef]
43. Ma, Z.-Z.; Ma, Y.; Liu, B.; Xu, L.; Jiao, H. A high-performance Co-MOF non-enzymatic electrochemical sensor for glucose detection. *New J. Chem.* **2021**, *45*, 21350–21358. [CrossRef]
44. Abrori, S.A.; Trisno, M.L.A.; Aritonang, R.A.; Anshori, I.; Nugraha; Suyatman; Yuliarto, B. Synthesis and characterization of metal-organic framework (MOF) CoBTC as a non-enzymatic electrochemical biosensor for glucose. *IOP Conf. Ser. Mater. Sci. Eng.* **2021**, *1045*, 012006. [CrossRef]
45. Pan, W.; Zheng, Z.; Wu, X.; Gao, J.; Liu, Y.; Yuan, Q.; Gan, W. Facile synthesis of 2D/3D hierarchical NiCu bimetallic MOF for non-enzymatic glucose sensor. *Microchem. J.* **2021**, *170*, 106652. [CrossRef]
46. Xu, Z.; Wang, Q.; Hui, Z.; Zhao, S.; Zhao, Y.; Wang, L. Carbon cloth-supported nanorod-like conductive Ni/Co bimetal MOF: A stable and high-performance enzyme-free electrochemical sensor for determination of glucose in serum and beverage. *Food Chem.* **2021**, *349*, 129202.
47. Liu, Q.; Chen, J.; Yu, F.; Wu, J.; Liu, Z.; Peng, B. Multifunctional book-like CuCo-MOF for highly sensitive glucose detection and electrocatalytic oxygen evolution. *New J. Chem.* **2021**, *45*, 16714–16721. [CrossRef]
48. Bagheri, A.R.; Aramesh, N. Towards the room-temperature synthesis of covalent organic frameworks: A mini-review. *J. Mater. Sci.* **2021**, *56*, 1116–1132. [CrossRef]
49. Wei, Z.; Zhu, W.; Li, Y.; Ma, Y.; Wang, J.; Hu, N.; Suo, Y.; Wang, J. Conductive leaflike cobalt metal-organic framework nanoarray on carbon cloth as a flexible and versatile anode toward both electrocatalytic glucose and water oxidation. *Inorg. Chem.* **2018**, *57*, 8422–8428. [CrossRef]
50. Li, P.; Bai, Y.; Zhang, G.; Guo, X.; Meng, X.; Pang, H. Surface-halogen-introduced 2D NiCo bimetallic MOFs via a modulation method for elevated electrochemical glucose sensing. *Inorg. Chem. Front.* **2022**, *9*, 5853–5861. [CrossRef]
51. Song, D.; Wang, L.; Qu, Y.; Wang, B.; Li, Y.; Miao, X.; Yang, Y.; Duan, C. A High-Performance Three-Dimensional Hierarchical Structure MOF-Derived NiCo LDH Nanosheets for Non-Enzymatic Glucose Detection. *J. Electrochem. Soc.* **2019**, *166*, B1681–B1688. [CrossRef]
52. Lin, C.; Du, Y.; Wang, S.; Wang, L.; Song, Y. Glucose oxidase@Cu-hemin metal-organic framework for colorimetric analysis of glucose. *Mater. Sci. Eng. C* **2021**, *118*, 111511. [CrossRef] [PubMed]
53. Chen, C.; Wang, Y.-L.; Lin, X.; Ma, S.-H.; Cao, J.-T.; Liu, Y.-M. Cu-MOFs/GOx Bifunctional Probe-Based Synergistic Signal Amplification Strategy: Toward Highly Sensitive Closed Bipolar Electrochemiluminescence Immunoassay. *ACS Appl. Mater. Interfaces* **2023**, *15*, 22959–22966. [CrossRef] [PubMed]
54. Hu, S.; Lin, Y.; Teng, J.; Wong, W.-L.; Qiu, B. In situ deposition of MOF-74(Cu) nanosheet arrays onto carbon cloth to fabricate a sensitive and selective electrocatalytic biosensor and its application for the determination of glucose in human serum. *Microchim. Acta* **2020**, *187*, 670. [CrossRef] [PubMed]
55. Zhao, Z.; Kong, Y.; Lin, X.; Liu, C.; Liu, J.; He, Y.; Yang, L.; Huang, G.; Mei, Y. Oxide nanomembrane induced assembly of a functional smart fiber composite with nanoporosity for an ultra-sensitive flexible glucose sensor. *J. Mater. Chem. A* **2020**, *8*, 26119–26129. [CrossRef]
56. Phan, P.T.; Hong, J.; Tran, N.; Le, T.H. The properties of microwave-assisted synthesis of metal-organic frameworks and their applications. *Nanomaterials* **2023**, *13*, 352. [CrossRef] [PubMed]
57. Vaitsis, C.; Sourkouni, G.; Argirusis, C. Metal organic frameworks (MOFs) and ultrasound: A review. *Ultrason. Sonochem.* **2019**, *52*, 106–119. [CrossRef]
58. Zeraati, M.; Alizadeh, V.; Kazemzadeh, P.; Safinejad, M.; Kazemian, H.; Sargazi, G. A new nickel metal organic framework (Ni-MOF) porous nanostructure as a potential novel electrochemical sensor for detecting glucose. *J. Porous Mater.* **2022**, *29*, 257–267. [CrossRef]
59. Friščić, T. New opportunities for materials synthesis using mechanochemistry. *J. Mater. Chem.* **2010**, *20*, 7599–7605. [CrossRef]
60. Koyappayil, A.; Yeon, S.-H.; Chavan, S.G.; Jin, L.; Go, A.; Lee, M.-H. Efficient and rapid synthesis of ultrathin nickel-metal organic framework nanosheets for the sensitive determination of glucose. *Microchem. J.* **2022**, *179*, 107462. [CrossRef]
61. Al-Kutubi, H.; Gascon, J.; Sudhölter, E.J.R.; Rassaei, L. Electrosynthesis of Metal–Organic Frameworks: Challenges and Opportunities. *ChemElectroChem* **2015**, *2*, 462–474. [CrossRef]

62. de Carvalho, M.H.; de Araújo, H.D.A.; da Silva, R.P.; dos Santos Correia, M.T.; de Freitas, K.C.S.; de Souza, S.R.; Barroso Coelho, L.C.B. Biosensor Characterization from Cratylia mollis Seed Lectin (Cramoll)-MOF and Specific Carbohydrate Interactions in an Electrochemical Model. *Chem. Biodivers.* **2022**, *19*, e202200515. [CrossRef]
63. Shahrokhian, S.; Ezzati, M.; Hosseini, H. Fabrication of a sensitive and fast response electrochemical glucose sensing platform based on co-based metal-organic frameworks obtained from rapid in situ conversion of electrodeposited cobalt hydroxide intermediates. *Talanta* **2020**, *210*, 120696. [CrossRef]
64. Wang, X.; Wang, Y.; Ying, Y. Recent advances in sensing applications of metal nanoparticle/metal–organic framework composites. *TrAC Trends Anal. Chem.* **2021**, *143*, 116395. [CrossRef]
65. Xiang, W.; Zhang, Y.; Lin, H.; Liu, C.-J. Nanoparticle/metal–organic framework composites for catalytic applications: Current status and perspective. *Molecules* **2017**, *22*, 2103. [CrossRef] [PubMed]
66. Zhang, Y.; Wei, X.; Gu, Q.; Zhang, J.; Ding, Y.; Xue, L.; Chen, M.; Wang, J.; Wu, S.; Yang, X.; et al. Cascade amplification based on PEI-functionalized metal–organic framework supported gold nanoparticles/nitrogen–doped graphene quantum dots for amperometric biosensing applications. *Electrochim. Acta* **2022**, *405*, 139803. [CrossRef]
67. Li, Y.; Li, J.-J.; Zhang, Q.; Zhang, J.-Y.; Zhang, N.; Fang, Y.-Z.; Yan, J.; Ke, Q. The multifunctional BODIPY@Eu-MOF nanosheets as bioimaging platform: A ratiometric fluorescencent sensor for highly efficient detection of F^-, H_2O_2 and glucose. *Sens. Actuators B Chem.* **2022**, *354*, 131140.
68. Liu, Y.; Shi, W.-J.; Lu, Y.-K.; Liu, G.; Hou, L.; Wang, Y.-Y. Nonenzymatic Glucose Sensing and Magnetic Property Based on the Composite Formed by Encapsulating Ag Nanoparticles in Cluster-Based Co-MOF. *Inorg. Chem.* **2019**, *58*, 16743–16751. [CrossRef]
69. Zhang, Y.; Zhou, Y.; Zhao, Y.; Liu, C.-J. Recent progresses in the size and structure control of MOF supported noble metal catalysts. *Catal. Today* **2016**, *263*, 61–68. [CrossRef]
70. Zhai, X.; Cao, Y.; Sun, W.; Cao, S.; Wang, Y.; He, L.; Yao, N.; Zhao, D. Core-shell composite N-doped-Co-MOF@polydopamine decorated with Ag nanoparticles for nonenzymatic glucose sensors. *J. Electroanal. Chem.* **2022**, *918*, 116491. [CrossRef]
71. Xu, T.; Zhang, Y.; Liu, M.; Wang, H.; Ren, J.; Tian, Y.; Liu, X.; Zhou, Y.; Wang, J.; Zhu, W.; et al. In-situ two-step electrodeposition of alpha-CD-rGO/Ni-MOF composite film for superior glucose sensing. *J. Alloys Compd.* **2022**, *923*, 166418. [CrossRef]
72. Jin, X.; Li, G.; Xu, T.; Su, L.; Yan, D.; Zhang, X. Ruthenium-based Conjugated Polymer and Metal-organic Framework Nanocomposites for Glucose Sensing. *Electroanalysis* **2021**, *33*, 1902–1910. [CrossRef]
73. Chen, S.; Liu, D.; Song, N.; Wang, C.; Lu, X. Promoting non-enzymatic electrochemical sensing performance toward glucose by the integration of conducting polypyrrole with metal-organic framework. *Compos. Commun.* **2022**, *30*, 101074. [CrossRef]
74. Arif, D.; Hussain, Z.; Sohail, M.; Liaqat, M.A.; Khan, M.A.; Noor, T. A non-enzymatic electrochemical sensor for glucose detection based on Ag@TiO_2@metal-organic framework (ZIF-67) nanocomposite. *Front. Chem.* **2020**, *8*, 573510. [CrossRef]
75. Hu, S.; Jiang, Y.; Wu, Y.; Guo, X.; Ying, Y.; Wen, Y.; Yang, H. Enzyme-Free Tandem Reaction Strategy for Surface-Enhanced Raman Scattering Detection of Glucose by Using the Composite of Au Nanoparticles and Porphyrin-Based Metal-Organic Framework. *ACS Appl. Mater. Interfaces* **2020**, *12*, 55324–55330. [CrossRef]
76. Ma, Z.-Z.; Wang, Y.-S.; Liu, B.; Jiao, H.; Xu, L. A Non-Enzymatic Electrochemical Sensor of Cu@Co-MOF Composite for Glucose Detection with High Sensitivity and Selectivity. *Chemosensors* **2022**, *10*, 416. [CrossRef]
77. Duraisamy, S.R.; Liu, W.-L.; Huang, H.-Y.; Lin, C.-H. Immobilization of protein on nanoporous metal-organic framework materials. *Comments Inorg. Chem.* **2015**, *35*, 331–349.
78. Kempahanumakkagari, S.; Kumar, V.; Samaddar, P.; Kumar, P.; Ramakrishnappa, T.; Kim, K.H. Biomolecule-embedded metal-organic frameworks as an innovative sensing platform. *Biotechnol. Adv.* **2018**, *36*, 467–481. [CrossRef]
79. Mu, Z.; Wu, S.; Guo, J.; Zhao, M.; Wang, Y. Dual Mechanism Enhanced Peroxidase-like Activity of Iron-Nickel Bimetal-Organic Framework Nanozyme and Its Application for Biosensing. *ACS Sustain. Chem. Eng.* **2022**, *10*, 2984–2993. [CrossRef]
80. Zhou, Z.; Chao, H.; He, W.; Su, P.; Song, J.; Yang, Y. Boosting the activity of enzymes in metal-organic frameworks by a one-stone-two-bird enzymatic surface functionalization strategy. *Appl. Surf. Sci.* **2022**, *586*, 152815. [CrossRef]
81. Shen, H.; Shi, H.; Yang, Y.; Song, J.; Ding, C.; Yu, S. Highly efficient synergistic biocatalysis driven by stably loaded enzymes within hierarchically porous iron/cobalt metal-organic framework via biomimetic mineralization. *J. Mater. Chem. B* **2022**, *10*, 1553–1560. [CrossRef]
82. Du, Y.; Jia, X.; Zhong, L.; Jiao, Y.; Zhang, Z.; Wang, Z.; Feng, Y.; Bilal, M.; Cui, J.; Jia, S. Metal-organic frameworks with different dimensionalities: An ideal host platform for enzyme@MOF composites. *Coord. Chem. Rev.* **2022**, *454*, 214327. [CrossRef]
83. Li, S.-F.; Chen, Y.; Wang, Y.-S.; Mo, H.-L.; Zang, S.-Q. Integration of enzyme immobilization and biomimetic catalysis in hierarchically porous metal-organic frameworks for multi-enzymatic cascade reactions. *Sci. China Chem.* **2022**, *65*, 1122–1128. [CrossRef]
84. Goud, B.S.; Shin, G.; Vattikuti, S.V.P.; Mameda, N.; Kim, H.; Koyyada, G.; Kim, J.H. Enzyme-integrated biomimetic cobalt metal-organic framework nanozyme for one-step cascade glucose biosensing via tandem catalysis. *Biochem. Eng. J.* **2022**, *188*, 108669. [CrossRef]
85. Nadar, S.S.; Vaidya, L.; Rathod, V.K. Enzyme embedded metal organic framework (enzyme–MOF): De novo approaches for immobilization. *Int. J. Biol. Macromol.* **2020**, *149*, 861–876. [CrossRef] [PubMed]
86. Chen, T.; Zhang, A.; Cheng, Y.; Zhang, Y.; Fu, D.; Liu, M.; Li, A.; Liu, J. A molecularly imprinted nanoreactor with spatially confined effect fabricated with nano-caged cascaded enzymatic system for specific detection of monosaccharides. *Biosens. Bioelectron.* **2021**, *188*, 113355. [CrossRef]

87. Wang, B.; Luo, Y.; Gao, L.; Liu, B.; Duan, G. High-performance field-effect transistor glucose biosensors based on bimetallic Ni/Cu metal-organic frameworks. *Biosens. Bioelectron.* **2021**, *171*, 112736. [CrossRef]
88. Zhao, Z.; Huang, Y.; Liu, W.; Ye, F.; Zhao, S. Immobilized glucose oxidase on boronic acid-functionalized hierarchically porous MOF as an integrated nanozyme for one-step glucose detection. *ACS Sustain. Chem. Eng.* **2020**, *8*, 4481–4488. [CrossRef]
89. Xu, W.; Jiao, L.; Yan, H.; Wu, Y.; Chen, L.; Gu, W.; Du, D.; Lin, Y.; Zhu, C. Glucose oxidase-integrated metal-organic framework hybrids as biomimetic cascade nanozymes for ultrasensitive glucose biosensing. *ACS Appl. Mater. Interfaces* **2019**, *11*, 22096–22101. [CrossRef]
90. Zhu, N.; Gu, L.; Wang, J.; Li, X.; Liang, G.; Zhou, J.; Zhang, Z. Novel and Sensitive Chemiluminescence Sensors Based on 2D-MOF Nanosheets for One-Step Detection of Glucose in Human Urine. *J. Phys. Chem. C* **2019**, *123*, 9388–9393. [CrossRef]
91. Zhong, N.; Gao, R.; Shen, Y.; Kou, X.; Wu, J.; Huang, S.; Chen, G.; Ouyang, G. Enzymes-Encapsulated Defective Metal-Organic Framework Hydrogel Coupling with a Smartphone for a Portable Glucose Biosensor. *Anal. Chem.* **2022**, *94*, 14385–14393. [CrossRef] [PubMed]
92. Zhang, S.; Zhao, W.; Zeng, J.; He, Z.; Wang, X.; Zhu, Z.; Hu, R.; Liu, C.; Wang, Q. Wearable non-invasive glucose sensors based on metallic nanomaterials. *Mater. Today Bio* **2023**, *20*, 100638. [CrossRef] [PubMed]
93. Chen, J.; Ma, Q.; Li, M.; Chao, D.; Huang, L.; Wu, W.; Fang, Y.; Dong, S. Glucose-oxidase like catalytic mechanism of noble metal nanozymes. *Nat. Commun.* **2021**, *12*, 3375. [CrossRef]
94. Yogesh, M.C.; Satish, B.J.; Padamaja, N.P.; Vikas, V.M.; Jayavant, L.G.; Chandrakant, D.L. Metal oxide-based composites in nonenzymatic electrochemical glucose sensors. *Ind. Eng. Chem. Res.* **2021**, *60*, 18195–18217.
95. Gonçalves, J.M.; Martins, P.R.; Rocha, D.P.; Matias, T.A.; Julião, M.S.S.; Munoz, R.A.A.; Angnes, L. Recent trends and perspectives in electrochemical sensors based on MOF-derived materials. *J. Mater. Chem. C* **2021**, *9*, 8718–8745. [CrossRef]
96. Lu, X.F.; Fang, Y.; Luan, D.; Lou, X.W.D. Metal–organic frameworks derived functional materials for electrochemical energy storage and conversion: A mini review. *Nano Lett.* **2021**, *21*, 1555–1565. [CrossRef] [PubMed]
97. Xiao, W.; Cheng, M.; Liu, Y.; Wang, J.; Zhang, G.; Wei, Z.; Li, L.; Du, L.; Wang, G.; Liu, H. Functional metal/carbon composites derived from metal–organic frameworks: Insight into structures, properties, performances, and mechanisms. *ACS Catal.* **2023**, *13*, 1759–1790. [CrossRef]
98. Zhao, C.; Tang, X.; Zhao, J.; Cao, J.; Jiang, Z.; Qin, J. MOF derived core-shell CuO/C with temperature-controlled oxygen-vacancy for real time analysis of glucose. *J. Nanobiotechnol.* **2022**, *20*, 507. [CrossRef] [PubMed]
99. Zhang, X.; Zhang, Y.; Guo, W.; Wan, K.; Zhang, T.; Arbiol, J.; Zhao, Y.-Q.; Xu, C.-L.; Xu, M.; Fransaer, J. A yolk-albumen-shell structure of mixed Ni-Co oxide with an ultrathin carbon shell for high-sensitivity glucose sensors. *Mater. Adv.* **2020**, *1*, 908–917. [CrossRef]
100. Archana, V.; Xia, Y.; Fang, R.; Kumar, G.G. Hierarchical CuO/NiO-Carbon Nanocomposite Derived from Metal Organic Framework on Cello Tape for the Flexible and High Performance Nonenzymatic Electrochemical Glucose Sensors. *ACS Sustain. Chem. Eng.* **2019**, *7*, 6707–6719. [CrossRef]
101. Li, L.; Liu, Y.; Ai, L.; Jiang, J. Synthesis of the crystalline porous copper oxide architectures derived from metal-organic framework for electrocatalytic oxidation and sensitive detection of glucose. *J. Ind. Eng. Chem.* **2019**, *70*, 330–337. [CrossRef]
102. Chu, D.; Li, F.; Song, X.; Ma, H.; Tan, L.; Pang, H.; Wang, X.; Guo, D.; Xiao, B. A novel dual-tasking hollow cube NiFe2O4-NiCo-LDH@rGO hierarchical material for high preformance supercapacitor and glucose sensor. *J. Colloid Interface Sci.* **2020**, *568*, 130–138. [CrossRef]
103. Xiao, L.; Yang, K.; Duan, J.; Zheng, S.; Jiang, J. The nickel phosphate rods derived from Ni-MOF with enhanced electrochemical activity for non-enzymatic glucose sensing. *Talanta* **2022**, *247*, 123587. [CrossRef]
104. Zhu, Q.; Hu, S.; Zhang, L.; Li, Y.; Carraro, C.; Maboudian, R.; Wei, W.; Liu, A.; Zhang, Y.; Liu, S. Reconstructing hydrophobic ZIF-8 crystal into hydrophilic hierarchically-porous nanoflowers as catalyst carrier for nonenzymatic glucose sensing. *Sens. Actuators B Chem.* **2020**, *313*, 128031. [CrossRef]
105. Zhang, Y.; Huang, Y.; Gao, P.; Yin, W.; Yin, M.; Pu, H.; Sun, Q.; Liang, X.; Fa, H.-b. Bimetal-organic frameworks MnCo-MOF-74 derived Co/MnO@HC for the construction of a novel enzyme-free glucose sensor. *Microchem. J.* **2022**, *175*, 107097. [CrossRef]
106. Vignesh, A.; Vajeeston, P.; Pannipara, M.; Al-Sehemi, A.G.; Xia, Y.; Kumar, G.G. Bimetallic metal-organic framework derived 3D hierarchical NiO/Co_3O_4/C hollow microspheres on biodegradable garbage bag for sensitive, selective, and flexible enzyme-free electrochemical glucose detection. *Chem. Eng. J.* **2022**, *430*, 133157. [CrossRef]
107. Jia, H.; Shang, N.; Feng, Y.; Ye, H.; Zhao, J.; Wang, H.; Wang, C.; Zhang, Y. Facile preparation of Ni nanoparticle embedded on mesoporous carbon nanorods for non-enzymatic glucose detection. *J. Colloid Interface Sci.* **2021**, *583*, 310–320. [CrossRef]
108. Abrori, S.A.; Septiani, N.L.W.; Nugraha; Anshori, I.; Suyatman; Suendo, V.; Yuliarto, B. Metal-organic-framework FeBDC-derived Fe_3O_4 for non-enzymatic electrochemical detection of glucose. *Sensors* **2020**, *20*, 4891. [CrossRef]
109. Wang, L.; Yang, Y.; Wang, B.; Duan, C.; Li, J.; Zheng, L.; Li, J.; Yin, Z. Bifunctional three-dimensional self-supporting multistage structure CC@MOF-74(NiO)@NiCo LDH electrode for supercapacitors and non-enzymatic glucose sensors. *J. Alloys Compd.* **2021**, *885*, 160899. [CrossRef]
110. Zhang, D.; Wang, Z.; Li, J.; Hu, C.; Zhang, X.; Jiang, B.; Cao, Z.; Zhang, J.; Zhang, R. MOF-derived $ZnCo_2O_4$ porous micro-rice with enhanced electro-catalytic activity for the oxygen evolution reaction and glucose oxidation. *RSC Adv.* **2020**, *10*, 9063–9069. [CrossRef]

111. Li, G.; Xie, G.; Chen, D.; Gong, C.; Chen, X.; Zhang, Q.; Pang, B.; Zhang, Y.; Li, C.; Hu, J.; et al. Facile synthesis of bamboo-like Ni_3S_2@NCNT as efficient and stable electrocatalysts for non-enzymatic glucose detection. *Appl. Surf. Sci.* **2022**, *585*, 152683. [CrossRef]
112. Zhang, C.; Zhang, J.-R.; Han, C.; Zhang, Y.-H.; Yang, S.-H.; Meng, W. Synthesis of Porous CuO Based on an Etching Strategy and Application in Electrochemical Glucose Sensing. *Chin. J. Inorg. Chem.* **2021**, *37*, 2249–2259.
113. Zhang, D.; Zhang, X.; Bu, Y.; Zhang, J.; Zhang, R. Copper cobalt sulfide structures derived from MOF precursors with enhanced electrochemical glucose sensing properties. *Nanomaterials* **2022**, *12*, 1394. [CrossRef] [PubMed]
114. Wang, F.; Hu, J.; Liu, Y.; Yuan, G.; Zhang, S.; Xu, L.; Xue, H.; Pang, H. Turning coordination environment of 2D nickel-based metal-organic frameworks by pi-conjugated molecule for enhancing glucose electrochemical sensor performance. *Mater. Today Chem.* **2022**, *24*, 100885. [CrossRef]
115. Sobhanie, E.; Roshani, A.; Hosseini, M. Microfluidic systems with amperometric and voltammetric detection and paper-based sensors and biosensors. In *Carbon Nanomaterials-Based Sensors*, 1st ed.; Manjunatha, J.G., Hussain, C.M., Eds.; Elsevier: Amsterdam, The Netherlands, 2022; pp. 275–287.
116. Ünlüer, Ö.B.; Ghorbani-Bidkorbeh, F.; Keçili, R.; Hussain, C.M. Future of the modern age of analytical chemistry: Nanominiaturization. In *Handbook on Miniaturization in Analytical Chemistry*, 1st ed.; Hussain, C.M., Ed.; Elsevier: Amsterdam, The Netherlands, 2020; pp. 277–296.
117. Venton, B.J.; DiScenza, D.J. Voltammetry. In *Electrochemistry for Bioanalysis*, 1st ed.; Patel, B., Ed.; Elsevier: Amsterdam, The Netherlands, 2021; pp. 27–50.
118. Foley, M.P.; Du, P.; Griffith, K.J.; Karty, J.A.; Mubarak, M.S.; Raghavachari, K.; Peters, D.G. Electrochemistry of substituted salen complexes of nickel(II): Nickel(I)-catalyzed reduction of alkyl and acetylenic halides. *J. Electroanal. Chem.* **2010**, *647*, 194–203. [CrossRef]
119. Nandhini, C.; Arul, P.; Huang, S.-T.; Tominaga, M.; Huang, C.-H. Electrochemical sensing of dual biomolecules in live cells and whole blood samples: A flexible gold wire-modified copper-organic framework-based hybrid composite. *Bioelectrochemistry* **2023**, *152*, 108434. [CrossRef] [PubMed]
120. Xue, Z.; Jia, L.; Zhu, R.-R.; Du, L.; Zhao, Q.-H. High-performance non-enzymatic glucose electrochemical sensor constructed by transition nickel modified Ni@Cu-MOF. *J. Electroanal. Chem.* **2020**, *858*, 113783. [CrossRef]
121. Lu, M.; Deng, Y.; Li, Y.; Li, T.; Xu, J.; Chen, S.W.; Wang, J. Core-shell MOF@MOF composites for sensitive nonenzymatic glucose sensing in human serum. *Anal. Chim. Acta* **2020**, *1110*, 35–43. [CrossRef]
122. Niu, X.; Li, X.; Pan, J.; He, Y.; Qiu, F.; Yan, Y. Recent advances in non-enzymatic electrochemical glucose sensors based on non-precious transition metal materials: Opportunities and challenges. *RSC Adv.* **2016**, *6*, 84893–84905. [CrossRef]
123. Kim, S.E.; Muthurasu, A. Metal-organic framework–assisted bimetallic Ni@Cu microsphere for enzyme-free electrochemical sensing of glucose. *J. Electroanal. Chem.* **2020**, *873*, 114356. [CrossRef]
124. Kim, S.E.; Muthurasu, A. Highly Oriented Nitrogen-doped Carbon Nanotube Integrated Bimetallic Cobalt Copper Organic Framework for Non-enzymatic Electrochemical Glucose and Hydrogen Peroxide Sensor. *Electroanalysis* **2021**, *33*, 1333–1345. [CrossRef]
125. Zoski, C.G. *Handbook of Electrochemistry*; Elsevier Science: Amsterdam, The Netherlands, 2006.
126. Franklin, R.K.; Martin, S.M.; Strong, T.D.; Brown, R.B. Chemical Sensors. In *Comprehensive Microsystems*, 1st ed.; Gianchandani, Y.B., Tabata, O., Zappe, H., Eds.; Elsevier: Oxford, UK, 2008; pp. 433–461.
127. Uzak, D.; Atiroglu, A.; Atiroglu, V.; Cakiroglu, B.; Ozacar, M. Reduced Graphene Oxide/Pt Nanoparticles/Zn-MOF-74 Nanomaterial for a Glucose Biosensor Construction. *Electroanalysis* **2020**, *32*, 510–519. [CrossRef]
128. Cheng, X.; Zhou, J.; Chen, J.; Xie, Z.; Kuang, Q.; Zheng, L. One-step synthesis of thermally stable artificial multienzyme cascade system for efficient enzymatic electrochemical detection. *Nano Res.* **2019**, *12*, 3031–3036. [CrossRef]
129. Batista Deroco, P.; Giarola, J.d.F.; Wachholz Júnior, D.; Arantes Lorga, G.; Tatsuo Kubota, L. Paper-based electrochemical sensing devices. In *Comprehensive Analytical Chemistry*, 1st ed.; Merkoçi, A., Ed.; Elsevier: Amsterdam, The Netherlands, 2020; Volume 89, pp. 91–137.
130. Simões, F.R.; Xavier, M.G. Electrochemical Sensors. In *Nanoscience and Its Applications*, 1st ed.; Da Róz, A.L., Ferreira, M., de Lima Leite, F., Oliveira, O.N., Eds.; William Andrew Publishing: Norwich, NY, USA, 2016; pp. 155–178.
131. Song, Y.; Xu, M.; Gong, C.; Shen, Y.; Wang, L.; Xie, Y.; Wang, L. Ratiometric electrochemical glucose biosensor based on GOD/AuNPs/Cu-BTC MOFs/macroporous carbon integrated electrode. *Sens. Actuators B Chem.* **2018**, *257*, 792–799. [CrossRef]
132. Clark, L.C., Jr.; Lyons, C. Electrode systems for continuous monitoring in cardiovascular surgery. *Ann. N. Y. Acad. Sci.* **1962**, *102*, 29–45. [CrossRef] [PubMed]
133. Adeloju, S.B. AMPEROMETRY. In *Encyclopedia of Analytical Science*, 2nd ed.; Worsfold, P., Townshend, A., Poole, C., Eds.; Elsevier: Oxford, UK, 2005; pp. 70–79.
134. Chu, C.; Rao, S.; Ma, Z.; Han, H. Copper and cobalt nanoparticles doped nitrogen-containing carbon frameworks derived from CuO-encapsulated ZIF-67 as high-efficiency catalyst for hydrogenation of 4-nitrophenol. *Appl. Catal. B* **2019**, *256*, 117792. [CrossRef]
135. Zhai, Z.; Wang, J.; Sun, Y.; Hao, X.; Niu, B.; Xie, H.; Li, C. MOFs/nanofiber-based capacitive gas sensors for the highly selective and sensitive sensing of trace SO_2. *Appl. Surf. Sci.* **2023**, *613*, 155772. [CrossRef]

136. Mo, G.; Zheng, X.; Ye, N.; Ruan, Z. Nitrogen-doped carbon dodecahedron embedded with cobalt nanoparticles for the direct electro-oxidation of glucose and efficient nonenzymatic glucose sensing. *Talanta* **2021**, *225*, 121954. [CrossRef]
137. Wang, L.; Miao, X.; Qu, Y.; Duan, C.; Wang, B.; Yu, Q.; Gao, J.; Song, D.; Li, Y.; Yin, Z. Rattle-type Au@NiCo LDH hollow core-shell nanostructures for nonenzymatic glucose sensing. *J. Electroanal. Chem.* **2020**, *858*, 113810. [CrossRef]
138. Zhang, L.; Wang, J.; Ren, X.; Zhang, W.; Zhang, T.; Liu, X.; Du, T.; Li, T.; Wang, J. Internally extended growth of core–shell NH_2-MIL-101(Al)@ZIF-8 nanoflowers for the simultaneous detection and removal of Cu(II). *J. Mater. Chem. A* **2018**, *6*, 21029–21038. [CrossRef]
139. Heller, A.; Feldman, B. Electrochemical Glucose Sensors and Their Applications in Diabetes Management. *Chem. Rev.* **2008**, *108*, 2482–2505. [CrossRef] [PubMed]
140. Witkowska Nery, E.; Kundys, M.; Jeleń, P.S.; Jönsson-Niedziółka, M. Electrochemical Glucose Sensing: Is There Still Room for Improvement? *Anal. Chem.* **2016**, *88*, 11271–11282. [CrossRef] [PubMed]
141. Shu, Y.; Su, T.; Lu, Q.; Shang, Z.; Xu, Q.; Hu, X. Highly Stretchable Wearable Electrochemical Sensor Based on Ni-Co MOF Nanosheet-Decorated Ag/rGO/PU Fiber for Continuous Sweat Glucose Detection. *Anal. Chem.* **2021**, *93*, 16222–16230. [CrossRef]
142. Sun, L.; Hendon, C.H.; Park, S.S.; Tulchinsky, Y.; Wan, R.; Wang, F.; Walsh, A.; Dincă, M. Is iron unique in promoting electrical conductivity in MOFs? *Chem. Sci.* **2017**, *8*, 4450–4457. [CrossRef] [PubMed]
143. Xuan, X.; Qian, M.; Pan, L.; Lu, T.; Han, L.; Yu, H.; Wan, L.; Niu, Y.; Gong, S. A longitudinally expanded Ni-based metal–organic framework with enhanced double nickel cation catalysis reaction channels for a non-enzymatic sweat glucose biosensor. *J. Mater. Chem. B* **2020**, *8*, 9094–9109. [CrossRef]
144. Chen, Y.; Tian, Y.; Zhu, P.; Du, L.; Chen, W.; Wu, C. Electrochemically Activated Conductive Ni-Based MOFs for Non-enzymatic Sensors Toward Long-Term Glucose Monitoring. *Front. Chem.* **2020**, *8*, 602752. [CrossRef]
145. Sun, Y.; Li, Y.; Wang, N.; Xu, Q.Q.; Xu, L.; Lin, M. Copper-based Metal-organic Framework for Non-enzymatic Electrochemical Detection of Glucose. *Electroanalysis* **2018**, *30*, 474–478. [CrossRef]
146. Hu, Q.; Qin, J.; Wang, X.F.; Ran, G.Y.; Wang, Q.; Liu, G.X.; Ma, J.P.; Ge, J.Y.; Wang, H.Y. Cu-Based Conductive MOF Grown in situ on Cu Foam as a Highly Selective and Stable Non-Enzymatic Glucose Sensor. *Front. Chem.* **2021**, *9*, 786970. [CrossRef]
147. Kim, K.; Kim, S.; Lee, H.N.; Park, Y.M.; Bae, Y.-S.; Kim, H.-J. Electrochemically derived CuO nanorod from copper-based metal-organic framework for non-enzymatic detection of glucose. *Appl. Surf. Sci.* **2019**, *479*, 720–726. [CrossRef]
148. Wang, Z.; Liu, T.; Yu, Y.; Asif, M.; Xu, N.; Xiao, F.; Liu, H. Coffee Ring-Inspired Approach toward Oriented Self-Assembly of Biomimetic Murray MOFs as Sweat Biosensor. *Small* **2018**, *14*, e1802670. [CrossRef]
149. Ezzati, M.; Shahrokhian, S.; Hosseini, H. In Situ Two-Step Preparation of 3D NiCo-BTC MOFs on a Glassy Carbon Electrode and a Graphitic Screen Printed Electrode as Nonenzymatic Glucose-Sensing Platforms. *ACS Sustain. Chem. Eng.* **2020**, *8*, 14340–14352. [CrossRef]
150. Zhang, L.; Wang, N.; Cao, P.; Lin, M.; Xu, L.; Ma, H. Electrochemical non-enzymatic glucose sensor using ionic liquid incorporated cobalt-based metal-organic framework. *Microchem. J.* **2020**, *159*, 105343. [CrossRef]
151. Chen, C.; Xu, H.; Zhan, Q.; Zhang, Y.; Wang, B.; Chen, C.; Tang, H.; Xie, Q. Preparation of novel HKUST-1-glucose oxidase composites and their application in biosensing. *Microchim. Acta* **2022**, *190*, 10. [CrossRef] [PubMed]
152. Kumar, R.; Chauhan, S. Nano/micro-scaled materials based optical biosensing of glucose. *Ceram. Int.* **2022**, *48*, 2913–2947. [CrossRef]
153. Guo, J.; Liu, Y.; Mu, Z.; Wu, S.; Wang, J.; Yang, Y.; Zhao, M.; Wang, Y. Label-free fluorescence detection of hydrogen peroxide and glucose based on the Ni-MOF nanozyme–induced self-ligand emission. *Microchim. Acta* **2022**, *189*, 219. [CrossRef]
154. Zhong, X.; Xia, H.; Huang, W.; Li, Z.; Jiang, Y. Biomimetic metal-organic frameworks mediated hybrid multi-enzyme mimic for tandem catalysis. *Chem. Eng. J.* **2020**, *381*, 122758. [CrossRef]
155. Shi, M.-Y.; Xu, M.; Gu, Z.-Y. Copper-based two-dimensional metal-organic framework nanosheets as horseradish peroxidase mimics for glucose fluorescence sensing. *Anal. Chim. Acta* **2019**, *1079*, 164–170. [CrossRef]
156. Ning, D.; Liu, Q.; Wang, Q.; Du, X.-M.; Ruan, W.-J.; Li, Y. Luminescent MOF nanosheets for enzyme assisted detection of H_2O_2 and glucose and activity assay of glucose oxidase. *Sens. Actuators B Chem.* **2019**, *282*, 443–448. [CrossRef]
157. Cui, Y.; Chen, F.; Yin, X.-B. A ratiometric fluorescence platform based on boric-acid-functional Eu-MOF for sensitive detection of H_2O_2 and glucose. *Biosens. Bioelectron.* **2019**, *135*, 208–215. [CrossRef]
158. Li, D.; Zhang, S.; Feng, X.; Yang, H.; Nie, F.; Zhang, W. A novel peroxidase mimetic Co-MOF enhanced luminol chemiluminescence and its application in glucose sensing. *Sens. Actuators B Chem.* **2019**, *296*, 126631. [CrossRef]
159. Jiang, G.; Yang, Z.; Zhu, K.; Zong, S.; Wu, L.; Wang, Z.; Cui, Y. Universal peroxidase-like strategy for sensitive glucose detection in complex matrix. *Nano Res.* **2023**, *16*, 1141–1148. [CrossRef]
160. Guo, L.; Chen, S.; Yu, Y.-L.; Wang, J.-H. A Smartphone Optical Device for Point-of-Care Testing of Glucose and Cholesterol Using Ag NPs/UiO-66-NH_2-Based Ratiometric Fluorescent Probe. *Anal. Chem.* **2021**, *93*, 16240–16247. [CrossRef] [PubMed]
161. Pan, Y.; Pang, Y.; Shi, Y.; Zheng, W.; Long, Y.; Huang, Y.; Zheng, H. One-pot synthesis of a composite consisting of the enzyme ficin and a zinc(II)-2-methylimidazole metal organic framework with enhanced peroxidase activity for colorimetric detection for glucose. *Microchim. Acta* **2019**, *186*, 213. [CrossRef]
162. Ilacas, G.C.; Basa, A.; Nelms, K.J.; Sosa, J.D.; Liu, Y.; Gomez, F.A. Paper-based microfluidic devices for glucose assays employing a metal-organic framework (MOF). *Anal. Chim. Acta* **2019**, *1055*, 74–80. [CrossRef]

163. Hong, J.I.; Chang, B.Y. Development of the smartphone-based colorimetry for multi-analyte sensing arrays. *Lab Chip* **2014**, *14*, 1725–1732. [CrossRef]
164. Wu, T.; Gao, X.-J.; Ge, F.; Zheng, H.-G. Metal–organic frameworks (MOFs) as fluorescence sensors: Principles, development and prospects. *CrystEngComm* **2022**, *24*, 7881–7901. [CrossRef]
165. Gui, R.; Jin, H.; Bu, X.; Fu, Y.; Wang, Z.; Liu, Q. Recent advances in dual-emission ratiometric fluorescence probes for chemo/biosensing and bioimaging of biomarkers. *Coord. Chem. Rev.* **2019**, *383*, 82–103. [CrossRef]
166. Hirano, T.; Matsuhashi, C. A stable chemiluminophore, adamantylideneadamantane 1,2-dioxetane: From fundamental properties to utilities in mechanochemistry and soft crystal science. *J. Photochem. Photobiol. C* **2022**, *51*, 100483. [CrossRef]
167. Zheng, Z.; Cong, S.; Gong, W.; Xuan, J.; Li, G.; Lu, W.; Geng, F.; Zhao, Z. Semiconductor SERS enhancement enabled by oxygen incorporation. *Nat. Commun.* **2017**, *8*, 1993. [CrossRef]
168. Sun, H.; Yu, B.; Pan, X.; Zhu, X.; Liu, Z. Recent progress in metal–organic frameworks-based materials toward surface-enhanced Raman spectroscopy. *Appl. Spectrosc. Rev.* **2022**, *57*, 513–528. [CrossRef]
169. Xue, Q.; Li, Z.; Wang, Q.; Pan, W.; Chang, Y.; Duan, X. Nanostrip flexible microwave enzymatic biosensor for noninvasive epidermal glucose sensing. *Nanoscale Horiz.* **2020**, *5*, 934–943. [CrossRef] [PubMed]
170. Bandodkar, A.J.; Gutruf, P.; Choi, J.; Lee, K.; Sekine, Y.; Reeder, J.T.; Jeang, W.J.; Aranyosi, A.J.; Lee, S.P.; Model, J.B.; et al. Battery-free, skin-interfaced microfluidic/electronic systems for simultaneous electrochemical, colorimetric, and volumetric analysis of sweat. *Sci. Adv.* **2019**, *5*, eaav3294. [CrossRef]
171. Zhao, H.; Su, R.; Teng, L.; Tian, Q.; Han, F.; Li, H.; Cao, Z.; Xie, R.; Li, G.; Liu, X.J.N. Recent advances in flexible and wearable sensors for monitoring chemical molecules. *Nanoscale* **2022**, *14*, 1653–1669. [CrossRef] [PubMed]
172. Sun, H.; Saeedi, P.; Karuranga, S.; Pinkepank, M.; Ogurtsova, K.; Duncan, B.B.; Stein, C.; Basit, A.; Chan, J.C.N.; Mbanya, J.C.; et al. IDF Diabetes Atlas: Global, regional and country-level diabetes prevalence estimates for 2021 and projections for 2045. *Diabetes Res. Clin. Pr.* **2022**, *183*, 109119. [CrossRef] [PubMed]
173. Shao, Z.; Chen, J.; Xie, Q.; Mi, L. Functional metal/covalent organic framework materials for triboelectric nanogenerator. *Coord. Chem. Rev.* **2023**, *486*, 215118. [CrossRef]
174. Barsiwal, S.; Babu, A.; Khanapuram, U.K.; Potu, S.; Madathil, N.; Rajaboina, R.K.; Mishra, S.; Divi, H.; Kodali, P.; Nagapuri, R.; et al. ZIF-67-Metal-Organic-Framework-Based Triboelectric Nanogenerator for Self-Powered Devices. *Nanoenergy Adv.* **2022**, *2*, 291–302. [CrossRef]
175. Yang, T.; Yu, X.; Ma, N.; Wu, R.; Li, H. An autonomous channel deep learning framework for blood glucose prediction. *Appl. Soft Comput.* **2022**, *120*, 108636. [CrossRef]
176. Zhu, T.; Kuang, L.; Daniels, J.; Herrero, P.; Li, K.; Georgiou, P. IoMT-Enabled Real-Time Blood Glucose Prediction With Deep Learning and Edge Computing. *IEEE Internet Things J.* **2023**, *10*, 3706–3719. [CrossRef]
177. Wu, K.; He, X.; Wang, J.; Pan, T.; He, R.; Kong, F.; Cao, Z.; Ju, F.; Huang, Z.; Nie, L. Recent progress of microfluidic chips in immunoassay. *Front. Bioeng. Biotechnol.* **2022**, *10*, 1112327. [CrossRef]
178. Martinez, A.W.; Phillips, S.T.; Butte, M.J.; Whitesides, G.M. Patterned Paper as a Platform for Inexpensive, Low-Volume, Portable Bioassays. *Angew. Chem. Int. Ed.* **2007**, *119*, 1340–1342. [CrossRef]
179. Liu, H.; Crooks, R.M. Paper-Based Electrochemical Sensing Platform with Integral Battery and Electrochromic Read-Out. *Anal. Chem.* **2012**, *84*, 2528–2532. [CrossRef] [PubMed]
180. Choi, J.; Bandodkar, A.J.; Reeder, J.T.; Ray, T.R.; Turnquist, A.; Kim, S.B.; Nyberg, N.; Hourlier-Fargette, A.; Model, J.B.; Aranyosi, A.J.; et al. Soft, Skin-Integrated Multifunctional Microfluidic Systems for Accurate Colorimetric Analysis of Sweat Biomarkers and Temperature. *ACS Sens.* **2019**, *4*, 379–388. [CrossRef] [PubMed]

chemosensors

MDPI

Review

Recent Advances in Electrochemiluminescence Emitters for Biosensing and Imaging of Protein Biomarkers

Lei Yang [1] and Jinghong Li [1,2,3,4,*]

1 Key Laboratory of Bioorganic Phosphorus Chemistry & Chemical Biology, Department of Chemistry, Center for Bioanalytical Chemistry, Tsinghua University, Beijing 100084, China; ly1994@mail.tsinghua.edu.cn
2 New Cornerstone Science Laboratory, Shenzhen 518054, China
3 Beijing Life Science Academy, Beijing 102206, China
4 Center for BioAnalytical Chemistry, Hefei National Laboratory of Physical Science at Microscale, University of Science and Technology of China, Hefei 230026, China
* Correspondence: jhli@mail.tsinghua.edu.cn

Abstract: Electrochemiluminescence (ECL) is a light-emitting process triggered by the high energy redox between electrochemically oxidized and reduced luminophores or some coreactive intermediate radicals, representing a blooming hot topic over decades with a wide variety of bioanalytical applications. Due to the superb sensitivity, ultralow background noise, specificity, ease of integration, and real-time and in situ analysis, ECL has been developed as a convenient and versatile technique for immunodiagnostics, nucleic acid analysis, and bioimaging. Discovering highly-efficient ECL emitters has been a promising subject that will benefit the development of sensitive bioanalytical methods with prominent potential prospects. To date, the interdisciplinary integrations of electrochemistry, spectroscopy, and nanoscience have brought up the continuous emergences of novel nanomaterials which can be flexibly conjugated with specific bio-recognition elements as functional ECL emitters for bioassays. Therefore, a critical overview of recent advances in developing highly-efficient ECL emitters for ultrasensitive detection of protein biomarkers is presented in this review, where six kinds of the most promising ECL nanomaterials for biosensing and imaging of various disease-related protein biomarkers are separately introduced with references to representative works. Finally, this review discusses the ongoing opportunities and challenges of ECL emitters in developing advanced bioassays for single-molecule analysis and spatiotemporally resolved imaging of protein biomarkers with future perspectives.

Keywords: electrochemiluminescence; emitter; biosensing; imaging; protein biomarkers

Citation: Yang, L.; Li, J. Recent Advances in Electrochemiluminescence Emitters for Biosensing and Imaging of Protein Biomarkers. *Chemosensors* **2023**, *11*, 432. https://doi.org/10.3390/chemosensors11080432

Academic Editor: Ilaria Rea

Received: 29 June 2023
Revised: 29 July 2023
Accepted: 1 August 2023
Published: 4 August 2023

1. Introduction

Electrochemiluminescence (ECL) is an electrically stimulated chemiluminescencent event that involves the excitation of luminophores through efficient redox reactions along with light emissions [1–3]. Benefiting from the unification of the high sensitivity of chemiluminescence and spatiotemporal controllability of electrochemistry, ECL has been developed as a powerful technique that not only needs no external light excitation but also perfectly avoids high-power laser irradiation and auto-photoluminescence with a high signal-to-noise ratio and almost zero background signal [4,5]. The ECL principles have been explored for over half a century since the 1960s, which can be generally categorized into two types: annihilation and coreactant ECL mechanisms [6,7]. Both of these two mechanisms require the efficient conversion of electrical energy into radiative energy that can be further exported as readable signals by ECL analyzers [8]. Usually, coreactant ECL performs stronger ECL emission in narrower potential windows than the self-annihilation pathway, which has promoted the application scope of the ECL technique from fundamental research to clinical diagnosis. Till now, coreactant ECL systems have stimulated the growing development of commercialized ECL instruments and facilities as powerful diagnostic tools in medical institutions.

As well known, ECL emitters with high luminous efficiencies play crucial roles in ECL analytical systems since their qualities directly dominated the signal transduction efficiencies of the developed biosensors [9]. With the developments of nanoscience and nanotechnology, more and more attractive nanomaterial-based ECL emitters have been developed for bioassays, which greatly enriched the existing pool of ECL luminophores where classic ECL reagents including highly ECL-active ruthenium complexes, luminol with its derivates and iridium complexes are the mainstream choices for commercialization and fundamental researches. Nanomaterials with merits of large specific surface areas, high conductivities, tunable optical properties, and excellent electrochemical activities have opened new possibilities for the explorations of newly patterned mechanism models, constructions of ultrasensitive biosensors, and novel single-cell bioimaging probes for ECL bioanalysis. Those nanomaterials with outstanding quantum efficiencies can be directly fabricated and employed as promising ECL emitters for signal transduction or image probing. For those nanomaterials with large specific surface areas or porous nanostructures, such as mesoporous silica nanospheres (MSNs) [10–13], multiwall carbon nanotubes [14–16], graphene oxide [17–20], Ti_3C_2 MXene [21–23], and Au nanoparticles (NPs) [24], they can be utilized as the carriers to stabilize luminophores and developed as ECL emitters for biosensors. Unlike self-annihilation ECL pathway, a proper coreactant reagent is essentially required for coreactant ECL since it can generate abundant electro-active intermediates to promote the production of excited states, thus achieving stronger ECL intensity [25–27]. Instead of being carriers, nanomaterials with excellent electrical conductivity, electrochemical and catalytic activities can be utilized as coreaction accelerators to facilitate electron-transfer during the oxidation/reduction of coreactants to produce more radical intermediates, thus obtaining better signal amplification and higher sensitivity [28–30].

By virtue of the unique combination of ECL emitters with biorecognition elements such as antibody, nanobody, DNA, and aptamer, various ECL transducers have been established for ECL immunoassay, nucleic acid analysis, and cell imaging [7,31–33]. Biomarkers, considered as quantifiable indicators for some certain biological states of human bodies, hold enormous potential in diagnosing various severe diseases to human health [34,35]. Protein-based biomarkers are recognized as golden indicators for diagnosing various cancers such as carcinoembryonic antigen (CEA) [36–38], prostate-specific antigen (PSA) [39–41], alpha fetoprotein (AFP) [42,43], mucin 1 (MUC1) [44–46], and cytokeratin 19 fragments (CYFRA 21-1) [47–49]. Benefiting from the remarkable intrinsic features of ECL technique, more and more novel ECL nanomaterials have been diversely prepared and biofunctionalized as novel ECL emitters for detecting protein biomarkers of different diseases [33,50].

This review aims to provide a fresh summarization of key advances in the delicate fabrication of different kinds of novel nanomaterial-based ECL emitters along with developed biosensing and imaging strategies for analyzing protein biomarkers from cancers and non-cancer diseases. First, the recently reported novel nanomaterial-based ECL emitters were classified into different categories including quantum dots (QDs), metallic nanocrystals (MeNCs), metal–organic frameworks (MOFs), covalent organic frameworks (COFs), hydrogen-bonded organic frameworks (HOFs), and polymer dots (Pdots). Then, the most advanced sensing methods including self-enhanced ECL, aggregation-induced ECL (AIECL), ratiometric assay, ECL-resonance energy transfer (ECL-RET), or quenching-typed ECL biosensors are summarized, followed by a brief but informative sketch of recent advancements in ECL microscopy (ECLM) techniques for membrane imaging of protein biomarkers. In the final section, the ongoing challenges and opportunities are carefully addressed with future perspectives for readers as well. Though this review was delicately organized as comprehensively as possible, the readers are still suggested to refer to other excellent reviews and those unintentionally uninvolved works out of this review to deepen the understanding of current advances and future trends in the ECL domain.

2. Nanomaterial-Based ECL Emitters

To date, the family of ECL emitters has been expanded from classical ECL reagents such as $Ru(bpy)_3^{2+}$ and luminol to advanced nanomaterial-based emitters such as QDs, MeNCs, MOFs, COFs/HOFs, and PdotsQDs, MeNCs, MOFs, COFs, HOFs, and Pdots. It is worth noticing that the above six kinds of nanomaterial-based ECL emitters can be recategorized into two major types: inorganic and organic ECL emitters. Obviously, QDs and MeNCs belong to the inorganic ECL emitters while the other four kinds including MOFs, COFs, HOFs, and Pdots belong to the organic ECL emitters. As for the latter type of organic ECL emitters, more specific types can be classified considering the complexities in their structural components during preparation. For instance, the ECL emission in MOFs can be further categorized into three main pathways, such as guest-encapsulation in MOFs induced by ECL, ligand-induced ECL in MOFs, and metal ion-induced ECL in MOFs [51–53]. In this section, efforts are made to cover every sub-type of each kind of nanomaterial as much as possible along with representative and elegant examples that have been recently published and mostly cited to date.

2.1. Quantum Dot-Based ECL Emitters

Quantum dots (QDs), as one type of zero-dimensional nanomaterials, have been variously developed as attractive participators for photocatalysis, cell imaging, light-emitting diodes, and biosensors by virtue of their structural, optical and electrical properties, high photoluminescence quantum yields (QY), excellent luminesce stability, ease of hybridizing with other components, and so on [54–57]. The first nonaqueous ECL study of silicon QDs was reported in 2002 [54]. After that, a wide variety of III-V and II-VI QDs including Cd-based QDs, Mo/W/Sn-based QDs, carbon QDs, and perovskite QDs gradually emerged as promising ECL luminophores. As previously reviewed by Zhou et al. [55], the ECL mechanism for QDs normally involves high-energy electron transfer along with the formation of radicals to generate pure ECL emission via the self-annihilation and coreactant pathways. Cd-based QDs include CdSe [56,57], CdS [58,59], CdTe [60–62], and core-shell CdTe@ZnS QDs [63]; CdSe@ZnS [64] have been exploited as ECL emitters in the last decade since this family usually owns narrow emission and size-dependent ECL with high stability in the aqueous phase and an ease of surface modification with other components. The hybridization of Cd-based pure QDs with other beneficial nanomaterials have been investigated for promoted ECL behaviors. For example, Liu et al. [65] recently prepared CdS QDs hybridized porphyrinic Zr-MOFs (PCN-224/CdS QDs) as cathodic ECL emitters for Sa-16S rDNA detection using $K_2S_2O_8$ as the coreactant. It was claimed that the electro-active PCN-224 with large specific surface area not only provided abundant sites to anchor CdS QDs stably but also functioned as a coreaction accelerator to facilitate the reduction of $S_2O_8^{2-}$ for ECL amplification, thus achieving a high detection sensitivity. However, Cd-based QDs showed limited applicability due to the environmental toxicity, complex preparation process, and poor biocompatibility [66]. To satisfy the demand of biological applications, outer-capping by a biocompatible shell on the external surface of Cd-based QDs has been developed as a doable solution to improve the biocompatibility. Pan et al. [67] prepared SiO_2-capped CdTe QDs hybrids (SiO_2@CdTe) as ECL emitters for biosensor construction. The SiO_2@CdTe showed a lowered cytotoxicity due to the encapsulation of the biocompatible SiO_2 shell, which endowed the biosensor a better performance for biological detection. Otherwise, some attractive QDs such as SnS_2 [68,69] MoS_2 [70], EuS [71], AgBr [72], SnO_2 [73], and black phosphorus (BP) [74,75] QDs containing no heavy-metal elements have also been developed as ideal alternatives for Cd-based QDs. As previously reported, Lei et al. [69] developed a green hydrothermal method to prepare heavy metal-free SnS_2 QDs which showed favorable cathodic ECL emission using $S_2O_8^{2-}$ as the coreactant. To achieve stronger ECL signals of SnS_2 QDs, Li et al. [76] used polyethyleneimine (PEI) functionalized titanium dioxide hollow spheres (THS) to stabilize large amounts of SnS_2 QDs, the formed hybrids exhibited stronger ECL emission for ultrasensitive detection of chloramphenicol. Hua et al. [71] first revealed the cathodic

ECL property of EuS QDs whose emission can be specifically quenched by Hg^{2+}, based on which, a facile and sensitive Hg^{2+} sensor was developed for seafood monitoring. Recently, Yu et al. [74] discovered a dual excited states-dominated ECL mechanism of the BPQDs prepared with an average size of 4 nm (Figure 1A). The surface oxidation defects of BPQDs were effectively passivated by externally modified arginine (Arg) moieties. It was highlighted that the electron transition channels of BPQDs can be modulated by those Arg moieties to generate significantly enhanced ECL emission using $K_2S_2O_8$ or $N_2H_4 \cdot H_2O$ as the coreactant (Figure 1B). Moreover, Arg moieties endowed BPQDs with greatly improved biocompatibility, thus BPQDs were applied as ECL emitters for the sensitive analysis of integrin inhibitors on cell surfaces, resulting in a well-evaluated inhibiting efficiency. New synthetic strategies have been exploited to fabricate ECL-active multinary QDs such as AgGaInS QDs [77], $CuInS_2$/ZnS QDs [78], and ZnAgInS/ZnS [79] through efficient interfacial engineering on the interior inorganic structures and inorganic–organic interfaces, which also injected new insights for the developments and regulations of more highly-efficient QDs in ECL bioassays.

Benefiting from the low toxicity, good biocompatibility, sturdy chemical inertness, and easy of functionalization, carbon-based QDs including carbon QDs (CQDs), graphene QDs (GQDs), Ti_3C_2 MXene QDs, and graphitic carbon nitride QDs (g-CNQDs) have gained considerable attention as a class of promising ECL emitters for bioassays [66,80]. With the aim of improving the ECL efficiency of pure CQDs, nitrogen (N) doping was observed as an efficient strategy to lower the oxidation potential of CQDs with a significantly boosted anodic ECL emission. Chen et al. [81] prepared N-doped CQDs by a solvothermal method (Figure 1C). It was found that after lots of N atoms were doped into the graphitic cores of CQDs, which promoted positive shift of highest occupied molecular orbital (HOMO) to an upper energy level, thus achieving a 2.5-fold increased ECL efficiency at a rather low excitation potential of +0.11 V. Meanwhile, efforts such as the co-doping of multi-atoms have also been devoted to enhancing the ECL efficiency of pristine CQDs. Chen et al. [82] prepared biocompatible and water-dispersive nitrogen/sulfur (N/S) co-doped CQDs (NS-CQDs) by using cysteine and tryptophan as the precursors, which exhibited high photoluminescence QY of 73% for ECL application. In addition, Liu et al. [83] prepared highly electronegative fluorine/nitrogen (F/N) co-doped CQDs (FN-CQDs) via a one-step solvothermal method for HIV DNA fragment detection (Figure 1D). Since phosphorus (P) possesses a larger atom radius with a higher electron-donating ability than N, it can potentially be used as a dopant for CQDs to improve the ECL efficiency. Therefore, Yang et al. [84] developed the P/N co-doped CQDs (P/N CQDs) as efficient ECL emitters which showed eight-fold enhanced ECL intensity than N-doped CQDs for the sensitive detection of the mutant BRAF gene.

2.2. Metal Nanocluster-Based ECL Emitters

Metal nanoclusters (MeNCs), entailing a metal core consisting of a few atoms protected by a peripheral ligand shell, have become a scientific interest in the ECL domain over the last decade due to the merits of their molecule-like structure, intense quantum size effect, tunable luminescence, stable electronic and physicochemical properties, good electrocatalytic activity, and low biotoxicity [85–87]. Chemical reduction and electrochemical reduction have been the most used preparation methods for ECL-active MeNCs which usually involve the formation of a metallic core capped by a shell made of thiol ligands such as glutathione (GSH), 3-mercaptopropionic acid (MPA), and dithiothreitol (DTT) [87]. In 2009, the pioneering work on MeNCs was conducted by Díez et al. [88] who first revealed the tunable ECL emission of silver nanoclusters (Ag NCs). Since then, the fundamental knowledge of synthetic procedures and optimized ECL emission by ligand engineering on MeNCs has been gradually enriched. To date, monometallic, bimetallic. and MeNC-derived composites have become the three most investigated categories in the exploration of novel ECL-active MeNCs. Au NCs are the most well-known, popular, and representative monometallic MeNCs among such hot topics in the ECL realm. How to improve the dispersion, stability, and quantum yield of Au NCs in the water phase is still a challenging issue. To embrace this challenge, Wang et al. [89] prepared self-enhanced Au NCs by the covalent conjugation of co-reactive *N,N*-diethylethylenediamine (DEDA) with Au NCs (Au-LA-DEDA). The well-matched redox activity between DEDA and Au NCs in one hybrid structure gave rise to significantly boosted ECL emission due to the shortened electron and mass transportations. Li et al. [90] and Fang et al. [91] successively investigated the cathodic and anodic ECL properties of BSA-stabilized Au NCs in aqueous media. Notably, the efficient external encapsulation of a layer of BSA shell endowed the pristine Au NCs with a well-improved water dispersity and stability for aqueous bioassay. In addition, Kim et al. [58] prepared glutathione-stabilized Au NCs which not only owned good dispersion and stability in water but also generated strong near-infrared ECL (NIR-ECL) emission for bioassays. Meanwhile, Peng et al. [59] prepared the methionine-capped Au NCs which performed strong anodic NIR-ECL emission with a dramatically increased ECL quantum yield of 66%. After that, the NIR-ECL Au NCs including Au_{18} [92], Au_{21} [93], and Au_{38} [94] NCs were comprehensively investigated by Ding's group. For example, Hesari et al. [93] prepared two $Au_{21}(SR)_{15}$ NCs of $Au_{21}{}^{hcp,0}$ and $Au_{21}{}^{fcc,0}$ whose electrochemistry behaviors and ECL mechanisms with tri-*n*-propylamine (TPrA) were studied in detail (Figure 2) which was quite informative for understanding the basic characteristics and ECL mechanisms of Au NCs with wide scopes of NIR-ECL applications.

The controllable formation of bimetallic NCs (BNCs) was reported as an effective strategy for ECL optimization of individual metallic counterparts due to the triggered synergetic effects [95]. To date, Pt [96], Pd [97], and Ag [98,99] have already been doped with Au NCs to form BNCs with boosted ECL emission and lower biological background interferences for biosensing than pristine Au NCs. Chen et al. [98] prepared rod-shape bimetallic $Au_{12}Ag_{13}$ NCs which exhibited 10-fold enhanced self-annihilation ECL efficiency and 400-fold enhanced coreactant ECL efficiency using $Ru(bpy)_3{}^{2+}$-TPrA as the reference. It was claimed that the ECL enhancements were attributed to the stabilization of charges on the LUMO orbital by the 13th Ag atom at the central position which made $Au_{12}Ag_{13}$ NCs more rigid in their core to reduce the nonradiative decays. Guo et al. [100] prepared well-confined ECL emitters of Au–Ag bimetallic NCs (Au–Ag NCs) capped by GSH stabilizers whose ECL emission was linearly and selectively enhanced by the exogenously added spermine, thus making Au–Ag NCs perfect ECL emitters for sensitive detection of spermine in practical applications.

Though ligand engineering or formation, bimetallic crystallization can effectively regulate the ECL efficiency of MeNCs; it is still challenging to endow them with a uniform distribution on substrates to decrease self-aggregation induced ECL quenching [101]. Therefore, conductive nanomaterials with large specific surface areas are employed as carriers to anchor MeNCs in a uniformly distributed manner; this has become a promising option to eliminate the ACQ effect. Previously reported works showed that Au NCs or Pt NCs can be anchored on the graphene oxides (GO) to act as luminous substrates for ECL detections [102,103]. Nie et al. [101] employed the GSH-protected Au NCs as building blocks that can assemble with Zn^{2+} to form the ECL-active Au NCs@ZIF-8 MOFs. It should be noted that the ordered distribution of Au NCs in MOF structure triggered a 10-fold stronger ECL emission than individual Au NCs in the water phase, thus a quenching-typed biosensor was constructed for the label-free detection of rutin with high sensitivity.

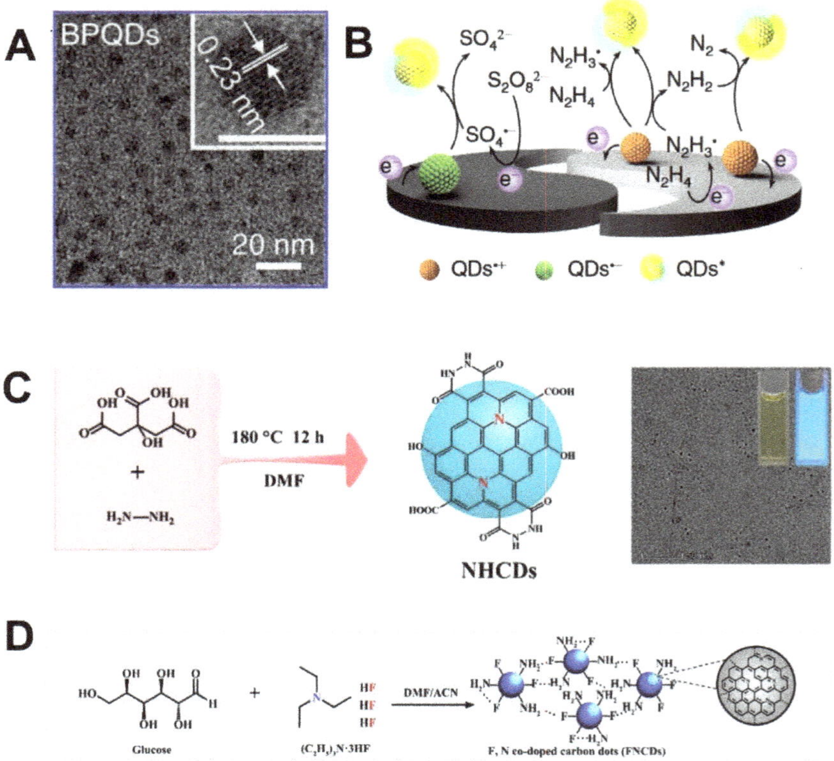

Figure 1. (**A**) Transmission electron microscopy image and high-resolution (inset) image (scale bar: 5 nm) of BPQDs; and the (**B**) cathodic and anodic ECL mechanisms for Arg-modified BPQDs/GCE using the $K_2S_2O_8$ and $N_2H_4{\cdot}H_2O$ as coreactants. Reprinted with permission from Ref. [74], Copyright 2022 Springer Nature. (**C**) Preparation of blue NHCDs. Reprinted with permission from Ref. [81], Copyright 2020 American Chemical Society. (**D**) Preparation of the FNCDs. Reprinted with permission from Ref. [83], Copyright 2023 Elsevier.

2.3. MOFs-Based ECL Emitters

Metal–organic frameworks (MOFs), as an emerging class of crystalline networks with high surface area and porosity, are formed by the ordered assembly of organic ligands with inorganic metal ions which have become promising candidates in chemical sensing, energy storage, adsorption and separation, catalysis, and luminescence [104,105]. Most of the members in the MOFs family can be further classified into three tailorable spatial dimensions (1D, 2D, and 3D) in terms of framework topologies and intricate skeletons [106]. To date, MOFs with different dimensions have been employed as carriers or scaffolds to stabilize ECL luminophores including $Ru(bpy)_3^{2+}$, luminol, and its derivatives as emitters for ECL biosensing applications [107]. To date, the feasibilities of UiO-66 [108], UiO-66-NH_2 [109], ZIF-8 [101], UiO-67 [110–112], MIL-101 [113], MOF-5 [114], PCN-222 [115], and PCN-777 [116] as carriers or scaffolds have been exploited; among these, Zr-based MOFs yield higher popularity due to their desirable physicochemical stabilities, porous features, and electrochemical activities [106]. Meanwhile, some Fe-/Co-based MOFs with enzyme-mimicing properties can catalyze the generation of ROSs for signal amplification of H_2O_2-based ECL systems [117–119]. In-depth studies were focused on how to conquer the unnecessary leakage of ECL luminophores from MOFs during real applications. One is to use those ECL luminophores as ligands to fabricate MOFs while the other one is to use

ECL-active metal ions to assemble with proper ligands. Until now, $Ru(bpy)_3^{2+}$ [110] and the derivatives of 9,10-di(pcarboxyphenyl)anthracene (DPA) [120], perylene [121], pyrene [122], porphyrins, and tetrakis (4-carboxyphenyl) porphyrin (TCPP) [123,124] have been served as ECL-active ligands to prepare MOFs. As a brave attempt, Zhu et al. [120] hired two different kinds of ligands to design a highly ECL emissive mixed-ligand MOF (m-MOF) in Figure 3A. These two ligands including DPA as a ECL luminophore and 1,4-diazabicyclo [2.2.2] octane ($D\text{-}H_2$) as the perfectly matched coreactant were well confined in the same reticular structure which was attributed to achieve a 26.5-fold ECL amplification via a newly demonstrated intra-reticular charge transition pathway (Figure 3B). Meanwhile, along with vigorous developments in aggregation science, aggregation induced emission (AIE) has injected new vitality into the preparation of highly-emissive ECL probes to avoid the unnecessary aggregation-caused quenching (ACQ) effect [125]. Reports also show that some planar AIEgens can serve as ligands whose intramolecular motions can be confined inside of rigid crystal MOF nanostructures to generate much stronger ECL emissions than their monomers, defined as aggregation induced ECL (AIECL) [126]. Notably, the AIE star molecule of tetraphenylethene (TPE) as well as its various derivatives of 1,1,2,2-tetra(4-carboxylbiphenyl)ethylene (H_4TCBPE) [45,125] and 1,1,2,2-tetrakis(4-(pyridin-4-yl)phenyl)-ethene (TPPE) [127] have all been explored as AIECL ligands to fabricate ECL-active MOFs for biosensing. As an attempt, Li et al. [125] prepared a green emissive Zr-MOF using the H_4TCBPE as AIECL organic ligands which was proved to perform an efficient coordination-triggered ECL (CT-ECL) emission that can be quenched by the $AuPd@SiO_2$ in a signal-off mode for sensitive immunoassays of neuron-specific enolase (NSE). A possible AIECL mechanism was mainly due to the restriction of intramolecular motions of H_4TCBPE in the rigid MOF matrix which efficiently reduced non-radiative decay for high ECL emission [126].

Recently, self-luminous lanthanide MOFs (Ln-MOFs) with unique antenna effects, a strong energy-transfer capability, and tunable luminescence properties are considered as promising ECL emitters for developing solid-state biosensors [107]. The pioneering studies on the luminescent Ln complexes were carried out by Bard et al. [128] in 1996. Ln^{3+} with a strong positive charge and large radius can collaborate with strongly absorbing organic ligands to display variable coordination configuration, thus increasing the structural tailor-ability and application scope of Ln-MOFs [129,130]. The so-called antenna effect in Ln-MOFs could be triggered when organic ligands absorb photons and act as an antenna to produce triplet states (T1) through intersystem crossing (ISC), sensitizing Ln^{3+} ions for ECL emission [48]. Different ECL-active lanthanide elements, including those already reported ones, namely Eu- [47,107,131–133], Ce- [45], Tb- [134–136], La- [137], and Yb- [138] based MOFs, have already been investigated for ECL applications. Until now, different Eu-based MOFs with excellent ECL properties were fabricated by sensitizing the Eu^{3+} with different ligands [47,107,131–133]. For example, Zhao et al. [47] prepared a mixed valence Eu-MOF by the coordination of $Sr(HCOO)_2$ ligands with Eu^{2+}/Eu^{3+} (Figure 3C). The antenna effect of the Eu-MOFs guaranteed the effective sensitization of $Sr(HCOO)_2$ toward Eu^{2+}/Eu^{3+}, thus generating high ECL efficiency by transferring the energy from the $4f^65d^1$ level of Eu^{2+} to the $^5D_0$ level of Eu^{3+} (Figure 3D). It is also quite innovative to design ECL-active Ln-MOFs with a self-triggered signal-amplification function by equipping some positive accelerators into the MOF structures. In addition, since the redox pair of Cu^{2+}/Cu^+ can efficiently catalyze the reduction in coreactants of $S_2O_8^{2-}$ for ECL amplification, a hollow and porous Cu^{2+} doped Tb-MOF (Cu:Tb-MOF) was thus prepared by Wang et al. [136] as self-enhanced ECL emitters for sensitive immunoassays of pro-gastrin-releasing peptide (ProGRP) which indicated that introducing a redox-active pair such as Cu^{2+}/Cu^+, Ce^{3+}/Ce^{4+}, or Co^{3+}/Co^{4+} might be a potential and doable strategy to regulate and improve the ECL efficiency of Ln-MOFs for biosensing application.

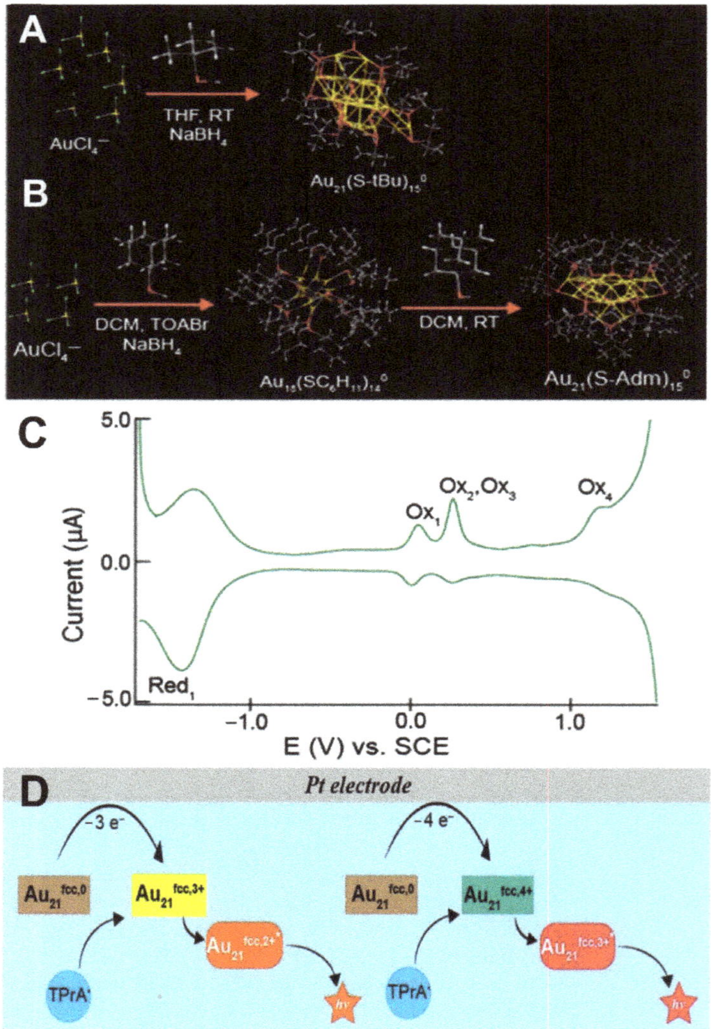

Figure 2. (**A**) One-step synthesis of $Au_{21}^{hcp,0}$. (**B**) Synthesis route of $Au_{21}^{fcc,0}$. (**C**) Differential pulse voltammograms (DPVs) of a 0.10 mM solution of $Au_{21}^{fcc,0}$ in dichloromethane containing 0.1 M $TBAPF_6$. (**D**) Anodic ECL mechanism of $Au_{21}^{fcc,0}$ on a Pt electrode. Reprinted with permission from Ref. [93], Copyright 2021 American Chemical Society.

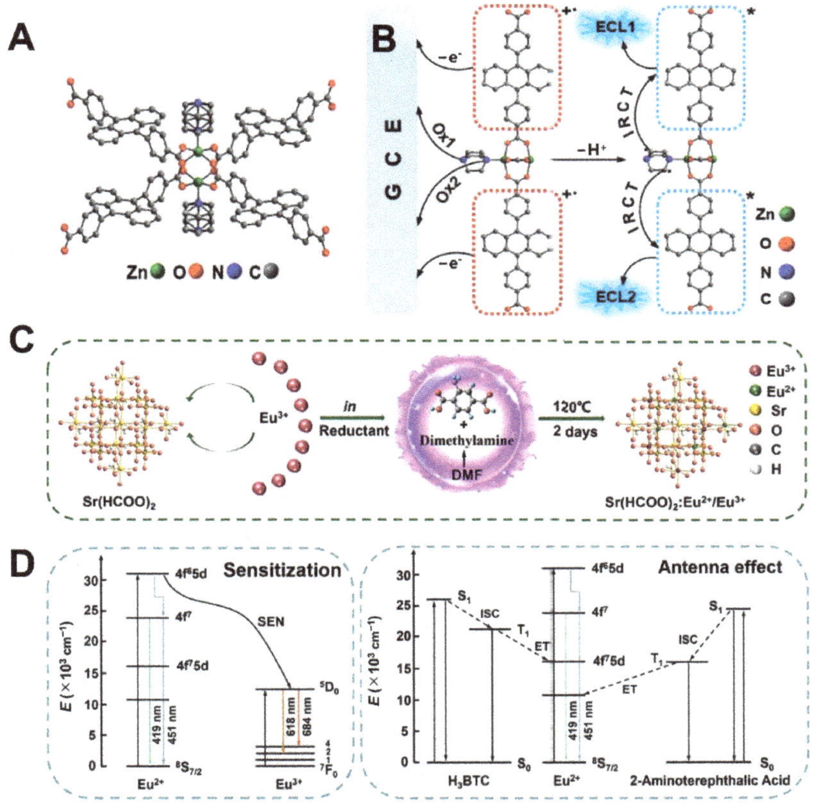

Figure 3. (**A**) Paddlewheel unit and (**B**) stepwise ECL mechanism of m-MOF via a dual intrareticular oxidation pathway. Reprinted with permission from Ref. [120], Copyright 2021 Springer. (**C**) The synthesis route of $Sr(HCOO)_2$:Eu^{2+}/$Eu^{3+}$and (**D**) simple models for the antenna effect of ligand to Eu^{2+} and the sensitization of Eu^{2+} to Eu^{3+}. Reprinted with permission from Ref. [47], Copyright 2023 Elsevier.

2.4. COFs/HOF-Based ECL Emitters

Though MOFs have gained considerable attention in the ECL domain, more and more concerns have been aroused from this hot topic, especially the underlying ECL quenching problem induced by the used metal ions or the ACQ effect of used planar ECL ligands [139,140]. By removing metal ions, covalent organic frameworks (COFs) have developed as a new class of fascinating metal-free crystalline porous ECL emitters which are attracting wide attention by virtue of their ultrahigh surface areas, favorable biocompatibility, adjustable porosity, and tunable physicochemical or luminescence activities that are on a par with MOFs [104,141]. Previous reports claimed that the ACQ effect of traditional planar luminophores in coordination polymers can be reduced or even eliminated by conjugation with AIEgens which will realize the ACQ-to-AIE transformation to obtain dramatically enhanced ECL emission [142]. Zhang et al. [143] reported a pyrene-based sp^2 carbon-conjugated covalent organic framework (COF) nanosheet (Py-sp^2c-CON) which was constructed by the C=C polycondensation of 2,2′-(1,4-phenylene)diacetonitrile (PDAN) and tetrakis(4-formylphenyl)pyrene (TFPPy). Topological attachment of TFPPy to PDAN in COFs can efficiently decrease the ACQ effect to fulfill an enhanced ECL efficiency. In addition, Luo et al. [144] designed the donor–acceptor (D–A) COFs with triphenylamine and triazine units which functioned by the tunable intrareticular charge transfer (IRCT) in Figure 4A. The IRCT dominated dual oxidation was proved to be an efficient ECL mechanism which resulted in an approximate 123-fold ECL enhancement compared to the benzene-based COFs (Figure 4B). Meanwhile, Li et al. [145] developed a scalable method to design Olefin-linked donor- and acceptor-conjugated COFs where different donors and acceptors were restricted in different electron configurations, forming highly transferred intramolecular electronic networks for stronger ECL emission in the presence of no exogenous poisonous coreactants (Figure 4C). Cui et al. [146] assembled an electron-withdrawing monomer of 2,4,6-trimethylbenzene-1,3,5-tricarbonitrile (TBTN) into the COFs structures conjugated with electron-donating olefin molecules which endowed the non-ECL emissive TBTN monomer with a favorable ECL activity through efficient intramolecular electron-transfer mechanism. Similarly, Li et al. [147] used olefin-linked COFs to fabricate a highly aligned array substrate where donors and acceptors were co-crystallized and stacked into COFs structure. By using 2,4,6-trimethylbenzene-1,3,5-tricarbonitrile (TBTN) as the acceptor, tunable ECL emissions of different non-ECL donors were triggered in the formed electron freely-transported networks. For realizing better electrochemical behavior, the poor conductivity of COFs were carefully dealt with in the following studies. For instance, Zhang et al. [148] prepared a conductive COF by using two large π-conjugated planar ligands of 2,3,6,7,10,11-hexahydroxytriphenylene (HHTP) and 2,3,6,7,10,11-hexaaminotriphenylene (HATP) as dual ECL luminophores. The formed conductive porous framework of HHTP-HATP-COF performed significantly improved electronic conductivity and accelerated the mass–transport ratio, resulting in strong cathodic ECL emission along with outstanding aqueous stability.

In addition to MOFs or COFs, hydrogen-bonded organic frameworks (HOFs) are another vibrant class of crystalline porous ECL emitters based on the non-covalent hydrogen-bonding interaction. The building motifs of HOFs usually contain scaffolds covered with abundant hydrogen-bonding interacting sites and some linkers to form rigid HOF backbones [149–151]. Due to the high electrochemical activity and π–π interaction propensity in those HOFs networks, the IRCT efficiency for amplifying ECL emissions can be guaranteed by efficient electron coupling effect [151]. For example, Hou et al. [151] reported two HOFs (HOF-100 and HOF-101) which were separately prepared by using 1,3,6,8-tetracarboxy pyrene (TCPY) and 1,3,6,8-tetra(4-carboxylphenyl)pyrene (TCPPY) as two building blocks through multiple π–π and hydrogen bond interactions. Owing to more efficient electron coupling effect, HOF-101 showed a 440-fold enhancement in ECL intensity compared with HOF-100, which indicated that establishing a tunable electron-coupled ECL system might light up a new way for investigating the intricate emission mechanism of HOFs. Like MOFs, HOFs can also be utilized as scaffolds to stabilize ECL luminophores to achieve signal-amplification. For instance, Shen et al. [152] prepared an ECL-active HOF by stabilizing abundant isoluminol molecules onto a self-assembled 2,4,6-tris(4-carboxyphenyl)-1,3,5-triazine (TATB) structure. TATB can catalyze the production of highly active oxygen-containing radicals ($OH^{\bullet}$ and $O_2{}^1$ radicals) for 23.4-fold signal amplification. Though the ECL studies on HOFs are still at the initial stage, it can be envisioned that more related investigations on ECL-active HOFs will be conducted in the future.

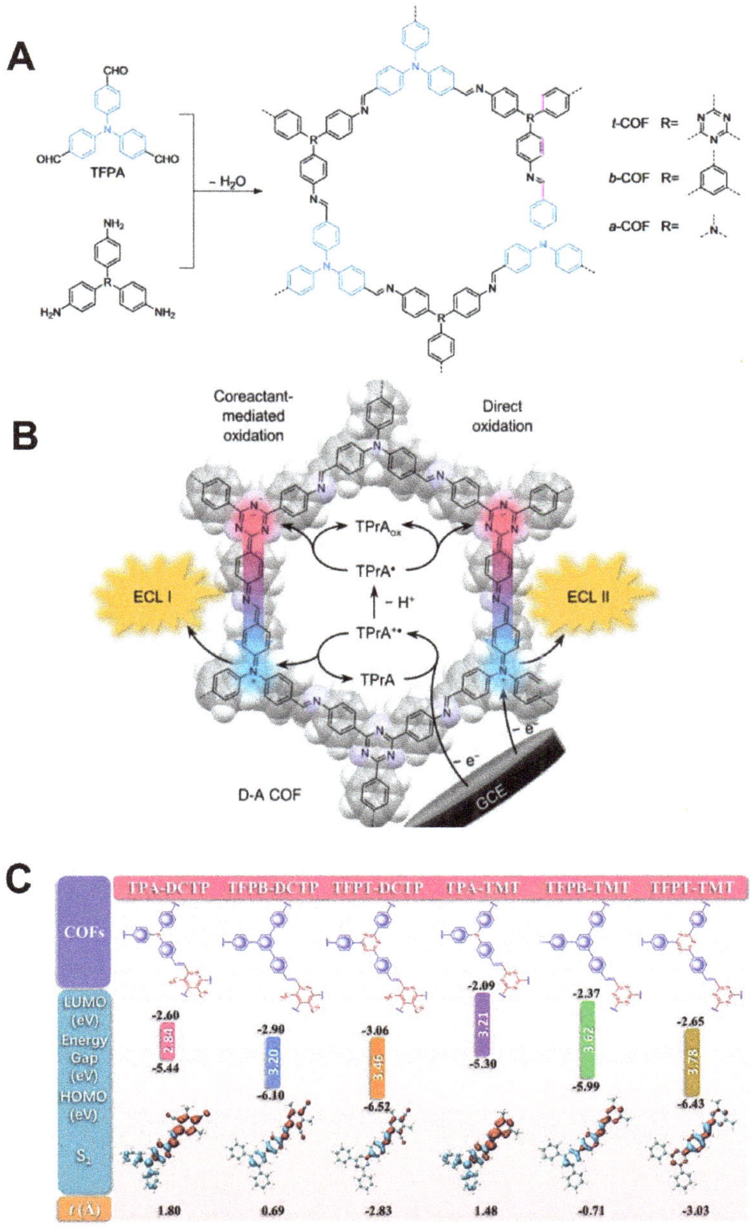

Figure 4. (**A**) Synthesis of three TFPA-based COFs and (**B**) schematic illustration of ECL mechanisms of COFs via the intrareticular charge transfer pathway. Reprinted with permission from Ref. [144], Copyright 2021 Springer Nature. (**C**) DFT calculations of COFs prepared by different organic building blocks with information on their corresponding HOMOs, LUMOs, calculated energy gaps, and *t* indexes. Reprinted with permission from Ref. [145], Copyright 2021 Springer Nature.

2.5. Polymer Dot-Based ECL Emitters

Since the first study of polymer dots (Pdots) by Bard et al. [153] in 2008, organic semiconductor Pdots as a kind of conjugated polymer nanoparticles (CNPs) have attracted extensive attention for their excellent fluorescence behaviors, bright luminescence, desirable biocompatibility, and superb photostability [154]. Direct polymerization of low molecular weight monomers and post-polymerization of high molecular weight polymers have been recognized as two major synthetic procedures for Pdots [155]. Post-polymerization methods including microemulsion and reprecipitation are the most used techniques for Pdots preparation nowadays. Compared to those QDs-based ECL emitters, Pdots have superiorities of excellent biocompatibility and easy functionalization with versatile synthetic strategies, which making them a promising type of ECL emitter for biosensing and imaging [156]. Nevertheless, strong ECL emission for most Pdots can be only observed in organic solvents and their excitation potentials are high in the aqueous, thus their applications in aqueous bioassays are restricted [157]. To deal with this challenge, Dai et al. [158] fabricated hydrophilic Pdots by using triton X-100 as a capping regent to encapsulate poly[2-methoxy-5-(2-ethylhexyloxy)-1,4-phenylenevinylene] (MEH-PPV) via a reprecipitation method. The resulting Pdots performed good aqueous ECL emission via anodic, cathodic, and annihilation routes. As expanded research, electron-withdrawing cyano (CN) groups were attached to the MEH-PPV by Feng et al. [159] to act as block copolymers for preparing Pdots. The obtained CN-PPV Pdots performed significantly enhanced coreaction and annihilation ECL efficiencies in aqueous media. To date, some commercially available polyfluorene polymers such as poly (9,9-di-n-octylfluorenyl-2,7-diyl) (PFO) and poly[(9,9-dioctylindole-2,7-diyl)-co-(1,4-benzo-{2,1′-3}-thiadiazole)] (PFBT) have been used to prepare highly ECL-active Pdots for biosensing [160–162] and PFO Pdots [163,164]. Moreover, $Ru(bpy)_3{}^{2+}$, luminol, and other traditional ECL regents have also been employed as luminous units to fabricate ECL-active Pdots. For instance, 1354 $Ru(bpy)_3{}^{2+}$ molecules were averagely doped into a carboxylated conjugated polymer by Feng et al. [165] to obtain a highly ECL-active RuPdot for detecting single-nucleotide polymorphism. Then, a potential/color-resolved ECL bioassay was established by Wang et al. [166] using the luminol-doped polymer dots (L-Pdots) and diethylamine-coupled Pdots (N-Pdots) as dual ECL emitters for the simultaneous analysis of multiplex microRNAs. Considering the indispensable role that coreactant plays in ECL generation, it can be envisioned that high ECL efficiency can also be fulfilled by conjugating coreactant-active moieties with Pdots in shortened electron-transfer in one structure with less consumed radiative energy for ECL enhancement. More importantly, this strategy needs no exogenous coreactants which can be both eco-friendly and cost-effective. For the very first time, Wang et al. [167] fabricated a coreactant-free TEA-PFBT Pdot by attaching two tertiary amine (TEA) groups to the side chain of PFBT, which generated 132-fold boosted ECL emission through a dual intramolecular electron transition pathway without any exogenous coreactants, which emphasized the advances of fabricating coreactant-free ECL Pdots with extraordinary luminous efficiencies for bioassays.

Another interesting topic in this direction is the combination of conjugated polymer backbones with attractive AIEgens to prepare AIE-active Pdots [168,169]. As shown in Figure 5A, an AIE-active Pdot was synthesized by Wang et al. [168] through a Suzuki reaction between the boron ketoiminate (BKM) monomers and the AIEgen of TPE followed by nanoprecipitation of poly(styreneco-maleic anhydride) (PSMA). The obtained AIE-active Pdot was used as an ECL emitter which showed sensitive response to $UO_2{}^{2+}$ in water environments. As shown in Figure 5B, Zhang et al. [169] incorporated AIE-active TPE moieties into the benzothiadiazole-based Pdots prepared by poly[4-(4-(2,2-bis(4-(octyloxy)phenyl)-1-phenylvinyl)phenyl)benzo-[c] [1,2,5] thiadiazole]. Benefiting from the AIECL property of TPE moieties, the prepared Pdots showed favorable sensing performances for nucleic acid detection. Notably, most of the reported methods for Pdots require nanoprecipitation of copolymers like PSMA with ECL-active moieties through weak noncovalent interactions. One should be concerned with the instability of the formed Pdots because of the possible desorption of those ECL-active moieties [170]. To further promote the advancement and application of Pdots-involved ECL sensing systems, covalent functionalization or some specific interactions with high affinities are expected to be introduced prior to or after the formation of Pdots, achieving better ECL stability for better sensing performances [171,172].

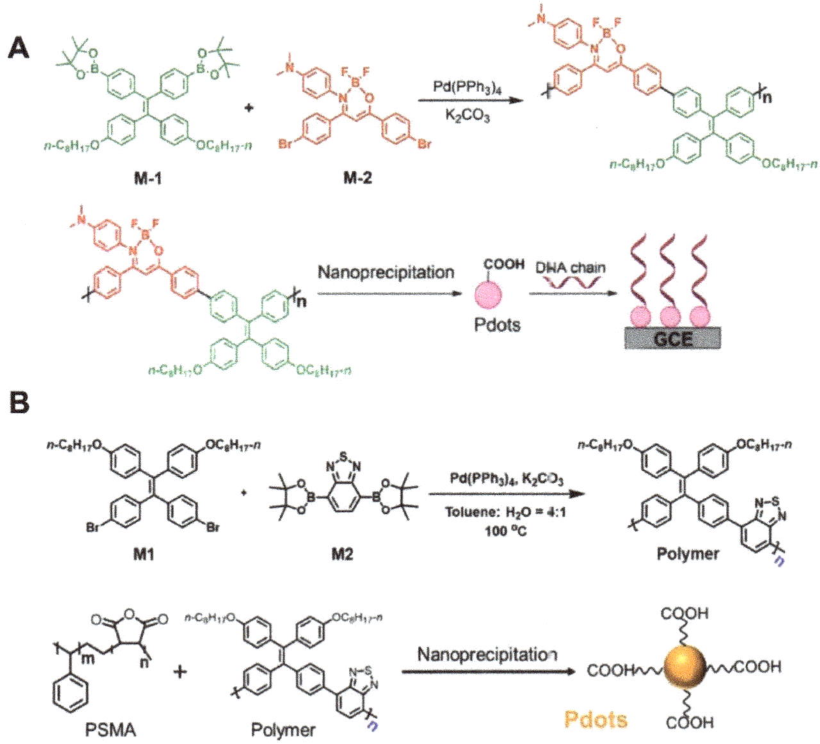

Figure 5. (**A**) Synthetic routes of AIE-active TPE-Pdots for electrode modification. Reprinted with permission from Ref. [168] Copyright 2020 Wiley. (**B**) Synthesis route of AIE-active iridium(III)-Pdots. Reprinted with permission from Ref. [169] Copyright 2021 American Chemical Society.

3. Application of ECL Emitters for the Analysis of Protein-Based Biomarkers

Biomarker, a kind of substance (gene, protein, polysaccharide, etc.) with valuable medical or biological significance, is generated from a normal physiological process or a biochemical pathway in the human body [34]. Pathologically related biomarker can give doctors insightful information about the health status of the body with minimal interference and less effort or warnings that a possible disease might be evolving, which is of great importance for patients to get the medical cure with efficient control of the disease in the early stages [173,174]. Biomarkers can be classified into imaging and non-imaging types. The former type is usually required in the X-rays or MIR imaging of diseases while the latter usually functions as a measurable indicator (protein, nucleic acid, small metabolite, cytogenetic, or cytokinetic factors) in body fluids or tissues that is closely relate to a certain disease. Among those types, protein-based biomarkers have been recognized as golden indicators for diagnosing various diseases in the early stages with trustful prognostic values [175–177]. As for the point-of-care detection of protein biomarkers with low abundance, great challenges are ahead due to the limited sensitivity of traditional assays, high background signals make it difficult to capture the accurate information of target protein rapidly in crude and complex samples without highly expensive and sophisticated instrumentations. ECL is now considered one of the most promising analytical techniques, owning superior sensitivity and ultralow background signals for biological applications with an ease of operation and rapid responses in real-time. With the aim of improving detection sensitivity and the reliability and accuracy in the diagnosis of protein biomarkers, a variety of promising and highly-efficient ECL nanomaterials were fabricated and employed as emitters of biosensors over the past decades.

3.1. Protein-Based Cancer Biomarkers

Protein biomarkers such as alpha fetoprotein (AFP), prostate-specific antigen (PSA), carcinoembryonic antigen (CEA), mucin 1 (MUC1), and cytokeratin 19 fragment (CYFRA 21-1), as recently studied by using advanced ECL emitters and biosensing strategies [173,175]. CEA as a broad-spectrum protein biomarker of malignant tumors is commonly related to potential pancreatic, colorectal, or breast cancer in human bodies [170]. To date, ECL biosensors have been developed as highly-sensitive tools for the clinical diagnosis of CEA-indicated cancers. For instance, the Zn^{2+}-doped Au NCs (Zn^{2+}-AIE-Au NCs) with AIE activity were prepared by Gao et al. [178] using mercaptopropionic acid as a capping regent. The Zn^{2+}-AIE-Au NCs as monochromatic (FWHM < 40 nm) emitters with high ECL efficiency were further integrated with the dual-stabilizer-capped $CuInS_2$@ZnS NCs to establish a spectrum-resolved ECL biosensor for multiplexing bioassay in real samples. In the assay, CEA was sensitively determined with an LOD of 0.3 pg/mL. Then, Zhang et al. [179] reported a band-edge effect-induced ECL immunosensor using the silica inverse opal photonic crystals (SIOPCs) for CEA incubation. The radiative transition probability was increased due to the formation of high density states at the band-edge in SIOPCs, which enhanced the photon extraction during propagation and achieved strong ECL outputs for CEA detection with an LOD of 0.032 pg/mL. To guarantee high accuracy, establishing a potential/wavelength- based ratiometric ECL assay is usually a preferred choice since it can provide self-calibration to decrease systematic errors and improve the detection reliability for CEA detection. As previously reported, Huang et al. [180] developed a dual-potential ECL immunoassay by using the highly ECL emissive luminol and CdS QDs as anodic and cathodic emitters, respectively. The assay showed ratio-metric ECL emission at −1.5 V and +0.3 V (vs. Ag/AgCl) using H_2O_2 as the coreactant which realized sensitive and accurate detection for CEA with an LOD of 0.62 pg/mL. With concern to the possible interferences between the two different ECL emitters in conventional ratiometric biosensors, Shang at al. [38] designed a potential-resolved ECL immunoassay by utilizing single emitters of luminol as the cathodic and the anodic emitter (−0.2 to 0.45 V, vs. Ag/AgCl). In this ratiometric assay, Pd NCs modified substrates could quench the anodic ECL (I_{anodic}) and amplify the anodic ECL ($I_{cathodic}$) of luminol −0.2 to 0.45 V, thus the detection results ($I_{cathodic}/I_{anodic}$) can be self-calibrated to endow the biosensor a stronger reliability for CEA detection with an ultralow LOD of 87.1 ag/mL. Obviously, ratiometric-typed ECL biosensors still hold great potential for the ultrasensitive detection of low-abundant protein biomarkers in complex samples in the future.

Recently, bipolar electrode (BPE) device as one of the most attractive electronic conductors based on bipolar electrochemistryhas attracted extensive scientific interest in the field of biosensing for protein biomarkers [181]. Li et al. [182] constructed a gold nanowire array-based BPE system using $Ru(bpy)_3{}^{2+}$-wrapped SiO_2 nanoparticles (Ru@SiO_2) as ECL emitters. The facile BPE sensing device realized sensitive detection of AFP expressed by the HepG2 cell surface with an LOD of 6.71 pg/cell. Meanwhile, Yu et al. [183] developed a near-infrared ECL immunoassay by using the biocompatible and environmentally friendly methionine-stabilized Au NCs as ECL emitters for AFP detection. The assay performed a good selectivity for AFP in complex samples with an ultralow LOD of 1 fg/mL, which was much lower than those reported ECL biosensors.

PSA, as a single-chain glycoprotein, is regarded as the most representative indicator for the early diagnosis of prostate cancer, the second most fatal cancer for men [184]. Qin et al. [185] developed a dual-quenching ECL immunoassay for PSA detection on a solid-state platform fabricated by the modification of an AIE-active layer of 6-aza-2-thiothymine capped Au NCs (ATT-Au NCs). The ECL emission of ATT-Au NCs can be regulated by the CeO_2-PEI@Ag probe via resonance energy transfer, thus achieving a sensitive detection of PSA with an LOD of 2.2 fg/mL lower than previously reported works. Wang et al. [184] designed a label-free PSA aptasensor based on the luminol-functionalized molecularly imprinted polymers (MIP) of electropolymerized dopamine (DA) film. Relying on the specific interaction of PSA with aptamer-covered imprinted cavities, the aptasensor exhibited gradually decreased ECL signals along with increased PSA concentration in the linear range of 5 pg/mL–50 ng/mL, achieving a low LOD of 3.0 pg/mL with acceptable stability and reproducibility. Considering the limited sensitivity of the label-free sensing format, as a following work, Wang et al. [40] further developed an ECL quenching strategy by using the carboxylated graphitic carbon nitride nanosheet as the donor and Ru@SiO_2 as the acceptor. Through the efficient electron-transfer mediated quenching mechanism, high sensitivity was achieved for PSA determination along with a rather low LOD of 1.2 fg/mL. In addition, an efficient ECL-RET aptasensor for PSA detection was constructed by Zhao et al. [186] using the Cys-$[Ru(dcbpy)_3]^{2+}$ as ECL emitters whose emission could be flexibly quenched by the ZnS QDs loaded mesoporous silica nanocontainers (ZnS@SiO_2) via a controlled releasing strategy (Figure 6). Due to the efficient ECL-RET between Cys-$[Ru(dcbpy)_3]^{2+}$ and ZnS@SiO_2, the detection sensitivity and accuracy for PSA detection were desirable, showing a low LOD of 1.01 fg/mL. Besides, other promising highly ECL-emissive NCs such as dual-stabilizer-capped InP/ZnS NCs and Ag-Ga-In-S NCs have also been fabricated as ECL emitters for PSA detection by Zou's group [77,187], which highly indicated the promising application potential of multinary NCs in the clinical diagnosis of protein biomarkers.

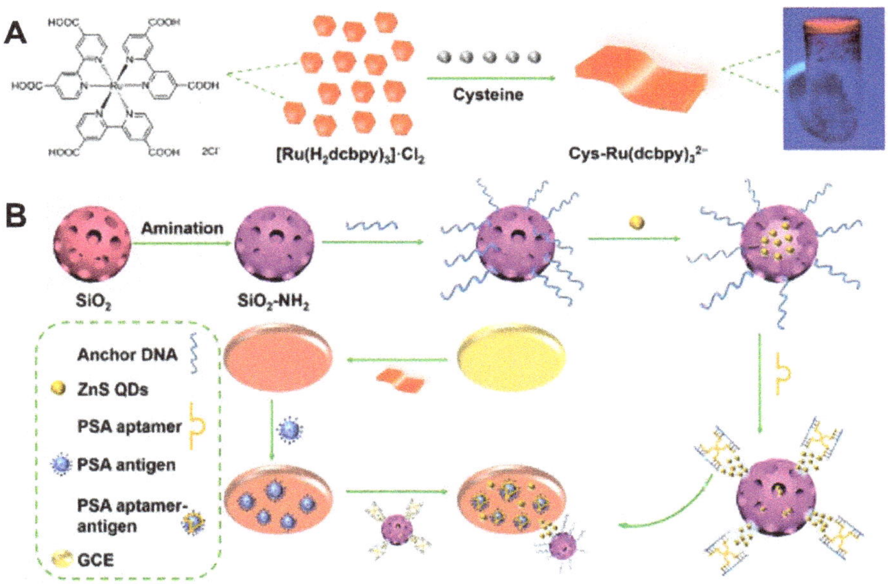

Figure 6. (**A**) Preparation procedures of Cys-$[Ru(dcbpy)_3]^{2+}$ and ZnS@SiO_2-ssDNA/apta. (**B**) Fabrication process of constructed aptasensor for PSA detection. Reprinted with permission from Ref. [186], Copyright 2023 American Chemical Society.

Lung cancers, including small cell lung cancer (SCLC) and non-small cell lung cancer (NSCLC), have become the primary worldwide cause of cancer mortality with the fastest growing incidence rate nowadays. Biomarkers for lung cancer including CEA, CYFRA 21-1 squamous cell carcinoma antigen (SCCA), and neuron-specific enolase (NSE) have exhibited tremendous potential as trustful indicators for lung cancer diagnosis in early stages [170]. To date, various ECL biosensing systems have been established for NSE detection [114,188–190]. For example, Dong et al. [114] prepared a $Ru(bpy)_3{}^{2+}$ functionalized Zr-MOF (Ru-MOF-5) as a dual-signal ratiometric ECL platform, achieving efficient dual-potential ECL emission for detecting NSE lower to 41 fg/mL along with self-calibrated detection accuracy. Furthermore, Han et al. [190] developed a sensitive NSE immunoassay by utilizing the nitrogen-doped CQDs (N-CQDs) as ECL emitters. The sensitivity of this assay was enhanced by introducing the coreaction accelerator of molybdenum sulfide/ferric oxide (MoS_2@Fe_2O_3) nanocomposites as an incubation substrate, achieving an ultralow LOD of 6.30 fg/mL with favorable performance. CYFRA 21-1 is a trustful protein biomarker for non-small cell lung cancer (NSCLC) which holds great reference value in the early stages of NSCLC. Recently, great efforts have been devoted into developing CYFRA 21-1 immunoassays with higher sensitivity and accuracy [191–193]. For example, Yang et al. [193] fabricated an AIE-active biocompatible ECL emitter by encapsulating *fac*-tris(2-phenylpyridine)iridium(III) complexes [Ir(ppy)$_3$] into an apoferritin (apoFt) cavity for CYRFA 21-1 detection based on a conductive and electroactive substrate of Fe_2N and Au NP-loaded rGO (Fe_2N/rGO/Au). Li et al. [194] proposed a BPE-assisted ECL immunoassay based on the glucose oxidase and horseradish peroxidase loaded ZIF-67@CaO_2 composites. ZIF-67@CaO_2 can in situ produce coreactants of O_2 and H_2O_2 to promote ECL reaction of luminol to realize the sensitive detection of CYFRA 21-1 with favorable performance.

3.2. Protein-Based Non-Cancer Biomarkers

Instead of focusing on cancer biomarkers, ECL biosensing systems have been developed for various protein biomarkers of other major diseases such as Alzheimer's disease (AD), cardiovascular diseases, sepsis, and so on [108,136,195–197]. AD has become a worldwide brain disease that threatened the physical health of the aged [197]. How to protect them from the AD attack and obtain timely medical treatment in the early stage is of great importance. Amyloid-β protein ($A\beta_{1-42}$) was recognized as the most critical and predictive protein biomarker for diagnosing AD. ECL-based biosensing strategies have been developed for the specific and sensitive analysis of $A\beta_{1-42}$ [27,108,198,199]. Recently, Qin et al. [200] proposed dual-wavelength ratiometric ECL immunoassays for $A\beta_{1-42}$ analysis based on twoECL emitters including Ru@TiO_2@Au (620 nm) and AuNPs-modified g-C_3N_4 (460 nm) with different emission wavelengths. By virtue of the ECL-RET between these two emitters, sensitive detection of $A\beta_{1-42}$ was achieved with an ultralow LOD of 2.6 fg/mL in cerebrospinal fluid. Also, Tan et al. [198] developed an advanced ECL aptasensing system by combining affinity screening of aptamers and size screening of silica nanochannels. The sensing mechanism was operated by the steric hindering formed by the specific interaction of $A\beta_{1-42}$ on the aptamer in arrayed nanochannels, which blocked the mass transport of luminol and led to ECL reduction in corresponding to the $A\beta_{1-42}$ concentration. Recently, Alzheimer-associated neuronal thread protein (AD7c-NTP) was found in the CSF, blood, and urine of AD patients and can be regarded as a new and trustful biomarker for AD diagnosis [201]. However, only enzyme-linked immunosorbent assays (ELISA) were used for AD7c-NTP detection. To obtain higher sensitivity than ELISA, the superior ECL technique was more suitable for ultra-trace analysis of AD7c-NTP. Liang et al. [202] recently constructed a dry chemistry-based BPE device for the ECL assay of AD7c-NTP. This device was composed of a screen-printed fiber chip filled with a Ru(II)-poly-L-lysine complex for immunoreaction where self-enhanced ECL detection can be completed in 6 min with an LOD of 0.15 pg/mL in human urine samples.

Heart diseases including acute myocardial infarction (AMI) and heart failure (HF) have comparable mortality rates to that of cancers and have become the most concerning class of disease that severely threatens the body health of human beings all around the world [203]. C-reactive protein (CRP) and cardiac troponin I (cTnI) are two common protein biomarkers for AMI and cardiovascular diseases [204]. Yang et al. [205] synthesized a series of novel multicolor iridium(III) complexes, among which the adap-based Ir(III) complexes exhibited a good binding affinity with β-cyclodextrin (β-CD) due to the strong hydrophobic interaction and acted as ECL emitters for aqueous-phase detection of CRP with an LOD of 72 pg/mL. Hong et al. [206] developed an ECL combined lateral flow immunoassay (ECL-LFI) which was suitable for full-range analysis of physiological CRP levels. This assay involved the usage of $Ru(bpy)_3{}^{2+}$-labeled AuNPs as CRP probes which plays an important role in the test line of the strip to achieve a sensitive and rapid read-out with a low LOD of 4.6 pg/mL in the clinical diagnosis of AMI. Guo et al. [207] designed an ECL emitter-free and disposable biosensor based on the quenched cathodic signal of porous TiO_2 microspheres by the ferrocene labeled Y-shaped probes. This efficient signal "off-on" assay showed good linearity in 100 fg/mL–100 ng/mL with a low LOD of 30.1 fg/mL for cTnI detection in human serum. Heart failure, as the result of the systolic or diastolic dysfunction of the heart, is usually regarded as the terminal stage of many cardiovascular diseases [196]. *N*-terminal pro-brain natriuretic peptide (NT-proBNP) has been established as a key biomarker which owns prognostic importance for heart failure and other cardiovascular diseases [170]. Luo et al. [195] proposed a simple and sensitive ECL-sensing device by using luminol functionalized AuNPs/ZIF-67 as an ECL emitter which can achieve sensitive NT-proBNP detection lower to 0.74 pg/mL with good specificity and stability. Ji et al. [208] designed a dual-signal ECL immunoassay by using the $Ru(bpy)_3^{2+}$@HKUST-1 as anodic ECL emitters and $Ce_2Sn_2O_7$ nanocubes as the cathodic emitters in which the detection accuracy for NT-proBNP was well improved, benefiting from the self-calibration of anodic and cathodic ECL responses (Figure 7).

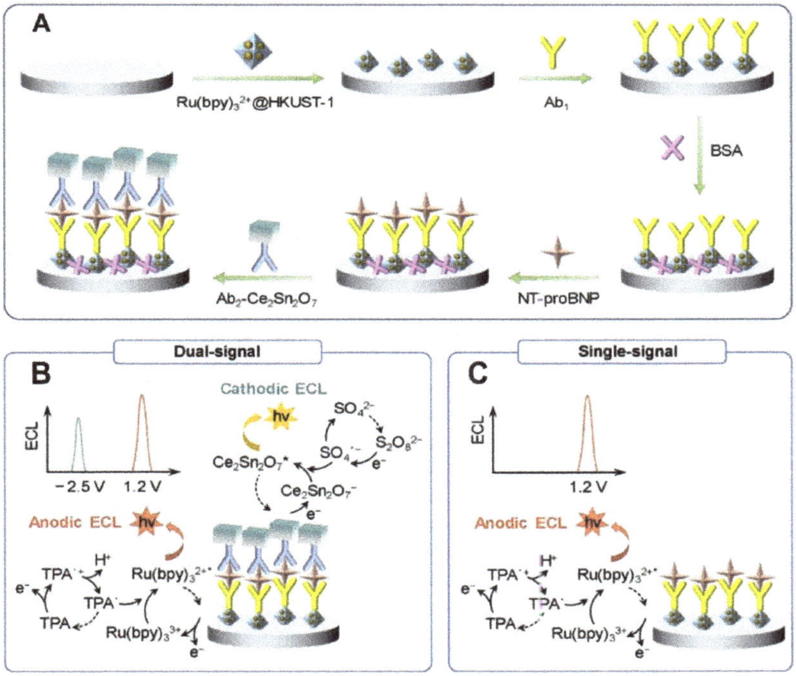

Figure 7. (**A**) Schematic illustration of a dual-signal ECL immunosensor for NT-proBNP detection. (**B**) Dual-signal and (**C**) single-signal ECL mechanisms. Reprinted with permission from Ref. [208] Copyright 2023 Royal Society of Chemistry.

Sepsis is a systemic response to infection most often caused by bacteria and results in life-threatening inflammation throughout the body. Procalcitonin (PCT) secreted by thyroid C cells has shown a strong correlation with the severity of sepsis, which is recognized as a representative biomarker in the early diagnosis of sepsis. Xu et al. [209] developed a quenching-typed ECL immunosensor based on ferrocene carboxylic acid (Fca) quenching the ECL of the $Ru(bpy)_3^{2+}$ and TPrA co-encapsulated ECL SiO_2 NPs. In this assay, good sensitivity for PCT detection was achieved by an electron-transfer mediated quenching mechanism which contributed to the low LOD of 0.85 pg/mL. Instead of the quenching method, ECL-RET was also recognized as a highly sensitive technique for PCT detection. Song et al. [210] designed an ECL-RET immunosensor by employing the carbon nanotubes

(CNT) and Au-functionalized graphitic carbon nitride ($g\text{-}C_3N_4$–CNT@Au) as the ECL donor and CuO nanospheres-polydopamine (PDA) layer (CuO@PDA) as the receptor, respectively. High RET efficiency from $g\text{-}C_3N_4$–CNT@Au to CuO@PDA was occurred for sensitive detection of PCT lower to 25.7 fg/mL in human serum. In addition, Wang et al. [211] prepared a dual-emitting ECL immunosensor by employing MnO_2 nanoflowers (NFs) and ZnS QDs as dual ECL emitters with $S_2O_8^{2-}$ as the coreactant which exhibited a good linearity in the range of 0.1 pg/mL to 100 ng/mL with an LOD of 0.033 pg/mL for PCT detection in serum. Most recently, Song et al. [212] designed a ternary ECL sensing system using lipoic acid (LA) capped Ag NCs as cathodic ECL emitters in $S_2O_8^{2-}$ containing electrolytes. In this assay, ligand-to-metal charge transfer between LA and Ag NCs induced ideal ECL emission which endowed the PCT immunosensor a low LOD of 3.56 fg/mL.

3.3. ECL Imaging

In recent years, ECLM as a fascinating optical microscopy technique has gained dramatically increased scientific interest in spatiotemporally resolved imaging of single NP or cell indicated by some specifically triggered ECL processes at the electrode surface [96,213]. Compared with traditional optical microscopy, ECLM owns superior features includingspatiotemporally controlled high resolution and high throughput r bioimaging of various membrane molecules at single-cell level [5,214]. More importantly, ECLM needs no extrinsic illumination source, which can eliminate those potential local photothermal effects and background noises for high-quality imaging [215]. Uniquely from serum analysis, specific visualization of protein biomarkers on cell membranes provides paramount information for elucidating molecular mechanisms in a wide variety of diseases [216]. To date, ECLM protocols have been gradually established to evaluate expression levels of protein biomarkers on cell membranes. Lu et al. [217] designed an efficient ECL nanoreactor for direct visualization of single membrane epithelial cell-adhesion molecule (EpCAM) proteins on heterogeneous MCF-7 cells by employing the $Ru(bpy)_3^{2+}$-loaded nanoporous zeolite nanoparticles (Ru@zeolite) as imaging probes (Figure 8). In the assay, nanoconfinement-enhanced ECL emission of $Ru(bpy)_3^{2+}$-TPrA was triggered in every single nanoreactor, which was conducive to achieving high-resolution ECL imaging of membranal biomarkers including AFP, PSA, and CEA. Liu et al. [10] proposed an ECL imaging strategy by using $Ru(bpy)_3^{2+}$-doped silica/Au NPs (RuDSNs/AuNPs) as probeswhich could create surface-confinement-controlled ECL enhancement for realizing single-molecule imaging of a protein biomarker of cytokeratin 19 at the electrode surface and cellular membrane. Liu et al. [218] developed a potential-resolved ECL imaging method for the diagnosis of apoptosis at the single-cell level. In the method, the epidermal growth factor receptor and phosphatidylserine were visualized on the surfaces of normal and cancer cell samples using $g\text{-}C_3N_4$ and Au@L012 as dual imaging probes. The assay not only offered a reliable imaging technique for the detection of various membrane biomarkers at the single-cell level, but also contributed to the guidance of rational drug applications for medical treatments. As well known, the addition of exogenous coreactants such as H_2O_2 or TPrA might induce potential cytotoxicity, thus making it quite urgent to design coreactant-free ECL emitters for cell imaging [219]. Wang et al. [167] designed coreactant-embedded polymer dots (TEA-Pdots) for in situ ECL imaging of single membrane protein without exogenous coreactant via an efficient intramolecular electron transfer mechanism. In addition to fabricating coreactant-embedded ECL emitters, developing biocompatible ECL coreactants is another doable and promising option. Chen et al. [219] prepared a guanine-rich DNA-aptamer complex that can not only function as the biocompatible coreactant for $Ru(bpy)_3^{2+}$ emission via a catalytic pathway but also specifically interact with overexpressed CEA on cell membranes for ECL imaging. Moreover, some it has been proved that high resolution for ECL imaging of membrane proteins can also be achieved without the direct utilization highly emissive probes. The pioneering work was reported by Zhang et al. [220]. They established capacitance-based ECL microscopy for the high-resolution imaging of low abundance CEA on the cellular plasma membrane without using any imaging probes. The system was dominated by the principle of locally dropped capacitance upon the binding of target proteins to the surface or cellular membrane without using any ECL emitters which opened up a brand-new avenue for efficient ECL bioimaging of membrane protein biomarkers in a label-free manner.

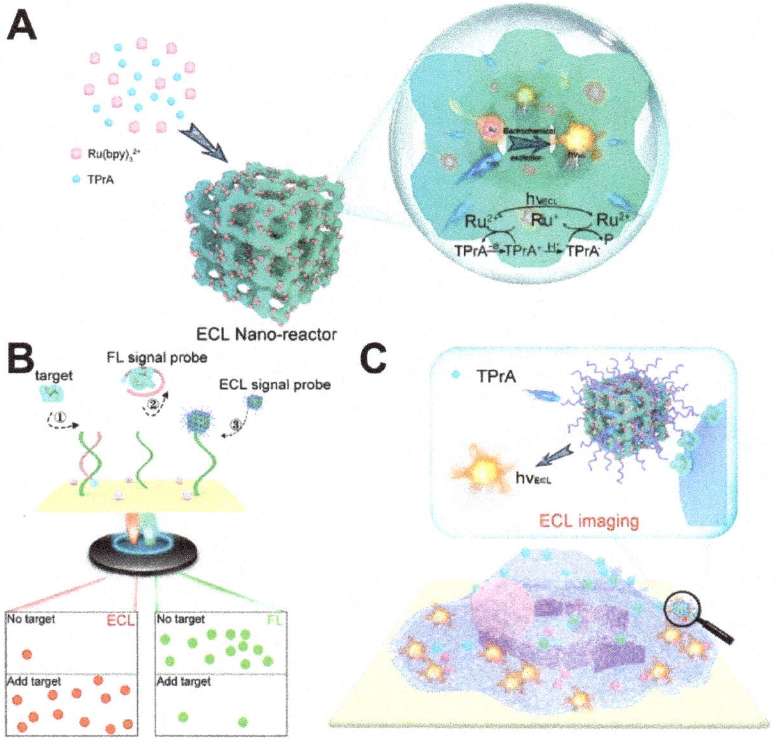

Figure 8. (**A**) Scheme for the nanoconfinement-enhanced ECL reaction at single nanoreactors. (**B**) Dual-signal (FL and ECL) imaging procedures ①–③ for individual protein analysis. (**C**) ECL imaging of a single protein on the cell membrane. Reprinted with permission from Ref. [217] Copyright 2023 American Chemical Society.

Despite that some advanced progress has been achieved in ECL imaging of protein biomarkers at the single-cell level, ECLM can also be employed as a powerful technique for other biological applications such as photodynamic therapy (PDT) and cell-adhesion visualizations. For instance, Chen et al. [221] established an ECL-driven PDT system which was dominated by the ECL-RET mechanism using $Ru(bpy)_3^{2+}$ as the donor and photo-sensitizer of chlorin e6 (Ce 6) as the acceptor. The mechanism revealed that the ECL emission of $Ru(bpy)_3^{2+}$ could sensitize the surrounding O_2 to produce ROSs to kill cancer cells. Moreover, ECLM has also been applied to investigate the cell matrix adhesion process in a visual view [222,223]. Ding et al. [222] reported a label-free strategy to image the cell-matrix adhesion of living cells on a silica nanochannel membrane-based electrode at the single-cell level. The ECL reaction was triggered between freely diffused $Ru(bpy)_3^{2+}$ and coreactants, which showed a direct visualization of different adhesion sites of living cells, thus offering more precise information for imaging cellular orientations during the collective migrations in a label-free manner. Moreover, Ding et al. [223] proposed a surface-sensitive ECLM imaging method to evaluate possible impacts of different chemically modified substrates on cell adhesions. As observed from the statistical analysis of spatial distribution, area, and strength of cell adhesion, Arg–Gly–Asp (RGD) modified electrodes can mediate robust cell matrix interactions. This work showcased that ECLM can be a qualified tool to analyze other cell behaviors during cancer metastasis on different microenvironments with high sensitivity and resolution.

4. Summary and Perspectives

Applying a highly-efficient emitter is the core of an ECL analytical system to obtain the high sensitivity and accuracy for target detection. Along with the interdisciplinary integration of electrochemistry, spectroscopy, and nanotechnology, various nanomaterial-based ECL emitters were developed and injected numerous vitalities into ECL bioassays with significantly improved analytical performances. In this review, we mainly focused on the recently reported fabrication of some attractive ECL emitters including QDs, MeNCs, MOFs, COFs/HOFs, and Pdots and their advances in sensitive ECL sensing and spatiotemporally resolved imaging of protein-based biomarkers. Through decent designation, high ECL efficiencies can be easily achieved among these ECL emitters. As for the preparation methods, QDs, MOFs, COFs, HOFs, and Pdots often require complicated organic synthetic procedures to obtain needed ligands or hydrothermal processes with high temperatures and pressures. Water stability issues for QDs, MeNCs, and MOFs should also be carefully concerned for aqueous application. Those nanosized QDs, MeNCs, and Pdots possess greater potential for high-resolution bioimaging than MOFs and COFs/ HOFs with larger sizes. Instead, MOFs, COFs/HOFs with large specific areas and electrochemical activity can be flexibly employed to develop promising ECL solid-sate devices for point-of-care application. In order to aim for higher-level advancements, opportunities and challenges are considered and addressed in this section to obtain a clear understanding of development trends in the future.

Obtaining extraordinary ECL efficiency via a newly established excited-state generation mechanism is becoming one of the most promising directions for exploiting high-quality ECL emitters. As for those above-mentioned QDs, MeNCs, MOFs, COFs, HOFs, and Pdots, most of their triplet excitons were consumed via nonradiative decay; only ~25% of the generated singlet excitons were decayed radiatively with a low ECL efficiency. Paving a new way to promote the utilization rate of triplet excitons with a proper theoretical model is urgently demanded. Some advanced efforts in the antenna effect, thermal activation of delayed luminescence, and triplet–triplet annihilation on this topic have been gradually exploited with significantly increased ECL efficiencies which injected new insights of the investigations on new ECL mechanisms.

Considering that most of the ECL bioassays are developed following the sandwich-typed format, efficient bioactivity maintenance of biomolecules in this format is of great importance. Interferences from the sensing substrate, probe, and electrolytes including pH and ionic strength toward the bioactivity of used biorecognition molecules like antibodies or aptamers should be carefully evaluated to guarantee the accuracy and practicability of immunosensors. Therefore, constructing more biocompatible interfaces on ECL emitters and substrates with vigorous and anti-interferent biorecognition molecules is becoming another attracting direction in which more endeavors are suggested to be devoted to establishing more robust specific interactions to capture the detection targets effectively. In this way, we believe that the developed ECL biosensor will perform better specificity, reproducibility, and reliability for practical applications.

In addition, a variety of advanced imaging methods combined with promising ECL emitters were displayed for membrane protein sensing at a single-cell level. No one can deny that the ECL imaging technique combined with immunoassays for the high-throughput detection of membrane biomarkers will become one of the most attractive microscopic branches in the future. Though some grounding breakthroughs have been achieved, no one can deny that the conventional $Ru(bpy)_3{}^{2+}$-TPrA system is mostly used for ECL imaging. To obtain stronger intensity, $Ru(bpy)_3{}^{2+}$ molecules are usually stabilized in a biocompatible carrier like SiO_2 NPs to act as imaging probe while the cell-incubated substrate is still immersed into electrolytes containing TPrA whose biotoxicity on cell viability is inevitable. Obviously, designing biocompatible coreactant is a doable idea to embrace the challenge. Also, preparing self-enhanced ECL emitters by implanting or modifying the coreactant-active moieties with luminophores can be regarded as an ideal alternative which needs no exogenous coreactant to directly contact the live cells. It can be envisioned that coreactant-free sensing strategies will present great application potential toward the development of eco-friendly ECL immunoassays or bioimaging in the future.

To summarize, the fabrication and application of novel and highly-efficient ECL emitters for biosensing and imaging are still incredibly exciting. They enriched the resources of synthetic routes for different kinds of nanomaterials, wide applications for various disease-related protein biomarkers, and advanced imaging techniques with high spatial-temporal resolution. ECL biosensors with advantages of being miniaturized, portable, easy to integrate with other devices, are still highly encouraged to create a bridge between lab research and commercialization. We believe that the remarkable ECL technique will open more scientific possibilities for the establishment of sophisticated mechanism models, achieving widely expanded applications in biosensing and imaging with favorable social and economic benefits.

Author Contributions: Writing—original draft preparation, L.Y.; Writing—review and editing, supervision, funding acquisition, J.L. All authors have read and agreed to the published version of the manuscript.

Funding: This work was financially supported by the National Key Research and Development Program of China (No. 2021YFA1200104), New Cornerstone Science Foundation, National Natural Science Foundation of China (No. 22027807, No. 22034004), Strategic Priority Research Program of Chinese Academy of Sciences (No. XDB36000000), Special support from China Postdoctoral Science Foundation (2022TQ0171), and Tsinghua-Vanke Special Fund for Public Health and Health Discipline Development (No. 2022Z82WKJ003).

Institutional Review Board Statement: Not applicable.

Informed Consent Statement: Not applicable.

Data Availability Statement: Not applicable.

Conflicts of Interest: The authors declare no competing financial interest.

References

1. Hu, L.; Xu, G. Applications and trends in electrochemiluminescence. *Chem. Soc. Rev.* **2010**, *39*, 3275–3304. [CrossRef] [PubMed]
2. Yang, X.; Hang, J.; Qu, W.; Wang, Y.; Wang, L.; Zhou, P.; Ding, H.; Su, B.; Lei, J.; Guo, W.; et al. Gold Microbeads Enabled Proximity Electrochemiluminescence for Highly Sensitive and Size-Encoded Multiplex Immunoassays. *J. Am. Chem. Soc.* **2023**, *145*, 16026–16036. [CrossRef]
3. Wang, Y.; Ding, J.; Zhou, P.; Liu, J.; Qiao, Z.; Yu, K.; Jiang, J.; Su, B. Electrochemiluminescence Distance and Reactivity of Coreactants Determine the Sensitivity of Bead-Based Immunoassays. *Angew. Chem. Int. Ed.* **2023**, *62*, e202216525. [CrossRef]
4. Miao, W. Electrogenerated Chemiluminescence and Its Biorelated Applications. *Chem. Rev.* **2008**, *108*, 2506–2553. [CrossRef]
5. Voci, S.; Goudeau, B.; Valenti, G.; Lesch, A.; Jović, M.; Rapino, S.; Paolucci, F.; Arbault, S.; Sojic, N. Surface-Confined Electrochemiluminescence Microscopy of Cell Membranes. *J. Am. Chem. Soc.* **2018**, *140*, 14753–14760. [CrossRef] [PubMed]
6. Visco, R.E.; Chandross, E.A. Electroluminescence in Solutions of Aromatic Hydrocarbons. *J. Am. Chem. Soc.* **1964**, *86*, 5350–5351. [CrossRef]
7. Han, Q.; Wang, C.; Liu, P.; Zhang, G.; Song, L.; Fu, Y. Three kinds of porphyrin dots as near-infrared electrochemiluminescence luminophores: Facile synthesis and biosensing. *Chem. Eng. J.* **2021**, *421*, 129761. [CrossRef]
8. Nasiri Khonsari, Y.; Sun, S. Recent trends in electrochemiluminescence aptasensors and their applications. *Chem. Commun.* **2017**, *53*, 9042–9054. [CrossRef]
9. Li, Z.; Huang, X.; Liu, H.; Luo, F.; Qiu, B.; Lin, Z.; Chen, H. Electrochemiluminescence Biosensor for Hyaluronidase Based on the Adjustable Electrostatic Interaction between the Surface-Charge-Controllable Nanoparticles and Negatively Charged Electrode. *ACS Sens.* **2022**, *7*, 2012–2019. [CrossRef]
10. Liu, Y.; Zhang, H.; Li, B.; Liu, J.; Jiang, D.; Liu, B.; Sojic, N. Single Biomolecule Imaging by Electrochemiluminescence. *J. Am. Chem. Soc.* **2021**, *143*, 17910–17914. [CrossRef]
11. Yang, X.; Yuan, R.; Chai, Y.; Zhuo, Y.; Mao, L.; Yuan, S. $Ru(bpy)_3^{2+}$-doped silica nanoparticles labeling for a sandwich-type electrochemiluminescence immunosensor. *Biosens. Bioelectron.* **2010**, *25*, 1851–1855. [CrossRef]

12. Dong, Y.-P.; Chen, G.; Zhou, Y.; Zhu, J.-J. Electrochemiluminescent Sensing for Caspase-3 Activity Based on $Ru(bpy)_3^{2+}$-Doped Silica Nanoprobe. *Anal. Chem.* **2016**, *88*, 1922–1929. [CrossRef] [PubMed]
13. Valenti, G.; Rampazzo, E.; Bonacchi, S.; Petrizza, L.; Marcaccio, M.; Montalti, M.; Prodi, L.; Paolucci, F. Variable Doping Induces Mechanism Swapping in Electrogenerated Chemiluminescence of $Ru(bpy)_3^{2+}$ Core–Shell Silica Nanoparticles. *J. Am. Chem. Soc.* **2016**, *138*, 15935–15942. [CrossRef] [PubMed]
14. Yue, Q.; Li, X.; Fang, J.; Li, M.; Zhang, J.; Zhao, G.; Cao, W.; Wei, Q. Oxygen Free Radical Scavenger PtPd@PDA as a Dual-Mode Quencher of Electrochemiluminescence Immunosensor for the Detection of AFB1. *Anal. Chem.* **2022**, *94*, 11476–11482. [CrossRef] [PubMed]
15. Huang, J.; Xiang, Y.; Li, J.; Kong, Q.; Zhai, H.; Xu, R.; Yang, F.; Sun, X.; Guo, Y. A novel electrochemiluminescence aptasensor based on copper-gold bimetallic nanoparticles and its applications. *Biosens. Bioelectron.* **2021**, *194*, 113601. [CrossRef]
16. Lin, Y.; Cao, J.; Li, X.; Zhang, X.; Zhang, J.; Lin, Z. A novel molecularly imprinted electrochemiluminescence sensor based on a $Ru(bpy)_3^{2+}$/MWCNTs/nano-TiO_2-Nafion electrode for the detection of bisphenol A. *Anal. Methods* **2016**, *8*, 7445–7452. [CrossRef]
17. Ju, J.; Ding, Q.; Xie, J.; Li, G. Label-free electrochemiluminescence immunosensor for mucoprotein 1 using a graphene oxide-$Ru(Bpy)_3^{2+}$-polyaniline nanocomposite. *Anal. Lett.* **2023**, 1–12. [CrossRef]
18. Lin, J.; Wu, H.; Lu, L.; Sun, Z.; Zhang, Y.; Dang, F.; Qian, L. Porous graphene containing immobilized Ru(II) tris-bipyridyl for use in electrochemiluminescence sensing of tripropylamine. *Microchim. Acta* **2016**, *183*, 1211–1217. [CrossRef]
19. Yang, W.; Zhang, G.; Ni, J.; Wang, Q.; Lin, Z. From signal amplification to restrained background: Magnetic graphene oxide assisted homogeneous electrochemiluminescence aptasensor for highly sensitive detection of okadaic acid. *Sens. Actuators B Chem.* **2021**, *327*, 128872. [CrossRef]
20. Gao, W.; Chen, Y.; Xi, J.; Zhang, A.; Chen, Y.; Lu, F.; Chen, Z. A novel electrochemiluminescence sensor based on $Ru(bpy)_3^{2+}$ immobilized by graphene on glassy carbon electrode surface via in situ wet-chemical reaction. *Sens. Actuators B Chem.* **2012**, *171–172*, 1159–1165. [CrossRef]
21. Yao, B.; Zhang, J.; Fan, Z.; Ding, Y.; Zhou, B.; Yang, R.; Zhao, J.; Zhang, K. Rational Engineering of the DNA Walker Amplification Strategy by Using a $Au@Ti_3C_2@PEI\text{-}Ru(dcbpy)_3^{2+}$ Nanocomposite Biosensor for Detection of the SARS-CoV-2 RdRp Gene. *ACS Appl. Mater. Interfaces* **2021**, *13*, 19816–19824. [CrossRef]
22. Xing, H.; Xia, H.; Fan, Y.; Xue, Y.; Peng, C.; Ren, J.; Li, J.; Wang, E. A Solid-State Electrochemiluminescence Sensor Based on Novel Two-Dimensional Ti_3C_2 MXene. *ChemElectroChem* **2021**, *8*, 1858–1863. [CrossRef]
23. Fang, Y.; Yang, X.; Chen, T.; Xu, G.; Liu, M.; Liu, J.; Xu, Y. Two-dimensional titanium carbide (MXene)-based solid-state electrochemiluminescent sensor for label-free single-nucleotide mismatch discrimination in human urine. *Sens. Actuators B Chem.* **2018**, *263*, 400–407. [CrossRef]
24. Wang, Y.; Shan, D.; Wu, G.; Wang, H.; Ru, F.; Zhang, X.; Li, L.; Qian, Y.; Lu, X. A novel "dual-potential" ratiometric electrochemiluminescence DNA sensor based on enhancing and quenching effect by G-quadruplex/hemin and Au-Luminol bifunctional nanoparticles. *Biosens. Bioelectron.* **2018**, *106*, 64–70. [CrossRef] [PubMed]
25. Gao, X.; Ren, X.; Ai, Y.; Li, M.; Zhang, B.; Zou, G. Luminophore-Surface-Engineering-Enabled Low-Triggering-Potential and Coreactant-Free Electrochemiluminescence for Protein Determination. *Anal. Chem.* **2023**, *95*, 6948–6954. [CrossRef] [PubMed]
26. He, Y.; Du, J.; Luo, J.; Chen, S.; Yuan, R. Coreactant-free electrochemiluminescence biosensor for the determination of organophosphorus pesticides. *Biosens. Bioelectron.* **2020**, *150*, 111898. [CrossRef]
27. Xie, J.; Yang, G.; Tan, X.; Yuan, R.; Chen, S. Coreactant-free electrochemiluminescence of polyfluorene nanoparticle coupling double quencher for β-amyloid$_{1\text{-}42}$ detection. *Talanta* **2023**, *258*, 124398. [CrossRef]
28. Song, X.; Zhao, L.; Luo, C.; Ren, X.; Wang, X.; Yang, L.; Wei, Q. Bioactivity-protective electrochemiluminescence sensor using CeO_2/Co_4N heterostructures as highly effective coreaction accelerators for ultrasensitive immunodetection. *Sens. Actuators B Chem.* **2022**, *355*, 131158. [CrossRef]
29. Wu, F.-F.; Zhou, Y.; Zhang, H.; Yuan, R.; Chai, Y.-Q. Electrochemiluminescence Peptide-Based Biosensor with Hetero-Nanostructures as Coreaction Accelerator for the Ultrasensitive Determination of Tryptase. *Anal. Chem.* **2018**, *90*, 2263–2270. [CrossRef]
30. Yang, G.; Zhang, Y.; Zhao, J.; He, Y.; Yuan, R.; Chen, S. Dual-emitting Iridium nanorods combining dual-regulating coreaction accelerator Ag nanoparticles for electrochemiluminescence ratio determination of amyloid-β oligomers. *Biosens. Bioelectron.* **2022**, *216*, 114629. [CrossRef]
31. Ma, C.; Cao, Y.; Gou, X.; Zhu, J.-J. Recent Progress in Electrochemiluminescence Sensing and Imaging. *Anal. Chem.* **2020**, *92*, 431–454. [CrossRef]
32. Husain, R.A.; Barman, S.R.; Chatterjee, S.; Khan, I.; Lin, Z.-H. Enhanced biosensing strategies using electrogenerated chemiluminescence: Recent progress and future prospects. *J. Mater. Chem. B* **2020**, *8*, 3192–3212. [CrossRef] [PubMed]
33. Qi, H.; Zhang, C. Electrogenerated Chemiluminescence Biosensing. *Anal. Chem.* **2020**, *92*, 524–534. [CrossRef] [PubMed]
34. Perera, G.S.; Ahmed, T.; Heiss, L.; Walia, S.; Bhaskaran, M.; Sriram, S. Rapid and Selective Biomarker Detection with Conductometric Sensors. *Small* **2021**, *17*, 2005582. [CrossRef]
35. Liu, R.; Ye, X.; Cui, T. Recent Progress of Biomarker Detection Sensors. *Research* **2020**, *2020*, 7949037. [CrossRef] [PubMed]
36. Shang, L.; Wang, X.; Zhang, W.; Jia, L.-P.; Ma, R.-N.; Jia, W.-L.; Wang, H.-S. A dual-potential electrochemiluminescence sensor for ratiometric detection of carcinoembryonic antigen based on single luminophor. *Sens. Actuators B Chem.* **2020**, *325*, 128776. [CrossRef]

37. Luo, W.; Ye, Z.; Ma, P.; Wu, Q.; Song, D. Preparation of a disposable electrochemiluminescence sensor chip based on an MXene-loaded ruthenium luminescent agent and its application in the detection of carcinoembryonic antigens. *Analyst* **2022**, *147*, 1986–1994. [CrossRef]
38. Shang, L.; Shi, B.-J.; Zhang, W.; Jia, L.-P.; Ma, R.-N.; Xue, Q.-W.; Wang, H.-S. Ratiometric Electrochemiluminescence Sensing of Carcinoembryonic Antigen Based on Luminol. *Anal. Chem.* **2022**, *94*, 12845–12851. [CrossRef]
39. Huang, B.; Liu, X.-P.; Chen, J.-S.; Mao, C.; Niu, H.-L.; Jin, B.-K. Electrochemiluminescence immunoassay for the prostate-specific antigen by using a CdS/chitosan/g-C_3N_4 nanocomposite. *Microchim. Acta* **2020**, *187*, 155. [CrossRef]
40. Wang, Y.; Feng, D.; Kan, X. The combination of highly efficient resonance energy transfer in one nanocomposite and ferrocene-quenching for ultrasensitive electrochemiluminescence bioanalysis. *Biosens. Bioelectron.* **2022**, *210*, 114347. [CrossRef]
41. Zheng, L.; Guo, Q.; Yang, C.; Wang, J.; Xu, X.; Nie, G. Electrochemiluminescence and photoelectrochemistry dual-signal immunosensor based on $Ru(bpy)_3{}^{2+}$-functionalized MOF for prostate-specific antigen sensitive detection. *Sens. Actuators B Chem.* **2023**, *379*, 133269. [CrossRef]
42. Chen, H.; Huang, J.; Zhang, R.; Yan, F. Dual-mode electrochemiluminescence and electrochemical sensor for alpha-fetoprotein detection in human serum based on vertically ordered mesoporous silica films. *Front. Chem.* **2022**, *10*, 1023998. [CrossRef] [PubMed]
43. Qin, D.; Jiang, X.; Mo, G.; Feng, J.; Deng, B. Boron nitride quantum dots as electrochemiluminescence coreactants of rGO@Au@Ru–SiO_2 for label-free detection of AFP in human serum. *Electrochim. Acta* **2020**, *335*, 135621. [CrossRef]
44. Liang, Z.; Liu, Y.; Zhang, Q.; Guo, Y.; Ma, Q. The high luminescent polydopamine nanosphere-based ECL biosensor with steric effect for MUC1 detection. *Chem. Eng. J.* **2020**, *385*, 123825. [CrossRef]
45. Huang, W.; Hu, G.-B.; Yao, L.-Y.; Yang, Y.; Liang, W.-B.; Yuan, R.; Xiao, D.-R. Matrix Coordination-Induced Electrochemiluminescence Enhancement of Tetraphenylethylene-Based Hafnium Metal–Organic Framework: An Electrochemiluminescence Chromophore for Ultrasensitive Electrochemiluminescence Sensor Construction. *Anal. Chem.* **2020**, *92*, 3380–3387. [CrossRef]
46. Huang, W.; Wang, Y.; Liang, W.-B.; Hu, G.-B.; Yao, L.-Y.; Yang, Y.; Zhou, K.; Yuan, R.; Xiao, D.-R. Two Birds with One Stone: Surface Functionalization and Delamination of Multilayered $Ti_3C_2T_x$ MXene by Grafting a Ruthenium(II) Complex to Achieve Conductivity-Enhanced Electrochemiluminescence. *Anal. Chem.* **2021**, *93*, 1834–1841. [CrossRef]
47. Zhao, L.; Wang, B.; Wang, C.; Fan, D.; Liu, X.; Wei, Q.; Ju, H.; Wu, D. Dual-strategy ECL biosensor based on rare Eu(II,III)-MOF as probe with antenna effect and sensitization for CYFRA 21-1 trace analysis. *Sens. Actuators B Chem.* **2023**, *377*, 133101. [CrossRef]
48. Jian, L.; Wang, X.; Hao, L.; Liu, Y.; Yang, H.; Zheng, X.; Feng, W. Electrochemiluminescence immunosensor for cytokeratin fragment antigen 21-1 detection using electrochemically mediated atom transfer radical polymerization. *Microchim. Acta* **2021**, *188*, 115. [CrossRef]
49. Xue, J.; Yang, L.; Du, Y.; Ren, Y.; Ren, X.; Ma, H.; Wu, D.; Ju, H.; Li, Y.; Wei, Q. Electrochemiluminescence sensing platform based on functionalized poly-(styrene-co-maleicanhydride) nanocrystals and iron doped hydroxyapatite for CYFRA 21-1 immunoassay. *Sens. Actuators B Chem.* **2020**, *321*, 128454. [CrossRef]
50. Li, J.; Yang, H.; Cai, R.; Tan, W. Ultrahighly Sensitive Sandwich-Type Electrochemical Immunosensor for Selective Detection of Tumor Biomarkers. *ACS Appl. Mater. Interfaces* **2022**, *14*, 44222–44227. [CrossRef]
51. Li, C.; Yang, J.; Xu, R.; Wang, H.; Zhang, Y.; Wei, Q. Progress and Prospects of Electrochemiluminescence Biosensors Based on Porous Nanomaterials. *Biosensors* **2022**, *12*, 508. [CrossRef] [PubMed]
52. Chen, X.; Liu, Y.; Wang, B.; Liu, X.; Lu, C. Understanding role of microstructures of nanomaterials in electrochemiluminescence properties and their applications. *TrAC Trends Anal. Chem.* **2023**, *162*, 117030. [CrossRef]
53. Padmakumari Kurup, C.; Abdullah Lim, S.; Ahmed, M.U. Nanomaterials as signal amplification elements in aptamer-based electrochemiluminescent biosensors. *Bioelectrochemistry* **2022**, *147*, 108170. [CrossRef]
54. Zou, G.; Ju, H. Electrogenerated Chemiluminescence from a CdSe Nanocrystal Film and Its Sensing Application in Aqueous Solution. *Anal. Chem.* **2004**, *76*, 6871–6876. [CrossRef]
55. Zhou, J.; Yang, Y.; Zhang, C. Toward Biocompatible Semiconductor Quantum Dots: From Biosynthesis and Bioconjugation to Biomedical Application. *Chem. Rev.* **2015**, *115*, 11669–11717. [CrossRef]
56. Zhang, X.; Zhang, B.; Miao, W.; Zou, G. Molecular-Counting-Free and Electrochemiluminescent Single-Molecule Immunoassay with Dual-Stabilizers-Capped CdSe Nanocrystals as Labels. *Anal. Chem.* **2016**, *88*, 5482–5488. [CrossRef]
57. Liu, S.; Zhang, X.; Yu, Y.; Zou, G. A Monochromatic Electrochemiluminescence Sensing Strategy for Dopamine with Dual-Stabilizers-Capped CdSe Quantum Dots as Emitters. *Anal. Chem.* **2014**, *86*, 2784–2788. [CrossRef]
58. Lin, D.; Wu, J.; Yan, F.; Deng, S.; Ju, H. Ultrasensitive Immunoassay of Protein Biomarker Based on Electrochemiluminescent Quenching of Quantum Dots by Hemin Bio-Bar-Coded Nanoparticle Tags. *Anal. Chem.* **2011**, *83*, 5214–5221. [CrossRef]
59. Fan, Y.; Liu, Z.; Wang, J.; Cui, C.; Hu, L. An "off–on" electrochemiluminescence aptasensor for determination of lincomycin based on CdS QDs/carboxylated g-C3N4. *Microchim. Acta* **2023**, *190*, 11. [CrossRef]
60. Chen, P.; Liu, Z.; Liu, J.; Liu, H.; Bian, W.; Tian, D.; Xia, F.; Zhou, C. A novel electrochemiluminescence aptasensor based CdTe QDs@NH_2-MIL-88(Fe) for signal amplification. *Electrochimica Acta* **2020**, *354*, 136644. [CrossRef]
61. Zhu, H.-Y.; Ding, S.-N. Dual-signal-amplified electrochemiluminescence biosensor for microRNA detection by coupling cyclic enzyme with CdTe QDs aggregate as luminophor. *Biosens. Bioelectron.* **2019**, *134*, 109–116. [CrossRef] [PubMed]
62. Zhao, J.; He, Y.; Tan, K.; Yang, J.; Chen, S.; Yuan, R. Novel Ratiometric Electrochemiluminescence Biosensor Based on BP-CdTe QDs with Dual Emission for Detecting MicroRNA-126. *Anal. Chem.* **2021**, *93*, 12400–12408. [CrossRef] [PubMed]

63. Wang, J.; Han, H.; Jiang, X.; Huang, L.; Chen, L.; Li, N. Quantum Dot-Based Near-Infrared Electrochemiluminescent Immunosensor with Gold Nanoparticle-Graphene Nanosheet Hybrids and Silica Nanospheres Double-Assisted Signal Amplification. *Anal. Chem.* **2012**, *84*, 4893–4899. [CrossRef] [PubMed]
64. Li, Z.; Lin, Z.; Wu, X.; Chen, H.; Chai, Y.; Yuan, R. Highly Efficient Electrochemiluminescence Resonance Energy Transfer System in One Nanostructure: Its Application for Ultrasensitive Detection of MicroRNA in Cancer Cells. *Anal. Chem.* **2017**, *89*, 6029–6035. [CrossRef] [PubMed]
65. Liu, Y.; Wang, F.; Ge, S.; Zhang, L.; Zhang, Z.; Liu, Y.; Zhang, Y.; Ge, S.; Yu, J. Programmable T-Junction Structure-Assisted CRISPR/Cas12a Electrochemiluminescence Biosensor for Detection of Sa-16S rDNA. *ACS Appl. Mater. Interfaces* **2023**, *15*, 617–625. [CrossRef]
66. Yang, E.; Zhang, Y.; Shen, Y. Quantum dots for electrochemiluminescence bioanalysis—A review. *Anal. Chim. Acta* **2022**, *1209*, 339140. [CrossRef]
67. Pan, D.; Chen, K.; Zhou, Q.; Zhao, J.; Xue, H.; Zhang, Y.; Shen, Y. Engineering of CdTe/SiO_2 nanocomposites: Enhanced signal amplification and biocompatibility for electrochemiluminescent immunoassay of alpha-fetoprotein. *Biosens. Bioelectron.* **2019**, *131*, 178–184. [CrossRef]
68. Shen, C.; Li, Y.; Li, Y.; Wang, S.; Li, Y.; Tang, F.; Wang, P.; Liu, H.; Li, Y.; Liu, Q. A double reaction system induced electrochemiluminescence enhancement based on SnS_2 QDs@MIL-101 for ultrasensitive detection of CA242. *Talanta* **2022**, *247*, 123575. [CrossRef]
69. Lei, Y.-M.; Zhou, J.; Chai, Y.-Q.; Zhuo, Y.; Yuan, R. SnS_2 Quantum Dots as New Emitters with Strong Electrochemiluminescence for Ultrasensitive Antibody Detection. *Anal. Chem.* **2018**, *90*, 12270–12277. [CrossRef]
70. Zhao, M.; Chen, A.-Y.; Huang, D.; Chai, Y.-Q.; Zhuo, Y.; Yuan, R. MoS_2 Quantum Dots as New Electrochemiluminescence Emitters for Ultrasensitive Bioanalysis of Lipopolysaccharide. *Anal. Chem.* **2017**, *89*, 8335–8342. [CrossRef]
71. Hua, Q.; Tang, F.; Wang, X.; Li, M.; Gu, X.; Sun, W.; Luan, F.; Tian, C.; Zhuang, X. Electrochemiluminescence sensor based on EuS nanocrystals for ultrasensitive detection of mercury ions in seafood. *Sens. Actuators B Chem.* **2022**, *352*, 131075. [CrossRef]
72. Jiang, D.; Wei, M.; Du, X.; Qin, M.; Shan, X.; Wang, W.; Chen, Z. Ultrasensitive near-infrared aptasensor for enrofloxacin detection based on wavelength tunable AgBr nanocrystals electrochemiluminescence emission triggered by O-terminated Ti_3C_2 MXene. *Biosens. Bioelectron.* **2022**, *200*, 113917. [CrossRef] [PubMed]
73. Lei, Y.-M.; Zhuo, Y.; Guo, M.-L.; Chai, Y.-Q.; Yuan, R. Pore Confinement-Enhanced Electrochemiluminescence on SnO_2 Nanocrystal Xerogel with NO_3^- As Co-Reactant and Its Application in Facile and Sensitive Bioanalysis. *Anal. Chem.* **2020**, *92*, 2839–2846. [CrossRef] [PubMed]
74. Yu, S.; Du, Y.; Niu, X.; Li, G.; Zhu, D.; Yu, Q.; Zou, G.; Ju, H. Arginine-modified black phosphorus quantum dots with dual excited states for enhanced electrochemiluminescence in bioanalysis. *Nat. Commun.* **2022**, *13*, 7302. [CrossRef]
75. Chen, J.; Wang, Q.; Liu, X.; Chen, X.; Wang, L.; Yang, W. Black phosphorus quantum dots as novel electrogenerated chemiluminescence emitters for the detection of Cu^{2+}. *Chem. Commun.* **2020**, *56*, 4680–4683. [CrossRef]
76. Li, P.; Yu, J.; Zhao, K.; Deng, A.; Li, J. Efficient enhancement of electrochemiluminescence from tin disulfide quantum dots by hollow titanium dioxide spherical shell for highly sensitive detection of chloramphenicol. *Biosens. Bioelectron.* **2020**, *147*, 111790. [CrossRef]
77. Fu, L.; Fu, K.; Gao, X.; Dong, S.; Zhang, B.; Fu, S.; Hsu, H.-Y.; Zou, G. Enhanced Near-Infrared Electrochemiluminescence from Trinary Ag–In–S to Multinary Ag–Ga–In–S Nanocrystals via Doping-in-Growth and Its Immunosensing Applications. *Anal. Chem.* **2021**, *93*, 2160–2165. [CrossRef]
78. Long, X.; Zhang, F.; He, Y.; Hou, S.; Zhang, B.; Zou, G. Promising Anodic Electrochemiluminescence of Nontoxic Core/Shell $CuInS_2$/ZnS Nanocrystals in Aqueous Medium and Its Biosensing Potential. *Anal. Chem.* **2018**, *90*, 3563–3569. [CrossRef]
79. Wang, C.; Liu, L.; Liu, X.; Chen, Y.; Wang, X.; Fan, D.; Kuang, X.; Sun, X.; Wei, Q.; Ju, H. Highly-sensitive electrochemiluminescence biosensor for NT-proBNP using MoS_2@Cu_2S as signal-enhancer and multinary nanocrystals loaded in mesoporous UiO-66-NH_2 as novel luminophore. *Sens. Actuators B Chem.* **2020**, *307*, 127619. [CrossRef]
80. Jiang, X.; Wang, H.; Shen, Y.; Hu, N.; Shi, W. Nitrogen-doped Ti_3C_2 MXene quantum dots as novel high-efficiency electrochemiluminescent emitters for sensitive mucin 1 detection. *Sens. Actuators B Chem.* **2022**, *350*, 130891. [CrossRef]
81. Chen, A.; Liang, W.; Wang, H.; Zhuo, Y.; Chai, Y.; Yuan, R. Anodic Electrochemiluminescence of Carbon Dots Promoted by Nitrogen Doping and Application to Rapid Cancer Cell Detection. *Anal. Chem.* **2020**, *92*, 1379–1385. [CrossRef]
82. Chen, Y.; Lin, J.; Zhang, R.; He, S.; Ding, Z.; Ding, L. Electrochemiluminescence of water-dispersed nitrogen and sulfur doped carbon dots synthesized from amino acids. *Analyst* **2021**, *146*, 5287–5293. [CrossRef]
83. Liu, J.; Zhang, Y.; Yuan, R.; Chai, Y. Fluorine-nitrogen co-doped carbon dots with stable and strong electrochemiluminescence as an emitter for ultrasensitive detection of HIV-DNA fragment. *Sens. Actuators B Chem.* **2023**, *379*, 133260. [CrossRef]
84. Yang, E.; Ning, Z.; Yin, F.; Fang, Z.; Chen, M.; Zhang, M.; Xu, W.; Zhang, Y.; Shen, Y. Surface plasmon-enhanced electrochemiluminescence of P, N-doped carbon dots for ultrasensitive detection of BRAF gene. *Sens. Actuators B Chem.* **2022**, *369*, 132288. [CrossRef]
85. Chen, S.; Jiang, L.; Yin, B.; Du, W.; Wang, X.; Liu, Y.; Chen, S.; Zhu, M. Bright Near-Infrared Circularly Polarized Electrochemiluminescence from Au_9Ag_4 Nanoclusters. *Chem. Sci.* **2023**, *14*, 7304–7309. [CrossRef]
86. Cao, Y.; Zhou, J.-L.; Ma, Y.; Zhou, Y.; Zhu, J.-J. Recent progress of metal nanoclusters in electrochemiluminescence. *Dalton Trans.* **2022**, *51*, 8927–8937. [CrossRef] [PubMed]

87. Han, S.; Zhao, Y.; Zhang, Z.; Xu, G. Recent Advances in Electrochemiluminescence and Chemiluminescence of Metal Nanoclusters. *Molecules* **2020**, *25*, 5208. [CrossRef] [PubMed]
88. Díez, I.; Pusa, M.; Kulmala, S.; Jiang, H.; Walther, A.; Goldmann, A.S.; Müller, A.H.E.; Ikkala, O.; Ras, R.H.A. Color Tunability and Electrochemiluminescence of Silver Nanoclusters. *Angew. Chem. Int. Ed.* **2009**, *48*, 2122–2125. [CrossRef]
89. Wang, T.; Wang, D.; Padelford, J.W.; Jiang, J.; Wang, G. Near-Infrared Electrogenerated Chemiluminescence from Aqueous Soluble Lipoic Acid Au Nanoclusters. *J. Am. Chem. Soc.* **2016**, *138*, 6380–6383. [CrossRef]
90. Li, L.; Liu, H.; Shen, Y.; Zhang, J.; Zhu, J.-J. Electrogenerated Chemiluminescence of Au Nanoclusters for the Detection of Dopamine. *Anal. Chem.* **2011**, *83*, 661–665. [CrossRef]
91. Fang, Y.-M.; Song, J.; Li, J.; Wang, Y.-W.; Yang, H.-H.; Sun, J.-J.; Chen, G.-N. Electrogenerated chemiluminescence from Au nanoclusters. *Chem. Commun.* **2011**, *47*, 2369–2371. [CrossRef] [PubMed]
92. Hesari, M.; Ding, Z. Efficient Near-Infrared Electrochemiluminescence from Au_{18} Nanoclusters. *Chem. Eur. J.* **2021**, *27*, 14821–14825. [CrossRef] [PubMed]
93. Hesari, M.; Ding, Z. Identifying Highly Photoelectrochemical Active Sites of Two Au_{21} Nanocluster Isomers toward Bright Near-Infrared Electrochemiluminescence. *J. Am. Chem. Soc.* **2021**, *143*, 19474–19485. [CrossRef] [PubMed]
94. Hesari, M.; Ma, H.; Ding, Z. Monitoring single Au_{38} nanocluster reactions via electrochemiluminescence. *Chem. Sci.* **2021**, *12*, 14540–14545. [CrossRef] [PubMed]
95. Li, D.; Sun, X.; Shao, N.; Zhang, G.; Li, S.; Zhou, H.; Wu, J.; Tian, Y. Crystal structure, optical properties and electrochemiluminescence of Cu(I), Ag(I) and Au(I) complexes that contain the cyanoacetic acid triphenylamine ligand. *Polyhedron* **2015**, *93*, 17–22. [CrossRef]
96. Zhu, M.-J.; Pan, J.-B.; Wu, Z.-Q.; Gao, X.-Y.; Zhao, W.; Xia, X.-H.; Xu, J.-J.; Chen, H.-Y. Electrogenerated Chemiluminescence Imaging of Electrocatalysis at a Single Au-Pt Janus Nanoparticle. *Angew. Chem. Int. Ed.* **2018**, *57*, 4010–4014. [CrossRef]
97. Qian, H.; Jiang, D.; Li, G.; Gayathri, C.; Das, A.; Gil, R.R.; Jin, R. Monoplatinum Doping of Gold Nanoclusters and Catalytic Application. *J. Am. Chem. Soc.* **2012**, *134*, 16159–16162. [CrossRef]
98. Chen, S.; Ma, H.; Padelford, J.W.; Qinchen, W.; Yu, W.; Wang, S.; Zhu, M.; Wang, G. Near Infrared Electrochemiluminescence of Rod-Shape 25-Atom AuAg Nanoclusters That Is Hundreds-Fold Stronger Than That of $Ru(bpy)_3$ Standard. *J. Am. Chem. Soc.* **2019**, *141*, 9603–9609. [CrossRef]
99. Tang, Y.; Xu, J.; Xiong, C.; Xiao, Y.; Zhang, X.; Wang, S. Enhanced electrochemiluminescence of gold nanoclusters via silver doping and their application for ultrasensitive detection of dopamine. *Analyst* **2019**, *144*, 2643–2648 [CrossRef]
100. Guo, Y.; Tian, L.; Wu, J.; Wu, Y.; Liu, Y.; Du, J.; Lu, X. A Highly Sensitive Electrochemiluminescence Spermine Biosensor Based on Au−Ag Bimetallic Nanoclusters. *Electroanalysis* **2021**, *33*, 2016–2024. [CrossRef]
101. Nie, Y.; Tao, X.; Zhang, H.; Chai, Y.; Yuan, R. Self-Assembly of Gold Nanoclusters into a Metal–Organic Framework with Efficient Electrochemiluminescence and Their Application for Sensitive Detection of Rutin. *Anal. Chem.* **2021**, *93*, 3445–3451. [CrossRef] [PubMed]
102. Luo, S.; Xiao, H.; Yang, S.; Liu, C.; Liang, J.; Tang, Y. Ultrasensitive detection of pentachlorophenol based on enhanced electrochemiluminescence of Au nanoclusters/graphene hybrids. *Sens. Actuators B Chem.* **2014**, *194*, 325–331. [CrossRef]
103. Yang, Y.; Wu, W.; Wang, Q.; Xiao, H.; Kuang, Y.; Liu, C. Novel anodic electrochemiluminescence system of Pt nanocluster/graphene hybrids for ultrasensitive detection of Cu^{2+}. *J. Electroanal. Chem.* **2016**, *772*, 73–79. [CrossRef]
104. Guo, C.; Duan, F.; Zhang, S.; He, L.; Wang, M.; Chen, J.; Zhang, J.; Jia, Q.; Zhang, Z.; Du, M. Heterostructured hybrids of metal–organic frameworks (MOFs) and covalent–organic frameworks (COFs). *J. Mater. Chem. A* **2022**, *10*, 475–507. [CrossRef]
105. Wang, Z.; Liu, L.; Li, Z.; Goyal, N.; Du, T.; He, J.; Li, G.K. Shaping of Metal–Organic Frameworks: A Review. *Energy Fuels* **2022**, *36*, 2927–2944. [CrossRef]
106. Hwang, J.; Ejsmont, A.; Freund, R.; Goscianska, J.; Schmidt, B.V.K.J.; Wuttke, S. Controlling the morphology of metal–organic frameworks and porous carbon materials: Metal oxides as primary architecture-directing agents. *Chem. Soc. Rev.* **2020**, *49*, 3348–3422. [CrossRef]
107. Wang, Y.; Zhao, G.; Chi, H.; Yang, S.; Niu, Q.; Wu, D.; Cao, W.; Li, T.; Ma, H.; Wei, Q. Self-Luminescent Lanthanide Metal–Organic Frameworks as Signal Probes in Electrochemiluminescence Immunoassay. *J. Am. Chem. Soc* **2021**, *143*, 504–512. [CrossRef]
108. Dong, X.; Zhao, G.; Li, X.; Fang, J.; Miao, J.; Wei, Q.; Cao, W. Electrochemiluminescence immunosensor of "signal-off" for β-amyloid detection based on dual metal-organic frameworks. *Talanta* **2020**, *208*, 120376. [CrossRef]
109. Huang, W.; Hu, G.-B.; Liang, W.-B.; Wang, J.-M.; Lu, M.-L.; Yuan, R.; Xiao, D.-R. Ruthenium(II) Complex-Grafted Hollow Hierarchical Metal–Organic Frameworks with Superior Electrochemiluminescence Performance for Sensitive Assay of Thrombin. *Anal. Chem.* **2021**, *93*, 6239–6245. [CrossRef]
110. Zhu, X.; Xing, H.; Xue, Y.; Li, J.; Wang, E.; Dong, S. Atom-Anchoring Strategy with Metal–Organic Frameworks for Highly Efficient Solid-State Electrochemiluminescence. *Anal. Chem.* **2021**, *93*, 9628–9633. [CrossRef]
111. Zhao, G.; Wang, Y.; Li, X.; Dong, X.; Wang, H.; Du, B.; Cao, W.; Wei, Q. Quenching Electrochemiluminescence Immunosensor Based on Resonance Energy Transfer between Ruthenium (II) Complex Incorporated in the UiO-67 Metal–Organic Framework and Gold Nanoparticles for Insulin Detection. *ACS Appl. Mater. Interfaces* **2018**, *10*, 22932–22938. [CrossRef] [PubMed]
112. Dong, X.; Zhao, G.; Liu, L.; Li, X.; Wei, Q.; Cao, W. Ultrasensitive competitive method-based electrochemiluminescence immunosensor for diethylstilbestrol detection based on $Ru(bpy)_3^{2+}$ as luminophor encapsulated in metal–organic frameworks UiO-67. *Biosens. Bioelectron.* **2018**, *110*, 201–206. [CrossRef] [PubMed]

113. Wang, C.; Zhang, N.; Wei, D.; Feng, R.; Fan, D.; Hu, L.; Wei, Q.; Ju, H. Double electrochemiluminescence quenching effects of Fe_3O_4@PDA-Cu_XO towards self-enhanced $Ru(bpy)_3{}^{2+}$ functionalized MOFs with hollow structure and its application to procalcitonin immunosensing. *Biosens. Bioelectron.* **2019**, *142*, 111521. [CrossRef]
114. Dong, X.; Du, Y.; Zhao, G.; Cao, W.; Fan, D.; Kuang, X.; Wei, Q.; Ju, H. Dual-signal electrochemiluminescence immunosensor for Neuron-specific enolase detection based on "dual-potential" emitter $Ru(bpy)_3{}^{2+}$ functionalized zinc-based metal-organic frameworks. *Biosens. Bioelectron.* **2021**, *192*, 113505. [CrossRef] [PubMed]
115. Li, P.; Luo, L.; Cheng, D.; Sun, Y.; Zhang, Y.; Liu, M.; Yao, S. Regulation of the Structure of Zirconium-Based Porphyrinic Metal–Organic Framework as Highly Electrochemiluminescence Sensing Platform for Thrombin. *Anal. Chem.* **2022**, *94*, 5707–5714. [CrossRef] [PubMed]
116. Hu, G.-B.; Xiong, C.-Y.; Liang, W.-B.; Zeng, X.-S.; Xu, H.-L.; Yang, Y.; Yao, L.-Y.; Yuan, R.; Xiao, D.-R. Highly Stable Mesoporous Luminescence-Functionalized MOF with Excellent Electrochemiluminescence Property for Ultrasensitive Immunosensor Construction. *ACS Appl. Mater. Interfaces* **2018**, *10*, 15913–15919. [CrossRef]
117. Wang, Z.; Jiang, X.; Yuan, R.; Chai, Y. *N*-(aminobutyl)-*N*-(ethylisoluminol) functionalized Fe-based metal-organic frameworks with intrinsic mimic peroxidase activity for sensitive electrochemiluminescence mucin1 determination. *Biosens. Bioelectron.* **2018**, *121*, 250–256. [CrossRef]
118. Wang, C.; Zhang, N.; Li, Y.; Yang, L.; Wei, D.; Yan, T.; Ju, H.; Du, B.; Wei, Q. Cobalt-based metal-organic frameworks as co-reaction accelerator for enhancing electrochemiluminescence behavior of *N*-(aminobutyl)-*N*-(ethylisoluminol) and ultrasensitive immunosensing of amyloid-β protein. *Sens. Actuators B Chem.* **2019**, *291*, 319–328. [CrossRef]
119. Wang, S.; Zhao, Y.; Wang, M.; Li, H.; Saqib, M.; Ge, C.; Zhang, X.; Jin, Y. Enhancing Luminol Electrochemiluminescence by Combined Use of Cobalt-Based Metal Organic Frameworks and Silver Nanoparticles and Its Application in Ultrasensitive Detection of Cardiac Troponin I. *Anal. Chem.* **2019**, *91*, 3048–3054. [CrossRef]
120. Zhu, D.; Zhang, Y.; Bao, S.; Wang, N.; Yu, S.; Luo, R.; Ma, J.; Ju, H.; Lei, J. Dual Intrareticular Oxidation of Mixed-Ligand Metal–Organic Frameworks for Stepwise Electrochemiluminescence. *J. Am. Chem. Soc.* **2021**, *143*, 3049–3053. [CrossRef]
121. Wang, J.-M.; Yao, L.-Y.; Huang, W.; Yang, Y.; Liang, W.-B.; Yuan, R.; Xiao, D.-R. Overcoming Aggregation-Induced Quenching by Metal−Organic Framework for Electrochemiluminescence (ECL) Enhancement: Zn-PTC as a New ECL Emitter for Ultrasensitive MicroRNAs Detection. *ACS Appl. Mater. Interfaces* **2021**, *13*, 44079–44085. [CrossRef] [PubMed]
122. Wu, J.; Wang, A.; Liu, P.; Hou, Y.; Song, L.; Yuan, R.; Fu, Y. Sulfur-functionalized zirconium(IV)-based metal-organic frameworks relieves aggregation-caused quenching effect in efficient electrochemiluminescence sensor. *Sens. Actuators B Chem.* **2020**, *321*, 128531. [CrossRef]
123. Wang, Y.; Shu, J.; Lyu, A.; Wang, M.; Hu, C.; Cui, H. Zn^{2+}-Modified Nonmetal Porphyrin-Based Metal–Organic Frameworks with Improved Electrochemiluminescence for Nanoscale Exosome Detection. *ACS Appl. Nano Mater.* **2023**, *6*, 4214–4223. [CrossRef]
124. Han, Q.; Wang, C.; Liu, P.; Zhang, G.; Song, L.; Fu, Y. Achieving synergistically enhanced dual-mode electrochemiluminescent and electrochemical drug sensors via a multi-effect porphyrin-based metal-organic framework. *Sens. Actuators B Chem.* **2021**, *330*, 129388. [CrossRef]
125. Li, J.; Jia, H.; Ren, X.; Li, Y.; Liu, L.; Feng, R.; Ma, H.; Wei, Q. Dumbbell Plate-Shaped AIEgen-Based Luminescent MOF with High Quantum Yield as Self-Enhanced ECL Tags: Mechanism Insights and Biosensing Application. *Small* **2022**, *18*, 2106567. [CrossRef]
126. Xiong, X.; Xiong, C.; Gao, Y.; Xiao, Y.; Chen, M.-M.; Wen, W.; Zhang, X.; Wang, S. Tetraphenylethylene-Functionalized Metal–Organic Frameworks with Strong Aggregation-Induced Electrochemiluminescence for Ultrasensitive Analysis through a Multiple Convertible Resonance Energy Transfer System. *Anal. Chem.* **2022**, *94*, 7861–7867. [CrossRef]
127. Xiao, S.; Wang, X.; Yang, C.; Jiang, Y.; Zhen, S.; Huang, C.; Li, Y. Electrochemiluminescence Resonance Energy Transfer System Based on Silver Metal–Organic Frameworks as a Double-Amplified Emitter for Sensitive Detection of miRNA-107. *Anal. Chem.* **2022**, *94*, 1178–1186. [CrossRef]
128. Richter, M.M.; Bard, A.J. Electrogenerated Chemiluminescence. 58. Ligand-Sensitized Electrogenerated Chemiluminescence in Europium Labels. *Anal. Chem.* **1996**, *68*, 2641–2650. [CrossRef]
129. Dong, H.; Liu, S.; Liu, Q.; Li, Y.; Xu, Z.; Li, Y.; Wei, Q. Mixed-Ligand-Regulated Self-Enhanced Luminous Eu-MOF as an ECL Signal Probe for an Oriented Antibody-Decorated Biosensing Platform. *Anal. Chem.* **2022**, *94*, 12852–12859. [CrossRef]
130. Dai, W.; Wang, X.; Chen, G.; Wang, X.; Hu, C.; Zhen, S.; Huang, C.; Li, Y. Facile synthesis of 2D Europium-metal organic frameworks nanosheets for highly efficient electrochemiluminescence in DNA detection. *Chem. Eng. J.* **2023**, *465*, 143037. [CrossRef]
131. Zhao, L.; Song, X.; Ren, X.; Fan, D.; Wei, Q.; Wu, D. Rare Self-Luminous Mixed-Valence Eu-MOF with a Self-Enhanced Characteristic as a Near-Infrared Fluorescent ECL Probe for Nondestructive Immunodetection. *Anal. Chem.* **2021**, *93*, 8613–8621. [CrossRef]
132. Zhao, L.; Song, X.; Ren, X.; Wang, H.; Fan, D.; Wu, D.; Wei, Q. Ultrasensitive near-infrared electrochemiluminescence biosensor derived from Eu-MOF with antenna effect and high efficiency catalysis of specific CoS2 hollow triple shelled nanoboxes for procalcitonin. *Biosens. Bioelectron.* **2021**, *191*, 113409. [CrossRef] [PubMed]
133. Zhao, L.; Wang, M.; Song, X.; Liu, X.; Ju, H.; Ai, H.; Wei, Q.; Wu, D. Annihilation luminescent Eu-MOF as a near-infrared electrochemiluminescence probe for trace detection of trenbolone. *Chem. Eng. J.* **2022**, *434*, 134691. [CrossRef]

134. Wang, C.; Han, Q.; Liu, P.; Zhang, G.; Song, L.; Zou, X.; Fu, Y. A Superstable Luminescent Lanthanide Metal Organic Gel Utilized in an Electrochemiluminescence Sensor for Epinephrine Detection with a Narrow Potential Sweep Range. *ACS Sens.* **2021**, *6*, 252–258. [CrossRef] [PubMed]
135. Zhou, L.; Yang, L.; Wang, C.; Jia, H.; Xue, J.; Wei, Q.; Ju, H. Copper doped terbium metal organic framework as emitter for sensitive electrochemiluminescence detection of CYFRA 21-1. *Talanta* **2022**, *238*, 123047. [CrossRef]
136. Wang, C.; Li, Z.; Ju, H. Copper-Doped Terbium Luminescent Metal Organic Framework as an Emitter and a Co-reaction Promoter for Amplified Electrochemiluminescence Immunoassay. *Anal. Chem.* **2021**, *93*, 14878–14884. [CrossRef]
137. Gao, H.; Wei, X.; Li, M.; Wang, L.; Wei, T.; Dai, Z. Co-Quenching Effect between Lanthanum Metal–Organic Frameworks Luminophore and Crystal Violet for Enhanced Electrochemiluminescence Gene Detection. *Small* **2021**, *17*, 2103424. [CrossRef]
138. Wang, X.Y.; Xiao, S.Y.; Jiang, Z.W.; Zhen, S.J.; Huang, C.Z.; Liu, Q.Q.; Li, Y.F. An ultrathin 2D Yb(III) metal-organic frameworks with strong electrochemiluminescence as a "on-off-on" platform for detection of picric acid and berberine chloride form. *Talanta* **2021**, *234*, 122625. [CrossRef]
139. Li, C.; Liu, L.; Kang, J.; Xiao, Y.; Feng, Y.; Cao, F.-F.; Zhang, H. Pristine MOF and COF materials for advanced batteries. *Energy Storage Mater.* **2020**, *31*, 115–134. [CrossRef]
140. Ren, X.; Liao, G.; Li, Z.; Qiao, H.; Zhang, Y.; Yu, X.; Wang, B.; Tan, H.; Shi, L.; Qi, X.; et al. Two-dimensional MOF and COF nanosheets for next-generation optoelectronic applications. *Coord. Chem. Rev.* **2021**, *435*, 213781. [CrossRef]
141. Freund, R.; Zaremba, O.; Arnauts, G.; Ameloot, R.; Skorupskii, G.; Dincă, M.; Bavykina, A.; Gascon, J.; Ejsmont, A.; Goscianska, J.; et al. The Current Status of MOF and COF Applications. *Angew. Chem. Int. Ed.* **2021**, *60*, 23975–24001. [CrossRef]
142. Lv, X.; Li, Y.; Cui, B.; Fang, Y.; Wang, L. Electrochemiluminescent sensor based on an aggregation-induced emission probe for bioanalytical detection. *Analyst* **2022**, *147*, 2338–2354. [CrossRef]
143. Zhang, J.-L.; Yang, Y.; Liang, W.-B.; Yao, L.-Y.; Yuan, R.; Xiao, D.-R. Highly Stable Covalent Organic Framework Nanosheets as a New Generation of Electrochemiluminescence Emitters for Ultrasensitive MicroRNA Detection. *Anal. Chem.* **2021**, *93*, 3258–3265. [CrossRef]
144. Luo, R.; Lv, H.; Liao, Q.; Wang, N.; Yang, J.; Li, Y.; Xi, K.; Wu, X.; Ju, H.; Lei, J. Intrareticular charge transfer regulated electrochemiluminescence of donor–acceptor covalent organic frameworks. *Nat. Commun.* **2021**, *12*, 6808. [CrossRef] [PubMed]
145. Li, Y.-J.; Cui, W.-R.; Jiang, Q.-Q.; Wu, Q.; Liang, R.-P.; Luo, Q.-X.; Qiu, J.-D. A general design approach toward covalent organic frameworks for highly efficient electrochemiluminescence. *Nat. Commun.* **2021**, *12*, 4735. [CrossRef] [PubMed]
146. Cui, W.-R.; Li, Y.-J.; Jiang, Q.-Q.; Wu, Q.; Luo, Q.-X.; Zhang, L.; Liang, R.-P.; Qiu, J.-D. Covalent Organic Frameworks as Advanced Uranyl Electrochemiluminescence Monitoring Platforms. *Anal. Chem.* **2021**, *93*, 16149–16157. [CrossRef]
147. Li, Y.-J.; Cui, W.-R.; Jiang, Q.-Q.; Liang, R.-P.; Li, X.-J.; Wu, Q.; Luo, Q.-X.; Liu, J.; Qiu, J.-D. Arousing Electrochemiluminescence Out of Non-Electroluminescent Monomers within Covalent Organic Frameworks. *ACS Appl. Mater. Interfaces* **2021**, *13*, 47921–47931. [CrossRef] [PubMed]
148. Zhang, J.-L.; Yao, L.-Y.; Yang, Y.; Liang, W.-B.; Yuan, R.; Xiao, D.-R. Conductive Covalent Organic Frameworks with Conductivity- and Pre-Reduction-Enhanced Electrochemiluminescence for Ultrasensitive Biosensor Construction. *Anal. Chem.* **2022**, *94*, 3685–3692. [CrossRef]
149. Qin, X.; Zhan, Z.; Ding, Z. Progress in electrochemiluminescence biosensors based on organic framework emitters. *Curr. Opin. Electrochem.* **2023**, *39*, 101283. [CrossRef]
150. Hisaki, I.; Xin, C.; Takahashi, K.; Nakamura, T. Designing Hydrogen-Bonded Organic Frameworks (HOFs) with Permanent Porosity. *Angew. Chem. Int. Ed.* **2019**, *58*, 11160–11170. [CrossRef] [PubMed]
151. Hou, H.; Wang, Y.; Wang, Y.; Luo, R.; Zhu, D.; Zhou, J.; Wu, X.; Ju, H.; Lei, J. Intrareticular electron coupling pathway driven electrochemiluminescence in hydrogen-bonded organic frameworks. *J. Mater. Chem. C* **2022**, *10*, 14488–14495. [CrossRef]
152. Shen, K.-Y.; Zhan, J.; Shen, L.; Xiong, Z.; Zhu, H.-T.; Wang, A.-J.; Yuan, P.-X.; Feng, J.-J. Hydrogen Bond Organic Frameworks as Radical Reactors for Enhancement in ECL Efficiency and Their Ultrasensitive Biosensing. *Anal. Chem.* **2023**, *95*, 4735–4743. [CrossRef] [PubMed]
153. Chang, Y.-L.; Palacios, R.E.; Fan, F.-R.F.; Bard, A.J.; Barbara, P.F. Electrogenerated Chemiluminescence of Single Conjugated Polymer Nanoparticles. *J. Am. Chem. Soc.* **2008**, *130*, 8906–8907. [CrossRef]
154. Bai, X.; Wang, K.; Chen, L.; Zhou, J.; Wang, J. Semiconducting polymer dots as fluorescent probes for in vitro biosensing. *J. Mater. Chem. B* **2022**, *10*, 6248–6262. [CrossRef] [PubMed]
155. Yuan, Y.; Hou, W.; Qin, W.; Wu, C. Recent advances in semiconducting polymer dots as optical probes for biosensing. *Biomater. Sci.* **2021**, *9*, 328–346. [CrossRef] [PubMed]
156. Feng, Y.; Wang, N.; Ju, H. Electrochemiluminescence biosensing and bioimaging with nanomaterials as emitters. *Sci. China Chem.* **2022**, *65*, 2417–2436. [CrossRef]
157. Wang, Z.; Guo, H.; Luo, Z.; Duan, Y.; Feng, Y. Low-Triggering-Potential Electrochemiluminescence from a Luminol Analogue Functionalized Semiconducting Polymer Dots for Imaging Detection of Blood Glucose. *Anal. Chem.* **2022**, *94*, 5615–5623. [CrossRef]
158. Dai, R.; Wu, F.; Xu, H.; Chi, Y. Anodic, Cathodic, and Annihilation Electrochemiluminescence Emissions from Hydrophilic Conjugated Polymer Dots in Aqueous Medium. *ACS Appl. Mater. Interfaces* **2015**, *7*, 15160–15167. [CrossRef]
159. Feng, Y.; Wang, N.; Ju, H. Highly Efficient Electrochemiluminescence of Cyanovinylene-Contained Polymer Dots in Aqueous Medium and Its Application in Imaging Analysis. *Anal. Chem.* **2018**, *90*, 1202–1208. [CrossRef]

160. Wang, N.; Feng, Y.; Wang, Y.; Ju, H.; Yan, F. Electrochemiluminescent Imaging for Multi-immunoassay Sensitized by Dual DNA Amplification of Polymer Dot Signal. *Anal. Chem.* **2018**, *90*, 7708–7714. [CrossRef]
161. Zhao, J.; Luo, J.; Liu, D.; He, Y.; Li, Q.; Chen, S.; Yuan, R. A coreactant-free electrochemiluminescence (ECL) biosensor based on in situ generating quencher for the ultrasensitive detection of microRNA. *Sens. Actuators B Chem.* **2020**, *316*, 128139. [CrossRef]
162. Luo, J.-H.; Li, Q.; Chen, S.-H.; Yuan, R. Coreactant-Free Dual Amplified Electrochemiluminescent Biosensor Based on Conjugated Polymer Dots for the Ultrasensitive Detection of MicroRNA. *ACS Appl. Mater. Interfaces* **2019**, *11*, 27363–27370. [CrossRef]
163. Liu, D.; Zhang, X.; Zhao, J.; Chen, S.; Yuan, R. An ultrasensitive sensing platform for microRNA-155 based on H_2O_2 quenched hydroxide-dependent ECL emission of PFO Pdots. *Biosens. Bioelectron.* **2020**, *150*, 111872. [CrossRef]
164. Yang, H.; Wang, H.; Xiong, C.; Chai, Y.; Yuan, R. Highly sensitive electrochemiluminescence immunosensor based on ABEI/H_2O_2 system with PFO dots as enhancer for detection of kidney injury molecule-1. *Biosens. Bioelectron.* **2018**, *116*, 16–22. [CrossRef]
165. Feng, Y.; Sun, F.; Wang, N.; Lei, J.; Ju, H. $Ru(bpy)_3{}^{2+}$ Incorporated Luminescent Polymer Dots: Double-Enhanced Electrochemiluminescence for Detection of Single-Nucleotide Polymorphism. *Anal. Chem.* **2017**, *89*, 7659–7666. [CrossRef] [PubMed]
166. Wang, N.; Chen, L.; Chen, W.; Ju, H. Potential- and Color-Resolved Electrochemiluminescence of Polymer Dots for Array Imaging of Multiplex MicroRNAs. *Anal. Chem.* **2021**, *93*, 5327–5333. [CrossRef] [PubMed]
167. Wang, N.; Gao, H.; Li, Y.; Li, G.; Chen, W.; Jin, Z.; Lei, J.; Wei, Q.; Ju, H. Dual Intramolecular Electron Transfer for In Situ Coreactant-Embedded Electrochemiluminescence Microimaging of Membrane Protein. *Angew. Chem. Int. Ed.* **2021**, *60*, 197–201. [CrossRef]
168. Wang, Z.; Pan, J.; Li, Q.; Zhou, Y.; Yang, S.; Xu, J.; Hua, D. Improved AIE-Active Probe with High Sensitivity for Accurate Uranyl Ion Monitoring in the Wild Using Portable Electrochemiluminescence System for Environmental Applications. *Adv. Funct. Mater.* **2020**, *30*, 2000220. [CrossRef]
169. Zhang, N.; Gao, H.; Jia, Y.-L.; Pan, J.-B.; Luo, X.-L.; Chen, H.-Y.; Xu, J.-J. Ultrasensitive Nucleic Acid Assay Based on AIE-Active Polymer Dots with Excellent Electrochemiluminescence Stability. *Anal. Chem.* **2021**, *93*, 6857–6864. [CrossRef]
170. Wang, C.; Liu, S.; Ju, H. Electrochemiluminescence nanoemitters for immunoassay of protein biomarkers. *Bioelectrochemistry* **2023**, *149*, 108281. [CrossRef] [PubMed]
171. Zhang, N.; Zhao, Z.-Y.; Gao, H.; Yu, Y.; Pan, J.-B.; Chen, H.-Y.; Xu, J.-J. An ultrasensitive electrochemiluminescence assay for nucleic acid detection based on carboxyl functionalized polymer dots. *J. Electroanal. Chem.* **2021**, *900*, 115743. [CrossRef]
172. Gao, H.; Zhang, N.; Hu, J.; Pan, J.-B.; Cheng, Y.-X.; Chen, H.-Y.; Xu, J.-J. Molecular Engineering of Polymer Dots for Electrochemiluminescence Emission. *ACS Appl. Nano Mater.* **2021**, *4*, 7244–7252. [CrossRef]
173. Khan, H.; Shah, M.R.; Barek, J.; Malik, M.I. Cancer biomarkers and their biosensors: A comprehensive review. *TrAC Trends Anal. Chem.* **2023**, *158*, 116813. [CrossRef]
174. Liu, X.; Zhao, S.; Tan, L.; Tan, Y.; Wang, Y.; Ye, Z.; Hou, C.; Xu, Y.; Liu, S.; Wang, G. Frontier and hot topics in electrochemiluminescence sensing technology based on CiteSpace bibliometric analysis. *Biosens. Bioelectron.* **2022**, *201*, 113932. [CrossRef]
175. Surinova, S.; Schiess, R.; Hüttenhain, R.; Cerciello, F.; Wollscheid, B.; Aebersold, R. On the Development of Plasma Protein Biomarkers. *J. Proteome Res.* **2011**, *10*, 5–16. [CrossRef]
176. Schwarzenbach, H.; Hoon, D.S.B.; Pantel, K. Cell-free nucleic acids as biomarkers in cancer patients. *Nat. Rev. Cancer* **2011**, *11*, 426–437. [CrossRef]
177. Binotti, W.W.; Bayraktutar, B.; Ozmen, M.C.; Cox, S.M.; Hamrah, P. A Review of Imaging Biomarkers of the Ocular Surface. *Eye Contact Lens Sci. Clin. Pract.* **2020**, *46*, S84–S105. [CrossRef]
178. Gao, X.; Liu, X.; Zeng, Y.; Zhang, Q.; Zhang, B.; Zou, G. Spectrum-Resolved Electrochemiluminescence to Multiplex the Immunoassay and DNA Probe Assay. *Anal. Chem.* **2022**, *94*, 15801–15808. [CrossRef] [PubMed]
179. Zhang, X.; Wang, Z.-J.; Wang, X.; Zhang, Y.-H.; Qu, J.; Ding, S.-N. Band-Edge Effect-Induced Electrochemiluminescence Signal Amplification Based on Inverse Opal Photonic Crystals for Ultrasensitive Detection of Carcinoembryonic Antigen. *Anal. Chem.* **2022**, *94*, 9919–9926. [CrossRef]
180. Huang, Y.; Lei, J.; Cheng, Y.; Ju, H. Ratiometric electrochemiluminescent strategy regulated by electrocatalysis of palladium nanocluster for immunosensing. *Biosens. Bioelectron.* **2016**, *77*, 733–739. [CrossRef] [PubMed]
181. Wu, M.-S.; Yuan, D.-J.; Xu, J.-J.; Chen, H.-Y. Electrochemiluminescence on bipolar electrodes for visual bioanalysis. *Chem. Sci.* **2013**, *4*, 1182–1188. [CrossRef]
182. Li, X.; Qin, X.; Tian, Z.; Wang, K.; Xia, X.; Wu, Y.; Liu, S. Gold Nanowires Array-Based Closed Bipolar Nanoelectrode System for Electrochemiluminescence Detection of α-Fetoprotein on Cell Surface. *Anal. Chem.* **2022**, *94*, 7350–7357. [CrossRef] [PubMed]
183. Yu, L.; Zhang, Q.; Kang, Q.; Zhang, B.; Shen, D.; Zou, G. Near-Infrared Electrochemiluminescence Immunoassay with Biocompatible Au Nanoclusters as Tags. *Anal. Chem.* **2020**, *92*, 7581–7587. [CrossRef]
184. Wang, Y.; Kan, X. Sensitive and selective "signal-off" electrochemiluminescence sensing of prostate-specific antigen based on an aptamer and molecularly imprinted polymer. *Analyst* **2021**, *146*, 7693–7701. [CrossRef]
185. Qin, D.; Meng, S.; Wu, Y.; Mo, G.; Deng, B. Aggregation-induced electrochemiluminescence resonance energy transfer with dual quenchers for the sensitive detection of prostate-specific antigen. *Sens. Actuators B Chem.* **2022**, *367*, 132176. [CrossRef]
186. Zhao, L.; Song, X.; Fan, D.; Liu, X.; Wang, H.; Wei, Q.; Wu, D. Highly Efficient Signal On/Off Electrochemiluminescence Gel Aptasensor Based on a Controlled Release Strategy for the Sensitive Detection of Prostate Specific Antigen. *Anal. Chem.* **2023**, *95*, 5695–5701. [CrossRef]

187. Fu, L.; Gao, X.; Dong, S.; Jia, J.; Xu, Y.; Wang, D.; Zou, G. Coreactant-Free and Direct Electrochemiluminescence from Dual-Stabilizer-Capped InP/ZnS Nanocrystals: A New Route Involving n-Type Luminophore. *Anal. Chem.* **2022**, *94*, 1350–1356. [CrossRef] [PubMed]
188. Yang, L.; Wu, T.; Du, Y.; Zhang, N.; Feng, R.; Ma, H.; Wei, Q. PEGylation Improved Electrochemiluminescence Supramolecular Assembly of Iridium(III) Complexes in Apoferritin for Immunoassays Using 2D/2D MXene/TiO_2 Hybrids as Signal Amplifiers. *Anal. Chem.* **2021**, *93*, 16906–16914. [CrossRef]
189. Yang, L.; Du, Y.; Fan, D.; Zhang, Y.; Kuang, X.; Sun, X.; Wei, Q. Facile Encapsulation of Iridium(III) Complexes in Apoferritin Nanocages as Promising Electrochemiluminescence Nanodots for Immunoassays. *Anal. Chem.* **2021**, *93*, 11329–11336. [CrossRef]
190. Han, Y.; Jia, Y.; Du, Y.; Li, Y.; Ren, X.; Ma, H.; Wu, D.; Kuang, X.; Fan, D.; Wei, Q. Controlled Growth of MoS_2 on Dendritic Ferric Oxide to Enhance Electrochemiluminescence of Nitrogen-Doped Carbon Quantum Dots for Sensitive Immunoassay. *Anal. Chem.* **2023**, *95*, 6655–6663. [CrossRef]
191. Du, Y.; Xue, J.; Sun, X.; Wu, D.; Liu, X.; Ju, H.; Yang, L.; Wei, Q. Oxygen Vacancy-Enhanced Electrochemiluminescence Sensing Strategy Using Luminol Thermally Encapsulated in Apoferritin as a Transducer for Biomarker Immunoassay. *Anal. Chem.* **2020**, *92*, 8472–8479. [CrossRef]
192. Xue, J.; Jia, Y.; Yang, L.; Feng, J.; Wu, D.; Ren, X.; Du, Y.; Ju, H.; Wei, Q. Etching Triangular Silver Nanoparticles by Self-generated Hydrogen Peroxide to Initiate the Response of an Electrochemiluminescence Sensing Platform. *Anal. Chem.* **2020**, *92*, 14203–14209. [CrossRef] [PubMed]
193. Yang, L.; Sun, X.; Wei, D.; Ju, H.; Du, Y.; Ma, H.; Wei, Q. Aggregation-Induced Electrochemiluminescence Bioconjugates of Apoferritin-Encapsulated Iridium(III) Complexes for Biosensing Application. *Anal. Chem.* **2021**, *93*, 1553–1560. [CrossRef] [PubMed]
194. Li, X.; Du, Y.; Wang, H.; Ma, H.; Wu, D.; Ren, X.; Wei, Q.; Xu, J.-J. Self-Supply of H_2O_2 and O_2 by Hydrolyzing CaO_2 to Enhance the Electrochemiluminescence of Luminol Based on a Closed Bipolar Electrode. *Anal. Chem.* **2020**, *92*, 12693–12699. [CrossRef] [PubMed]
195. Luo, Q.; Qian, X.; Mi, X.; Tu, Y. A novel electrochemiluminescent immunosensor for the detection of NT-proBNP based on a Au/ZIF-67 nanocomposite. *J. Electroanal. Chem.* **2022**, *912*, 116260. [CrossRef]
196. Yang, L.; Jia, Y.; Wu, D.; Zhang, Y.; Ju, H.; Du, Y.; Ma, H.; Wei, Q. Synthesis and Application of CeO_2/SnS_2 Heterostructures as a Highly Efficient Coreaction Accelerator in the Luminol–Dissolved O_2 System for Ultrasensitive Biomarkers Immunoassay. *Anal. Chem.* **2019**, *91*, 14066–14073. [CrossRef]
197. Jia, Y.; Yang, L.; Feng, R.; Ma, H.; Fan, D.; Yan, T.; Feng, R.; Du, B.; Wei, Q. $MnCO_3$ as a New Electrochemiluminescence Emitter for Ultrasensitive Bioanalysis of β-Amyloid$_{1-42}$ Oligomers Based on Site-Directed Immobilization of Antibody. *ACS Appl. Mater. Interfaces* **2019**, *11*, 7157–7163. [CrossRef]
198. Tan, R.; Wang, Y.; Mi, X.; Li, H.; Tu, Y. A dual-screening electrochemiluminescent aptasensor based on a mesoporous silica nano-sieve for specific detection of amyloid-β monomer. *Sens. Actuators B Chem.* **2022**, *352*, 131065. [CrossRef]
199. Ali, A.; Zhao, J.; Khan, M.S.; Wang, H.; Ren, X.; Hu, L.; Manzoor, R.; Wu, D.; Wei, Q. Electrochemiluminescence detection for β-amyloid$_{1-42}$ oligomers using silver nanoparticle decorated CuS@CoS_2 double shelled nanoboxes as dual-quencher. *Sens. Actuators B Chem.* **2021**, *329*, 129155. [CrossRef]
200. Qin, D.; Meng, S.; Wu, Y.; Mo, G.; Jiang, X.; Deng, B. Design of a Dual-Wavelength Ratiometric Electrochemiluminescence Immunosensor for Sensitive Detection of Amyloid-β Protein in Human Serum. *ACS Sustain. Chem. Eng.* **2021**, *9*, 7541–7549. [CrossRef]
201. Xu, M.-R.; Dai, R.-F.; Wei, Q.-Q.; Wang, J.; Feng, Y.-Y.; Hu, Y. Urinary AD7c-NTP Evaluates Cognition Impairment and Differentially Diagnoses AD and MCI. *Am. J. Alzheimers Dis. Dement.* **2022**, *37*, 15333175221111152. [CrossRef]
202. Liang, Y.; Xue, K.; Shi, Y.; Zhan, T.; Lai, W.; Zhang, C. Dry Chemistry-Based Bipolar Electrochemiluminescence Immunoassay Device for Point-of-Care Testing of Alzheimer-Associated Neuronal Thread Protein. *Anal. Chem.* **2023**, *95*, 3434–3441. [CrossRef] [PubMed]
203. Domenico, T.; Rita, A.; Giacomo, S.; Diego, A.; Thelma, P.; Mariana, G.; Giampaolo, N.; Francesco, N.; Maria, G.; Francesco, F.; et al. Salivary biomarkers for diagnosis of acute myocardial infarction: A systematic review. *Int. J. Cardiol.* **2023**, *371*, 54–64. [CrossRef] [PubMed]
204. Gumanova, N.G.; Bogdanova, N.L.; Metelskaya, V.A.; Tarasov, V.I.; Kots, A.Y.; Kutsenko V.A.; Kontsevaya, A.V.; Drapkina, O.M. Serum biomarkers, including nitric oxide metabolites (NOx), for prognosis of cardiovascular death and acute myocardial infarction in an ESSE-RF case–control cohort with 6.5-year follow up. *Sci. Rep.* **2022**, *12*, 18177. [CrossRef] [PubMed]
205. Yang, X.; Xu, Y.; Huang, X.; Hang, J.; Guo, W.; Dai, Z. Multicolor Iridium(III) Complexes with Host–Guest Recognition Motifs for Enhanced Electrochemiluminescence and Modular Labeling. *Anal. Chem.* **2023**, *95*, 4543–4549. [CrossRef]
206. Hong, D.; Kim, K.; Jo, E.-J.; Kim, M.-G. Electrochemiluminescence-Incorporated Lateral Flow Immunosensors Using $Ru(bpy)_3^{2+}$-Labeled Gold Nanoparticles for the Full-Range Detection of Physiological C-Reactive Protein Levels. *Anal. Chem.* **2021**, *93*, 7925–7932. [CrossRef]
207. Guo, X.-M.; Zhao, M.-L.; Liang, W.-B.; Yang, X.; Yuan, R.; Zhuo, Y. Programmable Y-Shaped Probes with Proximity Bivalent Recognition for Rapid Electrochemiluminescence Response of Acute Myocardial Infarction. *ACS Sens.* **2022**, *7*, 3208–3215. [CrossRef]

208. Ji, Y.; He, S.; Chen, Y.; Zhang, P.; Sun, J.; Li, Y.; Kuang, K.; Jia, N. A sensitive dual-signal electrochemiluminescence immunosensor based on $Ru(bpy)_3^{2+}$ @HKUST-1 and $Ce_2Sn_2O_7$ for detecting the heart failure biomarker NT-proBNP. *J. Mater. Chem. B* **2023**, *11*, 2754–2761. [CrossRef]
209. Xu, P.; Zhang, Y.; Li, X.; Ren, X.; Fan, D.; Wang, H.; Wei, Q.; Ju, H. Electrochemiluminescence immunosensor based on ferrocene functionalized ZIF-8 quenching the electrochemiluminescence of $Ru(bpy)_3^{2+}$-doped silica nanoparticles embodied *N*-butyldiethanolamine. *Sens. Actuators B Chem.* **2021**, *329*, 129101. [CrossRef]
210. Song, C.; Li, X.; Hu, L.; Shi, T.; Wu, D.; Ma, H.; Zhang, Y.; Fan, D.; Wei, Q.; Ju, H. Quench-Type Electrochemiluminescence Immunosensor Based on Resonance Energy Transfer from Carbon Nanotubes and Au-Nanoparticles-Enhanced *g*-C_3N_4 to CuO@Polydopamine for Procalcitonin Detection. *ACS Appl. Mater. Interfaces* **2020**, *12*, 8006–8015. [CrossRef]
211. Wang, N.; Yang, J.; Luo, Z.; Qin, D.; Wu, Y.; Deng, B. A dual-emitting immunosensor based on manganese dioxide nanoflowers and zinc sulfide quantum dots with enhanced electrochemiluminescence performance for the ultrasensitive detection of procalcitonin. *Analyst* **2023**, *148*, 2122–2132. [CrossRef] [PubMed]
212. Song, X.; Zhao, L.; Zhang, N.; Liu, L.; Ren, X.; Ma, H.; Kuang, X.; Li, Y.; Luo, C.; Wei, Q. Ultrasensitive Electrochemiluminescence Biosensor with Silver Nanoclusters as a Novel Signal Probe and α-Fe_2O_3–Pt as an Efficient Co-reaction Accelerator for Procalcitonin Immunoassay. *Anal. Chem.* **2023**, *95*, 1582–1588. [CrossRef] [PubMed]
213. Zhu, H.; Jiang, D.; Zhu, J.-J. High-resolution imaging of catalytic activity of a single graphene sheet using electrochemiluminescence microscopy. *Chem. Sci.* **2021**, *12*, 4794–4799. [CrossRef]
214. Knezevic, S.; Bouffier, L.; Liu, B.; Jiang, D.; Sojic, N. Electrochemiluminescence microscopy: From single objects to living cells. *Curr. Opin. Electrochem.* **2022**, *35*, 101096. [CrossRef]
215. Zhou, Y.; Dong, J.; Zhao, P.; Zhang, J.; Zheng, M.; Feng, J. Imaging of Single Bacteria with Electrochemiluminescence Microscopy. *J. Am. Chem. Soc.* **2023**, *145*, 8947–8953. [CrossRef]
216. Zhang, H.; Liu, Y.; Yao, M.; Han, W.; Zhang, S. Cathodic Electrochemiluminesence Microscopy for Imaging of Single Carbon Nanotube and Nucleolin at Single Tumor Cell. *Anal. Chem.* **2023**, *95*, 570–574. [CrossRef]
217. Lu, Y.; Huang, X.; Wang, S.; Li, B.; Liu, B. Nanoconfinement-Enhanced Electrochemiluminescence for in Situ Imaging of Single Biomolecules. *ACS Nano* **2023**, *17*, 3809–3817. [CrossRef]
218. Liu, G.; Jin, B.-K.; Ma, C.; Chen, Z.; Zhu, J.-J. Potential-Resolved Electrochemiluminescence Nanoprobes for Visual Apoptosis Evaluation at Single-Cell Level. *Anal. Chem.* **2019**, *91*, 6363–6370. [CrossRef]
219. Chen, Y.; Gou, X.; Ma, C.; Jiang, D.; Zhu, J.-J. A Synergistic Coreactant for Single-Cell Electrochemiluminescence Imaging: Guanine-Rich ssDNA-Loaded High-Index Faceted Gold Nanoflowers. *Anal. Chem.* **2021**, *93*, 7682–7689. [CrossRef] [PubMed]
220. Zhang, J.; Jin, R.; Jiang, D.; Chen, H.-Y. Electrochemiluminescence-Based Capacitance Microscopy for Label-Free Imaging of Antigens on the Cellular Plasma Membrane. *J. Am. Chem. Soc.* **2019**, *141*, 10294–10299. [CrossRef]
221. Chen, M.; Xu, C.; Zhao, W.; Chen, H.; Xu, J. Single Cell Imaging of Electrochemiluminescence-Driven Photodynamic Therapy. *Angew. Chem. Int. Ed.* **2022**, *61*, e2021174. [CrossRef]
222. Ding, H.; Guo, W.; Su, B. Imaging Cell-Matrix Adhesions and Collective Migration of Living Cells by Electrochemiluminescence Microscopy. *Angew. Chem. Int. Ed.* **2020**, *59*, 449–456. [CrossRef] [PubMed]
223. Ding, L.; Ding, H.; Zhou, P.; Xi, L.; Su, B. Surface-Sensitive Imaging Analysis of Cell–Microenvironment Interactions by Electrochemiluminescence Microscopy. *Anal. Chem.* **2022**, *94*, 10885–10892. [CrossRef] [PubMed]

chemosensors MDPI

Review

Near-Infrared-II Fluorophores for In Vivo Multichannel Biosensing

Feng Ren [1], Tuanwei Li [1], Tingfeng Yao [1], Guangcun Chen [1,2], Chunyan Li [1,2] and Qiangbin Wang [1,2,*]

[1] CAS Key Laboratory of Nano-Bio Interface, Division of Nanobiomedicine and i-Lab, Suzhou Institute of Nano-Tech and Nano-Bionics, Chinese Academy of Sciences, Suzhou 215123, China; fren2021@sinano.ac.cn (F.R.); twli2020@sinano.ac.cn (T.L.); tfyao2022@sinano.ac.cn (T.Y.); gcchen2011@sinano.ac.cn (G.C.); cyli2012@sinano.ac.cn (C.L.)

[2] School of Nano-Tech and Nano-Bionics, University of Science and Technology of China, Hefei 230026, China

* Correspondence: qbwang2008@sinano.ac.cn

Abstract: The pathological process involves a range of intrinsic biochemical markers. The detection of multiple biological parameters is imperative for providing precise diagnostic information on diseases. In vivo multichannel fluorescence biosensing facilitates the acquisition of biochemical information at different levels, such as tissue, cellular, and molecular, with rapid feedback, high sensitivity, and high spatiotemporal resolution. Notably, fluorescence imaging in the near-infrared-II (NIR-II) window (950–1700 nm) promises deeper optical penetration depth and diminished interferential autofluorescence compared with imaging in the visible (400–700 nm) and near-infrared-I (NIR-I, 700–950 nm) regions, making it a promising option for in vivo multichannel biosensing toward clinical practice. Furthermore, the use of advanced NIR-II fluorophores supports the development of biosensing with spectra-domain, lifetime-domain, and fluorescence-lifetime modes. This review summarizes the versatile designs and functions of NIR-II fluorophores for in vivo multichannel biosensing in various scenarios, including biological process monitoring, cellular tracking, and pathological analysis. Additionally, the review briefly discusses desirable traits required for the clinical translation of NIR-II fluorophores such as safety, long-wavelength emission, and clear components.

Keywords: NIR-II fluorophore; multichannel biosensing; spectra-domain mode; lifetime-domain mode; fluorescence-lifetime mode

Citation: Ren, F.; Li, T.; Yao, T.; Chen, G.; Li, C.; Wang, Q. Near-Infrared-II Fluorophores for In Vivo Multichannel Biosensing. *Chemosensors* **2023**, *11*, 433. https://doi.org/10.3390/chemosensors11080433

Academic Editor: Cynthia M. Dupureur

Received: 28 June 2023
Revised: 20 July 2023
Accepted: 2 August 2023
Published: 4 August 2023

1. Introduction

Fluorescence biosensing is a widely employed technique in the field of life sciences due to its ability to visualize multiplexed physiological and pathological information with rapid feedback, high sensitivity, and high spatiotemporal resolution [1–3]. Multichannel biosensing is particularly significant for the development of biomedicine in both fundamental research and clinical practice [4,5]. It is capable of monitoring multiple processes or quantitatively/qualitatively detecting biomarkers, providing accurate and valuable diagnostic information on tissue, cellular, and molecular features [6,7]. Nevertheless, multichannel fluorescence biosensing in the visible (400–700 nm) and first near-infrared (NIR-I, 700–950 nm) region encounters poor photon penetration depth in mammalian tissues due to the substantial photon scattering/absorption and the interferential autofluorescence (Figure 1a) [8–10]. In contrast, fluorescence imaging in the second near-infrared window (NIR-II, 950–1700 nm) can offer deeper penetration depth and better clarity owing to the suppressed photon scattering (us'$\propto\lambda$-α; u: scattering coefficient; λ: wavelength; and α: constant) and reduced tissue autofluorescence (inversely proportional to wavelength) compared with the visible and NIR-I regions (Figure 1b,c) [11–14]. Thus, the advancement of new NIR-II fluorescent probes is necessary to support the development of deep-tissue and multichannel fluorescence biosensing.

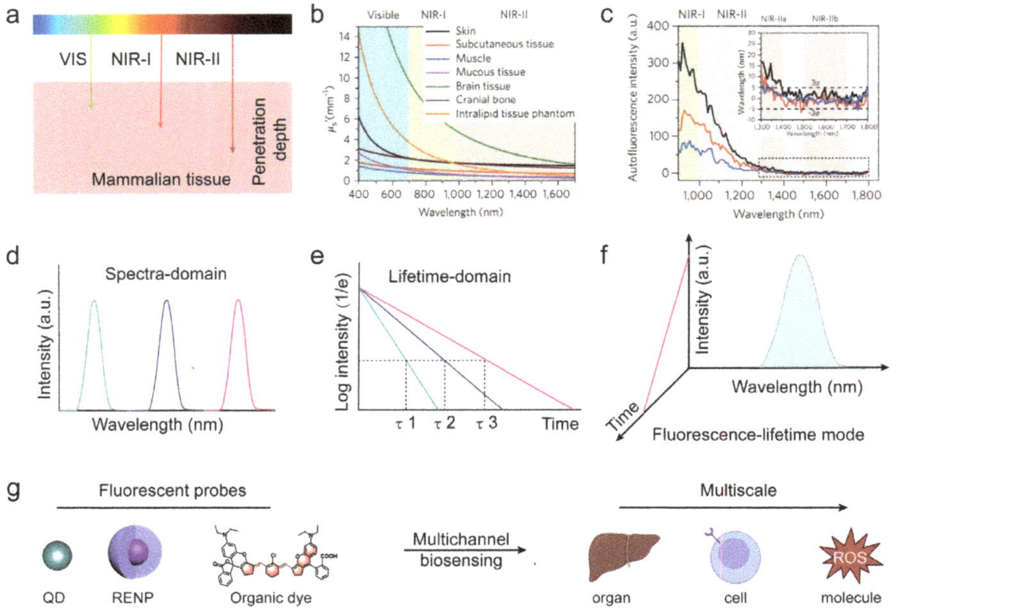

Figure 1. NIR-II fluorophores for deep-tissue multichannel biosensing in versatile modes at a multiscale level: (**a**–**c**) light–tissue interactions including tissue penetration depth of light (visible, NIR-I, NIR-II), tissue scattering spectra in 400–1700 nm, and tissue autofluorescence: (**a**) reproduced with permission from [10], copyright 2020, Wiley-VCH; (**b**,**c**) reproduced with permission from [8], copyright 2017, Springer Nature; (**d**–**f**) schematic illustration of multichannel biosensing on the spectra-domain, lifetime-domain, fluorescence-lifetime modes; (**g**) schematic illustration of multiple fluorescent probes (QD, RENP, and organic dye) for biosensing at the multiscale level (organ, cell, molecule).

NIR-II fluorophores, mainly including organic dyes [15,16], quantum dots (QDs) [17,18], and rare-earth-based nanoparticles (RENPs) [19,20], have been developed for multichannel fluorescence imaging using spectra-domain, lifetime-domain, and fluorescence-lifetime modes (Figure 1d–f). The fabrication of these fluorophores, which exhibit nonoverlap emissions–excitations or distinguishable fluorescence lifetimes, is crucial for enabling multichannel biosensing applications [21,22]. To facilitate the qualitative and quantitative analysis of niduses, organic dyes may be designed to control their chemical groups and conjugation units, thereby regulating their fluorescence emissions–excitations and responsiveness to pathological environments [23,24]. QDs may be tuned via the regulation of their size and chemical structure (e.g., core–shell structure and heteroatom doping) to alter their fluorescence emissions [17,25]. RENPs can be modified by doping with various sensitizers/activators and by altering their core–shell structures (e.g., interlayer and shell thickness), resulting in changes in their fluorescence emissions–excitations and lifetimes [19,26]. These fluorescent probes, possessing unique fluorescence and lifetime properties, can be utilized as a versatile toolbox to fulfill the demands of multiscale and multichannel biosensing in a variety of scenarios (Figure 1g).

This review focuses on the advanced development of in vivo multichannel biosensing in the NIR-II window for improvement in biomedical diagnostics. We highlight the multifaceted designs and properties of NIR-II fluorophores for in vivo multichannel biosensing, which can be utilized in a variety of modes, namely the spectra-domain mode (utilizing cut-off excitation–emission wavelengths), the lifetime-domain mode (using differential fluorescence lifetimes), and the fluorescence-lifetime mode (utilizing the decoding of fluorescence signals via the time-gating technique). We also highlight a selection of their distinctive bioapplication scenarios, including biological process monitoring, cellular tracking, and pathological analysis. Ultimately, we conclude by discussing the merits of multichannel biosensing modes, enabled by the use of versatile NIR-II fluorophores, as well as the outlook for potential challenges and opportunities in multichannel biosensing.

2. Near-Infrared-II Fluorophores

Since the discovery of single-walled carbon nanotubes (SWNTs) with NIR-II fluorescence by the Dai group in 2009 [27], NIR-II fluorophores have flourished. To deploy fluorescence multiplexing for in vivo multichannel biosensing, NIR-II fluorophores demand tunable optical properties in the NIR-II window. Typical fluorophores (QDs, RENPs, and organic dyes) with tunable NIR-II fluorescence are described in subsequent sections.

2.1. QDs

As a class of semiconductor nanocrystals with radii smaller than the excitation Bohr radius, the size and chemical structure of QDs can be easily tailored to tune their fluorescence emission. Lead sulfide (PbS) QDs have size-dependent fluorescence throughout the entire NIR-II region owing to their narrow bandgap and large Bohr radii [28,29]. Bruns et al. designed a series of InAs-based core–shell-structured QDs with narrow and tunable NIR-II emission. Notably, the PLQY (photoluminescence quantum yield) of the core–shell QDs could reach up to 30% in aqueous media compared with PbS QDs (4%) [17]. In addition to shell wrapping, the Wang group reported a cation exchange strategy to boost the PL efficiency of Ag_2E (E = Se and Te) QDs. With the incorporation of gold (Au) atoms into the crystal lattice, alloyed silver gold selenide (AgAuSe) QDs achieved an absolute PLQY of up to 65.3% at 978 nm, and Au:Ag_2Te achieved a calculated PLQY of 6.2% at 1490 nm [11,30]. These versatile NIR-II QDs provide a toolbox for fluorescence multiplexing.

2.2. RENPs

The large quantum numbers of 4f electron configurations provide lanthanide ions (Ln^{3+}) with a wealth of optical properties. By simply changing the dopants in RENPs, their emission wavelength can be effectively tuned over the NIR-II region. For instance, Naczynski et al. found that the typical Yb^{3+}-sensitized Ln^{3+} (Ln = Er, Ho, Tm, and Pr) in RENPs could produce differential emissions (Ho^{3+}: 1185 nm; Pr^{3+}: 1310 nm; Tm^{3+}: 1475 nm; and Er^{3+}: 1525 nm) after absorbing 980 nm photons [31]. In comparison, for a system highly doped with Er^{3+}, RENPs possess multiwavelength excitations (980 nm, 808 nm, and 658 nm) and NIR-II emission (1550 nm) after inert shell coating to alleviate surface and concentration quenching effect [32]. In addition to the tunable multiwavelength excitation–emission, the long luminescence lifetime is also one of the intrinsic properties of RENPs. Tailoring the structural parameters of core–shell RENPs, including the concentrations of dopants, the thicknesses of interlayer and inert shell and the crystal phase of host matrixes could effectively tune the NIR-II lifetime of Ln^{3+} activators with different proportions [33]. These feasible pathways to achieve NIR-II RENPs with highly distinguishable fluorescence wavelengths and lifetimes favor the development of advanced multiplexing modes.

2.3. Organic Dyes

Organic dyes with NIR-II fluorescence have been synthesized based on cyanine dyes (D-π-A) or donor–acceptor–donor (D-A-D) structures by tailoring their chemical groups or conjugation units. Cyanine dyes typically consist of two heterocyclic end groups connected by a length-tunable polymethine chain [34]. Lengthening the polymethine chain or modifying heterocycle can endow cyanine dyes (Flav7, IR-1061, IR-26, and FD-1080) with tunable NIR-II fluorescence by enhancing the π-conjugate strength [35,36]. The first NIR-II organic dye based on the D-A-D structure, CH1055, was reported by the Dai group, which is composed of a benzobisthia diazole acceptor and a triphenylamine donor [37].

Although organic dyes generally face the problem of aggregation-caused quenching (ACQ) in biological environments, mainly due to the strong intramolecular π–π interactions [38], Tang et al. have discovered and explored a type of luminogens with aggregation-induced emissions (AIEgens) since 2001 [39,40]. Tailoring the electron donors (D) and electron acceptors (A) of AIEgens can extend their emissions to the NIR-II window, such as D-A-structured 2TT-*o*C26B molecules [41,42]. More prominently, AIEgens are capable of multiphoton absorption, enabling multiphoton fluorescence bioimaging to decipher deep-tissue biological structures via NIR-II excitation [43–45]. Typical approaches such as tailoring D-A structures and expanding the conjugation length could enhance the multiphoton absorption cross-sections of AIEgens [46]. In addition, the highly ordered assembly of organic dyes provides an effective avenue to achieve longer wavelengths beyond 1500 nm. J-aggregates with head–tail molecular stacking of organic dyes resulted in strong red-shifted absorption–emission (approximately 300 nm) and boosted fluorescence intensity of FD-1080 compared with that of the dye monomer [14]. These versatile approaches to obtaining NIR-II organic dyes with brightness and differential excitation–emission wavelengths are beneficial for the development of fluorescence multiplexing methods.

3. Spectra-Domain Multichannel Biosensing

The implementation of in vivo multichannel fluorescence biosensing requires a meticulous selection of nonoverlap excitation–emission wavelengths for the use of NIR-II fluorophores. To achieve simultaneous visualization/tracking of multiple analytes at the tissue or cellular level, fluorophores with varying excitation–emission properties are necessary to bind specific targets. On the other hand, in order to quantitatively measure analytes at the molecular level, ratiometric fluorophores with multiple excitation–emission wavelengths are typically required to enable selective responsiveness. The rational designs and bioapplications of these NIR-II fluorophores will be elucidated in-depth in subsequent sections.

3.1. Excitation–Emission Multiplexed Biosensing

In order to facilitate the visualization of multiple pathological or physiological parameters at the tissue or cellular level, it is essential to use fluorophores with nonoverlap excitation–emission spectra that enable to simultaneously tag and visualize them in the NIR-II imaging window (Figure 2a,b). Recently, Cosco et al. tuned the absorption properties of flavylium polymethine dyes utilizing flavylium heterocycles, which allowed for the real-time excitation multiplexing of living mice at the tissue level using NIR-II emitters (MeOFlav7: 980 nm excitation; JuloFlav7: 1064 nm excitation) [36]. However, the long-term retention of these molecules has the potential to induce toxicity in major organs. To overcome this limitation, Yao et al. designed a series of highly efficient renal clearance and long blood circulation aza-boron-dipyrromethene (aza-BODIPY) NIR-II macromolecular fluorophores (FBP790: 730 nm excitation, 950 nm long-pass optical filter; FBP1025: 980 nm excitation, 1200 nm long-pass optical filter) and applied them for excitation multiplexed imaging of tumors and surrounding vessels [47]. Nevertheless, these NIR-II organic fluorophores are restricted to the emission wavelength range of 1000–1500 nm, thereby limiting their utility in achieving high-resolution NIR-IIb fluorescence imaging (1500–1700 nm) with millimeter-scale penetration depth and micron-level resolution [48–50]. Although AIEgens (DCBT dots) with multiphoton absorption could obtain fine structures of brain vasculatures of mice and macaque by using three-photon fluorescence bioimaging with NIR-IIb excitations (1550 nm), their progress in the field of excitation–emission multiplexed biosensing remains sluggish [51–53].

To date, only a handful of NIR-IIb fluorophores have been developed for multichannel fluorescence biosensing, including rare-earth-based nanoprobes and colloidal QDs [54,55]. Given the intrinsic location of 4f energy levels of lanthanides, Er(III) is the mainstream activator generating NIR-IIb emission (around 1530 nm) under varied excitation sources (650 nm, 808 nm, and 980 nm) [32,56–58]. To expand the range of small-molecule lanthanide complexes for biosensing, Wang et al. developed a hybrid Er(III)-bacteriochlorin complex (EB766) with 760 nm excitation and bright luminescence at 1530 nm [59]. Compared with typical co-doped lanthanide nanoparticles (ErNPs: $NaYF_4$:20%Yb,2%Er@$NaYF_4$), the nonoverlap excitations of EB766 (760 nm) and ErNPs (930 nm) facilitate robust NIR-IIb multiplexed biosensing for deciphering multiple tissue structures (such as lymph nodes and lymph vessels in jaw and footpads) with superior signal-to-background ratios (SNRs: 3.92–22.79) (Figure 2c). Alongside lanthanides, colloidal QDs, such as PbS QDs with a narrow bandgap and large Bohr radius, offer tunable emission coverage of the NIR-II region via size control [60,61]. By synthesizing NIR-IIa and NIR-IIb fluorescent PbS/CdS core–shell QDs functionalized with Gr-1 and CD-1b antibodies, respectively, Yu et al. exploited this emission multiplexing strategy for in vivo colocalization of myeloid-derived suppressor cells (MDSCs). By connecting the unique emission properties of these QDs to the targeted MDSCs, this approach holds significant potential for enhancing immunotherapy (Figure 2d) [21]. Overall, these results demonstrate that the excitation–emission multiplexing approach is highly suitable for a real-time diagnosis of tissue lesions and therapeutic feedback at the cellular level.

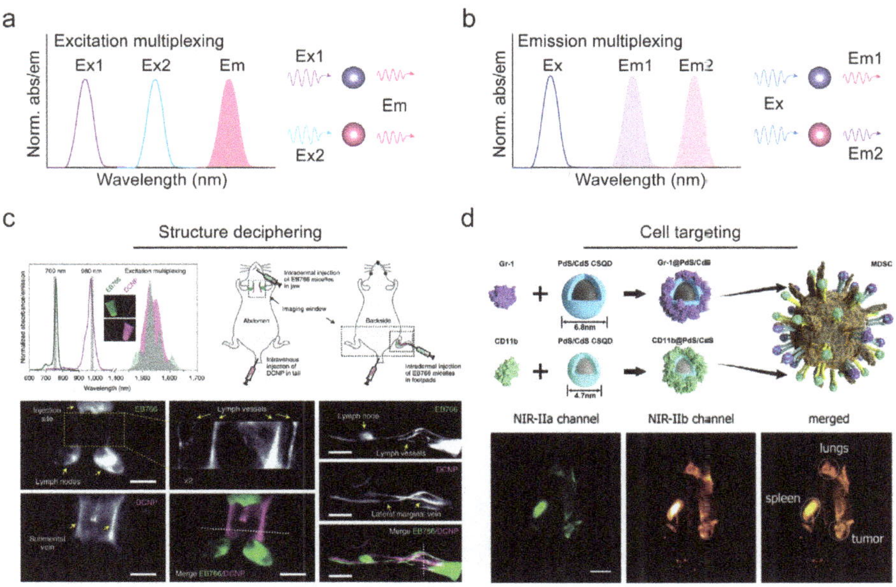

Figure 2. Excitation–emission multiplexed biosensing: (**a**,**b**) schematic illustration of NIR-II fluorophores enabled excitation–emission multiplexing; (**c**) excitation multiplexing biosensing to decipher multiple tissues of living mice using NIR-IIb lanthanide-based probes (EB766: 760 nm excitation, 1530 nm emission; DCNP: 980 nm excitation, 1530 emission). Reproduced with permission from [59], copyright 2021, Springer Nature; (**d**) emission multiplexing to target MDSCs using size-varied PbS/CdS QDs. Reproduced with permission from [21], copyright 2019, American Chemical Society.

3.2. Ratiometric Fluorescence Biosensing

In addition to multichannel biosensing utilizing multiple fluorophores, fluorophores with the capacity to emit or excite at different wavelengths have the potential to enable the quantitative measurement of pathological parameters at the molecular level. For this purpose, the fluorophores must possess an emission–excitation pattern that responds solely to the analyte (such as reactive oxygen species, enzymes, and pH), while the other remains inert and acts as a reference signal (Figure 3a,b). To achieve this, Lan et al. designed a series of responsive NIR-II probes (enzyme-responsive Rap-N, ROS-responsive Rap-R, and pH-responsive Py-H; ratiometric emission: 945 nm/1010 nm) based on the Py-2 molecular platform for ratiometric fluorescence biosensing [24]. These single-component NIR-II dyes have the potential to selectively visualize and quantitatively measure enzymes and small molecules in living mice with significant ratiometric fluorescence signals (Fl900LP/Fl1000LP). Additionally, NIR-II organic nanosensors for ratiometric fluorescence imaging can be designed based on intramolecular Förster resonance energy transfer (FRET). To create the NIR-II ratiometric fluorescent dyes, Yu et al. covalently linked an asymmetric aza-BODIPY with an $ONOO^-$-responsive meso-thiocyanine, known as the aBOP-IR1110 (ratiometric emission: 950 nm/1110 nm). This process results in $ONOO^-$-altered intramolecular FRET, which generates a linear ratiometric response (Fl950LP/Fl1100LP) [62]. The aBOP-IR1110 can withstand biological media, thereby preventing spectral shift and distortion and facilitating the dynamic monitoring of oxidative stress in traumatic brain injury and evaluating therapeutic efficiency with high in vivo sensing accuracy. However, the currently available organic ratiometric fluorophores have a limited emission wavelength that falls short of 1500 nm, along with a small Stokes shift, thereby limiting their usability in deep-tissue biosensing.

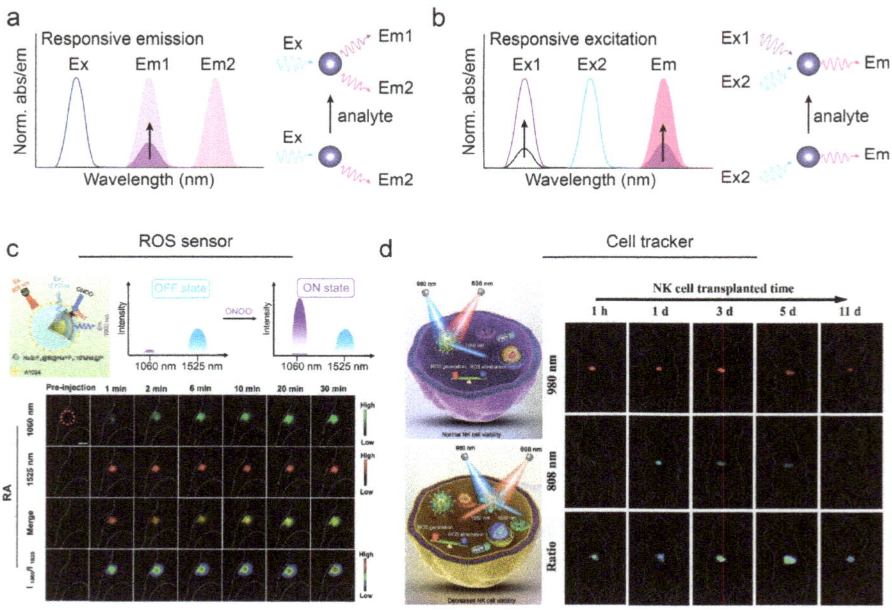

Figure 3. Ratiometric fluorescence biosensing: (**a**,**b**) schematic illustration of NIR-II fluorescent nanocomposites enabled ratiometric fluorescence biosensing; (**c**) emission-responsive NIR-II hybrid nanosensor for ratiometric fluorescence imaging of inflammation using A1094-RENPs hybrid nanosensor. Reproduced with permission from [63], copyright 2022, American Chemical Society; (**d**) excitation-responsive NIR-II hybrid nanosensor for ratiometric fluorescence imaging of NK cell viability using IR780-RENPs hybrid nanosensor. Reproduced with permission from [64], copyright 2021, Wiley-VCH.

Hybrid ratiometric nanosensors offer a variety of options for emission wavelength and responsive substances, thus enabling quantitative in vivo biosensing. Sun et al. developed a NIR-II ratiometric nanocomposite by coupling the dual-emission (1060 nm and 1525 nm) RENPs ($NaErF_4@NaYF_4@NaYF_4$:10%Nd@$NaYF_4$) with $ONOO^-$-responsive A1094 organic dyes (absorbance:1094 nm). The FRET between them allowed for the rapid and sensitive in vivo detection of $ONOO^-$ levels (Figure 3c) [63]. Since ROS/RNS levels are highly correlated with the activation and viability of immune cells, in vivo molecular analysis can provide valuable feedback during cancer immunotherapy [65,66]. Liao et al. developed a NIR-II ratiometric nanocomposite by coating dual-excitation (ex: 808 nm and 980 nm; em: 1525 nm) RENPs with IR786 dyes. The degradation of ROS-responsive dyes solely activates the 808 nm excitation channel through the absorption competition-induced emission effect (ACIE) [64]. Therefore, the hybrid nanosensor can evaluate the cell viability of natural killer (NK) cells by measuring the excess generation of ROS, while simultaneously tracking the NK cells via the stable signal excited at 980 nm (Figure 3d). Such intracellular ROS-induced ratiometric NIR-II fluorescence biosensing has the potential to pave the way for in vivo companion diagnostics in cancer immunotherapy.

3.3. Spectra-Domain Multichannel Biosensing in Various Scenarios

In the field of oncology, mapping the heterogeneity of primary and metastatic tumors is vital for precision medicine [4]. Detecting the metastasis of cancer cells and the migration of immune cells is essential from a fundamental research perspective, as it can help identify the most effective therapies [67,68]. To study cellular behavior in live animals, intravital microscopic multiplexing is essential [69]. Wang et al. developed a cell tracking probe (CT1530) for intracellular delivery, comprising the cell-penetrating peptide HIV-TAT-conjugated EB766–bovine serum albumin (BSA) complex [59]. Using fluorescence signals (green, 1100–1300 nm) from Cy7.5 phospholipid micelles to highlight the cerebrovascular system, the researchers found that cancer cells were arrested in vessel bifurcations through physical occlusion via the nonoverlap fluorescence (red, 1400–1600 nm) from CT1530-labeled 143B cells (Figure 4a). On the other hand, evaluating the infiltration of immune cells into tumors is crucial for determining therapeutic efficacy. Hao et al. developed a NIR-II emission multiplexing strategy to visualize the recruitment of NK cells into the tumor via the programmable administration of Ag_2Se-QD-based nanodrug (em: 1350 nm) containing SDF-1α (stromal-cell-derived factor-1α, the chemokine of NK-92 cells) and Tat-Ag_2S QDs-labeled NK-92 cells (em: 1050 nm) (Figure 4b) [70]. This multichannel biosensing strategy has significant implications not only for precision medicine but also for individual therapeutic schedules and companion diagnostics [4,71]

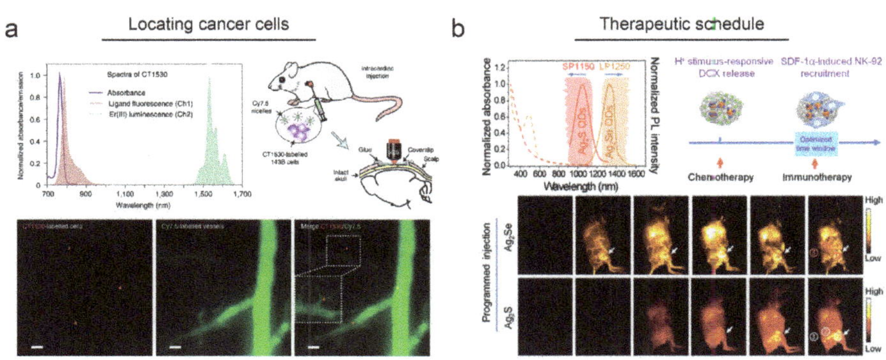

Figure 4. *Cont.*

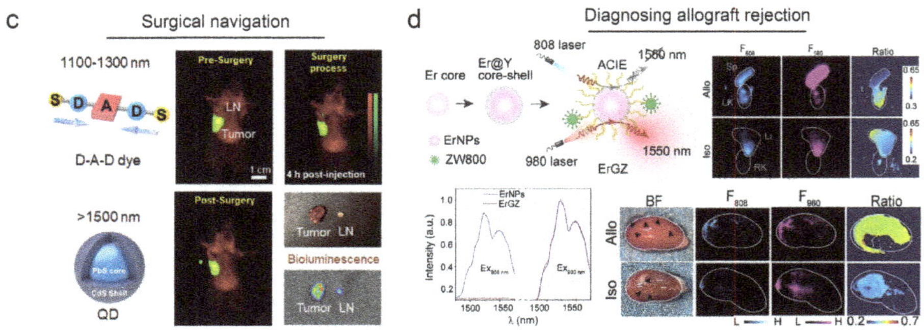

Figure 4. Spectra-domain multichannel biosensing in versatile scenarios: (**a**) intravital multiplexed imaging of cancer cell metastases in mouse brain. Reproduced with permission from [59], copyright 2021, Springer Nature; (**b**) NIR-II fluorescence multiplexed imaging to program the chemotherapy and immunotherapy against breast cancer. Reproduced with permission from [70], copyright 2018, Wiley-VCH; (**c**) NIR-II emission multiplexing imaging for surgical navigation using QDs and D-A-D dye. Reproduced with permission from [72], copyright 2020, Wiley-VCH; (**d**) NIR-IIb ratiometric imaging for allograft rejection diagnosis using granzyme-B-responsive ErGZ nanosensor. Reproduced with permission from [73], copyright 2023, Wiley-VCH.

In clinical practice, a first-in-human study has demonstrated the potential of intraoperative NIR-II fluorescence imaging in guiding liver tumor surgery [5]. In addition to primary tumors theranostics, detecting and monitoring cancer metastasis is critical for tumor staging, therapeutic decision making, and prognosis. For instance, in the case of breast cancer, multiorgan metastasis often leads to a median survival rate of fewer than two years for patients. Lymph nodes (LNs) are frequently the primary site of tumor cell dissemination, which can disrupt the immune microenvironment [74]. To enable precise surgical resection, Tian et al. developed an emission multiplexing approach that simultaneously labels the metastatic tumor (IR-FD dye) and tumor metastatic proximal LNs (PbS/CdS QDs) (Figure 4c) [72]. Meanwhile, diagnosing allograft rejection to improve the immune management of transplant recipients in the early stages is also essential, as transplantation can cause severe postoperative complications [75,76]. Chen et al. developed a responsive NIR-II fluorescent nanosensor by linking ErNPs ($NaErF_4@NaYF_4$: 980/808 nm excitation; 1550 nm emission) with ZW800 dye, allowing for the ratiometric biosensing of granzyme B, which is overexpressed in recipients' T cells during the onset of allograft rejection, in contrast to the gold-standard biopsy (Figure 4d) [73]. This strategy could also be applied to in situ monitoring of tissue regeneration. Pei et al. integrated 3D-printed bioactive glass scaffolds with a responsive NIR-II fluorescent nanosensor, enabling in situ monitoring of early inflammation, angiogenesis, and implant degradation during mouse skull repair [77]. These results illustrate the potential of spectral-domain fluorescence multiplexing in clinical translation.

4. Lifetime-Domain Multichannel Biosensing

As fluorescence imaging involves real-time light irradiation, it inevitably leads to tissue scattering and autofluorescence. However, the use of cut-off wavelengths allows for fluorescence multiplexing through optical filters, while simultaneously filtering out scattering light. Although tissue autofluorescence is significantly attenuated in the NIR-II region, this often compromises the contrast and sensitivity of fluorescence imaging [78,79]. To achieve background-free multiplexed imaging, manipulating the fluorescence lifetime of fluorophores matters. Fluorophores with longer lifetimes than tissue autofluorescence (0.1–5 ns) can provide a luminescence signal even after the laser is turned off (Figure 5a) [9,80]. This time-gating technique can eliminate interference from the excitation light to improve the contrast of fluorescence imaging [81]. To enable this lifetime-domain multichannel biosensing, it is essential that the fluorophores have a tunable lifetime distinct from that of tissue autofluorescence, which is mainly limited to RENPs, known as t-dots (Figure 5b).

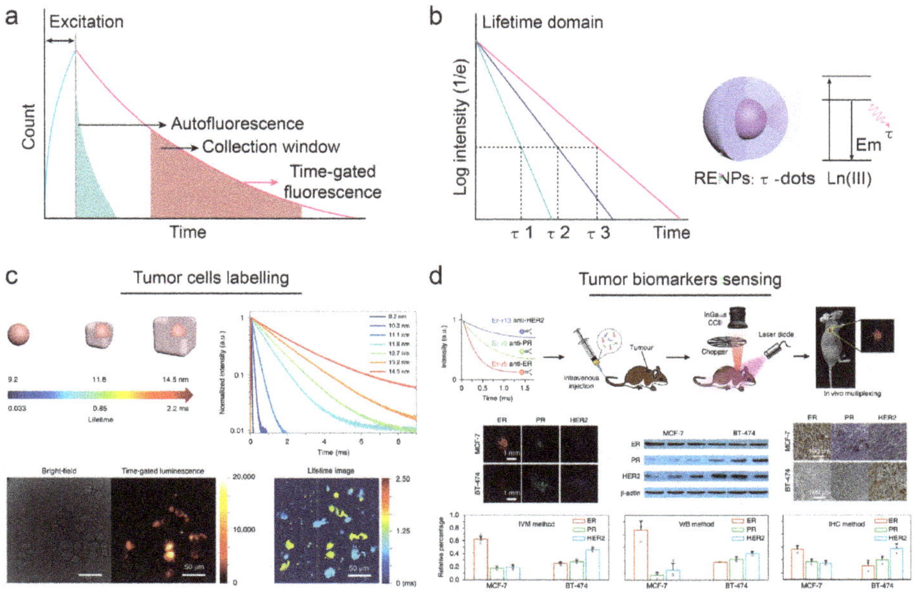

Figure 5. Lifetime-domain multichannel biosensing: (**a**) fluorophores with long lifetimes enable background-free fluorescence imaging through time-gated gear; (**b**) schematic illustration of fluorescence-lifetime-tunable RENPs; (**c**) fluorescence-lifetime-tunable $NaYbF_4@CaF_2$ by modulating shell thickness for tagging Hela cells. Reproduced with permission from [82], copyright 2019, Springer Nature; (**d**) lifetime-engineered ErNPs with versatile antibodies (ER, PR, and HER2) binging for in vivo lifetime-domain multiplexing to identify tumor subtype in living mice. Reproduced with permission from [33], copyright 2018, Springer Nature.

RENPs boast a flexible fluorescence lifetime ranging from microseconds to milliseconds, which can be easily tuned based on various structural parameters such as dopants, their concentrations, and relay–inert layer thickness [83–85]. Gu et al. achieved a highly efficient conversion of pulsed excitation into long-decaying NIR luminescence without energy loss by using a NIR light transducer ($NaYbF_4@CaF_2$ core–shell RENPs) [82]. The optically inert CaF_2 shell encapsulating the Yb(III) signal transducers allowed for the tunable lifetime of RENPs spanning from 0.033 to 2.2 ms (emission: 980 nm) by adjusting the shell thickness (Figure 5c). Unlike fluorescence imaging, lifetime-domain multichannel imaging demonstrated the ability to distinguish two different types of nanoparticles with unique lifetimes (0.84 and 1.45 ms, respectively) utilized for labeling HeLa cells (Figure 5c). Meanwhile, Fan et al. systematically modulated the energy relay in ErNPs (Er-doped RENPs) to enable fluorescence imaging with tunable lifetimes, spanning three orders of magnitude (from microseconds to milliseconds) with 1525 nm emission [33]. Moreover, by conjugating lifetime-discriminating ErNPs (Er-τ5, Er-τ9, and Er-τ13) with antibodies that target various breast cancer receptors, including estrogen receptor (ER), progesterone receptor (PR), and human epidermal growth factor receptor 2 (HER2), in vivo fluorescence-lifetime multiplexing was utilized to identify the precise tumor subtypes in living mice (Figure 5d).

5. Fluorescence-Lifetime Multichannel Biosensing

Spectra-domain multiplexing is often hindered by errors in penetration depth that arise from wavelength variations, and lifetime-domain multiplexing lacks the ability to acquire fluorescence signals in real time during imaging because of the time-consuming fluorescence signal acquisition protocol [86]. Nevertheless, the flexibility of alternating between these two fluorescence imaging modes offers a groundbreaking solution to overcome the limitations of each technique. In the fluorescence channel, emissions with the same wavelength can be distinguished via a time-resolved method or dynamic variations in fluorescence intensity can be detected. In comparison, in the lifetime channel, the time-gating technique acquires delayed luminescence signals accordingly. By seamlessly integrating these imaging channels, we can visualize pathological environments in real time without any penetration-depth errors. This cutting-edge strategy unlocks new prospects for the advancement of multichannel biosensing applications.

Immunotherapy is a pivotal treatment for malignant tumors that has demonstrated significant clinical outcomes. However, it still faces several formidable challenges, including low response rates and immune-related adverse events [87,88]. Therefore, predicting and monitoring the therapeutic response to immunotherapy is of paramount importance to achieve better patient outcomes [89,90]. Since dual emission with the same wavelength can be separated by the time-gating technique to allow for the colocalization of multiple targets, avoiding the penetration-depth error (Figure 6a) [86], Zhong et al. developed a two-plex NIR-IIb imaging approach for in vivo visualization of immune responses using long-lived ErNPs-aPDL1 (lifetime: 4.2 ms) to tag tumor cells and short-lived PbS-aCD8 QDs (lifetime: 46 μs) to tag CD8+ T cells, respectively (Figure 6b) [91]. The infiltration of CD8+ T cells could be visualized instantly, which is promising for predicting the extent of immune activation [92]. Therefore, this approach could serve as a companion diagnostic for immunotherapy in clinical practice [71].

In contrast to dual-NIR-IIb emission at the same wavelength, the coherent recovery of fluorescence and lifetime allows for real-time in vivo localization and measurement through two channels at the same penetration depth (Figure 6c). Zhao et al. devised a TME (tumor microenvironment) nanosensor for hepatocellular carcinoma (HCC) detection by using β-$NaYF_4$@$NaYF_4$:1%Nd NPs and MY-1057 cyanine dyes, which respond to $ONOO^-$-induced variation in FRET between them (Figure 6d) [79]. Furthermore, the authors developed a luminescence resonance energy transfer (LRET) toolbox consisting of amorphous manganese oxide (MnO_x) and versatile lanthanide nanoparticles for quantifying intratumor glutathione (GSH) levels in vivo [93]. The results of a blind study showed that the quantified intratumor GSH levels of different types of tumors were highly consistent with the commercial kit results. In addition to the hybrid nanosensor, Chang et al. prepared a novel glutathione-capped copper–indium–selenium (CISe) nanotube with a NIR-II emission (1010–1130 nm) for phosphorescence imaging [94]. Notably, phosphorescence is a type of photoluminescence related to fluorescence that is featured with long-lived luminescence and large Stokes shifts. Unlike dual-channel signal recovery, CISe nanotubes have the unique ability to self-assemble in an acidic physiological environment (pH: 5.5–6.5) and switch from short-lived fluorescence to phosphorescence, allowing for the precise and accurate identification of tumor lesions through fluorescence-lifetime dual-channel imaging. (Figure 7). These results strongly indicate that a logical switch between imaging modes based on fluorescence and lifetime channels holds great promise for in vivo, real-time identification and quantification of a wide range of biomarkers.

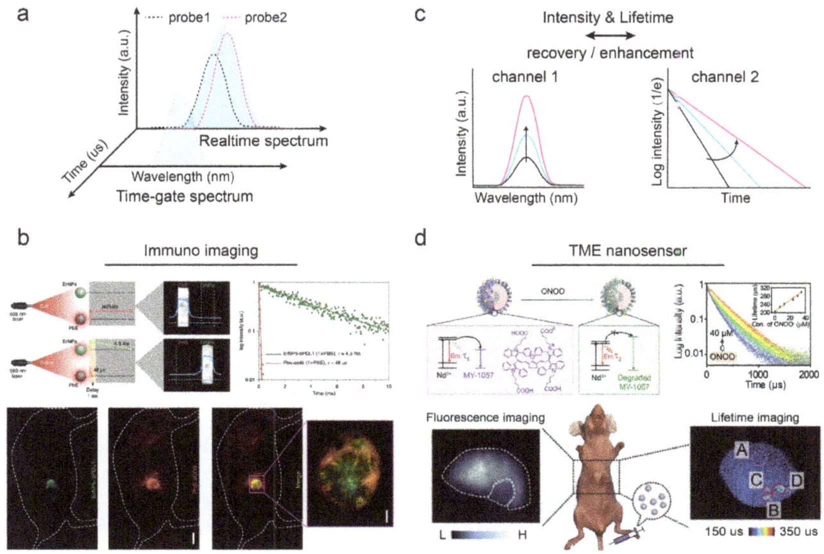

Figure 6. Fluorescence-lifetime multichannel biosensing: (**a**) schematic diagram of NIR-II signals separation through time-resolved technique; (**b**) NIR-II fluorescence-lifetime dual-channel imaging for monitoring immunotherapy using ErNPs (em:980 nm; ex: 1530 nm) and PbS/CdS QDs (em: 808/980 nm; em: 1600 nm). Reproduced with permission from [91], copyright 2019, Springer Nature; (**c**) schematic illustration of a responsive nanosensor with luminescence-lifetime synergistic recovery; (**d**) TME responsive ($ONOO^-$) nanosensor (DCNPs@MY-1057) with luminescence-lifetime dual-channel imaging for HCC liver tumor diagnosis. Reproduced with permission from [79], copyright 2020, Wiley-VCH.

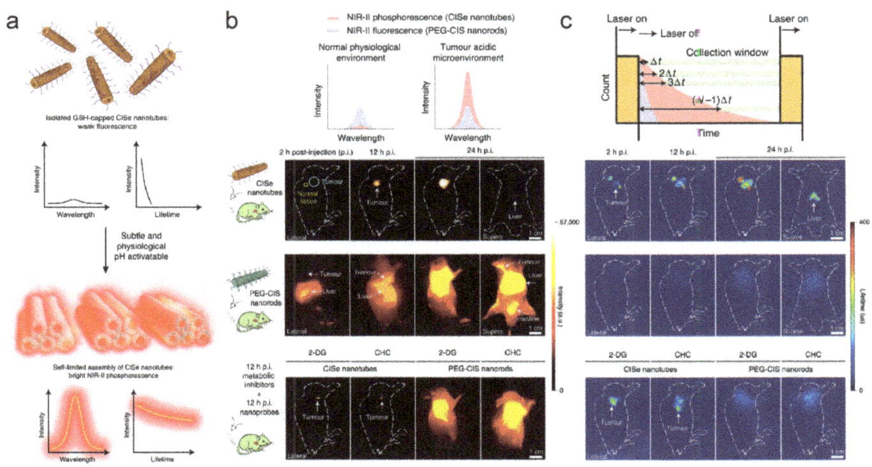

Figure 7. CISe nanotubes as a pH-responsive NIR-II phosphorescent probe for fluorescence-lifetime biosensing: (**a**) schematic illustration of emission and lifetime enhancement of CISe NIR-II phosphorescent probe upon pH responsiveness; (**b**) NIR-II fluorescence imaging and (**c**) time-resolved NIR-II imaging of living mice bearing 143B osteosarcoma cancer xenografts after intravenous administration of CISe nanotubes or PEG-CIS nanorods (dosage: 20 mg kg^{-1}). Reproduced with permission from [94], copyright 2021, Springer Nature.

6. Conclusions and Outlook

In conclusion, this review provides an overview of NIR-II fluorophores for multichannel fluorescence biosensing in two sections: imaging modes and biosensing applications. Based on the current development of NIR-II fluorophores, we divided multichannel biosensing into spectra-domain, lifetime-domain, and fluorescence-lifetime modes, and discussed their respective application scenarios (Table 1). With their respective merits, they can be applied in diverse scenarios. In vivo spectra-domain multiplexing can visualize pathological processes in real time; lifetime-domain multiplexing can pinpoint multiple targets with higher contrast; and fluorescence-lifetime multichannel biosensing can provide real-time and high-contrast imaging results. Multichannel NIR-II fluorescence biosensing is promising for noninvasive in vivo diagnosis with precision. Despite these impressive developments in multichannel fluorescence biosensing with the use of advanced NIR-II fluorophores, some fields are still worth exploring for their application in clinical practice.

Table 1. Summary of NIR-II fluorophores for multichannel biosensing.

NIR-II Sensor	Ex. (nm)	Em. (nm)	Mode	Application	Ref
Er(III)–bacteriochlorin complexes	766	1530	Spectra domain	Multiplexing of tissues and tumor cells	[59]
PbS/CdS QDs	808	1100&1600	Spectra domain	Multiplexing of MDSCs	[21]
$NaErF_4@NaYF_4@NaYF_4$:10%Nd $@NaYF_4$@A1094	808	1060&1525	Spectra domain	Ratiometric imaging of RNS	[63]
DCNP@IR760s	808&980	1550	Spectra domain	Ratiometric imaging of ROS to track NK cell viability	[64]
Ag_2S&Ag_2Se QDs	808	1050&1350	Spectra domain	Multiplexing-guided therapeutic schedule	[70]
PbS/CdS QDs &IR-FD dye	808	1100&1600	Spectra domain	Multiplexing-guided surgical navigation	[72]
$NaErF_4@NaYF_4$ @ZW800	808&980	1550	Spectra domain	Ratiometric imaging of GzB to diagnose allograft rejection	[73]
$NaYbF_4@CaF_2$	920	980	Lifetime domain	Tumor cells labeling	[82]
$NaGdF_4@NaGdF_4$:Yb,Er@$NaYF_4$:Yb@$NaNdF_4$:Yb	808	1525	Lifetime domain	Tumor biomarkers sensing	[33]
$NaYbF_4$:Er@$NaYF_4$& PbS QDs	980&808	1530&1600	Fluorescence lifetime	Tumor cells labeling	[91]
$NaYF_4@NaYF_4$:Nd@MY-1057	808	1060	Fluorescence lifetime	HCC tumor detection	[79]

First of all, the potential biotoxicity of nanoprobes induced by long-term retention often lacks systematic investigation. Currently, most researchers focus on the rational design of NIR-II fluorophores to achieve distinctive functions for specific application scenarios. However, large-sized and charged nanoprobes can be captured using the reticuloendothelial system and remain in live mammalians for an extended period of time (1 month or longer) after biosensing, which may cause biosafety issues [20,95]. Therefore, designing long-circulating and renal-clearable NIR-II fluorophores for multichannel biosensing could be a way forward. Renal-clearable NIR-I nanosensors, such as golden nanoclusters or polyfluorophore nanoprobes, are applicable for in vivo detection and humoral diagnosis in real time at a cellular level [96–98], which poses fewer concerns on the biosafety of these fluorophores during clinical implementation.

In addition to metabolizability, NIR-II fluorophores with long-wavelength emission in the NIR-IIb/NIR-IIc subwindow (NIR-IIb: 1500–1700 nm; NIR-IIc:1700–1880 nm) are favorable for deep-tissue biosensing with high resolution and contrast [99,100]. In terms of biosensing in the fluorescence-lifetime mode, fluorescent probes with responsiveness to stimulants in lifetime channels are still limited to the 950–1300 nm range [79,93]. Striking a delicate balance between the luminescence intensity and the lifetime of fluorescent probes is crucial in determining their effective use in practical applications. To meet this challenge, improving the luminescence intensity of NIR-IIb/NIR-IIc fluorophores while up- or downregulating the fluorescence lifetime might be the solution.

Last but not least, simplifying the components of NIR-II fluorophores can pave the way for clinical implementation. Hybrid nanoprobes remain the mainstream design for in vivo ratiometric quantification of various chemical substances (ROS, RNS, GSH, etc.), either in the form of chemical conjugation or physicochemical enclosure [54,62,64,66,73,77,79,93,101]. However, they may encounter dissociation or the release of components in physiological environments, posing significant biosafety risks. Our group has reported a single-component NIR-II fluorophore (LC-1250) for ratiometric fluorescence imaging of H_2O_2 to diagnose gastroenteritis with high sensitivity under physiological conditions (pH 1–2 in the stomach) [23], exemplifying that the ingenious design of organic dyes is key. Moreover, by endowing renal-clearable functionality, extending to longer emission wavelengths, or even incorporating portable devices [102], new fluorescent probes based on organic dyes could be developed for multichannel biosensing in future clinical practice.

Author Contributions: Conceptualization, writing—original draft preparation, F.R.; writing—review and editing, F.R., T.L. and T.Y.; visualization, G.C., C.L. and Q.W.; supervision, G.C., C.L. and Q.W.; project administration, Q.W.; funding acquisition, Q.W. All authors have read and agreed to the published version of the manuscript.

Funding: This research was funded by the National Natural Science Foundation of China (Grant Nos. 22201297, 21934007, 22127808, 22271308, 22174158, and 22177128); the National Key Research and Development Program (2021YFF0701804); the Chinese Academy of Sciences (Grant Nos. XDB36000000, ZDBS-LY-SLH021, and YJKYYQ20200036); the China Postdoctoral Science Foundation (Grant 2022M712327); the Science and Technology Project of Suzhou (Nos. SZS201904 and SJC2021001); the Natural Science Foundation of Jiangsu Province (Grant Nos. BK20222016, BE2022753, and BE2022745); and Jiangsu Funding Program for Excellent Postdoctoral Talent.

Institutional Review Board Statement: Not applicable.

Informed Consent Statement: Not applicable.

Data Availability Statement: There were no new data created or analyzed in this study. Data sharing does not apply to this article.

Acknowledgments: The authors thank researchers at Suzhou NIR-Optics Technology Co., Ltd., for their suggestions on the future development of in vivo NIR-II biosensing.

Conflicts of Interest: The authors declare no conflict of interest.

References

1. Li, C.; Wang, Q. Challenges and opportunities for intravital near-infrared fluorescence imaging technology in the second transparency window. *ACS Nano* **2018**, *12*, 9654–9659. [CrossRef]
2. Weissleder, R.; Pittet, M.J. Imaging in the era of molecular oncology. *Nature* **2008**, *452*, 580–589. [CrossRef]
3. Wang, K.; Du, Y.; Zhang, Z.; He, K.; Cheng, Z.; Yin, L.; Dong, D.; Li, C.; Li, W.; Hu, Z.; et al. Fluorescence image-guided tumour surgery. *Nat. Rev. Bioeng.* **2023**, *1*, 161–179. [CrossRef]
4. Andreou, C.; Weissleder, R.; Kircher, M.F. Multiplexed imaging in oncology. *Nat. Biomed. Eng.* **2022**, *6*, 527–540. [CrossRef]
5. Hu, Z.; Fang, C.; Li, B.; Zhang, Z.; Cao, C.; Cai, M.; Su, S.; Sun, X.; Shi, X.; Li, C.; et al. First-in-human liver-tumour surgery guided by multispectral fluorescence imaging in the visible and near-infrared-I/II windows. *Nat. Biomed. Eng.* **2020**, *4*, 259–271. [CrossRef] [PubMed]

6. Fan, Y.; Wang, S.; Zhang, F. Optical multiplexed bioassays for improved biomedical diagnostics. *Angew. Chem. Int. Ed.* **2019**, *58*, 13208–13219. [CrossRef] [PubMed]
7. Ma, T.; Hou, Y.; Zeng, J.; Liu, C.; Zhang, P.; Jing, L.; Shangguan, D.; Gao, M. Dual-ratiometric target-triggered fluorescent probe for simultaneous quantitative visualization of tumor microenvironment protease activity and pH in vivo. *J. Am. Chem. Soc.* **2018**, *140*, 211–218. [CrossRef] [PubMed]
8. Hong, G.S.; Antaris, A.L.; Dai, H.J. Near-infrared fluorophores for biomedical imaging. *Nat. Biomed. Eng.* **2017**, *1*, 0010. [CrossRef]
9. Chen, Y.; Wang, S.; Zhang, F. Near-infrared luminescence high-contrast in vivo biomedical imaging. *Nat. Rev. Bioeng.* **2023**, *1*, 60–78. [CrossRef]
10. Chen, G.; Cao, Y.; Tang, Y.; Yang, X.; Liu, Y.; Huang, D.; Zhang, Y.; Li, C.; Wang, Q. Advanced near-infrared light for monitoring and modulating the spatiotemporal dynamics of cell functions in living systems. *Adv. Sci.* **2020**, *7*, 1903783. [CrossRef]
11. Yang, H.; Huang, H.; Ma, X.; Zhang, Y.; Yang, X.; Yu, M.; Sun, Z.; Li, C.; Wu, F.; Wang, Q. Au-doped Ag_2Te quantum dots with bright NIR-IIb fluorescence for in situ monitoring of angiogenesis and arteriogenesis in a hindlimb ischemic model. *Adv. Mater.* **2021**, *33*, e2103953. [CrossRef] [PubMed]
12. Hong, G.S.; Robinson, J.T.; Zhang, Y.J.; Diao, S.; Antaris, A.L.; Wang, Q.B.; Dai, H.J. In vivo fluorescence imaging with Ag_2S quantum dots in the second near-infrared region. *Angew. Chem. Int. Ed.* **2012**, *51*, 9818–9821. [CrossRef] [PubMed]
13. Li, C.; Chen, G.; Zhang, Y.; Wu, F.; Wang, Q. Advanced fluorescence imaging technology in the near-infrared-II window for biomedical applications. *J. Am. Chem. Soc.* **2020**, *142*, 14789–14804. [CrossRef] [PubMed]
14. Sun, C.X.; Li, B.H.; Zhao, M.Y.; Wang, S.F.; Lei, Z.H.; Lu, L.F.; Zhang, H.X.; Feng, L.S.; Dou, C.R.; Yin, D.R.; et al. J-aggregates of cyanine dye for NIR-II in vivo dynamic vascular imaging beyond 1500 nm. *J. Am. Chem. Soc.* **2019**, *141*, 19221–19225. [CrossRef]
15. Yang, Y.; Sun, C.; Wang, S.; Yan, K.; Zhao, M.; Wu, B.; Zhang, F. Counterion-paired bright heptamethine fluorophores with NIR-II excitation and emission enable multiplexed biomedical imaging. *Angew. Chem. Int. Ed.* **2022**, *61*, e202117436.
16. Lu, L.; Li, B.; Ding, S.; Fan, Y.; Wang, S.; Sun, C.; Zhao, M.; Zhao, C.X.; Zhang, F. NIR-II bioluminescence for in vivo high contrast imaging and in situ ATP-mediated metastases tracing. *Nat. Commun.* **2020**, *11*, 4192. [CrossRef]
17. Bruns, O.T.; Bischof, T.S.; Harris, D.K.; Franke, D.; Shi, Y.; Riedemann, L.; Bartelt, A.; Jaworski, F.B.; Carr, J.A.; Rowlands, C.J.; et al. Next-generation in vivo optical imaging with short-wave infrared quantum dots. *Nat. Biomed. Eng.* **2017**, *1*, 0056. [CrossRef]
18. Du, Y.P.; Xu, B.; Fu, T.; Cai, M.; Li, F.; Zhang, Y.; Wang, Q.B. Near-infrared photoluminescent Ag_2S quantum dots from a single source precursor. *J. Am. Chem. Soc.* **2010**, *132*, 1470–1471. [CrossRef]
19. Zhu, X.; Liu, X.; Zhang, H.; Zhao, M.; Pei, P.; Chen, Y.; Yang, Y.; Lu, L.; Yu, P.; Sun, C.; et al. High-fidelity NIR-II multiplexed lifetime bioimaging with bright double interfaced lanthanide nanoparticles. *Angew. Chem. Int. Ed.* **2021**, *60*, 23545–23551. [CrossRef]
20. Ma, Z.; Wang, F.; Zhong, Y.; Salazar, F.; Li, J.; Zhang, M.; Ren, F.; Wu, A.M.; Dai, H. Cross-link-functionalized nanoparticles for rapid excretion in nanotheranostic applications. *Angew. Chem. Int. Ed.* **2020**, *59*, 20552–20560. [CrossRef]
21. Yu, G.T.; Luo, M.Y.; Li, H.; Chen, S.; Huang, B.; Sun, Z.J.; Cui, R.; Zhang, M. Molecular targeting nanoprobes with non-Overlap emission in the second near-infrared window for in vivo two-color colocalization of immune cells. *ACS Nano* **2019**, *13*, 12830–12839. [CrossRef] [PubMed]
22. Zhu, X.; Wang, X.; Zhang, H.; Zhang, F. Luminescence lifetime imaging based on lanthanide nanoparticles. *Angew. Chem. Int. Ed.* **2022**, *61*, e202209378. [CrossRef]
23. Li, T.; Cao, K.; Yang, X.; Liu, Y.; Wang, X.; Wu, F.; Chen, G.; Wang, Q. An oral ratiometric NIR-II fluorescent probe for reliable monitoring of gastrointestinal diseases in vivo. *Biomaterials* **2023**, *293*, 121956. [CrossRef] [PubMed]
24. Lan, Q.; Yu, P.; Yan, K.; Li, X.; Zhang, F.; Lei, Z. Polymethine molecular platform for ratiometric fluorescent probes in the second near-infrared window. *J. Am. Chem. Soc.* **2022**, *144*, 21010–21015. [CrossRef] [PubMed]
25. Yu, M.; Yang, X.; Zhang, Y.; Yang, H.; Huang, H.; Wang, Z.; Dong, J.; Zhang, R.; Sun, Z.; Li, C.; et al. Pb-doped Ag_2Se quantum dots with enhanced photoluminescence in the NIR-II window. *Small* **2021**, *17*, e2006111. [CrossRef]
26. Ortgies, D.H.; Tan, M.; Ximendes, E.C.; Del Rosal, B.; Hu, J.; Xu, L.; Wang, X.; Martin Rodriguez, E.; Jacinto, C.; Fernandez, N.; et al. Lifetime-encoded infrared-emitting nanoparticles for in vivo multiplexed imaging. *ACS Nano* **2018**, *12*, 4362–4368. [CrossRef]
27. Welsher, K.; Liu, Z.; Sherlock, S.P.; Robinson, J.T.; Chen, Z.; Daranciang, D.; Dai, H. A route to brightly fluorescent carbon nanotubes for near-infrared imaging in mice. *Nat. Nanotechnol.* **2009**, *4*, 773–780. [CrossRef]
28. Weidman, M.C.; Beck, M.E.; Hoffman, R.S.; Prins, F.; Tisdale, W.A. Monodisperse, air-stable PbS nanocrystals via precursor stoichiometry control. *ACS Nano* **2014**, *8*, 6363–6371. [CrossRef]
29. Hines, M.A.; Scholes, G.D. Colloidal PbS nanocrystals with size-tunable near-infrared emission: Observation of post-synthesis self-narrowing of the particle size distribution. *Adv. Mater.* **2003**, *15*, 1844–1849. [CrossRef]
30. Yang, H.; Li, R.; Zhang, Y.; Yu, M.; Wang, Z.; Liu, X.; You, W.; Tu, D.; Sun, Z.; Zhang, R.; et al. Colloidal alloyed quantum dots with enhanced photoluminescence quantum yield in the NIR-II window. *J. Am. Chem. Soc.* **2021**, *143*, 2601–2607. [CrossRef]
31. Naczynski, D.J.; Tan, M.C.; Zevon, M.; Wall, B.; Kohl, J.; Kulesa, A.; Chen, S.; Roth, C.M.; Riman, R.E.; Moghe, P.V. Rare-earth-doped biological composites as in vivo shortwave infrared reporters. *Nat. Commun.* **2013**, *4*, 2199. [CrossRef]
32. Johnson, N.J.; He, S.; Diao, S.; Chan, E.M.; Dai, H.; Almutairi, A. Direct evidence for coupled surface and concentration quenching dynamics in lanthanide-doped nanocrystals. *J. Am. Chem. Soc.* **2017**, *139*, 3275–3282. [CrossRef]

33. Fan, Y.; Wang, P.; Lu, Y.; Wang, R.; Zhou, L.; Zheng, X.; Li, X.; Piper, J.A.; Zhang, F. Lifetime-engineered NIR-II nanoparticles unlock multiplexed in vivo imaging. *Nat. Nanotechnol.* **2018**, *13*, 941–946. [CrossRef] [PubMed]
34. Bricks, J.L.; Kachkovskii, A.D.; Slominskii, Y.L.; Gerasov, A.O.; Popov, S.V. Molecular design of near infrared polymethine dyes: A review. *Dyes Pigments* **2015**, *121*, 238–255. [CrossRef]
35. Li, B.; Lu, L.; Zhao, M.; Lei, Z.; Zhang, F. An efficient 1064 nm NIR-II excitation fluorescent molecular dye for deep-tissue high-resolution dynamic bioimaging. *Angew. Chem. Int. Ed.* **2018**, *57*, 7483–7487. [CrossRef]
36. Cosco, E.D.; Spearman, A.L.; Ramakrishnan, S.; Lingg, J.G.P.; Saccomano, M.; Pengshung, M.; Arus, B.A.; Wong, K.C.Y.; Glasl, S.; Ntziachristos, V.; et al. Shortwave infrared polymethine fluorophores matched to excitation lasers enable non-invasive, multicolour in vivo imaging in real time. *Nat. Chem.* **2020**, *12*, 1123–1130. [CrossRef] [PubMed]
37. Antaris, A.L.; Chen, H.; Cheng, K.; Sun, Y.; Hong, G.S.; Qu, C.R.; Diao, S.; Deng, Z.X.; Hu, X.M.; Zhang, B.; et al. A small-molecule dye for NIR-II imaging. *Nat. Mater.* **2016**, *15*, 235–242. [CrossRef] [PubMed]
38. Bouit, P.A.; Aronica, C.; Toupet, L.; Guennic, B.L.; Andraud, C.; Maury, O. Continuous symmetry breaking induced by ion pairing effect in heptamethine cyanine dyes: Beyond the cyanine limit. *J. Am. Chem. Soc.* **2010**, *132*, 4328–4335. [CrossRef]
39. Luo, J.; Xie, Z.; Lam, J.W.; Cheng, L.; Chen, H.; Qiu, C.; Kwok, H.S.; Zhan, X.; Liu, Y.; Zhu, D.; et al. Aggregation-induced emission of 1-methyl-1,2,3,4,5-pentaphenylsilole. *Chem. Commun.* **2001**, *18*, 1740–1741. [CrossRef]
40. Li, Y.Y.; Liu, S.J.; Ni, H.W.; Zhang, H.; Zhang, H.Q.; Chuah, C.; Ma, C.; Wong, K.S.; Lam, J.W.Y.; Kwok, R.T.K.; et al. ACQ-to-AIE transformation: Tuning molecular packing by regioisomerization for two-photon NIR bioimaging. *Angew. Chem. Int. Ed.* **2020**, *59*, 12822–12826. [CrossRef]
41. Sheng, Z.; Guo, B.; Hu, D.; Xu, S.; Wu, W.; Liew, W.H.; Yao, K.; Jiang, J.; Liu, C.; Zheng, H.; et al. Bright aggregation-induced-emission dots for targeted synergetic NIR-II fluorescence and NIR-I photoacoustic imaging of orthotopic brain tumors. *Adv. Mater.* **2018**, *30*, e1800766. [CrossRef]
42. Li, Y.; Cai, Z.; Liu, S.; Zhang, H.; Wong, S.T.H.; Lam, J.W.Y.; Kwok, R.T.K.; Qian, J.; Tang, B.Z. Design of AIEgens for near-infrared IIb imaging through structural modulation at molecular and morphological levels. *Nat. Commun.* **2020**, *11*, 1255. [CrossRef] [PubMed]
43. Hong, Y.; Geng, W.; Zhang, T.; Gong, G.; Li, C.; Zheng, C.; Liu, F.; Qian, J.; Chen, M.; Tang, B.Z. Facile access to far-red fluorescent probes with through-space charge-transfer effects for in vivo two-photon microscopy of the mouse cerebrovascular system. *Angew. Chem. Int. Ed.* **2022**, *61*, e202209590. [CrossRef] [PubMed]
44. Qin, W.; Alifu, N.; Lam, J.W.Y.; Cui, Y.; Su, H.; Liang, G.; Qian, J.; Tang, B.Z. Facile synthesis of efficient luminogens with AIE features for three-photon fluorescence imaging of the brain through the intact skull. *Adv. Mater.* **2020**, *32*, e2000364. [CrossRef] [PubMed]
45. Qian, J.; Zhu, Z.; Leung, C.W.; Xi, W.; Su, L.; Chen, G.; Qin, A.; Tang, B.Z.; He, S. Long-term two-photon neuroimaging with a photostable AIE luminogen. Biomed. *Opt. Express* **2015**, *6*, 1477–1486. [CrossRef]
46. Mandal, A.K.; Sreejith, S.; He, T.H.; Maji, S.K.; Wang, X.J.; Ong, S.L.; Joseph, J.; Sun, H.D.; Zhao, Y.L. Three-photon-excited luminescence from unsymmetrical cyanostilbene aggregates: Morphology tuning and targeted bioimaging. *ACS Nano* **2015**, *9*, 4796–4805. [CrossRef]
47. Yao, C.; Chen, Y.; Zhao, M.; Wang, S.; Wu, B.; Yang, Y.; Yin, D.; Yu, P.; Zhang, H.; Zhang, F. A bright, renal-clearable NIR-II brush macromolecular probe with long blood circulation time for kidney disease bioimaging. *Angew. Chem. Int. Ed.* **2022**, *61*, e202114273. [CrossRef]
48. Wang, F.; Qu, L.; Ren, F.; Baghdasaryan, A.; Jiang, Y.; Hsu, R.; Liang, P.; Li, J.; Zhu, G.; Ma, Z.; et al. High-precision tumor resection down to few-cell level guided by NIR-IIb molecular fluorescence imaging. *Proc. Natl. Acad. Sci. USA* **2022**, *119*, e2123111119. [CrossRef]
49. Zhang, M.X.; Yue, J.Y.; Cui, R.; Ma, Z.R.; Wan, H.; Wang, F.F.; Zhu, S.J.; Zhou, Y.; Kuang, Y.; Zhong, Y.T.; et al. Bright quantum dots emitting at ~1600 nm in the NIR-IIb window for deep tissue fluorescence imaging. *Proc. Natl. Acad. Sci. USA* **2018**, *115*, 6590–6595. [CrossRef]
50. Ren, F.; Liu, H.H.; Zhang, H.; Jiang, Z.L.; Xia, B.; Genevois, C.; He, T.; Allix, M.; Sun, Q.; Li, Z.; et al. Engineering NIR-IIb fluorescence of Er-based lanthanide nanoparticles for through-skull targeted imaging and imaging-guided surgery of orthotopic glioma. *Nano Today* **2020**, *34*, 100905. [CrossRef]
51. Zhang, H.; Fu, P.; Liu, Y.; Zheng, Z.; Zhu, L.; Wang, M.; Abdellah, M.; He, M.; Qian, J.; Roe, A.W.; et al. Large-depth three-photon fluorescence microscopy imaging of cortical microvasculature on nonhuman primates with bright AIE probe in vivo. *Biomaterials* **2022**, *289*, 121809. [CrossRef]
52. He, M.; Li, D.; Zheng, Z.; Zhang, H.; Wu, T.; Geng, W.; Hu, Z.; Feng, Z.; Peng, S.; Zhu, L.; et al. Aggregation-induced emission nanoprobe assisted ultra-deep through-skull three-photon mouse brain imaging. *Nano Today* **2022**, *45*, 101536. [CrossRef]
53. Zheng, Z.; Zhang, H.; Cao, H.; Gong, J.; He, M.; Gou, X.; Yang, T.; Wei, P.; Qian, J.; Xi, W.; et al. Intra- and intermolecular synergistic engineering of aggregation-induced emission luminogens to boost three-photon absorption for through-skull brain imaging. *ACS Nano* **2022**, *16*, 6444–6454. [CrossRef] [PubMed]
54. Wang, T.; Chen, Y.; He, Z.; Wang, X.; Wang, S.; Zhang, F. Molecular-based FRET nanosensor with dynamic ratiometric NIR-IIb fluorescence for real-time in vivo imaging and sensing. *Nano Lett.* **2023**, *23*, 4548–4556. [CrossRef]
55. Wang, F.; Wan, H.; Ma, Z.; Zhong, Y.; Sun, Q.; Tian, Y.; Qu, L.; Du, H.; Zhang, M.; Li, L.; et al. Light-sheet microscopy in the near-infrared II window. *Nat. Methods* **2019**, *16*, 545–552. [CrossRef]

56. Dong, H.; Sun, L.D.; Yan, C.H. Local structure engineering in lanthanide-doped nanocrystals for tunable upconversion emissions. *J. Am. Chem. Soc.* **2021**, *143*, 20546–20561. [CrossRef]
57. Zhong, Y.; Ma, Z.; Zhu, S.; Yue, J.; Zhang, M.; Antaris, A.L.; Yuan, J.; Cui, R.; Wan, H.; Zhou, Y.; et al. Boosting the down-shifting luminescence of rare-earth nanocrystals for biological imaging beyond 1500 nm. *Nat. Commun.* **2017**, *8*, 737. [CrossRef] [PubMed]
58. Pei, P.; Chen, Y.; Sun, C.; Fan, Y.; Yang, Y.; Liu, X.; Lu, L.; Zhao, M.; Zhang, H.; Zhao, D.; et al. X-ray-activated persistent luminescence nanomaterials for NIR-II imaging. *Nat. Nanotechnol.* **2021**, *16*, 1011–1018. [CrossRef]
59. Wang, T.; Wang, S.; Liu, Z.; He, Z.; Yu, P.; Zhao, M.; Zhang, H.; Lu, L.; Wang, Z.; Wang, Z.; et al. A hybrid erbium(III)-bacteriochlorin near-infrared probe for multiplexed biomedical imaging. *Nat. Mater.* **2021**, *20*, 1571–1578. [CrossRef] [PubMed]
60. Bakueva, L.; Gorelikov, I.; Musikhin, S.; Zhao, X.S.; Sargent, E.H.; Kumacheva, E. PbS quantum dots with stable efficient luminescence in the NIR spectral range. *Adv. Mater.* **2004**, *16*, 926–929. [CrossRef]
61. McDonald, S.A.; Konstantatos, G.; Zhang, S.; Cyr, P.W.; Klem, E.J.; Levina, L.; Sargent, E.H. Solution-processed PbS quantum dot infrared photodetectors and photovoltaics. *Nat. Mater.* **2005**, *4*, 138–142. [CrossRef]
62. Yu, P.; Yan, K.; Wang, S.; Yao, C.; Lei, Z.; Tang, Y.; Zhang, F. NIR-II dyad-doped ratiometric nanosensor with enhanced spectral fidelity in biological media for in vivo biosensing. *Nano Lett.* **2022**, *22*, 9732–9740. [CrossRef] [PubMed]
63. Sun, Z.; Huang, H.; Zhang, R.; Yang, X.; Yang, H.; Li, C.; Zhang, Y.; Wang, Q. Activatable rare earth near-infrared-II fluorescence ratiometric nanoprobes. *Nano Lett.* **2021**, *21*, 6576–6583. [CrossRef] [PubMed]
64. Liao, N.; Su, L.; Zheng, Y.; Zhao, B.; Wu, M.; Zhang, D.; Yang, H.; Liu, X.; Song, J. In vivo tracking of cell viability for adoptive natural killer cell-based immunotherapy by ratiometric NIR-II fluorescence imaging. *Angew. Chem. Int. Ed.* **2021**, *60*, 20888–20896. [CrossRef] [PubMed]
65. Cui, D.; Li, J.; Zhao, X.; Pu, K.; Zhang, R. Semiconducting polymer nanoreporters for near-infrared chemiluminescence imaging of immunoactivation. *Adv. Mater.* **2020**, *32*, e1906314. [CrossRef]
66. Ramesh, A.; Kumar, S.; Brouillard, A.; Nandi, D.; Kulkarni, A. A nitric oxide (NO) nanoreporter for noninvasive real-time imaging of macrophage immunotherapy. *Adv. Mater.* **2020**, *32*, e2000648. [CrossRef]
67. Huang, Y.; Snuderl, M.; Jain, R.K. Polarization of tumor-associated macrophages: A novel strategy for vascular normalization and antitumor immunity. *Cancer Cell* **2011**, *19*, 1–2. [CrossRef]
68. Huang, Y.; Kim, B.Y.S.; Chan, C.K.; Hahn, S.M.; Weissman, I.L.; Jiang, W. Improving immune-vascular crosstalk for cancer immunotherapy. *Nat. Rev. Immunol.* **2018**, *18*, 195–203. [CrossRef]
69. Pittet, M.J.; Garris, C.S.; Arlauckas, S.P.; Weissleder, R. Recording the wild lives of immune cells. *Sci. Immunol.* **2018**, *3*, eaaq0491. [CrossRef]
70. Hao, X.; Li, C.; Zhang, Y.; Wang, H.; Chen, G.; Wang, M.; Wang, Q. Programmable chemotherapy and immunotherapy against breast cancer guided by multiplexed fluorescence imaging in the second near-infrared window. *Adv. Mater.* **2018**, *30*, e1804437. [CrossRef]
71. Lucero, M.Y.; Chan, J. Photoacoustic imaging of elevated glutathione in models of lung cancer for companion diagnostic applications. *Nat. Chem.* **2021**, *13*, 1248–1256. [CrossRef]
72. Tian, R.; Ma, H.; Zhu, S.; Lau, J.; Ma, R.; Liu, Y.; Lin, L.; Chandra, S.; Wang, S.; Zhu, X.; et al. Multiplexed N IR-II probes for lymph node-invaded cancer detection and imaging-guided surgery. *Adv. Mater.* **2020**, *32*, e1907365. [CrossRef] [PubMed]
73. Chen, Y.; Pei, P.; Yang, Y.; Zhang, H.; Zhang, F. Noninvasive early diagnosis of allograft rejection by a granzyme B protease responsive NIR-II bioimaging nanosensor. *Angew. Chem. Int. Ed.* **2023**, *62*, e202301696. [CrossRef] [PubMed]
74. Kantamneni, H.; Zevon, M.; Donzanti, M.J.; Zhao, X.; Sheng, Y.; Barkund, S.R.; McCabe, L.H.; Banach-Petrosky, W.; Higgins, L.M.; Ganesan, S.; et al. Surveillance nanotechnology for multi-organ cancer metastases. *Nat. Biomed. Eng.* **2017**, *1*, 993–1003. [CrossRef] [PubMed]
75. Duneton, C.; Winterberg, P.D.; Ford, M.L. Activation and regulation of alloreactive T cell immunity in solid organ transplantation. *Nat. Rev. Nephrol.* **2022**, *18*, 663–676. [CrossRef] [PubMed]
76. Cherry, C.; Maestas, D.R.; Han, J.; Andorko, J.I.; Cahan, P.; Fertig, E.J.; Garmire, L.X.; Elisseeff, J.H. Computational reconstruction of the signalling networks surrounding implanted biomaterials from single-cell transcriptomics. *Nat. Biomed. Eng.* **2021**, *5*, 1228–1238. [CrossRef]
77. Pei, P.; Hu, H.; Chen, Y.; Wang, S.; Chen, J.; Ming, J.; Yang, Y.; Sun, C.; Zhao, S.; Zhang, F. NIR-II ratiometric lanthanide-dye hybrid nanoprobes doped bioscaffolds for in situ bone repair monitoring. *Nano Lett.* **2022**, *22*, 783–791. [CrossRef]
78. Del Rosal, B.; Benayas, A. Strategies to overcome autofluorescence in nanoprobe-driven in vivo fluorescence imaging. *Small Methods* **2018**, *2*, 1800075. [CrossRef]
79. Zhao, M.; Li, B.; Wu, Y.; He, H.; Zhu, X.; Zhang, H.; Dou, C.; Feng, L.; Fan, Y.; Zhang, F. A tumor-microenvironment-responsive lanthanide-cyanine FRET sensor for NIR-II luminescence-lifetime in situ Imaging of hepatocellular carcinoma. *Adv. Mater.* **2020**, *32*, e2001172. [CrossRef]
80. Becker, W. Fluorescence lifetime imaging—Techniques and applications. *J. Microsc.* **2012**, *247*, 119–136. [CrossRef]
81. Del Rosal, B.; Ortgies, D.H.; Fernandez, N.; Sanz-Rodriguez, F.; Jaque, D.; Rodriguez, E.M. Overcoming autofluorescence: Long-lifetime infrared nanoparticles for time-gated in vivo imaging. *Adv. Mater.* **2016**, *28*, 10188–10193. [CrossRef]
82. Gu, Y.; Guo, Z.; Yuan, W.; Kong, M.; Liu, Y.; Liu, Y.; Gao, Y.; Feng, W.; Wang, F.; Zhou, J.; et al. High-sensitivity imaging of time-domain near-infrared light transducer. *Nat. Photon.* **2019**, *13*, 525–531. [CrossRef]

83. Li, H.; Tan, M.; Wang, X.; Li, F.; Zhang, Y.; Zhao, L.; Yang, C.; Chen, G. Temporal multiplexed in vivo upconversion imaging. *J. Am. Chem. Soc.* **2020**, *142*, 2023–2030. [CrossRef]
84. Wu, L.; Jia, M.; Li, D.; Chen, G. Shell engineering on thermal sensitivity of lifetime-based NIR nanothermometers for accurate temperature measurement in murine internal liver organ. *Nano Lett.* **2023**, *23*, 2862–2869. [CrossRef] [PubMed]
85. Lu, Y.; Zhao, J.; Zhang, R.; Liu, Y.; Liu, D.; Goldys, E.M.; Yang, X.; Xi, P.; Sunna, A.; Lu, J.; et al. Tunable lifetime multiplexing using luminescent nanocrystals. *Nat. Photon.* **2013**, *8*, 32–36. [CrossRef]
86. Qiu, X.; Zhou, Q.; Zhu, X.; Wu, Z.; Feng, W.; Li, F. Ratiometric upconversion nanothermometry with dual emission at the same wavelength decoded via a time-resolved technique. *Nat. Commun.* **2020**, *11*, 4. [CrossRef] [PubMed]
87. Ribas, J.D.W. Cancer immunotherapy using checkpoint blockade. *Science* **2018**, *359*, 1350–1355. [CrossRef]
88. June, C.H.; O'Connor, R.S.; Kawalekar, O.U.; Ghassemi, S.; Milone, M.C. CAR T cell immunotherapy for human cancer. *Science* **2018**, *359*, 1361–1365. [CrossRef]
89. Bensch, F.; van der Veen, E.L.; Lub-de Hooge, M.N.; Jorritsma-Smit, A.; Boellaard, R.; Kok, I.C.; Oosting, S.F.; Schroder, C.P.; Hiltermann, T.J.N.; van der Wekken, A.J.; et al. (89)Zr-atezolizumab imaging as a non-invasive approach to assess clinical response to PD-L1 blockade in cancer. *Nat. Med.* **2018**, *24*, 1852–1858. [CrossRef]
90. Zou, W.; Wolchok, J.D.; Chen, L. PD-L1 (B7-H1) and PD-1 pathway blockade for cancer therapy: Mechanisms, response biomarkers, and combinations. *Sci. Transl. Med.* **2016**, *8*, 328rv4. [CrossRef]
91. Zhong, Y.; Ma, Z.; Wang, F.; Wang, X.; Yang, Y.; Liu, Y.; Zhao, X.; Li, J.; Du, H.; Zhang, M.; et al. In vivo molecular imaging for immunotherapy using ultra-bright near-infrared-IIb rare-earth nanoparticles. *Nat. Biotechnol.* **2019**, *37*, 1322–1331. [CrossRef] [PubMed]
92. Ma, X.; Zhang, M.J.; Wang, J.; Zhang, T.; Xue, P.; Kang, Y.; Sun, Z.J.; Xu, Z. Emerging biomaterials imaging antitumor immune response. *Adv. Mater.* **2022**, *34*, e2204034. [CrossRef] [PubMed]
93. Zhao, M.; Zhuang, H.; Zhang, H.; Li, B.; Ming, J.; Chen, X.; Chen, M. A LRET nanoplatform consisting of lanthanide and amorphous manganese oxide for NIR-II luminescence lifetime imaging of tumor redox status. *Angew. Chem. Int. Ed.* **2022**, *61*, e202209592. [CrossRef] [PubMed]
94. Chang, B.; Li, D.; Ren, Y.; Qu, C.; Shi, X.; Liu, R.; Liu, H.; Tian, J.; Hu, Z.; Sun, T.; et al. A phosphorescent probe for in vivo imaging in the second near-infrared window. *Nat. Biomed. Eng.* **2021**, *6*, 629–639. [CrossRef]
95. Liu, J.; Wang, P.; Zhang, X.; Wang, L.; Wang, D.; Gu, Z.; Tang, J.; Guo, M.; Cao, M.; Zhou, H.; et al. Rapid degradation and high renal clearance of Cu_3BiS_3 nanodots for efficient cancer diagnosis and photothermal therapy in vivo. *ACS Nano* **2016**, *10*, 4587–4598. [CrossRef]
96. He, S.; Cheng, P.; Pu, K. Activatable near-infrared probes for the detection of specific populations of tumour-infiltrating leukocytes in vivo and in urine. *Nat. Biomed. Eng.* **2023**, *7*, 281–297. [CrossRef]
97. Huang, J.; Chen, X.; Jiang, Y.; Zhang, C.; He, S.; Wang, H.; Pu, K. Renal clearable polyfluorophore nanosensors for early diagnosis of cancer and allograft rejection. *Nat. Mater.* **2022**, *21*, 598–607. [CrossRef]
98. Loynachan, C.N.; Soleimany, A.P.; Dudani, J.S.; Lin, Y.; Najer, A.; Bekdemir, A.; Chen, Q.; Bhatia, S.N.; Stevens, M.M. Renal clearable catalytic gold nanoclusters for in vivo disease monitoring. *Nat. Nanotechnol.* **2019**, *14*, 883–890. [CrossRef]
99. Wang, F.; Ren, F.; Ma, Z.; Qu, L.; Gourgues, R.; Xu, C.; Baghdasaryan, A.; Li, J.; Zadeh, I.E.; Los, J.W.N.; et al. In vivo non-invasive confocal fluorescence imaging beyond 1,700 nm using superconducting nanowire single-photon detectors. *Nat. Nanotechnol.* **2022**, *17*, 653–660. [CrossRef]
100. Chang, Y.; Chen, H.; Xie, X.; Wan, Y.; Li, Q.; Wu, F.; Yang, R.; Wang, W.; Kong, X. Bright Tm^{3+}-based downshifting luminescence nanoprobe operating around 1800 nm for NIR-IIb and c bioimaging. *Nat. Commun.* **2023**, *14*, 1079. [CrossRef]
101. Zhao, M.; Wang, J.; Lei, Z.; Lu, L.; Wang, S.; Zhang, H.; Li, B.; Zhang, F. NIR-II pH sensor with a FRET adjustable transition point for in situ dynamic tumor microenvironment visualization. *Angew. Chem. Int. Ed.* **2021**, *60*, 5091–5095. [CrossRef] [PubMed]
102. Li, S.; Zhang, H.; Huang, Z.; Jia, Q. Spatially confining copper nanoclusters in porous ZrO_2 for fluorescence/colorimetry/smartphone triple-mode detection of metoprolol tartrate. *Biosens. Bioelectron.* **2023**, *231*, 115290. [CrossRef] [PubMed]

chemosensors

Article

$NaBH_4$-Mediated Co-Reduction Synthesis of Glutathione Stabilized Gold/Silver Nanoclusters for Detection of Magnesium Ions

Weiwei Chen [1], Yiying Chen [1], Xianhu Zhu [2], Miaomiao Xu [1], Zhihao Han [2], Lianhui Wang [2] and Lixing Weng [1,*]

[1] School of Geographic and Biologic Information, Nanjing University of Posts and Telecommunications, Nanjing 210023, China; chenww@njupt.edu.cn (W.C.); 1021173611@njupt.edu.cn (Y.C.); xumiaomiao@njupt.edu.cn (M.X.)

[2] State Key Laboratory for Organic Electronics and Information Displays, Jiangsu Key Laboratory for Biosensors, Institute of Advanced Materials (IAM), Jiangsu National Synergetic Innovation Center for Advanced Materials (SICAM), Nanjing University of Posts & Telecommunications, Nanjing 210023, China; 1221066727@njupt.edu.cn (X.Z.); 17698067082@163.com (Z.H.); iamlhwang@njupt.edu.cn (L.W.)

* Correspondence: lxweng@njupt.edu.cn

Abstract: The content of magnesium ions (Mg^{2+}) in drinking water is relatively high and the excessive Mg^{2+} ingestion may lead to pathological lesions in the human body system. At present, the detection of Mg^{2+} still relies on costly devices or/and complex organic fluorescence probes. To solve this problem, this work proposed a $NaBH_4$-mediated co-reduction strategy for the synthesis of glutathione-stabilized bimetallic AuAg nanoclusters (GSH@AuAg NCs) with performance recognition to Mg^{2+}. The preparation of GSH@AuAg NCs was simple and rapid and could be performed at mild conditions. The reaction parameters and sampling orders were optimized to understand the formation mechanism of GSH@AuAg NCs. The GSH@AuAg NCs exhibited a sensitive "light on" fluorescence response to Mg^{2+} due to the re-molding of the interfacial physicochemical environment following the Mg^{2+} coordination, which affected the surface charge transfer process, and thus led to a novel method for fluorescence detection of Mg^{2+} with admirable selectivity for Mg^{2+}. The proposed method showed a detection limit of 0.2 μM, and its practical utility for the detection of Mg^{2+} in a real sample of purified drinking water was also demonstrated, confirming its practicability in monitoring the Mg^{2+} concentration in drinking water.

Keywords: bimetallic AuAg nanoclusters; coordination; glutathione; $NaBH_4$; magnesium ions

Citation: Chen, W.; Chen, Y.; Zhu, X.; Xu, M.; Han, Z.; Wang, L.; Weng, L. $NaBH_4$-Mediated Co-Reduction Synthesis of Glutathione Stabilized Gold/Silver Nanoclusters for Detection of Magnesium Ions. *Chemosensors* **2023**, *11*, 435. https://doi.org/10.3390/chemosensors11080435

Academic Editor: Manuel Algarra

Received: 13 July 2023
Revised: 1 August 2023
Accepted: 4 August 2023
Published: 5 August 2023

1. Introduction

Quantum-sized metal nanoclusters (MNCs) that only contain a few dozen atoms possess plenteous optical properties, such as outstanding fluorescence, size-dependent fluorescence, and ligand-tunable fluorescence, and thus hold great promise in fluorescent sensing applications [1–5]. Among MNCs, bio-liganded Au and Ag NCs prepared by easy one-pot synthesis have received the most attention due to their intrinsic luminescence properties [5–8]. To obtain the nanoclusters with superior stable and strong fluorescence, different bio-ligands such as peptides, proteins, and DNA, etc., have been used as the protective shell to wrap the core of metal atoms [8–10]. Besides, glutathione (GSH, l-γ-glutamyl-l-cysteinyl-glycine) containing one thiol (-SH) and two carboxyl (-COOH) functional groups has been considered as the most adopted stabilizer since it can protect nanoclusters more efficiently [9,10]. However, GSH-capped Au and Ag NCs still have some problems in practical applications, which are caused by low quantum yields (QYs) and less-than-perfect photon stability [10,11]. To solve these problems, some proteins with large cavity and abundant functional groups have been combined with GSH to act as protective ligands [11]. Extra efforts have also been invested in the preparation of Au/Ag bimetallic NCs (AuAg NCs). The Au and Ag can be easily mingled without changing the lattice parameter of AuAg NCs due to their similar sizes to the Femi wavelength of an electron (0.7 nm). In addition, the interaction of Au and Ag can modify the electronic structure of alloy, surface composition, and deficiency, sharing distinct advantages of highly enhanced fluorescence and photon-stability over monometallic nanoclusters as the synergetic effect [11–13]. In this work, a $NaBH_4$-mediated co-reduction strategy was proposed for the synthesis of GSH@AuAg NCs with excellent fluorescence properties.

To prepare GSH@AuAg NCs, an appropriate chemical environment must be provided to match the assembly of GSH ligand and Au and Ag precursors, which inevitably brings rigor-reactive conditions such as strong alkali, high temperature, and prolonged reaction time [11,14]. Hence, there is a particular need to provide a simple and rapid method for GSH@AgAu NCs synthesis. Considering the mild conditions of $NaBH_4$ reduction for obtaining Au and Ag NCs [15], a one-pot synthesis method by simply mixing of certain proportion of $AuCl^{4-}$ and Ag^+ with GSH, and then $NaBH_4$, was developed (Scheme 1A), which could form GSH@AgAuNCs with strong fluorescent emission within a matter of minutes. Moreover, the GSH@AgAuNCs showed a sensitive "turn-on" fluorescence response to magnesium ions (Mg^{2+}) (Scheme 1B), thus providing a simple method for Mg^{2+} detection.

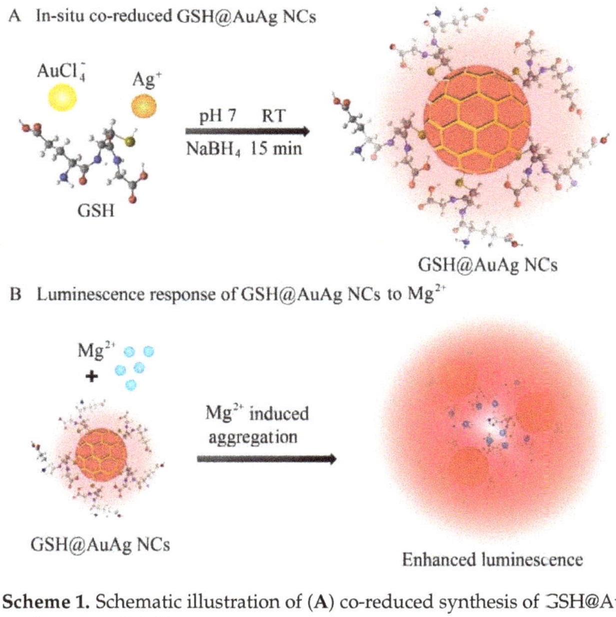

Scheme 1. Schematic illustration of (**A**) co-reduced synthesis of GSH@AuAg NCs and, (**B**) detection mechanism of Mg^{2+}.

Mg^{2+} is the most abundant intracellular divalent cation and plays a significant physiological role in numerous cellular processes [16]. Moreover, excessive Mg^{2+} ingestion may lead to pathological lesions in the cardiovascular, nervous, urinary, and hematopoietic systems [17,18]. However, most methods for detecting Mg^{2+} still need expensive instrumentations, such as atomic absorption spectroscopy (AAS) and inductively coupled plasma-mass spectrometry (ICP-MS), and complex organic fluorescent probes [18–20]. It is valuable to design a simpler and more readily applicable sensor for Mg^{2+}. This work provided a mild and easy preparation method for GSH@AgAuNCs with admirable sensitivity and selectivity for fluorescent detection of Mg^{2+} ions. The practical utility of GSH@AgAuNCs in monitoring the Mg^{2+} in a real sample of drinking water was also demonstrated.

2. Materials and Methods

2.1. Materials and Reagents

Silver nitrate ($AgNO_3$), glutathione (GSH), and sodium borohydride ($NaBH_4$) were purchased from Sigma-Aldrich (Shanghai, China). Chloroauric acid ($HAuCl_4$) was obtained from Energy Chemical Co., Ltd. (Shanghai, China). Sodium hydroxide (NaOH) and other metal salts were products of Shanghai Aladdin Biochemical Technology Co., Ltd.(Shanghai, China) Tyrosine (Try), arginine (Arg), lysine (Lys), proline (Pro), glycine (Gly), phenylalanine (Phe), histidine (His), aspartate (Asp), glucose (Glu), and cysteine (Cys). K^+, Na^+, Ca^{2+}, Mg^{2+}, Cr^{3+}, Cu^{2+}, Zn^{2+}, Ni^{2+}, Co^{2+}, Cd^{2+}, Mn^{2+}, Zr^{4+}, Al^{3+}, Fe^{2+}, and Fe^{3+} solutions were prepared with KCl, NaCl, $CaCl_2{\cdot}2H_2O$, $MgCl_2$, $CrCl_3$, $CuCl_2{\cdot}2H_2O$, $ZnCl_2$, $NiSO_4{\cdot}6H_2O$, $CoCl_2{\cdot}6H_2O$, $CdCl_2{\cdot}5H_2O$, $MnCl_2{\cdot}4H_2O$, $AlCl_3{\cdot}6H_2O$, $FeSO_4{\cdot}7H_2O$, and $FeCl_3{\cdot}6H_2O$, respectively. All reagents were of analytical reagent grade and used as received without further purification. Ultrapure water with a resistivity of 18.2 $M\Omega{\cdot}cm^{-1}$ obtained from a Millipore purification system was used for the experiments.

2.2. Characterization

Fluorescence spectra were recorded on a FluoroMax Plus spectrophotometer (Shi madzu, Kyoto, Japan). The UV-vis-NIR absorption spectra were recorded on a UV-3600 spectrophotometer (Shimadzu, Kyoto, Japan). Transmission electron microscopic (TEM) images were collected with a JEM-2100 transmission electron microscope (JEOL Ltd., Tokyo, Japan). The morphologies and structures of GSH@AuAg NCs were characterized by high-resolution TEM (HRTEM) (FEI Talos F200X). Fluorescence lifetimes of GSH@AuAg NCs were collected by an Edinburgh FLS920 spectrofluorometer (Edinburgh, Livington, England). X-ray photoelectron spectroscopy (XPS) was performed on PHI 5000 VersaProbe with A1 Kα (hυ = 1486.6 eV) X-ray source (Ulvac-Phi, Chigasaki, Japan).

The samples used for the TEM and AFM characterization were prepared as follows. For the TEM analysis, the copper wire mesh was dipped into a solution of freshly prepared GSH@AuAg NCs and then naturally dried after removal. For AFM sampling, a solution of freshly prepared GSH@AuAg NCs was deposited on the surface of a clean mica sheet. After being rapidly dried in air, they were imaged by AFM. In addition, purified GSH@AuAg NCs were used for XPS and FTIR spectral analysis. The actual photographs were taken with a digital camera under daylight. Fluorescence photographs of GSH@AuAg NCs were obtained with UV light by a Tanon MINI Space 1000 gel imager.

2.3. Synthesis of GSH@AuAg NCs

Firstly, 50 μL $HAuCl_4$ (10 mM) was added to 758 μL ultrapure water in a 1.5 mL centrifuge tube and mixed with 50 μL $AgNO_3$ (10 mM) to form flocculent precipitation. After stirring vigorously for 1 min, 50 μL GSH (20 mM) was added under vigorous stirring at room temperature for 5 min. Seventy-five μL $NaBH_4$ (1 mM) dissolved in 0.25 M NaOH was then add to the mixture under ice water. Finally, 17 μL NaOH (0.25 M) was added to adjust the pH to 7.5. The GSH@AuAg NCs could be formed within 15 min at room temperature (25 °C), which were washed using an Amicon Ultra-15 centrifugal filter with a molecular weight cutoff of 10 kDa (Merck Millipore, Billerica, MA, USA), and then suspended in 1 mL of ultrapure water and reserved at 4 °C.

2.4. Selective and Sensitive Detection of Mg^{2+}

After 100 μL GSH@AuAg NCs dispersion was diluted with 200 μL ultrapure water and added with 5.0 μL Mg^{2+} solution or sample to incubate at room temperature for 5 min, the fluorescent emission spectrum was recorded at the excitation wavelength of 468 nm.

The selectivity of the proposed method was in terms of fluorescence intensity variation of GSH@AuAg NCs in the presence of metal ions (Mg^{2+}, Ca^{2+}, Na^+, K^+, Al^{3+}, Zr^{4+}, Zn^{2+}, Mn^{2+}, Cd^{2+}, Cr^{3+}, Fe^{2+}, Fe^{3+}, Co^{2+}, Ni^{2+}, Cu^{2+}). One hundred μL GSH@AuAg NCs dispersion was diluted with 200 μL ultrapure water and then added with 50 μM metal ions (final concentration) to incubate at room temperature for 5 min, and the fluorescent emission spectra were recorded at the excitation wavelength of 468 nm.

To detect Mg^{2+} in purified drinking water, 100 μL of GSH@AuAg NCs were mixed with 200 μL purified drinking water which was taken from the C'est bon bottled purified drinking water. Then, three sets of specific concentrations of Mg^{2+} were mixed with the resulting solution and the resulting solution was incubated at room temperature for 5 min. Finally, the content of Mg^{2+} in purified drinking water was calculated by emission spectra.

3. Results and Discussion

3.1. Synthesis and Characterization of GSH@AuAg NCs

The reductant $NaBH_4$ was introduced in the mixture of $HAuCl_4$, $AgNO_3$, and GSH to accelerate the metallic nucleation and growth, which moderated the restricted synthesis conditions. After GSH@AuAg NCs were formed, the rate of $NaBH_4$ reduction was limited by harmonizing reaction parameters, such as solution pH, amounts of reactants, and temperature. Therefore, the reaction conditions, such as the molar ratio of $HAuCl_4$ to $AgNO_3$ (Au/Ag), the concentrations of GSH and $NaBH_4$, and the pH were firstly optimized by detecting the fluorescence of produced GSH@AuAg NCs. The effect of Au/Ag ratio was examined in the presence of 0.5 mM $HAuCl_4$. The fluorescence intensity of GSH@AuAg NCs prepared at three pHs all increased gradually with the decreasing ratio of Au/Ag and reached a maximum intensity at the molar ratio of 1:1 for pH 7–8 and 11–12, and 1:1.5 for pH 9–10 (Figure 1A–C). More $AgNO_3$ resulted in the decrease of fluorescence intensity of GSH@AuAg NCs and the appearance of a shoulder peak around 560 nm, which should result from other distinct clusters. The equal concentrations of Au and Ag precursors for the brightest nanoclusters hinted at the co-doping of Au and Ag.

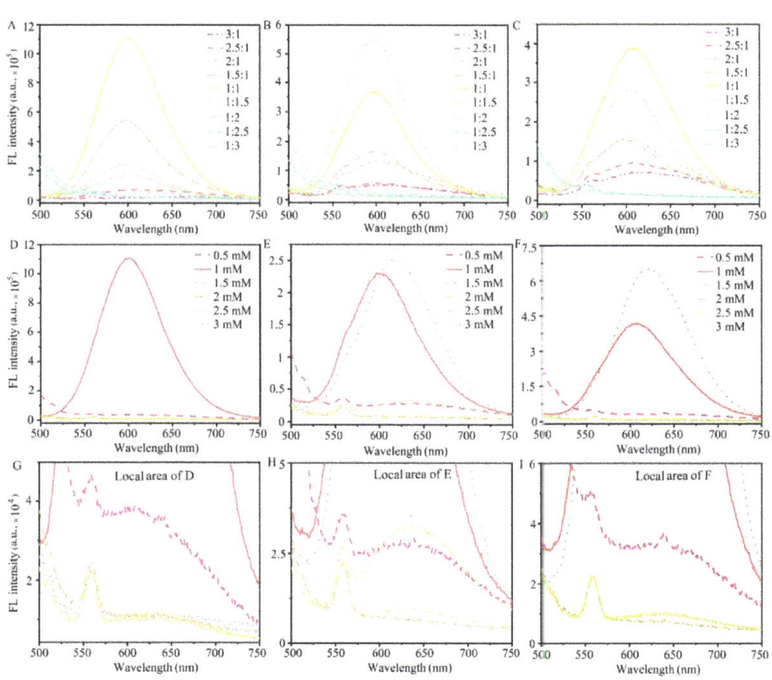

Figure 1. Condition optimization for GSH@AuAg NCs synthesis. (**A–C**) Molar ratio of Au to Ag at pH 7–8 (**A**), pH 9–10 (**B**), and pH 11–12 (**C**) in the presence of 0.5 mM Au, 1 mM GSH, and 0.75 mM $NaBH_4$. (**D–F**) Concentration of GSH at pH 7–8 (**D**), pH 9–10 (**E**), and pH 11–12 (**F**) in the presence of 0.5 mM Au and Ag precursors, and 0.75 mM $NaBH_4$. (**G–I**) G, H, and I were the partial dates from (**D–F**), respectively.

The effect of GSH ligand concentration was examined at different pHs with 0.5 mM Au and Ag precursors and 0.75 mM $NaBH_4$ to show the brightest GSH@AuAg NCs at 1.0 mM of GSH (Figure 1D,E). GSH at concentrations of 0.5 mM and 2–3 mM cannot obtain the GSH@AuAg NCs, where the fluorescence intensity of the solution was very weak (Figure 1G–I, the local amplified images of Figure 1D,E). The GSH@AuAg NCs synthesized at pH 7–8 showed much stronger fluorescence intensity than those obtained at higher pHs. Thus, pH 7.5 was selected for GSH@AuAg NCs synthesis.

The optimization of $NaBH_4$ concentration for GSH@AuAg NCs synthesis was performed at pH 7.5 with 0.5 mM Au and Ag precursors, and 1.0 mM GSH. The maximum fluorescence intensity occurred at 0.75 mM $NaBH_4$ (Figure S1 in SI). The optimized reaction time was 15 min, and a prolonged reaction time did not increase the fluorescence intensity of GSH@AuAg NCs (Figure S2 in SI).

As the first reported co-reduction synthesis of GSH@AuAg NCs mediated by $NaBH_4$, the growth of GSH@AuAg NCs was investigated by UV-vis and fluorescence spectroscopy to understand the complex reduction assembly of Au and Ag precursors and GSH ligands. First, to confirm the doping synthesis of GSH@AuAg NCs rather than other luminous oligomeric clusters, spectroscopic analysis was performed by preparing the NCs in the absence of each reactant. The GSH@AuAg NCs did not exhibit obvious plasmon resonance absorption in the range of 400–700 nm, while a clear absorption peak appeared when the synthesis was performed in the absence of GSH or Au precursors (Figure 2A). In addition, in the absence of the Ag precursor, the reaction solution was transparent and the UV-vis absorbance was close to zero, which was similar to that in the absence of $NaBH_4$ (Figure 2B, (3) and (5)) and could be attributed to the presence of GSH with two times higher concentration than $HAuCl_4$ to form stabilized Au (I)-GSH complexes, which limited the immediate reduction to Au (0) components [21,22]. In the absence of the Au precursor, an absorption peak appeared at 450 nm and the solution became a lilac color, which resulted from the formation of large Ag nanoparticles (AgNPs) (Figure 2B, (2)) due to the weaker stability of GSH and Ag^+ complexes. Interestingly, no fluorescence emission was observed in the absence of the Au or Ag precursor (Figure 2C), demonstrating that the fluorescence emission came from the bimetallic AuAg nanoclusters. In the absence of GSH, a shield peak occurred at 560 nm (Figure 2C, curve (4)), as observed in Figure 1, which eliminated the formation of the weak luminous metal (I)-GSH (M (I)-GSH) complexes. Besides, the mixture of Au and Ag precursors and GSH did not exhibit luminescence, which excluded the emission originating from the oligomeric metal (I)-GSH complex. The above results confirmed that all reactants are indispensable for the formation of GSH@AuAg NCs.

To understand the interaction of Au and Ag precursors with the GSH ligand, the ratio of GSH to Au precursor was changed while keeping the amount of GSH and total concentration of Au and Ag precursors at 1.0 mM to record the corresponding spectra. In the absence of the Ag precursor, an absorption peak appeared at 520 nm, and the solution color turned deep reddish brown (Figure 2D,E, (1)), indicating the formation of Au NPs, and the above protective function of GSH only occurred at a higher ratio of GSH to Au precursor. Interestingly, the absorption peak of Au NPs at 520 nm completely disappeared, and the solution became light brown–yellow after only 0.1 mM Ag precursor was added. More Ag precursor decreased the absorbance and the solution color became lighter till the presence of 0.5 mM Ag precursor (Figure 2D,E, (3)–(5)). The maximum fluorescence emission also occurred at 0.5 mM Au and Ag precursors (Figure 2F), demonstrating the synergism stability induced by co-doped Au and Ag and the formation of GSH@AuAg NCs.

The fluorescence of metal nanoclusters (MNCs) can be considered to originate from the aggregated M(I)-thiolate oligomers due to ligands inducing the ligand-to-metal charge transfer (LMCT) or ligand–metal–metal charge transfer (LMMCT) [21,22]. At high ratios of GSH to Au precursor, no Au (0) component was formed due to the protective effect of GSH ligand; thus, the nanoclusters could not be generated. While introducing Ag to decentralize the shielding of GSH as forming the M(I)–GSH complex, the $NaBH_4$ initially reduced the Ag (I) to Ag (0) as a less weak protection against reducing Ag (I) to Ag (0), as the intertwined M(I)–GSH complexes made the Au(I) ions close to the Ag (0) surface prompt the formation of bimetallic clusters [22].

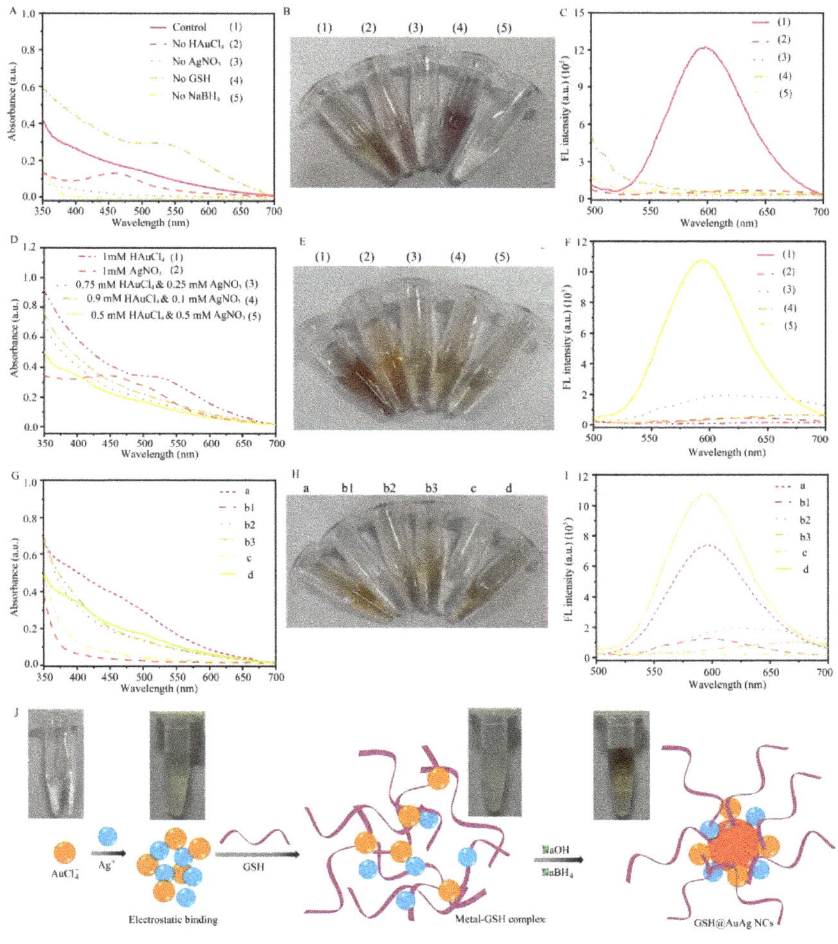

Figure 2. Growth process of GSH@AuAg NCs. (**A**,**D**,**G**) UV-vis absorption spectra, (**A**,**E**,**H**) photos and (**C**,**F**,**I**) FL spectra of the mixtures (1) 0.5 mM $HAuCl_4$, 0.5 mM $AgNO_3$, 1 mM GSH, and 0.75 mM $NaBH_4$ as control, and (2–5) (1) without the presence of $HauCl_4$ (2), $AgNO_3$ (3), GSH (4), and $NaBH_4$ (5) for (**A**–**C**); and 1 mM GSH, 0.75 mM $NaBH_4$, and (1) 1 + 0, (2) 0 + 1, (3) 0.75 + 0.25, (4) 0.9 + 0.1, and (5) 0.5 + 0.5 mM Au + Ag precursors for (**D**–**F**); the reaction time is 15 min. (**G**–**I**) Effects of feeding orders: (a) pre-mixed GSH and Au precursor + Ag precursor and then $NaBH_4$, (b,c) pre-reduced Au (b) or Ag (c) precursor with $NaBH_4$ and GSH for 5 min + another precursor at (b1) 0.5 + 0.5, (b2) 0.75 + 0.25, and (b3) 0.9 + 0.1 mM Au + Ag precursors, (d) following the schematic order shown in (**J**).

To further understand the assembly of Au and Ag precursors and GSH ligand, the effect of the feeding pattern was examined to elucidate the formulation of GSH@AuAg NCs. The premixing of the GSH ligand with the Au precursor and later, the introduction of the Ag precursor resulted in the growth of nanoclusters, which exhibited weaker luminosity and higher absorbance from 400 to 600 nm (Figure 2G–I, curve a), compared with the feeding mode that pre-mixed Au and Ag precursors, and then added GSH (Figure 2G–I, curve d), which showed the maximum luminosity, thus could be considered as an optimized sampling procedure (Figure 2J). In the optimized sampling procedure, the yellow mixture of Au and Ag precursors was attributed to the formation of multiple complexes by the electrostatic interaction; the solution became muddy white upon addition of GSH due to the formation of a mingled network of the Metal–GSH complex, and the addition of NaOH and $NaBH_4$ produced brown–yellow GSH@AuAg NCs.

Another sampling procedure was premixing the GSH ligand with the Au precursor and then, $NaBH_4$ to react for 5 min, then adding Ag precursor for another 10 min (Figure 2G–I, b). The absorbance of the final solution was extremely low at 0.5 mM Au and Ag precursors (Figure 2G, curve b1), which was different from that of the optimized sampling order (Figure 2G, curve d). This demonstrated the influence of the sampling order on the growth of GSH@AuAg NCs. In addition, further increasing the concentrations of the Au precursor, both the color and absorption spectra of the ultimate solution were close to that of the optimized sampling procedure, indicating the formation of nanoclusters (Figure 2G, curves b2 and b3). These results confirmed that the molar ratios of GSH to metal precursors significantly affected the assembly of Au, Ag, and GSH. Other sampling procedures, such as premixing the GSH ligand with the Ag precursor, then $NaBH_4$ to react for 5 min and then adding the Au precursor, were also examined (Figure 2G–I, c). After 5 min $NaBH_4$ reduction, the solution turned a lilac color, indicating the formation of large AgNPs as mentioned above. After introducing the Au precursor (0.5 mM), the solution color gradually decayed to colorless, and the absorbance was also significantly reduced (Figure 2G,H, curve c), suggesting the dissolution of AgNPs due to the interaction among Au, Ag, and GSH. The significantly weak emission intensity demonstrated the failure of the GSH@AuAg NCs assembly.

The TEM and AFM images and corresponding particle size distribution of GSH@AuAg NCs formed with the optimized sampling procedure showed spherical morphology and well dispersion with an average diameter of about 2.35 nm (Figure 3A–E). The clear lattice fringes were observed with an interspacing of 0.22 nm (inset in Figure 3B), corresponding to the d-spacing of the crystal plane of face-centered cubic Au (111) [23,24]. The GSH@AuAg NCs have also been characterized by fluorescence spectroscopy. As shown in Figure 3F, the maximum excitation and emission wavelengths of AuAg NCs were 468 nm and 600 nm, respectively. The GSH@AuAg NCs were yellow–brown under the visible light (inset in Figure 3F of a-1) and emitted a bright red fluorescence under UV irradiation (inset in Figure 3F of b-1), indicating forming high luminous nanoclusters. XPS spectroscopy was then used to confirm the elemental composition of the GSH@AuAg NCs and the valence states of Au and Ag (Figure S3 in SI, Figure 3D,E). The survey spectrum in Figure S3 confirmed that GSH@AuAg NCs were composed of Au 4f, Ag 3d, carbon (C 1s), sulfur (S 2p), oxygen (O 1s), and nitrogen (N 1s). The peaks of C, O, S, and N were attributed to GSH, indicating that the GSH was capped with the AuAg NCs. XPS spectroscopy was then used to confirm the elemental composition of the GSH@AuAg NCs and the valence states of Au and Ag (Figure 3D,E). The Au 4f XPS spectrum displayed a peak at 88.1 eV (Au $4f_{5/2}$) and a splitting peak at 84.4 eV (Au $4f_{7/2}$). The latter could be deconvoluted into two distinct components with the binding energies centered at 84.0 and 84.6 eV, assigned to Au (0) and Au (I), respectively. The predominant Au species in the GSH@AuAg NCs was identified as Au (I) (~62.5%). The Ag 3d pattern exhibited two peaks at 374.1 eV (Ag $3d_{3/2}$) and 368.1 eV (Ag $3d_{5/2}$), and the latter was deconvoluted into Ag (I) at 368.4 eV and Ag (0) at 367.7 eV (Figure 3E). The predominant Ag species in the GSH@AuAg NCs was identified as Ag (I) (~63.7%), indicating the successful doped synthesis of GSH@AuAg NCs [25,26]. FT-IR studies of GSH-AuAg NCs also demonstrated the presence of GSH. As depicted in Figure 3I, GSH (black line) has a set of characteristic IR bands such as the characteristic bands at 1535 cm^{-1} for the antisymmetric stretching of COO^- and 1646 cm^{-1} for the vibration of C=O/N-H [27,28], which can also be seen in the FT-IR spectrum of the GSH@AuAg NCs (red line), while disappearing with reduced synthesized AuAg NCs without GSH (blue line), further indicating the presence of GSH capping the AuAg NCs.

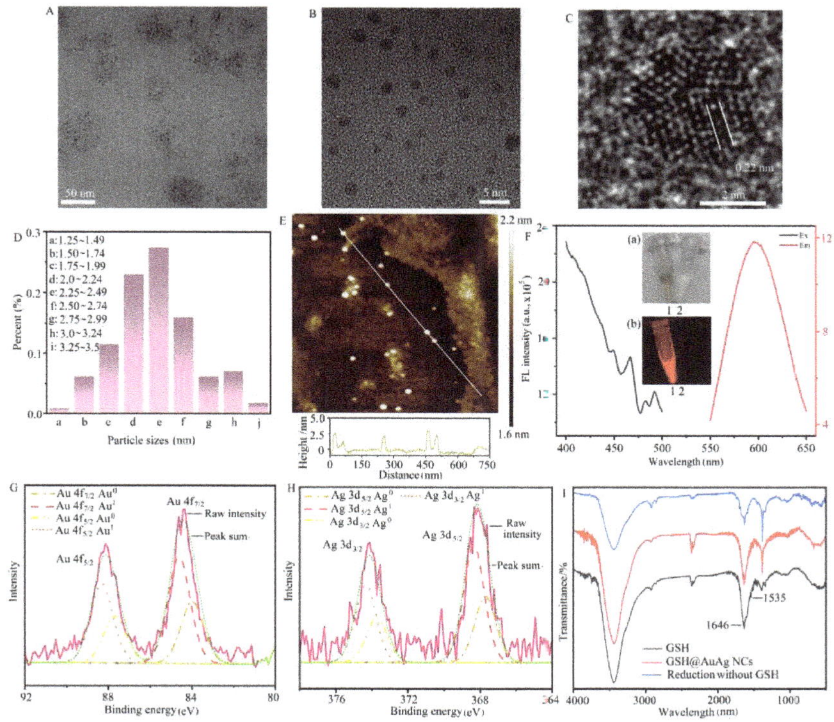

Figure 3. Characterization of GSH@AuAg NCs prepared using 0.5 mM $HauCl_4$, 0.5 mM $AgNO_3$, 1 mM GSH, and 0.75 mM $NaBH_4$ at pH 7.5 and 25 °C within 15 min. (**A**) TEM, (**B**,**C**) HRTEM, (**D**) Particle-size distribution histogram; (**E**) AFM images of GSH@AuAg NCs. (**F**) Fluorescence excitation and emission spectra of GSH@AuAg NCs, and the inset showed the photograph of GSH@AuAg NCs (1) and water (2) under visible light (a) and UV light from gel imager (b). (**G**) Au and (**H**) Ag XPS spectra of GSH@AuAg NCs. (**I**) FT-IR spectra of GSH, GSH@AuAg NCs, and reduced synthesized AuAg NCs without GSH.

3.2. Optimization of pH for Recognition of GSH@AuAg NCs to Mg^{2+}

MNCs, including AuAg NCs, have been reported as showing fancy numerous responses to different metal ions [7,23,29]. To provide a new fluorescence probe for Mg^{2+}, the recognition conditions of GSH@AuAg NCs to Mg^{2+} were optimized. Firstly, the effect of pH on the fluorescence of GSH@AuAg NCs dispersion was examined (Figure 4A). The fluorescence was negligible at pHs lower than 6.5, and quickly enhanced with the increasing pH from 6.5 to 7.5 (Figure 4B). At different pHs, the responses of GSH@AuAg NCs to several metal ions were also examined. As shown in Figure 4C, the fluorescence of GSH@AuAg NCs could be completely quenched by Cu^{2+} ions in the pH range of 6.0–10, while Mg^{2+} exhibited obvious boosting effect at different pHs. Besides, Ca^{2+} exhibited a relatively weaker FL enhancement, and Zn^{2+} could quench the fluorescence at low pH and increase the fluorescence at pHs more than 9.0. Mg^{2+} and Ca^{2+} are the second main-group elements and possess a similar coordinated interaction with GSH, leading to the FL enhancement. In order to achieve Mg^{2+} detection, the optimum pH was chosen at 7.0, at which the FL enhancement of Ca^{2+} was the weakest, and the effect of Zn^{2+} was also relatively low.

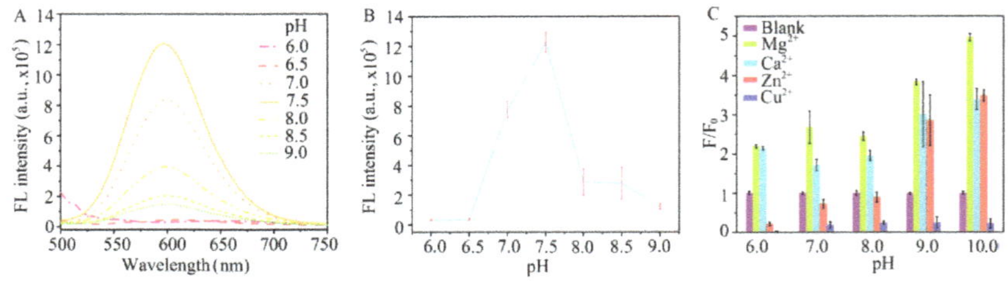

Figure 4. Effect of pH on FL response. (**A**) FL spectra and (**B**) intensity of GSH@AuAg NCs at different pHs. (**C**) FL responses to 50 μM metal ions.

3.3. Mechanism of Mg^{2+}-Mediated Fluorescence Enhancement of GSH@AuAg NCs

The TEM images of GSH@AuAg NCs showed the large aggregates with the size around 100 nm (Figure 5A), which was significantly larger than that of GSH@AuAg NCs (Figure 3), and could be attributed to the neutralization of negatively charged GSH@AuAg NCs by Mg^{2+} binding with -COO^- in GSH and/or the chelation of Mg^{2+} to carbonyl, hydroxyl, and other electron-donating groups [30,31]. The neutralized GSH@AuAg NCs weakened the dispersal stability of the GSH@AuAg NCs due to the loss of electrostatic repulsion. The aggregates induced by the chelation of Mg^{2+} ion deeply changed the surface metal–ligand [Au (1)–GSH] motifs state, which remolded the ligand conformation and strengthened the aurophilic interaction of the oligomeric GSH–$[Au (I)–GSH]_x$ motifs [8,21,31].

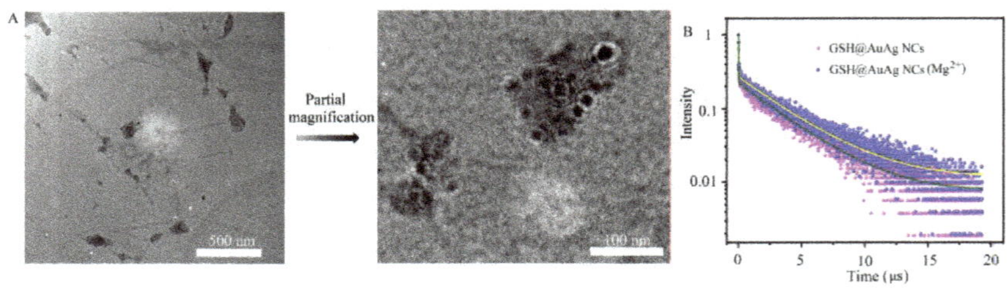

Figure 5. Mechanism of Mg^{2+}-induced FL enhancement. (**A**) TEM image of Mg^{2+}-induced aggregates of GSH@AuAg NCs. (**B**) FL lifetime of GSH@AuAg NCs before (green) and after (yellow) incubating with Mg^{2+}.

To further understand the interaction of Mg^{2+} with GSH@AuAg NCs, the average fluorescent lifetime of GSH@AuAg NCs was examined upon addition of Mg^{2+} (Figure 5B). The fluorescent lifetime increased from 3.29 μs to 3.45 μs, while the fluorescent QY increased from 2.47 % to 7.16 % with the ethanol solution of rhodamine 6G as a reference. The large stokes shift (130 nm) and long fluorescence lifetime of GSH@AuAg NCs in their excited-state decay indicated that luminescence originated from the ligand-to-metal charge transfer (LMCT) or ligand-to-metal–metal charge transfer (LMMCT) [21,22]. The increase of fluorescent QY and the prolongation of fluorescent lifetime upon the addition of Mg^{2+} indicated that the process of charge transfer underwent a transformation after Mg^{2+} chelation, further confirming the reconstructed local surface physicochemical environment.

3.4. Sensing Performance towards Mg^{2+}

In the optimization experimental conditions, we performed an Mg^{2+} assay as shown in Figure 6A. The emission intensity of GSH@AuAg NCs increased with the increasing Mg^{2+} concentration. The plot of fluorescence intensity vs the concentration showed a good linearity with an R^2 of 0.989 over the range from 0.2 μM to 1 μM (Figure 6B). The linear regression equation was $F = 177\ c\ (\mu M) + 3.848 \times 10^5$. To evaluate the specificity of the fluorometric detection of Mg^{2+}, 14 types of metal ions were chosen to be tested under the same experimental conditions as those for Mg^{2+}. As shown in Figure 6C, other common metal ions except Ca^{2+} showed a negligible effect on Mg^{2+} detection. The fluorescence quenching of GSH@AuAg NCs by the coexisting ions did not affect the detection of Mg^{2+} (Figure 6D), demonstrating the admirable specificity of GSH@AuAg NCs towards Mg^{2+} and the potential application of the proposed method in the quality testing of the Mg^{2+} amount.

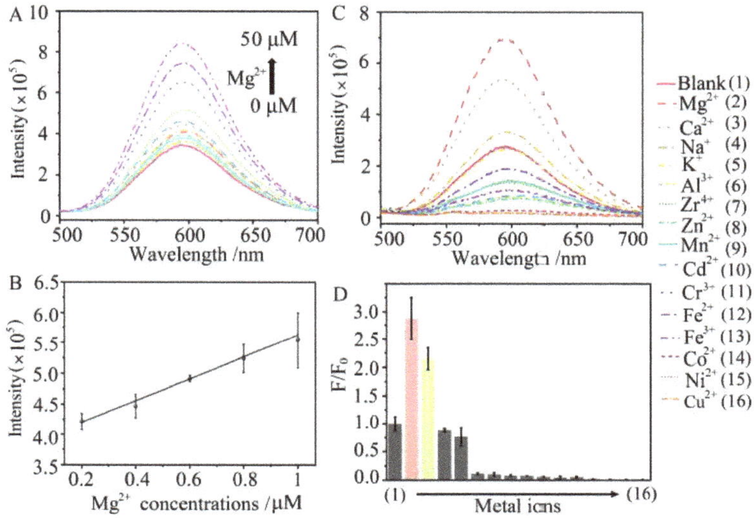

Figure 6. Sensing performance of GSH@AuAg NCs toward Mg^{2+}. (**A**) FL spectra of GSH@AuAg NCs after incubating with different concentrations of Mg^{2+}. (**B**) Plot of FL intensity of GSH@AuAg NCs vs Mg^{2+} concentration. (**C**) FL spectra of GSH@AuAg NCs after incubating with different metal ions. (**D**) FL response of GSH@ AuAg NCs to different metal ions.

To demonstrate the practical analytical application of the GSH@AuAg NCs, Mg^{2+} was spiked into purified drinking water at different concentrations. Recovery results were presented in Table S1, which showed the recoveries ranging from 94.7% to 97.4%, indicating an applicable Mg^{2+} detection in real drinking water. Unlike purified drinking water, the other dairy using water, such as the ground water and domestic treated water (tap water), tended to have higher concentrations of Mg^{2-} that can affect people's normal life. To achieve specific identification of Mg^{2+} in dairy using water, some issues, such as the interference from other heavy metal ions, pH-induced changes in the specificity for Mg^{2+}, and the signal crosstalk from Ca^{2+}, had to be overcame. The designed GSH@AuAg NCs had several advantages that may have a positive impact on Mg^{2+} quantitative identification. Of all the common metal ions, only Mg^{2+} and Ca^{2+} turn on the fluorescence intensity of the GSH@AuAg NCs, demonstrating its anti-jamming performance for the most heavy metal ions. Then, as known, the pH of the ground water and tap water may change dynamically according to different environment backgrounds, which the fluorescence intensity of GSH@AuAg NCs would decrease in an acidic or alkaline condition, but the Mg^{2+}-mediated turn-on fluorescence intensity of GSH@AuAg NCs was very stable at various pH values from 6 to 10. In addition, some heavy metal ions (Zn^{2+}) can also enhance the fluorescence intensity of GSH@AuAg NCs at alkaline condition. Yet, the concentrations of most heavy metal ions (less than micro-molar) were much lower than that of Mg^{2+} (hundreds of micro moles to millimoles) in dairy using water [32,33], which mean the heavy metal ions may not interface the Mg^{2+} measurement even under different pH conditions. Ultimately, the only drawback was the crosstalk of Ca^{2+}-mediated turn-on emission from GSH@AuAg NCs. The Ca^{2+} concentration in water was several times higher than that of Mg^2 [33]. The pretreatment of the water sample by settlement or filtration adsorption of Ca^{2+} may solve this problem, and there are many techniques for removing Ca^{2+} from water [34,35], which may eliminate the effect of Ca^{2+} crosstalk on the accurate quantifying of Mg^{2+}. By taking this into account, the specificity of Mg^{2+} in real-water samples can be achieved.

4. Conclusions

The facile, gentle, and fast synthesis of $NaBH_4$-reduced GSH@AuAg NCs with good dispersivity has been successfully achieved by pre-mixing Au and Ag precursors and adding GSH in the mixture to form a metal–GSH complex, followed with $NaBH_4$ addition for the reduction of metal ions at pH 7.5. The co-doping of Au and Ag at the molar ratio of 1:1 results in the synergism stability of the NCs. The GSH@AgAu NCs show a special enhancement of fluorescence response to magnesium ions (Mg^{2+}) at pH 7.0, which has been designed for the first nanoclusters-based fluorescent sensing of Mg^{2+}. The proposed method exhibits acceptable selectivity for Mg^{2+}detection, and practical utility in monitoring the Mg^{2+} content in daily drinking water.

Supplementary Materials: The following supporting information can be downloaded at: https://www.mdpi.com/article/10.3390/chemosensors11080435/s1, Figure S1: Optimization of NaBH4 concentration for GSH@AuAg NCs preparation; Figure S2: Fluorescent spectra of GSH@AuAg NCs obtained at different times; Figure S3: The survey spectra of AuAg NCs; Table S1: Detection of Mg^{2+} in purified drinking water samples.

Author Contributions: Conceptualization, W.C.; methodology, W.C. and Y.C.; validation, Y.C., X.Z. and Z.H.; investigation, Y.C., X.Z. and Z.H.; resources, L.W. (Lixing Weng); data curation, Y.C.; writing—original draft preparation, Y.C.; writing—review and editing, W.C., M.X. and L.W. (Lixing Weng); supervision, L.W. (Lianhui Wang); funding acquisition, W.C. and L.W. (Lixing Weng). All authors have read and agreed to the published version of the manuscript.

Funding: This research was financially supported by the National Natural Science Foundation of China (22004068), "the Belt and Road" Innovation Cooperation Project of Jiangsu (BZ2022011), the research fund from Nanjing University of Posts and Telecommunications (XK0320921144) and self-funding projects from State Key Laboratory of Analytical Chemistry for Life Science, Nanjing University (SKLACLS2207).

Institutional Review Board Statement: Not applicable.

Informed Consent Statement: Not applicable.

Data Availability Statement: Not applicable.

Conflicts of Interest: The authors declare no conflict of interest.

References

1. Yao, Q.; Chen, T.; Yuan, X.; Xie, J. Toward Total Synthesis of Thiolate-Protected Metal Nanoclusters. *Acc. Chem. Res.* **2018**, *51*, 1338–1348. [CrossRef]
2. Jia, H.; Yu, S.; Yang, L.; Wei, Q.; Ju, H. Near-Infrared Electrochemiluminescence of Dual-Stabilizer-Capped Au Nanoclusters for Immunoassays. *ACS Appl. Nano Mater.* **2021**, *4*, 2657–2663. [CrossRef]
3. Jia, H.; Yang, L.; Fan, D.; Kuang, X.; Sun, X.; Wei, Q.; Ju, H. Cobalt ion doping to improve electrochemiluminescence emisssion of gold nanoclusters for sensitive NIR biosensing. *Sens. Actuators B Chem.* **2022**, *367*, 132304. [CrossRef]
4. Srinivasulu, Y.G.; Goswami, N.; Yao, Q.; Xie, J. High-Yield Synthesis of AIE-type Au22(SG)18 Nanoclusters through Precursors Engineering and Its pH-Dependent Size Transformation. *J. Phys. Chem. C* **2021**, *125*, 4066–4076. [CrossRef]
5. Mastracco, P.; Gonzalez-Rosell, A.; Evans, J.; Bogdanov, P.; Copp, S.M. Chemistry-Informed Machine Learning Enables Discovery of DNA-Stabilized Silver Nanoclusters with Near-Infrared Fluorescence. *ACS Nano* **2022**, *16*, 16322–16331. [CrossRef] [PubMed]
6. Aires, A.; Sousaraei, A.; Moller, M.; Cabanillas-Gonzalez, J.; Cortajarena, A.L. Boosting the Photoluminescent Properties of Protein-Stabilized Gold Nanoclusters through Protein Engineering. *Nano Lett.* **2021**, *21*, 9347–9353. [CrossRef]
7. Li, J.; Peng, G.; Yu, Y.; Lin, B.; Zhang, L.; Guo, M.; Cao, Y.; Wang, Y. Cu^{2+}-mediated turn-on fluorescence biosensor based on DNA-templated silver nanoclusters for label-free and sensitive detection of adenosine triphosphate. *Microchim. Acta* **2023**, *190*, 41. [CrossRef]

8. Zanetti-Polzi, L.; Charchar, P.; Yarovsky, I.; Corni, S. Origins of the pH-Responsive Photoluminescence of Peptide-Functionalized Au Nanoclusters. *ACS Nano* **2022**, *16*, 20129–20140. [CrossRef]
9. Li, X.; Luo, J.; Jiang, X.; Yang, M.; Rasooly, A. Gold nanocluster-europium(III) ratiometric fluorescence assay for dipicolinic acid. *Microchim. Acta* **2021**, *188*, 26. [CrossRef]
10. Mi, W.; Tang, S.; Guo, S.; Li, H.; Shao, N. In situ synthesis of red fluorescent gold nanoclusters with enzyme-like activity for oxidative stress amplification in chemodynamic therapy. *Chin. Chem. Lett.* **2022**, *33*, 1331–1336. [CrossRef]
11. Mi, W.; Tang, S.; Jin, Y.; Shao, N. Au/Ag Bimetallic Nanoclusters Stabilized by Glutathione and Lysozyme for Ratiometric Sensing of H_2O_2 and Hydroxyl Radicals. *ACS Appl. Nano Mater.* **2021**, *4*, 1586–1595. [CrossRef]
12. Yin, M.-M.; Chen, W.-Q.; Hu, Y.-J.; Liu, Y.; Jiang, F.-L. Rapid preparation of water-soluble Ag@Au nanoclusters with bright deep-red emission. *Chem. Commun.* **2022**, *58*, 2492–2495. [CrossRef] [PubMed]
13. Cui, L.; Li, C.; Chen, B.; Huang, H.; Xia, Q.; Li, X.; Shen, Z.; Ge, Z.; Wang, Y. Surface functionalized red fluorescent dual-metallic Au/Ag nanoclusters for endoplasmic reticulum imaging. *Microchim. Acta* **2022**, *187*, 66. [CrossRef] [PubMed]
14. Li, Y.; Deng, Y.; Zhou, X.; Hu, J. A label-free turn-on-off fluorescent sensor for the sensitive detection of cysteine via blocking the Ag^+-enhancing glutathione-capped gold nanoclusters. *Talanta* **2018**, *179*, 742–752. [CrossRef]
15. Farrag, M.; Tschurl, M.; Heiz, U. Chiral Gold and Silver Nanoclusters: Preparation, Size selection, and Chiroptical Properties. *Chem. Mater.* **2013**, *25*, 862–870. [CrossRef]
16. Pontes, M.H.; Yeom, J.; Groisman, E.A. Reducing ribosome biosynthesis promotes translation during low Mg^{2+} stress. *Mol. Cell* **2016**, *64*, 480–492. [CrossRef]
17. Kim, D.Y.; Shinde, S.; Ghodake, G. Colorimetric detection of magnesium (II) ions using tryptophan functionalized gold nanoparticles. *Sci. Rep.* **2017**, *7*, 3966. [CrossRef] [PubMed]
18. Li, L.; Ding, Y.; Zhang, C.; Xian, H.; Chen, S.; Dai, G.; Wang, X.; Ye, C. Ratiometric Fluorescence Detection of Mg^{2+} based on Regulating Crown-Ether Modified Annihilators for Triplet-Triplet Annihilation Upconversion. *J. Phys. Chem. B* **2022**, *126*, 3276–3282. [CrossRef]
19. Gruskos, J.J.; Zhang, G.; Buccella, D. Visualizing Compartmentalized Cellular Mg^{2+} on Demand with Small-Molecule Fluorescent Sensors. *J. Am. Chem. Soc.* **2016**, *138*, 14639–14649. [CrossRef]
20. Fujii, T.; Shindo, Y.; Hotta, K.; Citterio, D.; Nishiyama, S.; Suzuki, K.; Oka, K. Design and Synthesis of a FlAsH-Type Mg^{2+} Fluorescent Probe for Specific Protein Labeling. *J. Am. Chem. Soc.* **2014**, *136*, 2374–2381. [CrossRef] [PubMed]
21. Goswami, N.; Yao, Q.; Luo, Z.; Li, J.; Chen, T.; Xie, J. Luminescent Metal Nanoclusters with Aggregation-Induced Emission. *J. Phys. Chem. Lett.* **2016**, *7*, 962–975. [CrossRef] [PubMed]
22. Luo, Z.; Yuan, X.; Yu, Y.; Zhang, Q.; Leong, D.T.; Lee, J.Y.; Xie, J. From Aggregation-Induced Emission of Au(I)-Thiolate Complexes to Ultrabright Au (0) @Au (I)-Thiolate Core-Shell Nanoclusters. *J. Am. Chem. Soc.* **2012**, *134*, 16662–16670. [CrossRef] [PubMed]
23. Wu, H.; Xie, R.; Hao, Y.; Pang, J.; Gao, H.; Qu, F.; Tian, M.; Guo, C.; Mao, B.; Chai, F. Portable smartphone-integrated AuAg nanoclusters electrospun membranes for multivariate fluorescent sensing of Hg^{2+}, Cu^{2+} and L-histidine in water and food samples. *Food Chem.* **2023**, *418*, 135961. [CrossRef]
24. Zhang, X.-L.; Li, X.; Li, X.-T.; Gao, Y.; Feng, F.; Yang, G.-J. Electrochemiluminescence sensor for pentoxifylline detection using Au nanoclusters@graphene quantum dots as an amplified electrochemiluminescence luminophore. *Sens. Actuators B Chem.* **2019**, *282*, 927–935. [CrossRef]
25. Jia, H.; Yang, L.; Dong, X.; Zhou, L.; Wei, Q.; Ju, H. Cysteine Modification of Glutathione-Stabilized Au Nanoclusters to Red-Shift and Enhance the Electrochemiluminescence for Sensitive Bioanalysis. *Anal. Chem.* **2022**, *94*, 2313–2320. [CrossRef] [PubMed]
26. Zhang, B.; Chen, L.; Zhang, M.; Deng, C.; Yang, X. A Gold-silver bimetallic nanocluster-based fluorescent probe for cysteine detection in milk and apple. *Spectrochim. Acta A Mol. Biomol. Spectrosc.* **2022**, *278*, 121345. [CrossRef]
27. Chandran, N.; Janardhanan, P.; Bayal, M.; Unniyampurath, U.; Pilankatta, R.; Nair, S.S. Label Free, Nontoxic Cu-GSH NCs as a Nanoplatform for Cancer Cell Imaging and Subcellular pH Monitoring Modulated by a Specific Inhibitor: Bafilomycin A1. *ACS Appl. Bio Mater.* **2020**, *3*, 1245–1257. [CrossRef]
28. Li, C.; Zhang, X.; Guo, Y.; Seidi, F.; Shi, X.; Xiao, H. Naturally Occurring Exopolysaccharide Nanoparticles: Formation Process and Their Application in Glutathione Detection. *ACS Appl. Mater. Interfaces* **2021**, *13*, 19756–19767. [CrossRef]
29. Tahir, F.L.F., Jr.; Aucelio, R.Q.; Cremona, M.; Padilha, J.S.; Margheri, G.; Zaman, Q.; Concas, G.C.; Gisbert, M.; Ali, S.; Toloza, C.A.T.; et al. Quenching of the Photoluminescence of Gold Nanoclusters Synthesized by Pulsed Laser Ablation in Water upon Interaction with Toxic Metal Species in Aqueous Solution. *Chemosensors* **2023**, *11*, 118. [CrossRef]
30. Sadhanala, H.K.; Aryal, S.; Sharma, K.; Orpaz, Z.; Michaeli, S.; Gedanken, A. Nitrogen-doped carbon dots as a highly selective and sensitive fluorescent probe for sensing Mg^{2+} ions in aqueous solution, and their application in the detection and imaging of intracellular Mg^{2+} ions. *Sens. Actuators B Chem.* **2022**, *366*, 131958. [CrossRef]
31. Chang, H.; Karan, N.S.; Shin, K.; Bootharaju, M.S.; Nah, S.; Chae, S.I.; Baek, W.; Lee, S.; Kim, J.; Son, Y.J.; et al. Highly Fluorescent Gold Cluster Assembly. *J. Am. Chem. Soc.* **2021**, *143*, 326–334. [CrossRef] [PubMed]
32. Yi, L.; Hong, Y.; Weng, L.; Zhu, Y. Measurement of free heavy metal ion concentrations in soils using Donnan membrane technique. *Chin. J. Geohem.* **2005**, *24*, 184–188.
33. Panhwar, A.H.; Kazi, T.G.; Afrid, H.I.; Shaikh, H.R.; Arain, S.A.; Arain, S.S.; Brahman, K.D. Evaluation of Calcium and Magnesium in Scalp Hair Samples of Population Consuming Different Drinking Water: Risk of Kidney Stone. *Biol. Trace Elem. Res.* **2013**, *156*, 67–73. [CrossRef]

34. Olsen, J.E.; Kero, I.T.; Engh, T.A.; Tranell, G. Model of Silicon Refining During Tapping: Removal of Ca, Al, and Other Selected Element Groups. *Metall. Mater. Trans. B* **2017**, *48*, 870–877. [CrossRef]
35. Xu, Y.; Xiang, S.; Zhou, H.; Wang, G.; Zhang, H.; Zhao, H. Intrinsic Pseudocapacitive Affinity in Manganese Spinel Ferrite Nanospheres for High-Performance Selective Capacitive Removal of Ca^{2+} and Mg^{2+}. *ACS Appl. Mater. Interfaces* **2021**, *13*, 38886–38896. [CrossRef]

chemosensors

Review

Advances in the Application of Nano-Enzymes in the Electrochemical Detection of Reactive Oxygen Species: A Review

Rongwei Gao and Shujuan Bao *

Institute for Clean Energy and Advanced Materials, School of Materials and Energy, Southwest University, Chongqing 400715, China; rongwei.gao@kuleuven.be
* Correspondence: baoshj@swu.edu.cn; Tel.: +86-23-68254943 or +86-023-68254372

Abstract: Reactive oxygen species (ROS) play an important role in maintaining human health and are recognized as indicators of oxidative stress linked to various conditions such as neurodegenerative and cardiovascular diseases, as well as cancer. Consequently, detecting ROS levels in biological systems is crucial for biomedical and analytical research. Electrochemical approaches offer promising opportunities for ROS determination due to their exceptional sensitivity, speed, and simplicity of equipment. This review covers studies using advanced electrochemical nanozyme sensors for detecting ROS in biological samples that were published over the last ten years, from 2013 to 2023. Emphasis is placed on the sensor materials and different types of modifiers employed for selective ROS detection. Furthermore, a comprehensive analysis of the sensors' selectivity was performed.

Keywords: reactive oxygen species (ROS); nanozymes; electrochemistry sensor; detection; biological objects

Citation: Gao, R.; Bao, S. Advances in the Application of Nano-Enzymes in the Electrochemical Detection of Reactive Oxygen Species: A Review. *Chemosensors* **2023**, *11*, 440. https://doi.org/10.3390/chemosensors11080440

Academic Editor: Marcello Mascini

Received: 3 July 2023
Revised: 29 July 2023
Accepted: 2 August 2023
Published: 7 August 2023

1. Introduction

Over the past few decades, substantial progress has been achieved in the study of ROS. Traditionally, ROS have been considered as oxygen derivatives generated through complete or incomplete oxidation processes [1]. Figure 1 illustrates that a significant portion of molecular oxygen utilized by humans is involved in oxidation reactions within mitochondria. In this context, a four-electron transfer to the O_2 molecule yields the formation of two water (H_2O) molecules. However, one-, two-, or three-electron reduction can also generate intermediate ROS, including hydrogen peroxide (H_2O_2), hydroxyl radical (•OH), superoxide anion ($O_2^{\bullet-}$), singlet oxygen ($^1O_2$), and others [2]. ROS primarily govern numerous signal transduction pathways within the human body by directly interacting with proteins, transcription factors, and genes, leading to structural modifications and modulation of their functions. Moderate increases in ROS levels typically promote cell metabolism and differentiation, whereas prolonged and excessive ROS production can induce oxidative damage, such as lipid peroxidation, protein degradation, and DNA fragmentation [3,4]. Consequently, the assessment of ROS levels in biological samples such as blood serum is a crucial aspect of the research of medical and analytical fields.

As the field of ROS detection advances, it is essential to address certain limitations. Notably, different ROS possess unique intrinsic properties, such as lifetimes, diffusion rates, and sources of generation, which can introduce challenges and potential inaccuracies in measurements. Moreover, the inherently limited and fluctuating levels of reactive oxygen species (ROS) at their sites of production pose significant challenges for accurate detection methods, particularly when applied to living cells. However, electrochemical methods have emerged as a promising approach for ROS detection in biological samples, offering several advantages, including simplicity of equipment, high sensitivity and selectivity, rapid response times, and the ability to fabricate micro- and nano-sized sensors for intracellular ROS determination [5,6].

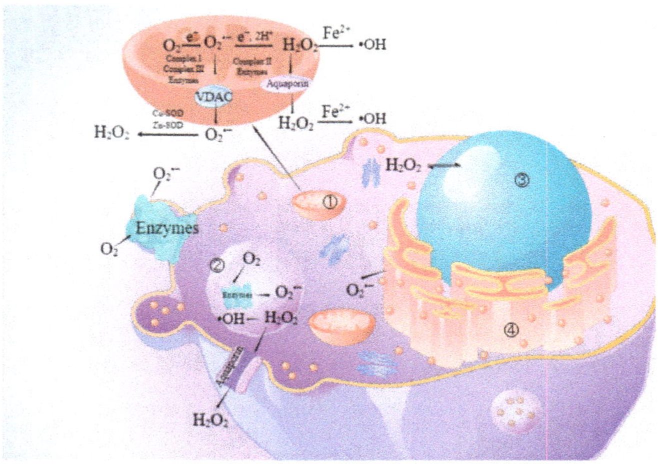

Figure 1. The generation of reactive oxygen species in biological systems. (By Figdraw). ① Mitochondrion; ② cytosol; ③ nuclei; ④ endoplasmic reticulum.

An excellent electrochemical biosensor comprises highly selective and active sites integrated into a stable carrier with excellent electrical conductivity. The integration of natural enzyme-based electrochemical biosensors represent a compelling approach that harnesses the inherent specificity of enzymes alongside the enhanced sensitivity and rapid response characteristics of electrochemical techniques. However, the instability of biological enzymes has hindered their development [7,8]. Since their emergence, nanomaterials have revolutionized the field by providing unprecedented possibilities for designing catalytic interfaces with exceptional activity and selectivity. Moreover, the integration of carbon/metal-based nanozymes, which exhibit high electron conductivity, with catalytic interfaces has demonstrated enhanced interfacial electron-transfer kinetics, thereby enabling the development of high-performance assays. The exceptional conductivity and facile functionalization capabilities of nanozymes, either through conductive matrices or redox-relay units, facilitate efficient electrocatalytic activation. As a result, nanozymes present greater promise for electrochemical sensing applications when compared to many naturally occurring enzymes [9–11]. However, despite the exciting stability and activity demonstrated by nanozymes, achieving selectivity in complex living systems remains a major challenge. This comprehensive review encompasses recent advancements in electrochemical techniques utilized for the detection of reactive oxygen species, including H_2O_2, $O_2^{\bullet-}$, and •OH, employing nanozymes as sensing elements. It goes beyond a mere exploration of trends, delving into the selectivity challenges encountered by electrochemical nanozyme sensors. We systematically analyze the challenges that researchers may encounter when employing specific approaches. With ongoing advancements in the sensitivity and selectivity of in vivo ROS biosensors, we hold an optimistic view that the field of ROS research will experience significant advancements.

2. Selection and Construction of Nanozymes with Highly Active Sites for Electrochemical Detection of ROS

From a biological perspective, the active site of a natural enzyme is defined as the region where substrate molecules undergo specific chemical reactions with a significant reduction in activation energy. It serves as the critical component responsible for enzyme function. Consequently, the most effective approach to achieving comparable catalytic performance to natural enzymes is to mimic their active sites. To date, there has been a growing utilization of diverse nanomaterials in the construction of electrochemical nanozymes for the detection of reactive oxygen species (ROS). This can be attributed to the distinctive physicochemical characteristics exhibited by nanomaterials, including their diminutive size, expansive surface area, and remarkable reactivity. Prominent examples of such nanomaterials encompass noble-metal-based materials, transition-metal-based materials, carbon-based materials, metal–organic framework-based materials, and various other emerging nanomaterials. These materials have garnered significant attention in the field of electrochemical nanozymes due to their unique properties and potential applications in ROS detection.

2.1. Carbon-Based Nanozymes

Nano-carbon materials, including carbon nanotubes, graphene, and other related structures, exhibit exceptional characteristics such as high strength, outstanding electrical conductivity, efficient charge transfer capabilities at high current densities, low resistance, substantial surface area, and remarkable chemical stability. These inherent properties make them highly desirable for various applications, particularly in the field of electrochemical sensing and nanozyme-based systems.

Detection of $O_2^{\bullet-}$. The most commonly used carbon material is graphene. A. Olean-Oliveira et al. recently reported on the development of a nonenzymatic chemiresistor sensor for the detection of superoxide radicals, utilizing an azo-polymer in conjunction with reduced graphene oxide (rGO) employed as the resistive platform for the sensor application [12]. The sensor platform was fabricated by sequentially depositing poly (azo-Bismarck Brown Y) and reduced graphene oxide films using a layer-by-layer assembly technique (Figure 2A). The resulting nanocomposite film demonstrated intriguing synergistic properties, combining the redox properties of the azo-polymer with the excellent electronic conductivity and stability of graphene. Real-time impedance measurements (chrono-impedance) using the poly(azo-BBY)/rGO sensor exhibited a linear relationship between the real impedance and the concentration of superoxide anions (ranging from 0.12 to 2.6 mM), with a detection limit of 81.0 μM. Xuan Cai et al. further introduced a novel approach for fabricating enzyme-mimicking metal-free catalysts, specifically for the electrochemical detection of $O_2^{\bullet-}$, incorporating phosphate groups into a graphene-based foam [13] (Figure 2B). This was achieved through a template-free hydrothermal process, involving the treatment of graphene oxide (GO) with varying amounts of phytic acid (PA) to obtain a three-dimensional porous graphene-based foam (PAGF). The characterization results confirmed the successful fabrication of the sensors, which were effectively employed for the determination of $O_2^{\bullet-}$ released by cells, showcasing exceptional performance in the dynamic monitoring of cellular $O_2^{\bullet-}$ levels.

Mesoporous carbon materials possess a significant number of edge-plane-like defective sites, which effectively facilitate electron transfer to analytes and enhance the electrochemical activity at the electrode interfaces. The presence of mesoporous channels within the carbon shells of Hollow Mesoporous Carbon Spheres (HMCSs) offers advantageous mass transport and/or charge transfer properties between the sensors and analytes. Li Liu et al. successfully developed an enzyme- and metal-free electrochemical method with remarkable sensitivity for detecting $O_2^{\bullet-}$. This method utilized a screen-printed carbon electrode (SPCE) that was modified with nitrogen-doped hollow mesoporous carbon spheres (N-HMCSs) [14]. As shown in Figure 2C, the electrochemical reduction of $O_2^{\bullet-}$ that takes place on the surface of the modified electrodes is represented by the following equation:

$$O_2^{\bullet-} + 2H^+ + e^- = H_2O_2 \quad (1)$$

The chronoamperometric responses of the N-HMCSs/SPCE towards $O_2^{\bullet-}$ at −0.15 V were recorded. The current exhibited distinct variations upon the addition of $O_2^{\bullet-}$, demonstrating a proportional relationship with the concentration of $O_2^{\bullet-}$ up to 480 mM. Based on the working area of the electrodes (0.071 cm^2), the sensitivity of the N-HMCSs/SPCE was calculated to be 1493.2 $mA \cdot cm^{-2}\ mM^{-1}$. Furthermore, the limit of detection (LOD) for $O_2^{\bullet-}$ was determined to be 2.2 μM.

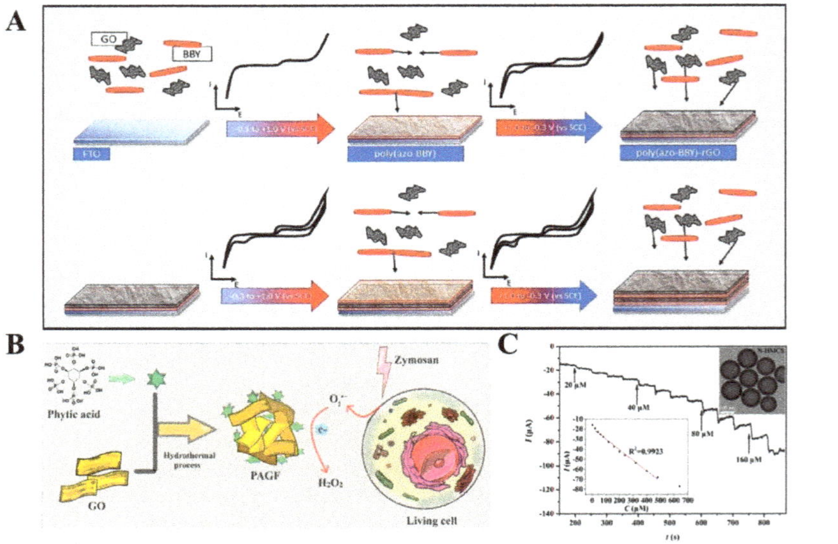

Figure 2. (**A**) Schematic illustration of the construction of a sensor utilizing a poly(azo-BBY)-rGO film [12], Copyright: Elsevier B.V. (**B**) Illustration of the synthesis of PAGF materials and their application in detecting cell-released superoxide anions [13], Copyright: Elsevier B.V. (**C**) Current–time response of N-HMCS/SPCE with successive injection of $O_2^{\bullet-}$ into 0.1 M deoxidized PBS (pH 7.4) at -0.15 V [14], Copyright: Elsevier B.V.

Detection of H_2O_2. The generation of H_2O_2 was first discovered in 1966, and since then, a great deal of work has been performed to investigate the production of hydrogen peroxide and its important role in the body. According to some works, H_2O_2, an essential and potent oxidant, plays a pivotal role in various biological processes, encompassing intercellular signaling, immune cell recruitment, and modulation of cellular morphology and differentiation. Its significance lies in its simplicity, importance, and remarkable oxidative capabilities, which contribute to the intricate mechanisms underlying fundamental biological functions. Despite its lower oxidant power compared to $O_2^{\bullet-}$, H_2O_2 is recognized as an extremely potent cytotoxic agent. However, in a groundbreaking study conducted by J. Q. Tian et al. in 2013, it was demonstrated for the first time that ultrathin graphitic carbon nitride (g-C_3N_4) nanosheets possess exceptional electrocatalytic properties, making them a cost-effective, environmentally friendly, and highly efficient catalyst for the reduction of hydrogen peroxide [15]. Another noteworthy contribution in this field was made by the research group led by J. Bai [16] who developed H_2O_2 sensors utilizing carbon dots (CDs) and multi-walled carbon nanotubes (MWCNTs). Notably, the CDs/MWCNTs/GCE sensor exhibited a significant synergistic effect, leading to enhanced performance, including an improved LOD of 0.25 μM.

H_2O_2 detection using carbon-based nanomaterials often relies on the electrochemical sensing principle, where H_2O_2 undergoes a redox reaction at the surface of the carbon nanomaterial, leading to measurable electrical signals. As represented by the following equation:

$$H_2O_2 + 2H^+ + 2e^- = 2H_2O \quad (2)$$

2.2. *Noble-Metal-Based Nanozymes*

Noble metal nanomaterials, such as gold nanoparticles (AuNPs) and silver nanoparticles (AgNPs), have captured significant interest within the scientific community due to their remarkable stability, high conductivity, facile preparation, expansive specific surface area, and outstanding biocompatibility. Notably, emerging research has highlighted the crucial influence of nanoparticle size, shape, and distribution on their electrocatalytic activity. These findings underscore the significance of tailoring these parameters to optimize the electrochemical performance of noble metal nanomaterials.

Detection of $O_2^{\bullet-}$. W. Z. Fan. et al. successfully obtained uniformly dispersed silver nanoparticles (AgNPs) by pyrolyzing a novel silver-based metal–organic framework, as reported in their study [17]. Specifically, the AgNPs@C nanocomposites were synthesized through thermal treatment of the Ag-based metal–organic frameworks, utilizing silver as the metal center and benzimidazole as the organic ligand. An electrochemical response towards the reduction of $O_2^{\bullet-}$ was observed in the obtained AgNPs@C nanocomposites, exhibiting an ultra-wide linear range (3.032×10^{-13} to 5.719×10^{-5} M) and an exceptionally low detection limit (1.011×10^{-13} M), as depicted in Figure 3A. Furthermore, the fabricated sensor was successfully employed for real-time detection of $O_2^{\bullet-}$ released from HeLa cells under both normal and oxidative stress conditions. Based on the experimental results, the authors presented the reduction mechanism of $O_2^{\bullet-}$ on the AgNPs@C/GCE as:

$$2O_2^{\bullet-} + 2H^+ \rightarrow H_2O_2 + O_2 \tag{3}$$

$$H_2O_2 \overset{AgNPs}{\rightarrow} H_2O + 1/2O_2 \tag{4}$$

$$O_2 + 2H^+ + 2e^- \overset{electrode}{\rightarrow} H_2O_2 \tag{5}$$

As shown in Equation (3), $O_2^{\bullet-}$ rapidly decomposes into H_2O_2 and O_2 in aqueous solution due to its inherent instability. It is well-known that Ag nanoparticles possess catalytic activity towards the decomposition of H_2O_2 [18]. Consequently, it was observed that AgNPs exhibit enhanced catalytic efficacy in the decomposition of H_2O_2 as depicted in Equation (4). Subsequently, the liberated oxygen originating from Equations (3) and (4) undergoes diffusion in the vicinity of the electrode, allowing for its detection through reduction on the modified electrode, as illustrated in Equation (5). Other common methods involve the combination of nanoparticles and carbon materials. For instance, T. D. Wu. et al. fabricated a nanocomposite (NCF-Ag) comprising nitrogen-doped cotton carbon fiber and silver nanoparticles (Figure 3B). The fabricated sensor exhibited remarkable electrochemical performance in the identification of $O_2^{\bullet-}$, demonstrating notable attributes such as a low limit of detection (LOD) of 2.53×10^{-14} M, an extensive linear detection range spanning from 7.59×10^{-14} to 7.22×10^{-5} M, and exceptional selectivity [19]. In a separate investigation, a nanocomposite comprising silver nanoparticles and multi-walled carbon nanotubes (AgNPs/MWNTs) was employed as an efficient electrode material, enabling sensitive detection of superoxide anions.

Detection of H_2O_2. D. Z. Zhu [20] drew inspiration from self-assembling peptide nanofibers (PNFs) and successfully designed and synthesized a novel hybrid material called PtNWs-PNFs/GO, which consists of biomimetic graphene-supported ultrafine platinum nanowires (PtNWs) integrated with PNFs. The controllable self-assembly process allowed the PNFs to act as a bridge between the GO nanosheets and PtNWs. The presence of PtNWs, known for their high catalytic activity, imparts remarkable electrochemical properties to the PtNWs-PNFs/GO hybrid-based sensor. The sensor demonstrates an extended linear detection range of 0.05 mM to 15 mM and a low detection limit of 0.0206 mM. Euna Ko et al. [21] immobilized bimetallic Au and Pt nanoparticles on the surface of agarose microbeads through chemical means, resulting in a hybrid nanostructure termed Au@PtNP/GO (Figure 4). The synergistic effect between the bimetallic nanoparticles and GO conferred strong peroxidase-like catalytic activity to the hybrid nanostructure, particularly towards the 3,3′,5,5′-tetramethylbenzidine (TMB) substrate in the presence of H_2O_2. This hybrid nanostructure enables dual applications of colorimetric and electrochemical detection. Upon the introduction of the TMB substrate solution containing H_2O_2, the catalytic oxidation of TMB takes place. Consequently, on the electrode surfaces, the oxidized TMB is subsequently subjected to electrochemical reduction, thereby leading to the achievement of an expanded detection range for H_2O_2 spanning from 1 μM to 3 mM, along with a reduced lower limit of detection quantified as 1.62 μM. Furthermore, the developed point-of-care (POC) devices exhibited accurate determination of H_2O_2, demonstrating strong repeatability and reproducibility in real sample testing using artificial urine.

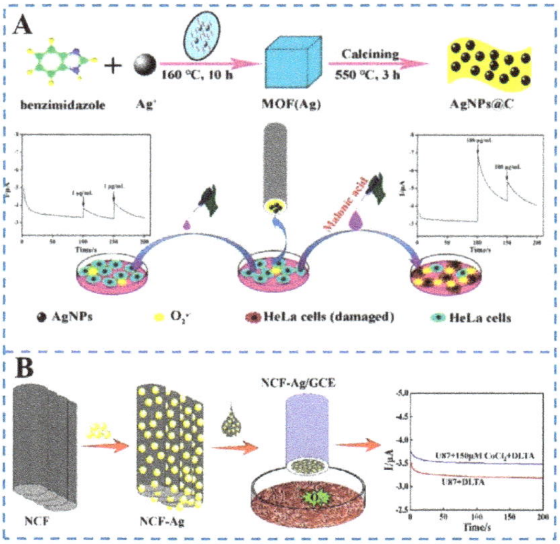

Figure 3. (**A**) A sensitive superoxide anion sensor constructed with Ag-based metal–organic frameworks, which can directly detect $O_2^{\bullet-}$ released from cells [17]. Copyright: Wiley Analytical Science. (**B**) Schematic illustration of the preparation and application of the NCF-Ag/GCE [19]. Copyright: Elsevier B.V.

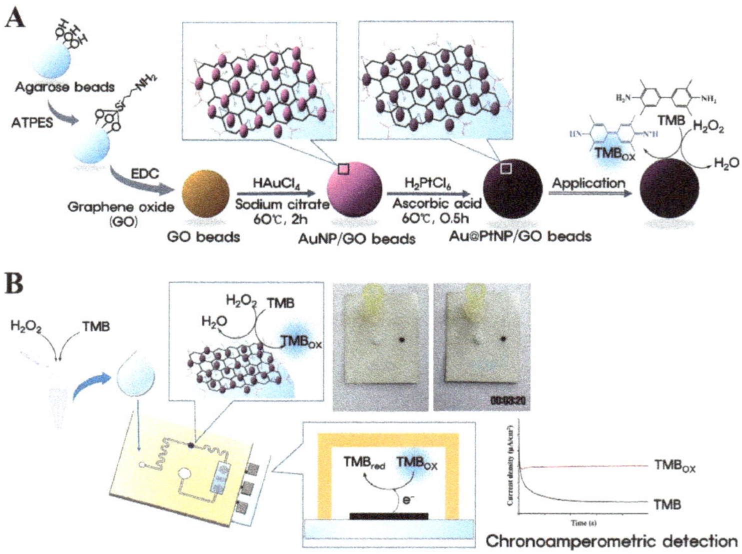

Figure 4. Schematic diagram of (**A**) the preparation of Au@PtNP/GO microbeads and (**B**) H_2O_2 detection on electrochemical POC devices with Au@PtNP/GO nanozymes [21]. Copyright: Elsevier B.V.

Detection of •OH. As is commonly known, among all the free radicals, hydroxyl radicals (•OH) exhibit exceptional reactivity and pose significant dangers due to their nanosecond-scale lifetime. Consequently, the electrochemical detection of hydroxyl radicals presents a formidable challenge, necessitating rapid response times, ultra-high sensitivity, and low detection limits. Regrettably, the current literature offers limited insight into the electrochemical detection of hydroxyl radicals, and the majority of developed sensors have not been employed for the analysis of biological samples. In this regard, two groups of modifier materials have emerged as potential candidates for •OH detection: thiol self-assembled monolayers (SAM) and CeOx nanoclusters.

As shown in Figure 5, X. W. Xu et al. [22] have successfully developed a novel electrochemical sensor utilizing a self-assembled nanoporous gold layer (NPGL) modified with 6-(Ferrocenyl) hexanethiol (6-FcHT) on a glassy carbon electrode (6-FcHT/NPGL/GE). This sensor demonstrates remarkable sensitivity and selectivity in detecting the release of •OH from living cells. The enhanced sensitivity can be attributed to the unique porous architecture of NPGL, which significantly increases the electrode's surface area and facilitates rapid electron transport during electrochemical reactions. Moreover, NPGL offers abundant active binding sites for the efficient assembly of the •OH capture agent (6-FcHT), ensuring excellent selectivity.

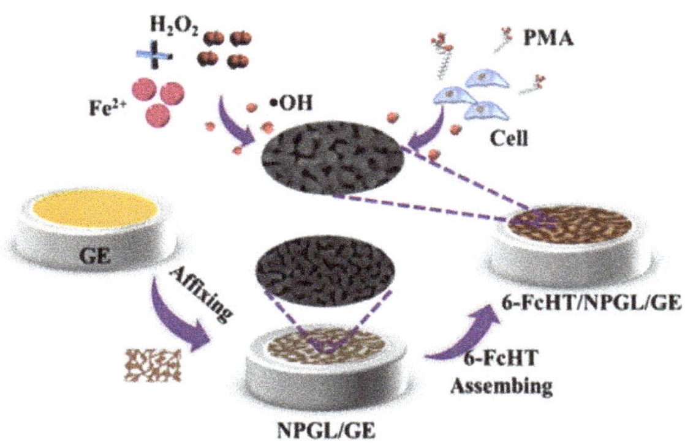

Figure 5. The schematic for preparing 6-FcHT/NPGL/GE [22]. Copyright: Elsevier B.V.

In comparison, when 6-FcHT was solely immobilized on a glassy carbon electrode (6-FcHT/GE), the sensitivity for •OH detection was measured at 0.0305 mA nM^{-1} with a detection limit of 0.133 nM within the linear range of 0.4 nM to 70 nM. After implementing the NPGL modification, the sensitivity of 6-FcHT/NPGL/GE towards •OH increased substantially to 0.1364 mA nM^{-1}, while the detection limit significantly decreased to 0.316 pM. Furthermore, the linear detection range was effectively extended from 1 pM to 100 nM. Importantly, the sensor exhibits additional merits in terms of reproducibility, repeatability, and stability, making it suitable for direct electrochemical detection of •OH in HepG2 cells.

2.3. *Transition-Metal-Based Nanozymes*

2.3.1. Metallic Oxides

Metal oxides are considered to be highly significant and extensively studied solid catalysts due to their exceptional structural characteristics, including a large surface area, semiconducting behavior, and facile synthesis. These unique properties make metal oxides attractive candidates for the development of enzyme-less sensors with enhanced effectiveness. Extensive research has been conducted to explore the potential of metal oxides in this regard.

Detection of $O_2^{\bullet-}$. Y. L. Liu et al. [23] synthesized Co_3O_4 nanoparticles (NPs) with an average diameter of about 30 nm by utilizing ZIF-9 as a template (Figure 6A). The resulting electrode exhibited excellent electrochemical activity for the determination of $O_2^{\bullet-}$. This enhancement in performance can be attributed to the introduction of multi-walled carbon nanotubes (MWCNTs) as a substrate, which enhances the electron-conductive property. The three-dimensional architectures of metal–organic frameworks (MOFs) utilized here served as a template material and underwent a conversion process to yield Co_3O_4@CMWCNTs. These hybrid structures possess a high specific surface area and well-defined mesopores, facilitating enhanced contact with $O_2^{\bullet-}$ species. A. L. Ding and F. Liu et al. [24] conducted an investigation into the synthesis of MnO nanocomposites integrated within cellulose nanofiber (CNF) aerogels for the purpose of determining $O_2^{\bullet-}$. The process involved the synthesis of MnO nanoparticles through the calcination of potassium permanganate ($KMnO_4$) embedded within bacterial cellulose (BC) hydrogels, followed by the carbonization of these hydrogels to yield CNF aerogels with embedded MnO nanoparticles. The resulting sensor demonstrated a linear amperometric response characterized by a considerable sensitivity of 76.2×10^{-3} μA cm^{-2} M^{-1} and a low detection limit of 1.2×10^{-9} M within a concentration range of 5.0×10^{-9} M to 2.5×10^{-6} M. This novel approach provides a promising strategy for $O_2^{\bullet-}$ sensing, utilizing well-designed nanocomposite structures embedded within CNF aerogels.

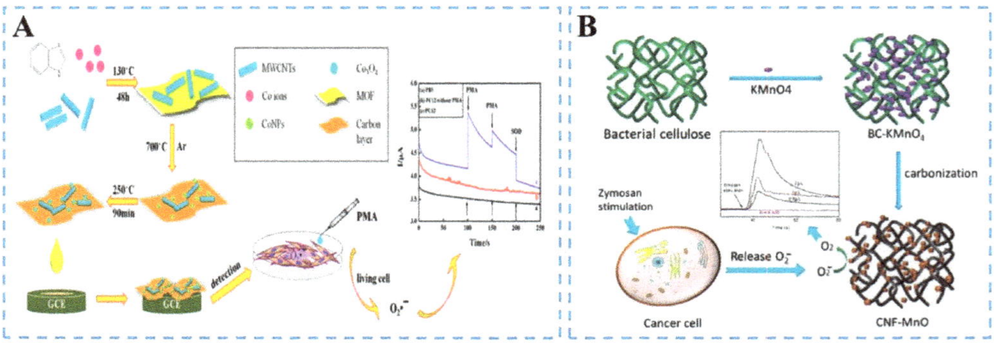

Figure 6. (**A**) Schematic illustration of preparation and application of Co_3O_4 nanocrystals [23]. Copyright: Elsevier B.V. (**B**) Schematic illustration of the preparation of 3D MnO−CNF nanocomposites [24]. Copyright: Wiley.

Detection of H_2O_2. Gaseous hydrogen peroxide plays a significant role as a biomarker linked to severe medical conditions like lung cancer and asthma. The conventional approach for monitoring exhaled breath biomarkers involves the utilization of exhaled breath condensate along with standard analytical techniques. In this case, Ursa Klun et al. [25] presented a novel H_2O_2-gas-sensing approach, depicted in Figure 7. The aqueous polyacrylate gel electrolyte containing Cu (II) ions served as a sensing material that facilitated the accumulation and stabilization of the gaseous analyte. Through redox interaction with Cu (II) ions, the H_2O_2 was rapidly and sensitively detected. Notably, the gas sensor developed by Klun et al. demonstrated successful and rapid detection of gaseous H_2O_2 within only 2 min of accumulation under ambient conditions. The sensor demonstrated favorable sensitivity in the lower concentration range, and it exhibited a broad linear response across the investigated concentration range of 10–100 mg m^{-3}. This research breakthrough not only advances the field of H_2O_2 detection but also expands the potential applications in emerging fields such as explosive detection, environmental monitoring, and occupational health and safety. By providing a fast, sensitive, and reliable method for H_2O_2 gas sensing, this work opens doors for various practical applications and contributes to the advancement of detection techniques in diverse fields.

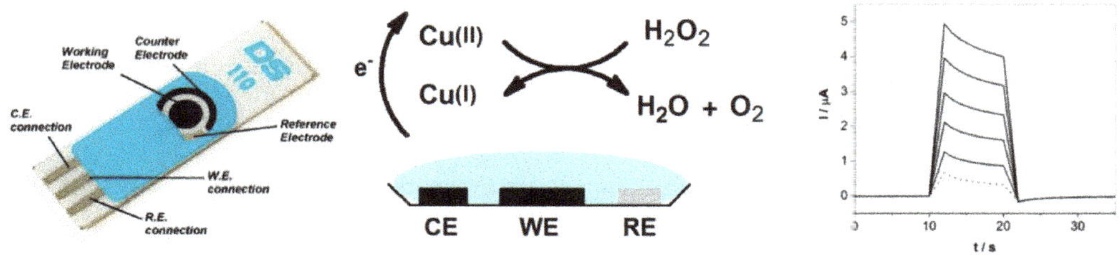

Figure 7. Drop Sens screen-printed carbon electrode and schematic presentation of sensor processes [25]. Copyright: Elsevier B.V.

Detection of •OH. The prevailing approach for the catalytic reduction of hydroxyl radicals (•OH) involves the utilization of cerium oxide (CeOx) nanoclusters. H. Ghaedamini's group [26] proposed a novel electrochemical sensor that incorporated a composite of cerium oxide nanoclusters, gold nanoparticles (AuNPs), and a highly conductive carbon material for the detection of •OH. The fabrication process involved the deposition of AuNPs onto carbon (Au/Carbon), followed by a controlled surface reaction to decorate the AuNP surface with CeOx nanoclusters to scavenge •OH. The synergistic effect of CeOx and AuNPs resulted in enhanced electrochemical signals and excellent •OH detection capability of the sensor. Duanghathaipornsuk et al. [27] conducted a research study to examine the impact of size and content variations of cerium oxide nanoparticles (CeNPs) on a composite sensor developed for the detection of hydroxyl radicals (•OH). Their findings revealed that composite sensors containing 50 wt% CeNPs displayed the most significant responses in the detection of •OH.

In summary, although CeOx-based electrochemical sensors display robust catalytic activity, their electrical conductivity is limited and often necessitates the incorporation of carbon materials with superior electrical conductivity. Additionally, achieving selectivity in practical testing remains a significant challenge. We posit that the issue of accurately determining •OH in biological samples should receive heightened attention in future investigations. Furthermore, in our perspective, an area of great interest lies in the development of non-toxic and readily available modifiers that exhibit high selectivity toward the hydroxyl radical.

2.3.2. Transition Metal Phosphates

Transition metal phosphates have garnered significant attention as a class of noble-metal-free materials investigated for use in electrochemical sensors. These materials are particularly appealing due to several advantageous characteristics. Firstly, transition metal phosphates are low-cost, making them economically viable for large-scale production and widespread application. Additionally, they are abundantly available in the Earth's crust, ensuring a sustainable and readily accessible supply.

Detection of $O_2^{\bullet-}$. Manganese phosphate ($Mn_3(PO_4)_2$) has gained considerable attention in the field of electrochemical sensors due to its biocompatible nature, making it suitable for biomedical applications [28]. Moreover, $Mn_3(PO_4)_2$ holds pivotal importance in material sciences owing to its unique catalytic and electronic properties. However, the performance of electrochemical sensors based on transition metal phosphate nanozymes is significantly influenced by the microstructure and size of the materials. To address this challenge, various strategies have been explored. M. Q. Wang [29] synthesized nanostructured $Mn_3(PO_4)_2$ hollow spheres with tunable pore structures using a micro-emulsion method, as depicted in Figure 8. The resulting nanostructured $Mn_3(PO_4)_2$ hollow spheres exhibited a high surface area and porous structure, enabling exceptional catalytic performance towards the detection of $O_2^{\bullet-}$. In another study, A. L. Ding et al. [30] developed a graphene/DNA/$Mn_3(PO_4)_2$ biomimetic enzyme. The inclusion of graphene as a carrier proved highly beneficial in augmenting the catalytic activity of $Mn_3(PO_4)_2$ nanoparticles, while DNA adsorbed on the graphene surface facilitated the growth of $Mn_3(PO_4)_2$. The resulting graphene/DNA/$Mn_3(PO_4)_2$ nanozyme exhibited outstanding electrochemical performance, notably reducing the response time and significantly enhancing the sensor's sensitivity towards $O_2^{\bullet -}$. The sensor achieved a low detection limit of 1.67 nM and a sensitivity of 3.54 mA mM^{-1}. Similarly, Y. Wang et al. [31] employed chitosan to confine ultra-small $Mn_3(PO_4)_2$ particles, demonstrating excellent biocompatibility. They further constructed an in situ detection chip for the electrochemical sensing of $O_2^{\bullet -}$ from murine breast tumor cells (4T1), achieving a low detection limit of 9.4 nM and good selectivity. X. Cai et al. [32] introduced a carbon-mediated approach for the rapid synthesis of transition metal phosphates with precise control over their shape and size. This strategy encompassed the assembly of melamine, phytic acid, and transition metal ions on the carbon substrate's surface.

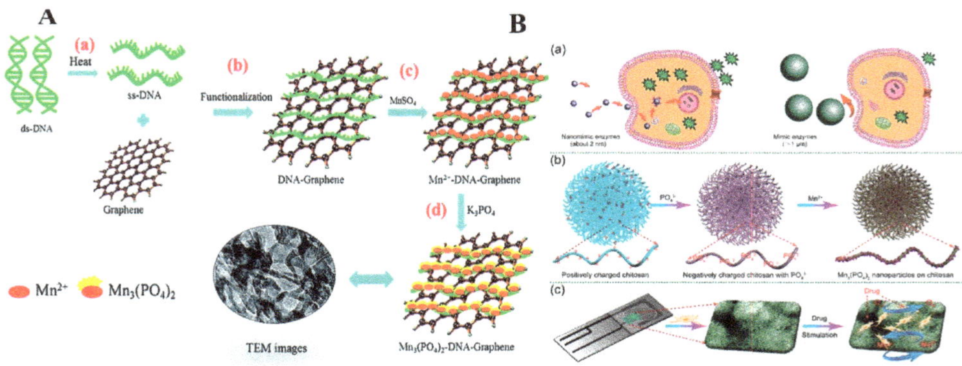

Figure 8. (**A**) Schematic illustration of the preparation procedure for graphene/DNA/$Mn_3(PO_4)_2$ [30]. (**a**) The solution was annealed at 95 °C for 15 min to produce single-stranded DNA (ssDNA); (**b**) ssDNA mixed with 15 mL of 1 mg mL^{-1} graphene; (**c**) The mixture was dispersed into 0.1 M $MnSO_4$ solution; (**d**) 0.1 M K_3PO_4 solution were dropwisely added under stirring and kept at room temperature for 30 min. Copyright: Royal Society of Chemistry. (**B**) (**a**) Nanomimic enzymes in immediate cell environment or inside the cells (left image) induce oxidative-stress-related cytotoxicity and smaller damage of larger mimic enzymes in immediate cell environment (right image); (**b**) the mechanism of growth of nano-$Mn_3(PO_4)_2$ on chitosan; (**c**) preparation of the chip device in 3D cell adsorption environment and the oxidation mechanism [31]. Copyright: Elsevier B.V.

In summary, manganese phosphate-based biomimetic sensors exhibit high selectivity towards $O_2^{\bullet -}$, a long shelf-life (approximately 30 days), an easy modification procedure, low cost, and a low nanomolar-level detection limit. However, these sensors suffer from the disadvantage of a large size due to the rapid formation of manganese phosphate, which hinders rapid and sensitive in vivo detection. Additionally, their poor electrical conductivity poses a challenge. In addition, other phosphates commonly used to detect superoxide anions are cobalt phosphate ($Co_3(PO_4)_2$) [33], iron phosphate ($FePO_4$) [34], nickel phosphate ($NiPO_4$NRs) [35], etc.

Detection of H_2O_2. The Ag/$FePO_4$ nanozymes were prepared by D.J. Rao group [36] using a modified silver mirror reaction at the gas–liquid interface. The $FePO_4$ nanospheres possessed a high surface-to-volume ratio and displayed a negative surface charge, enabling a large surface area for loading more Ag nanoparticles (NPs) and enhancing the catalytic performance towards the reduction of H_2O_2. Electrochemical investigations of the sensor demonstrated excellent analytical performance, with a wide linear range from 3.0×10^{-5} to 1.1×10^{-2} mol·L^{-1} and a low detection limit of 4.7 μmol·L^{-1}. The sensor also exhibited acceptable reproducibility and anti-interference ability. L. J. Peng et al. [37] synthesized the $Co_3(PO_4)_2 \cdot 8H_2O$ nanozyme in situ through direct reaction between cobalt (II) and phosphate. The researchers confirmed that the $Co_3(PO_4)_2 \cdot 8H_2O$ nanozyme exhibited peroxidase-like activity. The investigation further revealed that the peroxidase-like activity followed Michaelis–Menten kinetics, with a Michaelis constant (Km) of 0.073 mM, indicating a higher affinity towards H_2O_2 compared to natural horseradish peroxidase (HRP) and some other peroxidase nanozymes. The mechanistic exploration experiments confirmed that the hydroxyl radical (•OH) was the primary active species involved in the catalytic reaction of $Co_3(PO_4)_2{\cdot}8H_2O$. This research introduced a novel nanozyme that can be conveniently synthesized in situ, thereby eliminating the requirement for harsh, laborious, and time-consuming synthesis and purification procedures.

2.4. *Metal–Organic-Framework-Based Nanozymes*

Metal–organic frameworks (MOFs) have garnered significant attention due to their unique properties, such as a high specific surface area, tunable pore structure, and exposed activity. Comprising metal nodes and organic ligands, MOFs have emerged as versatile materials with diverse applications in the fields of electrochemistry, fluorescence, colorimetry, photo-electrochemistry, and electrochemiluminescence sensing. Their exceptional properties make them well-suited for these applications, enabling the development of advanced sensing platforms and devices. The high specific surface area of MOFs facilitates efficient analyte adsorption and interaction, while the tunable pore structure allows for the selective trapping and recognition of target molecules. Additionally, the exposed activity of MOFs contributes to their enhanced performance in various sensing modalities. As a result, MOFs have demonstrated great potential as promising materials for the development of next-generation sensing technologies.

Detection of $O_2^{\bullet-}$. Y. H. Zhang et al. [38] conducted a study aimed at enhancing the sensing performance of $O_2^{\bullet-}$ by developing a straightforward one-step strategy for the morphology-controllable synthesis of a manganese–organic framework (Mn-MOF) (Figure 9). Interestingly, they achieved the synthesis of Mn-MOF nanoparticles, asymmetric nano-lollipops, and nanorods with homogeneous components by carefully adjusting the solvent ratios and regulating the initial precursor concentrations. Subsequently, the authors found that the Mn-MOF nano-lollipops exhibited superior $O_2^{\bullet-}$-sensing capabilities compared to other nanostructures due to their larger active surface areas, which can be attributed to the excellent dispersibility provided by the asymmetric structure, as well as the accelerated electron transfer rate facilitated by the stem structure. By utilizing the Mn-MOF nano-lollipops for $O_2^{\bullet-}$ detection, a high sensitivity of 105 $\mu A\ cm^{-2}{\cdot}\mu M^{-1}$ was achieved, enabling the successful real-time and in situ detection of $O_2^{\bullet-}$ released from living cells. This research not only provides valuable insights into the solvents engineered morphologies of other MOF nanomaterials but also advances the understanding and potential applications of Mn-MOF nanostructures in sensing technologies.

Detection H_2O_2. MOF-based electrochemical sensors have gained significant attention for the detection of H_2O_2. Gao et al. [39] synthesized a Pt-nanoparticle-modified metalloporphyrin MOF (Pt@PMOF(Fe)) with multienzyme activity, which was utilized to construct an electrochemical H_2O_2 sensor. Another study by Z. Q. Wei's group [40] involved the synthesis of a 3D Co-based Zeolitic Imidazolate Framework (3D ZIF-67) for H_2O_2 detection. X. Liu's group [41] reported the development of a nickel metal–organic framework nanosheet array on Ti-mesh (Ni-MOF/TM) as an enzyme-free electrochemical sensing platform for H_2O_2 determination. X. L. Yang et al. [42] selected MIL-47(V) as an electrocatalyst to explore the feasibility of electrochemical sensing of H_2O_2. Nevertheless, these conventional approaches were executed in the single-readout mode, which makes them vulnerable to false-positive or false-negative outcomes arising from external interferences, such as the intricacies of the biological milieu, non-standardized testing protocols, and discrepancies among operators or testing environments. These limitations pose substantial challenges to the accuracy of single-readout analytical methods and impose constraints on their practical applications in disease diagnosis.

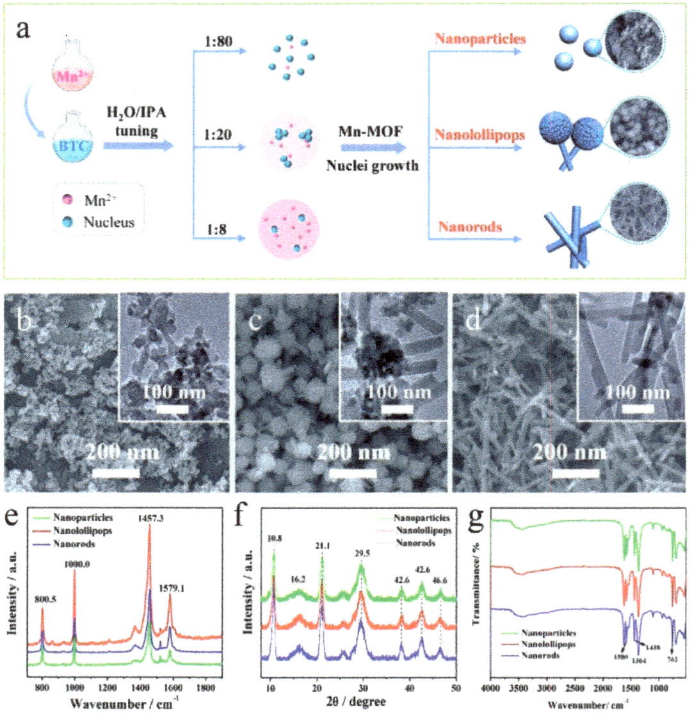

Figure 9. (**a**) Scheme of solutions of ratio-controlled Mn-MOF nanoparticles, nanoparticles, and nanorods. SEM images of Mn-MOF prepared in ratio of 1:80 (**b**), 1:20 (**c**), and 1:8 (**d**); insets are the corresponding TEM images. (**e**) Raman spectra and (**f**) XRD patterns of Mn-MOF nanoparticles, Mn-MOF nanorods, and Mn-MOF nanoparticles. (**g**) FTIR spectra of different Mn-MOF nanomaterials [38]. Copyright: Elsevier B.V.

To address this issue, K. Yu et al. [43] proposed a novel approach using a portable colorimetric and electrochemical dual-mode sensor for the detection of cell-secreted H_2O_2 and H_2S, based on the MOF-818 nanozyme. As depicted in Figure 10, the trinuclear copper centers in MOF-818 catalyze the formation of •OH from H_2O_2, leading to the oxidation of the substrate 3,3′,5,5′-tetramethylbenzidine (TMB) and the production of blue oxTMB. The authors developed a smartphone sensing system by capturing the Hue-Saturation Value (HSV) of the reaction solution using a "Color Identifier" App. Furthermore, MOF-818 exhibited outstanding electrocatalytic activity in H_2O_2 reduction, as evidenced by its reduction peak potential of 0.08 V vs. Ag/AgCl. This exceptional performance enabled the integration of a smartphone and a mini electrochemical analyzer, leading to the establishment of an ultra-sensitive sensing system for H_2O_2. The inherent advantage of a dual-modal assay lies in the fact that the two signals obtained can be utilized to calibrate the assay results and effectively mitigate false-positive/negative outcomes. Additionally, analytical methods featuring dual readout signals can readily adapt to different analytical conditions, catering to the diverse needs of assay tasks. This innovative dual-mode sensing strategy represents a significant advancement in MOF-based electrochemical sensors and holds great promise for enhancing their accuracy and practical applicability in disease diagnosis and beyond.

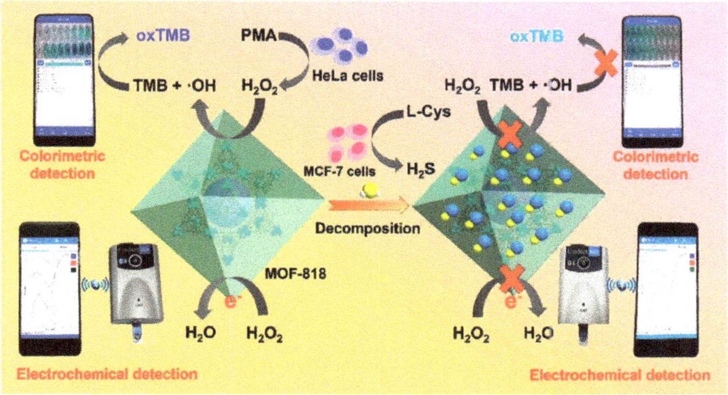

Figure 10. Illustration of MOF-818 nanozyme-based colorimetric and electrochemical dual-mode smartphone sensing platform for in situ detection of H_2O_2 and H_2S released from living cells [43]. Copyright: Elsevier B.V.

To sum up, within recent decades, numerous nanozymes were applied as electrochemical biosensors. However, we also can find that most of studies are focused on the detection of superoxide anions (Table 1) and H_2O_2 (Table 2). The elements commonly used to construct the active sites of nanozymes for detection of ROS are mainly Au, Ag, Pt, Ni, Co, Mn, Fe, Cu, Ti, and Ce. Specially, due to its special chemical properties, CeOx has an excellent antioxidant capacity and catalytic performance, so it has great application potential in detecting ROS. On the one hand, CeOx, as a potent antioxidant, is able to reversibly change its oxidation state in the presence of ROS through the redox cycle, thereby converting excess ROS into harmless substances and protecting cells from oxidative stress. On the other hand, CeOx nanozymes can enhance their selectivity and sensitivity to ROS through surface modifications or composite materials, so as to achieve accurate detection and monitoring of ROS concentration. With the in-depth study of the mechanism of ROS in vivo and the continuous optimization of the properties of nanozyme materials, it is believed that nanozymes will play an important role in the fields of life science, medicine, and environmental studies, providing new solutions for the prevention, diagnosis, and treatment of ROS-related diseases.

Table 1. Main analytical features and application of superoxide anion electrochemical sensors.

Electrode	Linear Range (μM)	LOD (nM)	Sensitivity ($\mu A\ cm^{-2}\ mM^{-1}$)	Stability (Days)	Application Potential (V)	Ref.
AgNPs@C/GCE	7.422×10^{-4}–0.5719	1.011×10^{-4}	-	7	−0.7	[17]
Co_3O_4@CMWCNTs/GCE	5×10^{-9}–10	1.6767×10^{-9}	-	-	-	[23]
Mn-MPSA-HCS/SPCE	0–1257.4	1.25	224	-	0.75	[32]
Ni(PO4)NRs/C-MWCNTs/GCE	1–80	97	5.67×10^4	25	−0.3	[35]
2D-mNC@CeO_2/SPCEs	8–536	179	401.4	20	−0.5	[44]
PAMAM-Au/GCE	3.69×10^{-5}–37.2	0.0123	-	-	−0.7	[45]
AgNPs-MC/GCE	1.68×10^{-3}–30.6	0.012	-	15	−0.5	[46]
Co-NPs–NG/GCE	1.67×10^{-3}–0.575	1.67	628.86	-	0.9	[47]
$Co_3(PO_4)_2$/I-rGO/GCE	2.4×10^{-3}–2.195	2.4	177.14	30	0.6	[48]
Mn-MPSA-MWCNTs/SPCE	0–1817	127	77.47	30	0.7	[49]

Table 2. Summary of amperometric sensors using nanozymes for the detection of H_2O_2.

Electrode	Linear Range (μM)	LOD (nM)	Sensitivity (μA cm^{-2} mM^{-1})	Stability (Days)	Application Potential (V)	Ref.
Fe SAs-N/C/GCE	764–9664	340	22.1	-	−0.05	[50]
Cu@Cu_2O/GCE	2–860	460	1855.53	-	−0.5	[51]
rGO/Au-NPs/GCE	25–3000	6.55	0.0641	-	−0.8	[52]
Bi_2S_3/g-C_3N_4/GCE	0.5–950	78	1011	7	0.26	[53]
Pt-LEPG/GCE	0.01×10^{-3}–0.029	0.65	575.75	14	0.5	[54]
CoHCF-NSp's /GCE	2–1130	2.1	329	-	0.8	[55]
Ag/PNA/GCE	1–3000	0.972	1844.76	10	−0.42	[56]
RGO–Pt NPs/GCE	0.5–3475	0.2	459	14	−0.08	[57]
Ag-Au/RGO/TiO_2/GCE	10–30,000	3	-	-	-	[58]
MOF-Au@Pt/GCE	0.8–3000	86	24.14	-	−0.12	[59]
LDH/PPy-Ag/GCE	30–800	280	257.64	30	−0.3	[60]
CuO-CeO_2/MXene/GCE	5–100	1.67	84.44	-	−0.3	[61]
Ag-Cu nanoalloys/GCE	2000–961,000	152	-	-	−0.07	[62]
AgNPs/2D Zn-MOFs/GCE	5–70,000	1.67×10^3	358.7	6	−0.55	[63]
$BiVO_4$/TiO_2/GCE	5–400	5×10^3	3014	90	0.5	[64]
$NiCo_2S_4$/rGO/GCE	25–11,250	190	118.5	14	−0.45	[65]
Pt/C-CeO_2/GCE	10–30,000	2×10^3	185.4	15	−0.4	[66]
Co_3N NW/TM/GCE	2–28	1×10^3	139.9	30	−0.7	[67]
MnO_2/Ta/GCE	1–2	60	1111.09	-	−1.21	[68]

3. Selective Challenges of Nanozymes for Electrochemical Detection of ROS

As mentioned in Section 2, with the development of nano-materials, more and more nanozymes have been designed to detection ROS. Although significant progress has been made in the development of biosensors, many of them still face challenges when it comes to practical applications. The selectivity of biosensors becomes a crucial factor when operating in real biological samples, as the presence of various electro-active species, including ascorbate and O_2, can lead to significant interference. Consequently, improving the specificity of nanozymes has been a focus of research efforts. Many researchers have also tried various methods to enhance the specificity of nanozymes, such as the K. L. Fan group [69] who employed a strategy to mimic the enzymatic microenvironment of natural peroxidase enzymes. They achieved this by introducing histidine residues onto the surface of Fe_3O_4 nanoparticles (Figure 11). Compared with unmodified Fe_3O_4, the imidazole of the histidine side chain can form hydrogen bond with hydrogen peroxide, so the modified Fe_3O_4 has an at least 10 times stronger affinity for hydrogen peroxide. However, the selective challenges of nanozymes and the electrochemical method itself still needs to be taken into account. In addition, R. W. Gao et al. also put forward the need to comprehensively consider the view of sensing chips; they believe that only the high activity and selectivity of the recognition factor is not enough, and it is also necessary to consider the adhesion of the recognition factor to the electrode and the rapid transfer and collection of electrical signals.

3.1. *Selective Challenge of the Same Nanozymes*

First and foremost, despite the significant advancements in the catalytic activity of nanozymes, their selectivity still falls short compared to natural enzymes. As previously discussed, the selectivity of a nanozyme can vary due to factors such as its preparation method, structure, size, and so on (Table 3). For instance, D. Q. Xu's group [70] achieved a rapid and sensitive response to H_2O_2 using a silver-nanoparticle-decorated carbon nanotube (AgNPs-MWCNT) composite, with a linear range of 1 to 1000 μM and LOD of 0.38 μM (S/N = 3). However, X. H. Liu et al. reported a nanocomposite composed of AgNPs/MWNTs utilized as an efficient electrode material for the sensitive detection superoxide anions. Secondly, the utilization of synergistic effects through the combination of different nanomaterials undoubtedly enhances sensor performance. However, it is important to consider the inherent variations in nanomaterial production, which can significantly impact the reproducibility of sensors across different laboratories. Incorporating multiple nanomaterials further amplifies these variations, thereby posing challenges to sensor reproducibility. Thirdly, the range of nanomaterial types is vast, and new nanomaterials are constantly being synthesized. The traditional approach of discovering nanocomposites with high catalytic activity by combining various nanomaterials is time-consuming and labor-intensive.

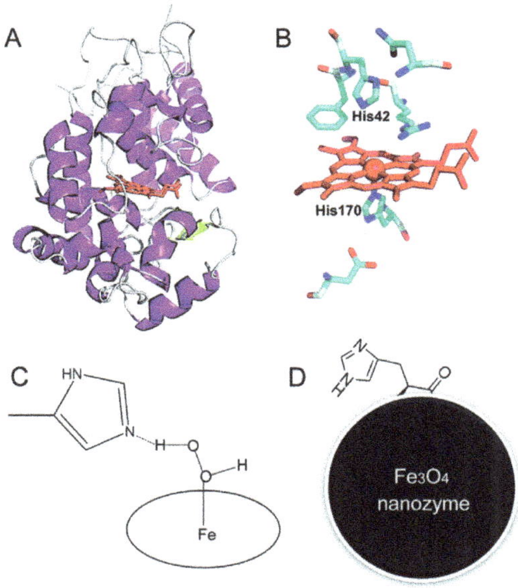

Figure 11. Architecture of the active site in HRP and comparison with the histidine-modified Fe_3O_4 nanozyme. (**A**) Protein structure of HRP (PDB entry 1HCH); (**B**) architecture of active site in HRP; (**C**) H bond between histidine residual and H_2O_2 in the initial state of catalysis of HRP; (**D**) enhancement of Fe_3O_4 nanozyme activity by histidine modification [69]. Copyright: Royal Society of Chemistry.

Table 3. Summary of electrochemical sensors using the same nanozymes for the detection of different analytes.

Nanozyme	Analyte	Linear Range (μM)	LOD (nM)	Sensitivity ($\mu A\ cm^{-2}\ mM^{-1}$)	Ref.
AgNPs-MWCNT	H_2O_2	1–1000	380	2556	[70]
AgNPs-MWCNT	$O_2^{\bullet-}$	3.65×10^{-7}–5.59×10^{-4}	0.1192	80.22	[71]
CeO_2	H_2O_2	0.001–0.125	0.4	141.96	[72]
CeO_2	$O_2^{\bullet-}$	8–536	179	401.4	[44]
MnO_2	H_2O_2	25×10^{-3}–2	5	3261	[73]
MnO_2	oxalate	7.8–250	910	-	[74]
Pt/TiO_2 nanotube	H_2O_2	0–20	400	40	[75]
Pt-TiO_2 NPs	NO	0.01–17,790	2.47	7.81	[76]
PBA/UiO-66/NF	H_2O_2	50–3500	0.02	1903	[77]
PBA/UiO-66/NF	Glucose	200–450	0.28	22,800	[77]

Addressing these challenges requires concerted efforts in the field of nanozyme research. Strategies aimed at improving the selectivity of nanozymes, developing standardized protocols for nanomaterial synthesis, and exploring novel approaches for efficient discovery of high-performance nanocomposites are crucial steps toward overcoming these limitations.

3.2. Selective Challenge of Electrochemical Technology

In recent years, several remarkable electrochemical studies have been published, highlighting the attractiveness of electrochemical technology as a tool for the determination of important small molecules and biomarkers. This technique offers advantages such as rapid detection, small device size, portability, and the ability for continuous monitoring. However, the issue of limited selectivity arises when current signals are mixed, compromising the accuracy of measurements. In an effort to address this challenge, V. Kumar et al. [78] developed a hybrid alginate–polyacrylamide hydrogel through in situ self-assembly of a reduced graphene oxide–cerium oxide nanocomposite (rGO-CeO_2) and cytochrome c (Cyt c) for electrochemical detection of reactive oxygen species (ROS). The hydrogel platform provided a significantly enlarged electroactive surface area, enhancing the reactivity of Cyt c. Moreover, the integration of the rGO-CeO_2 nanocomposite into the hydrogel improved both the electrochemical signals and structural stability of the sensor.

However, it is worth noting from Figure 12 that when the potential exceeds 0.75 V, all three species (H_2O_2, •OH, and $O_2^{\bullet-}$) undergo oxidation. Since their oxidation potentials are very similar, the resulting current signals become mixed, further emphasizing the necessity of developing sensors with improved sensitivity and selectivity for comprehensive investigations. To overcome these limitations, future research efforts should focus on the development of sensors that can offer enhanced sensitivity and selectivity. Such advancements will enable more precise and accurate studies in the field of electrochemical sensing, paving the way for applications in various areas of science and technology.

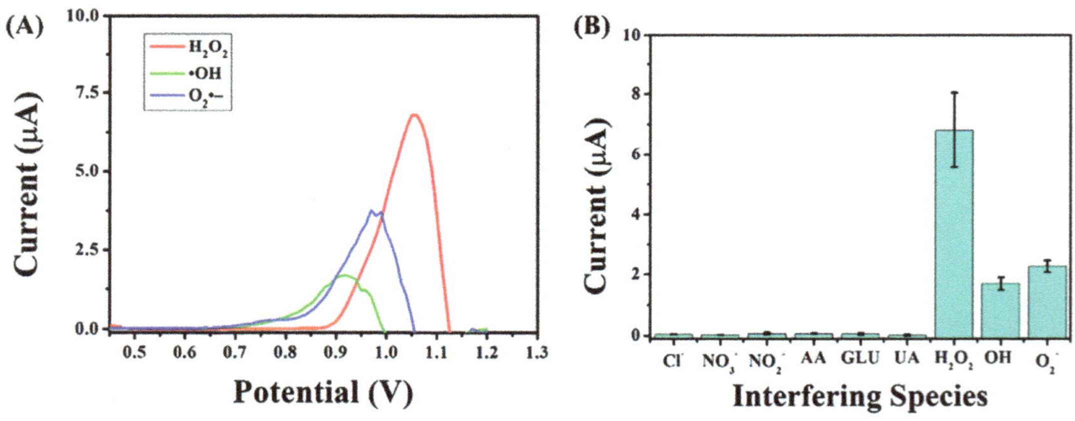

Figure 12. (**A**) Representative DPV responses after simultaneous addition of H_2O_2, •OH, and $O_2^{\bullet-}$ (30 μM) on rGO–CeO_2@Cyt *c* hydrogel/SPE. (**B**) Selective DPV responses of rGO–CeO_2@Cyt *c* hydrogel/SPE towards ROS over various interfering species, as labeled [78]. Copyright: Royal Society of Chemistry.

4. Conclusions and Future Perspectives

In this comprehensive review, we have discussed the recent advances in the application of nanomaterial-based electrochemical sensors for the detection of reactive oxygen species (ROS), specifically focusing on H_2O_2, •OH, and $O_2^{\bullet-}$. Despite the concerted efforts of numerous research groups, the field of electrochemical ROS detection and quantification still faces significant challenges that need to be addressed.

Firstly, as this review of recent studies has shown, most electrochemical sensors for detection ROS are focused on H_2O_2 and $O_2^{\bullet-}$, which is due to their half-life being longer than that of •OH (the half-life of H_2O_2, $O_2^{\bullet-}$, and •OH are 10–20 h, 6 –10 min, and 10 s, respectively). Thus, they are more easily detected by electrochemical methods. At the same time, we also notice that the materials and determination process for the detection of H_2O_2 and $O_2^{\bullet-}$ are similar. However, it is important to note that the determination of •OH requires distinct approaches due to the unique characteristics of this free radical. Therefore, the development of sensors with improved sensitivity and selectivity is necessary to investigate the impact of •OH in mammalian cells and its association with oxidative-stress-induced diseases.

The second crucial issue pertains to enhancing the sensitivity and selectivity of electrochemical sensors for ROS detection, particularly in the context of medical applications. The electrochemical technique shows immense promise for real-time detection utilizing micro- or even nanoscale electrodes, paving the way for the advancement of in vivo sensors capable of real-time monitoring of live cells, irrespective of their type or location. This necessitates sensors with enhanced sensitivity and selectivity. In this regard, nanozymes can greatly benefit from interdisciplinary collaborations, including advanced information technology such as machine learning. Prior to the synthesis of nanomaterials, machine learning algorithms can be employed to simulate and identify candidate materials that meet specific criteria, resulting in the establishment of a library of catalytically active materials with high activity, stability, and repeatability. Furthermore, the application of machine learning algorithms enables training using extracted features from cyclic voltametric curves of known concentrations of target analytes, thus facilitating precise concentration calculations.

Thirdly, when developing nanozyme-based electrochemical sensors, the potential for industrialization and commercialization should be taken into consideration. Emerging technologies like the Internet of Things can be employed for real-time monitoring and automated control of the production process.

In conclusion, the remarkable ability of nanozyme-based electrochemical sensors for ROS detection underscores their potential for occupying a significant market share in the commercial application of sensors as the synthesis of nanozymes continues to mature. With ongoing improvements in the sensitivity and selectivity of in vivo ROS biosensors, the field of ROS research stands to benefit significantly, advancing our understanding of oxidative-stress-related phenomena.

Author Contributions: Writing—original draft preparation: R.G.; Supervision, S.B. All authors have read and agreed to the published version of the manuscript.

Funding: This work is financially supported by National Natural Science Foundation of China (Grant number: 21972111).

Institutional Review Board Statement: Not applicable.

Informed Consent Statement: Not applicable.

Data Availability Statement: Not applicable.

Conflicts of Interest: The authors declare no conflict of interest.

References

1. Sies, H.; Jones, D.P. Reactive oxygen species (ROS) as pleiotropic physiological signalling agents. *Nat. Rev. Mol. Cell Biol.* **2020**, *21*, 363–383. [CrossRef] [PubMed]
2. Halliwell, B. Reactive Species and Antioxidants. Redox Biology Is a Fundamental Theme of Aerobic Life. *Plant Physiol.* **2006**, *141*, 312–322. [CrossRef] [PubMed]
3. Rahal, A.; Kumar, A.; Singh, V.; Yadav, B.; Tiwari, R.; Chakraborty, S.; Dhama, K. Oxidative stress, prooxidants, and antioxidants: The interplay. *Biomed. Res. Int.* **2014**, *2014*, 761264. [CrossRef] [PubMed]
4. Sayre, L.M.; Perry, G.; Smith, M.A. Oxidative Stress and Neurotoxicity. *Chem. Res. Toxicol.* **2008**, *21*, 172–188. [CrossRef]
5. Geraskevich, A.V.; Solomonenko, A.N.; Dorozhko, E.V.; Korotkova, E.I.; Barek, J. Electrochemical Sensors for the Detection of Reactive Oxygen Species in Biological Systems: A Critical Review. *Crit. Rev. Anal. Chem.* **2022**, 1–33. [CrossRef]
6. Duanghathaipornsuk, S.; Farrell, E.J.; Alba-Rubio, A.C.; Zelenay, P.; Kim, D.S. Detection Technologies for Reactive Oxygen Species: Fluorescence and Electrochemical Methods and Their Applications. *Biosensors* **2021**, *11*, 30. [CrossRef]
7. Gao, R.; Yang, X.; Yang, Q.; Wu, Y.; Wang, F.; Xia, Q.; Bao, S.-J. Design of an amperometric glucose oxidase biosensor with added protective and adhesion layers. *Microchim. Acta* **2021**, *188*, 312. [CrossRef]
8. Kaur, J.; Choudhary, S.; Chaudhari, R.; Jayant, R.; Joshi, A. *Enzyme-Based Biosensors*; ResearchGate: Berlin, Germany, 2019; pp. 211–240.

9. Wang, X.; Dong, S.; Wei, H. Recent Advances on Nanozyme-based Electrochemical Biosensors. *Electroanalysis* **2022**, *35*, 38–39. [CrossRef]
10. Wang, H.; Wan, K.; Shi, X. Recent Advances in Nanozyme Research. *Adv. Mater.* **2019**, *31*, e1805368. [CrossRef]
11. Campuzano, S.; Pedrero, M.; Yáñez-Sedeño, P.; Pingarrón, J.M. Nanozymes in electrochemical affinity biosensing. *Microchim. Acta* **2020**, *187*, 423. [CrossRef]
12. Olean-Oliveira, A.; Pacheco, J.C.; Seraphim, P.M.; Teixeira, M.F.S. Synergistic effect of reduced graphene oxide/azo-polymer layers on electrochemical performance and application as nonenzymatic chemiresistor sensors for detecting superoxide anion radicals. *J. Electroanal. Chem.* **2019**, *852*, 113520. [CrossRef]
13. Cai, X.; Chen, H.L.; Wang, Z.X.; Sun, W.Q.; Shi, L.B.; Zhao, H.L.; Lan, M.B. 3D graphene-based foam induced by phytic acid: An effective enzyme-mimic catalyst for electrochemical detection of cell-released superoxide anion. *Biosens. Bioelectron.* **2019**, *123*, 101–107. [CrossRef]
14. Liu, L.; Zhao, H.L.; Shi, L.B.; Lan, M.B.; Zhang, H.W.; Yu, C.Z. Enzyme- and metal-free electrochemical sensor for highly sensitive superoxide anion detection based on nitrogen doped hollow mesoporous carbon spheres. *Electrochim. Acta* **2017**, *227*, 69–76. [CrossRef]
15. Tian, J.Q.; Liu, Q.; Ge, C.J.; Xing, Z.C.; Asiri, A.M.; Al-Youbi, A.O.; Sun, X.P. Ultrathin graphitic carbon nitride nanosheets: A low-cost, green, and highly efficient electrocatalyst toward the reduction of hydrogen peroxide and its glucose biosensing application. *Nanoscale* **2013**, *5*, 8921–8924. [CrossRef]
16. Bai, J.; Sun, C.H.; Jiang, X.E. Carbon dots-decorated multiwalled carbon nanotubes nanocomposites as a high-performance electrochemical sensor for detection of H_2O_2 in living cells. *Anal. Bioanal. Chem.* **2016**, *408*, 4705–4714. [CrossRef] [PubMed]
17. Fan, W.Z.; Liu, X.H.; Wu, J.S.; Liu, Q.; Ding, L.; Liu, X.H. Development of a Novel Silver-based Sensing Platform for Detecting Superoxide Anion Released from HeLa Cells Directly. *Electroanalysis* **2022**, *34*, 987–994. [CrossRef]
18. Welch, C.M.; Banks, C.E.; Simm, A.O.; Compton, R.G. Silver nanoparticle assemblies supported on glassy-carbon electrodes for the electro-analytical detection of hydrogen peroxide. *Anal. Bioanal. Chem.* **2005**, *382*, 12–21. [CrossRef]
19. Wu, T.D.; Li, L.; Jiang, X.C.; Liu, F.X.; Liu, Q.; Liu, X.H. Construction of silver-cotton carbon fiber sensing interface and study on the protective effect of antioxidants on hypoxia-induced cell damage. *Microchem. J.* **2020**, *159*, 105345. [CrossRef]
20. Zhu, D.Z.; He, P.; Kong, H.; Yang, G.Z.; Luan, X.; Wei, G. Biomimetic graphene-supported ultrafine platinum nanowires for colorimetric and electrochemical detection of hydrogen peroxide. *J. Mater. Chem. B* **2022**, *10*, 9216–9225. [CrossRef] [PubMed]
21. Ko, E.; Van-Khue, T.; Son, S.E.; Hur, W.; Choi, H.; Seong, G.H. Characterization of Au@PtNP/GO nanozyme and its application to electrochemical microfluidic devices for quantification of hydrogen peroxide. *Sens. Actuators B-Chem.* **2019**, *294*, 166–176. [CrossRef]
22. Xu, Y.W.; Wang, D.Y.; Zhang, Y.; Zhang, J.J.; Jiang, S.Y.; Liang, W.C.; Zhu, T.T.; Ye, B.C. A novel electrochemical sensor for determination of hydroxyl radicals in living cells by coupling nanoporous gold layer with self-assembled 6-(Ferrocenyl) hexanethiol. *Anal. Chim. Acta* **2020**, *1096*, 69–75. [CrossRef] [PubMed]
23. Liu, Y.L.; Wei, H.W.; Jiang, X.C.; Guo, H.X.; Liu, X.H. Synthesis of metal-organic frameworks derived nanocomposites for superoxide anion radical sensing and cell monitoring upon oxidative stress. *J. Electroanal. Chem.* **2018**, *820*, 51–59. [CrossRef]
24. Ding, A.L.; Liu, F.; Zheng, J.S.; Chen, J.C.; Li, C.M.; Wang, B. Synthesis of Manganese Oxide Embedded Carbon Nanofibers as Effective Biomimetic Enzymes for Sensitive Detection of Superoxide Anions Released from Living Cells. *Macromol. Mater. Eng.* **2018**, *303*, 1800079. [CrossRef]
25. Klun, U.; Zorko, D.; Stojanov, L.; Mirceski, V.; Jovanovski, V. Amperometric sensor for gaseous H_2O_2 based on copper redox mediator incorporated electrolyte. *Sens. Actuators Rep.* **2023**, *5*, 100144. [CrossRef]
26. Ghaedamini, H.; Alba-Rubio, A.C.; Kim, D.S. A Novel Electrochemical Sensor Based on a Cerium Oxide/Gold/Carbon Nanocomposite for the Detection of Hydroxyl Free Radicals. *J. Electrochem. Soc.* **2023**, *170*, 047510. [CrossRef]
27. Duanghathaipornsuk, S.; Alateeq, F.A.O.; Kim, S.S.; Kim, D.-S.; Alba-Rubio, A.C. The effects of size and content of cerium oxide nanoparticles on a composite sensor for hydroxyl radicals detection. *Sens. Actuators B Chem.* **2020**, *321*, 128467. [CrossRef]
28. Barnese, K.; Gralla, E.B.; Cabelli, D.E.; Selverstone Valentine, J. Manganous Phosphate Acts as a Superoxide Dismutase. *J. Am. Chem. Soc.* **2008**, *130*, 4604–4606. [CrossRef]
29. Wang, M.Q.; Ye, C.; Bao, S.J.; Xu, M.W. Controlled synthesis of $Mn_3(PO_4)_2$ hollow spheres as biomimetic enzymes for selective detection of superoxide anions released by living cells. *Microchim. Acta* **2017**, *184*, 1177–1184. [CrossRef]
30. Ding, A.L.; Wang, B.; Ma, X.Q.; Diao, J.L.; Zheng, J.S.; Chen, J.C.; Li, C.M. DNA-induced synthesis of biomimetic enzyme for sensitive detection of superoxide anions released from live cell. *Rsc Adv.* **2018**, *8*, 12354–12359. [CrossRef]
31. Wang, Y.; Wang, D.; Sun, L.H.; Xue, P.; Wang, M.Q.; Lu, Z.S.; Wang, F.; Xia, Q.Y.; Xu, M.W.; Bao, S.J. Constructing high effective nano-$Mn_3(PO_4)_2$-chitosan in situ electrochemical detection interface for superoxide anions released from living cell. *Biosens. Bioelectron.* **2019**, *133*, 133–140. [CrossRef]
32. Cai, X.; Wang, Z.X.; Zhang, H.H.; Li, Y.F.; Chen, K.C.; Zhao, H.L.; Lan, M.B. Carbon-mediated synthesis of shape-controllable manganese phosphate as nanozymes for modulation of superoxide anions in HeLa cells. *J. Mater. Chem. B* **2019**, *7*, 401–407. [CrossRef] [PubMed]
33. Wang, M.Q.; Ye, C.; Bao, S.J.; Xu, M.W.; Zhang, Y.; Wang, L.; Ma, X.Q.; Guo, J.; Li, C.M. Nanostructured cobalt phosphates as excellent biomimetic enzymes to sensitively detect superoxide anions released from living cells. *Biosens. Bioelectron.* **2017**, *87*, 998–1004. [CrossRef] [PubMed]

34. Wang, Y.; Wang, M.Q.; Lei, L.L.; Chen, Z.Y.; Liu, Y.S.; Bao, S.J. $FePO_4$ embedded in nanofibers consisting of amorphous carbon and reduced graphene oxide as an enzyme mimetic for monitoring superoxide anions released by living cells. *Microchim. Acta* **2018**, *185*, 140. [CrossRef]
35. Cui, M.; Zhao, H.Y.; Wen, X.F.; Li, N.; Ren, J.J.; Zhang, C. Facile synthesis of nickel phosphate nanorods as biomimetic enzyme with excellent electrocatalytic activity for highly sensitive detection of superoxide anion released from living cells. *J. Pharm. Biomed. Anal.* **2022**, *212*, 114653. [CrossRef]
36. Rao, D.J.; Zhang, J.; Zheng, J.B. Synthesis of silver nanoparticles-decorated $FePO_4$ nanosphere at a gas-liquid interface for the electrochemical detection of Hydrogen peroxide. *J. Chem. Sci.* **2016**, *128*, 839–847. [CrossRef]
37. Peng, L.J.; Zhou, H.Y.; Zhang, C.Y.; Yang, F.Q. Study on the peroxidase-like activity of cobalt phosphate and its application in colorimetric detection of hydrogen peroxide. *Colloids Surf. A-Physicochem. Eng. Asp.* **2022**, *647*, 129031. [CrossRef]
38. Zhang, Y.; Guo, C.X.; Du, H.; Wang, X.; Liu, L.; Li, C.M. Solvent-engineered morphologies of Mn-MOF toward ultrasensitive sensing cell superoxide. *Electrochim. Acta* **2022**, *431*, 141147. [CrossRef]
39. Ling, P.; Cheng, S.; Chen, N.; Qian, C.; Gao, F. Nanozyme-Modified Metal-Organic Frameworks with Multienzymes Activity as Biomimetic Catalysts and Electrocatalytic Interfaces. *ACS Appl. Mater. Interfaces* **2020**, *12*, 17185–17192. [CrossRef]
40. Wei, Z.Q.; Li, W.; Yang, H.; Li, T.; He, S.N.; Wang, Y.; Hu, Y.L. Synthesis of 3D Co-based Zeolitic Imidazolate Framework and Application as Electrochemical Sensor for H_2O_2 Detection. *Int. J. Electrochem. Sci.* **2022**, *17*, 221132. [CrossRef]
41. Liu, X.; Xiang, M.H.; Zhang, X.Y.; Li, Q.; Liu, X.Y.; Zhang, W.J.; Qin, X.; Qu, F.L. An Enzyme-free Electrochemical H_2O_2 Sensor Based on a Nickel Metal-organic Framework Nanosheet Array. *Electroanalysis* **2022**, *34*, 369–374. [CrossRef]
42. Yang, X.; Qiu, W.; Gao, R.; Wang, Y.; Bai, Y.; Xu, Z.; Bao, S.-J. MIL-47(V) catalytic conversion of H_2O_2 for sensitive H_2O_2 detection and tumor cell inhibition. *Sens. Actuators B Chem.* **2022**, *354*, 131201. [CrossRef]
43. Yu, K.; Li, M.J.; Chai, H.N.; Liu, Q.; Hai, X.; Tian, M.W.; Qu, L.J.; Xu, T.L.; Zhang, G.Y.; Zhang, X.J. MOF-818 nanozyme-based colorimetric and electrochemical dual-mode smartphone sensing platform for in situ detection of H_2O_2 and H_2S released from living cells. *Chem. Eng. J.* **2023**, *451*, 138321. [CrossRef]
44. Wang, Z.; Zhao, H.; Chen, K.; Zhou, F.; Magdassi, S.; Lan, M. Two-dimensional mesoporous nitrogen-rich carbon nanosheets loaded with CeO_2 nanoclusters as nanozymes for the electrochemical detection of superoxide anions in HepG2 cells. *Biosens Bioelectron* **2022**, *209*, 114229. [CrossRef] [PubMed]
45. Liu, F.X.; Yu, R.; Wei, H.W.; Wu, J.S.; He, N.; Liu, X.H. Construction of a novel electrochemical sensing platform to investigate the effect of temperature on superoxide anions from cells and superoxide dismutase enzyme activity. *Anal. Chim. Acta* **2022**, *1198*, 339561. [CrossRef] [PubMed]
46. Liu, F.X.; Jiang, X.C.; He, N.; Yu, R.; Xue, Z.H.; Liu, X.H. Electrochemical investigation for enhancing cellular antioxidant defense system based on a superoxide anion sensor. *Sens. Actuators B-Chem.* **2022**, *368*, 132190. [CrossRef]
47. Zou, Z.; Shi, Z.Z.; Yuan, C.S.; Tang, C.Y.; Wu, C.; Liang, T.T.; Tang, K.L.; Chen, H.; Yang, H.B.; Li, C.M. Steric shelter-free cobalt nanoparticle-based high-sensitive biomimetic superoxide anion sensor. *Mater. Chem. Front.* **2021**, *5*, 6860–6864. [CrossRef]
48. Zou, Z.; Chen, J.; Shi, Z.Z.; Yuan, C.S.; Zhou, G.D.; Liu, Q.; Chen, H.; Zeng, Q.X.; Liang, T.T.; Tang, K.L.; et al. Cobalt Phosphates Loaded into Iodine-Spaced Reduced Graphene Oxide Nanolayers for Electrochemical Measurement of Superoxide Generated by Cells. *ACS Appl. Nano Mater.* **2021**, *4*, 3631–3638. [CrossRef]
49. Cai, X.; Shi, L.; Sun, W.; Zhao, H.; Li, H.; He, H.; Lan, M. A facile way to fabricate manganese phosphate self-assembled carbon networks as efficient electrochemical catalysts for real-time monitoring of superoxide anions released from HepG2 cells. *Biosens. Bioelectron.* **2018**, *102*, 171–178. [CrossRef]
50. Liang, Y.; Zhao, P.; Zheng, J.L.; Chen, Y.Y.; Liu, Y.Y.; Zheng, J.; Luo, X.G.; Huo, D.Q.; Hou, C.J. Fe Single-Atom Electrochemical Sensors for H_2O_2 Produced by Living Cells. *ACS Appl. Nano Mater.* **2022**, *5*, 11852–11863. [CrossRef]
51. Li, H.X.; Jiang, L.L.; Shao, D.; Wu, C.S.; Gao, Y.J.; Yang, Z.Q.; Yang, Z.J. Facile synthesis of Cu@Cu_2O aerogel for an effective electrochemical hydrogen peroxide sensor. *Chin. J. Anal. Chem.* **2022**, *50*, 100060. [CrossRef]
52. Patella, B.; Buscetta, M.; Di Vincenzo, S.; Ferraro, M.; Aiello, G.; Sunseri, C.; Pace, E.; Inguanta, R.; Cipollina, C. Electrochemical sensor based on rGO/Au nanoparticles for monitoring H_2O_2 released by human macrophages. *Sens. Actuators B-Chem.* **2021**, *327*, 128901. [CrossRef]
53. Othmani, A.; Derbali, M.; Kalfat, R.; Touati, F.; Dhaouadi, H. A novel 1D/2D Bi_2S_3/g-C_3N_4 core-shell nanocomposite as a highly performing H_2O_2 non-enzymatic electrochemical sensor. *J. Mater. Res. Technol. -JmrT* **2021**, *15*, 5762–5775. [CrossRef]
54. Lu, Z.W.; Wu, L.; Dai, X.X.; Wang, Y.Y.; Sun, M.M.; Zhou, C.L.; Du, H.J.; Rao, H.B. Novel flexible bifunctional amperometric biosensor based on laser engraved porous graphene array electrodes: Highly sensitive electrochemical determination of hydrogen peroxide and glucose. *J. Hazard. Mater.* **2021**, *402*, 123774. [CrossRef]
55. Banavath, R.; Srivastava, R.; Bhargava, P. Nanoporous Cobalt Hexacyanoferrate Nanospheres for Screen-Printed H_2O_2 Sensors. *ACS Appl. Nano Mater.* **2021**, *4*, 5564–5576. [CrossRef]
56. Saidu, F.K.; Joseph, A.; Varghese, E.V.; Thomas, G.V. Silver nanoparticles-embedded poly(1-naphthylamine) nanospheres for low-cost non-enzymatic electrochemical H_2O_2 sensor. *Polym. Bull.* **2020**, *77*, 5825–5846. [CrossRef]
57. Zhang, Y.Y.; Bai, X.Y.; Wang, X.M.; Shiu, K.K.; Zhu, Y.L.; Jiang, H. Highly Sensitive Graphene-Pt Nanocomposites Amperometric Biosensor and Its Application in Living Cell H_2O_2 Detection. *Anal. Chem.* **2014**, *86*, 9459–9465. [CrossRef]
58. Han, L.; Cui, S.F.; Deng, D.M.; Li, Y.Y.; Yan, X.X.; He, H.B.; Luo, L.Q. Synthesis of Ag-Au/Reduced Graphene Oxide/TiO_2 Nanocomposites: Application as a Non-enzymatic Amperometric H_2O_2 Sensor. *Curr. Anal. Chem.* **2020**, *16*, 485–492. [CrossRef]

59. Wang, H.F.; Chen, W.X.; Chen, Q.Y.; Liu, N.; Cheng, H.J.; Li, T. Metal-organic framework (MOF)-Au@Pt nanoflowers composite material for electrochemical sensing of H_2O_2 in living cells. *J. Electroanal. Chem.* **2021**, *897*, 115603. [CrossRef]
60. Zhang, K.; Zeng, H.Y.; Wang, M.X.; Li, Z. Non-enzymatic H_2O_2 electrochemical sensor based on NiAl-LDH/PPy-Ag composite. *Ionics* **2022**, *28*, 5561–5570. [CrossRef]
61. Zhou, K.W.; Li, Y.; Zhuang, S.J.; Ren, J.; Tang, F.; Mu, J.L.; Wang, P. A novel electrochemical sensor based on CuO-CeO_2/MXene nanocomposite for quantitative and continuous detection of H_2O_2. *J. Electroanal. Chem.* **2022**, *921*, 116655. [CrossRef]
62. Shafa, M.; Ahmad, I.; Hussain, S.; Asif, M.; Pan, Y.; Zairov, R.; Alothman, A.A.; Ouladsmane, M.; Ullah, Z.; Ullah, N.; et al. Ag-Cu nanoalloys: An electrochemical sensor for H_2O_2 detection. *Surf. Interfaces* **2023**, *36*, 102616. [CrossRef]
63. Chen, S.Y.; Xie, Y.X.; Guo, X.J.; Sun, D.P. Self-supporting electrochemical sensors for monitoring of cell-released H_2O_2 based on metal nanoparticle/MOF nanozymes. *Microchem. J.* **2022**, *181*, 107715. [CrossRef]
64. Derbali, M.; Othmani, A.; Kouass, S.; Touati, F.; Dhaouadi, H. $BiVO_4/TiO_2$ nanocomposite: Electrochemical sensor for hydrogen peroxide. *Mater. Res. Bull.* **2020**, *125*, 110771. [CrossRef]
65. Wang, M.; Ma, J.W.; Guan, X.L.; Peng, W.C.; Fan, X.B.; Zhang, G.L.; Zhang, F.B.; Li, Y. A novel H_2O_2 electrochemical sensor based on $NiCo_2S_4$ functionalized reduced graphene oxide. *J. Alloys Compd.* **2019**, *784*, 827–833. [CrossRef]
66. Uzunoglu, A.; Ipekci, H.H. The use of CeO_2-modified Pt/C catalyst inks for the construction of high-performance enzyme-free H_2O_2 sensors. *J. Electroanal. Chem.* **2019**, *848*, 113302. [CrossRef]
67. Xie, F.Y.; Cao, X.Q.; Qu, F.L.; Asiri, A.M.; Sun, X.P. Cobalt nitride nanowire array as an efficient electrochemical sensor for glucose and H_2O_2 detection. *Sens. Actuators B-Chem.* **2018**, *255*, 1254–1261. [CrossRef]
68. Vijayalakshmi, K.; Renitta, A.; Alagusundaram, K.; Monamary, A. Novel two-step process for the fabrication of MnO_2 nanostructures on tantalum for enhanced electrochemical H_2O_2 detection. *Mater. Chem. Phys.* **2018**, *214*, 431–439. [CrossRef]
69. Fan, K.L.; Wang, H.; Xi, J.Q.; Liu, Q.; Meng, X.Q.; Duan, D.M.; Gao, L.Z.; Yan, X.Y. Optimization of Fe_3O_4 nanozyme activity via single amino acid modification mimicking an enzyme active site. *Chem. Commun.* **2017**, *53*, 424–427. [CrossRef]
70. Xu, D.Q.; Hou, B.B.; Qian, L.S.; Zhang, X.J.; Liu, G.D. Non-Enzymatic Electrochemical Sensor Based on Sliver Nanoparticle-Decorated Carbon Nanotubes. *Molecules* **2019**, *24*, 3411. [CrossRef]
71. Liu, X.; Ran, M.; Liu, G.; Liu, X.; Xue, Z.; Lu, X. A sensitively non-enzymatic amperometric sensor and its application in living cell superoxide anion radical detection. *Talanta* **2018**, *186*, 248–255. [CrossRef]
72. Rajendran, S.; Manoj, D.; Suresh, R.; Vasseghian, Y.; Ghfar, A.A.; Sharma, G.; Soto-Moscoso, M. Electrochemical detection of hydrogen peroxide using micro and nanoporous CeO_2 catalysts. *Environ. Res.* **2022**, *214*, 113961. [CrossRef]
73. Shu, Y.; Xu, J.; Chen, J.; Xu, Q.; Xiao, X.; Jin, D.; Pang, H.; Hu, X. Ultrasensitive electrochemical detection of H_2O_2 in living cells based on ultrathin MnO_2 nanosheets. *Sens. Actuators B Chem.* **2017**, *252*, 72–78. [CrossRef]
74. Gan, Y.; Hu, N.; He, C.; Zhou, S.; Tu, J.; Liang, T.; Pan, Y.; Kirsanov, D.; Legin, A.; Wan, H.; et al. MnO_2 nanosheets as the biomimetic oxidase for rapid and sensitive oxalate detection combining with bionic E-eye. *Biosens. Bioelectron.* **2019**, *130*, 254–261. [CrossRef] [PubMed]
75. Leonardi, S.G.; Aloisio, D.; Donato, N.; Russo, P.A.; Ferro, M.C.; Pinna, N.; Neri, G. Amperometric Sensing of H_2O_2 using Pt-TiO_2/Reduced Graphene Oxide Nanocomposites. *Chemelectrochem* **2014**, *1*, 617–624. [CrossRef]
76. Patra, D.C.; Deka, N.; Dash, A.; Khan, S.A.; Misra, T.K.; Mondal, S.P. Noninvasive Electrochemical Nitric Oxide Detection in Human Saliva Using Pt and TiO_2 Nanoparticle Composite Electrodes. *ACS Appl. Electron. Mater.* **2023**, *5*, 832–845. [CrossRef]
77. Jiang, Q.; Wang, J.; Liu, T.; Ying, S.; Kong, Y.; Chai, N.; Yi, F.-Y. UiO-66-Derived PBA Composite as Multifunctional Electrochemical Non-Enzymatic Sensor Realizing High-Performance Detection of Hydrogen Peroxide and Glucose. *Inorg. Chem.* **2023**, *62*, 7014–7023. [CrossRef] [PubMed]
78. Kumar, V.; Sachdev, A.; Matai, I. Self-assembled reduced graphene oxide-cerium oxide nanocomposite@cytochrome c hydrogel as a solid electrochemical reactive oxygen species detection platform. *New J. Chem.* **2020**, *44*, 11248–11255. [CrossRef]

chemosensors

Perspective

Linkage Pathways of DNA–Nanoparticle Conjugates and Biological Applications

Shan Huang and Jun-Jie Zhu *

State Key Laboratory of Analytical Chemistry for Life Science, School of Chemistry and Chemical Engineering, Nanjing University, Nanjing 210023, China; huangshan@njtech.edu.cn
* Correspondence: jjzhu@nju.edu.cn; Tel.: +86-025-89687204

Abstract: DNA–nanoparticle conjugates have extraordinary optical and catalytic properties that have attracted great interest in biosensing and biomedical applications. Combining these special qualities has made it possible to create extremely sensitive and selective biomolecule detection methods, as well as effective nanopharmaceutical carriers and therapy medications. In particular, inorganic nanoparticles, such as metal nanoparticles, metal–organic framework nanoparticles, or upconversion nanoparticles with relatively inert surfaces can easily bind to DNA through covalent bonds, ligand bonds, electrostatic adsorption, biotin–streptavidin interactions and click chemistry to form DNA–nanoparticle conjugates for a broad range of applications in biosensing and biomedicine due to their exceptional surface modifiability. In this review, we summarize the recent advances in the assembly mechanism of DNA–nanoparticle conjugates and their biological applications. The challenges of designing DNA–nanoparticle conjugates and their further applications are also discussed.

Keywords: DNA; inorganic nanoparticles; biosensing; biomedical

Citation: Huang, S.; Zhu, J.-J. Linkage Pathways of DNA–Nanoparticle Conjugates and Biological Applications. *Chemosensors* **2023**, *11*, 444. https://doi.org/10.3390/chemosensors11080444

Academic Editor: Léon Reubsaet

Received: 26 June 2023
Revised: 31 July 2023
Accepted: 7 August 2023
Published: 10 August 2023

1. Introduction

With its simple structure and complicated functions, DNA is a crucial building block in the creation of the blueprint for life and is employed extensively in biosensing and biomedicine [1,2]. However, the instability of the DNA structure in living organisms limits the efficient performance of its functions. Therefore, the construction of nanoscale models by employing DNA as a building block for improving its stability has been widely studied [3,4]. Nevertheless, the high price of its assembly also restricts its application for a wide range of purposes. In the past decades, DNA nanotechnology has rapidly evolved from structural DNA nanotechnology alone to assembly DNA nanotechnology in order to propose more effective methods to enhance the stability of DNA applied in living organisms [5–7].

Among them, the assembly of DNA on the surface of nanomaterials as a carrier and changing the presence state of DNA from the solution phase to interfacial connection can be a promising approach to solve the confusion of DNA instability in living organisms [8,9]. Incorporating DNA molecules to decorate inorganic nanoscale components brings up new possibilities for creating nanoparticles with distinctive applications in the realms of biosensing and biomedical applications [10]. The flexibility and adaptability of DNA-based platforms allow for the construction of complex structures from nanoparticles utilizing DNA-programmable interparticle interactions [11,12]. To date, in order to further expand the fields of its application, the investigation stages for the functionalization of DNA on the surface of nanoparticles can be classified into the following aspects:

(1) Searching for diversely functional DNA sequences The chemical activity of DNA grows into new areas outside of storing and transferring genetic information in the field of functional DNA nanotechnology. The two main representative categories of functional DNAs that are generated by in vitro selection with particular binding

affinities and catalytic capabilities are aptamers and DNAzymes [13,14]. Aptamers are chosen by a procedure called systematic evolution of ligands by exponential enrichment (SELEX), and the selection process has evolved from in vitro to in vivo. Aptamers are widely employed in the assembly of sensitive biosensors, the construction of bioimaging agents, and targeted therapeutics [15]. An additional category of valuable DNA molecules with catalytic activity is DNAzymes. RNA-cleaving DNAzymes are of particular fascination due to their quick reaction time and simplicity in application in life [16]. DNAzymes act by binding to specific metal ions as catalytic cofactors and are ideal for functionalized nanoparticle surfaces [17].

(2) Development of inorganic nanoparticles with facile modification of DNA on the surface. By adding DNA nanotechnology into nanoparticle research, precise geometric construction of nanoparticles and change in surface properties have been described [18,19]. The best candidates for carrying smart DNA walkers/motors, activatable aptamers, and DNAzyme-based systems that respond to stimuli for biosensing, bioimaging, and biomedical applications are nanoparticles with a substantial surface area, favorable biocompatibility, outstanding stability, and beneficial physical and chemical properties, such as gold nanoparticles (AuNPs), upconversion nanomaterials (UCNPs) and so on [20–22].

(3) Exploring the way DNA connects to nanoparticles. The facile modification properties of DNA encourage the interface engineering of nanoparticles based on several theories. Mirkin was the first to suggest that sulfurizing DNA and attaching it to AuNPs would result in the development of AuNPs-linked DNA [23,24]. The coordinated binding between DNA base structure and rare earth elements can also connect DNA to UCNPs [25–29]. The interface between DNA molecules and nanoparticles also involves additional covalent and noncovalent connecting techniques, such as the biotin-Streptavidin interaction and click chemistry [30,31]. Thereafter, electrostatic contact is another often employed technique for affixing DNA molecules to positively charged nanoparticles on the surface due to the negative charge of the phosphate backbone on the surface of the DNA structure [32]. Electrostatic interactions exist in various forms, such as hydrogen bonding, hydrophobic contacts, van der Waals forces, and ionic bonding, DNA–nanoparticle conjugates assembled with the above forms also demonstrate excellent stability and extremely promising applications. For instance, the mechanism of DNA's denaturation and rehybridization is employed to create hydrogen bonds between base pairs that are complementary on adjacent DNA strands, resulting in the formation of interconnected structures. Then, silicate nanodiscs are utilized to create extra network points by luring electrostatic interactions with the DNA backbone [33]. This improves the mechanical elasticity of the hydrogel formulation and achieves the release of the loaded drug dexamethasone, realizing the conjugate's ability to treat osteoporosis disease.

The functionalized connection of DNA on the surface of nanoparticles can not only stabilize DNA but also combine the unique physicochemical properties of nanoparticles themselves to realize the superposition function after the double unit compound (Figure 1). In recent years, the multifaceted applications of DNA–nanoparticle conjugates in biosensing and biomedicine are also summarized and introduced, but the articles focusing on the mode of connection between DNA and nanoparticles and thus summarizing the state-of-the-art of their applications in biosensing and biomedical have not yet been reported. Consequently, we outline many tried and true methods for DNA functionalization on various surface ligands and nanoparticles and provide a summary and overview of the biological applications of these methods over the past five years.

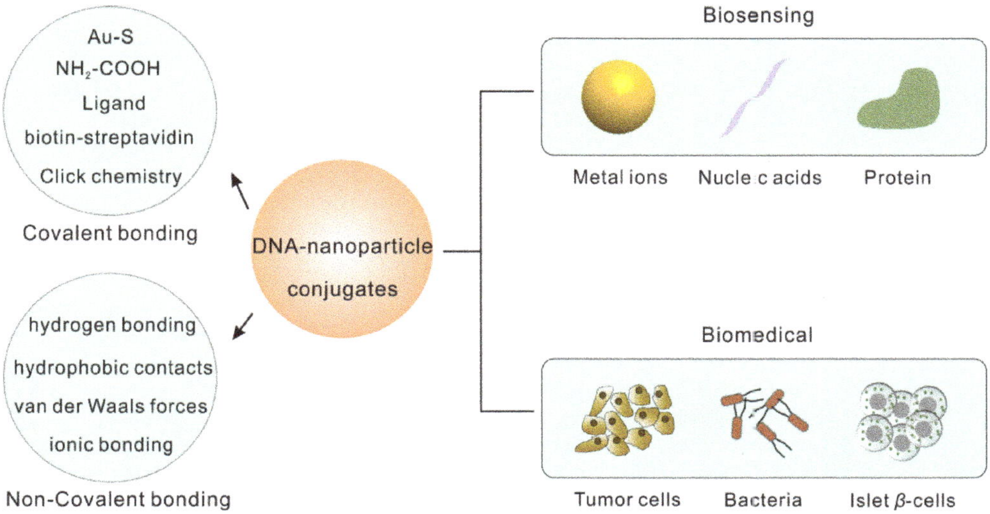

Figure 1. Schematic of the classes of DNA-conjugated nanoparticles and their applications.

2. DNA–Nanoparticle Conjugates Design

Fabricating the physico-chemical properties of inorganic nanoparticles with DNA attached to produce novel structures is particularly exciting because these materials' physicochemical properties differ greatly from those of their bulk structure [34]. Their small size and high surface-to-volume ratio give rise to these novel features, which may be tailored by varying their chemical makeup, size, and shape [35]. The approaches employed to achieve this goal can be classified as top-down or bottom-up. In contrast to bottom-up approaches, which rely on the organization of building blocks (such as molecules, atoms, and nanoparticles) to form larger structures and have broader structural versatility, top-down approaches employ external sources to design nanoscale features on the surface of the body but are more expensive to synthesize [36].

In this section, we summarize several approaches for the bottom-up synthesis of DNA–nanoparticle conjugates, employing DNA as a linker and inorganic nanomaterials as structural units.

2.1. Covalent Bonding

Covalent bonding to nanoparticles by exploiting the chemical modifiability of DNA and the properties of the bases themselves is the key method of DNA–nanoparticle conjugate formation [12]. Among the chemically modified DNA, phosphorothioate DNA, amino-modified DNA, biotin-modified DNA, and alkyne or azide-modified DNA are the primary representatives.

The technique that is employed the most frequently is to modify DNA with -SH functional groups on the surface of plasma metal nanoparticles and then to ensure that DNA and nanoparticles contact closely by means of covalent metal–SH interactions [37]. It is worth noting that end-labeled thiol is the main anchor for linking DNA to AuNPs because of strong Au–thiol interactions. For example, SH-functionalized DNA sequences are covalently attached to the surface of AuNPs to construct a structurally stable thermal sensor for the highly sensitive in situ detection of exosomal miRNAs without RNA extraction or target amplification (Figure 2a) [38]. It has been demonstrated that DNA-functionalized gold nanoprobes linked by Au–sulfur bonds exhibit remarkable stability in exosomes, ensuring the proper operation of the sensing probes. This stable structure is also employed in a number of nanoprobes constructed based on this metal–SH bonding method [39]. However, another method of functionalizing DNA on the surface of AuNPs, namely polyadenine (poly A) DNA functionalized AuNPs, has received increasing attention due to the high cost of thiols [40]. Metal ions can undergo coordination reactions with the amino acid side chains of adenine bases, thus completing the attachment of functionalized DNA strands to AuNPs. This more time-efficient method of conjugation can facilitate the further development of this method in the future for several applications [41].

Covalent cross-linking and phosphate group coordination are the two primary techniques for ligating functionalized DNA on the surface of UCNPs [42]. Covalent cross-linking is often accomplished by modifying functional molecules. For example, as shown in Figure 2b, transferring UCNPs from the organic phase to the aqueous phase and modifying the surface with functional modification of -COOH, covalent cross-linking can occur with the DNA strands with modified -NH_2 at the ends to generate -NH-CO- bond, thus realizing the functionalization of DNA on the surface of UCNPs [43]. In addition, Ln^{3+}-based UCNPs provide a surface rich in multiple functional groups through the coordination of Ln^{3+} with electron-rich groups, including -NH_2, -COOH, $SO_4{}^{2-}$, and phosphate group $PO_4{}^{3-}$. The phosphate group on the DNA backbone can easily replace -COOH because the coordination between $PO_4{}^{3-}$ and Ln^{3+} is stronger than that between -COOH and Ln^{3+}, enabling the coordination bonding between the DNA and UCNPs to complete the assembly of stable nanoprobes [44]. Biotin-modified DNA can be attached to the surface of nanoparticles by covalent bonds between biotin and streptavidin (biotin–avidin), which is also a connecting method for preparing stable nanosensor molecules (Figure 2c) [45–48]. The above two types of modification of functional molecules on nanoparticles can extend the forms of covalently attached DNA molecules and provide more possibilities for the formation of stable functionalized sensing probes.

Recently, the attachment of alkyne-modified DNA to the surface of nanoparticles utilizing a click reaction is extensively performed, which is based on a copper-catalyzed azide–alkyne cyclization reaction [49–51]. This is because the click reaction, which is a quick and effective chemical reaction, takes place in mild conditions and does not require a catalyst to produce a high yield of product. For example, a "click chemistry" approach allows for the preparation of stable Janus nanoparticles containing both PEG and azide–DNA to form AuNP dimers under target responders and conditions for in situ SERS detection and imaging analysis of cancer cells [52]. By adding an azide group and an alkyne group to the two DNA strands in order to bind the two AuNPs together, the efficiency of the assembly may also be guaranteed (Figure 2d) [53]. In addition to improving the stability of the probe, the high efficiency of click chemistry can be exploited to enhance the enrichment of DNA on the surface of nanoparticles, allowing further improvement of the sensing sensitivity of the target detectors [54]. For other nanoparticles, such as metal–organic framework (MOF) materials, DNA can also be attached to the surface of the nanoparticles by click chemistry, acting as a reaction switch for the target and further extending the biological applications of DNA–nanoparticle conjugates [55].

2.2. Non-Covalent Bonding

Hydrogen bonds, hydrophobic contacts, van der Waals forces, and ionic bonds are non-covalent bonds collectively known as electrostatic adsorption that can bring DNA closer to nanoparticles and form DNA–nanoparticle conjugates [56–58].

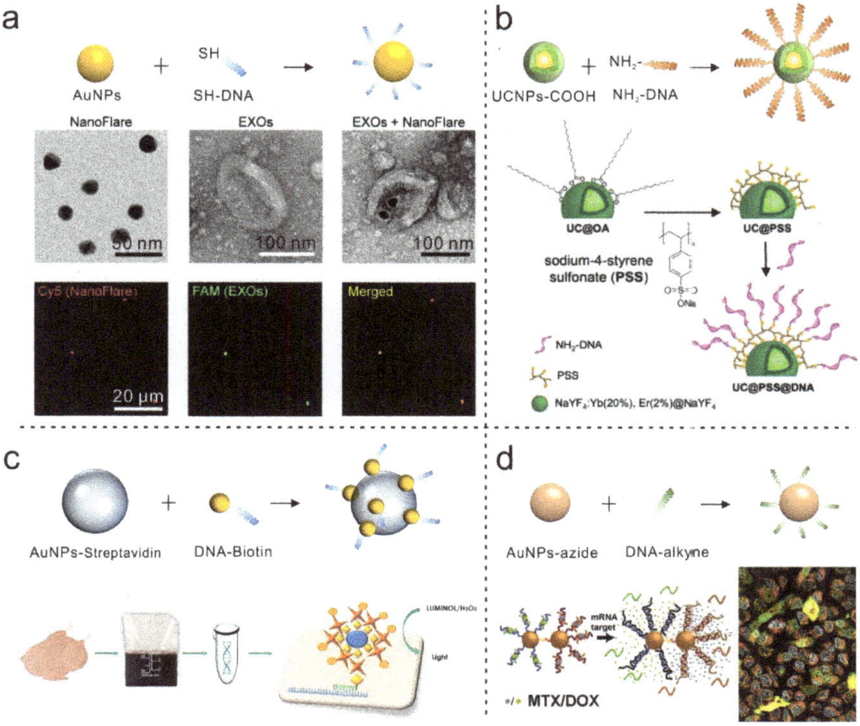

Figure 2. Schematic diagram of covalent bonding of DNA–nanoparticle conjugates. (**a**) Performance of nanoflares for miRNA detection and transportation of nanoflares into EXOs. Reproduced from Ref. [38] with permission. Copyright 2020 American Chemical Society. (**b**) Presentation of the UCNP-DNA bioconjugation in a schematic Reproduced from Ref. [43] with permission. Copyright 2022 Wiley-VCH GmbH. (**c**) Schematic diagram of the construction of biotinylated DNA and biotinylated nanoparticles connected by streptavidin to complete the sensor. Reproduced from Ref. [45] with permission. Copyright 2021 ELSEVIER. (**d**) Illustration of a multiplexed nanoparticle dimer that was joined via click chemistry using a DIBO-azide. Reproduced from Ref. [53] with permission. Copyright 2018 American Chemical Society.

Two DNA aptamer sequences are assembled on the surface of UCNPs by electrostatic adsorption, and the photoactivation ability of UCNPs is utilized to overcome the electrostatic adsorption and release the loaded two DNA aptamer strands after reaching a specific location, completing ultrasensitive imaging monitoring of single molecules on cell membranes (Figure 3a) [59]. Similarly, as shown in Figure 3b, aptamer DNA strands with redox activation ability can be electrostatically adsorbed on AuNPs with targeting ability to construct DNA–nanoparticle conjugates with gating effects that enable simultaneous spatial imaging of ATP and glutathione (GSH) in mitochondria [60]. This information is capable of being utilized to build pH probes with great temporal and spatial accuracy for tracking various physiological and pathological processes, which is equally crucial (Figure 3c) [61]. A variety of DNA sequences that have been detached from the nanoparticles and are reassembled into functional structures on their own can be loaded in addition to a single DNA sequence (Figure 3d). The sensitivity of DNA–nanoparticle conjugates for intracellular imaging monitoring of mRNA can be efficiently boosted with this technique [62]. All of the above work is illustrating the significance of interchangeable connections in biosensing applications.

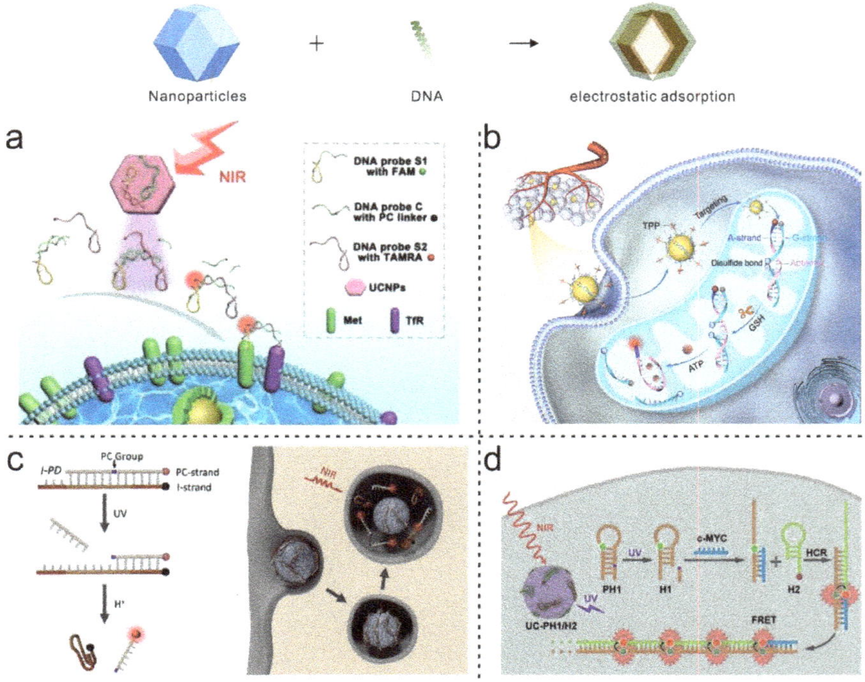

Figure 3. Schematic diagram of non-covalent bonding of DNA–nanoparticle conjugates. (**a**) A schematic illustration: DNA-UCNPs nanoprobes for detecting receptor dimers. Reproduced from Ref. [57] with permission. Copyright 2021 American Chemical Society. (**b**) Redox-activatable DNA nanodevice design schematic for AND-gated imaging of ATP and GSH in mitochondria. Reproduced from Ref. [58] with permission. Copyright 2021 American Chemical Society. (**c**) Diagram of the pH-sensing UV-activated DNA probe, I-PD. Reproduced from Ref. [59] with permission. Copyright 2020 American Chemical Society. (**d**) NIR light-initiated hybridization chain reaction (HCR) for spatiotemporally resolved mRNA imaging and signal amplification in living cells. Reproduced from Ref. [60] with permission. Copyright 2019 Wiley-VCH GmbH.

Notably, the triggering conditions for the above DNA–nanoparticle conjugates to work are diverse, including near-infrared light irradiation, pH stimulation, and the presence of a target, and all of these triggering effects are stronger than the force of electrostatic adsorption. Therefore, the conversion of functional switches can be successfully accomplished by taking advantage of the diverse and realistic physiological environment in organisms, as well as the strength of the forces acting, which is also a major advantage of non-covalent bond binding.

3. Biological Applications

This section will go over several notable applications of DNA–nanoparticle conjugates in biological and biomedical applications, such as sensing, bioimaging, and therapy (gene therapy, photothermal therapy, photodynamic therapy, and synergistic therapy). The link between functional DNA sequences and nanoparticles is highlighted in this section, with a focus on the personalized creation of DNA–nanoparticle conjugates with particular activities.

3.1. Biosensing

MicroRNAs (miRNAs) in mammalian cells can be imaged in situ spatiotemporally to reveal their structure and biological functions. The employing of DNA–nanoparticle conjugates as imaging tools for miRNAs in situ has tremendous potential. To achieve intracellular miRNA imaging and accurate quantification in living cells without disrupting the cell membrane, some work has proposed a near-infrared (NIR) light-activated nanoprobe that can be utilized for highly sensitive in situ controllable miRNA imaging in living cells (Figure 4a). The NIR-activated nanoprobe employs a UCNP that acts as an NIR-UV sensor, triggering the following on the surface of the nanoparticle attached to the photocleavage of the dumbbell DNA probe. The structural alteration of the dumbbell probe induces a catalytic hairpin assembly of the target miRNA, through which an amplified fluorescence signal can be read in situ [63]. It is worth noting here that the photodegradable n-nitrobenzyl-protected DNA architecture is the architecture most extensively practiced as a functional DNA shell layer [64]. The same work successfully created an intelligent system based on NIR light-initiated DNA walkers for precise spatiotemporal regulation of living cells utilizing two DNA–nanoparticle conjugates (UCNPs and AuNPs) (Figure 4b). UCNPs are one of them, and they operate as DNA probe carriers, converting NIR into UV light to activate the precursor probes. Regarding high-resolution spatiotemporal imaging of intracellular miRNAs, AuNPs are employed as carriers of ATP-driven DNA walkers [65]. The device is exceedingly stable and has a very low fluorescence background. Meanwhile, it is highly desirable that DNA-based molecular circuits can perform complex information processing in biological systems. Li's group reports a conceptual approach to constructing photonic nanocircuits that enable DNA molecular computation with superior spatial accuracy in vitro and in vivo. Upon remote activation of spatially restricted NIR-light inputs, two cancer biomarker inputs, ATP and miRNA, can sequentially trigger conformational changes in the DNA circuit via structural switch aptamers and toe point-mediated strand exchange, leading to the release of signal output and the formation of more accurate in vivo imaging data of the organism [66]. Alternatively, multiple intracellular miRNAs can be exploited for simultaneous triggering, enabling spatially and temporally controlled on-demand precision imaging [67].

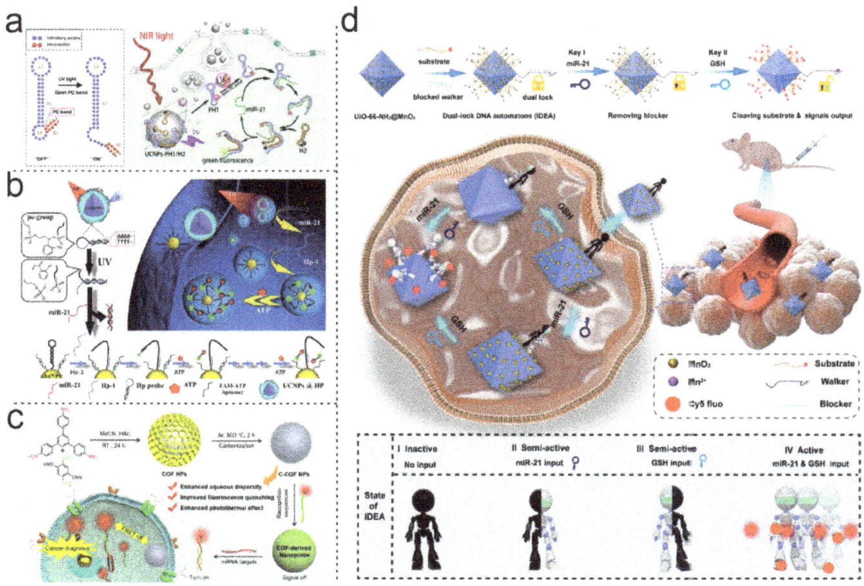

Figure 4. DNA–nanoparticle conjugates for biosensing applications. (**a**) The proposed controllable miRNA imaging nanoprobe's general principle. Reproduced from Ref. [63] with permission. Copyright 2020 American Chemical Society. (**b**) Schematic of the photoactivatable DNA walker system for

spatiotemporally resolved miRNA sensing. Reproduced from Ref. [65] with permission. Copyright 2021 American Chemical Society. (**c**) Schematic illustration of the preparation of carbonized COF-based nanoprobes for cancer cell imaging. Reproduced from Ref. [68] with permission. Copyright 2021 American Chemical Society. (**d**) Diagram of IDEAs for precise tumor imaging. Reproduced from Ref. [69] with permission. Copyright 2023 American Chemical Society.

In addition to UCNPs as carriers for DNA, covalent organic frameworks (COF) and MOF are also ideal core materials with excellent physicochemical properties. The work leads to the remarkable conclusion that carbonized COFs (C-COFs) can significantly improve the fluorescence burst efficiency and water stability of nanoscale COFs (Figure 4c). The probes prepared by the physisorption of dye-labeled DNA recognition sequences onto C-COFs for cell imaging can effectively illuminate biomarkers (survivin and TK1 mRNA) in living cells [68]. DNA-functionalized ZrMOF@MnO_2 exhibits a significantly enhanced fluorescence signal only when miRNA and GSH from tumor cells coexist, enabling accurate differentiation between tumor cells and healthy cells (Figure 4d) [69].

It is essential to involve a variety of enzymes in addition to miRNA, ATP, and GSH as regulatory molecules in the organism. An efficient type of DNA nanomachine was designed by utilizing two enzymes in the organism, telomerase (TE) and purine-free/pyrimidine-free endonuclease 1 (APE1) (Figure 5a). The walker of this nanomachine moves along a TE-regenerated trajectory, generating multiple amplified signals by which APE1 can be imaged in situ, providing a new paradigm for the development of more applicable and efficient DNA nanomachines [70]. In parallel, DNA–nanoparticle conjugates (E-SNAs) can be manipulated for nucleoplasmic translocation of proteins with cancer cell selectivity. E-SNAs are constructed by programmable design schemes, based on modules of aptamers that carry enzyme response units at pre-designed sites and are further combined with SNA nanotechnology (Figure 5b). E-SNAs are able to efficiently and specifically regulate the cytoplasmic–nuclear shuttle of RelA proteins, while remaining inactive in normal cells due to insufficient enzyme expression, thus accurately localizing the target protein within the cell [71]. The linkage pathway and applications of the DNA–nanoparticle conjugates from the above work are shown in Table 1.

Table 1. Summarizes the types of linkages and biosensing applications of DNA–nanoparticle conjugates.

DNA Structure	Nanoparticle's Type	Connection Type	Analyte	LOD	Application	Ref.
DNA dumbbell structure	UCNPs	Allotropic bond	miRNA-21	Single cell	In situ imaging	[62]
Hairpin DNA structure	UCNPs	Allotropic bond	miRNA-21	Single cell	In situ imaging	[64]
Double-stranded	UCNPs	Allotropic bond	ATP, miRNA-21	Single cell	In vivo imaging	[65]
Triangle structure	UCNPs	Allotropic bond	miRNA-21, miRNA-373, miRNA-155	Single cell	In situ imaging	[66]
Single strand	COFs	Electrostatic adsorption	mRNA	Single cell	Cancer diagnosis	[67]
Double-strands	ZrMOF@MnO_2	Covalent bonding	miRNA-21, GSH	Single cell	In vivo imaging	[68]
Double-Strands	AuNPs	Covalent bonding	RelA protein	Single cell	In situ imaging	[69]
Double-Strands	AuNPs	Covalent bonding	APE1 enzyme	Single cell	In situ imaging	[70]

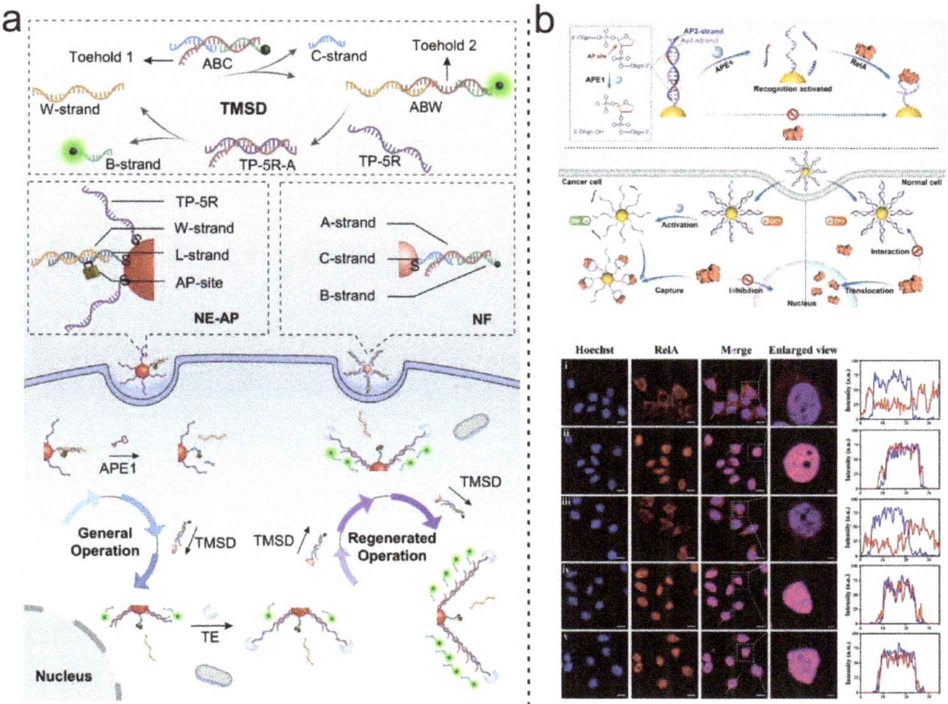

Figure 5. (**a**) Diagram of the TE-activatable regenerative DNA nanomachine (NEAP/NF) for sensing of intracellular APE1. Reproduced from Ref. [70] with permission. Copyright 2023 Wiley-VCH GmbH. (**b**) Design of the enzyme-operated SNA (E-SNA) for cancer cell-selective regulation of cytoplasmic-to-nuclear translocation of RelA protein. Scale bar, 2 µm. Reproduced from Ref. [71] with permission. Copyright 2023 Wiley-VCH GmbH.

3.2. Biomedical

Precise application of DNA–nanoparticle conjugates for imaging targets in living organisms is a prerequisite for accurate biomedical treatment [72–75]. The introduction of small interfering RNA (siRNA) in the DNA shell layer is an effective therapeutic approach to regulate the expression of target genes. The multifunctional three-dimensional (3D) DNA shell contains a degradation-resistant Y-shaped backbone (hardened triangular sticky DNA brick) and is programmed to lay flat on the siRNA-packed AuNPs. After the alignment of the aptamer on the outer surface, siRNA-encapsulated DNA–nanoparticle conjugates with excellent biocompatibility are obtained, siRNA/Ap-CS. In this, the siRNA is internally encapsulated in the 3D DNA shell, ensuring that it is not degraded by the enzymes in the outermost layer of the 3D DNA shell (Figure 6a). The conjugate-loaded siRNA can be released by endogenous miRNA and gene silencing in tumor cells, leading to apoptosis and enabling gene therapy [76]. Next, a freshly developed approach utilizes a simple, versatile, and inexpensive platform (AuNPs-DNA) to control the release of cholesterol-coupled oligonucleotides [77]. Nanoclusters formed by AuNPs-DNA combine the platform into a dual-release system that allows the delivery of a hydrophobic drug with zero-order kinetics followed by the rapid release of cholesterol-coupled DNA for therapeutic purposes and holds promise for future applications in downregulating the expression of b-galactosidase in *E. coli* (Figure 6b).

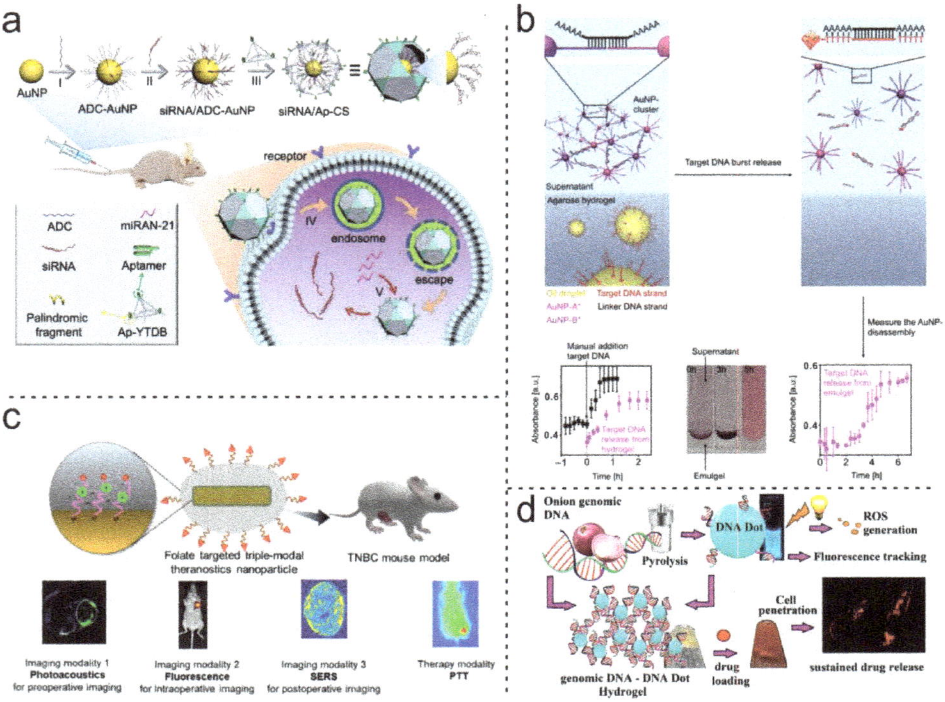

Figure 6. DNA–nanoparticle conjugates for biomedical applications. (**a**) Building a gold nanoparticle with a multi-purpose 3D DNA self-assembled multilayer core/shell nanostructure (siRNA/Ap-CS) and employing it to deliver siRNA to tumor cells with precision and control their release. Reproduced from Ref. [76] with permission. Copyright 2021 Nature. (**b**) Schematic of the DNA-AuNP-system and the disassembly of AuNP aggregates induced manually and by releasing the target DNA. Reproduced from Ref. [77] with permission. Copyright 2021 Nature. (**c**) Schematic representation of a TMNP and different optical imaging and therapy modalities offered by the TMNPs for in vivo applications. Reproduced from Ref. [78] with permission. Copyright 2022 American Chemical Society. (**d**) Schematic of synthesis and in vivo application of DNA hydrogel. Reproduced from Ref. [79] with permission. Copyright 2023 American Chemical Society.

For in vivo imaging and treatment of organisms, such as triple-negative breast cancer, early and precise identification, and treatment is critical for better disease management and increased life expectancy. On the basis of these needs, innovative biocompatible tri-modal nanoprobes (TMNPs) provide optical imaging employing photoacoustic, fluorescence, and surface-enhanced Raman scattering (SERS), as well as photothermal therapy (PTT) exploiting near-infrared (NIR) light [78]. Based on the modifiability of DNA, positively charged NIR fluorophores were designed and screened to obtain optimal fluorescence emission and SERS signals. As seen in the experimental results, selective exposure of tumors to the NIR laser showed effective thermal tissue ablation without causing systemic toxicity, creating a treatment platform with excellent imaging–treatment integration (Figure 6c). In addition to synthetic inorganic nanoparticles as carriers, plant-sourced DNA–nanoparticles (DNA dots) created from onion genomic DNA (gDNA) are without any chemicals and exhibit superior biocompatibility (Figure 6d). DNA dots further form stimuli-responsive hydrogels with multifunctional properties by self-assembly with hybridization-mediated precursor gDNA. Its main functional type of strand is contributed by surface-suspended DNA strands resulting from incomplete carbonization during annealing. DNA dot hybrid hydrogels proved to be an excellent drug delivery vehicle for on-demand reactions by tracking slow release through the intrinsic fluorescence of DNA dots and by normal visible light photoexcitation [79]. The linkage pathway and applications of the DNA–nanoparticle conjugates from the above work are shown in Table 2.

Table 2. Summarizes the types of linkages and biomedical applications of DNA–nanoparticle conjugates.

DNA Structure	Nanoparticle's Type	Connection Type	Treatment Modalities	Application	Ref.
Y-shaped backbone-rigidified triangular DNA	AuNPs	Covalent bonding	Gene silencing	Cancer therapy	[75]
Cholesterol-conjugated DNA	AuNPs	Conjugate connection	Gene silencing	Drug release calculation	[76]
Single DNA	AuNRs	Covalent bonding	Phototherapy	Triple-negative breast cancer therapy	[77]
Biomass DNA	DNA dots	Conjugate connection	Photoactivated ROS Generation	Cancer therapy	[78]

4. Conclusions and Outlook

DNA–nanoparticle conjugates have attracted widespread attention in biosensing, bioimaging, drug delivery, and photoactivated biotherapeutics due to their high water solubility, functionalized surface modifications, and excellent physicochemical properties. In this review, we summarize the various assembly principles of DNA–nanoparticle conjugates and the recent research progress in the fields of bioimaging, photothermal, photodynamic therapy, gene therapy, etc. Although DNA–nanoparticle conjugates have made great progress in these aspects, many problems still exist in practical applications: (1) The broadest functional DNA sequences currently screened and applied include DNA aptamers and DNAzyme, followed by complex DNA structures introduced for the purpose of signal amplification, such as HCR, catalytic hairpin assembly (CHA), Y-types, and other steric structures. The aforementioned functional DNAs can be applied in a variety of biomedical fields; therefore, designing and screening more kinds of functional DNA sequences is an effective means to develop novel DNA–nanoparticle conjugates and expand their applications; (2) Exploring more diverse methods of surface modification of nanoparticles and assembling multiple types of functionalized DNA sequences on them. It is expected that the physicochemical properties of the abundant nanocarriers can be fully utilized to organically combine them with functionalized DNA to achieve the accumulation of dual functions and maximize the synergistic functions; (3) To develop more types of DNA shell layer and nanoparticle binding methods, and to introduce more abundant and on-demand controlled covalent bonding modes into the conjugate binding to form more diversified and stable DNA–nanoparticle conjugates. We believe that the successive development of DNA–nanoparticle conjugates based on the above-mentioned richer species will lead to breakthroughs in solving major problems in biosensing and biomedical fields, benefiting more research fields in the near future.

Author Contributions: Conceptualization, methodology, writing—original draft preparation, writing—review and editing, S.H.; visualization, J.-J.Z.; funding acquisition, J.-J.Z. All authors have read and agreed to the published version of the manuscript.

Funding: The authors greatly appreciate the financial support from the National Natural Science Foundation of China (21834004), State Key Laboratory of Analytical Chemistry for Life Science (SKLACLS2305).

Institutional Review Board Statement: Not applicable.

Informed Consent Statement: Not applicable.

Data Availability Statement: Not applicable.

Conflicts of Interest: The authors declare no conflict of interest.

References

1. Ashworth, C. DNA nanotechnology: Building big with DNA bricks. *Nat. Rev. Mater.* **2018**, *3*, 17092. [CrossRef]
2. Hu, Q.; Li, H.; Wang, L.; Gu, H.; Fan, C. DNA Nanotechnology-Enabled Drug Delivery Systems. *Chem. Rev.* **2019**, *119*, 6459–6506. [CrossRef] [PubMed]
3. Rothemund PW, K. Folding DNA to create nanoscale shapes and patterns. *Nature* **2006**, *440*, 297–302. [CrossRef] [PubMed]
4. Knappe, G.A.; Wamhoff, E.-C.; Bathe, M. Functionalizing DNA origami to investigate and interact with biological systems. *Nat. Rev. Mater.* **2023**, *8*, 123–138. [CrossRef] [PubMed]
5. Saliba, D.; Luo, X.; Rizzuto, F.J.; Sleiman, H.F. Programming rigidity into size-defined wireframe DNA nanotubes. *Nanoscale* **2023**, *15*, 5403–5413. [CrossRef]
6. Yutong Huang, F.; Kumar Lat, P.; Sen, D. Unusual Paradigm for DNA–DNA Recognition and Binding: "Socket-Plug" Complementarity. *J. Am. Chem. Soc.* **2023**, *145*, 3146–3157. [CrossRef]
7. Zhang, P.; Ouyang, Y.; Zhuo, Y.; Chai, Y.; Yuan, R. Recent Advances in DNA Nanostructures Applied in Sensing Interfaces and Cellular Imaging. *Anal. Chem.* **2023**, *95*, 407–419. [CrossRef]
8. Wang, W.; Yu, S.; Huang, S.; Bi, S.; Han, H.; Zhang, J.R.; Lu, Y.; Zhu, J.J. Bioapplications of DNA nanotechnology at the solid-liquid interface. *Chem. Soc. Rev.* **2019**, *48*, 4892–4920. [CrossRef]
9. Ahmadi-Sangachin, E.; Bazzi, F.; Xu, G.; Hosseini, M. Biosensing using DNA-based structures integrated with nanosheets. *Microchem. J.* **2023**, *191*, 108779. [CrossRef]
10. Shaikh, S.; Younis, M.; Yuan, L. Functionalized DNA nanostructures for bioimaging. *Coord. Chem. Rev.* **2022**, *469*, 214648. [CrossRef]
11. Chandler, M.; Jain, S.; Halman, J.; Hong, E.; Dobrovolskaia, M.A.; Zakharov, A.V.; Afonin, K.A. Artificial Immune Cell, AI-cell, a New Tool to Predict Interferon Production by Peripheral Blood Monocytes in Response to Nucleic Acid Nanoparticles. *Small* **2022**, *18*, 2204941. [CrossRef] [PubMed]
12. De Fazio, A.F.; Misatziou, D.; Baker, Y.R.; Muskens, O.L.; Brown, T.; Kanaras, A.G. Chemically modified nucleic acids and DNA intercalators as tools for nanoparticle assembly. *Chem. Soc. Rev.* **2021**, *50*, 13410–13440. [CrossRef] [PubMed]
13. Kong, R.M.; Zhang, X.B.; Chen, Z.; Tan, W. Aptamer-assembled nanomaterials for biosensing and biomedical applications. *Small* **2011**, *7*, 2428–2436. [CrossRef] [PubMed]
14. Breaker, R.R.; Joyce, G.F. A DNA enzyme that cleaves RNA. *Chem. Biol.* **1994**, *1*, 223–229. [CrossRef]
15. Fu, X.; Peng, F.; Lee, J.; Yang, Q.; Zhang, F.; Xiong, M.; Kong, G.; Meng, H.M.; Ke, G.; Zhang, X.B. Aptamer-Functionalized DNA Nanostructures for Biological Applications. *Top. Curr. Chem.* **2020**, *378*, 21. [CrossRef]
16. Zhu, Y.; Wu, J.; Zhou, Q. Functional DNA sensors integrated with nucleic acid signal amplification strategies for non-nucleic acid targets detection. *Biosens. Bioelectron.* **2023**, *230*, 115282. [CrossRef]
17. Su, J.; Sun, C.; Du, J.; Xing, X.; Wang, F.; Dong, H. RNA-Cleaving DNAzyme-Based Amplification Strategies for Biosensing and Therapy. *Adv. Healthcare Mater.* **2023**, 2300367. [CrossRef]
18. Zhao, Y.; Shi, L.; Kuang, H.; Xu, C. DNA-Driven Nanoparticle Assemblies for Biosensing and Bioimaging. *Top. Curr. Chem.* **2020**, *378*, 18. [CrossRef]
19. Hu, Q.; Wu, J.; Chen, L.; Lou, X.; Xia, F. Recent Development of DNA-modified AIEgen Probes for Biomedical Application. *Chem. Res. Chin. Univ.* **2021**, *37*, 66–72. [CrossRef]
20. Ohta, S.; Glancy, D.; Chan, W.C.W. DNA-controlled dynamic colloidal nanoparticle systems for mediating cellular interaction. *Science* **2016**, *351*, 841–845. [CrossRef]
21. Jia, R.; Wang, Y.; Ma, W.; Huang, J.; Sun, H.; Chen, B.; Cheng, H.; He, X.; Wang, K. Activatable Dual Cancer-Related RNA Imaging and Combined Gene-Chemotherapy through the Target-Induced Intracellular Disassembly of Functionalized DNA Tetrahedron. *Anal. Chem.* **2022**, *94*, 5937–5945. [CrossRef] [PubMed]
22. Huang, S.; Song, Y.; He, Z.; Zhang, J.-R.; Zhu, J.-J. Self-assembled nanomaterials for biosensing and therapeutics: Recent advances and challenges. *Analyst* **2021**, *146*, 2807–2817. [CrossRef] [PubMed]
23. Cutler, J.I.; Auyeung, E.; Mirkin, C.A. Spherical nucleic acids. *J. Am. Chem. Soc.* **2012**, *134*, 1376–1391. [CrossRef]
24. Callmann, C.E.; Vasher, M.K.; Das, A.; Kusmierz, C.D.; Mirkin, C.A. In Vivo Behavior of Ultrasmall Spherical Nucleic Acids. *Small* **2023**, *19*, 2300097. [CrossRef]
25. Li, L.L.; Wu, P.; Hwang, K.; Lu, Y. An exceptionally simple strategy for DNA-functionalized up-conversion nanoparticles as biocompatible agents for nanoassembly, DNA delivery, and imaging. *J. Am. Chem. Soc.* **2013**, *135*, 2411–2414. [CrossRef]
26. Huang, L.J.; Yu, R.Q.; Chu, X. DNA-functionalized upconversion nanoparticles as biosensors for rapid, sensitive, and selective detection of Hg($^{2+}$) in complex matrices. *Analyst* **2015**, *140*, 4987–4990. [CrossRef] [PubMed]
27. Ge, H.; Wang, D.; Pan, Y.; Guo, Y.; Li, H.; Zhang, F.; Zhu, X.; Li, Y.; Zhang, C.; Huang, L. Sequence-Dependent DNA Functionalization of Upconversion Nanoparticles and Their Programmable Assemblies. *Angew. Chem. Int. Ed.* **2020**, *59*, 8133–8137. [CrossRef]
28. Zhang, Y.; Zhang, Y.; Song, G.; He, Y.; Zhang, X.; Liu, Y.; Ju, H. A DNA-Azobenzene Nanopump Fueled by Upconversion Luminescence for Controllable Intracellular Drug Release. *Angew. Chem. Int. Ed.* **2019**, *58*, 18207–18211. [CrossRef]
29. Zhang, Y.; Chen, W.; Zhang, Y.; Zhang, X.; Liu, Y.; Ju, H. A Near-Infrared Photo-Switched MicroRNA Amplifier for Precise Photodynamic Therapy of Early-Stage Cancers. *Angew. Chem. Int. Ed.* **2020**, *59*, 21454–21459. [CrossRef]

30. Lu, J.; Chen, Y.; Liu, D.; Ren, W.; Lu, Y.; Shi, Y.; Piper, J.; Paulsen, I.; Jin, D. One-Step Protein Conjugation to Upconversion Nanoparticles. *Anal. Chem.* **2015**, *87*, 10406–10413. [CrossRef]
31. Wu, D.; Yang, K.; Zhang, Z.; Feng, Y.; Rao, L.; Chen, X.; Yu, G. Metal-free bioorthogonal click chemistry in cancer theranostics. *Chem. Soc. Rev.* **2022**, *51*, 1336–1376. [CrossRef]
32. Li, M.; Zhao, J.; Chu, H.; Mi, Y.; Zhou, Z.; Di, Z.; Zhao, M.; Li, L. Light-Activated Nanoprobes for Biosensing and Imaging. *Adv. Mater.* **2019**, *31*, 1804745. [CrossRef]
33. Basu, S.; Pacelli, S.; Feng, Y.; Lu, Q.; Wang, J.; Paul, A. Harnessing the Noncovalent Interactions of DNA Backbone with 2D Silicate Nanodisks To Fabricate Injectable Therapeutic Hydrogels. *ACS Nano* **2018**, *12*, 9866–9880. [CrossRef]
34. Baig, N.; Kammakakam, I.; Falath, W. Nanomaterials: A review of synthesis methods, properties, recent progress, and challenges. *Mater. Adv.* **2021**, *2*, 1821–1871. [CrossRef]
35. Zhao, H.; Lee, Y.; Han, M.; Sharma, B.K.; Chen, X.; Ahn, J.-H.; Rogers, J.A. Nanofabrication approaches for functional three-dimensional architectures. *Nano Today* **2020**, *30*, 100825. [CrossRef]
36. Liddle, J.A.; Gallatin, G.M. Nanomanufacturing: A Perspective. *ACS Nano* **2016**, *10*, 2995–3014. [CrossRef] [PubMed]
37. Carnerero, J.M.; Jimenez-Ruiz, A.; Castillo, P.M.; Prado-Gotor, R. Covalent and Non-Covalent DNA–Gold-Nanoparticle Interactions: New Avenues of Research. *ChemPhysChem* **2017**, *18*, 17–33. [CrossRef]
38. Zhao, J.; Liu, C.; Li, Y.; Ma, Y.; Deng, J.; Li, L.; Sun, J. Thermophoretic Detection of Exosomal microRNAs by Nanoflares. *J. Am. Chem. Soc.* **2020**, *142*, 4996–5001. [CrossRef]
39. Wang, W.-J.; Li, J.-J.; Rui, K.; Gai, P.-P.; Zhang, J.-R.; Zhu, J.-J. Sensitive Electrochemical Detection of Telomerase Activity Using Spherical Nucleic Acids Gold Nanoparticles Triggered Mimic-Hybridization Chain Reaction Enzyme-Free Dual Signal Amplification. *Anal. Chem.* **2015**, *87*, 3019–3026. [CrossRef]
40. Zhu, D.; Li, J.; Wang, L.; Li, Q.; Wang, L.; Song, B.; Zhou, R.; Fan, C. Hydrophobic collapse-driven nanoparticle coating with poly-adenine adhesives. *Chem. Commun.* **2021**, *57*, 3801–3804. [CrossRef]
41. Ye, Y.; Hou, S.; Wu, X.; Cheng, X.; He, S. Freeze-Driven Adsorption of Poly-A DNA on Gold Nanoparticles: From a Stable Biointerface to Plasmonic Dimers. *Langmuir* **2022**, *38*, 4625–4632. [CrossRef] [PubMed]
42. Zhang, D.; Peng, R.; Liu, W.; Donovan, M.J.; Wang, L.; Ismail, I.; Li, J.; Li, J.; Qu, F.; Tan, W. Engineering DNA on the Surface of Upconversion Nanoparticles for Bioanalysis and Therapeutics. *ACS Nano* **2021**, *15*, 17257–17274. [CrossRef] [PubMed]
43. Francés-Soriano, L.; Estebanez, N.; Pérez-Prieto, J.; Hildebrandt, N. DNA-Coated Upconversion Nanoparticles for Sensitive Nucleic Acid FRET Biosensing. *Adv. Funct. Mater.* **2022**, *32*, 2201541. [CrossRef]
44. Yang, Z.; Loh, K.Y.; Chu, Y.-T.; Feng, R.; Satyavolu, N.S.R.; Xiong, M.; Nakamata Huynh, S.M.; Hwang, K.; Li, L.; Xing, H.; et al. Optical Control of Metal Ion Probes in Cells and Zebrafish Using Highly Selective DNAzymes Conjugated to Upconversion Nanoparticles. *J. Am. Chem. Soc.* **2018**, *140*, 17656–17665. [CrossRef] [PubMed]
45. Vizzini, P.; Manzano, M.; Farre, C.; Meylheuc, T.; Chaix, C.; Ramarao, N.; Vidic, J. Highly sensitive detection of Campylobacter spp. In chicken meat using a silica nanoparticle enhanced dot blot DNA biosensor. *Biosens. Bioelectron.* **2021**, *171*, 112689. [CrossRef]
46. D'Agata, R.; Palladino, P.; Spoto, G. Streptavidin-coated gold nanoparticles: Critical role of oligonucleotides on stability and fractal aggregation. *Beilstein J. Nanotechnol.* **2017**, *8*, 1–11. [CrossRef]
47. Li, M.; Zhang, T.; Zhang, Y. Ultrasensitive electrochemical sensing platform for miRNA-21 detection based on manganese dioxide-gold nanoparticle nanoconjugates coupled with hybridization chain reaction and horseradish peroxidase signal amplification. *Analyst* **2023**, *148*, 2180–2188. [CrossRef]
48. Chen, M.; Song, Z.; Yang, X.; Song, Z.; Luo, X. Antifouling peptides combined with recognizing DNA probes for ultralow fouling electrochemical detection of cancer biomarkers in human bodily fluids. *Biosens. Bioelectron.* **2022**, *206*, 114162. [CrossRef]
49. van der Meer, S.B.; Loza, K.; Wey, K.; Heggen, M.; Beuck, C.; Bayer, P.; Epple, M. Click Chemistry on the Surface of Ultrasmall Gold Nanoparticles (2 nm) for Covalent Ligand Attachment Followed by NMR Spectroscopy. *Langmuir* **2019**, *35*, 7191–7204. [CrossRef]
50. Yan, W.; Zhong, Z.; Ma, J.; Rujiralai, T. Highly sensitive colorimetric sensing of copper(ii) ions based on "CLICK-17" DNAzyme-catalyzed azide modified gold nanoparticles and alkyne capped dsDNA cycloaddition. *RSC Adv.* **2021**, *11*, 24196–24205. [CrossRef]
51. Hu, L.; Takezawa, Y.; Shionoya, M. CuII-mediated DNA base pairing of a triazole-4-carboxylate nucleoside prepared by click chemistry. *Chem. Commun.* **2023**, *59*, 892–895. [CrossRef] [PubMed]
52. Zhang, X.; Ge, Y.; Liu, M.; Pei, Y.; He, P.; Song, W.; Zhang, S. DNA-Au Janus Nanoparticles for In Situ SERS Detection and Targeted Chemo-photodynamic Synergistic Therapy. *Anal. Chem.* **2022**, *94*, 7823–7832. [CrossRef]
53. Kyriazi, M.-E.; Giust, D.; El-Sagheer, A.H.; Lackie, P.M.; Muskens, O.L.; Brown, T.; Kanaras, A.G. Multiplexed mRNA Sensing and Combinatorial-Targeted Drug Delivery Using DNA-Gold Nanoparticle Dimers. *ACS Nano* **2018**, *12*, 3333–3340. [CrossRef]
54. Damavandi, F.; Wang, W.; Shen, W.-Z.; Cetinel, S.; Jordan, T.; Jovel, J.; Montemagno, C.; Wong, G.K.-S. Enrichment of low abundance DNA/RNA by oligonucleotide-clicked iron oxide nanoparticles. *Sci. Rep.* **2021**, *11*, 13053. [CrossRef] [PubMed]
55. Zhang, P.; Fischer, A.; Ouyang, Y.; Wang, J.; Sohn, Y.S.; Nechushtai, R.; Pikarsky, E.; Fan, C.; Willner, I. Aptamer-modified DNA tetrahedra-gated metal–organic framework nanoparticle carriers for enhanced chemotherapy or photodynamic therapy. *Chem. Sci.* **2021**, *12*, 14473–14483. [CrossRef] [PubMed]
56. Zhao, J.; Gao, J.; Xue, W.; Di, Z.; Xing, H.; Lu, Y.; Li, L. Upconversion Luminescence-Activated DNA Nanodevice for ATP Sensing in Living Cells. *J. Am. Chem. Soc.* **2018**, *140*, 578–581. [CrossRef]

57. Morzy, D.; Tekin, C.; Caroprese, V.; Rubio-Sánchez, R.; Di Michele, L.; Bastings, M.M.C. Interplay of the mechanical and structural properties of DNA nanostructures determines their electrostatic interactions with lipid membranes. *Nanoscale* **2023**, *15*, 2849–2859. [CrossRef]
58. Hao, Y.; Li, Y.; Song, L.; Deng, Z. Flash Synthesis of Spherical Nucleic Acids with Record DNA Density. *J. Am. Chem. Soc.* **2021**, *143*, 3065–3069. [CrossRef]
59. Zhao, X.; Han, Q.; Na, N.; Ouyang, J. Spatiotemporally Controlled DNA Nanoclamps: Single-Molecule Imaging of Receptor Protein Oligomerization. *Anal. Chem.* **2021**, *93*, 14514–14520. [CrossRef]
60. Chai, X.; Fan, Z.; Yu, M.-M.; Zhao, J.; Li, L. A Redox-Activatable DNA Nanodevice for Spatially-Selective, AND-Gated Imaging of ATP and Glutathione in Mitochondria. *Nano Lett.* **2021**, *21*, 10047–10053. [CrossRef]
61. Zhao, J.; Li, Y.; Yu, M.; Gu, Z.; Li, L.; Zhao, Y. Time-Resolved Activation of pH Sensing and Imaging in Vivo by a Remotely Controllable DNA Nanomachine. *Nano Lett.* **2020**, *20*, 874–880. [CrossRef]
62. Chu, H.; Zhao, J.; Mi, Y.; Zhao, Y.; Li, L. Near-Infrared Light-Initiated Hybridization Chain Reaction for Spatially and Temporally Resolved Signal Amplification. *Angew. Chem. Int. Ed.* **2019**, *58*, 14877–14881. [CrossRef] [PubMed]
63. Zhao, X.; Zhang, L.; Gao, W.; Yu, X.; Gu, W.; Fu, W.; Luo, Y. Spatiotemporally Controllable MicroRNA Imaging in Living Cells via a Near-Infrared Light-Activated Nanoprobe. *ACS Appl. Mater. Interfaces* **2020**, *12*, 35958–35966. [CrossRef]
64. O'Hagan, M.P.; Duan, Z.; Huang, F.; Laps, S.; Dong, J.; Xia, F.; Willner, I. Photocleavable Ortho-Nitrobenzyl-Protected DNA Architectures and Their Applications. *Chem. Rev.* **2023**, *123*, 6839–6887. [CrossRef] [PubMed]
65. Ye, M.; Kong, Y.; Zhang, C.; Lv, Y.; Cheng, S.; Hou, D.; Xian, Y. Near-Infrared Light Controllable DNA Walker Driven by Endogenous Adenosine Triphosphate for in Situ Spatiotemporal Imaging of Intracellular MicroRNA. *ACS Nano* **2021**, *15*, 14253–14262. [CrossRef] [PubMed]
66. Mi, Y.; Zhao, J.; Chu, H.; Li, Z.; Yu, M.; Li, L. Upconversion Luminescence-Controlled DNA Computation for Spatiotemporally Resolved, Multiplexed Molecular Imaging. *Anal. Chem.* **2021**, *93*, 2500–2509. [CrossRef]
67. Zhao, T.; Gao, Y.; Wang, J.; Cui, Y.; Niu, S.; Xu, S.; Luo, X. From Passive Signal Output to Intelligent Response: "On-Demand" Precise Imaging Controlled by Near-Infrared Light. *Anal. Chem.* **2021**, *93*, 12329–12336. [CrossRef]
68. Gao, P.; Shen, X.; Liu, X.; Cui, B.; Wang, M.; Wan, X.; Li, N.; Tang, B. Covalent Organic Framework-Derived Carbonous Nanoprobes for Cancer Cell Imaging. *ACS Appl. Mater. Interfaces* **2021**, *13*, 41498–41506. [CrossRef]
69. Wang, L.; Wang, K.; Wang, X.; Niu, R.; Chen, X.; Zhu, Y.; Sun, Z.; Yang, J.; Liu, G.; Luo, Y. Intelligent Dual-Lock Deoxyribonucleic Acid Automatons Boosting Precise Tumor Imaging. *ACS Appl. Mater. Interfaces* **2023**, *15*, 3826–3838. [CrossRef]
70. Zhang, Q.; Wang, Y.; Wang, W.; Min, Q.; Zhang, J.R.; Zhu, J.J. A Telomerase-Assisted Strategy for Regeneration of DNA Nanomachines in Living Cells. *Angew. Chem. Int. Ed.* **2023**, *62*, e202213884. [CrossRef]
71. Liu, Q.; Huang, Y.; Li, L.; Li, Z.; Li, M. Endogenous Enzyme-Operated Spherical Nucleic Acids for Cell-Selective Protein Capture and Localization Regulation. *Angew. Chem. Int. Ed.* **2023**, *62*, e202214958. [CrossRef]
72. Li, L.; Xing, H.; Zhang, J.; Lu, Y. Functional DNA Molecules Enable Selective and Stimuli-Responsive Nanoparticles for Biomedical Applications. *Acc. Chem. Res.* **2019**, *52*, 2415–2426. [CrossRef] [PubMed]
73. Liu, Q.; Wu, B.; Li, M.; Huang, Y.; Li, L. Heterostructures Made of Upconversion Nanoparticles and Metal–Organic Frameworks for Biomedical Applications. *Adv. Sci.* **2022**, *9*, 2103911. [CrossRef]
74. Xu, R.; Li, Y.; Zhu, C.; Liu, D.; Yang, Y.R. Cellular Ingestible DNA Nanostructures for Biomedical Applications. *Adv. NanoBiomed Res.* **2023**, *3*, 2200119. [CrossRef]
75. Tunc, C.U.; Culha, M. Gold nanoparticles conjugated DNA-tile nanomaterials for simultaneous delivery of morpholino antisense oligonucleotides and doxorubicin. *J. Drug Delivery Sci. Technol.* **2022**, *74*, 103546. [CrossRef]
76. Xue, C.; Hu, S.; Gao, Z.-H.; Wang, L.; Luo, M.-X.; Yu, X.; Li, B.-F.; Shen, Z.; Wu, Z.-S. Programmably tiling rigidified DNA brick on gold nanoparticle as multi-functional shell for cancer-targeted delivery of siRNAs. *Nat. Commun.* **2021**, *12*, 2928. [CrossRef]
77. Tebcharani, L.; Akter, N.; Fan, D.; Lieleg, O.; Gibbs, J.M.; Boekhoven, J. Hydrolyzable emulsions as a dual release platform for hydrophobic drugs and DNA. *Chem. Commun.* **2023**, *59*, 8099. [CrossRef]
78. Pal, S.; Koneru, J.K.; Andreou, C.; Rakshit, T.; Rajasekhar, V.K.; Wlodarczyk, M.; Healey, J.H.; Kircher, M.F.; Mondal, J. DNA-Functionalized Gold Nanorods for Perioperative Optical Imaging and Photothermal Therapy of Triple-Negative Breast Cancer. *ACS Appl. Nano Mater.* **2022**, *5*, 9159–9169. [CrossRef]
79. Shankar, R.; Nayak, S.; Singh, S.; Sen, A.; Kumar, N.; Bhushan, R.; Aggarwal, M.; Das, P. Simultaneous Sustained Drug Delivery, Tracking, and On-Demand Photoactivation of DNA–Hydrogel Formulated from a Biomass-Derived DNA Nanoparticle. *ACS Appl. Bio Mater.* **2023**, *6*, 1556–1565. [CrossRef] [PubMed]

chemosensors

Article

Visual Measurement of Fumonisin B_1 with Bipolar Electrodes Array-Based Electrochemiluminescence Biosensor

Longsheng Jin, Huihui Yu, Weishuai Liu, Ziying Xiao, Haijian Yang, Bing Jin and Meisheng Wu *

Department of Chemistry, College of Sciences, Nanjing Agricultural University, 1 Weigang, Nanjing 210095, China
* Correspondence: wumeisheng@njau.edu.cn

Abstract: Fumonisin B_1 (FB_1) is a toxin produced by the metabolism of Fusarium oxysporum, which can cause serious effects on the nervous, respiratory, digestive, and reproductive systems of humans or animals; it is known as one of the highly toxic epidemic contaminants. Herein, we report the visual inspection of FB_1 using bipolar electrodes (BPEs) with an array-based electrochemiluminescence (ECL) platform. The sensor consists of a PDMS cover and a glass substrate containing an array of 10 ITO electrodes. A specific sensing interface was constructed on the cathode of the BPE, which could modulate the ECL reactions that occurred at the anode of BPEs. To amplify the ECL signal, methylene blue (MB)-encapsulated Zr-MOFs (MB@Zr-MOFs) were synthesized and immobilized on the cathode of the BPE, which could amplify the ECL signal at the anode. By coupling the cyclic amplification effect of the DNA walker and nicking endonuclease (Nb.BbvCI), the biosensor can realize the visual measurement of FB_1 in the range of 5×10^{-5}~0.5 ng/mL. In addition, the developed biosensor was used to monitor the concentration of FB_1 in maize and peanut samples. The recoveries were in the range of 99.2%~110.6%, which demonstrated the good accuracy of the designed BPE-ECL biosensor for FB_1 assay in food samples.

Keywords: fumonisin B_1; bipolar electrode-electrochemiluminescence; visualization; Zr-MOFs

Citation: Jin, L.; Yu, H.; Liu, W.; Xiao, Z.; Yang, H.; Jin, B.; Wu, M. Visual Measurement of Fumonisin B_1 with Bipolar Electrodes Array-Based Electrochemiluminescence Biosensor. *Chemosensors* **2023**, *11*, 451. https://doi.org/10.3390/chemosensors11080451

Academic Editor: Danila Moscone

Received: 10 July 2023
Revised: 6 August 2023
Accepted: 10 August 2023
Published: 12 August 2023

1. Introduction

There has been growing attention to the contamination of foodstuffs caused by mycotoxins with the rapid development of agriculture and animal husbandry [1]. Mycotoxins are widely found in moldy agricultural products and their processed products, which can enter human and animal bodies through the food chain and thus cause a variety of diseases [2,3]. It is estimated by the Food and Agriculture Organization of the United Nations that 25% of global agricultural crops are contaminated by mycotoxins [2]. Among various mycotoxins, fumonisin B_1 (FB_1) has been identified as one of the most toxic mycotoxins, capable of disrupting sphingolipid metabolism, destroying the immune system, and increasing the risks of cancer. It is produced by *Fusarium* species and contaminates different grains such as maize, wheat, rice, and sorghum [4]. In addition, a considerable proportion of FB_1 contamination is attributed to inappropriate storage conditions. The tolerable level established by the European Union for cereals and maize is 2 μg/kg and 5 μg/kg, respectively [5]. A survey performed by González-Arias on 43 rice samples and 25 maize samples revealed that FB_1 wasn't detected in polished rice samples; however, both field maize and market maize samples were contaminated with FB_1 [6]. Therefore, the development of sensitive analytical devices for the measurement of FB_1 in agricultural products is crucial for point-of-care diagnosis [7].

Up to now, versatile sensitive biosensors have been explored for accurate detection of mycotoxins using electrochemical [8,9] and optical transduction mechanisms [10–12]. Among them, electrochemiluminescence (ECL) as a sensitive optical analytical technique that combines the properties of electrochemical and luminescence has attracted great attention due to its unique advantages, such as high sensitivity and spatial controllability [13–16].

Additionally, the intensive ECL signal could easily be observed by the naked eye, which enabled visual readouts of targets. ECL biosensors, especially sensors based on bipolar electrode (BPE) arrays, have been quickly developed in recent years. By transducing the faradic current into an optical signal, the BPE-ECL platform based on an electrode array can realize the simultaneous screening of multiple targets in a single device using simple optical equipment (e.g., smartphone and digital camera) and then analyzed with appropriate software [17–20].

To achieve sensitive detection, various powerful signal amplification methods such as nucleic acid-assisted amplification [21], enzyme catalysis amplification strategies [22,23], nanomaterial-assisted methods [24,25], and electroactive species-based technologies [26,27] have been integrated into the BPE-ECL device to realize sensitive measurement. For example, the introduction of electroactive substance (thionine)-covalent organic frameworks at the cathode of the BPE could reduce the external voltage for driving the ECL reactions at the anode of the BPE and greatly amplify the ECL intensity [28]. Furthermore, some researchers reported the integration of multiple amplification strategies in order to achieve remarkable enhancement of ECL sensitivity [25,29]. Ge et al. reported a paper-based BPE-ECL device using the synergistic effects of the excellent catalytic activity of AuPd NPs and the hybridization chain reaction to realize the detection of miRNA-155 with a low detection limit of 0.67 pM [25].

Herein, a novel ECL device based on a 10 BPE array was prepared for the simultaneous measurement of mycotoxin. The ECL signal emitted from the anodes of the BPE could be simultaneously recorded by a CCD camera. To improve the sensitivity of the biosensor, methylene blue (MB) was embedded in zirconium-based metal–organic frameworks (Zr-MOFs). Zr-MOFs are an ideal carrier for the immobilization of electroactive dyes and biomolecules due to their unique properties, such as water stability, good stability, large surface area, and tunable surface chemistry, making them an excellent candidate in the construction of biosensors [30]. When the cathode of the BPE was exposed to MB@Zr-MOFs, the reduction current could be greatly improved due to the accumulation of a large number of MB molecules, leading to a significant enhancement of the ECL signal due to the charge neutrality of the BPE. In the presence of the target, the DNA walker was activated, which led to the continuous release of MB from the electrode surface with the assistance of nicking endonuclease. As a result, the BPE array-based ECL biosensor with dual-signal amplification strategies exhibited a good detection performance for screening of FB_1 measurement. Additionally, the biosensor successfully evaluated FB_1 contamination in corn and peanut samples with satisfactory stability and recovery. The portable BPE sensors array based on ECL imaging opens up a new avenue toward simultaneous screening of multiple mycotoxins in a sample.

2. Materials and Methods

2.1. Reagents

Multi-wall carbon nanotubes (MWCNTs) with a diameter of 10–20 nm (10–30 μm in length, purity 95%) were supplied by Nanjing XFNANO Materials Tech Co., Ltd. Polydimethylsiloxane (PDMS), Nanjing, China; the monomer and curing reagent (Sylgard 184) were provided by Dow Corning (Midland, MI, USA). The screen stencil (hole diameter of 0.5 mm) was acquired from a nearby shop. Dibutylaminoethanol (DBAE) was purchased from Shanghai Energy Chemical Co., Ltd. (Shanghai, China). $ZrCl_4$, $HAuCl_4$, dodecanoic acid, and 6-mercapto-1-hexanol (MCH) were purchased from Aladdin. N, N-dimethylformamide (DMF) was purchased from Guangzhou Jinhua Chemical Reagent Co., Ltd. (Guangzhou, China). 2-aminoterephthalic acid (H_2N-BDC) was purchased from Shanghai Maclin Biochemical Technology Co., Ltd. (Shanghai, China). Screen printable etching paste (TP-005-4HC) was acquired from ShenZhen LaiYuan Technology Developing Co., Ltd. (Shenzhen, China). Tris (2-carboxyethyl) phosphine hydrochloride (TCEP), Tris(2,2′-bipyridyl)ruthenium-(II) dichloride hexahydrate ($Ru(bpy)_3Cl_2\cdot6H_2O$), fumonisin B_1 (FB_1), ochratoxin A (OTA), aflatoxin G1 (AFG_1), aflatoxin B_1 (AFB_1), and zearalenone

(ZEA) were provided by Sigma-Aldrich, St. Louis, MO, USA. An amount of 10 mM PBS (phosphate buffer solution, pH 7.4) was supplied by Jiangsu Kaiji Biotechnology Co., Ltd. (Nanjing, China). All of the DNA oligonucleotides were purchased from Shanghai Shenggong Biological Co., Ltd. (Shanghai, China); they are listed below.

DNA walker: 5′-SH-$(CH_2)_6$-TTTTTTTTTTTTTTTTTTTTTTTTTTTTTTTTTTTTTTT TGAATAAGCTGGTATCCTCAG C-3′

Aptamer: 5′–TGAGGATACCAGCTTATTCAATTAATCGCATTACCTTATACCAGCTT ATTCAATTACGTCTGCACATACCAGCTTATTCAATTAGATAGTAAGTGCAATCT-3′

Capture DNA (cDNA): 5′-SH-$(CH_2)_6$-TTTTTT**GC*TGAGG**TT-$(CH_2)_6$-NH_2-3′

2.2. Instruments

An MS23 CCD camera (Guangzhou Mingmei Technology Co., Ltd. Guangzhou, China) was used to capture ECL images. The UV–vis spectra were measured using an Ensight multi-mode plate reader (Perkin Elmer, Waltham, MA, USA). Transmission electron microscopy (TEM) was acquired using an FEI Tecnai G2 F20 apparatus. Two MWCNTs/PDMS fibers (diameter 1 mm) were used as driving electrodes and connected to a power supply.

2.3. Synthesis of Zr-MOFs and MB@Zr-MOFs

Zr-MOFs were synthesized according to the literature with minor modifications [31,32]. Briefly, 0.048 g zirconium chloride and 1.44 g dodecanoic acid were dissolved in 12 mL of DMF and ultrasonicated for 25 min to obtain a clarified solution, then 0.0186 g 2-aminoterephthalic acid (H_2N-BDC) was added and ultrasonicated again for 5 min. After that, the mixture was transferred to a hydrothermal reactor and the reaction was continued for 48 h in an oven at 120 °C. The obtained product was centrifuged at 8000 r/min for 5 min, washed 3 times with DMF and ethanol, and then dried at 60 °C to obtain Zr-MOFs (yellow powder).

For the preparation of MB@Zr-MOFs, 15 mg Zr-MOFs was mixed with 5 mL of methylene blue (1 mM) solution and shaken for 24 h, after which the product was centrifuged for 5 min at 8000 r/min. Finally, the sediment was washed with PBS several times to remove the excess MB and re-dispersed in 5 mL of PBS.

2.4. Preparation of ITO Array Electrodes

An array of bipolar electrodes was prepared using the screen-printed method according to our previous study [33]. Firstly, the pattern of the electrode array (10 electrodes, 0.15 cm × 1 cm) was designed using Adobe Illustrator and then sent to a local printing shop to fabricate a template with 200-mesh screens. Then, the screen printable etching paste was patterned on the ITO substrate (10 cm × 8 cm) with the help of the template. The patterned ITO substrate was heated to 120 °C and kept for 2 h to etch the ITO layer underneath the etching glue. After washing with water, the electrode was sonicated in isopropanol solution and ultrapure water for 15 min to remove the etching glue and other impurities from the electrode surface. The cleaned ITO was dried at 60 °C and stored in a clean environment before use.

2.5. Preparation of Au/ITO Array Electrodes

Au NPs were deposited on one pole of the ITO array electrode using a bipolar electrodeposition method [34]. An open BPE electrochemical cell was prepared as follows. A PDMS slice was prepared by mixing the monomer and curing agent in a ratio of 10:1. After degassing, the mixture was poured into a flat glass substrate and heated at 80 °C for 1 h. The PDMS slice was peeled off from the glass substrate and a rectangular reservoir (9 cm × 2 cm) was fabricated on the PDMS slice using a knife. Then, the PDMS slice was attached to the ITO electrode array surface. Two Pt wires were placed at the two ends of the reservoirs to serve as driving electrodes and connected to the external power supply. An amount of 3 mL of $HAuCl_4$ solution was introduced and a constant voltage was applied for 1 min to trigger the simultaneous deposition of Au NPs on the cathode pole of the BPE. The Au/ITO electrode array was then washed with ultrapure water and dried at 60 °C.

2.6. Preparation of DNA Mix

The DNA mix contains MB@Zr-MOFs/cDNA and double-stranded DNA (dsDNA). MB@Zr-MOFs/cDNA was prepared by using glutaraldehyde as a linker molecule. An amount of 5 mL of MB@Zr-MOFs was treated with 2.5% glutaraldehyde under continuous shaking for 4 h. The glutaralized MB@Zr-MOFs were washed with PBS three times and dispersed in 5 mL of PBS. An amount of 75 μL of cDNA (20 μM) was pretreated with 75 μL TCEP (10 mM) at 37 °C for 1 h to cleave the disulfide bonds and then conjugated on glutaraldehyde-chemical MB@Zr-MOFs (150 μL) by incubating at 37 °C for 4 h. The obtained MB@Zr-MOFs/cDNA was stored at 4 °C before use.

A total of 15 μL of DNA walker (20 μM) was reacted with 15 μL of TCEP (10 mM) at 37 °C for 1 h. Afterward, 15 μL of aptamer (20 μM) and 55 μL of PBS solution were introduced. The mixture was heated to 95 °C for 5 min, cooled naturally to room temperature, and then stored at 4 °C for 30 min to obtain dsDNA.

The DNA mix was prepared by introducing 300 μL of the above-prepared MB@Zr-MOFs/cDNA solution, 50 μL of TCEP (10 mM), and 50 μL of PBS solution to the dsDNA solution.

2.7. Fabrication of ECL Biosensor

Ten holes with a diameter of 2 mm were punched on a PDMS slice to act as reservoirs. Then, the PDMS slice was covered on the Au/ITO electrode array surface to construct a specific biosensing interface for FB_1. The distance between each hole was specially designed so that each electrode was located in one reservoir.

A total of 20 μL of DNAmix solution was added into the reservoirs and incubated at 37 °C for 3 h. The obtained DNAmix/Au/ITO electrode was carefully washed with PBS. Then, the unreacted sites of the Au/ITO electrode were blocked by 20 μL of MCH (0.1 mM) at 37 °C for 1 h. After rinsing with PBS, the modified electrode was incubated with 20 μL of mixture solution containing Nb.BbvCI (10 U/20 μL), different concentrations of FB_1, and 1× Cutsmart buffer solution at 37 °C for 3 h, followed by washing with PBS. The PDMS slice was removed, and the electrode was stored at 4 °C before ECL measurement.

2.8. ECL Imaging

A PDMS with two reservoirs (9 cm × 1.75 cm) was attached to the ITO BPE array to fabricate a closed BPE-ECL device for ECL imaging (Figure S1). The distance between these two reservoirs was 5 mm. MWCNTs-PDMS fibers with a diameter of 1 mm were glued on the PDMS slice to serve as driving electrodes. They were prepared according to our previous work [35] and connected to a CHI660E electrochemical workstation.

The reservoir containing the anode of the BPE array was filled with 2 mL of $Ru(bpy)_3^{2+}$ (5 mM)/DBAE (50 mM) solution, and the cathode of the BPE was immersed in PBS. ECL images were captured using a microscope equipped with a CCD in a dark environment with an exposure time of 10 s. ECL intensity was analyzed using Image-Pro Plus 6.0. ECL–potential curves were obtained using an MPI-E electrochemiluminescence analyzer (Xi'An Remax Electronic Science &Technology Co. Ltd., Xi'an, China).

2.9. Real Sample Detection

Corn and peanuts were purchased from local markets and pretreated following the reported literature [36]. Samples were grounded into powder, filtered through a 20-mesh sieve, and then dispersed into methanol. Different concentrations of FB_1 standard solutions were spiked into 1 mL of sample (1 g)/methanol solution. After the complete evaporation of solvent at room temperature, the sample was extracted with methanol/water (2:8, v/v) solution by ultrasonication for 15 min. The suspension was collected by centrifugation and diluted into 50 mL with PBS. The concentration of FB_1 was then measured with the fabricated biosensor.

3. Results and Discussion

3.1. Principle of the BPE- ECL Biosensors Array for FB_1 Detection

Scheme 1 illustrates the fabrication process and detection mechanism of the BPE-ECL biosensor array. cDNA was labeled with MB@Zr-MOFs (Scheme 1A). Aptamer was hybridized with a DNA walker to form double-stranded DNA (dsDNA), which was then mixed with cDNA/MB@Zr-MOFs to form a DNA mix solution (Scheme 1B). The cathodes of the ITO BPE were modified with Au NPs (Scheme 1C) through bipolar electrodeposition. The large surface area of Au NPs allows more DNA to be immobilized on the electrode surface; meanwhile, the excellent conductivity of Au NPs could amplify the ECL signal at the anode of the BPE. Then, the specific sensing interface detection of FB_1 was constructed on the Au NPs' surface. A DNA mix solution containing an aptamer–DNA walker duplex and MB@Zr-MOFs-labeled cDNA was immobilized on Au NPs. Owing to the accumulation of plenty of MB molecules on the cathode of the BPE, the reduction of MB can cause the enhancement of the ECL signal of $Ru(bpy)_3^{2+}$/DBAE at the anode. In the presence of FB_1, the competitive binding to aptamer between FB_1 and the DNA walker forced the release of aptamer from the electrode surface, which initiated the walking process of the DNA walker. The hybridization of DNA walker with neighboring cDNA produced specific recognition sites for Nb. BbvCI nicking endonuclease cleaved cDNA into two short DNA fragments, causing the liberation of MB@Zr-MOFs-labeled DNA fragment from the cathode. As a result, the cyclic walking of the DNA walker and the continuous cleavage of cDNA could greatly improve the sensitivity of the biosensor.

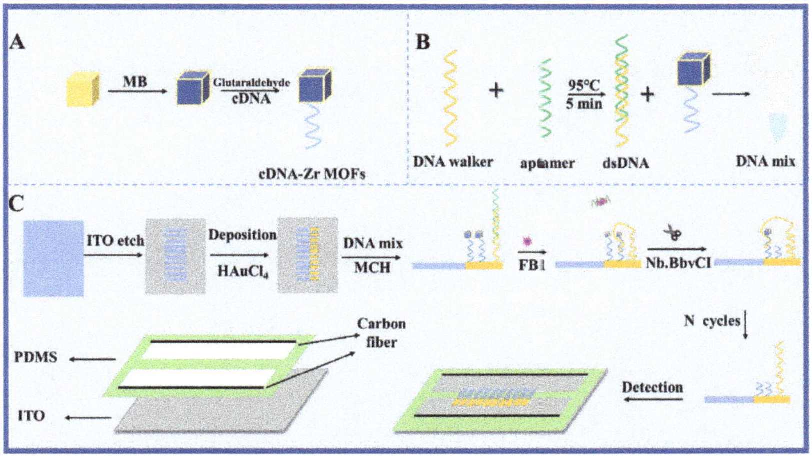

Scheme 1. Schematic diagram of (**A**) the preparation process of cDNA-Zr MOFs, (**B**) synthesis process of DNA mix, and (**C**) visual bipolar electrode-ECL array electrode for FB_1 detection.

3.2. Characterization of MB@Zr-MOFs

The morphology of MB@Zr-MOFs was studied using transmission electron microscopy (TEM). Figure 1A showed that the diameter of Zr-MOFs was about 70–80 nm, which is in agreement with the literature [26]. After the loading of MB (Figure 1B), the structure of the nanoparticles was not significantly changed. Meanwhile, the color of the MOFs turned from light yellow to blue (inset of Figure 1C), indicating the successful encapsulation of MB molecules in the Zr-MOFs. The UV-vis absorption spectra of Zr-MOFs and MB@Zr-MOFs were displaced in Figure 1C. Zr-MOFs have no obvious peak in the range from 400 to 800 nm (curve A). The MB solution exhibited a strong absorption peak at around 661 nm. The MB@Zr-MOFs (curve C) also displayed an obvious peak at the same position as the MB molecules (curve B).

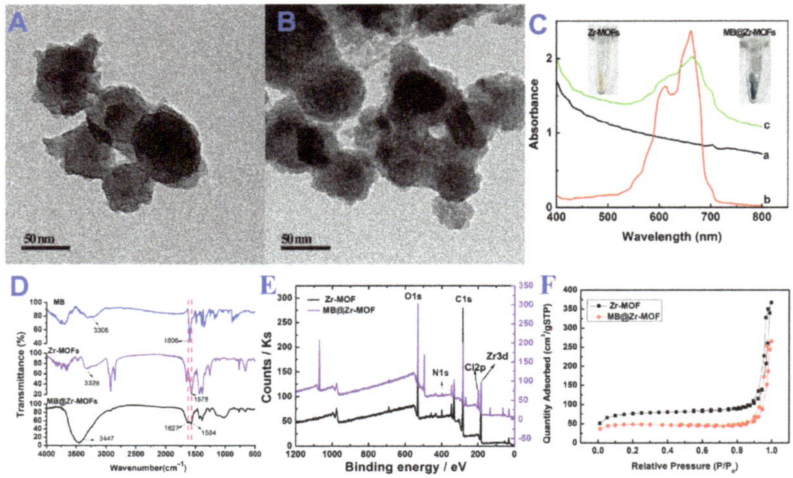

Figure 1. TEM images of Zr-MOFs (**A**) and MB@Zr-MOFs (**B**). (**C**) UV-vis absorption spectra of Zr-MOFs (a), MB (b), and MB@Zr-MOFs (c). Inset were the optical images of Zr-MOFs and MB@Zr-MOFs solution. (**D**) FT-IR spectra of Zr-MOFs, MB and MB@Zr-MOFs. (**E**) XPS spectra of Zr-MOF and MB@Zr-MOFs. (**F**) nitrogen adsorption–desorption isotherm of Zr-MOFs and MB@Zr-MOFs.

FT-IR analysis was then conducted to obtain the characteristic functional groups in MB@Zr-MOFs (Figure 1D). MB exhibited characteristic peaks at 3308 cm^{-1} and 1606 cm^{-1}, corresponding to the N-H stretching vibration and phenyl ring vibration. For Zr-MOFs, the peaks at 3328 cm^{-1} and 1578 cm^{-1} were related to the N-H stretching vibration and Zr-C=N MOF's stretching vibration [37], respectively. When MB molecules were embedded into the Zr-MOFs (MB@Zr-MOFs), the characteristic peaks at 3447 cm^{-1} corresponding to the N-H stretching vibration could be observed. Additionally, the peak caused by the Zr-C=N MOF's stretching vibration also appeared at 1564 cm^{-1}.

The chemical composition of Zr-MOF and MB@Zr-MOF was analyzed using X-ray photoelectron spectroscopy (XPS). As can be seen from the full survey spectra in Figure 1E that both Zr-MOF and MB@Zr-MOF have C1s, O1s, N1s, and Zr3d peaks. Because of the combination of MB on Zr-MOF, a new peak corresponding to Cl2p appeared in the spectrum of MB@Zr-MOF. We could see from the high-resolution Zr3d spectrum of MB@Zr-MOF in Figure S2 that the Zr 3d peak could be divided into two peaks at 182.4 and 184.8 eV, corresponding to Zr $3d_{5/2}$ and Zr $3d_{3/2}$.

To further detect the specific surface area and porosity of Zr-MOFs and MB@Zr-MOFs, nitrogen adsorption–desorption isotherm analysis was carried out. The Brunauer–Emmett–Teller (BET) surface area and the pore volume of Zr-MOFs were 230.04 m^2/g and 0.5568 m^3/g, respectively. After the combination of MB molecules, the surface area and pore volume reduced to 138.09 m^2/g and 0.4012 m^3/g, respectively.

3.3. Feasibility of the BPE-ECL Device for FB_1 Assay

The electrochemical performance of MB@Zr-MOFs and the feasibility of MB@Zr-MOFs-mediated ECL enhancement in the BPE system were then investigated using cyclic voltammetry (CV) and ECL. Figure 2A shows the cyclic voltammograms obtained in a three-electrode system using the prepared Au NPs/ITO electrode as the working electrode. As can be seen, both MB and MB@Zr-MOFs displayed a pair of reversible redox peaks at ca. 0.169 and –0.196 V. It indicated that MB@Zr-MOFs underwent similar two-electron redox reaction as MB [38]. The electrochemical reduction of MB@Zr-MOFs inducing an ECL amplification effect in the BPE system was also studied. Curve a in Figure 2B shows the ECL–voltage curve obtained by introducing Zr-MOFs solution into the cathodic reservoir of the BPE. As can be seen, the ECL intensity was improved when the voltage exceeded

2.5 V. When MB@Zr-MOFs were added (curve b), the onset voltage for driving the ECL reaction reduced to 2.15 V; meanwhile, the ECL intensity was improved by 3.7-fold. It confirmed that the reduction of MB@Zr-MOFs on the cathode of the BPE could mediate the ECL signal emitted from the anode of the BPEs array. The corresponding ECL–time curve is displayed in the inset of Figure 2B. The relative standard deviation (RSD) of the ECL intensity under continuous 10 cycles of CV sweep was 2.04%, suggesting good ECL stability of the prepared ITO electrode array.

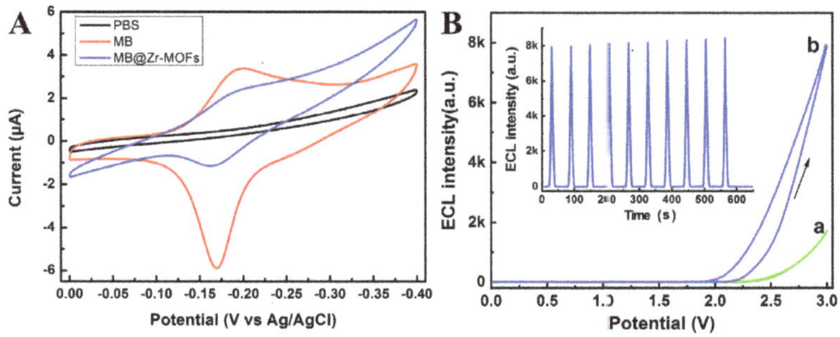

Figure 2. (**A**) Cyclic voltammograms obtained on Au NPs/ITO electrodes in the presence of PBS, MB, and MB@Zr-MOFs. (**B**) ECL–voltage curves were obtained by introducing $Ru(bpy)_3^{2+}$/DBAE into the anodic cell. The cathodic cell was filled with Zr-MOFs (a) and MB@Zr-MOFs (b), respectively. Inset was the ECL–time curve recorded by filling the cathodic cell with MB@Zr-MOFs.

The immobilization of MB@Zr-MOFs/cDNA on Au NPs/ITO BPE was then assessed by CV (Figure 3A). there was no noticeable peak when the Au NPs/ITO electrode was immersed in PBS (black line). After treating the Au NPs/ITO electrode with DNA mix, a pair of peaks appeared at –0.239 V and –0.175 V, ascribing to the oxidation and reduction of MB. It confirmed that MB@Zr-MOFs/cDNA was successfully assembled on the Au NPs/ITO electrode. The peaks vanished after subsequent incubation with FB_1 and Nb.BbvCI, demonstrating the recognition of the modified electrode toward FB_1 and the cleavage of cDNA by Nb.BbvCI.

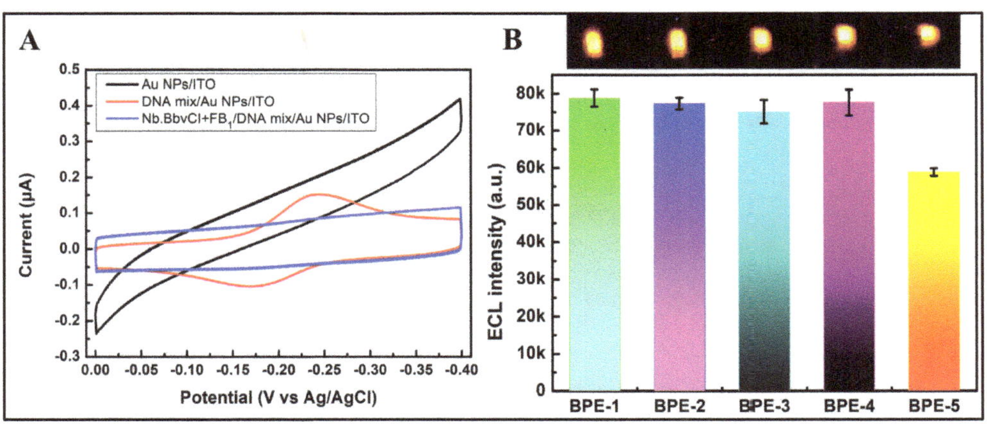

Figure 3. (**A**) Cyclic voltammograms obtained on different electrodes in PBS. (**B**) ECL images and the corresponding ECL intensity obtained on Au NPs/ITO BPE (BPE-1), DNA mix/Au NPs/ITO BPE (BPE-2), FB_1/DNA mix/Au NPs/ITO BPE (BPE-3), Nb.BbvCI/DNA mix/Au NPs/ITO BPE (BPE-4), and Nb.BbvCI/FB_1/DNA mix/Au NPs/ITO BPE (BPE-5), respectively.

The feasibility of the developed BPE-ECL biosensor for visual analysis of FB_1 is displayed in Figure 3B. A strong ECL signal could be observed on the anode of Au NPs/ITO BPE (BPE-1). When the DNA mix was immobilized on the electrode surface, no significant ECL signal could be observed. This is because although the reduction of MB@Zr-MOFs on Au NPs/ITO BPE could amplify the ECL signal, the increased resistance of the electrode caused by the immobilization of DNA would inhibit the ECL signal. When FB_1 (0.5 ng/mL) was added, the ECL signal changed slightly (BPE-3). When the DNA mix/Au NPs/ITO BPE was treated with Nb.BbvCI in the absence of FB_1 (BPE-4), the ECL signal also exhibited no obvious change, suggesting that Nb.BbvCI cannot digest cDNA without FB_1. When FB_1/DNA mix/Au NPs/ITO BPE was incubated with Nb.BbvCI (BPE-5), the ECL intensity demonstrated a sharp decrement, confirming that the cleavage of MB@Zr-MOFs-labeled cDNA by Nb.BbvCI was activated by FB_1 and the feasibility of the developed biosensor for FB_1 measurement.

3.4. Optimization of Experimental Conditions

To achieve sensitive measurement of FB_1, the experimental conditions, such as the deposition time of Au NPs and the incubation time of FB_1 and Nb.BbvCI, were optimized. Figure S3 shows the optical images of the ITO electrode array captured during the deposition of Au NPs under different voltages and times. As can be seen, when the external voltage was 6 V, the color of the ITO electrode changed apparently due to the reduction of $HAuCl_4$ on the cathode of the BPE. As the deposition time increased, the color of the electrode turned darker. With the increase in external voltage from 6 to 9 V, the coverage area of the ITO electrode by Au NPs increased from 15.24% to 36.30%. Therefore, we chose 9 V for the preparation of Au/ITO BPE.

Figure 4 shows the effect of $HAuCl_4$ concentration on the ECL behavior of the BPE array. The ECL images were obtained by applying different voltages to trigger the ECL reactions on the BPE array. The ECL intensity increased significantly with increasing $HAuCl_4$ concentration (1 mM to 8 mM). However, when the concentration of $HAuCl_4$ reached 8 mM, the thick coating layer of gold can easily fall off from the ITO surface, resulting in poor reproducibility of the biosensor. Therefore, to improve the performance of the biosensor, 5 mM $HAuCl_4$ was selected for the construction of Au/ITO BPE.

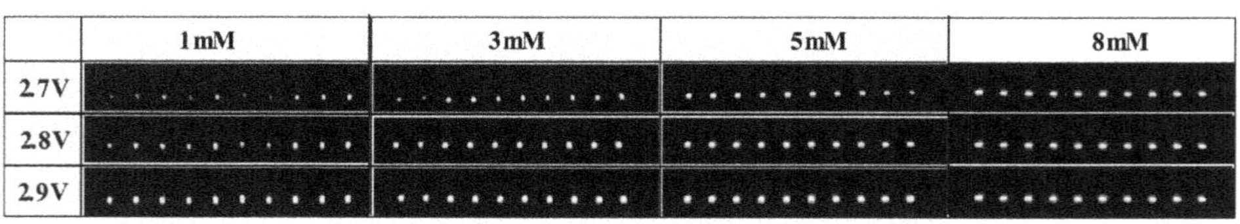

Figure 4. The effect of $HAuCl_4$ concentration on the ECL behavior of BPE array.

Figure 4 also reveals that the ECL intensity exhibited a sharp enhancement with the increase in voltage. Figure S4A shows the ECL stability of the developed Au/BPE array in 200 s, and that the voltage for triggering the ECL reactions was 2.9 V. Stable and intensive ECL signals could be observed on each electrode. Figure S4B demonstrates that the average ECL intensity at different electrodes on one array reached a plateau in 60 s. Additionally, the RSD of the ECL intensities in one array was below 9.2%. These results suggest the good stability and reproducibility of the BPE array.

Then, we optimized the reaction time of the DNA mix/Au/ITO electrode and the mixture of FB_1 and Nb.BbvCI. Figure 5 shows that the ECL signal intensity decreased with increasing incubation time (0 to 2.5 h), which is caused by the cleavage of cDNA/MB@Zr-MOFs on the electrode surface by Nb.BbvCI. After 2.5 h, the ECL intensity reached a stable value, indicating that the cleavage reaction was completed. Therefore, 2.5 h was chosen for the subsequent experiments.

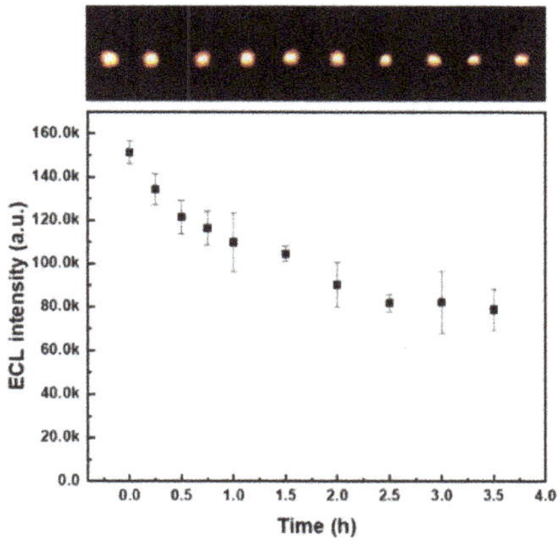

Figure 5. Optimization of the incubation time of FB_1.

3.5. Calibration Curve for FB_1 Measurement

Under optimized conditions, the designed biosensor was then used for a quantitative assay of FB_1. Figure 6A shows that the brightness of the ECL signal decreased as the concentration of FB_1 increased. The ECL signal intensity was linear in the range of 5×10^{-5}~0.5 ng/mL with the logarithm of the FB_1 concentration (R^2 = 0.9995). The linear equation was I = 45,894–17,409 logC(ng/mL), indicating that the prepared biosensor can be applied in the quantitative detection of FB_1.

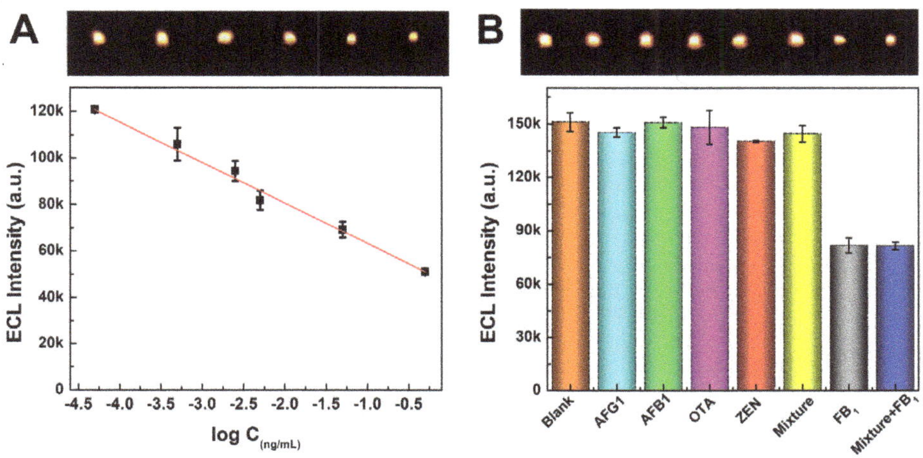

Figure 6. (**A**) ECL response and the calibration curve of the BPE biosensor for sensing FB_1. (**B**) Selectivity evaluation of the BPE biosensor on different mycotoxins.

The specificity of the prepared sensor for FB_1 detection was conducted by measuring the ECL signal of the biosensor in the presence of interference mycotoxins. As shown in Figure 6B, the presence of AFG_1, AFB_1, OTA, ZEN, and the mixture could not cause a significant change in ECL intensity compared to that of the blank sample. When the mixture containing all these interfering toxins and FB_1 was introduced, the ECL intensity of the biosensor decreased sharply. Additionally, the ECL intensity was very close to that obtained in the presence of FB_1. All of these results suggested the superior selectivity of the proposed biosensor for FB_1 analysis.

3.6. Application of the BPE Array Biosensor in Real Samples

To testify the applicability of the developed biosensor for FB_1 measurement in real samples, maize and peanut samples were grounded and different concentrations of FB_1 standard solution were spiked. After extracting by solvent, the suspension was collected using sonication and analyzed using the BPE array. As shown in Table 1, the recoveries were in the range of 99.2%~110.6%, and the RSD was from 1.8% to 9.5%, indicating the acceptable accuracy of the as-prepared biosensor for mycotoxin measurement in real samples.

Table 1. Detection of FB_1 in actual samples (n = 3).

Samples	Added (pg/mL)	Founded (ng/mL)	RSD (%)	Recovery (%)
Corn	0	Not detected		-
	0.500	0.5207 ± 0.0104	2.0	104.1
	2.500	2.490 ± 0.085	3.4	99.6
	5.000	5.184 ± 0.454	8.7	103.7
Peanut	0	Not detected		-
	0.500	0.4958 ± 0.0091	1.8	99.2
	2.500	2.646 ± 0.251	9.5	105.9
	5.000	5.530 ± 0.806	7.3	110.6

4. Conclusions

In the present work, a novel ECL biosensor based on an ITO BPEs array was constructed for visual analysis of FB_1 in food samples. By introducing MB molecules-decorated Zr-MOFs at the cathode of the BPE, the ECL signal produced at the anode of the BPE could be significantly amplified due to the neutrality of the BPE. Meanwhile, in the presence of the target, a DNA walker was activated, which triggered the continuous release of MB@Zr-MOFs from the electrode surface, causing the significant quenching effect of the ECL signal. As a result, the developed biosensor demonstrated a broad linear range and a low detection limit for FB_1 measurement. Additionally, this work provides a new concept for screening multiple targets. Owing to the high sensitivity of the portable BPE-ECL device, the biosensor might be commercialized by merging smartphone photography and analysis technologies, as well as better manufacturing procedures for larger-scale production.

Supplementary Materials: The following supporting information can be downloaded at: https://www.mdpi.com/article/10.3390/chemosensors11080451/s1, Figure S1: BPEs array-ECL device; Figure S2: high-resolution Zr 3d XPS spectra of MB@Zr-MOFs; Figure S3: optical images of ITO electrode array during the deposition of Au NPs; Figure S4: ECL images captured on ITO BPEs array under different time and the average ECL intensity–time curve.

Author Contributions: Conceptualization, L.J., H.Y. (Huihui Yu), W.L., Z.X., H.Y. (Haijian Yang) and B.J.; data curation, L.J. and H.Y. (Huihui Yu); formal analysis, L.J., H.Y. (Huihui Yu), W.L., Z.X., H.Y. (Haijian Yang) and B.J.; methodology, L.J., H.Y. (Huihui Yu), W.L., Z.X., H.Y. (Haijian Yang), B.J. and M.W.; resources, L.J. and H.Y. (Huihui Yu); software, L.J.; validation, L.J. and H.Y. (Huihui Yu); visualization, L.J., investigation, W.L., Z.X., H.Y. (Haijian Yang) and B.J.; writing—original draft preparation, L.J. and H.Y. (Huihui Yu); writing—review and editing, L.J., W.L., Z.X., H.Y. (Haijian Yang), B.J. and M.W.; funding acquisition, M.W.; project administration, M.W.; supervision, M.W. All authors have read and agreed to the published version of the manuscript.

Funding: This research was funded by the Qinglan Project of Jiangsu Province of China and the National Natural Science Foundation of China (No. 21675087).

Institutional Review Board Statement: Not applicable.

Informed Consent Statement: Not applicable.

Data Availability Statement: Not applicable.

Conflicts of Interest: The authors declare no conflict of interest.

References

1. Yang, C.; Song, G.; Lim, W. Effects of mycotoxin-contaminated feed on farm animals. *J Hazard. Mater.* **2020**, *389*, 122087. [CrossRef] [PubMed]
2. Gao, Z.C.; Luo, K.X.; Zhu, Q.X.; Peng, J.H.; Liu, C.; Wang, X.Y.; Li, S.J.; Zhang, H.Y. The natural occurrence, toxicity mechanisms and management strategies of Fumonisin B1: A review. *Environ. Pollut.* **2023**, *320*, 121065. [CrossRef]
3. Lumsangkul, C.; Chiang, H.I.; Lo, N.W.; Fan, Y.K.; Ju, J.C. Developmental Toxicity of Mycotoxin Fumonisin B-1 in Animal Embryogenesis: An Overview. *Toxins* **2019**, *11*, 114. [CrossRef]
4. Hao, K.; Suryoprabowo, S.; Hong, T.; Song, S.; Liu, L.; Zheng, Q.; Kuang, H. Immunochromatographic strip for ultrasensitive detection of fumonisin B-1. *Food Agric. Immunol.* **2018**, *29*, 699–710. [CrossRef]
5. He, D.; Wu, Z.; Cui, B.; Xu, E. Aptamer and gold nanorod-based fumonisin B1 assay using both fluorometry and SERS. *Microchim. Acta* **2020**, *187*, 215. [CrossRef]
6. Molina-Pintor, I.B.; Ruiz-Arias, M.A.; Guerrero-Flores, M.C.; Rojas-Garcia, A.E.; Barron-Vivanco, B.S.; Medina-Diaz, I.M.; Bernal-Hernandez, Y.Y.; Ortega-Cervantes, L.; Rodriguez-Cervantes, C.H.; Ramos, A.J.; et al. Preliminary survey of the occurrence of mycotoxins in cereals and estimated exposure in a northwestern region of Mexico. *Int. J. Environ. Health Res.* **2022**, *32*, 2271–2285. [CrossRef] [PubMed]
7. Arumugam, T.; Ghazi, T.; Chuturgoon, A.A. Molecular and epigenetic modes of Fumonisin B-1 mediated toxicity and carcinogenesis and detoxification strategies. *Crit. Rev. Toxicol.* **2021**, *51*, 76–94. [CrossRef] [PubMed]
8. Han, Z.; Tang, Z.; Jiang, K.; Huang, Q.; Meng, J.; Nie, D.; Zhao, Z. Dual-target electrochemical aptasensor based on co-reduced molybdenum disulfide and Au NPs (rMoS2-Au) for multiplex detection of mycotoxins. *Biosens. Bioelectron.* **2020**, *150*, 111894. [CrossRef]
9. De Rycke, E.; Foubert, A.; Dubruel, P.; Bol'hakov, O.I.; De Saeger, S.; Beloglazova, N. Recent advances in electrochemical monitoring of zearalenone in diverse matrices. *Food Chem.* **2021**, *353*, 129342. [CrossRef]
10. Jiang, F.; Li, P.; Zong, C.; Yang, H. Surface-plasmon-coupled chemiluminescence amplification of silver nanoparticles modified immunosensor for high-throughput ultrasensitive detection of multiple mycotoxins. *Anal. Chim. Acta* **2020**, *1114*, 58–65. [CrossRef]
11. Zong, C.; Jiang, F.; Wang, X.; Li, P.; Xu, L.; Yang, H. Imaging sensor array coupled with dual-signal amplification strategy for ultrasensitive chemiluminescence immunoassay of multiple mycotoxins. *Biosens. Bioelectron.* **2021**, *177*, 112998. [CrossRef]
12. Jiang, P.; Luo, L.; Liu, X.; Zhao, W.; Bi, X.; Luo, L.; Li, L.; You, T. A potential-resolved ratiometric electrochemiluminescence aptasensor for Pb^{2+}: Gold nanoclusters and amino-terminated perylene derivative as both emitters and resonance energy transfer donor-acceptor pair. *Sens. Actuators B Chem.* **2023**, *386*, 133758. [CrossRef]
13. Luo, L.; Liu, X.; Bi, X.; Li, L.; You, T. Dual-quenching effects of methylene blue on the luminophore and co-reactant: Application for electrochemiluminescent-electrochemical ratiometric zearalenone detection. *Biosens. Bioelectron.* **2023**, *222*, 114991. [CrossRef] [PubMed]
14. Wang, Q.; Xiong, C.; Li, J.; Deng, Q.; Zhang, X.; Wang, S.; Chen, M.-M. High-performance electrochemiluminescence sensors based on ultra-stable perovskite quantum dots@ZIF-8 composites for aflatoxin B1 monitoring in corn samples. *Food Chem.* **2023**, *410*, 135325. [CrossRef] [PubMed]
15. Wei, M.; Wang, C.; Xu, E.; Chen, J.; Xu, X.; Wei, W.; Liu, S. A simple and sensitive electrochemiluminescence aptasensor for determination of ochratoxin A based on a nicking endonuclease-powered DNA walking machine. *Food Chem.* **2019**, *282*, 141–146. [CrossRef]
16. Xi, M.; Wu, Z.; Luo, Z.; Ling, L.; Xu, W.; Xiao, R.; Wang, H.; Fang, Q.; Hu, L.; Gu, W.; et al. Water Activation for Boosting Electrochemiluminescence. *Angew. Chem. Int. Ed. Engl.* **2023**, *62*, e202302166. [CrossRef]
17. Qin, X.; Gao, J.; Jin, H.-J.; Li, Z.-Q.; Xia, X.-H. Closed Bipolar Electrode Array for Optical Reporting Reaction-Coupled Electrochemical Sensing and Imaging. *Chem. A Eur. J.* **2022**, *29*, e202202687.
18. Liu, Y.; Cheng, Q.-Y.; Gao, H.; Chen, H.-Y.; Xu, J.-J. Microfluidic Gradient Culture Arrays for Cell Pro-oxidation Analysis Using Bipolar Electrochemiluminescence. *Anal. Chem.* **2023**, *95*, 8376–8383. [CrossRef]
19. Anderson, T.J.; Defnet, P.A.; Zhang, B. Electrochemiluminescence (ECL)-Based Electrochemical Imaging Using a Massive Array of Bipolar Ultramicroelectrodes. *Anal. Chem.* **2020**, *92*, 6748–6755. [CrossRef]
20. Liu, Y.; Zhang, N.; Pan, J.-B.; Song, J.; Zhao, W.; Chen, H.-Y.; Xu, J.-J. Bipolar Electrode Array for Multiplexed Detection of Prostate Cancer Biomarkers. *Anal. Chem.* **2022**, *94*, 3005–3012. [CrossRef]
21. Zhang, R.; Liang, Y.; Su, Y.; Lai, W.; Zhang, C. Cloth-based closed bipolar electrochemiluminescence DNA sensors (CCBEDSs): A new class of electrochemiluminescence gene sensors. *J. Lumin.* **2021**, *238*, 118209. [CrossRef]
22. Li, X.; Du, Y.; Wang, H.; Ma, H.; Wu, D.; Ren, X.; Wei, Q.; Xu, J.-J. Self-Supply of H_2O_2 and O_2 by Hydrolyzing CaO_2 to Enhance the Electrochemiluminescence of Luminol Based on a Closed Bipolar Electrode. *Anal. Chem.* **2020**, *92*, 12693–12699. [CrossRef] [PubMed]
23. Yang, X.; Wang, Y.; Wang, L.; Zhu, J.; Zhao, J.; Zong, H.; Chen, C. Bipolar electrode ratiometric electrochemiluminescence biosensing analysis based on boron nitride quantum dots and biological release system. *Biosens. Bioelectron.* **2021**, *191*, 113393. [CrossRef] [PubMed]
24. Khoshfetrat, S.M.; Bagheri, H.; Mehrgardi, M.A. Visual electrochemiluminescence biosensing of aflatoxin M1 based on luminol-functionalized, silver nanoparticle-decorated graphene oxide. *Biosens. Bioelectron.* **2018**, *100*, 382–388. [CrossRef]

25. Wang, F.; Fu, C.; Huang, C.; Li, N.; Wang, Y.; Ge, S.; Yu, J. Paper-based closed Au-Bipolar electrode electrochemiluminescence sensing platform for the detection of miRNA-155. *Biosens. Bioelectron.* **2020**, *150*, 111917. [CrossRef] [PubMed]
26. Chen, B.; Tao, Q.; OuYang, S.; Wang, M.; Liu, Y.; Xiong, X.; Liu, S. Biocathodes reducing oxygen in BPE-ECL system for rapid screening of E. coli O157:H7. *Biosens. Bioelectron.* **2023**, *221*, 114940. [CrossRef]
27. Che, Z.-Y.; Wang, X.-Y.; Ma, X.; Ding, S.-N. Bipolar electrochemiluminescence sensors: From signal amplification strategies to sensing formats. *Coord. Chem. Rev.* **2021**, *446*, 214116. [CrossRef]
28. Jia, Y.-L.; Xu, C.-H.; Li, X.-Q.; Chen, H.-Y.; Xu, J.-J. Visual analysis of Alzheimer disease biomarker via low-potential driven bipolar electrode. *Anal. Chim. Acta* **2023**, *1251*, 340980. [CrossRef]
29. Ge, S.; Zhao, J.; Wang, S.; Lan, F.; Yan, M.; Yu, J. Ultrasensitive electrochemiluminescence assay of tumor cells and evaluation of H_2O_2 on a paper-based closed-bipolar electrode by in-situ hybridization chain reaction amplification. *Biosens. Bioelectron.* **2018**, *102*, 411–417. [CrossRef]
30. Zhang, S.; Rong, F.; Guo, C.; Duan, F.; He, L.; Wang, M.; Zhang, Z.; Kang, M.; Du, M. Metal–organic frameworks (MOFs) based electrochemical biosensors for early cancer diagnosis in vitro. *Coord. Chem. Rev.* **2021**, *439*, 213948. [CrossRef]
31. He, B.S.; Dong, X.Z. Nb. BbvCI powered DNA walking machine-based Zr-MOFs-labeled electrochemical aptasensor using Pt@AuNRs/Fe-MOFs/PEI-rGO as electrode modification material for patulin detection. *Chem. Eng. J.* **2021**, *405*, 126642. [CrossRef]
32. He, B.; Dong, X. Hierarchically porous Zr-MOFs labelled methylene blue as signal tags for electrochemical patulin aptasensor based on ZnO nano flower. *Sens. Actuators B Chem.* **2019**, *294*, 192–198. [CrossRef]
33. Chen, J.; Zhang, J.; Qiao, J.; Wu, M. Closed bipolar electrochemical biosensor based on ohmic loss mechanism for noncontact measurements. *J. Electroanal. Chem.* **2020**, *860*, 113873. [CrossRef]
34. Yu, H.H.; Yang, H.J.; Liu, W.S.; Jin, L.S.; Jin, B.; Wu, M.S. Novel electrochemiluminescence biosensor of fumonisin B1 detection using MWCNTs-PDMS flexible bipolar electrode. *Talanta* **2023**, *257*, 124379. [CrossRef]
35. Li, Y.C.; Zheng, C.R.; Liu, S.; Huang, L.; Fang, T.S.; Li, J.X.Z.; Xu, F.; Li, F. Smart Glove Integrated with Tunable MWNTs/PDMS Fibers Made of a One-Step Extrusion Method for Finger Dexterity, Gesture, and Temperature Recognition. *ACS Appl. Mater. Interfaces* **2020**, *12*, 23764–23773. [CrossRef]
36. Ren, C.C.; Li, H.M.; Lu, X.T.; Qian, J.; Zhu, M.Y.; Chen, W.; Liu, Q.; Hao, N.; Li, H.N.; Wang, K. A disposable aptasensing device for label-free detection of fumonisin B1 by integrating PDMS film-based micro-cell and screen-printed carbon electrode. *Sens. Actuators B Chem.* **2017**, *251*, 192–199. [CrossRef]
37. Al-Hazmi, G.A.A.; El-Zahhar, A.A.; El-Desouky, M.G.; El-Bindary, M.A.; El-Bindary, A.A. Adsorption of industrial dye onto a zirconium metal-organic framework: Synthesis, characterization, kinetics, thermodynamics, and DFT calculations. *J. Coord. Chem.* **2022**, *75*, 1203–1229. [CrossRef]
38. Mollarasouli, F.; Asadpour-Zeynali, K.; Campuzano, S.; Yáñez-Sedeño, P.; Pingarrón, J.M. Non-enzymatic hydrogen peroxide sensor based on graphene quantum dots-chitosan/methylene blue hybrid nanostructures. *Electrochimica Acta* **2017**, *246*, 303–314. [CrossRef]

chemosensors

Article

Development and Application of an Electrochemical Sensor with 1,10-Phenanthroline-5,6-dione-Modified Electrode for the Detection of *Escherichia coli* in Water

Yining Fan [1,2,†], Yanran Liu [1,2,†], Guanyue Gao [1,2,*], Hanxin Zhang [1,2] and Jinfang Zhi [1,2]

1 Key Laboratory of Photochemical Conversion and Optoelectronic Materials, Technical Institute of Physics and Chemistry, Chinese Academy of Sciences, Beijing 100190, China; zhi-mail@mail.ipc.ac.cn (J.Z.)
2 University of Chinese Academy of Sciences, Beijing 100049, China
* Correspondence: gaoguanyue@mail.ipc.ac.cn
† These authors contributed equally to this work.

Abstract: The routine monitoring of bacterial populations is crucial for ensuring water quality and safeguarding public health. Thus, an electrochemical sensor based on a 1,10-phenanthroline-5,6-dione-modified electrode was developed and explored for the detection of *E. coli*. The modified electrode exhibited enhanced NADH oxidation ability at a low potential of 0.1 V, which effectively eliminated the interference from other redox compounds in bacteria. The sensitivity for NADH was 0.222 μA/μM, and the limit of detection was 0.0357 μM. Upon cell lysis, the intracellular NADH was released, and the concentration of *E. coli* was determined through establishing the relationship between the oxidation current signal and NADH concentration. The performance of the electrochemical sensor in the detection of NADH and *E. coli* suspensions was validated using the WST-8 colorimetric method. The blank recovery experiment in real water samples exhibited good accuracy, with recovery rates ranging from 89.12% to 93.26% and relative standard deviations of less than 10%. The proposed electrochemical sensor realized the detection of *E. coli* without the usage of biomarkers, which provides a promising approach for the broad-spectrum detection of microbial contents in complex water environments.

Keywords: electrochemical sensor; 1,10-phenanthroline-5,6-dione; NADH; *Escherichia coli*; rapid detection

Citation: Fan, Y.; Liu, Y.; Gao, G.; Zhang, H.; Zhi, J. Development and Application of an Electrochemical Sensor with 1,10-Phenanthroline-5,6-dione-Modified Electrode for the Detection of *Escherichia coli* in Water. *Chemosensors* **2023**, *11*, 458. https://doi.org/10.3390/chemosensors11080458

Academic Editor: Marcello Mascini

Received: 29 June 2023
Revised: 4 August 2023
Accepted: 10 August 2023
Published: 15 August 2023

1. Introduction

Biodiverse and complex water ecosystems provide favorable environments for a wide range of microorganisms, resulting in significant microbial populations within aquatic environments. However, the spreading of pathogenic microorganisms through water can lead to severe epidemics and infectious diseases [1]. Consequently, strict regulations have been established to restrict the microbial contents in various water sources, such as drinking water, surface water, and groundwater, in order to safeguard public health [2]. Recent advancements in water quality assessment methodologies have contributed to the progressive refinement of the standards concerning microbial indicators. The key microbiological indicators utilized in evaluating water quality predominantly include the total number of colonies, coliform groups, fecal coliform groups, and other pathogenic bacteria. The water standards have adopted analytical methods like flat colony counting, multitube fermentation, filter membrane, and turbidimetric counting to assessing microbial contents through enumerating colonies or measuring optical density [3]. However, these traditional culture methods require specific culture media; precise control over experimental conditions such as temperature, pH, and oxygen levels; as well as specialized instruments and skilled personnel. Additionally, their lengthy culture times and poor anti-interference ability have further limited their application in real detection scenarios. Furthermore, with the development of immunoassay and bioassay technologies, bioanalytical techniques, represented by enzyme-linked immunosorbent assay (ELISA) and polymerase chain reaction (PCR), have offered sensitive approaches for bacterial detection with high accuracy and specificity [4]. Nevertheless, their application is constrained by their high cost and inability to meet the rapid detection requirements for multiple targets in water.

In addition to conventional standard methods, optical techniques, including fluorescence analysis [5], surface plasmon resonance [6,7], surface-enhanced Raman scattering [8,9], and electrochemical sensing, can selectively identify and sensitively detect microorganisms through the assistance of endogenous biomarkers such as antibodies/antigens, aptamers, DNA, etc. Among them, electrochemical biosensors have emerged as a promising candidate due to their fast response, ease of miniaturization, and cost-effectiveness [10–12]. Via incorporation with biomarkers, the electrochemical biosensors have achieved the sensitive detection of DNA [13], toxins [14], cancer cells [15], and bacteria [16]. Ju et al. developed a sensitive and specific electrochemical biosensor based on target-induced aptamer displacement and achieved the direct detection of *E. coli* O111 with a limit of detection (LOD) of 112 CFU/mL [16]. Güner et al. developed an electrochemical immunosensor by modifying a poly-pyrrole/Au/MWCNT/chitosan modified electrode with *E. coli* monoclonal antibodies [17]. Zou et al. employed a "sandwich" mode to capture *E. coli* with antibodies and bind them to mediators. The resulting differential pulse voltammetry signal was proportional to the number of *E. coli*, enabling an LOD of 10 CFU/mL [18]. In recent years, innovative immunosignal amplification strategies, including 3D walkers, clustered regularly interspaced short palindromic repeats (CRISPR) technology [19,20], and ring-mediated isothermal amplification, have been integrated into the development of electrochemical sensors. Zhang et al. employed a 3D DNA walking molecule amplification strategy in the detection of *E. coli*, achieving an LOD of 20 CFU/mL [21].

Electrochemical sensors can also achieve indirect bacterial detection through the detection of intracellular endogenous substances. Kwon et al. detected the reduction current of 4-methoxyphenyl-β-d-galactopyranoside to 4-methoxyphenol under the catalysis of the β-galactosidase present in *E. coli* and realized the detection of *E. coli* with an LOD of 2×10^3 CFU/mL [22]. Currently, most electrochemical biosensors rely on bio-immune and DNA analysis, which possess high specificity [23–25], but this characteristic also leads to constraints in the detection of multiple microorganisms. The use of nonbiological identification methods to achieve broad-spectrum detection of microorganisms is still relatively rare.

In present study, an electrochemical sensor was fabricated using 1,10-phenanthroline-5,6-dione as a redox mediator, and its potential for quantitatively detecting NADH and *E. coli* was investigated. The performance and anti-interference ability of the as-prepared electrochemical sensor were evaluated. To determine the accuracy of the electrochemical sensor in quantifying *E. coli* concentration and intracellular NADH content, a WST-8 colorimetric method was employed. Additionally, three real river water samples were collected, and a blank recovery experiment was conducted to assess the capability of the electrochemical sensor in detecting *E. coli* concentrations. The primary objective of this work was to determine the *E. coli* population using NADH as a signal indicator, thereby creating a feasible and broad-spectrum method for the detection of microorganisms in complex water samples.

2. Materials and Methods

2.1. Reagents

Escherichia coli (ATCC 25922) was provided by China General Microbiological Culture Collection Center. The 1,10-phenanthroline-5,6-dione (PD), nicotinamide adenine dinucleotide (NADH), thionine (TH), reduced flavin adenine dinucleotide ($FADH_2$), agar, tri (hydroxymethyl) amino methane hydrochloride (Tris-HCl), and ethylene diamine tetraacetic acid (EDTA) were all purchased from Beijing Lanyi Chemical Products Co., Ltd., Beijing, China. Multiwalled carbon nanotubes (MWCNTs) were provided by XFNANO Co., Ltd., Nanjing, China. Triton X-100 and NAD^+/NADH assay kits were purchased from Beyotime

Biotechnology Co., Ltd., Shanghai, China. We purchased 0.1 M phosphate buffer solution (PBS) from Sinopharm Chemical Reagent Co., Ltd., Shanghai, China, and 5% Nafion™ 117 solution was purchased from Merck, Darmstadt, Germany. Beef extract and peptone were purchased from Aoboxing biotech company, Beijing, China. All chemicals were used without further purification. Milli-Q water (R > 18.0 MΩ•cm) was used to prepare all solutions.

Three real water samples were collected from Puhe River in Shanxi Province, Tangwang River in Heilongjiang Province, and Xiaoyun River in Beijing. All the real water samples were filtered through a 13 mm×0.22 μm filtration membrane to remove the sediments before electrochemical detection.

2.2. Apparatus

Cyclic voltammetry (CV) and chronoamperometric measurements were carried out with an electrochemical workstation (CHI 660E). Field-emission scanning electron microscopy (FESEM, Quanta FEG 250, Oregon, OR, USA) with an operating voltage of 10 kV was used to characterize the morphology of the modified electrode. The concentration of *E. coli* in the suspension was measured with a UV spectrophotometer (SECOMAM UVIKONXL, Domont, France) at 600 nm, recorded as OD_{600}. OD_{600} refers to the light absorption value of a bacterial solution at a wavelength of 600 nm, which is used to characterize bacterial concentrations. The higher the bacterial concentration, the higher the absorbance [26,27].

2.3. Preparation of Electrochemical Sensor

We dissolved 100 mg of MWCNTs in 10 mL of deionized water, which we then dispersed ultrasonically for 30 min. Then, 100 μL of 5% Nafion solution was added and mixed ultrasonically for 5 min to prepare the MWCNTs–Nafion solution. We dropped 15 μL of as-prepared MWCNTs–Nafion solution on the surface of a glassy carbon electrode (GCE), which was dried at room temperature to form MWCNTs-Nafion@GCE. Then, 5 mg of PD was dissolved in 5 mL ethanol solution and sonicated for 10 min to prepare PD solution; 10 μL of PD solution was dropped on the MWCNTs-Nafion@GCE and dried for 2 h. The modified electrode was rinsed with deionized water and dried at room temperature to obtain PD-MWCNTs-Nafion@GCE, which was stored in a refrigerator at 4 °C until use.

Similarly, the TH-modified electrode was prepared according to same procedure. A total of 10 μL of 0.3 mM TH aqueous solution was dropped on MWCNTs-Nafion@GCE and dried for 2 h. Then, the modified electrode was rinsed with deionized water to remove excess TH and dried at room temperature to obtain TH-MWCNTs-Nafion@GCE.

2.4. Culture of E. coli

A small aliquot of *E. coli* was placed in 100 mL of liquid culture medium that contained 5 g/L beef extract, 10 g/L peptone, and 5 g/L sodium chloride. Then, the inoculated culture medium was cultured aerobically in a rotary shaker in a 37 °C water bath for 16 h. The cultured bacterial suspension was centrifuged at 6000 rpm for 5 min and washed twice with PBS solution to fully remove the culture medium. The concentration of *E. coli* suspension was adjusted using a UV spectrophotometer and diluted to different concentrations ranging from OD_{600} = 0.2 to OD_{600} = 2.6.

The number of *E. coli* in bacterial suspension with varied OD_{600} values was further determined using the flat colony counting method. Solid agar culture plates were prepared as follows: The agar liquid medium (beef extract 5 g/L, peptone 10 g/L, sodium chloride 5 g/L, and agar 15 g/L) was first fully mixed and sterilized at 121 °C. The liquid agar medium was then cooled to 46 °C and poured into the plates and solidified at room temperature. *E. coli* suspensions with different OD_{600} values were fully diluted, then 1 mL of dilution was dropped into the sterile agar plate and cultured at 37 °C for 24 h. Three parallel groups were set for each OD_{600} value. The number of *E. coli* colonies was observed and counted with a colony counter, and the bacterial concentration in each OD_{600} value is presented as colony-forming units per milliliter (CFU/mL).

2.5. Detection of NADH Using As-Prepared Electrochemical Sensor

A three-electrode electrochemical cell was used to determine the content of NADH. The as-prepared PD-MWCNTs-Nafion@GCE was used as the working electrode, a Pt wire was used as the counter electrode, and Ag/AgCl (sat' KCl) was used as the reference electrode. During the detection process, the modified electrode was initially activated in 10 mL of 0.1 M PBS solution (pH = 7.0) under 900 rpm stirring. The applied potential was 0.1 V vs. Ag/AgCl (sat' KCl). Standard NADH solutions with varied concentrations were added into the electrochemical cell once the current was steady, and the current values were measured 30 s after each addition. The calibration curve was obtained by plotting the current changes (ΔI) against the corresponding NADH concentrations.

2.6. Detection of E. coli *Using As-Prepared Electrochemical Sensor*

The *E. coli* concentration was determined by quantitively analyzing their intercellular NADH content. Therefore, cells lysis was conducted to release the intracellular NADH in order to detect *E. coli* centration. We mixed 5 mL of *E. coli* suspension with different OD_{600} values with 0.1 M Tris-HCl, 0.02 M EDTA, and 0.05% Triton X-100, and sonicated in an ice bath for 15 min to prepare *E. coli* lysate.

The concentration of *E. coli* was determined via chronoamperometry. The current response of the as-prepared PD-MWCNTs-Nafion@GCE in 10 mL of *E. coli* lysate was recorded at the applied potential of 0.1 V. The steady-state current was recorded after 100 s, and the current–concentration relationship was established by detecting *E. coli* suspension with various OD_{600} values.

2.7. Detection of E. coli *Using Colorimetric Method*

The intracellular concentration of NADH was also verified using the WST-8 colorimetric method with the following procedure: We prepared 20 μL of 0, 0.25, 0.5, 1, 2, 4, 6, 8, and 10 μM standard NADH solutions, which we added into a 96-well plate. Then, 20 μL of *E. coli* lysate with varying OD_{600} values was also added into the 96-well plate for detection. Next, 90 μL of PBS and 10 μL of 1-methoxy-5-methylphenazinium methyl sulfate chromogenic solution were added into each sample, which were then incubated in the dark for 45 min. The absorbance intensity was measured at a wavelength of 450 nm with a microplate reader (TECAN Infinite50, Männedorf, Switzerland). According to the established absorbance intensity–NADH concentration calibration curve, the NADH concentrations in *E. coli* lysate were determined.

3. Results and Discussion

3.1. The Detection Principles of As-Prepared Electrochemical Sensor

In this work, a novel electrochemical approach was developed for the sensitive detection of *E. coli* based on the quantification of intracellular NADH content. NADH is a crucial product of microbial carbohydrate metabolism and plays a vital role in the microbial respiratory electron transfer processes. Studies have demonstrated that the number of NADH molecules within each bacterial cell remains relatively constant, typically ranging from 10^6 to 10^8, making NADH a stable and universally present endogenous substance in bacterial systems [28]. However, it is important to identify the specific potential at which NADH can undergo oxidation to coenzyme I in the electrochemical detection of NADH. This oxidation process typically occurs at high oxidation potentials, resulting in the concurrent oxidation of other intracellular electroactive substances and subsequently the generation of interference signals. Nevertheless, previous research has indicated that PD is an ideal mediator capable of oxidizing NADH at a lower potential, thus enabling selective and accurate detection [29].

In the design of electrochemical sensors, MWCNTs are widely employed as electrode-modification materials, which exhibit remarkable electrochemical stability and electrical conductivity [30]. Nafion, as a type of perfluorosulfonic acid polymer, exhibits good thermal stability and high proton conductivity. It has been revealed that the interaction between the hydrophobic sidewalls of MWCNTs and the main chain of Nafion fluorocarbons facilitates the uniform dispersion of MWCNTs within Nafion, effectively enhancing Nafion's ability to immobilize MWCNTs on the electrode [31]. Consequently, the integration of these materials can enhance electrode conductivity while simultaneously improving the immobilization of the PD mediator on the electrode surface. Therefore, we developed a PD-mediated electrochemical sensor to detect NADH at low potential, subsequently realizing the accurate quantification of *E. coli* concentrations. The electrode preparation and *E. coli* detection process are depicted in Scheme 1. The GCE was first modified with MWCNTs–Nafion solution to form a polymer film. Then, PD mediator was drop casted on the MWCNTs-Nafion@GCE, and the PD-modified electrode was prepared. In the *E. coli* detection process, the bacterial suspension was incubated with cell lysis buffer to fully release the intracellular NADH. The NADH content was determined using the as-prepared PD-modified electrode at a suitable potential, and the current response was recorded.

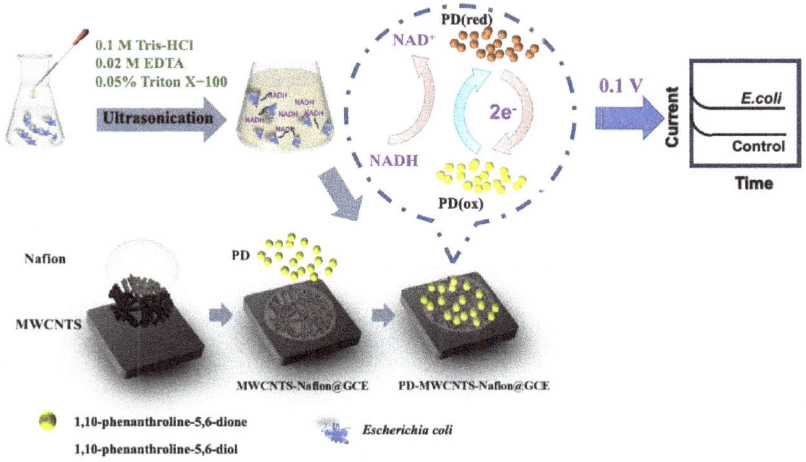

Scheme 1. Schematic representation of the preparation and detection process of electrochemical sensor.

3.2. Surface Characterization of PD-Modified Electrode

The morphologies of the PD particles, MWCNTs, and modified electrodes were characterized using a field-emission scanning electron microscope, the results are shown in Figure 1. Needle-like clusters of PD particles with different shapes and sizes were observed, indicating the agglomeration of PD particles on the GCE surface (Figure 1a). The morphology of the MWCNTs is shown in Figure 1b, which had an average tube length of 8 μm. The as-prepared PD-MWCNTs-Nafion@GCE was uniformly covered by a smooth and porous film, signifying the successful deposition of the Nafion film onto the GCE (Figure 1c). The presence of large particles on the electrode surface corresponded to agglomerated PD, while the tubular structures indicated that the MWCNTs were partially embedded within the Nafion film. The above results demonstrated the effective immobilization of both PD and MWCNTs on the electrode surface through Nafion.

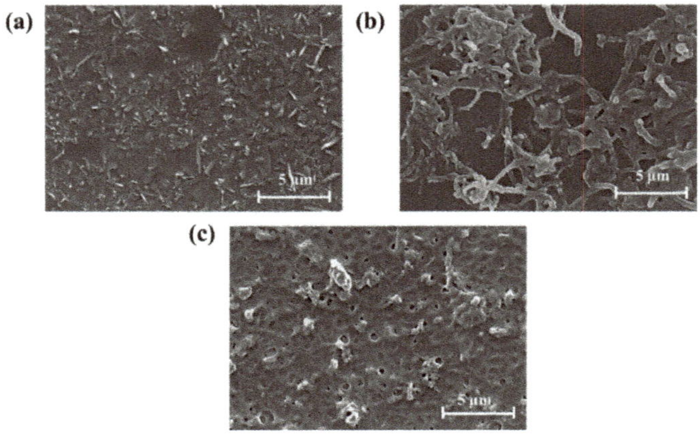

Figure 1. FESEM images of different modified electrodes: (**a**) PD@GCE; (**b**) MWCNTs@GCE; (**c**) PD-MWCNTs-Nafion@GCE.

3.3. Electrochemical Characterization of PD-Modified Electrode

The electrochemical properties of the modified electrodes were firstly examined via cyclic voltammetry, as shown in Figure 2a. In PBS solution, PD-MWCNTs-Nafion@GCE exhibited oxidation and reduction peaks at 0.05 V and −0.13 V (Figure 2a, red dashed line), representing the oxidation and reduction of PD mediator. Notably, the PD-modified electrode displayed good reversibility with a ΔEp value of 0.18 V. In the absence of the PD mediator, no redox peak was observed for MWCNTs-Nafion@GCE, owing to the inherent stability of MWCNTs and Nafion polymer (Figure 2a, blue dashed line).

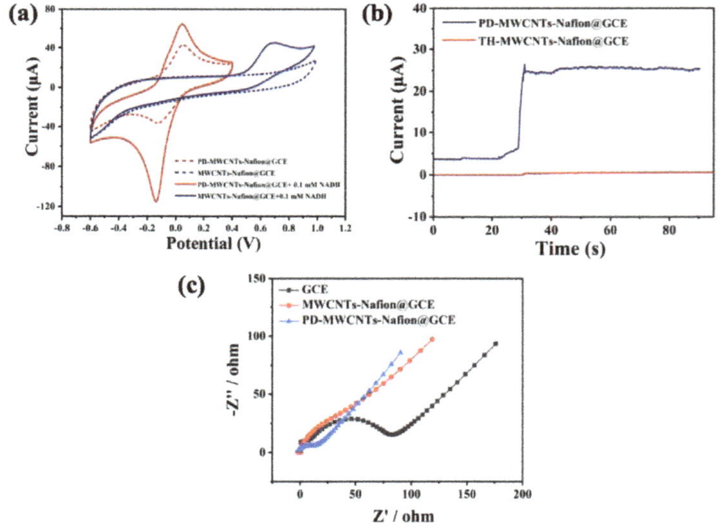

Figure 2. (**a**) Cyclic voltammetry curves of PD-MWCNTs-Nafion@GCE and MWCNTs-Nafion@GCE before and after the addition of 0.1 mM NADH, scan rate 0.1 V/s; (**b**) i–t curves of TH-MWCNTs-Nafion@GCE and PD-MWCNTs-Nafion@GCE in 0.1 mM NADH at a potential of 0.1 V (vs. Ag/AgCl); (**c**) Nyquist plots of bare GCE and modified electrodes.

Moreover, the electrochemical reaction between the modified electrode and NADH was also investigated. In 0.1 mM NADH solution, the MWCNTs-Nafion@GCE electrode displayed a distinct oxidation peak at approximately 0.6 V (Figure 2a, blue line), indicating that NADH undergoes oxidation and decomposition at a high potential on MWCNTs-Nafion@GCE. It is worth noting that many substances, such as uric acid and ascorbic acid, can also undergo direct oxidation at a potential of 0.6 V. Hence, NADH detection at this potential suffers from poor anti-interference ability. In contrast, the PD-MWCNTs-Nafion@GCE exhibited a much more sensitive response to NADH (Figure 2a, red line). The oxidation of NADH was achieved at a much lower potential of 0.05 V, and the oxidation peak current of the modified electrode increased by 20 μA after 0.1 mM NADH addition. This decrease in oxidation potential offers several advantages: On the one hand, the decrease in oxidation potential can effectively avoid the electrode fouling caused by high potential oxidation and the signal interference caused by the oxidation reactions of other substances in water samples. On the other hand, the enhanced redox signal response can further improve the sensitivity in NADH and *E. coli* detection. This improved performance of the PD-MWCNTs-Nafion@GCE electrode can be attributed to the reactions between the PD mediator and NADH. As shown in Equations (1) and (2), PD catalyzes the oxidative decomposition of NADH and is reduced to 1,10-phenanthroline-5,6-diol; then, 1,10-phenanthroline-5,6-diol is oxidized back to PD under the applied oxidation potential and generates an oxidation current.

$$\text{PD(ox)} + 2\text{NADH} \rightarrow \text{PD(red)} + 2\text{NAD}^+ \quad (1)$$

$$\text{PD(red)} \rightarrow \text{PD(ox)} + 2\text{H}^+ + 2\text{e}^- \quad (2)$$

In order to verify the superiority of PD as a mediator in the detection of NADH, we also considered the electrochemical performance of a thionine-modified electrode. TH is a widely used mediator in the construction of microbial biosensors and microbial fuel cells, which possesses good electron transfer ability in bacterial metabolic processes [32]. The chronoamperometric responses of modified electrodes utilizing different mediators were assessed in the detection of NADH. As depicted in Figure 2b, the TH-modified electrode only exhibited a feeble oxidation current ($\Delta I = 0.5$ μA) upon the addition of 0.1 mM NADH (Figure 2b, red line). In comparison, the PD-modified electrode demonstrated a distinct current response ($\Delta I = 21.6$ μA, Figure 2b, blue line) because TH reacts with NADH at a negative potential, which leads to the low oxidation efficiency at 0.1 V. Therefore, PD is a more favorable mediator in the detection of NADH at low potentials.

The changes in the modified electrode's interfacial properties were characterized using electrochemical impedance spectroscopy (EIS). The impedance spectra of different electrodes were recorded in 0.1 M KCl solution containing 5.0 mM $[Fe(CN)_6]^{3-}/[Fe(CN)_6]^{4-}$ at a frequency ranging from 10^5 to 10^{-1} Hz. As depicted in Figure 2c, the Nyquist plot of bare GCE appears as an oblique line at low frequencies and a semicircle at high frequencies, with an R_{ct} of 90 Ω. After the modification of MWCNTs−Nafion, the R_{ct} value remained at the same level. It could be attributed to the combined effect of the high conductivity of carbon nanotubes and the electrostatic repulsion of Nafion membranes against $[Fe(CN)_6]^{3-}/[Fe(CN)_6]^{4-}$ mediators. Notably, the R_{ct} value decreased to 15 Ω after the adsorption of PD, which demonstrated the effect of the PD mediator in facilitating the electron transfer process of PD-MWCNTs-Nafion@GCE.

3.4. The Performance of As-Prepared Electrochemical Sensor in NADH Detection

To validate the feasibility of as-prepared electrochemical sensor in the detection of NADH, chronoamperometric measurements were performed by consecutively adding NADH at varying concentration gradients. As shown in Figure 3a, when 10 μM NADH was added, the diffusion equilibrium was reached within 10 s under stirring, and the steady-state current was obtained. The current change−NADH concentration relationships between 10–100 μM and 0.25–8 μM were established according to the amperiometric results. As shown in Figure 3b, the as-prepared electrochemical sensor exhibited excellent linear relationships in both the higher and lower concentration ranges. The correlation coefficients were above 0.99, indicating high accuracy in NADH detection. The sensitivity for NADH was 0.222 μA/μM, and the LOD value was 0.0357 μM. The LOD was calculated using 3 S_b/m, where m is the slope of calibration curve at low concentration, and S_b is standard deviation from three blank experiments. Compared with previously reported sensors (Table 1), the present electrochemical sensor has a relatively low detection limit. In addition, our method could be operated at a relatively low voltage, which can effectively prevent interference from other electroactive substances.

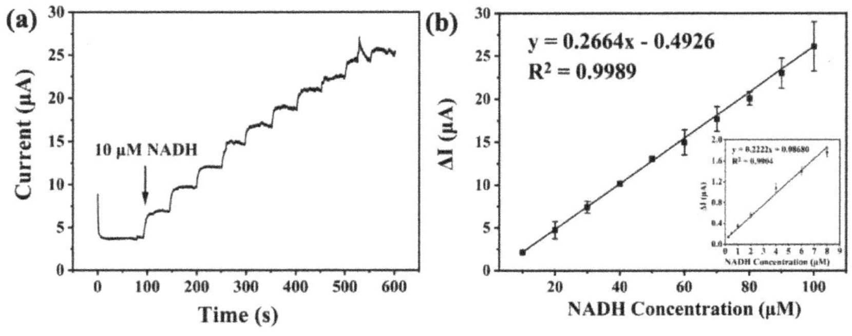

Figure 3. (**a**) i–t curves of PD-MWCNTs-Nafion@GCE in the detection of NADH, E = 0.1 V (vs. Ag/AgCl); (**b**) calibration curves for NADH.

Table 1. Comparison of electrochemical sensors in the detection of NADH.

Sensor	Detection Potential (V)	LOD (μM)	Linear Range (μM)	Sensitivity (μA/μM)	Ref.
CS–DA-MWCNTs-COOH/Au	+0.25 vs. SCE	0.012	0.1–600	0.009	[33]
PAA-MWNTs/GCE	+0.13 vs. Ag/AgCl	1	4–100	0.094	[34]
ERGO–PTH/GC	+0.4 vs. SCE	0.1	10–3900	0.143	[35]
CHIT/MWNTs/GC	+0.4 vs. SCE	0.3	0.8–1600	/	[36]
POA/GR@GCE	+0.045 vs. SCE	1.3	0.166–1.772	47.1	[37]
PD-MWCNTs-Nafion@GCE	+0.1 vs. Ag/AgCl	0.0357	0.25–8, 10–100	0.222	This work

The bacterial respiratory metabolism is a complex process involving cytochrome c, cytochrome a, $FADH_2$, and various enzymes involved in respiratory electron transfer, such as NADH dehydrogenase, coenzyme Q, succinate dehydrogenase, etc. Thus, the anti-interference experiment was conducted to evaluate the potential impact of active intracellular redox substances on the oxidation current signal of NADH. Among these substances, coenzyme Q exhibited an anodic peak at approximately −0.1 V and cytochrome c at −0.26 V for MWCNTs-modified GCE [38,39], which do not interfere with the NADH oxidation signal. In contrast, $FADH_2$, which participates in the ATP conversion process together with NADH, is an electron donor similar to NADH, which may potentially interfere with the oxidation of NADH. Therefore, an interference experiment was conducted to examine whether $FADH_2$ generates an oxidation current under the same experimental conditions. A potential of 0.1 V was applied on the modified electrode, and chronoamperometry was performed in 0.1 M PBS solution. The result is shown in Figure 4a: no significant current increase was observed after the addition of 0.1 mM $FADH_2$, indicating that $FADH_2$ does not interfere with the redox reaction of PD and NADH at a potential of 0.1 V. This result was also verified via the cyclic voltammetry of PD-MWCNTs-Nafion@GCE in the presence of $FADH_2$. The CV curve displayed distinct redox peaks for $FADH_2$ at −0.3 V and −0.5 V, respectively, which suggested the low redox potential of $FADH_2$ on the modified electrode (Figure 4b). The PD mediator was not involved in the oxidation of $FADH_2$ at a potential of 0.1 V. Therefore, the electrochemical sensor showed nonresponse to $FADH_2$ during the detection process. In summary, the low oxidation potential of PD and NADH effectively eliminates the interference of active intracellular redox substances during the detection process, ensuring the specificity and accuracy of the as-prepared electrochemical sensor in the detection of NADH.

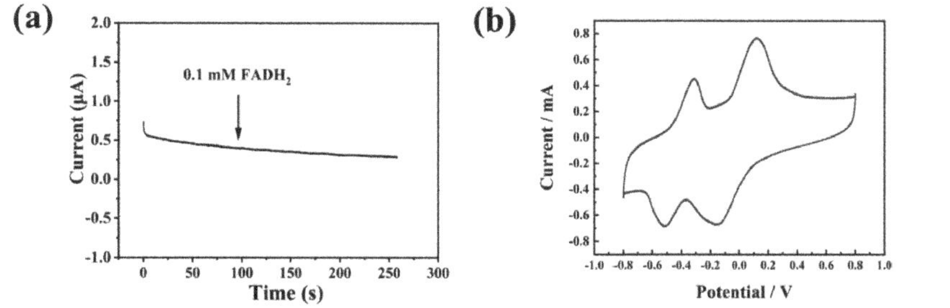

Figure 4. (**a**) i–t curve of PD-MWCNTs-Nafion@GCE upon the addition of 0.1 mM $FADH_2$ (in 0.1 M PBS), E = 0.1 V (vs. Ag/AgCl). (**b**) Cyclic voltammetry of PD-MWCNTs-Nafion@GCE in 0.1 M PBS solution containing $FADH_2$ at the scan rate of 0.1 V/s.

3.5. Detection of E. coli by As-Prepared Electrochemical Sensor

In water quality assessments, the amount of *E. coli* and the total number of bacterial colonies are crucial microbial parameters. Therefore, the applicability of the as-prepared electrochemical sensor in detecting *E. coli* was investigated. Chronoamperometry was employed to detect *E. coli* lysates with varying concentrations. The chronoamperometric response of the electrochemical sensor to *E. coli* lysates with OD_{600} values of 0, 0.4, 0.77, 1.2, 1.64, and 2.0 is shown in Figure 5a. The steady-state currents increased with rising concentrations of *E. coli* suspension. The steady-state current of the *E. coli* suspension with OD_{600} values of 0.4 and 0.77 were 0.32 μA and 0.37 μA, respectively, while the steady-state current of OD_{600} = 2.0 reached 1.53 μA. It is worth noting that the relationship between the current response and OD_{600} values was found to be nonlinear. In order to explore the underlying reason for this phenomenon, the numbers of *E. coli* colonies under different OD_{600} values from 0.2 to 2.6 were detected using the plate colony counting method. The relationship of the current response and *E. coli* colony numbers is illustrated in Figure 5b; the current–*E. coli* colony numbers conformed to a linear relationship at low bacterial concentrations, which indicated that the NADH concentrations in bacterial lysates increased linearly with *E. coli* colony numbers. However, the current response slowly plateaued as *E. coli* colony numbers further increased. This could be attributed to the insufficient cell lysis and nonexhaustive release of intracellular NADH as the bacterial content further increased. The LOD of the as-prepared electrochemical sensor for *E. coli* was determined using $3S_b/m$, where m is the slope at low concentrations, and S_b is the standard deviation for three blank measurements. The LOD value for *E. coli* was 1.12×10^{10} CFU/mL. Therefore, the as-prepared sensor can accurately detect the NADH content in *E. coli* solutions, and the detection signal is positively correlated with the *E. coli* concentration, which further proves the suitability of the electrochemical sensor for the detection of *E. coli*.

To further validate the accuracy of the as-prepared electrochemical sensor in the quantification of NADH content in *E. coli* solutions, three *E. coli* lysate samples with OD_{600} values of 0.2, 1.0, and 1.5 were selected for analysis. The NADH concentrations in these bacterial lysates were determined with both the electrochemical sensor and the WST-8 colorimetric method. To determine the NADH concentration in unknown samples via the WST-8 colorimetric method, the calibration curve between absorbance intensity and NADH concentration was established. As shown in Figure 6a, the absorbance intensity and NADH concentration conformed to a good linear relationship. Subsequently, the NADH contents of *E. coli* lysates with OD_{600} values of 0.2, 1.0, and 1.5 were detected following the same procedure. To eliminate solution turbidity from interfering with the absorbance intensity, the supernatants of the *E. coli* lysates were obtained through centrifugation. By comparing the detected absorbance intensity values with the calibration curve, the NADH concentrations in *E. coli* lysates were determined to be 0.965 μM, 2.194 μM, and 5.429 μM for OD_{600} values of 0.2, 1.0, and 1.5, respectively.

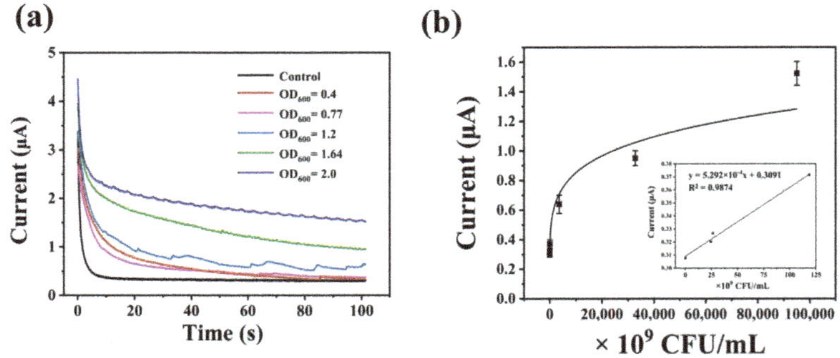

Figure 5. (**a**) i–t curves of as-prepared electrochemical sensor in the detection of *E. coli* in solution with different OD_{600} values, E = 0.1 V (vs. Ag/AgCl); (**b**) relationship of current response with *E. coli* colony numbers.

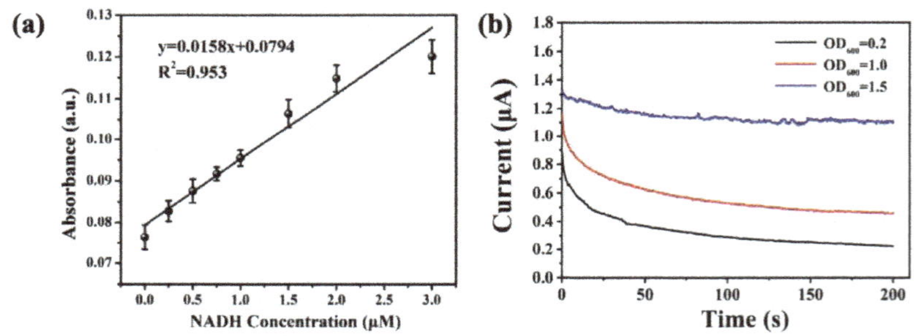

Figure 6. (**a**) Linear curve of NADH concentration and absorbance intensity obtained with WST-8 colorimetric method; (**b**) i–t curves of as-prepared electrochemical sensor in the detection of *E. coli* solution with different OD_{600} values, E = 0.1 V (vs. Ag/AgCl).

The NADH contents of the aforementioned *E. coli* lysate samples were also assessed using the electrochemical sensor. As depicted in Figure 6b, the ΔI observed for the three *E. coli* lysates with OD_{600} values of 0.2, 1.0, and 1.5 were 0.221 μA, 0.452 μA, and 1.110 μA, respectively. By comparing these current responses to the established current–NADH concentration relationship (Figure 3b), the NADH concentrations were determined to be 1.008 μM, 2.120 μM, and 5.294 μM, respectively. The relative standard deviations (RSDs) between these two methods were 4.46%, 3.37%, and 2.48%. These results demonstrated that the proposed electrochemical sensor has excellent accuracy in NADH detection hen compared with the WST-8 colorimetric method.

3.6. *E. coli* Detection in Real Water Samples

The applicability of the electrochemical sensor in detecting *E. coli* concentrations in real water was evaluated through a blank recovery experiment. Three real water samples were collected from cities, villages, and rural regions across different provinces. To eliminate the interference by intrinsic microorganisms, we employed the WST-8 technique to determine the NADH concentration in the real water samples. The result indicated that the NADH concentrations were lower than the detection limit. Hence, the contents of native microbes in the real water samples were extremely low and could be disregarded. Therefore, a blank recovery experiment was performed to investigate the feasibility of the developed electrochemical sensor for *E. coli* detection with real water samples. The same amount of *E. coli* suspension with OD_{600} = 0.2 was added into these real water samples, and the current responses and colorimetric changes were recorded. The NADH concentration in the *E. coli* suspension was determined to be 0.0965 μM using the WST-8 colorimetric method. The detected NADH concentrations and recovery rates of the three water samples determined with electrochemical sensor are listed in Table 2. The recovery rates of the NADH concentrations detected with the electrochemical sensors ranged from 89.12% to 93.26%. The RSD values between two method were less than 10%. The standard deviation for the as-prepared electrochemical sensor was 0.002 μM. These results confirm the good accuracy of the electrochemical sensor in quantifying NADH contents, thereby demonstrating the capability of the electrochemical sensor based on a PD mediator to detect the number of *E. coli* colonies in real water.

Table 2. Results of NADH blank recovery tests in three water samples.

Water Sample	NADH Concentration Detected with Electrochemical Sensor (μM)	Recovery Rate (%)	RSD (%)
Pu River	0.086	89.12	8.26
Tangwang River	0.090	93.26	2.08
Xueqing River	0.088	91.19	4.45

4. Conclusions

In this study, we successfully developed an electrochemical sensor utilizing 1,10-phenanthroline-5,6-dione as a mediator, which demonstrated good capabilities in both NADH detection and *E. coli* determination. The experimental results demonstrated that the utilization of the PD mediator significantly reduced the applied potential, leading to improved anti-interference ability. There was good linear relationship between the NADH concentration and current response of the as-prepared electrochemical sensor. Furthermore, the applicability of as-prepared electrochemical sensor in bacterial detection was verified using *E. coli* as a model organism. The results of comparative analysis with the WST-8 colorimetric method revealed that the electrochemical sensor exhibited good accuracy in detecting both *E. coli* suspensions and real water samples. Even though our study only investigated the performance in the detection of *E. coli*, as NADH widely exists in various bacterial species, the as-prepared sensor may also be feasible for the detection of total colony numbers in water samples, offering a promising approach for the broad-spectrum detection of microbial contents in complex water samples without the usage of biomarkers.

Author Contributions: Conceptualization, Y.L., G.G. and J.Z.; methodology, Y.F. and Y.L.; validation, Y.F. and Y.L.; formal analysis, Y.F. and Y.L.; investigation, Y.L. and Y.F.; data curation, Y.F. and G.G.; writing—original draft preparation, Y.L., G.G., H.Z. and J.Z.; writing—review and editing, G.G. and J.Z.; visualization, Y.F.; supervision, G.G. and J.Z.; funding acquisition, G.G. and J.Z. All authors have read and agreed to the published version of the manuscript.

Funding: This research was funded by the National Key Research and Development Program, grant number 2022YFB3304003; and the Presidential Foundation of Technical Institute of Physics and Chemistry, grant number E2A8F301.

Institutional Review Board Statement: Not applicable.

Informed Consent Statement: Not applicable.

Data Availability Statement: The data are available upon request to the corresponding author.

Conflicts of Interest: The authors declare no conflict of interest.

References

1. Guo, X.D.; Han, J.Z.; Li, J.Y.; Wang, Z.B.; Zhang, Z.H.; Kang, X.; Zhu, W.; Deng, H.B. Water hazard detection: A 20-year review. *J. Terramechanics* **2023**, *105*, 53–66. [CrossRef]
2. Ribeiro, V.H.A.; Moritz, S.; Rehbach, F.; Reynoso-Meza, G. A novel dynamic multi-criteria ensemble selection mechanism applied to drinking water quality anomaly detection. *Sci. Total Environ.* **2020**, *749*, 142368. [CrossRef]

3. Zulkifli, S.N.; Rahim, H.A.; Lau, W.J. Detection of contaminants in water supply: A review on state-of-the-art monitoring technologies and their applications. *Sens. Actuators B Chem.* **2018**, *255*, 2657–2689. [CrossRef] [PubMed]
4. Nnachi, R.C.; Sui, N.; Ke, B.W.; Luo, Z.H.; Bhalla, N.; He, D.P.; Yang, Z.G. Biosensors for rapid detection of bacterial pathogens in water, food and environment. *Environ. Int.* **2022**, *166*, 107357. [CrossRef] [PubMed]
5. Huang, Y.R.; Chen, W.J.; Chung, J.; Yin, J.; Yoon, J. Recent progress in fluorescent probes for bacteria. *Chem. Soc. Rev.* **2021**, *50*, 7725–7744. [CrossRef]
6. Syal, K.; Iriya, R.; Yang, Y.Z.; Yu, H.; Wang, S.P.; Haydel, S.E.; Chen, H.Y.; Tao, N.J. Antimicrobial Susceptibility Test with Plasmonic Imaging and Tracking of Single Bacterial Motions on Nanometer Scale. *ACS Nano* **2016**, *10*, 845–852. [CrossRef]
7. Li, Y.S.; Zhu, J.W.; Zhang, H.Y.; Liu, W.M.; Ge, J.C.; Wu, J.S.; Wang, P.F. High sensitivity gram-negative bacteria biosensor based on a small-molecule modified surface plasmon resonance chip studied using a laser scanning confocal imaging-surface plasmon resonance system. *Sens. Actuators B Chem.* **2018**, *259*, 492–497. [CrossRef]
8. Andrei, C.C.; Moraillon, A.; Lau, S.; Felidj, N.; Yamakawa, N.; Bouckaert, J.; Larquet, E.; Boukherroub, R.; Ozanam, F.; Szunerits, S.; et al. Rapid and sensitive identification of uropathogenic Escherichia coli using a surface-enhanced-Raman-scattering-based biochip. *Talanta* **2020**, *219*, 121174. [CrossRef]
9. Tadesse, L.F.; Ho, C.S.; Chen, D.H.; Arami, H.; Banaei, N.; Gambhir, S.S.; Jeffrey, S.S.; Saleh, A.A.E.; Dionne, J. Plasmonic and Electrostatic Interactions Enable Uniformly Enhanced Liquid Bacterial Surface-Enhanced Raman Scattering (SERS). *Nano Lett.* **2020**, *20*, 7655–7661. [CrossRef]
10. Xu, D.K.; Chen, H.Y. Electrochemical Detection for Arrayed Electrode Chips. *Prog. Chem.* **2009**, *21*, 2379–2387.
11. Wu, J.; Liu, H.; Chen, W.; Ma, B.; Ju, H. Device integration of electrochemical biosensors. *Nat. Rev. Bioeng.* **2023**, *1*, 346–360. [CrossRef] [PubMed]
12. Chen, Y.H.; Gao, X.L.; Liu, G.S.; Zhu, R.T.; Yang, W.L.; Li, Z.S.; Liu, F.M.; Zhou, K.C.; Yu, Z.M.; Wei, Q.P.; et al. Correlation of the role of boron concentration on the microstructure and electrochemical properties of diamond electrodes. *Funct. Diam.* **2022**, *1*, 197–204. [CrossRef]
13. Dong, X.Y.; Zhao, W.W.; Sun, G.B.; Xu, J.J.; Chen, H.Y. An Electrochemical DNA Biosensor Based on Gold Nanofilm and Stable Y Junction Structure. *Acta Chim. Sin.* **2012**, *70*, 1457–1463. [CrossRef]
14. Tong, P.; Zhang, L.; Xu, J.J.; Chen, H.Y. Simply amplified electrochemical aptasensor of Ochratoxin A based on exonuclease-catalyzed target recycling. *Biosens. Bioelectron.* **2011**, *29*, 97–101. [CrossRef]
15. Du, D.; Ju, H.X.; Zhang, X.J.; Chen, J.; Cai, J.; Chen, H.Y. Electrochemical immunoassay of membrane P-glycoprotein by immobilization of cells on gold nanoparticles modified on a methoxysilyl-terminated butyrylchitosan matrix. *Biochemistry* **2005**, *44*, 11539–11545. [CrossRef] [PubMed]
16. Luo, C.H.; Lei, Y.N.; Yan, L.; Yu, T.X.; Li, Q.; Zhang, D.C.; Ding, S.J.; Ju, H.X. A Rapid and Sensitive Aptamer-Based Electrochemical Biosensor for Direct Detection of *Escherichia coli* O111. *Electroanalysis* **2012**, *24*, 1186–1191. [CrossRef]
17. Guner, A.; Cevik, E.; Senel, M.; Alpsoy, L. An electrochemical immunosensor for sensitive detection of Escherichia coli O157:H7 by using chitosan, MWCNT, polypyrrole with gold nanoparticles hybrid sensing platform. *Food Chem.* **2017**, *229*, 358–365. [CrossRef]
18. Zou, Y.J.; Liang, J.; She, Z.; Kraatz, H.B. Gold nanoparticles-based multifunctional nanoconjugates for highly sensitive and enzyme-free detection of *E. coli* K12. *Talanta* **2019**, *193*, 15–22. [CrossRef] [PubMed]
19. Li, C.; Liu, C.J.; Liu, R.; Wang, Y.X.; Li, A.Y.; Tian, S.; Cheng, W.; Ding, S.J.; Li, W.T.; Zhao, M.; et al. A novel CRISPR/Cas14a-based electrochemical biosensor for ultrasensitive detection of *Burkholderia pseudomallei* with PtPd@PCN-224 nanoenzymes for signal amplification. *Biosens. Bioelectron.* **2023**, *225*, 115098. [CrossRef]
20. Li, L.L.; Yu, S.Q.; Wu, J.; Ju, H.X. Regulation of Target-activated CRISPR/Cas12a on Surface Binding of Polymer Dots for Sensitive Electrochemiluminescence DNA Analysis. *Anal. Chem.* **2023**, *95*, 7396–7402. [CrossRef]
21. Zhang, J.L.; Liu, W.J.; Li, J.H.; Lu, K.Q.; Wen, H.R.; Ren, J.L. Rapid bacteria electrochemical sensor based on cascade amplification of 3D DNA walking machine and toehold-mediated strand displacement. *Talanta* **2022**, *249*, 123646. [CrossRef] [PubMed]
22. Kwon, J.; Kang, H.Y.; Yang, H. Permeabilization-free beta-galactosidase-induction-based electrochemical detection of Escherichia coli. *Sens. Actuators B Chem.* **2021**, *337*, 129768. [CrossRef]
23. Ertl, P.; Mikkelsen, S.R. Electrochemical biosensor array for the identification of microorganisms based on lectin-lipopolysaccharide recognition. *Anal. Chem.* **2001**, *73*, 4241–4248. [CrossRef]
24. Johnson, B.J.; Delehanty, J.B.; Lin, B.; Ligler, F.S. Immobilized proanthocyanidins for the capture of bacterial lipopolysaccharides. *Anal. Chem.* **2008**, *80*, 2113–2117. [CrossRef]
25. Castle, L.M.; Schuh, D.A.; Reynolds, E.E.; Furst, A.L. Electrochemical Sensors to Detect Bacterial Foodborne Pathogens. *ACS Sens.* **2021**, *6*, 1717–1730. [CrossRef]
26. Myers, J.A.; Curtis, B.S.; Curtis, W.R. Improving accuracy of cell and chromophore concentration measurements using optical density. *BMC Biophys.* **2013**, *6*, 4. [CrossRef] [PubMed]
27. Beal, J.; Farny, N.G.; Haddock-Angelli, T.; Selvarajah, V.; Baldwin, G.S.; Buckley-Taylor, R.; Gershater, M.; Kiga, D.; Marken, J.; Sanchania, V.; et al. Robust estimation of bacterial cell count from optical density. *Commun. Biol.* **2020**, *3*, 512. [CrossRef]
28. Wos, M.L.; Pollard, P.C. Cellular nicotinamide adenine dinucleotide (NADH) as an indicator of bacterial metabolic activity dynamics in activated sludge. *Water Sci. Technol.* **2009**, *60*, 783–791. [CrossRef]
29. Mao, X.Y.; Wu, Y.H.; Xu, L.L.; Cao, X.J.; Cui, X.J.; Zhu, L.D. Electrochemical biosensors based on redox carbon nanotubes prepared by noncovalent functionalization with 1,10-phenanthroline-5,6-dione. *Analyst* **2011**, *136*, 293–298. [CrossRef]

30. Dube, I.; Jimenez, D.; Fedorov, G.; Boyd, A.; Gayduchenko, I.; Paranjape, M.; Barbara, P. Understanding the electrical response and sensing mechanism of carbon-nanotube-based gas sensors. *Carbon* **2015**, *87*, 330–337. [CrossRef]
31. Wang, T.T.; Zhao, D.L.; Guo, X.F.; Correa, J.; Riehl, B.L.; Heineman, W.R. Carbon Nanotube-Loaded Nafion Film Electrochemical Sensor for Metal Ions: Europium. *Anal. Chem.* **2014**, *86*, 4354–4361. [CrossRef] [PubMed]
32. Chen, Y.F.; Wang, D.C.; Liu, Y.R.; Gao, G.Y.; Zhi, J.F. Redox activity of single bacteria revealed by electrochemical collision technique. *Biosens. Bioelectron.* **2021**, *176*, 112914. [CrossRef] [PubMed]
33. Ge, B.; Tan, Y.; Xie, Q.; Ma, M.; Yao, S. Preparation of chitosan–dopamine-multiwalled carbon nanotubes nanocomposite for electrocatalytic oxidation and sensitive electroanalysis of NADH. *Sens. Actuators B Chem.* **2009**, *137*, 547–554. [CrossRef]
34. Liu, A.; Watanabe, T.; Honma, I.; Wang, J.; Zhou, H. Effect of solution pH and ionic strength on the stability of poly(acrylic acid)-encapsulated multiwalled carbon nanotubes aqueous dispersion and its application for NADH sensor. *Biosens. Bioelectron.* **2006**, *22*, 694–699. [CrossRef]
35. Li, Z.; Huang, Y.; Chen, L.; Qin, X.; Huang, Z.; Zhou, Y.; Meng, Y.; Li, J.; Huang, S.; Liu, Y.; et al. Amperometric biosensor for NADH and ethanol based on electroreduced graphene oxide–polythionine nanocomposite film. *Sens. Actuators B Chem.* **2013**, *181*, 280–287. [CrossRef]
36. Zhai, X.; Wei, W.; Zeng, J.; Gong, S.; Yin, J. Layer-by-Layer Assembled Film Based on Chitosan/Carbon Nanotubes, and its Application to Electrocatalytic Oxidation of NADH. *Microchim. Acta* **2006**, *154*, 315–320. [CrossRef]
37. Sangamithirai, D.; Narayanan, V.; Muthuraaman, B.; Stephen, A. Investigations on the performance of poly(o-anisidine)/graphene nanocomposites for the electrochemical detection of NADH. *Mater. Sci. Eng. C* **2015**, *55*, 579–591. [CrossRef]
38. Arthisree, D.; Devi, K.S.S.; Devi, S.L.; Meera, K.; Joshi, G.M.; Senthil Kumar, A. A hydrophobic coenzyme Q10 stabilized functionalized-MWCNT modified electrode as an efficient functional biomimetic system for the electron-transfer study. *Colloids Surf. A Physicochem. Eng. Asp.* **2016**, *504*, 53–61. [CrossRef]
39. Eguílaz, M.; Gutiérrez, A.; Rivas, G. Non-covalent functionalization of multi-walled carbon nanotubes with cytochrome c: Enhanced direct electron transfer and analytical applications. *Sens. Actuators B Chem.* **2016**, *225*, 74–80. [CrossRef]

chemosensors

MDPI

Review

Recent Advances in Rational Design and Engineering of Signal-Amplifying Substrates for Surface-Enhanced Raman Scattering-Based Bioassays

Song Gao, Zhanchen Guo and Zhen Liu *

State Key Laboratory of Analytical Chemistry for Life Science, School of Chemistry and Chemical Engineering, Nanjing University, Nanjing 210023, China; song.gao@smail.nju.edu.cn (S.G.); zcguo@smail.nju.edu.cn (Z.G.)
* Correspondence: zhenliu@nju.edu.cn

Abstract: In recent decades, surface-enhanced Raman spectroscopy (SERS) has become a powerful detection scheme for many applications, particularly bioassays, due to its unique strengths, such as its ultrasensitive performance. Due to the development of various SERS substrates, more SERS-based bioassays with improved sensitivity and reproducibility have been designed and manufactured. SERS is able to provide the intrinsic vibration information of molecules through the unique Raman fingerprint to enable direct detection and quantitation. Meanwhile, with the assistance of Raman-active labels, biomolecules, like proteins and nucleic acids, can be detected by the immunosandwich assay. In this review, we focus on the rational design and engineering of signal-enhancing substrates for SERS-based bioassays. Those substrates are classified into two categories, i.e., nanoparticles in colloidal suspension and nanostructures on a solid support. Each category is discussed in detail with stress on their biomedical application potential. Afterward, we summarize the SERS-based assays of proteins, nucleic acids, and viruses, for which both label-free and labeled approaches play important roles. Finally, we present the remaining challenges in the field of SERS-based bioassays and sketch out promising directions for future development.

Keywords: Raman spectroscopy; SERS substrate; rational design; bioassays; high-sensitivity analysis

Citation: Gao, S.; Guo, Z.; Liu, Z. Recent Advances in Rational Design and Engineering of Signal-Amplifying Substrates for Surface-Enhanced Raman Scattering-Based Bioassays. *Chemosensors* **2023**, *11*, 461. https://doi.org/10.3390/chemosensors11080461

Academic Editor: Rui Martins

Received: 25 July 2023
Revised: 10 August 2023
Accepted: 13 August 2023
Published: 16 August 2023

1. Introduction

Surface-enhanced Raman scattering (SERS) has emerged as a powerful analytical technique across many application domains. It combines Raman spectroscopy with nanostructured metal surfaces to achieve significant enhancement of the Raman signal. The conventional Raman spectroscopy, discovered by Raman and Krishnan in 1928 [1], has served as an analytical tool for a wide range of applications. Raman scattering provides insights into molecular structures by revealing their vibrational spectra. It provides compound-specific information and molecular-level fingerprinting without the need for complex instruments. This characteristic makes it promising for sensing various analyte molecules, and it has found extensive use in biological applications [2–5]. For example, it has been employed in identifying cancer cells [6], investigating biomolecular structures [7], and diagnosing diseases and pathologies [8,9]. Despite its utility, Raman scattering intensity is typically weak, as the process suffers from low efficiency. Upon light-matter interactions, only one in every 10^8 photons would undergo an inelastic scattering as estimated [10,11]. This limitation has hindered its application in trace analysis, necessitating methods to enhance the Raman signal. Fortunately, Fleishmann, Hendra, and McQuilian first observed unexpectedly strong Raman signals in 1974 from pyridine adsorbed on roughened silver electrodes [12]. Later, in 1977, Jeanmaire and Van Duyne discovered that placing a Raman-active species on a roughened noble metal surface significantly amplified the Raman signal, surpassing the limitations of conventional Raman scattering [13]. This phenomenon, known as surface-enhanced Raman scattering, has revolutionized the sensitivity of Raman spectroscopy. Now SERS is one of the key technologies to dramatically amplify the Raman scattering signal in order to bring several advantages, including ultrahigh sensitivity, less susceptibility to sample environment, rapid readout speed, and the possibility of on-site or field detection [14].

Bioassays are crucial tools in biological and pharmaceutical research, providing valuable information about the role and effects of substances in living organisms. They facilitate the evaluation of biological activity [15], assist early disease diagnostics [16], aid in drug discovery and development [17], ensure quality control [18], assess toxicity [19], monitor environmental impact [20], and contribute to expanding our understanding of living systems [21]. Both labeled and label-free bioassays have gained prominence as powerful tools for studying and analyzing biological systems. Labeled bioassays involve the use of molecular probes or labels, such as fluorescent dyes and radioisotopes, to detect and quantify the biomolecules of interest that have been specifically captured. These labels enable the direct measurement and visualization of the target, providing specific and sensitive results. Label-free bioassays, on the other hand, rely on the direct measurement of intrinsic properties of the captured biomolecule, such as mass, absorbance, and Raman response, without the use of exogenous labels. In general, labeled assays provide high sensitivity and versatility, but require sample manipulation and may introduce biases. Conversely, label-free assays offer minimal sample processing and real-time analysis but may lack sensitivity and pose challenges in data interpretation. As technology continues to advance, the development of hybrid strategies and improved detection methods may bridge the gap between these two bioassay paradigms, enabling researchers to harness the strengths of both approaches in their scientific investigations.

Conventional bioassays mainly rely on the use of absorbance, fluorescence, and chemiluminescence for detection. However, SERS-based bioassays have garnered significant interest in recent decades due to the unique advantages over these detection schemes. One of the primary advantages is suitability for multiplexing [22]. Raman bands in SERS have narrower widths as compared with fluorescence, enabling simultaneous measurement of multiple analytes within a single detection. Another advantage is the high enhancement factor (EF) in SERS, which ranges from 10^6 to 10^{14} [23–25], making SERS an exceptionally sensitive tool. The high sensitivity of SERS is particularly favorable for the detection of low-abundance substances of biological importance in complex real samples. By combining the specific binding capabilities of antibodies with the enhanced sensitivity of SERS, SERS-based bioassays provide a powerful tool for the detection and quantification of target analytes.

Raman signal-enhancing substrates play a vital role in SERS, enabling ultrasensitive detection and characterization of biological analytes. The rational design of SERS substrates involves tailoring their properties at the nanoscale to optimize the enhancement factor and ensure high reproducibility. Since there have been numerous publications devoted to the design and fabrication of substrates for SERS measurement, several review papers have well-surveyed the advances in this area [26–30]. A critical review is still needed on the progress of SERS-based bioassays, particularly from the view of the exquisite design, performance improvement, and real-world application potential.

In this review, we survey recent advances in SERS-based bioassays with a focus on the rational design and engineering of signal-enhancing substrates. For convenience of comparison and understanding, we classify the substrates into two categories, i.e., nanoparticles in colloidal suspension and nanostructures on a solid support. Each category is discussed in detail with stress on their biomedical application potential. SERS-based bioassays have been applied to various clinical samples. We further summarize the SERS-based assays of proteins, nucleic acids, and viruses, for which both label-free and labeled approaches play important roles. Finally, we suggest other challenges in the field of SERS-based bioassays and sketch out promising directions for future development.

2. Basic Theory of SERS

It is now generally accepted that the Raman scattering signal is amplified through a combination of electromagnetic (EM) enhancement mechanism, and chemical (CM) enhancement mechanism. In brief, electromagnetic enhancement arises from the excitation of localized surface plasmons, which are collective oscillations of conduction electrons on the metal surface. These plasmons generate intense electromagnetic fields at the metal surface, leading to increased Raman scattering intensity. The chemical enhancement, on the other hand, results from the charge transfer between the molecule and the metal surface, further enhancing the Raman signal. Chemical enhancement is highly dependent on the specific molecule–substrate interactions and can vary for different analytes.

2.1. Electromagnetic Enhancement

The underlying principle of SERS involves the amplification of the incident and scattered electromagnetic fields when a nanostructured surface of the metal is irradiated with the frequency of localized surface plasmons resonance (LSPR) on the metal surface. This phenomenon is explained by considering a metal nanosphere subjected to an external electric field (Figure 1A). The oscillating electric field (E_0) of the incident light interacts with the metal nanoparticle, inducing a polarization of charge and creating a dipolar localized surface plasmon resonance. This polarization generates an induced dipole moment (μ_{ind}) determined by the metal polarizability (α_{met}) and the incident electric field ($E_0(\omega_{inc})$) [31].

$$\mu_{ind} = \alpha_{met} E_0(\omega_{inc}) \tag{1}$$

As a Raman scattering arises on a molecule, a dipole moment is first created by the incident light. After being scattered, the photons are detected as the Raman signal. SERS follows a similar process but with the enhancement of the local electromagnetic field due to hotspots on the metal surface. In SERS, the inelastic scattering of E_0 generates an enhanced local electric field (E_{loc}) near the metal surface. The interaction between the local electric field and the molecule adsorbed on the surface creates an induced dipole moment (μ_{ind}) determined by the molecular polarizability (α_{mol}) and the enhanced local electric field ($E_{loc}(\omega_{inc})$).

$$\mu_{ind} = \alpha_{mol} E_{loc}(\omega_{inc}) \tag{2}$$

According to the vibrational spectrum theory in quantum mechanics, the presence of inelastic scattering for a vibrating molecule is explained by the incident electric field (E_{inc}) and the eigenvalue of angular frequency (ω_{vib}) of the vibrating molecule. This inelastic scattering gives rise to three dipole components: Rayleigh scattering $\mu_{ind}(\omega_{inc})$, Stokes scattering $\mu_{ind}(\omega_{inc} - \omega_{vib})$, and anti-Stokes scattering $\mu_{ind}(\omega_{inc} + \omega_{vib})$ (Figure 1B). The resonance frequency of surface plasmons on the metal surface determines the enhancement of the scattered Stokes field. Taking into account the incident and scattered field intensities, the overall SERS enhancement intensity can be described by an equation that involves the local electric field ($E_{inc}(\omega_{inc})$) at the incident frequency (ω_{inc}) and the electric field ($E(\omega_s)$) at the Stokes-shifted frequency ($\omega_s = \omega_{inc} - \omega_{vib}$) [31].

$$I_{SERS} = I_{inc}(\omega_{inc}) I(\omega_s) \tag{3}$$

$$= |E_{inc}(\omega_{inc})|^2 |E(\omega_s)|^2 \tag{4}$$

When the electric field values at the incident frequency ($E_{inc}(\omega_{inc})$) and the Stokes shifted frequency ($E(\omega_s)$) are in proximity, the resulting enhancement derived from the electromagnetic mechanism follows the relationship where the SERS enhancement factor is directly proportional to the fourth power of the induced electric field enhancement value ($E(\omega_{inc})$).

$$I_{SERS} = |E(\omega_{inc})|^4 \tag{5}$$

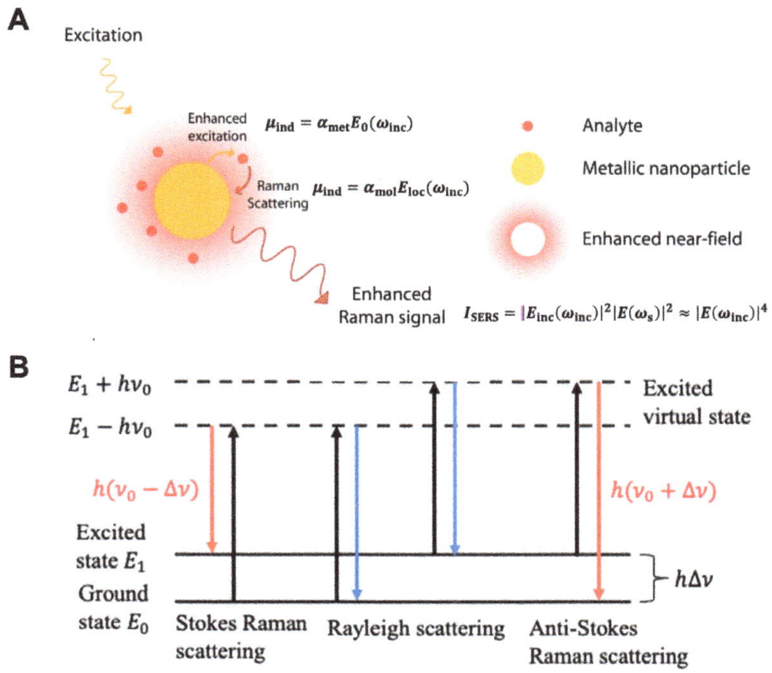

Figure 1. (**A**) Principal of surface-enhanced Raman scattering (reproduced under the terms of CC BY 4.0 from Ref. [32], copyright 2015, the authors). (**B**) Rayleigh scattering and Raman scattering energy-level diagram (reproduced under the terms of CC BY 4.0 from Ref. [33], copyright 2022, the authors).

2.2. Chemical Enhancement

The chemical effect is another significant mechanism contributing to SERS enhancement, and it requires direct contact between the SERS substrate and the analyte. This effect involves the interaction of the adsorbate and the surface to form a complex through electronic coupling. During this interaction, electrons on the substrate surface transfer from the Fermi level to the lowest unoccupied molecular orbital of the analyte molecule. This process leads to the creation of charge transfer intermediates, which demonstrate higher Raman cross-sections in comparison to the free molecule. If the incident photon at frequency ω_{inc} resonates with the charge transfer transition of the formed complex, the scattered Stokes intensity contains information about the vibrational state of the molecule. In other words, the Raman spectrum of the analyte molecule is enhanced when it forms a chemical bond or interacts strongly with the SERS-active metal surface.

The magnitude of the chemical enhancement effect is typically in the range of 10^0 to 10^2, which is weaker compared to the electromagnetic enhancement. However, the chemical effect can provide additional information about the chemical interactions between the analyte molecule and the metal surface, allowing for a more detailed characterization of the molecular species.

3. Substrates in Colloidal Suspension

SERS substrates designed for practical applications and problem-solving purposes have been extensively explored. These substrates in colloids can be broadly categorized into two groups: dispersed particles and aggregated systems. Dispersed particle systems in solution are relatively simple to prepare, making them widely used. Furthermore, advancements in producing increasingly uniform particles have improved the reproducibility of SERS measurements. Surface functionalization of colloidal particles is also often straightforward, allowing for a wide range of targets and applications [34–36]. However, these systems suffer from some drawbacks, with colloidal stability being a primary concern, as colloidal systems tend to be unstable and prone to irreversible aggregation, particularly in complex environments.

Highly enhancing SERS substrates often consist of aggregated or assembled nanoparticles. While yielding significant SERS enhancement factors, reproducibility can sometimes be challenging due to the difficulty in controlling aggregation. Controlled particle assembly offers a solution to this issue, allowing for precise nanoengineering and reducing the irreproducibility associated with aggregated particles. The outcome of the assembly process is influenced by surface functionalization and nanoparticle geometry.

3.1. Dispersed Particles

With the advancements in nanoscience and nanotechnology, the synthesis of metal nanoparticles (with precise control over their sizes, shapes, and structures) has been widely explored. Numerous synthesis methods have been developed to create increasingly intricate nanostructures, including nanospheres [37], nanostars [38–40], nanotriangles [41], nanorods [42–44], nanocubes [34,45–47], core–shell particles [48,49], and others (Figure 2A). Higher order structures in SERS allow for increased control over regions with high field enhancement and enable optimization of plasmon modes for specific experiments. Chemical reduction methods have been used to achieve uniform structural control of nanoparticles. For example, using the seed-mediated growth method, gold nanospheres with successive and tunable diameters can be prepared, bringing size-dependent SERS properties [37,50].

Surfactants play a crucial role in the conventional synthesis process of nanoparticles, and the development of new types of surfactants can enable simplified and controllable synthetic methods for nanostructures, expanding their applications. Zhang et al. involved the natural compound epigallocatechin gallate (EGCG) as a reducing agent to synthesize different morphologies of gold nanoparticles (AuNPs) by changing the solution pH values [51]. The addition of halogen ions not only alleviates laser damage to nanoparticles but also promote the formation of a 3D hotspot, leading to a noticeable SERS effect that brings about high reproducibility and stability. In another study, bromide ions were introduced to induce anisotropic growth when synthesizing gold nanocubes (AuNCs) [45]. As bromide ions favorably adsorb onto the (100) facet, in combination with cetyltrimethylammonium chloride (CTAC), a difference in the accessibility of gold precursors to different facets is made, leading to distinct growth rates between [100] and other facets. At low bromide densities, the adsorbed bromide ions do not completely block the (100) facet, producing AuNCs with round corners. However, when an adequate amount of facet-directing agents is provided, preferential binding to the (100) facet maximizes the growth-rate difference between the (100) facet and the other two facets, resulting in AuNCs with sharp corners (Figure 2B). This straightforward and generally applicable synthetic strategy enables the precise control of the size and corner sharpness of gold nanocubes in a high yield. Similarly, iodide ions are used in the synthesis of gold nanotriangles [41]. Highly symmetric gold nanostars are enclosed by high-index facets to grow the hexagonal pyramidal arms, which are regulated by alkyl amines, like dimethylamine (DMA), methylamine, ethylamine, butylamine, and octylamine as shape-control agents [38]. Due to their high symmetry, Au nanostars demonstrate superior single-particle SERS performance in terms of both intensity and reproducibility when compared to asymmetric Au nanostars. In addition to chemical methods, laser-induced morphological remodeling of gold nanorods has been achieved by regulating the intensity of illumination and density of surfactant, resulting in different aspect ratios and ultranarrow local surface plasmon resonance bands [43]. This laser-controlled method enables the synthesis of gold nanorods with an ultranarrow LSPR band.

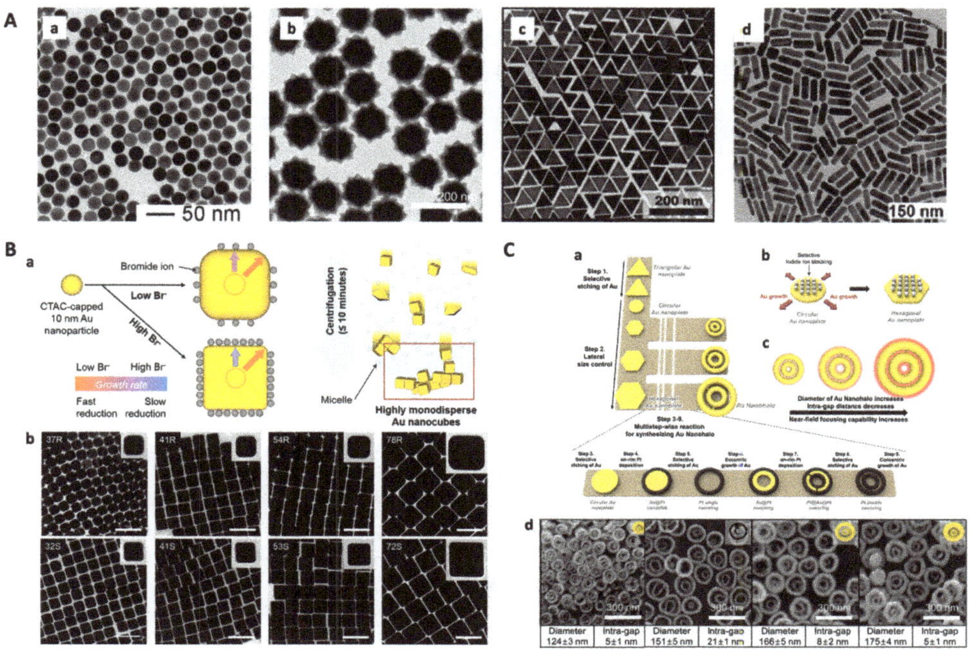

Figure 2. Representative substrates of dispersed nanoparticles in colloidal suspension. (**A**) Transmission electron microscope (TEM) images of controllable gold nanoparticles with different shapes: (**a**) nanospheres (reproduced with permission from Ref. [37], copyright 2013, WILEY-VCH Verlag GmbH & Co. KGaA, Weinheim, Germany), (**b**) nanostars (reproduced with permission from Ref. [38], copyright 2015, American Chemical Society), (**c**) nanotriangles (reproduced with permission from Ref. [41], copyright 2021, American Chemical Society), (**d**) nanorods (reproduced with permission from Ref. [42], copyright 2019, American Chemical Society). (**B**) (**a**) Schematic illustration representing the synthesis of corner-sharpness-controlled AuNCs with varied bromide densities followed by refinement by centrifugation-driven depletion-induced flocculation in surfactant micelle solutions; (**b**) TEM images of different sizes of round-cornered (R) and sharp-cornered (S) AuNCs (reproduced with permission from Ref. [45], copyright 2018, American Chemical Society). (**C**) (**a**) Schematic illustration representing the synthetic pathway of size-controlled Au nanohalos (Au double nanorings), including selective etching of Au, lateral size control, on-rim Pt deposition, selective etching of Au, eccentric growth of Au, and concentric growth of Au; (**b**) iodide ions blocking the top and bottom flat facets of circular Au nanoplates allow us to obtain diameter-controlled hexagonal Au nanoplates during the lateral size control step; (**c**) near-field focusing capability of Au nanohalos varies as a function of the diameter and intragap distance of Au nanohalos; (**d**) TEM images of Au nanohalos with different diameters and intra-gaps (reproduced with permission from Ref. [52], copyright 2022, American Chemical Society).

To achieve maximum near-field focusing around nanoscale metal particles, more complicated morphologies like gold double nanorings have been developed (Figure 2C) [52]. These nanorings consist of two concentric rings with nanoscale gaps between them. The circular intra-nanogaps create hot halos, enabling single-particle SERS when analytes are present within these gaps. The synthesis of Au nanohalos is based on a novel strategy utilizing hexagonal Au nanoplates as a template. By controlling the ratio of Au ions to iodide ions while maintaining the thickness, the lateral size of the nanoplates can be precisely controlled. Through multiple sequential steps, the initially triangular Au nanoplates are transformed into Au nanohalos with near-field focusing capabilities. The estimated enhancement factor achieved by these structures is remarkably high, ranging from 1.1×10^9 to 3.6×10^9, and the results are highly reproducible. Another approach involves the evolution of chirality by adding chiral molecules during the synthesis, promoting asymmetric growth and inducing the chiroptical response of nanostructures [53]. This method allows for the preparation of single plasmonic gold nanoparticles with a significant uniform chiral gap, aiding in precise morphology and property control for chiral nanomaterials.

Overall, these advancements in the synthesis of reproducible nanostructures using various methods, including chemical reduction, surfactant-assisted fabrication, laser control, and chiral shape modifiers, offer opportunities for precise control over the morphology and optical properties of nanoparticles. These advances have implications for a wide range of applications, including SERS-based detection, where reproducibility, sensitivity, and reliability are critical.

3.2. Coupling of Particles

Multimeric nanostructures, such as dimers and trimers, provide inter-nanogap structures containing hotspots that exhibit intense electromagnetic field enhancement between adjacent particles. These hotspots are responsible for the extraordinary enhancement of Raman signals in SERS. He et al. introduced a dual biomimetic recognition-driven plasmonic nanogap-enhanced Raman scattering (DBR-PNERS) strategy for highly sensitive protein fingerprinting and quantification [54] (Figure 3A). This approach utilizes a pair of protein terminal epitope-imprinted plasmonic nanoantennas (PNAs) designed to bind to the N- and C-terminals of the target protein. An ultrathin imprinting layer (approximately 5 nm) is engineered on the PNAs, preserving the plasmonic signal enhancement effect of the nanoantennas and overcoming the limitations of larger antibodies. When both PNAs are present in a sample containing the target protein, specific binding occurs, forming a nanogap where the target protein is located. This nanogap provides a well-defined hot spot for Raman signal amplification, enabling precise molecular information readout and quantification of the target protein. Additionally, an internal standard (IS), in the form of a Raman-active small molecule, was incorporated into the nanoantenna, enabling a ratiometric assay for accurate and reliable quantification. Compared to existing approaches, DBR-PNERS exhibited several significant merits, including fingerprinting, robust quantitation, ultrahigh sensitivity, minimal sample consumption, and so on.

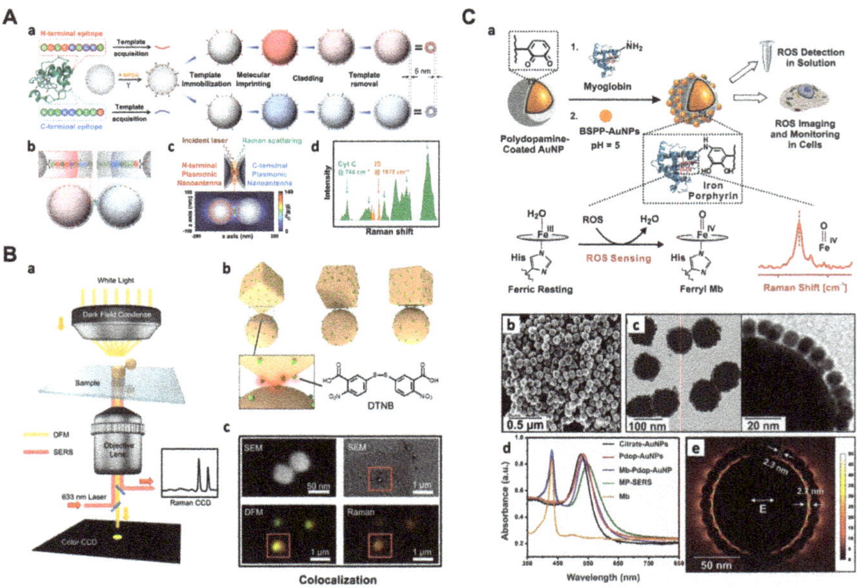

Figure 3. Representative substrates of coupled nanoparticles in colloidal suspension. (**A**) DBR-PNERS strategy for ultrasensitive protein fingerprinting and quantitation. (**a**) Preparation of terminal epitope-

imprinted PNAs; (**b**) protein-targeting DBR for fabrication of plasmonic nanogap; (**c**) protein-targeting EM enhancement for PNERS; (**d**) representative Raman spectrum for detection and molecular fingerprinting (reproduced with permission from Ref. [54], copyright 2022, American Chemical Society). (**B**) Single-particle SERS measurements of AANs with DTNB Raman dye on the surface. (**a**) Dark-field microscopy (DFM)-Raman integrated apparatus for measuring the structure-correlated optical and plasmonic properties of individual AANs; (**b**) schematic of AANs at different gap morphology with DTNB Raman dye on the surface; (**c**) the colocalization of scanning electron microscope (SEM), DFM characterization and Raman mapping of an on-vertex AAN (reproduced with permission from Ref. [55], copyright 2021, Wiley-VCH GmbH). (**C**) (**a**) Schematic representation for the synthesis of myoglobin and polydopamine-engineered SERS nanoprobes and SERS-based bio-detection chemistry of ROS at the porphyrin center of myoglobin; (**b**) SEM image of MP-SERS nanoprobes; (**c**) TEM images of MP-SERS nanoprobes; (**d**) UV-Vis spectra of citrate-AuNPs, pdop-AuNPs, Mb-functionalized pdop-AuNPs, MP-SERS nanoprobes, and Mb; (**e**) 3D-FEM simulation image of MP-SERS nanoprobe showing several hot spots (reproduced with permission from Ref. [56], copyright 2017, WILEY-VCH Verlag GmbH & Co. KGaA, Weinheim).

Recent progress in programmable materials has led to significant advancements in the field. Specifically, well-designed DNA nanostructures, including DNA origami, offer a highly programmable method for creating precise nanopatterns with nanoscale addressing capabilities [57,58]. This allows for the construction of architectures with precisely defined geometries [59–62]. Compared to nanoantenna gaps where molecules are adsorbed randomly on the surface, target molecules can be placed specifically in hotspots due to the nanogaps precisely controlled by self-assembled DNA nanostructures [63], providing single molecule resolution [64]. Even the anisotropic functionalization of plasmonic gold nanostructures with relative spatial directionality and sequence asymmetry is under precise control [65]. Moreover, the fabrication processes involved in DNA nanostructures are more cost-effective compared to traditional top-down lithographic techniques [66].

Taking advantage of these features, several SERS substrates have been developed with reliable signals. For example, Zhan et al. constructed plasmonic bowtie nanostructures using a DNA origami-based bottom-up assembly strategy, enabling precise control over the geometrical configuration of the bowtie with an approximate 5 nm gap [67]. A single Raman probe was accurately positioned at the gap of the bowtie, resulting in an exceptionally high enhancement factor of 10^9 so that a single-molecule level SERS detection is possible. Similarly, a DNA-based synthetic method for gold–silver core–shell nanodumbbells (GSNDs) was developed, offering good structural reproducibility and precise nanometer-level control over the size of the nanogaps across multiple particles [63]. The method involves preparing DNA-modified AuNPs by controlling the stoichiometric ratio of two thiolated DNA sequences. Each particle is modified with a maximum of one linking DNA strand, while densely modifying them with protecting DNA strands to maximize dimer yield and minimize higher multimeric structures. Subsequently, Ag nanoshells are formed on the Au dimers using polyvinylpyrrolidone(PVP)-based delivery of Ag precursors, enabling precise control and reproducibility of the interparticle gaps at the nanometer scale. Moreover, Niu et al. developed a DNA origami-based nanoprinting (DOBNP) strategy to transfer the essential DNA strands with predefined sequences and positions to the surface of AuNCs [55] (Figure 3B). This approach facilitated the controlled assembly of AuNC–AuNP nanostructures (AANs) with specific geometry and composition. The anchoring of a single dye molecule in hotspot regions results in a significant enhancement of the electromagnetic field, leading to a stronger amplification of the SERS signal.

Alternatively, nanoparticles of varying sizes or shapes can be assembled onto a central particle to form core-satellite assembled structures [56,69–72]. For instance, Kumar et al. developed myoglobin and polydopamine(pdop)-engineered SERS nanoprobes (MP-SERS) for real-time and quantitative detection of reactive oxygen species (ROS) levels in living cells [56] (Figure 3C). The probes consisted of 80 nm gold nanoparticles as a core, a thin pdop layer as a spacer, Mb as a ROS-responsive Raman reporter, and 10 nm AuNPs as satellite particles. The plasmonic coupling between the core and satellites created a highly enhanced electromagnetic field in the interstitial sites, leading to nanogap-enhanced Raman scattering. The probes exhibited high sensitivity to ROS, good biocompatibility, and were successfully internalized into cells.

4. Substrates on a Solid Support

Compared to traditional colloid-based systems, solid-supported SERS substrates offer numerous advantages. Nanostructures embedded on the surface exhibit high uniformity, provide compatibility with ultra-low volumes, enable single-nanoparticle measurements, and facilitate the preparation process. Similar to colloidal systems, various solid-supported substrates have been prepared. Colloidal nanoparticles, assemblies, or aggregates can be deposited onto surfaces [73–75]. A chemical layer can be used to functionalize surfaces and direct the adsorption of colloidal particles. [74,76]. By adjusting nanoparticle concentration and deposition time, the particle density on the surface can be optimized. These substrates offer similar advantages to colloidal systems while enabling comprehensive optical and structural characterization of individual particles during usage, which is particularly advantageous for single-molecule studies.

Apart from surface self-assembly methods, nanolithography is a commonly employed technique for fabricating surfaces with highly enhancing metal nanostructures. Colloidal lithography [77], and electron beam lithography (EBL) [78–80], are among the most prevalent lithographic methods used in SERS applications. These methods vary in terms of ease and cost of fabrication. Although lithography allows for the preparation of various highly sensitive structures, one drawback of deposited metal nanoparticles is their vulnerability to annealing, as structural changes in the nanoparticles can lead to shifts in LSPR position and alter SERS capabilities, sometimes even during the course of a measurement. Advancements in synthesis and nanolithography methods have contributed to higher field enhancements, improved spatial control, and greater uniformity and reproducibility.

4.1. Bottom-Up Strategies

The bottom-up approach, mainly relying on self-assembly, offers a straightforward and high-throughput method for preparing SERS substrates. This approach involves organizing small nanoparticles into arrays with ordered nanostructures, and various techniques have been developed for this purpose. One method is oblique deposition, where nanoparticles are deposited at an angle to the substrate surface [81]. Another method is electro-displacement, which utilizes electrostatic forces to assemble nanoparticles into desired patterns [82].

Liquid substrate-based methods have gained popularity in bottom-up self-assembly in recent years. These methods allow for the formation of highly ordered nanostructures without complex instruments and additional materials. One example is the nanocapillary pumping model developed by Ge et al. [83]. In this model, a membrane structure with nanoparticles is subjected to solvent evaporation, then the surface of the film undergoes deformation, bringing the nanoparticles closer to each other and creating a substrate with hotspots in a large area. The pressure difference among gaps induces the movement of target molecules towards the smaller nanoparticle gaps, resulting in molecules trapped by the hotspots (Figure 4A). The detection limit is significantly reduced in this nanocapillary pump model. However, one limitation of this approach is the inability to control the nanoparticle array forming in a monolayer. Nanoparticles may obtain accumulated during the solvent evaporation, leading to plasmon resonance intensity of the substrate being non-uniform, reducing the reproducibility and reliability of SERS measurements. To address the challenges of non-uniform nanoparticle assembly and interference from non-adsorptive molecules in liquid SERS analysis, researchers have developed self-assembly methods depending on two-phase substrates and ternary regulations. In the two-phase substrate approach, gold nanoparticles tend to assemble at the interface of oil/water due to reduced electrostatic repulsion [84,85]. This leads to the formation of a large-scale self-assembled monolayer with controlled spacing on a nanometer scale (Figure 4B). Such self-assembled monolayer provides a uniform distribution of hotspots, enhancing the performance of substrates. This method has been applied in various practical detection scenarios. For example, a liquid-like 3D plasmonic array was formed at the oil/water interface, enabling sensitive SERS analysis of different analytes [86]. The liquid interfacial plasmonic platform broke through the limitations of low molecular affinity and achieved sub-ppb level sensitivity. Researchers successfully detected polycyclic aromatic hydrocarbons and distinguished different molecular configurations using this approach.

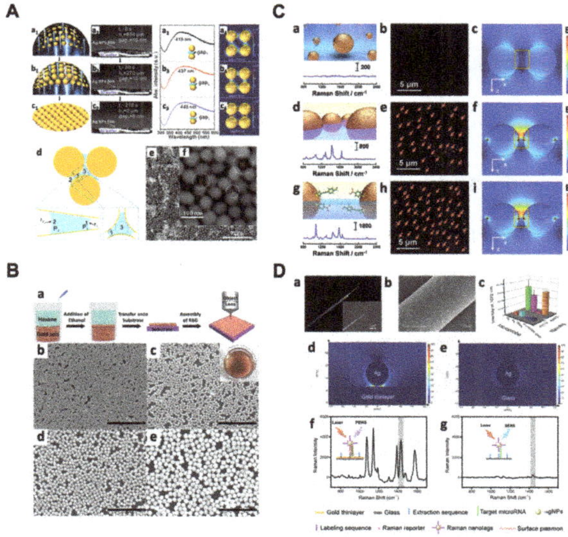

Figure 4. Representative substrates on a solid support using bottom-up strategies. (**A**) Formation of the nanocapillary pumping model; (a_1–a_4) schematic diagram of the assembled nanoparticle film, high-speed camera picture, in situ UV spectrum, and simulation of electromagnetic enhancement between nanoparticles with different gaps in the initial wet stage of the monolayer nanoparticle film; (b_1–b_4) formation of the effective hot-spot stage of the monolayer nanoparticle film; (c_1–c_4) final dry stage of the monolayer nanoparticle film; (**d**) principle causing the solvent to move toward the smaller gaps of the nanocapillary pumping model; (**e**,**f**) SEM image of the assembled monolayer silver nanoparticle film in the final dry stage (reproduced with permission from Ref. [83], copyright 2021, American Chemical Society). (**B**) (**a**) A scheme for fabricating and transferring Au nanoparticle monolayers from the water/hexane interface and an illustration of the SERS experiment; (**b**–**e**) SEM and optical images (inset) of the Au nanoparticle monolayer obtained through the water/oil interface by adjusting the addition of ethanol into different Au solutions with diameters of (**b**) 30 nm, (**c**) 60 nm, (**d**) 90 nm, and (**e**) 120 nm. Scale bar: 1 μm (reproduced with permission from Ref. [85], copyright 2016, WILEY-VCH Verlag GmbH & Co. KGaA, Weinheim). (**C**) Schematic and SERS spectrum of (**a**) AuNPs in water phase, (**d**) AuNPs at OA/W interface, and (**g**) AuNPs/RP1+MPBA at OA/W interface; (**b**,**e**,**h**) the corresponding dark field microscope image; (**c**,**f**,**i**) FDTD-calculated electromagnetic enhancement distribution for SERS in the xz-plane. The black dashed line indicates the OA/W interface (reproduced with permission from Ref. [87], copyright 2022, Wiley-VCH GmbH). (**D**) (**a**) Photo and SEM images of an extraction microprobe; (**b**) SEM image showing immunosandwiches formed on an extraction microprobe after extracting target protein from a single cell and being labeled with Raman nanotags; (**c**) dependence of Raman intensity on the combination of extraction microprobe and nanotag; electric-field intensity simulated by finite-difference time-domain (FDTD) under (**d**) PERS mode and (**e**) SERS mode; representative Raman spectra for (**f**) the PERS mode and (**g**) SERS mode, and insertions show the schematic of the principles of PERS and SERS modes (reproduced with permission from Ref. [88], copyright 2016, WILEY-VCH Verlag GmbH & Co. KGaA, Weinheim and under the terms of CC BY-NC 3.0 from Ref. [89], copyright 2018, the authors).

In the ternary regulation approach, charged noble metal nanoparticles are directly assembled under an electrostatic drive to form 2D thin arrays at the liquid/liquid interface [90]. The resulting nanoarray exhibits long-term stability and even distribution of hotspots. When transferred to a solid substrate, this nanoarray achieved single-cell Raman fingerprinting and discrimination of eight different bacteria species. However, controlling nanogaps at the molecular scale using interfacial tension and electrostatic forces can be challenging, leading to inaccuracies in trace substance analysis. Non-adsorptive molecules from complex biological systems may bring about interference in biological applications. To overcome these challenges, a novel interfacial SERS platform was developed using a functionalized AuNP array formed at the liquid/liquid interface [87]. This platform was achieved through ternary regulations involving double recognition of rigid molecular probes, uniform distribution of hotspots in a consistent manner, and the SERS readout in a silent region (Figure 4C). This platform allows real-time quantification, addressing signal interference and increasing the reliability of the SERS analysis. The reproducibility of the SERS analysis is improved by the fixed nanogap formed by the rigid structure of the probe. The use of specific Raman vibrations as signal output reduces interference from substances embedded in the brain.

A technique known as shell-isolated nanoparticle-enhanced Raman spectroscopy (SHINERS) has garnered significant attention and was introduced and developed by the Tian group [49,91]. In SHINERS, a thin dielectric layer is applied over the surface of metal nanoparticles, effectively creating a masking effect with a different material interface. This dielectric spacer layer, when sufficiently thin, allows molecules bound to the surface to experience a different surface chemistry while still benefiting from the significant field enhancement provided by the nearby metallic surface [92]. Silica is the most commonly used material for the shell in SHINERS, as it allows for the functionalization of gold and silver nanoparticles using silane chemistry, facilitating shell growth. This offers several advantages, including the ability to use different surface chemistry types for targeted adsorption in sensing applications, improved stability, and enhanced biocompatibility. Additionally, the use of insulating materials to create pinhole-free shells eliminates the occurrence of chemical enhancement resulting from charge transfer [49]. This altered interface between the shell and certain biological molecules or catalysts can lead to minimal perturbation of the electronic structure of the target caused by the underlying metal surface.

The Liu group developed a simple but effective approach named plasmon-enhanced Raman scattering (PERS) to provide ultrasensitive detection at the single-molecule level [88]. With the combination of gold substrate and silver nanoparticles, hotspots were generated with a 1–2 orders of magnitude higher signal compared to the conventional SERS mode with gold nanoparticles on a glass substrate (Figure 4D). Gold thin-layer-coated glass of a slightly lower enhancement was used as an alternative to save costs. AgNPs are uniformly dispersed on the Au layer, providing better reproducibility while maintaining high sensitivity compared with the coupling between particles. More importantly, the fabrication of substrates is simple and straight, which is conducive to the practical application. By the integration with microprobes for immunoaffinity extraction, a powerful tool for probing low-abundance proteins in single living cells has emerged. Depending on the interaction between the microprobe and the target, different bioassays, like the plasmonic immunosandwich assay (PISA) [88] and plasmonic affinity sandwich assay (PASA) [89], have been proposed and accomplished, empowering the analysis of low-abundance species, such as proteins and RNA in single cells.

4.2. *Top-Down Strategies*

Top-down lithography techniques realize uniform hotspots on SERS substrates in a wide range with high precision. Electron beam lithography (EBL), a maskless technology that directly writes micro- and nanopatterns using an electron beam sensitive material, has shown compatibility with mass production manufacturing and excellent signal reproducibility [93,94]. For instance, on single-crystal Si substrates, nanopore arrays fabricated using two-layer EBL demonstrated high sensitivity to the increased intensity of scattered light from enhanced incident electromagnetic radiation [95]. Focused ion beam (FIB) lithography, as another branch of top-down lithography, can focus the ion beams into tiny sizes using electrostatic lenses to achieve nanometer-scale positioning accuracy [96]. Etching techniques, which use physical or chemical methods to selectively remove unwanted moieties, enable high-throughput fabrication of substrates with desired nanostructures. Polymer nanopillars etched with Ar plasma from a polyethylene terephthalate (PET) substrate are subsequently deposited with Au, resulting in large-scale, highly reproducible SERS-active nanopillar arrays [97]. Chemical etching techniques can also be used to precisely regulate nanometer spacing. Porous nano-web structures fabricated by Lee et al. demonstrated a controllable gap between the nanoring core and the nano-web, confining the electromagnetic field in the nanogap [98]. The resulting structure demonstrated single-particle SERS dependent on the gap distance, being an ideal platform for SERS measurement for its structural integrity and stability, open accessibility of analytes, incident polarization independence, and reproducibility of SERS signals.

The combination of multiple nanofabrication technologies is emerging as a development trend in the SERS substrate preparation. Flauraud et al. utilized EBL and dry etching to fabricate lithographed funneled traps and auxiliary sidewalls on a solid substrate [99], directing the capillary assembly of Au nanorods, and achieving simultaneous control over their positions, orientations, and interparticle distances at the nanometer level. Multi-technique combination for biocompatible SERS-active materials is often used to enhance reliability. Li et al. fabricated digital nanopillar arrays with a combination of EBL, physical vapor deposition of Au, and selective reactive ion etching (RIE) to reveal the pillar structure [100] (Figure 5A). Four kinds of antibodies were conjugated to the array of pillars to selectively recognize cytokines, which were labeled with different Raman reporters to be visualized in color. The digital nanopillar SERS assay achieved highly specific and sensitive real-time cytokine detection down to the attomolar level, showing promise for advancing personalized medicine and predicting the higher risk of developing severe immune toxicities.

Colloidal lithography (also known as nanosphere lithography) involves monodispersed nanospheres as templates for the fabrication of SERS substrates in a wide range. By a chemical route, a colloidal template-induced precursor solution dipping strategy was used for the formation of periodic semi-hollow sphere arrays [101]. This array serves as a substrate for magnetron sputtering deposition of Au, resulting in novel micro/nanostructured arrays. These arrays offer diverse performance for different requirements. By involving a combination of molecular self-assembly, colloidal nanosphere lithography, and physical peeling, Luo et al. reported a fabrication process for fabricating large-area arrays of triangular nanogaps with adjustable widths ranging from ~10 to ~3 nm, exhibiting remarkably high measured enhancement factors of up to 10^8 compared to a thin gold film [102] (Figure 5B). Colloidal lithography can also form large-area nanostructures in parallel at low-cost on flexible substrates. An in situ chemical patterning technique is proposed based on plasmonic nanochemistry [103] (Figure 5C). A large-area-ordered Au nanohole array (NHA) film was prepared using colloidal lithography, where a polystyrene (PS) microsphere array was etched by the RIE process, and deposited with Au film. Via the wet chemical method, Ag nanoparticles were grown in situ from the edge to the center of the holes. After peeling off the Au NHA film, an ordered and large-area Ag nanoparticle assembly array (NAA) was left on the substrate. The region-specific Ag NAAs demonstrated outstanding SERS performance for quantitative detection and Raman imaging, with applications in multilevel encryption and anti-counterfeiting labels. This technique is also suitable for flexible substrates, such as polydimethylsiloxane (PDMS) and curved glass, allowing for extensive applications.

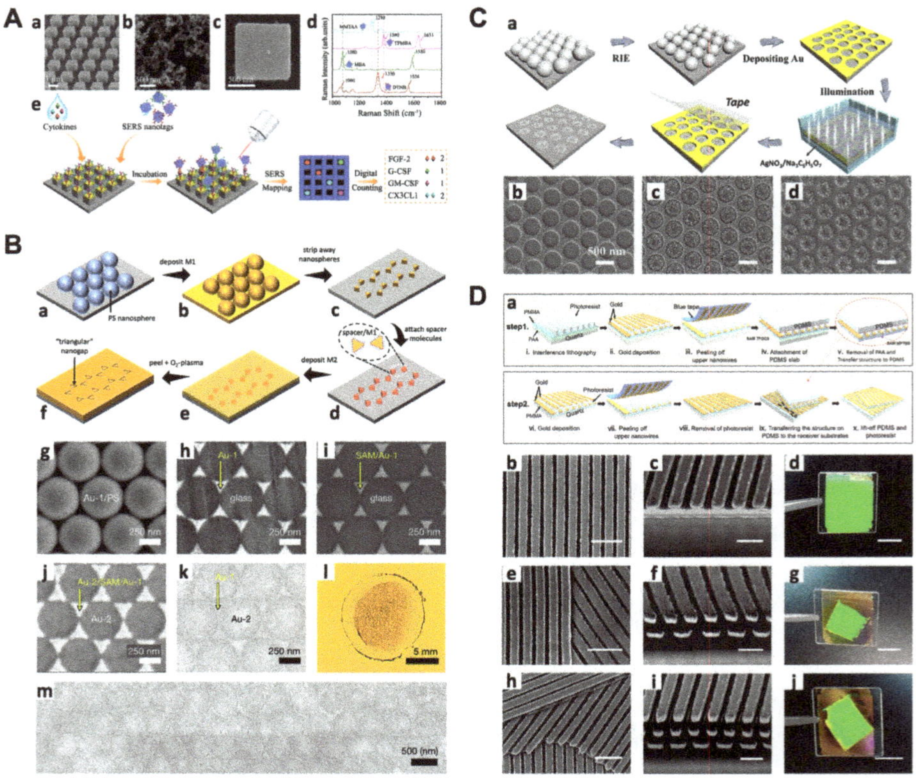

Figure 5. Representative substrates on a solid support using top-down strategies. (**A**) Digital single-molecule nanopillar SERS platform for parallel counting of four types of cytokines. SEM images of (**a**) pillar array side view, (**b**) nanoboxes, and (**c**) a single nanobox on the top of a pillar; (**d**) SERS spectra of nanoboxes conjugated with Raman reporters; (**e**) workflow for multiplex counting of cytokines (reproduced under the terms of CC BY 4.0 from Ref. [100], copyright 2021, the authors). (**B**) (**a–f**) Fabrication procedure for triangular nanogap arrays; (**g–m**) images of N = 1 Au/Au TNG arrays at various stages in the fabrication procedure (reproduced under the terms of CC BY 4.0 from Ref. [102], copyright 2022, the authors). (**C**) (**a**) Schematic of the fabrication process; SEM images of (**b**) Au NHAs, (**c**) Au NHAs with Ag NPs selectively growing in the interior of the holes, (**d**) Ag NP-NAAs after removing the NHA (reproduced with permission from Ref. [103], copyright 2021, Wiley-VCH GmbH). (**D**) (**a**) Schematics of the procedure for PAA-assisted nanotransfer printing to assemble 3D nanostructures layer-by-layer; (**b–j**) top- (**b**,**e**,**h**) and side-view (**c**,**f**,**i**) SEM images of the twisted triple-layer-grating nanostructures and their corresponding optical photographs (**d**,**g**,**j**); (**b–d**) the bottom nanograting layer; (**e–g**) double-layer 3D nanostructure with a twist angle $\theta = 30°$; (**h–j**) triple-layer 3D nanostructure with a twist angle between the neighboring nanogratings $\theta = 30°$. Scale bar: 1 µm (**b**,**e**,**h**), 500 nm (**c**,**f**,**i**), and 1 cm (**d**,**g**,**j**) (reproduced with permission from Ref. [104], copyright 2019, American Chemical Society).

In addition to traditional lithography methods, unconventional strategies, such as nanoprinting, have been explored for the convenient fabrication of 3D nanostructures. Nanotransfer printing technology has been developed to transfer large-area and crack-free 3D multilayer nanostructures onto flexible substrates [104] (Figure 5D). This approach improves the uniformity and reproducibility of SERS substrates. Tape imprinting methods have been utilized to create lotus-shaped nanoarray structures with ultra-sensitive characteristics for detection [105]. Combination approaches, such as colloidal nanosphere lithography with metal-assisted chemical etching, can produce nanopillar array SERS platforms with stable and reproducible Raman signals [106]. These platforms exhibit strong sensitivity and have applications in extended monitoring of cell surfaces and live cell analysis.

5. Applications

Conventional methods, such as bacterial culture, polymerase chain reaction (PCR), luminescence, and microarray are commonly utilized in the biomedical field to test clinical analytes. However, these methods have drawbacks including lengthy processing times, limited portability, insufficient multiplexing capabilities, and low sensitivity. In contrast, SERS provides rapid, on-site analysis with the ability to perform fingerprint multiplexing and achieve high sensitivity. Consequently, point-of-care (POC) SERS approaches have been developed and implemented to overcome the limitations of conventional methods. These approaches have been applied to various clinical samples, encompassing proteins, nucleic acids, and viruses.

5.1. Detection of Proteins

Protein biomarkers play a crucial role in disease diagnostics and treatment monitoring. The development of SERS-based bioassays for protein analysis has gained significant concern [107,108]. One approach is to incorporate SERS detection into lateral flow immunoassays (LFAs) using disposable and portable test strips [109–111]. In these SERS-LFA assays, ultrabright SERS nanotags conjugated with antibodies are used for sandwich recognition of the target protein in clinical samples. For example, a rapid SERS-LFA assay has been developed for the detection of human chorionic gonadotropin (hCG), a hormone associated with pregnancy [111]. This assay achieved fast detection within 2–5 seconds with a limit of detection (LOD) of approximately 1.6 $mIU{\cdot}mL^{-1}$ using a portable Raman reader. Compared to traditional raster-scanning Raman microscopy, the use of line illumination significantly improved detection speed and sensitivity, making it 15 times more sensitive than commercial LFAs. In addition to SERS-LFA assays, the integration of digital microfluidics (DMF) with SERS-based immunoassays has been reported for protein detection [112] (Figure 6A). This approach involves a sandwich immunoassay in which magnetic beads coated with antibodies and antibody-functionalized SERS tags are used to capture and label the antigens for sensitive detection. The automation capability of DMF simplifies the assay procedure and minimizes the risk of exposure to hazardous samples.

The DMF-SERS method was utilized for the quantitative detection of H5N1, an avian influenza virus. This approach demonstrated outstanding sensitivity (LOD of 74 $pg{\cdot}mL^{-1}$) and selectivity for H5N1 detection, with a shorter assay time (< 1 h) and reduced reagent consumption (~30 μL) compared to the standard ELISA method. By combining DMF with a portable Raman spectrometer, this SERS offers a low reagent consumption and minimized exposure risk for hazardous samples, making it a promising tool for the diagnosis of infectious diseases.

Low-copy-number proteins that are expressed fewer than 1000 molecules per cell play vital roles in various essential cellular processes, yet accurately quantifying and understanding the functions of low-copy-number proteins at a single-cell level is challenging. Advanced techniques, such as the single-cell plasmonic immunosandwich assay (scPISA), are being developed to enhance the sensitivity and enable quantitative analysis of these proteins [113]. It combines in vivo microextraction for specific enrichment of target proteins from individual cells, and PERS for highly sensitive detection of low-copy-number proteins (Figure 6B). This approach holds great promise for advancing our understanding of cellular heterogeneity and the functions of low-copy-number proteins in biological processes and disease states [114,115]. PERS-based bioassays are not limited to a single-cell analysis. Depending on the principle of PERS, immunoassays that utilize antibodies or antibody mimics like molecular imprinting polymers (MIPs) and aptamers are effective in protein detection [116–120]. For example, Xing et al. presented an approach called the dual MIP-based plasmonic immunosandwich assay (duMIP-PISA) for the sensitive and specific detection of protein biomarkers in complex biological samples [118]. By preparing a C-terminal epitope-imprinted self-assembled gold nanoparticle monolayer and N-terminal epitope-imprinted Raman-responsive Ag@SiO_2 nanoparticles, the target protein was specifically captured and labeled sequentially. The formation of the MIP-protein-MIP sandwich-like complexes results in a significantly enhanced SERS signal. The duMIP-PISA approach offers several advantages over regular enzyme-linked immunosorbent assay (ELISA), including a simpler procedure, faster speed, lower sample volume requirement, and wider linear range, making it a promising tool for various important applications, particularly in disease diagnosis.

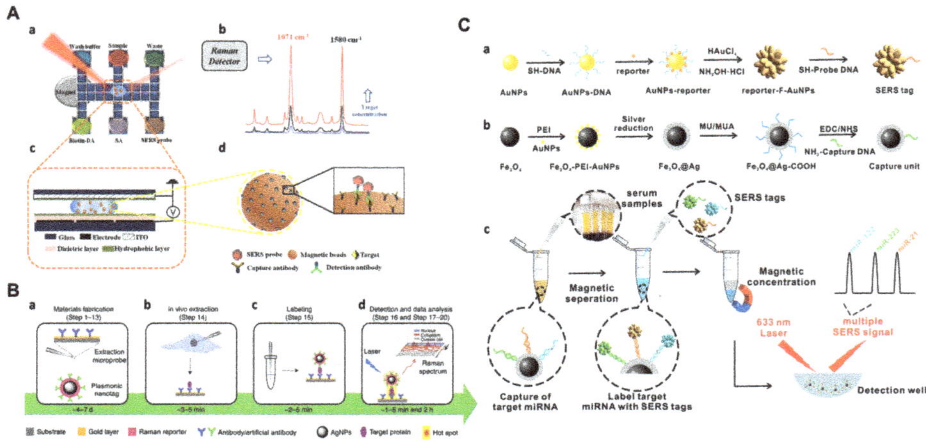

Figure 6. Representative applications of SERS bioassays for detection of proteins and nucleic acids. (**A**) Schematic illustration of SERS-based immunoassay with digital microfluidics. (**a**) Illustration of DMF-SERS method and bottom plate of DMF chip; (**b**) two characteristic Raman peaks of 4-MBA at 1071 and 1580 cm^{-1}; (**c**) side view of DMF chip containing a droplet with magnetic beads; (**d**) immunocomplex functionalized with SERS tags on magnetic beads (reproduced with permission from Ref. [112], copyright 2018, American Chemical Society). (**B**) Overview of the scPISA procedure for the determination of low-copy-number proteins in single living cells. (**a**) Fabrication of the materials used for scPISA, including affinity ligand-functionalized gold-based extraction microprobe and affinity ligand-modified silver-based plasmonic nanotag; (**b**) in vivo extraction by an affinity extraction microprobe precisely inserted into a single living cell under test; (**c**) in vitro labeling of target protein captured from the single cell with plasmonic nanotag and the formation of extraction microprobe/protein/plasmonic nanotag sandwich-like immunocomplexes on the microprobe surface; (**d**) Raman signal readout by plasmonic detection and data analysis (reproduced with permission from Ref. [113], copyright 2021, the authors, under exclusive license to Springer Nature Limited). (**C**) Schematic illustration of multiplex miRNA assay via the SERS sandwich strategy. R6G-, CV-, and 4-ATP-encoded fractal Au nanoparticles were utilized as SERS tags, and Ag-coated magnetic nanoparticles were utilized as the capture substrate. (**a**) Schematic processes of synthesizing SERS tag; (**b**) design and synthesis of the capture substrate; (**c**) detection procedure of multiple miRNAs based on the capture substrate/miRNA/SERS tag sandwich structure (reproduced with permission from Ref. [121], copyright 2021, American Chemical Society).

5.2. Detection of Nucleic Acids

Nucleic acids, including RNA, DNA, and miRNA, play a crucial role as genetic material and serve as important biomarkers in various applications. Conventional methods for nucleic acid detection, such as real-time quantitative PCR (RT-qPCR) and northern blotting, often suffer from limitations such as low sensitivity, lack of portability, long assay times, and extensive sample preparations. To overcome these challenges, researchers have explored alternative detection methods, especially optical and electrochemical techniques.

Among these, SERS has become a promising strategy for nucleic acids, leading to the development of various SERS-based detection methods.

One particular class of nucleic acids, miRNA, consists of an average length of only 22 nucleotides. The short length and high sequence homology of miRNAs pose challenges for specific detection and amplification using conventional methods like PCR. However, miRNAs have raised potential as biomarkers of cancers since they play a critical role in protein post-transcriptional regulation. To address these limitations, many strategies based on SERS for miRNA detection have been proposed [121–127]. For example, a magnetically assisted sandwich-type SERS-based biosensor was manufactured [121], capable of ultrasensitive and multiplex detection of three hepatocellular carcinoma-related miRNA biomarkers (miRNA-122, miRNA-223, and miRNA-21) (Figure 6C). The biosensor consists of a magnetic capture substrate to capture DNA-conjugated Ag-coated magnetic nanoparticles (AgMNPs) and SERS tags to probe DNA-conjugated DNA-engineered fractal gold nanoparticles (F-AuNPs). This assay exhibited excellent selectivity, specificity, and high accuracy in multiplexed detection of actual human serum samples and liver cancer patient serum assay, even in the presence of other miRNAs. Remarkably, the LOD for the three miRNAs were exceptionally low, with values of 349 aM for miRNA-122, 374 aM for miRNA-223, and 311aM for miRNA-21. Additionally, a technique named PASA was employed for the detection of nucleic acids [128] and their modifications [129] for several key advantages, including ultrahigh sensitivity, fast analysis time, and a minimal sample volume requirement. Zhang et al. developed a rapid and highly sensitive method called the AuNP-decorated Ag@SiO_2 nanocomposite-based PASA for the detection of circulating miRNAs in human serum [128]. Using this method, researchers successfully quantified miR-21 in human serum with a LOQ of 10^{-14} M and differentiated a breast cancer patient from a healthy individual. The AuNP-decorated Ag@SiO_2 nanocomposite and PASA method can be easily adapted for the detection of other miRNAs and circulating tumor DNA, making it a promising tool for cancer diagnosis.

In addition to the detection of miRNA, longer nucleic acids, such as RNA and DNA, have also been successfully detected using SERS-based methods [130–138]. For example, a "lab-in-a-stick" portable device that integrates a SERS-based bioassay and the washing process has enabled the direct detection of pathogen RNA [133]. Two signal enhancement levels were utilized to achieve the sensitivity required for direct detection. Each target sequence was tagged with an ultrabright SERS-encoded nanorattle, providing ultrahigh SERS signals. These tagged target sequences were then concentrated into a focused spot for detection using hybridization sandwiches with magnetic microbeads. Through this approach, the device achieved direct detection of synthetic targets with a LOD of 200 fM. SERS devices have also been developed for the detection of DNA. DNA mutation pattern profiling is crucial for the classification of cancer types and plays a fundamental role in advancing precision medicine. Wu et al. developed an amplification-free SERS biochip that allows for the direct and simultaneous identification of multiple point mutations in tumor cells [135]. By integrating the SERS spectra encoding technique with a supervised learning algorithm, a panel of nucleotide mixtures can be effectively distinguished based on their mutation profiles. The SERS sensor is incorporated into a microfluidic chip, enabling one-step multiplex analysis within 40 min.

5.3. Detection of Viruses

The detection of viruses is crucial for early diagnosis and treatment and monitoring disease progression, especially in the context of recent pandemic and epidemic outbreaks like COVID-19. SERS-based detection methods for viruses have been developed, both in label-free and labeled approaches, offering rapid and reliable detection capabilities.

In label-free SERS detection, Yeh et al. presented a portable microfluidic platform named VIRRION (virus capture with rapid Raman spectroscopy detection and identification), which contains carbon nanotube arrays with differential filtration porosity, allowing for rapid enrichment and optical identification of viruses [139] (Figure 7A). The platform employs a multi-virus capture component in conjunction with SERS, enabling real-time enrichment and identification of different emerging strains or unknown viruses. Following viral capture and detection on the chip, the viruses remain viable and become purified within the microdevice. This enables subsequent in-depth characterizations using various conventional methods. The technology successfully enriched and identified rhinovirus, influenza virus, and parainfluenza viruses, maintaining the stoichiometric viral proportions even when multiple types of viruses were present, simulating co-infection. The process of viral capture and detection took only a few minutes and resulted in a 70-fold enrichment enhancement, with a detection limit of 10^2 $EID_{50}\cdot mL^{-1}$ and a virus specificity of 90%, making it a promising tool for real-time tracking and monitoring of viral outbreaks.

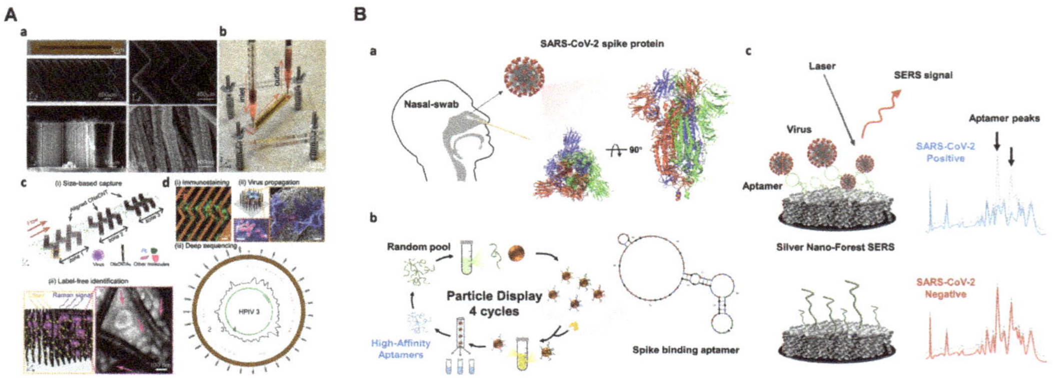

Figure 7. Representative applications of SERS bioassays for detection of viruses. (**A**) Design and working principle of VIRRION for effective virus capture and identification. (**a**) Photograph and SEM images of aligned CNTs exhibiting herringbone patterns decorated with gold nanoparticles; (**b**) picture showing assembled VIRRION device, processing a blood sample; (**c**) illustration of (i) size-based capture and (ii) in situ Raman spectroscopy for label-free optical virus identification; (**d**) on-chip virus analysis and enrichment for NGS, (i) on-chip immunostaining for captured H5N2, (ii) on-chip viral propagation through cell culture, and (iii) genomic sequencing and analysis of human parainfluenza virus type 3 (HPIV 3) (reproduced under the terms of CC BY 4.0 from Ref. [139], copyright 2020, the authors). (**B**) Design strategy for the label-free SERS-based aptasensor platform for SARS-CoV-2. (**a**) Targeted aptamer screening against spike (S) protein for detecting SARS-CoV-2 from clinical samples. Side and top view of the trimeric S protein on the surface of SARS-CoV-2 (PDB: 6VXX); (**b**) the particle display aptamer discovery process, in which solution-phase aptamer library molecules are converted to monoclonal aptamer particles, incubated with fluorescently-labeled S protein, and then subjected to fluorescence-activated cell sorting (FACS) to enrich library molecules with a strong affinity for this target; (**c**) the aptamer is then conjugated onto a silver nanoforest (SNF) substrate for the detection of SARS-CoV-2 and the intrinsic aptamer peaks shift in response to the conformational changes triggered by the S protein binding to the aptamer (reproduced with permission from Ref. [140], copyright 2023, Elsevier B.V.).

In the labeled SERS detection, typical immunoassays have been employed for virus detection [141–146]. For example, a highly sensitive and quantitative SERS-based LFA strip was developed for the simultaneous detection of influenza A H1N1 virus and human adenovirus (HAdV) [144]. Fe_3O_4@Ag nanoparticles served as magnetic SERS nanotags, which were conjugated with dual-layer Raman dye molecules and target virus-capture antibodies, allowing for specific recognition and magnetic enrichment of target viruses in the solution, as well as SERS detection of the viruses on the strip. The magnetic SERS strip can directly be used for real biological samples without requiring any sample pretreatment steps. The LOD achieved for H1N1 and HAdV were 50 and 10 $pfu\cdot mL^{-1}$, respectively, which were significantly higher compared to the standard colloidal gold strip method. This

SERS-based LFA strip is a potential tool for early detection of virus infections for its easy operation, rapidity, stability, and high throughput.

SERS also provides great potential for fast-screening detection of SARS-CoV-2 virus and its variants [140,147–153]. Park et al. developed an ultra-sensitive label-free aptasensor based on SERS for the universal detection of SARS-CoV-2 variants of concern [140] (Figure 7B). Two DNA aptamers with high affinity for binding to the spike protein of SARS-CoV-2 were first identified through high-throughput screening. The dissociation constants (k_D) of the aptamers are 1.47 ± 0.30 nM and 1.81 ± 0.39 nM, respectively. By combining these aptamers with a silver nanoforest, this ultra-sensitive SERS platform achieves a detection limit at the attomolar (10^{-18} M) level using a recombinant trimeric spike protein. Furthermore, the intrinsic properties of the aptamer make the approach label-free, eliminating the need for a Raman tag. Clinical samples with variants of concern, like the wild-type, delta, and omicron variants, were successfully detected. In another study, a hand-held breathalyzer based on SERS for the rapid identification of individuals infected with COVID-19 was developed, which is capable of providing results in under 5 minutes and demonstrates high sensitivity (>95%) and specificity across a cohort of 501 participants [152]. These developments hold promise as a powerful tool for the rapid and reliable detection of SARS-CoV-2 variants, facilitating effective control measures against the ongoing pandemic.

While both label-free and labeled SERS bioassays have been developed for the detection of viruses, there is still a need for diverse functions, such as virus capture and enrichment, and minimized sample exposure, apart from sensitive, specific, and rapid detection. Such advancements would be highly desirable for real-world applications.

6. Conclusions and Future Perspectives

SERS has emerged as a very powerful and promising analytical technique, and its capabilities continue to advance primarily through improvements in substrates. The sensitivity, compatibility with probed molecules, and measurement reproducibility are determined by the substrate. These substrates often involve well-controlled particles and their tunable couplings to achieve optimal enhancement. While SERS-based bioassays have already been widely used in biomolecular detection, the rational design of substrates and applied methodologies discussed here will continue to drive advancements toward higher sensitivity and reproducibility for clinical applications. The labeled SERS bioassays using SERS nanotags for biomolecule detection offer a sensitive method with enhanced specificity through the use of targeting moieties on the nanoconstructs. One notable advantage of SERS nanotags is the ability to design and synthesize Raman-reporter labels with unique spectral peaks, enabling multiplex detection by avoiding overlap. These nanoprobes hold potential for early disease diagnosis, where simultaneous detection of multiple biomarkers at low concentrations is crucial.

Additionally, the implementation of artificial intelligence (AI) can greatly enhance the analysis of complex samples for which the involuted Raman spectra are often hard to distinguish, enabling disease diagnosis without the labeling of biomarkers [154–158]. For more related information, please refer to Ref. [159,160]. In brief, acquired Raman spectra in complex media can be processed using spectral unmixing algorithms to estimate concentration profiles, leading to improved quantification By "learning" the fingerprint of Raman spectra and identifying specific characteristics from massive data, deep learning algorithms can automate analysis processes, outperforming human-operated analysis and potentially enabling more efficient and faster disease diagnosis. The combination of SERS signals with data processing software, especially involving AI, is an active area of research, offering potential improvements in accuracy and reliability.

While the SERS technique offers high sensitivity and intrinsic molecular fingerprint information, its reproducibility has historically been a challenge, limiting its widespread application. Recent advancements in nanotechnology have provided researchers with the means to create more uniform and reproducible SERS substrates, both at the micro/nanoscale and macroscopic levels. However, achieving precise nanogap regulation for specific recognition remains a critical requirement, especially in complex measurement environments. SERS analysis in vivo still pose both opportunities and challenges. One pressing need is to fix the nanogap between nanoparticles at a stable distance on a molecular level to achieve simultaneous reproducibility and selectivity in detection. Framework nucleic acids hold promise as they can integrate chemical recognition elements and provide a stable structure. By altering Raman reporters and refining molecule structures, it is possible to minimize fluctuations in the SERS signal. In addition, it is worth noting that some delicate methods, due to the complexity of their designs, may have a relatively high coefficient of variation in analyzing complex samples. Therefore, developing simple yet effective methods is of great importance for practical use in real-world applications.

Another significant challenge in advancing SERS is the limited availability of sensitive, affordable, and portable Raman spectrometers. Confocal Raman microscopy, despite its sensitivity, is often of cumbersome size and complex structure, limiting its usage as a portable device. Moreover, since only one micro area can be acquired at a time with Raman microscopy, bias may result from the microscopic heterogeneity, lacking representation at the overall level. To overcome this hurdle, extensive research is needed to develop miniaturized spectrometers with higher sensitive and Raman-stabilized lasers that can be seamlessly integrated into on-chip platforms. This development would be crucial in realizing the vision of portable and accessible SERS devices.

Author Contributions: Conceptualisation, S.G., Z.G. and Z.L; writing—original draft preparation, S.G. and Z.L.; writing—review and editing, S.G., Z.G. and Z.L.; visualization, S.G.; supervision, Z.L.; funding acquisition, Z.L. All authors have read and agreed to the published version of the manuscript.

Funding: This research was funded by the Key Grant (21834003) and the Key Scientific Instrumentation Grant (21627810) from the National Natural Science Foundation of China.

Institutional Review Board Statement: Not applicable.

Informed Consent Statement: Not applicable.

Data Availability Statement: Data sharing not applicable.

Conflicts of Interest: The authors declare no conflict of interest.

References

1. Raman, C.V.; Krishnan, K.S. A new type of secondary radiation. *Nature* **1928**, *121*, 501–502. [CrossRef]
2. Zong, C.; Xu, M.; Xu, L.J.; Wei, T.; Ma, X.; Zheng, X.S.; Hu, R.; Ren, B. Surface-enhanced Raman spectroscopy for bioanalysis: reliability and challenges. *Chem. Rev.* **2018**, *118*, 4946–4980. [CrossRef]
3. Howes, P.D.; Chandrawati, R.; Stevens, M.M. Colloidal nanoparticles as advanced biological sensors. *Science* **2014**, *346*, 1247390. [CrossRef] [PubMed]
4. Farka, Z.; Jurik, T.; Kovar, D.; Trnkova, L.; Skládal, P. Nanoparticle-based immunochemical biosensors and assays: Recent advances and challenges. *Chem. Rev.* **2017**, *117*, 9973–10042. [CrossRef]
5. Lane, L.A.; Qian, X.; Nie, S. SERS nanoparticles in medicine: from label-free detection to spectroscopic tagging. *Chem. Rev.* **2015**, *115*, 10489–10529. [CrossRef]
6. Kallaway, C.; Almond, L.M.; Barr, H.; Wood, J.; Hutchings, J.; Kendall, C.; Stone, N. Advances in the clinical application of Raman spectroscopy for cancer diagnostics. *Photodiagn. Photodyn. Ther.* **2013**, *10*, 207–219. [CrossRef] [PubMed]
7. Krafft, C.; Popp, J. The many facets of Raman spectroscopy for biomedical analysis. *Anal. Bioanal. Chem.* **2015**, *407*, 699–717. [CrossRef]
8. Stone, N.; Matousek, P. Advanced transmission Raman spectroscopy: A promising tool for breast disease diagnosis. *Cancer Res.* **2008**, *68*, 4424–4430. [CrossRef] [PubMed]
9. Movasaghi, Z.; Rehman, S.; Rehman, I.U. Raman spectroscopy of biological tissues. *Appl. Spectrosc. Rev.* **2007**, *42*, 493–541. [CrossRef]
10. Mosca, S.; Conti, C.; Stone, N.; Matousek, P. Spatially offset Raman spectroscopy. *Nat. Rev. Methods Primers* **2021**, *1*, 21. [CrossRef]
11. Ellis, D.I.; Cowcher, D.P.; Ashton, L.; O'Hagan, S.; Goodacre, R. Illuminating disease and enlightening biomedicine: Raman spectroscopy as a diagnostic tool. *Analyst* **2013**, *138*, 3871–3884. [CrossRef]
12. Fleischmann, M.; Hendra, P.J.; McQuillan, A.J. Raman spectra of pyridine adsorbed at a silver electrode. *Chem. Phys. Lett.* **1974**, *26*, 163–166. [CrossRef]

13. Jeanmaire, D.L.; Van Duyne, R.P. Surface Raman spectroelectrochemistry: Part I. Heterocyclic, aromatic, and aliphatic amines adsorbed on the anodized silver electrode. *J. Electroanal. Chem. Interfacial Electrochem.* **1977**, *84*, 1–20. [CrossRef]
14. Ye, J.; Chen, Y.; Liu, Z. A boronate affinity sandwich assay: an appealing alternative to immunoassays for the determination of glycoproteins. *Angew. Chem., Int. Ed.* **2014**, *53*, 10386–10389. [CrossRef]
15. Ahn, K.C.; Zhao, B.; Chen, J.; Cherednichenko, G.; Sanmarti, E.; Denison, M.S.; Lasley, B.; Pessah, I.N.; Kültz, D.; Chang, D.P.; et al. In vitro biologic activities of the antimicrobials triclocarban, its analogs, and triclosan in bioassay screens: receptor-based bioassay screens. *Environ. Health Perspect.* **2008**, *116*, 1203–1210. [CrossRef]
16. Fan, Y.; Wang, S.; Zhang, F. Optical Multiplexed Bioassays for Improved Biomedical Diagnostics. *Angew. Chem. Int. Ed.* **2019**, *58*, 13208–13219. [CrossRef] [PubMed]
17. Kerns, E.H.; Di, L.; Carter, G.T. In vitro solubility assays in drug discovery. *Curr. Drug Metab.* **2008**, *9*, 879–885. [CrossRef] [PubMed]
18. Johnson, I.; Hutchings, M.; Benstead, R.; Thain, J.; Whitehouse, P. Bioassay selection, experimental design and quality control/assurance for use in effluent assessment and control. *Ecotoxicology* **2004**, *13*, 437–447. [CrossRef]
19. Sprague, J. Measurement of pollutant toxicity to fish I. Bioassay methods for acute toxicity. *Water Res.* **1969**, *3*, 793–821. [CrossRef]
20. Schuetzle, D.; Lewtas, J. Bioassay-directed chemical analysis in environmental research. *Anal. Chem.* **1986**, *58*, 1060A–1075A. [CrossRef]
21. Rampersad, S.N. Multiple applications of Alamar Blue as an indicator of metabolic function and cellular health in cell viability bioassays. *Sensors* **2012**, *12*, 12347–12360. [CrossRef] [PubMed]
22. Laing, S.; Jamieson, L.E.; Faulds, K.; Graham, D. Surface-enhanced Raman spectroscopy for in vivo biosensing. *Nat. Rev. Chem.* **2017**, *1*, 0060. [CrossRef]
23. Francis, M.K.; Sahu, B.K.; Bhargav, P.B.; Balaji, C.; Ahmed, N.; Das, A.; Dhara, S. Ag nanowires based SERS substrates with very high enhancement factor. *Physica E Low Dimens. Syst. Nanostruct.* **2022**, *137*, 115080. [CrossRef]
24. Xu, S.; Man, B.; Jiang, S.; Wang, J.; Wei, J.; Xu, S.; Liu, H.; Gao, S.; Liu, H.; Li, Z.; et al. Graphene/Cu nanoparticle hybrids fabricated by chemical vapor deposition as surface-enhanced Raman scattering substrate for label-free detection of adenosine. *ACS Appl. Mater. Interfaces* **2015**, *7*, 10977–10987. [CrossRef] [PubMed]
25. Chen, S.; Li, X.; Zhao, Y.; Chang, L.; Qi, J. Graphene oxide shell-isolated Ag nanoparticles for surface-enhanced Raman scattering. *Carbon* **2015**, *81*, 767–772. [CrossRef]
26. Wang, X.; Huang, S.C.; Hu, S.; Yan, S.; Ren, B. Fundamental understanding and applications of plasmon-enhanced Raman spectroscopy. *Nat. Rev. Phys.* **2020**, *2*, 253–271. [CrossRef]
27. Langer, J.; Jimenez de Aberasturi, D.; Aizpurua, J.; Alvarez-Puebla, R.A.; Auguié, B.; Baumberg, J.J.; Bazan, G.C.; Bell, S.E.; Boisen, A.; Brolo, A.G.; et al. Present and future of surface-enhanced Raman scattering. *ACS Nano* **2019**, *14*, 28–117. [CrossRef]
28. Karthick Kannan, P.; Shankar, P.; Blackman, C.; Chung, C.H. Recent advances in 2D inorganic nanomaterials for SERS sensing. *Adv. Mater.* **2019**, *31*, 1803432. [CrossRef] [PubMed]
29. Shi, L.; Zhang, L.; Tian, Y. Rational Design of Surface-Enhanced Raman Scattering Substrate for Highly Reproducible Analysis. *Anal. Sens.* **2023**, *3*, e202200064. [CrossRef]
30. Lee, H.K.; Lee, Y.H.; Koh, C.S.L.; Phan-Quang, G.C.; Han, X.; Lay, C.L.; Sim, H.Y.F.; Kao, Y.C.; An, Q.; Ling, X.Y. Designing surface-enhanced Raman scattering (SERS) platforms beyond hotspot engineering: emerging opportunities in analyte manipulations and hybrid materials. *Chem. Soc. Rev.* **2019**, *48*, 731–756. [CrossRef]
31. Schlücker, S. Surface-Enhanced raman spectroscopy: Concepts and chemical applications. *Angew. Chem. Int. Ed.* **2014**, *53*, 4756–4795. [CrossRef]
32. Boujday, S.; Lamy de la Chapelle, M.; Srajer, J.; Knoll, W. Enhanced vibrational spectroscopies as tools for small molecule biosensing. *Sensors* **2015**, *15*, 21239–21264. [CrossRef]
33. Ding, H.; Hu, D.J.J.; Yu, X.; Liu, X.; Zhu, Y.; Wang, G. Review on all-fiber online raman sensor with hollow core microstructured optical fiber. *Photonics* **2022**, *9*, 134. [CrossRef]
34. Jorgenson, E.; Ianoul, A. Biofunctionalization of plasmonic nanoparticles with short peptides monitored by SERS. *J. Phys. Chem. B* **2017**, *121*, 967–974. [CrossRef] [PubMed]
35. Chen, Y.; Xianyu, Y.; Jiang, X. Surface modification of gold nanoparticles with small molecules for biochemical analysis. *Acc. Chem. Res.* **2017**, *50*, 310–319. [CrossRef]
36. He, J.; Unser, S.; Bruzas, I.; Cary, R.; Shi, Z.; Mehra, R.; Aron, K.; Sagle, L. The facile removal of CTAB from the surface of gold nanorods. *Colloids Surf.* **2018**, *163*, 140–145. [CrossRef] [PubMed]
37. Zheng, Y.; Zhong, X.; Li, Z.; Xia, Y. Successive, Seed-Mediated Growth for the Synthesis of Single-Crystal Gold Nanospheres with Uniform Diameters Controlled in the Range of 5–150 nm. *Part. Part. Syst. Charact.* **2014**, *31*, 266–273. [CrossRef]
38. Niu, W.; Chua, Y.A.A.; Zhang, W.; Huang, H.; Lu, X. Highly symmetric gold nanostars: crystallographic control and surface-enhanced Raman scattering property. *J. Am. Chem. Soc.* **2015**, *137*, 10460–10463. [CrossRef]
39. Khoury, C.G.; Vo-Dinh, T. Gold nanostars for surface-enhanced Raman scattering: synthesis, characterization and optimization. *J. Phys. Chem. C* **2008**, *112*, 18849–18859. [CrossRef] [PubMed]
40. Childs, A.; Vinogradova, E.; Ruiz-Zepeda, F.; Velazquez-Salazar, J.J.; Jose-Yacaman, M. Biocompatible gold/silver nanostars for surface-enhanced Raman scattering. *J. Raman Spectrosc.* **2016**, *47*, 651–655. [CrossRef]

41. Scarabelli, L.; Liz-Marzan, L.M. An extended protocol for the synthesis of monodisperse gold nanotriangles. *ACS Nano* **2021**, *15*, 18600–18607. [CrossRef]
42. Khanal, B.P.; Zubarev, E.R. Chemical transformation of nanorods to nanowires: Reversible growth and dissolution of anisotropic gold nanostructures. *ACS Nano* **2019**, *13*, 2370–2378. [CrossRef] [PubMed]
43. González-Rubio, G.; Díaz-Núñez, P.; Rivera, A.; Prada, A.; Tardajos, G.; González-Izquierdo, J.; Bañares, L.; Llombart, P.; Macdowell, L.G.; Alcolea Palafox, M.; et al. Femtosecond laser reshaping yields gold nanorods with ultranarrow surface plasmon resonances. *Science* **2017**, *358*, 640–644. [CrossRef] [PubMed]
44. Deng, R.; Qu, H.; Liang, L.; Zhang, J.; Zhang, B.; Huang, D.; Xu, S.; Liang, C.; Xu, W. Tracing the therapeutic process of targeted aptamer/drug conjugate on cancer cells by surface-enhanced Raman scattering spectroscopy. *Anal. Chem.* **2017**, *89*, 2844–2851. [CrossRef] [PubMed]
45. Park, J.E.; Lee, Y.; Nam, J.M. Precisely shaped, uniformly formed gold nanocubes with ultrahigh reproducibility in single-particle scattering and surface-enhanced Raman scattering. *Nano Lett.* **2018**, *18*, 6475–6482. [CrossRef]
46. Matteini, P.; Cottat, M.; Tavanti, F.; Panfilova, E.; Scuderi, M.; Nicotra, G.; Menziani, M.C.; Khlebtsov, N.; de Angelis, M.; Pini, R. Site-selective surface-enhanced Raman detection of proteins. *ACS Nano* **2017**, *11*, 918–926. [CrossRef]
47. Guo, P.; Sikdar, D.; Huang, X.; Si, K.J.; Xiong, W.; Gong, S.; Yap, L.W.; Premaratne, M.; Cheng, W. Plasmonic core–shell nanoparticles for SERS detection of the pesticide thiram: size-and shape-dependent Raman enhancement. *Nanoscale* **2015**, *7*, 2862–2868. [CrossRef]
48. Jana, D.; Gorunmez, Z.; He, J.; Bruzas, I.; Beck, T.; Sagle, L. Surface enhanced Raman spectroscopy of a Au@Au core–shell structure containing a spiky Shell. *J. Phys. Chem. C* **2016**, *120*, 20814–20821. [CrossRef]
49. Li, J.F.; Zhang, Y.J.; Ding, S.Y.; Panneerselvam, R.; Tian, Z.Q. Core–shell nanoparticle-enhanced Raman spectroscopy. *Chem. Rev.* **2017**, *117*, 5002–5069. [CrossRef]
50. Zhang, P.; Li, Y.; Wang, D.; Xia, H. High–Yield Production of Uniform Gold Nanoparticles with Sizes from 31 to 577 nm via One-Pot Seeded Growth and Size-Dependent SERS Property. *Part. Part. Syst. Charact.* **2016**, *33*, 924–932. [CrossRef]
51. Zhang, D.; Tang, L.; Chen, J.; Tang, Z.; Liang, P.; Huang, Y.; Cao, M.; Zou, M.; Ni, D.; Chen, J.; et al. Controllable self-assembly of SERS hotspots in liquid environment. *Langmuir* **2021**, *37*, 939–948. [CrossRef] [PubMed]
52. Shin, J.; Lee, S.; Yoo, S.; Jung, I.; Lee, S.; Kim, J.; Son, J.; Kim, J.E.; Kim, J.M.; Nam, J.M.; et al. Enormous enhancement in single-particle surface-enhanced Raman scattering with size-controllable Au double nanorings. *Chem. Mater.* **2022**, *34*, 2197–2205. [CrossRef]
53. Cho, N.H.; Byun, G.H.; Lim, Y.C.; Im, S.W.; Kim, H.; Lee, H.E.; Ahn, H.Y.; Nam, K.T. Uniform chiral gap synthesis for high dissymmetry factor in single plasmonic gold nanoparticle. *ACS Nano* **2020**, *14*, 3595–3602. [CrossRef] [PubMed]
54. He, H.; Zhou, L.; Guo, Z.; Li, P.; Gao, S.; Liu, Z. Dual Biomimetic Recognition-Driven Plasmonic Nanogap-Enhanced Raman Scattering for Ultrasensitive Protein Fingerprinting and Quantitation. *Nano Lett.* **2022**, *22*, 9664–9671. [CrossRef] [PubMed]
55. Niu, R.; Song, C.; Gao, F.; Fang, W.; Jiang, X.; Ren, S.; Zhu, D.; Su, S.; Chao, J.; Chen, S.; et al. DNA Origami-Based Nanoprinting for the Assembly of Plasmonic Nanostructures with Single-Molecule Surface-Enhanced Raman Scattering. *Angew. Chem. Int. Ed.* **2021**, *60*, 11695–11701. [CrossRef] [PubMed]
56. Kumar, S.; Kumar, A.; Kim, G.H.; Rhim, W.K.; Hartman, K.L.; Nam, J.M. Myoglobin and polydopamine-engineered Raman nanoprobes for detecting, imaging, and monitoring reactive oxygen species in biological samples and living cells. *Small* **2017**, *13*, 1701584. [CrossRef] [PubMed]
57. Hong, F.; Zhang, F.; Liu, Y.; Yan, H. DNA origami: scaffolds for creating higher order structures. *Chem. Rev.* **2017**, *117*, 12584–12640. [CrossRef]
58. Liu, N.; Liedl, T. DNA-assembled advanced plasmonic architectures. *Chem. Rev.* **2018**, *118*, 3032–3053. [CrossRef]
59. Liu, W.; Halverson, J.; Tian, Y.; Tkachenko, A.V.; Gang, O. Self-organized architectures from assorted DNA-framed nanoparticles. *Nat. Chem.* **2016**, *8*, 867–873. [CrossRef]
60. Liu, Z.; Pei, H.; Zhang, L.; Tian, Y. Mitochondria-targeted DNA nanoprobe for real-time imaging and simultaneous quantification of Ca^{2+} and pH in neurons. *ACS Nano* **2018**, *12*, 12357–12368. [CrossRef]
61. Huang, X.; Zhao, W.; Chen, X.; Li, J.; Ye, H.; Li, C.; Yin, X.; Zhou, X.; Qiao, X.; Xue, Z.; et al. Gold nanoparticle-bridge array to improve DNA hybridization efficiency of SERS sensors. *J. Am. Chem. Soc.* **2022**, *144*, 17533–17539. [CrossRef] [PubMed]
62. Niu, R.; Gao, F.; Wang, D.; Zhu, D.; Su, S.; Chen, S.; YuWen, L.; Fan, C.; Wang, L.; Chao, J. Pattern recognition directed assembly of Plasmonic gap nanostructures for single-molecule SERS. *ACS Nano* **2022**, *16*, 14622–14631. [CrossRef]
63. Fang, W.; Jia, S.; Chao, J.; Wang, L.; Duan, X.; Liu, H.; Li, Q.; Zuo, X.; Wang, L.; Wang, L.; et al. Quantizing single-molecule surface-enhanced Raman scattering with DNA origami metamolecules. *Sci. Adv.* **2019**, *5*, eaau4506. [CrossRef] [PubMed]
64. Simoncelli, S.; Roller, E.M.; Urban, P.; Schreiber, R.; Turberfield, A.J.; Liedl, T.; Lohmuller, T. Quantitative single-molecule surface-enhanced Raman scattering by optothermal tuning of DNA origami-assembled plasmonic nanoantennas. *ACS Nano* **2016**, *10*, 9809–9815. [CrossRef] [PubMed]
65. Liu, B.; Song, C.; Zhu, D.; Wang, X.; Zhao, M.; Yang, Y.; Zhang, Y.; Su, S.; Shi, J.; Chao, J.; et al. DNA-Origami-based assembly of anisotropic plasmonic gold nanostructures. *Small* **2017**, *13*, 1603991. [CrossRef] [PubMed]
66. Tanwar, S.; Haldar, K.K.; Sen, T. DNA origami directed Au nanostar dimers for single-molecule surface-enhanced Raman scattering. *J. Am. Chem. Soc.* **2017**, *139*, 17639–17648. [CrossRef]

67. Zhan, P.; Wen, T.; Wang, Z.g.; He, Y.; Shi, J.; Wang, T.; Liu, X.; Lu, G.; Ding, B. DNA origami directed assembly of gold bowtie nanoantennas for single-molecule surface-enhanced Raman scattering. *Angew. Chem. Int. Ed.* **2018**, *57*, 2846–2850. [CrossRef]
68. Lim, D.K.; Jeon, K.S.; Kim, H.M.; Nam, J.M.; Suh, Y.D. Nanogap-engineerable Raman-active nanodumbbells for single-molecule detection. *Nat. Mater.* **2010**, *9*, 60–67. [CrossRef]
69. Gandra, N.; Abbas, A.; Tian, L.; Singamaneni, S. Plasmonic planet–satellite analogues: hierarchical self-assembly of gold nanostructures. *Nano Lett.* **2012**, *12*, 2645–2651. [CrossRef]
70. Li, A.; Tang, L.; Song, D.; Song, S.; Ma, W.; Xu, L.; Kuang, H.; Wu, X.; Liu, L.; Chen, X.; et al. A SERS-active sensor based on heterogeneous gold nanostar core–silver nanoparticle satellite assemblies for ultrasensitive detection of aflatoxinB1. *Nanoscale* **2016**, *8*, 1873–1878. [CrossRef]
71. Li, Q.; Ge, X.; Ye, J.; Li, Z.; Su, L.; Wu, Y.; Yang, H.; Song, J. Dual ratiometric SERS and photoacoustic core–satellite nanoprobe for quantitatively visualizing hydrogen peroxide in inflammation and cancer. *Angew. Chem. Int. Ed.* **2021**, *60*, 7323–7332. [CrossRef] [PubMed]
72. Hu, C.; Hu, Y.; Fan, C.; Yang, L.; Zhang, Y.; Li, H.; Xie, W. Surface-enhanced Raman spectroscopic evidence of key intermediate species and role of NiFe dual-catalytic center in water oxidation. *Angew. Chem. Int. Ed.* **2021**, *60*, 19774–19778. [CrossRef] [PubMed]
73. Rodríguez-Lorenzo, L.; Alvarez-Puebla, R.A.; Pastoriza-Santos, I.; Mazzucco, S.; Stéphan, O.; Kociak, M.; Liz-Marzán, L.M.; García de Abajo, F.J. Zeptomol detection through controlled ultrasensitive surface-enhanced Raman scattering. *J. Am. Chem. Soc.* **2009**, *131*, 4616–4618. [CrossRef] [PubMed]
74. Su, Q.; Ma, X.; Dong, J.; Jiang, C.; Qian, W. A reproducible SERS substrate based on electrostatically assisted APTES-functionalized surface-assembly of gold nanostars. *ACS Appl. Mater. Interfaces* **2011**, *3*, 1873–1879. [CrossRef] [PubMed]
75. Fortuni, B.; Fujita, Y.; Ricci, M.; Inose, T.; Aubert, R.; Lu, G.; Hutchison, J.A.; Hofkens, J.; Latterini, L.; Uji-i, H. A novel method for in situ synthesis of SERS-active gold nanostars on polydimethylsiloxane film. *Chem. Commun.* **2017**, *53*, 5121–5124. [CrossRef]
76. Lee, J.; Hua, B.; Park, S.; Ha, M.; Lee, Y.; Fan, Z.; Ko, H. Tailoring surface plasmons of high-density gold nanostar assemblies on metal films for surface-enhanced Raman spectroscopy. *Nanoscale* **2014**, *6*, 616–623. [CrossRef] [PubMed]
77. Zrimsek, A.B.; Henry, A.I.; Van Duyne, R.P. Single molecule surface-enhanced Raman spectroscopy without nanogaps. *J. Phys. Chem. Lett.* **2013**, *4*, 3206–3210. [CrossRef]
78. Das, G.; Chirumamilla, M.; Toma, A.; Gopalakrishnan, A.; Zaccaria, R.P.; Alabastri, A.; Leoncini, M.; Di Fabrizio, E. Plasmon based biosensor for distinguishing different peptides mutation states. *Sci. Rep.* **2013**, *3*, 1792. [CrossRef]
79. Chirumamilla, M.; Toma, A.; Gopalakrishnan, A.; Das, G.; Zaccaria, R.P.; Krahne, R.; Rondanina, E.; Leoncini, M.; Liberale, C.; De Angelis, F.; et al. 3D nanostar dimers with a sub-10-nm gap for single-/few-molecule surface-enhanced Raman scattering. *Adv. Mater.* **2014**, *26*, 2353–2358. [CrossRef]
80. Gopalakrishnan, A.; Chirumamilla, M.; De Angelis, F.; Toma, A.; Zaccaria, R.P.; Krahne, R. Bimetallic 3D nanostar dimers in ring cavities: recyclable and robust surface-enhanced Raman scattering substrates for signal detection from few molecules. *ACS Nano* **2014**, *8*, 7986–7994. [CrossRef]
81. Liu, Y.; Wu, H.; Ma, L.; Zou, S.; Ling, Y.; Zhang, Z. Highly stable and active SERS substrates with Ag–Ti alloy nanorods. *Nanoscale* **2018**, *10*, 19863–19870. [CrossRef] [PubMed]
82. Han, W.; Stepula, E.; Philippi, M.; Schlücker, S.; Steinhart, M. Evaluation of 3D gold nanodendrite layers obtained by templated galvanic displacement reactions for SERS sensing and heterogeneous catalysis. *Nanoscale* **2018**, *10*, 20671–20680. [CrossRef] [PubMed]
83. Ge, M.; Li, P.; Zhou, G.; Chen, S.; Han, W.; Qin, F.; Nie, Y.; Wang, Y.; Qin, M.; Huang, G.; et al. General surface-enhanced Raman spectroscopy method for actively capturing target molecules in small gaps. *J. Am. Chem. Soc.* **2021**, *143*, 7769–7776. [CrossRef] [PubMed]
84. Kim, K.; Han, H.S.; Choi, I.; Lee, C.; Hong, S.; Suh, S.H.; Lee, L.P.; Kang, T. Interfacial liquid-state surface-enhanced Raman spectroscopy. *Nat. Commun.* **2013**, *4*, 2182. [CrossRef] [PubMed]
85. Si, S.; Liang, W.; Sun, Y.; Huang, J.; Ma, W.; Liang, Z.; Bao, Q.; Jiang, L. Facile fabrication of high-density sub-1-nm gaps from Au nanoparticle monolayers as reproducible SERS substrates. *Adv. Funct. Mater.* **2016**, *26*, 8137–8145. [CrossRef]
86. Tian, L.; Su, M.; Yu, F.; Xu, Y.; Li, X.; Li, L.; Liu, H.; Tan, W. Liquid-state quantitative SERS analyzer on self-ordered metal liquid-like plasmonic arrays. *Nat. Commun.* **2018**, *9*, 3642. [CrossRef] [PubMed]
87. Shi, L.; Liu, M.; Zhang, L.; Tian, Y. A liquid interfacial SERS platform on a nanoparticle array stabilized by rigid probes for the quantification of norepinephrine in rat brain microdialysates. *Angew. Chem. Int. Ed.* **2022**, *61*, e202117125.
88. Liu, J.; Yin, D.; Wang, S.; Chen, H.Y.; Liu, Z. Probing Low-Copy-Number Proteins in a Single Living Cell. *Angew. Chem. Int. Ed.* **2016**, *55*, 13215–13218. [CrossRef]
89. Liu, J.; Wen, Y.; He, H.; Chen, H.Y.; Liu, Z. Probing cytoplasmic and nuclear microRNAs in single living cells via plasmonic affinity sandwich assay. *Chem. Sci.* **2018**, *9*, 7241–7246. [CrossRef]
90. Tian, T.; Yi, J.; Liu, Y.; Li, B.; Liu, Y.; Qiao, L.; Zhang, K.; Liu, B. Self-assembled plasmonic nanoarrays for enhanced bacterial identification and discrimination. *Biosens. Bioelectron.* **2022**, *197*, 113778. [CrossRef]
91. Li, J.F.; Huang, Y.F.; Ding, Y.; Yang, Z.L.; Li, S.B.; Zhou, X.S.; Fan, F.R.; Zhang, W.; Zhou, Z.Y.; Wu, D.Y.; et al. Shell-isolated nanoparticle-enhanced Raman spectroscopy. *Nature* **2010**, *464*, 392–395. [CrossRef] [PubMed]

92. Tian, X.D.; Liu, B.J.; Li, J.F.; Yang, Z.L.; Ren, B.; Tian, Z.Q. SHINERS and plasmonic properties of Au Core SiO_2 shell nanoparticles with optimal core size and shell thickness. *J. Raman Spectrosc.* **2013**, *44*, 994–998. [CrossRef]
93. Duan, H.; Hu, H.; Kumar, K.; Shen, Z.; Yang, J.K. Direct and reliable patterning of plasmonic nanostructures with sub-10-nm gaps. *ACS Nano* **2011**, *5*, 7593–7600. [CrossRef] [PubMed]
94. Tu, M.; Xia, B.; Kravchenko, D.E.; Tietze, M.L.; Cruz, A.J.; Stassen, I.; Hauffman, T.; Teyssandier, J.; De Feyter, S.; Wang, Z.; et al. Direct X-ray and electron-beam lithography of halogenated zeolitic imidazolate frameworks. *Nat. Mater.* **2021**, *20*, 93–99. [CrossRef] [PubMed]
95. Scherrer, D.; Vogel, D.; Drechsler, U.; Olziersky, A.; Sparr, C.; Mayor, M.; Lörtscher, E. Monitoring Solid-Phase Reactions in Self-Assembled Monolayers by Surface-Enhanced Raman Spectroscopy. *Angew. Chem. Int. Ed.* **2021**, *60*, 17981–17988. [CrossRef]
96. Schröder, T.; Trusheim, M.E.; Walsh, M.; Li, L.; Zheng, J.; Schukraft, M.; Sipahigil, A.; Evans, R.E.; Sukachev, D.D.; Nguyen, C.T.; et al. Scalable focused ion beam creation of nearly lifetime-limited single quantum emitters in diamond nanostructures. *Nat. Commun.* **2017**, *8*, 15376. [CrossRef] [PubMed]
97. Park, S.G.; Xiao, X.; Min, J.; Mun, C.; Jung, H.S.; Giannini, V.; Weissleder, R.; Maier, S.A.; Im, H.; Kim, D.H. Self-assembly of nanoparticle-spiked pillar arrays for plasmonic biosensing. *Adv. Funct. Mater.* **2019**, *29*, 1904257. [CrossRef]
98. Lee, S.; Lee, S.; Son, J.; Kim, J.M.; Lee, J.; Yoo, S.; Haddadnezhad, M.; Shin, J.; Kim, J.; Nam, J.M.; et al. Web-above-a-Ring (WAR) and Web-above-a-Lens (WAL): Nanostructures for Highly Engineered Plasmonic-Field Tuning and SERS Enhancement. *Small* **2021**, *17*, 2101262. [CrossRef]
99. Flauraud, V.; Mastrangeli, M.; Bernasconi, G.D.; Butet, J.; Alexander, D.T.; Shahrabi, E.; Martin, O.J.; Brugger, J. Nanoscale topographical control of capillary assembly of nanoparticles. *Nat. Nanotechnol.* **2017**, *12*, 73–80. [CrossRef]
100. Li, J.; Wuethrich, A.; Sina, A.A.; Cheng, H.H.; Wang, Y.; Behren, A.; Mainwaring, P.N.; Trau, M. A digital single-molecule nanopillar SERS platform for predicting and monitoring immune toxicities in immunotherapy. *Nat. Commun.* **2021**, *12*, 1087. [CrossRef]
101. Zhang, H.; Liu, M.; Zhou, F.; Liu, D.; Liu, G.; Duan, G.; Cai, W.; Li, Y. Physical deposition improved SERS stability of morphology controlled periodic micro/nanostructured arrays based on colloidal templates. *Small* **2015**, *11*, 844–853. [CrossRef]
102. Luo, S.; Mancini, A.; Wang, F.; Liu, J.; Maier, S.A.; de Mello, J.C. High-throughput fabrication of triangular nanogap arrays for surface-enhanced Raman spectroscopy. *ACS Nano* **2022**, *16*, 7438–7447. [CrossRef] [PubMed]
103. Guan, Y.; Ai, B.; Wang, Z.; Chen, C.; Zhang, W.; Wang, Y.; Zhang, G. In situ chemical patterning technique. *Adv. Funct. Mater.* **2022**, *32*, 2107945. [CrossRef]
104. Zheng, C.; Shen, Y.; Liu, M.; Liu, W.; Wu, S.; Jin, C. Layer-by-layer assembly of three-dimensional optical functional nanostructures. *ACS Nano* **2019**, *13*, 5583–5590. [CrossRef] [PubMed]
105. Wang, Y.; Zhang, M.; Feng, L.; Dong, B.; Xu, T.; Li, D.; Jiang, L.; Chi, L. Tape-imprinted hierarchical lotus seedpod-like arrays for extraordinary surface-enhanced Raman spectroscopy. *Small* **2019**, *15*, 1804527. [CrossRef] [PubMed]
106. Lin, D.; Wu, Z.; Li, S.; Zhao, W.; Ma, C.; Wang, J.; Jiang, Z.; Zhong, Z.; Zheng, Y.; Yang, X. Large-area Au-nanoparticle-functionalized Si nanorod arrays for spatially uniform surface-enhanced Raman spectroscopy. *ACS Nano* **2017**, *11*, 1478–1487. [CrossRef]
107. Kamil Reza, K.; Wang, J.; Vaidyanathan, R.; Dey, S.; Wang, Y.; Trau, M. Electrohydrodynamic-induced SERS immunoassay for extensive multiplexed biomarker sensing. *Small* **2017**, *13*, 1602902. [CrossRef]
108. Lai, Y.; Schlücker, S.; Wang, Y. Rapid and sensitive SERS detection of the cytokine tumor necrosis factor alpha (tnf-α) in a magnetic bead pull-down assay with purified and highly Raman-active gold nanoparticle clusters. *Anal. Bioanal. Chem.* **2018**, *410*, 5993–6000. [CrossRef]
109. Sánchez-Purrà, M.; Carré-Camps, M.; de Puig, H.; Bosch, I.; Gehrke, L.; Hamad-Schifferli, K. Surface-enhanced Raman spectroscopy-based sandwich immunoassays for multiplexed detection of Zika and Dengue viral biomarkers. *ACS Infect. Dis.* **2017**, *3*, 767–776. [CrossRef]
110. Hu, S.W.; Qiao, S.; Pan, J.B.; Kang, B.; Xu, J.J.; Chen, H.Y. A paper-based SERS test strip for quantitative detection of Mucin-1 in whole blood. *Talanta* **2018**, *179*, 9–14. [CrossRef]
111. Tran, V.; Walkenfort, B.; König, M.; Salehi, M.; Schlücker, S. Rapid, quantitative, and ultrasensitive point-of-care testing: A portable SERS reader for lateral flow assays in clinical chemistry. *Angew. Chem. Int. Ed.* **2019**, *58*, 442–446. [CrossRef] [PubMed]
112. Wang, Y.; Ruan, Q.; Lei, Z.C.; Lin, S.C.; Zhu, Z.; Zhou, L.; Yang, C. Highly sensitive and automated surface enhanced Raman scattering-based immunoassay for H5N1 detection with digital microfluidics. *Anal. Chem.* **2018**, *90*, 5224–5231. [CrossRef]
113. Liu, J.; He, H.; Xie, D.; Wen, Y.; Liu, Z. Probing low-copy-number proteins in single living cells using single-cell plasmonic immunosandwich assays. *Nat. Protoc.* **2021**, *16*, 3522–3546. [CrossRef] [PubMed]
114. Xie, D.; Wen, Y.; Chen, J.; Guo, Z.; Li, P.; Liu, Z. Probing Protein 4′-Phosphopantetheinylation in Single Living Cells. *Anal. Chem.* **2023**, *95*, 7229–7236. [CrossRef] [PubMed]
115. Wen, Y.; Zhao, J.; He, H.; Zhao, Q.; Liu, Z. Multiplexed single-cell plasmonic immunoassay of intracellular signaling proteins enables non-destructive monitoring of cell fate. *Anal. Chem.* **2021**, *93*, 14204–14213. [CrossRef] [PubMed]
116. Muhammad, P.; Tu, X.; Liu, J.; Wang, Y.; Liu, Z. Molecularly imprinted plasmonic substrates for specific and ultrasensitive immunoassay of trace glycoproteins in biological samples. *ACS Appl. Mater. Interfaces* **2017**, *9*, 12082–12091. [CrossRef] [PubMed]

117. Zhou, L.; Wang, Y.; Xing, R.; Chen, J.; Liu, J.; Li, W.; Liu, Z. Orthogonal dual molecularly imprinted polymer-based plasmonic immunosandwich assay: a double characteristic recognition strategy for specific detection of glycoproteins. *Biosens. Bioelectron.* **2019**, *145*, 111729. [CrossRef]
118. Xing, R.; Wen, Y.; Dong, Y.; Wang, Y.; Zhang, Q.; Liu, Z. Dual molecularly imprinted polymer-based plasmonic immunosandwich assay for the specific and sensitive detection of protein biomarkers. *Anal. Chem.* **2019**, *91*, 9993–10000. [CrossRef]
119. Pang, J.; Li, P.; He, H.; Xu, S.; Liu, Z. Molecularly imprinted polymers outperform lectin counterparts and enable more precise cancer diagnosis. *Chem. Sci.* **2022**, *13*, 4589–4597. [CrossRef]
120. Guo, Z.; Zhang, Q.; Xing, R.; Liu, Z. Molecularly imprinted and cladded polymers for constructing a portable plasmonic immunoassay for peptides in biofluids. *Chem. Commun.* **2023**, *59*, 3075–3078. [CrossRef]
121. Wu, J.; Zhou, X.; Li, P.; Lin, X.; Wang, J.; Hu, Z.; Zhang, P.; Chen, D.; Cai, H.; Niessner, R.; et al. Ultrasensitive and simultaneous SERS detection of multiplex microRNA using fractal gold nanotags for early diagnosis and prognosis of hepatocellular carcinoma. *Anal. Chem.* **2021**, *93*, 8799–8809. [CrossRef] [PubMed]
122. Zhou, W.; Tian, Y.F.; Yin, B.C.; Ye, B.C. Simultaneous surface-enhanced Raman spectroscopy detection of multiplexed microRNA biomarkers. *Anal. Chem.* **2017**, *89*, 6120–6128. [CrossRef] [PubMed]
123. Ma, D.; Huang, C.; Zheng, J.; Tang, J.; Li, J.; Yang, J.; Yang, R. Quantitative detection of exosomal microRNA extracted from human blood based on surface-enhanced Raman scattering. *Biosens. Bioelectron.* **2018**, *101*, 167–173. [CrossRef] [PubMed]
124. Lee, J.U.; Kim, W.H.; Lee, H.S.; Park, K.H.; Sim, S.J. Quantitative and specific detection of exosomal miRNAs for accurate diagnosis of breast cancer using a surface-enhanced Raman scattering sensor based on plasmonic head-flocked gold nanopillars. *Small* **2019**, *15*, 1804968. [CrossRef] [PubMed]
125. Pang, Y.; Wang, C.; Lu, L.; Wang, C.; Sun, Z.; Xiao, R. Dual-SERS biosensor for one-step detection of microRNAs in exosome and residual plasma of blood samples for diagnosing pancreatic cancer. *Biosens. Bioelectron.* **2019**, *130*, 204–213. [CrossRef] [PubMed]
126. Jiang, S.; Li, Q.; Wang, C.; Pang, Y.; Sun, Z.; Xiao, R. In situ exosomal MicroRNA determination by target-triggered SERS and Fe_3O_4@ TiO_2-based exosome accumulation. *ACS Sens.* **2021**, *6*, 852–862. [CrossRef] [PubMed]
127. Zhou, H.; Zhang, J.; Li, B.; Liu, J.; Xu, J.J.; Chen, H.Y. Dual-mode SERS and electrochemical detection of miRNA based on popcorn-like gold nanofilms and toehold-mediated strand displacement amplification reaction. *Anal. Chem.* **2021**, *93*, 6120–6127. [CrossRef]
128. Zhang, Q.; Liu, J.; Dong, Y.; Li, W.; Xing, R.; Ma, Y.; Liu, Z. Gold nanoparticle-decorated Ag@SiO_2 nanocomposite-based plasmonic affinity sandwich assay of circulating microRNAs in human serum. *ACS Appl. Nano Mater.* **2019**, *2*, 3960–3970. [CrossRef]
129. Xie, D.; Wen, Y.; Chen, J.; Lu, H.; He, H.; Liu, Z. Probing Queuosine Modifications of Transfer RNA in Single Living Cells via Plasmonic Affinity Sandwich Assay. *Anal. Chem.* **2022**, *94*, 12828–12835. [CrossRef]
130. Zhou, Q.; Zheng, J.; Qing, Z.; Zheng, M.; Yang, J.; Yang, S.; Ying, L.; Yang, R. Detection of circulating tumor DNA in human blood via DNA-mediated surface-enhanced Raman spectroscopy of single-walled carbon nanotubes. *Anal. Chem.* **2016**, *88*, 4759–4765. [CrossRef]
131. Fu, X.; Cheng, Z.; Yu, J.; Choo, P.; Chen, L.; Choo, J. A SERS-based lateral flow assay biosensor for highly sensitive detection of HIV-1 DNA. *Biosens. Bioelectron.* **2016**, *78*, 530–537. [CrossRef]
132. Wu, L.; Xiao, X.; Chen, K.; Yin, W.; Li, Q.; Wang, P.; Lu, Z.; Ma, J.; Han, H. Ultrasensitive SERS detection of Bacillus thuringiensis special gene based on Au@ Ag NRs and magnetic beads. *Biosens. Bioelectron.* **2017**, *92*, 321–327. [CrossRef] [PubMed]
133. Ngo, H.T.; Freedman, E.; Odion, R.A.; Strobbia, P.; De Silva Indrasekara, A.S.; Vohra, P.; Taylor, S.M.; Vo-Dinh, T. Direct detection of unamplified pathogen RNA in blood lysate using an integrated lab-in-a-stick device and ultrabright SERS nanorattles. *Sci. Rep.* **2018**, *8*, 4075. [CrossRef]
134. Zhang, Y.; Wang, Z.; Wu, L.; Zong, S.; Yun, B.; Cui, Y. Combining multiplex SERS nanovectors and multivariate analysis for in situ profiling of circulating tumor cell phenotype using a microfluidic chip. *Small* **2018**, *14*, 1704433. [CrossRef] [PubMed]
135. Wu, L.; Teixeira, A.; Garrido-Maestu, A.; Muinelo-Romay, L.; Lima, L.; Santos, L.L.; Prado, M.; Diéguez, L. Profiling DNA mutation patterns by SERS fingerprinting for supervised cancer classification. *Biosens. Bioelectron.* **2020**, *165*, 112392. [CrossRef] [PubMed]
136. Liu, Y.; Lyu, N.; Rajendran, V.K.; Piper, J.; Rodger, A.; Wang, Y. Sensitive and direct DNA mutation detection by surface-enhanced Raman spectroscopy using rational designed and tunable plasmonic nanostructures. *Anal. Chem.* **2020**, *92*, 5708–5716. [CrossRef] [PubMed]
137. Moitra, P.; Chaichi, A.; Hasan, S.M.A.; Dighe, K.; Alafeef, M.; Prasad, A.; Gartia, M.R.; Pan, D. Probing the mutation independent interaction of DNA probes with SARS-CoV-2 variants through a combination of surface-enhanced Raman scattering and machine learning. *Biosens. Bioelectron.* **2022**, *208*, 114200. [CrossRef] [PubMed]
138. Choi, J.H.; Shin, M.; Yang, L.; Conley, B.; Yoon, J.; Lee, S.N.; Lee, K.B.; Choi, J.W. Clustered regularly interspaced short palindromic repeats-mediated amplification-free detection of viral DNAs using surface-enhanced Raman spectroscopy-active nanoarray. *ACS Nano* **2021**, *15*, 13475–13485. [CrossRef]
139. Yeh, Y.T.; Gulino, K.; Zhang, Y.; Sabestien, A.; Chou, T.W.; Zhou, B.; Lin, Z.; Albert, I.; Lu, H.; Swaminathan, V.; et al. A rapid and label-free platform for virus capture and identification from clinical samples. *Proc. Natl. Acad. Sci. USA* **2020**, *117*, 895–901. [CrossRef]

140. Park, K.S.; Choi, A.; Kim, H.J.; Park, I.; Eom, M.S.; Yeo, S.G.; Son, R.G.; Park, T.I.; Lee, G.; Soh, H.T.; et al. Ultra-sensitive label-free SERS biosensor with high-throughput screened DNA aptamer for universal detection of SARS-CoV-2 variants from clinical samples. *Biosens. Bioelectron.* **2023**, *228*, 115202. [CrossRef]
141. Kamińska, A.; Witkowska, E.; Winkler, K.; Dzięcielewski, I.; Weyher, J.L.; Waluk, J. Detection of Hepatitis B virus antigen from human blood: SERS immunoassay in a microfluidic system. *Biosens. Bioelectron.* **2015**, *66*, 461–467. [CrossRef] [PubMed]
142. Sun, Y.; Xu, L.; Zhang, F.; Song, Z.; Hu, Y.; Ji, Y.; Shen, J.; Li, B.; Lu, H.; Yang, H. A promising magnetic SERS immunosensor for sensitive detection of avian influenza virus. *Biosens. Bioelectron.* **2017**, *89*, 906–912. [CrossRef] [PubMed]
143. Camacho, S.A.; Sobral-Filho, R.G.; Aoki, P.H.B.; Constantino, C.J.L.; Brolo, A.G. Zika immunoassay based on surface-enhanced Raman scattering nanoprobes. *ACS Sens.* **2018**, *3*, 587–594. [CrossRef] [PubMed]
144. Wang, C.; Wang, C.; Wang, X.; Wang, K.; Zhu, Y.; Rong, Z.; Wang, W.; Xiao, R.; Wang, S. Magnetic SERS strip for sensitive and simultaneous detection of respiratory viruses. *ACS Appl. Mater. Interfaces* **2019**, *11*, 19495–19505. [CrossRef] [PubMed]
145. Chen, H.; Park, S.G.; Choi, N.; Moon, J.I.; Dang, H.; Das, A.; Lee, S.; Kim, D.G.; Chen, L.; Choo, J. SERS imaging-based aptasensor for ultrasensitive and reproducible detection of influenza virus A. *Biosens. Bioelectron.* **2020**, *167*, 112496. [CrossRef] [PubMed]
146. Zhuang, J.; Zhao, Z.; Lian, K.; Yin, L.; Wang, J.; Man, S.; Liu, G.; Ma, L. SERS-based CRISPR/Cas assay on microfluidic paper analytical devices for supersensitive detection of pathogenic bacteria in foods. *Biosens. Bioelectron.* **2022**, *207*, 114167. [CrossRef] [PubMed]
147. Yadav, S.; Sadique, M.A.; Ranjan, P.; Kumar, N.; Singhal, A.; Srivastava, A.K.; Khan, R. SERS based lateral flow immunoassay for point-of-care detection of SARS-CoV-2 in clinical samples. *ACS Appl. Bio Mater.* **2021**, *4*, 2974–2995. [CrossRef]
148. Sitjar, J.; Liao, J.D.; Lee, H.; Tsai, H.P.; Wang, J.R.; Liu, P.Y. Challenges of SERS technology as a non-nucleic acid or-antigen detection method for SARS-CoV-2 virus and its variants. *Biosens. Bioelectron.* **2021**, *181*, 113153. [CrossRef]
149. Liu, H.; Dai, E.; Xiao, R.; Zhou, Z.; Zhang, M.; Bai, Z.; Shao, Y.; Qi, K.; Tu, J.; Wang, C.; et al. Development of a SERS-based lateral flow immunoassay for rapid and ultra-sensitive detection of anti-SARS-CoV-2 IgM/IgG in clinical samples. *Sens. Actuators B* **2021**, *329*, 129196. [CrossRef]
150. Zhang, M.; Li, X.; Pan, J.; Zhang, Y.; Zhang, L.; Wang, C.; Yan, X.; Liu, X.; Lu, G. Ultrasensitive detection of SARS-CoV-2 spike protein in untreated saliva using SERS-based biosensor. *Biosens. Bioelectron.* **2021**, *190*, 113421. [CrossRef]
151. Chen, H.; Park, S.G.; Choi, N.; Kwon, H.J.; Kang, T.; Lee, M.K.; Choo, J. Sensitive detection of SARS-CoV-2 using a SERS-based aptasensor. *ACS Sens.* **2021**, *6*, 2378–2385. [CrossRef] [PubMed]
152. Leong, S.X.; Leong, Y.X.; Tan, E.X.; Sim, H.Y.F.; Koh, C.S.L.; Lee, Y.H.; Chong, C.; Ng, L.S.; Chen, J.R.T.; Pang, D.W.C.; et al. Noninvasive and point-of-care surface-enhanced Raman scattering (SERS)-based breathalyzer for mass screening of coronavirus disease 2019 (COVID-19) under 5 min. *ACS Nano* **2022**, *16*, 2629–2639. [CrossRef]
153. Peng, Y.; Lin, C.; Li, Y.; Gao, Y.; Wang, J.; He, J.; Huang, Z.; Liu, J.; Luo, X.; Yang, Y. Identifying infectiousness of SARS-CoV-2 by ultra-sensitive SnS2 SERS biosensors with capillary effect. *Matter* **2022**, *5*, 694–709. [CrossRef] [PubMed]
154. Shin, H.; Oh, S.; Hong, S.; Kang, M.; Kang, D.; Ji, Y.g.; Choi, B.H.; Kang, K.W.; Jeong, H.; Park, Y.; et al. Early-stage lung cancer diagnosis by deep learning-based spectroscopic analysis of circulating exosomes. *ACS Nano* **2020**, *14*, 5435–5444. [CrossRef] [PubMed]
155. Huang, Z.; Siddhanta, S.; Zheng, G.; Kickler, T.; Barman, I. Rapid, Label-free Optical Spectroscopy Platform for Diagnosis of Heparin-Induced Thrombocytopenia. *Angew. Chem. Int. Ed.* **2020**, *59*, 5972–5978. [CrossRef] [PubMed]
156. Xie, Y.; Su, X.; Wen, Y.; Zheng, C.; Li, M. Artificial intelligent label-free SERS profiling of serum exosomes for breast cancer diagnosis and postoperative assessment. *Nano Lett.* **2022**, *22*, 7910–7918. [CrossRef]
157. Shu, W.; Zhang, M.; Zhang, C.; Li, R.; Pei, C.; Zeng, Y.; Zhao, L.; Zhao, J.; Wan, J. An Alloy Platform of Dual-Fingerprints for High-Performance Stroke Screening. *Adv. Funct. Mater.* **2023**, *33*, 2210267. [CrossRef]
158. Huang, L.; Sun, H.; Sun, L.; Shi, K.; Chen, Y.; Ren, X.; Ge, Y.; Jiang, D.; Liu, X.; Knoll, W.; et al. Rapid, label-free histopathological diagnosis of liver cancer based on Raman spectroscopy and deep learning. *Nat. Commun.* **2023**, *14*, 48. [CrossRef]
159. Lussier, F.; Thibault, V.; Charron, B.; Wallace, G.Q.; Masson, J.F. Deep learning and artificial intelligence methods for Raman and surface-enhanced Raman scattering. *TrAC Trends Anal. Chem.* **2020**, *124*, 115796. [CrossRef]
160. Luo, R.; Popp, J.; Bocklitz, T. Deep learning for Raman spectroscopy: A review. *Analytica* **2022**, *3*, 287–301. [CrossRef]

chemosensors

Review

Aptamer–Molecularly Imprinted Polymer Multiple-Recognition System: Construction and Application

Kangping Ning, Yingzhuo Shen, Yao Yao, Wenzheng Xie, Cheng Ma * and Qin Xu *

School of Chemistry and Chemical Engineering, Yangzhou University, Yangzhou 225002, China; 19859937580nkp@sina.com (K.N.); dx120220073@stu.yzu.edu.cn (Y.S.); dayaoyao11@163.com (Y.Y.); 17855508916@163.com (W.X.)

* Correspondence: chengma@yzu.edu.cn (C.M.); xuqin@yzu.edu.cn (Q.X.)

Abstract: Molecularly imprinted polymers (MIPs) and aptamers (Apts) are widely used in substance detection due to their specific recognition abilities. However, both of them have limitations in terms of stability or sensitivity. Therefore, an increasingly employed strategy is to combine MIPs and aptamers to form mixed components for detecting various substances, such as viruses, bacteria, proteins, heavy-metal ions, and hormones. The aim of this review is to provide a comprehensive summary of the scientific research conducted on the construction and application of aptamer–MIP multiple-recognition components in the past five years. It also aims to analyze their research and development strategies, construction mechanisms, advantages, and potential applications, as well as limitations and current challenges that need to be addressed.

Keywords: molecularly imprinted polymer; aptamer; multiple recognition; detection

Citation: Ning, K.; Shen, Y.; Yao, Y.; Xie, W.; Ma, C.; Xu, Q. Aptamer–Molecularly Imprinted Polymer Multiple-Recognition System: Construction and Application. *Chemosensors* **2023**, *11*, 465. https://doi.org/10.3390/chemosensors11080465

Academic Editor: Ambra Giannetti

Received: 7 July 2023
Revised: 2 August 2023
Accepted: 16 August 2023
Published: 18 August 2023

1. Introduction

The molecular recognition theory has been widely applied in various fields, including separation [1], sensing [2], material manufacturing [3], disease diagnosis [4], and drug delivery [5], due to its significant practical value. Nature provides a plethora of biometric pairs, such as glycans and glycan-binding proteins [6], enzymes and substrates [7], and antigens and antibodies [8]. They are widely used for detecting target molecules because of their outstanding specific recognition and binding ability. However, the identification of target molecules during application is often hindered by environmental interference. Additionally, their practical application is hampered by natural molecular instability, high costs, modification complexities, and limitations on mild conditions. Therefore, it is crucial to develop molecular recognition systems with enhanced properties.

In recent years, molecularly imprinted polymers have emerged as a cost-effective and highly stable alternative [9] to traditional molecular recognition systems, offering superior reusability and performance. Molecular imprinting is a biomimetic technology that utilizes template molecules and crosslinking agents to form three-dimensional cavity structures [10] with specific selection and high affinity through polymerization. Based on the "key-and-lock" principle [11], the obtained cavity can be used for the selective binding of targeted molecules, thus achieving high sensitivity. MIPs have been successfully applied in various scientific research fields, such as environmental pollutant adsorption, biomimetic sensing [12], and solid-phase extraction. However, the complexity of actual samples may result in the retention of large molecules like proteins on the surface of MIPs, leading to the blockage of specific recognition sites. Additionally, solutions still need to be sought for the limited analytical window [13] and limitations posed by non-polar environments.

Aptamers are short RNA/DNA oligonucleotides that specifically bind to targeted complementary molecules [14]. These molecules were first reported by Ellington [15] and Gold [16] in 1990. Aptamers possess several advantages, including high specificity, facile modification, small size, non-immunogenicity [17], and high affinity. As a potential substitute for natural antibodies, aptamers have been widely used in biosensors, clinical diagnosis and therapy, imaging technology, and cellular immunity. Particularly in the field of sensors, aptamer-based biosensors exhibit superior biocompatibility and reproducible functions [18] compared to other types of sensors, such as natural antibody sensors. However, the presence of nucleases everywhere can cause the rapid degradation of aptamers [19], leading to the loss of their three-dimensional structure and function. Meanwhile, there is a need to improve the thermal stability of aptamers and enhance the diversity of the functional groups they possess [20–22].

To address the functional limitations of MIPs and aptamers when used alone, researchers have proposed a hybrid detection system that combines both materials to achieve inherent affinity [23]. The aptamer–MIP hybrid creates a "best of both worlds" scenario. When the aptamer is made to be polymerizable, it can become the recognition part of a MIP by incorporating it into the polymer matrix via polymerizable groups on the aptamer. In this way, the high affinity and specificity of the aptamer are preserved whilst imparting the robustness and added shape specificity generated by the MIPs. The presence of the polymer should protect the aptamer from environmental degradation and potentially widen the scope of use of aptamers for recognition. An aptamer could also be used as an additional recognition monomer and further amplify the sensing signal of the system as a marker. By combining the specificity of MIPs and aptamers, a hybrid can have high sensitivity, a low limit of detection, high stability, and excellent detection results. The successful development of aptamer–MIP hybrid materials would pave the way for a new generation of MIPs and for their use to become more mainstream, offering significant commercial gain. In 2014, Bai and Spivak [24] successfully verified this idea by using the molecularly imprinted aptamer multiple-recognition strategy to construct an efficient virus detection sensor. Since then, more researchers have developed a plethora of hybrids [25,26] by combining the advantages of MIPs and aptamers, which have been effectively employed in detecting antibiotics, viruses, hormones, and other molecules. This review summarizes the preparation methods of the multiple-recognition imprinting layer and MIP–aptamer, as well as comprehensively reviews the application of MIP–aptamer recognition systems in various sensor applications (Figure 1). Furthermore, it discusses and analyzes current shortcomings and future prospects for this technology. This review is expected to provide researchers with a clearer understanding of the advantages of multiple-recognition systems based on molecular imprinting and aptamers while encouraging them to seek solutions for future challenges.

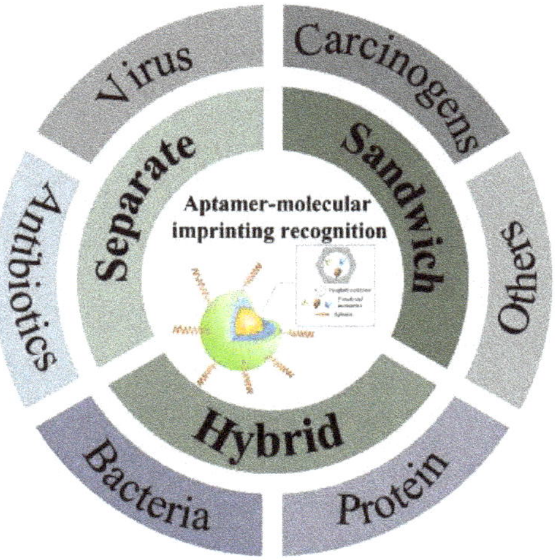

Figure 1. Overview of different types of aptamer–MIP multiple-recognition systems.

2. Acquisition of Molecularly Imprinted Layer

In the aptamer–MIP multiple-recognition system, molecular-imprinting sites play a crucial role in recognition. These sites refer to three-dimensional imprinting cavities obtained through the elution of polymerized imprinting molecules, and the number of these sites depends on the effectiveness of the polymerization process. Currently, thermal polymerization, photopolymerization, the sol–gel method, and electrochemistry are commonly used for polymerizing molecularly imprinted films. Thermal polymerization is generally carried out by heating in a water bath with methacrylic acid (MAA) as a functional monomer, EDGMA (ethylene glycol di-methacrylate) as a crosslinking agent, and AIBN (azodiisobutyronitrile) as an initiator. The severe limitation of the method is the requirement for an organic solvent. This presents obvious drawbacks to their use in environmental and biological applications. Photopolymerization, on the other hand, is typically performed by triggering free radicals using either a high-pressure mercury lamp or a purple interconnector to complete the polymerization of the imprinted layer. The photopolymerization method is relatively easy to operate; however, it has the following disadvantages: long polymerization time, expensive photoinitiators, and poor surface regeneration ability. The sol–gel method possesses numerous significant advantages, such as being simple and low-cost, requiring mild conditions, and providing products with high thermal and mechanical stability [27]. The sol–gel imprinting process occurs by dissolving a metal oxide precursor or metal halides ($M(OR)_n$ or MX_n, with M = Si, Al, Ti, etc.) in a low-molecular-weight solvent medium using a catalyst (acid, base, or ions such as F^-), followed by a hydrolysis (water) and polycondensation step [28]. Silicon alkoxides have been regarded as the preferred molecular precursors for the sol–gel process because they have moderate reactivity but the chemical stability of the Si–C bond is high. During the molecular-imprinting process, alkoxysilane units (Si–OR) are transformed into silanol (Si–OH) with the release of alcohol (R–OH), and then the condensation reaction occurs with the formed silanol group. Both the hydrolysis and condensation reactions will continue simultaneously, leading to the formation of a three-dimensional network. However, the low capacity hinders its wide application. Electrically driven polymerization methods are mainly used for the fabrication of MIP films directly on a conducting substrate. Electropolymerization involves placing a pre-treated electrode into a solution containing a template and specific functional monomers, which undergo electrolysis to generate free radicals under the influence of an electric current. Polymerization is then completed in the presence of template molecules. The mechanism is simply explained as an alternation between chemical and electrode reaction steps. A radical cation is most likely formed during the electro-oxidation step, and then the radical reacts with the monomer and the protonated dimer of the radical formed. These steps follow one after another to form the polymer [29]. The electropolymerization method is low-cost and allows for adjusting the thickness and morphology of the imprinted film while also exhibiting excellent reproducibility. Although there are numerous advantages, the absence of a crosslinker during the electropolymerization process results in fewer recognition sites and poor recognition ability [30].

3. Multiple-Recognition Systems Based on Aptamer and MIPs

3.1. Sandwich Type

In a sandwich assay, the target is bound between a capture antibody and a detection antibody. The capture antibody is immobilized on a surface, while the detection antibody (conjugated to an enzyme or fluorophore label) is applied as the last step before quantitation. Because two antibodies against the same antigen are used, this method is flexible and sensitive. However, it is not always easy or possible to have pairs of antibodies that work well in this type of assay. As described in the introduction section, MIPs and aptamers can be obtained at a large scale using simple methods. They are good candidates for antibodies. The sandwich type consists of a MIP/target/aptamer [31], which combines the specific recognition capabilities of MIPs and aptamers to achieve signal amplification [32]. The MIP material acts as a pre-concentration part for the targets, while the aptamer is responsible for signaling the presence of the target on the MIP platform. The costs and time of MIP and aptamer production are substantially low. Meanwhile, incorporating nanoparticles into this approach serves to independently amplify the signal intensity following secondary recognition, thereby achieving stable and enhanced sensing of signal output. Based on this strategy, Li Tang et al. [33] developed a sensor for the ultrasensitive visual detection of EV71 (Enterovirus 71), which is a plus-stranded RNA virus that causes hand-foot-mouth disease and central nervous system infection [34]. The fabrication process of the sensor is shown in Figure 2. Firstly, Fe_3O_4-coated CD (carbon quantum dot) carriers were constructed, and MIPs were prepared using the sol–gel method with APTES (3-aminopropyltriethoxysilane) and TEOS (ethyl silicate) as functional monomers and crosslinkers. Then, the aptamers were introduced into the imprinting layer. The surface of the imprinting layer was modified with EV71 aptamers, and phenolphthalein was coated on the surface of ZIF-8 (Zeolitic Imidazolate Framework-8). When EV71 was captured by imprinted particles and combined with ZIF-8, the fluorescence signal was quenched, achieving the first signal amplification. Then, when the pH of the solution was adjusted to 12, ZIF-8 decomposed and released loaded phenolphthalein molecules, turning the solution system red and achieving the second signal amplification. Efficient visual detection can be achieved within 20 min of sensor application, with fluorescence and visual detection limits reaching 8.33 fM and 2.08 pM, respectively. The repeatability and stability of this sensor are commendable. After conducting five repeated determination experiments on the same sensor, the fluorescence signal maintains 82% of the initial value. Additionally, even after being stored for 6 months, the sensor still maintains a satisfactory color-rendering function, with an absorbance level at 91.9% of the initial value. Compared with traditional virus detection methods, which are time-consuming, expensive, and insensitive, this sensor allows for low-cost and highly accurate clinical diagnosis from the beginning. The research team's work is expected to be used in future virus-screening efforts to reduce medical staff workload and clinical diagnosis costs.

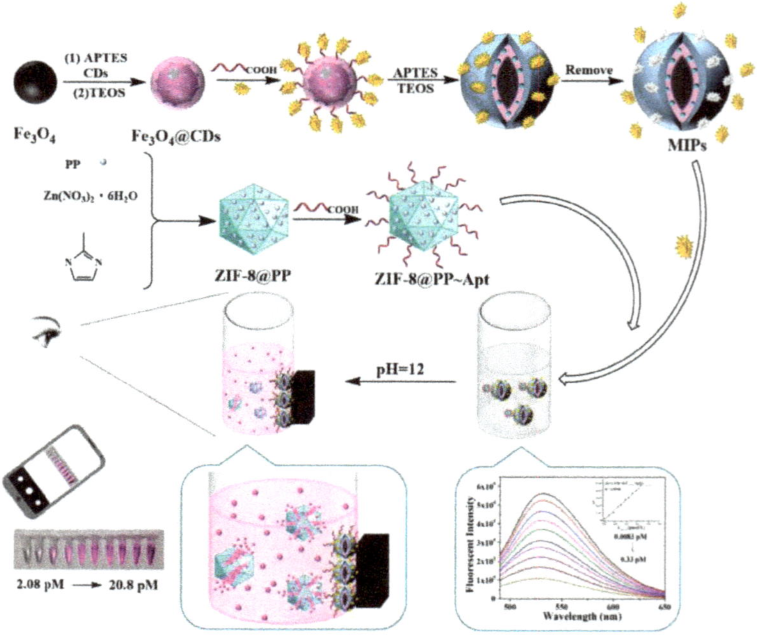

Figure 2. Construction principle and flow diagram of the virus sensor. Reprinted with permission from ref. [33].

Exosomes, which are extracellular vesicles composed of nanoscale lipid bilayers, have emerged as promising biomarkers for early cancer diagnosis and monitoring due to their ability to provide rich molecular information similar to that of parental cells. However, the lack of efficient methods for separating and capturing intact exosomes has hindered the widespread application of exosome analysis platforms in clinical practice. To address this issue, Liao's team [35] developed a more sensitive "on" fluorescent MIP sensor for detecting lysozyme, which is one of the innate immunity proteins carried by the intact exosome. The captured lysozyme was detected through selective fluorescence induced by aptamer-mediated aggregation. A linear relationship between the lysozyme content and fluorescence intensity was obtained. The specific working process is illustrated in Figure 3, where, in the first step, both lysozyme and characteristic exosomes from the sample are added to the surface polymerization system. After the reaction of the template and crosslinker, customized complementary cavities are formed. In the second step, AIMIP ("Artificial intelligence" imprinted polymers) particles selectively capture and enrich targeted analytes. In the third step, after magnetic collection is completed, the aptamers are incubated with a MIP. Utilizing a sandwich induction strategy, the sensor selectively captures AIIP target molecules and modulates the spin of water-soluble AIE (aggregation-induced emission) fluorescence agents through conformational changes in DNA aptamers. The fluorescence effects are enhanced, enabling the specific detection of lysozyme and exosomes within complex proteins through selective "on" fluorescent lighting. This technology has been preliminarily applied to the detection of clinical serum samples. The fluorescence intensity showed a strong linear correlation with the exosome concentration (0.9823), indicating a significant relationship between the two variables. Moreover, the detection limit was approximately 1.3×10^6 capsules/mL, demonstrating high sensitivity and accurate quantification capabilities. Additionally, the sensor exhibited good reproducibility, with coefficients of variation of 2.10×10^{-7} and 2.38×10^{-5}, corresponding to percentages of 6% and 4.5%, respectively. Furthermore, the recovery rate in repeated tests reached as high as 107%, highlighting its potential application in future medical fields.

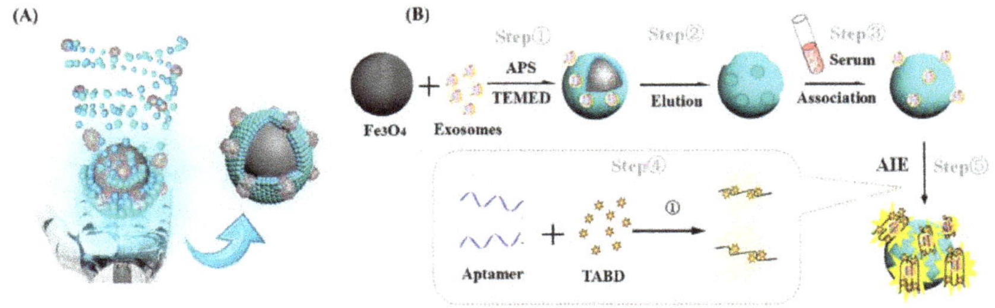

Figure 3. (**A**) Conceptual graph of self-assembly of AIMIPs; (**B**) schematic of the preparation of AIMIPs and AIMIP separation integrated aptamer/AIE fluorescent sensors for exosomes. Reprinted with permission from ref. [35].

In conclusion, the use of a MIP and an aptamer in the sandwich system greatly improves the specific recognition ability of the target and realizes the secondary amplification of the signal, which would improve the sensitivity of the sensor. However, the susceptibility of the aptamer to thermal, chemical, and enzymatic degradation is still the main bottleneck limiting its wide application because of the lack of protective shells around aptamers. Further work should be conducted to improve the stability of aptamers.

3.2. Hybrid Type

Unlike the sandwich method, this hybrid strategy involves directly embedding the aptamer into the molecularly imprinted structure. After the hybrid structure is formed, the molecularly imprinted polymer acts as a protective scaffold for the aptamer molecule, confining it to a specific conformation and effectively separating it from potential interference from DNA enzymes present in the substrate. At the same time, the aptamer can bind to the target molecule specifically to make up for the deficiency of the sensitivity of the molecular imprinting itself. This enhances the stability and specificity of aptamer-mediated recognition. The interaction forces, such as a hydrogen bond, between the target molecule and MIP greatly improve the affinity between the recognition system and the target molecule.

This strategy has been successfully applied to the detection of antibiotics, such as amoxicillin (AMOX). In 2021, Lu's group [36] developed a sensor that specifically detects amoxicillin using a co-deposition technique and a dopamine electropolymerization technique. The construction scheme of the sensor was as follows: (1) a large-surface-area conductive sensing platform was constructed by electrodepositing AuNPs (Au nanoparticles) on the electrode surface; (2) the aptamer–amoxicillin complex was fixed on the electrode surface with the Au-S bond; (3) the molecular-imprinting layer was prepared by dopamine electropolymerization; (4) the imprinted cavity was obtained after elution. CV (cyclic voltammetry) and DPV (differential pulse voltammetry) techniques were used for characterization and quantification by using $Fe(CN)_6{}^{3-}/Fe(CN)_6{}^{4-}$ as probes. Compared to the curve depicted in Figure 4, Curve a presents a pair of standard REDOX peaks. Curve b shows a significant increase in the peak current, indicating that AuNPs/ZnO-rGO can accelerate the electron transfer rate on the electrode surface. In contrast, Curve c exhibits a decrease in peak value due to the formation of a self-assembled insulation layer on the electrode surface, which renders the molecular-imprinting layer non-conductive and impedes electron transfer. Curve d shows the peak current after electropolymerizing dopamine to obtain the blotting layer. Because the imprinting layer is non-conductive, the charge transfer of the probe is blocked, resulting in a significant decrease in the peak current. However, after the removal of template molecules, leaving behind an imprinting cavity, aptamer–AMOX probes can reach the electrode surface through pores and complete electron transfer, resulting in an increase in the peak value for Curve e. Finally, upon the completion of the recognition process and the occupation of imprinting sites, charge transfer is hindered again, leading to a reduction in the peak current for Curve f. During the process of sensor construction, the author utilized an aptamer as a functional monomer, combining its recognition ability with a molecular-imprinting layer to reduce non-specific recognition in complex systems. Additionally, it was discovered that the optimal mole ratio of the aptamer to the template molecule is 1:2. By using this ratio to construct the sensor system, the best detection effect can be achieved, with a detection limit of 3.3×10^{-15} M. Three electrodes were selected under identical conditions to construct a MIEAS for the determination of AMOX at equimolar concentrations, yielding an RSD of 2.99% upon repeated experimentation. Furthermore, in order to assess sensor stability, response signals from the same electrode were measured after 30 days and found to be 95.1% of their initial values. These results demonstrate that the constructed MIEAS exhibits excellent repeatability and stability. However, in order to optimize the utilization of this sensor system, further research on electrode material development is necessary due to its poor selectivity.

Taking Krishnan's work [37] as an example, hemophilia B, a genetic disease that can cause cardiovascular disease and paralysis, is often clinically diagnosed by the human coagulation factor IX protein (FIX). The authors designed the MIP embedded with the aptamer as a biomimetic biosensor, which can quickly and sensitively recognize FIX specifically. The author simultaneously utilized the resilience of the molecularly imprinted polymer layer to mitigate biological contamination on the electrode surface. The preliminary work of the research and development process is shown in Figure 5. (a) Human coagulation FIX-imprinted polymer was prepared by free radical initiation polymerization, and an aptamer–target complex was obtained by incubating the aptamer and imprinted template. (b) The working electrode was fabricated through the process of aluminum deposition. Then, the working electrode was modified with carbon nanomaterials and metal nanomaterials, and the aptamer–target complex was fixed on the modified electrode. After the modification, the performance of the electrode was greatly improved in terms of conductivity, sensitivity, and affinity. The amperometry method was used for the detection of FIX. Compared to a traditional Apt sensor without MIPs, it was discovered that the MIP-incorporated sensor has a 3-fold higher sensitivity and a 40 fM detection limit. However, experimental data show that the aptamer–MIP hybrid AIDE's (Archimedean Interdigitated Sensor) detection capability is limited to the decreasing current response as surface mass loads are increased. Based on the author's research, it can be inferred that potential advancements for this strategy primarily involve incorporating nanomaterials, eliminating impurities from the electrode surface [38], introducing additional affinity detection monomers to enhance sensitivity, and enabling the detection of various substances through the modification of a single component with the same synthesis approach.

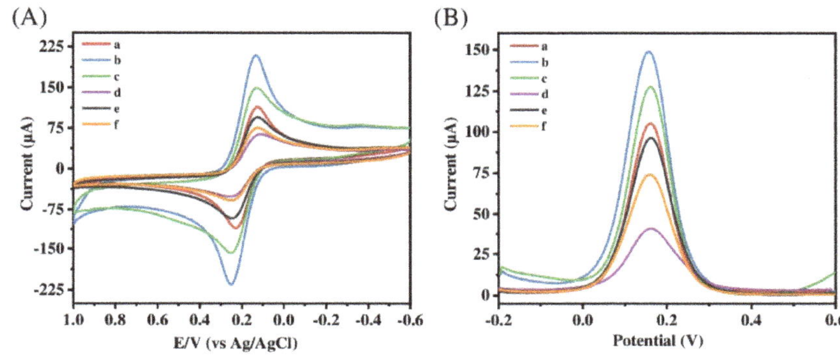

Figure 4. (**A**) CV and (**B**) DPV for (a) bare GCE, (b) AuNPs/ZnO-rGO/GCE, (c) Apt-AMOX/AuNPs/ZnO-rGO/GCE, (d) MIP-Apt/AuNPs/ZnO-rGO before elution, (e) MIP-Apt/AuNPs/ZnO-rGO after elution, and (f) MIP-Apt/AuNPs/ZnO-rGO after adsorption in 5 mM $[Fe(CN)_6]^{3-/4-}$ and 0.1 M KCl (CV scan rate: 100 $mV \cdot s^{-1}$, scan range: −0.2–+0 6 V; DPV scan range: −0.2–+0.6 V). Reprinted with permission from ref. [36].

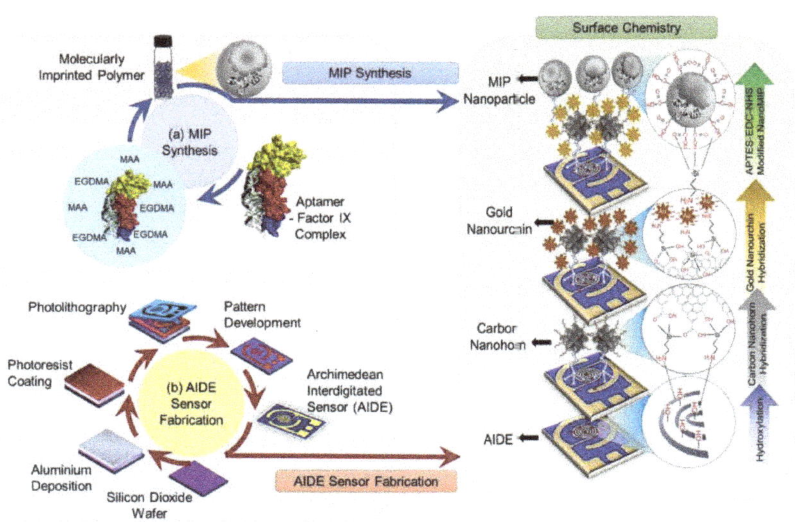

Figure 5. Schematic illustration of the overall research. It comprised of MIP synthesis (**a**), AIDE sensor fabrication by photolithography process (**b**). Reprinted with permission from ref. [37].

For the hybrid assay, the affinity and specificity of the prepared aptamer–MIP hybrid toward the target were improved dramatically as compared with the aptamer or MIP alone.

However, the hybrid process is still based on a trial-and-error process. The monomer in the imprinting process may not be ideal, leading to a low imprinting factor. Rational computer modeling would help to improve the aptamer–MIP hybrid preparation and make the hybrid more efficient.

3.3. *Recapture–Detection–Separate Sensing Strategy*

Unlike the sandwich and hybrid modes, the recapture–detection–separate sensing strategy incorporates a microfluidic control system with miniature valves on the chip. By regulating the opening and closing of the valves, it detects signals produced by target molecules as they flow through narrow channels. In this system, the molecularly imprinted polymer and aptamer function as two distinct recognition elements. Once captured and eluted by the MIP site, analytes are transferred to the aptamer for secondary capture. A furan sensor is taken as an example, as furan is an insecticide widely used in agriculture, but its residue poses a serious threat to human health. Although the commonly used methods for detecting furan include chromatography [39,40] and electrochemical methods [41,42], due to the operational complexity and poor targeting selectivity of these two technologies, there is a general trend to develop a new type of sensor. Under such requirements, Li et al. [43] developed an interesting design idea by etching a microfluidic channel, and two functional regions were etched on polydimethylsiloxane (PDMS). These two functional areas serve as recognition areas for the MIP and aptamer. In this sensing process, the target furan molecule is initially transported through the microchannel and subsequently captured by the MIP recognition site. Following elution, the furan is conveyed to the aptamer's functional region, where it undergoes a secondary capture event facilitated by DNA aptamer binding. The Pt-Au-Ag/AgCl three-electrode system generates a current signal at the second capture, while GO (graphene oxide)-AuNPs amplify the detected signal. This research group realized the high-sensitivity detection of furan in complex samples, and the detection limit was up to 67 pM. After optimizing the chip preparation process, this strategy is expected to expand to the detection of other harmful substances and promote the development process in the field of sensors. In this method, molecular imprinting and the aptamer work independently without interference. However, the production process of the working chip for microfluidics is often costly, and the construction of microchannels and recognition units remains a technical problem that needs improvement. The unoccupied imprinting sites on the chip are prone to binding with aptamers and generating interference signals. Therefore, future research should focus on improving controllability.

4. Applications of MIP–Aptamer Multiple-Recognition Systems

4.1. *Virus*

Viruses are microorganisms with the smallest and simplest structures, yet their types are complex. Most viruses exhibit strong infectivity and can cause life-threatening diseases in severe cases. Although vaccines are available for most viruses, an outbreak can still result in significant damage to the economy and public health, as seen during the three-year pandemic caused by the novel severe acute respiratory syndrome coronavirus 2 (SARS-CoV-2) worldwide [44]. Enhanced virus detection is a crucial element in managing and resolving epidemics. Currently, the enzyme-linked immunosorbent assay [2] and the isolation culture and detection of virus strains [45] are often used for virus-related detection, but these methods can be expensive and susceptible to complex substrates, resulting in low sensitivity and poor selectivity. To overcome these shortcomings, Chen S.Y. [46] designed a molecularly imprinted polymer–aptamer sensor for the non-autofluorescence detection of the H5N1 virus (Influenza A virus). In this work, the author fixed three functional monomers, including acrylamide, to the modified magnetic Fe_3O_4 and used AIBN to initiate polymerization. The H5N1 virus template was embedded into the polymer, and the subsequent elution of viral molecules yielded an imprinted cavity with selective recognition for the H5N1 virus, serving as the identification probe. Additionally, the aptamer was immobilized onto amino-modified ZGO (Zn_2GeO_4:Mn^{2+}) via amide bonding to serve as a secondary recognition probe. When the target H5N1 virus particle is introduced into the recognition system, it is captured by the imprinting chamber to form MIP-H5N1. Furthermore, the cavity selectively captures ZGO-H5N1 Apt from the solution, resulting in the formation of a magnetic sandwich structure; however, nanoparticles without a sandwich structure remain in the solution system. As the concentration of virus molecules increases, there is a continuous decrease in the residual aptamer amount, leading to changes in the persistent photoluminescence (PL) signal. The correlation curve between the PL signal and template molecule concentration was obtained. Through these processes, a sandwich sensor with a detection limit of 1.16 fM was constructed.

Hepatitis B virus is a common blood-borne pathogen that can lead to death from cirrhosis and liver cancer in carriers [47]. To detect this virus, Chen S.Y. [48] introduced a resonance light-scattering sensor in 2021 based on molecular imprinting and an aptamer for the early diagnosis of hepatitis B virus (HBV) using the same sandwich strategy, as shown in Figure 6. The imprinting factor of the sensor reached 7.56, surpassing other molecularly imprinted sensors in virus detection. The sensing system employs a molecular-imprinting layer supported by carbon spheres as probe 1 and utilizes a specific aptamer as probe 2. Simultaneously, glucose is introduced during the construction of the imprinting vector to generate hydroxyl and carboxyl groups on the carrier's surface, thereby enhancing hydrogen bonding and improving the binding capacity of the imprinting layer toward the target virus molecules. The successful binding of the imprinted probe to the target molecule generates a resonant light-scattering response. Upon the double binding of the aptamer to the target molecule, this light signal is further enhanced, as depicted in Figure 7, demonstrating excellent detection efficacy. The authors' study indicates that an imprinting layer of 25 nm thickness made with 2.5 $\mu L\ mL^{-1}$ TEOS yields the optimal signal strength and imprinting effect. When the imprinting layer is too thin, it results in a loose network structure for the constructed imprinting cavity, which lacks specificity. Meanwhile, the thickness of the imprinting layer should be carefully considered during the construction of the dual-recognition system to prevent excessive crosslinking, as an excessively thick layer hinders the removal of target molecules from the imprinting cavity, thereby impeding secondary signal generation by aptamers. Only an appropriate thickness of the imprinting layer can ensure sufficient recognition sites for aptamer binding. The ratio of MIPs and SiO_2@Apt was also optimized. A dosage ratio of MIPs and SiO_2@Apt of 1:2 and a dosage of MIPs of 80 ng/mL^{-1} were used. In general, this sensing approach holds significant implications for the timely detection of a wide range of viruses.

Hepatitis C virus (HCV) is also a causative agent of liver disease. In Ghanbari's work [49], the aptamer was fixed on a multi-wall carbon-nanotube–chitosan nanocomposite (MWCNT-CHIT), and after the imprinting layer was obtained by dopamine electropolymerization, the HCV core antigen sensor was prepared. When the aptamer binds to the target molecule, the peak current is gradually increased by the DPV method, and then the antigen changes conformation to overcome the electrostatic effect during the process of separation from the electrode surface, so the current density changes with the antigen concentration. The method combines the high stability of the nanocomposite platform, the specific recognition capability provided by the MIP–aptamer, and the excellent sensitivity of the hybrid receptor to successfully detect HCV core antigens in human serum, maintaining an initial response of 95% and a relative standard deviation of 5.4% after 50 cycles, with good repeatability. These findings demonstrate the effectiveness of this method in detecting HCV and its potential for widespread application in analyzing complex samples.

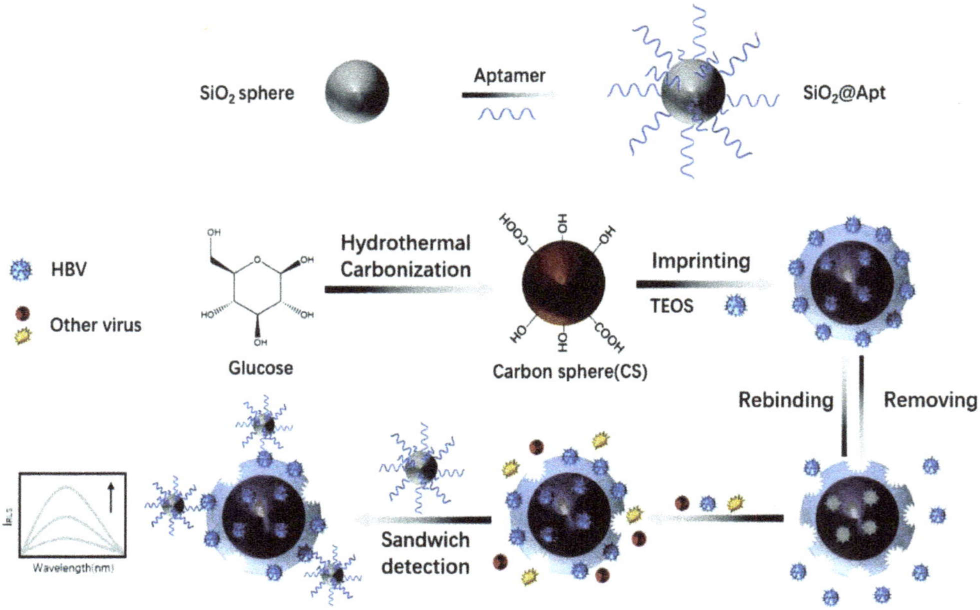

Figure 6. Principle of the preparation of virus–MIPs. Reprinted with permission from ref. [48].

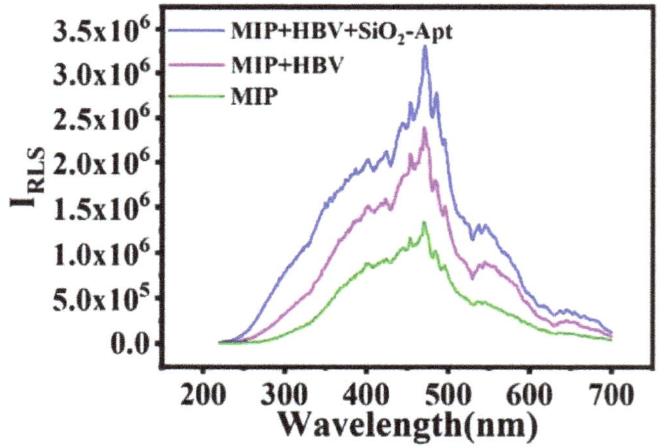

Figure 7. Detection possibility verification: MIP, MIP + HBV, and MIP + HBV + SiO_2-Apt. Reprinted with permission from ref. [48].

4.2. Carcinogens

As one of the most lethal diseases, cancer is caused by various carcinogenic factors, among which aflatoxin is included. Aflatoxin B1 (AFB1) [50] is the representative, which has the highest yield, the highest toxicity, and the strongest carcinogenicity. Once AFB1 enters the food chain, it can result in significant economic losses and pose a grave threat to human health through biological accumulation. In 2022 and 2023, Roushani [51] and Chi Hai [52] put forward a summary of the work on the multiple-recognition strategy for AFB1 detection. In Roushani's study, a novel electrochemical biological device with a detection limit of 12.0 pg/L was designed by hybridizing MIPs and aptamers on a glassy carbon electrode modified with Cu_2O NCs (nanocubes) with a large surface area. The experimental data were measured by the impedance method. In Chi's work, the recognition probe MIP/PC (porous carbon) and signal probe CdTe/ZnS-Apt were used, which have good matching and stable fluorescence signals. In this detection process, the intensity of the fluorescence signal is determined by the synthesis ratio of CdTe/ZnS and Apt, and the concentration is insufficient to bind to all modified aptamers in the sensor, resulting in fluorescence interference. When the optimal concentration of CdTe/ZnS was 0.8 mg/mL and the Apt content was 10 nmol/L, the fluorescence intensity reached its maximum efficacy. The results show that there is satisfactory compatibility and synergy between the aptamer and the molecular-imprinting layer, so the detection of AFB1 is realized with a very low detection limit of 4.0 pg/mL^{-1}. The simultaneous determination of 10 ng/mL samples using five identical sensors has a relative standard deviation of 4.1%. In addition, after three times of repeated use, the sensor system showed a fluorescence signal reduction of only 9.7%, which indicates that the sensor has great commercial potential in the food industry.

Ochratoxin (OTA) is another mycotoxin that has attracted worldwide attention. The toxicity, carcinogenicity, and teratogenicity of OTA seriously harm human health [53]. In view of the fact that the aptamer–MIP materials developed in the past have not been applied to the detection of OTA, Lyu [54] first proposed a novel aptamer–MIP material for OTA-specific recognition in 2019. In this study, APT-OTA complexes were obtained by mixing an aptamer with a concentration of 60 μmolL^{-1} with template molecules, and then the raw materials, such as the monomer, crosslinker, initiator, and complex, were placed in the purple linkage instrument to complete photopolymerization. The hydrogen bonding between the monomer AMPS (2-acrylamide-2-methylpropanesulfonic acid) and template OTA and the specific affinity of the aptamer greatly increased the adsorption effect. Based on these advantages, the material was successfully applied to actual beer samples, and the OTA-sensitive LOD and LOQ were measured to be 0.07 ng/mL^{-1} and 0.14 ng/mL^{-1}, respectively, realizing the specific enrichment and high-sensitivity detection of OTA. To evaluate the stability of the aptamer–MIP monolithic column, changes in flow velocity were measured in five mobile phases with different polarities. The linear correlation coefficients R^2 of these five groups of data are in the range of 0.9908~0.9985, indicating that the measurement precision and accuracy are high. Furthermore, after one month of use, the recovery rate of aptamer–MIP was found to be 87.3%, demonstrating excellent mechanical stability and reproducibility. However, as the components have an impact on the overall pore structure and permeability of the aptamer–MIP, the next step is to optimize the composition of the polymeric mixture in order to comprehensively enhance the performance of this multiple-recognition material.

There are heavy-metal ions [55] with potent carcinogenicity, among which the bivalent chromium ion is a representative. Upon entering the human body, cadmium will affect the formation of human bone cells, leading to deformation and pain and inducing cancer and even death in patients in severe cases [56]. To achieve the effective detection of Cd^{2+}, Li [57] devised a fluorescence-quenching sensor. The team fixed SN-CQD/Au (carbon quantum dots co-doped with sulfur and nitrogen atoms) fluorescent particles on the surface of the indium tin oxide glass electrode, and then they combined the self-assembled aptamer–Cd^{2+} complex with fluorescent particles and obtained the molecular-imprinting sites by electropolymerization. Two signal units, molecular imprinting and the aptamer, are used to show two signals in the detection. For electrical signals, when Cd^{2+} is captured by the imprinting cavity formed, the diffusion channel will be blocked due to the density of the MIP structure, and the charge transfer will be blocked, showing a high impedance value. The fluorescence response signal is caused by the quenching of Cd^{2+}. When the aptamer binds to the gold nanoparticles, it will show a strong fluorescence signal, but because the polymerized MIP layer on the electrode surface will cover the original structure after binding Cd^{2+}, resulting in the partial absorption and refraction of the fluorescence, it will show weakened signal intensity. Based on the specific affinity of MIP and the aptamer in the sensor for Cd^{2+}, interference and non-specific recognition are greatly reduced. The actual measurement results show that the method has a good resolution, and the detection limit is 1.2×10^{-12} mol/L^{-1}. To ensure the reproducibility and stability of the sensors, a relative standard deviation of 2.65% was achieved for 5.0×10^{-9} mol/L^{-1} Cd^{2+} using nine sensors under identical conditions, while a relative standard deviation of 3.14% was obtained for fluorescence signals when utilizing the same sensors. Furthermore, after storing the sensor in an environment at 4 °C for twenty days, there was a decrease in the fluorescence signal of 7.48% compared to its initial value; these results demonstrate that this study's sensor system exhibits exceptional performance. However, due to the unsatisfactory number of aptamers and target metal ions, the sensor must overcome this limitation in order to improve its adsorption capacity.

4.3. Antibiotics

Antibiotics are either metabolites or synthetic analogs produced by microorganisms. The most common classes of antibiotics include beta-lactams, quinolones, and aminoglycosides. Antibiotics are utilized to treat infections caused by pathogenic bacteria and have widespread use in clinical medicine. However, due to widespread use and even abuse, bacteria gradually adapt to the environment where antibiotics exist and develop resistance, which has become a potential threat to global health [58,59]. Lincomycin is a kind of amide drug with an antibacterial effect against bacteria, especially Gram-positive bacteria, and also protozoans bioactive [60]. Li [61] proposed a new method for detecting lincomycin based on the energy resonance transfer between the Au-GO nanomaterial and C-dots. In the absence of target molecules, C-dots exhibit an enhanced light signal after energy absorption. After the introduction of the target molecule, the aptamer and MIP compete to bind with lincomycin within the sensor system. Due to the structural change in the aptamer after binding with lincomycin, the electron supply and energy transfer between GO-Au and C-dots are hindered, resulting in a decrease in the electrochemical luminescence signal strength in the monitoring system. Through this signal change, the trace detection of lincomycin can be directly detected. At the same time, this experiment shows that when the adsorption time is 8 min and the solution pH is maintained at 8.0, the current intensity is the best. Based on this signal change, the method can achieve multiple identifications with high resolution, and the detection limit is 1.6×10^{-13} mol/L.

Kanamycin (KAN) is an aminoglycoside antibiotic commonly employed for the treatment of Gram-positive bacterial infections. However, its neurotoxicity and nephrotoxicity necessitate the development of an efficient sensor for monitoring kanamycin levels. In response to this societal need, Bi [62] proposed an electrochemical detection strategy in 2019. In this study, Fe_3O_4 material was loaded with gold nanoparticles and modified with a KAN aptamer. The imprinted cavity formed by pre-polymerization on the electrode surface selectively captures and recognizes kanamycin molecules in the substrate, followed by the specific binding of the aptamer to kanamycin during sensor application. The sensor system goes through two processes of capture and binding to generate electrical signals, which are quickly detected by cyclic voltammetry. The presence of sufficient binding sites on the electrode surface enables the sensor to have a wide linear range. Furthermore, when the molar ratio of the template molecule to the functional monomer is 1:2, the physical properties and performance of the material are the best, the current response is the largest, and the recovery rate is 96.5% to 101.5% when applied to the detection of groundwater samples. For the stability and repeatability of the sensor system, the relative standard deviation of the corresponding current obtained by the determination of 100 nM KAN by the DPV method is 3.0%. At the same time, the detached electrode was stored at a low temperature for 2 days and measured again. After 7 repeated cycles, the relative deviation was only 1.8% compared with the initial peak value. These data confirm that the stability and repeatability of the research work are excellent.

Later in 2022, there was also a new testing strategy for tetracycline (TC), another threat in the antibiotic family. Tetracycline is a natural antibiotic produced by actinomycetes [63] and has important applications in aquaculture and other fields due to its antibacterial activity. However, the accumulation of tetracycline in the environment inevitably leads to the emergence of drug-resistant strains. Ma [64] reported a magnetic material for separating and enriching tetracycline. As shown in Figure 8, the authors utilized aptamers and β-CD (β-cyclodextrin) as replacements for traditional functional monomers and employed free radical polymerization to construct molecular-imprinting recognition sites on the outer layer of the material. In practical applications, the functional monomer modified on the material works cooperatively with the imprinting chamber to selectively adsorb tetracycline. The magnetic properties of the material enable the adsorbed magnet to remove tetracycline from the substrate, thereby achieving TC enrichment. Compared with other molecularly imprinted materials, the material has a core–shell structure, which ensures ecological friendliness and non-toxicity. This effectively isolates the aptamer from nucleases in the environment and prevents inactivation due to enzymatic hydrolysis. Due to these advantages, the material has a blot factor of 7.6 and a detection limit of 1.0 μg/L^{-1} by UV spectroscopy, which shows the special potential for the trace detection of TC in complex samples. In the selection of a material for testing, reusability and stability are crucial parameters for evaluating its practicality. The adsorption capacity of the material was assessed after six cycles of use, yielding a result of 84.6% compared to the initial treatment. This demonstrates the excellent reusability of the material. Furthermore, the sustained adsorption effectiveness, even after repeated usage, indicates that the constructed imprinted cavity remains intact and exhibits remarkable stability. Additionally, TC was detected in five samples with varying concentrations under identical experimental conditions, resulting in an RSD value of 0.81% (n = 5), confirming the method's exceptional reproducibility and stability. In summary, this study presents a novel approach for utilizing molecularly imprinting technology in environmental protection and has promising applications for detecting and enriching other hazardous substances in the environment.

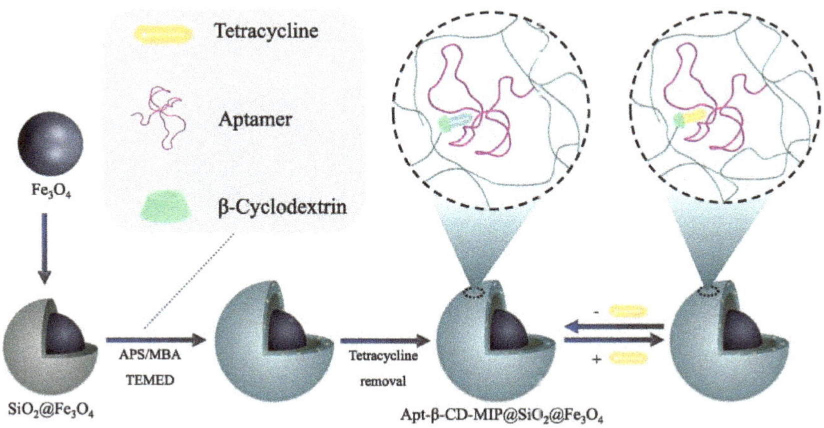

Figure 8. Schematic illustration of the Apt-β-CD-MIP@SiO_2@Fe_3O_4 preparation and its application to detect TC. Reprinted with permission from ref. [64].

4.4. Bacteria

As the most abundant organisms in nature, bacteria are widely distributed in soil and water. On the one hand, bacteria are closely related to human production and life, such as food production, digestion and absorption, battery manufacturing, and antibiotic extraction. On the other hand, these prokaryotes pose a serious threat to human health due to their diversity of infectivity and transmission routes. *Pseudomonas aeruginosa* is an opportunistic bacterium that is one of the main sources of infection in hospitals. It is more common in postoperative patients, and infected individuals may experience a range of uncomfortable symptoms, including death in severe cases [65]. Therefore, it is imperative to develop a rapid method for detecting *Pseudomonas aeruginosa*. In Sarabaegi's work [66], they designed an electrochemical sensor for multiple molecular recognition. The preparation process is shown in Figure 9: (1) gold nanoparticles are deposited on the surface of the glassy carbon electrode; (2) the aptamer is modified on the electrode surface, and the molecularly imprinted polymerization layer is obtained by the electrochemical polymerization of dopamine; (3) the eluting layer obtains the cavity and adds the target for detection. In order to observe the cavity obtained after elution, FE-SEM (Field-Emission Scanning Electron Microscope) was used to analyze the morphology and structure of the electrode. According to the scanning images, the removal of template molecules made the surface of the molecular-imprinting layer rougher, and a large number of imprinting cavities were observed, indicating that the elution process was successful, and identification sites with a satisfactory number and specific surface area were obtained. During the construction of the recognition system, when gold nanoparticles are deposited on the exposed glassy carbon electrode, these nanoparticles increase the electrode area, and the current peak value is increased. Then, when the aptamer is modified on the electrode surface, the carboxylate group in the aptamer generates a repulsive force to reduce the current peak value. Finally, when the molecular-imprinting site captures the target molecule and undergoes desorption, the peak current is enhanced again, and by observing the changes in this series of currents, the direct detection of *Pseudomonas aeruginosa* can be realized. The detection limit of the sensor is 1 CFU/mL^{-1}.

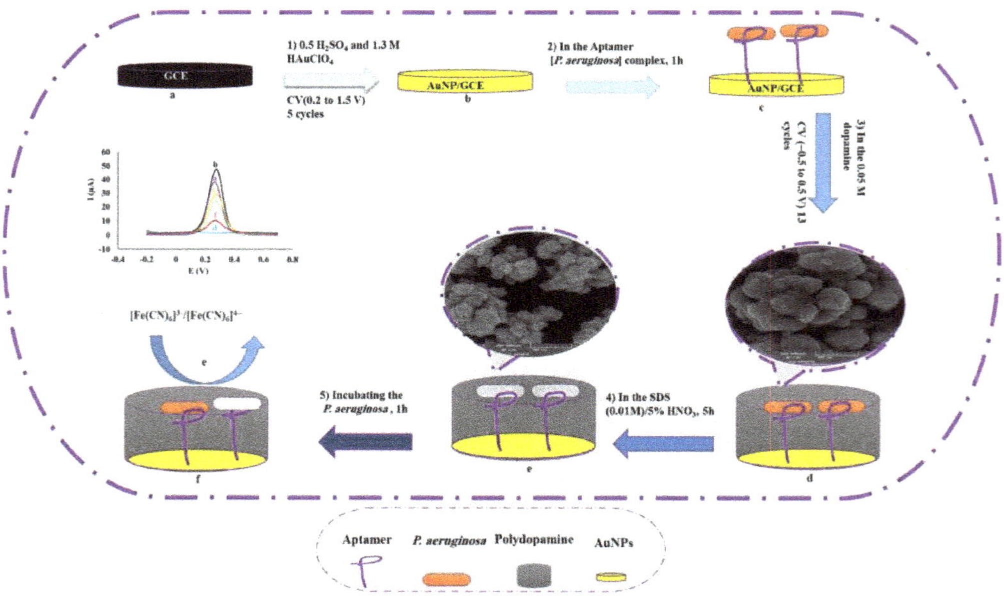

Figure 9. Schematic diagram of aptamer–MIP preparation for detection of *P. aeruginosa*. Reprinted with permission from ref. [66].

Staphylococcus aureus is a prevalent pathogenic bacterium that exhibits robust antibacterial resistance and frequently induces pneumonia and sepsis in humans [67,68]. The simple sensor constructed by El-Wekil using electrical analysis technology shows outstanding application potential [69]. In this study, El-Wekil immobilized aptamers onto composite materials via Au-S bonds and electropolymerized imprinting layers on the material surface using o-phenylenediamine as a functional monomer in the presence of *Staphylococcus aureus*. As the recognition sites formed are complementary to the target molecules, the binding of *S. aureus* to the imprinted cavity and aptamer can result in a decrease in the peak current on the sensing electrode, thereby achieving the effective separation of *S. aureus* from complex matrices. In this study, the MIP–aptamer dual-identification system was successfully implemented for the first time to identify Staphylococcus aureus. The linear range of the sensor is 101 to 107 CFU mL^{-1}, and the detection limit LOD is 1 CFU mL^{-1}. Given its robust detection capabilities and high recovery rates, we anticipate widespread adoption of this technology in the future.

4.5. Protein

Proteins are substances that possess specific spatial conformations, participate in the formation of vital components within the human body, and serve as primary mediators of organismal life processes. Given the functional diversity exhibited by human proteins, characteristic protein detection can effectively facilitate disease diagnosis and screening. For example, the medical diagnosis of Alzheimer's disease is often based on amyloid-beta oligomers (AβOs) [70]. In 2020, You Min [31] proposed a diagnostic strategy that could be used as an alternative to these methods. As shown in Figure 10, the author introduced silver and silicon dioxide nanoparticles, followed by the self-assembly of the aptamer on silica particles, thereby accomplishing the construction of a sandwich structure on the MIP membrane. Once the imprinted cavity captures AβOs, the captured object is transferred to the aptamer for secondary binding, resulting in electron transfer and electrical signal transmission within the three-dimensional imprinted cavity. This method combines site-specific binding with aptamer affinity toward targeted molecules, enabling highly sensitive electrochemical signals to be obtained even when a small number of AβOs enter the detection system. During the construction process of this detection system, the number of imprinted cavities is directly determined by the concentration of the template, which, in turn, determines the number of captured template molecules. Only when an appropriate concentration of template molecules is used can a stable current response signal be generated. Additionally, it is important to consider the concentration of the AβO-specific aptamer added to the system. When its concentration exceeds 5 μM, excessive aptamers will inhibit the transmission of electrons and weaken the current signal. It was determined that the combination of 1 mg/mL^{-1} Aβ1-42 oligomer and 5μM aptamer is the best concentration choice and shows the best electrochemical signal. The sensor based on this strategy has successfully achieved the microdetection of AβOs in human serum. It exhibits good linearity within the concentration range of 5 pg/mL^{-1} to 10 ng/mL^{-1} and a detection limit of 1.22 pg/mL^{-1}.

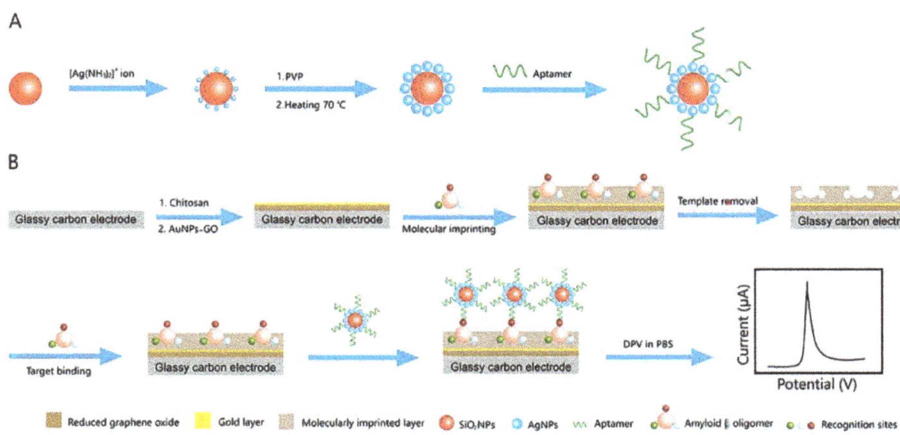

Figure 10. Schematic illustration of (**A**) the preparation of the SiO_2@Ag–aptamer composite and (**B**) the fabrication of the MIP-based antibody-free biosensor and the electrochemical detection of AβO via a sandwich-type assay. Reprinted with permission from ref. [31].

Similar strategies have also been applied in the detection of troponin (cTnI), which is a marker of myocardial injury and necrosis and is often used as a diagnostic basis for myocardial infarction [71]. In Mokhtari's study [72], a non-immune dual-recognition approach was employed to fabricate a sensing platform for cTnI detection. After immobilizing the cTnl aptamer on the electrode surface, the imprinted polymer was obtained by the electropolymerization of methylene blue monomers around the aptamer. During the detection process, cTnl was selectively adsorbed by the hybrid receptor of MIPs and the aptamer due to their selective affinity and cavity effects, respectively. The determination of the adsorbed cTnl was achieved through differential pulse voltammetry. The detection limit of the sensor platform was 1.04 pM (2.61×10^{-5} μg/mL), and the RSD of the sensor system designed in this study was 5.37% for cTnl in a protein mixture containing cardiac troponin, and several electrochemical values obtained in this study demonstrate that the novel hybrid probe developed by the authors possesses superior performance.

In addition, Wang [73] successfully detected alpha-fetoprotein (AFP) and insulin using a nanoprobe. The team utilized static electricity between gold nanoparticles and the abundant amino groups, as well as Au-S bonds, to increase the fixed number of aptamers on the substrate. Additionally, a synergistic interaction between the imprinted cavity and aptamer lowered the detection limit. When insulin is used as the template molecule, the detection limit reaches approximately 0.5 ng/mL^{-1}, whereas when AFP serves as the template molecule, this magnetic nanoprobe can detect AFP in samples ranging from 1000 ng/mL^{-1} to 20 ng/mL^{-1}. It is worth mentioning that this study demonstrates the controllability of preparing an imprinting layer through the sol–gel method, as evidenced by the direct relationship between the consumption time and thickness change in the SiO_2 coating. The optimal coating time of 90 min results in the highest specific recognition ability. These findings highlight the significant potential of our magnetic nanoprobe for the efficient detection of disease serum protein markers.

To expand the detection range, Yang [74] developed a novel electrochemiluminescence (ECL) "signal on" sensor platform for the efficient detection of thrombin, a bioactive protein that plays a crucial role in blood clotting and has significant clinical implications in the treatment of various diseases. In this detection system, the author employed molecularly imprinted nanocavities to bind target molecules to thrombin aptamers. Simultaneously, the synergistic effect between polyetherimide (PEI) and $Ru(bpy)_3{}^{2+}$ (RuNP) is utilized to expedite electron transfer, thereby enhancing the signal strength of the sensing platform by sevenfold compared with conventional methods. The sensor platform exhibits excellent stability and detection efficacy, with the ECL sensor signal strength remaining robust even after ten cycles of testing at concentrations of 10^{-10} M and 10^{-12} M, yielding corresponding RSDs of 2.0972% and 2.1509%, respectively. Following thirty days of low-temperature storage, the ECL signal strength remains at approximately 90% of its initial value, while the detection limit is as low as 1.73×10^{-15} M, indicating significant potential for application in food safety monitoring. It is worth mentioning that previous molecularly imprinted aptamer sensors are the signal closure type, but the signal flow type was used in this study, which provides a new research direction for the construction of a new small-molecule detection platform in the future.

4.6. Others

In view of the remarkable benefits and immense potential of the multiple-recognition strategy based on aptamer–MIPs, researchers have increasingly adopted this approach. In addition to well-known molecular categories, even relatively obscure substances such as cortisol, bisphenol A [75], and histamine can be detected using this method.

Histamine (HIS) is a hydrophilic biogenic amine that is produced by bacterial fermentation and is commonly found in food. Studies have shown that high levels of histamine in food can indirectly lead to Gram-negative bacterial infections. In order to realize highly sensitive detection of HIS in canned tuna and human serum, Mahmoud [76] developed a nanosensor. Based on molecular-imprinting polymerization technology and DNA aptamer affinity characteristics, the MIP-apt/AuNPs/cCNTs/GCE architecture was fabricated by the surface modification of gold nanoparticles and carboxylated carbon nanotubes on the glassy carbon electrode, followed by the covalent immobilization of the thiolate aptamer and the electropolymerization of phenylenediamine. The system was applied to the detection of plasma and canned tuna. The detection limit was 0.11 nmol, the recovery rate was 96.2~105.2%, and the RSD% was 95.3~104.4%. These data confirmed that this detection system has higher accuracy and precision for the analysis of HIS in complex components.

Cortisol is a steroid hormone involved in the regulation of human metabolism. In the medical field, cortisol detection is frequently utilized for the diagnosis and treatment of depression and post-traumatic stress disorder. Due to the similarity in detecting cortisol levels between saliva and blood samples, Cheng [77] proposed a sensing method with a detection limit as low as 3.3×10^{-13} M. In this study, a bare glassy carbon electrode was initially coated with a dispersion of graphene and carbon quantum dots. Subsequently, the aptamers were immobilized on the electrode surface through electrostatic interaction. Finally, chitosan was utilized as a raw material to form an imprinting layer, and cortisol molecules were eluted to obtain the imprinting cavity. In this experiment, the authors' success can be attributed to two factors. Firstly, the combination of graphene and carbon quantum dots increases the number of active sites, resulting in a more sensitive detection method. Secondly, optimization of the concentration of the template molecule (1μM) and the molar ratio of the template molecule to aptamer (1:2) further enhances sensitivity. The developed sensor is reported to outperform other methods for detecting cortisol.

As a steroid hormone, progesterone is often used for oral contraception. However, excessive progesterone released into the environment will enter the human body along with water circulation. In order to achieve the trace analysis of this substance, an ultrasensitive sensor platform was established based on SnO_2-Gr and AuNPs by Huang Yan [78]. The construction process of the sensor is as follows. Firstly, SnO_2-Gr is modified on the exposed glassy carbon electrode to increase the conductivity and specific surface area, then gold nanoparticles are electrodeposited, and aptamers are fixed by Au-S bonds. Finally, the imprinted cavity is obtained by electropolymerization and the elution process. The sensor platform exhibits excellent performance with a low detection limit (1.73×10^{-15} M) while achieving a relative standard deviation (RSD) of 2.79% at a concentration of 10^{-10} M using this hybrid system. Furthermore, the DPV current was monitored every 10 days after storage at low temperatures, and even after four cycles, the DPV value remained at an impressive 93.19% of its initial value, demonstrating the exceptional repeatability and stability of the developed sensor. Thus, it possesses immense potential for effective detection in the field of food safety.

Table 1 lists the MIP layer formation methods, recognition systems, and detection methods of aptamer–MIP hybrid systems in recent years. The results show that aptamer–MIP hybrids can be used as superior recognition elements for the detection of various targets.

Table 1. Summary of polymerization methods, recognition systems, and detection methods of aptamer–MIP hybrid systems.

Electrochemical Polymerization								
Target	**Potential Range**	**Monomers**	**Polymerization Time**	**Scan Rate**	**Recognition Systems**	**Detection Methods**	**Year**	**Ref.**
Acrylamide	−0.4–0.9 V	o-Phenanthroline	15 cycles	100 mV·s^{-1}	Hybrid	DPV	2023	[79]
Progesterone	−0.2–0.6 V	p-Aminothiophenol	15 cycles	150 mV·s^{-1}	Hybrid	DPV	2023	[78]
Lysozyme	−0.4–1.2 V	Methylene blue	20 cycles	50 mV·s^{-1}	Hybrid	DPV	2022	[80]
SARS-CoV-2 virus	−0.5–0.5 V	Dopamine	15 cycles	50 mV·s^{-1}	Hybrid	EIS	2022	[81]
Amoxicillin	−0.5–1.0 V	Dopamine	15 cycles	75 mV·s^{-1}	Hybrid	DPV	2022	[36]
Prostate-specific antigen	−0.5–0.5 V	Dopamine	12 cycles	20 mV·s^{-1}	Hybrid	EIS	2021	[82]
Pseudomonas aeruginosa	−0.5–0.5 V	Dopamine	13 cycles	20 mV·s^{-1}	Hybrid	DPV	2021	[66]
Trypsin	−0.5–0.5 V	Dopamine	15 cycles	20 mV·s^{-1}	Hybrid	DPV	2021	[83]
Ractopamine	−0.5–0.5 V	Dopamine	13 cycles	20 mV·s^{-1}	Hybrid	EIS	2020	[84]
Cardiac troponin I	−0.4–1.2 V	Methylene blue	20 cycles	50 mV·s^{-1}	Hybrid	DPV	2020	[72]
Kanamycin	0–0. 8 V	3-Aminophenylboronic acid	20 cycles	50 mV·s^{-1}	Sandwich	DPV	2020	[62]
Chloramphenicol	0–1.2 V	Risorcinol	14 cycles	100 mV·s^{-1}	Hybrid	EIS	2019	[85]
Chlorpyrifos	0–1.0 V	o-Dihydroxybenzene o-Phenylenediamine	10 cycles	50 mV·s^{-1}	Hybrid	DPV	2018	[86]
Thermal Polymerization								
Target	Monomers	Crosslinking agent	Temperature	Time	Recognition systems	Detection methods	Year	Ref.
Human blood clotting factor IX protein	MAA	EGDMA	40 °C	24 h	Hybrid	UV-Vis	2022	[37]
H5N1 Virus	AA MAA	MBA	65 °C	6 h	Sandwich	Fluorimetry	2022	[46]
Cytochrome C	MAA	MBA	37 °C	12 h	Hybrid	Fluorimetry	2018	[87]
Carbofuran	MAA	MBA	50 °C	2 h	Microfluidic	DPV	2018	[43]
Kanamycin	MAA	MBAA	40 °C	6 h	Hybrid	Fluorimetry	2018	[88]

Table 1. *Cont.*

Sol–Gel Method								
Target	Monomers	Crosslinking agent	Solvent	Catalyst	Recognition systems	Detection methods	Year	Ref.
Aflatoxin B1	APTES	TEOS	Ethanol	$NH_3 \cdot H_2O$	Sandwich	Fluorimetry	2023	[49]
Malachite green	APTES	TEOS	Ultrapure water	$NH_3 \cdot H_2O$	Hybrid	Fluorimetry UV-Vis	2023	[89]
Virus enterovirus 71	APTES	TEOS	Ultrapure water	$NH_3 \cdot H_2O$	Sandwich	Fluorimetry UV-Vis	2022	[33]
Photopolymerization								
Target	Monomers	Crosslinking agent	Illuminating source	Irradiation time	Recognition systems	Detection methods	Year	Ref.
Cadmium (II)	L-Alanine	N-hydroxysuccinimide	UV	30 min	Hybrid	Fluorimetry	2019	[57]

EDGMA—ethylene glycol dimethacrylate; AA—acrylic acid; MAA—methacrylic acid; MBA—methylene diacrylamide; MBAA—5,5′-methylenedianthranilic acid; APTES—3-aminopropyltriethoxysilane; TEOS—ethyl silicate; EIS—electrochemical impedance spectroscopy.

5. Conclusions

In summary, we review the work conducted in recent years on multiple-recognition systems utilizing MIPs and aptamers. Aptamers and molecular imprinting have made remarkable contributions to sensing as highly favorable detection elements. Numerous researchers have demonstrated that combining the specific recognition ability of molecular-imprinting technology with the affinity of aptamers can significantly enhance detection effectiveness. This recognition strategy finds applications in various fields, including medical treatment, environmental monitoring, and agricultural production. In comparison to a single-recognition-element approach, a multi-recognition strategy offers improved selectivity, sensitivity, stability, and reusability. Nevertheless, it still encounters numerous challenges.

5.1. Limitations

In the process of constructing a multiple-recognition system, the principle of the recognition mechanism is not perfect, the polymerization process is uncontrollable, the repeatability is unstable, and conformational changes will occur in the process of imprinting sites and the aptamer-specific recognition and capture of target molecules. In addition, the identification of large water-soluble biomacromolecules remains problematic because conformational integrity cannot be guaranteed during synthesis. These challenges hinder the wide application of multiple-recognition strategies.

5.2. Challenges

The method requires further refinement, including finding the best molecular-imprinting film thickness, optimizing the molar concentration ratio of the template molecule to the aptamer, enhancing the electron transfer effect on the electrode, finding a suitable electrode modifier to multiply the signal, finding a method to freely switch monomers to detect different or similar substances, and improving the commercial application range of the technology. This strategy has shown great promise in environmental monitoring, disease prevention, and medical diagnosis and treatment, and we believe that this approach, as researchers continue to optimize it, will open up a broad avenue for the analysis of multi-component complex components.

Author Contributions: Conceptualization, K.N. and Q.X.; writing—original draft preparation, writing—review and editing, K.N., Y.S., Y.Y., W.X., C.M. and Q.X.; visualization, K.N. and Q.X. All authors have read and agreed to the published version of the manuscript.

Funding: This research was funded by NSFC (22076161, 21675140 and 21705141), the Talent Support Program of Yangzhou University, Yangzhou University Interdisciplinary Research Foundation for Chemistry Discipline of Targeted Support (yzuxk202009), the project funded by the PAPD and TAPP.

Institutional Review Board Statement: Not applicable.

Informed Consent Statement: Not applicable.

Data Availability Statement: Not applicable.

Conflicts of Interest: The authors declare no conflict of interest.

Abbreviations

Templates	
EV71	Enterovirus 71
AMOX	Amoxicillin
FIX	Factor IX protein
SARS-CoV-2	Severe acute respiratory syndrome coronavirus 2
H5N1 virus	Influenza A virus
HBV	Hepatitis B virus
HCV	Hepatitis C virus

AFB1	Aflatoxin B1
OTA	Ochratoxin
KAN	Kanamycin
TC	Tetracycline
AβOs	Amyloid-beta oligomers
cTnI	Troponin
AFP	Alpha-fetoprotein
HIS	Histamine
Functional monomers	
Apt	Aptamer
AA	Acrylic acid
MAA	Methacrylic acid
MBA	Methylene diacrylamide
MBAA	5,5′-Methylenedianthranilic acid
APTES	3-Aminopropyltriethoxysilane
AMPS	2-Acrylamide-2-methylpropanesulfonic acid
β-CD	β-Cyclodextrin
Crosslinking agent	
EDGMA	Ethylene glycol dimethacrylate
APTES	3-Aminopropyltriethoxysilane
TEOS	Ethyl silicate
Patterns	
AIMIP	"Artificial intelligence" imprinted polymers
AIE	Aggregation-induced emission
AIDE	Archimedean Interdigitated Sensor
Base	
ZIF-8	Zeolitic Imidazolate Framework-8
PDMS	Polydimethylsiloxane
PC	Porous carbon
MWCNT-CHIT	Multi-wall carbon-nanotube–chitosan nanocomposite
Surface modification	
MIPs	Molecularly imprinted polymers
CDs	Carbon quantum dots
AuNPs	Au nanoparticles
GO	Graphene oxide
ZGO	$Zn_2GeO_4:Mn^{2+}$
NCs	Nanocubes
SN-CQD/Au	Carbon quantum dots co-doped with sulfur and nitrogen atoms
Others	
AIBN	Azodiisobutyronitrile
CV	Cyclic voltammetry
DPV	Differential pulse voltammetry
PL	Persistent luminescence
FE-SEM	Field-Emission Scanning Electron Microscope
ECL	Electrochemiluminescence
PEI	Polyetherimide
RuNP	$Ru(bpy)_3{}^{2+}$
EIS	Electrochemical impedance spectroscopy

References

1. Cheng, Y.; Zhao, X.; Zhang, Q.; Li, X.; Wei, Z. Constructing imprinted reticular structure in molecularly imprinted hybrid membranes for highly selective separation of acteoside. *Sep. Purif. Technol.* **2022**, *298*, 121572. [CrossRef]
2. Zhou, J.; Ni, Y.; Wang, D.; Fan, B.; Zhu, X.; Zhou, J.; Hu, Y.; Li, L.; Li, B. Development of a Competitive Enzyme-Linked Immunosorbent Assay Targeting the-p30 Protein for Detection of Antibodies against African Swine Fever Virus. *Viruses* **2023**, *15*, 154. [CrossRef] [PubMed]
3. Wang, Y.; Xu, Y.; Gao, R.; Tian, X.; Heinlein, J.; Hussain, S.; Pfefferle, L.D.; Chen, X.; Zhang, X.; Hao, Y. Strategic design and fabrication of lightweight sesame ball-like hollow double-layer hybrid magnetic molecularly imprinted nanomaterials for the highly specific separation and recovery of tetracycline from milk. *Green Chem.* **2022**, *24*, 8036–8045. [CrossRef]

4. Feng, D.; Ren, M.; Miao, Y.; Liao, Z.; Zhang, T.; Chen, S.; Ye, K.; Zhang, P.; Ma, X.; Ni, J.; et al. Dual selective sensor for exosomes in serum using magnetic imprinted polymer isolation sandwiched with aptamer/graphene oxide based FRET fluorescent ignition. *Biosens. Bioelectron.* **2022**, *207*, 114112. [CrossRef] [PubMed]
5. Kakkar, V.; Narula, P. Role of molecularly imprinted hydrogels in drug delivery—A current perspective. *Int. J. Pharm.* **2022**, *625*, 121883. [CrossRef]
6. Gao, C.; Wei, M.; McKitrick, T.R.; McQuillan, A.M.; Heimburg-Molinaro, J.; Cummings, R.D. Glycan Microarrays as Chemical Tools for Identifying Glycan Recognition by Immune Proteins. *Front. Chem.* **2019**, *7*, 833. [CrossRef]
7. Fang, Y.Z.; Jiang, L.; He, Q.; Cao, J.; Yang, B. Deubiquitination complex platform: A plausible mechanism for regulating the substrate specificity of deubiquitinating enzymes. *Acta Pharm. Sin. B* **2023**, *13*, 2295–2816. [CrossRef]
8. Kachhawa, P.; Mishra, S.; Jain, A.K.; Tripura, C.; Joseph, J.; Radha, V.; Chaturvedi, N. Antigen-Antibody Interaction-Based GaN HEMT Biosensor for C3G Detection. *IEEE Sens. J.* **2022**, *22*, 6256–6262. [CrossRef]
9. Wang, W.; Wang, X.; Cheng, N.; Luo, Y.; Lin, Y.; Xu, W.; Du, D. Recent advances in nanomaterials-based electrochemical (bio)sensors for pesticides detection. *TrAC Trends Anal. Chem.* **2020**, *132*, 116041. [CrossRef]
10. Cui, M.; Che, Z.; Gong, Y.; Li, T.; Hu, W.; Wang, S. A graphdiyne-based protein molecularly imprinted biosensor for highly sensitive human C-reactive protein detection in human serum. *Chem. Eng. J.* **2022**, *431*, 133455. [CrossRef]
11. Carballido, L.; Karbowiak, T.; Cayot, P.; Gerometta, M.; Sok, N.; Bou-Maroun, E. Applications of molecularly imprinted polymers and perspectives for their use as food quality trackers. *Chem* **2022**, *8*, 2330–2341. [CrossRef]
12. Yarman, A.; Scheller, F.W. Coupling Biocatalysis with Molecular Imprinting in a Biomimetic Sensor. *Angew. Chem.-Int. Ed.* **2013**, *52*, 11521–11525. [CrossRef] [PubMed]
13. Rampey, A.M.; Umpleby, R.J.; Rushton, G.T.; Iseman, J.C.; Shah, R.N.; Shimizu, K.D. Characterization of the imprint effect and the influence of imprinting conditions on affinity, capacity, and heterogeneity in molecularly imprinted polymers using the Freundlich isotherm-affinity distribution analysis. *Anal. Chem.* **2004**, *76*, 1123–1133. [CrossRef]
14. Aljohani, M.M.; Cialla-May, D.; Popp, J.; Chinnappan, R.; Al-Kattan, K.; Zourob, M. Aptamers: Potential Diagnostic and Therapeutic Agents for Blood Diseases. *Molecules* **2022**, *27*, 383. [CrossRef]
15. Ellington, A.D.; Szostak, J.W. In vitro selection of RNA molecules that bind specific ligands. *Nature* **1990**, *346*, 818–822. [CrossRef]
16. Tuerk, C.; Gold, L. Systematic evolution of ligands by exponential enrichment: RNA ligands to bacteriophage T4 DNA polymerase. *Science* **1990**, *249*, 505–510. [CrossRef]
17. Kaur, H.; Bruno, J.G.; Kumar, A.; Sharma, T.K. Aptamers in the Therapeutics and Diagnostics Pipelines. *Theranostics* **2018**, *8*, 4016–4032. [CrossRef] [PubMed]
18. Nxele, S.R.; Nkhahle, R.; Nyokong, T. The composites of asymmetric Co phthalocyanines-graphitic carbon nitride quantum dots-aptamer as specific electrochemical sensors for the detection of prostate specific antigen. *J. Am. Chem. Soc.* **2021**, *900*, 115730. [CrossRef]
19. Poma, A.; Brahmbhatt, H.; Pendergraff, H.M.; Watts, J.K.; Turner, N.W. Generation of Novel Hybrid Aptamer-Molecularly Imprinted Polymeric Nanoparticles. *Adv. Mater.* **2015**, *27*, 750–758. [CrossRef]
20. Keefe, A.D.; Pai, S.; Ellington, A. Aptamers as therapeutics. *Nat. Rev. Drug Discov.* **2010**, *9*, 537–550. [CrossRef]
21. Pfeiffer, F.; Rosenthal, M.; Siegl, J.; Ewers, J.; Mayer, G. Customised nucleic acid libraries selection for enhanced aptamer and performance. *Curr. Opin. Biotechnol.* **2017**, *48*, 111–118. [CrossRef] [PubMed]
22. Rothlisberger, P.; Hollenstein, M. Aptamer chemistry. *Adv. Drug Deliv. Rev.* **2018**, *134*, 3–21. [CrossRef] [PubMed]
23. Bai, W.; Gariano, N.A.; Spivak, D.A. Macromolecular Amplification of Binding Response in Superaptamer Hydrogels. *J. Am. Chem. Soc.* **2013**, *135*, 6977–6984. [CrossRef] [PubMed]
24. Bai, W.; Spivak, D.A. A Double-Imprinted Diffraction-Grating Sensor Based on a Virus-Responsive Super-Aptamer Hydrogel Derived from an Impure Extract. *Angew. Chem.-Int. Ed.* **2014**, *53*, 2095–2209. [CrossRef] [PubMed]
25. Liu, X.; Ren, J.; Su, L.; Gao, X.; Tang, Y.; Ma, T.; Zhu, L.; Li, J. Novel hybrid probe based on double recognition of aptamer-molecularly imprinted polymer grafted on upconversion nanoparticles for enrofloxacin sensing. *Biosens. Bioelectron.* **2017**, *87*, 203–208. [CrossRef]
26. Zhou, Q.Q.; Xu, Z.G.; Liu, Z.M. Molecularly Imprinting-Aptamer Techniques and Their Applications in Molecular Recognition. *Biosensors* **2022**, *12*, 576. [CrossRef]
27. Moein, M.M.; Abdel-Rehim, A.; Abdel-Rehim, M. Recent Applications of Molecularly Imprinted Sol-Gel Methodology in Sample Preparation. *Molecules* **2019**, *24*, 2889. [CrossRef]
28. Gutiérrez-Climente, R.; Clavié, M.; Dumy, P.; Mehdi, A.; Subra, G. Sol-gel process: The inorganic approach in protein imprinting. *J. Mater. Chem. B* **2021**, *9*, 2155–2178. [CrossRef]
29. Malitesta, C.; Mazzotta, E.; Picca, R.A.; Poma, A.; Chianella, I.; Piletsky, S.A. MIP sensors—The electrochemical approach. *Anal. Bioanal. Chem.* **2012**, *402*, 1827–1846. [CrossRef]
30. Palladino, P.; Bettazzi, F.; Scarano, S. Polydopamine: Surface coating, molecular imprinting, and electrochemistry-successful applications and future perspectives in (bio)analysis. *Anal. Bioanal. Chem.* **2019**, *411*, 4327–4338. [CrossRef]
31. You, M.; Yang, S.; An, Y.; Zhang, F.; He, P. A novel electrochemical biosensor with molecularly imprinted polymers and aptamer-based sandwich assay for determining amyloid-beta oligomer. *J. Electroanal. Chem.* **2020**, *862*, 114017. [CrossRef]
32. Ocana, C.; del Valle, M. Three different signal amplification strategies for the impedimetric sandwich detection of thrombin. *Anal. Chim. Acta* **2016**, *912*, 117–124. [CrossRef] [PubMed]

33. Tang, L.; Liang, K.; Wang, L.; Chen, C.; Cai, C.; Gong, H. Construction of an Ultrasensitive Molecularly Imprinted Virus Sensor Based on an "Explosive" Secondary Amplification Strategy for the Visual Detection of Viruses. *Anal. Chem.* **2022**, *94*, 13879–13888. [CrossRef] [PubMed]
34. Ma, H.Y.; Lu, C.Y.; Tsao, K.C.; Shih, H.M.; Cheng, A.L.; Huang, L.M.; Chang, L.Y. Association of EV71 3C polymorphisms with clinical severity. *J. Microbiol. Immunol. Infect.* **2018**, *51*, 608–613. [CrossRef]
35. Liao, Z.; Peng, J.; Chen, S.; Zhang, P.; Chen, H.; Feng, D.; Zhang, T.; Ye, K.; Deng, Y.; Dong, Y.; et al. Sensitive fluorescent sensor for the fuzzy exosomes in serum based on the exosome imprinted polymer sandwiched with aggregation induced emission. *Sens. Actuators B-Chem.* **2022**, *358*, 131182. [CrossRef]
36. Lu, H.; Huang, Y.; Cui, H.; Li, L.; Ding, Y. A molecularly imprinted electrochemical aptasensor based on zinc oxide and co-deposited gold nanoparticles/reduced graphene oxide composite for detection of amoxicillin. *Microchim. Acta* **2022**, *189*, 421. [CrossRef]
37. Krishnan, H.; Gopinath, S.C.; Arshad, M.M.; Zulhaimi, H.I.; Anbu, P.; Subramaniam, S. Molecularly imprinted polymer enhances affinity and stability over conventional aptasensor for blood clotting biomarker detection on regimented carbon nanohorn and gold nanourchin hybrid layers. *Sens. Actuators B-Chem.* **2022**, *363*, 131842. [CrossRef]
38. Shahdost-fard, F.; Roushani, M. Impedimetric detection of trinitrotoluene by using a glassy carbon electrode modified with a gold nanoparticle@fullerene composite and an aptamer-imprinted polydopamine. *Microchim. Acta* **2017**, *184*, 3997–4006. [CrossRef]
39. Ho, I.P.; Yoo, S.J.; Tefera, S. Determination of furan levels in coffee using automated solid-phase microextraction and gas chromatography/mass spectrometry. *J. AOAC Int.* **2005**, *88*, 574–576.
40. Hashemi-Moghaddam, H.; Ahmadifard, M. Novel molecularly-imprinted solid-phase microextraction fiber coupled with gas chromatography for analysis of furan. *Talanta* **2016**, *150*, 148–154. [CrossRef]
41. Dounin, V.; Veloso, A.J.; Schulze, H.; Bachmann, T.T.; Kerman, K. Disposable electrochemical printed gold chips for the analysis of acetylcholinesterase inhibition. *Anal. Chim. Acta* **2010**, *669*, 63–67. [CrossRef] [PubMed]
42. Mariyappan, V.; Keerthi, M.; Chen, S.M. Highly Selective Electrochemical Sensor Based on Gadolinium Sulfide Rod-Embedded RGO for the Sensing of Carbofuran. *J. Agric. Food Chem.* **2021**, *69*, 2679–2688. [CrossRef]
43. Li, S.; Li, J.; Luo, J.; Xu, Z.; Ma, X. A microfluidic chip containing a molecularly imprinted polymer and a DNA aptamer for voltammetric determination of carbofuran. *Microchim. Acta* **2018**, *185*, 295. [CrossRef] [PubMed]
44. Sullivan, M.V.; Allabush, F.; Flynn, H.; Balansethupathy, B.; Reed, J.A.; Barnes, E.T.; Robson, C.; O'Hara, P.; Milburn, L.J.; Bunka, D.; et al. Highly Selective Aptamer-Molecularly Imprinted Polymer Hybrids for Recognition of SARS-CoV-2 Spike Protein Variants. *Glob. Chall.* **2021**, *12*, 4394–4405. [CrossRef] [PubMed]
45. Abe-Chayama, H.; Hayes, C.N.; Chayama, K. Pan-genotypic cell culture system for propagation of hepatitis C virus clinical isolates. *Hepatology* **2016**, *64*, 1356–1358. [CrossRef]
46. Chen, S.; Cai, G.; Gong, X.; Wang, L.; Cai, C.; Gong, H. Non-autofluorescence Detection of H5N1 Virus Using Photochemical Aptamer Sensors Based on Persistent Luminescent Nanoparticles. *ACS Appl. Mater. Interfaces* **2022**, *14*, 46964–46971. [CrossRef]
47. Shih, C.; Yang, C.C.; Choijilsuren, G.; Chang, C.H.; Liou, A.T. Hepatitis B Virus. *Trends Microbiol.* **2018**, *26*, 386–387. [CrossRef]
48. Chen, S.; Luo, L.; Wang, L.; Chen, C.; Gong, H.; Cai, C. A sandwich sensor based on imprinted polymers and aptamers for highly specific double recognition of viruses. *Analyst* **2021**, *146*, 3924–3932. [CrossRef]
49. Ghanbari, K.; Roushani, M. A nanohybrid probe based on double recognition of an aptamer MIP grafted onto a MWCNTs-Chit nanocomposite for sensing hepatitis C virus core antigen. *Sens. Actuators B Chem.* **2018**, *258*, 1066–1071. [CrossRef]
50. Wang, Q.; Yang, Q.L.; Wu, W. Progress on Structured Biosensors for Monitoring Aflatoxin B1 From Biofilms: A Review. *Front. Microbiol.* **2020**, *11*, 408. [CrossRef]
51. Roushani, M.; Farokhi, S.; Rahmati, Z. Development of a dual-recognition strategy for the aflatoxin B1 detection based on a hybrid of aptamer-MIP using a Cu_2O NCs/GCE. *Microchem. J.* **2022**, *178*, 107328. [CrossRef]
52. Chi, H.; Liu, G. A fluorometric sandwich biosensor based on molecular imprinted polymer and aptamer modified CdTe/ZnS for detection of aflatoxin B1 in edible oil. *Food Sci. Technol.* **2023**, *180*, 114726. [CrossRef]
53. Khalil, O.A.A.; Hammad, A.A.; Sebaei, A.S. Aspergillus flavus and Aspergillus ochraceus inhibition and reduction of aflatoxins and ochratoxin A in maize by irradiation. *Toxicon* **2021**, *198*, 111–120. [CrossRef] [PubMed]
54. Lyu, H.; Sun, H.; Zhu, Y.; Wang, J.; Xie, Z.; Li, J. A double-recognized aptamer-molecularly imprinted monolithic column for high-specificity recognition of ochratoxin A. *Anal. Chim. Acta* **2020**, *1103*, 97–105. [CrossRef]
55. Aralekallu, S.; Palanna, M.; Hadimani, S.; Prabhu C.P., K.; Sajjan, V.A.; Thotiyl, M.O.; Sannegowda, L.K. Biologically inspired catalyst for electrochemical reduction of hazardous hexavalent chromium. *Dalton Trans.* **2020**, *49*, 15061–15071. [CrossRef]
56. Ma, Y.; Ran, D.; Shi, X.; Zhao, H.; Liu, Z. Cadmium toxicity: A role in bone cell function and teeth development. *Sci. Total Environ.* **2021**, *769*, 144646. [CrossRef]
57. Li, S.; Ma, X.; Pang, C.; Tian, H.; Xu, Z.; Yang, Y.; Lv, D.; Ge, H. Fluorometric aptasensor for cadmium(II) by using an aptamer-imprinted polymer as the recognition element. *Microchim. Acta* **2019**, *186*, 823. [CrossRef]
58. Hauser, A.R.; Mecsas, J.; Moir, D.T. Beyond antibiotics: New therapeutic approaches for bacterial infections. *Int. J. Antimicrob. Agents* **2017**, *50*, S19–S20. [CrossRef]
59. Hoeksema, M.; Brul, S.; Kuile, B.H.T. Influence of Reactive Oxygen Species on De Novo Acquisition of Resistance to Bactericidal Antibiotics. *Antimicrob. Agents Chemother.* **2018**, *62*, 17. [CrossRef]

60. Spizek, J.; Rezanka, T. Lincosamides: Chemical structure, biosynthesis, mechanism of action, resistance, and applications. *Biochem. Pharmacol.* **2017**, *133*, 20–28. [CrossRef]
61. Li, S.; Liu, C.; Yin, G.; Zhang, Q.; Luo, J.; Wu, N. Aptamer-molecularly imprinted sensor base on electrogenerated chemiluminescence energy transfer for detection of lincomycin. *Biosens. Bioelectron.* **2017**, *91*, 687–691. [CrossRef] [PubMed]
62. Bi, H.; Wu, Y.; Wang, Y.; Liu, G.; Ning, G.; Xu, Z. A molecularly imprinted polymer combined with dual functional Au@Fe_3O_4 nanocomposites for sensitive detection of kanamycin. *J. Electroanal. Chem.* **2020**, *870*, 114216. [CrossRef]
63. Sanchez, A.R.; Rogers, R.S.; Sheridan, P.J. Tetracycline and other tetracycline-derivative staining of the teeth and oral cavity. *Int. J. Dermatol.* **2004**, *43*, 709–715. [CrossRef] [PubMed]
64. Ma, Y.; Liao, X.; Zhao, Y.; Qiu, L.; Yao, Y.; Wang, S.; Yang, X.; Hu, X. Fabrication of magnetic molecularly imprinted polymers based on aptamers and ss-cyclodextrin for synergistic recognition and separation of tetracycline. *Anal. Chim. Acta* **2022**, *1236*, 340572. [CrossRef] [PubMed]
65. Paprocka, P.; Durnaś, B.; Mańkowska, A.; Król, G.; Wollny, T.; Bucki, R. *Pseudomonas aeruginosa* Infections in Cancer Patients. *Pathogens* **2022**, *11*, 679. [CrossRef]
66. Sarabaegi, M.; Roushani, M. Rapid and sensitive determination of *Pseudomonas aeruginosa* by using a glassy carbon electrode modified with gold nanoparticles and aptamer-imprinted polydopamine. *Microchem. J.* **2021**, *168*, 106388. [CrossRef]
67. Jarneborn, A.; Mohammad, M.; Engdahl, C.; Hu, Z.; Na, M.; Ali, A.; Jin, T. Tofacitinib treatment aggravates Staphylococcus aureus septic arthritis, but attenuates sepsis and enterotoxin induced shock in mice. *Sci. Rep.* **2020**, *10*, 10891. [CrossRef]
68. Cai, R.; Yin, F.; Zhang, Z.; Tian, Y.; Zhou, N. Functional chimera aptamer and molecular beacon based fluorescent detection of Staphylococcus aureus with strand displacement-target recycling amplification. *Anal. Chim. Acta* **2019**, *1075*, 128–136. [CrossRef]
69. El-Wekil, M.M.; Halby, H.M.; Darweesh, M.; Ali, M.E.; Ali, R. An innovative dual recognition aptasensor for specific detection of Staphylococcus aureus based on Au/Fe_3O_4 binary hybrid. *Sci. Rep.* **2022**, *12*, 12502. [CrossRef]
70. De Oliveira, J.; Kucharska, E.; Garcez, M.L.; Rodrigues, M.S.; Quevedo, J.; Moreno-Gonzalez, I.; Budni, J. Inflammatory Cascade in Alzheimer's Disease Pathogenesis: A Review of Experimental Findings. *Cells* **2021**, *10*, 2581. [CrossRef]
71. Roos, A.; Edgren, G. Using historical cardiac troponins to identify patients at a high risk of myocardial infarction. *Heart* **2023**, *109*, 127–133. [CrossRef]
72. Mokhtari, Z.; Khajehsharifi, H.; Hashemnia, S.; Solati, Z.; Azimpanah, R.; Shahrokhian, S. Evaluation of molecular imprinted polymerized methylene blue/aptamer as a novel hybrid receptor for Cardiac Troponin I (cTnI) detection at glassy carbon electrodes modified with new biosynthesized ZnONPs. *Sens. Actuators B-Chem.* **2020**, *320*, 108316. [CrossRef]
73. Wang, Z.; Fang, X.; Sun, N.; Deng, C. A rational route to hybrid aptamer-molecularly imprinted magnetic nanoprobe for recognition of protein biomarkers in human serum. *Anal. Chim. Acta* **2020**, *1128*, 1–10. [CrossRef] [PubMed]
74. Yang, C.; Tian, Y.; Wang, B.; Guo, Q.; Nie, G. "Signal-on" molecularly imprinting-aptamer electrochemiluminescence platform for ultrasensitive detection of thrombin. *Sens. Actuators B-Chem.* **2021**, *338*, 129870. [CrossRef]
75. Ensafi, A.A.; Amini, M.; Rezaei, B. Molecularly imprinted electrochemical aptasensor for the attomolar detection of bisphenol A. *Microchim. Acta* **2018**, *185*, 265. [CrossRef]
76. Mahmoud, A.M.; Alkahtani, S.A.; Alyami, B.A.; El-Wekil, M.M. Dual-recognition molecularly imprinted aptasensor based on gold nanoparticles decorated carboxylated carbon nanotubes for highly selective and sensitive determination of histamine in different matrices. *Anal. Chim. Acta* **2020**, *1133*, 58–65. [CrossRef]
77. Yu, C.; Li, L.; Ding, Y.; Liu, H.; Cui, H. Molecularly imprinted electrochemical aptasensor based on functionalized graphene and nitrogen-doped carbon quantum dots for trace cortisol assay. *Analyst* **2022**, *147*, 744–752. [CrossRef]
78. Huang, Y.; Ye, D.; Yang, J.; Zhu, W.; Li, L.; Ding, Y. Dual recognition elements for selective determination of progesterone based on molecularly imprinted electrochemical aptasensor. *Anal. Chim. Acta* **2023**, *1264*, 341288. [CrossRef]
79. Ali, R.; El-Wekil, M.M. A dual-recognition-controlled electrochemical biosensor for selective and ultrasensitive detection of acrylamide in heat-treated carbohydrate-rich food. *Food Chem.* **2023**, *413*, 135666. [CrossRef]
80. Beiki, T.; Najafpour-Darzi, G.; Mohammadi, M.; Shakeri, M.; Boukherroub, R. Fabrication of a novel electrochemical biosensor based on a molecular imprinted polymer-aptamer hybrid receptor for lysozyme determination. *Anal. Bioanal. Chem.* **2023**, *415*, 899–911. [CrossRef]
81. Rahmati, Z.; Roushani, M. SARS-CoV-2 virus label-free electrochemical nanohybrid MIP-aptasensor based on Ni-3(BTC)(2) MOF as a high-performance surface substrate. *Microchim. Acta* **2022**, *189*, 287–296. [CrossRef] [PubMed]
82. Wang, Y.Y.; Kan, X.W. Sensitive and selective "signal-off" electrochemiluminescence sensing of prostate-specific antigen based on an aptamer and molecularly imprinted polymer. *Analyst* **2021**, *146*, 7693–7701. [CrossRef] [PubMed]
83. Roushani, M.; Zalpour, N. Impedimetric ultrasensitive detection of trypsin based on hybrid aptamer-2DMIP using a glassy carbon electrode modified by nickel oxide nanoparticle. *Microchem. J.* **2021**, *172*, 106955. [CrossRef]
84. Roushani, M.; Ghanbarzadeh, M.; Shahdost-Fard, F. Fabrication of an electrochemical biodevice for ractopamine detection under a strategy of a double recognition of the aptamer/molecular imprinting polymer. *Bioelectrochemistry* **2021**, *138*, 107722. [CrossRef] [PubMed]
85. Roushani, M.; Rahmati, Z.; Hoseini, S.J.; Fath, R.H. Impedimetric ultrasensitive detection of chloramphenicol based on aptamer MIP using a glassy carbon electrode modified by 3-ampy-RGO and silver nanoparticle. *Colloids Surf. B Biointerfaces* **2019**, *183*, 110451. [CrossRef]

86. Roushani, M.; Nezhadali, A.; Jalilian, Z. An electrochemical chlorpyrifos aptasensor based on the use of a glassy carbon electrode modified with an electropolymerized aptamer-imprinted polymer and gold nanorods. *Microchim. Acta* **2018**, *185*, 551. [CrossRef]
87. Tan, J.; Guo, M.; Tan, L.; Geng, Y.; Huang, S.; Tang, Y.; Su, C.; Lin, C.C.; Liang, Y. Highly efficient fluorescent QDs sensor for specific detection of protein through double recognition of hybrid aptamer-molecular imprinted polymers. *Sens. Actuators B Chem.* **2018**, *274*, 627–635. [CrossRef]
88. Geng, Y.; Guo, M.; Tan, J.; Huang, S.; Tang, Y.; Tan, L.; Liang, Y. A fluorescent molecularly imprinted polymer using aptamer as a functional monomer for sensing of kanamycin. *Sens. Actuators B-Chem.* **2018**, *268*, 47–54. [CrossRef]
89. Duan, N.; Chen, X.; Lin, X.; Ying, D.; Wang, Z.; Yuan, W.; Wu, S. Paper-based fluorometric sensing of malachite green using synergistic recognition of aptamer-molecularly imprinted polymers and luminescent metal–organic frameworks. *Sens. Actuators B Chem.* **2023**, *384*, 133665. [CrossRef]

chemosensors

Review

Recent Advances in Wearable Sensors for the Monitoring of Sweat: A Comprehensive Tendency Summary

Zhe Xing [1,2,†], Jianan Hui [1,†], Bo Lin [1,2], Zhenhua Wu [1,*] and Hongju Mao [1,2,*]

1 State Key Laboratory of Transducer Technology, Shanghai Institute of Microsystem and Information Technology, Chinese Academy of Sciences, Shanghai 200050, China; xingzhe22@mails.ucas.ac.cn (Z.X.); jiananhui2@mail.sim.ac.cn (J.H.); linbo201@mails.ucas.ac.cn (B.L.)
2 Center of Materials Science and Optoelectronics Engineering, University of Chinese Academy of Sciences, Beijing 100049, China
* Correspondence: wuzhx@mail.sim.ac.cn (Z.W.); hjmao@mail.sim.ac.cn (H.M.)
† These authors contributed equally to this work.

Abstract: Sweat, as a biofluid that is easy to extract and contains a variety of biomarkers, can provide various types of physiological information for health monitoring. In recent years, research on wearable sensors for sweat sensing has been emerging continuously. Wearable sweat sensing will probably become an alternative method to traditional chemical analysis. This is due to its advantages of portability, non-invasiveness, comfort, and continuous monitoring. Since the inception of this research field, wearable sweat sensors have achieved significant development in terms of materials, structures, systems, and application directions. Research interests are gradually evolving from single biomarker detection to the pursuit of multi-channel, multi-modal system-level architecture. The analysis of physiological signals has also developed from single signal characterization to omics analysis using multiple physiological information sources. Based on the changes mentioned above, this paper mainly introduces the latest researches of wearable sweat sensors from the aspects of strategy, architecture, material, system, data processing, etc., and tries to summarize the trends of sweat sensors. Finally, this paper analyzes the challenges faced by the sensing platform and possible methods for optimization.

Keywords: health monitoring; sweat analysis; wearable device; sensing platform; non-invasive detection

Citation: Xing, Z.; Hui, J.; Lin, B.; Wu, Z.; Mao, H. Recent Advances in Wearable Sensors for the Monitoring of Sweat: A Comprehensive Tendency Summary. *Chemosensors* **2023**, *11*, 470. https://doi.org/10.3390/chemosensors11090470

Academic Editor: Vardan Galstyan

Received: 29 June 2023
Revised: 27 July 2023
Accepted: 21 August 2023
Published: 23 August 2023

1. Introduction

In the last few years, there has been an increasing focus on personal health and disease prevention. Monitoring human physiological signals is a crucial approach to achieving this goal. One common situation is that individuals often experience discomfort or specific symptoms prior to seeking medical attention, which can result in delayed treatment. Traditional detection techniques typically depend on trained professionals, numerous sample extractions, complex and expensive instruments, and lengthy sample analysis durations. Moreover, these techniques are often invasive and rely upon subjective diagnostic results from healthcare professionals. As such, they may be inadequate to meet the growing demand for convenient and accessible detection. In contrast, wearable sensors offer several advantages, including portability, low cost, comfort, and non-invasiveness. These devices enable in situ detection and wireless data transmission, overcoming many of the limitations of traditional methods. Therefore, wearable sensors have the potential to serve as an alternative approach to detection and have attracted significant research interest. So far, both physical and chemical sensors have made notable advancements [1–3]. Wearable chemical sensors are capable of detecting a wide range of biomarkers, such as metabolites, nutrients, drugs and so on. Through the use of transducers and data analysis, these biomarkers can be converted into practical information for use in various scenarios, including daily life, sports performance, clinical diagnosis, and rehabilitation. Biomarkers exist in a wide range of biofluids, e.g., blood, interstitial fluid, sweat, tears, saliva, etc. Blood has always been the gold standard for biochemical detection, while other non-invasively accessible biofluids also contain many of the same components as blood. Thus, interstitial fluid, sweat, tears and saliva have become the primary target biofluids for wearable sensors. Many studies on tear monitoring have utilized contact lenses for glucose detection, which may cause discomfort for the wearer [2]. The composition of saliva is highly similar to that of blood, and sample extraction is relatively simple. However, a disadvantage of saliva monitoring is that it is difficult to deal with the noise interference introduced by eating and the oral environment [4]. Conversely, sweat is easy to extract and the sample is less disturbed by the environment. It also contains similar biomarkers to blood, including metabolites, electrolytes, proteins, hormones, and micronutrients. Consequently, there are compelling reasons to use sweat as a complementary tool for standard analytical diagnosis [5], and sweat sensors have always been a popular area in the field of wearable biosensor research. Since Gao et al. published their fully integrated sweat sensor [6], many wearable sensors with different shapes and principles have emerged, including tattoos, wristbands, patches, and rings [7]. After several years of development, the focus of wearable sensor research has shifted significantly from its initial direction. Wearable sensors were initially designed to simply capture and process signals produced by the human body. However, these devices are now increasingly focused on biochemical analysis to obtain more detailed information about health conditions and are utilizing omics research to provide a more comprehensive picture of human health. Regarding sensor structure and fabrication, with the development of flexible electronics, research has focused more on sensing materials, structural design, and system integration to achieve multichannel and multimodal sensing. This article focuses on showing the latest research results in the field of wearable sweat sensors, according to their classification of applications, strategies, morphology, materials, structures and challenges, respectively, and by giving examples of the latest specific research in recent years, in order to summarize the development trend of wearable sweat sensors in recent years. First, the new sweat sensors for different target analytes are introduced according to the application scenarios, which mainly serve as an overview. Then, several modification methods of sweat sensors are discussed. The next section explains the detection strategy used in sweat sensors and some purposeful optimization research. After that, the development of sweat sensor forms and the latest trends are analyzed by introducing innovative progress in materials, structure and systems. Subsequently, the current application of machine learning in sensor signal processing is introduced. Next, challenges and potential feasible solutions in the current sweat sensor research are listed. The last section summarizes the full article.

2. Target Analytes/Application Scenarios for Sweat Sensor

The main components of sweat are ions/electrolytes, small molecules (ethanol, cortisol, urea, lactic acid, etc.) and proteins (neuropeptides and cytokines) [8]. The choice of target analyte for a sweat sensor is determined not only by the application scenario, but also by reference to existing physiological studies.

2.1. Biochemical Analytes

2.1.1. Metabolites

Glucose: Enzyme-based glucose sensors have advanced from the first generation (electron transfer based on hydrogen peroxide involved in redox reactions) to the second generation (media-based electron transfer) [9], and the third generation (electron transfer directly between the enzyme and the electrode without any mediator) [10]. Since the first media-based amperometric sensor [11] was reported, the second generation of glucose sensors have made great progress, but they still have the disadvantages of relative toxicity, solubility of the medium and insufficient system stability [12], which has driven the birth of the third-generation glucose sensor. In the field of wearables, each generation of sensors is currently being applied, and the second and third generations are more prevalent, mainly used in the management (prevention, screening, detection and treatment) of type II diabetes [13].

Lactate: Lactate reflects metabolic, renal and cardiorespiratory health. During exercise, sweat lactate concentration increases with blood lactate concentration [14], so sweat lactate testing can replace blood testing to some extent. Lactate dehydrogenase (LDH) or lactate oxidase (LOx/LOD) are commonly used for selective identification, and the addition of Prussian blue [9] (a good electron transfer medium) for lactate detection can reduce the reaction potential and improve selectivity [15–20].

Alcohol: Sweat alcohol concentration correlates with blood alcohol concentration [21] and blood ethanol content (BAC) can be determined by detecting transdermal alcohol concentration (TAC) using wearable sweat sensors [22]. Not limited to liquid phase testing, alcohol detection can take advantage of alcohol's volatility. Khemtonglang et al. [23] built a smart wristband for sweat alcohol detection based on an MOX gas sensor, which can detect drunk driving with 94.66% accuracy and can integrate an IoT-based alarm system to facilitate law enforcement.

Cortisol: Cortisol is a glucocorticoid whose concentration in sweat is associated with changes in psychological or physiological stress and is regulated by the HPA axis [24]. Cortisol secretion can help the body maintain homeostasis, but excessive secretion can lead to disease (depression, anxiety, etc.), so real-time monitoring is necessary. Common detection methods include antibodies, DNA aptamers, and molecularly imprinted polymers (MIPs) [25]. Torrente-Rodriguez et al. [26] developed a wearable portable sensing platform for the real-time monitoring of sweat cortisol dynamics, and reported the first cortisol diurnal cycle and dynamic stress response curve constructed by human sweat according to the monitoring results (Figure 1a). Compared with the subjective stress monitoring method of a traditional questionnaire survey, it provides a data-based non-invasive real-time stress monitoring method, and demonstrates the good correlation between the hormonal circadian cycle and stress curve, which is helpful for further improving stress-management methods.

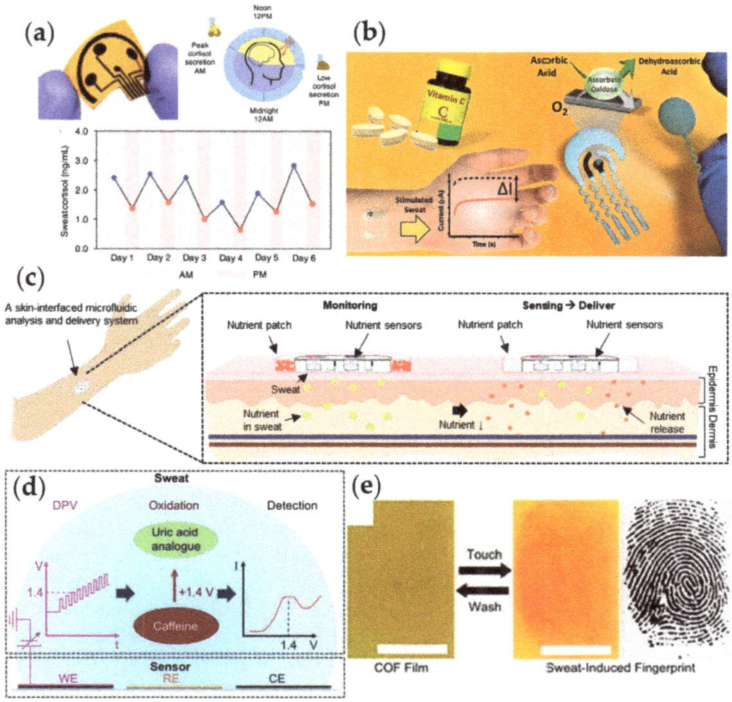

Figure 1. Novel wearable sweat sensors for multiple analytes. (**a**) A novel cortisol sensor and circadian rhythm study. Reprinted from ref. [26], Copyright (2020), with permission from Elsevier. (**b**) An enzymatic biosensor that monitors vitamin C in sweat. Reprinted from ref. [27], Copyright (2020), with permission from American Chemical Society. (**c**) A sweat sensor that monitors multiple nutrients simultaneously, allowing for nutrient replenishment. Reprinted from ref. [28] Copyright (2021), with permission from John Wiley and Sons. (**d**) Sweat sensor to detect methylxanthine drugs (caffeine) using differential pulsed voltammetry. Reprinted from ref. [29] Copyright 2018, with permission from John Wiley and Sons. (**e**) A sweat-sensing patch for fingerprint liveness detection measurement. Reprinted from ref. [30], Copyright (2023), with permission from Springer Nature.

Electrolyte: Electrolyte concentration can reflect body fluid homeostasis, and monitoring sweat can help ensure timely replenishment and prevent excessive loss. Electrolytes can be measured through electrochemical methods using ion-selective electrodes or colorimetrical methods, with specific strategies tailored to their needs. Among the common electrolytes, sweat chloride can reflect cystic fibrosis, so detecting chloride concentration has been studied for a long time [31]. Zinc ions are early diagnostic biomarkers for muscle stress and fatigue [32] and play a critical role in gene expression and immune function [33]; magnesium ions are essential cations for central nervous system function [34], are involved in neuroendocrine metabolic processes, and can reflect stress levels and sleep status [33]. The detection of these ions can be achieved using sweat-sensing patches [35]. Sodium ion concentrations vary greatly during sweating [36] and Hashimoto et al. used this to propose a real-time colorimetric sensor that expresses electrolyte concentration changes in sweat using sodium chloride as the reference analyte [37]. In the field of electrolyte detection, there have been many studies based on sodium and potassium ions, such as a wearable platform reported by Porvano et al. [38] that can simultaneously detect sodium and potassium ions in sweat, and can detect other target ions by swapping components.

Changes in pH are associated with many diseases or abnormalities, including infection, asthma, diabetes, kidney disease, lung disease and heart disease. pH sensors mostly use potentiometric methods for signal measurement [39–41].

2.1.2. Biomolecules

Amino acid: Amino acids can be used to synthesize proteins, which are vital nutrients for the functioning of the immune and nervous systems. Additionally, proteins contribute to muscle and bone growth and are fundamental substances for human life activities. Yang et al. [42] proposed a wearable sweat sensor for uric acid and tyrosine. It can detect exercise performance after protein ingestion and determine the presence of gout through sweat analysis after the ingestion of purine-rich foods. The results can also be correlated with serum measurements in healthy subjects and ventilated patients. Further research could extend the sensor's capabilities to detect cardiovascular disease, type II diabetes and kidney disease.

Vitamin: As essential micronutrients, vitamins generally cannot be produced by the human body and must be obtained through a sensible diet. Take vitamin C, for example, which can boost immunity, promote iron absorption and, if not consumed in sufficient quantities, can lead to iron deficiency anemia, weakened immunity and even scurvy. Sempionatto et al. [27] reported an enzymatic biosensor for the detection of vitamin C in sweat, which holds promise for the detection of nutrient levels and the provision of personalized nutritional solutions (Figure 1b). Ascorbate oxidase (AAOx)-based printable tattoo electrodes used in the device are highly selective and support the detection of real-time changes in vitamin C for timely assessment of vitamin intake and dietary changes.

Wang et al. [43] designed a wearable electrochemical sensor that continuously monitors amino acid and vitamin levels in sweat. The sensor can assess the risk of metabolic syndrome by correlating amino acid levels in serum and sweat, thus facilitating the adjustment of nutritional health status.

After assessing nutritional levels, further supplementation may be necessary. In addition to improving dietary habits and taking oral supplements, the integration of a nutrient release function into the sensor, similar to a drug delivery module, could also be considered. Kim et al. [28] reported on a miniature sensor system targeting several essential nutrients (vitamin C, calcium, zinc and iron) in sweat and verified the dynamic correlation between the concentration of these nutrients in sweat and the corresponding concentration in blood based on human study results (Figure 1c). During the wearing period, nutrients can also be continuously delivered to the body through transdermal patches that are directly integrated with microfluidic structures. Once colorimetric results indicate a nutrient imbalance, the block layer can be removed to initiate the release process. The efficiency of the drug delivery module has been verified and is comparable to that of oral administration.

2.1.3. Hazardous Substances/Drugs

Drug testing is important for stimulant control and precision medicine, and sweat is an attractive biofluid that holds promise as an alternative to traditional invasive (blood draw) assays [29]. The sweat sensor proposed by Tai et al. [29] in 2018 used differential pulsed voltammetry to detect methylxanthine drugs like caffeine (Figure 1d). The device's ability to measure caffeine levels under different conditions can greatly assist clinicians in adjusting dosages, monitoring patient compliance and understanding pharmacokinetics.

Nicotine is a highly addictive drug and excessive use can lead to heart, brain and respiratory disease, and even lung cancer [44]. Conversely, studies of nicotine in the treatment of disease have also shown that NIC can be used to treat ulcerative colitis, Alzheimer's disease and Parkinson's disease [44]. Based on the above applications, Magesh et al. [45] developed an electrochemical sensor based on MXene ($Ti_3C_2T_x$)/palladium hydroxide-supported carbon (Pearlman's catalyst) composite (MXene/$Pd(OH)_2$/C) to detect nicotine levels in human sweat. The sensor was validated using sweat samples from smokers and can achieve high sensitivity and selectivity.

2.2. Physical Indicators

2.2.1. Sweat Rate

Sweat rate monitoring is important for screening for physical abnormalities such as hyperhidrosis, thermoregulatory and heat stress-related disorders, mental stress and neurological disorders, and to help prevent dehydration and heat stroke. Sweat rate can be measured using colorimetric (detecting the amount of sweat) and flow methods. Apart from the traditional methods of measuring sweat rate (described above), many new ideas have been developed recently. Kwon et al. [46] used a thermal actuator and thermistor-based system for flow measurement, requiring only a short straight channel for measurement. Hashimoto et al. [37] chose to direct sweat through a short microfluidic pathway in the form of droplets for easy quantification. Momose et al. [47] did not use flow to measure sweat rate, but integrated a capacitive humidity sensor into a ventilation capsule for monitoring. Dautta et al. [48] converted flow to admittance, using the change in admittance when the electrode comes into contact with sweat to measure flow. After the sweat enters the spiral microfluidic channel embedded in the electrode, the admittance value increases as the contact area grows, and the real-time change in sweat rate can be obtained from the admittance data over time. The article also mentions that the real-time monitoring of sweat rate using wearable devices can also help study the relationship between local and systemic sweat rate.

Heat stroke has become a serious global problem. Children, the elderly and manual laborers are particularly susceptible to heat stroke in hot and humid weather. Timely hydration and cooling can effectively reduce the risk of heat stroke. Based on the study of the original sweat meter [49,50], Momose et al. [47] developed an early warning system for the risk of heat stroke using a portable sweat meter. In addition to capacitive moisture sensors, a smartphone-based warning system was developed. The system can provide an early warning of heat stroke risk at the time when the sweat rate changes from negative to positive in the second order, and remind the wearer to rehydrate according to the sweat volume.

The measurement of athletes' physiological data in training relies on accurate, real-time monitoring. While monitoring exercise performance in a laboratory environment requires wired connections and computer analysis, wearables enable non-invasive, continuous sensing and wireless data transmission. Measurements from wearable sweat sensors promise to provide a real-time reflection of an athlete's physical condition, preventing injury, dehydration or cramps, while helping healthcare professionals/coaches tailor hydration, nutrition and rehabilitation programs to each athlete [51]. The quest for lightness and comfort in wearable sports devices has led to a focus on simplifying sensor structures. For example, Liu et al. [52] developed a paper-based sweat sensor using pH and H_2O_2 test strips to meet the needs of sports scenarios. The colorimetric patch sensor has a simple, lightweight, and thin structure, as well as a low cost and strong stretchability. Its minimal impact on human activities makes it suitable for monitoring in motion. Additionally, dynamic sweat measurement can be achieved by integrating smartphone image processing.

2.2.2. Fingerprint

Early studies have used sweat pores to reflect fingerprint patterns to facilitate fingerprint analysis and pore-dysfunction diagnosis; for example, Lee et al. [53] used water-induced materials to reflect sweat pore distribution. More recent work in this area includes the development of a covalent organic framework for live fingerprint detection by Hao et al. [30] This framework can distinguish human fingerprint patterns based on sweat response and is unresponsive to artificial false fingerprints (Figure 1e). This study exploits the essential difference between living and fake fingerprints, i.e., the secretion of sweat. Sweat hole detection by materials is used to complete biometrics, eliminating the need for complex feature extraction algorithms used in previous recognition steps [54–58]. This approach is expected to improve the reliability and security of existing fingerprint-recognition technologies. The material can be re-used with ethanol cleaning, ensuring practicality and reusability. Sweat biomarkers have also been mentioned as potentially useful for determining biological age [59].

Table 1 provides a list of common target analytes in sweat and their intended applications. Although the discussion above is separated by different biomarkers, current wearable sweat sensors are often not designed for a single biomarker. A significant number of them also detect body temperature, heart rate, gait, and other indicators simultaneously. This is because the physical state of human body is related to numerous metabolites and small molecules. It is clear that testing each biomarker individually is not as efficient as simultaneously detecting multiple indicators to characterize health status. At the same time, advances in wearable sensor technology allow for the further expansion of device detection capabilities. Multiplex detection also enables the discovery of unknown biomarkers associated with specific pathological conditions.

Table 1. Common target analytes in sweat and related diseases or applications.

Category	Analyte	Target	Applications	References
Biochemical analyte	Metabolites	Glucose	Preventing and treating type II diabetes	[13]
		Lactate	Checking metabolic, kidney, and cardiopulmonary health	[9,14]
		Alcohol	Detecting drunk driving	[21,22]
		Cortisol	Measuring stress changes to prevent depression and anxiety	[24,26]
	Biomolecules	Electrolyte	Maintaining human fluid homeostasis, detecting cystic fibrosis (Cl)	[35–38,60]
		pH	Checking for infection, asthma, metabolism, pH balance, and wound healing	[39–41]
		Amino acid	Demonstrating exercise performance after protein intake, detecting gout and cardiovascular disease	[28,42]
		Vitamin C	Revealing the level of human nutrition, reminding subjects to supplement in time to maintain immunity	[27,28]
	Hazardous substances or drugs	Caffeine	Helping doctors adjust dosage, observe prescription adherence, and understand pharmacokinetics	[29]
		Nicotine	Revealing the risk of heart, brain and respiratory diseases, and even lung cancer	[44,45]
Physical indicators		Sweat rate	Avoiding dehydration and heat stroke, detecting mental stress and neurological disorders, and improving athletic performance	[46,47,51,52]
		Fingerprint	Biometrics	[30,53]
		Age		[59]

3. Modification of Sweat Sensors

3.1. Antibody

Electrochemical immunosensors detect target analytes through the specific binding of antibodies to antigens. In recent studies, immunosensors were commonly used to detect cortisol levels in sweat. Cortisol, as an antigen, interacts with antibodies and inhibits electron transfer, thereby reducing the current response [61]. Santiago et al. [62] adopted this principle to characterize different sweat cortisol concentrations using cyclic voltammetry.

3.2. DNA

DNAs commonly used to modify electrodes include single-stranded DNA (ssDNA) and DNA-based molecular pendulums. The former is specifically identified using DNA hybridization or signal probe displacement. The latter uses specific binding to target molecules to induce changes in shape and probe electrode spacing to produce recognition signals [63]. When the analyte binds to the DNA on the electrode, the structure and dynamics of the electrode change, affecting the efficiency of electron transfer [64]. DNA-modified electrodes employing electrochemical impedance spectroscopy can provide the continuous real-time detection of cortisol [65].

3.3. Artificial Receptors

3.3.1. Aptamer

When binding to the target molecule, the aptamer undergoes a conformational conversion. When bound to the electrode, the aptamer's identification process causes electron transfer, resulting in a current response. The molecular structure of the aptamers affect the current response, and the distribution density of the aptamers also affects sensor performance [66]. Singh et al. [67] developed a pseudo-junction aptamer-based microfluidic sensor. Through multiple comparative experiments on different aptamers with different hybridization regions, they identified a pseudo-knot aptamer with high sensitivity and low variability. However, aptamers attached to sensing electrodes are not completely stable because sweat contains a small amount of DNase-I, which can degrade aptamers with long-term exposure [68,69]. Surface sealants such as actin and 2ME are widely used to inhibit the activity of DNase-I [70,71]. This study verified the protective effect of 2ME + actin on aptamers.

3.3.2. Molecularly Imprinted Polymer

Molecularly imprinted polymers (MIPs) are specific receptors formed by combining functional monomers with imprinted molecules to form a polymer and then dissociating them, which is widely exploited in biochemical sensing due to its high sensitivity and selectivity. However, the classical MIP structure requires washing after use, which makes it difficult to reuse and can be detrimental to continuous sensing. Wang et al. [43] incorporated functional groups into the polymerized structure of laser-engraved graphene (LEG) after the polymerization of functional monomers to obtain a renewable special acceptor that can be regenerated after binding to amino acids at a constant potential applied by the working electrode.

The lifespan of such sweat sensors is largely determined by the adhesion duration and the activity of the modifier. Therefore, numerous studies have focused on two areas: developing more stable attachment methods to prevent modifier detachment in liquid environments, and designing modifiers that can be regenerated and slow the rate of activity loss. The ultimate goal of these studies is to reduce the frequency of sensor replacement and increase the duration of a single usage.

4. Sweat Sensing Strategies

Typical wearable sweat sensors can be divided into electrochemical and optical sensors. In the last few years, devices using other methods have also been developed.

4.1. Electrochemical Sensing Strategies

The advantages of electrochemical technology are high selectivity, high sensitivity, rapid response, small sample requirements and reproducibility. Electrochemical methods combined with electronic systems can convert biomarker concentrations into electronic signals to accurately quantify physiological parameters. Electrochemical methods are used not only in measurement but also in device characterization.

4.1.1. Potentiometric Sensing

Potentiometric sensors are based on the Nernst equation and measure the electrolyte concentration using the potential difference between the reference electrode and the working electrode [1]. The target of potentiometric sensors includes diverse ions and the working electrode is ion-selective. During selective ion identification, a potential change is induced and converted into a voltage signal by the transducer.

4.1.2. Amperometric Sensing

Amperometry quantifies the transfer of electrons during reactions at electrodes. Generally, when the reactants undergo oxidation or reduction reactions, there are electron gains and losses. After applying an appropriate operating potential, electron transfer occurs between the sensing electrode and the reaction environment, resulting in a recordable current. The intensity of the current reflects the concentration of the target analyte. Amperometric sensors combined with enzymes are commonly used for detecting small organic molecules [72], such as lactate [18], glucose [73] and ethanol [22]. During sensor operation, amperometric sensors often use the three-electrode system. This is necessary because after electrons are transferred from the solution to the electrode, the counter electrode supplies electrons to the solution, forming a current loop with the working electrode.

Enzyme activity is influenced by the external environment (temperature, pH, etc.) [74] and, therefore, affects the performance of the sensor. The sweat environment varies from individual to individual [74,75] so the results of the sensor must be calibrated. Previous studies have solved the correction of the sensor for temperature and pH changes [9,74,75]. Recently, Assaduzzaman et al. [9] performed pH and T corrections for the sensor using chronoamperometry to detect sweat glucose and lactose, used the chronoamperometric response of the sensor in artificial sweat to obtain the corresponding T and pH effects, and recalculated the target concentration according to the slope and intercept of the calibration curve in the reference state.

Amperometry has a faster response time, wider detection range, and higher sensitivity than potentiometry. However, amperometry is more susceptible to interference and its detection is often less stable than potentiometry. In addition, amperometry relies on the application of a potential or current, which is not required for potentiometry.

4.1.3. Voltammetric Sensing

In voltammetry, a continuously varying potential is applied to the working electrode and the response current signal is detected to record the current–potential curve and then analyze the composition of a solution. Depending on the potential scanning mode, voltammetry can be divided into cyclic voltammetry, differential pulse voltammetry, and square-wave pulse voltammetry. Voltammetry also involves electron transfer during detection, making three-electrode systems prevalent. It is worth noting that voltammetry may not be as selective as potentiometry and amperometry because different electroactive molecules may have similar reaction potentials [63].

Cyclic voltammetry (CV) can monitor electron transfer behavior, which can be used for electrochemical deposition to prepare electrodes [39] and characterize intrinsic electrochemical properties [76] to obtain the optimum electrode working potential. Magesh et al. [44] verified the selectivity and sensitivity of the sensing electrode to the target (nicotine) in buffer solution using CV. Square-Wave Voltammetry (SWV) is one of the fastest and most sensitive voltammetric techniques for sweat detection [77], because the square-wave voltage eliminates charging current and provides a higher detection threshold for substances such as caffeine [29] or heavy metal ions [78]. When the square wave in SWV is replaced by a pulse, it becomes differential pulse voltammetry. DPV is also suitable for detecting substances at low concentrations and can also characterize redox reactions, such as the binding of MIP templates to target molecules [43].

4.1.4. Impedance/Electrochemical Impedance Spectroscopy

Polymer composites impart moisture absorption to the sensor and the adsorbed moisture can change the dielectric and electrical properties of the electrode. Based on this principle, sweat composition can be monitored via the impedance principle, and the impedance change caused by sweat adsorption reflects the change in sweat composition or volume [79]. Changes in fluid impedance can serve as an indicator of infection detection to aid wound healing [80,81].

Electrochemical impedance spectroscopy (EIS) is an alternating current-based electrochemical measurement technique. Compared to cyclic voltammetry, which scans the entire potential range at a given rate, EIS has the advantage of measuring at different set potentials [82]. However, it requires more complex data-processing methods than amperometry and potentiometry. The input–response relationship reflects the complex impedance characteristics of the sensing electrode, and the impedance modulus at a given frequency contains resistance and capacitance information, revealing the specific binding process of the receptor to the electrode [83–87]. Ganguly et al. [65] calibrated the sweat cortisol sensor using EIS to obtain a response curve that reliably reflects the concentration of synthetic sweat cortisol. Veeralingam et al. [88] utilized interdigitated electrodes to provide an external AC field to measure skin impedance at a frequency of 10 kHz to indicate skin hydration. The data obtained from EIS can be optimized into reliable output data using machine learning.

4.1.5. Others

Yin. L et al. [89] developed a stretchable epidermal sensor platform. An integrated electrochromic display displays the sampled signal from the electrochemical sensor in real time. An enzyme-modified electrode can be used to detect sweat pH, electrolytes, lactic acid, etc. For visualization, the sampled signal is converted to digital by the MCU and connected to ten addressable ECD pixels containing the PSS electrolyte. Compared to conventional electrochemical sensors with complex electronic systems, the patch has a lower energy consumption, is lighter and conforms to the skin. Compared to colorimetric sensors, there is no need to use a smartphone or camera to convert the results [89–94] and the detection results are more intuitive.

4.2. *Optical Sensing Strategies*

In 2014, Lee et al. [53] combined fluorescence and colorimetry to fabricate a sweat pore distribution sensor using the conjugated polymer poly(PCDA-Cs), pioneering the field of optical methods for sweat detection. The advantage of the optical method is that the detection results can be displayed in a relatively intuitive way. In recent years, more research has chosen to incorporate cameras and smartphones to record test results and perform image processing to achieve higher sensitivity. The use of smartphones avoids the use of laboratory instruments for analysis, and results can be analyzed without the need for a specialist. In addition, optical methods can partially overcome the shortcomings of enzymatic electrochemical sensors (enzyme consumption, temperature instability). The disadvantages of optical methods are limited resolution, sensitivity and detection limits, and it is difficult to distinguish small changes under naked-eye conditions. However, it is more suitable for the concept of dichotomy or semi-quantitative detection [95].

4.2.1. Colorimetric Sensing

In wearable sweat colorimetric sensors, the presence of the target analyte in sweat causes a visible color change in the reagent. Typically, there are two common ways of colorimetric sensing: one is to fix the colorimetric reagent in the designated chamber, a colorimetric reaction will occur after sweat flows in, and the colorimetric reaction will be analyzed according to the color changes; the second is to keep the colorimetric reagent in a fixed position, and the reagent will be taken away to produce a color block of a certain length after sweat flows in, and the composition will be analyzed according to the length. Most existing colorimetric sensors incorporate microfluidic systems to allow in situ sweat detection [91,96,97]. Methyl orange and bromocresol green can be used for the fluorescent detection of pH and enzymes for the detection of metabolites such as lactate. Promphet et al. [98] have shown that colorimetric detection can be achieved by co-depositing these reagents or substrates with chitosan and sodium carboxymethylcellulose on substrates.

Colorimetry has long been favored for its simplicity and low cost. From the beginning to the present, patch has been the mainstream form of the colorimetric sensor. For example, Yue et al. [99] presented a representative patch sensor that can use colorimetry to display a variety of physiological data (lactic acid, uric acid, pH, glucose and sweat rate) at the same time (Figure 2a).

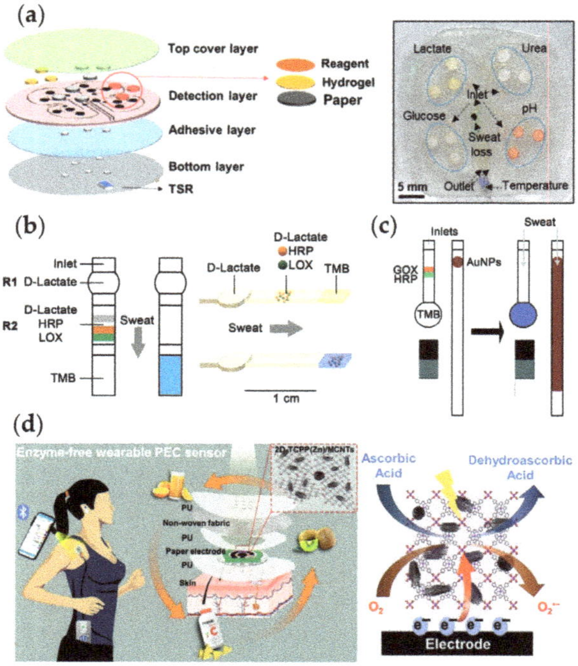

Figure 2. Sweat sensing devices using different strategies. (**a**) A colorimetric sensor that can display multiple physical and chemical data simultaneously. Reprinted from ref. [99], Copyright (2022), with permission from American Chemical Society. (**b**) A sweat sensing patch for fingerprint liveness-detection measurement. Reprinted from ref. [100], Copyright (2021), with permission from American Chemical Society. (**c**) A filter paper-based glucose colorimetric sensor. Reproduced from ref. [101], Copyright (2021) with permission from the Royal Society of Chemistry. (**d**) A two-dimensional material-based photoelectrochemical sensor for Vitamin C detection. Reprinted from ref. [102], Copyright (2023) with permission from Elsevier.

However, colorimetric sensors have the disadvantage of low sensitivity and a high lower limit of detection. Vaquer et al. [100] reported an enzyme-based lactate colorimetric sensor with an adjustable dynamic range (Figure 2b). The sensor also detects sweat volume at the same rate. The dynamic range of the sensor is modified by adding a competitive inhibitor (D-lactate) to different reservoirs. Sweat volume sensors measure the volume that is colored by gold nanoparticles (AuNPs) as they flow with sweat.

Using a similar structure, another study by the same group proposed a filter paper-based colorimetric blood glucose sensor. This sensor, with a lower detection limit, was used to prevent hypoglycemia during exercise [101]. Sweat brings glucose oxidase GOx and horseradish peroxidase HRP into the monitoring area to produce colorimetric reactions, and reagent partitioning can avoid spontaneous reactions between reagents (Figure 2c). By increasing the pore size of the filter paper, sensitivity and signal can be boosted. In addition, using a round tip to enlarge the sample capacity can also improve the signal.

volume [79]. Changes in fluid impedance can serve as an indicator of infection detection to aid wound healing [80,81].

Electrochemical impedance spectroscopy (EIS) is an alternating current-based electrochemical measurement technique. Compared to cyclic voltammetry, which scans the entire potential range at a given rate, EIS has the advantage of measuring at different set potentials [82]. However, it requires more complex data-processing methods than amperometry and potentiometry. The input–response relationship reflects the complex impedance characteristics of the sensing electrode, and the impedance modulus at a given frequency contains resistance and capacitance information, revealing the specific binding process of the receptor to the electrode [83–87]. Ganguly et al. [65] calibrated the sweat cortisol sensor using EIS to obtain a response curve that reliably reflects the concentration of synthetic sweat cortisol. Veeralingam et al. [88] utilized interdigitated electrodes to provide an external AC field to measure skin impedance at a frequency of 10 kHz to indicate skin hydration. The data obtained from EIS can be optimized into reliable output data using machine learning.

4.1.5. Others

Yin. L et al. [89] developed a stretchable epidermal sensor platform. An integrated electrochromic display displays the sampled signal from the electrochemical sensor in real time. An enzyme-modified electrode can be used to detect sweat pH, electrolytes, lactic acid, etc. For visualization, the sampled signal is converted to digital by the MCU and connected to ten addressable ECD pixels containing the PSS electrolyte. Compared to conventional electrochemical sensors with complex electronic systems, the patch has a lower energy consumption, is lighter and conforms to the skin. Compared to colorimetric sensors, there is no need to use a smartphone or camera to convert the results [89–94] and the detection results are more intuitive.

4.2. *Optical Sensing Strategies*

In 2014, Lee et al. [53] combined fluorescence and colorimetry to fabricate a sweat pore distribution sensor using the conjugated polymer poly(PCDA-Cs), pioneering the field of optical methods for sweat detection. The advantage of the optical method is that the detection results can be displayed in a relatively intuitive way. In recent years, more research has chosen to incorporate cameras and smartphones to record test results and perform image processing to achieve higher sensitivity. The use of smartphones avoids the use of laboratory instruments for analysis, and results can be analyzed without the need for a specialist. In addition, optical methods can partially overcome the shortcomings of enzymatic electrochemical sensors (enzyme consumption, temperature instability). The disadvantages of optical methods are limited resolution, sensitivity and detection limits, and it is difficult to distinguish small changes under naked-eye conditions. However, it is more suitable for the concept of dichotomy or semi-quantitative detection [95].

4.2.1. Colorimetric Sensing

In wearable sweat colorimetric sensors, the presence of the target analyte in sweat causes a visible color change in the reagent. Typically, there are two common ways of colorimetric sensing: one is to fix the colorimetric reagent in the designated chamber, a colorimetric reaction will occur after sweat flows in, and the colorimetric reaction will be analyzed according to the color changes; the second is to keep the colorimetric reagent in a fixed position, and the reagent will be taken away to produce a color block of a certain length after sweat flows in, and the composition will be analyzed according to the length. Most existing colorimetric sensors incorporate microfluidic systems to allow in situ sweat detection [91,96,97]. Methyl orange and bromocresol green can be used for the fluorescent detection of pH and enzymes for the detection of metabolites such as lactate. Promphet et al. [98] have shown that colorimetric detection can be achieved by co-depositing these reagents or substrates with chitosan and sodium carboxymethylcellulose on substrates.

4.2.2. Fluorescence Sensing

Fluorescence methods require an excitation light source and anti-optical interference equipment. The mechanism of this method is more complex than the color method, but the sensitivity and selectivity are better. A recent example is the fluorescence sensor proposed by Xu et al. [103] for the detection of Cl in sweat. Under UV irradiation, fluorescence signals can be recorded and warnings given during exercise.

Luminescent materials can also be used for fluorescence detection. Carbon quantum dots (CQDs) emit different colors of light under ultraviolet light irradiation and have pH-responsive properties after treatment. When the pH changes, the functional groups of the CQDs are ionized and affect the degree of deprotonation, causing the fluorescence color and intensity to change with different pH stimuli. Doping CQDs can be made into pH-sensitive sensing materials. Wei et al. [104] obtained a good pH response (4~7) by using N-doped CQDs (yellow fluorescence) and hydrogel impregnation doping, and a proof of concept was performed for sweat detection during exercise. In addition, biomass-derived CQDs can be extracted from natural wood fibers, which reduces environmental pollution during production and still achieves excellent fluorescence performance [105].

4.2.3. Others

Image acquisition and analysis using smartphones in optical methods is a common means, and smartphones play the role of receiver in this process. Furthermore, smartphones also have an excitation function that can send data, and act as a light source, sound source, etc. The photoelectrochemical (PEC) method combines the advantages of electrochemical methods (simple structure, fast response) and optical methods (low noise, high sensitivity). Yan et al. [102] proposed a photoelectrochemical sensor based on two-dimensional zinc porphyrinic MOF nanosheets/multi-walled carbon nanotubes (2D-TCPP(Zn)/MCNTs). The material (2D-TCPP(Zn)/MCNTs) used in the device stably generates hole-electron pairs under photo-stimulation, promoting vitamin C oxidation through their photoelectric conversion capability (Figure 2d). Driven by smartphone light, the 2D-TCPP(Zn)/MCNTs electrodes exhibit an enhanced cathode photocurrent response to vitamin C and can send sensing data back to the phone. In addition, polymer-based laser-induced graphene and In-doped CdS composites, capable of photocurrent sensing, can be used for Cu^{2+} detection [106]. Photoelectrochemical methods can also be employed for the label-free detection of lactate molecules [107].

There are many other optical principles that are applicable for sweat monitoring. Nan et al. [108] fabricated a sweat cortisol sensor using the localized surface plasmon resonance (LSPR) properties of gold nanoparticles (AuNPs) to detect the cortisol concentration from changes in extinction peaks in LSPR spectra.

For non-biochemical indicators such as sweat rate, colorimetric sensing is most commonly used. In addition, measurements can be made using flow sensors. Some flowmeters operate indirectly using thermal principles (thermal actuator and thermal sensor [46]). The work of Jose et al. [79] combined the impedance principle (fluid electrical properties) and the transient planar source (TPS, liquid heat flow rate) principle to design two coupled sensors to jointly measure sweat composition and fluid volume. The same idea is reflected in the work of Mei et al. [109]. This work integrates both electrochemical and colorimetric sensors, combining the advantages of both.

Innovative sensing strategies are expected to make a significant impact in real-world applications, provided that more research is conducted to test their stability and reliability. In addition, when selecting a sensing strategy for the device, it is necessary to consider not only the continuous detection time, sensitivity, detection limit, and other parameters of the strategy, but also the effect of the sensing environment and excitation conditions on the results. For example, optical sensors in low-light conditions require special image-processing algorithms, and underwater environments require a better sealing of the sensor. Table 2 summarizes the main features, advantages, and disadvantages of the above mainstream sensing strategies.

Table 2. A comparison of different sensing strategies.

Strategy	Overview	Advantages	Disadvantages	Refs
Potentiometry	Potential change is induced and converted into a voltage signal by the transducer	• Simplicity • Stability	• Only suitable for detecting electrically charged substances • Accuracy difficult to guarantee at low concentrations	[1,110]
Amperometry	Amperometry quantifies the transfer of electrons during reactions at electrodes	• Fast response time • Wider detection limits • Higher sensitivity	• Sensing performance is affected by enzyme activity • Applied voltage required	[72,74]
Voltammetry	A continuously varying potential is applied to the working electrode and the response current signal is detected to record the current–potential curve and then analyze the composition of a solution	• Multiple analytes can be analyzed simultaneously using voltage scans on the same two electrodes • Higher detection limit than chronocurrent method	• Different electroactive molecules may have similar reaction potentials, so voltammetry may not be as selective as voltage and amperometric methods	[63,111]
Electrochemical impedance spectroscopy	The impedance varies with sweat composition, and the impedance modulus at a given frequency contains resistance and capacitance information, revealing the specific binding process of the receptor to the electrode	• Ability to measure at different selected potentials compared to cyclic voltammetry • No need to couple electrochemical inactives to redox labels	• Requiring complex subsequent data processing and high time costs	[112–114]
Colorimetry	The colorimetric reaction of sweat with the reagent reflects the concentration of the target analytes	• Simple operation • Low cost	• Lower resolution makes it difficult to quantify the indicator • Relatively high detection limit	[96,100,101]
Fluorescence sensing	Under light excitation, the fluorescence signal changes depending on the target analyte	• Better sensitivity and selectivity than colorimetric methods	• Requiring excitation light source and anti-light interference equipment, the mechanism is more complex	[103,104]

5. Materials, Structure and System Composition of Sweat Sensor

5.1. Sensing Materials

5.1.1. Common Materials for Electrodes

Common counter-electrode materials include gold, platinum and carbon. The reference electrode is typically Ag/AgCl. The working electrode is divided into enzymatic reaction electrode, molecular selection electrode [115] and ion-selective electrode. When the sensitive membrane of the ion selective electrode contacts the solution, an electrical potential is generated at the interface that is directly related to the corresponding ion concentration. For example, the multiplexed sensor proposed by Mei et al. integrates five electrochemical sensors detecting lactate, glucose, pH, Cl^- and urea, of which the detection of lactate and glucose uses enzyme-modified electrodes, using current changes to detect concentration; pH, Cl^- and urea are sensed by electrodes modified with PANI, Ag treated with $FeCl_3$ and NH_4^+ ion-selective membranes, respectively. As mentioned above, sensing materials are also gradually developing in the direction of multifunctional multi-modulus in the form of arrays. For example, the multiplexed sensor proposed by Mei et al. [109] integrates five electrochemical sensors, lactate, glucose, pH, Cl^- and urea, of which the detection of lactate and glucose uses enzyme-modified electrodes that rely on current changes to detect concentration; pH, Cl^- and urea are sensed by electrodes modified with PANI, Ag treated with $FeCl_3$ and NH_4^+ ion-selective membranes, respectively (Figure 3a). As mentioned above, sensing materials are helping to improve devices in a multifunctional and multimodal way in the form of arrays.

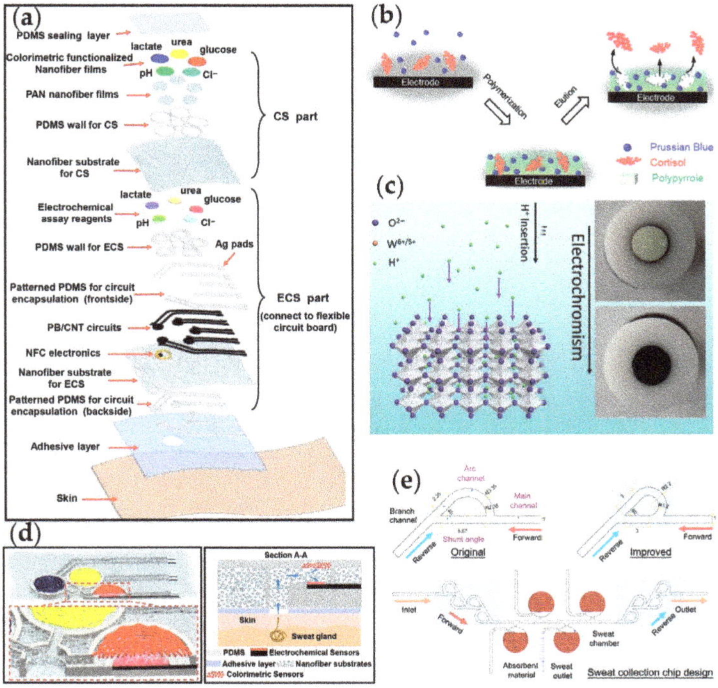

Figure 3. The role of newly developed materials in sweat sensors, and the demonstration of novel microfluidic channels. (**a**) An electrochemical sensor and electrode modification method for the simultaneous detection of lactate, glucose, pH, Cl^- and urea. Reprinted from ref. [109], Copyright (2023) with permission from Elsevier. (**b**) A molecularly imprinted polymer combined with a redox probe touch-based cortisol sensor. Reprinted from ref. [116], Copyright (2021) with permission from John Wiley and Sons. (**c**) A pH sensor based on the inorganic material WO_3. Reprinted from ref. [117], Copyright (2021) with permission from John Wiley and Sons. (**d**) A nanofiber-based microfluidic analysis system that collects and transports sweat using a highly porosity, hydrophilic nanofiber membrane. Reprinted from ref. [109], Copyright (2023) with permission from Elsevier. (**e**) An improved design of the Tesla valve. Reprinted from ref. [118]. Copyright (2022) with permission from Elsevier.

Recent studies have focused on the repeatability of electrode materials in wearable sensor electrodes [109,119]. The application of reusable electrodes reduces the need for replacement, thereby improving the user experience.

5.1.2. Nanomaterials

As materials science continues to advance, the application of nanomaterials has become increasingly sophisticated. Due to their unique properties, such as porosity, rigidity, and impedance, as well as their customizable electrical and wetting properties, nanomaterials have been widely used in the fabrication of sensing materials, substrates, and other components to enhance sensor performance. Lin et al. [120] utilized N-GODs to anchor a PANI matrix as a sensing electrode, resulting in improved sensitivity and stability while avoiding the cracking commonly associated with rigid electrodes. Furthermore, the modification of electrodes with metal nanomaterials can optimize sensing performance. For example, the modification of a Prussian blue electrode with gold nanoparticles (AuNPs) yielded a larger amperometric response [121]. In the fabrication of electrochemical electrodes, metal nanomaterials are often combined with redox graphene (rGO) [122,123], which serves as a supporting substrate to enhance sensitivity. Due to its high electron mobility and large active surface area, this composite often leads to significant improvements in the sensitivity and accuracy of sensors. Additionally, such nanocomposites generally exhibit excellent electrocatalytic activity [124,125]. Yu et al. [126] developed a CuxO-nanoflake (NFs)/Cu nanoparticles (NPs) nanocomposite as a sensing material. The sensor was enriched with more active sites, a low detection limit, high selectivity, and resistance to bending and twisting. Assaduzzaman et al. [9] developed a flexible, multifunctional patch based on laser-induced graphene (LIG) combined with hybrid nano-porous carbon (H-NPC). Sensors based on nanomaterials exhibit excellent electrochemical activity, reliability and a wide detection range.

5.1.3. Two-Dimensional Materials

Graphene: As a monolayer two-dimensional material, Graphene has long been of interest to researchers. Special variants of graphene, such as graphene oxide and reduced graphene oxide, are widely used in wearable sweat sensors. Their high surface-area-to-volume ratio provides a large number of reaction sites, excellent electrocatalytic performance and high electron mobility. Graphene is often used as a substrate for electrodes and is combined with other materials or modification methods to enhance the overall performance of the electrode. After modification with graphene, the sensing electrode shows high sensitivity, good selectivity and flexibility, enabling the detection of small-molecule concentrations such as glucose [127,128], lactate [9], cortisol [26] and uric acid [129]. Wang, Y et al. [130] synthesized reduced graphene oxide (RGO) in situ on cotton fabric (CF). This imparted softer physical properties to the graphene material while achieving a fast response and excellent selectivity. In addition, the electrodes can be fabricated by engraving graphene films into specific patterns using laser-engraving techniques. Wang, M et al. [43] fabricated an electrode based on laser-engraved graphene (LEG) and Molecularly Imprinted Polymers (MIP) that can be repeatedly regenerated in situ. This electrode can be used to assess amino acid intake levels [131]. LEG can likewise be modified with nanomaterials (e.g., AuNPs) for highly sensitive electrochemical detection. According to the research conducted by Tu et al. [132], the deposition of AuNPs via pulsed potential results in their uniform distribution throughout the mesoporous graphene structure. This enhances their catalytic ability and provides a large number of specific binding sites. Laser-induced graphene (LIG), on the other hand, produces graphene directly on polymers using lasers without needing a film [133]. Feng et al. [134] demonstrated the use of LIG and single-walled carbon nanotubes (SWCNT) in the development of a sweat sensor. The LIG fabrication process is both rapid, taking less than a minute, and inexpensive. Additionally, laser-scribed graphene (LSG) has also been shown to have potential for use in chemical sensing [135]. The emergence of LEG and LIG has made the large-scale fabrication of graphene devices increasingly promising.

MXene: MXene is a novel material discovered by Gogosti in 2011 [136]. It is a two-dimensional transition metal carbide and nitride with a tunable structure and can exhibit excellent electrical, optical, catalytic and mechanical properties when appropriately modified [137]. The formation of composites with MXene and other materials, such as Pt and polymers, can result in enhanced performance, making it suitable for use in wearable devices. $Ti_3C_2T_x$ was the first synthesized MXene material due to its low formation energy and in wearable sweat-sensor studies; $Ti_3C_2T_x$ is a commonly used MXene material [138]. $Ti_3C_2T_x$ exhibits high electrical conductivity, excellent hydrophilicity, and a large number of active sites and terminal functional groups. These properties reduce signal interference and promote the adsorption of gas and water molecules, making $Ti_3C_2T_x$ well-suited for use in the manufacture of sensors [139]. Sharifuzzaman et al. [140] reported a catalyst composed of MXene/fluoropolymer-derived porous TiO_2 nanomaterials. On the other hand, Magesh et al. [45] developed a carbon (Pearlman catalyst) using a MXene/palladium hydroxide-loaded composite (MXene/$Pd(OH)_2$/C) for electrochemical sensors, achieving good selectivity, sensitivity and stability. Li et al. [141] used MXene to develop a fully optimized glucose sensor. Pt/MXene nanomaterials were fabricated using Pt nanoparticles for glucose detection on MXene nanosheets, yielding a wide linear range under neutral conditions. The sensor substrate was a conductive hydrogel synthesized from MXene and polyvinyl alcohol (PVA), providing reliable stability to the sensor.

5.1.4. Hydrogel

Hydrogel-based flexible sensors possess numerous properties such as compatibility, self-healing, and adhesion [142]. Due to their water content of up to 90%, they exhibit excellent biocompatibility, electrical conductivity, and flexibility. The use of hydrogel as a sensor substrate or electrode material can improve the sensing performance. The elastic hydrogel system results in a skin-like base that can significantly improve the wearability and mechanical deformability of the device. It has been used in applications such as tissue engineering [143,144], drug delivery carriers [145], soft robots [146,147], and wound dressings [148,149]. Wang et al. [150] used TPU film to combine hydrogel and conductive material to provide the sensor with excellent mechanical and electrical properties at the same time, verifying the ability of hydrogel to improve comfort.

Gao et al. [151] used porous hydrogel instead of liquid electrolyte on sensor electrodes, maintaining better conductivity while achieving improved swelling and biocompatibility. A recent study showed that glycerin gel can also achieve alternative functions while providing anti-bending and sweat absorbing effects [127]. As a new type of gel, noble metal aero-/hydrogels possess the properties of both aerogel materials and noble metal nanoparticles [152,153]. Their highly branched, fused nanowire structure imparts inherent self-healing and flexible properties, making them promising for flexible sensing [154,155]. Li et al. [156] designed a bis-structured Pt-Ni hydrogel that achieved the aforementioned excellent properties and remained stable for two months. Chen et al. [129] reported a wearable biosensing platform based on metal aerogels using three-dimensional porous bis-structured aerogels (N-rGO/Au DAs) consisting of gold nanowires and N-doped graphene nanosheets as working electrodes. The high surface area and porous structure, as well as the synergistic effect of Au and N-rGO aerogels, enable better catalytic and electron-transfer properties than mono aerogels, facilitating non-enzymatic uric acid sensors.

5.1.5. Organic Electrochemical Transistors

An organic electrochemical transistor (OECT) is a low-cost, highly sensitive device that responds to physical stimuli, chemical functionalization, and shape changes. The OECT is composed of an ion-permeable organic semiconductor channel (OSC) connecting the source and drain electrodes. Its operation involves the conversion of the ionic current between the electrodes into a modulation of the polymer film conductivity, allowing for the detection of electrolyte solution concentrations. Poly(3,4-ethylenedioxythiophene):poly(styrenesulfonate) (PEDOT:PSS) is commonly used as a conductive pathway between the source and drain of the OECT due to its ability to amplify signals, ease of miniaturization, and excellent biocompatibility [157,158]. PEDOT:PSS-based OECTs exhibit high transconductance ($\approx$2 mS) at low voltages (<0.5 V) [159], making them promising for use in biosensing applications.

However, these OECTs are sensitive to ionic strength [160] and the PEDOT:PSS channel accepts all cation insertions from the analyte, resulting in a lack of selectivity [161]. The selectivity of OECTs for analytes in sweat can be improved through the use of an ion-selective membrane (ISM), which permits only specific ions to pass through. In 2014, Sessolo et al. [162] first proposed the incorporation of an ISM into an OECT for the detection of K^+, inspiring subsequent research in the field of wearable sweat sensors. Coppedè et al. [163] later proposed an OECT using an ISM for the detection of Ca^{2+}, with the device being fabricated by functionalizing textile fibers. It is worth noting that many recent studies on OECT sweat sensors have incorporated textiles as a platform. Textiles are easy to prepare and can be integrated into clothing, and the detection of target analytes is not limited to ions. Tao et al. [164] developed a fabric-based OECT lactate sensor that enables PEDOT:PSS to exhibit high transconductance in both depletion and accumulation modes. Fang et al. [165] addressed the issue of insufficient transconductance in fiber OECTs by modifying carbon nanotube (CNT) fiber grids. Qing et al. [166] further integrated fiber OECTs with thermoelectric fabrics, using a manufacturing substrate composed of cotton/poly(3,4-ethylenedioxythiophene):poly(styrenesulfonate)/dimethylsulfoxide/(3-glycidyloxypropyl)trimethoxysilane (PDG) yarn, which is lightweight, strong, and sweat-resistant.

5.1.6. Other Materials

There are numerous reports on fiber-based wearable electronics, with many studies focusing on the application of fibers for physical sensing, such as pressure [167] and thermal [168] sensing. Instead, Wu et al. [169] focused on biochemical parameters to develop a flexible microelectronic fiber that can be woven into textile substrates. Detection of UA oxidation or Na^+ concentration can be achieved by modifying ion-sensitive membranes or nanoparticles.

Garcia-Rey et al. [5] conducted a proof-of-concept study on alginate-based biological systems to evaluate the efficacy of the colorimetric determination of sweat glucose. Alginate beads were obtained by integrating enzyme assays, consisting of GOx, HRP, and TMB, into the alginate scaffold. The porosity of the beads enabled highly sensitive glucose detection, and a similar system has been used in previous studies to calibrate and detect lactate in sweat [170]. Furthermore, another report stated that the integration of TiO_2 nanotubes into alginate can provide enhanced hydrophilicity, further improving the sensing performance of this material [171]. The touch-based cortisol sensor reported by Tang et al. [116] embeds Prussian blue redox probes within a molecularly imprinted polymer (imprinted polypyrrole, PPy, film) network (Figure 3b). The selective binding of cortisol to the imprinted PPy film impedes PB charge transfer, resulting in a rapid change in current.

Tang et al. [117] reported a WO_3-based pH sensor. According to the article, conventional pH sensing methods primarily utilize organic materials, which carry the risk of biological toxicity (Figure 3c). As an alternative, WO_3 is an inorganic and non-toxic material that is sensitive to hydrogen ions. However, its exchange rate is very low. The embedding of lattice H^+ transforms WO_3 from a monoclinic phase to a cubic phase (H_xWO_3), enhancing its H+ exchange capacity and reducing its resistance, thereby improving detection sensitivity and response speed.

The use of composite sensing materials is becoming an increasingly popular trend in the research into wearable sweat sensors. While individual materials may have specific limitations, composite materials can be specifically designed to balance the strengths and weaknesses of different materials to achieve the desired functionality. Nanomaterials are also gradually becoming significant in this field. The various materials mentioned above can be nanosized or combined with nanomaterials to produce interfaces or electrodes with a large number of active sites, a large surface area, and good mechanical properties, thereby optimizing sensor performance.

5.2. Forms and Structure

5.2.1. Development of Sweat Sensor Forms

In the early years of wearable sensor research, various forms of sensors, including tattoos, wristbands, and patches, have been reported. In 2012, Professor Joseph Wang's research team reported a tattoo-based electrochemical sensor [172]. Based on this research, many applications, including the detection of ammonium [173], lactate [75], and sodium [110], have been developed in vitro and attached to the skin using transfer technology. Tattoo sensors integrated with carbon fiber exhibit excellent immunity to disturbances such as stretching, bending, and skin abrasion, and have better electrical performance than screen-printed electrodes. In addition, these sensors can be combined with artistic tattoos, which has attracted the attention of many researchers. In practice, tattoo-based sensors are more like a new type of electrode. While research on tattoo-based sensors has seen significant improvements, their popularity in subsequent research has gradually declined. Katseli et al. [7] proposed a 3D-printed electrochemical ring for sweat glucose detection. The complex fabrication process of wearable sensors is simplified by embedding conductive plastic electrodes in a ring holder and using a dual-extrusion 3D printer and commercially available filaments in a one-step printing process. In addition to the above research, many sweat sensors are now manufactured in patch form, including readable disposable sensor patches and system-based patches.

Currently, the research focus has gradually shifted, first from single-parameter to multi-parameter detection. The human body system is complex and constantly in a dynamically changing environment. As such, it is difficult to fully express the physiological health status of the human body by interpreting a single parameter. Consequently, more research is now focused on multi-parameter detection, including metabolites and electrolytes [174], sweat analytes and sweat rate [175], providing data on multiple biomarkers for analysis.

From single-component components to system-level studies, Gao et al. [6,176] proposed a sensor array capable of multiplexed sweat composition analysis, in 2016. The array was connected to integrated circuits and flexible circuit boards to simultaneously detect sweat metabolites, such as lactic acid and glucose, and electrolyte ions, including pH, calcium, sodium, and potassium. This study aimed to propose a design scheme for wearable sweat sensors from the perspectives of system integration, signal calibration, and multiplexing. This groundbreaking work provided researchers with a more complete system-level design.

Multi-function sweat sensors broaden the information collection range of single-channel sensors, and further expansion (e.g., combining physical signal detection such as acoustics) is expected to provide a more comprehensive picture of human health information. Sempionatto et al. [177] report a multiplexed sensor capable of not only detecting multiple biomarkers in sweat, but also integrating an ultrasound transducer to detect blood pressure and heart rate. One of the issues that must be considered when multiplexing integrated sensors is to prevent signal crosstalk from affecting the reliability of the data from each sensor. A solution to the signal crosstalk problem was proposed that spatially separates components at optimal distances and prevents crosstalk between acoustic and electrochemical transducers with solid-state ultrasound and sensing hydrogel layers.

While the concept of multiplexed detection and system design research is on the rise, making sensor systems lighter is also an important issue in the development of complex sensors. Gao et al. [178] proposed a lightweight flexible sensor based on their previous research. To achieve this, the researchers aimed to make materials that fit the body lighter, such as paper-based and textile sensors.

A disadvantage of using sweat as a detection target is the small sample size available at rest. To overcome this limitation, flexible sensors with integrated sweat-stimulation capabilities are being developed. Sam et al. [179] used iontophoresis electrodes to deliver agonists to stimulate sweat gland secretion, which has gradually become the main method for sweat induction. Research into natural sweat production is also increasing due to its non-invasive nature.

Research on sweat sensors does not stop at simply completing the detection function. Researchers also expect them to have therapeutic capabilities. Lee et al. [128] used microneedles to design sweat sensors capable of transcutaneous drug delivery, adding medical value to wearable devices.

As wearable sweat sensors move closer to practical applications, portability and miniaturization are inevitable trends for clinical and everyday consumer markets [180,181]. One important method for achieving device miniaturization is to replace traditional standard discrete components with integrated circuits. Wang et al. [182] reported an ultra-small wearable sweat sensing system with a size of only 1.5 cm × 1.5 cm. Its core processing chip, the MS02 chip, integrates functional circuits such as amplification and filtering, which greatly reduces the overall system area while maintaining sensor performance.

5.2.2. Structure of Sweat Sensor

Wearable sweat sensors can be roughly divided into two categories based on whether they have a sweat transport structure: patch and microfluidic-based devices. The former undergoes electron transfer or optical reactions directly upon contact with sweat, while the latter collects or transports sweat to electrodes or reagents through microfluidic channels for detection. Taking electrochemical detection as an example, when the detection target is relatively simple, sweat can be brought directly into contact with the electrode to achieve rapid real-time monitoring. However, when multiple analytes are involved, the lack of sweat sample becomes more noticeable, making it difficult to supply multiple electrodes for simultaneous analysis. Timed sampling with an absorbent patch or a regular sweat collector can solve this problem, but it increases the effort required [183]. In addition, changes in sweat rate over time and across body parts, combined with factors such as epidermal contaminants and sweat evaporation, can easily cause fluctuations in the results of sweat sensor patches [184]. By incorporating microfluidic channels, sweat can be collected and distributed to different sensing electrodes on demand, overcoming problems such as inadequate sweat sample volume, instability, and sweat evaporation. The effects of contamination can also be reduced with specially designed channels.

Microfluidics: Microfluidic systems have the advantages of simple structure, strong ductility, and storage and transport capabilities. By designing the size (e.g., capillary [115], width variation), structure (e.g., serpentine structure, fiber, reservoir distribution), and other properties of the microfluidic channels, they can acquire specific capabilities. Examples include designing numerous pathways for the multi-modal analysis of sweat [96] and building attractive fluid dynamics and reservoirs for rapid filling [185]. Microfluidics have long been used in wearable sweat sensors as channels for the rapid sampling, collection, and analysis of sweat. Microfluidic sensing patches can be mass-produced using techniques like Roll-to-Roll (R2R) printing [186], a low-cost rotary screen printing process. Biocompatible and stretchable materials including PDMS [90] (polydimethylsiloxane) and PET (polyethylene terephthalate) are commonly used to construct microfluidic channels, allowing sensors to be integrated anywhere on the body [187].

As mentioned above, the structure of microfluidic channels can be customized. However, a more complex microfluidic system does not necessarily result in better performance. For instance, the flowmeter-based sweat rate sensor reported by Kwon et al. [46] relies on a short, straight microchannel that is thermally coupled to sweat to accurately measure temperature differences and obtain flow values. Microfluidic channels can also be designed as three-dimensional structures, allowing the vertical integration of multiple modules for sampling, transport, sensing, and storage [188].

The microfluidic channel is a semi-open system that depends on capillary action to transport sweat during collection. As such, it is important to control the fluid flow within the channel. Shi et al. [118] proposed the addition of forward and reverse Tesla valves to the inlet and outlet to prevent inlet backflow, and reduce contact between the outlet and the environment (Figure 3e). This design is highly suitable for low-flow-rate conditions, such as sweat monitoring.

With the ongoing advancement of materials science, new microfluidic systems have been developed to address the limitations of traditional hollow microfluidic channels, such as contamination, miniaturization, and diffusion of chemical agents. Currently, there are studies exploring the combination of nanofibers with microfluidic channel materials. Mei et al. [109] reported a nanofiber-based microfluidic analysis system (NFMAS). This system utilizes highly porous and hydrophilic polyimide/sodium dodecyl sulfate (PI/SDS) nanofiber films as a microfluidic network substrate for collecting and conducting sweat (Figure 3d). The system features both electrochemical and colorimetric sensing modules, which are separated by nanofibers. These sensors exhibit excellent mechanical properties and resistance to contamination. In subsequent research, the same group continued to apply this nanofiber microfluidic technology to fabricate flexible sensors [189], with the goal of better improvement.

Another approach is to use paper-based microfluidics. By cutting paper into a channel shape, sweat can be transported by capillary action [190,191]. Also, paper can be folded to form three-dimensional structures, allowing the creation of more complex channels. Liang et al. [192] demonstrated a three-dimensional paper-based microfluidic electrochemical device (3D-PMED). The device is constructed by prefabricating a pattern on cellulose paper and folding it into a five-layer stacked structure containing a sweat collector, vertical channel, transverse channel, electrode layer, and sweat evaporator. Table 3 presents a comparison of currently commercially available microfluidic sweat wearable sensors with those from recent research.

Table 3. A comparison of microfluidic sweat sensors on the market and in the latest research.

Device	Localization	Indicators	Mechanism	Features	Ref
Nix Hydration Biosensor produced by Nix (Boston, MA, USA)	Biceps brachii	Sweat rate, electrolyte loss rate, and sweat composition	An electronic pod clips onto the single-use sweat patch and wirelessly transmits the test data to a cell phone, watch or bike computer.	• Real-time tracking of fluid and electrolyte loss during workouts • Calculate the Nix Index™ for a given workout in the future based on the weather forecast • 36 h of battery life • The patch weighs less than 0.5 ounces and is similar in size to a Garmin dial	[193]
AbsolusSweat P1 produced by Shenzhen Refresh Biosensor Technology Co., Ltd. (Shenzhen, China)	Chest, forehead and upper arm	Glucose, potassium, sodium, sweat volume, sweat rate	The sweat analysis terminal is connected to the sensing patch, which analyzes the sweat and transmits the data to the app.	• Pre-hydration plan based on historical data • Advice on supplementation during exercise • Personalized Recovery Plan After Exercise • Timely alerts to prevent injuries	[194]
Gx Sweat Patch produced by Gatorade, PepsiCo Inc. (New York, NY, USA)	Inner left forearm	Sweat rate, fluid loss, and sodium loss	Scanning patches, processing and analyzing images using Gx applications.	• Gx app records sweat data as a sweat profile. No patch is required for workouts in conditions similar to the profile. • Targeted advice on hydration and nutrition • The patch measures 2.5 in. × 1.5 in. and weighs 1.25 oz. • Valid for 20 to 120 min	[195]

Table 3. *Cont.*

Device	Localization	Indicators	Mechanism	Features	Ref
An on-skin platform for wireless monitoring	Inner forearm	Sweat rate, sweat loss and temperature	The platform's electronic system contains a thermal actuator and thermistor, which utilizes the temperature difference to calculate the flow rate and, thus, the sweat rate.	• Short and straight microfluidic structure • Expansion modules (colorimetric patches) can be integrated into the platform to measure other biomarkers	[46]
An epidermal wearable microfluidic patch	Inner forearm	Cl^- in sweat, Ca^{2+} in interstitial fluid (ISF)	The patch uses iontophoresis to extract sweat and ISF, detecting both ions simultaneously in a three-dimensional microfluidic channel.	• Three-dimensionally structured microfluidic channels for efficient sampling and storage of fluids	[188]
Nanofiber-based microfluidic systems with integrated dual-mode sensing arrays	Wrist	Lactate, urea, glucose, pH, Cl^-	The system direct sweat to different microchannels and chambers for electrochemical and colorimetric detection, respectively.	• nanofiber-based microfluidic network (NFMN) • Dual-mode sensing arrays (electrochemical array and colorimetric array)	[109]
An integrated three-dimensional paper-based microfluidic electrochemical device (3D-PMED)	Forearm	Potassium	Sweat is driven through the capillary tube into the vertical channel, where it flows and completes the detection.	• Five-layer three-dimensional structure composed of folded patterned cellulose paper • Flow driven by sweat evaporation	[192]

5.3. *System and Module*

5.3.1. Sweat Collection

There are several common methods for collecting sweat using wearable sensing systems. One approach is to use external stimuli, such as iontophoresis, to induce the release of sweat from the skin. Another approach is to collect sweat produced naturally during high-intensity exercise or to absorb sweat from the skin under resting conditions [190].

Iontophoresis: Iontophoresis involves the use of a pair of positive and negative electrodes. The anode releases pilocarpine or other long-acting sweat-inducing drugs, such as carbachol [196], to induce sweating without significant stimulation. The anode used in iontophoresis is often positioned in the center of the measuring electrodes [197] to enable real-time sweat sampling or detection. However, due to the mixing of sweat and gel and the lack of dynamic sweat sampling, this method can result in a limited number of effective samples and reduced measurement accuracy [43]. By combining iontophoresis with microfluidics [188], induced sweat can be collected and directly transferred to the sensing electrode, resulting in a more stable and quantifiable sample. Bolat et al. [196] reported on a specialized wearable, single-step sensing platform that integrates iontophoresis with microfluidics. The iontophoresis electrode is positioned directly around the PDMS microchannel, stimulating sweat glands near the microfluidic inlet. This allows secreted sweat to precisely enter the channel and be delivered to the sensor. A similar approach has been used by Wang et al. [43] in their sweat sensing patches.

Hydrophilic/Hydrophobic Substrate: Zhang et al. [97] developed a stretchable, colorimetric sweat sensor that utilizes the large wettability gradient between a superhydrophobic substrate and superhydrophilic assays to collect sweat (Figure 4a). Superhydrophilic assays are composed of elastomeric nanofiber textiles modified with SiO_2 nanoparticles, rather than the more commonly used filter papers. Superhydrophobic substrates are created by modifying the surface of elastomer substrates with nanoparticles. This elastic substrate, featuring a wetting gradient and favorable comfortability, is well-suited for non-invasive and efficient sweat collection.

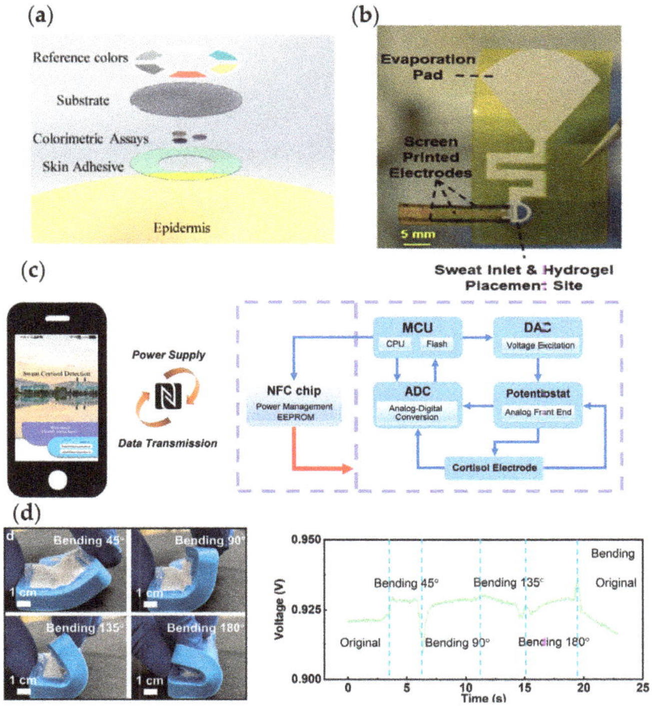

Figure 4. Development of different system modules for wearable sweat sensors. (**a**) A sweat sensor using a superhydrophobic substrate and superhydrophilic assays capable of collecting sweat using an ultra-large wettability gradient. Reprinted from ref. [97], Copyright (2021), with permission from American Chemical Society. (**b**) A sweat sensor based on a specially designed paper channel. The evaporation pad after bending and increasing the area can delay channel saturation and increase the rate of sweat evaporation. Reprinted from ref. [190], Copyright (2021), with permission from American Chemical Society. (**c**) A sweat sensor that uses NFC for wireless communication and function. To start the measurement, the mobile phone should be held close to the sensor. Reprinted from ref. [198], Copyright (2021), with permission from Elsevier. (**d**) A flexible sweat-activated battery that uses soft silicone as a package. Reprinted from ref. [199], Copyright (2021), with permission from John Wiley and Sons.

Hydrogel for Natural Sweat Extraction: Skin-conforming hydrogels have a hydrophilic interface. The hydrogel that adheres directly to the skin has the ability to reduce Laplace pressure, which prevents sweat from forming on the skin surface. In addition, water-based patches are an effective matrix for metabolite extraction, and the proper thickness of hydrogel patches can substantially improve the sensor's capability to collect biomarkers [76]. Lin et al. [76] developed a wearable hydrogel patch that leverages the above advantages to quickly collect natural sweat from the hands. Tang et al. [116] designed a sensor that can rapidly detect cortisol through fingertip contact, using a highly permeable, porous PVA hydrogel. Its detection results can be obtained within 3.5 min of fingertip contact. Patches utilizing PVA hydrogel also exhibit excellent resistance to staining [106].

Sweat Transport and Management: When continuously monitoring sweat, it is important to consider how to prevent the accumulation of old sweat, as residual sweat can saturate the device and affect its long-term performance. Saha et al. [190] introduced an osmotic wearable for lactate sensing in sweat (OWLSS) that is capable of continuous monitoring, utilizing a specially designed paper channel to effectively manage sweat (Figure 4b). Adding an evaporation pad to the channel and creating a serpentine channel path to increase the distance sweat must travel before reaching the evaporation pad can help delay channel saturation. Increasing the area of the evaporation pad also ensures that sweat evaporates faster than it flows into the channel under all conditions, allowing for continuous monitoring.

5.3.2. Signal Transmission and Power Management

Wearable sweat sensors require real-time monitoring over extended periods of time, necessitating the use of wireless signal transmission. Technologies such as Bluetooth, NFC, and RFID are commonly employed in the electronic systems of wearable sweat sensors due to their advantages of miniaturization and low power consumption. The communication and control of wearable sensors consume energy, particularly in the signal processing and wireless transmission circuits, such as those utilizing Bluetooth or radio frequency transmission [200–202]. However, the traditional battery power supply method hinders the continuous operation of the sensor on the skin due to its difficulty in deformation, poor battery life, and the presence of harmful substances. To address the issue of energy consumption, new battery materials suitable for use in wearable devices have been developed. In their study, Liu et al. [203] mentioned the use of super-resilient hard carbon nanofabrics (s-HCNFs) as carbonaceous anode materials. This high-capacity material exhibits excellent mechanical stability under twisting and folding, and its properties can be further enhanced by designing it as a 3D micro- or nanolayered structure. These properties make s-HCNFs well-suited for use in the fabrication of flexible wearable devices. Moreover, researchers are also applying scientific research results such as self-generation, wireless power supply, and flexible batteries to integrate them into wearable sensors.

NFC: The advantages of NFC include low power consumption, few components, and small size. Only small coils and NFC chips are required to achieve wireless signal transmission [204]. NFC chips and sensors can be magnetically coupled to form a stable system [90]. The modulated antenna can also provide wireless energy supply [35]. Cheng et al. [198] developed a sweat cortisol sensor that uses NFC for wireless communication and power (Figure 4c). When detection is required, the NFC antenna is powered after the smartphone approaches the patch to power the circuit and begin measurement. The performance of the NFC antenna is also verified under different radii of curvature to ensure that the NFC chip can still provide stable power when the patch is bent.

Biofuels and Sweat-Activated Batteries: Sweat has the potential to provide electricity support for epidermal electronics as a sustainable bioenergy source [205,206]. Biofuel cells provide power based on the bioenergy in sweat. Yu et al. [200] report on a lactate biofuel cell based on full perspiration-powered electronic skin (PPES). The enzyme-catalyzed batteries exhibit a higher current density and stability. The electrolytes in sweat (Na^+, K^+, Cl^-) can also be used for electrochemical energy storage. Chen et al. [207] designed an array of sector electrodes (using copper and zinc thin films) to create a sweat-based energy generator based on redox reactions. Manjakkal et al. [208] report a supercapacitor using PEDOT: PSS as the active electrode and current collector. This material has been used in sensitive electrodes of electrochemical sensors [6] due to its high conductivity, large voltage window, and good mechanical properties.

In response to the need for flexibility, Liu et al. [199] developed a flexible sweat-activated battery that can be directly integrated onto the skin (Figure 4d). Soft silicone is used for fixing and loading functional materials as a package housing. Two Nylon fabric bags containing electrolyte are used to stabilize the output voltage under various deformation conditions.

Triboelectric Nanogenerator: Since the invention of triboelectric nanogenerators (TENGs), many sensor platforms incorporating TENGs have been reported [209–211] due to their advantages of a high output voltage, power, low cost, and ease of fabrication [212–215]. In sweat-sensing applications, methods for optimizing TENGs are also being explored. Zahed et al. [216] used a PVDF/Co_3O_4 nanofiber-based TENG and a miniature Halbach magnet array for EMG to improve its electrical energy conversion efficiency. Baro et al. [217] proposed a textile-based sweat sensor incorporating a ZnO-based unipolar friction nanogenerator (STENG). Their study experimentally verified that hydrated salts in sweat promote an increase in conduction band electrons in ZnO, allowing the integrated STENG to output higher voltages. TENGs can also be integrated with flexible supercapacitors (SCs) to form a distributed power system, further enhancing the power-management capabilities of wearable devices [218]. The construction of self-charging power systems (SCPS) for TENG-SCs requires consideration of two key factors. Firstly, the AC power output by the TENG cannot be directly fed into the SC and must be converted using a rectifier. Secondly, it is difficult to share electrodes between the TENG and the SC, and alternative solutions such as shared substrates or packages are typically used [219]. These can include fabrics [220], thin films [221], carbon cloth [222,223], hydrogels [224], and others. The integration of TENGs with energy-storage technologies has led to the development of uninterrupted power supply (UPS) systems for wearable devices [225]. These systems provide a viable solution for the continuous power supply of wearable devices.

Solar Power: Singh et al. [226] installed a solar panel on top of a sweat sensor in the form of a smart watch to collect energy. A circular solar panel with a diameter of 2.2 inches was mounted on the top of the device as an energy source to power the integrated circuit of the smart watch. This combination provides a new idea for the power supply scheme of wearable devices.

Inertial Energy Harvesters: Inertial energy harvesters can convert the kinetic energy generated by human motion into electrical energy to power devices. These harvesters contain a proof mass that is excited by human motion, with the inertial energy of the motion being converted into electrical energy through piezoelectric, electrostatic, or electromagnetic methods [227,228]. To address the power density limitation imposed by the volume of the proof mass, Cai et al. [229] proposed an alternative approach. They utilized a planetary structure with a power-generating unit consisting of a base, coils, rotor, and magnets as an eccentric mass, thereby avoiding the use of a proof mass. In walking tests, the power output of this design reached 1.84 mW at the wrist and 2.95 mW at the ankle. The output of energy harvesters is influenced by factors such as the frequency of human movement and the placement of the device. To address this, Hoareau et al. [230] proposed a method for determining the optimal placement of cantilevered piezoelectric generators (PEGs). This method uses electromechanical modeling and accelerometer data to identify the optimal power source for cantilevered PEGs. According to the scenario presented in the article, the optimal placement for the device is along the axis normal to the surface of the right hand. Additionally, extremity segments such as the feet, hands, and forearms were identified as better energy sources. Beach et al. [231] compared the amount of energy harvested from the wrist, hip, ankle, and foot. Based on their analysis of the spectral content, optimal harvester parameters, and the average power output, it was concluded that the foot is able to provide more power, utilize a larger area, and exhibit a better broadband response during walking. Additionally, the foot was found to be least affected by changes in walking speed. To further improve energy harvesting, Sandhu et al. [232] proposed combining solar and kinetic energy harvesting as both energy and activity sensors.

Thermoelectricity: Thermoelectric systems convert the temperature difference (ΔT) between two interfaces into electricity. The ΔT between the human body and the environment is generally stable, making thermoelectric generators (TEGs) a promising power for providing a stable energy supply. However, TEGs require a high temperature difference [233] or a booster [234] to provide sufficient voltage, which unfortunately reduces the flexibility of portable devices and increases the complexity of the system. To solve the issues mentioned above, fiber-based TEGs can be structurally optimized to increase the output voltage [235,236]. The all-fiber thermoelectric sensing device proposed by Qing et al. [166] integrates the fiber-based organic electrochemical transistor (FOECT) and thermoelectric fiber (TEF). Such a power supply has sufficient output and does not require energy storage devices.

Despite promising research into self-powered modules, both biofuel cells and energy harvesters still face challenges in terms of energy efficiency, cell size, and durability. As a result, the dominance of lithium batteries remains difficult to challenge. A long-term vision and sustained efforts are required to fully transition from traditional rigid batteries to flexible power sources suitable for wearable systems.

6. Applying Machine Learning to Sweat Sensing

The trend toward multi-sensor integration is becoming mainstream, allowing in situ monitoring to provide multiple data sets simultaneously. Not only biochemical analytes, but also physical signals are rich in information. For more complex monitoring results, the right data analysis tool is critical. Machine learning has been shown in numerous medical studies [63] to be an effective tool for dealing with large and complex data sets. Machine learning is currently widely used in data analysis, image processing, etc. It can perform feature extraction, pattern recognition, and other operations. Consequently, machine learning can be combined with wearable sweat sensors to use sensor data to create personal physical health portraits.

In particular, machine learning can help classify detection results. Support vector machines (SVM) [237], decision trees [238,239], k-nearest neighbor algorithms (KNN) [240], and other methods can help with the classification of data. Machine learning can evaluate the trend of physiological indicators based on historical data and provide the current state, such as determining the rise and fall of cortisol over time according to the change in sweat cortisol concentration [241]. The data collected by the sensor is pre-processed and can be used as training data for machine learning. Sabilla et al. [242] used Principal Component Analysis (PCA) to preprocess sensor data for dimensionality reduction. Then, three machine learning methods (SVM, decision tree, and KNN) were tested to discriminate individual gender based on sweat composition. After comparison, it was found that SVM could complete the judgment with the best accuracy rate (94.12%).

Machine learning can also predict the trend of indicator changes and the probability of events occurring, as well as detect the occurrence of events. Regression methods are often used to predict changes in physiological parameters. Sankhala et al. [243] used machine learning methods to optimize the signal obtained from a sweat sensor. Data sets included impedance signals, skin temperature, and perspiration (%RH). The sensor converted the discrete input into glucose concentration, interpolated it to build a continuous signal, and then used a decision tree regression algorithm to meet the standard RMSE. The algorithm was tested on three subjects and verified to correctly capture glucose trends. The algorithm predicts the percentage of body weight loss (%BWL) using five parameters: heart rate (HR), core temperature (core T), whole-body sweat rate (WBSR), regional sweat rate (RSR), and sweat sodium ion content. The sweat sensing platform developed by Lafaye et al. [244] also uses machine learning algorithms (linear regression, LR) for prediction. The algorithm predicts the percentage of body weight loss (%BWL) using five parameters: heart rate (HR), core temperature (core T), whole-body sweat rate (WBSR), regional sweat rate (RSR), and sweat sodium ion content. Studies have shown that multimodal analysis using machine learning prediction can effectively increase the detection limit of sensing systems [245]. In addition to its predictive capabilities, machine learning can also identify events that have already occurred. For example, it can be applied in the detection of dehydration state [246] and has the ability to estimate the amount of sweating [247] throughout the body.

For optical sensing, image processing algorithms can be used to quantify colorimetric or fluorescence results. Baker et al. [195,248] developed smartphone applications using digital image processing algorithms. After acquiring colorimetric images of wearable microfluidic patches, the algorithm could perform real-time analysis under different directions and lighting conditions, and provide quantified sweat rate and chloride concentration. This approach can be applied in recreational sports testing.

7. Challenges and Solutions

7.1. Contamination

Sweat sensors must fit closely to the human skin. However, oils and proteins from sweat, skin, beverages, and food debris outside the body can easily remain in the device, reducing its sensing performance. There are many solutions to the problem of fouling residues in electrochemical sensors, such as fluorination and electrode layering [249] or chemical modification [250]. In the field of colorimetric sensors, this problem is relatively ignored [251].

In response to the contamination problem of textile sensors, a fluorine-containing polymer can be introduced on their surface to change the wettability of the fabrics [252]. Based on this, Zhao et al. [251] studied the contamination treatment of fabric colorimetric sensors and proposed a solution. Using the piCVD method, the colorimetric array is encapsulated with a thin conformal coating of fluorinated hydrophobic polymer (Figure 5a). The coating prevents dirt from adhering to the surface of the array. Sensors produced with this technique are protected against contamination from various liquids, such as milk, coffee, and olive oil. In the same study, to overcome the interference of fabric texture on image processing, image-sharpening technology was applied during processing to ensure the accuracy of sensor identification of different pH values.

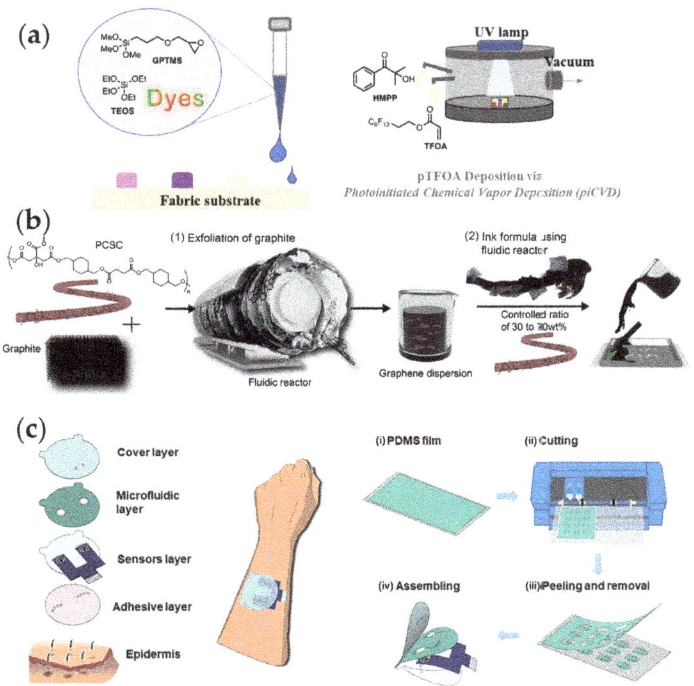

Figure 5. (**a**) Contamination treatment for fabric sensors. Using piCVD with fluorinated hydrophobic polymers as conformal coatings prevents dirt from remaining on the surface. Reprinted from ref. [251], Copyright (2023), with permission from Elsevier. (**b**) Fabrication process of the self-healing P-Gr ink and its screen-printing application. Reprinted from ref. [253], Copyright (2023), with permission from Elsevier. (**c**) Exploded view of the device fabricated by CAP technology and schematics of the fabrication process including four steps. Reprinted from ref. [121], Copyright (2022), with permission from Elsevier.

One approach to managing contamination is to treat it after it enters the system, through processes such as cleaning and retention prevention. Another effective option is to block contaminants at the point of sweat collection. Yang et al. [191] proposed a pH sensor with a paper-based sandwich structure based on oil control paper. After control experiments (compared with monolayer filter paper) and sample analysis, the paper-based sensor has obvious resistance to triglycerides and squalene, and can filter more than 90% of sebum in sweat. The simplicity of this microchannel manufacturing process, requiring only scissors and paper, further demonstrates the potential for practical applications. Huang et al. [254] integrated a microfilter at the entrance of the microchannel, which blocks skin debris through ten continuously arranged micropillars, thereby realizing the filtering function.

7.2. *Interference*

Waterproof: The operating environment of a wearable sweat sensor is determined by the wearer. If the wearer is exposed to rain, splashes, or underwater activities, the device's water resistance, adhesive stability, and resistance to contamination may be more demanding. Reeder et al. [255] previously proposed a waterproof microfluidic patch sensor. The design included a low-permeability polymer shell, a water-resistant microchannel inlet and outlet, conformal technology to reduce the effects of shear stress, underwater bonding materials, and other targeted proposals to improve water resistance.

Optical Crosstalk: Photoelectrochemical sweat sensor measurements can be affected by incomplete sweat coverage and are highly dependent on light intensity. As a result, they are susceptible to interference from vibrations caused by human movement and battery depletion. Amphoteric Bi_2O_3 has been reported to overcome these challenges [256].

Flow Rate Crosstalk: Based on current research, Komkova et al. [257] investigated the effect of sweat flow on sensor parameters, such as sensitivity. They made recommendations for addressing this issue, including flow correction, potentiostat control, and pulse detection.

7.3. *Self-Healing Ability*

Wearable devices must remain attached to the human epidermis during daily activities. These activities can cause deformation of the substrate and impacts to the device. Both deformation and impacts can potentially cause damage to the device. As a result, if the device possesses self-healing capabilities, it could significantly improve the lifespan, reliability, and stability of the wearable device. In this direction, there have been some new explorations, and one option is to use composite gel materials. Wang et al. [142] developed a hydrogel patch that detects sweat in situ without sweat collection. Polyvinyl alcohol and sucrose are crosslinked to form hydrogel, the self-healing adhesive properties of which are derived from hydrogen bonds between molecules. Its self-healing ability is also used to enable the autonomous assembling of colorimetric modules and substrates. Son et al. [253] prepared a P-Gr ink using a self-healing bio-based elastomer matrix incorporated with graphene (Figure 5b). This ink was used to print a conductive electrode with self-healing capabilities. These developments represent potential avenues for further research.

7.4. *Technology*

The complexity and high cost of manufacturing processes are among the major factors limiting the market availability of wearable sensing devices. To reduce production costs, existing methods such as paper-based fabrication, 3D printing, laser engraving, inkjet printing, and integrated circuit processes have been employed. In addition, researchers have developed new technologies specifically for wearable devices. These include the vacuum filtration transfer printing method (PS-VFTP), inspired by the photosensitive seal (PS) proposed by Hao et al. [258], and the cut and paste (CAP) technology used by Sun et al. [121] in the fabrication of wearable devices (Figure 5c). Further exploration is necessary to identify manufacturing processes that are universally applicable to the production of wearable sweat sensors. Figure 6 provides an overview of the current challenges and potential solutions in the development of wearable sweat sensors.

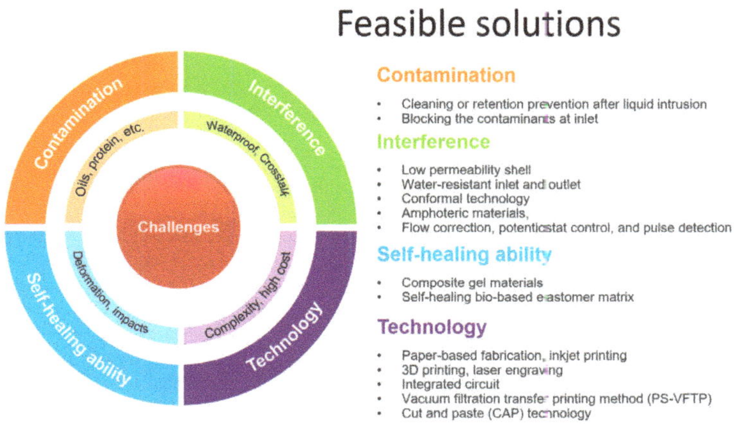

Figure 6. A summary diagram of the current challenges in wearable sweat-sensor research, along with a list of feasible solutions.

8. Conclusions

Wearable sweat sensors have drawn increasing attention due to their portability, flexibility, non-invasiveness, and ability to provide continuous real-time monitoring. Studies in this multidisciplinary research area span a range of theoretical and engineering disciplines, including biomedical principles, materials science, chemistry, physics, and electrical engineering. Distinctive advances have been made in each of these directions. This paper provides an overview of recent advances in biomarkers, sensing strategies, materials, and system architectures for wearable sweat sensors. It also summarizes current challenges and potential solutions in this field, with the aim of providing researchers with a comprehensive perspective and targets for future development. Since their inception, wearable sweat sensors have made significant advances in both form and function for practical applications. Current research is focused on the development of full-system sensors—devices capable of independently performing the entire process of sensing, signal processing and transmission, power supply, and data analysis. This represents the product form closest to practical application. In the future, it is expected that wearable sweat-sensor systems will continue to improve and new types of sensors will emerge. All of these innovative efforts will hopefully make the detection of human physiological signals more scientific and feasible.

Author Contributions: Writing—original draft, review and editing, Z.X.; writing—original draft, review and editing, J.H.; writing—review and editing, B.L.; writing—review and editing, Z.W.; supervision, writing—review and editing, H.M. All authors have read and agreed to the published version of the manuscript.

Funding: This research was funded by National Key R&D Program of China, grant number 2022YFF1202700, National Natural Science Foundation of China (No. 62231025 and 61971410), Program of Science and Technology Commission of Shanghai Municipality (No. 23141901100), Shanghai Municipal Science and Technology Major Project (No. ZD2021CY001), and Shanghai Engineer & Technology Research Center of Internet of Things for Respiratory Medicine (20DZ2254400).

Institutional Review Board Statement: Not applicable.

Informed Consent Statement: Not applicable.

Data Availability Statement: No new data were created or analyzed in this study. Data sharing is not applicable to this article.

Conflicts of Interest: The authors declare no conflict of interest.

References

1. Bariya, M.; Nyein, H.Y.Y.; Javey, A. Wearable sweat sensors. *Nat. Electron.* **2018**, *1*, 160–171. [CrossRef]
2. Kim, J.; Campbell, A.S.; de Avila, B.E.; Wang, J. Wearable biosensors for healthcare monitoring. *Nat. Biotechnol.* **2019**, *37*, 389–406. [CrossRef] [PubMed]
3. Shrivastava, S.; Trung, T.Q.; Lee, N.-E. Recent progress, challenges, and prospects of fully integrated mobile and wearable point-of-care testing systems for self-testing. *Chem. Soc. Rev.* **2020**, *49*, 1812–1866. [CrossRef] [PubMed]
4. Mani, V.; Beduk, T.; Khushaim, W.; Ceylan, A.E.; Timur, S.; Wolfbeis, O.S.; Salama, K.N. Electrochemical sensors targeting salivary biomarkers: A comprehensive review. *Trends Anal. Chem.* **2021**, *135*, 116164. [CrossRef]
5. Garcia-Rey, S.; Gil-Hernandez, E.; Basabe-Desmonts, L.; Benito-Lopez, F. Colorimetric Determination of Glucose in Sweat Using an Alginate-Based Biosystem. *Polymers* **2023**, *15*, 1218. [CrossRef]
6. Gao, W.; Emaminejad, S.; Nyein, H.Y.Y.; Challa, S.; Chen, K.; Peck, A.; Fahad, H.M.; Ota, H.; Shiraki, H.; Kiriya, D.; et al. Fully integrated wearable sensor arrays for multiplexed in situ perspiration analysis. *Nature* **2016**, *529*, 509–514. [CrossRef]
7. Katseli, V.; Economou, A.; Kokkinos, C. Smartphone-Addressable 3D-Printed Electrochemical Ring for Nonenzymatic Self-Monitoring of Glucose in Human Sweat. *Anal. Chem.* **2021**, *93*, 3331–3336. [CrossRef]
8. Sonner, Z.; Wilder, E.; Heikenfeld, J.; Kasting, G.; Beyette, F.; Swaile, D.; Sherman, F.; Joyce, J.; Hagen, J.; Kelley-Loughnane, N.; et al. The microfluidics of the eccrine sweat gland, including biomarker partitioning, transport, and biosensing implications. *Biomicrofluidics* **2015**, *9*, 031301. [CrossRef]
9. Asaduzzaman, M.; Zahed, M.A.; Sharifuzzaman, M.; Reza, M.S.; Hui, X.; Sharma, S.; Shin, Y.D.; Park, J.Y. A hybridized nano-porous carbon reinforced 3D graphene-based epidermal patch for precise sweat glucose and lactate analysis. *Biosens. Bioelectron.* **2022**, *219*, 114846. [CrossRef]
10. Pullano, S.A.; Greco, M.; Bianco, M.G.; Foti, D.; Brunetti, A.; Fiorillo, A.S. Glucose biosensors in clinical practice: Principles, limits and perspectives of currently used devices. *Theranostics* **2022**, *12*, 493–511. [CrossRef]
11. Cass, A.E.G.; Davis, G.; Francis, G.D.; Hill, H.A.O.; Aston, W.J.; Higgins, I.J.; Plotkin, E.V.; Scott, L.D.L.; Turner, A.P.F. Ferrocene-mediated enzyme electrode for amperometric determination of glucose. *Anal. Chem.* **1984**, *56*, 667–671. [CrossRef] [PubMed]
12. Papanikolaou, E.; Simos, Y.V.; Spyrou, K.; Tzianni, E.I.; Vezyraki, P.; Tsamis, K.; Patila, M.; Tigas, S.; Prodromidis, M.I.; Gournis, D.P.; et al. Is graphene the rock upon which new era continuous glucose monitors could be built? *Exp. Biol. Med.* **2023**, *248*, 14–25. [CrossRef]
13. Khor, S.M.; Choi, J.; Won, P.; Ko, S.H. Challenges and Strategies in Developing an Enzymatic Wearable Sweat Glucose Biosensor as a Practical Point-Of-Care Monitoring Tool for Type II Diabetes. *Nanomaterials* **2022**, *12*, 221. [CrossRef] [PubMed]
14. Choi, Y.M.; Lim, H.; Lee, H.-N.; Park, Y.M.; Park, J.-S.; Kim, H.-J. Selective Nonenzymatic Amperometric Detection of Lactic Acid in Human Sweat Utilizing a Multi-Walled Carbon Nanotube (MWCNT)-Polypyrrole Core-Shell Nanowire. *Biosensors* **2020**, *10*, 111. [CrossRef] [PubMed]
15. Kim, J.; Valdés-Ramírez, G.; Bandodkar, A.J.; Jia, W.; Martinez, A.G.; Ramírez, J.; Mercier, P.; Wang, J. Non-invasive mouthguard biosensor for continuous salivary monitoring of metabolites. *Analyst* **2014**, *139*, 1632–1636. [CrossRef] [PubMed]
16. Terse-Thakoor, T.; Punjiya, M.; Matharu, Z.; Lyu, B.; Ahmad, M.; Giles, G.E.; Owyeung, R.; Alaimo, F.; Shojaei Baghini, M.; Brunyé, T.T.; et al. Thread-based multiplexed sensor patch for real-time sweat monitoring. *NPJ Flexible Electron.* **2020**, *4*, 18. [CrossRef]
17. Sempionatto, J.R.; Nakagawa, T.; Pavinatto, A.; Mensah, S.T.; Imani, S.; Mercier, P.; Wang, J. Eyeglasses based wireless electrolyte and metabolite sensor platform. *Lab Chip* **2017**, *17*, 1834–1842. [CrossRef]
18. Gillan, L.; Teerinen, T.; Suhonen, M.; Kivimäki, L.; Alastalo, A. Simultaneous multi-location wireless monitoring of sweat lactate trends. *Flexible Printed Electron.* **2021**, *6*, 034003. [CrossRef]
19. Nagamine, K.; Mano, T.; Shiwaku, R.; Furusawa, H.; Matsui, H.; Kumaki, D.; Tokito, S. An L-lactate Biosensor Based on Printed Organic Inverter Circuitry and with a Tunable Detection Limit. *Sens. Mater.* **2019**, *31*, 1205. [CrossRef]
20. Komkova, M.A.; Eliseev, A.A.; Poyarkov, A.A.; Daboss, E.V.; Evdokimov, P.V.; Eliseev, A.A.; Karyakin, A.A. Simultaneous monitoring of sweat lactate content and sweat secretion rate by wearable remote biosensors. *Biosens. Bioelectron.* **2022**, *202*, 113970. [CrossRef]
21. Xu, L.; Zhou, Z.; Fan, M.; Fang, X. Advances in wearable flexible electrochemical sensors for sweat monitoring: A mini-review. *Int. J. Electrochem. Sci.* **2023**, *18*, 13–19. [CrossRef]
22. Gamella, M.; Campuzano, S.; Manso, J.; de Rivera, G.G.; Lopez-Colino, F.; Reviejo, A.J.; Pingarron, J.M. A novel non-invasive electrochemical biosensing device for in situ determination of the alcohol content in blood by monitoring ethanol in sweat. *Anal. Chim. Acta* **2014**, *806*, 1–7. [CrossRef] [PubMed]
23. Khemtonglang, K.; Chaiyaphet, N.; Kumsaen, T.; Chaiyachati, C.; Chuchuen, O. A Smart Wristband Integrated with an IoT-Based Alarming System for Real-Time Sweat Alcohol Monitoring. *Sensors* **2022**, *22*, 6435. [CrossRef]
24. Kogler, L.; Müller, V.I.; Chang, A.; Eickhoff, S.B.; Fox, P.T.; Gur, R.C.; Derntl, B. Psychosocial versus physiological stress—Meta-analyses on deactivations and activations of the neural correlates of stress reactions. *NeuroImage* **2015**, *119*, 235–251. [CrossRef] [PubMed]
25. Ok, J.; Park, S.; Jung, Y.H.; Kim, T.I. Wearable and Implantable Cortisol Sensing Electronics for Stress Monitoring. *Adv. Mater.* **2023**, 2211595. [CrossRef]

26. Torrente-Rodriguez, R.M.; Tu, J.; Yang, Y.; Min, J.; Wang, M.; Song, Y.; Yu, Y.; Xu, C.; Ye, C.; IsHak, W.W.; et al. Investigation of cortisol dynamics in human sweat using a graphene-based wireless mHealth system. *Matter* **2020**, *2*, 921–937. [CrossRef]
27. Sempionatto, J.R.; Khorshed, A.A.; Ahmed, A.; De Loyola, E.S.A.N.; Barfidokht, A.; Yin, L.; Goud, K.Y.; Mohamed, M.A.; Bailey, E.; May, J.; et al. Epidermal Enzymatic Biosensors for Sweat Vitamin C: Toward Personalized Nutrition. *ACS Sens.* **2020**, *5*, 1804–1813. [CrossRef]
28. Kim, J.; Wu, Y.; Luan, H.; Yang, D.S.; Cho, D.; Kwak, S.S.; Liu, S.; Ryu, H.; Ghaffari, R.; Rogers, J.A. A Skin-Interfaced, Miniaturized Microfluidic Analysis and Delivery System for Colorimetric Measurements of Nutrients in Sweat and Supply of Vitamins Through the Skin. *Adv. Sci.* **2022**, *9*, e2103331. [CrossRef]
29. Tai, L.C.; Gao, W.; Chao, M.; Bariya, M.; Ngo, Q.P.; Shahpar, Z.; Nyein, H.Y.Y.; Park, H.; Sun, J.; Jung, Y.; et al. Methylxanthine Drug Monitoring with Wearable Sweat Sensors. *Adv. Mater.* **2018**, *30*, e1707442. [CrossRef]
30. Hao, Q.; Ren, X.R.; Chen, Y.; Zhao, C.; Xu, J.; Wang, D.; Liu, H. A sweat-responsive covalent organic framework film for material-based liveness detection and sweat pore analysis. *Nat. Commun.* **2023**, *14*, 578. [CrossRef]
31. Choi, D.-H.; Li, Y.; Cutting, G.R.; Searson, P.C. A wearable potentiometric sensor with integrated salt bridge for sweat chloride measurement. *Sens. Actuators B* **2017**, *250*, 673–678. [CrossRef]
32. Mondal, S.; Subramaniam, C. Point-of-Care, Cable-Type Electrochemical Zn^{2+} Sensor with Ultrahigh Sensitivity and Wide Detection Range for Soil and Sweat Analysis. *ACS Sustain. Chem. Eng.* **2019**, *7*, 14569–14579 [CrossRef]
33. Lopresti, A.L. The Effects of Psychological and Environmental Stress on Micronutrient Concentrations in the Body: A Review of the Evidence. *Adv. Nutr.* **2020**, *11*, 103–112. [CrossRef]
34. Botturi, A.; Ciappolino, V.; Delvecchio, G.; Boscutti, A.; Viscardi, B.; Brambilla, P. The Role and the Effect of Magnesium in Mental Disorders: A Systematic Review. *Nutrients* **2020**, *12*, 1661. [CrossRef] [PubMed]
35. Rose, D.P.; Ratterman, M.E.; Griffin, D.K.; Hou, L.; Kelley-Loughnane, N.; Naik, R.R.; Hagen, J.A.; Papautsky, I.; Heikenfeld, J.C. Adhesive RFID Sensor Patch for Monitoring of Sweat Electrolytes. *IEEE Trans. Biomed. Eng.* **2015**, *62*, 1457–1465. [CrossRef] [PubMed]
36. Buono, M.J.; Claros, R.; Deboer, T.; Wong, J. Na^+ secretion rate increases proportionally more than the Na^+ reabsorption rate with increases in sweat rate. *J. Appl. Physiol.* **2008**, *105*, 1044–1048. [CrossRef]
37. Hashimoto, Y.; Ishihara, T.; Kuwabara, K.; Amano, T.; Togo, H. Wearable Microfluidic Sensor for the Simultaneous and Continuous Monitoring of Local Sweat Rates and Electrolyte Concentrations. *Micromachines* **2022**, *13*, 575. [CrossRef]
38. Pirovano, P.; Dorrian, M.; Shinde, A.; Donohoe, A.; Brady, A.J.; Moyna, N.M.; Wallace, G.; Diamond, D.; McCaul, M. A wearable sensor for the detection of sodium and potassium in human sweat during exercise. *Talanta* **2020**, *219*, 121145. [CrossRef]
39. Yoon, E.S.; Park, H.J.; Kil, M.S.; Kim, J.; Lee, K.G.; Choi, B.G. Preparation of nanopillar array electrode of iridium oxide for high performance of pH sensor and its real-time sweat monitoring. *Bull. Korean Chem. Soc.* **2023**, *44*, 528–535. [CrossRef]
40. Cui, X.; Bao, Y.; Han, T.; Liu, Z.; Ma, Y.; Sun, Z. A wearable electrochemical sensor based on beta-CD functionalized graphene for pH and potassium ion analysis in sweat. *Talanta* **2022**, *245*, 123481. [CrossRef]
41. Chen, L.; Chen, F.; Liu, G.; Lin, H.; Bao, Y.; Han, D.; Wang, W.; Ma, Y.; Zhang, B.; Niu, L. Superhydrophobic Functionalized Ti(3)C(2)T(x) MXene-Based Skin-Attachable and Wearable Electrochemical pH Sensor for Real-Time Sweat Detection. *Anal. Chem.* **2022**, *94*, 7319–7328. [CrossRef] [PubMed]
42. Yang, Y.; Song, Y.; Bo, X.; Min, J.; Pak, O.S.; Zhu, L.; Wang, M.; Tu, J.; Kogan, A.; Zhang, H.; et al. A laser-engraved wearable sensor for sensitive detection of uric acid and tyrosine in sweat. *Nat. Biotechnol.* **2020**, *38*, 217–224. [CrossRef] [PubMed]
43. Wang, M.; Yang, Y.; Min, J.; Song, Y.; Tu, J.; Mukasa, D.; Ye, C.; Xu, C.; Heflin, N.; McCune, J.S.; et al. A wearable electrochemical biosensor for the monitoring of metabolites and nutrients. *Nat. Biomed. Eng.* **2022**, *6*, 1225–1235. [CrossRef] [PubMed]
44. Karthika, A.; Karuppasamy, P.; Selvarajan, S.; Suganthi, A.; Rajarajan, M. Electrochemical sensing of nicotine using CuWO(4) decorated reduced graphene oxide immobilized glassy carbon electrode. *Ultrason. Sonochem.* **2019**, *55*, 196–206. [CrossRef]
45. Magesh, V.; Sundramoorthy, A.K.; Ganapathy, D.; Atchudan, R.; Arya, S.; Alshgari, R.A.; Aljuwayid, A.M. Palladium Hydroxide (Pearlman's Catalyst) Doped MXene (Ti(3)C(2)Tx) Composite Modified Electrode for Selective Detection of Nicotine in Human Sweat. *Biosensors* **2022**, *13*, 54. [CrossRef]
46. Kwon, K.; Kim, J.U.; Deng, Y.; Krishnan, S.R.; Choi, J.; Jang, H.; Lee, K.; Su, C.-J.; Yoo, I.; Wu, Y.; et al. An on-skin platform for wireless monitoring of flow rate, cumulative loss and temperature of sweat in real time. *Nat. Electron.* **2021**, *4*, 302–312. [CrossRef]
47. Momose, H.; Takasaka, M.; Watanabe-Asaka, T.; Hayashi, M.; Maejima, D.; Kawai, Y.; Ohhashi, T. Heatstroke risk informing system using wearable perspiration ratemeter on users undergoing physical exercise. *Sci. Rep.* **2023**, *13*, 416. [CrossRef]
48. Dautta, M.; Ayala-Cardona, L.F.; Davis, N.; Aggarwal, A.; Park, J.; Wang, S.; Gillan, L.; Jansson, E.; Hietala, M.; Ko, H.; et al. Tape-Free, Digital Wearable Band for Exercise Sweat Rate Monitoring. *Adv. Mater. Technol.* **2023**, *8*, 2201187. [CrossRef]
49. Sakaguchi, M.; Mori, K.; Yokochi, H.; Nakashima, K.; Ohhashi, T. A new apparatus for continuous recording of sweating rate by use of a hygrometer. *Iyodenshi Seitai Kogaku* **1988**, *26*, 213–217.
50. Toshio, O.; Masao, S.; Takao, T. Human perspiration measurement. *Physiol. Meas.* **1998**, *19*, 449.
51. Seshadri, D.R.; Li, R.T.; Voos, J.E.; Rowbottom, J.R.; Alfes, C.M.; Zorman, C.A.; Drummond, C.K. Wearable sensors for monitoring the physiological and biochemical profile of the athlete. *NPJ Digit. Med.* **2019**, *2*, 72. [CrossRef]
52. Liu, D.; Liu, Z.; Feng, S.; Gao, Z.; Chen, R.; Cai, G.; Bian, S. Wearable Microfluidic Sweat Chip for Detection of Sweat Glucose and pH in Long-Distance Running Exercise. *Biosensors* **2023**, *13*, 157. [CrossRef]

53. Lee, J.; Pyo, M.; Lee, S.H.; Kim, J.; Ra, M.; Kim, W.Y.; Park, B.J.; Lee, C.W.; Kim, J.M. Hydrochromic conjugated polymers for human sweat pore mapping. *Nat. Commun.* **2014**, *5*, 3736. [CrossRef]
54. Marasco, E.; Ross, A. A Survey on Antispoofing Schemes for Fingerprint Recognition Systems. *ACM Comput. Surv.* **2014**, *47*, 28. [CrossRef]
55. Agarwal, R.; Jalal, A.S.; Arya, K.V. A review on presentation attack detection system for fake fingerprint. *Mod. Phys. Lett. B* **2020**, *34*, 2030001. [CrossRef]
56. Ding, B.; Wang, H.; Chen, P.; Zhang, Y.; Liang, R.; Liu, Y. Subcutaneous sweat pore estimation from optical coherence tomography. *IET Image Process* **2021**, *15*, 3267–3280. [CrossRef]
57. Chengsheng, Y.; Sun, X.; Rui, L. Fingerprint liveness detection based on multi-scale LPQ and PCA. *China Commun.* **2016**, *13*, 60–65. [CrossRef]
58. Lu, M.Y.; Chen, Z.Q.; Sheng, W.G. A Pore-Based Method for Fingerprint Liveness Detection. In Proceedings of the 2015 International Conference on Computer Science and Applications (CSA), Wuhan, China, 20–22 November 2015; pp. 77–81.
59. Niederberger, C.; Vermeersch, A.; Davidhi, F.; Ewald, C.Y.; Havenith, G.; Goldhahn, J.; Dincer, C.; Brasier, N. Wearable sweat analysis to determine biological age. *Trends Biotechnol.* **2023**, *41*, 1113–1116. [CrossRef] [PubMed]
60. Kim, J.; de Araujo, W.R.; Samek, I.A.; Bandodkar, A.J.; Jia, W.; Brunetti, B.; Paixão, T.R.L.C.; Wang, J. Wearable temporary tattoo sensor for real-time trace metal monitoring in human sweat. *Electrochem. Commun.* **2015**, *51*, 41–45. [CrossRef]
61. Laochai, T.; Yukird, J.; Promphet, N.; Qin, J.; Chailapakul, O.; Rodthongkum, N. Non-invasive electrochemical immunosensor for sweat cortisol based on L-cys/AuNPs/ MXene modified thread electrode. *Biosens. Bioelectron.* **2022**, *203*, 114039. [CrossRef] [PubMed]
62. Santiago, E.; Poudyal, S.S.; Shin, S.Y.; Yoon, H.J. Graphene Oxide Functionalized Biosensor for Detection of Stress-Related Biomarkers. *Sensors* **2022**, *22*, 558. [CrossRef] [PubMed]
63. Sempionatto, J.R.; Lasalde-Ramirez, J.A.; Mahato, K.; Wang, J.; Gao, W. Wearable chemical sensors for biomarker discovery in the omics era. *Nat. Rev. Chem.* **2022**, *6*, 899–915. [CrossRef] [PubMed]
64. Xiao, Y.; Lai, R.Y.; Plaxco, K.W. Preparation of electrode-immobilized, redox-modified oligonucleotides for electrochemical DNA and aptamer-based sensing. *Nat. Protoc.* **2007**, *2*, 2875–2880. [CrossRef]
65. Ganguly, A.; Lin, K.C.; Muthukumar, S.; Prasad, S. Autonomous, Real-Time Monitoring Electrochemical Aptasensor for Circadian Tracking of Cortisol Hormone in Sub-microliter Volumes of Passively Eluted Human Sweat. *ACS Sens.* **2021**, *6*, 63–72. [CrossRef] [PubMed]
66. Singh, N.K.; Chung, S.; Sveiven, M.; Hall, D.A. Cortisol Detection in Undiluted Human Serum Using a Sensitive Electrochemical Structure-Switching Aptamer over an Antifouling Nanocomposite Layer. *ACS Omega* **2021**, *6*, 27888–27897. [CrossRef] [PubMed]
67. Singh, N.K.; Chung, S.; Chang, A.Y.; Wang, J.; Hall, D.A. A non-invasive wearable stress patch for real-time cortisol monitoring using a pseudoknot-assisted aptamer. *Biosens. Bioelectron.* **2023**, *227*, 115097. [CrossRef]
68. Baker, L.B. Physiology of sweat gland function: The roles of sweating and sweat composition in human health. *Temperature* **2019**, *6*, 211–259. [CrossRef]
69. Patterson, M.J.; Galloway, S.D.R.; Nimmo, M.A. Variations in Regional Sweat Composition in Normal Human Males. *Exp. Physiol.* **2000**, *85*, 869–875. [CrossRef]
70. Humpolicek, P.; Kasparkova, V.; Saha, P.; Stejskal, J. Biocompatibility of polyaniline. *Synthetic Metals* **2012**, *162*, 722–727. [CrossRef]
71. Robergs, R.A.; Ghiasvand, F.; Parker, D. Biochemistry of exercise-induced metabolic acidosis. *Am. J. Physiol. Regul. Integr. Comp. Physiol* **2004**, *287*, R502–R516. [CrossRef]
72. Bilbao, E.; Garate, O.; Rodríguez Campos, T.; Roberti, M.; Mass, M.; Lozano, A.; Longinotti, G.; Monsalve, L.; Ybarra, G. Electrochemical Sweat Sensors. *Chemosensors* **2023**, *11*, 244. [CrossRef]
73. Yang, J.; Gong, X.; Chen, S.; Zheng, Y.; Peng, L.; Liu, B.; Chen, Z.; Xie, X.; Yi, C.; Jiang, L. Development of Smartphone-Controlled and Microneedle-Based Wearable Continuous Glucose Monitoring System for Home-Care Diabetes Management. *ACS Sens.* **2023**, *8*, 1241–1251. [CrossRef] [PubMed]
74. Wiorek, A.; Parrilla, M.; Cuartero, M.; Crespo, G.A. Epidermal Patch with Glucose Biosensor: pH and Temperature Correction toward More Accurate Sweat Analysis during Sport Practice. *Anal. Chem.* **2020**, *92*, 10153–10161. [CrossRef] [PubMed]
75. Jia, W.; Bandodkar, A.J.; Valdés-Ramírez, G.; Windmiller, J.R.; Yang, Z.; Ramírez, J.; Chan, G.; Wang, J. Electrochemical Tattoo Biosensors for Real-Time Noninvasive Lactate Monitoring in Human Perspiration. *Anal. Chem.* **2013**, *85*, 6553–6560. [CrossRef] [PubMed]
76. Lin, P.H.; Sheu, S.C.; Chen, C.W.; Huang, S.C.; Li, B.R. Wearable hydrogel patch with noninvasive, electrochemical glucose sensor for natural sweat detection. *Talanta* **2022**, *241*, 11. [CrossRef] [PubMed]
77. Das, R.; Nag, S.; Banerjee, P. Electrochemical Nanosensors for Sensitization of Sweat Metabolites: From Concept Mapping to Personalized Health Monitoring. *Molecules* **2023**, *28*, 1259. [CrossRef]
78. Luo, D.; Sun, H.; Li, Q.; Niu, X.; He, Y.; Liu, H. Flexible Sweat Sensors: From Films to Textiles. *ACS Sens.* **2023**, *8*, 465–481. [CrossRef]
79. Jose, M.; Oudebrouckx, G.; Bormans, S.; Veske, P.; Thoelen, R.; Deferme, W. Monitoring Body Fluids in Textiles: Combining Impedance and Thermal Principles in a Printed, Wearable, and Washable Sensor. *ACS Sens.* **2021**, *6*, 896–907. [CrossRef]

80. Ciani, I.; Schulze, H.; Corrigan, D.K.; Henihan, G.; Giraud, G.; Terry, J.G.; Walton, A.J.; Pethig, R.; Ghazal, P.; Crain, J.; et al. Development of immunosensors for direct detection of three wound infection biomarkers at point of care using electrochemical impedance spectroscopy. *Biosens. Bioelectron.* **2012**, *31*, 413–418. [CrossRef]
81. Farrow, M.J.; Hunter, I.S.; Connolly, P. Developing a Real Time Sensing System to Monitor Bacteria in Wound Dressings. *Biosensors* **2012**, *2*, 171–188. [CrossRef]
82. Wang, S.; Zhang, J.; Gharbi, O.; Vivier, V.; Gao, M.; Orazem, M.E. Electrochemical impedance spectroscopy. *Nat. Rev. Methods Primers* **2021**, *1*, 41. [CrossRef]
83. Daniels, J.S.; Pourmand, N. Label-Free Impedance Biosensors: Opportunities and Challenges. *Electroanalysis* **2007**, *19*, 1239–1257. [CrossRef] [PubMed]
84. Randviir, E.P.; Banks, C.E. Electrochemical impedance spectroscopy: An overview of bioanalytical applications. *Anal. Methods* **2013**, *5*, 1098–1115. [CrossRef]
85. Pänke, O.; Balkenhohl, T.; Kafka, J.; Schäfer, D.; Lisdat, F. Impedance spectroscopy and biosensing. *Adv. Biochem. Eng./Biotechnol.* **2008**, *109*, 195–237. [PubMed]
86. Mandler, D. Fritz Scholz (Ed.): Electroanalytical methods. Guide to experiments and applications, 2nd ed. *Anal. Bioanal. Chem.* **2010**, *398*, 2771–2772. [CrossRef]
87. Bard, A.J.; Faulkner, L.R.; White, H.S. *Electrochemical Methods: Fundamentals and Applications*. Wiley: New York, NY, USA, 2022.
88. Veeralingam, S.; Khandelwal, S.; Badhulika, S. AI/ML-Enabled 2-D—RuS_2 Nanomaterial-Based Multifunctional, Low Cost, Wearable Sensor Platform for Non-Invasive Point of Care Diagnostics. *IEEE Sens. J.* **2020**, *20*, 8437–8444. [CrossRef]
89. Yin, L.; Cao, M.; Kim, K.N.; Lin, M.; Moon, J.-M.; Sempionatto, J.R.; Yu, J.; Liu, R.; Wicker, C.; Trifonov, A.; et al. A stretchable epidermal sweat sensing platform with an integrated printed battery and electrochromic display. *Nat. Electron.* **2022**, *5*, 694–705. [CrossRef]
90. Bandodkar, A.J.; Gutruf, P.; Choi, J.; Lee, K.; Sekine, Y.; Reeder, J.T.; Jeang, W.J.; Aranyosi, A.J.; Lee, S.P.; Model, J.B.; et al. Battery-free, skin-interfaced microfluidic/electronic systems for simultaneous electrochemical, colorimetric, and volumetric analysis of sweat. *Sci. Adv.* **2019**, *5*, eaav3294. [CrossRef]
91. Choi, J.; Bandodkar, A.J.; Reeder, J.T.; Ray, T.R.; Turnquist, A.; Kim, S.B.; Nyberg, N.; Hourlier-Fargette, A.; Model, J.B.; Aranyosi, A.J.; et al. Soft, Skin-Integrated Multifunctional Microfluidic Systems for Accurate Colorimetric Analysis of Sweat Biomarkers and Temperature. *ACS Sens.* **2019**, *4*, 379–388. [CrossRef]
92. Li, Z.; Suslick, K.S. Portable Optoelectronic Nose for Monitoring Meat Freshness. *ACS Sens.* **2016**, *1*, 1330–1335. [CrossRef]
93. Gao, L.; Zhang, Y.; Malyarchuk, V.; Jia, L.; Jang, K.-I.; Chad Webb, R.; Fu, H.; Shi, Y.; Zhou, G.; Shi, L.; et al. Epidermal photonic devices for quantitative imaging of temperature and thermal transport characteristics of the skin. *Nat. Commun.* **2014**, *5*, 4938. [CrossRef] [PubMed]
94. Araki, H.; Kim, J.; Zhang, S.; Banks, A.; Crawford, K.E.; Sheng, X.; Gutruf, P.; Shi, Y.; Pielak, R.M.; Rogers, J.A. Materials and Device Designs for an Epidermal UV Colorimetric Dosimeter with Near Field Communication Capabilities. *Adv. Funct. Mater.* **2017**, *27*, 1604465. [CrossRef]
95. Promphet, N.; Ummartyotin, S.; Ngeontae, W.; Puthongkham, P.; Rodthongkum, N. Non-invasive wearable chemical sensors in real-life applications. *Anal. Chim. Acta* **2021**, *1179*, 338643. [CrossRef]
96. Koh, A.; Kang, D.; Xue, Y.; Lee, S.; Pielak, R.M.; Kim, J.; Hwang, T.; Min, S.; Banks, A.; Bastien, P.; et al. A soft, wearable microfluidic device for the capture, storage, and colorimetric sensing of sweat. *Sci. Transl. Med.* **2016**, *8*, ra165–ra366. [CrossRef]
97. Zhang, K.; Zhang, J.; Wang, F.; Kong, D. Stretchable and Superwettable Colorimetric Sensing Patch for Epidermal Collection and Analysis of Sweat. *ACS Sens.* **2021**, *6*, 2261–2269. [CrossRef]
98. Promphet, N.; Rattanawaleedirojn, P.; Siralertmukul, K.; Soatthiyanon, N.; Potiyaraj, P.; Thanawattano, C.; Hinestroza, J.P.; Rodthongkum, N. Non-invasive textile based colorimetric sensor for the simultaneous detection of sweat pH and lactate. *Talanta* **2019**, *192*, 424–430. [CrossRef] [PubMed]
99. Yue, X.; Xu, F.; Zhang, L.; Ren, G.; Sheng, H.; Wang, J.; Wang, K.; Yu, L.; Wang, J.; Li, G.; et al. Simple, Skin-Attachable, and Multifunctional Colorimetric Sweat Sensor. *ACS Sens.* **2022**, *7*, 2198–2208. [CrossRef]
100. Vaquer, A.; Baron, E.; de la Rica, R. Wearable Analytical Platform with Enzyme-Modulated Dynamic Range for the Simultaneous Colorimetric Detection of Sweat Volume and Sweat Biomarkers. *ACS Sens.* **2021**, *6*, 130–136 [CrossRef]
101. Vaquer, A.; Baron, E.; de la Rica, R. Detection of low glucose levels in sweat with colorimetric wearable biosensors. *Analyst* **2021**, *146*, 3273–3279. [CrossRef]
102. Yan, T.; Zhang, G.; Yu, K.; Chai, H.; Tian, M.; Qu, L.; Dong, H.; Zhang, X. Smartphone light-driven zinc porphyrinic MOF nanosheets-based enzyme-free wearable photoelectrochemical sensor for continuous sweat vitamin C detection. *Chem. Eng. J.* **2023**, *455*, 140779. [CrossRef]
103. Xu, X.Y.; Yan, B. A fluorescent wearable platform for sweat Cl- analysis and logic smart-device fabrication based on color adjustable lanthanide MOFs. *J. Mater. Chem. C* **2018**, *6*, 1863–1869. [CrossRef]
104. Wei, X.; Li, J.; Hu, Z.; Wang, C.; Gao, Z.; Cao, Y.; Han, J.; Li, Y. Carbon Quantum Dot/Chitosan-Derived Hydrogels with Photo-stress-pH Multiresponsiveness for Wearable Sensors. *Macromol. Rapid Commun.* **2023**, *44*, 2200928. [CrossRef] [PubMed]
105. Tao, X.; Liao, M.; Wu, F.; Jiang, Y.; Sun, J.; Shi, S. Designing of biomass-derived carbon quantum dots@polyvinyl alcohol film with excellent fluorescent performance and pH-responsiveness for intelligent detection. *Chem. Eng. J.* **2022**, *443*, 136442. [CrossRef]

106. Zhang, S.; Liu, Y.; Wang, J.; Liu, Z. A Laser-Induced Photoelectrochemical Sensor for Natural Sweat Cu^{2+} Detection. *Chemosensors* **2022**, *10*, 169. [CrossRef]
107. Han, D.; Li, X.; Liang, Z.; Zhao, B.; Wu, Z.; Han, F.; Han, D.; Niu, L. Label-free photoelectric sensor for lactic acid determination in human sweat. *Chin. Chem. Lett.* **2023**, *34*, 107722. [CrossRef]
108. Nan, M.; Darmawan, B.A.; Go, G.; Zheng, S.; Lee, J.; Kim, S.; Lee, T.; Choi, E.; Park, J.O.; Bang, D. Wearable Localized Surface Plasmon Resonance-Based Biosensor with Highly Sensitive and Direct Detection of Cortisol in Human Sweat. *Biosensors* **2023**, *13*, 184. [CrossRef]
109. Mei, X.; Yang, J.; Liu, J.; Li, Y. Wearable, nanofiber-based microfluidic systems with integrated electrochemical and colorimetric sensing arrays for multiplex sweat analysis. *Chem. Eng. J.* **2023**, *454*, 140248. [CrossRef]
110. Bandodkar, A.J.; Molinnus, D.; Mirza, O.; Guinovart, T.; Windmiller, J.R.; Valdes-Ramirez, G.; Andrade, F.J.; Schoning, M.J.; Wang, J. Epidermal tattoo potentiometric sodium sensors with wireless signal transduction for continuous non-invasive sweat monitoring. *Biosens. Bioelectron.* **2014**, *54*, 603–609. [CrossRef]
111. Qiao, Y.; Qiao, L.; Chen, Z.; Liu, B.; Gao, L.; Zhang, L. Wearable Sensor for Continuous Sweat Biomarker Monitoring. *Chemosensors* **2022**, *10*, 273. [CrossRef]
112. Muñoz, J.; Montes, R.; Baeza, M. Trends in electrochemical impedance spectroscopy involving nanocomposite transducers: Characterization, architecture surface and bio-sensing. *TrAC Trends Anal. Chem.* **2017**, *97*, 201–215. [CrossRef]
113. Lisdat, F.; Schafer, D. The use of electrochemical impedance spectroscopy for biosensing. *Anal. Bioanal. Chem.* **2008**, *391*, 1555–1567. [CrossRef] [PubMed]
114. Huang, J.; Zhang, Y.; Wu, J. Review of non-invasive continuous glucose monitoring based on impedance spectroscopy. *Sens. Actuators A* **2020**, *311*, 112103. [CrossRef]
115. Parlak, O.; Keene, S.T.; Marais, A.; Curto, V.F.; Salleo, A. Molecularly selective nanoporous membrane-based wearable organic electrochemical device for noninvasive cortisol sensing. *Sci. Adv.* **2018**, *4*, eaar2904. [CrossRef] [PubMed]
116. Tang, W.; Yin, L.; Sempionatto, J.R.; Moon, J.M.; Teymourian, H.; Wang, J. Touch-Based Stressless Cortisol Sensing. *Adv. Mater.* **2021**, *33*, e2008465. [CrossRef]
117. Tang, Y.; Gan, S.; Zhong, L.; Sun, Z.; Xu, L.; Liao, C.; Lin, K.; Cui, X.; He, D.; Ma, Y.; et al. Lattice Proton Intercalation to Regulate WO_3-Based Solid-Contact Wearable pH Sensor for Sweat Analysis. *Adv. Funct. Mater.* **2021**, *32*, 2107653. [CrossRef]
118. Shi, H.H.; Cao, Y.; Zeng, Y.N.; Zhou, Y.N.; Wen, W.H.; Zhang, C.X.; Zhao, Y.L.; Chen, Z. Wearable tesla valve-based sweat collection device for sweat colorimetric analysis. *Talanta* **2022**, *240*, 10. [CrossRef]
119. Zhai, Q.; Yap, L.W.; Wang, R.; Gong, S.; Guo, Z.; Liu, Y.; Lyu, Q.; Wang, J.; Simon, G.P.; Cheng, W. Vertically Aligned Gold Nanowires as Stretchable and Wearable Epidermal Ion-Selective Electrode for Noninvasive Multiplexed Sweat Analysis. *Anal. Chem.* **2020**, *92*, 4647–4655. [CrossRef]
120. Lin, Y.-C.; Rinawati, M.; Chang, L.-Y.; Wang, Y.-X.; Wu, Y.-T.; Yen, Y.-H.; Chen, K.-J.; Ho, K.-C.; Yeh, M.-H. A non-invasive wearable sweat biosensor with a flexible N-GQDs/PANI nanocomposite layer for glucose monitoring. *Sens. Actuators B* **2023**, *383*, 133617. [CrossRef]
121. Sun, T.; Hui, J.; Zhou, L.; Lin, B.; Sun, H.; Bai, Y.; Zhao, J.; Mao, H. A low-cost and simple-fabricated epidermal sweat patch based on "cut-and-paste" manufacture. *Sens. Actuators B* **2022**, *368*, 132184. [CrossRef]
122. Tao, T.; Zhou, Y.; Ma, M.; He, H.; Gao, N.; Cai, Z.; Chang, G.; He, Y. Novel graphene electrochemical transistor with ZrO_2/rGO nanocomposites functionalized gate electrode for ultrasensitive recognition of methyl parathion. *Sens. Actuators B* **2021**, *328*, 128936. [CrossRef]
123. Tao, T.; Gao, N.; He, H.; Zhou, R.; Tu, B.; Cai, Z.; Chang, G.; He, Y.; Ji, X. Au-PEDOT/rGO nanocomposites functionalized graphene electrochemical transistor for ultra-sensitive detection of acetaminophen in human urine. *Anal. Chim. Acta* **2022**, *1191*, 339306. [CrossRef] [PubMed]
124. Thanh, T.D.; Balamurugan, J.; Lee, S.H.; Kim, N.H.; Lee, J.H. Effective seed-assisted synthesis of gold nanoparticles anchored nitrogen-doped graphene for electrochemical detection of glucose and dopamine. *Biosens. Bioelectron.* **2016**, *81*, 259–267. [CrossRef] [PubMed]
125. Gao, N.; He, C.; Ma, M.; Cai, Z.; Zhou, Y.; Chang, G.; Wang, X.; He, Y. Electrochemical co-deposition synthesis of Au-ZrO_2-graphene nanocomposite for a nonenzymatic methyl parathion sensor. *Anal. Chim. Acta* **2019**, *1072*, 25–34. [CrossRef] [PubMed]
126. Yu, Z.; Wu, H.; Xu, Z.; Yang, Z.; Lv, J.; Kong, C. Wearable Noninvasive Glucose Sensor Based on Cu_xO NFs/Cu NPs Nanocomposites. *Sensors* **2023**, *23*, 695. [CrossRef]
127. Gao, N.; Zhou, R.; Tu, B.; Tao, T.; Song, Y.; Cai, Z.; He, H.; Chang, G.; Wu, Y.; He, Y. Graphene electrochemical transistor incorporated with gel electrolyte for wearable and non-invasive glucose monitoring. *Anal. Chim. Acta* **2023**, *1239*, 340719. [CrossRef]
128. Lee, H.; Choi, T.K.; Lee, Y.B.; Cho, H.R.; Ghaffari, R.; Wang, L.; Choi, H.J.; Chung, T.D.; Lu, N.; Hyeon, T.; et al. A graphene-based electrochemical device with thermoresponsive microneedles for diabetes monitoring and therapy. *Nat. Nanotechnol.* **2016**, *11*, 566–572. [CrossRef]
129. Chen, Y.; Li, G.; Mu, W.; Wan, X.; Lu, D.; Gao, J.; Wen, D. Nonenzymatic Sweat Wearable Uric Acid Sensor Based on N-Doped Reduced Graphene Oxide/Au Dual Aerogels. *Anal. Chem.* **2023**, *95*, 3864–3872. [CrossRef]
130. Wang, Y.; Chen, F.; Ye, J.; Liu, H.; Zhang, T.; Li, Z. Reduced graphene oxide cotton fabric based on copper nanowires for flexible non-enzyme glucose sensor. *Cellulose* **2023**, *30*, 5131–5143. [CrossRef]

131. Mukasa, D.; Wang, M.; Min, J.; Yang, Y.; Solomon, S.A.; Han, H.; Ye, C.; Gao, W. A Computationally Assisted Approach for Designing Wearable Biosensors toward Non-Invasive Personalized Molecular Analysis. *Adv. Mater.* **2023**, 2212161. [CrossRef]
132. Tu, J.; Min, J.; Song, Y.; Xu, C.; Li, J.; Moore, J.; Hanson, J.; Hu, E.; Parimon, T.; Wang, T.Y.; et al. A wireless patch for the monitoring of C-reactive protein in sweat. *Nat. Biomed. Eng.* **2023**, *14*, 15. [CrossRef]
133. Huang, L.; Su, J.; Song, Y.; Ye, R. Laser-Induced Graphene: En Route to Smart Sensing. *Nanomicro Lett* **2020**, *12*, 157. [CrossRef] [PubMed]
134. Feng, J.; Jiang, Y.; Wang, K.; Li, J.; Zhang, J.; Tian, M.; Chen, G.; Hu, L.; Zhan, Y.; Qin, Y. An Energy-Efficient Flexible Multi-Modal Wireless Sweat Sensing System Based on Laser Induced Graphene. *Sensors* **2023**, *23*, 4818. [CrossRef] [PubMed]
135. Wang, Y.; Guo, H.; Yuan, M.; Yu, J.; Wang, Z.; Chen, X. One-step laser synthesis platinum nanostructured 3D porous graphene: A flexible dual-functional electrochemical biosensor for glucose and pH detection in human perspiration. *Talanta* **2023**, *257*, 124362. [CrossRef] [PubMed]
136. Naguib, M.; Kurtoglu, M.; Presser, V.; Lu, J.; Niu, J.; Heon, M.; Hultman, L.; Gogotsi, Y.; Barsoum, M.W. Two-dimensional nanocrystals produced by exfoliation of Ti_3AlC_2. *Adv. Mater.* **2011**, *23*, 4248–4253. [CrossRef]
137. VahidMohammadi, A.; Rosen, J.; Gogotsi, Y. The world of two-dimensional carbides and nitrides (MXenes). *Science* **2021**, *372*, eabf1581. [CrossRef]
138. Zhao, Q.-N.; Zhang, Y.-J.; Duan, Z.-H.; Wang, S.; Liu, C.; Jiang, Y.-D.; Tai, H.-L. A review on $Ti_3C_2T_x$-based nanomaterials: Synthesis and applications in gas and humidity sensors. *Rare Met.* **2020**, *40*, 1459–1476. [CrossRef]
139. Zhang, Y.; Wang, L.; Zhang, N.; Zhou, Z. Adsorptive environmental applications of MXene nanomaterials: A review. *RSC Adv.* **2018**, *8*, 19895–19905. [CrossRef]
140. Sharifuzzaman, M.; Zahed, M.A.; Reza, M.S.; Asaduzzaman, M.; Jeong, S.; Song, H.; Kim, D.K.; Zhang, S.; Park, J.Y. MXene/Fluoropolymer-Derived Laser-Carbonaceous All-Fibrous Nanohybrid Patch for Soft Wearable Bioelectronics. *Adv. Funct. Mater.* **2023**, *33*, 2208894. [CrossRef]
141. Li, Q.-F.; Chen, X.; Wang, H.; Liu, M.; Peng, H.-L. Pt/MXene-Based Flexible Wearable Non-Enzymatic Electrochemical Sensor for Continuous Glucose Detection in Sweat. *ACS Appl. Mater. Interfaces* **2023**, *15*, 13290–13298. [CrossRef]
142. Wang, L.; Xu, T.; He, X.; Zhang, X. Flexible, self-healable, adhesive and wearable hydrogel patch for colorimetric sweat detection. *J. Mater. Chem. C* **2021**, *9*, 14938–14945. [CrossRef]
143. Yang, C.; Suo, Z. Hydrogel ionotronics. *Nat. Rev. Mater.* **2018**, *3*, 125–142. [CrossRef]
144. Tringides, C.M.; Vachicouras, N.; de Lázaro, I.; Wang, H.; Trouillet, A.; Seo, B.R.; Elosegui-Artola, A.; Fallegger, F.; Shin, Y.; Casiraghi, C.; et al. Viscoelastic surface electrode arrays to interface with viscoelastic tissues. *Nat. Nanotechnol.* **2021**, *16*, 1019–1029. [CrossRef] [PubMed]
145. Daly, A.C.; Riley, L.; Segura, T.; Burdick, J.A. Hydrogel microparticles for biomedical applications. *Nat. Rev. Mater.* **2020**, *5*, 20–43. [CrossRef] [PubMed]
146. Xu, Z.; Zhou, F.; Yan, H.; Gao, G.; Li, H.; Li, R.; Chen, T. Anti-freezing organohydrogel triboelectric nanogenerator toward highly efficient *Flex.* human-machine interaction at −30 °C. *Nano Energy* **2021**, *90*, 106614. [CrossRef]
147. Duan, X.; Yu, J.; Zhu, Y.; Zheng, Z.; Liao, Q.; Xiao, Y.; Li, Y.; He, Z.; Zhao, Y.; Wang, H.; et al. Large-Scale Spinning Approach to Engineering Knittable Hydrogel Fiber for Soft Robots. *ACS Nano* **2020**, *14*, 14929–14938. [CrossRef]
148. Wang, L.; Zhou, M.; Xu, T.; Zhang, X. Multifunctional hydrogel as wound dressing for intelligent wound monitoring. *Chem. Eng. J.* **2022**, *433*, 134625. [CrossRef]
149. Liang, Y.; He, J.; Guo, B. Functional Hydrogels as Wound Dressing to Enhance Wound Healing. *ACS Nano* **2021**, *15*, 12687–12722. [CrossRef]
150. Wang, L.; Wang, J.; Fan, C.; Xu, T.; Zhang, X. Skin-like hydrogel-elastomer based electrochemical device for comfortable wearable biofluid monitoring. *Chem. Eng. J.* **2023**, *455*, 140609. [CrossRef]
151. Gao, N.; Cai, Z.; Chang, G.; He, Y. Non-invasive and wearable glucose biosensor based on gel electrolyte for detection of human sweat. *J. Mater. Sci.* **2023**, *58*, 890–901. [CrossRef]
152. Sarkar, R.; Farghaly, A.A.; Arachchige, I.U. Oxidative Self-Assembly of Au/Ag/Pt Alloy Nanoparticles into High-Surface Area, Mesoporous, and Conductive Aerogels for Methanol Electro-oxidation. *Chem. Mater.* **2022**, *34*, 5874–5887. [CrossRef]
153. Wang, C.; Gao, W.; Wan, X.; Yao, B.; Mu, W.; Gao, J.; Fu, Q.; Wen, D. In situ electrochemical synthesis of Pd aerogels as highly efficient anodic electrocatalysts for alkaline fuel cells. *Chem. Sci.* **2022**, *13*, 13956–13965. [CrossRef] [PubMed]
154. Du, R.; Joswig, J.-O.; Fan, X.; Hübner, R.; Spittel, D.; Hu, Y.; Eychmüller, A. Disturbance-Promoted Unconventional and Rapid Fabrication of Self-Healable Noble Metal Gels for (Photo-)Electrocatalysis. *Matter* **2020**, *2*, 908–920. [CrossRef] [PubMed]
155. Gao, W.; Wen, D. Recent advances of noble metal aerogels in biosensing. *View* **2021**, *2*, 20200124. [CrossRef]
156. Li, G.; Wang, C.; Chen, Y.; Liu, F.; Fan, H.; Yao, B.; Hao, J.; Yu, Y.; Wen, D. Dual Structural Design of Platinum-Nickel Hydrogels for Wearable Glucose Biosensing with Ultrahigh Stability. *Small* **2023**, *19*, e2206868. [CrossRef] [PubMed]
157. Keene, S.T.; Fogarty, D.; Cooke, R.; Casadevall, C.D.; Salleo, A.; Parlak, O. Wearable Organic Electrochemical Transistor Patch for Multiplexed Sensing of Calcium and Ammonium Ions from Human Perspiration. *Adv Healthcare Mater.* **2019**, *8*, e1901321. [CrossRef]
158. Aerathupalathu Janardhanan, J.; Chen, Y.L.; Liu, C.T.; Tseng, H.S.; Wu, P.I.; She, J.W.; Hsiao, Y.S.; Yu, H.H. Sensitive Detection of Sweat Cortisol Using an Organic Electrochemical Transistor Featuring Nanostructured Poly(3,4-Ethylenedioxythiophene) Derivatives in the Channel Layer. *Anal. Chem.* **2022**, *94*, 7584–7593. [CrossRef]

159. Khodagholy, D.; Rivnay, J.; Sessolo, M.; Gurfinkel, M.; Leleux, P.; Jimison, L.H.; Stavrinidou, E.; Herve, T.; Sanaur, S.; Owens, R.M.; et al. High transconductance organic electrochemical transistors. *Nat. Commun.* **2013**, *4*, 2133. [CrossRef]
160. Wang, N.; Liu, Y.; Fu, Y.; Yan, F. AC Measurements Using Organic Electrochemical Transistors for Accurate Sensing. *ACS Appl. Mater. Interfaces* **2018**, *10*, 25834–25840. [CrossRef]
161. Li, Y.; Cui, B.; Zhang, S.; Li, B.; Li, J.; Liu, S.; Zhao, Q. Ion-Selective Organic Electrochemical Transistors: Recent Progress and Challenges. *Small* **2022**, *18*, e2107413. [CrossRef]
162. Sessolo, M.; Rivnay, J.; Bandiello, E.; Malliaras, G.G.; Bolink, H.J. Ion-selective organic electrochemical transistors. *Adv. Mater.* **2014**, *26*, 4803–4807. [CrossRef]
163. Coppedè, N.; Giannetto, M.; Villani, M.; Lucchini, V.; Battista, E.; Careri, M.; Zappettini, A. Ion selective textile organic electrochemical transistor for wearable sweat monitoring. *Org. Electron.* **2020**, *78*, 105579. [CrossRef]
164. Tao, Y.; Zhu, R.; Hao, P.; Jiang, W.; Li, M.; Liu, Q.; Yang, L.; Wang, Y.; Wang, D. Textile-based dual-mode organic electrochemical transistors for lactate biosensing. *Mater. Sci. Eng. B* **2023**, *290*, 116356. [CrossRef]
165. Fang, Y.; Feng, J.; Shi, X.; Yang, Y.; Wang, J.; Sun, X.; Li, W.; Sun, X.; Peng, H. Coaxial fiber organic electrochemical transistor with high transconductance. *Nano Res.* **2023**. [CrossRef]
166. Qing, X.; Chen, H.; Zeng, F.; Jia, K.; Shu, Q.; Wu, J.; Xu, H.; Lei, W.; Liu, D.; Wang, X.; et al. All-Fiber Integrated Thermoelectrically Powered Physiological Monitoring Biosensor. *Adv. Fiber Mater.* **2023**, *5*, 1025–1036. [CrossRef]
167. Qu, Y.; Nguyen-Dang, T.; Page, A.G.; Yan, W.; Das Gupta, T.; Rotaru, G.M.; Rossi, R.M.; Favrod, V.D.; Bartolomei, N.; Sorin, F. Superelastic Multimaterial Electronic and Photonic Fibers and Devices via Thermal Drawing. *Adv. Mater.* **2018**, *30*, 1707251. [CrossRef]
168. Zhang, T.; Wang, Z.; Srinivasan, B.; Wang, Z.; Zhang, J.; Li, K.; Boussard-Pledel, C.; Troles, J.; Bureau, B.; Wei, L. Ultraflexible Glassy Semiconductor Fibers for Thermal Sensing and Positioning. *ACS Appl. Mater. Interfaces* **2019**, *11*, 2441–2447. [CrossRef]
169. Wu, J.; Sato, Y.; Guo, Y. Microelectronic fibers for multiplexed sweat sensing. *Anal. Bioanal. Chem.* **2023**, *415*, 4307–4318. [CrossRef]
170. Garcia-Rey, S.; Ojeda, E.; Gunatilake, U.B.; Basabe-Desmonts, L.; Benito-Lopez, F. Alginate Bead Biosystem for the Determination of Lactate in Sweat Using Image Analysis. *Biosensors* **2021**, *11*, 379. [CrossRef]
171. Gunatilake, U.B.; Garcia-Rey, S.; Ojeda, E.; Basabe-Desmonts, L.; Benito-Lopez, F. TiO_2 Nanotubes Alginate Hydrogel Scaffold for Rapid Sensing of Sweat Biomarkers: Lactate and Glucose. *ACS Appl. Mater. Interfaces* **2021**, *13*, 37734–37745. [CrossRef]
172. Windmiller, J.R.; Bandodkar, A.J.; Valdes-Ramirez, G.; Parkhomovsky, S.; Martinez, A.G.; Wang, J. Electrochemical sensing based on printable temporary transfer tattoos. *Chem. Commun.* **2012**, *48*, 6794–6796. [CrossRef]
173. Guinovart, T.; Bandodkar, A.J.; Windmiller, J.R.; Andrade, F.J.; Wang, J. A potentiometric tattoo sensor for monitoring ammonium in sweat. *Analyst* **2013**, *138*, 7031–7038. [CrossRef] [PubMed]
174. Anastasova, S.; Crewther, B.; Bembnowicz, P.; Curto, V.; Ip, H.M.; Rosa, B.; Yang, G.Z. A wearable multisensing patch for continuous sweat monitoring. *Biosens. Bioelectron.* **2017**, *93*, 139–145. [CrossRef] [PubMed]
175. Nyein, H.Y.Y.; Tai, L.C.; Ngo, Q.P.; Chao, M.; Zhang, G.B.; Gao, W.; Bariya, M.; Bullock, J.; Kim, H.; Fahad, H.M.; et al. A Wearable Microfluidic Sensing Patch for Dynamic Sweat Secretion Analysis. *ACS Sens.* **2018**, *3*, 944–952. [CrossRef]
176. Nyein, H.Y.; Gao, W.; Shahpar, Z.; Emaminejad, S.; Challa, S.; Chen, K.; Fahad, H.M.; Tai, L.C.; Ota, H.; Davis, R.W.; et al. A Wearable Electrochemical Platform for Noninvasive Simultaneous Monitoring of $Ca^{(2+)}$ and pH. *ACS Nano* **2016**, *10*, 7216–7224. [CrossRef] [PubMed]
177. Sempionatto, J.R.; Lin, M.; Yin, L.; De la Paz, E.; Pei, K.; Sonsa-Ard, T.; de Loyola Silva, A.N.; Khorshed, A.A.; Zhang, F.; Tostado, N.; et al. An epidermal patch for the simultaneous monitoring of haemodynamic and metabolic biomarkers. *Nat. Biomed. Eng.* **2021**, *5*, 737–748. [CrossRef]
178. Gao, W.; Nyein, H.Y.Y.; Shahpar, Z.; Fahad, H.M.; Chen, K.; Emaminejad, S.; Gao, Y.; Tai, L.-C.; Ota, H.; Wu, E.; et al. Wearable Microsensor Array for Multiplexed Heavy Metal Monitoring of Body Fluids. *ACS Sens.* **2016**, *1*, 866–874. [CrossRef]
179. Emaminejad, S.; Gao, W.; Wu, E.; Davies, Z.A.; Yin Yin Nyein, H.; Challa, S.; Ryan, S.P.; Fahad, H.M.; Chen, K.; Shahpar, Z.; et al. Autonomous sweat extraction and analysis applied to cystic fibrosis and glucose monitoring using a fully integrated wearable platform. *Proc. Natl. Acad. Sci. USA* **2017**, *114*, 4625–4630. [CrossRef]
180. Kim, J.; Gutruf, P.; Chiarelli, A.M.; Heo, S.Y.; Cho, K.; Xie, Z.; Banks, A.; Han, S.; Jang, K.-I.; Lee, J.W.; et al. Miniaturized Battery-Free Wireless Systems for Wearable Pulse Oximetry. *Adv. Funct. Mater.* **2017**, *27*, 1604373. [CrossRef]
181. Chen, H.; Bao, S.; Lu, C.; Wang, L.; Ma, J.; Wang, P.; Lu, H.; Shu, F.; Oetomo, S.B.; Chen, W. Design of an Integrated Wearable Multi-Sensor Platform Based on Flexible Materials for Neonatal Monitoring. *IEEE Access* **2020**, *8*, 23732–23747. [CrossRef]
182. Wang, J.; Wang, L.; Li, G.; Yan, D.; Liu, C.; Xu, T.; Zhang, X. Ultra-Small Wearable Flexible Biosensor for Continuous Sweat Analysis. *ACS Sens.* **2022**, *7*, 3102–3107. [CrossRef]
183. He, X.; Xu, T.; Gu, Z.; Gao, W.; Xu, L.-P.; Pan, T.; Zhang, X. Flexible and Superwettable Bands as a Platform toward Sweat Sampling and Sensing. *Anal. Chem.* **2019**, *91*, 4296–4300. [CrossRef] [PubMed]
184. Steijlen, A.S.M.; Jansen, K.M.B.; Bastemeijer, J.; French, P.J.; Bossche, A. Low-Cost Wearable Fluidic Sweat Collection Patch for Continuous Analyte Monitoring and Offline Analysis. *Anal. Chem.* **2022**, *94*, 6893–6901. [CrossRef]
185. Martin, A.; Kim, J.; Kurniawan, J.F.; Sempionatto, J.R.; Moreto, J.R.; Tang, G.; Campbell, A.S.; Shin, A.; Lee, M.Y.; Liu, X.; et al. Epidermal Microfluidic Electrochemical Detection System: Enhanced Sweat Sampling and Metabolite Detection. *ACS Sens.* **2017**, *2*, 1860–1868. [CrossRef] [PubMed]

186. Nyein, H.Y.Y.; Bariya, M.; Kivimäki, L.; Uusitalo, S.; Liaw, T.S.; Jansson, E.; Ahn, C.H.; Hangasky, J.A.; Zhao, J.; Lin, Y.; et al. Regional and correlative sweat analysis using high-throughput microfluidic sensing patches toward decoding sweat. *Sci. Adv.* **2019**, *5*, eaaw9906. [CrossRef]
187. Huang, X.; Li, J.; Liu, Y.; Wong, T.; Su, J.; Yao, K.; Zhou, J.; Huang, Y.; Li, H.; Li, D.; et al. Epidermal self-powered sweat sensors for glucose and lactate monitoring. *Bio-Des. Manuf.* **2021**, *5*, 201–209. [CrossRef]
188. Paul Kunnel, B.; Demuru, S. An epidermal wearable microfluidic patch for simultaneous sampling, storage, and analysis of biofluids with counterion monitoring. *Lab Chip* **2022**, *22*, 1793–1804. [CrossRef] [PubMed]
189. Mei, X.; Yang, J.; Yu, X.; Peng, Z.; Zhang, G.; Li, Y. Wearable molecularly imprinted electrochemical sensor with integrated nanofiber-based microfluidic chip for in situ monitoring of cortisol in sweat. *Sens. Actuators B* **2023**, *381*, 133451. [CrossRef]
190. Saha, T.; Songkakul, T.; Knisely, C.T.; Yokus, M.A.; Daniele, M.A.; Dickey, M.D.; Bozkurt, A.; Velev, O.D. Wireless Wearable Electrochemical Sensing Platform with Zero-Power Osmotic Sweat Extraction for Continuous Lactate Monitoring. *ACS Sens.* **2022**, *7*, 2037–2048. [CrossRef]
191. Yang, M.; Sun, N.; Lai, X.; Wu, J.; Wu, L.; Zhao, X.; Feng, L. Paper-Based Sandwich-Structured Wearable Sensor with Sebum Filtering for Continuous Detection of Sweat pH. *ACS Sens.* **2023**, *8*, 176–186. [CrossRef]
192. Liang, B.; Cao, Q.; Mao, X.; Pan, W.; Tu, T.; Fang, L.; Ye, X. An Integrated Paper-Based Microfluidic Device for Real-Time Sweat Potassium Monitoring. *IEEE Sens. J.* **2021**, *21*, 9642–9648. [CrossRef]
193. Revolutionary Sweat Science: The First Biosensor to Analyze Sweat and Provide Endurance Athletes with Personalized Hydration Data-Scientifically Validated and Delivered in Real Time. Available online: https://nixbiosensors.com/pages/product#patch (accessed on 25 July 2023).
194. AbsolutSweat P1. Available online: http://www.refresh.cc/product/sweat (accessed on 25 July 2023).
195. Baker, L.B.; Seib, M.S.; Barnes, K.A.; Brown, S.D.; King, M.A.; De Chavez, P.J.D.; Qu, S.K.; Archer, J.; Wolfe, A.S.; Stofan, J.R.; et al. Skin-Interfaced Microfluidic System with Machine Learning-Enabled Image Processing of Sweat Biomarkers in Remote Settings. *Adv. Mater. Technol.* **2022**, *7*, 13. [CrossRef]
196. Bolat, G.; De la Paz, E.; Azeredo, N.F.; Kartolo, M.; Kim, J.; de Loyola, E.S.A.N.; Rueda, R.; Brown, C.; Angnes, L.; Wang, J.; et al. Wearable soft electrochemical microfluidic device integrated with iontophoresis for sweat biosensing. *Anal. Bioanal. Chem.* **2022**, *414*, 5411–5421. [CrossRef] [PubMed]
197. Kim, J.; Jeerapan, I.; Imani, S.; Cho, T.N.; Bandodkar, A.; Cinti, S.; Mercier, P.P.; Wang, J. Noninvasive Alcohol Monitoring Using a Wearable Tattoo-Based Iontophoretic-Biosensing System. *ACS Sens.* **2016**, *1*, 1011–1019. [CrossRef]
198. Cheng, C.; Li, X.; Xu, G.; Lu, Y.; Low, S.S.; Liu, G.; Zhu, L.; Li, C.; Liu, Q. Battery-free, wireless, and flexible electrochemical patch for in situ analysis of sweat cortisol via near field communication. *Biosens. Bioelectron.* **2021**, *172*, 112782. [CrossRef] [PubMed]
199. Liu, Y.; Huang, X.; Zhou, J.; Yiu, C.K.; Song, Z.; Huang, W.; Nejad, S.K.; Li, H.; Wong, T.H.; Yao, K.; et al. Stretchable Sweat-Activated Battery in Skin-Integrated Electronics for Continuous Wireless Sweat Monitoring. *Adv. Sci.* **2022**, *9*, e2104635. [CrossRef]
200. Yu, Y.; Nassar, J.; Xu, C.; Min, J.; Yang, Y.; Dai, A.; Doshi, R.; Huang, A.; Song, Y.; Gehlhar, R.; et al. Biofuel-powered soft electronic skin with multiplexed and wireless sensing for human-machine interfaces. *Sci. Robot.* **2020**, *5*, eaaz7946. [CrossRef]
201. Chen, J.; Huang, Y.; Zhang, N.; Zou, H.; Liu, R.; Tao, C.; Fan, X.; Wang, Z.L. Micro-cable structured textile for simultaneously harvesting solar and mechanical energy. *Nat. Energy* **2016**, *1*, 16138. [CrossRef]
202. Suarez, F.; Nozariasbmarz, A.; Vashaee, D.; Öztürk, M.C. Designing thermoelectric generators for self-powered wearable electronics. *Energy Environ. Sci.* **2016**, *9*, 2099–2113. [CrossRef]
203. Liu, S.; Kang, L.; Zhang, J.; Jun, S.C.; Yamauchi, Y. Carbonaceous Anode Materials for Non-aqueous Sodium- and Potassium-Ion Hybrid Capacitors. *ACS Energy Lett.* **2021**, *6*, 4127–4154. [CrossRef]
204. Anabtawi, N.; Freeman, S.; Ferzli, R. A fully implantable, NFC enabled, continuous interstitial glucose monitor. In Proceedings of the 2016 IEEE-EMBS International Conference on Biomedical and Health Informatics (BHI), Las Vegas, NV, USA, 24–27 February 2016; pp. 612–615.
205. Escalona-Villalpando, R.A.; Ortiz-Ortega, E.; Bocanegra-Ugalde, J.P.; Minteer, S.D.; Ledesma-Garcia, J.; Arriaga, L.G. Clean energy from human sweat using an enzymatic patch. *J. Power Sources* **2019**, *412*, 496–504. [CrossRef]
206. Bandodkar, A.J.; You, J.M.; Kim, N.H.; Gu, Y.; Kumar, R.; Mohan, A.M.V.; Kurniawan, J.; Imani, S.; Nakagawa, T.; Parish, B.; et al. Soft, stretchable, high power density electronic skin-based biofuel cells for scavenging energy from human sweat. *Energy Environ. Sci.* **2017**, *10*, 1581–1589. [CrossRef]
207. Chen, Y.; Xue, Y.; Liu, W.; Li, S.; Wang, X.; Zhou, W.; Zhang, G.; Liu, K.; Zhang, H.; Zhao, Y.; et al. Untethered artificial muscles powered by wearable sweat-based energy generator. *Nano Today* **2023**, *49*, 101765. [CrossRef]
208. Manjakkal, L.; Pullanchiyodan, A.; Yogeswaran, N.; Hosseini, E.S.; Dahiya, R. A Wearable Supercapacitor Based on Conductive PEDOT:PSS-Coated Cloth and a Sweat Electrolyte. *Adv. Mater.* **2020**, *32*, e1907254. [CrossRef] [PubMed]
209. Qin, Y.; Mo, J.; Liu, Y.; Zhang, S.; Wang, J.; Fu, Q.; Wang, S.; Nie, S. Stretchable Triboelectric Self-Powered Sweat Sensor Fabricated from Self-Healing Nanocellulose Hydrogels. *Adv. Funct. Mater.* **2022**, *32*, 2201846. [CrossRef]
210. Song, Y.; Min, J.H.; Yu, Y.; Wang, H.B.; Yang, Y.R.; Zhang, H.X.; Gao, W. Wireless battery-free wearable sweat sensor powered by human motion. *Sci. Adv.* **2020**, *6*, 10. [CrossRef]
211. Gai, Y.; Wang, E.; Liu, M.; Xie, L.; Bai, Y.; Yang, Y.; Xue, J.; Qu, X.; Xi, Y.; Li, L.; et al. A Self-Powered Wearable Sensor for Continuous Wireless Sweat Monitoring. *Small Methods* **2022**, *6*, 2200653. [CrossRef]

212. Lin, Z.; Chen, J.; Yang, J. Recent Progress in Triboelectric Nanogenerators as a Renewable and Sustainable Power Source. *J. Nanomater.* **2016**, *2016*, 5651613. [CrossRef]
213. Hu, Y.; Wang, Z.L. Recent progress in piezoelectric nanogenerators as a sustainable power source in self-powered systems and active sensors. *Nano Energy* **2015**, *14*, 3–14. [CrossRef]
214. Briscoe, J.; Dunn, S. Piezoelectric nanogenerators—A review of nanostructured piezoelectric energy harvesters. *Nano Energy* **2015**, *14*, 15–29. [CrossRef]
215. Hinchet, R.; Seung, W.; Kim, S.-W. Recent Progress on Flexible Triboelectric Nanogenerators for SelfPowered Electronics. *ChemSusChem* **2015**, *8*, 2327–2344. [CrossRef]
216. Zahed, M.A.; Rana, S.M.S.; Sharifuzzaman, M.; Jeong, S.; Pradhan, G.B.; Song, H.S.; Park, J.Y. A Hybrid Nanogenerator-Driven Self-Powered Wearable Perspiration Monitoring System. In Proceedings of the 2023 IEEE 36th International Conference on Micro Electro Mechanical Systems (MEMS), Munich, Germany, 15–19 January 2023; pp. 732–735.
217. Baro, B.; Khimhun, S.; Das, U.; Bayan, S. ZnO based triboelectric nanogenerator on textile platform for wearable sweat sensing application. *Nano Energy* **2023**, *108*, 108212. [CrossRef]
218. Lu, Y.; Wu, T.; Ma, Z.; Mi, Y.; Zhao, Z.; Liu, F.; Cao, X.; Wang, N. Integration of Flexible Supercapacitors with Triboelectric Nanogenerators: A Review. *Batteries* **2023**, *9*, 281. [CrossRef]
219. Pu, X.; Hu, W.; Wang, Z.L. Toward Wearable Self-Charging Power Systems: The Integration of Energy-Harvesting and Storage Devices. *Small* **2018**, *14*, 1702817. [CrossRef] [PubMed]
220. Zhao, J.; Cong, Z.; Hu, J.; Lu, H.; Wang, L.; Wang, H.; Malyi, O.I.; Pu, X.; Zhang, Y.; Shao, H.; et al. Regulating zinc electroplating chemistry to achieve high energy coaxial fiber Zn ion supercapacitor for self-powered textile-based monitoring system. *Nano Energy* **2022**, *93*, 106893. [CrossRef]
221. Wang, L.; Wu, H.; Zhai, X.; Shi, J.; Zhou, Q.; Li, H.; Wan, J. $Ti_3C_2T_x$ MXene/dopamine-modified polypyrrole flexible composite electrodes with application in energy storage devices. *J. Alloys Compd.* **2023**, *946*, 169347. [CrossRef]
222. Huang, Y.; Wang, L.; Li, X.; Yang, X.; Lü, W. Washable All-in-One Self-Charging Power Unit Based on a Triboelectric Nanogenerator and Supercapacitor for Smart Textiles. *Langmuir* **2023**, *39*, 8855–8864. [CrossRef] [PubMed]
223. Zhu, Z.; Liang, X.; Luo, H.; Wang, L.; Gao, Y.; Li, X.; Yang, X.; Lü, W. Flexible self-powered energy systems based on H_2O/Ni^{2+} intercalated $Ni_xV_2O_5{\cdot}nH_2O$. *Chem. Eur. J.* **2023**, e202301583. [CrossRef] [PubMed]
224. Wang, Z.; Yao, S.; Wang, S.; Liu, Z.; Wan, X.; Hu, Q.; Zhao, Y.; Xiong, C.; Li, L. Self-powered energy harvesting and implantable storage system based on hydrogel-enabled all-solid-state supercapacitor and triboelectric nanogenerator. *Chem. Eng. J.* **2023**, *463*, 142427. [CrossRef]
225. Mi, Y.; Lu, Y.; Wang, X.; Zhao, Z.; Cao, X.; Wang, N. From Triboelectric Nanogenerator to Uninterrupted Power Supply System: The Key Role of Electrochemical Batteries and Supercapacitors. *Batteries* **2022**, *8*, 215. [CrossRef]
226. Singh, J.; Ning, B.; Lee, P.; Liu, L. A Solar-Driven Wearable Multiplexed Bio-Sensing System For Noninvasive Healthcare Monitoring In Sweat. In Proceedings of the 2023 IEEE 36th International Conference on Micro Electro Mechanical Systems (MEMS), Munich, Germany, 15–19 January 2023; pp. 440–443.
227. Roundy, S.; Rantz, R.; Xue, T.; Halim, M.A. Inertial Energy Harvesting for Wearables. In Proceedings of the 2018 IEEE SENSORS, New Delhi, India, 28–31 October 2018; pp. 1–4.
228. Mitcheson, P.D.; Yeatman, E.M.; Rao, G.K.; Holmes, A.S.; Green, T.C. Energy Harvesting From Human and Machine Motion for Wireless Electronic Devices. *Proc. IEEE* **2008**, *96*, 1457–1486. [CrossRef]
229. Cai, M.; Liao, W.H. High-Power Density Inertial Energy Harvester Without Additional Proof Mass for Wearables. *IEEE Internet Things J.* **2021**, *8*, 297–308. [CrossRef]
230. Hoareau, D.; Jodin, G.; Laaraibi, A.-r.A.; Prioux, J.; Razan, F. Available Kinetic Energy Sources on the Human Body during Sports Activities: A Numerical Approach Based on Accelerometers for Cantilevered Piezoelectric Harvesters. *Energies* **2023**, *16*, 2695. [CrossRef]
231. Beach, C.; Casson, A.J. Inertial Kinetic Energy Harvesters for Wearables: The Benefits of Energy Harvesting at the Foot. *IEEE Access* **2020**, *8*, 208136–208148. [CrossRef]
232. Sandhu, M.M.; Khalifa, S.; Geissdoerfer, K.; Jurdak, R.; Portmann, M.; Kusy, B. FusedAR: Energy-Positive Human Activity Recognition Using Kinetic and Solar Signal Fusion. *IEEE Sens. J.* **2023**, *23*, 12411–12426. [CrossRef]
233. Yang, Y.; Lin, Z.-H.; Hou, T.; Zhang, F.; Wang, Z.L. Nanowire-composite based flexible thermoelectric nanogenerators and self-powered temperature sensors. *Nano Res.* **2012**, *5*, 888–895. [CrossRef]
234. Kim, J.; Khan, S.; Wu, P.; Park, S.; Park, H.; Yu, C.; Kim, W. Self-charging wearables for continuous health monitoring. *Nano Energy* **2021**, *79*, 105419. [CrossRef]
235. Sun, T.; Zhou, B.; Zheng, Q.; Wang, L.; Jiang, W.; Snyder, G.J. Stretchable fabric generates electric power from woven thermoelectric fibers. *Nat. Commun.* **2020**, *11*, 572. [CrossRef]
236. Lee, J.A.; Aliev, A.E.; Bykova, J.S.; de Andrade, M.J.; Kim, D.; Sim, H.J.; Lepró, X.; Zakhidov, A.A.; Lee, J.-B.; Spinks, G.M.; et al. Woven-Yarn Thermoelectric Textiles. *Adv. Mater.* **2016**, *28*, 5038–5044. [CrossRef]
237. Boubin, M.; Shrestha, S. Microcontroller Implementation of Support Vector Machine for Detecting Blood Glucose Levels Using Breath Volatile Organic Compounds. *Sensors* **2019**, *19*, 2283. [CrossRef]

238. Tomasik, J.; Han, S.Y.S.; Barton-Owen, G.; Mirea, D.-M.; Martin-Key, N.A.; Rustogi, N.; Lago, S.G.; Olmert, T.; Cooper, J.D.; Ozcan, S.; et al. A machine learning algorithm to differentiate bipolar disorder from major depressive disorder using an online mental health questionnaire and blood biomarker data. *Transl. Psychiatry* **2021**, *11*, 41. [CrossRef]
239. Sardesai, A.U.; Tanak, A.S.; Subramaniam, K.; Striegel, D.A.; Schully, K.L.; Clark, D.V.; Muthukumar, S.; Prasad, S. An approach to rapidly assess sepsis through multi-biomarker host response using machine learning algorithm. *Sci. Rep.* **2021**, *11*, 16905. [CrossRef] [PubMed]
240. Kallapur, B.; Ramalingam, K.; Bastian, B.; Mujib, A.; Sarkar, A.; Sethuraman, S. Quantitative estimation of sodium, potassium and total protein in saliva of diabetic smokers and nonsmokers: A novel study. *J. Nat. Sci. Biol. Med.* **2013**, *4*, 341–345. [PubMed]
241. Shahub, S.; Upasham, S.; Ganguly, A.; Prasad, S. Machine learning guided electrochemical sensor for passive sweat cortisol detection. *Sens. Bio-Sens. Res.* **2022**, *38*, 100527. [CrossRef]
242. Sabilla, I.A.; Cahyaningtyas, Z.A.; Sarno, R.; Al Fauzi, A.; Wijaya, D.R.; Gunawan, R. Classification of Human Gender from Sweat Odor using Electronic Nose with Machine Learning Methods. In Proceedings of the 2021 IEEE Asia Pacific Conference on Wireless and Mobile (APWiMob), Bandung, Indonesia, 8–10 April 2021; pp. 109–115.
243. Sankhala, D.; Sardesai, A.U.; Pali, M.; Lin, K.C.; Jagannath, B.; Muthukumar, S.; Prasad, S. A machine learning-based on-demand sweat glucose reporting platform. *Sci. Rep.* **2022**, *12*, 2442. [CrossRef] [PubMed]
244. Lafaye, C.; Rovira, M.; Demuru, S.; Wang, S.; Kim, J.; Kunnel, B.P.; Besson, C.; Fernandez-Sanchez, C.; Serra-Graells, F.; Margarit-Taule, J.M.; et al. Real-time smart multisensing wearable platform for monitoring sweat biomarkers during exercise. In Proceedings of the 2022 IEEE Biomedical Circuits and Systems Conference (BioCAS), Taipei, China, 13–15 October 2022; pp. 173–177.
245. Kammarchedu, V.; Butler, D.; Ebrahimi, A. A machine learning-based multimodal electrochemical analytical device based on eMoSx-LIG for multiplexed detection of tyrosine and uric acid in sweat and saliva. *Anal. Chim. Acta* **2022**, *1232*, 340447. [CrossRef]
246. Wang, S.; Lafaye, C.; Saubade, M.; Besson, C.; Margarit-Taule, J.M.; Gremeaux, V.; Liu, S.C. Predicting Hydration Status Using Machine Learning Models From Physiological and Sweat Biomarkers During Endurance Exercise: A Single Case Study. *IEEE J. Biomed. Health Inform.* **2022**, *26*, 4725–4732. [CrossRef] [PubMed]
247. Pavlov, K.; Perchik, A.; Tsepulin, V.; Megre, G.; Nikolaev, E.; Volkova, E.; Park, J.; Chang, N.; Lee, W.; Kim, J.Y. Sweat Loss Estimation Solution for Smartwatch. In Proceedings of the 2022 IEEE-EMBS International Conference on Wearable and Implantable Body Sensor Networks (BSN), Ioannina, Greece, 27–30 September 2022; pp. 1–4.
248. Baker, L.B.; Model, J.B.; Barnes, K.A.; Anderson, M.L.; Lee, S.P.; Lee, K.A.; Brown, S.D.; Reimel, A.J.; Roberts, T.J.; Nuccio, R.P.; et al. Skin-interfaced microfluidic system with personalized sweating rate and sweat chloride analytics for sports science applications. *Sci. Adv.* **2020**, *6*, eabe3929. [CrossRef]
249. Hang, T.; Xiao, S.; Yang, C.; Li, X.; Guo, C.; He, G.; Li, B.; Yang, C.; Chen, H.-j.; Liu, F.; et al. Hierarchical graphene/nanorods-based H2O2 electrochemical sensor with self-cleaning and anti-biofouling properties. *Sens. Actuators B* **2019**, *289*, 15–23. [CrossRef]
250. Jolly, P.; Miodek, A.; Yang, D.-K.; Chen, L.-C.; Lloyd, M.D.; Estrela, P. Electro-Engineered Polymeric Films for the Development of Sensitive Aptasensors for Prostate Cancer Marker Detection. *ACS Sens.* **2016**, *1*, 1308–1314. [CrossRef]
251. Zhao, P.; Patamia, E.D.; Andrew, T.L. Strategies to combat the fouling and surface texture issues associated with fabric-based colorimetric sensors. *Sens. Actuators B* **2023**, *377*, 133099. [CrossRef]
252. Song, Q.; Zhao, R.; Liu, T.; Gao, L.; Su, C.; Ye, Y.; Chan, S.Y.; Liu, X.; Wang, K.; Li, P.; et al. One-step vapor deposition of fluorinated polycationic coating to fabricate antifouling and anti-infective textile against drug-resistant bacteria and viruses. *Chem. Eng. J.* **2021**, *418*, 129368. [CrossRef] [PubMed]
253. Gyu Son, S.; Jun Park, H.; Kim, S.-M.; Jin Kim, S.; Sik Kil, M.; Jeong, J.-M.; Lee, Y.; Eom, Y.; Yeon Hwang, S.; Park, J.; et al. Ultra-fast self-healable stretchable bio-based elastomer/graphene ink using fluid dynamics process for printed wearable sweat-monitoring sensor. *Chem. Eng. J.* **2023**, *454*, 140443. [CrossRef]
254. Huang, X.; Liu, Y.; Park, W.; Li, J.; Ma, J.; Yiu, C.K.; Zhang, Q.; Li, J.; Wu, P.; Zhou, J.; et al. Intelligent Soft Sweat Sensors for the Simultaneous Healthcare Monitoring and Safety Warning. *Adv. Healthcare Mater.* **2023**, *12*, e2202846. [CrossRef] [PubMed]
255. Reeder, J.T.; Choi, J.; Xue, Y.; Gutruf, P.; Hanson, J.; Liu, M.; Ray, T.; Bandodkar, A.J.; Avila, R.; Xia, W.; et al. Waterproof, electronics-enabled, epidermal microfluidic devices for sweat collection, biomarker analysis, and thermography in aquatic settings. *Sci. Adv.* **2019**, *5*, eaau6356. [CrossRef] [PubMed]
256. Dong, B.; Zhang, X.; Cao, L.; Jiang, X.; Wang, F. Anti-interference monitoring of sweat pH: A new sensing mechanism based on the p–n transition potential of a flexible Bi2O3 photoelectrode. *J. Mater. Chem. C* **2023**, *11*, 2074–2081. [CrossRef]
257. Komkova, M.A.; Poyarkov, A.A.; Eliseev, A.A.; Eliseev, A.A. Mass transport limitations for electrochemical sensing in low-flux excretory fluids. *Biosens. Bioelectron.* **2023**, *227*, 115148. [CrossRef]
258. Hao, J.; Zhu, Z.; Hu, C.; Liu, Z. Photosensitive-Stamp-Inspired Scalable Fabrication Strategy of Wearable Sensing Arrays for Noninvasive Real-Time Sweat Analysis. *Anal. Chem.* **2022**, *94*, 4547–4555. [CrossRef]

chemosensors

Article

Electrochemical Etching-Assisted Fabrication of Quantum Tunneling Sensing Probes with Controlled Nanogap Width

Bangrui Shao [1,†], Qiuxiang He [1,†], Tao Jiang [1], Biaofeng Zeng [1], Cuifang Kuang [1], Xu Liu [1] and Longhua Tang [1,2,*]

1 State Key Laboratory of Modern Optical Instrumentation, College of Optical Science and Engineering, Zhejiang University, Hangzhou 310027, China
2 Institute of Quantum Sensing, Interdisciplinary Centre for Quantum Information, Zhejiang University, Hangzhou 310027, China
* Correspondence: lhtang@zju.edu.cn
† These authors contributed equally to this work.

Abstract: Quantum tunneling electrical probes, consisting of a pair of nanoelectrodes with a gap width of less than 5 nm, can be used as a robust electrical sensing platform for the detection of various nanoscale objects. To achieve this, stable and gap-width-controllable electrodes are essential. Although various methods, including lithography and electrochemical strategies, have been proposed for the fabrication of tunneling electrodes, the ability to precisely control the gap width and ensure reproducibility is still lacking. Here, we report a feedback-controlled electrochemical etching approach to fabricate the tunneling electrodes with a controlled nanogap. The connected nanoelectrodes, derived from a dual-barrel nanopipette, were subjected to a controlled electrochemical etching process from a short-circuited state to a tunneling gap. The resulting tunneling electrodes exhibited solvent-response current–voltage electrical behavior, which was well fitted with the Simons model, indicating the formation of tunneling electrodes. Overall, a success rate of more than 60% could be achieved to obtain the tunneling gaps. Furthermore, to verify the function of tunneling electrodes, we used the etched-tunneling electrodes for free-diffusing protein detection, showing the potential of etched-tunneling electrodes as single-molecule sensors.

Keywords: quantum tunneling probe; nanogap width electrode; single-molecule detection

Citation: Shao, B.; He, Q.; Jiang, T.; Zeng, B.; Kuang, C.; Liu, X.; Tang, L. Electrochemical Etching-Assisted Fabrication of Quantum Tunneling Sensing Probes with Controlled Nanogap Width. *Chemosensors* **2023**, *11*, 480. https://doi.org/10.3390/chemosensors11090480

Academic Editor: Christos Kokkinos

Received: 30 June 2023
Revised: 14 August 2023
Accepted: 29 August 2023
Published: 1 September 2023

1. Introduction

Single-molecule detection methods based on electrical signals can achieve direct, real-time, and label-free monitoring of the dynamic processes of molecular events at the single-molecule level [1]. It provides a platform for exploring the detailed information of chemical and biological reaction processes, including intermediate/transient states and stochastic processes, which usually drown in traditional ensemble-average experimental methods that are crucial to chemistry, biology, and medicine research [2,3]. Among these newly developed detection technologies, the method based on the quantum tunneling effect has excellent performance in the field of single-molecule detection due to its advantages of high spatial resolution, ultra-high sensitivity, and easy integration [4].

The quantum tunneling effect refers to the transport of microscopic particles such as electrons across the potential barrier, including a nanogap, a molecule, a nano-sized insulating layer, etc. It becomes increasingly pronounced as the gap width decreases to sub-5 nm [5]. Quantum tunneling sensors, which are composed of a pair of electrodes with a gap of sub-5 nm, have recently gained considerable attention due to their label-free direct in situ sensing ability in liquid environments and the potential for on-chip integrated platforms [4–7]. When a bias voltage is applied to the electrodes, electrons transport across the gap due to the tunneling effect, generating a tunneling current. The current is dependent on the applied bias, the gap width, and the medium between electrodes [4,8]. In a typical sensing system, if a specific molecule enters the tunneling region, the tunneling current will be modulated, producing a characteristic electrical signal. The acquired signal is normally superimposed on the tunneling current, with the featured current amplitude and dwell time. By extracting and analyzing these features, specific information about the target molecule can be revealed, including the sub-molecular structure and electronic structure [4].

The fabrication of tunneling electrodes with controllable gaps is the key to quantum sensing applications. In recent years, a variety of techniques have been developed for this purpose, including scanning tunneling microscope (STM) [9,10], breaking junctions [11–14], electrodeposition [5,15], electromigration [16], and advanced lithography [17–19]. According to the main methods used for nanogap formation, the techniques can be divided into three main categories, including additive, subtractive, and splitting methods. The additive methods, such as electrodeposition and shadow mask deposition, mainly form nanogaps by depositing electrode materials. In subtractive methods like electron beam lithography (EBL) and focused ion beam (FIB) lithography, the electrode material is removed to create a nanogap. For splitting methods including mechanically controllable break junctions (MCBJs) and crack-defined break junctions, the nanogap is generated by the fracture of electrode material. Typically, these methods have their own priorities in fabricating the tunneling gaps and cannot combine the advantages of controllability, reproducibility, stability, and mass production. For instance, shadow mask deposition produces nanogap electrodes by depositing electrode materials and local blocking suspended masks, while the grain-like growth phenomenon of metallic electrode material will cause instability and uncontrollability of electrode gap width [20]. FIB lithography generates sub-5 nm nanogaps by directly applying ion beams to electrode materials without the need for a resistance layer or a mask. It has the advantages of being maskless and having high reliability, but the ion sputtering process may cause contamination of the electrode surface. Besides the high cost, the long time consumed also influences reproducibility [18,19,21]. MCBJ allows for the formation of tunneling electrodes with an adjustable gap width by controlling the piezo actuator and is not suitable for large-scale preparation [11,21]. Moreover, most of these fabrication methods require state-of-the-art facilities, instruments, and sophisticated processes, which leads to high costs [21–23]. Therefore, it is essential to develop new strategies for fabricating tunneling electrodes with a more controllable gap width and reproducibility. Meanwhile, it is worth noting that a combination of manufacturing techniques is usually utilized to create nanogaps. Recently, some novel hybrid methods have been proposed to fabricate gap-width-controllable electrodes [24–26]. For instance, in situ adjustable metal nanogaps can be achieved by either a lateral expandable piezoelectric sheet or a stretchable membrane [26]. These nanogaps with angstrom resolution were developed into in-plane molecular break junctions and successfully applied to measure single-molecule conductance.

Recently, we reported feedback-controlled electrochemical deposition for efficiently fabricating standalone tunneling probes, which can be used directly in solution for single-molecule detection [5]. However, this electrodeposition approach tends to form narrow gaps, making it difficult to control the gap width from the atomic contact scale to a wider scale. Here, we demonstrate an electrochemical etching-assisted method for fabricating the tunneling electrodes with stable nanogaps, which can be used either independently or as a complementary method to the electrodeposition approach. Although electrochemical etching has been widely applied in STM probe fabrication with various materials such as gold (Au) [27–29], platinum (Pt) [30], and tungsten (W) [31], few studies have explored its capability to directly form the tunneling gap. We used a mixed solution of potassium chloride (KCl) and potassium thiosulfate ($K_2S_2O_3$) as the electrolyte and a double-electrode potentiostatic electrochemical setup as the etching strategy. We achieved a success rate of approximately 60% in fabricating the tunneling probes. We further applied our tunneling electrodes to single-molecule detection. We conducted free diffusion detection on two different protein molecules (bovine serum albumin and glucose oxidase) and obtained evident single-molecule tunneling signals. Data analysis shows that there are differences in the distribution of conductance and signal dwell time between the two proteins. The results indicate that our etched-tunneling electrodes have reliable and sensitive sensing capability at the single-molecular level in liquid environments.

2. Materials and Methods

2.1. Chemicals and Materials

The Au-plating solution, consisting of 4.4 mmol/L NH_4AuSO_3 dissolved in 52 mmol/L $(NH_4)_2SO_3$, was purchased from Tianyue (Shenzhen, China). Potassium thiosulfate ($K_2S_2O_3$), with a purity of 98%, was purchased from Chentong Biochemical Technology Co., Ltd. (Hangzhou, China). All the other reagents not mentioned above used for device fabrication and characterization in this study were purchased from Sigma Aldrich (Burlington, MA, USA). All aqueous solutions should be prepared using ultrapure deionized water (18.2 MΩ·cm at 25 °C). Dual-barrel theta-shaped quartz capillaries (Friedrich & Dimmock, Inc. (Millville, NJ, USA), 1.2 mm outside diameter (OD), 0.90 mm inside diameter (ID), 75 mm long) were used for fabricating tunneling electrodes. Copper wires were 0.25 mm in diameter and purchased from Goodfellow (Shanghai, China). Silver wires (Agar Scientific, Stansted, UK, 0.25 mm OD) were made into silver/silver chloride (Ag/AgCl) electrodes in 1 mol/L potassium chloride (KCl) aqueous solution through an electrochemical workstation.

2.2. Apparatus and Characterization

The puller we used for pulling quartz capillaries was the P-2000 laser puller from Sutter Instrument (Novato, CA, USA). All electrochemical processes in this work were conducted on a CHI760c electrochemical workstation (Chenhua Instruments Shanghai Co., Ltd., Shanghai, China). All tunneling current measurements were performed using a MultiClamp 700B (Molecular Devices, San Jose, CA, USA) operated in voltage-clamp mode. The recorded signals were filtered using a four-pole Bessel filter at 10 kHz and digitized using an Axon Digidata 1550B (Molecular Devices, San Jose, CA, USA). Data analysis was performed using custom-written MATLAB code developed in-house. During tunneling current measurements, the tunneling electrode and the amplifier headstage were placed in a Faraday cage to shield external electrical noise. The setup was placed on stable support to minimize vibrational interference. In addition, short cables and electrode connecting wires were selected in priority to minimize electrical noise. The amplifier, Faraday cage, and supporting platform were connected to the same grounding point. Morphological characterization of the electrode tip was performed by a Zeiss ULTRA-55 field-emission scanning electron microscope (Oberkochen, Germany) under an operating voltage of 10 kV.

2.3. Fabrication of Dual-Barrel Au-Deposited Nanoelectrodes

The Au-deposited probes, consisting of two proximally addressable Au nanoelectrodes in a standalone double-barrel nanopipette, were obtained according to our previous work [5]. Briefly, a theta-shaped dual-barrel quartz capillary was first laser-pulled into a sharp nanopipette using a P-2000 laser puller with a custom two-line program (First line, Heat = 850, Filament = 4, Velocity = 30, Delay = 160, Pull = 100. Second line, Heat = 860, Filament = 3, Velocity = 20, Delay = 128, Pull = 160). The two nanopores formed at the probe tip were 20~60 nm in diameter and separated by a septum of about 20 nm. Next, both barrels were filled with carbon by the pyrolytic decomposition of butane, resulting in two coplanar carbon nanoelectrodes at the tip of the pipette. Copper wires were inserted from the end of the quartz capillary, in close contact with the deposited carbon, to provide the electrical connection to the instrumentation. Carbon etching was followed and formed a pit on the carbon electrode surface. This is beneficial for the subsequent electrochemical deposition of gold and improves the stability of the final prepared tunneling electrode. Then, gold was electrochemically deposited on the carbon electrodes in a bipotentiostatic configuration until it formed the connected electrodes (>4.75 μS in conductance).

3. Results and Discussion

3.1. Electrochemical Etching Tunneling Electrodes Fabrication

The tunneling probes were obtained with an electrochemical etching strategy, as shown in Figure 1a. At first, the dual-barrel Au-deposited nanoelectrodes were fabricated via our previously reported method [5]. The detailed fabrication procedure has been shown above (see Section 2). Subsequently, the Au deposited on the electrodes was controllably removed to generate the tunneling gap in a double-electrode electrochemical etching configuration. In the potentiostatic electrochemical experimental setup, both of the connected electrodes were immersed in the etching electrolyte as the working electrodes (WE1 and WE2), and the silver/silver chloride electrode was used as the counter electrode and the reference electrode (CE and RE). In principle, under the presence of the etching agent potassium chloride (KCl), Au is dissolved in $AuCl_2^-$ by reacting with Cl^- at a positive potential [32]. It should be noted that the formed $AuCl_2^-$ will be partially transformed to Au (0) and $AuCl_4^-$ in solution. The Au reduced by disproportionation will adhere to the electrode surface, roughening the electrode surface or even reconnecting the etched-open electrode. In order to increase the controllability of the etching process and to obtain a smooth electrode surface, we introduced the solvent potassium thiosulfate ($K_2S_2O_3$) into the etching solution, which could efficiently react with the disproportionated Au and form the $Au_2S_2O_3$ complex (Equation (1)) [15].

$$2Au + S_2O_3^{2-} \rightleftharpoons Au_2S_2O_3 + 2e^- \quad (1)$$

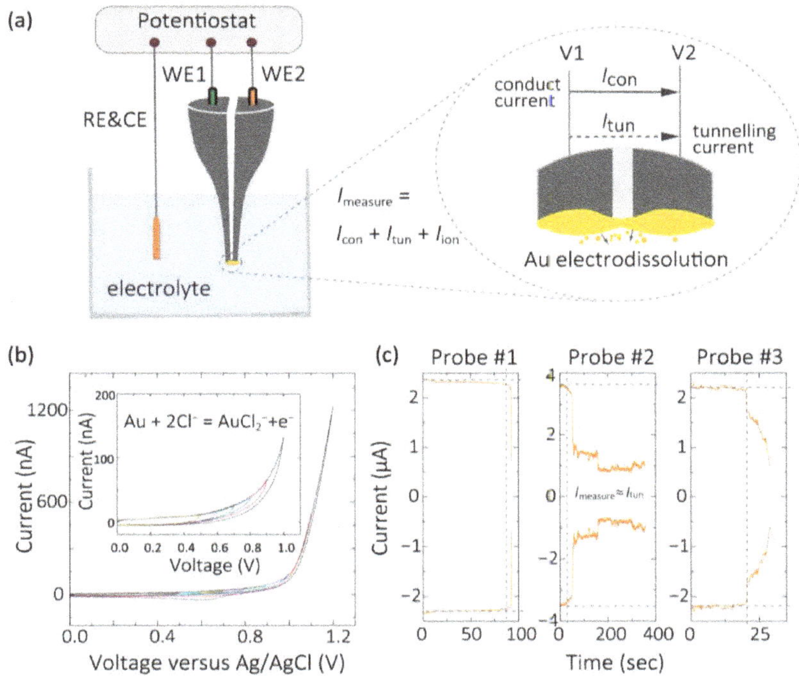

Figure 1. Fabrication of tunneling electrodes by electrochemical etching. (**a**) Schematic diagram of experimental setup. A double-electrode potentiostatic electrochemical system is used for Au etching. The potentials applied to working electrode 1 (WE1) and working electrode 2 (WE2) are 1.1 V and 1.2 V, respectively. A local enlargement of the etching process at the electrode tip is shown on the right. (**b**) Redox curves measured in 1 mol/L potassium chloride (KCl) aqueous solution and 2 mol/L potassium thiosulfate ($K_2S_2O_3$) aqueous solution with a volume ratio of 4:1. (**c**) Three typical current–time (*I*–*t*) curves measured during the etching process.

We further investigated the chemical composition of the etching electrolyte. The etching rate increases with the increase in the concentration of KCl. In order to control the etching rate in an appropriate range, we determined that a mixed solution containing 1 mol/L KCl aqueous solution and 2 mol/L $K_2S_2O_3$ aqueous solution with a volume ratio of 4:1 is optimal.

The potentials we applied to the two working electrodes were selected based on the redox curve, as shown in Figure 1b. During the etching process, when the voltage exceeds 0.6 V, Au is oxidized to Au^+ on the anode. In addition, the reaction becomes intense as the voltage increases. When the voltage reaches 1.4 V, the current oscillates due to the continuous dissolution and disproportionation reduction in Au [32], which increases the roughness of the electrode surface and makes the etching process more uncontrollable. Hence, a voltage range of 0.9 V to 1.2 V is preferred for etching. In order to obtain real-time feedback information from the current–time (*I*–*t*) curve and help determine the timing of etching cessation, a bias of 100 mV is set between two working electrodes. Due to the exponential correlation between tunneling current and the tunneling gap distance, the measured current containing tunneling current will sharply decrease as the gap broadens. Therefore, if a sharp decrease in measured current is observed, the etching can be stopped, which demonstrates the formation of the tunneling electrodes. Accordingly, we selected etching potentials of 1.1 V and 1.2 V for working electrode 1 (WE1) and working electrode 2 (WE2), respectively. The etching rate can be adjusted by changing the potentials of WE1 and WE2 during the etching process.

The real-time current monitoring at working electrodes revealed the etching process, as shown by the typical current curves shown in Figure 1c. Generally, the current measured at working electrodes is composed of conduction current ($I_{conduct}$, on the order of 10^{-6} A, under the set working potential, the same below), tunneling current ($I_{tunneling}$, on the order of 10^{-9} A~10^{-6} A), and ion current (I_{ion}, on the order of 10^{-10} A), such that $I_{measure} = I_{conduct} + I_{tunneling} + I_{ion}$ for WE1 and $I_{measure} = (-I_{conduct}) + (-I_{tunneling}) + I_{ion}$ for WE2. The positive and negative signs represent the direction of the current. The conduction current is generated by the directional flow of electrons between the two electrodes when they contact, and the tunneling current occurs when the electrode gap enters the tunneling range. The ion current is related to the directional flow of ions in an electrolyte solution. Due to the fact that the ion current is much smaller than the former two parts of the measured current, the currents of WE1 and WE2 exhibit symmetric changes in the curve. During the etching process, the current measured decreased with time. At the start of the etching process, the two electrodes are connected, and the conduction current is dominant under the potential difference applied to the two working electrodes. As the etching process progresses, when the electrodes are etched apart, the conduction current almost goes down to zero and the tunneling current becomes the dominant factor, which depends exponentially on the gap width. To obtain tunneling electrodes, the etching is stopped at this stage. If the etching continues, more Au will be etched away, resulting in wider gap-width electrodes.

3.2. Characterization of Electrochemical Etching Electrodes

The schematic diagrams of the structure of connected electrodes, tunneling electrodes, and open electrodes are shown in Figure 2a. We performed current–voltage (*I*–*V*) characterization on three types of electrodes separately. All the *I*–*V* curves shown in Figure 2b had the same scale of horizontal and vertical coordinates. The connected electrodes had a large conductance (>4.75 μS), which showed up on the curve as a straight line with a very high slope. The tunneling junction showed a classical S-shape on the curve, which is the feature that represents tunneling, and the conductance typically ranged in 0.1~2 μS. The open electrode had a gap greater than 3 nm between electrodes, and the curve was almost a horizontal line on the order of 10^{-10} A.

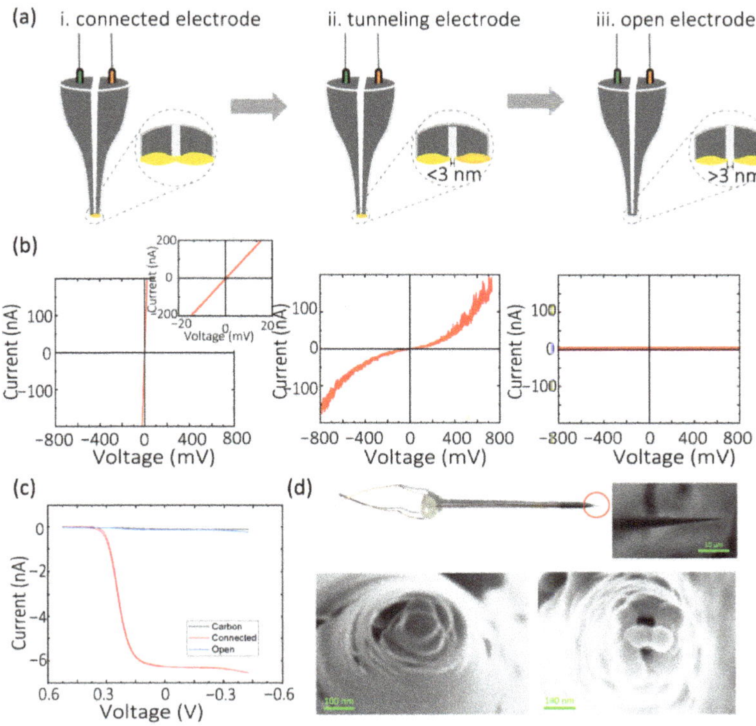

Figure 2. Characterization of electrochemical etching electrodes. (**a**) Schematics of connected electrodes, tunneling electrodes, and open electrodes. (**b**) *I–V* curves of connected electrodes, tunneling electrodes, and open electrodes. (**c**) The steady-state current curves of the same electrode in the carbon electrode state, connected electrode state, and etched-open electrode state. (**d**) The optical image of the tunneling electrode and SEM images of connected electrodes or open electrodes. Inset: the magnified SEM image of the electrode tip (inside the red circle).

Steady-state voltammetry is a reliable electrochemical method used to estimate the size of ultrasmall electrodes. This method calculates the effective radius of the electrode from the steady-state current of the electrode obtained by measuring the cyclic voltammetry curve of the redox pair. Assuming the electrode is hemispherical and the RG value (RG = rg/a, where rg is the radius of the insulator around the electrode and a is the effective radius of the electrode) is greater than 10, the effective radius can be estimated by steady-state current using Equation (2) [33].

$$I_{\text{lim}} = 2\pi n_e a F D c \quad (2)$$

where I_{lim} is the steady-state current, n_e is the number of electrons transferred in electrochemical reactions, F is the Faraday constant, and $D = 7.2 \times 10^{-6}$ cm^2 s^{-1} is the diffusion coefficient of $[Fe(CN)_6]^{3-}$. C is the concentration of the redox mediator, and a is the electrode radius. Here, we obtained the voltammograms in 10 mmol/L $K_3[Fe(CN)_6]$ using the standard three-electrode system. Figure 2c shows typical curves of the same electrode in the carbon electrode state, connected state, and etched-open electrode state. The estimated effective radius of carbon electrodes, connected electrodes, and open electrodes is 23 nm, 1540 nm, and 25 nm, respectively. The size of the etched-open electrode is similar to the initial carbon electrode size, indicating that the gold on the electrode surface can be effectively etched away. The tips of the electrodes we fabricated are semi-elliptical in shape, with RG

1.5~2. Hence, the radius calculated by steady-state current is only an approximation of the actual electrode radius. The optical image of the tunneling electrode and SEM images of connected electrodes and open electrodes are shown in Figure 2d. The upper-right inset shows an enlarged SEM image of the tunneling electrode tip (inside the red circle). There is a lot of Au attached to the tip of the connected electrodes, and it has a relatively large area. As for open electrodes, we can see that Au on the tip has been etched away, exposing theta-shaped holes (about 50~100 nm in diameter).

We further characterized the etched-tunneling electrodes by *I–V* curves and fitted them with the Simons model. To give an intuitive comparison, we present the typical *I–V* curves of the connected, tunneling, and open electrodes in one graph, as shown in Figure 3a. The area on the yellow background is the typical curve of connected electrodes. The etched-open electrodes are displayed in the gray background area, and the remaining regions represent the etched-tunneling electrodes. The *I–V* curves of the tunneling electrodes are fitted by the Simons model, and the fitted curve is represented by solid lines in the graph. Generally, the *I–V* curves of electrodes with a smaller gap width tend to have steeper slopes and take on a more linear shape. The gap width can be obtained by fitting curves using the Simons model, and the gap width distribution of tunneling electrodes is shown in the Supplementary Materials (Figure S1). In total, 29 standalone etched-tunneling electrodes were fabricated with gap widths ranging from 1 nm to 3.5 nm.

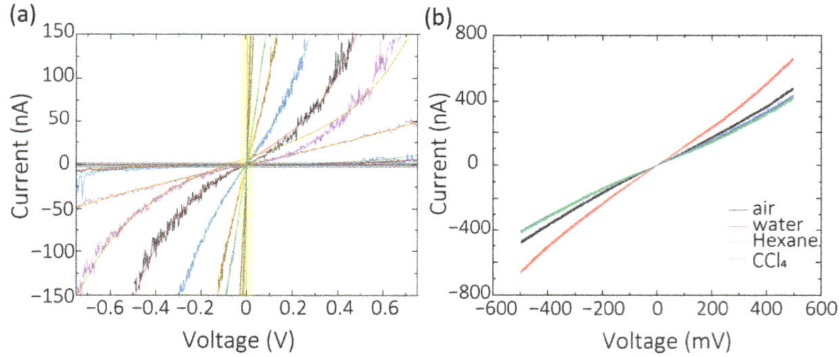

Figure 3. Graph of statistical results of electrodes fabricated by electrochemical etching. (**a**) Typical *I–V* curves of connected electrodes (yellow background), tunneling electrodes, and etched-open electrodes (gray background) measured in air. (**b**) *I–V* curves of the same etched-tunneling electrodes in air, DI water, tetrachloromethane (CCl_4), and hexane. The curves from top to below in panel b: water, air, hexane and CCl_4.

To further confirm the functionality of the etched-tunneling electrodes, *I–V* curves were measured in different materials with different barrier heights, including air, DI water, tetrachloromethane (CCl_4), and hexane. The data for all four substances were collected by the same electrode. We measured 30 *I–V* curves for each material. When switching electrodes from one material to another, we first dipped the electrodes in ethanol for 5 min to dissolve the residue and then kept them in the air for 2 min to evaporate the ethanol, which can reduce the impact caused by the residue on the electrode. The average *I–V* curves fitted by the Simons model are shown in Figure 3b. According to Equation (3) [34], there is an exponential dependence between the square root of barrier height and the tunneling current.

$$I_{\text{tunnel}} \sim \frac{V_{\text{bias}}}{z} \exp\left(-A\sqrt{\Phi_{\text{eff}} z}\right) \tag{3}$$

where V_{bias} is the bias potential between electrodes, A is the effective tunneling area, z is the gap width, and Φ_{eff} is the effective barrier height. That means, while using the same electrode and under the same voltage, the tunneling current is negatively correlated with the barrier height. From this point of view, the order of barrier height of hexane, DI water, and CCl_4 determined by *I–V* curves is consistent with the previous results [35], that is, $\phi_{CCl4} > \phi_{hexane} > \phi_{water}$. It is worth mentioning that the Simons model fitting method can also be used to calculate the medium barrier height. However, due to cross-contamination of the electrodes, the fitting results are not accurate enough, so we chose the former method for analysis.

3.3. Single-Molecule Detection Using Etched-Tunneling Electrodes

In order to confirm the function of the etched-tunneling junctions, we applied the obtained quantum tunneling electrodes to single-molecule detection. Single-molecule tunneling detection methods can be divided into two categories based on whether the molecule is fixed to the electrodes: single-molecule junctions (SMJs) and molecular-free diffusion. SMJ methods bridge the molecules between two separate electrodes by chemical modification and obtain molecular conductance and other information by measuring the current. However, this method requires specific molecular modification systems, and the complexity of modification and measurements leads to low reproducibility and stability. For methods based on molecular free diffusion, the amplitude of tunneling current is related to the intrinsic properties of molecules and the interaction between electrodes and molecules [36,37]. We conducted free-diffusing single-molecule signal measurements on two protein molecules, bovine serum albumin (BSA) and glucose oxidase (GOD). BSA is the main component in bovine serum and is widely used in biochemical laboratories, with a molecular size of approximately $4 \times 4 \times 10$ nm^3. GOD is widely present in animals, plants, and microorganisms and is an important industrial enzyme in the food industry. Its molecular size is about $7 \times 5.5 \times 8$ nm^3. The size of both proteins is much larger than the gap width, so both of them cannot fully diffuse into the tunneling region.

BSA in 0.01 mol/L phosphate buffered saline (PBS, pH 7.4) with a concentration of 10 nmol/L was measured using an electrode with a gap distance of 2 nm by applying a constant voltage of 100 mV. The typical *I–t* curve and signal peak shape measured are shown in Figure 4a. The signal height that reflects the conductance of the electrode itself is about 5 nA. The pulse signal appearing indicates that BSA molecules were in or near the tunneling region, affecting electron transport and generating characteristic signals related to molecular characteristics in the measured current. The low-amplitude signals were attributed to the adsorption and desorption of small molecules by a single electrode. While the higher amplitude signals were likely caused by protein molecules bridging and binding to both electrodes, followed by desorption. For studying the conductivity of individual protein molecules, the higher-amplitude signals formed by bridging are believed to be effective, while low-amplitude adsorption signals are ignored. A statistical analysis was conducted on the signal curve with a time length of 120 s. The distribution of characteristic signal peaks and dwell time obtained is shown in Figure 4b. The signal peaks are concentrated around 4 nA, with a dwell time mostly less than 5 ms. The same experiments were conducted on 10 nmol/L GOD in a 0 01 mol/L PBS solution with an electrode with a 2.1 nm gap distance, and the typical current–time curve and signal peak shape are shown in Figure 4c. The signal height was about 2 nA. It is worth noting that the instability of the voltage applied during the measurement process can lead to a baseline offset. The signals with an apparently longer duration (>10 ms) in the figure are believed to be caused by the voltage instability between the two electrodes. Baseline fitting should be used to remove the influence of such signals during analysis. After statistical analysis, the distribution of peaks and dwell time of the signals is shown in Figure 4d, with a peak of about 5 nA and dwell time mostly below 5 ms. Figure 4e shows the histogram of the conductance distribution of BSA and GOD. The graph shows that the conductance of BSA is about 38 nS and the conductance of GOD is about 50 nS, which means the two protein molecules can be distinguished by the conductance distribution. The magnitude of the measured values of BSA conductivity is consistent with the data reported in previous reports [38,39], indicating the reliability of this method. In addition, the measured tunneling current signals of GOD obtained from the same electrode under different bias voltages

(50~200 mV) were analyzed, and a predominately linear current dependence on voltage was observed with a slope of 25.77 ± 1.54 nS (Figure S2). Figure 4f shows the dwell time histograms of BSA and GOD, indicating differences in the dwell time distribution of the two proteins. The experiment results indicate the potential to differentiate proteins directly through specific tunneling signal characteristics and demonstrate the feasibility of applying this method to drug screening and biomarker detection.

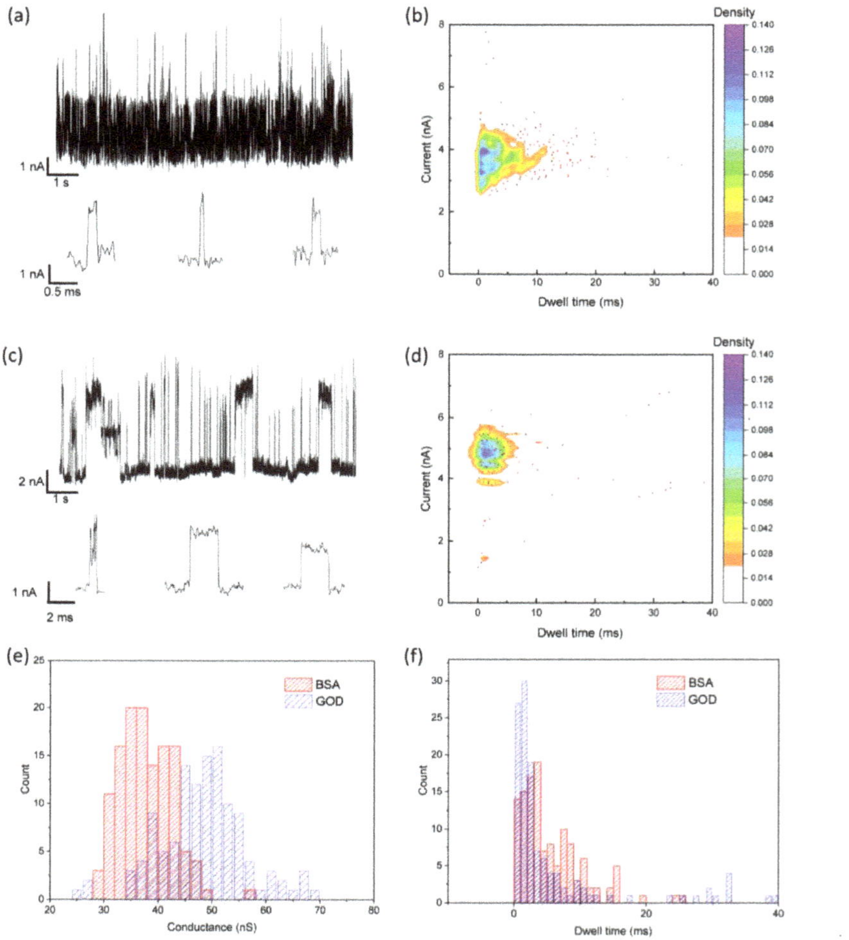

Figure 4. Free-diffusing protein detection using the etched-tunneling electrodes. (**a**) The *I–t* curve and typical signal peak figure of bovine serum albumin protein (BSA). (**b**) Scatter plot of peak amplitude and dwell time distribution of BSA detection signals. (**c**) The *I–t* curve and typical signal peak figure of glucose oxidase (GOD). (**d**) Scatter plot of peak amplitude and dwell time distribution of GOD detection signals. (**e**) Histograms of conductance distribution for BSA and GOD. (**f**) Histograms of dwell time distribution for BSA and GOD.

4. Conclusions

In summary, we proposed and experimentally confirmed that electrochemical etching can be used independently for the fabrication of tunneling electrodes or as a complement to the electrodeposition method. We used various methods to characterize the obtained electrode, demonstrating the feasibility of the scheme. Electrochemical etching combined with electrodeposition greatly increases the success rate of electrode fabrication and improves electrode reusability. We verified the tunneling performance of the obtained electrodes by measuring the barrier heights of different substances. Moreover, we have demonstrated that the tunneling electrodes fabricated by electrochemical etching can be used in free diffusing protein molecule detection via tunneling current, which verifies the function of the fabricated quantum tunneling probes with nanogap width and shows a new strategy for developing quantum tunneling sensors.

Supplementary Materials: The following supporting information can be downloaded at: https://www.mdpi.com/article/10.3390/chemosensors11090480/s1, Figure S1: Histogram of gap width distribution for etched-tunneling electrodes; Figure S2: Tunneling current measurement of GOD under different bias voltages.

Author Contributions: Conceptualization, Q.H. and L.T.; methodology, B.S., Q.H., T.J. and L.T.; validation, B.S., Q.H. and B.Z.; formal analysis, B.S. and Q.H.; investigation, B.S. and Q.H.; resources, T.J. and B.Z.; writing—original draft preparation, B.S. and Q.H.; writing—review and editing, T.J., L.T., C.K. and X.L.; supervision, L.T., C.K. and X.L.; project administration, B.S., Q.H. and L.T.; funding acquisition, L.T. All authors have read and agreed to the published version of the manuscript.

Funding: This research was funded by the National Natural Science Foundation of China (grant nos. 62127818), the Natural Science Foundation of Zhejiang Province (grant no. LR22F050003), and the Fundamental Research Funds for the Central Universities.

Institutional Review Board Statement: Not applicable.

Informed Consent Statement: Not applicable.

Data Availability Statement: Any additional data in support of the findings of this study are available from the corresponding author upon reasonable request.

Conflicts of Interest: The authors declare no conflict of interest.

References

1. Gu, C.; Jia, C.; Guo, X. Single-Molecule Electrical Detection with Real-Time Label-Free Capability and Ultrasensitivity. *Small Methods* **2017**, *1*, 1700071. [CrossRef]
2. Wang, Y.; Shan, X.; Tao, N. Emerging Tools for Studying Single Entity Electrochemistry. *Faraday Discuss.* **2016**, *193*, 9–39. [CrossRef] [PubMed]
3. Mathwig, K.; Aartsma, T.J.; Canters, G.W.; Lemay, S.G. Nanoscale Methods for Single-Molecule Electrochemistry. *Annu. Rev. Anal. Chem.* **2014**, *7*, 383–404. [CrossRef]
4. Albrecht, T. Electrochemical Tunneling Sensors and Their Potential Applications. *Nat. Commun.* **2012**, *3*, 829. [CrossRef] [PubMed]
5. Tang, L.; Nadappuram, B.P.; Cadinu, P.; Zhao, Z.; Xue, L.; Yi, L.; Ren, R.; Wang, J.; Ivanov, A.P.; Edel, J.B. Combined Quantum Tunneling and Dielectrophoretic Trapping for Molecular Analysis at Ultra-Low Analyte Concentrations. *Nat. Commun.* **2021**, *12*, 913. [CrossRef]
6. Di Ventra, M.; Taniguchi, M. Decoding DNA, RNA and Peptides with Quantum Tunneling. *Nat. Nanotech* **2016**, *11*, 117–126. [CrossRef]
7. Xin, N.; Guan, J.; Zhou, C.; Chen, X.; Gu, C.; Li, Y.; Ratner, M.A.; Nitzan, A.; Stoddart, J.F.; Guo, X. Concepts in the Design and Engineering of Single-Molecule Electronic Devices. *Nat. Rev. Phys.* **2019**, *1*, 211–230. [CrossRef]
8. Simmons, J.G. Generalized Formula for the Electric Tunnel Effect between Similar Electrodes Separated by a Thin Insulating Film. *J. Appl. Phys.* **1963**, *34*, 1793–1803. [CrossRef]
9. Xu, B.; Tao, N.J. Measurement of Single-Molecule Resistance by Repeated Formation of Molecular Junctions. *Science* **2003**, *301*, 1221–1223. [CrossRef]
10. Li, X.-M.; Wang, Y.-H.; Seng, J.-W.; Zheng, J.-F.; Cao, R.; Shao, Y.; Chen, J.-Z.; Li, J.-F.; Zhou, X.-S.; Mao, B.-W. Z-Piezo Pulse-Modulated STM Break Junction: Toward Single-Molecule Rectifiers with Dissimilar Metal Electrodes. *ACS Appl. Mater. Interfaces* **2021**, *13*, 8656–8663. [CrossRef]
11. Tsutsui, M.; Shoji, K.; Taniguchi, M.; Kawai, T. Formation and Self-Breaking Mechanism of Stable Atom-Sized Junctions. *Nano Lett.* **2008**, *8*, 345–349. [CrossRef] [PubMed]
12. Xiang, D.; Jeong, H.; Lee, T.; Mayer, D. Mechanically Controllable Break Junctions for Molecular Electronics. *Adv. Mater.* **2013**, *25*, 4845–4867. [CrossRef]
13. Vrouwe, S.A.G.; Van Der Giessen, E.; Van Der Molen, S.J.; Dulic, D.; Trouwborst, M.L.; Van Wees, B.J. Mechanics of Lithographically Defined Break Junctions. *Phys. Rev. B* **2005**, *71*, 035313. [CrossRef]

14. Pan, R.; Yang, Y.; Wang, Y.; Li, S.; Liu, Z.; Su, Y.; Quan, B.; Li, Y.; Gu, C.; Li, J. Nanocracking and Metallization Doubly Defined Large-Scale 3D Plasmonic Sub-10 Nm-Gap Arrays as Extremely Sensitive SERS Substrates. *Nanoscale* **2018**, *10*, 3171–3180. [CrossRef] [PubMed]
15. Missault, N.; Schwarzacher, W. Bridging the Nanogap: Au Electrodeposition at Parallel Electrodes with NM-Scale Spacing. *J. Electrochem. Soc.* **2021**, *168*, 082511. [CrossRef]
16. Naitoh, Y.; Tani, Y.; Koyama, E.; Nakamura, T.; Sumiya, T.; Ogawa, T.; Misawa, G.; Shima, H.; Sugawara, K.; Suga, H.; et al. Single-Molecular Bridging in Static Metal Nanogap Electrodes Using Migrations of Metal Atoms. *J. Phys. Chem. C* **2020**, *124*, 14007–14015. [CrossRef]
17. Manheller, M.; Trellenkamp, S.; Waser, R.; Karthäuser, S. Reliable Fabrication of 3 Nm Gaps between Nanoelectrodes by Electron-Beam Lithography. *Nanotechnology* **2012**, *23*, 125302. [CrossRef]
18. Li, H.; Wani, I.H.; Hayat, A.; Jafri, S.H.M.; Leifer, K. Fabrication of Reproducible Sub-5 Nm Nanogaps by a Focused Ion Beam and Observation of Fowler-Nordheim Tunneling. *Appl. Phys. Lett.* **2015**, *107*, 103108. [CrossRef]
19. Nagase, T.; Gamo, K.; Kubota, T.; Mashiko, S. Direct Fabrication of Nano-Gap Electrodes by Focused Ion Beam Etching. *Thin Solid. Film.* **2006**, *499*, 279–284. [CrossRef]
20. Sun, L.F.; Chin, S.N.; Marx, E.; Curtis, K.S.; Greenham, N.C.; Ford, C.J.B. Shadow-Evaporated Nanometre-Sized Gaps and Their Use in Electrical Studies of Nanocrystals. *Nanotechnology* **2005**, *16*, 631–634. [CrossRef]
21. Dubois, V.; Bleiker, S.J.; Stemme, G.; Niklaus, F. Scalable Manufacturing of Nanogaps. *Adv. Mater.* **2018**, *30*, 1801124. [CrossRef] [PubMed]
22. Li, T.; Hu, W.; Zhu, D. Nanogap Electrodes. *Adv. Mater.* **2010**, *22*, 286–300. [CrossRef] [PubMed]
23. He, Q.; Tang, L. Sub-5 Nm Nanogap Electrodes towards Single-Molecular Biosensing. *Biosens. Bioelectron.* **2022**, *213*, 114486. [CrossRef] [PubMed]
24. Jeong, H.; Li, H.B.; Domulevicz, L.; Hihath, J. An On-Chip Break Junction System for Combined Single-Molecule Conductance and Raman Spectroscopies. *Adv. Funct. Mater.* **2020**, *30*, 2000615. [CrossRef]
25. Zhang, S.; Guo, C.; Ni, L.; Hans, K.M.; Zhang, W.; Peng, S.; Zhao, Z.; Guhr, D.C.; Qi, Z.; Liu, H.; et al. In-Situ Control of on-Chip Angstrom Gaps, Atomic Switches, and Molecular Junctions by Light Irradiation. *Nano Today* **2021**, *39*, 101226. [CrossRef]
26. Zhao, X.; Zhang, X.; Yin, K.; Zhang, S.; Zhao, Z.; Tan, M.; Xu, X.; Zhao, Z.; Wang, M.; Xu, B.; et al. In Situ Adjustable Nanogaps and In-Plane Break Junctions. *Small Methods* **2023**, *7*, 2201427. [CrossRef] [PubMed]
27. Yang, B.; Kazuma, E.; Yokota, Y.; Kim, Y. Fabrication of Sharp Gold Tips by Three-Electrode Electrochemical Etching with High Controllability and Reproducibility. *J. Phys. Chem. C* **2018**, *122*, 16950–16955. [CrossRef]
28. Wang, X.; Cui, Y.; Ren, B. Fabrication of Au tips for tip-enhanced Raman spectroscopy. *Chem. J. Chin. Univ.-Chin.* **2007**, *28*, 522–525.
29. Ren, B.; Picardi, G.; Pettinger, B. Preparation of Gold Tips Suitable for Tip-Enhanced Raman Spectroscopy and Light Emission by Electrochemical Etching. *Rev. Sci. Instrum.* **2004**, *75*, 837–841. [CrossRef]
30. Takami, T.; Kitamura, R.; Hiramoto, T.; Oki, S.; Yoshioka, K.; Aoyama, Y. Automatic Shutdown System of Alternating Current Electrochemical Etching for the Preparation of a Platinum/Iridium Tip for Scanning Tunneling Microscopy and the Investigation of the Byproduct of Platinum Chloride Particles. *Jpn. J. Appl. Phys.* **2019**, *58*, SIIC05. [CrossRef]
31. Tahmasebipour, G.; Hojjat, Y.; Ahmadi, V.; Abdullah, A. Optimization of STM/FIM Nanotip Aspect Ratio Based on the Taguchi Method. *Int. J. Adv. Manuf. Technol.* **2009**, *44*, 80–90. [CrossRef]
32. Mao, B.; Ren, B.; Cai, X.; Xiong, L. Electrochemical Oscillatory Behavior Under a Scanning Electrochemical Microscopic Configuration. *J. Electroanal. Chem.* **1995**, *394*, 155–160. [CrossRef]
33. Penner, R.M.; Heben, M.J.; Lewis, N.S. Preparation and Electrochemical Characterization of Conical and Hemispherical Ultramicroelectrodes. *Anal. Chem.* **1989**, *61*, 1630–1636. [CrossRef]
34. Hugelmann, M.; Schindler, W. Tunnel Barrier Height Oscillations at the Solid/Liquid Interface. *Surf. Sci.* **2003**, *541*, L643–L648. [CrossRef]
35. Prokopuk, N.; Son, K.-A.; Waltz, C. Electron Tunneling through Fluid Solvents. *J. Phys. Chem. C* **2007**, *111*, 6533–6537. [CrossRef]
36. Taniguchi, M. Single-Molecule Analysis Methods Using Nanogap Electrodes and Their Application to DNA Sequencing Technologies. *Bull. Chem. Soc. Jpn.* **2017**, *90*, 1189–1210. [CrossRef]
37. Zwolak, M.; Di Ventra, M. Electronic Signature of DNA Nucleotides via Transverse Transport. *Nano Lett.* **2005**, *5*, 421–424. [CrossRef]
38. Ron, I.; Sepunaru, L.; Itzhakov, S.; Belenkova, T.; Friedman, N.; Pecht, I.; Sheves, M.; Cahen, D. Proteins as Electronic Materials: Electron Transport through Solid-State Protein Monolayer Junctions. *J. Am. Chem. Soc.* **2010**, *132*, 4131–4140. [CrossRef]
39. Ron, I.; Pecht, I.; Sheves, M.; Cahen, D. Proteins as Solid-State Electronic Conductors. *Acc. Chem. Res.* **2010**, *43*, 945–953. [CrossRef]

chemosensors

MDPI

Review

Recent Advances in Functionalization Strategies for Biosensor Interfaces, Especially the Emerging Electro-Click: A Review

Feiyu Wang, Yiwen Xie, Weijie Zhu and Tianxiang Wei *

School of Environment, Nanjing Normal University, Nanjing 210023, China; 212502033@njnu.edu.cn (F.W.); 222502041@njnu.edu.cn (Y.X.); 222512016@njnu.edu.cn (W.Z.)
* Correspondence: weitianxiang@njnu.edu.cn

Abstract: The functionalization of biosensor interfaces constitutes a crucial aspect of biosensing systems, as it directly governs key characteristics, including sensitivity, selectivity, accuracy, and rapidity. Among the diverse range of functionalization strategies available for biosensor interfaces, the click reaction has emerged as an exceptionally straightforward and stable approach for modifying electrodes and sensing films. Notably, the electro-click reaction enables the reagent-free functionalization of the biosensing interface, offering significant advantages, such as high speed, selectivity, and minimal pollution. Consequently, this strategy has garnered substantial attention and is widely regarded as a promising avenue for enhancing biosensor interface functionalization. Within this comprehensive review, we commence by presenting the latest advancements in functionalized biosensor interfaces, organizing the regulatory strategies into distinct categories based on the mediators employed, ranging from nanomaterials to biomolecules. Subsequently, we provide a comprehensive summary with an emphasis on recently developed electro-click strategies for functionalizing electrochemical and optical biosensor interfaces, covering both principles and applications. It is our anticipation that gaining a profound understanding of the principles and applications underlying electro-click strategies for biosensor interface functionalization will facilitate the design of highly selective and sensitive biosensor systems for diverse domains, such as clinical, pharmaceutical, environmental, and food analyses.

Keywords: biosensor; interface; functionalization; electro-click; electrochemistry

Citation: Wang, F.; Xie, Y.; Zhu, W.; Wei, T. Recent Advances in Functionalization Strategies for Biosensor Interfaces, Especially the Emerging Electro-Click: A Review. *Chemosensors* **2023**, *11*, 481. https://doi.org/10.3390/chemosensors11090481

Academic Editor: Camelia Bala

Received: 26 June 2023
Revised: 12 August 2023
Accepted: 24 August 2023
Published: 1 September 2023

1. Introduction

Biosensors represent a distinct branch within the broader domain of analytical sensors, with a specific emphasis on incorporating biometric entities into the detection mode [1]. These remarkable devices house miniature analytical sensing systems that seamlessly integrate biometric components, thereby showcasing micro-level analyte specificity. By leveraging this capability, biosensors can proficiently identify a diverse array of analytes, including proteins, cells, and viruses within living organisms, as well as track heavy-metal pollutants and biochemical waste within the environment encompassing air, water, and soil [2,3]. The accurate detection of these analytes assumes paramount importance in disease surveillance and treatment and environmental monitoring [4]. Typical biosensors contain two basic functional units: recognition elements and signal transducers [5]. The selectivity of a biosensor hinges upon the inherent properties of the recognition element, while its sensitivity relies upon the signal transducer. The fusion of these two components facilitates the attainment of stringent requirements for both high selectivity and sensitivity. The analysis process of a biosensor is usually carried out on the sensor interface. It is at this interface that the core functionality and defining characteristics of the biosensor, such as the sensitivity, selectivity, accuracy, and speed, are directly determined. Accordingly, the development of innovative interfaces stands as the foundational approach for enhancing the biosensor performance. Hence, the functionalization of biosensor interfaces assumes a vital role within the biosensing system.

Numerous studies have focused on the development of novel and improved materials for the functionalization of biosensor interfaces [6,7]. In addition, researchers have proposed more controllable, safe, and convenient strategies to enhance the efficiency of functionalization. In 2001, Professor Sharpless and his team introduced the concept of "Click Chemistry" along with a series of reactions aligned with this concept, marking its first introduction to the scientific community [8]. Despite being a relatively recent concept, click chemistry has gained widespread adoption in various fields, including biomolecular labeling and detection, biomolecular modification, drug lead-compound discovery, drug delivery, polymer modification, fluorescence imaging, CRISPR sgRNA synthesis, target gene labeling, and more. Notably, click chemistry has found significant application in the development of highly controllable, selective, and sensitive biosensing platforms [9–15]. It is worth mentioning that the 2022 Nobel Prize in Chemistry was jointly awarded to American chemist Carolyn Bertozzi, Danish chemist Morten Meldal, and American chemist Barry Sharpless for their contributions to the development of click chemistry and biological orthogonal chemistry.

In recent years, electrochemical reactions have garnered increasing attention due to their rapid reaction kinetics, relatively straightforward reaction conditions, and minimal environmental impact. This has led to the emergence of the electro-click method. Compared to traditional click reactions, the reagent-free nature of the electro-click method makes it a good candidate for functionalization strategies employed in biosensor interfaces [16]. In this paper, we present a comprehensive review of recent advancements in biosensor interface functionalization strategies, with a particular emphasis on the electro-click functionalization strategy. It is anticipated that this review will positively contribute to the development of more precise and high-performance biosensors through the utilization of the electro-click strategy.

2. Functionalization Technologies of Biosensor Interfaces

As previously mentioned, the fundamental components of typical biosensors are recognition elements and signal transducers. To imbue biosensors with specific functions and enhance their performance, it becomes necessary to functionalize the biosensing interface. In this context, nanostructures, macromolecules, small molecules, and cells assume crucial roles as regulators mediating the process of functionalization. Various strategies exist to modify these regulators onto the biosensor interface, including physical adsorption, covalent binding, redox reactions, electrostatic self-assembly, and more. These functionalization strategies empower biosensors to achieve efficient and specific target recognition, minimize background noise, and amplify sensing signals. In the subsequent discussion, we will delve into recent research advancements in this field.

2.1. *Nanostructured Biosensor Interface*

The integration of nanomaterials into the biosensor interface results in a nanostructured biosensor interface, thereby altering the chemical binding profile and composition of the sensing interface. This integration facilitates the attainment of unique detection modes, enhances sensitivity and specificity, and imparts novel properties to the entire biosensor system [17]. The interface between nanomaterials and biomolecules has played a pivotal role in the development of biosensors optimized for diverse objectives and applications.

2.1.1. Carbon Nanomaterials

Carbon nanomaterials, owing to their exceptional electrical conductivity, high stability, and ease of functionalization, have found extensive utilization in various biosensor interfaces [18].

Graphene Nanomaterials

Graphene, with its large specific surface area and excellent conductivity, has emerged as an ideal sensing surface that can be functionalized by grafting different functional groups [19]. As a monolayer graphite with intriguing physical and chemical properties [20], graphene has been extensively employed in the construction of nanostructured biosensor interfaces, as highlighted in many research reviews [21–24].

Graphene oxide (GO) utilizes its high specific surface area to accommodate a greater number of recognition elements or employs its unique sp^2 structure to specifically adsorb corresponding molecules. Moreover, it enhances the performance of transducers by facilitating electron transfer through its high conductivity [25]. GO, which can be readily dispersed in water [19] and prepared in large quantities through the stripping of graphite under acidic and oxidizing conditions [26], possesses abundant carboxyl, hydroxyl, and epoxy groups on its base surface. Additionally, it exhibits a nanoscale domain comprising crystalline carbon or highly oxidized regions, allowing for the differential adsorption of biomolecules with varying affinities in different regions. Upon the successful preparation of the GO interface, fluorescence-based determination represents a commonly employed method for target detection. This method relies on the interference of external substances with the interaction between dye-labeled biomolecules and GO [27]. Iwe et al. [28] introduced a fluorescence-based method for DNA determination, utilizing GO, exonuclease III (Exo III), and specially designed fluorophore-labeled hairpin probes (HP1 and HP2). This approach capitalizes on the differential binding ability of GO towards hairpin DNA probes and single nucleotides (Figure 1a). Based on this principle of binding difference, many GO-based biosensor interfaces have been constructed for the detection of DNA or DNA-related biomolecules [29–31]. Affected by DNA hybridization [32], changes in voltage, current, or impedance may accompany DNA hybridization in GO-structured biosensor interfaces. Zhao et al. [6] synthesized a controllable flower-like Pt–graphene oxide (PtNFS-GO) structure via the layer-by-layer electrostatic self-assembly method and used it for biosensing research on DNA damage marker 8-hydroxy-2′-deoxyguanosine (8-OHdG). The PtNFS-GO structure, which resulted from the improved combination facilitated via electrostatic self-assembly, demonstrated high conductivity and exhibited an excellent electrochemical biosensor performance for the oxidation state of 8-OHdG.

The reduction from GO to reduced graphene oxide (rGO) introduces a higher density of defects, resulting in enhanced electrochemical activity compared to GO. This property proves particularly advantageous for the development of electrochemical biosensors [33]. Furthermore, rGO inherits the unique morphological structure and characteristics suitable for sensing applications. Cao et al. [34] reported on the combination of rGO and exonuclease III, which enables signal amplification in electrochemical impedance spectroscopy for DNA detection. By employing enzyme-assisted target recycling, the biosensing system transitions from a high-impedance state, in which ssDNA probes are directly adsorbed onto rGO, to a low-impedance state generated by the continuous desorption of target-probe ssDNA hybrid products and ssDNA probe digestion (Figure 1b). This approach allows for efficient interfacial tuning, and the change in the electron-transfer resistance becomes more pronounced after the removal of the ssDNA probe. As a result, changes in impedance signals can be measured to sensitively detect ssDNA targets with a remarkably low detection limit (LOD) of 10 aM. In recent practical applications, Ali et al. [35] reported an advanced biosensing platform based on rGO nanomaterials capable of the rapid detection of COVID-19 antibodies. They employed rGO nanosheets, immobilized specific viral antigens on their surfaces, and integrated the electrode with a microfluidic device. When antibodies were introduced to the electrode surface, they selectively bound to the antigens, and the binding event was detected by monitoring the impedance spectrum.

The examples above show that graphene nanomaterials and their derivatives play important roles in biosensor interfaces.

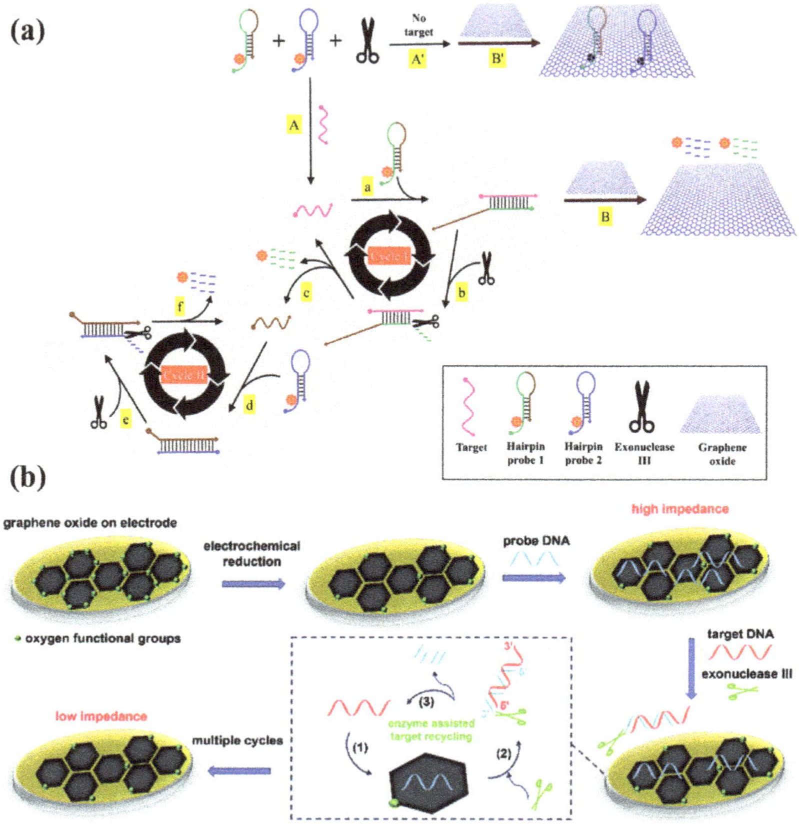

Figure 1. (**a**) Schematic of the detection strategy. On addition of GO without the presence of the target DNA (path A′), HP1 and HP2 are adsorbed onto the GO surface, giving rise to a low background signal (path B′). Upon the addition of the target DNA (path A), it hybridizes and forms a duplex structure with HP1 (a). The HP1-target DNA complex in cycle I allows Exo III to digest HP1 from the 3′ blunt end (b) to free the part of HP1 complementary to HP2, release the fluorophore, and regenerate the target DNA (c). In cycle II, the released fragment of HP1 binds with HP2 (d) and activates the enzyme again (e) to digest HP2, free the fluorophore, and release a fragment of HP1 to repeat the cycle (f). The subsequent addition of GO (path B) cannot quench the fluorescence of FAM, leading to the production of a strong fluorescence signal Reprinted with permission from [28], copyright 2019, *Microchimica Acta*. (**b**) Schematic representation of the fabrication of DNA impedimetric biosensor and its detection through enzyme-assisted target recycling. After the target DNA binds to the probe DNA on the electrode (path 1), exonuclease III is added to selectively hydrolyze the DNA probe in the double-stranded structure (path 2), and the released target DNA will bind to another probe to initiate another round of exonuclease III digestion (path 3). Reprinted with permission from [34], copyright 2019, *Analyst*.

Porous Carbon

Porous carbon possesses notable advantages, including a high specific surface area and superior electron-transfer performance, making it a promising candidate for biosensor applications [36]. In recent years, there has been significant interest in porous carbon derived from biomaterials due to its ease of preparation through low-cost precursor carbonization. Guan et al. [37] prepared a self-powered biosensor using lactate oxidase-modified porous-carbon film. This porous-carbon film exhibits the ability to absorb sweat and generate electricity by harnessing the heat from natural sweat evaporation. The output voltage of the biosensor increases in proportion to the concentration of lactic acid present in sweat. Furthermore, the surface enzymatic reaction alters the zeta potential of the carbon, thereby influencing the output voltage. Gao et al. [38] utilized a luminescent reagent, luminol, for the in situ reduction in chloroauric acid on the nanopores of porous carbon. By combining this approach with an enzyme-cycle and chain-replacement mechanism, they achieved the ultra-sensitive detection of mucin1, an important tumor biomarker. The incorporation of Au-Lum nanoparticles (NPs) within the pores and hollow interiors of porous-carbon-accelerated electron transfer, resulting in excellent luminescence properties of the composite (Figure 2a). Tian et al. [39] designed an accurate biosensor based on n-doped porous-carbon-containing Fe (Fe/N-C) nanocomposites for the detection of H_2O_2 in real water samples and renal epithelial 293T cells. The Fe/N-C nanocomposites substantially enhanced the electron-transfer process and catalytic activity of the sensor, enabling the detection of trace amounts of H_2O_2. This development holds significance for various applications, such as monitoring H_2O_2 levels in water samples and studying its impact on renal epithelial cells.

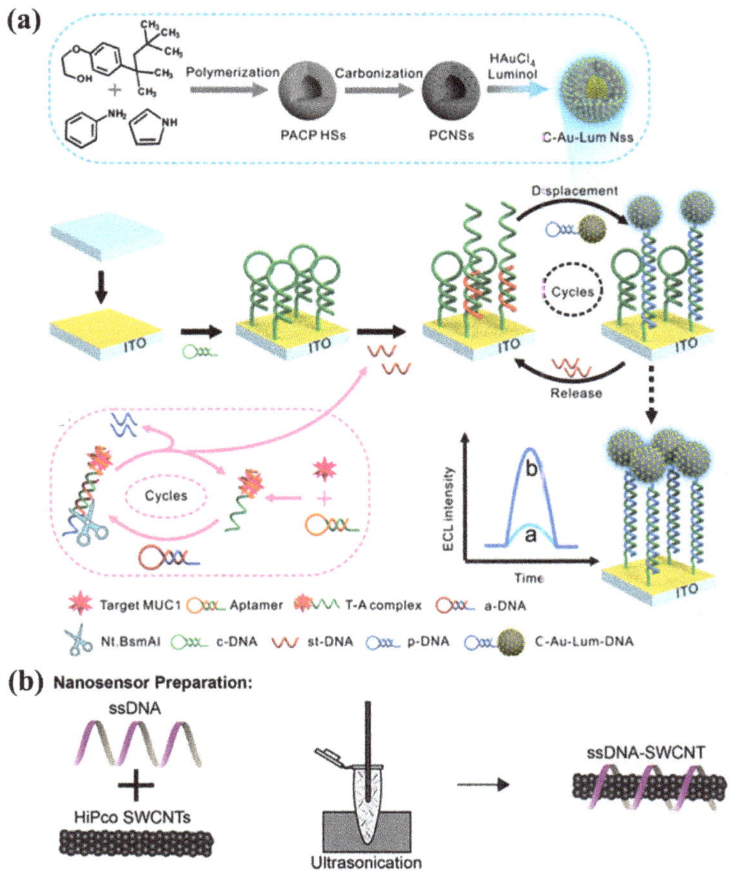

Figure 2. (**a**) The schematic of ECL biosensors for MUC1 detection. After further integrating with proximity-initiated secondary target DNA strand displacement (**a**), the ECL signal quality was significantly higher than that before replacement (**b**). Reprinted with permission from [38], copyright 2019, *Nanoscale*. (**b**) Nanosensor preparation via probe-tip sonicating SWCNTs in the presence of ssDNA followed by ultracentrifugation of the resultant dispersion. Reprinted with permission from [40], copyright 2021, *Advanced Functional Materials*.

Carbon Nanotubes

Carbon nanotubes, resembling rolled-up graphene cylinders, exhibit remarkable tensile strength, high conductivity, and excellent electrocatalytic abilities [41]. Single-walled carbon nanotubes (SWCNTs) consist of a single graphene cylinder, while multi-walled carbon nanotubes (MWCNTs) are composed of multiple layers [42,43]. Notably, through a comprehensive analysis of the existing literature, Gooding et al. [44] observed that SWCNTs outperformed their MWCNT counterparts in terms of electroanalytical performance. Although SWCNT-based electrodes expose a larger number of surface oxides to reactants and facilitate faster electron transfer, the reaction primarily occurs in a non-oriented manner, with the sidewalls predominantly immersed in the solution. This arrangement hampers the transmission and detection of the charge [36]. Consequently, current research predominantly focuses on SWCNTs. Exploiting the excellent light stability and fluorescence properties of SWCNTs in the near-infrared (NIR) range, which lies outside the autofluorescence region of chlorophyll, Lew et al. [45] developed a detection platform utilizing a pair of DNA-coated SWCNT probes. This ratio measurement platform enabled the in vivo detection of endogenous hydrogen peroxide in plants. The functionalized biosensing interface not only holds potential for monitoring H_2O_2 levels in human tumors, but also finds application in monitoring the therapeutic response of pancreatic ductal adenocarcinoma (PDAC) cells to tumors in vitro and in vivo, as well as in evaluating the efficacy of chemotherapy drugs (Figure 2b), as demonstrated by Bhattacharya et al. [46]. Safaee et al. [40] employed optical core–shell microfiber textiles incorporating SWCNTs for the real-time optical monitoring of the hydrogen peroxide concentration in in vitro wounds, enabling the tracking of the wound-healing progress. The versatility of the SWCNT biosensor extends to other applications depending on the utilization of wrapped single-stranded DNA. For instance, Harvey et al. [47] selected this implantable optical biosensor to swiftly quantify the exposure of doxorubicin in living tissues. Furthermore, Salem et al. [48] devised a sensor capable of distinguishing Cu(II), Cd(II), Hg(II), and Pb(II) at a concentration of 100 μM.

2.1.2. Polymer and Bio-Nanomaterials

Conducting polymers possessing favorable electrochemical properties, nanostructural morphology, and biological-coupling functionality play a crucial role in polymer biosensor systems. To reduce reliance on traditional nanomaterials, Meng et al. [49] employed tetrabutylammonium perchlorate as a soft template to fabricate a bifunctional poly(3,4-ethylenedioxythiophene) (PEDOT) interface with an adjustable 3D nanofiber network and carboxylic acid groups. This was achieved by controlling the copolymerization of 3,4-ethylenedioxythiophene(EDOT) and EDOT-COOH monomers (Figure 3a). The newly developed Bio-Nano-PEDOT-COOH interface allows for the coupling of various biorecognition molecules through the carboxylic acid groups, enabling the development of advanced all-polymer biosensors. Expanding on this work, Zhao et al. [50] demonstrated the sensitive and specific monitoring of 17β-estradiol (E2) by modifying electrodeposited PEDOT-graphene oxide (GO) with Au@Pt nanocrystals (Au@Pt). The PEDOT-GO nanocomposite film was polymerized in situ on a glassy carbon electrode using cyclic voltammetry, followed by the synthesis of Au@Pt on the conductive polymer. This provided a platform for aptamer immobilization. With the addition of E2, the differential pulse voltammetry signal gradually decreased due to hindered electron transfer at the E2–aptamer complex interface. Additionally, Zhu et al. [51] investigated the self-assembly of size-controlled tetrahedral framework nucleic acids (FNAs) on a microfluidic microchannel interface, allowing the elevation of DNA probes from the interface to construct a nanoscale three-dimensional reaction space. The highly ordered orientation, configuration, and density of the DNA probes within this three-dimensional reaction space optimize the reaction kinetics in molecular recognition processes. The FNA-designed interface was successfully applied to the one-stop detection of Escherichia coli O157:H7 (*E. coli* O157:H7), achieving a bacterial detection efficiency of 10 CFU/mL with excellent selectivity and precision. In recent advancements, the design and development of polymer-based biosensing platforms for detection have been realized through novel analytical and scientific methodologies.

2.1.3. Other Nanomaterials

Besides the graphene nanomaterials and polymer and bio-nanomaterials mentioned earlier, various other nanomaterials, such as metal nanomaterials and metal oxide nanomaterials, have also been explored [52,53]. Zhou et al. [54] introduced a novel class of chameleon DNA-template silver nanoclusters (AgNCs) capable of switching fluorescence colors among red, orange, and yellow in response to Mg^{2+}, non-fluorescent auxiliary AgNC and complementary DNA. Based on this principle, a ratiometric fluorescence analysis platform was developed for the detection of target DNA. The yellow fluorescence of the probe increased, and the red fluorescence weakened as the amount of target DNA decreased. Regarding metal oxide nanomaterials, Fan et al. [55] discovered that MnO_2 nanosheets (NSs) can effectively quench the fluorescence of highly fluorescent Scopoletin (SC) while enhancing the fluorescence of non-fluorescent Amplex Red (AR) through an oxidation reaction. Upon the addition of glutathione (GSH), MnO_2 was reduced to Mn^{2+} and lost the oxide properties (Figure 3b). At this stage, the fluorescence of the SC intensified, and the AR was quenched. The biosensor system monitors GSH levels by detecting changes in the fluorescence signals of SC and AR.

Another commonly employed nanomaterial is MoS_2, a layered 2D transition metal dihalide exhibiting unique structural, physicochemical, optical, and biological properties [56]. Yan et al. [57] developed an aptasensor using hierarchical MoS_2 nanostructuring and SiO_2 nano-signal amplification. In this approach, hierarchical MoS_2 nanostructures served as functional interfaces, enhancing the accessibility between molecules and improving the efficiency of DNA hybridization. Simultaneously, the SiO_2 nanoprobes, combined with electroactive labels and DNA probes, amplified the electrochemical signal. This biosensing system enabled the simultaneous detection of prostate-specific antigen (PSA) and sarcosine, two prostate cancer (PCa) biomarkers.

Overall, nanomaterials play a pivotal role in the functionalization strategies of biosensing interfaces. Table 1 provides a comparison of nanostructure-functionalized biosensor interfaces.

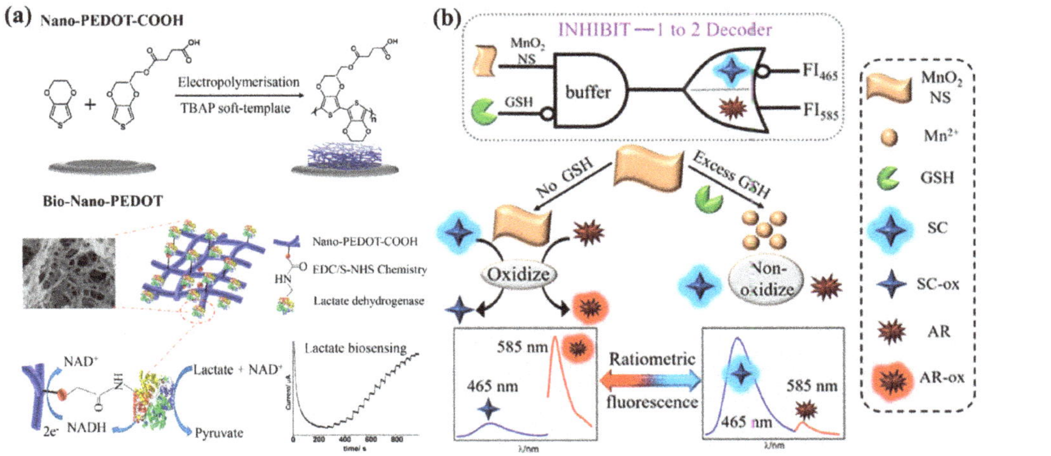

Figure 3. (**a**) Schematic of 3D Nano-PEDOT-COOH network preparation via copolymerization of EDOT and EDOT-COOH monomers using TBAP as a soft template; Bio-Nano-PEDOT; Nano-PEDOT-COOH bio-conjugation with lactate dehydrogenase via EDC/S–NHS chemistry for lactate biosensing. Reprinted with permission from [49], copyright 2022, *Biosensors and Bioelectronics*. (**b**) Schematic operations of the MnO_2 NS-based ratiometric fluorescent sensor for GSH based on two fluorescent substrates. Reprinted with permission from [55], copyright 2017, *ACS Appl Mater Interfaces*.

Table 1. Comparison of nanostructure-functionalized biosensor interfaces.

Functionalized Goal	Functionalized Strategy	Nanostructure	Principle	Effect	Reference
Fluorescence analysis detection of hemochromatosis protein gene	Direct adsorption	GO	π-stacking interactions	Ensures a very low background signal	[28]
Specific detection of biomarker-8-hydroxy-2′-deoxyguanosine	Layer–layer electrostatic self-assembly	GO	Electrostatic interaction	Improves the electrocatalytic performance of Pt nanoparticles	[6]
Detection of ultrasensitive target DNA	Direct adsorption	rGO	π-stacking interactions	Enlarges impedimetric signals	[34]
Detection of COVID-19 antibodies within seconds	Amidation reaction	rGO	Covalent bond	Enhances the transport of diffusing species in an electrochemical cell	[35]
Building of a self-powered wearable-lactate analyzer	Direct adsorption	Porous carbon	π-stacking interactions	Soaks up environmental thermal energy to generate electricity	[37]
Achievement of an ultrasensitive ECL biosensor	In situ reduction	Porous carbon	Electron transfer	Increases the mass transfer of reagents; accelerates the electron transport	[38]
Development of a highly sensitive and selective ECL * sensor based on the noble metal-free electrode	Pyrolytic process	Porous carbon	Breaking and polymerization of molecular bonds	Facilitates rapid electron transfer; improves the conductivity of the electrode	[39]
Real-time, in situ biochemical measurement of H_2O_2 in plants	DNA wrapping	SWNT	π-stacking interactions	An ideal probe for in vivo plant applications	[45]
Detection of H_2O_2 to determine the response to tumor therapy	DNA wrapping	SWNT	π-stacking interactions	Determines dynamic alteration of hydrogen peroxide in tumor; evaluates the effectiveness of chemotherapeutics	[46]
Continuous monitoring of ROS through a wearable diagnostic platform to prevent chronic and pathogenic infections	DNA wrapping	SWNT	π-stacking interactions	In situ measurements of peroxide in wounds	[40]
Use of implantable optical nanosensors to rapidly quantificate doxorubicin in living tissues	DNA wrapping	SWNT	π-stacking interactions	Quantifies doxorubicin exposure to tissues within living organisms	[47]
Fixing of near-infrared fluorescent SWCNT sensor on the paper substrate for sensing	DNA wrapping	SWNT	π-stacking interactions	Immobilizates SWCNTs onto paper substrates; distinguishes metal-ion contaminants	[48]

Table 1. *Cont.*

Functionalized Goal	Functionalized Strategy	Nanostructure	Principle	Effect	Reference
Fabrication of a Bio-Nano-PEDOT *-based biosensor for lactate detection	EDC/S–NHS chemistry	PEDOT-COOH	Chemical coupling	Low charge-transfer resistance; high transduction activity towards the co-enzyme NADH	[49]
Preparation of an electrochemical aptasensor electrodeposited of PEDOT *-GO coupled with Au@Pt	EDC/S–NHS chemistry	PEDOT-COOH	Chemical coupling	Fabricates a simple, label-free electrochemical aptasensor to detect estradiol	[50]
Development of an integrated "one-stop" microfluidic biosensor	Complementary base pairing	Framework nucleic acids	Hydrogen bond	One-stop detection of E.Coli O157:H7 with capture, release, enrichment, cell culture, and antimicrobial susceptibility testing	[51]
Detection of dual PCa * biomarkers, PSA and sarcosine	DNA wrapping	DNA-SiO_2	π-stacking interactions	Enhances the diagnostic performance of PCa *	[57]

* ECL: electrochemiluminescence; PEDOT: poly (3,4-ethylenedioxythiophene); PCa: prostate cancer.

2.2. Small-Molecule-Mediated Interfacial Regulation

Small molecules offer several advantages, including good biocompatibility, clear structures, and easy modification, making them versatile for biosensor functionalization [58]. For instance, Cui et al. [59] utilized dopamine as a functional monomer to synthesize graphdiyne (GDY) with high biocompatibility and conductivity through hydrogen bonding and multipoint electrostatic attraction. They incorporated GDY into C-reactive molecular-imprinted polymers (C-MIPs), achieving high sensitivity and selective recognition of human C-reactive protein (Figure 4a). Salimian et al. [60] employed polyethylene glycol to modify the sensing interface, effectively preventing non-specific protein adsorption and enabling the sensitive and specific detection of cancer biomarkers in serum. Similarly, Zhang et al. [61] employed 6-mercapto-1-hexanol to modify DNA probes and immobilized them on the gold-nanoparticle interface to analyze DNA methylation through current changes.

Fluorescent probes based on small molecules have also found widespread application in the detection of crucial biological analyses [62]. These probes undergo changes in luminescence intensity or emission wavelength through various sensing mechanisms. Among them, electrochemiluminescence (ECL), also known as electrogenerated chemiluminescence, stands out as a superior method for biosensing compared to other photoexcitation spectroscopy techniques. ECL involves the generation of free radicals on the electrode surface, which undergo high-energy electron-transfer reactions to form excited states and emit light [63]. Consequently, ECL does not require excitation light and exhibits minimal background signals [64]. Currently, luminol serves as the most commonly used organic-small-molecule ECL emitter. Zhao et al. [65] developed a paper-based dual-mode detection platform to detect Pb^{2+} based on the oxidation reaction initiated by horseradish peroxidase (HRP) in the presence of H_2O_2. Upon the addition of Pb^{2+} to the interface, cleaved oligonucleotide fragments linked to HRP-functionalized Au nanocubes penetrated into the cellulose, quenching the ECL signals of CDs and quantum dots via resonance energy transfer [63] while enhancing the ECL intensity generated by luminol catalyzed by H_2O_2 (Figure 4b). On this basis, Guo et al. [66] further advanced the field by using reactive oxygen species (ROS) instead of H_2O_2 to develop a new luminol-ROS ECL system for GSH detection. The dissolved oxygen was reduced to superoxide radicals (O_2^-) by atomized gold-loaded 2D VO_2 nanobelts (Au/VO_2), which combined with the loaded luminol to promote the ECL. The ECL resonance energy transfer (ECL-RET) between the hollow SiO_2 nanospheres and luminol resulted in a significant decrease in the ECL-signal response. In the presence of GSH, the effective redox reaction between SiO_2 and GSH restored the ECL signal.

Due to distinct luminescence mechanisms and low electroluminescence efficiency in the aqueous phase, most small organic molecules are unsuitable for recognition and response analysis, limiting their application in biosensing [67]. Therefore, further research is needed to explore the interfacial regulation mediated by organic electroluminescent small molecules. Table 2 shows a comparison of small-molecule-functionalized biosensor interfaces.

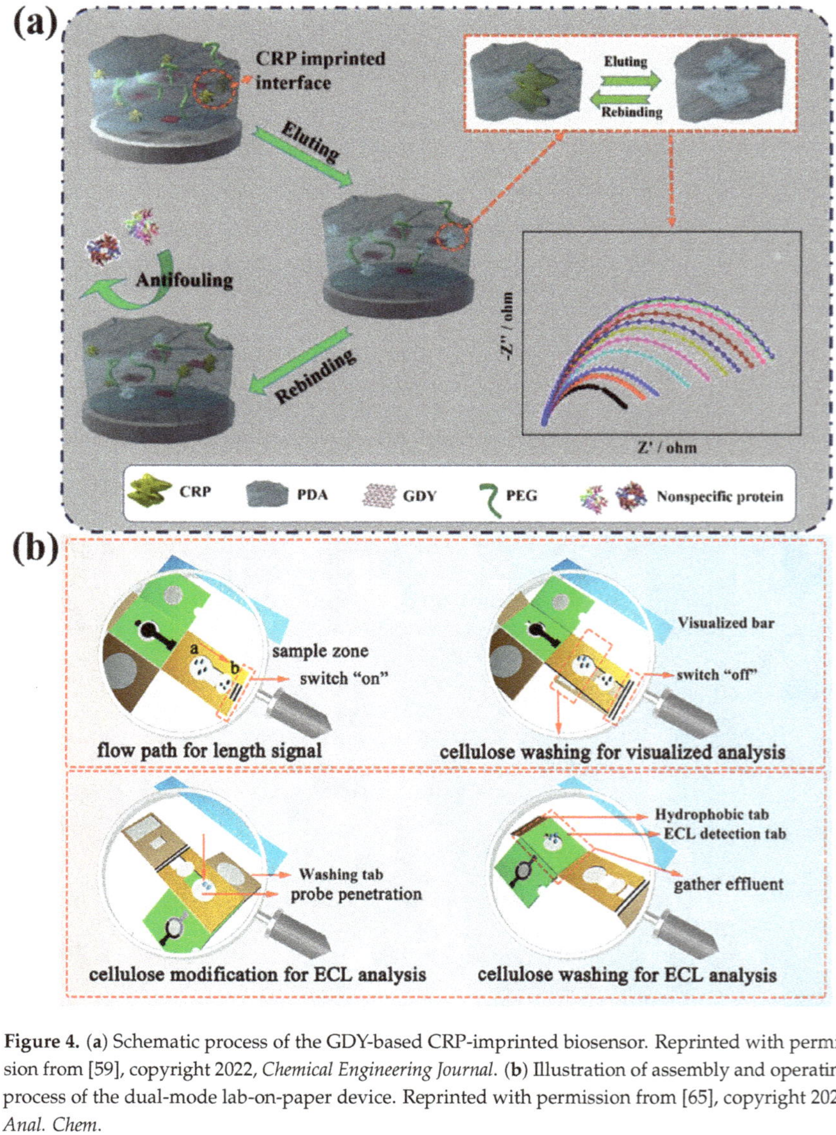

Figure 4. (**a**) Schematic process of the GDY-based CRP-imprinted biosensor. Reprinted with permission from [59], copyright 2022, *Chemical Engineering Journal*. (**b**) Illustration of assembly and operating process of the dual-mode lab-on-paper device. Reprinted with permission from [65], copyright 2020, *Anal. Chem*.

Table 2. Comparison of small-molecule-functionalized biosensor interfaces.

Functionalized Goal	Functionalized Strategy	Small Molecule	Principle	Effect	Reference
Construction of a highly sensitive protein MIP biosensor	MIT *	Dopamine	Hydrogen bond, multipoint electrostatic attraction	Achieves the highly sensitive and selective recognition of human C-reactive protein	[59]
Development of an electrocatalytically amplified assay for analysis of HER-2 *	Reaction of Au with mercapto groups	Polyethylene glycol	Au-S bond	Achieves sensitive and specific sensing of cancer biomarkers in serum	[60]
Preparation of an electrochemical sensor via the co-assembling of DNA probe and 6-mercapto-1-hexanol onto a gold electrode	Reaction of Au with mercapto groups	6-mercapto-1-hexanol	Au-S bond	Analyzes dynamic DNA methylation process	[61]
Integration of visual readings and ratiometric ECL analysis on paper substrates	Direct adsorption	Luminol	Au-N bond, electrostatic interaction	Obtains accurate monitoring performance for H_2O_2	[65]
Preparation of a new luminol-ROS ECL system to detect GSH *	Direct adsorption	Luminol	Au-N bond, electrostatic interaction	Greatly promotes the ECL emission	[66]
Forster resonance energy transfer based ratiometric imaging of lysosomal HOCl *	Redox reaction	Non-fluorescent spirothioether unit	Electron transfer	Exhibits distinct biochemical properties for facile monitoring of HOCl via conventional flow cytometry	[68]
Exhibition of high detection sensitivity to $OONO^-$ *	Chemical reaction	Cyanine 3, cyanine 5	Condensation reaction	Produces a ratiometric fluorescence signal	[69]
Development of "smart" noninvasive bioimaging probes for trapping specific enzyme activities	Substitution reaction	β-galactosidase	Nucleophilic substitution	Real-time fluorescence quantification; capture of in vivo and in situ β-GAL activity	[70]
Use of new fluorophore (azulene) to prepare an effective two-photon fluorescent probe	Inversion of internal charge transfer	Boronate	Electron transfer	Detects reactive oxygen species; has good cell penetration	[71]
Development of fluorescent probes that can show different modes of fluorescence signals fo distinct concentrations	Substitution reaction	2,4-dinitrobenzenesulfonate, the chloro group	Nucleophilic substitution	Shows the signal in the low-concentration range of thiols and the ratio response to high-concentration thiols	[72]
A conformationally induced "off–on" tyrosine kinase cell membrane fluorescent sensor	Linker group connectivity	Sunitinib, pyrene	Hydrogen bond	Enables fluorescence microscopy imaging of receptor protein tyrosine kinases in the cell membranes of living cells	[73]

Table 2. *Cont.*

Functionalized Goal	Functionalized Strategy	Small Molecule	Principle	Effect	Reference
Design of a mitochondria-specific coumarin pyrrolidinium-derived fluorescence probe	Linker group connectivity	7-diethylamino-coumarin moiety	Hydrogen bond	Allows real-time ratio imaging of HCN in living cells	[74]

* MIT: molecular-imprinting technology; HER-2: human epidermal growth factor receptor-2; GSH: glutathione; HOCl: hypochlorous acid; $OONO^-$: peroxynitrite.

2.3. *Biomacromolecule-Mediated Interfacial Regulation*

In response to the growing demand for biosensing, the development of biomacromolecule-mediated sensing interfaces with precise molecular recognition, controllable signal amplification, and enhanced sensing performance has become imperative. Various biomacromolecules, including enzymes, DNA, proteins, and cells, have been employed to modify the sensing interface, enabling the creation of highly sensitive and selective biosensors [32].

2.3.1. Enzyme-Based Interfaces

Enzymes, in particular, are widely utilized components in biosensor interfaces, leveraging their ability to detect targets with high specificity through the catalytic conversion of the analyte or enzyme inhibition [7]. To achieve a superior sensing performance, it is crucial to immobilize efficient biological enzymes onto the sensing interface while preserving their biological properties. Consequently, numerous immobilization methods have been developed, ranging from membrane encapsulation and physical adsorption to covalent binding [75].

For example, Gu et al. [76] employed a membrane encapsulation method to modify the electrode surface with glucose oxidase (GOx) for glucose sensing (Figure 5a). They utilized a 3D porous $Ti_3C_2T_x$·Mxene–graphene (MG) hybrid film with an adjustable porous structure, which provided a highly hydrophilic microenvironment for GOx immobilization. Dhanjai et al. [77] employed chitosan (CHI) as a binder for the physical adsorption of glucose oxidase (GOx) onto a cobalt oxide-loaded mesoporous carbon framework (Co_3O_4 @ MCF), enabling highly selective glucose detection upon the modification of a glassy carbon electrode. Levien et al. [7] covalently linked GO_x and acetylcholinesterase (AChE) onto the surface of plasma-polymerized (pp) hydrogels to fabricate an electrochemical biosensor interface. They utilized glucose as the substrate for GOx and eserine as the AChE inhibitor to validate the practicality of the biosensing approach.

2.3.2. DNA-Based Interfaces

DNA, a genetic biological macromolecule, plays a crucial role in regulating life activities with remarkable accuracy. Additionally, DNA possesses the unique attribute of sequence programmability, allowing for the incorporation of desired functions through well-designed sequences [78]. Currently, DNA modifications can be broadly categorized into chain DNA, DNA nanomaterials, and framework DNA (FDNA) [79–81]. For instance, Su et al. [80] investigated the impact of probe DNA on the detection performance of adenosine triphosphate aptamer (ATPA) and adenosine triphosphate (ATP) in the presence of hairpin DNA and double-stranded DNA (DsDNA). Subsequently, Zhu et al. [82] separated the two and developed two aptamer sensors based on hairpin DNA (HDNA) and linear single-stranded DNA (ssDNA) for the detection of aflatoxin B1 and diethyl phthalate.

In the DNA/nano-functionalized electrochemical sensing interface, thiol-modified DNA is immobilized on the gold-electrode surface through strong gold–thiol interactions [81]. Miao et al. [83] proposed an electrochemical biosensor utilizing DNA-modified Fe_3O_4 @ Au magnetic NPs (Figure 5b). The three DNA probes contain specific

mismatched base pairs, and the presence of different heavy-metal ions promotes hybridization with distinct DNA probes, enabling the detection of corresponding electrochemical species on the magnetic nanoparticle surface. Han et al. [84] anchored methylene blue-labeled polyadenine DNA onto a gold electrode to construct an electrochemical biosensor interface for the detection of the COVID-19 virus.

Comparatively, FDNA offers a higher probe density on the electrode surface compared to DNA nanomaterials [79]. Su et al. [85] covalently coupled DNA tetrahedrons to the carbon surface and applied them for the detection of various bioactive molecules. Furthermore, Mao et al. [86] developed a three-dimensional pure DNA hydrogel serving as a scaffold for electron transfer. The DNA hydrogel incorporates an embedded electron mediator through binding, while DNAzyme is introduced at the hydrogel scaffold nodes to enable the long-range acquisition of DNAzyme catalytic signals.

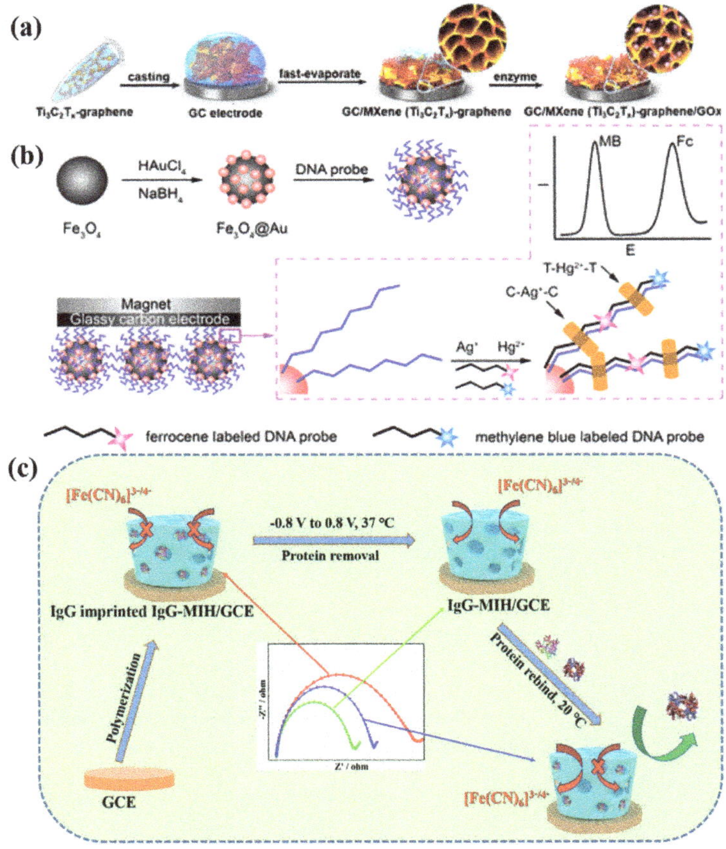

Figure 5. (**a**) Preparation of MG hybrid film for enzyme immobilization. Reprinted with permission from [76], copyright 2019, *ACS Applied Nano Materials*. (**b**) Preparation schematic diagram of DNA-modified Fe_3O_4@Au NPs and detection of Ag^+ and Hg^{2+}. Reprinted with permission from [83], copyright 2017, *ACS Appl Mater Interfaces*. (**c**) Schematic fabrication protocol of the electrochemical biosensor based on IgG-imprinted hydrogel. Reprinted with permission from [87], copyright 2021, *Sensors and Actuators B: Chemical*.

2.3.3. Protein-Based Interfaces

Proteins, as quintessential biological macromolecules, exhibit distinct binding affinities and sequence specificities towards DNA and RNA, as elucidated by Berezovski et al. [88].

For instance, Campuzano et al. [89] demonstrated that Tombusviral p19 protein forms dimers and selectively binds to short dsRNA, making it a valuable tool for identifying miRNAs upon hybridization with specific RNA probes. Rubio et al. [90] harnessed the concept of de novo-designed protein switches to create protein-based biosensors, enabling the reversal of information flow. Applying this approach, they developed a biosensor for the SARS-CoV-2 spike protein, incorporating a de novo-designed spike receptor-binding domain (RBD) binder4, with a remarkable limit of detection of 15 pM. Innovative techniques involving the combination of proteins and molecular-imprinting technology (MIT) have led to the development of protein-imprinted hydrogels (MIHs) [91]. Extensive efforts have been made in this area. Utilizing free-radical polymerization, a new type of protein-imprinting hydrogel was created by combining N,N′-methylenebis(acrylamide) cross-linkers, N,N′-dimethylaminoethyl methacrylate gas-sensitive monomers, and human serum albumin (HSA) as the template protein. This hydrogel exhibited unique self-recognition properties towards HSA protein [92]. Furthermore, Cui et al. [87] synthesized a protein-imprinted hydrogel using acrylamide as a functional monomer through free-radical polymerization. This hydrogel demonstrated high sensitivity and selectivity for detecting target immunoglobulins in complex biological samples (Figure 5c).

2.4. Cell-Based Interfaces

Cell-mediated biosensing interfaces have emerged as promising tools in biomedicine [93]. For example, Qi et al. [94] utilized marine-pathogen sulfate-reducing bacteria (SRB) to facilitate the synthesis of bio-imprinted membranes, leading to the formation of a biosensing interface. The electrochemical impedance method enabled the direct measurement of the bacterial concentration. Building upon this work, Jiang et al. [95] employed electropolymerization to create a Salmonella-imprinted membrane, resulting in the development of a biosensor interface capable of detecting Salmonella within a short span of 20 min. In addition to molecular-imprinting technology (MIT), red blood cell (RBC)-mediated biosensing interfaces have also witnessed significant advancements. Shete et al. [96] employed a reversible membrane-opening/resealing method to incorporate a designed nanosensor into RBCs, facilitating the detection of exchangeable Pb^{2+} concentrations over time and space. Furthermore, Chen et al. [97] immobilized gold-coated Fe_3O_4 core–shell nanocomposites on RBCs, enabling the use of RBCs as cell biosensors for detecting H_2O_2. As technology continues to progress, cell-mediated interface regulation strategies are becoming increasingly sophisticated. In 2018, Liu et al. [98] combined red cell membrane vesicles with near-infrared persistent luminescent nanophosphors (PLNPs) and employed mesoporous SiO_2 as a carrier to fabricate a biosensor interface for the in situ monitoring of tumor-growth inhibition. Although there are relatively fewer examples of cell-mediated biosensing interfaces compared to nanomaterials and other biomolecules, their development prospects remain highly promising. This is attributed to the superior biocompatibility of cells, which allows biosensors to play a crucial role in human medicine. Table 3 shows a comparison of biomacromolecule-functionalized biosensor interfaces.

Table 3. Comparison of biomacromolecule-functionalized biosensor interfaces.

Functionalized Goal	Functionalized Strategy	Biomacro-Molecule	Principle	Effect	Reference
Construction of a 3D porous $Ti_3C_2T_x$ MG * hybrid film for the determination of glucose in serum	Membrane encapsulation	GO_x	Wrapped	Enhances the stable fixation and retention of GO_x in the membrane	[76]
Preparation of advanced functional nanostructures based on Co_3O_4@MCF *	Physical adsorption	GO_x	Hydrogen bond	Highly selective detection of glucose	[77]

Table 3. *Cont.*

Functionalized Goal	Functionalized Strategy	Biomacro-Molecule	Principle	Effect	Reference
Study of the bio-interaction properties of PP hydrogel composed of HEMA * and DEAEMA * on SPE	Covalent binding	GO_x	Covalent bond	Optimizes the response of the sensor via different biomolecular ratios	[7]
Construction of a targeted induced hairpin-mediated biosensing interface	Reaction of Au with mercapto groups	Hairpin DNA	Au-S bond	Enhances the effect of probe DNA on the detection performance of ATPA * and ATP *	[80]
Development of a dual-ratiometric electrochemical apta-sensing strategy for the simultaneous detection of AFB1 * and OTA *	Complementary base pairing	ssDNA	Hydrogen bond	Greatly improves the assembly and recognition efficiency of the sensing interface	[82]
Development of an electrochemical biosensor based on DNA-modified Fe_3O_4 @ Au magnetic NPs for the detection of trace heavy-metal ions	Reaction of Au with mercapto groups	DNA	Au-S bond	Detects heavy-metal ions with no obvious interference at the same time; maintains the high sensitivity	[83]
Proposition of an electrochemical aptamer sensor based on CRISPR/CAS12a	Reaction of Au with mercapto groups	Polyadenine DNA	Au-S bond	Detects COVID-19 NPs * rapidly and is ultrasensitive	[84]
Preparation of an electrochemical biosensor by using a specific DNA skeleton-DNA tetrahedron	Covalent binding	DNA tetrahedron	Covalent bond	Detects a variety of bioactive molecules with high signal-to-noise ratio, sensitivity, and specificity.	[85]
Construction of an electrochemical biosensor by introducing DNAzyme with peroxidase-like activity into the junction of hydrogel	Embedding binding	DNAzyme	π–π conjugation, hydrophobic interaction	Overcomes the limitation of two-dimensional electrode, obtains long-distance catalytic signal of DNAzyme	[86]
Construction of a specific and long-acting antifouling biosensor interface based on protein-imprinted hydrogel	Complex	Template protein IgG *	Multiple-point electrostatic interaction, hydrogen bond	Detects target immunoglobulins in complex biological samples	[87]
Preparation of a new gas-sensitive-imprinted hydrogel via free-radical polymerization	Free-radical polymerization method	π bond is broken and start of poly-merization reaction	Human serum albumin (HSA)-template proteins	Shows unique self-recognition characteristics to HSA protein	[92]

* MG: Mxene–graphene; GO_x: glucose oxidase; MCF: mesoporous carbon framework; HEMA: hydroxyethyl methacrylate; DEAEMA: 2-(diethylamino)ethyl methacrylate; SPE: gold screen-printed electrodes; ATPA: adenosine triphosphate aptamer; ATP: adenosine triphosphate; AFB1: aflatoxin B1; OTA: ochratoxin A; NP: nucleocapsid protein; IgG: immunoglobulin G.

3. Electro-Click Chemistry for the Functionalization of Biosensor Interfaces

In 2001, Sharpless et al. [8] first proposed the concept of connecting small units with heteroatom connections (C-X-C) to generate matter. They hope to develop a set of powerful and selective "modules" that can work reliably in both large and small applications. The basis of this method is named "click chemistry". In simple terms, click chemistry is a selective assembly of two molecular building blocks under mild reaction conditions. It has the characteristics of high yield, harmless by-products, and easy separation via non-chromatographic methods [14].

For click chemistry, carbon–heteroatom-bond-formation reactions are the most common examples, including cycloadditions of unsaturated species, nucleophilic substitution chemistry, carbonyl chemistry of the "non-aldol" type, and additions to carbon–carbon multiple bonds [8]. Then, researchers from the same group reported that the reaction rate and regioselectivity can be improved by using Cu(I) to catalyze the Huisgen dipolar cycloaddition in 2002 [99]. A few years later, Sharpless et al. [100] and Rodionov et al. [101] further found that Cu(I)-catalyzed alkyne–azide cycloaddition (CuAAC) can effectively form 1-4-disubstituted 1,2,3-triazole bonds in water and organic solvents (Figure 6). Today, CuAAC has become the most representative cycloaddition reaction, compatible with a wide range of solvents, pH values, and temperatures [102]. Of course, there are also the thiol–ene reaction, Michael addition reaction, imine and oxime reaction, and Diels–Alder cycloaddition reaction [103]. Click chemistry is considered an effective strategy to immobilize biomolecules while maintaining their biological activities [104–106]. This review will take the CuAAC click reaction as an example to introduce the development in recent years.

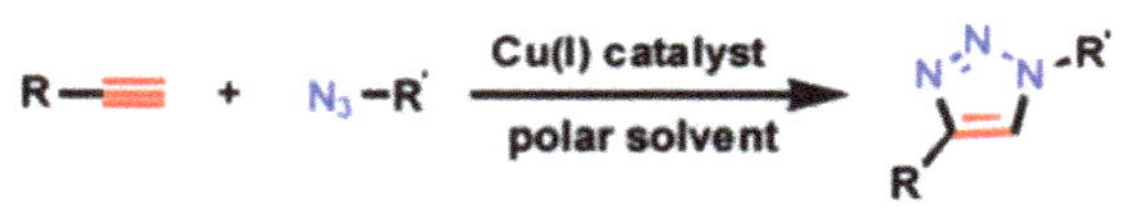

Figure 6. CuAAC process. Reprinted with permission from [107], copyright 2008, *J. Am. Chem. Soc.*

3.1. Traditional Click Technology

Because CuAAC is an addition reaction of azides and alkynes catalyzed by Cu(I), Cu(I) can be directly added to the reaction environment or in situ generated by the reaction of Cu(II) with reducing agents [9]. Although Cu(I) can be directly added to the reaction system, the development of stable Cu(I) catalysts is of great significance for the application of CuAAc due to the poor stability of Cu(I). Yang et al. [108] prepared $Cu_{2-x}S_ySe_{1-y}$ nanoparticles, which have a large number of Cu vacancies, and Cu(II) and Cu(I) coexist, to catalyze the click chemistry of CuAAC (Figure 7a). Because glutathione can stabilize Cu(I), Zhang et al. [104] successfully prepared a novel nanocatalyst containing abundant and stable Cu (I). Therefore, the reduction in Cu(II) by a reducing agent is also a common method of the CuAAC click reaction. Guerrero et al. [105] reported the first electrochemical immunosensor for the detection of CXCL7 (chemokine (C-X-C motif) ligand 7) as an autoimmune marker via the click reaction of azide-functionalized MWCNTs and ethynyl-IgG on screen-printed carbon electrodes using ascorbic acid to reduce the Cu(II) (Figure 7b). Then, Xue et al. [109] used specific antigen–antibody recognition to trigger the in situ reduction in Cu MOF, thereby generating a powerful click catalyst for the CuAAC click reaction.

However, as a catalyst, Cu(I) has certain toxicity, and some Cu-catalyzed fixed viruses or oligonucleotide chains will degrade [111]. Bertozzi et al. [112] first proposed copper-free click chemistry, which eliminates the adverse effects of Cu(I), simplifies the experimental steps, and becomes an alternative CuAAC click reaction. This is an alternative method to activate alkynes for the catalyst-free [3+2] cycloaddition reaction with azides. Then, Xiang et al. [110] reported a biosensor system based on azide-co-functionalized graphene oxide (GO-N_3) and carbon dots (CDs). Carbon-dot-labeled DNA (CD-DNA) binds to GO-N_3 by copper-free click chemistry and quenches the fluorescence of CDs by fluorescence resonance energy transfer (FRET) (Figure 7c). After adding carcinoembryonic antigen (CEA), CEA binds to the aptamer and fluorescence recovers. The latest research progress was proposed by Yang et al. [113]. They designed and constructed a nanopore complex containing unnatural amino acid, which is a conical rigid complex formed by octameric Mycobacterium smegmatis porin A (MspA). Through the strain-promoted azide–alkyne cycloaddition reaction, single-stranded DNA composed of 40 thymines (poly-T(40)) and lysozyme molecules with dibenzocyclooctyne (DBCO) were successfully connected to the functionalized nanopore interface, and it was found that it exhibited a new signal distribution pattern and the signal was significantly enhanced. When combined with different oligosaccharide substrates, the corresponding signals will change significantly. The prepared functionalized interface provides a new direction for single-molecule sensing. We can see that click chemistry is also an attractive technique in the biomedical field. However, there are only a few papers on the latest application of click chemistry in the biomedical field [15].

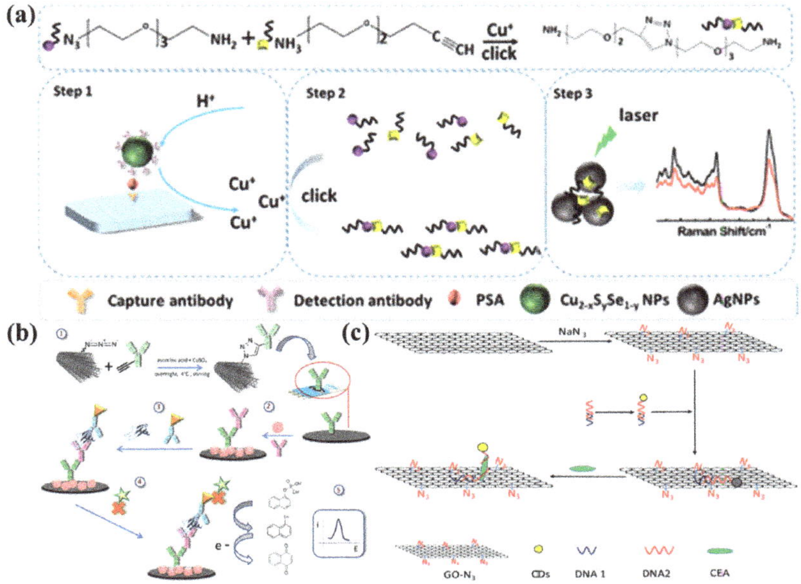

Figure 7. (**a**) $Cu_{2-x}S_ySe_{1-y}$-NP-catalyzed click chemistry for SERS immunoassay of PSA detection. Reprinted with permission from [108], copyright 2018, *Anal Chem*. (**b**) Schematic display of the different steps involved in the preparation of the immunosensor for the determination of CXCL7. Reprinted with permission from [105], copyright 2019. *Journal of Electroanalytical Chemistry*. (**c**) Schematic illustration of CEA detection system based on click chemistry and FRET between GO-N_3 and CD -DNA. Reprinted with permission from [110], copyright 2020, *Anal Chim Acta*.

3.2. Electro-Click Technology

In the CuAAC click reaction, Cu(I) salts or the in situ reduction in Cu(II) are commonly employed to introduce a catalytic amount of Cu(I). Various methods can be used to generate the required Cu(I) ions, including microwave [114], ultrasound [115], UV irradiation [116], and electrochemical reactions [117]. The electrochemical approach enables the reagent-free functionalization of biosensing interfaces, known as electro-click [16]. Devaraj et al. [118] demonstrated that catalytic Cu(I) ions can be generated by applying a mild reduction potential (−300 mV vs. Ag/AgCls) at the interface in contact with the Cu(II) solution. For example, Lesniewski et al. [119] deposited gold nanoparticles modified with terminal alkynyl groups onto azide-functionalized glassy carbon surfaces. The generation of the Cu catalyst can be precisely controlled by adjusting the treatment time, allowing control over the surface coverage by altering the deposition time. Villalba et al. [120] achieved the catalytic assembly of multilayers through the covalent bonding of poly(acrylic acid) (PAA) multilayers with alkynyl (PAAalk) and azide (PAAaz) groups. Recently, Yamamoto et al. [121] utilized electro-click chemistry to modify boron-doped diamond (BDD) electrodes. The electro-click strategy enables precise control over the amount of Cu(I) catalyst and the loading of azide-terminated ferrocene on alkyne-terminated BDD electrodes. Compared with the chemical synthesis method, the electro-click approach offers advantages such as faster reaction kinetics, environmental friendliness, and the absence of chemical residues. Furthermore, the assessment technique inherent to electro-click chemistry is inherently uncomplicated, affording the direct interpretation of the reaction through the analysis of the cyclic voltammetry profile.

These reasons contribute to the ongoing popularity of the electro-click strategy.

3.3. Electro-Click Strategies for the Functionalization of Biosensor Interfaces

Before constructing biosensors, careful attention must be given to the functionalization strategies of the biosensor interfaces in order to meet diverse sensing requirements. The resulting interface should not only immobilize the biometric element in a stable manner, but also offer a selective and sensitive response. In this section, we will discuss the functionalization strategies for electrochemical biosensor interfaces and optical biosensor interfaces.

3.3.1. Electro-Click Strategies for the Functionalization of Electrochemical Biosensor Interfaces

Electrochemical biosensors have long been extensively utilized across various fields. These biosensors can generate sensing signals by monitoring changes in impedance, current, and dielectric properties when the target analyte is formed on the electrode surface. Numerous strategies exist for the modification of interfaces in electrochemical biosensors, encompassing approaches such as self-assembled monolayers, covalent immobilization, and electro-click methodologies, among others [122,123]. Concurrently, a diverse array of electrochemical biosensor variants is available. Notably, owing to its electrochemically mediated modifications and notable versatility, the electro-click strategy finds predominant utility within the domain of electrochemical biosensors. The functional modification strategies based on electro-click for electrochemical biosensor electrodes can be categorized into four parts: electrografting, electropolymerization, electrodeposition, and bipolar electrode methods.

Electrografting

The most commonly employed approach for modifying biosensing interfaces is electrografting [124,125]. Electro-click-mediated electrografting plays a crucial role in preparing controllable nanostructures, enhancing the sensor performance, and expanding the sensor functionalities [126]. In simple terms, azide and alkyne are grafted onto the biosensing interface and modifier, respectively, and then connected by the CuAAC electro-click reaction. For instance, Sciortino et al. [127] achieved a network structure of organic/inorganic hybrid products containing alkynes and polyethylene glycol azides with bifunctional groups through the electro-click reaction on an F-SnO_2 electrode. The composition and loading of these hybrid films can be adjusted, demonstrating the versatility of such hybrid coatings. Fenoy et al. [126] reported the synthesis of azide-derived organic electrochemical transistor monomers and applied them in a CuAAC reaction with a dibenzocyclooctyne-grafted thrombin-specific HD22 aptamer to achieve specific thrombin recognition. Also, Guerrero et al. [128] utilized ethynylated IgG bonded to azide-modified multi-walled carbon nanotube (MWCNT) electrodes to construct an electrochemical immunosensor for the cytokine interleukin 1b (IL-1b). The detection limit of this immunosensor was significantly improved to 5.2 pg/mL compared to the commercial kit. Effective coupling methods between interfaces are vital for achieving both sensitivity and selectivity. Therefore, we employed cucurbituril- and azide-grafted graphene oxide to develop a novel functional nanomaterial as a biosensor interface for the detection of VEGF165 protein (Figure 8a) [129]. This interface demonstrated the simultaneous acquisition of sensitivity and selectivity, with a dynamic detection range of from 10 fg/mL to 1 ng/mL and a detection limit of 8 fg/mL. Electro-click-mediated electrografting enables the preparation of multipoint-specific electrode functionalization for multi-target biosensors on various materials. The examples discussed above clearly illustrate the potential of electro-click-mediated electrografting in expanding the capabilities of biosensor interfaces [130].

Electropolymerization

Electropolymerization (EP) refers to the network polymerization of solution components into a film when applying electrical stimulation through cyclic voltammetry (CV) [131]. This unique film structure enables the immobilization of functional molecules on or within the coating, making it suitable for combining with electro-click chemistry to create new functional biosensing interfaces. Rydzek et al. [131] developed self-assembled functional polymer films based on aniline and naphthalene using a one-pot method that involved simultaneous EP and electro-click functionalization. CV facilitated the oxidation of 4-azidoaniline and the reduction in Cu(II) ions, allowing for the simultaneous polymerization of the former and the CuAAC click reaction. For electropolymerization on conductive substrates, electrochemically mediated surface-initiated atom-transfer radical polymerization (SI-eATRP) offers a convenient approach to control polymer growth with low catalyst concentrations [132]. Wu et al. [133] achieved the fixation of sparse monomolecular membranes of ethynylphenyl on a carbon matrix through the electroreduction in aryldiazonium ions. They subsequently polymerized Poly(N-isopropylacrylamide) (PNIPAM) from the substrate in a one-pot solution containing an azide-derived initiator, PNIPAM, and a Cu catalyst (Figure 8b). The electro-reduction from Cu(II) to Cu(I) facilitated both the click reaction and SI-eATRP on the surface of the ethynylphenyl. This method is simple, convenient, and combines the advantages of electro-click chemistry and SI-eATRP.

Electrodeposition

Electrodeposition, as an electrochemical synthesis method based on solution-based synthesis, allows for the synthesis of different material types and the manipulation of various synthetic variables affecting morphology. It also enables the fabrication of uniform and tightly joined multi-junction electrodes through continuous multilayer deposition [134]. This feature proves particularly beneficial for synthesizing and functionalizing biosensing interfaces.

Conductive hydrogels, among various synthetic materials, have garnered attention from researchers. These hydrogels consist of conductive polymer networks formed by combining hydrogels with inherently conductive polymers. They offer a combination of biocompatibility, flexibility, high diffusivity, and conductivity [135,136]. Hydrogels can be prepared in specific forms on various substrates to suit the desired purpose. Hu et al. [137] described a method for coating chitosan hydrogel on a conductive surface using the electro-click-mediated electrodeposition technique. Chitosan was functionalized with azide or alkynyl groups, and a cathodic potential was applied to a gold chip to reduce Cu(II) ions to Cu(I) ions, catalyzing the CuAAC click reaction between the alkynyl chitosan and azide chitosan to form a conjugated chitosan hydrogel attached to the gold surface (Figure 8c). Experiments demonstrated the encapsulation of biomolecules within this hydrogel for biosensing purposes. Similarly, Choi et al. [138] electrodeposited PVA-based hydrogel on an indium tin oxide glass electrode using a Cu(I)-catalyzed CuAAC reaction with the reduction in Cu(II) ions. They found that embedding carbon nanotubes within the hydrogel enabled the deposition of thicker films due to the larger electrochemically active area provided by the carbon nanotubes.

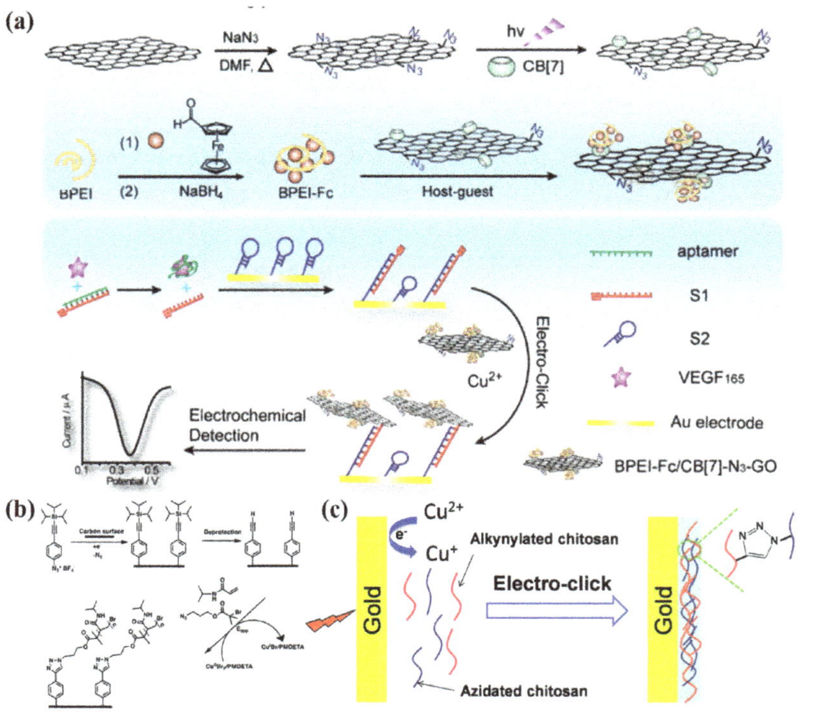

Figure 8. (**a**) Schematic of the BPEI-Fc/CB [7]-N_3-GO composite preparation and representation of the electro-click biosensing platform for VEGF165 analysis based on the composition. Reprinted with permission from [129], copyright 2017, *Anal Chem.* (**b**) Strategy for the preparation of H-Eth-Ar monolayers and grafting PNIPAM. Reprinted with permission from [133], copyright 2019, *ChemElectroChem*. (**c**) Schematic illustrating the "electro-click" chitosan hydrogel on a gold chip triggered by a negative potential that reduces Cu^{2+} to Cu^{+}. Reprinted with permission from [137], copyright 2014, *RSC Advances*.

Bipolar Electrodes

Bipolar electrodes (BPEs) can simultaneously perform anodic and cathodic reactions, making them a versatile electrochemical-reaction-driving mode and suitable as radio electrodes [139]. BPEs do not require electrical connection and can accommodate conductive materials of any shape and size, and even multiple materials at once (Figure 9a) [140]. These characteristics align well with electro-click chemistry (Figure 9b). The potential gradient at the BPE interface has been successfully utilized as a controllable template for forming molecular- or polymer-gradient materials, making it suitable for biomimetic materials or biosensor analysis equipment [141]. A study was conducted on the gradient modification of azide-functionalized conductive-polymer film using electrogenerated Cu(I) species on a bipolar electrode through an electroshock reaction in the presence of terminal alkyne [142]. This allowed the introduction of various gradient functions, such as gradient hydrophobicity/hydrophilicity and visible labeling, on the poly(3,4-ethylenedioxythiophene) (PEDOT) film to prepare the biosensing interface. In subsequent work, Zhou et al. [143] introduced a bipolar electrolytic micelle disruption (BEMD) system for shaping organic films. They created a U-shaped bipolar electrolysis system with an S-shaped potential gradient on the BPE, enabling the formation of a gradient film containing various organic compounds. This approach provided a novel idea and method for modifying organic membranes using the CuAAC electro-click reaction.

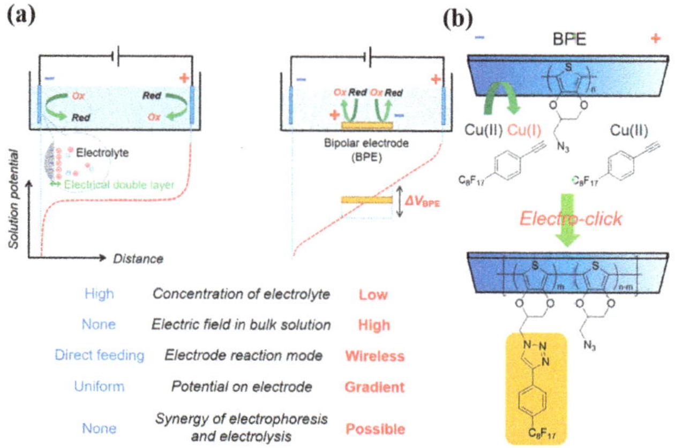

Figure 9. (**a**) Comparison of electrolytic systems for conventional and bipolar electrolysis. Red dotted line represents an ideal electric field generated between driving electrodes. (**b**) Scheme for the electro-click reaction of an azido-functionalized polymer film on a BPE with an alkyne derivative using an electrogenerated Cu(I) species. Reprinted with permission from [140], copyright 2019, *Acc Chem Res*.

Applications of Electro-Click Strategies in Electrochemical Biosensor Functionalization with Submicron Resolution

Numerous techniques are available for effecting functional alterations in biosensing interfaces, including single-molecule modification/deposition [144]. However, these methods seldom permit intricate modifications at the micron scale. Furthermore, they often exhibit constrained control over interface surfaces or necessitate costly equipment [145]. In stark contrast, a notable facet of electro-click chemistry, which entails electrochemical reactions, is its spatial selectivity. This distinctive attribute facilitates the precise functionalization of nanogaps, gradients, and microelectrodes, all achieved with submicron precision.

The utilization of NPs as fundamental building blocks has catalyzed the emergence of a novel category of nano-building materials, thereby illuminating a promising avenue for nano-gap functionalization. In a pioneering endeavor, Rydezk et al. [146] orchestrated a nanodevice with a high aspect ratio, selectively focused through covalent self-assembly via spatially arranged Cu(I)-catalyzed electro-click reactions. Employing dendritic iron oxide nanoparticles endowed with azide and alkyne functionalities as building blocks, they achieved a spatially orchestrated electro-click network structure. This method of NP functionalization proves highly adaptable and can be extrapolated to diverse nanoparticle varieties featuring clickable moieties. As such, the electro-click approach emerges as a sanguine instrument for seamlessly integrating covalent NPs into nanodevices, thereby beckoning forth prospects for biosensors and granular electronic devices [147].

The manipulation of physicochemical property gradients—marked by gradual spatial and temporal transitions—can substantially amplify the efficacy of catalyst and drug design, introducing novel analytical avenues of considerable import in solution and interface functionalization. Krabbenborg et al. [148] harnessed electro-click chemistry to fabricate a solution gradient, thereby enabling the high-throughput control and monitoring of the surface reactivity across spatial and temporal dimensions. By virtue of the solution gradient, finely adjustable and inherently predictable variations in the spatial concentration of chemically active species are achieved, engendering micron-scale surface gradients. In analogous fashion, Nicosia et al. [149] harnessed the CuAAC reaction to generate a monolayer on a glass surface flanked by platinum electrodes, effectively manipulating the gradient of the Cu(I) solution to effect microelectrode interface functionalization.

Beyond gradient-driven surface modification, the repertoire of electro-click chemistry extends to the direct functionalization of microelectrodes. Hansen et al. [150] directly functionalized microelectrodes with PEDOT-N_3, showcasing a universally applicable and convenient strategy for diverse conductive-polymer functionalization. Notwithstanding the aforementioned applications, numerous prospects for submicron-resolution applications await exploration.

3.3.2. Electro-Click Strategies for the Functionalization of Optical Biosensor Interfaces

In recent times, the augmentation of optical biosensors using the electro-click approach has gained prominence. This development is attributed to the understanding that electron dynamics can arise not exclusively from direct electrochemical processes, but also from alternative energy excitations [151]. Consequently, as investigations into the interplay between photons and electrons advance, the electro-click strategy has found application within the realm of optical sensing [142,152]. Optical biosensors rely on measuring changes in the optical properties of the transducer surface when the analyte and recognition element form a complex [153]. They can be categorized as direct optical biosensors or indirect optical biosensors. While there are limited examples of optical biosensing interfaces prepared using electro-click chemistry, luminescent products prepared through electro-click have been identified. It is expected that similar functionalization strategies will be applied in the near future to prepare corresponding optical biosensors.

Direct optical biosensor signals depend on the complex formation on the biosensor interface. Shida et al. [142] used rhodamine-based acetylene as a visible indicator. After electroshock reaction on the bipolar electrode (Figure 10a), the polymer film was uniformly oxidized to the doped state, reducing the background absorption of PEDOT in the visible-light region. Consequently, the cathode part turned purple, allowing for naked-eye detection (Figure 10b). In addition to the existing direct optical sensor, Goll et al. [152] blended EDOT and branched 2,2′:3′,2′′-terthiophene (3T) to form an EDOT-N_3 film that facilitated functional modification through an electro-click reaction. By altering the mixing ratio of the two materials, the optical properties exhibited significant changes that could be detected via infrared spectroscopy (Figure 10c).

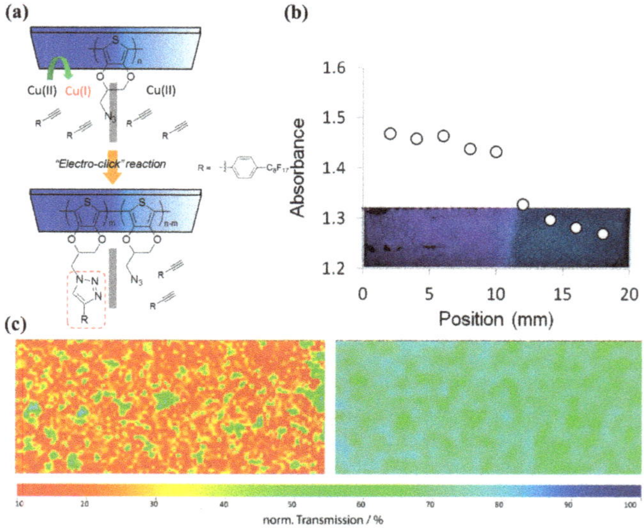

Figure 10. (**a**) Electro-click reaction of PEDOT-N_3 film and terminal alkyne using cathodically generated Cu(I) species. (**b**) Photograph of PEDOT-N_3 film gradually modified with rhodamine and the absorption profile (550 nm) across the film. Reprinted with permission from [142], copyright 2012, *ACS Macro Letters*. (**c**) IR mappings (bottom) of the azide band intensity at 2100 cm^{-1} of copolymer films deposited under potentiostatic control on gold in 0.1 M NBu_4PF_6/MeCN. Reprinted with permission from [152], copyright 2015, *Beilstein J Org Chem*.

Indirect optical biosensors commonly employ various labels, such as fluorophores, to detect and amplify signals. As early as 2008, Ku et al. [107] introduced the use of electro-click reactions to attach non-reactive fluorescent molecular patterns onto azide glass substrates (Figure 11a). This approach can be extended to enable the growth of any fluorescent molecule on an insulating substrate, forming a direct covalent bond beneath the microelectrode. The F-SnO_2 coating developed by Sciortino et al. [127] using the aforementioned electro-click reaction exhibited notable differences in fluorescence spectra when loaded with 4,4-Difluoro-8-(4-trimethylsilylethynylphenyl)-1,3,5,7-tetramethyl-2,6-diethyl-4-bora-3a,4a-diaza-s-indacene(BODIPY) (Figure 11b). This method also proves to be a viable approach for indirect optical biosensors. Coceancigh et al. [154] employed electrochemically assisted CuAAC to precisely control the inner surface of poly(ethylene terephthalate) (PET) track-etched pores. In this process, the ethynyl groups within the etched holes were first modified via the amidation of the surface-COOH groups. Subsequently, a solution containing copper (II) and azide-labeled fluorescent dye was introduced and sandwiched between gold electrodes (Figure 11c). Rydzek et al. [155] harnessed the electro-click reaction methodology to fabricate a thin film atop a gold electrode. Variations in the pH induced distinct electrostatic-repulsion patterns between PAA-Alk moieties, thereby governing the structural attributes of the resultant film. These disparities in the film structure were anticipated to manifest as discernible divergences in fluorescence responses. As a result, the fluorescent group could be directly immobilized onto the membrane. Furthermore, materials such as CNTs, CDs, and luminol exhibit light-specific properties [45,63,110]. If they can be utilized for modification, then a broader range of functional optical biosensor interfaces could be developed to meet diverse requirements.

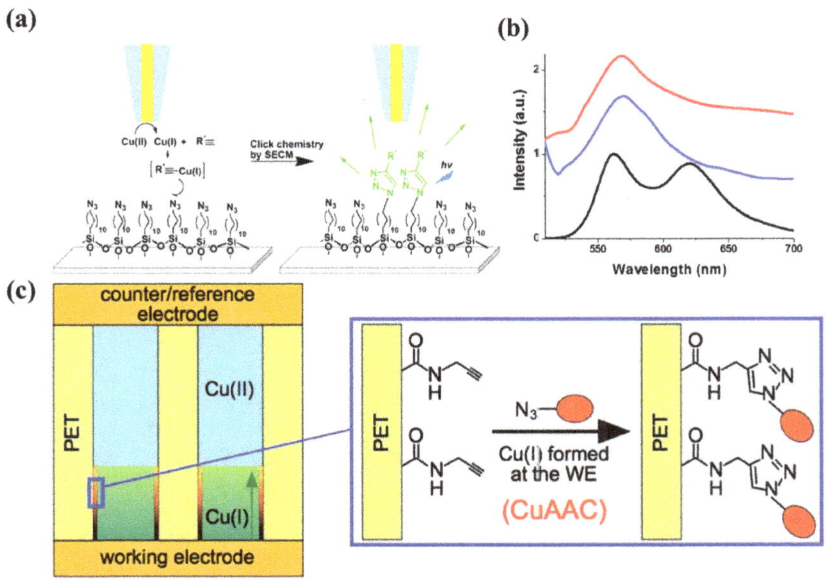

Figure 11. (**a**) Local reduction of Cu(II) to Cu(I) at a gold microelectrode (left) and immobilization of acetylene fluorophore derivatives onto a glass. Reprinted with permission from [107], copyright 2008, *Journal of the American Chemical Society*. (**b**) Fluorescence spectra in the dry state (λ_{exc} at 480 nm) of BODIPY (black line), drop-casted bodipy-loaded hybridosome (blue line), and an electro-clicked film based on BODIPY-loaded hybridosomes (red line). Reprinted with permission from [127], copyright 2018, *Phys Chem Chem Phys*. (**c**) Electrochemically controlled pore modification based on CuAAC. Reprinted with permission from [154], copyright 2017, *Langmuir*.

3.3.3. Limits of Electro-Click Strategies for the Functionalization of Biosensor Interfaces

Electro-click chemistry offers an expedient, versatile, and facile approach to biosensing interface functionalization; nevertheless, it is not without its imperfections. Several challenges persist, warranting researchers' dedicated efforts. Within the realm of biosensors, a paramount concern in the electro-click methodology is the retention of copper ions (Cu(II)/Cu(I)) during functionalization, which could potentially impede biomolecular activity. Cu(II) might catalyze the production of reactive oxygen species at the biosensing interface, engendering lipid peroxidation, protein oxidation, and DNA degradation [156]. Cu(I) is highly thiophilic and can directly damage the Fe-S-cluster-containing enzyme modifiers in the biosensing interface [157]. Schaaf et al. [158] ventured to ameliorate the Cu(II) influence via EDTA, yielding commendable outcomes. Contemporary endeavors, such as those by Cheng et al. [159], persist in deploying EDTA overdosing to tackle this conundrum. Beyond chemical mitigation, physical adsorption emerges as an effective countermeasure against Cu(II). In this vein, Liu et al. [160] prepared triethanolamine-GO nanosheets to robustly adsorb Cu(II), displaying an elevated equilibrium capacity even at higher initial Cu(II) concentrations.

Early forays into electro-click applications also confronted challenges in characterizing functionalized biosensing interfaces. Evolving scientific and technological advancements have enriched the characterization methodologies, enhancing precision and deepening researchers' appreciation for the electro-click paradigm. Presently, indirect characterization techniques, spanning SEM [104,119,127], TEM [104,127], XRD [104], BET [104], fluorescence [104,127], CV [119,120,127,128,131], and EIS [128], discerningly probe interfacial substance connections. Complementary to these indirect avenues, polarization-modulated infrared-reflectance-absorption spectroscopy (PM-IRRAS) [119,120], XPS [104,131,142], and EDS [104,127,142] have emerged as more direct means of elucidating reactions via element- or chemical-group alterations at electrode surfaces. It is firmly believed that future endeavors will engender a proliferation of characterization methods, augmenting researchers' arsenal for scrutinizing electro-click-functionalized biosensing interfaces.

The widespread adoption of the electro-click reaction rests on its attributes of expeditious reaction kinetics, ecological benignity, and minimal pollution. However, investigations into the recycling and reutilization of electro-click-functionalized biosensing interfaces have remained relatively sparse, casting a spotlight on a nascent concern for future inquiry. Encouragingly, select researchers have delved into recycling experiments concerning electro-click-functionalized biosensors. For instance, Fomo et al. [106] demonstrated the potential for electrode reuse by washing it with deionized water after sample detection, subsequently reemploying the electrode with negligible alteration in DPSV voltammograms, attesting to the sensor viability. Likewise, Qi et al. [117] exhibited a stable immunosensor retention for 15 days, with the subsequent detection revealing a commendable 88% retention of the initial steady-state current. These instances provide inspiring precedents. Thus, within the domain of electro-click-functionalized biosensing interfaces, the focus must expand beyond novel modification methodologies, encompassing heightened attention toward interface recyclability and, building upon this foundation, the exploration of straightforward recovery protocols. In general, the journey toward electro-click-modified-functionalized biosensing interfaces stretches ahead, necessitating persistent exploration and innovation.

4. Conclusions

In summary, we have comprehensively examined the different strategies for functionalizing biosensing interfaces, encompassing nanomaterials, small molecules, biomacromolecules, and cell-mediated interfacial regulations. We have also delved into the process of transforming identified elements into signals, imparting functionality to the sensor, and enhancing its performance.

Among the various functionalization strategies for biosensing interfaces, click chemistry stands out due to its advantages of rapidity, selectivity, and low pollution. We have elucidated the origins and principles of click chemistry, with CuAAC serving as an exemplar, highlighting its advantages. In the realm of electrochemical biosensors, the use of click-chemistry-based modifications on the sensing interface offers improved stability and convenience, leading to more discernible electrochemical signals. For optical biosensors, the modified sensing interface not only generates optical signals in response to the target analyte, but also amplifies the signal through the incorporated materials, thereby bolstering the reliability of the sensing outcomes. Within this domain, electro-click chemistry emerges as an environmentally friendly, swift, and convenient approach, surpassing traditional click chemistry in terms of functionalizing biosensing interfaces. We have summarized the research and application progress of electro-click chemistry in functionalizing electrochemical biosensors and optical sensors. While there may be fewer corresponding examples in optical biosensors, drawing from related literature, this review has identified potential examples that can be utilized for constructing optical biosensor interfaces.

This review provides a fresh perspective for the development of novel functional biosensing interface strategies. Moreover, the expansion of biosensing systems based on electro-click chemistry will contribute to the selection of materials and the advancement of strategies. Ultimately, this will broaden the applications of biosensors across diverse fields, such as chemistry, agriculture, food science, clinical diagnostics, pharmaceuticals, and environmental analysis.

Author Contributions: Conceptualization, T.W.; methodology, T.W.; investigation, F.W.; writing—original draft preparation, F.W., Y.X. and W.Z.; writing—review and editing, T.W., F.W., Y.X. and W.Z. All authors have read and agreed to the published version of the manuscript.

Funding: This work was supported by the National Natural Science Foundation of China (Grant No. 22074064).

Institutional Review Board Statement: Not applicable.

Informed Consent Statement: Not applicable.

Data Availability Statement: No new data were created.

Conflicts of Interest: The authors declare no conflict of interest.

Abbreviations

Acronym	Definition	Acronym	Definition
GO	Graphene oxide	ATPA	Adenosine triphosphate aptamer
Exo III	Exonuclease III	ATP	Adenosine triphosphate
HP1 and HP2	Hairpin probes	DsDNA	Double-stranded DNA
PtNFS-GO	Flower-like Pt-graphene oxide	HDNA	Hairpin DNA
8-OHdG	8-hydroxy-2′-deoxyguanosine	SsDNA	Single-stranded DNA
rGO	Reduced graphene oxide	NPs	Nanoparticles
LOD	Low detection limit	RBD	Receptor-binding domain
Fe/N-C	N-doped porous-carbon-containing Fe	HSA	Human serum albumin
ECL	Electrochemiluminescence	SRB	Sulfate-reducing bacteria
SWCNTs	Single-walled carbon nanotubes	RBC	Red blood cell
MWCNTs	Multi-walled carbon nanotubes	PLNPs	Persistent luminescent nanophosphors
NIR	Near-infrared	MCF	Mesoporous carbon framework
PDAC	Pancreatic ductal adenocarcinoma	HEMA	Hydroxyethyl methacrylate
EDOT	3,4-ethylenedioxythiophene	DEAEMA	2-(diethylamino)ethyl methacrylate
PEDOT	Poly(3,4-ethylenedioxythiophene)	SPE	Gold screen-printed electrodes
Au@Pt	Au@Pt nanocrystals	AFB1	Aflatoxin B1
FNAs	Framework nucleic acids	OTA	Ochratoxin A
E. coli O157:H7	Escherichia coli O157:H7	NP	Nucleocapsid protein
AgNCs	Silver nanoclusters	IgG	Immunoglobulin G
SC	Scopoletin	CuAAC	Cu(I)-catalyzed alkyne–azide cycloaddition
AR	Amplex Red	CXCL7	Chemokine (C-X-C motif) ligand 7

Acronym	Definition	Acronym	Definition
GSH	Glutathione	GO-N_3	Azide-co-functionalized graphene oxide
PSA	Prostate-specific antigen	CDs	Carbon dots
PCa	Prostate cancer	CDs-DNA	Carbon-dot-labeled DNA
GDY	Graphdiyne	FRET	Fluorescence resonance energy transfer
C-MIPs	C-reactive-molecular-imprinted polymers	CEA	Carcinoembryonic antigen
HRP	Horseradish peroxidase	MspA	Mycobacterium smegmatis porin A
ROS	Reactive oxygen species	DBCO	Dibenzocyclooctyne
O_2^-	Superoxide radicals	PAA	Poly(acrylic acid)
Au/VO_2	Gold-loaded 2D VO_2 nanobelts	PAAalk	Poly(acrylic acid) multilayers with alkynyl
RET	Resonance energy transfer	PAAaz	Poly(acrylic acid) multilayers with azide
MIT	Molecular-imprinting technology	BDD	Boron-doped diamond
HER-2	Human epidermal growth factor receptor 2	EP	Electropolymerization
HOCl	Hypochlorous acid	CV	Cyclic voltammetry
$OONO^-$	Peroxynitrite	SI-eATRP	Surface-initiated atom-transfer radical polymerization
GOx	Glucose oxidase	PNIPAM	Poly(N-isopropylacrylamide)
MG	Mxene–graphene	BPEs	Bipolar electrodes
CHI	Chitosan	PEDOT	Poly(3,4-ethylenedioxythiophene)
Co_3O_4 @ MCF	Cobalt oxide-loaded mesoporous carbon framework	BEMD	Bipolar electrolytic micelle disruption
AChE	Acetylcholinesterase	3T	2,2′:3′,2″-terthiophene
pp	Plasma-polymerized	PET	Poly(ethylene terephthalate)
FDNA	Framework DNA	BODIPY	4,4-Difluoro-8-(4-trimethylsilylethynylphenyl) -1,3,5,7-tetramethyl-2,6-diethyl-4-bora-3a, 4a-diaza-s-indacene

References

1. Walther, B.K.; Dinu, C.Z.; Guldi, D.M.; Sergeyev, V.G.; Creager, S.E.; Cooke, J.P.; Guiseppi-Elie, A. Nanobiosensing with graphene and carbon quantum dots: Recent advances. *Mater. Today* **2020**, *39*, 23–46.
2. Singh, S.; Kumar, V.; Dhanjal, D.S.; Datta, S.; Prasad, R.; Singh, J. Biological biosensors for monitoring and diagnosis. In *Microbial Biotechnology*; Springer: Berlin/Heidelberg, Germany, 2020; pp. 317–335.
3. Cheng, Z.; Wei, J.; Gu, L.; Zou, L.; Wang, T.; Chen, L.; Li, Y.; Yang, Y.; Li, P. DNAzyme-based biosensors for mercury (II) detection: Rational construction, advances and perspectives. *J. Hazard. Mater.* **2022**, *431*, 128606.
4. Gosai, A.; Khondakar, K.R.; Ma, X.; Ali, M.A. Application of Functionalized Graphene Oxide Based Biosensors for Health Monitoring: Simple Graphene Derivatives to 3D Printed Platforms. *Biosensors* **2021**, *11*, 384. [PubMed]
5. Bandodkar, A.J.; Wang, J. Non-invasive wearable electrochemical sensors: A review. *Trends Biotechnol.* **2014**, *32*, 363–371.
6. Zhao, Q.; Zhang, Q.; Sun, Y.; Liu, Y.; Lu, H.; Fan, X.; Wang, H.; Zhang, Y.; Wang, H. Design synthesis of a controllable flower-like Pt-graphene oxide architecture through electrostatic self-assembly for DNA damage biomarker 8-hydroxy-2′-deoxyguanosine biosensing research. *Analyst* **2018**, *143*, 3619–3627. [PubMed]
7. Levien, M.; Farka, Z.; Pastucha, M.; Skládal, P.; Nasri, Z.; Weltmann, K.-D.; Fricke, K. Functional plasma-polymerized hydrogel coatings for electrochemical biosensing. *Appl. Surf. Sci.* **2022**, *584*, 152511.
8. Kolb, H.C.; Finn, M.G.; Sharpless, K.B. Click Chemistry: Diverse Chemical Function from a Few Good Reactions. *Angew. Chem. Int. Ed.* **2001**, *40*, 2004–2021.
9. Hein, J.E.; Fokin, V.V. Copper-catalyzed azide–alkyne cycloaddition (CuAAC) and beyond: New reactivity of copper (I) acetylides. *Chem. Soc. Rev.* **2010**, *39*, 1302–1315.
10. Deng, Y.; Shavandi, A.; Okoro, O.V.; Nie, L. Alginate modification via click chemistry for biomedical applications. *Carbohydr. Polym.* **2021**, *270*, 118360.
11. Shi, W.; Tang, F.; Ao, J.; Yu, Q.; Liu, J.; Tang, Y.; Jiang, B.; Ren, X.; Huang, H.; Yang, W. Manipulating the Click Reactivity of Dibenzoazacyclooctynes: From Azide Click Component to Caged Acylation Reagent by Silver Catalysis. *Angew. Chem. Int. Ed.* **2020**, *132*, 20112–20116.
12. Yoon, H.Y.; Lee, D.; Lim, D.K.; Koo, H.; Kim, K. Copper-Free Click Chemistry: Applications in Drug Delivery, Cell Tracking, and Tissue Engineering. *Adv. Mater.* **2022**, *34*, 2107192.
13. Kondengadan, S.M.; Bansal, S.; Yang, C.; Liu, D.; Fultz, Z.; Wang, B. Click chemistry and drug delivery: A bird's-eye view. *Acta Pharm. Sin. B* **2022**, *13*, 1990–2016. [CrossRef]
14. Fitzgerald, P.R.; Paegel, B.M. DNA-encoded chemistry: Drug discovery from a few good reactions. *Chem. Rev.* **2020**, *121*, 7155–7177. [CrossRef]
15. Kim, E.; Koo, H. Biomedical applications of copper-free click chemistry: In vitro, in vivo, and ex vivo. *Chem. Sci.* **2019**, *10*, 7835–7851. [CrossRef]

16. Rydzek, G.; Ji, Q.; Li, M.; Schaaf, P.; Hill, J.P.; Boulmedais, F.; Ariga, K. Electrochemical nanoarchitectonics and layer-by-layer assembly: From basics to future. *Nano Today* **2015**, *10*, 138–167. [CrossRef]
17. Deo, K.A.; Jaiswal, M.K.; Abasi, S.; Lokhande, G.; Bhunia, S.; Nguyen, T.-U.; Namkoong, M.; Darvesh, K.; Guiseppi-Elie, A.; Tian, L. Nanoengineered Ink for Designing 3D Printable Flexible Bioelectronics. *ACS Nano* **2022**, *16*, 8798–8811. [CrossRef]
18. Zhang, W.; Zhu, S.; Luque, R.; Han, S.; Hu, L.; Xu, G. Recent development of carbon electrode materials and their bioanalytical and environmental applications. *Chem. Soc. Rev.* **2016**, *45*, 715–752.
19. Lopez, A.; Liu, J. Covalent and Noncovalent Functionalization of Graphene Oxide with DNA for Smart Sensing. *Adv. Intell. Syst.* **2020**, *2*, 2000123. [CrossRef]
20. Georgakilas, V.; Tiwari, J.N.; Kemp, K.C.; Perman, J.A.; Bourlinos, A.B.; Kim, K.S.; Zboril, R. Noncovalent functionalization of graphene and graphene oxide for energy materials, biosensing, catalytic, and biomedical applications. *Chem. Rev.* **2016**, *116*, 5464–5519.
21. Zhang, M.; Li, Y.; Su, Z.; Wei, G. Recent advances in the synthesis and applications of graphene–polymer nanocomposites. *Polym. Chem.* **2015**, *6*, 6107–6124. [CrossRef]
22. Liu, B.; Salgado, S.; Maheshwari, V.; Liu, J. DNA adsorbed on graphene and graphene oxide: Fundamental interactions, desorption and applications. *Curr. Opin. Colloid. Interface Sci.* **2016**, *26*, 41–49.
23. Szunerits, S.; Boukherroub, R. Graphene-based bioelectrochemistry and bioelectronics: A concept for the future? *Curr. Opin. Electrochem.* **2018**, *12*, 141–147. [CrossRef]
24. Vivaldi, F.M.; Dallinger, A.; Bonini, A.; Poma, N.; Sembranti, L.; Biagini, D.; Salvo, P.; Greco, F.; Di Francesco, F. Three-dimensional (3D) laser-induced graphene: Structure, properties, and application to chemical sensing. *ACS Appl. Mater. Interfaces* **2021**, *13*, 30245–30260. [CrossRef]
25. Wei, T.; Dai, Z.; Lin, Y.; Du, D. Electrochemical Immunoassays Based on Graphene: A Review. *Electroanalysis* **2016**, *28*, 4–12. [CrossRef]
26. Perrozzi, F.; Prezioso, S.; Ottaviano, L. Graphene oxide: From fundamentals to applications. *J. Phys. Condens. Matter* **2014**, *27*, 013002. [CrossRef]
27. Gao, L.; Lian, C.; Zhou, Y.; Yan, L.; Li, Q.; Zhang, C.; Chen, L.; Chen, K. Graphene oxide-DNA based sensors. *Biosens. Bioelectron.* **2014**, *60*, 22–29. [CrossRef] [PubMed]
28. Iwe, I.; Li, Z.; Huang, J. Graphene oxide and enzyme-assisted dual-cycling amplification method for sensitive fluorometric determination of DNA. *Mikrochim. Acta* **2019**, *186*, 716. [CrossRef]
29. Wei, T.; Chen, Y.; Tu, W.; Lan, Y.; Dai, Z. A phosphomolybdic acid anion probe-based label-free, stable and simple electrochemical biosensing platform. *Chem. Commun.* **2014**, *50*, 9357–9360. [CrossRef]
30. Shamsipur, M.; Molaei, K.; Molaabasi, F.; Hosseinkhani, S.; Taherpour, A.; Sarparast, M.; Moosavifard, S.E.; Barati, A. Aptamer-based fluorescent biosensing of adenosine triphosphate and cytochrome c via aggregation-induced emission enhancement on novel label-free DNA-capped silver nanoclusters/graphene oxide nanohybrids. *ACS Appl. Mater. Interfaces* **2019**, *11*, 46077–46089. [CrossRef] [PubMed]
31. Asefifeyzabadi, N.; Holland, T.E.; Sivakumar, P.; Talapatra, S.; Senanayake, I.M.; Goodson, B.M.; Shamsi, M.H. Sequence-Independent DNA Adsorption on Few-Layered Oxygen-Functionalized Graphene Electrodes: An Electrochemical Study for Biosensing Application. *Biosensors* **2021**, *11*, 273. [CrossRef] [PubMed]
32. Pena-Bahamonde, J.; Nguyen, H.N.; Fanourakis, S.K.; Rodrigues, D.F. Recent advances in graphene-based biosensor technology with applications in life sciences. *J. Nanobiotechnology* **2018**, *16*, 75. [PubMed]
33. Qiu, L.; Li, D.; Cheng, H.M. Structural Control of Graphene-Based Materials for Unprecedented Performance. *ACS Nano* **2018**, *12*, 5085–5092. [CrossRef]
34. Cao, S.H.; Li, L.H.; Wei, W.Y.; Feng, Y.; Jiang, W.L.; Wang, J.L.; Zhang, X.P.; Cai, S.H.; Chen, Z. A label-free and ultrasensitive DNA impedimetric sensor with enzymatic and electrical dual-amplification. *Analyst* **2019**, *144*, 4175–4179. [CrossRef] [PubMed]
35. Ali, M.A.; Hu, C.; Jahan, S.; Yuan, B.; Saleh, M.S.; Ju, E.; Gao, S.J.; Panat, R. Sensing of COVID-19 Antibodies in Seconds via Aerosol Jet Nanoprinted Reduced-Graphene-Oxide-Coated 3D Electrodes. *Adv. Mater.* **2021**, *33*, 2006647. [CrossRef]
36. Maduraiveeran, G.; Jin, W. Nanomaterials based electrochemical sensor and biosensor platforms for environmental applications. *Trends Environ. Anal.* **2017**, *13*, 10–23.
37. Guan, H.; Zhong, T.; He, H.; Zhao, T.; Xing, L.; Zhang, Y.; Xue, X. A self-powered wearable sweat-evaporation-biosensing analyzer for building sports big data. *Nano Energy* **2019**, *59*, 754–761. [CrossRef]
38. Gao, J.W.; Chen, M.M.; Wen, W.; Zhang, X.; Wang, S.; Huang, W.H. Au-Luminol-decorated porous carbon nanospheres for the electrochemiluminescence biosensing of MUC1. *Nanoscale* **2019**, *11*, 16860–16867. [CrossRef] [PubMed]
39. Tian, H.; Tan, B.; Dang, X.; Zhao, H. Enhanced Electrochemiluminescence Detection for Hydrogen Peroxide Using Peroxidase-Mimetic Fe/N-Doped Porous Carbon. *J. Electrochem. Soc.* **2019**, *166*, 1594–1601. [CrossRef]
40. Safaee, M.M.; Gravely, M.; Roxbury, D. A Wearable Optical Microfibrous Biomaterial with Encapsulated Nanosensors Enables Wireless Monitoring of Oxidative Stress. *Adv. Funct. Mater.* **2021**, *31*, 2006254. [CrossRef]
41. Saheb, A.; Janata, J.; Josowicz, M. Reference electrode for ionic liquids. *Electroanalysis* **2006**, *18*, 405–409. [CrossRef]
42. Iijima, S.; Ichihashi, T. Single-shell carbon nanotubes of 1-nm diameter. *Nature* **1993**, *363*, 603–605.
43. Iijima, S. Helical microtubules of graphitic carbon. *Nature* **1991**, *354*, 56–58. [CrossRef]

44. Gooding, J.J. Nanostructuring electrodes with carbon nanotubes: A review on electrochemistry and applications for sensing. *Electrochim. Acta* **2005**, *50*, 3049–3060.
45. Lew, T.T.S.; Koman, V.B.; Silmore, K.S.; Seo, J.S.; Gordiichuk, P.; Kwak, S.Y.; Park, M.; Ang, M.C.; Khong, D.T.; Lee, M.A.; et al. Real-time detection of wound-induced H_2O_2 signalling waves in plants with optical nanosensors. *Nat. Plants* **2020**, *6*, 404–415.
46. Bhattacharya, S.; Gong, X.; Wang, E.; Dutta, S.K.; Caplette, J.R.; Son, M.; Nguyen, F.T.; Strano, M.S.; Mukhopadhyay, D. DNA-SWCNT Biosensors Allow Real-Time Monitoring of Therapeutic Responses in Pancreatic Ductal Adenocarcinoma. *Cancer Res.* **2019**, *79*, 4515–4523. [PubMed]
47. Harvey, J.D.; Williams, R.M.; Tully, K.M.; Baker, H.A.; Shamay, Y.; Heller, D.A. An in Vivo Nanosensor Measures Compartmental Doxorubicin Exposure. *Nano Lett.* **2019**, *19*, 4343–4354.
48. Salem, D.P.; Gong, X.; Liu, A.T.; Akombi, K.; Strano, M.S. Immobilization and Function of nIR-Fluorescent Carbon Nanotube Sensors on Paper Substrates for Fluidic Manipulation. *Anal. Chem.* **2020**, *92*, 916–923.
49. Meng, L.; Turner, A.P.F.; Mak, W.C. Tunable 3D nanofibrous and bio-functionalised PEDOT network explored as a conducting polymer-based biosensor. *Biosens. Bioelectron.* **2020**, *159*, 112181.
50. Zhao, Z.; Chen, H.; Cheng, Y.; Huang, Z.; Wei, X.; Feng, J.; Cheng, J.; Mugo, S.M.; Jaffrezic-Renault, N.; Guo, Z. Electrochemical aptasensor based on electrodeposited poly (3, 4-ethylenedioxythiophene)-graphene oxide coupled with Au@ Pt nanocrystals for the detection of 17β-estradiol. *Microchim. Acta* **2022**, *189*, 178.
51. Zhu, F.; Bian, X.; Zhang, H.; Wen, Y.; Chen, Q.; Yan, Y.; Li, L.; Liu, G.; Yan, J. Controllable design of a nano-bio aptasensing interface based on tetrahedral framework nucleic acids in an integrated microfluidic platform. *Biosens. Bioelectron.* **2021**, *176*, 112943.
52. Jia, Y.; Yi, X.; Li, Z.; Zhang, L.; Yu, B.; Zhang, J.; Wang, X.; Jia, X. Recent advance in biosensing applications based on two-dimensional transition metal oxide nanomaterials. *Talanta* **2020**, *219*, 121308.
53. Fan, Y.; Liu, S.; Yi, Y.; Rong, H.; Zhang, J. Catalytic Nanomaterials toward Atomic Levels for Biomedical Applications: From Metal Clusters to Single-Atom Catalysts. *ACS Nano* **2021**, *15*, 2005–2037.
54. Zhou, W.; Saran, R.; Liu, J. Metal sensing by DNA. *Chem. Rev.* **2017**, *117*, 8272–8325.
55. Fan, D.; Shang, C.; Gu, W.; Wang, E.; Dong, S. Introducing Ratiometric Fluorescence to MnO_2 Nanosheet-Based Biosensing: A Simple, Label-Free Ratiometric Fluorescent Sensor Programmed by Cascade Logic Circuit for Ultrasensitive GSH Detection. *ACS Appl. Mater. Interfaces* **2017**, *9*, 25870–25877.
56. Yadav, V.; Roy, S.; Singh, P.; Khan, Z.; Jaiswal, A. 2D MoS(2)-Based Nanomaterials for Therapeutic, Bioimaging, and Biosensing Applications. *Small* **2019**, *15*, 1803706.
57. Yan, R.; Lu, N.; Han, S.; Lu, Z.; Xiao, Y.; Zhao, Z.; Zhang, M. Simultaneous detection of dual biomarkers using hierarchical MoS_2 nanostructuring and nano-signal amplification-based electrochemical aptasensor toward accurate diagnosis of prostate cancer. *Biosens. Bioelectron.* **2022**, *197*, 113797.
58. Han, H.-H.; Tian, H.; Zang, Y.; Sedgwick, A.C.; Li, J.; Sessler, J.L.; He, X.-P.; James, T.D. Small-molecule fluorescence-based probes for interrogating major organ diseases. *Chem. Soc. Rev.* **2021**, *50*, 9391–9429.
59. Cui, M.; Che, Z.; Gong, Y.; Li, T.; Hu, W.; Wang, S. A graphdiyne-based protein molecularly imprinted biosensor for highly sensitive human C-reactive protein detection in human serum. *Chem. Eng. J.* **2022**, *431*, 133455.
60. Salimian, R.; Kékedy-Nagy, L.; Ferapontova, E.E. Specific Picomolar Detection of a Breast Cancer Biomarker HER-2/neu Protein in Serum: Electrocatalytically Amplified Electroanalysis by the Aptamer/PEG-Modified Electrode. *ChemElectroChem* **2017**, *4*, 872–879.
61. Zhang, L.; Xiao, X.; Xu, Y.; Chen, D.; Chen, J.; Ma, Y.; Dai, Z.; Zou, X. Electrochemical assay for continuous monitoring of dynamic DNA methylation process. *Biosens. Bioelectron.* **2018**, *100*, 184–191. [PubMed]
62. Tian, X.; Murfin, L.C.; Wu, L.; Lewis, S.E.; James, T.D. Fluorescent small organic probes for biosensing. *Chem. Sci.* **2021**, *12*, 3406–3426.
63. Ma, X.; Gao, W.; Du, F.; Yuan, F.; Yu, J.; Guan, Y.; Sojic, N.; Xu, G. Rational Design of Electrochemiluminescent Devices. *Acc. Chem. Res.* **2021**, *54*, 2936–2945. [PubMed]
64. Ma, C.; Cao, Y.; Gou, X.; Zhu, J.J. Recent Progress in Electrochemiluminescence Sensing and Imaging. *Anal. Chem.* **2020**, *92*, 431–454. [PubMed]
65. Zhang, Y.; Xu, J.; Zhou, S.; Zhu, L.; Lv, X.; Zhang, J.; Zhang, L.; Zhu, P.; Yu, J. DNAzyme-Triggered Visual and Ratiometric Electrochemiluminescence Dual-Readout Assay for Pb(II) Based on an Assembled Paper Device. *Anal. Chem.* **2020**, *92*, 3874–3881.
66. Guo, J.; Xie, M.; Du, P.; Liu, Y.; Lu, X. Signal Amplification Strategy Using Atomically Gold-Supported VO(2) Nanobelts as a Co-reaction Accelerator for Ultrasensitive Electrochemiluminescent Sensor Construction Based on the Resonance Energy Transfer Platform. *Anal. Chem.* **2021**, *93*, 10619–10626.
67. Wu, K.; Zheng, Y.; Chen, R.; Zhou, Z.; Liu, S.; Shen, Y.; Zhang, Y. Advances in electrochemiluminescence luminophores based on small organic molecules for biosensing. *Biosens. Bioelectron.* **2023**, *223*, 115031.
68. Wu, X.; Li, Z.; Yang, L.; Han, J.; Han, S. A self-referenced nanodosimeter for reaction based ratiometric imaging of hypochlorous acid in living cells. *Chem. Sci.* **2013**, *4*, 460–467.
69. Jia, X.; Chen, Q.; Yang, Y.; Tang, Y.; Wang, R.; Xu, Y.; Zhu, W.; Qian, X. FRET-Based Mito-Specific Fluorescent Probe for Ratiometric Detection and Imaging of Endogenous Peroxynitrite: Dyad of Cy3 and Cy5. *J. Am. Chem. Soc.* **2016**, *138*, 10778–10781. [PubMed]

70. Gu, K.; Xu, Y.; Li, H.; Guo, Z.; Zhu, S.; Zhu, S.; Shi, P.; James, T.D.; Tian, H.; Zhu, W.H. Real-Time Tracking and In Vivo Visualization of beta-Galactosidase Activity in Colorectal Tumor with a Ratiometric Near-Infrared Fluorescent Probe. *J. Am. Chem. Soc.* **2016**, *138*, 5334–5340. [PubMed]
71. Murfin, L.C.; Weber, M.; Park, S.J.; Kim, W.T.; Lopez-Alled, C.M.; McMullin, C.L.; Pradaux-Caggiano, F.; Lyall, C.L.; Kociok-Kohn, G.; Wenk, J.; et al. Azulene-Derived Fluorescent Probe for Bioimaging: Detection of Reactive Oxygen and Nitrogen Species by Two-Photon Microscopy. *J. Am. Chem. Soc.* **2019**, *141*, 19389–19396.
72. Chen, H.; Tang, Y.; Ren, M.; Lin, W. Single near-infrared fluorescent probe with high- and low-sensitivity sites for sensing different concentration ranges of biological thiols with distinct modes of fluorescence signals. *Chem. Sci.* **2016**, *7*, 1896–1903.
73. Jiao, Y.; Yin, J.; He, H.; Peng, X.; Gao, Q.; Duan, C. Conformationally Induced Off-On Cell Membrane Chemosensor Targeting Receptor Protein-Tyrosine Kinases for in Vivo and in Vitro Fluorescence Imaging of Cancers. *J. Am. Chem. Soc.* **2018**, *140*, 5882–5885. [CrossRef]
74. Long, L.; Huang, M.; Wang, N.; Wu, Y.; Wang, K.; Gong, A.; Zhang, Z.; Sessler, J.L. A Mitochondria-Specific Fluorescent Probe for Visualizing Endogenous Hydrogen Cyanide Fluctuations in Neurons. *J. Am. Chem. Soc.* **2018**, *140*, 1870–1875. [CrossRef] [PubMed]
75. Sassolas, A.; Blum, L.J.; Leca-Bouvier, B.D. Immobilization strategies to develop enzymatic biosensors. *Biotechnol. Adv.* **2012**, *30*, 489–511. [PubMed]
76. Gu, H.; Xing, Y.; Xiong, P.; Tang, H.; Li, C.; Chen, S.; Zeng, R.; Han, K.; Shi, G. Three-Dimensional Porous $Ti_3C_2T_x$ MXene–Graphene Hybrid Films for Glucose Biosensing. *ACS Appl. Nano Mater.* **2019**, *2*, 6537–6545. [CrossRef]
77. Dhanjai; Balla, P.; Sinha, A.; Wu, L.; Lu, X.; Tan, D.; Chen, J. Co_3O_4 nanoparticles supported mesoporous carbon framework interface for glucose biosensing. *Talanta* **2019**, *203*, 112–121. [CrossRef]
78. Yao, C.; Ou, J.; Tang, J.; Yang, D. DNA Supramolecular Assembly on Micro/Nanointerfaces for Bioanalysis. *Acc. Chem. Res.* **2022**, *55*, 2043–2054. [CrossRef]
79. Li, F.; Li, Q.; Zuo, X.; Fan, C. DNA framework-engineered electrochemical biosensors. *Sci. China Life Sci.* **2020**, *63*, 1130–1141. [PubMed]
80. Su, S.; Ma, J.; Xu, Y.; Pan, H.; Zhu, D.; Chao, J.; Weng, L.; Wang, L. Electrochemical Analysis of Target-Induced Hairpin-Mediated Aptamer Sensors. *ACS Appl. Mater. Interfaces* **2020**, *12*, 48133–48139. [CrossRef]
81. Li, M.; Lv, M.; Wang, L.; Fan, C.; Zuo, X. Engineering electrochemical interface for biomolecular sensing. *Curr. Opin. Electrochem.* **2019**, *14*, 71–80.
82. Zhu, C.; Liu, D.; Li, Y.; Ma, S.; Wang, M.; You, T. Hairpin DNA assisted dual-ratiometric electrochemical aptasensor with high reliability and anti-interference ability for simultaneous detection of aflatoxin B1 and ochratoxin A. *Biosens. Bioelectron.* **2021**, *174*, 112654.
83. Miao, P.; Tang, Y.; Wang, L. DNA Modified Fe_3O_4@Au Magnetic Nanoparticles as Selective Probes for Simultaneous Detection of Heavy Metal Ions. *ACS Appl. Mater. Interfaces* **2017**, *9*, 3940–3947. [CrossRef]
84. Han, C.; Li, W.; Li, Q.; Xing, W.; Luo, H.; Ji, H.; Fang, X.; Luo, Z.; Zhang, L. CRISPR/Cas12a-Derived electrochemical aptasensor for ultrasensitive detection of COVID-19 nucleocapsid protein. *Biosens. Bioelectron.* **2022**, *200*, 113922.
85. Su, J.; Liu, W.; Chen, S.; Deng, W.; Dou, Y.; Zhao, Z.; Li, J.; Li, Z.; Yin, H.; Ding, X.; et al. A Carbon-Based DNA Framework Nano-Bio Interface for Biosensing with High Sensitivity and a High Signal-to-Noise Ratio. *ACS Sens.* **2020**, *5*, 3979–3987. [CrossRef]
86. Mao, X.; Mao, D.; Chen, T.; Jalalah, M.; Al-Assiri, M.S.; Harraz, F.A.; Zhu, X.; Li, G. DNA Hydrogel-Based Three-Dimensional Electron Transporter and Its Application in Electrochemical Biosensing. *ACS Appl. Mater. Interfaces* **2020**, *12*, 36851–36859. [CrossRef]
87. Cui, M.; Gong, Y.; Du, M.; Wang, K.; Li, T.; Zhu, X.; Wang, S.; Luo, X. An antifouling electrochemical biosensor based on a protein imprinted hydrogel for human immunoglobulin G recognition in complex biological media. *Sens. Actuators B Chem.* **2021**, *337*, 129820. [CrossRef]
88. Berezovski, M.; Krylov, S.N. Using DNA-binding proteins as an analytical tool. *J. Am. Chem. Soc.* **2003**, *125*, 13451–13454. [CrossRef]
89. Campuzano, S.; Pedrero, M.; Pingarrón, J.M. Viral protein-based bioanalytical tools for small RNA biosensing. *Trends Analyt. Chem.* **2016**, *79*, 335–343.
90. Quijano-Rubio, A.; Yeh, H.W.; Park, J.; Lee, H.; Langan, R.A.; Boyken, S.E.; Lajoie, M.J.; Cao, L.; Chow, C.M.; Miranda, M.C.; et al. De novo design of modular and tunable protein biosensors. *Nature* **2021**, *591*, 482–487.
91. Wei, Y.; Zeng, Q.; Hu, Q.; Wang, M.; Tao, J.; Wang, L. Self-cleaned electrochemical protein imprinting biosensor basing on a thermo-responsive memory hydrogel. *Biosens. Bioelectron.* **2018**, *99*, 136–141. [CrossRef]
92. Wei, Y.; Zeng, Q.; Huang, J.; Guo, X.; Wang, L.; Wang, L. Preparation of Gas-Responsive Imprinting Hydrogel and Their Gas-Driven Switchable Affinity for Target Protein Recognition. *ACS Appl. Mater. Interfaces* **2020**, *12*, 24363–24369. [CrossRef]
93. Kerry, R.G.; Ukhurebor, K.E.; Kumari, S.; Maurya, G.K.; Patra, S.; Panigrahi, B.; Majhi, S.; Rout, J.R.; del Pilar Rodriguez-Torres, M.; Das, G. A comprehensive review on the applications of nano-biosensor-based approaches for non-communicable and communicable disease detection. *Biomater. Sci.* **2021**, *9*, 3576–3602.
94. Qi, P.; Wan, Y.; Zhang, D. Impedimetric biosensor based on cell-mediated bioimprinted films for bacterial detection. *Biosens. Bioelectron.* **2013**, *39*, 282–288.

95. Jiang, H.; Jiang, D.; Liu, X.; Yang, J. A self-driven PET chip-based imprinted electrochemical sensor for the fast detection of Salmonella. *Sens. Actuators B Chem.* **2021**, *349*, 130785. [CrossRef]
96. Benson, V.S.S.a.D.E. Protein Design Provides Lead(II) Ion Biosensors for Imaging Molecular Fluxes around Red Blood Cells. *Biochemistry* **2009**, *48*, 462–470.
97. Chen, C.; Liu, Y.; Gu, H.-Y. Cellular biosensor based on red blood cells immobilized on Fe_3O_4 Core/Au Shell nanoparticles for hydrogen peroxide electroanalysis. *Microchim. Acta* **2010**, *171*, 371–376. [CrossRef]
98. Liu, J.M.; Zhang, D.D.; Fang, G.Z.; Wang, S. Erythrocyte membrane bioinspired near-infrared persistent luminescence nanocarriers for in vivo long-circulating bioimaging and drug delivery. *Biomaterials* **2018**, *165*, 39–47. [CrossRef]
99. Rostovtsev, V.V.; Green, L.G.; Fokin, V.V.; Sharpless, K.B. A stepwise huisgen cycloaddition process: Copper (I)-catalyzed regioselective "ligation" of azides and terminal alkynes. *Angew. Chem. Int. Ed.* **2002**, *114*, 2708–2711.
100. Rodionov, V.O.; Presolski, S.I.; Díaz Díaz, D.; Fokin, V.V.; Finn, M. Ligand-accelerated Cu-catalyzed azide– alkyne cycloaddition: A mechanistic report. *J. Am. Chem. Soc.* **2007**, *129*, 12705–12712. [CrossRef]
101. Chan, T.R.; Hilgraf, R.; Sharpless, K.B.; Fokin, V.V. Polytriazoles as copper (I)-stabilizing ligands in catalysis. *Org. Lett.* **2004**, *6*, 2853–2855. [CrossRef] [PubMed]
102. Yanez-Sedeno, P.; Gonzalez-Cortes, A.; Campuzano, S.; Pingarron, J.M. Copper(I)-Catalyzed Click Chemistry as a Tool for the Functionalization of Nanomaterials and the Preparation of Electrochemical (Bio)Sensors. *Sensors* **2019**, *19*, 2379. [CrossRef]
103. Nicosia, C.; Huskens, J. Reactive self-assembled monolayers: From surface functionalization to gradient formation. *Mater. Horiz.* **2014**, *1*, 32–45. [CrossRef]
104. Zhang, X.; Wu, Y.; Chen, J.; Yang, Y.; Li, G. Bioinspired Artificial "Clickase" for the Catalytic Click Immunoassay of Foodborne Pathogens. *Anal. Chem.* **2021**, *93*, 3217–3225. [CrossRef]
105. Guerrero, S.; Cadano, D.; Agüí, L.; Barderas, R.; Campuzano, S.; Yáñez-Sedeño, P.; Pingarrón, J.M. Click chemistry-assisted antibodies immobilization for immunosensing of CXCL7 chemokine in serum. *J. Electroanal. Chem.* **2019**, *837*, 246–253. [CrossRef]
106. Fomo, G.; Nwaji, N.; Nyokong, T. Low symmetric metallophthalocyanine modified electrode via click chemistry for simultaneous detection of heavy metals. *J. Electroanal. Chem.* **2018**, *813*, 58–66.
107. Genzer, J.; Bhat, R.R. Surface-bound soft matter gradients. *Langmuir* **2008**, *24*, 2294–2317. [PubMed]
108. Yang, L.; Gao, M.X.; Zou, H.Y.; Li, Y.F.; Huang, C.Z. Plasmonic $Cu_{2-x}S_ySe_{1-y}$ Nanoparticles Catalyzed Click Chemistry Reaction for SERS Immunoassay of Cancer Biomarker. *Anal. Chem.* **2018**, *90*, 11728–11733. [CrossRef] [PubMed]
109. Xue, L.; Yang, Y.; Wu, S.; Huang, Y.; Li, J.; Xiang, Y.; Li, G. In Situ Reduction of Porous Copper Metal-Organic Frameworks for Three-Dimensional Catalytic Click Immunoassay. *Anal. Chem.* **2020**, *92*, 2972–2978. [CrossRef]
110. Xiang, W.; Zhang, Z.; Weng, W.; Wu, B.; Cheng, J.; Shi, L.; Sun, H.; Gao, L.; Shi, K. Highly sensitive detection of carcinoembryonic antigen using copper-free click chemistry on the surface of azide cofunctionalized graphene oxide. *Anal. Chim. Acta* **2020**, *1127*, 156–162. [CrossRef]
111. Lutz, J.F. Copper-free azide-alkyne cycloadditions: New insights and perspectives. *Angew. Chem. Int. Ed.* **2008**, *47*, 2182–2184. [CrossRef]
112. Agard, N.J.; Prescher, J.A.; Bertozzi, C.R. A strain-promoted [3+2] azide– alkyne cycloaddition for covalent modification of biomolecules in living systems. *J. Am. Chem. Soc.* **2004**, *126*, 15046–15047. [CrossRef] [PubMed]
113. Yang, J.; Wang, K.; Zhang, S.; Zheng, X.; Cui, T.; Yang, X.; Liu, Y.; Lu, D.; Wang, Y.; Tian, X.; et al. Site-Specific Introduction of Bioorthogonal Handles to Nanopores by Genetic Code Expansion. *Angew. Chem. Int. Ed.* **2023**, *62*, 202216115. [CrossRef] [PubMed]
114. Toulemon, D.; Pichon, B.P.; Leuvrey, C.d.; Zafeiratos, S.; Papaefthimiou, V.; Cattoën, X.; Bégin-Colin, S. Fast assembling of magnetic iron oxide nanoparticles by microwave-assisted copper (I) catalyzed alkyne–azide cycloaddition (CuAAC). *Chem. Mater.* **2013**, *25*, 2849–2854. [CrossRef]
115. Cintas, P.; Barge, A.; Tagliapietra, S.; Boffa, L.; Cravotto, G. Alkyne–azide click reaction catalyzed by metallic copper under ultrasound. *Nat. Prot.* **2010**, *5*, 607–616. [CrossRef]
116. Tasdelen, M.A.; Yagci, Y. Light-induced click reactions. *Angew. Chem. Int. Ed.* **2013**, *52*, 5930–5938. [CrossRef] [PubMed]
117. Qi, H.; Li, M.; Zhang, R.; Dong, M.; Ling, C. Double electrochemical covalent coupling method based on click chemistry and diazonium chemistry for the fabrication of sensitive amperometric immunosensor. *Anal. Chim. Acta* **2013**, *792*, 28–34. [CrossRef]
118. Devaraj, N.K.; Dinolfo, P.H.; Chidsey, C.E.; Collman, J.P. Selective functionalization of independently addressed microelectrodes by electrochemical activation and deactivation of a coupling catalyst. *J. Am. Chem. Soc.* **2006**, *128*, 1794–1795. [CrossRef]
119. Lesniewski, A.; Matyjewicz, J.; Palys, B.; Niedziolka-Jonsson, J. Electroassisted click chemistry immobilisation of gold nanoparticles on a solid substrate. *Electrochem. Commun.* **2015**, *53*, 20–23. [CrossRef]
120. Villalba, M.; Bossi, M.; Pozo, M.D.; Calvo, E.J. Palladium Nanoparticles Embedded in a Layer-by-Layer Nanoreactor Built with Poly(Acrylic Acid) Using "Electro-Click Chemistry". *Langmuir* **2016**, *32*, 6836–6842. [CrossRef]
121. Yamamoto, T.; Akahori, M.; Natsui, K.; Saitoh, T.; Einaga, Y. Controlled decoration of boron-doped diamond electrodes by electrochemical click reaction (e–CLICK). *Carbon* **2018**, *130*, 350–354. [CrossRef]
122. Rydzek, G.; Jierry, L.; Parat, A.; Thomann, J.S.; Voegel, J.C.; Senger, B.; Hemmerle, J.; Ponche, A.; Frisch, B.; Schaaf, P.; et al. Electrochemically triggered assembly of films: A one-pot morphogen-driven buildup. *Angew. Chem. Int. Ed.* **2011**, *50*, 4374–4377. [CrossRef]

123. Hu, L.; Zhao, P.; Deng, H.; Xiao, L.; Qin, C.; Du, Y.; Shi, X. Electrical signal guided click coating of chitosan hydrogel on conductive surface. *RSC Adv.* **2014**, *4*, 13477. [CrossRef]
124. Cheng, C.; Oueslati, R.; Wu, J.; Chen, J.; Eda, S. Capacitive DNA sensor for rapid and sensitive detection of whole genome human herpesvirus-1 dsDNA in serum. *Electrophoresis* **2017**, *38*, 1617–1623. [CrossRef]
125. Castiello, F.R.; Porter, J.; Modarres, P.; Tabrizian, M. Interfacial capacitance immunosensing using interdigitated electrodes: The effect of insulation/immobilization chemistry. *Phys. Chem. Chem. Phys.* **2019**, *21*, 15787–15797. [CrossRef] [PubMed]
126. Cesbron, M.; Levillain, E.; Breton, T.; Gautier, C. Click chemistry: A versatile method for tuning the composition of mixed organic layers obtained by reduction of diazonium cations. *ACS Appl. Mater. Interfaces* **2018**, *10*, 37779–37782. [CrossRef]
127. Cernat, A.; Griveau, S.; Martin, P.; Lacroix, J.C.; Farcau, C.; Sandulescu, R.; Bedioui, F. Electrografted nanostructured platforms for click chemistry. *Electrochem. Commun.* **2012**, *23*, 141–144. [CrossRef]
128. Fenoy, G.E.; Hasler, R.; Quartinello, F.; Marmisolle, W.A.; Lorenz, C.; Azzaroni, O.; Bauerle, P.; Knoll, W. "Clickable" Organic Electrochemical Transistors. *JACS Au* **2022**, *2*, 2778–2790. [CrossRef] [PubMed]
129. Sciortino, F.; Rydzek, G.; Grasset, F.; Kahn, M.L.; Hill, J.P.; Chevance, S.; Gauffre, F.; Ariga, K. Electro-click construction of hybrid nanocapsule films with triggered delivery properties. *Phys. Chem. Chem. Phys.* **2018**, *20*, 2761–2770. [CrossRef]
130. Guerrero, S.; Agui, L.; Yanez-Sedeno, P.; Pingarron, J.M. Design of electrochemical immunosensors using electro-click chemistry. Application to the detection of IL-1beta cytokine in saliva. *Bioelectrochemistry* **2020**, *133*, 107484. [CrossRef]
131. Wei, T.; Dong, T.; Xing, H.; Liu, Y.; Dai, Z. Cucurbituril and Azide Cofunctionalized Graphene Oxide for Ultrasensitive Electro-Click Biosensing. *Anal. Chem.* **2017**, *89*, 12237–12243. [CrossRef] [PubMed]
132. Levrie, K.; Jans, K.; Vos, R.; Ardakanian, N.; Verellen, N.; Van Hoof, C.; Lagae, L.; Stakenborg, T. Multiplexed site-specific electrode functionalization for multitarget biosensors. *Bioelectrochemistry* **2016**, *112*, 61–66. [CrossRef]
133. Rydzek, G.; Terentyeva, T.G.; Pakdel, A.; Golberg, D.; Hill, J.P.; Ariga, K. Simultaneous electropolymerization and electro-click functionalization for highly versatile surface platforms. *ACS Nano* **2014**, *8*, 5240–5248. [CrossRef]
134. Chmielarz, P.; Fantin, M.; Park, S.; Isse, A.A.; Gennaro, A.; Magenau, A.J.; Sobkowiak, A.; Matyjaszewski, K. Electrochemically mediated atom transfer radical polymerization (eATRP). *Prog. Polym. Sci.* **2017**, *69*, 47–78. [CrossRef]
135. Wu, T.; Lankshear, E.R.; Downard, A.J. Simultaneous Electro-Click and Electrochemically Mediated Polymerization Reactions for One-Pot Grafting from a Controlled Density of Anchor Sites. *ChemElectroChem* **2019**, *6*, 5149–5154. [CrossRef]
136. Kang, D.; Kim, T.W.; Kubota, S.R.; Cardiel, A.C.; Cha, H.G.; Choi, K.S. Electrochemical Synthesis of Photoelectrodes and Catalysts for Use in Solar Water Splitting. *Chem. Rev.* **2015**, *115*, 12839–12887. [CrossRef]
137. Guiseppi-Elie, A. Electroconductive hydrogels: Synthesis, characterization and biomedical applications. *Biomaterials* **2010**, *31*, 2701–2716. [CrossRef]
138. Nederberg, F.; Trang, V.; Pratt, R.C.; Kim, S.-H.; Colson, J.; Nelson, A.; Frank, C.W.; Hedrick, J.L.; Dubois, P.; Mespouille, L. Exploring the versatility of hydrogels derived from living organocatalytic ring-opening polymerization. *Soft Matter* **2010**, *6*, 2006. [CrossRef]
139. Choi, E.J.; Shin, J.; Khaleel, Z.H.; Cha, I.; Yun, S.-H.; Cho, S.-W.; Song, C. Synthesis of electroconductive hydrogel films by an electro-controlled click reaction and their application to drug delivery systems. *Polym. Chem.* **2015**, *6*, 4473–4478. [CrossRef]
140. Fosdick, S.E.; Knust, K.N.; Scida, K.; Crooks, R.M. Bipolar electrochemistry. *Angew. Chem. Int. Ed.* **2013**, *52*, 10438–10456. [CrossRef]
141. Shida, N.; Zhou, Y.; Inagi, S. Bipolar Electrochemistry: A Powerful Tool for Electrifying Functional Material Synthesis. *Acc. Chem. Res.* **2019**, *52*, 2598–2608. [CrossRef] [PubMed]
142. Shida, N.; Ishiguro, Y.; Atobe, M.; Fuchigami, T.; Inagi, S. Electro-Click Modification of Conducting Polymer Surface Using Cu(I) Species Generated on a Bipolar Electrode in a Gradient Manner. *ACS Macro. Lett.* **2012**, *1*, 656–659. [CrossRef]
143. Zhou, Y.; Shida, N.; Tomita, I.; Inagi, S. Fabrication of Gradient and Patterned Organic Thin Films by Bipolar Electrolytic Micelle Disruption Using Redox-Active Surfactants. *Angew. Chem. Int. Ed.* **2021**, *60*, 14620–14629. [CrossRef]
144. Khan, A.; Khan, A.A.; Asiri, A.M.; Rub, M.A.; Azum, N.; Rahman, M.M.; Khan, S.B.; Ghani, S.A. A new trend on biosensor for neurotransmitter choline/acetylcholine—An overview. *Appl. Biochem. Biotechnol.* **2013**, *169*, 1927–1939. [CrossRef] [PubMed]
145. Wu, J.; Mao, Z.; Tan, H.; Han, L.; Ren, T.; Gao, C. Gradient biomaterials and their influences on cell migration. *Interface Focus* **2012**, *2*, 337–355. [CrossRef] [PubMed]
146. Rydzek, G.; Toulemon, D.; Garofalo, A.; Leuvrey, C.; Dayen, J.F.; Felder-Flesch, D.; Schaaf, P.; Jierry, L.; Begin-Colin, S.; Pichon, B.P.; et al. Selective Nanotrench Filling by One-Pot Electroclick Self-Constructed Nanoparticle Films. *Small* **2015**, *11*, 4638–4642. [CrossRef] [PubMed]
147. Quinton, D.; Maringa, A.; Griveau, S.; Nyokong, T.; Bedioui, F. Surface patterning using scanning electrochemical microscopy to locally trigger a "click" chemistry reaction. *Electrochem. Commun.* **2013**, *31*, 112–115. [CrossRef]
148. Krabbenborg, S.O.; Nicosia, C.; Chen, P.; Huskens, J. Reactivity mapping with electrochemical gradients for monitoring reactivity at surfaces in space and time. *Nat. Commun.* **2013**, *4*, 1667. [CrossRef]
149. Nicosia, C.; Krabbenborg, S.O.; Chen, P.; Huskens, J. Shape-controlled fabrication of micron-scale surface chemical gradients via electrochemically activated copper(i) "click" chemistry. *J. Mater. Chem. B* **2013**, *1*, 5417–5428. [CrossRef]
150. Hansen, T.S.; Daugaard, A.E.; Hvilsted, S.r.; Larsen, N.B. Spatially Selective Functionalization of Conducting Polymers by "Electroclick" Chemistry. *Adv. Mat.* **2009**, *21*, 4483–4486. [CrossRef]

151. Cui, X.; Wei, T.; Hao, M.; Qi, Q.; Wang, H.; Dai, Z. Highly sensitive and selective colorimetric sensor for thiocyanate based on electrochemical oxidation-assisted complexation reaction with Gold nanostars etching. *J. Hazard. Mater.* **2020**, *391*, 122217. [CrossRef]
152. Goll, M.; Ruff, A.; Muks, E.; Goerigk, F.; Omiecienski, B.; Ruff, I.; Gonzalez-Cano, R.C.; Lopez Navarrete, J.T.; Ruiz Delgado, M.C.; Ludwigs, S. Functionalized branched EDOT-terthiophene copolymer films by electropolymerization and post-polymerization "click"-reactions. *Beilstein. J. Org. Chem.* **2015**, *11*, 335–347. [CrossRef] [PubMed]
153. Saylan, Y.; Erdem, O.; Unal, S.; Denizli, A. An Alternative Medical Diagnosis Method: Biosensors for Virus Detection. *Biosensors* **2019**, *9*, 65. [CrossRef] [PubMed]
154. Ku, S.-Y.; Wong, K.-T.; Bard, A.J. Surface patterning with fluorescent molecules using click chemistry directed by scanning electrochemical microscopy. *J. Am. Chem. Soc.* **2008**, *130*, 2392–2393. [CrossRef] [PubMed]
155. Coceancigh, H.; Tran-Ba, K.H.; Siepser, N.; Baker, L.A.; Ito, T. Longitudinally Controlled Modification of Cylindrical and Conical Track-Etched Poly(ethylene terephthalate) Pores Using an Electrochemically Assisted Click Reaction. *Langmuir* **2017**, *33*, 11998–12006. [CrossRef]
156. Rydzek, G.; Polavarapu, P.; Rios, C.; Tisserant, J.-N.; Voegel, J.-C.; Senger, B.; Lavalle, P.; Frisch, B.; Schaaf, P.; Boulmedais, F.; et al. Morphogen-driven self-construction of covalent films built from polyelectrolytes and homobifunctional spacers: Buildup and pH response. *Soft Matter* **2012**, *8*, 10336. [CrossRef]
157. Chatterjee, A.K.; Chakraborty, R.; Basu, T. Mechanism of antibacterial activity of copper nanoparticles. *Nanotechnology* **2014**, *25*, 135101. [CrossRef]
158. Yang, M.; Jalloh, A.S.; Wei, W.; Zhao, J.; Wu, P.; Chen, P.R. Biocompatible click chemistry enabled compartment-specific pH measurement inside *E. coli*. *Nat. Commun.* **2014**, *5*, 4981. [CrossRef]
159. Cheng, X.; Liu, D.; Jin, Y.; Yang, M.; Xiang, J. Addressing Cu^{2+} interference for accurate aptamer-based biomarker determinations of Alzheimer's disease. *Anal. Sci.* **2022**, *38*, 317–322. [CrossRef]
160. Liu, G.; Gui, S.; Zhou, H.; Zeng, F.; Zhou, Y.; Ye, H. A strong adsorbent for Cu^{2+}: Graphene oxide modified with triethanolamine. *Dalton. Trans.* **2014**, *43*, 6977–6980. [CrossRef]

chemosensors

Article

Novel Electrochemical Sensor Based on MnO_2 Nanowire Modified Carbon Paper Electrode for Sensitive Determination of Tetrabromobisphenol A

Chunmao Zhu, Qi Wu, Fanshu Yuan, Jie Liu, Dongtian Wang and Qianli Zhang *

School of the Chemistry and Life Sciences, Suzhou University of Science and Technology, Suzhou 215009, China; zcm2468260200@163.com (C.Z.); wq1213857@163.com (Q.W.); yuanfanshute@126.com (F.Y.); jliu@mail.usts.edu.cn (J.L.); dtwang@mail.usts.edu.cn (D.W.)
* Correspondence: zqlmhb@mail.usts.edu.cn

Abstract: In this paper, a MnO_2 nanowire (MnO_2-NW) modified carbon paper electrode (CP) was developed as a novel electrochemical sensor for the sensitive determination of tetrabromobisphenol A (TBBPA). The MnO_2 nanowire was prepared by a hydrothermal synthesis method, and the morphology and structure of MnO_2 were characterized using scanning electron microscopy, X-ray diffraction and X-ray photoelectron spectroscopy. The electrochemical performance of TBBPA on MnO_2-NW/CP was investigated by cyclic voltammetry, and the result confirmed that MnO_2-NW/CP exhibited excellent sensitivity for the determination of TBBPA due to the high specific surface area and good electrical conductivity of the nanowire-like MnO_2. Moreover, the important electrochemical factors such as pH value, incubation time and modified material proportion were systematically studied to improve the determination sensitivity. The interferences from similar structure compounds on TBBPA have also been investigated. Under the optimal conditions, MnO_2-NW/CP displayed a linear range of 70~500 nM for TBBPA with a detection limit of 3.1 nM. This was superior to some electrochemical methods in reference. The work presents a novel and simple method for the determination of TBBPA.

Keywords: tetrabromobisphenol A; MnO_2; carbon paper; modified electrode

Citation: Zhu, C.; Wu, Q.; Yuan, F.; Liu, J.; Wang, D.; Zhang, Q. Novel Electrochemical Sensor Based on MnO_2 Nanowire Modified Carbon Paper Electrode for Sensitive Determination of Tetrabromobisphenol A. *Chemosensors* **2023**, *11*, 482. https://doi.org/10.3390/chemosensors11090482

Academic Editor: Gerd-Uwe Flechsig

Received: 26 June 2023
Revised: 17 July 2023
Accepted: 16 August 2023
Published: 1 September 2023

1. Introduction

Tetrabromobisphenol A (TBBPA), an organic compound with the chemical formula $C_{15}H_{12}Br_4O_2$, is widely used as a reactive flame retardant in electronic equipment, furniture, plastics and textile products [1,2]. TBBPA accounts for 80% of the global demand in Asia. As a result of its massive use, release from industrial products, as well as the migration of terrestrial and marine food chains, TBBPA is continuously released into the environment [3–5]. TBBPA is a potential persistent organic pollutant reported to be bioaccumulative and highly toxic. Chronic exposure to TBBPA can induce immune, reproductive and neurotoxic effects on animals [6–8]. In 2017, the World Health Organization's International Agency for Research on Cancer published a preliminary list of carcinogens for reference, and tetrabromobisphenol A is on the list of carcinogens in category 2A. Therefore, detecting TBBPA quickly and accurately is important for environmental monitoring and the protection of human health.

Several analytical methods have been reported for the detection of TBBPA, including chromatographic [9,10], spectroscopic [11,12] and immunoassay methods [13,14]. For example, Zhang et al. [10] developed a high-performance liquid chromatography method coupled with triple quadrupole mass spectrometry (HPLC-MS/MS) with atmospheric pressure chemical ionization (APCI) source for simultaneous detection of TBBPA and its ten derivatives in determining complicated environmental samples, including sewage sludge, river water and vegetable samples. Feng et al. [12] explored a fluorescent sensor by coating a molecularly imprinted polymer layer on CdTe quantum dots for the sensitive determination of TBBPA. Although these methods have good selectivity and high detection sensitivity, they are limited by the high costs of instrument maintenance, sophisticated operating skills, and extensive organic solvent consumption. Many researchers pay attention to electrochemical sensors due to their high efficiency, low cost, prominent sensitivity, and rapid detection speed [15–17]. Presently, some electrochemical sensors have been reported for the detection of TBBPA in the environment. Nonetheless, some TBBPA electrochemical sensors' utility is limited by the non-conductive film that forms on the electrode surface due to TBBPA's electrochemical oxidation, impairing detection stability [18].

The materials used for electrode modification play a pivotal role in mitigating the passivation induced by TBBPA oxidation products. The conception and advancement of nanomaterials have considerably influenced electrochemical sensors, with their properties largely being a function of their microstructure and form. A variety of nanomaterials, such as metal oxides, carbon-based materials, metal-organic frameworks, and nanocomposites, are being studied for TBBPA detection. Zhou et al. [19] prepared a conductive composite of carbon nanotubes@zeolitic imidazole framework-67 (CNTs@ZIF-67), which possesses an excellent adsorption capacity (92.12 mg g^{-1}) for TBBPA. The composite was used to modify an acetylene black electrode for the sensitive determination of TBBPA in spiked rain and pool water samples with the aid of perfluorodecanoic acid. The sensor is stable, reproducible, and has a linear range of 0.01–1.5 μM TBBPA concentration, with a 4.2 nM detection limit (at S/N = 3). Recently, carbon dots (CDs) have attracted a lot of attention due to their low toxicity, biocompatibility and good electrical conductivity, and are widely used in biosensors, photocatalysis, electrocatalysis and electrochemical sensors. Guo et al. [20] developed magnetic CDs composed of carbon dots (CDs) and Fe_3O_4 nanoparticles via an amination reaction. A glassy carbon electrode modified with magnetic CDs and cetyltrimethylammonium bromide (CTAB) was employed as an electrochemical sensor for TBBPA detection in beverages. Magnetic CDs facilitate TBBPA oxidation, and CTAB's hydrophobic effect can enrich TBBPA. The combined impact of Magnetic CDs and CTAB boosts electrochemical sensor performance, indicating a linear range between 1 and 1000 nM, a detection limit of 0.75 nM, and displaying benefits such as rapidity, superior sensitivity, and robust stability. MXene, an exceptional electrode modification material, is a two-dimensional transition metal carbide obtained by etching aluminum in Ti_3AlC_2 with hydrofluoric acid. The combination of precious metals nanoparticles and Mxene can take advantage of the extraordinary electrocatalytic properties of noble metal nanomaterials and the specific surface area and electrical conductivity of MXene. Shao et al. [21] developed a TBBPA electrochemical sensor by modifying it to a glassy carbon electrode with the MXene/Au nanocomposite. The finalized sensor exhibited excellent linearity for TBBPA concentrations between 0.05 and 10 nM. It had a detection limit of 0.0144 nM and successfully detected TBBPA in water, with recoveries ranging from 97.1% to 106% for the added standards. Lu et al. [22] reported that the composite of graphitic carbon nitride (g-C_3N_4) and N-butylpyridinium hexafluorophosphate (NBH) could promote the oxidation of TBBPA. The detection of TBBPA was fulfilled using a g-C_3N_4-NBH modified carbon paste electrode in the range of 1 nM to 30 nM and 30 nM to 500 nM with a limit of 0.4 nM.

Various metal oxides, including CuO, Fe_3O_4 and Fe_2O_3, have extensive applications in electrochemical sensors due to their distinct morphology [23–25]. Zhou and colleagues [23] produced different forms of CuO nanomaterials, namely nanostrips, nanowires, and microspheres. These were combined with graphene nanoplates and used to modify the surface of glassy carbon electrodes for electrochemical detection of both glucose and TBBPA. The composite made from CuO nanostrips demonstrated the most significant active surface area, the least charge transfer resistance, and the highest detection sensitivity. The designed electrochemical sensor offered sensitive TBBPA detection within a linear range of 5 to 600 nM, with an anticipated detection limit of 0.73 nM based on a signal-to-noise ratio of three. Luo et al. [24] developed an Fe_3O_4-activated biochar using surplus sludge, leading to the successful creation of an electrochemical sensor for TBBPA detection. The results from electrochemical tests suggested that the Fe_3O_4–activated biochar film possessed a more substantial active surface area, reduced charge transfer resistance, and increased TBBPA accumulation efficiency. The sensor exhibited solid linearity for TBBPA concentrations ranging from 5 to 1000 nM, a reasonably low detection limit of 3.2 nM, and was effectively utilized to determine TBBPA in water samples. Zhang et al. [25] synthesized different morphologies of Fe_2O_3 nanomaterials such as nanoplate, nanorod and flower-like by a hydrothermal method. The electrochemical activity of Fe_2O_3 nanomaterial modified electrodes was found to be closely related to Fe_2O_3's morphology. The composite of flower-like Fe_2O_3 (f-Fe_2O_3) and expanded graphite (EG) exhibited the largest electrochemical active area and the lowest electron transfer resistance. The f-Fe_2O_3/EG modified electrode displayed the best sensing performance of TBBPA with a detection limit of 1.23 nM.

Nano manganese dioxide (MnO_2) was widely used as electrode material owing to its extraordinary electrochemical performance, low cost, abundant storage, non-toxic nature and simplicity in preparation. Currently, MnO_2 nanomaterials show promising potential in solar cell devices, bioapplications, sensing, dye mineralization etc., because MnO_2 can be easily tuned into desired structure and morphology [26]. Different morphological MnO_2 nanoparticles prepared under specific conditions display varying electrochemical characteristics, boasting qualities such as excellent electrical conductivity, stability, and sensitivity [27–29]. Karami-Kolmoti et al. [28] exhibited a sensitive electrochemical sensor based on a manganese dioxide nanorods/graphene oxide nanocomposite for the determination of hydroquinone in water samples. Yu et al. [29] prepared novel manganese dioxide-graphene nanosheets (MnO_2-GNSs) by a one-step hydrothermal method. A MnO_2-GNSs modified glassy carbon electrode was employed for the sensitive detection of hydrogen peroxide. Yakubu et al. [30] introduced a novel competitive electrochemical immunosensor built on gold-palladium bimetallic nanoparticles for quick and sensitive detection of TBBPA in environmental water. The nanoparticles were successfully synthesized and modified with amine-functionalized nanoflower-like manganese oxide. Under the best conditions, the competitive sensors demonstrated superb performance, including commendable sensitivity (LOD, 0.10 ng/mL; S/N = 3) and satisfactory accuracy (recovery rate, 84–120%). This proposed method has been applied in the analysis of TBBPA from various water sources, demonstrating the significant potential for the sensitive detection of trace amounts of TBBPA in aquatic settings. However, a non-enzymatic electrochemical sensor for the detection of TBBPA based on a MnO_2 nanowire modified electrode has not been reported.

In this work, an easy-to-use and sensitive electrochemical sensor, a MnO_2 nanowire modified carbon paper electrode, was fabricated for the sensitive detection of TBBPA. Carbon paper (CP) is a kind of rigid material composed of a complex and tortuous fibrous structure with a resin [31]. It is well known that the electrochemical behavior of sensors depends on their surface properties. Compared with other conventional electrodes, carbon paper electrodes possess some unique characteristics, such as adjustable size, porosity, conductivity and mechanical strength [32,33]. In addition, the use of carbon paper electrodes eliminates the tedious electrode grinding of solid electrodes, making miniaturization and portability possible and advancing on-line and in-situ measurements for electrochemical testing. The effects of pH, the usage of MnO_2, and the enrichment time were also studied to enhance the detection performance of MnO_2-NW/CP for the detection of TBBPA.

2. Materials and Methods

2.1. Chemicals and Instruments

The carbon paper (CP) used in this research was purchased from Toray (Japan). TBBPA and Potassium ferricyanide ($K_3[Fe(CN)_6]$) were purchased from Aladdin Reagent Co. Ltd. (Shanghai, China). Potassium permanganate, sodium dihydrogen phosphate (NaH_2PO_4), dibasic sodium phosphate (Na_2HPO_4), N, N-Dimethylformamide (DMF), anhydrous ethanol and acetonitrile were purchased from Jiangsu Qiangsheng Functional Chemistry Co. Ltd. (Nanjing, China). All the chemicals were of analytical grade and used without further purification. All of the aqueous solutions were prepared with twice-distilled water throughout the whole experiment. The experimental temperatures were all at room temperature.

All electrochemical experiments, including differential pulse voltammetry (DPV) and cyclic voltammetry (CV), were carried out in an electrochemical workstation RST 3000 with a three-electrode system. CP or modified electrode was used as the working electrode, and Ag/AgCl and Pt wire were used as the reference and auxiliary electrodes, respectively.

2.2. Fabrication of Manganese Dioxide

Manganese dioxide was prepared by two methods.

Method 1: 0.728 g of potassium permanganate was added to a mixture solution of 80 mL H_2O and 15 mL ethanol and then stirred at 25 °C for 2 h. After the reaction was completed, the product was washed 3 times with deionized water and anhydrous ethanol and dried in an oven at 60 °C for 12 h [34].

Method 2: 0.45 g of potassium permanganate was dissolved in 30 mL of 0.4 M acetic acid solution, transferred to a 50 mL Teflon-lined stainless-steel autoclave (Anhui Kemi Instruments Co., Hefei, China) and kept at 140 °C for 12 h. After the reaction was completed, the product was washed 3 times with deionized water and anhydrous ethanol and dried in an oven at 60 °C for 12 h.

2.3. Preparation of Modified Electrodes

The CP was cut into 1 × 0.5 cm and ultrasonically cleaned in ethanol for five minutes. The geometric area of the carbon paper electrode in solution was controlled to be 0.5 × 0.5 cm^2 with the insulating adhesive, and the upper end of the carbon paper electrode was connected to the electrochemical workstation. To excite the electrochemical properties of the CP [35], the electrode was immersed in 0.1 M sulfuric acid solution and scanned for 50 cycles from −0.2 V to 1.0 V at 100 mV/s using the CV method. Then, the electrode was cleaned with deionized water and dried at room temperature.

A total of 4.0 mg of manganese dioxide was dispersed in 2.0 mL of DMF solution with the aid of an ultrasonic bath to obtain a suspension of 2 mg/mL of manganese dioxide. Next, 7.0 μL of the above suspension was dropped onto the surface of the treated carbon paper electrode and dried at room temperature.

3. Results and Discussion

3.1. Characterization of Manganese Dioxide

Scanning electron microscope (SEM) images were used to investigate the microstructures and morphologies of MnO_2 nanomaterials. Different microstructures and morphologies of MnO_2 nanomaterials were found when MnO_2 was prepared by different methods. As shown in Figure 1a, an irregular sheet structure was obtained when MnO_2 was synthesized by reducing potassium permanganate using ethanol as a reducing agent under normal temperature and pressure at pH 7 (Method 1). MnO_2 nanowire (Figure 1b) was observed when MnO_2 was prepared via a hydrothermal synthesis under acidic conditions (Method 2). The length of the nanowire is approximately 6–10 μm. The obvious different morphologies were expected to result in different electrochemical performances. The crystal structure of the MnO_2 nanowire (MnO_2-NW) was characterized by X-ray powder diffraction (XRD). In XRD patterns of MnO_2-NW (Figure 1c), the diffraction peaks at 12.78°,18.06°, 28.82°, 37.82° and 42.17° are indexed to crystal Surface of (110), (200), (310), (211) and (301), respectively, which are matched with the standard diffraction pattern of α-MnO_2 (JCPDS card No.44-0141). More importantly, no other impurities are found in these XRD patterns, revealing that as-synthesized MnO_2-NW is well crystallized and at high purity.

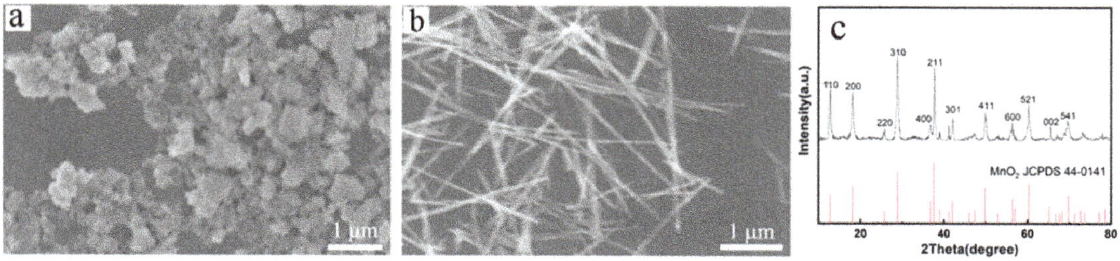

Figure 1. SEM of granular MnO_2 (**a**) and MnO_2-NW (**b**); XRD of MnO_2-NW (**c**).

X-ray photoelectron spectroscopy (XPS) was used to study the surface composition of MnO_2-NW (Figure 2). According to the full scan XPS spectrum (Figure 2a), Mn and O were observed on the surface of the sample of MnO_2-NW. In the high-resolution spectrum of O 1s (Figure 2b), the peak at 529.88 eV corresponded to lattice oxygen, and the appearance of a high-intensity peak indicates that a large amount of lattice oxygen (O latt) was present in MnO_2 nanowires. The characteristic peak of O abs at 531.49 eV was related to surface-adsorbed oxygen, hydroxyl groups, or defect-related oxygen species [36,37]. Figure 2c displays the XPS energy spectrum of Mn 3s. The valence state of Mn was usually determined based on the peak spacing of Mn 3s. The peak spacing of Mn 3s in MnO_2-NW was 4.70 eV, indicating that most of the Mn in the sample exists in the form of +4 valence. In the XPS spectra of Mn 2p (Figure 2d), the peaks at 642.47 eV and 641.04 eV correspond to Mn^{4+} and Mn^{3+}, respectively. The content of Mn^{3+} is relatively low, indicating that most of the Mn in MnO_2-NW samples exist in the form of +4 valence [38,39].

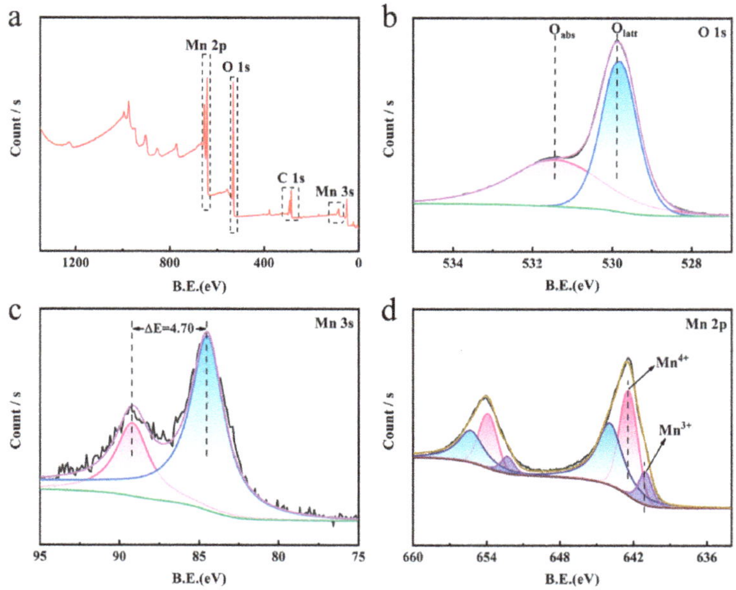

Figure 2. XPS figures of MnO_2-NW. (**a**) full spectrum, (**b**) O 1s, (**c**) Mn 3s, (**d**) Mn 2p.

3.2. Electrochemical Behavior of TBBPA on MnO_2-NW/CP

Figure 3a shows the cyclic voltammograms of 500 nM TBBPA on CP, MnO_2/CP (MnO_2 nanosheet), and MnO_2-NW/CP, respectively. The buffer solution was 0.10 M phosphate buffer solution (pH 6.5). As can be seen from Figure 3a, an irreversible oxidation peak at 0.60 V (vs. Ag/AgCl) was found to correspond to the irreversible oxidation of TBBPA at the working electrode surface. The comparison of the oxidation peak current of TBBPA on the above three electrodes is displayed in Figure 3b. It was shown that the oxidation peak current of TBBPA was slightly enhanced by modifying the nanosheet MnO_2 on CP. In addition, the electrochemical signal of TBBPA on MnO_2-NW/CP was approximately twice higher than that on the bare CP electrode. Furthermore, the oxidation potential shifted to 0.58 V. It is well understood that the electrochemical activity of nano MnO_2 is shape-dependent. Compared with the MnO_2 nanosheet, the MnO_2 nanowire can provide a higher surface area and more abundant surface oxygen, which can result in enhanced catalytic activity. The oxidation of TBBPA was obviously promoted when the CP electrode was modified by MnO_2-NW, which may be due to the larger specific surface area, better electrical conductivity and the oxidizing ability of MnO_2-NW. Hence, MnO_2-NW/CP was chosen for the detection of TBBPA.

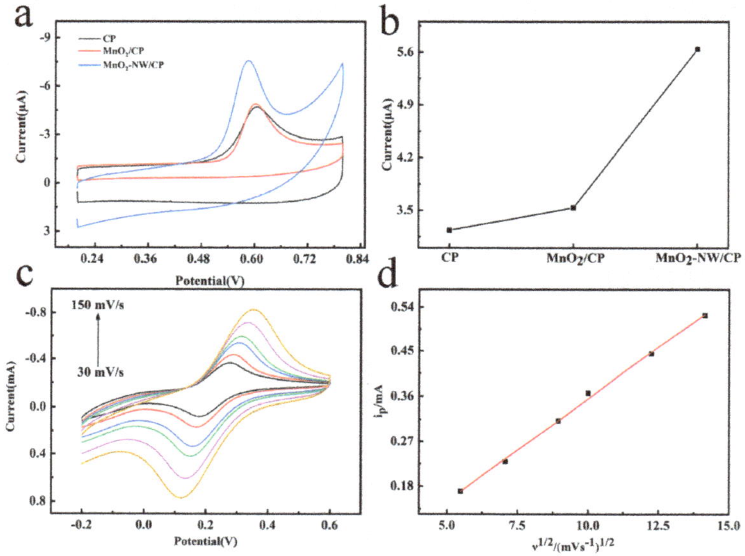

Figure 3. (**a**) CVs of 500 nM TBBPA on different electrodes; (**b**) oxidation current of TBBPA on different electrodes; (**c**) CVs of MnO_2–NW/CP in 2.5 mM $K_3[Fe(CN)_6]$ at different scan rates; (**d**) The linear plot of $K_3[Fe(CN)_6]$ ipa vs. $\upsilon^{1/2}$.

CVs of MnO_2-NW/CP in 0.1 M KCl solution containing 2.5 mM $K_3[Fe(CN)_6]$ at various scan rates were recorded in the range of 30–150 mV/s to investigate the effective electrochemical area of the MnO_2-NW/CP electrode. It was observed that the oxidation and reduction peaks of potassium ferricyanide increase with increasing scanning speed at the MnO_2-NW/CP electrode, and the potential difference between the oxidation and reduction peaks becomes larger. Diffusion-controlled behavior of $K_3[Fe(CN)_6]$ could be deduced from the linear relationship between oxidation peak currents i_p and the square root of scan rates $\upsilon^{1/2}$ with the regression equation of i_p = 0.04109 $\upsilon^{1/2}$ − 0.05549 (R^2 = 0.9976) (Figure 3d). The Randles–Sevcik equation was used to calculate the effective surface area (ESA) [40].

$$i_p = 0.4463nFAC\left(\frac{nFvD}{RT}\right)^{1/2} = 2.69 \times 10^5 n^{2/3} ACD^{1/2} v^{1/2} \quad (25\ ^\circ\text{C}) \tag{1}$$

where i_p (A) is the oxidation or reduction peak current; n is the total number of electrons exchanged in the redox reaction; A (cm^2) is the effective surface area (ESA); C (mol/cm^3) is the concentration of $K_3[Fe(CN)_6]$ (2.5 mM); D (cm^2/s) is the diffusion coefficient of $[Fe(CN)_6]^{3-}$ (6.5×10^{-6} cm^2/s); and v (V/s) is the scan rate. In our experiment, the slope was 0.04109, and the ESA value was calculated as 0.24 cm^2. The ESA value of the bare carbon paper electrode was determined using a similar process to be 0.026 cm^2. Thus, the ESA of MnO_2-NW/CP was about nine times greater than that of a bare carbon paper electrode, suggesting the excellent electrochemical properties of MnO_2-NW/CP.

Figure 4a shows CVs of MnO_2-NW/CP in 0.1 M pH 6.5 phosphate buffer solutions containing 500 nM TBBPA at different scan rates in the range of 25–150 mV/s. The peak current in CVs is coupled with the change of concentration gradient with time. Thus, the peak currents are obviously influenced by the scan rate. This is because the scan rate determines the time between the switching potential and the peak potential. It was shown that the oxidation peak currents of TBBPA kept increasing with the increase in the scan rate in the range of 25–150 mV/s. Figure 4b shows a plot of the TBBPA oxidation peak current versus sweep rate with the linear equation $i_p = 0.0975\upsilon + 0.00459$ ($R^2 = 0.9915$). Figure 4c is a plot of i_p versus $\upsilon^{1/2}$ with the linear equation $i_p = 1.67755\upsilon^{1/2}–6.61462$ ($R^2 = 0.9531$). The oxidation peak currents of TBBPA were background subtracted to correct for the larger charging current contribution at the higher scan rate. The oxidation peak current of TBBPA is proportional to the potential scan rate, which is in accordance with the formula of $i_p = (n^2F^2\Gamma AV)/(4RT)$. Moreover, the independence between i_p and the square root of the scan rate reveals that the oxidation of TBBPA on MnO_2-NW/CP was not controlled by the diffusing process. Thus, TBBPA oxidation on a MnO_2-NW/CP is an adsorption-controlled process. The oxidation peak potentials of TBBPA were found to shift toward positive potentials with the increase in scan rates, which is related to the irreversible oxidation process of TBBPA.

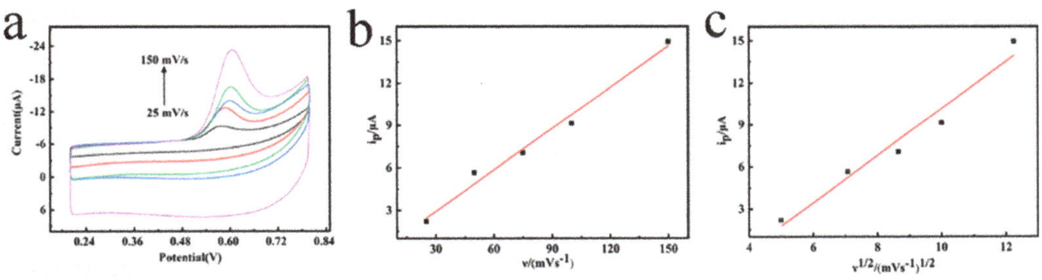

Figure 4. (**a**) CVs of 500 nM TBBPA in 0.1 M pH 6.5 PBS on MnO_2−NW/CP at different scan rates; (**b**)The linear plot of i_p vs. υ; (**c**)The linear plot of i_p vs. $\upsilon^{1/2}$.

3.3. Optimization of Experimental Conditions

DPV was chosen to prepare a sensitive sensor for the determination of TBBPA. The experiment parameters, including the pH value of the buffer solution, the enriching time and the amount of MnO_2-NW, were optimized by recording the electrochemical performance of TBBPA under different experiment conditions.

Figure 5a shows the oxidation peak currents and peak potentials of TBBPA at different pH conditions. It was found that the oxidation peak currents of TBBPA increased with increasing pH in the range of 5.5–6.5 and reached a maximum at pH 6.5. The pKa_1 and pKa_2 of TBBPA were reported to be 7.5 and 8.5. TBBPA was not dissociated when the pH was 6.5. The undissociated TBBPA can be adsorbed more easily on a MnO_2-NW/CP electrode than the dissociated form. This resulted in the maximum oxidation peak current of TBBPA at pH 6.5. Hence, pH 6.5 was chosen as the optimum pH condition in the subsequent experiments. In the pH range of 5.8–8.0, the oxidation peak potential of TBBPA negatively linearly shifted with an increase in pH value following the equations of $Ep = 0.978–0.0594$ pH. This shows that protons took part in the electrochemical reaction of TBBPA. The slope −0.0594 V/pH was close to the value obtained from the Nernst equation (−0.059 V/pH), which indicated that the number of electrons was equal to the number of protons during the oxidation of TBBPA [41]. Referring to the oxidation mechanism of TBBPA on Fe_3O_4@SiO_2@CDs-CTAB/GCE [20], the oxidation mechanism of TBBPA on MnO_2-NW/CP is deduced in Figure 6.

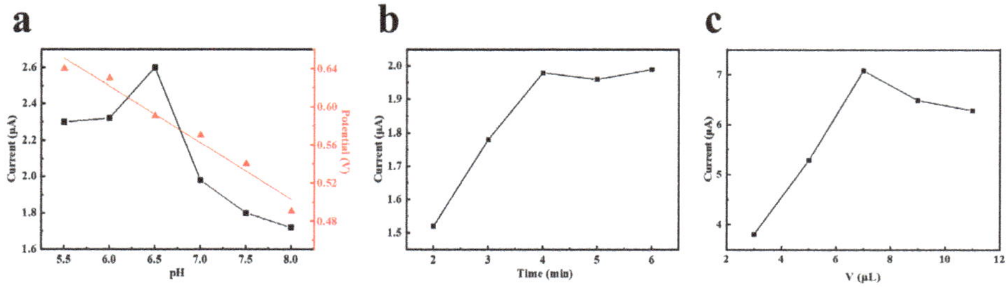

Figure 5. (**a**) The effect of pH on the oxidation peak current and potential of TBBPA; (**b**) The effect of enrichment time on the peak current of TBBPA; (**c**) The effect of MnO_2-NW amount on the peak current of TBBPA.

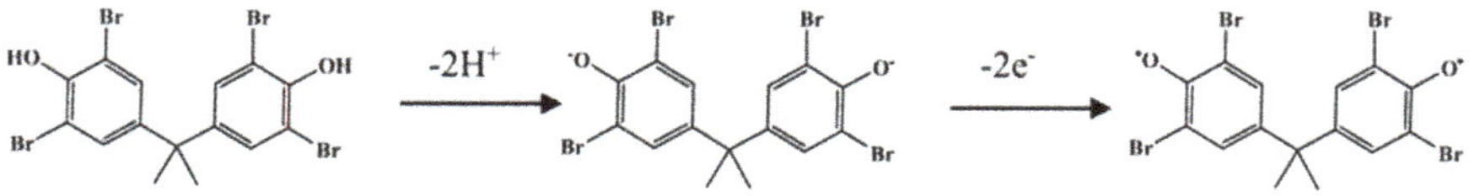

Figure 6. Electrochemical oxidation mechanism of TBBPA.

The effect of enrichment time on the oxidation peak current at the open circuit of TBBPA (500 nM) is shown in Figure 5b. When the enrichment time was less than 5 min, the peak current increased rapidly with the increase in enrichment time. This is because more TBBPA was adsorbed on the MnO_2-NW/CP electrode with increasing enrichment time. When the enrichment time was above 5 min, the oxidation peak current tended to be stable due to the adsorption saturation of TBBPA. Therefore, 5 min of enrichment time was chosen.

The effect of the modification amount of MnO_2-NW was investigated in Figure 5c. It was found that the oxidation current of TBBPA increased sharply when the modification amount of MnO_2-NW suspension increased from 3.0 to 7.0 µL. This might be a result of more catalyst-enhanced catalytic ability. However, when the modification amount was over 7.0 µL, the oxidation peak current decreased due to the deterioration of electrical conductivity. Therefore, 7.0 µL suspension of MnO_2-NW was selected as the optimal modification amount in this work.

3.4. Electrochemical Detection of TBBPA

DPV of MnO_2-NW/CP in different TBBPA concentrations was investigated under optimal experimental conditions (Figure 7). As shown in Figure 7a, the oxidation currents increased when TBBPA was added successively from 70 to 500 nM. The linear regression equation was $i_p = 0.00323c + 0.06304$, $R^2 = 0.9914$ (Figure 7b). The estimated detection limit is 3.1 nM based on a three-signal-to-noise ratio of 3. It could be seen that our electrochemical sensor exhibited a wide linear range and a low detection limit, which might be due to the high ESA of the MnO_2-NW/CP, with plenty of active sites. Our experimental results were compared with the literature reports in Table 1, indicating that this method has some favorable advantages.

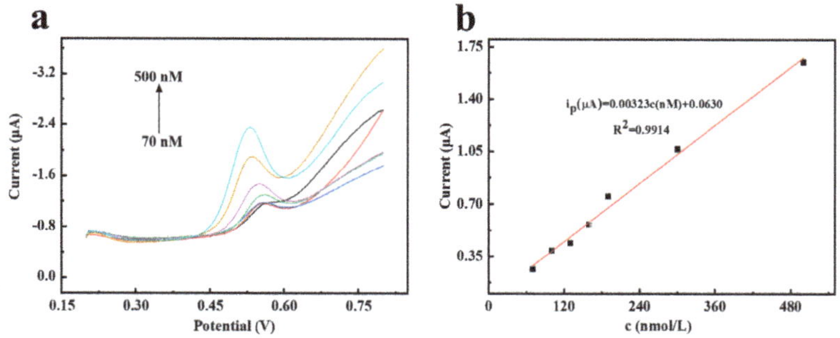

Figure 7. (**a**) DPVs of MnO_2−NW/CP in 0.1 M pH 6.5 PBS containing different concentrations of TBBPA; (**b**) The corresponding calibration plots between the peak currents and the concentrations of TBBPA.

Table 1. Comparison of the experimental results with other electrodes or methods.

Sensor Assembly	Method	LOD	Linear Range	Reference
CNTs@ZIF-67/PFDA/AB	DPV	4.2 nM	10~150 nM	[19]
AuNPs-PSSA/GCE	DPV	25 pM	0.1~10 nM	[42]
MI-TiO_2/Au/rGO	DPV, CV	0.51 nM	1.68~100 nM	[43]
CTAB-MnO_2/Pd	I-t	0.17 ng/mL	0~81 ng/mL	[44]
poly(PPBim-DS)/GCE	DPV	20 nM	0.05~10 μM	[39]
MnO_2-NW/CP	DPV	3.1 nM	70~500 nM	This work

Five parallel MnO_2-NW/CP electrodes were prepared for the detection of 500 nM TBBPA, and the detection relative standard deviation was 3.8%. One of the above five electrodes was further investigated five consecutive times, and the relative standard deviation was 1.9%. The results showed good reproducibility and precision of MnO_2-NW/CP.

3.5. Real Samples Analysis

To estimate the potential of a MnO_2-NW for the detection of TBBPA in practical applications, the fabricated electrochemical sensor was employed to detect TBBPA in lake water (Suzhou, China). Before detection, the water was filtered through a 0.45 μm membrane to remove suspended solids. No response to TBBPA was found in these real samples, indicating that the concentration of TBBPA was extremely low or that no TBBPA existed in these real samples. Then standard solutions of TBBPA were spiked into real sample solutions for recovery evaluation. The recoveries were in the range of 94.5% to 100.5%. The results indicated that MnO_2-NW/CP is reliable and could be applied to detect TBBPA in the actual water sample.

4. Conclusions

In this study, a cost-effective, simple fabricated and sensitive electrochemical sensor was prepared by modifying carbon paper with a MnO_2 nanomaterial. The performance of electrochemical sensors was proven to be closely related to the microscopic morphology of MnO_2. The fabricated MnO_2-NW/CP displayed attractive performance for the detection of TBBPA. The sensitivity and stability of MnO_2-NW/CP were satisfied, and the selectivity of MnO_2-NW/CP can be further enhanced by combining other materials, such as molecular imprinting technology. It is reasonable to believe that MnO_2 nanowire-based electrochemical sensors might be a promising method for the effective detection of TBPPA in environment monitoring.

Author Contributions: Conceptualization, Q.Z.; methodology, Q.Z. and C.Z.; software, C.Z. and Q.W.; validation, Q.W.; investigation, C.Z.; resources, Q.Z.; data curation, C.Z. and F.Y.; writing—original draft preparation, C.Z. and F.Y.; writing—review and editing Q.Z., J.L. and D.W.; funding acquisition, Q.Z. All authors have read and agreed to the published version of the manuscript.

Funding: This work was financially supported by the National Natural Science Foundation of China (No. 20905055), Postgraduate Research and Practice Innovation Program of Jiangsu Province (No. SJCX22_1564).

Institutional Review Board Statement: Not applicable.

Informed Consent Statement: Not applicable.

Data Availability Statement: Data is contained within the article.

Conflicts of Interest: The authors declare no conflict of interest.

References

1. Lee, S.; Song, G.J.; Kannan, K.; Moon, H.B. Occurrence of PBDEs and other alternative brominated flame retardants in sludge from wastewater treatment plants in Korea. *Sci. Total Environ.* **2014**, *470*, 1422–1429. [PubMed]
2. Shi, Z.X.; Zhang, L.; Li, J.G.; Wu, Y.N. Legacy and emerging brominated flame retardants in China: A review on food and human milk contamination, human dietary exposure and risk assessment. *Chemosphere* **2018**, *198*, 522–536. [CrossRef] [PubMed]
3. Shi, Z.X.; Zhang, L.; Zhao, Y.F.; Sun, Z.W.; Zhou, X.Q.; Li, J.G.; Wu, Y.N. Dietary exposure assessment of Chinese population to tetrabromobisphenol-A, hexabromocyclododecane and decabrominated diphenyl ether: Results of the 5th Chinese Total Diet Study. *Environ. Pollut.* **2017**, *229*, 539–547. [CrossRef] [PubMed]
4. Wluka, A.; Wozniak, A.; Wozniak, E.; Michalowicz, J. Tetrabromobisphenol A, terabromobisphenol S and other bromophenolic flame retardants cause cytotoxic effects and induce oxidative stress in human peripheral blood mononuclear cells (in vitro study). *Chemosphere* **2020**, *261*, 127705. [CrossRef]
5. Zuiderveen, E.A.R.; Slootweg, J.C.; de Boer, J. Novel brominated flame retardants—A review of their occurrence in indoor air, dust, consumer goods and food. *Chemosphere* **2020**, *255*, 126816.
6. Nakajima, A.; Saigusa, D.; Tetsu, N.; Yamakuni, T.; Tomioka, Y.; Hishinuma, T. Neurobehavioral effects of tetrabromobisphenol A, a brominated flame retardant, in mice. *Toxicol. Lett.* **2009**, *189*, 78–83. [CrossRef]
7. Feiteiro, J.; Mariana, M.; Cairrao, E. Health toxicity effects of brominated flame retardants: From environmental to human exposure. *Environ. Pollut.* **2021**, *285*, 117475.
8. Jagic, K.; Dvorscak, M.; Klincic, D. Analysis of brominated flame retardants in the aquatic environment: A review. *Arh. Hig. Rada Toksikol.* **2021**, *72*, 254–267. [PubMed]
9. Gao, W.; Li, G.L.; Liu, H.; Tian, Y.; Li, W.T.; Fa, Y.; Cai, Y.Q.; Zhao, Z.S.; Yu, Y.L.; Qu, G.B.; et al. Covalent organic frameworks with tunable pore sizes enhanced solid-phase microextraction direct ionization mass spectrometry for ultrasensitive and rapid analysis of tetrabromobisphenol A derivatives. *Sci. Total Environ.* **2021**, *764*, 144388. [CrossRef]
10. Zhang, S.; Liu, J.; Hou, X.; Zhang, H.; Zhu, Z.; Jiang, G. Sensitive method for simultaneous determination of TBBPA and its ten derivatives. *Talanta* **2023**, *264*, 124750.
11. Zeng, L.S.; Cui, H.R.; Chao, J.L.; Huang, K.; Wang, X.; Zhou, Y.K.; Jing, T. Colorimetric determination of tetrabromobisphenol A based on enzyme-mimicking activity and molecular recognition of metal-organic framework-based molecularly imprinted polymers. *Microchem. J.* **2020**, *187*, 142. [CrossRef]
12. Feng, J.; Tao, Y.; Shen, X.; Jin, H.; Zhou, T.; Zhou, Y.; Lee, Y.I. Highly sensitive and selective fluorescent sensor for tetrabromobisphenol-A in electronic waste samples using molecularly imprinted polymer coated quantum dots. *Microchem. J.* **2019**, *144*, 93–101. [CrossRef]
13. Fu, H.J.; Wang, Y.; Xiao, Z.L.; Wang, H.; Li, Z.F.; Shen, Y.D.; Lei, H.T.; Sun, Y.M.; Xu, Z.L.; Hammock, B. A rapid and simple fluorescence enzyme-linked immunosorbent assay for tetrabromobisphenol A in soil samples based on a bifunctional fusion protein. *Ecotoxicol. Environ. Saf.* **2020**, *188*, 109904. [CrossRef] [PubMed]
14. Li, Z.F.; Wang, Y.; Vasylieva, N.; Wan, D.B.; Yin, Z.H.; Dong, J.X.; Hammock, B. An Ultrasensitive Bioluminescent Enzyme Immunoassay Based on Nanobody/Nanoluciferase Heptamer Fusion for the Detection of Tetrabromobisphenol A in Sediment. *Anal. Chem.* **2020**, *92*, 10083–10090. [CrossRef]
15. Zenasni, M.; Quintero-Jaime, A.; Salinas-Torres, D.; Benyoucef, A.; Morallón, E. Electrochemical synthesis of composite materials based on titanium carbide and titanium dioxide with poly (N-phenyl-o-phenylenediamine) for selective detection of uric acid. *J. Electroanal. Chem.* **2021**, *895*, 115481. [CrossRef]
16. Zhang, Z.H.; Cai, R.; Long, F.; Wang, J. Development and application of tetrabromobisphenol A imprinted electrochemical sensor based on graphene/carbon nanotubes three-dimensional nanocomposites modified carbon electrode. *Talanta* **2015**, *134*, 435–442. [CrossRef]

17. Zhou, T.T.; Feng, Y.Q.; Zhou, L.X.; Tao, Y.; Luo, D.; Jing, T.; Shen, X.L.; Zhou, Y.K.; Mei, S.R. Selective and sensitive detection of tetrabromobisphenol-A in water samples by molecularly imprinted electrochemical sensor. *Sens. Actuators B Chem.* **2016**, *236*, 153–162. [CrossRef]
18. Wang, Y.Y.; Liu, G.S.; Hou, X.D.; Huang, Y.N.; Li, C.Y.; Wu, K.B. Assembling gold nanorods on a poly-cysteine modified glassy carbon electrode strongly enhance the electrochemical reponse to tetrabromobisphenol A *Microchim. Acta* **2015**, *183*, 689–696. [CrossRef]
19. Zhou, T.T.; Zhao, X.Y.; Xu, Y.H.; Tao, Y.; Luo, D.; Hu, L.Q.; Jing, T.; Zhou, Y.K.; Wang, P.; Mei, S.R. Electrochemical determination of tetrabromobisphenol A in water samples based on a carbon nanotubes@zeolitic imidazole framework-67 modified electrode. *RSC Adv.* **2020**, *10*, 2123–2132. [CrossRef]
20. Guo, J.H.; Zhou, B.Y.; Li, S.Y.; Tong, Y.Y.; Li, Z.; Liu, M.H.; Li, Y.H.; Qu, T.X.; Zhou, Q.X. Novel electrochemical sensor from magnetic carbon dots and cetyltrimethylammonium bromide for sensitive measurement of tetrabromobisphenol A in beverages. *Chemosphere* **2022**, *298*, 134326. [CrossRef]
21. Shao, Y.M.; Zhu, Y.; Zheng, R.; Wang, P.; Zhao, Z.Z.; An, J. Highly sensitive and selective surface molecularly imprinted polymer electrochemical sensor prepared by Au and MXene modified glassy carbon electrode for efficient detection of tetrabromobisphenol A in water. *Adv. Compos. Hybrid Mater.* **2022**, *5*, 3104–3116. [CrossRef]
22. Zhao, Q.; Zhou, H.; Wu, W.; Wei, X.; Jiang, S.; Zhou, T.; Lu, Q. Sensitive electrochemical detection of tetrabromobisphenol A based on poly (diallyldimethylammonium chloride) modified graphitic carbon nitride-ionic liquid doped carbon paste electrode. *Electrochim. Acta* **2017**, *254*, 214–222. [CrossRef]
23. Zhou, Q.; Zhang, Y.Y.; Zeng, T.; Wan, Q.J.; Yang, N.J. Morphology-dependent sensing performance of CuO nanomaterials. *Anal. Chim. Acta* **2021**, *1171*, 338663. [CrossRef] [PubMed]
24. Luo, S.X.; Yang, M.Z.; Wu, Y.H.; Li, J.; Qin, J.; Feng, F. A Low Cost Fe_3O_4-Activated Biochar Electrode Sensor by Resource Utilization of Excess Sludge for Detecting Tetrabromobisphenol A. *Micromachines* **2022**, *13*, 115. [CrossRef]
25. Chen, X.Y.; Zhang, Y.Y.; Li, C.; Li, C.; Zeng, T.; Wan, Q.J.; Li, Y.W.; Ke, Q.; Yang, N.J. Nanointerfaces of expanded graphite and Fe_2O_3 nanomaterials for electrochemical monitoring of multiple organic pollutants. *Electrochim. Acta* **2020**, *329*, 135118. [CrossRef]
26. Baral, A.; Satish, L.; Zhang, G.Y.; Ju, S.H.; Ghosh, M.K. A Review of Recent Progress on Nano MnO_2: Synthesis, Surface Modification and Applications. *J. Inorg. Organomet. Polym. Mater.* **2020**, *31*, 899–922. [CrossRef]
27. Yu, Y.; Liu, S.L.; Ji, J.; Huang, H.B. Amorphous MnO_2 surviving calcination: An efficient catalyst for ozone decomposition. *Catal. Sci. Technol.* **2019**, *9*, 5090–5099. [CrossRef]
28. Karami-Kolmoti, P.; Beitollahi, H.; Modiri, S. Electrochemical Sensor for Simple and Sensitive Determination of Hydroquinone in Water Samples Using Modified Glassy Carbon Electrode. *Biomedicines* **2023**, *11*, 1869. [CrossRef]
29. Huang, Z.N.; Liu, G.C.; Zou, J.; Jiang, X.Y.; Liu, Y.P.; Yu, J.G. A hybrid composite of recycled popcorn-shaped MnO_2 microsphere and Ox-MWCNTs as a sensitive non-enzymatic amperometric H_2O_2 sensor. *Microchem. J.* **2020**, *158*, 105215. [CrossRef]
30. Yakubu, S.; Xiao, J.X.; Gu, J.P.; Cheng, J.; Wang, J.; Li, X.S.; Zhang, Z. A competitive electrochemical immunosensor based on bimetallic nanoparticle decorated nanoflower-like MnO_2 for enhanced peroxidase-like activity and sensitive detection of Tetrabromobisphenol A. *Sens. Actuators B Chem.* **2020**, *325*, 128909. [CrossRef]
31. Torrinha, A.; Morais, S. Electrochemical (bio)sensors based on carbon cloth and carbon paper: An overview. *Trends Anal. Chem.* **2021**, *142*, 116324. [CrossRef]
32. Li, Y.H.; Xu, Q.Z.; Li, Q.Y.; Wang, H.Q.; Huang, Y.G.; Xu, C.W. Pd deposited on MWCNTs modified carbon fiber paper as high-efficient electrocatalyst for ethanol electrooxidation. *Electrochim. Acta* **2014**, *147*, 151–156. [CrossRef]
33. Kannan, P.; Maiyalagan, T.; Marsili, E.; Ghosh, S.; Niedziolka-Jonsson, J.; Jonsson-Niedziolka, M. Hierarchical 3-dimensional nickel-iron nanosheet arrays on carbon fiber paper as a novel electrode for non-enzymatic glucose sensing. *Nanoscale* **2016**, *8*, 843–855. [CrossRef]
34. Sun, H.Y.; Wang, C.X.; Xu, Y.J.; Dai, D.M.; Deng, X.Y.; Gao, H.T. A Novel Electrochemical Sensor Based on A Glassy Carbon Electrode Modified with GO/MnO_2 for Simultaneous Determination of Trace Cu(II) and Pb(II) in Environmental Water. *ChemistrySelect* **2019**, *4*, 11862–11871. [CrossRef]
35. Radulescu, M.C.; Bucur, M.P.; Bucur, B.; Radu, G.L. Ester flavorants detection in foods with a bienzymatic biosensor based on a stable Prussian blue-copper electrodeposited on carbon paper electrode. *Talanta* **2019**, *199*, 541–546. [CrossRef] [PubMed]
36. Tagsin, P.; Suksangrat, P.; Klangtakai, P.; Srepusharawoot, P.; Ruttanapun, C.; Kumnorkaew, P.; Pimanpang, S.; Amornkitbamrung, V. Electrochemical mechanisms of activated carbon, alpha-MnO_2 and composited activated carbon-alpha-MnO2 films in supercapacitor applications. *Appl. Surf. Sci.* **2021**, *570*, 151056. [CrossRef]
37. Cui, Y.; Song, H.K.; Shi, Y.Y.; Ge, P.X.; Chen, M.D.; Xu, L.L. Enhancing the Low-Temperature CO Oxidation over CuO-Based alpha-MnO_2 Nanowire Catalysts. *Nanomaterials* **2022**, *12*, 2083. [CrossRef]
38. Zhou, X.; Ye, X.X.; Wu, K.B.; Li, C.; Wang, Y.Y. Electrochemical sensing of terabromobisphenol A at a polymerized ionic liquid film electrode and the enhanced effects of anions. *Ionics* **2018**, *24*, 2843–2850. [CrossRef]
39. Ganesh, A.; Sivakumar, T.; Sankar, G. Biomass-derived porous carbon-incorporated MnO_2 composites thin films for asymmetric supercapacitor: Synthesis and electrochemical performance. *J. Mater. Sci. Mater. Electron.* **2022**, *33*, 14772–14783. [CrossRef]
40. Shen, T.Y.; Liu, T.C.; Mo, H.Q.; Yuan, Z.C.; Cui, F.; Jin, Y.X.; Chen, X.J. Cu-based metal-organic framework HKUST-1 as effective catalyst for highly sensitive determination of ascorbic acid. *RSC Adv.* **2020**, *10*, 22881–22890. [CrossRef]

41. Stradins, J.; Hasanli, B. Anodic voltammetry of phenol and benzenethiol derivatives.1. Influence of pH on electrooxidation potentials of substituted phenols and evaluation of pK(a) from anodic voltammetry data. *J. Electroanal. Chem.* **1993**, *353*, 57–69. [CrossRef]
42. Shen, J.; Bian, C.; Xia, S.H.; Wu, K.B. Poly(sulfosalicylic acid)-functionalized gold nanoparticles for the detection of tetrabromobisphenol A at pM concentrations. *J. Hazard. Mater.* **2020**, *388*, 121733. [CrossRef]
43. Li, Z.; Hu, J.Y.; Lou, Z.Z.; Zeng, L.X.; Zhu, M.S. Molecularly imprinted photoelectrochemical sensor for detecting tetrabromobisphenol A in indoor dust and water. *Microchem. J.* **2021**, *188*, 320. [CrossRef] [PubMed]
44. Yakubu, S.; Jia, B.Y.; Guo, Y.J.; Zou, Y.M.; Song, N.H.; Xiao, J.X.; Liang, K.L.; Bu, Y.Q.; Zhang, Z. Indirect competitive-structured electrochemical immunosensor for tetrabromobisphenol A sensing using CTAB-MnO_2 nanosheet hybrid as a label for signal amplification. *Anal. Bioanal. Chem.* **2021**, *413*, 4217–4226. [CrossRef] [PubMed]

chemosensors

MDPI

Review

Single-Atom Nanomaterials in Electrochemical Sensors Applications

Jinglin Fu and Yang Liu *

Department of Chemistry, Beijing Key Laboratory for Analytical Methods and Instrumentation, Kay Lab of Bioorganic Phosphorus Chemistry and Chemical Biology of Ministry of Education, Tsinghua University, Beijing 100084, China
* Correspondence: liu-yang@mail.tsinghua.edu.cn; Tel.: +86-10-62798137

Abstract: In recent years, the development of highly sensitive sensors has become a popular research topic. Some functional nanomaterials occupy an important position in the sensing field by virtue of their unique structures and catalytic properties, but there are still problems such as low sensitivity and poor specificity. Single-atom nanomaterials (SANs) show significant advantages in amplifying sensing signals and improving sensor interference resistance due to their high atomic utilization, structural simplicity, and homogeneity. They are expected to achieve high sensitivity and high specificity monitoring by modulating the active sites. In this review, the recent progress on SANs for electrochemical sensing applications was summarized. We first briefly summarize the features and advantages of single-atom catalysts. Then recent advances in the regulation of reaction sites in noble and non-noble metal-based SANs, including the introduction of defects in the carrier, other metal atoms, and ligand atoms, were highlighted. After that, the SANs for the construction of electrochemical, electrochemiluminescent (ECL), and photoelectrochemical (PEC) sensors and their applications in biochemical and environmental analysis were demonstrated. Finally, the future research aspect of SANs-based electrochemical sensing and the challenges of the SANs design and structure-properties revelation were illustrated, giving guidance on sensitive and accurate biosensing toward clinic diagnostic and environmental analysis.

Keywords: single-atom nanomaterial; electrochemical; sensor

Citation: Fu, J.; Liu, Y. Single-Atom Nanomaterials in Electrochemical Sensors Applications. *Chemosensors* **2023**, *11*, 486. https://doi.org/10.3390/chemosensors11090486

Academic Editor: Maria Luz Rodriguez-Mendez

Received: 30 June 2023
Revised: 2 August 2023
Accepted: 1 September 2023
Published: 3 September 2023

1. Introduction

Electrochemical analysis is a technique that involves the measurement of electrical signals from chemical reactions occurring at the electrode interface [1]. Because of the obvious advantages of high sensitivity, fast response time, simple operation, and miniaturization, electrochemical sensors have become an important analytical tool in areas such as environmental and biological samples [2]. In recent years, as researchers have been studying the physicochemical properties of nanomaterials, it has been possible to achieve precise modulation of their structure and size to a large extent, which has led to many critical breakthroughs in the utilization of nanomaterials as sensing interfaces for signal amplification and interfacial biomolecule recognition [3]. However, the extensive application of electrochemical sensors is still limited by their low sensitivity for the detection of trace substances, and their specificity in complex biological systems is yet to be improved. Since the chemical and physical features of nanomaterials are highly correlative to their intrinsic properties such as surface areas, numbers of active sites, morphologies, etc., precise control of their structures can not only improve catalytic activity and amplify signals but also facilitate specific identification of the analytes to be detected, thus promoting the performance of electrochemical sensors in clinical diagnosis, environmental analysis, and so on [4,5]. In these years, SANs have received much attention since their introduction due to their special geometry and electronic structure, and the precise metal sites of SANs cater to the challenges encountered in the sensitivity and specificity of electrochemical sensors [6].

SANs are distinguished by the isolated metal atoms dispersed on the support material and the absence of metal–metal bonds. SANs not only maximize atomic utilization and have unique electronic structures and atomic coordination environments, but more importantly, the simplicity and homogeneity of their structures facilitate the accurate identification and characterization of active sites, which can provide insight into the structure–activity relationship and make reasonable regulation of them possible [7,8]. These merits enable SANs to be effective materials for the development of sensors with high analytical performance and defined mechanisms. The presence of a large number of unsaturated, low-coordinated metal atoms in SANs facilitates signal amplification and also improves the selectivity of detection by adjusting the metal centers and their coordination environment, which is conducive to achieving sensitive detection of trace substances. By adjusting the structure of SANs, researchers have continuously optimized their sensitivity, selectivity, and stability in sensing, making them great candidates for development in analytical chemistry [1,9,10]. According to the classification of support materials, they broadly include metal–carbon-based materials [11–16], metal–metal oxides [17–19], metal–metal sulfides [20–22], and metal–metal substrates [23,24] (i.e., single-atom alloys) synthesized mainly through three strategies, including defect engineering strategy, spatial confinement strategy, and sacrificial template-assisted strategy (Figure 1).

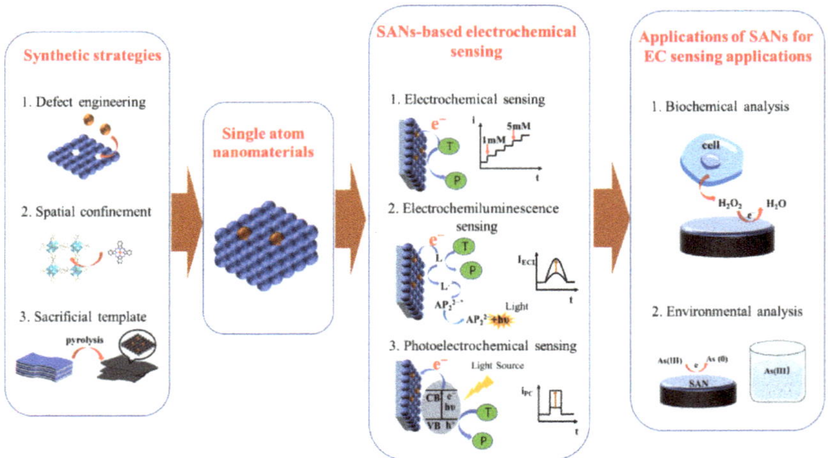

Figure 1. Schematic illustration of the main content of this review.

In this review, we systematically summarize recent advances in SANs for applications in catalytic performance optimization and electrochemical sensing. Firstly, the recent achievements in the modulation of the active sites, including the noble and non-noble metal-based SANs, are presented. The active centers of SANs are mostly the metallic atoms in them, which are not only affected by the metal species but also the numbers of the metal centers as well as their coordinative environment. In this way, the properties of the SANs can be modulated by the central metal atoms and their coordination environment. Then, the research progress of SANs in electrochemistry [25–27], electrochemiluminescence (ECL) [28,29], and photoelectrochemical (PEC) [30,31] sensing for environmental and biochemical sample analysis was highlighted, which has demonstrated large potential for in vivo and in vitro detection of biological samples as well as environmental samples. Finally, the challenges and opportunities for the application of SANs in the field of electrochemical sensing are illustrated (Figure 1).

2. Active Centers Modulation of SANs

The active sites are the catalytically active parts of the SAN, including the metal center and its coordination atoms, which directly affect the catalytic activity and selectivity of the

SAN. The metal centers of the SANs are generally transition metals. Most conventional electrocatalysts are based on noble metals, which have been widely used in catalysis due to their unique intrinsic properties and irreplaceable catalytic activity, such as Pt [18], Pd [32], Ru [33], Rh [34–36], Ir [37], Ag [35,38], Au [39], etc. In order to solve the problems of high cost and low reserves, a series of non-noble metal elements with highly active and economical alternatives have been developed in recent years, such as Fe [40,41], Co [42–44], Cu [45,46], Mn [47], Mo [48], Zn [49], Ni [50,51], etc. The transition metals have empty d or f orbitals and can form coordination bonds with substrate molecules, which facilitates the binding of analytes to SANs [52]. These metals can be dispersed on the supports as single or multiple atoms, and the specific recognition of the analyte can be achieved by modulation of the metal centers. The metal atoms in SANs are isolated and confined within the support materials and are given specific coordination environments. The differences in coordination environments result in distinct electronic structures and densities and thereby can affect the binding energy between substances and SANs as well as the adsorption energy of some reaction intermediates and products, which are important factors in determining catalytic activity and the specificity of sensing. Therefore, methods such as heteroatom doping and axial ligand modulation are also effective ways to modulate the sensing performance of SANs. The following section reviews the work on improving the catalytical activities of noble metal-based and non-noble metal-based SANs by modulating the active centers.

2.1. Noble Metal-Based SANs

Noble metals have been widely studied in the fields of catalysis and sensing due to their superior properties, but high costs and small reserves have severely limited their large-scale applications. The most significant advantage of SANs is the maximization of atomic utilization, which means less waste of active sites and the promise of achieving the same activity with reduced metal loading. Therefore, combining noble metal materials with single-atom strategies is undoubtedly the most effective way to reduce costs. By modifying single Pt atoms on carbon nitride nanorods, SA–Pt/g–C_3N_4–K with peroxidase (POD)-mimicking activity was obtained and used to construct H_2O_2 and antibiotic sensors [53]. After experimental investigation, the enzyme activity of SA–Pt/g–C_3N_4–K was found to be the highest at pH = 4 and 48 °C. Moreover, compared with the natural enzyme horseradish peroxidase (HRP), SA–Pt/g–C_3N_4–K can maintain catalytic stability over a wider temperature range. The POD activities of SA–Pt/g–C_3N_4–K, g–C_3N_4–K, and HRP were compared by colorimetric method using UV-vis absorption spectroscopy with 1 mM tetramethylbenzidine (TMB) and 1 mM H_2O_2 at pH = 4. It was found that SA–Pt/g–C_3N_4–K exhibited higher POD-like activity (Figure 2), which could be targeted to improve the sensitivity of detecting H_2O_2. More than just H_2O_2 sensors, Pt-based SANs can also be utilized for the detection of glucose. A Pt single-atom material on Cu@CuO core-shell nanowires (NWs) has been fabricated by Zhao et al [54]. Cu NWs were first synthesized and further oxidized to Cu@Cu_xO core-shell NWs by oxidation with H_2O_2, and then Pt atoms were anchored to Cu@CuO NWs by an impregnation method and were modified on a glassy carbon electrode (GCE) for glucose sensing. Pt_1/Cu@ CuO NWs can catalyze the oxidation of glucose and exhibit a lower onset potential and higher response current than Cu@CuO and Cu NWs. The introduction of the oxidation layer and single Pt atom improved the electron transfer ability and glucose adsorption ability of the nanomaterials, and this Pt_1/Cu@CuO NWs-based glucose sensor not only had high sensitivity but also good anti-interference ability and long-term stability. Theoretical calculations show that the excellent sensing performance of Pt_1/Cu@CuO NWs stems from the synergistic effect between single Pt atoms and Cu@CuO core-shell NWs, which results in the strong binding energy of glucose on the NWs. Since the doping of other elements (e.g., O, P, S, B, and Cl) in SANs with metal–N–C coordination structures (M–N–C) can modulate the electronic structure of the central metal atom through the electronegativity difference of the heteroatoms, this has become an effective strategy to improve the catalytic activity of SANs. A single-atom nanozyme (SAE) with unique Pt_1–N_3PS active centers has been designed by Chen et al. [55], and in this work, the direct atomization of Pt nanoparticles (NPs) into single atoms by reversing the thermal sintering process was first reported. This Pt SAN exhibited significant POD-like catalytic activity and kinetics that far surpassed those of Pt NPs. High-angle annular dark field-scanning transmission electron microscopy equipped with a spherical aberration corrector (AC HAADF-STEM) image provided a bright contrast of metal atoms against the low background, in which the atomically dispersed Pt sites could be clearly observed. No Pt–Pt peaks were observed in Fourier-transformed extended Pt L-edge X-ray absorption fine structure (EXAFS) spectra, which suggested that the Pt atoms exist as isolated single-atoms. Pt L_3-edge X-ray absorption near edge structure (XANES) curves indicated that the adsorption threshold of Pt single-atom nanozyme (Pt_{TS}-SAzyme) was located between those of PtO_2 and Pt foil, showing that the Pt species carry a positive charge. The EXAFS fitting analysis results revealed that a Pt atom is coordinated to N, S, and P with coordination numbers 2.5, 1, and 1, respectively. The experimental XANES spectrum of Pt_{TS}-SAzyme matched well with the calculated XANES spectra, further proving the existence of the Pt_1–N_3SP site (Figure 3A–D). The excellent catalytic performance for oxygen reduction reaction (ORR) of SANs has been investigated for the amplification of electrochemiluminescence (ECL) sensing signals to improve the sensitivity of detection [56]. Therefore, improving the ORR catalytic performance of SANs through ligand environment modulation also has an important role in the sensing field. Qin et al. [57] have reported a Ru-based SAN more durable than the 3D transition metal-based SANs, in which the second coordination shell of Ru centers was doped with S anions bonded to N. S anions in the second coordination shell modulated the energy barrier in the ORR reaction by adjusting the electronic structure of Ru centers, resulting in higher ORR catalytic activity than commercial Pt/C. The Bader charge analysis of RuN_4–S and RuN_4 without S revealed that the S coordination leads to a significant increase in the charge density of the Ru sites, reducing their binding energy to OOH*, O*, and OH*, which results in a higher ORR catalytic activity of Ru–SAN/SNC than Ru–SAN/NC (Figure 4A–D). RuN_4–S (S bonded to the N atom of the second shell layer) and RuN_3S (S bonded directly to the Ru atom of the first shell layer) were also compared, and the evaluation result showed that the binding energy of RuN_4–S is greater than that of Ru–N_3S, thus the RuN_4–S conformation is more easily formed. The introduction of other metal atoms in the active site can also optimize the sensing performance through synergistic effects. Catalysts with atomically dispersed Ru_3 sites were reported by Wu et al. to have higher electrocatalytic activity for uric acid (UA) than Ru SAN [58]. This was due to the optimized electronic structure of the multi-atom sites, which facilitates the adsorption of hydroxyl anion intermediates. Owing to the excellent catalytic activity for the oxidation of small biomolecules of Ru_3/NC, it could be used to construct electrochemical sensors with high sensitivity for UA detection in serum samples, which had broad practical application prospects.

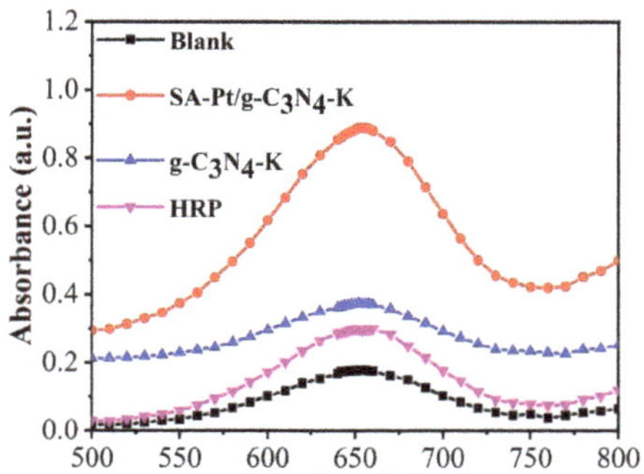

Figure 2. UV–vis absorption spectra of SA–Pt/g–C_3N_4–K, g–C_3N_4–K, and HRP with 1 mM TMB and 1 mM H_2O_2 at pH = 4 [53]. Copyright 2021 with permission from Elsevier.

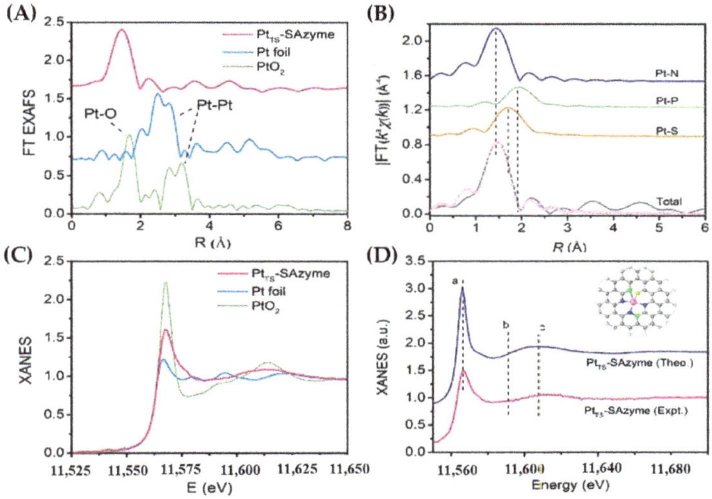

Figure 3. Atomic structural analysis of Pt$_{TS}$-SAzyme. (**A**) Fourier-transformed magnitudes of the experimental Pt L$_3$-edge EXAFS signals of Pt$_{TS}$-SAzyme, Pt foil, and PtO$_2$. (**B**) EXAFS fitting analysis of Pt$_{TS}$-SAzyme in R space. Curves from top to bottom are the Pt–N, Pt–P, and Pt–S three-body backscattering signals, the fitting curve total signal (pink line), and the experimental signal (gray line). (**C**) Pt L$_3$-edge XANES spectra of Pt$_{TS}$-SAzyme, Pt foil, and PtO$_2$. (**D**) Comparison between the experimental XANES spectra (pink line) and the theoretically simulated XANES spectra (blue line) of Pt$_{TS}$-SAzyme (a, b and c indicate the comparison of the shapes of the two spectra.) [55].

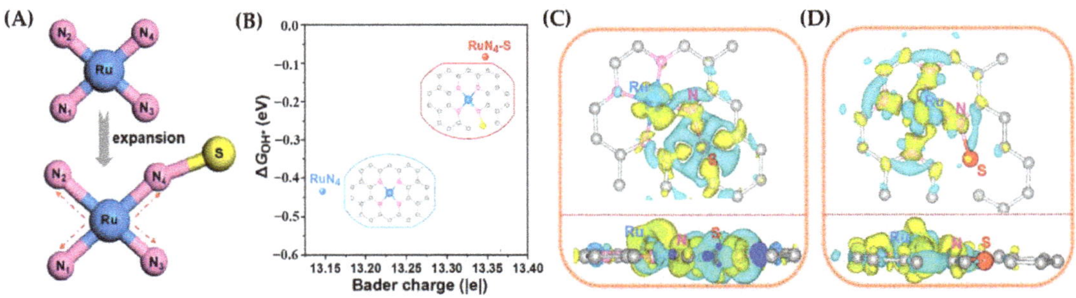

Figure 4. (**A**) Schematic atomic model of Ru–N bond length changes before and after S introduction. Color scheme: blue for Ru, yellow for S, pink for N, and gray for C. (**B**) Relationship between Bader charge and OH* binding energy of single-atom Ru in RuN$_4$ and RuN$_4$–S, respectively. Insert: the corresponding schematic models of the samples. Color scheme: blue for Ru, yellow for S, pink for N, and gray for C. Charge density difference between (**C**) S as the target and (**D**) Ru as the target in the RuN$_4$–S moiety. Color scheme: blue for Ru, red for S, pink for N, and gray for C. Yellow regions indicate charge accumulation, while cyan regions indicate charge depletion [57].

2.2. Non-Noble Metal-Based SANs

The high cost and limited availability of noble metals pose a significant challenge to sustainability. Non-noble metals are now widely used in various fields as cost-effective and sustainable alternatives to other commonly used, expensive transition metals, and their fabrication into single atoms can further amplify the advantages and improve their performance in analysis. As the most widely studied SAN, Fe SANs have demonstrated excellent catalytic performance and application potential, especially their ORR catalytic ability and enzyme-like activity. A nanozyme with single Fe atoms anchored on N-doped carbon nanotubes (CNT/FeNC) was proposed by Cheng et al. [59], in which the individual iron atoms were all surface atoms and had high structural similarity to the active center in natural PODs, thus possessing the most adequate enzyme-like reaction active centers and better POD-like activity. Its application as a signal element in the construction of a series of paper-based biosensors has successfully achieved the sensitive detection of H_2O_2, glucose, and ascorbic acid (AA), providing a novel and efficient signal element for the construction of future biosensors. The synergy of two non-noble metal atoms has also proven to be an effective way to improve detection sensitivity [60]. Using amphiphilic poly(vinyl alcohol) (PVA) aerogel as a substrate material for stabilizing metal single-atoms, Ma et al. [61] obtained a Zn/Mo dual single-atom nanomaterial supported on the macroscope aerogel (Zn/Mo DSAN-SMA) by soaking the aerogels with acetonitrile solutions of supramolecular coordination complexes/polyoxometalates (SCCs/POMs) followed by carbonization, which can be used as a novel nanozyme with ultra-long-term stability as a POD mimetic (Figure 5A). The constructed sensors can be used for the detection of glucose, AA, cholesterol, and H_2O_2 (Figure 5B). Theoretical calculations indicated that the Zn/Mo site is the main active center, and the synergistic effect between Zn and Mo atoms led to the superior activity of Zn/Mo DSAN-SMA. Achieving controlled synthesis and modulation of diatomic sites is essential for optimizing the performance of dual-atom nanomaterials (DANs). Due to the influence of geometry on the local electronic structure, the atomic sites at the edges are very different from those on the base surface of the supported materials in terms of electron density. The introduction of defects in the substrates is an effective way to increase the number of active sites at the edges while also facilitating material migration and the exposure of active sites. Kim et al. [62] designed a large number of defects on N-doped mesoporous carbon NPs using OH· generated by the decomposition of H_2O_2 under a hydrothermal process at 180 °C to etch the carbon-based support, and these defect edges provided anchor sites for Fe atoms (Figure 6A). Such edge sites exhibited significantly enhanced POD and oxidase (OXD)-like properties. Theoretical calculations suggested that the increased activity is due to the higher electron density of the N atoms at the edge sites, allowing for new reaction pathways at the edge sites (Figure 6B). Similar to noble metal-based SANs, modulating the metal coordination environment of non-noble metal-based SANs, such as the introduction of axial ligands or heteroatoms, can also affect the electronic structure of the active sites, thereby improving the catalytic activity as well as the specific recognition of the catalytic substrate [63,64], and the design of these SANs is also expected to be applied in the field of electrochemical sensors. A boron-doped Fe–N–C (FeBNC) nanozyme was developed by Jiao et al. [65] by mimicking the active sites of natural POD, in which the B atoms in the second coordination shell induced the electronic rearrangement of iron. The FeBNC nanozyme specifically enhanced the POD-like activity compared to the FeNC nanozyme. Axial nitrogen ligands play a crucial role in stabilizing the active structure and enhancing enzyme activity. By mimicking the structure of natural enzymes, a five-coordinated Fe-based single-atom nanozyme has been synthesized, displaying 7.64-fold higher POD-like activity than Fe–N_4 nanozyme [66] (Figure 7A). Theoretical calculations suggested that this reason is that Fe–N_5 molecules exhibit higher affinity for H_2O_2 and better activation ability (Figure 7B,C).

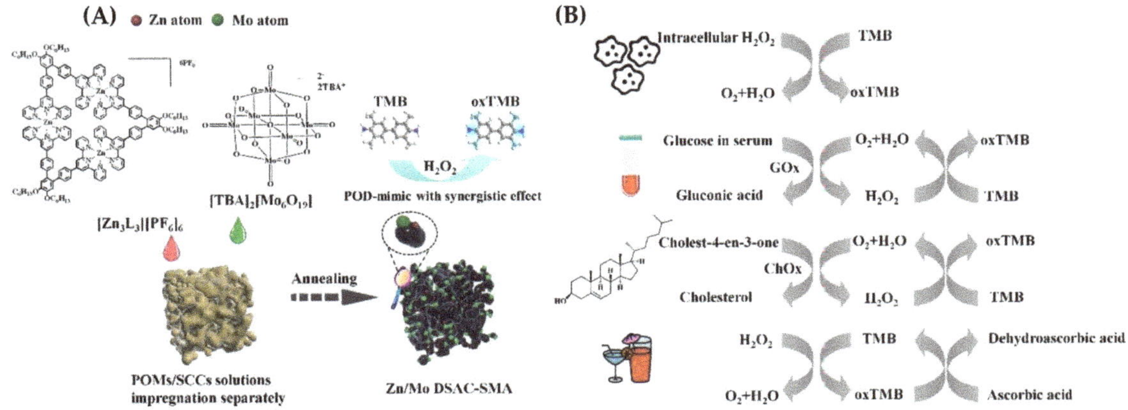

Figure 5. (**A**) Schematic illustration of the fabrication process of the confined Zn/Mo dual single-atom nanozyme loaded on PVA-based aerogel. (**B**) Versatile applications in intracellular H_2O_2 detection, glucose detection in serum, cholesterol determination, and AA in beverage detection [61]. Copyright 2022 with permission from Wiley.

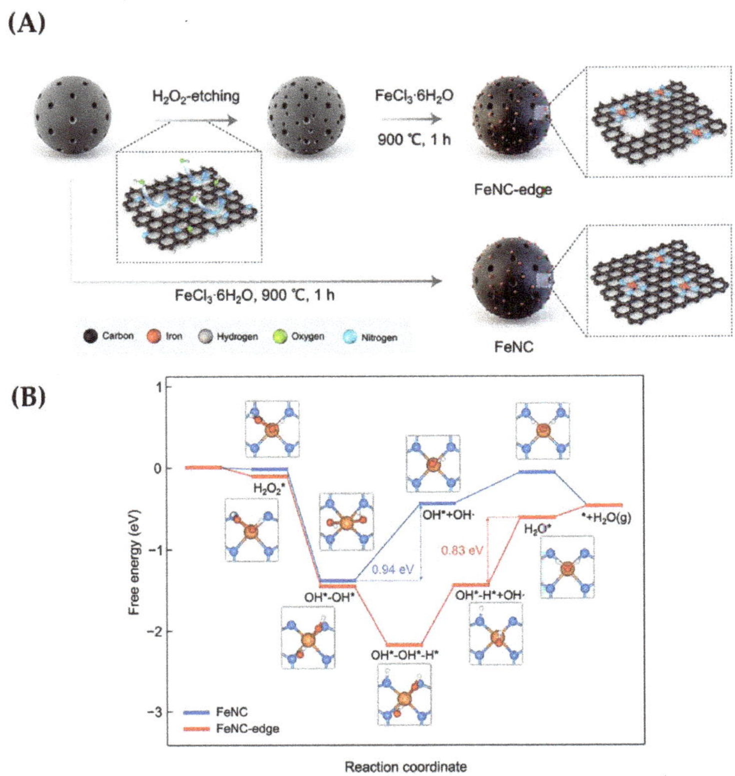

Figure 6. (**A**) Schematic illustration of FeNC-edge synthesis. (**B**) Free energy diagram for the POD-like reaction of FeNC and FeNC-edge. Color code: Fe, orange; nitrogen, blue; oxygen, red; H, white [62]. Copyright 2023 with permission from Wiley.

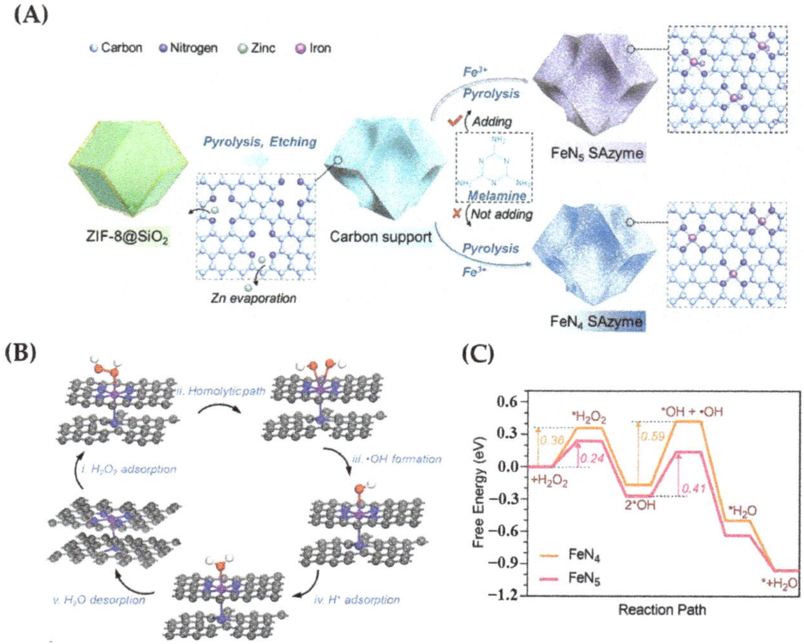

Figure 7. (**A**) Schematic illustration of the synthesis process for FeN_5 SAzyme. (**B**) Proposed catalytic mechanism for peroxidase-like reaction on FeN_5. (**C**) Corresponding free energy diagram for peroxidase-like reactions on FeN_5 and FeN_4 [66]. Copyright 2022 with permission from Wiley.

3. Types of SANs-Based Electrochemical Sensing

Integrating the virtues of SANs with electrochemical analytical methods has become a popular approach for constructing highly sensitive and selective analytical sensors, and the following section specifically describes the application of SANs in the fields of electrochemical sensing, electrochemiluminescent sensing, and photoelectrochemical sensing.

3.1. SANs-Based Electrochemical Sensing

Electrochemical analysis refers to a set of techniques used to study and analyze the behavior of chemical substances and processes, involving the measurement and interpretation of electrical properties such as current, potential, cuductivity, and so on, arising from chemical reactions occurring at the electrode interface. The use of SANs in electrochemical sensors offers numerous advantages, including enhanced sensitivity, selectivity, faster response times, stability, miniaturization, and energy efficiency because of the high catalytic activity and specificity of SANs. These characteristics make single-atom materials promising candidates for the development of advanced sensing devices with improved performance and wider practical applications.

The main class of electrochemical sensors constructed by SANs are those based on current analysis, that is, a certain voltage is applied and a corresponding current is generated by the redox reaction of the substance to be measured at the electrode through electrochemical catalysis, thus allowing quantitative analysis (Table 1). The amperometric method is currently a very popular method in electrochemical sensing. The four-electron pathway of ORR plays a very important role in electrochemical oxygen sensing, but the commonly used Pt/C catalysts for the analysis of oxygen in complex systems need to be improved due to the current influence of the H_2O_2 reduction reaction. A SAN with Co–N_4 as the active center was designed, which can effectively promote the four-electron ORR in a potential- and loading density-dependent manner under neutral conditions [67]. A hydrophobic hexamine monomolecular layer was electrochemically applied to the carbon fiber microelectrode (CFE) surface by CV scanning, and then the Co–N_4/C catalyst was adsorbed onto the modified CFE via hydrophobic interactions. The CFE was implanted as a working electrode (WE) in the right cortex of the rat brain, and oxygen was effectively detected by a typical amperometric response. The weak interaction between H_2O_2 and the active sites led to its highly selective sensing of oxygen, thus realizing real-time specific sensing of oxygen in vivo. Li et al. [68] prepared a Fe-based SAN using two-dimensional nitrogen-doped graphene as a support. It possessed a larger specific surface area to ensure high loading of active sites than 1D carbon substrates. And the distance between adjacent active centers in the catalyst matched the -O–O- bridge adsorption mode, allowing the H_2O_2 reduction reaction to occur in a 2-electron transfer path with higher catalytic activity compared to the 1-electron path occurring in the 1D NW-loaded SAN. Fe–SAN/NW was modified onto GCE and then Nafion solution was deposited onto the sensor surface to obtain an electrochemical sensor for remarkably sensitive detection of H_2O_2, which can also be used for in situ monitoring of H_2O_2 release from cells. H_2O_2 was catalyzed by the Fe active centers on the electrode surface for reduction to produce water, and the current was linearly related to the H_2O_2 concentration. The introduction of synergistic components in SANs can optimize the electronic and geometric structure of metal atoms, which enables efficient electrochemical sensing. Aiming to improve the POD activity of SAN, Fe single atomic sites with carbon-encapsulated Fe_3C crystals (Fe_3C@C/Fe–N–C) were synergized by Wei et al. [69] to enhance the adsorption of H_2O_2 molecules on Fe sites by transferring electrons from Fe_3C@C to single atomic Fe sites. It was used to construct a hydrogen peroxide electrochemical sensor with a high sensitivity of 1225 $\mu A/mM \cdot cm^2$ and a low detection limit of 0.26 μM. In addition to the amperometric method, differential pulse voltammetry (DPV) is one of the commonly used techniques, in which square pulses are applied to a linear potential sweep. An ultrasensitive dopamine (DA) electrochemical sensor has been constructed by doping Mn atoms on electrodeposited MoS_2 nanosheets (Mn–MoS_2) and using Mn–MoS_2 as WE [70]. DA was catalytically oxidized on the electrode surface, and the current measured using the DPV method increased with increasing DA concentration. In contrast to MoS_2, Mn–MoS_2 had higher selectivity and sensitivity for DA detection, with limits of detection (LOD) of 50 pM, 5 nM, and 50 nM in buffer, 10% serum, and artificial sweat, respectively. The Mn atoms took the place of Mo atoms in the MoS_2 lattice (Mn_{Mo}) or adsorb on Mo atoms (Mn_{topMo}), and the former was more energetically favorable compared to the latter. DA molecules were physisorbed on Mn_{Mo}, unlike chemisorption on Mn_{topMo}, where the former dominated at low DA concentrations and the latter dominated at high concentrations. There are also studies in which the linear sweep voltammetry (LSV) method was used for electrochemical quantification. Ding et al. [71] doped iron single atoms onto polypyrrole-derived carbon NW to synthesize Fe–N–C-based single-atomic site catalysts (Fe–SASC/NW), which featured Fe–N_x structures that can mimic the active sites of heme enzymes. The electrochemical sensor constructed by modifying Fe–SASC/NW onto the electrode can achieve highly sensitive detection of H_2O_2, and the LSV results demonstrated that the Fe–SASC/NW modified electrode can produce a linear current response to H_2O_2 with a linear concentration range from 5.0×10^{-10} M to 0.5 M and a LOD of 46.35×10^{-9} M.

In some studies, potentiometric analysis is used, i.e., the potential difference between the WE and the reference electrode in the open circuit is recorded and used as the output signal for the quantitative analysis of chemical substances. Pan et al. [72] coupled SAN with galvanic redox potentiometry (GRP) for the detection of H_2S in living mouse brains. (Figure 8) GRP is achieved by constructing a " galvanic cell " that spontaneously forms a redox process with the substance to be measured and by recording the open-circuit potential (OCP) without an applied polarization voltage. To avoid other coexisting neurochemicals with similar redox potentials from affecting the H_2S detection, they constructed an electrochemical sensor with high selectivity for H_2S by hydrophobically adsorbing hollow carbon spheres loaded with single Ni atoms onto the electrode to promote electrochemical H_2S

oxidation at very low potentials, thus driving spontaneous bipolarization of a single carbon fiber. Almost no current flowed in the circuit during the measurement, so the process did not electrically affect or interfere with the nervous system.

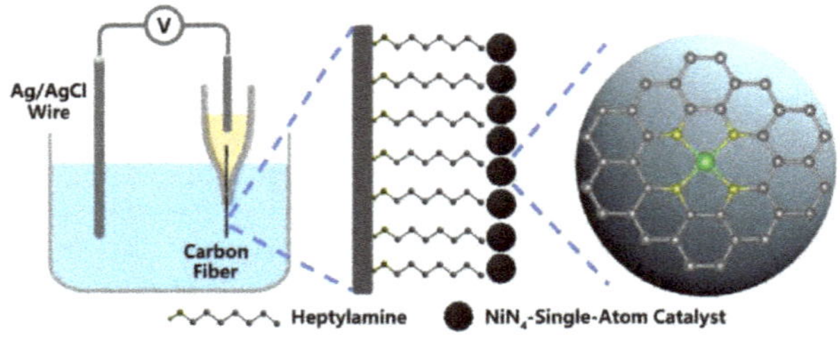

Figure 8. Schematic illustration of the NiN_4–SACs–GRP microsensor for H_2S sensing. The inner pole of the carbon fiber was immersed in an artificial cerebrospinal fluid (aCSF) solution containing 10 mM $K_3Fe(CN)_6$ and $K_4Fe(CN)_6$ (yellow). The outer pole of the carbon fiber was modified with NiN_4-SACs to catalyze the electrochemical oxidation of H_2S in bulk media (blue) [72]. Copyright © 2022 American Chemical Society.

Table 1. Summary of reported SANs-based electrochemical biosensors and applications.

SANs	Analyte	Sensitivity ($\mu A\ mM^{-1}$)	Linear Range (μM)	LOD (μM)	Reference
Fe–SASC/NW	H_2O_2	-	5×10^{-4}~5×10^5	0.04635	[71]
Fe_3C@C/Fe–N–C	H_2O_2	1225 cm^{-2}	1~6000	0.26	[69]
Cu_1/C_3N_4	H_2O_2	0.155	-	-	[73]
Fe–SASC/G [1]	H_2O_2	3214.28 cm^{-2} 1785.71 cm^{-2}	10~920 920~7020	0.2	[68]
Fe Sas–N/C	H_2O_2	86.99 32.66 19.91	1~54 54~764 764~9664	0.34	[74]
Ni–SAC	H_2O_2	-	2×10^{-5}~2.22×10^4	6.87×10^{-6}	[75]
Se SA/NC	H_2O_2	403.9 cm^{-2}	40~1.11×10^4	18	[76]
Fe_1Se_1/NC	H_2O_2	1508.6 cm^{-2}	20~1.3×10^4	11.5	[77]
Co–N–C-800	H_2O_2 DA	943.9 cm^{-2} 979.6 cm^{-2}	0.3~1.0×10^6 0.06~1200	0.13 0.04	[78]
Ni–MoS_2	DA	-	1×10^{-6}~1000	1×10^{-6}	[79]
Ru–Ala [2]–C_3N_4	DA UA	0.083 0.033	0.06~490 0.5~2135	0.02 0.17	[80]
Ru_3/NC	DA UA	58 24	0.01~200 0.05~1000	0.033 0.01	[58]
Fe–N_5–SAC	DA UA	2150 2740	0.005~500 0.01~480	7 27	[25]
Pt_1/Cu@CuO NWs	glucose	852.163 cm^{-2}	0.01~5.18	3.6	[54]
NCA [3]–Co	glucose	7.8 1	0.5~1000 1000~6000	0.1	[27]
Pt_1/Ni_6Co_1LDHs [4]/NG [5]	glucose	273.78 cm^{-2}	100~2180	10	[81]
$Pt_1/Ni(OH)_2$/NG	glucose	220.75 cm^{-2}	10~2180	-	[82]
Ti-MOF [6]–Pt	thrombin	-	4×10^{-6}~0.2	1.3×10^{-6}	[83]
SANb–BCN [7]	NB	480.37 156.64	2~100 100~600	0.7	[84]

[1] G—graphene; [2] Ala—alanine; [3] NCA—N-doped carbon aerogel; [4] LDHs—layered double hydroxides; [5] NG—nitrogen-doped graphene; [6] MOF—metal-organic framework; [7] SANb-BCN—single-atom niobium-doped boron–carbon–nitrogen nanotubes.

3.2. SANs-Based Electrochemiluminescence Sensing

ECL consists of an electrochemical initiation step and an optical readout step and is a luminescence process resulting from the relaxation of electronically excited products to the ground state following an electrochemical reaction. It has been widely used in biomedical fields because it does not require an excitation light source, giving it a lower background compared to other optical methods, and is potential and spatially controlled. Combining the high activity and large surface area of SANs, SANs offer the advantages of enhanced ECL signal intensity, tunable emission properties, high stability, fast response time, selectivity and specificity, integration with nanostructures, and miniaturization. In addition, the performances of SANs can be enhanced by their controlled atomic structure and strong bonding interactions; as a result, the sensitivity and selectivity of SANs-based ECL sensors can be improved for a wide range of applications, such as biomedical diagnostics, environmental monitoring, and chemical analysis.

The sensitivity of ECL sensing is significantly related to the efficiency of photon generation; therefore, SANs can act as co-reaction accelerators by virtue of their high catalytic activities to effectively promote the generation of excited luminescent substances at the surface of the electrode. Luminol, a classical luminescent, has a low oxidation potential, which reduces interference from other reactions. There have been many studies using SANs to catalyze the generation of reactive oxygen species (ROS) from the co-reactant to achieve signal amplification. Gu et al. [56] modified the electrode with Fe-based SAN containing Fe-N_4 active sites, which showed better enhancement of ECL efficiency than nitrogen-doped carbon and Fe_3O_4 NPs. (Figure 9A,B) To verify the ECL signal enhancement mechanism, they added isopropanol (IPA) or benzoquinone (BQ) as $\cdot OH$ and $O_2^{\cdot -}$ radical scavengers, respectively, to the Fe–N–C–luminol system, and the significantly decreased luminescence signal proved that the SAN promoted the ECL reaction through the radical pathway. FeN_4 active sites catalyzed the generation of ROS from dissolved oxygen at the electrode surface, amplifying the ECL signal of luminol. Exploiting the fact that antioxidants can eliminate free radicals to inhibit ECL, a Trolox sensor was constructed with a linear range of detection from 0.8 μM to 1.0 mM (Figure 9C). To further investigate the ECL mechanism, they later also designed two carbon-supported nickel SANs with Ni–N_4 and Ni–N_2O_2 for catalytic ORR, demonstrating four- and two-electron pathways, respectively [85]. The results showed that the Ni–N_4 active site had a better enhancement of the ECL signal and that $O_2^{\cdot -}$ was the main active intermediate species for the ECL reaction. The Ni–N_4/C–luminol ECL system was used to detect AA in the linear range of 70 μM to 350 nM. The plasma exciton effect, a collective oscillation of dense electrons, is capable of converting light energy into electronic excitation, and combining it with ECL is considered to be an efficient way to improve detection sensitivity. Au@SiO_2 and Fe–SAN were coupled by Bushira et al. [86] to enhance the ECL of the luminol-dissolved oxygen system by plasmon effect and construct sensitive and stable ECL sensors for the detection of DA, heme, and mercury (Hg^{2+}). The relationship between the cathodic ECL behavior of the luminol-oxygen system and the ORR electrocatalytic activity of SANs was examined by Xia et al. [87], and the results showed that the ECL intensity was positively correlated with the ORR catalytic activity. Two Fe-based SANs with different active sites, Fe–SAN and Fe–SAN(O), were designed, and enhanced luminescence signals were detected at the cathode without direct luminol electrochemical oxidation. Fe–SAN and Fe–SAN(O) generated electrocatalytic ORR via four-electron and two-electron pathways, respectively, and the difference in the electronic structure of the metal centers caused significant differences in the ECL signals. Among them, Fe-SAN with a $4e^-$-pathway tended to generate more kinds of ROS with stronger ECL intensity. This work achieved the tuning of the cathode ECL performance, and the ECL sensor constructed based on Fe–SAN had a detection limit of 0.10 nM for AA. Recently, some researchers have also employed SANs in the ECL system of $Ru(bpy)_3^{2+}/S_2O_8^{2-}$. Fe–SAN effectively activated $S_2O_8^{2-}$ to $SO_4^{\cdot -}$ and significantly enhanced the cathodic ECL emission of $Ru(bpy)_3^{2+}$ [88].

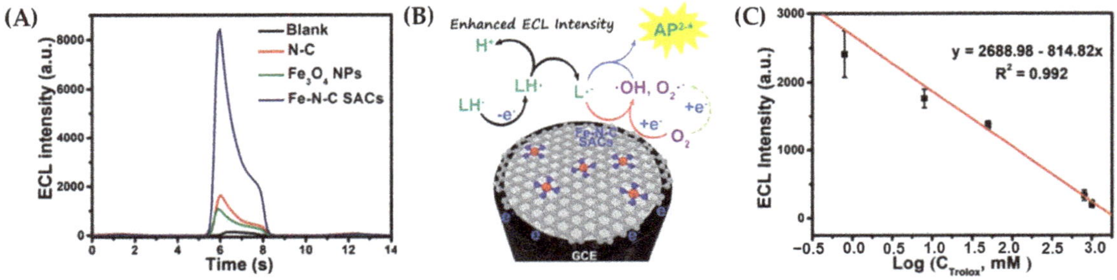

Figure 9. (**A**) ECL intensity of bare GCE and modified GCE with N–C, Fe_3O_4 NPs, and Fe–N–C SACs. (**B**) The mechanism of luminol–O_2 ECL systems with Fe–N–C SACs as coreactant accelerators. (**C**) The linear relationship between the Trolox concentration and ECL peak intensity [56]. Copyright 2019 with permission from Wiley.

SAN is not only able to act as a co-reaction accelerator to promote the generation of ROS but also as an ECL probe in some systems. Ma et al. [89] modified Ni SANs with polyethylene glycol (PEG) to enhance their hydrophilicity and promote catalyst dispersion in water, and the functionalized Ni SANs could be used as ECL probes to label biomolecules. They labeled phage recombinant cell-binding domains with PEGylated Ni SANs (Ni@PEG) for recognizing methicillin-resistant Staphylococcus aureus (MRSA) and modified porcine IgG as a capture molecule on GCE deposited with gold nanoparticle membranes to construct a sensor for MRSA with a detection limit of 25 CFU/mL. In addition to oxygen, SANs also have similar applications in the ECL system of Luminol/H_2O_2. By combining Cu–SAN and CdS quantum dots as a cathodic luminophore, Eu MOF-loaded isoluminol–Au NPs as an anodic luminophore, and modifying human epithelial protein 4 (HE4) Ab_2 and carbohydrate antigen 125 (CA125) Ab_2 on the two luminophores, respectively, a sandwich immunosensor was constructed for the simultaneous detection of two different markers [90]. CuSAN was used as a co-reaction accelerator to catalyze the generation of large amounts of $^{\cdot}OH$ and $O_2^{\cdot-}$ from H_2O_2 to promote luminescence.

3.3. SANs-Based Photoelectrochemical Sensing

PEC is composed of two processes: photoelectric conversion and electrochemical processing. Firstly, the PEC active material is excited by absorbing photons under light irradiation, and the photogenerated carriers generate photovoltage or photocurrent through charge transfer and transmission, thus realizing photoelectric conversion. Then the photogenerated carriers are transferred to the loaded electrode or solid–liquid interface, and charge exchange occurs at the interface, completing the redox reaction and transforming the chemical information into electrical signals [3]. The development of PEC materials with high photoactivity by using the huge surface active sites of SANs to trigger unique surface reactions is crucial to improving the analytical performance of PEC sensors [91]. Single metal atoms can effectively modulate the energy band and electronic structure of semiconductor frameworks, thus improving their corresponding light trapping and charge transport behaviors. At the same time, single atoms can significantly accelerate interfacial redox reactions due to their high catalytic activity and thus reduce the aggregation of charge carriers.

A Pt-based SAN has been synthesized by anchoring Pt atoms on the surface of hollow CdS (HCdS–Pt_1) [92], which was used as a PEC sensing platform. The introduction of Pt_1 increased the carrier density, leading to higher PEC activity in HCdS–Pt_1 compared to HCdS and HCdS–PtNPs. A biomolecular sensor was constructed by encapsulating HRP and glucose oxidase (GOx) in DNA flowers (HRP and GOx-DFs) as recognition elements and exploiting the phenomenon that target exosome-enriched HRP and GOx-DFs irreversibly bio-etch HCdS–Pt_1 in the presence of glucose, thereby causing changes in the photocurrent intensity. Since Pt is an ideal PEC photoactive material and its strong ability to provide electrons through chemical Pt–S interactions with CdS provides high photocurrent signal output, Qin et al. [93] prepared Pt single atoms dispersed on CdS nanorods (Pt SAs–CdS) that exhibited favorable ability for the separation of electron-hole pairs and constructed a prostate-specific antigen (PSA) sensing platform. The secondary antibodies (Ab_2) were labeled with CuO, and under acidic conditions, CuO NPs can dissociate into Cu^{2+} ions, which changed the PEC properties of Pt SAs–CdS photoelectrodes by reacting with them. Therefore, the addition of PSA led to a decrease in photocurrents, and the enzyme-free PEC immunosensor constructed using this principle was used to detect PSA in the linear range of 5 pg/mL to 10 ng/mL with a detection limit of 0.92 pg/mL. To improve the photoactivity and stability of pure CdS loaded with single atoms, CdS has been replaced with $Zn_{0.5}Cd_{0.5}S$, which can improve the photogenerated holes/electrons mobility, thus reducing the bulk hole-electron complex and the oxidation of divalent sulfide ions by photogenerated holes [94]. The constructed PEC sensor had a detection limit of 0.22 pg/mL for PSA. Except for Pt, the very commonly used Fe-based SAN can also improve PEC sensing performance by promoting interfacial reactions in a typical p-type semiconductor of Cu_2O. Fe SANs have also been integrated with $Cu_2O/Ti_3C_2T_x$ by Qin et al. [31] to construct a highly sensitive PEC biosensor by enhancing the ORR catalytic activity at the interface and thus the PEC signal (Figure 10). Fe SANs were also found to exhibit superior POD activity and could catalyze the oxidation of 4-chloro-1-naphthol (4-CN) on the photoelectrode surface to form insoluble precipitates and thus weaken the PEC signal. Since acetylcholinesterase (AChE) was able to catalyze the hydrolysis of acetylcholine (ACh) to form acetic acid, which led to a change in the pH value of the solution, thus affecting the POD activity of Fe SANs/$Ti_3C_2T_x$/Cu_2O, the PEC sensor constructed based on this principle provided highly sensitive detection of AChE activity and organophosphorus pesticides (OPs, AChE inhibitors). TiO_2, as a semiconductor with chemical stability and strong light absorption ability, when combined with gC_3N_4 to form a heterojunction, can achieve effective separation of carriers. Bott-Neto et al. [30] modified TiO_2 and graphitic carbon nitride anchored with nickel single atoms (Ni–gC_3N_4) to form heterojunctions on screen-printed carbon electrodes (SPCEs) and functionalized TiO_2 with electrodeposited aryl diazonium salts to anchor antibodies and facilitate the separation of charge carriers. A miniaturized 3D-printed PEC device that can detect PSA under visible LED light irradiation with good stability and a detection limit of 0.06 fg/mL was constructed using the synergy of these three materials.

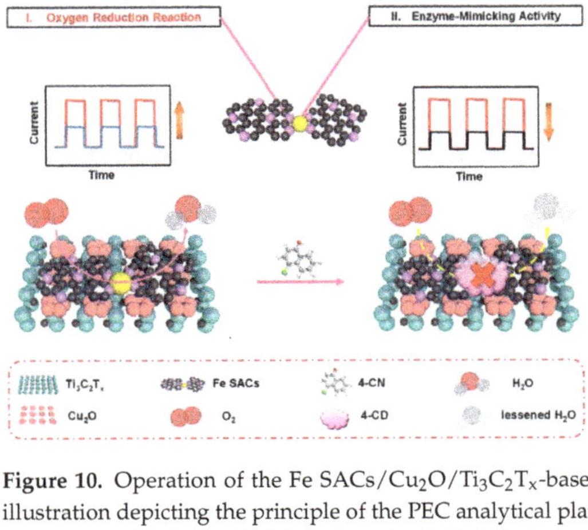

Figure 10. Operation of the Fe SACs/Cu_2O/$Ti_3C_2T_x$-based PEC analytical platform. Schematic illustration depicting the principle of the PEC analytical platform with the expected change in photocurrent profile due to the bifunctional Fe SACs [31]. Copyright © 2022 American Chemical Society.

4. Applications of SANs in Analytical Chemistry

SANs-based sensors have been used in a wide range of analytical applications. Some trace biological and environmental samples require high sensitivity for detection, and SANs have shown great applicational values in electrochemical sensing of biomolecules and environmental contaminants in vivo and in vitro with their powerful structural and performance advantages. The following section focuses on their application in environmental and biochemical analysis.

4.1. Applications of SANs in Biochemical Analysis

The detection of some signaling molecules in living organisms is of crucial importance for the prevention and diagnosis of diseases. The development of SANs has led to non-negligible progress in this research area of constructing highly sensitive biosensing platforms [27,71,78]. The analysis of endogenous substances generated by cells is a typical class of application. For example, NO, as an endogenous cellular substance associated with a variety of physiological and pathological conditions, requires high sensitivity and transient recording capability on the sensor. A Ni-based SAN has been designed by anchoring Ni atoms on nitrogen-doped hollow carbon spheres (Ni SANs/N–C) by Zhou et al. [95], which can effectively catalyze the electrochemical oxidation of NO, and constructed a stretchable electrochemical sensor by confining Ni SANs/N–C on a flexible dimethylsiloxane (PDMS) substrate (Figure 11A,B). This sensor had good biocompatibility, realized real-time detection of NO release from endothelial cells during drug and traction stimulation, and also provided a new idea for the design of sensing platforms for chemical signals in the environment of living cells. H_2O_2, an important representative of ROS, is generated by intracellular oxygen metabolism and plays a crucial role in stimulating cell proliferation, differentiation, and migration. The construction of sensors that can provide accurate, real-time detection of H_2O_2 produced by living cells has been an important topic. Liang et al. [74] constructed an electrochemical sensor for H_2O_2 detection with a detection limit of 0.34 μM by modifying Fe SAs–N/C prepared by high-temperature calcination carbonization of hemin@zeolitic imidazolate framework-8 (hemin@ZIF-8) on GCE and exploiting the POD activity of Fe–N_x active sites.

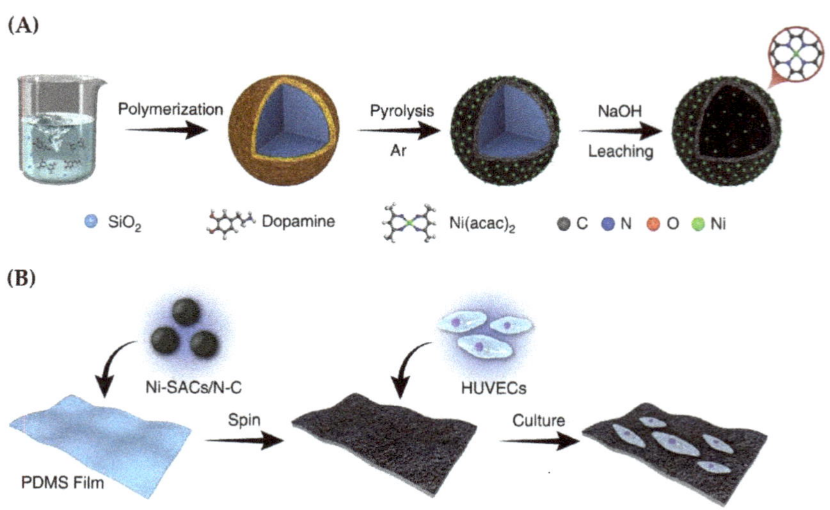

Figure 11. (**A**) Schematic illustration of the synthesis of Ni SACs/N–C. (**B**) Schematic illustration of the fabrication of Ni SACs/N–C-based stretchable sensors for NO sensing and HUVEC culturing [95]. Copyright © 2020, Nature, open access.

Achieving in vivo monitoring of neurochemicals and then studying brain function is crucial and challenging. An enzyme-based electrochemical sensor for glucose has been designed using the property that Co-based SAN can catalyze the oxidation of H_2O_2 at low potentials [96], which in turn enabled online monitoring of glucose in rat brain microdialysate. This sensor inherited the superiority of electrochemical analysis and SANs, with a good response to glucose and no interference from other electroactive substances. This work provided a new idea for in vivo analysis that can also be applied to the analysis of other chemicals. An effective method for the detection of H_2O_2 is to use a hydrogen peroxide reduction reaction (HPRR), but the detection process is often disturbed by ORR. To improve the selectivity of H_2O_2 detection. Gao et al. [73] dispersed single Cu atoms on mesoporous graphitic carbon nitride through the impregnation method, and the obtained Cu_1/C_3N_4 was proved to have better HPRR electrocatalytic performance than ORR in neutral media by theoretical calculation. Based on this, a microsensor with high selectivity for hydrogen peroxide was designed and implanted into the rat brain to achieve selective monitoring of H_2O_2 fluctuations in vivo.

In addition, electrochemical sensors based on SANs have also been used for some other biological small-molecule detection or immunoassays. To address the problem that the detection of UA in serum is difficult to achieve an ultra-wide linear range and an ultra-low detection limit, and the detection mechanism is unclear, an electrochemical UA sensor has been developed by Hu et al. [97], relying on a Co single-atom nanozyme (A–Co–NG) for the first time. These Co atoms in the prepared A–Co–NG nanozyme were coordinated to 3.4 N atoms on average in the form of Co^{2+}. The detection range of this A–Co–NG sensor was 0.4 to 41,950 μM, and the detection limit was 33.3 ± 0.024 nm, which was significantly better than the previously reported sensors based on various nanomaterials. This work provided excellent material for realizing a UA sensor with a wide detection range and low detection limit, which met the need for practical diagnosis and provided new ideas to guide the exploration of other biosensing processes. An electrochemical sensor that can detect DA and UA simultaneously was designed by Xie et al. [80] through dispersing Ru atoms on a C_3N_4 substrate, with linear ranges of 0.06 to 490 μM and 0.5 to 2135 μM for DA and UA, respectively, and detection limits of 20 and 170 nM, respectively. Furthermore, an immunosensor for the detection of PSA was also constructed by inducing PEC signal inhibition by CuO NPs-labeled sandwich immunocomplexes.

4.2. Applications of SANs in Environmental Analysis

SANs have also been applied to the detection of some heavy metal ions in environmental samples. For instance, an ultrasensitive electrochemical sensor for the heavy metal ion Pb^{2+} has been designed by Zhou et al. [98] through doping Mn atoms into MoS_2 nanosheets, in which Mn atoms took the place of some Mo atoms. The introduction of Mn atoms caused lattice destabilization and sulfur vacancies (V_S) on the one hand and phase changes on the other, adding another 1T-phase to Mn–MoS_2 compared with pure MoS_2 containing only 2H-phase. Defect- and phase-engineering enabled Mn–MoS_2 not only to have excellent electronic properties but also to form Pb–S bonds with lead ions, which significantly promoted in situ catalytic redox reactions. Yao et al. [99] synthesized a homogeneous dodecahedral N-doped carbon modified by a Fe–N–C SAE and modified it onto the gate electrode of the solution-gated graphene transistor (SGGT) to construct an electrochemical sensor for real-time monitoring of Hg^{2+} in environmental samples. (Figure 12) Combining the excellent electrocatalytic performance of Fe–N–C SAE with the high signal amplification efficiency of SGGT, this sensor has good sensitivity and selectivity for Hg^{2+} with a detection limit as low as 1 nM. Li et al. [100] realized that Co-based SANs have great superiority in the field of detection of trace-level As(III). The Co atoms anchored on the N-doped carbon substrate were active sites for catalyzing the reduction reaction of H_3AsO_3 with the formation of Co–O hybridization bonds, which resulted in a Co SAN that was more favorable than Co NPs in terms of both kinetics and thermodynamics. The As (III) electrochemical sensor constructed based on Co SAN had a good selectivity and a sensitivity of 11.44 μA ppb^{-1}. Some other environmental pollutants, such as nitrobenzene (NB) and hydroquinone (HQ), also have problems in terms of sensitivity and selectivity for detection, and the development of SAN provides a new idea for them. Four different scales of Mo-based nanostructures were prepared by Cong et al. [101], including Mo_2C NPs, Mo_2C nanodots, Mo nanoclusters, and Mo single atoms, on N, P, and O co-doped carbon substrates to compare their performance for electrochemical detection of HQ. The results showed that Mo single atoms exhibited the most sensitive results with a wide linear range (0.02 to 200 μM), a low detection limit (0.005 μM), and good anti-interference ability. Nb SAN has been prepared by Li et al. [84] using boron–carbon–nitrogen nanotubes as supports and modified onto GCE for the construction of electrochemical sensors for NB. Compared with bare GCE, SANb–BCN/ GCE for catalytic reduction reactions exhibited higher current intensity and could achieve detection limits as low as 0.70 mM, which was able to be used for the detection of NB in water samples. They also tested the anti-interference ability of the sensor using various inorganic and organic substances and found little effect on the detection signal, indicating a high specificity of the sensor for NB. Luo et al. [102] applied the ORR activity of the Ir SAN catalyst to electrochemical detection. They achieved the monitoring of AChE activity by exploiting the inhibitory effect of thiocholine (TCh) on the ORR activity of Ir–N_x sites, and the AChE–Ir SAN-based biosensor can be used to detect OPs in environmental samples. The linear range of OPs detection was 0.5 to 500 ng mL^{-1}, with a low detection limit of 0.17 ng mL^{-1}.

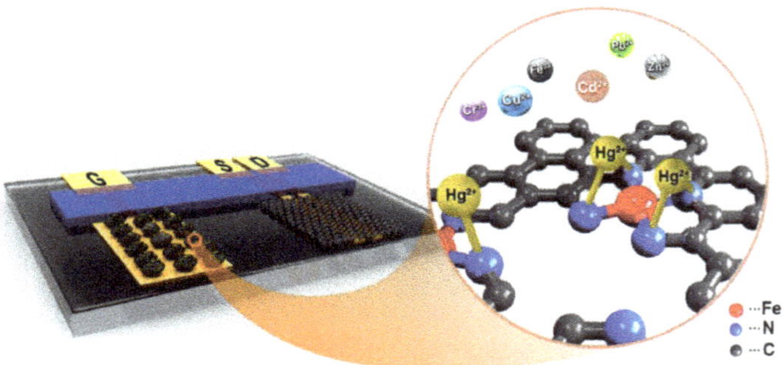

Figure 12. Schematic diagram of a solution-gated graphene transistor (SGGT) based on the Fe–N–C SAE and Cs/Au-modified gate electrode for the detection of Hg^{2+}. D, S, and G represent the drain, source, and gate electrodes, respectively [99]. Copyright © 2020 American Chemical Society.

5. Summary and Outlook

In this review, we first introduced the structural characteristics of SANs and the advantages that exist compared to conventional nanomaterials. Then, we highlight the progress made by researchers in recent years in tuning and optimizing the composition and structure of the active sites of SANs to improve catalytic performance, classified according to noble and non-noble metals, where the main approaches include the synergy of two or three metal atoms and the tuning of the coordination environment of metal atoms. Recent advances in the application of SANs in electrochemical sensing were also presented, including three types of sensing: electrochemical sensors, ECL sensors, and PEC sensors. Finally, the practical applications of SANs were presented according to the classification of biochemical and environmental analysis, which are promising for application in the sensitive detection of signaling molecules in living organisms and in vitro, as well as environmental pollutants. However, there are still some problems with SANs for applications in electrochemical sensors.

Firstly, SANs still face important challenges in their synthesis. Traditional synthesis methods still have difficulty achieving the desired metal loading, which hinders their large-scale application in practical production and is unfavorable to their application in the construction of highly sensitive sensors. There is a need to continue to develop new synthesis methods that should ensure high metal loading while avoiding the aggregation of metal atoms into clusters or NPs. The dispersion of the metal is also very importantly related to the interaction between the metal and the support, so further search for suitable supports and optimization of their surface properties to make them more suitable for anchoring single metal atoms are needed. In addition to this, the active sites of some SANs may be encapsulated in the supports, which greatly reduces the utilization of the atoms. Ultrathin, ultrasmall supports such as carbon dots and monolayer 2D nanomaterials appear to increase the exposure of metal atoms significantly over bulk materials.

Secondly, the performance of SANs for biosensing still needs to be further optimized. The catalytic activity and specificity of SANs are far less than those of biological enzymes, so the synthesis of SANs with better performance needs to be realized through the precise analysis of the electronic structure and coordination environment of metal atoms to understand the structure-function relationship. Moreover, SANs still have a series of problems in terms of biocompatibility, targeting, and stability that need to be considered for bioanalytical applications, especially in vivo analysis.

Third, the application of SANs in wearable sensors is yet to be developed. Wearable sensors are powerful tools to monitor human health, but one of the major problems they are currently facing is the fast energy consumption due to miniaturized batteries, which prevents continuous monitoring for a long period of time. The development of wearable self-powered sensors is the most effective way to address this problem, and the key to it is to design a continuous and efficient power supply. Due to the excellent performance of SANs in catalysis, some researchers have already used SANs to construct self-powered sensors, such as using a SAN as the ORR catalyst for zinc-air batteries, which provided long-term stability and high power density and can be utilized for sensitive detection of glucose when integrated with glucose oxidase [103]. Recent work has also combined a SAN with enzymatic activity with photoactive materials to construct highly active PEC fuel cells that can be used for long-term and sensitive electrochemical sensing [104]. Because fuel cells and metal-air batteries are efficient sources of power supply in self-powered systems, coupled with the fact that SANs have shown great potential in catalyzing ORR, one of the two half-reactions essential to these two novel energy storage and conversion systems, SANs will surely play an indispensable role for self-powered sensors in the future.

In the future, with a further understanding of the conformational relationships and catalytic mechanisms of SANs, we believe that SANs will have great potential in the field of sensing.

Funding: This work was financially supported by the National Natural Science Foundation of China (NO. 22174084).

Institutional Review Board Statement: The study did not require ethical approval.

Informed Consent Statement: This study did not involve humans.

Conflicts of Interest: The authors declare no conflict of interest.

References

1. Zheng, W. Single-Atom Materials as Electrochemical Sensors: Sensitivity, Selectivity, and Stability. *Anal. Sens.* **2022**, *3*, e20220007. [CrossRef]
2. Li, H.Y.; Qi, H.J.; Chang, J.F.; Gai, P.P.; Li, F. Recent progress in homogeneous electrochemical sensors and their designs and applications. *Trac-Trends Anal. Chem.* **2022**, *156*, 116712. [CrossRef]
3. Chen, Y.; Jiao, L.; Yan, H.; Xu, W.; Wu, Y.; Zheng, L.; Gu, W.; Zhu, C. Fe-N-C Single-Atom Catalyst Coupling with Pt Clusters Boosts Peroxidase-like Activity for Cascade-Amplified Colorimetric Immunoassay. *Anal. Chem.* **2021**, *93*, 12353–12359. [CrossRef] [PubMed]
4. Jiao, L.; Xu, W.; Wu, Y.; Wang, H.; Hu, L.; Gu, W.; Zhu, C. On the Road from Single-Atom Materials to Highly Sensitive Electrochemical Sensing and Biosensing. *Anal. Chem.* **2023**, *95*, 433–443. [CrossRef]
5. Chang, B.; Zhang, L.; Wu, S.; Sun, Z.; Cheng, Z. Engineering single-atom catalysts toward biomedical applications. *Chem. Soc. Rev.* **2022**, *51*, 3688–3734. [CrossRef] [PubMed]

6. Qiao, B.; Wang, A.; Yang, X.; Allard, L.F.; Jiang, Z.; Cui, Y.; Liu, J.; Li, J.; Zhang, T. Single-atom catalysis of CO oxidation using Pt_1/FeOx. *Nat. Chem.* **2011**, *3*, 634–641. [CrossRef] [PubMed]
7. Chen, Y.J.; Ji, S.F.; Chen, C.; Peng, Q.; Wang, D.S.; Li, Y.D. Single-Atom Catalysts: Synthetic Strategies and Electrochemical Applications. *Joule* **2018**, *2*, 1242–1264. [CrossRef]
8. Huang, Z.; Sun, X.; Wang, P.; Wan, H. Emerging single-atom catalysts in electrochemical biosensing. *View* **2023**, *4*, 20220058. [CrossRef]
9. Tajik, S.; Dourandish, Z.; Garkani Nejad, F.; Beitollahi, H.; Afshar, A.A.; Jahani, P.M.; Di Bartolomeo, A. Review—Single-Atom Catalysts as Promising Candidates for Electrochemical Applications. *J. Electrochem. Soc.* **2022**, *169*, 046504. [CrossRef]
10. Wang, A.Q.; Li, J.; Zhang, T. Heterogeneous single-atom catalysis. *Nat. Rev. Chem.* **2018**, *2*, 65–81. [CrossRef]
11. Wang, P.; Ren, Y.; Wang, R.; Zhang, P.; Ding, M.; Li, C.; Zhao, D.; Qian, Z.; Zhang, Z.; Zhang, L.; et al. Atomically dispersed cobalt catalyst anchored on nitrogen-doped carbon nanosheets for lithium-oxygen batteries. *Nat. Commun.* **2020**, *11*, 1576. [CrossRef]
12. Xia, C.; Qiu, Y.; Xia, Y.; Zhu, P.; King, G.; Zhang, X.; Wu, Z.; Kim, J.Y.T.; Cullen, D.A.; Zheng, D.; et al. General synthesis of single-atom catalysts with high metal loading using graphene quantum dots. *Nat. Chem.* **2021**, *13*, 887–894. [CrossRef]
13. Ran, L.; Li, Z.; Ran, B.; Cao, J.; Zhao, Y.; Shao, T.; Song, Y.; Leung, M.K.H.; Sun, L.; Hou, J. Engineering Single-Atom Active Sites on Covalent Organic Frameworks for Boosting CO_2 Photoreduction. *J. Am. Chem. Soc.* **2022**, *144*, 17097–17109. [CrossRef]
14. Qin, L.; Gan, J.; Niu, D.; Cao, Y.; Duan, X.; Qin, X.; Zhang, H.; Jiang, Z.; Jiang, Y.; Dai, S.; et al. Interfacial-confined coordination to single-atom nanotherapeutics. *Nat. Commun.* **2022**, *13*, 91. [CrossRef] [PubMed]
15. Zhang, E.; Tao, L.; An, J.; Zhang, J.; Meng, L.; Zheng, X.; Wang, Y.; Li, N.; Du, S.; Zhang, J.; et al. Engineering the Local Atomic Environments of Indium Single-Atom Catalysts for Efficient Electrochemical Production of Hydrogen Peroxide. *Angew. Chem. Int. Ed. Engl.* **2022**, *61*, e202117347. [CrossRef] [PubMed]
16. Yanan, S.; Xiaoguang, D.; Shaobin, W.; Qinyan, Y.; Baoyu, G.; Xing, X. Carbon-based single atom catalyst: Synthesis, characterization, DFT calculations. *Chin. Chem. Lett.* **2021**, *33*, 663–673. [CrossRef]
17. Chen, J.; Kang, Y.; Zhang, W.; Zhang, Z.; Chen, Y.; Yang, Y.; Duan, L.; Li, Y.; Li, W. Lattice-Confined Single-Atom Fe_1S_x on Mesoporous TiO_2 for Boosting Ambient Electrocatalytic N_2 Reduction Reaction. *Angew. Chem. Int. Ed. Engl.* **2022**, *61*, e202203022. [CrossRef] [PubMed]
18. Jiang, Z.; Tian, M.; Jing, M.; Chai, S.; Jian, Y.; Chen, C.; Douthwaite, M.; Zheng, L.; Ma, M.; Song, W.; et al. Modulating the Electronic Metal-Support Interactions in Single-Atom Pt_1-CuO Catalyst for Boosting Acetone Oxidation. *Angew. Chem. Int. Ed. Engl.* **2022**, *61*, e202200763. [CrossRef] [PubMed]
19. Kuai, L.; Liu, L.; Tao, Q.; Yu, N.; Kan, E.; Sun, N.; Liu, S.; Geng, B. High-Areal Density Single-Atoms/Metal Oxide Nanosheets: A Micro-Gas Blasting Synthesis and Superior Catalytic Properties. *Angew. Chem. Int. Ed. Engl.* **2022**, *61*, e202212338. [CrossRef] [PubMed]
20. Meza, E.; Diaz, R.E.; Li, C.W. Solution-Phase Activation and Functionalization of Colloidal WS_2 Nanosheets with Ni Single Atoms. *ACS Nano* **2020**, *14*, 2238–2247. [CrossRef] [PubMed]
21. Yang, Y.; Song, R.; Fan, X.; Liu, Y.; Kong, N.; Lin, H.; Li, Y. A mechanistic study of selective propane dehydrogenations on MoS_2 supported single Fe atoms. *Chin. Chem. Lett.* **2022**, *34*, 107257. [CrossRef]
22. Zhang, J.; Xu, X.; Yang, L.; Cheng, D.; Cao, D. Single-Atom Ru Doping Induced Phase Transition of MoS_2 and S Vacancy for Hydrogen Evolution Reaction. *Small Methods* **2019**, *3*, 1900653. [CrossRef]
23. Peng, Y.; Geng, Z.; Zhao, S.; Wang, L.; Li, H.; Wang, X.; Zheng, X.; Zhu, J.; Li, Z.; Si, R.; et al. Pt Single Atoms Embedded in the Surface of Ni Nanocrystals as Highly Active Catalysts for Selective Hydrogenation of Nitro Compounds. *Nano Lett.* **2018**, *18*, 3785–3791. [CrossRef] [PubMed]
24. Pei, G.X.; Liu, X.Y.; Yang, X.F.; Zhang, L.L.; Wang, A.Q.; Li, L.; Wang, H.; Wang, X.D.; Zhang, T. Performance of Cu-Alloyed Pd Single-Atom Catalyst for Semihydrogenation of Acetylene under Simulated Front-End Conditions. *Acs Catal.* **2017**, *7*, 1491–1500. [CrossRef]
25. Bushira, F.A.; Kitte, S.A.; Li, H.; Zheng, L.; Wang, P.; Jin, Y. Enzyme-like $Fe-N_5$ single atom catalyst for simultaneous electrochemical detection of dopamine and uric acid. *J. Electroanal. Chem.* **2022**, *904*, 115956. [CrossRef]
26. Liang, W.; Gao, M.; Li, Y.; Tong, Y.; Ye, B.C. Single-atom electrocatalysts templated by MOF for determination of levodopa. *Talanta* **2021**, *225*, 122042. [CrossRef]
27. Song, Y.Y.; He, T.; Zhang, Y.L.; Yin, C.Y.; Chen, Y.; Liu, Q.M.; Zhang, Y.; Chen, S.W. Cobalt single atom sites in carbon aerogels for ultrasensitive enzyme-free electrochemical detection of glucose. *J. Electroanal. Chem.* **2022**, *906*, 116024. [CrossRef]
28. Bushira, F.A.; Wang, P.; Wang, Y.; Hou, S.; Diao, X.; Li, H.; Zheng, L.; Jin, Y. Plasmon-Boosted Fe, Co Dual Single-Atom Catalysts for Ultrasensitive Luminol-Dissolved O_2 Electrochemiluminescence Detection of Prostate-Specific Antigen. *Anal. Chem.* **2022**, *94*, 9758–9765. [CrossRef]
29. Gu, W.; Wang, X.; Xi, M.; Wei, X.; Jiao, L.; Qin, Y.; Huang, J.; Cui, X.; Zheng, L.; Hu, L.; et al. Single-Atom Iron Enables Strong Low-Triggering-Potential Luminol Cathodic Electrochemiluminescence. *Anal. Chem.* **2022**, *94*, 9459–9465. [CrossRef]
30. Bott-Neto, J.L.; Martins, T.S.; Buscaglia, L.A.; Machado, S.A.S.; Oliveira, O.N., Jr. Photocatalysis of TiO_2 Sensitized with Graphitic Carbon Nitride and Electrodeposited Aryl Diazonium on Screen-Printed Electrodes to Detect Prostate Specific Antigen under Visible Light. *ACS Appl. Mater. Interfaces* **2022**, *14*, 22114–22121. [CrossRef] [PubMed]

31. Qin, Y.; Wen, J.; Wang, X.; Jiao, L.; Wei, X.; Wang, H.; Li, J.; Liu, M.; Zheng, L.; Hu, L.; et al. Iron Single-Atom Catalysts Boost Photoelectrochemical Detection by Integrating Interfacial Oxygen Reduction and Enzyme-Mimicking Activity. *ACS Nano* **2022**, *16*, 2997–3007. [CrossRef] [PubMed]
32. Zhao, L.; Qin, X.; Zhang, X.; Cai, X.; Huang, F.; Jia, Z.; Diao, J.; Xiao, D.; Jiang, Z.; Lu, R.; et al. A Magnetically Separable Pd Single-Atom Catalyst for Efficient Selective Hydrogenation of Phenylacetylene. *Adv. Mater.* **2022**, *34*, e2110455. [CrossRef] [PubMed]
33. Ji, K.; Xu, M.; Xu, S.M.; Wang, Y.; Ge, R.; Hu, X.; Sun, X.; Duan, H. Electrocatalytic Hydrogenation of 5-Hydroxymethylfurfural Promoted by a Ru_1Cu Single-Atom Alloy Catalyst. *Angew. Chem. Int. Ed. Engl.* **2022**, *61*, e202209849. [CrossRef]
34. Zheng, K.; Li, Y.; Liu, B.; Jiang, F.; Xu, Y.; Liu, X. Ti-doped CeO_2 Stabilized Single-Atom Rhodium Catalyst for Selective and Stable CO_2 Hydrogenation to Ethanol. *Angew. Chem. Int. Ed. Engl.* **2022**, *61*, e202210991. [CrossRef]
35. Zou, R.; Xie, R.; Peng, Y.; Guan, W.; Lin, Y.; Lu, C. Ag-O-Co Interface Modulation-Amplified Luminol Cathodic Electrogenerated Chemiluminescence. *Anal. Chem.* **2022**, *94*, 4813–4820. [CrossRef] [PubMed]
36. Li, X.; Wang, Z.; Tian, Y.; Li, X.; Cai, Q.; Zhao, J. Single-atom rhodium anchored on S-doped black phosphorene as a promising bifunctional electrocatalyst for overall water splitting. *Chin. Chem. Lett.* **2022**, *354*, 107812. [CrossRef]
37. Xu, J.; Xu, H.; Dong, A.; Zhang, H.; Zhou, Y.; Dong, H.; Tang, B.; Liu, Y.; Zhang, L.; Liu, X.; et al. Strong Electronic Metal-Support Interaction between Iridium Single Atoms and a WO_3 Support Promotes Highly Efficient and Robust CO_2 Cycloaddition. *Adv. Mater.* **2022**, *34*, e2206991. [CrossRef] [PubMed]
38. Shangguan, Y.; Zhou, Y.; Zheng, R.; Feng, X.; Ge, Q.; Wang, R.; Yang, D.; Wei, W.; Wu, X.; Lin, J.; et al. Bandgap engineering of tetragonal phase $CuFeS_2$ quantum dots via mixed-valence single-atomic Ag decoration for synergistic Cr(VI) reduction and RhB degradation. *Chin. Chem. Lett.* **2021**, *32*, 3450–3456. [CrossRef]
39. Yan, P.; Shu, S.; Shi, X.; Li, J. Promotion effect of Au single-atom support graphene for CO oxidation. *Chin. Chem. Lett.* **2022**, *33*, 4822–4827. [CrossRef]
40. Peng, L.; Yang, J.; Yang, Y.; Qian, F.; Wang, Q.; Sun-Waterhouse, D.; Shang, L.; Zhang, T.; Waterhouse, G.I.N. Mesopore-Rich Fe-N-C Catalyst with FeN_4-O-NC Single-Atom Sites Delivers Remarkable Oxygen Reduction Reaction Performance in Alkaline Media. *Adv. Mater.* **2022**, *34*, e2202544. [CrossRef]
41. Cai, X.; Ma, F.; Jiang, J.; Yang, X.; Zhang, Z.; Jian, Z.; Liang, M.; Li, P.; Yu, L. Fe-N-C single-atom nanozyme for ultrasensitive, on-site and multiplex detection of mycotoxins using lateral flow immunoassay. *J. Hazard. Mater.* **2023**, *441*, 129853. [CrossRef] [PubMed]
42. Du, Z.; Chen, X.; Hu, W.; Chuang, C.; Xie, S.; Hu, A.; Yan, W.; Kong, X.; Wu, X.; Ji, H.; et al. Cobalt in Nitrogen-Doped Graphene as Single-Atom Catalyst for High-Sulfur Content Lithium-Sulfur Batteries. *J. Am. Chem. Soc.* **2019**, *141*, 3977–3985. [CrossRef] [PubMed]
43. Sun, H.; Liu, J. Carbon-supported CoS_4-C single-atom nanozyme for dramatic improvement in CO_2 electroreduction to HCOOH: A DFT study combined with hybrid solvation model. *Chin. Chem. Lett.* **2022**, *34*, 108018. [CrossRef]
44. Cai, S.; Liu, J.; Ding, J.; Fu, Z.; Li, H.; Xiong, Y.; Lian, Z.; Yang, R.; Chen, C. Tumor-Microenvironment-Responsive Cascade Reactions by a Cobalt-Single-Atom Nanozyme for Synergistic Nanocatalytic Chemotherapy. *Angew. Chem. Int. Ed. Engl.* **2022**, *61*, e202204502. [CrossRef]
45. Wu, W.; Wang, Y.; Luo, L.; Wang, M.; Li, Z.; Chen, Y.; Wang, Z.; Chai, J.; Cen, Z.; Shi, Y.; et al. CO_2 Hydrogenation over Copper/ZnO Single-Atom Catalysts: Water-Promoted Transient Synthesis of Methanol. *Angew. Chem. Int. Ed. Engl.* **2022**, *61*, e202213024. [CrossRef]
46. Yitao, Z.; Lei, T. Towards catalytic reactions of Cu single-atom catalysts: Recent progress and future perspective. *Chin. Chem. Lett.* **2023**, 108571. [CrossRef]
47. Zhu, Y.; Wang, W.; Cheng, J.; Qu, Y.; Dai, Y.; Liu, M.; Yu, J.; Wang, C.; Wang, H.; Wang, S.; et al. Stimuli-Responsive Manganese Single-Atom Nanozyme for Tumor Therapy via Integrated Cascade Reactions. *Angew. Chem. Int. Ed. Engl.* **2021**, *60*, 9480–9488. [CrossRef] [PubMed]
48. Jiang, Y.; Sung, Y.; Choi, C.; Joo Bang, G.; Hong, S.; Tan, X.; Wu, T.S.; Soo, Y.L.; Xiong, P.; Meng-Jung Li, M.; et al. Single-Atom Molybdenum-N_3 Sites for Selective Hydrogenation of CO_2 to CO. *Angew. Chem. Int. Ed. Engl.* **2022**, *61*, e202203836. [CrossRef]
49. Xu, B.; Wang, H.; Wang, W.; Gao, L.; Li, S.; Pan, X.; Wang, H.; Yang, H.; Meng, X.; Wu, Q.; et al. A Single-Atom Nanozyme for Wound Disinfection Applications. *Angew. Chem. Int. Ed. Engl.* **2019**, *58*, 4911–4916. [CrossRef] [PubMed]
50. Zhou, S.; Zhao, Y.; Shi, R.; Wang, Y.; Ashok, A.; Heraly, F.; Zhang, T.; Yuan, J. Vacancy-Rich MXene-Immobilized Ni Single Atoms as a High-Performance Electrocatalyst for the Hydrazine Oxidation Reaction. *Adv. Mater.* **2022**, *34*, e2204388. [CrossRef]
51. Quanguo, J.; Yushuai, Q.; Yuqing, L.; Min, H.; Zhimin, A. Strain boosts CO oxidation on Ni single-atom-catalyst supported by defective graphene. *Chin. Chem. Lett.* **2022**, *34*, 107395. [CrossRef]
52. Ai, Y.; Hu, Z.N.; Liang, X.; Sun, H.b.; Xin, H.; Liang, Q. Recent Advances in Nanozymes: From Matters to Bioapplications. *Adv. Funct. Mater.* **2021**, *32*, 107395. [CrossRef]
53. Fan, Y.; Gan, X.; Zhao, H.; Zeng, Z.; You, W.; Quan, X. Multiple application of SAzyme based on carbon nitride nanorod-supported Pt single-atom for H_2O_2 detection, antibiotic detection and antibacterial therapy. *Chem. Eng. J.* **2022**, *427*, 131572. [CrossRef]
54. Zhao, Y.; Jiang, Y.; Mo, Y.; Zhai, Y.; Liu, J.; Strzelecki, A.C.; Guo, X.; Shan, C. Boosting Electrochemical Catalysis and Nonenzymatic Sensing Toward Glucose by Single-Atom Pt Supported on Cu@CuO Core-Shell Nanowires. *Small* **2023**, *19*, 2207240. [CrossRef] [PubMed]

55. Chen, Y.; Wang, P.; Hao, H.; Hong, J.; Li, H.; Ji, S.; Li, A.; Gao, R.; Dong, J.; Han, X.; et al. Thermal Atomization of Platinum Nanoparticles into Single Atoms: An Effective Strategy for Engineering High-Performance Nanozymes. *J. Am. Chem. Soc.* **2021**, *143*, 18643–18651. [CrossRef]
56. Gu, W.; Wang, H.; Jiao, L.; Wu, Y.; Chen, Y.; Hu, L.; Gong, J.; Du, D.; Zhu, C. Single-Atom Iron Boosts Electrochemiluminescence. *Angew. Chem. Int. Ed. Engl.* **2020**, *59*, 3534–3538. [CrossRef]
57. Qin, J.; Liu, H.; Zou, P.; Zhang, R.; Wang, C.; Xin, H.L. Altering Ligand Fields in Single-Atom Sites through Second-Shell Anion Modulation Boosts the Oxygen Reduction Reaction. *J. Am. Chem. Soc.* **2022**, *144*, 2197–2207. [CrossRef]
58. Wu, N.; Zhong, H.; Zhang, Y.; Wei, X.; Jiao, L.; Wu, Z.; Huang, J.; Wang, H.; Beckman, S.P.; Gu, W.; et al. Atomically dispersed Ru_3 site catalysts for electrochemical sensing of small molecules. *Biosens. Bioelectron.* **2022**, *216*, 114609. [CrossRef] [PubMed]
59. Cheng, N.; Li, J.C.; Liu, D.; Lin, Y.; Du, D. Single-Atom Nanozyme Based on Nanoengineered Fe-N-C Catalyst with Superior Peroxidase-Like Activity for Ultrasensitive Bioassays. *Small* **2019**, *15*, e1901485. [CrossRef]
60. Chen, X.; Wang, Y.; Feng, M.; Deng, D.; Xie, X.; Deng, C.; Khattak, K.N.; Yang, X. Dual-active-site Fe/Cu single-atom nanozymes with multifunctional specific peroxidase-like properties for S^{2-} detection and dye degradation. *Chin. Chem. Lett.* **2022**, *34*, 107969. [CrossRef]
61. Ma, C.B.; Xu, Y.; Wu, L.; Wang, Q.; Zheng, J.J.; Ren, G.; Wang, X.; Gao, X.; Zhou, M.; Wang, M.; et al. Guided Synthesis of a Mo/Zn Dual Single-Atom Nanozyme with Synergistic Effect and Peroxidase-like Activity. *Angew. Chem. Int. Ed. Engl.* **2022**, *61*, e202116170. [CrossRef] [PubMed]
62. Kim, K.; Lee, J.; Park, O.K.; Kim, J.; Kim, J.; Lee, D.; Paidi, V.K.; Jung, E.; Lee, H.S.; Lee, B.; et al. Geometric Tuning of Single-Atom FeN_4 Sites via Edge-Generation Enhances Multi-Enzymatic Properties. *Adv. Mater.* **2023**, *35*, 2207666. [CrossRef] [PubMed]
63. Ji, S.F.; Jiang, B.; Hao, H.G.; Chen, Y.J.; Dong, J.C.; Mao, Y.; Zhang, Z.D.; Gao, R.; Chen, W.X.; Zhang, R.F.; et al. Matching the kinetics of natural enzymes with a single-atom iron nanozyme. *Nat. Catal.* **2021**, *4*, 407–417. [CrossRef]
64. Bian, W.; Shen, X.; Tan, H.; Fan, X.; Liu, Y.; Lin, H.; Li, Y. The triggering of catalysis via structural engineering at atomic level: Direct propane dehydrogenation on Fe-N3P-C. *Chin. Chem. Lett.* **2023**, *34*, 107289. [CrossRef]
65. Jiao, L.; Xu, W.; Zhang, Y.; Wu, Y.; Gu, W.; Ge, X.; Chen, B.; Zhu, C.; Guo, S. Boron-doped Fe-N-C single-atom nanozymes specifically boost peroxidase-like activity. *Nano Today* **2020**, *35*, 100971. [CrossRef]
66. Xu, B.; Li, S.; Zheng, L.; Liu, Y.; Han, A.; Zhang, J.; Huang, Z.; Xie, H.; Fan, K.; Gao, L.; et al. A Bioinspired Five-Coordinated Single-Atom Iron Nanozyme for Tumor Catalytic Therapy. *Adv. Mater.* **2022**, *34*, e2107088. [CrossRef] [PubMed]
67. Wu, F.; Pan, C.; He, C.T.; Han, Y.; Ma, W.; Wei, H.; Ji, W.; Chen, W.; Mao, J.; Yu, P.; et al. Single-Atom $Co-N_4$ Electrocatalyst Enabling Four-Electron Oxygen Reduction with Enhanced Hydrogen Peroxide Tolerance for Selective Sensing. *J. Am. Chem. Soc.* **2020**, *142*, 16861–16867. [CrossRef] [PubMed]
68. Li, J.; Wu, C.; Yuan, C.; Shi, Z.; Zhang, K.; Zou, Z.; Xiong, L.; Chen, J.; Jiang, Y.; Sun, W.; et al. Single-Atom Iron Anchored on 2-D Graphene Carbon to Realize Bridge-Adsorption of O-O as Biomimetic Enzyme for Remarkably Sensitive Electrochemical Detection of H_2O_2. *Anal. Chem.* **2022**, *94*, 14109–14117. [CrossRef]
69. Wei, X.; Song, S.; Song, W.; Xu, W.; Jiao, L.; Luo, X.; Wu, N.; Yan, H.; Wang, X.; Gu, W.; et al. Fe_3C-Assisted Single Atomic Fe Sites for Sensitive Electrochemical Biosensing. *Anal. Chem.* **2021**, *93*, 5334–5342. [CrossRef] [PubMed]
70. Lei, Y.; Butler, D.; Lucking, M.C.; Zhang, F.; Xia, T.; Fujisawa, K.; Nakajima, T.G.; Silva, R.C.; Endo, M.; Terrones, H.; et al. Single-atom doping of MoS_2 with manganese enables ultrasensitive detection of dopamine: Experimental and computational approach. *Sci. Adv.* **2020**, *6*, eabc4250. [CrossRef] [PubMed]
71. Ding, S.; Lyu, Z.; Fang, L.; Li, T.; Zhu, W.; Li, S.; Li, X.; Li, J.C.; Du, D.; Lin, Y. Single-Atomic Site Catalyst with Heme Enzymes-Like Active Sites for Electrochemical Sensing of Hydrogen Peroxide. *Small* **2021**, *17*, e2100664. [CrossRef] [PubMed]
72. Pan, C.; Wu, F.; Mao, J.; Wu, W.; Zhao, G.; Ji, W.; Ma, W.; Yu, P.; Mao, L. Highly Stable and Selective Sensing of Hydrogen Sulfide in Living Mouse Brain with NiN_4 Single-Atom Catalyst-Based Galvanic Redox Potentiometry. *J. Am. Chem. Soc.* **2022**, *144*, 14678–14686. [CrossRef] [PubMed]
73. Gao, X.; Ma, W.; Mao, J.; He, C.T.; Ji, W.; Chen, Z.; Chen, W.; Wu, W.; Yu, P.; Mao, L. A single-atom $Cu-N_2$ catalyst eliminates oxygen interference for electrochemical sensing of hydrogen peroxide in a living animal brain. *Chem. Sci.* **2021**, *12*, 15045–15053. [CrossRef] [PubMed]
74. Liang, Y.; Zhao, P.; Zheng, J.L.; Chen, Y.Y.; Liu, Y.Y.; Zheng, J.; Luo, X.G.; Huo, D.Q.; Hou, C.J. Fe Single-Atom Electrochemical Sensors for H_2O_2 Produced by Living Cells. *Acs Appl. Nano Mater.* **2022**, *5*, 11852–11863. [CrossRef]
75. Li, Z.; Li, Y.; Chen, S.; Zha, Q.; Zhu, M. In-situ monitoring of hydrogen peroxide production at nickel single-atom electrocatalyst. *Chem. Eng. J.* **2023**, *460*, 141657. [CrossRef]
76. Qi, C.; Wang, W.; Dong, Y. Synthesis of Se single atoms on nitrogen-doped carbon as novel electrocatalyst for sensitive nonenzymatic sensing of hydrogen peroxide. *Anal. Bioanal. Chem.* **2023**, *415*, 5391–5401. [CrossRef]
77. Qi, C.; Luo, Y.; Dong, Y. Synergistic effects of Fe-Se dual single-atom sites for boosting electrochemical nonenzymatic H_2O_2 sensing. *Appl. Surf. Sci.* **2023**, *637*, 157900. [CrossRef]
78. Shu, Y.J.; Li, Z.J.; Yang, Y.; Tan, J.W.; Liu, Z.Y.; Shi, Y.H.; Ye, C.X.; Gao, Q.S. Isolated Cobalt Atoms on N-Doped Carbon as Nanozymes for Hydrogen Peroxide and Dopamine Detection. *Acs Appl. Nano Mater.* **2021**, *4*, 7954–7962. [CrossRef]
79. Sun, X.J.; Chen, C.; Xiong, C.; Zhang, C.M.; Zheng, X.S.; Wang, J.; Gao, X.P.; Yu, Z.Q.; Wu, Y.E. Surface modification of MoS_2 nanosheets by single Ni atom for ultrasensitive dopamine detection. *Nano Res.* **2022**, *16*, 917–924. [CrossRef]

80. Xie, X.; Wang, D.P.; Guo, C.; Liu, Y.; Rao, Q.; Lou, F.; Li, Q.; Dong, Y.; Li, Q.; Yang, H.B.; et al. Single-Atom Ruthenium Biomimetic Enzyme for Simultaneous Electrochemical Detection of Dopamine and Uric Acid. *Anal. Chem.* **2021**, *93*, 4916–4923. [CrossRef] [PubMed]
81. Long, B.; Cao, P.; Zhao, Y.; Fu, Q.; Mo, Y.; Zhai, Y.; Liu, J.; Lyu, X.; Li, T.; Guo, X.; et al. Pt_1/Ni_5Co_1 layered double hydroxides/N-doped graphene for electrochemical non-enzymatic glucose sensing by synergistic enhancement of single atoms and doping. *Nano Res.* **2022**, *16*, 318–324. [CrossRef]
82. Long, B.; Zhao, Y.; Cao, P.; Wei, W.; Mo, Y.; Liu, J.; Sun, C.J.; Guo, X.; Shan, C.; Zeng, M.H. Single-Atom Pt Boosting Electrochemical Nonenzymatic Glucose Sensing on $Ni(OH)_2$/N-Doped Graphene. *Anal. Chem.* **2022**, *94*, 1919–1924. [CrossRef] [PubMed]
83. Jiang, J.; Cai, Q.; Deng, M. Construction of Electrochemical Aptamer Sensor Based on Pt-Coordinated Titanium-Based Porphyrin MOF for Thrombin Detection. *Front. Chem.* **2021**, *9*, 812983. [CrossRef]
84. Li, M.; Peng, X.; Liu, X.; Wang, H.; Zhang, S.; Hu, G. Single-atom niobium doped BCN nanotubes for highly sensitive electrochemical detection of nitrobenzene. *RSC Adv.* **2021**, *11*, 28988–28995. [CrossRef] [PubMed]
85. Gu, W.; Wang, X.; Wen, J.; Cao, S.; Jiao, L.; Wu, Y.; Wei, X.; Zheng, L.; Hu, L.; Zhang, L.; et al. Modulating Oxygen Reduction Behaviors on Nickel Single-Atom Catalysts to Probe the Electrochemiluminescence Mechanism at the Atomic Level. *Anal. Chem.* **2021**, *93*, 8663–8670. [CrossRef]
86. Bushira, F.A.; Kitte, S.A.; Xu, C.; Li, H.; Zheng, L.; Wang, P.; Jin, Y. Two-Dimensional-Plasmon-Boosted Iron Single-Atom Electrochemiluminescence for the Ultrasensitive Detection of Dopamine, Hemin, and Mercury. *Anal. Chem.* **2021**, *93*, 9949–9957. [CrossRef] [PubMed]
87. Xia, H.; Zheng, X.; Li, J.; Wang, L.; Xue, Y.; Peng, C.; Han, Y.; Wang, Y.; Guo, S.; Wang, J.; et al. Identifying Luminol Electrochemiluminescence at the Cathode via Single-Atom Catalysts Tuned Oxygen Reduction Reaction. *J. Am. Chem. Soc.* **2022**, *144*, 7741–7749. [CrossRef] [PubMed]
88. Luo, Z.; Xu, W.; Wu, Z.; Jiao, L.; Luo, X.; Xi, M.; Su, R.; Hu, L.; Gu, W.; Zhu, C. Iron Single-Atom Catalyst-Enabled Peroxydisulfate Activation Enhances Cathodic Electrochemiluminescence of Tris(bipyridine)ruthenium(II). *Anal. Chem.* **2023**, *95*, 10762–10768. [CrossRef] [PubMed]
89. Ma, Y.; Zhang, Y.; Gao, J.; Ouyang, H.; He, Y.; Fu, Z. PEGylated Ni Single-Atom Catalysts as Ultrasensitive Electrochemiluminescent Probes with Favorable Aqueous Dispersibility for Assaying Drug-Resistant Pathogens. *Anal. Chem.* **2022**, *94*, 14047–14053. [CrossRef]
90. Tang, Y.; Liu, Y.; Xia, Y.; Zhao, F.; Zeng, B. Simultaneous Detection of Ovarian Cancer-Concerned HE4 and CA125 Markers Based on Cu Single-Atom-Triggered CdS QDs and Eu MOF@Isoluminol ECL. *Anal. Chem.* **2023**, *95*, 4795–4802. [CrossRef] [PubMed]
91. Liu, D.; Wan, X.; Kong, T.; Han, W.; Xiong, Y. Single-atom-based catalysts for photoelectrocatalysis: Challenges and opportunities. *J. Mater. Chem. A* **2022**, *10*, 5878–5888. [CrossRef]
92. Zeng, R.J.; Wang, W.J.; Cai, G.N.; Huang, Z.L.; Tao, J.M.; Tang, D.P.; Zhu, C.Z. Single-atom platinum nanocatalyst-improved catalytic efficiency with enzyme-DNA supermolecular architectures. *Nano Energy* **2020**, *74*, 104931. [CrossRef]
93. Qin, Y.; Wen, J.; Zheng, L.; Yan, H.; Jiao, L.; Wang, X.; Cai, X.; Wu, Y.; Chen, G.; Chen, L.; et al. Single-Atom-Based Heterojunction Coupling with Ion-Exchange Reaction for Sensitive Photoelectrochemical Immunoassay. *Nano Lett.* **2021**, *21*, 1879–1887. [CrossRef] [PubMed]
94. Li, B.; Guo, L.; Chen, M.; Guo, Y.; Ge, L.; Kwok, H.F. Single-atom Pt-anchored $Zn_{0.5}Cd_{0.5}S$ boosted photoelectrochemical immunoassay of prostate-specific antigen. *Biosens. Bioelectron.* **2022**, *202*, 114006. [CrossRef] [PubMed]
95. Zhou, M.; Jiang, Y.; Wang, G.; Wu, W.; Chen, W.; Yu, P.; Lin, Y.; Mao, J.; Mao, L. Single-atom $Ni\text{-}N_4$ provides a robust cellular NO sensor. *Nat. Commun.* **2020**, *11*, 3188. [CrossRef]
96. Hou, H.F.; Mao, J.J.; Han, Y.H.; Wu, F.; Zhang, M.N.; Wang, D.S.; Mao, L.Q.; Li, Y.D. Single-atom electrocatalysis: A new approach to in vivo electrochemical biosensing. *Sci. China-Chem.* **2019**, *62*, 1720–1724. [CrossRef]
97. Hu, F.X.; Hu, T.; Chen, S.; Wang, D.; Rao, Q.; Liu, Y.; Dai, F.; Guo, C.; Yang, H.B.; Li, C.M. Single-Atom Cobalt-Based Electrochemical Biomimetic Uric Acid Sensor with Wide Linear Range and Ultralow Detection Limit. *Nanomicro Lett.* **2020**, *13*, 7. [CrossRef] [PubMed]
98. Zhou, W.Y.; Li, S.S.; Xiao, X.Y.; Chen, S.H.; Liu, J.H.; Huang, X.J. Defect- and phase-engineering of Mn-mediated MoS_2 nanosheets for ultrahigh electrochemical sensing of heavy metal ions: Chemical interaction-driven in situ catalytic redox reactions. *Chem. Commun.* **2018**, *54*, 9329–9332. [CrossRef] [PubMed]
99. Yao, L.; Gao, S.; Liu, S.; Bi, Y.; Wang, R.; Qu, H.; Wu, Y.; Mao, Y.; Zheng, L. Single-Atom Enzyme-Functionalized Solution-Gated Graphene Transistor for Real-Time Detection of Mercury Ion. *ACS Appl. Mater. Interfaces* **2020**, *12*, 6268–6275. [CrossRef]
100. Li, P.H.; Yang, M.; Li, Y.X.; Song, Z.Y.; Liu, J.H.; Lin, C.H.; Zeng, J.; Huang, X.J. Ultra-Sensitive and Selective Detection of Arsenic (III) via Electroanalysis over Cobalt Single-Atom Catalysts. *Anal. Chem.* **2020**, *92*, 6128–6135. [CrossRef] [PubMed]
101. Cong, W.; Song, P.; Zhang, Y.; Yang, S.; Liu, W.; Zhang, T.; Zhou, J.; Wang, M.; Liu, X. Supramolecular confinement pyrolysis to carbon-supported Mo nanostructures spanning four scales for hydroquinone determination. *J. Hazard. Mater.* **2022**, *437*, 129327. [CrossRef] [PubMed]
102. Luo, X.; Luo, Z.; Wei, X.; Jiao, L.; Fang, Q.; Wang, H.; Wang, J.; Gu, W.; Hu, L.; Zhu, C. Iridium Single-Atomic Site Catalysts with Superior Oxygen Reduction Reaction Activity for Sensitive Monitoring of Organophosphorus Pesticides. *Anal. Chem.* **2022**, *94*, 1390–1396. [CrossRef] [PubMed]

103. Luo, X.; Yang, M.; Song, W.; Fang, Q.; Wei, X.; Jiao, L.; Xu, W.; Kang, Y.; Wang, H.; Wu, N.; et al. Neutral Zn-Air Battery Assembled with Single-Atom Iridium Catalysts for Sensitive Self-Powered Sensing System. *Adv. Funct. Mater.* **2021**, *31*, 2101193. [CrossRef]
104. Tan, R.; Qin, Y.; Liu, M.; Wang, H.; Su, R.; Xiao, R.; Li, J.; Hu, L.; Gu, W.; Zhu, C. Bifunctional Single-Atom Iron Cocatalysts Enable an Efficient Photoelectrochemical Fuel Cell for Sensitive Biosensing. *Adv. Funct. Mater.* **2023**, 2305673. [CrossRef]

chemosensors

MDPI

Review

Recent Progress in Electrochemical Aptasensors: Construction and Application

Renqiang Yuan, Jing Cai, Haojie Ma, Yi Luo, Lianhui Wang and Shao Su *

State Key Laboratory of Organic Electronics and Information Displays & Jiangsu Key Laboratory for Biosensors, Institute of Advanced Materials (IAM), Nanjing University of Posts and Telecommunications, 9 Wenyuan Road, Nanjing 210023, China; iamrqyuan@njupt.edu.cn (R.Y.); 1022233716@njupt.edu.cn (J.C.); 1022233718@njupt.edu.cn (H.M.); iamyluo@njupt.edu.cn (Y.L.); iamlhwang@njupt.edu.cn (L.W.)

* Correspondence: iamssu@njupt.edu.cn

Abstract: Electrochemical aptasensors have gained significant attention due to their exceptional sensitivity, selectivity, stability, and rapid response, combining the advantages of electrochemical techniques with the specific recognition ability of aptamers. This review aims to provide a comprehensive summary of the recent advances in electrochemical aptasensors. Firstly, the construction method and the advantages of electrochemical aptasensors are introduced. Subsequently, the review highlights the application progress of electrochemical aptasensors in detecting various chemical and biological molecules, including metal ions, small biological molecules, drugs, proteins, exosomes, tumor cells, bacteria, and viruses. Lastly, the prospects and challenges associated with electrochemical aptasensors are discussed.

Keywords: electrochemical aptasensor; aptamer; biochemical analysis; disease diagnosis

Citation: Yuan, R.; Cai, J.; Ma, H.; Luo, Y.; Wang, L.; Su, S. Recent Progress in Electrochemical Aptasensors: Construction and Application. *Chemosensors* **2023**, *11*, 488. https://doi.org/10.3390/chemosensors11090488

Academic Editor: Edward P. C. Lai

Received: 9 July 2023
Revised: 31 August 2023
Accepted: 31 August 2023
Published: 4 September 2023

1. Introduction

Electrochemical biosensors have been widely applied in many fields including environmental monitoring, food safety, disease diagnosis, agricultural engineering, and even public safety owing to their excellent advantages of high sensitivity, fast analysis speed, simple operation, high reproducibility, long-term stability, ease of miniaturization, and on-site/in situ analysis [1–3]. Nowadays, antibody-based bio-molecular recognition events are the general way to construct electrochemical biosensors for the recognition and detection of target molecules. For example, different electrochemical immunosensors have been developed for the analysis of SARS-CoV-2 [4], cardiac troponin I [5], carcinoembryonic antigen [6], etc. Though great advances in electrochemical immunosensors have been achieved, they still have some limitations in the application of environmental monitoring, small biological molecules, and drug metabolism analysis due to the specific recognition scope of antibodies. Therefore, it is necessary to develop other electrochemical biosensors to enable sensitive and selective detection of target molecules.

In general, an aptamer is a single-stranded DNA or RNA sequence, which is selected by systematic evolution of ligands by exponential enrichment (SELEX) in vitro, which utilizes protein-, cell-, and live animal-based selection processes [7]. Like antibodies, aptamers have high specificity and affinity for a wide range of target substances, including metal ions, biological small molecules, proteins, exosomes, cells, and microbial pathogens [8–10]. Therefore, it is considered as a promising alternative bio-molecular recognition element. More importantly, the unique nature of aptamers including easy synthesis, multiplexed functionalization, and higher stability than antibodies make them widely applicable in the field of analytical chemistry. Coupled with the electrochemical technique, the aptamer-based electrochemical biosensor (also called electrochemical aptasensor) has gradually become a research hotspot due to its excellent performance and wide application fields [11–14]. With the fast development of nanotechnology, the introduction of nanomaterials greatly improves the analytical performance of electrochemical aptasensors [15–17]. For example, Ahmadi and co-workers have constructed an electrochemical aptasensor based on gold nanoparticles (AuNPs)/nitrogen-doped carbon nano-onions (NCNOs) for the detection of *Staphylococcus aureus* (*S. aureus*) [18]. The synergetic effect of AuNPs and NCNOs effectively enhanced the active surface area and the conductivity of the electrode, leading to an outstanding performance of the designed aptasensor with a wide linear range (10–10^8 CFU/mL) and a low detection limit (3 CFU/mL). This biosensor also showed excellent repeatability, reproducibility, and long-term stability, allowing it to detect low levels of *S. aureus* in human serum samples. Additionally, to improve the sensitivity and feasibility of the photoelectrochemical (PEC) aptasensor, Ju et al. decorated $MgIn_2S_4$ nanoplates on a TiO_2 nanowire array (TiONA) to form $MgIn_2S_4$-TiONA heterojunctions for the construction of a visible light PEC aptasensing platform [19]. The designed PEC aptasensor showed excellent performance for adenosine triphosphate (ATP) detection due to the large surface area and strong absorption efficiency of the $MgIn_2S_4$-TiONA heterojunction. The outstanding detection performance has greatly promoted the development of nanomaterial-based electrochemical aptasensors in the analytical fields.

In recent years, great advances in electrochemical aptasensors have been achieved. In this review, we are aiming to give a comprehensive overview of the recent progresses in electrochemical aptasensors with some typical examples: (1) to summarize the immobilization methods of the aptamer as the key recognition unit; (2) to cover a variety of target substances for detection applications in different fields; (3) to put insightful comments on various detection strategies with high sensitivity, specificity, and selectivity and give the exploration experiences and underlying experimental regularity. Finally, the opportunities and challenges of electrochemical aptasensors in the future are also discussed, which will be beneficial to clinical applications or commercial transformations of scientific research.

2. Immobilizations of Aptamers on the Surface of Electrode

How to efficiently immobilize aptamers on the electrode surface is a key step in the construction of electrochemical aptasensors, which greatly affects the sensing performance and application fields [20,21]. In general, electrochemical indicator- (such as methylene blue, ferrocene), thiol-, biotin-, and carboxyl-labeled aptamers have been synthesized by companies, which are designed for sensing purposes. Nowadays, three methods including adsorption, biological recognition, and covalent bonds have been widely used to construct electrochemical aptasensors (Scheme 1). The advantages and disadvantages of these immobilization methods are listed in Table 1 [22–24].

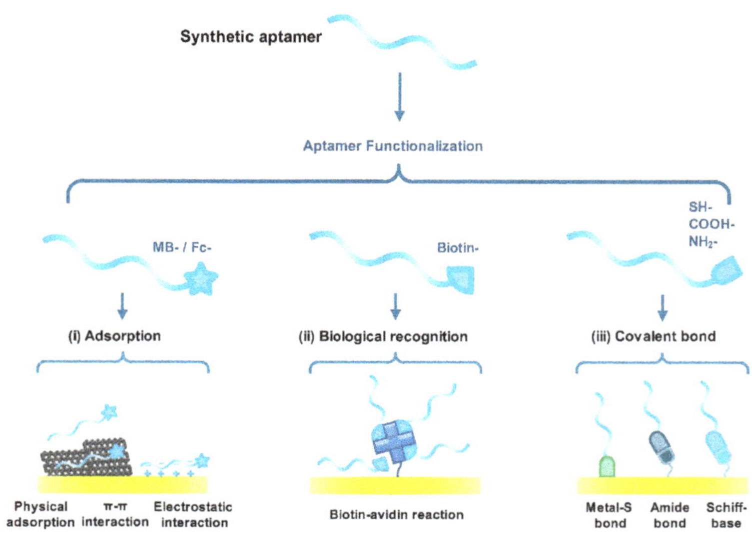

Scheme 1. A general scheme of the modification and immobilization of aptamers on sensing interfaces.

Table 1. Immobilization methods for the construction of electrochemical aptasensors.

Immobilization Methods	Technique Principles	Advantages	Disadvantages	References
adsorption method	electrostatic interaction, physical adsorption, π–π interaction	no chemical modification, simple, fast	easy detachment, low density, poor orientation	[25–27]
biological recognition	biotin–avidin reaction	mild conditions, high immobilization efficiency, orderly assembly	decrease in electrochemical signal	[28–30]
covalent bond	metal-S bond, amide bond, Schiff-base	good stability, high recognition activity, easy regeneration, adjustable process	complex chemical modification, introduced interferences	[31–33]

2.1. Adsorption Method

As a nucleic acid sequence, aptamers usually have negative charges due to their phosphate backbone. Therefore, aptamers easily adsorb on the surface of nanomaterials with positive charges via electrostatic reaction [10,34], such as metal–organic frameworks (MOFs) [35], mesoporous silica nanoparticles (MSNs) [36], and magnetic nanoparticles (MNPs) [37]. Moreover, aptamers also easily adsorb on the modified electrode surface or nanomaterials' surface through physical adsorption or π–π interaction, such as graphene [38], carbon nanotubes [39], single-layered MoS_2 [40], noble metal nanoparticles [41], etc. This immobilization method is simple, fast, and without chemical modification. However, this method has obvious shortcomings including easy detachments (sensitivity to pH, ionic strength, temperature, etc.), low density, and poor orientation on the sensing interfaces [26,27].

2.2. Biological Recognition

The biological recognition system has been recently employed to immobilize aptamers on the electrode surface and nanomaterials' surface through affinity labeling strategies, such as the biotin–streptavidin recognition reaction [42]. Generally, one streptavidin molecule has four identical subunits, which can specifically bind up to four biotin molecules. The binding constant (Ka) of biotin–streptavidin is about 10^{15} mol^{-1}, suggesting that this biological recognition is specific and efficient [43]. However, the poor conductivity of biotin and streptavidin molecules often leads to a decrease in the signal of electrochemical aptasensors, resulting in low detection sensitivity [28–30].

2.3. Covalent Bond

The covalent bond is a popular method for the construction of electrochemical aptasensors. To achieve this purpose, it usually requires the synthesized aptamer with specific functional groups such as thiol, amino, carboxylic acid, or phosphate groups [44,45]. Correspondingly, noble metal nanoparticles or various active groups including hydroxyl and carboxyl are introduced to the electrode surface. Through a chemical reaction, an aptamer can be efficiently and controllably immobilized on the electrode surface. This method for the construction of electrochemicals exhibits several advantages of good stability, high binding efficiency, easy regeneration, and a controllable process [31–33]. Meanwhile, this method required chemical modification, complicated steps, and introduced other interference factors [46,47].

Besides immobilization methods, experimental conditions are important factors to greatly affect the detection performance of electrochemical aptasensors, such as pH value, aptamer concentration, binding time between aptamer and target molecules, elution time, and temperature [48–50]. It should be noted that the experimental conditions are generally optimized according to the designed detection strategies and the properties of introduced nanomaterials and target molecules. Taking the pH value effect as an example, Liu et al. found that the electrochemical signal of the designed aptasensor reached a maximum value when the pH value was 7.0. Low and high pH could affect the binding efficiency of the aptamer and dopamine [51]. In another work, the optimized pH value of the developed aptasensor for Aflatoxin B1 detection was about 4.0, which is greatly dependent on the properties of the introduced titanium carbide/carboxylated graphene oxide-poly(4-vinyl pyridine) nanocomposites [52]. For the temperature effect, many electrochemical aptasensors are typically used at room temperature. Especially, electrochemical aptasensors for biological molecules' detection are usually used at 37 degrees Celsius, the optimized temperature, which is close to normal human body temperature [53,54]. Therefore, suitable experimental conditions could result in better detection performance.

3. Application of Electrochemical Aptasensors for the Detection of Different Target Molecules

3.1. Detection of Metal Ions

The detection of metal ions is particularly important in both environmental monitoring and biomedical diagnosis [55,56]. In recent years, with the acceleration of industrialization, environmental pollutants have become an increasingly serious problem. Among them, heavy metal ions, as a kind of pollutant, are difficult to degrade under natural conditions and can be accumulated through the food chain to harm human health. Therefore, it is an urgent problem to develop rapid, simple, and efficient sensors for qualitative and quantitative analysis of heavy metal ions in the environment.

Although traditional methods (such as atomic absorption spectroscopy, inductively coupled plasma mass spectrometry, high-performance liquid chromatography, etc.) have achieved high detection performance, the high cost of instruments and the need for specialized operators have limited their widespread application. Compared to traditional analytical methods, electrochemical methods show many advantages for heavy metal ions' detection due to their real-time detection capability, high sensitivity, fast response time, miniaturization, and integration [57,58]. To further improve detection selectivity in complex environments, electrochemical aptasensors are considered as promising detection methods for the determination of heavy metal ions [1,59]. For example, Zhang et al. designed an electrochemical aptasensor for trace heavy metal ions' detection based on the Fe(III)-based metal–organic framework (Fe-MOF) and mesoporous Fe_3O_4@C nanocapsules core-shell nanocomposite (denoted as Fe-MOF@mFe_3O_4@mC) [60]. The designed nanocomposites showed excellent electrochemical performance and high loading of aptamers, resulting in the high performance of this aptasensor for Pb^{2+} and As^{3+} detection. Under optimal conditions, the constructed aptasensor can detect as low as 2.27 pM Pb^{2+} and 6.73 pM As^{3+}, respectively. Besides the selective determination of metal ions, Gao et al. developed a dual signal-based electrochemical aptasensor for the simultaneous detection of Pb^{2+} and Hg^{2+} in environmental water samples by in situ assembling electrochemical signal tags (Cu^{2+}-Melamine and Nile blue) on the terminal of Pb^{2+}-binding aptamer (PBA) and Hg^{2+}-binding aptamer (HBA), respectively (Figure 1A) [61]. Once specifically capturing target Pb^{2+} and Hg^{2+}, the two electrochemical signal tags are far away from the electrode surface due to the rigidity of PBA-Pb^{2+} and HBA- Hg^{2+} secondary structures, leading to electrochemical "signal-off" with the increase in Cu^{2+}-Melamine and Nile blue. Based on this detection mechanism, the detection limits of the developed aptasensor for Pb^{2+} and Hg^{2+} analysis are 0.98 pM and 19 pM, respectively. Moreover, the designed aptasensor could achieve the successful determination of Pb^{2+} and Hg^{2+} in real water samples, presenting its latent capacity in the monitoring of heavy metal ions.

Metal ion concentration in the body also plays an important physiological role in maintaining the balance of intracellular and extracellular fluids, as well as the excitability of nerves and muscles [62,63]. It has been found that abnormal metal ion concentration is closely related to a variety of nervous system diseases [64]. For example, copper ions (Cu^{2+}), zinc ions (Zn^{2+}), iron ions (Fe^{3+}), and aluminum ions (Al^{3+}) are involved in the pathogenesis of Alzheimer's disease (AD) [65]. Therefore, it is significant to monitor these metal ions' concentrations for the early diagnosis of AD patients. For the sensitive and selective determination of Zn^{2+}, Zhao and co-workers designed an electrochemical aptasensor by fixing aptamers on the surface of bi-functional SiO_2-Pt@meso-SiO_2 core-shell nanoparticles [66]. The developed bi-functional nanoparticles can not only greatly enhance the conductivity of the electrode but also can pre-enrich Zn^{2+} onto the electrode surface. As a result, this aptasensor can determine the Zn^{2+} concentration as low as 65 pM. Moreover, this aptasensor can successfully detect Zn^{2+} in human blood and disrupted human cells. Similarly, Salehan and co-workers proposed an electrochemical aptasensor for potassium ion (K^+) analysis through electrochemical impedance spectroscopy [67]. They assembled the selected aptamer [$G_3(T_2AG_3)_3$] onto the surface of a polyaniline (PANI) modified electrode via electrostatic adsorption. In the presence of K^+, a single-strand aptamer folds into a G-quadruplex configuration, which blocks the electron transfer between the electrochemical indicator and electrode surface (Figure 1B). As a result, the charge transfer resistance (R_{ct}) increased with the increasing concentration of K^+. The linear relationship between ΔR_{ct} and the logarithm of K^+ concentration is plotted ranging from 10 pM to 60 μM with a detection limit of 3.7 pM. Moreover, this developed aptasensor can efficiently determine K^+ in serum samples, suggesting that it is a promising method for metal ion detection.

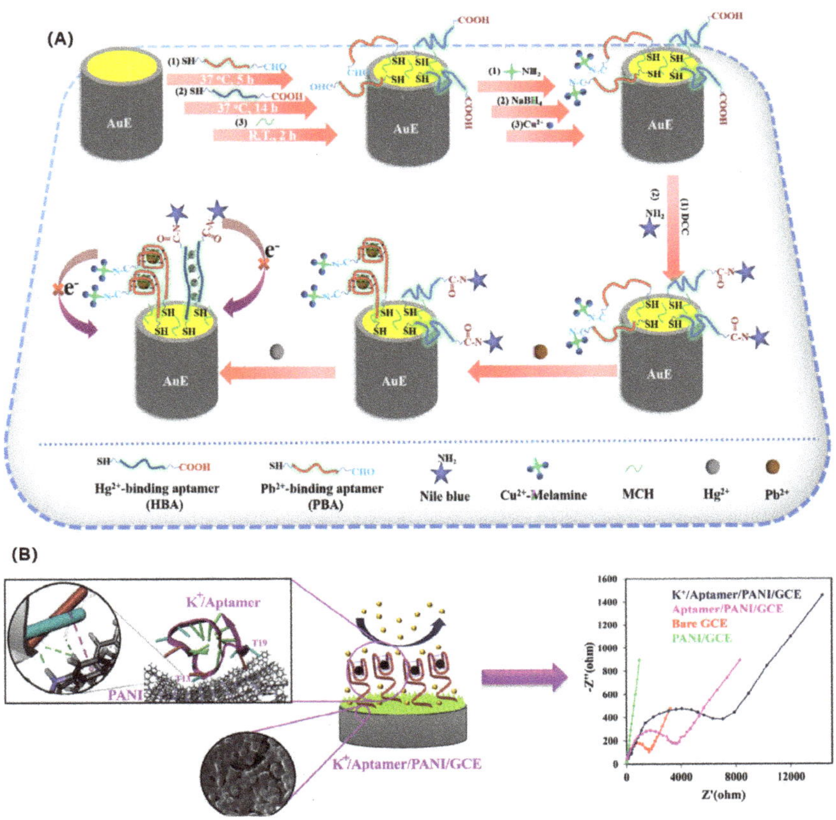

Figure 1. (**A**) Fabrication illustration of the dual signal−based electrochemical aptasensor for simultaneous analysis of Pb^{2+} and Hg^{2+}. (Reproduced with permission from Ref. [61] Copyright 2022, Elsevier). (**B**) Scheme of the impedimetric biosensor for the determination of potassium ions based on polyaniline/GCE and DNA G-quadruplex conformation. (Reproduced with permission from Ref. [67] Copyright 2022, Elsevier).

3.2. Detection of Pesticides

Over decades, the over-range and excessive dosage use of pesticides not only causes environmental pollution but also directly or indirectly threatens human health due to the long-term accumulation of residues. Nowadays, gas chromatography (GC), high-performance liquid chromatography (HPLC), GC-mass spectrometry (MS), and fluorescence and electrochemistry methods have been widely used to determine pesticides with accepted sensitivity, selectivity, and efficiency [68]. It is noted that some methods have disadvantages in the analysis of pesticides, such as low sensitivity, poor selectivity, and high cost [69]. Therefore, it is necessary to develop new methods for the sensitive and efficient detection of pesticides.

Electrochemical aptasensors have efficiently addressed these issues in pesticide detection. For example, Venegas and co-workers developed a label-free electrochemical aptasensor for the selective detection of carbendazim (CBZ) [70]. As shown in Figure 2A, the oxidized carbon nanotubes (CNT-COOH) were deposited on the surface of screen-printed carbon electrodes (SPCE) via physical adsorption. Then, amine-labeled aptamer ($AP\text{-}NH_2$) was assembled on the surface of SPCE via covalent immobilization. In the presence of CBZ, the corresponding aptamers specifically capture analytes to form secondary structures, causing a decrease in the electrochemical signal due to the blocking of electron transfer. According to this phenomenon, this proposed electrochemical aptasensor can detect values as low as 4.35 nM (0.83 ng/mL) CBZ, which is comparable to the HPLC method. Moreover, this aptasensor can efficiently determine CBZ in tomatoes, proving that the aptasensor can detect pesticides in food samples.

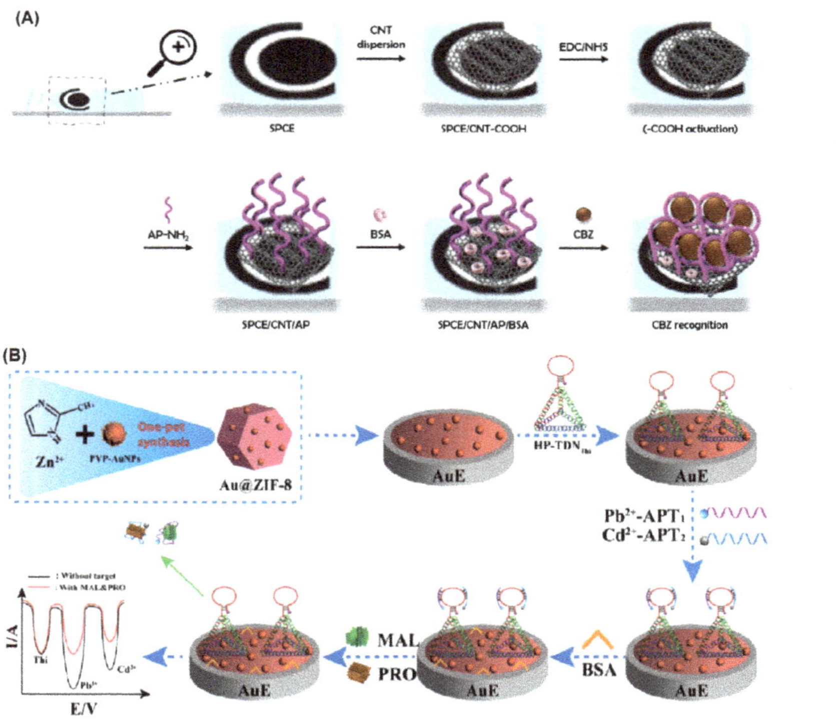

Figure 2. (**A**) The fabrication of a label-free electrochemical aptasensor for CBZ detection. (Reproduced with permission from Ref. [70] Copyright 2023, MDPI AG). (**B**) Scheme of the dual-ratiometric electrochemical aptasensing platform for the simultaneous detection of MAL and PRO. (Reproduced with permission from Ref. [71] Copyright 2023, Elsevier).

Electrochemical aptasensors also can simultaneously determine pesticides. Liu et al. designed a dual-ratiometric electrochemical aptasensor for the simultaneous detection of malathion (MAL) and profenofos (PRO) [71]. To obtain signal amplification, they firstly fabricated gold nanoparticle (AuNP)-encapsulated zeolitic imidazolate framework (ZIF-8) nanocomposites (Au@ZIF-8) to improve the immobilization amount of hairpin-tetrahedral DNA nanostructures (HP-TDN) and facilitate the electron transfer between the DNA probe and the electrode surface (Figure 2B). Then, metal ions (Pb^{2+} and Cd^{2+}) served as signal tracers to further amplify detection signal. Pb^{2+}-functionalized MAL aptamer (Pb^{2+}-APT1) and Cd^{2+}-functionalized PRO aptamer (Cd^{2+}-APT2) were co-assembled on the surface of thionine-labeled DNA nanostructures (HP-TDN_{Thi}) to specifically recognize MAL and PRO, respectively. With the addition of MAL and PRO, Pb^{2+}-APT_1 and Cd^{2+}-APT_2 dissociated from the surface of HP-TDN_{Thi}, leading to a decrease in both I_{Pb2+}/I_{Thi} and I_{Cd2+}/I_{Thi} oxidation current ratios. The detection limits of this aptasensor for MAL and PRO detection can reach 4.3 pg/mL and 13.3 pg/mL under optimized conditions, respectively. The obtained high performance suggested that this work may pave a way to develop aptasensors for multiple pesticides' detection in food safety and environmental monitoring.

3.3. Detection of Small Biological Molecules

Physiological metabolic activities of living organisms are closely related to various biological small molecules, such as neurotransmitters, adenosine triphosphate (ATP), nicotinamide adenine dinucleotide (NADH), hydrogen peroxide (H_2O_2), etc. [72–74]. Some small molecules including neurotransmitters and H_2O_2 can be directly detected by electrochemical sensors. However, direct detection is susceptible to interference from other electrochemical active molecules, resulting in low selectivity and sensitivity. Therefore, the biological recognition reaction is a promising tool to construct an electrochemical sensing platform for some biological small molecules' detection. According to this concept, electrochemical immunosensors and aptasensors have been widely developed to analyze ATP, dopamine (DA), uric acid (AA), 8-hydroxy-2′-deoxyguanosine (8-OHDG), etc. [75,76]. Compared with electrochemical immunosensors, electrochemical aptasensors attracted more and more attention due to their easy design, low cost, and controllable assembly.

Taking DA as an example, many electrochemical biosensors have been developed to directly determine it [77,78]. However, the co-existing AA, norepinephrine, adrenaline, and isoproterenol often affect the analytical performance of those designed biosensors. To improve the sensitivity and selectivity of electrochemical biosensors for dopamine detection, Zhang et al. used aptamer as a recognition unit, methylene blue-integrated m-PdNFs as a signal amplification unit, and Ce-MOF as an electrode-modified unit to construct a novel electrochemical aptasensor [79]. As shown in Figure 3A, a single-stranded DNA (S2) was assembled on the surface of the designed m-PdNFs-G4-MBs nanocomposites to obtain signal probes (m-PdNFs-S2-G4-MBs), which can offer electrochemical signals for detection. In the absence of DA, the anti-DA aptamer was hybridized with a single-stranded DNA (S1), which was further co-immobilized on the surface of the Ce-MOF nanocomposite-modified electrode. At this moment, the m-PdNFs-S2-G4-MBs signal probe cannot be assembled on the electrode surface via a DNA hybridization reaction. As a result, no electrochemical signal was obtained. In the presence of DA, DA preferentially binds with the aptamer to form the DA–aptamer complex, leading to the destruction of the aptamer-S1 double-strand structure. Therefore, the exposed S1 can be hybridized with the m-PdNFs-S2-G4-MBs signal probe to form double-stranded structures, generating significant electrochemical signals. Due to the extraordinary conductivity of Ce-MOF and the signal amplification of m-PdNFs-S2-G4-MBs, the designed electrochemical aptasensor achieved a low detection limit of 6 pM for DA detection. The practical DA detection result is highly consistent with the UPLC-MS method, suggesting that this aptasensor has high sensitivity and specificity.

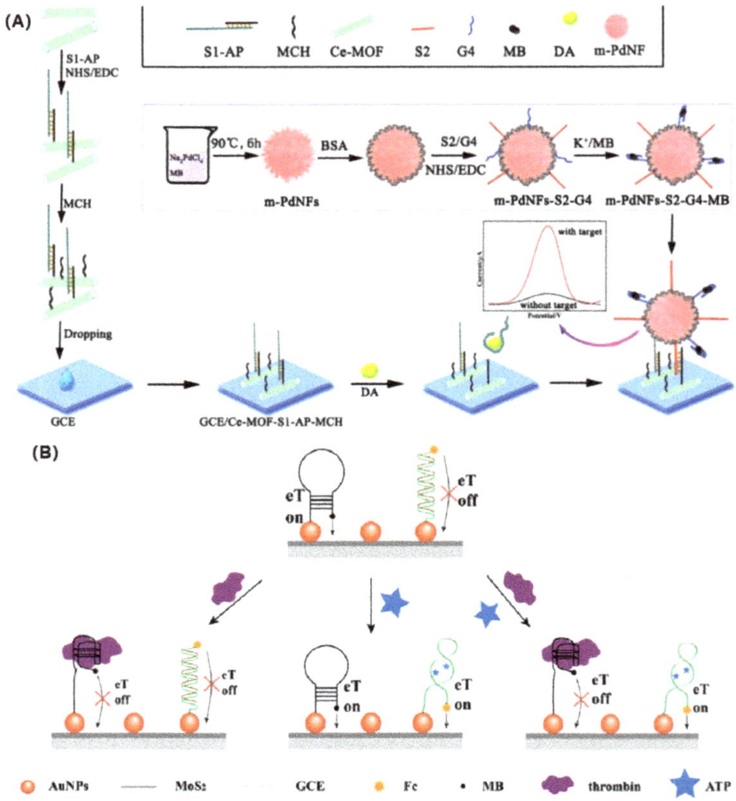

Figure 3. (**A**) Schematic diagram of a novel electrochemical aptasensor for dopamine detection based on MB-integrated m-PdNFs signal material. (Reproduced with permission from Ref. [79] Copyright 2022, Elsevier). (**B**) Schematic description of a dual-target electrochemical biosensor for ATP and thrombin detection based on DNA structural switching. (Reproduced with permission from Ref. [80] Copyright 2016, American Chemical Society).

Another example was given by Su et al. [80]. They constructed an electrochemical aptasensor for the simultaneous detection of ATP and thrombin (Figure 3B). They utilized the advantages of molybdenum disulfide (MoS_2) and noble metallic nanoparticles to synthesize MoS_2-based nanocomposites. After co-immobilizing anti-ATP and anti-thrombin aptamers on the surface of the MoS_2-based nanocomposite-modified electrode, a dual-target electrochemical aptasensor was constructed. Once the binding with ATP and thrombin is completed, the structures of two aptamers are changed, resulting in the labeled ferrocene (Fc) and methylene blue (MB) close to and far away from the electrode surface. Correspondingly, the electrochemical signal increased and decreased with the addition of ATP and thrombin, respectively. In other words, the designed aptasensor featured both "signal-on" and "signal-off" sensing mechanisms for the detection of ATP and thrombin, respectively. Under optimal conditions, this proposed aptasensor can simultaneously determine values as low as 0.74 nM ATP and 0.0012 nM thrombin. Furthermore, they employed the aptasensor to construct an "AND" logic gate platform for ATP and thrombin detection, which demonstrated a general methodology for the advancement of the highly sensitive and selective detection of various aptamer-specific binding targets.

3.4. Detection of Drug and Antibiotic Molecules

As we know, drugs and antibiotics greatly ensure the health of living organisms. However, the abuse of drugs and antibiotics, as well as their concentration-related toxicity, can endanger human health and even lives. Therefore, accurate, fast, and low-cost detection of drugs and antibiotics plays a vital role in disease diagnosis, treatment evaluation, and monitoring after operation.

An electrochemical aptasensor is also a powerful tool for the detection of drugs. For example, Chung et al. chose a de novo aptamer to develop an electrochemical aptasensor for the analysis of the anti-epileptic drug carbamazepine (CBZ) (Figure 4) [81]. The selected aptamer specifically binds with CBZ (K_d < 12 nM) to form a G-quadruplex structure, leading to the labeled MB close to the electrode surface. Correspondingly, the electron transfer between MB and the electrode is facilitated, increasing the electrochemical signal. It should be noted that the designed aptasensor has two functional modules: a 30 min assay for routine monitoring and a 5 min assay for rapid emergency testing. Under optimal conditions, this aptasensor exhibits a wide dynamic range (10 nM to 100 μM) and low detection limits of 1.25 nM and 1.82 nM for 5 min and 30 min analysis, respectively. The practical testing in the pricking blood sample (<50 μL) for CBZ determination proved the clinical applicability of this electrochemical aptasensor, suggesting that it has promising potential in point-of-care testing of personalized CBZ dosing. Using a similar sensing mechanism, cocaine, diclofenac, and amphetamine can be sensitively and selectively detected by electrochemical aptasensors without introducing other signal indicators [82–84].

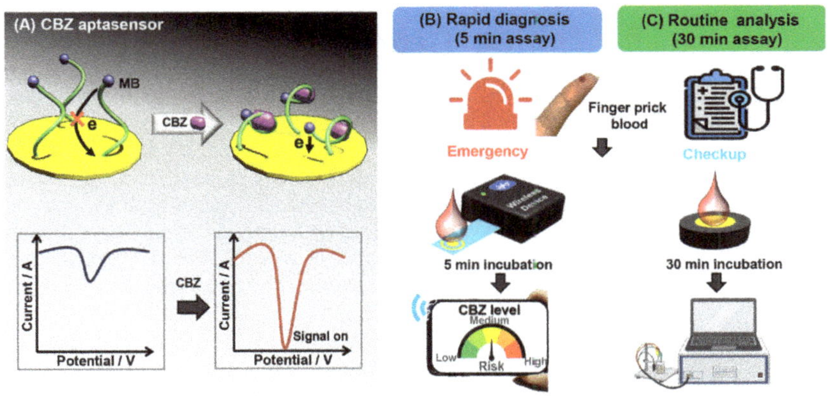

Figure 4. (**A**) Sensing schematic of the electrochemical aptasensor and its workflow for (**B**) 5 min rapid assay and (**C**) 30 min routine analysis for CBZ analysis. (Reproduced with permission from Ref. [81] Copyright 2022, American Chemical Society).

Besides drugs, electrochemical aptasensors are also widely employed to detect antibiotics [85]. A typical example is offered by Sun's group. They designed an electrochemical aptasensor for the ultrasensitive detection of neomycin by using a dual-signal amplification strategy [86]. In this sensing strategy, Fc-decorated multi-walled carbon nanotubes (MWCNTs) and the silica hybridized mesoporous ferroferric oxide nanoparticles ($Fe_3O_4@SiO_2$) were co-immobilized on the gold electrode surface to improve the electronic signal, which was used to load more RNA aptamers (Apt1) and facilitate electron transfer. Another RNA aptamer (Apt2) was assembled on the surface of gold nanoparticles to form a nanoprobe for signal amplification. In the presence of neomycin, a classical sandwiched structure was formed on the electrode surface, generating an obvious electrochemical signal. Due to the synergistic effect of different nanomaterials, this electrochemical aptasensor can detect 0.759 nM neomycin with high specificity and selectivity. Moreover, this aptasensor can efficiently determine neomycin in milk. Similarly, Emaminejad's, Wang's, Du's, and

Roushani's groups developed a series of electrochemical aptasensors for the detection of tobramycin [87], kanamycin [88], enrofloxacin [89], and streptomycin [90], respectively.

3.5. Detection of Proteins

Immunosensors coupled with different detection techniques have become universal tools for the detection of proteins, which have been widely applied in early diagnosis and clinical treatment of diseases [91]. However, antibody-based immunosensors have some disadvantages including the limited source of antibodies, high cost, disordered assembly on sensing interfaces, and the loss of biological activity [92], which greatly affect their detection performance and limit their further application. Therefore, electrochemical aptasensors have attracted more and more attention in protein detection due to their outstanding advantages, such as low cost, easy preparation, ordered assembly on the sensing interfaces, and anti-environment disturbance ability.

Highly efficient analytical approaches are of great significance for patients to realize early and rapid diagnosis, receive appropriate and timely treatment, and reduce mortality. Currently, C-reactive protein (CRP) has been identified as an angiocardiopathy special indicator for early diagnosis of cardiovascular disease with a high incidence rate [93]. Li and his colleagues have constructed an enzyme-free sandwich-type aptasensor for sensitive CRP detection based on $Mn_3(PO_4)_2$/CRP aptamer biomineralized nanosheets and polydopamine/Au nanospheres (PDA/Au NPs) (Figure 5A) [94]. The designed $Mn_3(PO_4)_2$/CRP aptamer biomineralized nanosheets are not only used as recognition units for capturing target CRP, but also serve as signal enhancers to improve electrochemical responses. Coupled with the signal amplification nanoprobe of PDA/Au NPs, the aptasensor exhibits excellent detection performance with a wide linear range (1 pg/mL–1 ng/mL) and a low detection limit (0.37 pg/mL). The repeatability, specificity, and reliability of this aptasensor are superior to that of the traditional immunoturbidimetry assay, suggesting that it has great prospects in clinical diagnostic applications of cardiovascular disease. Vascular endothelial growth factor (VEGF) is a key biomarker of diabetic retinopathy (DR), which can be also determined by electrochemical aptasensors [95]. For instance, Mei et al. designed a noninvasive electrochemical aptasensor to detect VEGF in tears [96]. They developed a cascade signal amplification strategy by combining a hybridization chain reaction (HCR) with CeO_2 nanoparticles to improve the analytical performance (Figure 5B). The introduced CeO_2 nanoparticles can efficiently catalyze H_2O_2 to generate electrochemical signals. Under optimal conditions, the aptasensor can detect 1 fg/mL~0.1 ng/mL VEGF with a detection limit of 0.27 fg/mL. Based on the acceptable stability and reproducibility, the proposed aptasensor can accurately analyze VEGF in tears comparable to the ELISA method, indicating that it is a potential tool for early diagnosis and monitoring of DR. Similarly, Li's, Hosseini's, and Gao's teams have constructed various electrochemical sensing platforms coupled with aptamers and a signal amplification strategy for different protein biomarkers' detection, including thrombin [97], tumor necrosis factor α [98], and tau protein [99], respectively.

Electrochemical aptasensors have also been widely used to monitor infectious diseases, such as SARS-CoV-2 [100], malaria [101], tuberculosis [102], etc. As shown in Figure 5C, Idili et al. have developed a simple electrochemical aptasensor for the detection of SARS CoV-2 S protein in undiluted samples (serum and artificial saliva) [103]. Typically, the added SARS CoV-2 S protein induced the change in the methylene blue derivative-labeled aptamer conformation, resulting in the electrochemical indicator close to the electrode surface. Correspondingly, a concentration-related electrochemical signal is generated for the qualitative and quantitative detection of SARS CoV-2 S protein. This electrochemical aptasensor is simple (single step) and fast (detection time < 5 min), suggesting that it is an ideal device for COVID-19 monitoring. Another typical example was offered by Rahmati et al. They constructed a label-free electrochemical aptasensor for the analysis of SARS-CoV-2 spike glycoprotein based on a copper hydroxide nanorod-modified screen-printed carbon electrode (SPCE) [104]. The obtained SARS-CoV-2 spike glycoprotein–

aptamer complex blocked the electron transfer, leading to a decrease in the electrochemical signal. Based on this phenomenon, the designed aptasensor can efficiently detect SARS-CoV-2 spike glycoprotein in buffer, labeled saliva samples, and actual clinical samples. It is noted that the analytical performance of this aptasensor is comparable to the PCR results, proving that it has great potential in rapid and on-site COVID-19 monitoring.

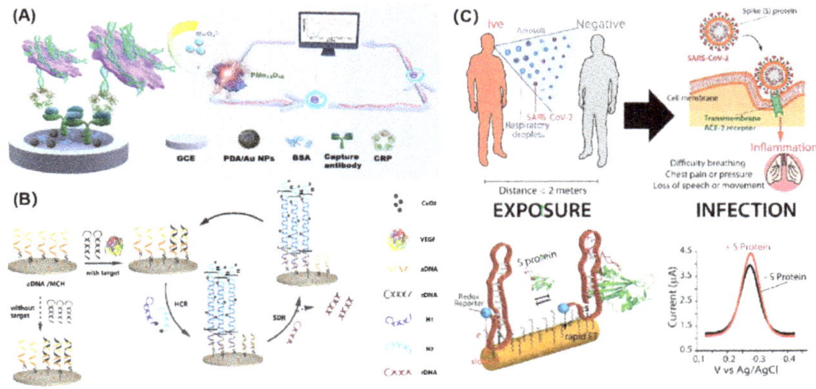

Figure 5. (**A**) The preparation of the $Mn_3(PO_4)_2$/aptamer hybrid nanosheet−based electrochemical aptasensor for CRP detection. (Reproduced with permission from Ref. [94] Copyright 2021, Elsevier). (**B**) Schematic diagram of electrochemical aptasensor coupled with signal amplification strategy for sensitive detection of VEGF. (Reproduced with permission from Ref. [96] Copyright 2021, Elsevier). (**C**) Schematic of electrochemical aptasensor for the rapid and efficient detection of the SARS-COV-2 spike protein. (Reproduced with permission from Ref. [103] Copyright 2021, American Chemical Society).

3.6. Detection of Exosomes

As nanoscale extracellular vesicles, exosomes play an important role in promoting tumor growth, migration, invasion, and information transfer between cells [105]. Therefore, exosomes are considered as excellent biomarkers for early diagnosis of cancer [106]. However, it is still a challenge to sensitively and selectively detect exosomes due to their low content in biological fluids during the early stages of cancer [107,108]. In recent years, electrochemical aptasensors have been gradually employed to detect exosomes due to their advantages. Coupled with signal amplification strategies, the analytical performance of electrochemical aptasensors can meet the need for exosome detection [109].

Zhang's group developed an electrochemical aptasensor for exosome detection based on a microelectrode and HCR amplification strategy [110]. They introduced EpCAM aptamers to selectively recognize and capture EpCAM-positive cancerous exosomes. Due to the HCR reaction, a large number of avidin–horseradish peroxidase (HRP) molecules were assembled on the microelectrode surface, which can catalyze the 3,3′,5,5′-tetramethylbenzidine (TMB)+H_2O_2 reaction strategy to generate a large electrochemical signal. Due to the synergetic effect of the enzyme catalytic reaction, HCR reaction, and microelectrode, this aptasensor has a detection limit as low as 5×10^2 exosomes per milliliter. Moreover, this aptasensor can successfully detect cancerous extracellular vesicles in serum samples of early- and late-stage lung cancer patients. Chang et al. used programmed cell death ligand 1 protein-positive (PD-L1+) exosome as a biomarker to evaluate the diagnosis of non-small cell lung cancer (NSCLC) [111]. They integrated the specific recognition ability of the aptamer, the excellent peroxidase-like catalytic activity of palladium–copper–boron alloy microporous nanospheres (PdCuB MNs), and the high conductivity of Au@$CuCl_2$ nanowires (NWs) to construct an electrochemical aptasensor for detecting PD-L1+ exosomes (Figure 6A). In the presence of PD-L1$^+$ exosomes, a classical sandwiched structure was formed on the electrode surface. The introduced PdCuB

MNs efficiently catalyze H_2O_2 to generate a significant electrochemical signal. According to this result, the aptasensor can detect 1×10^2–1×10^8 particles/mL exosomes with a low detection limit of 36 particles/mL. Results obtained from clinical samples suggested that this aptasensor has a potential application in the early diagnosis of NSCLC.

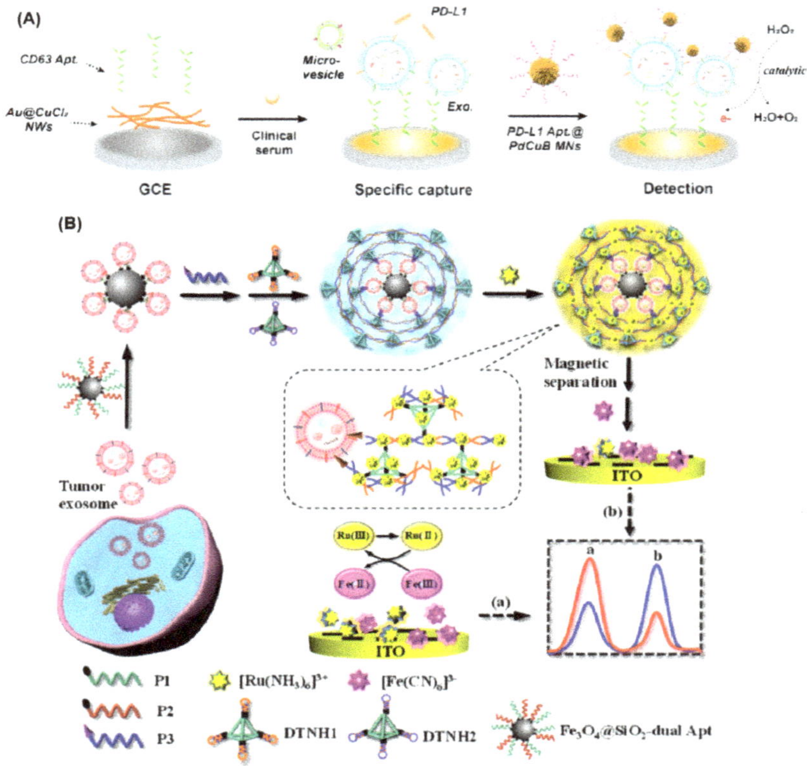

Figure 6. (**A**) Schematic illustration of a nanomaterial-based electrochemical aptasensor for PD-L1⁺ exosome analysis. (Reproduced with permission from Ref. [111] Copyright 2023, Springer Nature). (**B**) Schematic illustration of a ratiometric immobilization-free electrochemical aptasensor for precise capture and direct quantification of (a) in the absence and (b) in the presence of the tumor exosomes. (Reproduced with permission from Ref. [112] Copyright 2021, American Chemical Society).

Besides captured probes' immobilized mode, Li et al. developed a dual-aptamer recognition system for precise capture and direct analysis of tumor-specific exosomes without immobilization [112]. They assembled both CD63 and mucin 1 (MUC1) aptamers onto the surface of amino-functionalized Fe_3O_4@SiO_2 nanoparticles to construct dual-aptamer-modified nanoprobes (Fe_3O_4@SiO_2-dual Apt). As shown in Figure 6B, the dual-aptamer-modified nanoprobes can specifically capture target exosomes due to the recognition ability of two aptamers. After magnetic separation, cholesterol-labeled DNA probes (P3) can anchor on the surface of exosomes and activate the hyperbranched hybridization chain reaction, generating a magnetic bead–exosome–hyperbranched DNA superstructure (Fe_3O_4@SiO_2-exosome-HDS). This superstructure can adsorb a large number of $Ru(NH_3)_6^{3+}$(Ru (III)) molecules, resulting in a few unbound Ru (III) remaining in the supernatant after magnetic separation. Therefore, the redox reaction between Ru (II) and $[Fe(CN)_6]^{3-}$ (Fe (III)) is inhibited, leading to a significant increase in $I_{Fe(III)}/I_{Ru(III)}$. The ratio of the electrochemical signal was related to exosomes, which can be used to quantitatively and qualitatively analyze target exosomes. Consequently, this ratiometric dual-signal electrochemical aptasensor exhibits high reliability, accuracy, and easy manipulation, which can sensitively detect tumor exosomes in complex environments and human serum samples.

3.7. Detection of Tumor Cells

It is well known that the incidence rate and mortality of cancer have increased year by year, which has brought a huge threat to human health. Thus, the identification and detection of cancer cells in patients' tumor lesions and circulating blood is crucial for early clinical diagnosis, toxicity monitoring, and public health protection [113,114]. However, conventional clinical, imaging, and other diagnostic methods often have weak sensitivity and specificity due to the complex etiology of tumors, leading to misdiagnosis and missed diagnosis [115,116]. Although a variety of technologies including PCR-based methods [117], cell counting methods [118], and fluorescence determination [119] have been presented, many disadvantages of these methods have limited their wide application, such as expensive fluorescent labeling reagents, cumbersome processes, and the need for advanced instruments and professionals. Therefore, it is necessary to develop reliable, cost-effective, sensitive, and easy to be clinically popularized detection tools for high-precision diagnosis of cancer cells.

Electrochemical aptasensors have received widespread attention in the high-performance detection of tumor cells due to their significant advantages, such as high sensitivity, simplicity, rapid response, reusability, and low cost [120]. An impressive example was given by Cai et al. [121]. As shown in Figure 7A, a double-stranded DNA formed by a biotin-labeled aptamer and DNA walker was assembled on the surface of the streptavidin-modified magnetic microsphere to construct nanoprobe, which was used to recognize and capture target cancer cells. Once capturing target breast cancer MCF-7 cells, an MCF-7–aptamer complex was formed and a DNA walker was released from the surface of the nanoprobe. The released DNA walker (DNAzyme) repeatedly cleaved D-RNA with the help of Mg^{2+} to produce a large number of short nucleic acids, leading to a great decrease in electrochemical signals. With DNAzyme-assisted signal amplification, the designed aptasensor can analyze as low as 47 cells/mL MCF-7 cells. G-quadruplex DNAzyme is also widely used for signal amplification to construct electrochemical aptasensors for cancer cell detection. Chen and co-workers designed a super sandwich G-quadruplex DNAzyme to construct an electrochemical aptasensor for the highly sensitive detection of cancer cells [122]. In Figure 7B, amino-modified aptamers (S1) and their complementary DNA sequence (S2) were immobilized onto the surface of magnetic beads (Fe_3O_4 NPs) to form S2/S1/Fe_3O_4 conjugates, which were used to specifically capture K562 cells. Once encountering K562 cells, the double-stranded S1–S2 structure was dissociated due to the stronger affinity between K562 cells and S1. As a result, S2 was released from the surface of Fe_3O_4 NPs and hybridized with a capture probe (S3) on the electrode surface. The S2–S3 structure triggered a hybridization cascade between S4 and S5, resulting in the super sandwich structure. In the presence of Hemin, a large number of G-quadruplex DNAzymes were formed on the electrode surface. Finally, the G-quadruplex DNAzyme catalyzes the reduction of H_2O_2 by MB, leading to a dramatically amplified electrochemical signal. Under optimal conditions, the detection limit of the as-fabricated aptasensor was estimated to be 14 cells/mL K562 cells.

Electrochemical aptasensors are also extensively employed to detect circulating tumor cells (CTCs), which are of great significance for the clinical diagnosis, metastasis, and prognosis of tumors [123,124]. Shaegh et al. reported a novel microfluidic electrochemical aptasensor by coupling a microchip with a screen-printed gold electrode for detecting A549 cells, which is considered as a CTC model [125]. As shown in Figure 7C, the added A549 cells were specifically bound with the aptamer to form a complex on the electrode surface, which blocked the electron transfer between $[Fe(CN)_6]^{3-/4-}$ and the electrode surface. As a result, the electrochemical signal is decreased with the increasing concentration of A549 cells. According to this phenomenon, this biosensor can sensitively detect 50~5 × 10^5 cells/mL A549 cells with a low detection limit of 14 cells/mL. More importantly, this microfluidic aptasensor can effectively determine A549 cells in complex matrices (such as human serum), providing that this sensing platform can be used to detect low-abundance biomarkers obtained from liquid biopsies.

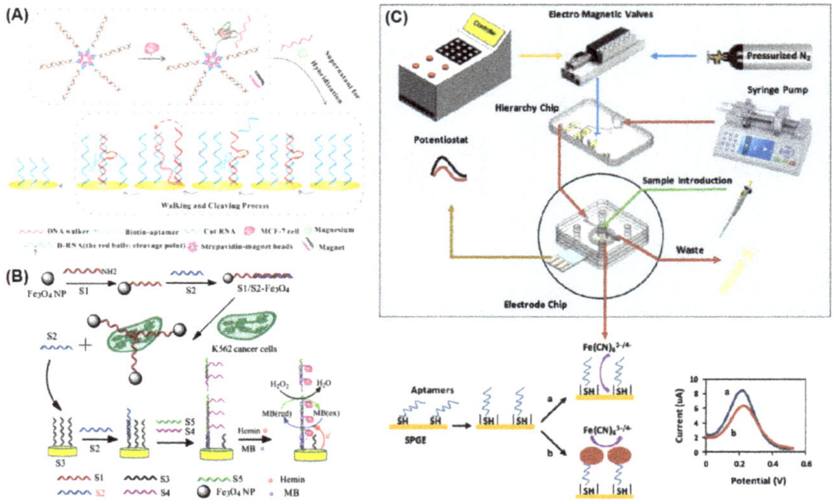

Figure 7. (**A**) Schematic illustration of the electrochemical aptasensor for MCF-7 cells' detection based on a DNA walker signal amplification strategy. (Reproduced with permission from Ref. [121] Copyright 2016, Elsevier). (**B**) Schematic representation of signal-amplified G-quadruplex DNAzyme for ultrasensitive detection of K562 cells. (Reproduced with permission from Ref. [122] Copyright 2015, Elsevier). (**C**) Schematic illustration of microfluidic electrochemical aptasensor and its electrochemical signal (a) in the absence of and (b) in the presence of A549 cells. (Reproduced with permission from Ref. [125] Copyright 2023, Elsevier).

3.8. Detection of Microbial Pathogens

Since the COVID-19 pandemic, infectious diseases have gained people's attention. Accurate and fast detection of pathogenic microorganisms is of great significance in controlling and preventing microbial infectious diseases. At present, direct smear microscopy, isolation and culture, serological reaction, and PCR are the main methods for the rapid identification and detection of pathogenic microorganisms [126]. However, some limitations of these methods exist, such as being time-consuming, requiring professional personnel and instruments, having low sensitivity, and so on. Therefore, it is urgent to develop simple, rapid, sensitive, and on-site methods for the detection of microbial pathogens.

To meet these requirements, a series of electrochemical aptasensors have been developed for the high-performance analysis of microbial pathogens due to their outstanding advantages. Bian et al. have constructed an electrochemical aptasensor for highly selective and ultrasensitive detection of single-cell levels of bacteria [127]. In Figure 8A, Au NPs modified with aptamers and 6-(Ferrocenyl) hexanethiol molecules (Au@Fc-Apt) and bacteria-imprinted polymer films (BIFs) were employed as the signal nanoprobe and the capture probe, respectively. In the presence of *Staphylococcus aureus* (*S. aureus*), a classical sandwiched structure was formed on the electrode surface. As a result, an obvious electrochemical signal is generated from the assembled Au@Fc-Apt signal nanoprobe, which can be used to qualitatively and quantitatively detect bacteria. In view of this, the aptasensor could detect single-cell level and 10 CFU mL^{-1} *S. aureus* in an ideal buffer and a complex milk sample, respectively. Moreover, this proposed aptasensor can efficiently distinguish *S. aureus* from the multiple coexisting interfering bacteria, suggesting that this aptasensor has outstanding selectivity.

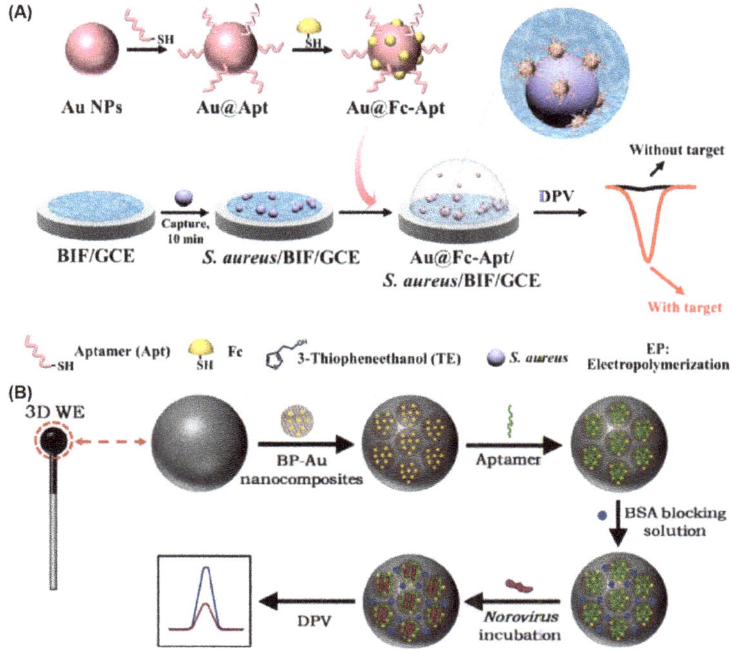

Figure 8. (**A**) Schematic preparation of the electrochemical aptasensor for highly selective and ultrasensitive detection of *S. aureus*. (Reproduced with permission from Ref. [127] Copyright 2023, American Chemical Society). (**B**) Workflow scheme for the electrode functionalization and aptasensing of norovirus based on BP-AuNCs. (Reproduced with permission from Ref. [128] Copyright 2022, Elsevier).

Besides bacteria detection, electrochemical aptasensors have also been widely used to detect viruses. For fast identification and detection of norovirus, Jiang et al. developed a 3D electrochemical aptasensor for sensitive detection of norovirus based on a movable spherical working electrode (WE) decorated with phosphorene–gold nanocomposites (BP-AuNCs) [128] (Figure 8B). The design of a movable spherical WE is beneficial for increasing surface area, simplifying sampling, and avoiding cross-contamination. The added norovirus specifically binds with aptamer and induces the change in aptamer structure, leading to the decrease in the electrochemical signal. As expected, the 3D electrochemical aptasensor can detect 1 ng/mL–10 μg/mL norovirus with a detection limit of 0.28 ng/mL. The recovery of this aptasensor for norovirus detection in spiked oyster samples ranges from 97.2% to 103.7%, suggesting that this aptasensor has a potential practical application. The proposed aptasensor for norovirus detection provides a facile, low-cost, highly sensitive, and selective platform, which is potentially applied in the fields of food safety and clinical diagnosis. Besides norovirus, avian influenza virus [129], SARS-CoV-2 virus [130], and dengue virus [131] have also been analyzed by electrochemical aptasensors.

4. Prospects and Challenges of Electrochemical Aptasensors

Nowadays, electrochemical aptasensors have achieved great advances in environmental monitoring, drug safety, biochemical analysis, disease diagnosis, and public safety due to their outstanding advantages of fast response, simple preparation, easy design, high specificity for chemical and biological molecules, excellent chemical stability, and low cost. Inspired by the above exciting progresses, integrating electrochemical aptasensors into portable, wearable, and implantable devices has become a popular trend for their application in POCT and real-time monitoring fields, which has the potential to completely change diagnosis and health management, especially in application scenarios for home users or remote areas where it is difficult to receive hospital nursing care [132]. In addition, with the fast development of nanotechnologies and miniaturization technologies, as well as the urgent need for various detection strategies, electrochemical aptasensors have broad applications and bright prospects in the future.

For achieving these goals, some challenges of electrochemical aptasensors should be solved: (1) For broad applications, more aptamers should be selected, which play a vital role in the detection of various chemical and biological molecules. Nowadays, hundreds of aptamers have been selected to specifically recognize target chemical/biological molecules. However, it is not enough for sensing/biosensing applications in many fields. Therefore, the selection of more aptamers with high binding affinity can promote the development of electrochemical aptasensors. (2) The integration of cutting-edge innovative technologies (e.g., DNAzyme, CRISPR, microfluidic chips, etc.) into electrochemical aptasensors is very prospective to design novel aptasensor devices, which can efficiently enlarge their application [133]. (3) The construction of suitable signal amplification strategies is a vital effect factor to construct high-performance electrochemical aptasensors. Up to now, several signal amplification strategies have been used to construct electrochemical aptasensors, such as loop-mediated isothermal amplification, rolling circle amplification (RCA), HCR, catalyzed hairpin assembly (CHA), enzyme-assisted signal amplification, and nanoprobe-based signal amplification, which can efficiently amplify the detection signal and reach lower detection limits [134]. Therefore, how to choose a signal amplification strategy greatly depends on the analytes and application scenarios. (4) For the POCT and real-time monitoring applications, it is still a challenge to construct a portable and intelligent electrochemical aptasensor by combining the detection instruments and signal communication technologies. To quickly keep up with global health issues, electrochemical aptasensors should gradually being clinically transformed and adapted to market demands. Therefore, the development of home-care, low-cost, real-time, and portable devices has become a popular and attractive research field [135,136]. In a word, electrochemical aptasensors will be an ideal sensing platform for sensitive, selective, real-time, and on-site detection of target molecules in complex environments.

Author Contributions: Conceptualization, R.Y. and S.S.; methodology, R.Y. and J.C.; formal analysis, J.C. and H.M.; investigation, R.Y., H.M., and Y.L.; writing—original draft preparation, R.Y.; writing—review and editing, S.S. and Y.L.; supervision, S.S. and L.W.; project administration, S.S. and L.W.; funding acquisition, L.W. and S.S. All authors have read and agreed to the published version of the manuscript.

Funding: This work is supported by the Leading-edge Technology Program of Jiangsu Natural Science Foundation (Grant No. BK20212012), the "Belt and Road" Innovation Cooperation Project of Jiangsu (Grant No. BZ2022011), the "Six Talents Peak" Foundation of Jiangsu Province (Grant No. SWYY-046), the Natural Science Foundation of the Jiangsu Higher Education Institutions (Grant No. 22KJA150003), and the Natural Science Research Start-up Foundation of Recruiting Talents of Nanjing University of Posts and Telecommunications (Grant No. NY222077).

Institutional Review Board Statement: Not applicable.

Informed Consent Statement: Not applicable.

Data Availability Statement: Not applicable.

Conflicts of Interest: The authors declare no conflict of interest.

References

1. Umapathi, R.; Ghoreishian, S.M.; Sonwal, S.; Rani, G.M.; Huh, Y.S. Portable electrochemical sensing methodologies for on-site detection of pesticide residues in fruits and vegetables. *Coord. Chem. Rev.* **2022**, *453*, 214305. [CrossRef]
2. Baranwal, J.; Barse, B.; Gatto, G.; Broncova, G.; Kumar, A. Electrochemical sensors and their applications: A review. *Chemosensors* **2022**, *10*, 363. [CrossRef]

3. Gong, L.; Feng, L.; Zheng, Y.; Luo, Y.; Zhu, D.; Chao, J.; Su, S.; Wang, L. Molybdenum disulfide-based nanoprobes: Preparation and sensing application. *Biosensors* **2022**, *12*, 87. [CrossRef] [PubMed]
4. Fabiani, L.; Saroglia, M.; Galata, G.; De Santis, R.; Fillo, S.; Luca, V.; Faggioni, G.; D'Amore, N.; Regalbuto, E.; Salvatori, P.; et al. Magnetic beads combined with carbon black -based screen-printed electrodes for COVID-19: A reliable and miniaturized electrochemical immunosensor for SARS-CoV-2 detection in saliva. *Biosens. Bioelectron.* **2021**, *171*, 112686. [CrossRef]
5. Feng, S.; Yan, M.; Xue, Y.; Huang, J.; Yang, X. Electrochemical immunosensor for cardiac troponin I detection based on covalent organic framework and enzyme-catalyzed signal amplification. *Anal. Chem.* **2021**, *93*, 13572–13579. [CrossRef]
6. Su, S.; Sun, Q.; Wan, L.; Gu, X.; Zhu, D.; Zhou, Y.; Chao, J.; Wang, L. Ultrasensitive analysis of carcinoembryonic antigen based on MoS2-based electrochemical immunosensor with triple signal amplification. *Biosens. Bioelectron.* **2019**, *140*, 77–82. [CrossRef]
7. Zhou, J.; Rossi, J. Aptamers as targeted therapeutics: Current potential and challenges. *Nat. Rev. Drug Discov.* **2017**, *16*, 181–202. [CrossRef]
8. Liu, L.; Wang, F.; Ge, Y.; Lo, P.K. Recent developments in aptasensors for diagnostic applications. *ACS Appl. Mater. Interfaces* **2021**, *13*, 9329–9358. [CrossRef]
9. He, Y.; He, G.; He, T. Specifically targeted transport of plasma membrane transporters: From potential mechanisms for regulating cell health or disease to applications. *Membranes* **2021**, *11*, 736. [CrossRef]
10. Atapour, A.; Khajehzadeh, H.; Shafie, M.; Abbasi, M.; Mosleh-Shirazi, S.; Kasaee, S.R.; Amani, A.M. Gold nanoparticle-based aptasensors: A promising perspective for early-stage detection of cancer biomarkers. *Mater. Today Commun.* **2022**, *30*, 103181. [CrossRef]
11. Parihar, A.; Singhal, A.; Kumar, N.; Khan, R.; Khan, M.A.; Srivastava, A.K. Next-generation intelligent Mxene-based electrochemical aptasensors for point-of-care cancer diagnostics. *Nano-Micro Lett.* **2022**, *14*, 100. [CrossRef]
12. Zaimbashi, R.; Tajik, S.; Beitollahi, H.; Torkzadeh-Mahani, M. Fabrication of a novel and ultrasensitive label-free electrochemical aptasensor based on gold nanostructure for detection of homocysteine. *Biosensors* **2023**, *13*, 244. [CrossRef]
13. Rozenblum, G.T.; Pollitzer, I.G.; Radrizzani, M. Challenges in electrochemical aptasensors and current sensing architectures using flat gold surfaces. *Chemosensors* **2019**, *7*, 57. [CrossRef]
14. Rahman, M.M. Progress in Electrochemical biosensing of SARS-CoV-2 Virus for COVID-19 management. *Chemosensors* **2022**, *10*, 287. [CrossRef]
15. Hou, Y.; Long, N.; Xu, Q.; Li, Y.; Song, P.; Yang, M.; Wang, J.; Zhou, L.; Sheng, P.; Kong, W. Development of a Nafion-MWCNTs and in-situ generated Au nanopopcorns dual-amplification electrochemical aptasensor for ultrasensitive detection of OTA. *Food Chem.* **2023**, *403*, 134375. [CrossRef]
16. Zhang, Z.; Karimi-Maleh, H. Label-free electrochemical aptasensor based on gold nanoparticles/titanium carbide MXene for lead detection with its reduction peak as index signal. *Adv. Compos. Hybrid Mater.* **2023**, *6*, 68. [CrossRef]
17. Su, S.; Sun, Q.; Gu, X.; Xu, Y.; Shen, J.; Zhu, D.; Chao, J.; Fan, C.; Wang, L. Two-dimensional nanomaterials for biosensing applications. *TrAC, Trends Anal. Chem.* **2019**, *119*, 115610. [CrossRef]
18. Sohouli, E.; Ghalkhani, M.; Zargar, T.; Joseph, Y.; Rahimi-Nasrabadi, M.; Ahmadi, F.; Plonska-Brzezinska, M.E.; Ehrlich, H. A new electrochemical aptasensor based on gold/nitrogen-doped carbon nano-onions for the detection of *Staphylococcus aureus*. *Electrochim. Acta* **2022**, *403*, 139633. [CrossRef]
19. Yang, L.; Liu, X.; Li, L.; Zhang, S.; Zheng, H.; Tang, Y.; Ju, H. A visible light photoelectrochemical sandwich aptasensor for adenosine triphosphate based on $MgIn_2S_4$-TiO_2 nanoarray heterojunction. *Biosens. Bioelectron.* **2019**, *142*, 111487. [CrossRef]
20. Mattarozzi, M.; Toma, L.; Bertucci, A.; Giannetto, M.; Careri, M. Aptamer-based assays: Strategies in the use of aptamers conjugated to magnetic micro- and nanobeads as recognition elements in food control. *Anal. Bioanal. Chem.* **2022**, *414*, 63–74. [CrossRef] [PubMed]
21. Shaver, A.; Arroyo-Curras, N. The challenge of long-term stability for nucleic acid-based electrochemical sensors. *Curr. Opin. Electrochem.* **2022**, *32*, 100902. [CrossRef] [PubMed]
22. Liu, Z.; Galli, F.; Janssen, K.G.H.; Jiang, L.; van der Linden, H.J.; de Geus, D.C.; Voskamp, P.; Kuil, M.E.; Olsthoorn, R.C.L.; Oosterkamp, T.H.; et al. Stable single-walled carbon nanotube-streptavidin complex for biorecognition. *J. Phys. Chem. C* **2010**, *114*, 4345–4352. [CrossRef]
23. Villalonga, A.; Pérez-Calabuig, A.M.; Villalonga, R. Electrochemical biosensors based on nucleic acid aptamers. *Anal. Bioanal. Chem.* **2020**, *412*, 55–72. [CrossRef] [PubMed]
24. Fortunati, S.; Rozzi, A.; Curti, F.; Giannetto, M.; Corradini, R.; Careri, M. Novel amperometric genosensor based on peptide nucleic acid (PNA) probes immobilized on carbon nanotubes-screen printed electrodes for the determination of trace levels of non-amplified DNA in genetically modified (GM) soy. *Biosens. Bioelectron.* **2019**, *129*, 7–14. [CrossRef] [PubMed]
25. Wang, L.; Wang, H.; Huang, S.; Wu, F.; Niu, X. Electrochemical sensor for detecting streptomycin in milk based on label-free aptamer chain and magnetic adsorption. *Food Chem.* **2023**, *403*, 134399. [CrossRef] [PubMed]
26. Acquah, C.; Danquah, M.K.; Yon, J.L.S.; Sidhu, A.; Ongkudon, C.M. A review on immobilised aptamers for high throughput biomolecular detection and screening. *Anal. Chim. Acta* **2015**, *888*, 10–18. [CrossRef] [PubMed]
27. Meirinho, S.G.; Dias, L.G.; Peres, A.M.; Rodrigues, L.R. Voltammetric aptasensors for protein disease biomarkers detection: A review. *Biotechnol. Adv.* **2016**, *34*, 941–953. [CrossRef]

28. Bharti, A.; Rana, S.; Dahiya, D.; Agnihotri, N.; Prabhakar, N. An electrochemical aptasensor for analysis of MUC1 using gold platinum bimetallic nanoparticles deposited carboxylated graphene oxide. *Anal. Chim. Acta* **2020**, *1097*, 186–195. [CrossRef] [PubMed]
29. Lucarelli, F.; Marrazza, G.; Turner, A.P.F.; Mascini, M. Carbon and gold electrodes as electrochemical transducers for DNA hybridisation sensors. *Biosens. Bioelectron.* **2004**, *19*, 515–530. [CrossRef]
30. Sassolas, A.; Leca-Bouvier, B.D.; Blum, L.J. DNA biosensors and microarrays. *Chem. Rev.* **2008**, *108*, 109–139. [CrossRef]
31. Jo, H.; Her, J.; Lee, H.; Shim, Y.B.; Ban, C. Highly sensitive amperometric detection of cardiac troponin I using sandwich aptamers and screen-printed carbon electrodes. *Talanta* **2017**, *165*, 442–448. [CrossRef] [PubMed]
32. Zhou, L.; Wang, M.H.; Wang, J.P.; Ye, Z.Z. Application of biosensor surface immobilization methods for aptamers. *Chin. J. Anal. Chem.* **2011**, *39*, 432–438. [CrossRef]
33. Diba, F.S.; Kim, S.; Lee, H.J. Amperometric bioaffinity sensing platform for avian influenza virus proteins with aptamer modified gold nanoparticles on carbon chips. *Biosens. Bioelectron.* **2015**, *72*, 355–361. [CrossRef]
34. Palecek, E.; Bartosik, M. Electrochemistry of nucleic acids. *Chem. Rev.* **2012**, *112*, 3427–3481. [CrossRef] [PubMed]
35. Song, Y.; Xu, M.; Liu, X.; Li, Z.; Wang, C.; Jia, Q.; Zhang, Z.; Du, M. A label-free enrofloxacin electrochemical aptasensor constructed by a semiconducting CoNi-based metal-organic framework (MOF). *Electrochim. Acta* **2021**, *368*, 137609. [CrossRef]
36. Tang, J.; Tang, D.; Niessner, R.; Knopp, D.; Chen, G. Hierarchical dendritic gold microstructure-based aptasensor for ultrasensitive electrochemical detection of thrombin using functionalized mesoporous silica nanospheres as signal tags. *Anal. Chim. Acta* **2012**, *720*, 1–8. [CrossRef]
37. Zhang, Y.; Xia, J.; Zhang, F.; Wang, Z.; Liu, Q. A dual-channel homogeneous aptasensor combining colorimetric with electrochemical strategy for thrombin. *Biosens. Bioelectron.* **2018**, *120*, 15–21. [CrossRef]
38. Qin, X.; Yin, Y.; Yu, H.; Guo, W.; Pei, M. A novel signal amplification strategy of an electrochemical aptasensor for kanamycin, based on thionine functionalized graphene and hierarchical nanoporous PtCu. *Biosens. Bioelectron.* **2016**, *77*, 752–758. [CrossRef]
39. Zhao, P.; Zheng, J.; Liang, Y.; Tian, F.; Peng, L.; Huo, D.; Hou, C. Functionalized carbon nanotube-decorated Mxene nanosheet-enabled microfluidic electrochemical aptasensor for carcinoembryonic antigen determination. *ACS Sustain. Chem. Eng.* **2021**, *9*, 15386–15393. [CrossRef]
40. Zhou, Y.; Li, F.; Wu, H.; Chen, Y.; Yin, H.; Ai, S.; Wang, J. Electrochemical aptasensing strategy for kanamycin detection based on target-triggered single-strand DNA adsorption on MoS_2 nanosheets and enzymatic signal amplification. *Sens. Actuators B* **2019**, *296*, 126664. [CrossRef]
41. Gu, C.; Yang, L.; Wang, M.; Zhou, N.; He, L.; Zhang, Z.; Du, M. A bimetallic (Cu-Co) Prussian Blue analogue loaded with gold nanoparticles for impedimetric aptasensing of ochratoxin A. *Microchim. Acta* **2019**, *186*, 343. [CrossRef]
42. Taskinen, B.; Zauner, D.; Lehtonen, S.I.; Koskinen, M.; Thomson, C.; Kahkonen, N.; Kukkurainen, S.; Maatta, J.A.E.; Ihalainen, T.O.; Kulomaa, M.S.; et al. Switchavidin: Reversible biotin-avidin-biotin bridges with high affinity and specificity. *Bioconjug. Chem.* **2014**, *25*, 2233–2243. [CrossRef]
43. Kim, D.S.; Park, H.J.; Park, J.E.; Shin, J.K.; Kang, S.W.; Seo, H.I.; Lim, G. MOSFET-type biosensor for detection of streptavidin-biotin protein complexes. *Sens. Mater.* **2005**, *17*, 259–268.
44. Balamurugan, S.; Obubuafo, A.; Soper, S.A.; Spivak, D.A. Surface immobilization methods for aptamer diagnostic applications. *Anal. Bioanal. Chem.* **2008**, *390*, 1009–1021. [CrossRef]
45. Thapa, K.; Liu, W.; Wang, R. Nucleic acid-based electrochemical biosensor: Recent advances in probe immobilization and signal amplification strategies. *Wiley Interdiscip. Rev. Nanomed. Nanobiotechnol.* **2022**, *14*, e1765. [CrossRef] [PubMed]
46. Du, Y.; Li, B.; Wang, E. "Fitting" makes "Sensing" simple: Label-free detection strategies based on nucleic acid aptamers. *Acc. Chem. Res.* **2013**, *46*, 203–213. [CrossRef] [PubMed]
47. Pilehvar, S.; Reinemann, C.; Bottari, F.; Vanderleyden, E.; Van Vlierberghe, S.; Blust, R.; Strehlitz, B.; De Wael, K. A joint action of aptamers and gold nanoparticles chemically trapped on a glassy carbon support for the electrochemical sensing of ofloxacin. *Sens. Actuators B* **2017**, *240*, 1024–1035. [CrossRef]
48. Ganguly, A.; Lin, K.C.; Muthukumar, S.; Prasad, S. Autonomous, real-time monitoring electrochemical aptasensor for circadian tracking of cortisol hormone in sub-microliter volumes of passively eluted human sweat. *ACS Sens.* **2021**, *6*, 63–72. [CrossRef] [PubMed]
49. Li, C.; Li, J.; Yang, X.; Gao, L.; Jing, L.; Ma, X. A label-free electrochemical aptasensor for sensitive myoglobin detection in meat. *Sens. Actuators B* **2017**, *242*, 1239–1245. [CrossRef]
50. Tu, C.; Dai, Y.; Xu, K.; Qi, M.; Wang, W.; Wu, L.; Wang, A. Determination of tetracycline in water and honey by iron (II, III)/aptamer-based magnetic solid-phase extraction with high-performance liquid chromatography analysis. *Anal. Lett.* **2019**, *52*, 1653–1669. [CrossRef]
51. Liu, S.; Xing, X.; Yu, J.; Lian, W.; Li, J.; Cui, M.; Huang, J. A novel label-free electrochemical aptasensor based on graphene-polyaniline composite film for dopamine determination. *Biosens. Bioelectron.* **2012**, *36*, 186–191. [CrossRef]
52. Guo, W.; Umar, A.; Algadi, H.; Albargi, H.; Ibrahim, A.A.; Cui, K.; Wang, L.; Pei, M.; Wang, Y. Design of a unique "ON/OFF" switch electrochemical aptasensor driven by the pH for the detection of Aflatoxin B1 in acid solutions based on titanium carbide/carboxylated graphene oxide- poly(4-vinyl pyridine)/aptamer composite. *Microchem. J.* **2021**, *169*, 106548. [CrossRef]

53. Sun, Y.; Jin, H.; Jiang, X.; Gui, R. Black phosphorus nanosheets adhering to thionine-doped 2D MOF as a smart aptasensor enabling accurate capture and ratiometric electrochemical detection of target microRNA. *Sens. Actuators B* **2020**, *309*, 127777. [CrossRef]
54. Wang, S.; Chen, S.; Shang, K.; Gao, X.; Wang, X. Sensitive electrochemical detection of cholesterol using a portable paper sensor based on the synergistic effect of cholesterol oxidase and nanoporous gold. *Int. J. Biol. Macromol.* **2021**, *189*, 356–362. [CrossRef] [PubMed]
55. Khizar, S.; Zine, N.; Jaffrezic-Renault, N.; Elaissari, A. Prospective analytical role of sensors for environmental screening and monitoring. *TrAC Trends Anal. Chem.* **2022**, *157*, 116751. [CrossRef]
56. Mohanty, S.; Ghosh, S.; Bal, B.; Das, A.P. A review of biotechnology processes applied for manganese recovery from wastes. *Rev. Environ. Sci. Bio/Technol.* **2018**, *17*, 791–811. [CrossRef]
57. Mohammed, M.Q.; Ismail, H.K.; Alesary, H.F.; Barton, S. Use of a Schiff base-modified conducting polymer electrode for electrochemical assay of Cd(II) and Pb(II) ions by square wave voltammetry. *Chem. Pap.* **2022**, *76*, 715–729. [CrossRef]
58. Alshawi, J.M.S.; Mohammed, M.Q.; Alesary, H.F.; Ismail, H.K.; Barton, S. Voltammetric determination of Hg^{2+}, Zn^{2+}, and Pb^{2+} ions using a PEDOT/NTA-modified electrode. *ACS Omega* **2022**, *7*, 20405–20419. [CrossRef] [PubMed]
59. Bansod, B.; Kumar, T.; Thakur, R.; Rana, S.; Singh, I. A review on various electrochemical techniques for heavy metal ions detection with different sensing platforms. *Biosens. Bioelectron.* **2017**, *94*, 443–455. [CrossRef] [PubMed]
60. Zhang, Z.; Ji, H.; Song, Y.; Zhang, S.; Wang, M.; Jia, C.; Tian, J.Y.; He, L.; Zhang, X.; Liu, C.S. Fe(III)-based metal-organic framework-derived core-shell nanostructure: Sensitive electrochemical platform for high trace determination of heavy metal ions. *Biosens. Bioelectron.* **2017**, *94*, 358–364. [CrossRef]
61. Gao, F.; Zhan, F.; Li, S.; Antwi-Mensah, P.; Niu, L.; Wang, Q. Dual signal-based electrochemical aptasensor for simultaneous detection of Lead(II) and Mercury(II) in environmental water samples. *Biosens. Bioelectron.* **2022**, *209*, 114280. [CrossRef]
62. Jomova, K.; Makova, M.; Alomar, S.Y.; Alwasel, S.H.; Nepovimova, E.; Kuca, K.; Rhodes, C.J.; Valko, M. Essential metals in health and disease. *Chem.-Biol. Interact.* **2022**, *367*, 110173. [CrossRef]
63. Xu, M.; Xing, J.; Yuan, B.; He, L.; Lu, L.; Chen, N.; Cai, P.; Wu, A.; Li, J. Organic small-molecule fluorescent probe-based detection for alkali and alkaline earth metal ions in biological systems. *J. Mater. Chem. B* **2023**, *11*, 3295–3306. [CrossRef]
64. Guo, F.; Zylinska, L.; Boczek, T. Role of metal ions in central nervous system: Physiology and pathophysiology. *Front. Cell. Neurosci.* **2022**, *16*, 1093224. [CrossRef]
65. Chen, W.T.; Liao, Y.H.; Yu, H.M.; Cheng, I.H.; Chen, Y.R. Distinct effects of Zn^{2+}, Cu^{2+}, Fe^{3+}, and Al^{3+} on Amyloid-beta stability, oligomerization, and aggregation. *J. Biol. Chem.* **2011**, *286*, 9646–9656. [CrossRef]
66. Li, Z.; Liu, M.; Fan, L.; Ke, H.; Luo, C.; Zhao, G. A highly sensitive and wide-ranged electrochemical zinc(II) aptasensor fabricated on core-shell SiO_2-Pt@meso-SiO_2. *Biosens. Bioelectron.* **2014**, *52*, 293–297. [CrossRef]
67. Salehan, P.; Ensafi, A.A.; Mousaabadi, K.Z.; Ghasemi, J.B.; Aghaee, E.; Rezaei, B. A theoretical and experimental study of polyaniline/GCE and DNA G-quadruplex conformation as an impedimetric biosensor for the determination of potassium ions. *Chemosphere* **2022**, *292*, 133460. [CrossRef] [PubMed]
68. Gruber, B.; David, F.; Sandra, P. Capillary gas chromatography-mass spectrometry: Current trends and perspectives. *TrAC Trends Anal. Chem.* **2020**, *124*, 115475. [CrossRef]
69. Fuyal, M.; Giri, B. A combined system of paper device and portable spectrometer for the detection of pesticide residues. *Food Anal. Methods* **2020**, *13*, 1492–1502. [CrossRef]
70. Venegas, C.J.; Rodríguez, L.; Sierra-Rosales, P. Selective Label-Free Electrochemical Aptasensor Based on Carbon Nanotubes for Carbendazim Detection. *Chemosenros* **2023**, *11*, 117. [CrossRef]
71. Li, J.S.; Yang, F.Z.; Chen, X.F.; Fang, H.G.; Zha, C.Y.; Huang, J.C.; Sun, X.; Ahmed, M.B.M.; Guo, Y.M.; Liu, Y. Dual-ratiometric aptasensor for simultaneous detection of malathion and profenofos based on hairpin tetrahedral DNA nanostructures. *Biosens. Bioelectron.* **2023**, *227*, 114853. [CrossRef]
72. Krishnan, S.K.; Singh, E.; Singh, P.; Meyyappan, M.; Nalwa, H.S. A review on graphene-based nanocomposites for electrochemical and fluorescent biosensors. *RSC Adv.* **2019**, *9*, 8778–8881. [CrossRef] [PubMed]
73. Sassetti, E.; Clausen, M.H.; Laraia, L. Small-molecule inhibitors of reactive oxygen species production. *J. Med. Chem.* **2021**, *64*, 5252–5275. [CrossRef] [PubMed]
74. Sinha, A.; Lu, X.; Wu, L.; Tan, D.; Li, Y.; Chen, J.; Jain, R. Voltammetric sensing of biomolecules at carbon based electrode interfaces: A review. *TrAC Trends Anal. Chem.* **2018**, *98*, 174–189. [CrossRef]
75. Shen, W.J.; Zhuo, Y.; Chai, Y.Q.; Han, J.; Li, E.K.; Yuan, R. An enzyme-free signal amplified strategy based on hollow platinum nanochains catalyzed oxidation of uric acid for electrochemical aptasensor construction. *Electrochim. Acta* **2014**, *143*, 240–246. [CrossRef]
76. Jia, L.P.; Wang, L.J.; Ma, R.N.; Shang, L.; Zhang, W.; Xue, Q.W.; Wang, H.S. An electrochemical aptasensor for the highly sensitive detection of 8-hydroxy-2′-deoxyguanosine based on the hybridization chain reaction. *Talanta* **2018**, *179*, 414–419. [CrossRef]
77. Su, S.; Sun, H.; Xu, F.; Yuwen, L.; Wang, L. Highly sensitive and selective determination of dopamine in the presence of ascorbic acid using gold nanoparticles-decorated MoS_2 nanosheets modified electrode. *Electroanalysis* **2013**, *25*, 2523–2529. [CrossRef]
78. Su, S.; Hao, Q.; Yan, Z.; Dong, R.; Yang, R.; Zhu, D.; Chao, J.; Zhou, Y.; Wang, L. A molybdenum disulfide@methylene blue nanohybrid for electrochemical determination of microRNA-21, dopamine and uric acid. *Microchim. Acta* **2019**, *186*, 607. [CrossRef]

79. Zhang, C.; You, X.; Li, Y.; Zuo, Y.; Wang, W.; Li, D.; Huang, S.; Hu, H.; Yuan, F.; Shao, F.; et al. A novel electrochemical aptasensor for serum dopamine detection based on methylene blue-integrated m-PdNFs signal material. *Sens. Actuators B* **2022**, *354*, 131233. [CrossRef]
80. Su, S.; Sun, H.; Cao, W.; Chao, J.; Peng, H.; Zuo, X.; Yuwen, L.; Fan, C.; Wang, L. Dual-target electrochemical biosensing based on DNA structural switching on gold nanoparticle-decorated MoS_2 nanosheets. *ACS Appl. Mater. Interfaces* **2016**, *8*, 6826–6833. [CrossRef]
81. Chung, S.; Singh, N.K.; Gribkoff, V.K.; Hall, D.A. Electrochemical carbamazepine aptasensor for therapeutic drug monitoring at the point of care. *ACS Omega* **2022**, *7*, 39097–39106. [CrossRef] [PubMed]
82. Du, Y.; Chen, C.; Yin, J.; Li, B.; Zhou, M.; Dong, S.; Wang, E. Solid-state probe based electrochemical aptasensor for cocaine: A potentially convenient, sensitive, repeatable, and integrated sensing platform for drugs. *Anal. Chem.* **2010**, *82*, 1556–1563. [CrossRef]
83. Derikvand, H.; Roushani, M.; Abbasi, A.R.; Derikvand, Z.; Azadbakht, A. Design of folding-based impedimetric aptasensor for determination of the nonsteroidal anti-inflammatory drug. *Anal. Biochem.* **2016**, *513*, 77–86. [CrossRef]
84. Soni, S.; Jain, U.; Burke, D.H.; Chauhan, N. A label free, signal off electrochemical aptasensor for amphetamine detection. *Surf. Interfaces* **2022**, *31*, 102023. [CrossRef]
85. Evtugyn, G.; Porfireva, A.; Tsekenis, G.; Oravczova, V.; Hianik, T. Electrochemical aptasensors for antibiotics detection: Recent achievements and applications for monitoring food safety. *Sensors* **2022**, *22*, 3684. [CrossRef]
86. Li, F.; Gao, X.; Wang, X.; Guo, Y.; Sun, X.; Yang, Q.; Zhang, Y. Ultrasensitive sandwich RNA-aptasensor based on dual-signal amplification strategy for highly sensitive neomycin detection. *Food Control* **2022**, *131*, 108445. [CrossRef]
87. Lin, S.; Cheng, X.; Zhu, J.; Wang, B.; Jelinek, D.; Zhao, Y.; Wu, T.Y.; Horrillo, A.; Tan, J.; Yeung, J.; et al. Wearable microneedle-based electrochemical aptamer biosensing for precision dosing of drugs with narrow therapeutic windows. *Sci. Adv.* **2022**, *8*, eabq4539. [CrossRef] [PubMed]
88. Huang, S.; Gan, N.; Zhang, X.; Wu, Y.; Shao, Y.; Jiang, Z.; Wang, Q. Portable fluoride-selective electrode as signal transducer for sensitive and selective detection of trace antibiotics in complex samples. *Biosens. Bioelectron.* **2019**, *128*, 113–121. [CrossRef]
89. Wang, M.; Hu, M.; Liu, J.; Guo, C.; Peng, D.; Jia, Q.; He, L.; Zhang, Z.; Du, M. Covalent organic framework-based electrochemical aptasensors for the ultrasensitive detection of antibiotics. *Biosens. Bioelectron.* **2019**, *132*, 8–16. [CrossRef]
90. Ghanbari, K.; Roushani, M. A novel electrochemical aptasensor for highly sensitive and quantitative detection of the streptomycin antibiotic. *Bioelectrochemistry* **2018**, *120*, 43–48. [CrossRef] [PubMed]
91. Filik, H.; Avan, A.A. Nanostructures for nonlabeled and labeled electrochemical immunosensors: Simultaneous electrochemical detection of cancer markers: A review. *Talanta* **2019**, *205*, 120153. [CrossRef]
92. Arshavsky-Graham, S.; Heuer, C.; Jiang, X.; Segal, E. Aptasensors versus immunosensors-Which will prevail? *Eng. Life Sci.* **2022**, *22*, 319–333. [CrossRef] [PubMed]
93. Xifre-Perez, E.; Ferre-Borrull, J.; Marsal, L.F. Oligonucleotic probes and immunosensors based on nanoporous anodic alumina for screening of diseases. *Adv. Mater. Technol.* **2022**, *7*, 2101591. [CrossRef]
94. Tan, X.; Sun, X.; Li, Y.; Zeng, Y.; Gong, J.; Wang, Z.; An, Y.; Li, H. Biomineralized $Mn_3(PO_4)_2$/aptamer nanosheets for enhanced electrochemical determination of C-reactive protein. *Sens. Actuators B* **2021**, *333*, 129510. [CrossRef]
95. Mei, C.; Zhang, Y.; Pan, L.; Dong, B.; Chen, X.; Gao, Q.; Xu, H.; Xu, W.; Fang, H.; Liu, S.; et al. A One-step electrochemical aptasensor based on signal amplification of metallo nanoenzyme particles for vascular endothelial growth factor. *Front. Bioeng. Biotechnol.* **2022**, *10*, 850412. [CrossRef] [PubMed]
96. Mei, C.; Pan, L.; Xu, W.; Xu, H.; Zhang, Y.; Li, Z.; Dong, B.; Ke, X.; McAlinden, C.; Yang, M.; et al. An ultrasensitive reusable aptasensor for noninvasive diabetic retinopathy diagnosis target on tear biomarker. *Sens. Actuators B* **2021**, *345*, 130398. [CrossRef]
97. Qing, M.; Sun, Z.; Wang, L.; Du, S.Z.; Zhou, J.; Tang, Q.; Luo, H.Q.; Li, N.B. CRISPR/Cas12a-regulated homogeneous electrochemical aptasensor for amplified detection of protein. *Sens. Actuators B* **2021**, *348*, 130713. [CrossRef]
98. Ghalehno, M.H.; Mirzaei, M.; Torkzadeh-Mahani, M. Electrochemical aptasensor for tumor necrosis factor alpha using aptamer-antibody sandwich structure and cobalt hexacyanoferrate for signal amplification. *J. Iran. Chem. Soc.* **2019**, *16*, 1783–1791. [CrossRef]
99. Wei, J.; Qiu, Z.; Yu, D.; Yin, Y.; Tang, Q.; Liao, X.; Zhang, G.; Liu, Z.; Gao, F. DNAzyme-driven tripedal DNA walker triggered hybridization chain reaction for label-free electrochemical detection of Alzheimer's tau protein. *Sens. Actuators B* **2023**, *384*, 133656. [CrossRef]
100. Biswas, G.C.; Choudhury, S.; Rabbani, M.M.; Das, J. A review on potential electrochemical point-of-care tests targeting pandemic infectious disease detection: COVID-19 as a reference. *Chemosensors* **2022**, *10*, 269. [CrossRef]
101. Liu, Y.; Tuleouva, N.; Ramanculov, E.; Revzin, A. Aptamer-based electrochemical biosensor for interferon gamma detection. *Anal. Chem.* **2010**, *82*, 8131–8136. [CrossRef] [PubMed]
102. Thakur, H.; Kaur, N.; Sabherwal, P.; Sareen, D.; Prabhakar, N. Aptamer based voltammetric biosensor for the detection of Mycobacterium tuberculosis antigen MPT64. *Microchim. Acta* **2017**, *184*, 1915–1922. [CrossRef]
103. Idili, A.; Parolo, C.; Alvarez-Diduk, R.; Merkoci, A. Rapid and efficient detection of the SARS-CoV-2 spike protein using an electrochemical aptamer-based sensor. *ACS Sens.* **2021**, *6*, 3093–3101. [CrossRef]

104. Rahmati, Z.; Roushani, M.; Hosseini, H.; Choobin, H. Label-free electrochemical aptasensor for rapid detection of SARS-CoV-2 spike glycoprotein based on the composite of $Cu(OH)_2$ nanorods arrays as a high-performance surface substrate. *Bioelectrochemistry* **2022**, *146*, 108106. [CrossRef]
105. Chen, Z.; Wang, X. The role and application of exosomes and their cargos in reproductive diseases: A systematic review. *Vet. Sci.* **2022**, *9*, 706. [CrossRef]
106. Pan, H.; Dong, Y.; Gong, L.; Zhai, J.; Song, C.; Ge, Z.; Su, Y.; Zhu, D.; Chao, J.; Su, S.; et al. Sensing gastric cancer exosomes with MoS_2-based SERS aptasensor. *Biosens. Bioelectron.* **2022**, *215*, 114553. [CrossRef]
107. Mei, K.; Yan, T.; Wang, Y.; Rao, D.; Peng, Y.; Wu, W.; Chen, Y.; Ren, M.; Yang, J.; Wu, S.; et al. Magneto-nanomechanical array biosensor for ultrasensitive detection of oncogenic exosomes for early diagnosis of cancers. *Small* **2023**, *19*, 2205445. [CrossRef] [PubMed]
108. Wu, Q.; Ding, Q.; Lin, W.; Weng, Y.; Feng, S.; Chen, R.; Chen, C.; Qiu, S.; Lin, D. Profiling of tumor cell-delivered exosome by surface enhanced raman spectroscopy-based biosensor for evaluation of nasopharyngeal cancer radioresistance. *Adv. Healthc. Mater.* **2022**, *12*, 2202482. [CrossRef] [PubMed]
109. Huang, R.; He, L.; Xia, Y.; Xu, H.; Liu, C.; Xie, H.; Wang, S.; Peng, L.; Liu, Y.; Liu, Y.; et al. A sensitive aptasensor based on a hemin/G-quadruplex-assisted signal amplification strategy for electrochemical detection of gastric cancer exosomes. *Small* **2019**, *15*, 1900735. [CrossRef]
110. Zhang, W.; Tian, Z.; Yang, S.; Rich, J.; Zhao, S.; Klingeborn, M.; Huang, P.H.; Li, Z.; Stout, A.; Murphy, Q.; et al. Electrochemical micro-aptasensors for exosome detection based on hybridization chain reaction amplification. *Microsyst. Nanoeng.* **2021**, *7*, 63. [CrossRef]
111. Chang, L.; Wu, H.; Chen, R.; Sun, X.; Yang, Y.; Huang, C.; Ding, S.; Liu, C.; Cheng, W. Microporous PdCuB nanotag-based electrochemical aptasensor with $Au@CuCl_2$ nanowires interface for ultrasensitive detection of PD-L1-positive exosomes in the serum of lung cancer patients. *J. Nanobiotechnol.* **2023**, *21*, 86. [CrossRef]
112. Yang, L.; Yin, X.; An, B.; Li, F. Precise capture and direct quantification of tumor exosomes via a highly efficient dual-aptamer recognition-assisted ratiometric immobilization-free electrochemical strategy. *Anal. Chem.* **2021**, *93*, 1709–1716. [CrossRef]
113. Tan, P.; Chen, X.; Zhang, H.; Wei, Q.; Luo, K. Artificial intelligence aids in development of nanomedicines for cancer management. *Semin. Cancer Biol.* **2023**, *89*, 61–75. [CrossRef] [PubMed]
114. Fitzgerald, R.C.; Antoniou, A.C.; Fruk, L.; Rosenfeld, N. The future of early cancer detection. *Nat. Med.* **2022**, *28*, 666–677. [CrossRef] [PubMed]
115. Xie, X.; Fu, C.C.; Lv, L.; Ye, Q.; Yu, Y.; Fang, Q.; Zhang, L.; Hou, L.; Wu, C. Deep convolutional neural network-based classification of cancer cells on cytological pleural effusion images. *Mod. Pathol.* **2022**, *35*, 609–614. [CrossRef] [PubMed]
116. de Beur, S.M.J.; Minisola, S.; Xia, W.B.; Abrahamsen, B.; Body, J.J.; Brandi, M.L.; Clifton-Bligh, R.; Collins, M.; Florenzano, P.; Houillier, P.; et al. Global guidance for the recognition, diagnosis, and management of tumor-induced osteomalacia. *J. Intern. Med.* **2023**, *293*, 309–328. [CrossRef]
117. Chen, X.; Zhou, F.; Li, X.; Yang, G.; Zhang, L.; Ren, S.; Zhao, C.; Deng, Q.; Li, W.; Gao, G.; et al. Folate receptor-positive circulating tumor cell detected by LT-PCR based method as a diagnostic biomarker for non-small cell lung cancer. *J. Clin. Oncol.* **2015**, *33*, 11032. [CrossRef]
118. Choi, H.; Kim, K.B.; Jeon, C.S.; Hwang, I.; Lee, S.; Kim, H.K.; Kim, H.C.; Chung, T.D. A label-free DC impedance-based microcytometer for circulating rare cancer cell counting. *Lab Chip* **2013**, *13*, 970–977. [CrossRef]
119. Yin, J.; He, X.; Wang, K.; Xu, F.; Shangguan, J.; He, D.; Shi, H. Label-free and turn-on aptamer strategy for cancer cells detection based on a DNA-silver nanocluster fluorescence upon recognition-induced hybridization. *Anal. Chem.* **2013**, *85*, 12011–12019. [CrossRef] [PubMed]
120. Kivrak, E.; Ince-Yardimci, A.; Ilhan, R.; Kirmizibayrak, P.B.; Yilmaz, S.; Kara, P. Aptamer-based electrochemical biosensing strategy toward human non-small cell lung cancer using polyacrylonitrile/polypyrrole nanofibers. *Anal. Bioanal. Chem.* **2020**, *412*, 7851–7860. [CrossRef]
121. Cai, S.; Chen, M.; Liu, M.; He, W.; Liu, Z.; Wu, D.; Xia, Y.; Yang, H.; Chen, J. A signal amplification electrochemical aptasensor for the detection of breast cancer cell via free-running DNA walker. *Biosens. Bioelectron.* **2016**, *85*, 184–189. [CrossRef] [PubMed]
122. Lu, C.Y.; Xu, J.J.; Wang, Z.H.; Chen, H.Y. A novel signal-amplified electrochemical aptasensor based on supersandwich G-quadruplex DNAzyme for highly sensitive cancer cell detection. *Electrochem. Commun.* **2015**, *52*, 49–52. [CrossRef]
123. Ring, A.; Nguyen-Sträuli, B.D.; Wicki, A.; Aceto, N. Biology, vulnerabilities and clinical applications of circulating tumour cells. *Nat. Rev. Cancer* **2023**, *23*, 95–111. [CrossRef] [PubMed]
124. Eslami-S, Z.; Cortés-Hernández, L.E.; Thomas, F.; Pantel, K.; Alix-Panabières, C. Functional analysis of circulating tumour cells: The KEY to understand the biology of the metastatic cascade. *Br. J. Cancer* **2022**, *127*, 800–810. [CrossRef] [PubMed]
125. Khaksari, S.; Ameri, A.R.; Taghdisi, S.M.; Sabet, M.; Bami, S.M.J.G.; Abnous, K.; Shaegh, S.A.M. A microfluidic electrochemical aptasensor for highly sensitive and selective detection of A549 cells as integrin $\alpha 6\beta 4$ -containing cell model via IDA aptamers. *Talanta* **2023**, *252*, 123781. [CrossRef]
126. Reis, H.J.; Wang, L.; Verano-Braga, T.; Pimenta, A.M.C.; Kalman, J.; Bogats, G.; Babik, B.; Vieira, L.B.; Teixeira, A.L.; Mukhamedyarov, M.A.; et al. Evaluation of post-surgical cognitive function and protein fingerprints in the cerebro-spinal fluid utilizing surface-enhanced laser desorption/ionization time-of-flight mass-spectrometry (SELDI-TOF MS) after coronary artery bypass grafting: Review of proteomic analytic tools and introducing a new syndrome. *Curr. Med. Chem.* **2011**, *18*, 1019–1037.

127. Lin, X.H.; Liu, P.P.; Yan, J.; Luan, D.L.; Sun, T.; Bian, X.J. Dual synthetic receptor-based sandwich electrochemical sensor for highly selective and ultrasensitive detection of pathogenic bacteria at the single-cell level. *Anal. Chem.* **2023**, *95*, 5561–5567. [CrossRef]
128. Jiang, H.; Sun, Z.; Zhang, C.; Weng, X. 3D-architectured aptasensor for ultrasensitive electrochemical detection of norovirus based on phosphorene-gold nanocomposites. *Sens. Actuators B* **2022**, *354*, 131232. [CrossRef]
129. Lee, I.; Kim, S.E.; Lee, J.; Woo, D.H.; Lee, S.; Pyo, H.; Song, C.S.; Lee, J. A self-calibrating electrochemical aptasensing platform: Correcting external interference errors for the reliable and stable detection of avian influenza viruses. *Biosens. Bioelectron.* **2020**, *152*, 112010. [CrossRef]
130. Rahmati, Z.; Roushani, M. SARS-CoV-2 virus label-free electrochemical nanohybrid MIP-aptasensor based on $Ni_3(BTC)_2$ MOF as a high-performance surface substrate. *Microchim. Acta* **2022**, *189*, 287. [CrossRef]
131. Rashid, S.; Nawaz, M.H.; Marty, J.L.; Hayat, A. Label free ultrasensitive detection of NS1 based on electrochemical aptasensor using polyethyleneimine aggregated AuNPs. *Microchem. J.* **2020**, *158*, 105285. [CrossRef]
132. Tu, J.; Torrente-Rodriguez, R.M.; Wang, M.; Gao, W. The era of digital health: A review of portable and wearable affinity biosensors. *Adv. Funct. Mater.* **2020**, *30*, 1906713. [CrossRef]
133. Wu, J.; Liu, H.; Chen, W.; Ma, B.; Ju, H. Device integration of electrochemical biosensors. *Nat. Rev. Bioeng.* **2023**, *1*, 346–360. [CrossRef] [PubMed]
134. Li, J.; Macdonald, J. Advances in isothermal amplification: Novel strategies inspired by biological processes. *Biosens. Bioelectron.* **2015**, *64*, 196–211. [CrossRef]
135. Kulkarni, M.B.; Ayachit, N.H.; Aminabhavi, T.M. Recent advancements in nanobiosensors: Current trends, challenges, applications, and future scope. *Biosensors* **2022**, *12*, 892. [CrossRef] [PubMed]
136. Li, S.; Zhang, H.; Zhu, M.; Kuang, Z.; Li, X.; Xu, F.; Miao, S.; Zhang, Z.; Lou, X.; Li, H.; et al. Electrochemical biosensors for whole blood analysis: Recent progress, challenges, and future perspectives. *Chem. Rev.* **2023**, *123*, 7953–8039. [CrossRef] [PubMed]

chemosensors

Review

Chemical Sensing and Analysis with Optical Nanostructures

Chenyu Dong, Yifan Wang, Xiaoyan Zhao, Jie Bian and Weihua Zhang *

College of Engineering and Applied Sciences, State Key Laboratory of Analytical Chemistry for Life Science, and Jiangsu Key Laboratory of Artificial Functional Materials, Nanjing University, Nanjing 210023, China
* Correspondence: zwh@nju.edu.cn

Abstract: Nanostructures and nanomaterials, especially plasmonic nanostructures, often show optical properties that conventional materials lack and can manipulate light, as well as various light–matter interactions, in both their near-field and far-field regions with a high efficiency. Thanks to these unique properties, not only can they be used to enhance the sensitivity of chemical sensing and analysis techniques, but they also provide a solution for designing new sensing devices and simplifying the design of analytical instruments. The earliest applications of optical nanostructures are surface-enhanced spectroscopies. With the help of the resonance field enhancement of plasmonic nanostructures, molecular signals, such as Raman, infrared absorption, and fluorescence can be significantly enhanced, and even single-molecule analysis can be realized. Moreover, the resonant field enhancements of plasmonic nanostructures are often associated with other effects, such as optical forces, resonance shifts, and photothermal effects. Using these properties, label-free plasmonic sensors, nano-optical tweezers, and plasmonic matrix-assisted laser desorption/ionization have also been demonstrated in the past two decades. In the last few years, the research on optical nanostructures has gradually expanded to non-periodic 2D array structures, namely metasurfaces. With the help of metasurfaces, light can be arbitrarily manipulated, leading to many new possibilities for developing miniaturized integrated intelligent sensing and analysis systems. In this review, we discuss the applications of optical nanostructures in chemical sensing and analysis from both theoretical and practical aspects, aiming at a concise and unified framework for this field.

Keywords: optical nanostructure; plasmonics; surface-enhanced spectroscopy; sensing; miniaturization

Citation: Dong, C.; Wang, Y.; Zhao, X.; Bian, J.; Zhang, W. Chemical Sensing and Analysis with Optical Nanostructures. *Chemosensors* **2023**, *11*, 497. https://doi.org/10.3390/chemosensors11090497

Academic Editor: Marco Pisco

Received: 25 June 2023
Revised: 4 September 2023
Accepted: 7 September 2023
Published: 9 September 2023

1. Introduction

Scientists have long dreamed of a rapid, simple, and low-cost analytical technique for tracing amounts of components in complex samples. This technique has broad applications in chemistry, material sciences, and life sciences but also presents a great challenge to researchers. Traditional methods are no longer applicable when faced with very small quantities of molecules or samples with complex nanostructures and nano-compositions (such as single cells, a droplet of blood, and complex polymer mixtures) due to the limitations of sample size and scale. New physical or chemical effects are required to transduce and amplify extremely weak sample information into readable physical signals. To meet this demand, many new advanced analytical techniques, particularly optical methods, have been developed in the last few decades thanks to breakthroughs in nano-optics.

Optical techniques are the most widely used techniques in chemical sensing and analysis because light–matter interactions are one of the most fundamental types of physical interactions in nature and form the basis of many analytical methods [1,2]. For example, the optical interactions with molecules' electronic states lead to UV-Vis spectroscopy, and interactions with the vibrational states lead to Raman spectroscopy and infrared spectroscopy. Using the optothermal effect, one can even vaporize and ionize molecules and subsequently measure their molecular weights. At the same time, with the development of the semiconductor industry, crucial components, e.g., lasers and photodetectors, have become cheap and easy to access today. All these factors have made optics a powerful, versatile, and convenient choice for chemical sensing and analysis. However, light–matter interactions are generally weak due to the fact that the size of molecules is several orders of magnitude smaller than the wavelength of light. This causes the sensitivity issue in optical measurements (Figure 1). Fortunately, the development of nano-optical technology makes it possible to circumvent this issue and obtain ultrasensitive analysis [3–5]. With an optical nanostructure, light can be resonantly enhanced, scattered, and absorbed, and one can therefore enhance local optical signals significantly, making ultrasensitive nano-analysis possible [6–8].

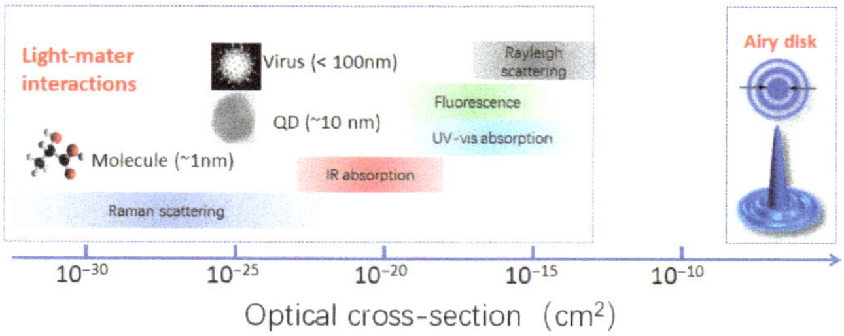

Figure 1. Optical cross-sections of typical light–matter interactions at the nanoscale.

To date, countless optical nanostructures, particularly metallic nanostructures, with different shapes and properties have been studied. The metallic nanostructures are also known as plasmonic nanostructures because they support collective oscillations of electrons (so-called plasmon polaritons). These plasmon oscillations can be divided into two categories: (1) surface plasmon resonance (SPR), which propagates along the surface of metal surfaces, and (2) localized surface plasmon resonance (LSPR), which cannot [9–11]. Both are tightly confined at the surfaces of structures and, in particular, with LSPR, light can be confined to sub-10 nm "hot spots" adjacent to metal nanostructures. This extreme light confinement can also lead to strong local enhancement of the electric fields and consequently various light–matter interactions, making it very useful in various analysis techniques (Figure 2). For example, when a molecule is placed in a "hot spot", its signal (e.g., Raman scattering, fluorescence, and absorption) can be increased by several orders of magnitude, and even single-molecule Raman detection can be achieved [12,13]. The presence of the molecule also shifts the plasmon resonance, and in other words, the plasmonic nanostructure can also be used as a high-performance index nanosensor [14–17]. Moreover, thanks to their strong optothermal effects, plasmonic nanostructures can even be used as the ion source for mass spectrometry (MS) [18–20]. Besides the plasmonic structure, it was reported that dielectric micro-resonators can also be used for detecting chemicals or viruses by monitoring the induced resonance shifts, which is similar to the LSPR sensors [21–25]. However, due to the absence of free electrons, these dielectric micro-resonators cannot confine or absorb light like their metallic counterparts.

In recent years, researchers started to realize that one can achieve arbitrary regulation of the wavefront of light by patterning separated nanostructures into a planar 2D device, which is also known as a metasurface [26,27]. A variety of designs have been demonstrated, including periodic arrays, gradient structures, and even more complex designs. With these structures, one can conduct sensing, imaging, spectroscopy, and even polarimetry with good quality and efficiency [28–35]. This not only opens the door towards new sensor designs but also makes the miniaturization of conventional analysis equipment possible.

Today, optical nanostructures have become one of the most dynamic research areas in chemical sensing and analysis, and there have been quite a few review articles on this topic [36–38]. Most of them are mainly focused on the applications related to field confinement and enhancement, such as surface-enhanced spectroscopies and LSPR-based sensors, but other important topics, such as photothermal effect-related applications, miniaturized sensors, and metasurface-based polarimetric techniques, are not covered. We, therefore, write this review to sort out the various basic properties of nano-optical structures, further discuss their applications in chemical analysis, from conventional surface-enhanced spectroscopies and index sensing to more complex phenomena like laser-induced desorption and ionization, and then cover the application of metasurfaces for circular dichroism (CD) measurement, as shown in Figure 2.

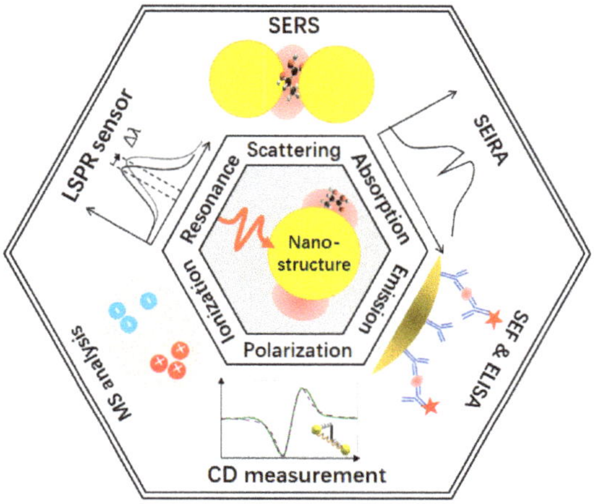

Figure 2. Chemical sensing and analysis with optical nanostructures.

2. Optical Properties of Nanostructures

In this section, we will give a concise review of the theory of the localized plasmon resonance of subwavelength nanostructures. Associated optical effects, such as field confinement, field enhancement, and local heating, which are crucial for chemical sensing and analysis, will also be discussed. At the end of this section, some of the basic ideas of metasurfaces will also be introduced based on the discussion of the local behaviors of subwavelength optical nanostructures.

2.1. Localized Plasmon Resonance and Field Enhancement

The simplest model for understanding plasmon resonance phenomena is spherical nanoparticles [11]. When the nanoparticle's size is much smaller than the wavelength of light, it can be treated as a dipole,

$$\mathbf{p} = \varepsilon_0 \alpha \mathbf{E}. \tag{1}$$

Here, α is the polarizability, and **E** is the external excitation field. The light scattering cross-section and absorption cross-section of this structure then become:

$$A_{sca} = \frac{k^4}{6\pi}|\alpha|^2 \tag{2}$$

$$A_{abs} = kIm(\alpha) \tag{3}$$

For a deep subwavelength particle, α has a simple form:

$$\alpha = 4\pi r^3 \frac{\varepsilon - \varepsilon_0}{\varepsilon + 2\varepsilon_0} \tag{4}$$

Interestingly, for metallic materials, the real part of the ε is negative when the frequency is below their bulk plasma resonance frequencies, which typically lie within the optical spectral range. As a result, the denominator of the α can approach zero, resulting in a strong resonance of the free electrons (i.e., localized plasmon resonances). This leads to the strong scattering and absorption of light by the nanostructures.

The above plasmon resonances are not limited to spherical particles and can occur for any metallic nanoparticles with any shape [39,40]. Theoretically, the scattering problem of electromagnetic fields can be rigorously described and solved using the Lippmann–Schwinger equation [41]:

$$\boldsymbol{E}(\boldsymbol{r}) = \boldsymbol{E}_{ext}(\boldsymbol{r}) + \int_{cavity} d\boldsymbol{r}' \boldsymbol{G}_0(\boldsymbol{r}, \boldsymbol{r}') \cdot \Delta\varepsilon(\boldsymbol{r}') k_0{}^2 \boldsymbol{E}(\boldsymbol{r}') \tag{5}$$

where $\Delta\varepsilon = \varepsilon_c - \varepsilon_0$ is the permittivity contrast between the scatter (nanocavity in this work) and the background medium, E_{ext} is the external field, G_0 is the free-space Green's tensor, and k_0 is the wavenumber in the background medium. In previous works, the author of this work demonstrated that under the quasi-static approximation, a nanoparticle always has a complete set of orthogonal eigenmodes, $|\boldsymbol{E}_i\rangle$. The scattering fields E, obtained under excitation field E_0, can be written as a linear superposition of this set of eigenmodes (Figure 3):

$$\mathbf{E} = \sum_i a_i \mathbf{E}_i \tag{6}$$

$$a_i = \frac{s}{s - s_i} \langle \mathbf{E}_i | \mathbf{E}_0 \rangle \tag{7}$$

Here, $s = \varepsilon_0/(\varepsilon_c - \varepsilon_0)$ is a material-related parameter, and s_i are the eigenvalues.

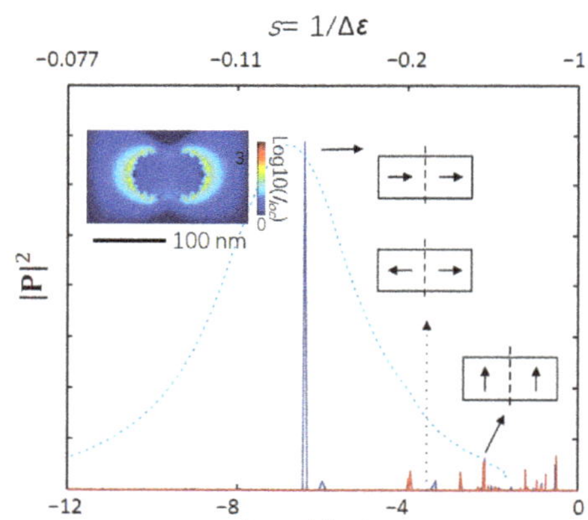

Figure 3. Eigenmodes of a plasmonic nanostructure. The arrows depict the symmetry of the eigenmodes. Inset is the field distribution of the dominant dipole mode.

Equation 5 can be numerically used after being discretized, and interestingly, it can be shown that s_i are always between -1 and 0 [41]. Considering that the permittivity of the local environment, ε_0, is positive (e.g., $\varepsilon_0 = 1$ for vacuum), it means that a negative permittivity is always required to reach the resonance condition of a subwavelength nanostructure. In the optical regime, only metals exhibit negative dielectric constants due to the existence of free electrons [8]. This explains why only metal nanostructures show extraordinary optical properties in the optical regime.

In the quasi-electrostatic model described above, the field enhancement effect is determined by two factors, (1) resonance enhancement and (2) mode distribution, that is, the field distribution of the eigenmode $|\mathbf{E}_i\rangle$. The first factor is determined by the intrinsic properties of the material. At the resonance wavelength, the real part of the denominator of $s/(s - s_i)$ will be zero, and it is evident that the smaller the imaginary part of the dielectric constant of the material is, the greater the resonance enhancement will be. For noble metals, e.g., Ag and Au, the resonance factors can be greater than ten or even reach hundreds. The second factor is determined by the geometry of the nanostructure. It is well known that large field enhancement is often associated with some specific features, such as sharp corners (i.e., lightning rod effect) and nanometer gaps [42]. In practice, by combining both the resonance and geometric effects, the field enhancement factor, $g = |\mathbf{E}| / |\mathbf{E}_0|$, can reach >100 times at some "hot spots". If there are molecules present in the "hot spot", their excitation rate will be increased by a factor of g^2, which is more than 10,000 times.

The above theoretical description can also be understood with the following physical picture. At the resonance frequency, a plasmon nanostructure can collect propagating light in free space from a cross-section larger than its geometric size and squeeze the light into nanometer "hot spots". This process is very similar to the function of radio antennas, which can collect the radio waves in the free space into their near field efficiently. This is why plasmon nanostructures are often called optical antennas or plasmonic nano-antennas [43–45].

2.2. Enhancement of the Near-Field Scattering and Fluorescence Emission

In antenna theory, it is known that radio antennas are bidirectional devices. They not only collect the far-field signals to their near field but also broadcast signals from their near-field to the far-field region [46]. Similarly, plasmon nanostructures can also greatly increase the rate of Raman scattering, as well as the fluorescence radiation of molecules in their near-field range. The former can be understood using the reciprocity principle in light scattering, and to understand the latter, one needs to consider the change in the local density of states caused by the antenna.

Let us first examine the process of Raman scattering. We consider a molecule located at $\mathbf{r}_1$ in the near field of a metallic particle, and the scattered signal $\mathbf{E}$ at $\mathbf{r}_2$ in the far field can then be described using Green's function $\mathbf{G}(\mathbf{r}_2, \mathbf{r}_1)$, as shown in Figure 4a. To excite the Raman signal, we let the emission of a dipole at $\mathbf{r}_2$ be the incident light, and at $\mathbf{r}_1$, the local excitation can be described by $\mathbf{G}(\mathbf{r}_1, \mathbf{r}_2)$, as shown in Figure 4b. In most cases, the reciprocal principle holds [46,47]:

$$\mathbf{G}(\mathbf{r}_1, \mathbf{r}_2) = \mathbf{G}(\mathbf{r}_2, \mathbf{r}_1) \tag{8}$$

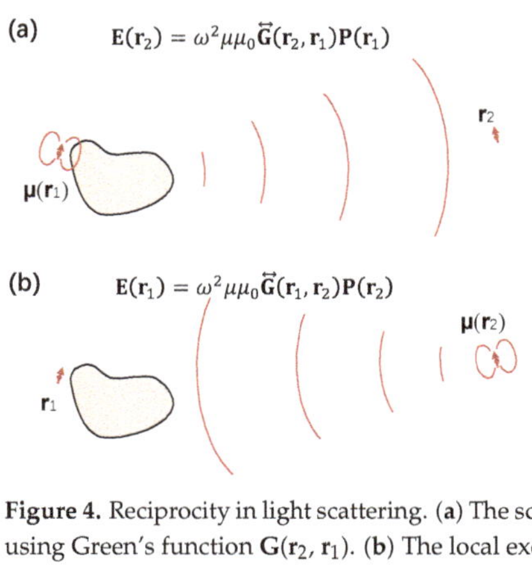

Figure 4. Reciprocity in light scattering. (**a**) The scattered signal $\mathbf{E}$ at $\mathbf{r}_2$ in the far field can be described using Green's function $\mathbf{G}(\mathbf{r}_2, \mathbf{r}_1)$. (**b**) The local excitation at $\mathbf{r}_1$ can be described by $\mathbf{G}(\mathbf{r}_1, \mathbf{r}_2)$.

In other words, if exaction light is enhanced by g^2 times due to the presence of the particle, in the emission process from $\mathbf{r}_1$ to $\mathbf{r}_2$, the signal will be enhanced by g^2 times again. Therefore, the total enhancement of the Raman signal is g^4, which is called the fourth power law in surface-enhanced Raman spectroscopy.

On the other hand, the enhancement of the emission process of fluorescence is very different because it is linked with the change in the lifetime of the electronic states, which can no longer be described by the theory of classical electrodynamics. To describe the emission process, Fermi's Golden Rule is used [3,6]. That is, the rate of transition (radiation) of molecules is completely determined by the local density of state (LDOS), ρ:

$$\gamma_{i-f} = \frac{2\pi}{\hbar}|\langle f|H'|i\rangle|^2\rho(E_f) \tag{9}$$

Interestingly, although state density is a quantum concept, it can be derived from the Green's function, which describes the classical electromagnetic fields [8,48,49], namely:

$$\rho(\mathbf{r},\omega) = \frac{2\omega}{\pi c^2}Im\{Tr[\mathbf{G}(\mathbf{r},\mathbf{r},\omega)]\} \tag{10}$$

When a plasmonic nanostructure is present, the LDOS can be significantly enhanced. This can be calculated using Equation (10), where Green's tensor can be directly calculated using iterative techniques developed by Martin and colleagues [50]. The LDOS can also be obtained using Equation (9), which states that the LDOS is proportional to the emission rate and, consequently, the power of a dipole source. One can therefore calculate the LDOS by simulating the emission properties of a dipole source using normal numerical solvers for electromagnetic fields, e.g., the finite-difference time-domain (FDTD) and finite element method (FEM).

Fluorescence signals can be affected by many factors, such as temperature, pH, concentration, and so on. In addition to the emission rate, the fluorescence signal can also be influenced by the quantum yield. The presence of plasmonic nanoparticles may lower the quantum yield due to the material losses of metal, and this can even quench the fluorescence signals despite the enhancement [51,52], particularly for the dye molecules in the visible spectral range, whose intrinsic quantum yields are often close to 100% [53].

It is also worth noting that surface-enhanced fluorescence (SEF) experiments are normally only performed with weak excitation light. When the excitation becomes strong, the excitation rate will reach the emission rate, and that will saturate the system. In this case, the local field enhancement will not influence the signal intensity anymore.

2.3. Plasmonic Trapping and Sensing

Another important effect related to the local field enhancement is the significant increase in the field gradient, which has been widely used for optical trapping and sensing in the last two decades. At a "hot spot", the intensity gradient of light can be enhanced by more than three orders of magnitude thanks to both the enhancement and spatial confinement of fields [54]. This will greatly enhance the gradient forces (also known as the dielectrophoresis effect in chemistry), leading to non-negligible attractions for nanoparticles and surrounding molecules. With the plasmon-enhanced trapping forces, individual nanoparticles and even biomolecules can be trapped under milliwatt-scale laser illumination, allowing researchers to enrich molecules at "hot spots" and investigate them further with surface-enhanced spectroscopy techniques [17,55,56].

Interestingly, if we treat the nanoparticle as an optical nano-resonator without radiation losses, we can link the trapping effect with the optical sensing techniques together. When a molecule or nanoparticle is trapped, the energy of the system will become lower, and consequently, the resonance frequency of the plasmonic nanostructure will be shifted. This property makes the plasmonic nanostructure a high-performance nanosensor that is extremely sensitive to external analytes [55]. Today, this effect has been widely used for label-free detections of local index changes induced by the analytes in the surrounding environment.

Using perturbation theory, the above resonance frequency change can be described with an explicit formula [57]:

$$\Delta\omega_i = -\alpha_{NP}\frac{d\omega_i}{d\varepsilon_{ca}}\frac{|\mathbf{E}(\mathbf{r}_{NP})|^2}{\int_{cavity} d\mathbf{r}|\mathbf{E}(\mathbf{r})|^2} \tag{11}$$

This is similar to the case of dielectric microcavities, whose resonance frequency shift can be written as:

$$\Delta\omega_i = -\frac{\omega_i}{2}\frac{\int d\mathbf{r}\Delta\varepsilon(\mathbf{r})|\mathbf{E}(\mathbf{r})^{\ 2}}{\int d\mathbf{r}\varepsilon(\mathbf{r})|\mathbf{E}(\mathbf{r})|^2} \tag{12}$$

In both cases, the resonance frequency shift is proportional to both the local light intensity and the refractive index change induced by the trapped nano-objects. Therefore, a larger enhancement will always lead to a better sensitivity of the LSPR-based sensors.

Meanwhile, there are also differences between Equations (11) and (12). In Equation (12), the resonance shift is inversely proportional to $\int d\mathbf{r}\varepsilon(\mathbf{r})|\mathbf{E}(\mathbf{r})|^2$, the integral of the electric field energy over the whole space, while in Equation (11), the integral is limited to the space inside the cavity. This is because, for the resonance mode of a plasmonic nanocavity, the integral $\int d\mathbf{r}\varepsilon(\mathbf{r})|\mathbf{E}(\mathbf{r})|^2 = 0$ is due to the negative permittivity of the metal.

2.4. Photothermal Effect

Large field enhancement means strong optical absorption. From Equation (3), it can be seen that the absorption cross-section of a nanoparticle can also be enhanced by resonance. In the case of gold and silver nanoparticles, their absorption interface can even be greater than their physical size. Meanwhile, the heat capacity and dissipation rate of plasmonic nanoparticles are often very low. For example, in the case of porous nanostructures where all nanopores are aligned along the z direction, the effective thermal capacity C_{eff} and conductivity along the z direction $K_{eff,zz}$ and x-y direction $K_{eff,xy}$ can be written as

$$C_{eff} = \phi C_1 + (1-\phi)C_2 \tag{13}$$

$$\frac{K_{eff,zz}}{K_m} = 1 + (\frac{K_1 - K_m}{K_m})\phi \tag{14}$$

$$\frac{K_{eff,xy}}{K_m} = 1 + \frac{2\phi}{A_1 - \phi + A_2(0.30584\phi^4 + 0.013363\phi^8)} \tag{15}$$

where C_1, C_m, K_1, and K_m are the thermal capacity and conductance of the filler and matrix, respectively, ϕ is the volume fraction of the filler, $A_1 = \frac{K_1+K_m}{K_1-K_m}$, and $A_2 = \frac{K_2+K_m}{K_2-K_m}$. Consider that the filler is air and its thermal capacity and conductivity are close to zero. When ϕ is close to 1, the thermal capacity and conductivity of the porous material will be very small. It is therefore possible to obtain unexpectedly high temperatures under a relatively mild excitation using specially designed plasmonic nanostructures [18]. This provides a new solution for energy-intensive processes, e.g., water desalination and sewage treatment [19]. In the chemical analysis, it has been demonstrated that the high temperature can be used to improve the efficiency of laser-induced desorption and ionization [58].

2.5. Metasurfaces and Light Manipulation

The plasmon resonance effect not only provides a means for the enhancement/regulation of the local field strength but also enables the manipulation of the local phase and polarization of the optical field [26,27]. In other words, it is possible to manipulate light arbitrarily at any point in space. Based on this idea, Capasso and coworkers proposed the concept of metasurfaces, which use arrays of plasmon nanostructures to regulate wavefronts at every point in space. This idea greatly expands the design freedom of optical devices and provides a new path for the development of optical sensing devices and analytical instruments [31,32].

In light manipulation, precise control of phase is crucial. There are two main methods: resonance tuning and geometric phase. The former uses the resonance behavior of the optical nanostructure to tune the local phase of light [26]. By changing the resonance across the working frequency, one can adjust the phase of the scattered light, achieving a 180-degree phase shift. The method is simple and efficient but is wavelength-dependent and often causes undesired dispersions. Another method is the geometric phase [59–61]. We know that linearly polarized light is the superposition of left-handed and right-handed circular polarized components. After a simple calculation, one can find that the phase of the left and right circular polarization components is linearly related to the direction of the linearly polarized light. In other words, the phase of circularly polarized light can be controlled by simply manipulating the polarization direction of the field using anisotropic nanostructures. The method is solely determined by the geometric parameters of the nanostructure and is independent of wavelength but has relatively low conversion efficiency. In addition to the above methods, other phase control methods were also demonstrated, e.g., the waveguide method in dielectric metasurfaces. Using the above method, traditional bulky optical systems can be replaced with an ultra-thin planar structure. This opens up many new exciting possibilities for future instrument design [27].

3. Surface-Enhanced Spectroscopies

In this section, we will give a brief review of the surface-enhanced spectroscopies, which are direct applications of large field enhancement of plasmonic nanostructures. They are also the most important and widely studied topics in nanostructure-based chemical analysis.

3.1. Surface-Enhanced Raman Scattering

Raman spectroscopy is commonly known as the fingerprint of molecules in analytical chemistry, but Raman signals are commonly extremely weak, preventing them from being used in many applications. Because of this, SERS attracted a lot of attention after it was first reported in the 1970s [62–64]. In particular, Nie and Kneipp's groups independently reported single-molecule Raman measurements from some "hot spots" of Ag nanoparticles [12,13], and for the first time, people were capable of performing structural analysis for individual molecules in an ambient environment. Encouraged by this, many researchers carried out research over the last two decades to develop SERS into a quantitative, robust, and reliable ultrasensitive analytical method. However, it was found that SERS had a series of limitations due to its own mechanism [65].

One of the major challenges in SERS is the fabrication of high-performance SERS substrates, which is a long-standing issue in the field. In the early stages, SERS is often performed with rough metal surfaces or metallic nanoparticle aggregates, in which "hot spots" can be randomly formed in the nanogaps between particles (Figure 5a) [66]. Later, the concept of nano-antennas was introduced [43], and "hot spots" can be designed and fabricated using laterally coupled structures in a "controllable" fashion (Figure 5c) [43,67–69]. However, until around 2010, "hot spots" on SERS substrates were always sparse and random. This sparse random "hot spots" problem stems from the fact that the enhancement factor of a "hot spot" is extremely sensitive to its local geometry. The SERS performance can be very different even if two nanostructures are almost identical under electron microscopy. For example, one of the authors of this work demonstrated that even 1 nm surface roughness can lead to a one order of magnitude change in the Raman signal [70]. Moreover, nanometer features are unstable under illumination due to local photothermal effects [71]. It is therefore almost impossible to create "hot spots" in a repeatable fashion because it is extremely difficult to control the detailed features at the nanoscale using conventional nanofabrication techniques.

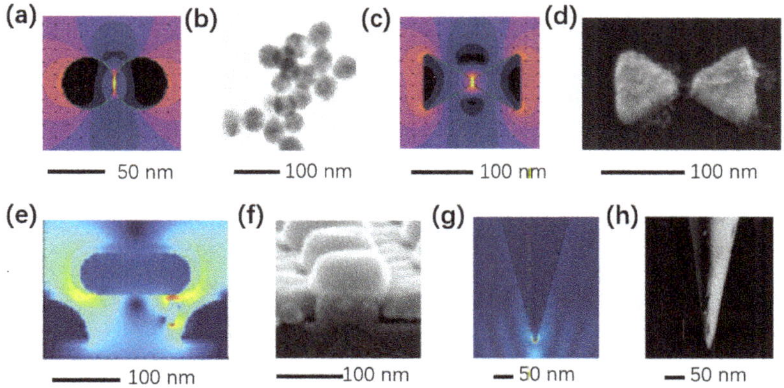

Figure 5. Different types of "hot spots" in surface-enhanced spectroscopies. (**a**,**b**) The field enhancement and SEM image of a nanoparticle cluster [72]. (**c**,**d**) Plasmonic bow-tie nano-antenna. (**e**,**f**) Vertically coupled plasmonic nanocavity array [73]. (**g**,**h**) Metallic nanotip [72].

To solve this problem, various vertically coupled designs were proposed to circumvent the fabrication challenge faced by laterally coupled structures (Figure 5e). Currently, deposition and etching techniques allow us to control the thickness of a layer at sub-nanometer precision. It is, therefore, possible to construct vertically coupled nanostructures with extremely high precision [74–76]. To date, various vertically coupled structures with a high density and large area uniformity of "hot spots" have been reported, and with these structures, SERS has gradually become a quantitative analytical tool [77].

Recently, some successful strategies that partly overcome this problem have been reported. One work reported that alkanethiolate ligand-regulated Ag nanoparticle films can be used to achieve quantitative SERS measurements down to the single-molecule level [78]. Another showed that arrays of weakly coupled Ag nanohelices achieved both homogeneous and strong near-field enhancements [79]. These works allow reproducible SERS detection over a large area with excellent uniformity and a high Raman enhancement factor.

Another challenge in SERS is that the SERS signals from "hot spots" are very unstable and often contaminated. The main reason is that the local metal structures of "hot spots" are very "active" due to the photothermal effect, as well as photocatalysis processes. To avoid the above problems, researchers have developed ultra-thin protective layer technology for SERS substrates, which greatly improves the stability of signals [80].

In addition to top-down microfabrication technology, there are also many interesting progresses with bottom-up methods. One promising approach is the combination of chemistry and microfluidics technology, which creates clusters of particles in large numbers in solutions that stabilize the total SERS signal during measurements [81]. This method is based on microfluidic chips, so it can be easily combined with pretreatment functions, i.e., separation and enrichment, making it a powerful tool for real-world applications.

In recent years, there has been growing interest in exploring novel SERS-active substrates to enhance the Raman signals of target molecules. One work reports Mo2C as a highly sensitive semiconductor substrate [82], and another uses hat-shaped MoS2 films to separate two layers of metal nanoparticles and exhibits superior SERS capability [83]. These recent works collectively contribute to the understanding of 2D films as SERS-active substrates and showcase their potential as promising candidates for sensitive molecular detection and spectroscopic analyses.

In addition, new measurement techniques for SERS were also developed to meet the request of different application scenarios. One example is the surface-enhanced spatially offset Raman spectroscopy (SESORS) [84]. By introducing a spatial offset between the excitation and detection optical paths, it allows selective detection of the Raman signal from deep tissues [85,86]. Another important SERS measurement technique is the fiber-integrated SERS substrate. In recent decades, different design and fabrication strategies were developed. The flexibility of the fiber system extended the detection to environments that cannot be reached by traditional substrates, such as tissues, cells, and harsh environments [87,88].

Today, SERS is not only used for the detection of trace amounts of chemicals but has also been applied in complex systems, for example, the detection and photothermal sterilization of bacteria [89] and intracellular detections using single-particle SERS. In situ real-time detection and tracing of various chemicals in cells are essential for understanding the basic processes of life. The current main method involves using gene editing technology to make specific proteins have luminescent properties and observing and analyzing them under fluorescence microscopy. While achieving great success, fluorescence methods also have limitations. They require labeling and also face difficulties in detecting small molecules. But SERS does not suffer from these issues. It enhances the signal of any molecule which is absorbed on nanoparticles, making SERS a suitable tool for the detection of small molecules, especially metabolism-related small molecules [90–94]. For more relevant developments in this field, one can read a related review article by Ren [37].

The above ability to detect chemical substances in complex systems shows the great potential of SERS in medical diagnosis [95,96]. For example, SERS has been proven to be a sensitive and effective method for detecting cancer [97–100]. Other applications include the detection of HIV and COVID-19 [95]. Researchers have demonstrated a reliable detection of COVID-19 based on soft SERS substrates [101].

SERS can also be used as a powerful tool for food safety and environmental detection. In the past two decades, there have been significant advances in these fields, which have been well summarized in recent review articles [102–105].

When talking about SERS, one also needs to mention tip-enhancement Raman spectroscopy (TERS) (Figure 5g), which is the combination of SERS and scanning probe microscopy. TERS utilizes a sharp metallic tip as a single "hot spot". By scanning the "hot spot" and collecting SERS signals point-by-point, one can map the chemical information of a sample at nanometer resolution with single-molecule sensitivity [106,107]. This provides researchers with a powerful tool for understanding the composition of various chemical compositions and chemical reactions at interfaces [71,108–110]. Despite its huge successes, today, the application of TERS is still limited, and it takes a lot of effort to obtain a high-quality TERS image. One of the main reasons is the reliability of the probes used. The fabrication of high-performance, chemically stable probes that are not easily damaged or contaminated is still a challenging issue in this field.

Another important issue with TERS is that it requires a complex and expensive optical system to excite and collect the Raman signal from the tip. This system needs to be well integrated with a scanning probe microscope, and the precision of aligning the optical beam with the tip-end needs to be subwavelength. To address this issue, one way is to integrate the metallic probe with waveguides, which can guide the excitation light to the tip apex and collect the signal from the metallic tip to far-field detectors. For example, Liu and his colleagues developed the nanowire-fiber integrated tip, which can be directly used with a commercial scanning tunneling microscope without any additional supporting optical system [111]. This greatly simplified the design of TERS systems.

3.2. Surface-Enhanced Absorption Spectroscopies

UV-Vis absorption spectroscopy and IR absorption spectroscopy are the two most widely used spectroscopic tools in analytical laboratories. In particular, IR spectroscopy, like Raman spectroscopy, is the fingerprint spectroscopy for chemical analysis, and the information of molecular functional groups can be accurately obtained through the analysis of characteristic absorption peaks, which has many applications in material science, surface science, and other fields.

Conventionally, surface-enhanced infrared absorption spectroscopy (SEIRA) is mainly performed with metal plasmon antennas using the local field enhancement effect [67,112,113].

However, in the infrared region, metals tend to be perfect conductors, and their performance is close to that of traditional antennas. The signal enhancement caused by the resonance of the materials discussed earlier is not significant as in the case of the visible or near-IR regime. To this end, similar to the development of substrates for SERS, new vertical antenna-based coupled designs were introduced, which greatly improved the performance of the enhanced spectrum [114].

Interestingly, it has been found that graphene provides excellent properties in the IR region that traditional metal structures do not have. For graphene, the concentration of its conductive electrons is relatively low compared with metals, and its plasmon frequency is therefore in the IR regime instead of the visible range. Moreover, graphene is atomically thin, and this causes strong field enhancement at the edges, which are extremely sharp. Thanks to the above properties, graphene substrates have received a lot of attention in the field of SEIRA [115,116]. However, compared with SERS, the sensitivity of SEIRA is relatively low and cannot reach the single-molecule level. In addition, optical components are expensive for the IR regime. This limits the application of SEIRA.

There are also reports on surface-enhanced UV-Vis absorption spectroscopy, but the number is much lower compared with SERIA and SERS. This is because of several different factors, mainly the limitation of materials, the high price of optical equipment, and the lack of structural information in the ultraviolet spectral region. Moreover, the intrinsic absorption section of organic molecules in the UV-Vis regime is relatively large, and in most cases, the sensitivity already meets the requirements.

3.3. *Surface-Enhanced Fluorescence*

Surface-enhanced fluorescence (SEF) also has a long history, similar to the case of SERS. However, because the signal intensity of commonly used fluorescent molecules is much stronger than that of Raman scattering and infrared absorption, single-molecule fluorescence detection and tracing at room temperature can be achieved without any enhancement, and surface-enhanced fluorescence has received far less attention in chemical analysis than SERS and SEIRA.

This situation changed considerably after near-IR (NIR) dyes became popular in bioimaging in bio-/chem-sensing. Compared with fluorescence techniques in the visible range, near-infrared dyes offer a larger penetration depth in tissue imaging and lower background thanks to their low autofluorescence background. These properties make NIR dyes very popular for in vivo bioimaging and enzyme-linked immunosorbent assays (ELISA). However, the absorption cross-section of NIR dyes is much smaller than that in the visible or UV spectral range, and the fluorescence yield is often less than 10%. These properties bring up the issue of low signal again in a way similar to Raman spectroscopy, and it is therefore important to find ways to enhance the fluorescence of NIR dyes.

To address this, Chou's group developed a vertically coupled antenna-based technology [74,117], which can enhance the average fluorescence by thousands of times, and at "hot spots", the fluorescence signal can be enhanced by up to six orders of magnitude [73,118]. Using this method, the detection limit of the Ebola virus was successfully pushed to sub-fM [119]. This technology is compatible with existing ELISA detection equipment, and using nanoimprinting technology, plasmon substrates themselves can be prepared in large quantities. It is believed that this high-performance SEF-based technique will play an increasingly important role in the field of high-sensitivity detection, especially in the early diagnosis of major diseases.

4. Index Sensing and Laser-Induced Ionization with Plasmonic Nanostructures

In this section, we review the applications of plasmonic nanostructures based on more complex effects, such as optical forces and photothermal effects.

4.1. Refractive Index Sensing

As aforementioned, the resonance frequency of a plasmon resonance mode can be shifted by the presence of external analytes due to the works of optical forces. This effect makes plasmonic nanostructures an important tool for index sensing in chemistry (Table 1). Researchers have invested great enthusiasm in developing different types of refractive index sensors, and different reading and processing methods were also developed. Today, thin-film-based SPR sensors (Figure 6a) have become the gold standard for many applications in bio-/chem-analysis with a sensitivity down to 10^{-7} RIU, and they have been well documented by review works by different groups. We therefore only focus on individual nanostructure-based sensors as well as array structures, which are less covered.

Table 1. Comparison between SPR, LSPR, and ELISA.

	SPR	LSPR	ELISA
Label-free	Yes	Yes	No
Sensitivity	10^{-6} nm/RIU	10^{-2} nm/RIU	10^{-18} M [119]
Single-molecule detection	No	Yes [120]	Yes
Detection mode	Absorption wavelength/angle, imaging	Scattering, extinction, imaging	Fluorescence, imaging
Spatial resolution	10 μm	1 μm	1 μm
Real-time detection	Yes	Yes	No
Multiplexing	Yes	Yes	Yes

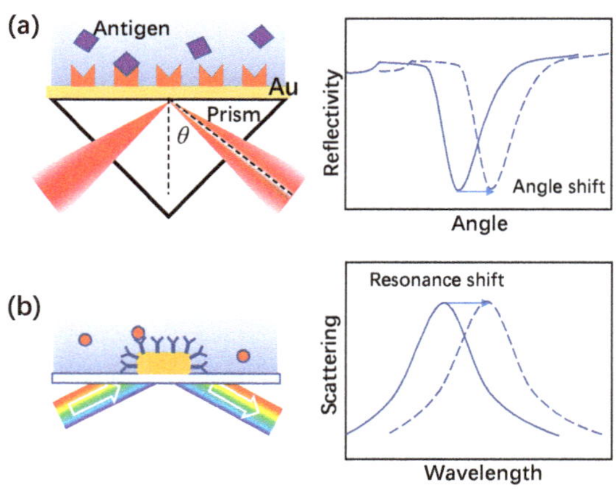

Figure 6. Principle of SPR sensor (**a**) and LSPR sensor (**b**).

Figure 6 shows the principle of this type of sensor. For a plasmonic nanoparticle, when a single molecule or viral particle is adsorbed on it, the resonance wavelength will undergo a step-like shift due to a change of the system's energy (Figure 6b), allowing us to probe a single particle or even single molecule in its local environment [17,121,122]. Array structures were also reported. They are mainly used for measuring the average concentration of solutions via continuous resonance shifts induced by global index changes [123].

Compared with conventional sensors, individual nanostructure-based LSRP sensors offer better sensitivity and spatial information. In recent decades, many interesting results have been reported using individual nanostructure-based LSPR sensors. For example, using separated plasmonic nanorods, researchers demonstrated the detection of single protein molecules [120]. When further combined with plasmon nano-antennas and traditional whispering-gallery-mode microcavities, the sensitivity can be further improved [124–126]. These above methods are based on the displacement detection of formants, which requires sophisticated spectroscopic equipment. Interestingly, it has been reported that by using nanopore design, optical detection of single molecules can be achieved by measuring the transmission intensity directly without requiring sophisticated spectroscopic equipment [127–129].

Due to their small size, plasmonic nanoparticles can even be used to detect local chemical information in living cells. This is because their scattering spectra can still be clearly recorded with dark-field imaging spectrographic instruments even after being injected into tissues [130,131]. Thanks to the high sensitivity and stability of plasmonic nanoparticles, they are expected to play an increasingly important role in in situ real-time analysis for life science in the future.

LSPR sensors can also be integrated with optical fibers [132,133]. In fact, fibers can be used for index sensing themselves using their own optical modes, such as the lossy-mode resonance (LMR)-based sensing technique [134–138]. The integration of LSPR can improve the sensitivity of the fiber-based sensors and further enable the combination of the SERS technique.

Thanks to their high sensitivity and ease of use, SPR and LSPR are now widely used in a variety of applications. Compared to ELISA, they can detect dynamic processes, which makes SPR technology particularly important in measuring molecular binding processes.

4.2. Photothermal Effects and Their Applications in Mass Spectrometry

As aforementioned, plasmonic nanostructures often exhibit strong photothermal effects. With their large light absorption cross-section and nanoscale volume, plasmonic nanostructures can be instantly heated to hundreds or even thousands of degrees when excited by laser pulses. This high temperature can be used to melt, desorb, and even break chemical bonds of the sample, making plasmonic nanostructures an interesting choice for building ion sources for MS measurements.

In MS, laser-induced desorption/ionization (LDI) is one of the most important ionization techniques for biomolecules [139,140]. It often uses small organic molecules as an assisting material to improve the efficiency of LDI and soften the ionization process of biomolecules. In the matrix-assisted LDI (MALDI) technique, the matrix molecules are ionized first, immediately after being hit by laser pulses, and then they transfer their charges to the target sample molecules. Because the energy is mainly absorbed by the matrix molecules, it does not destroy the structure of the organic analyte and can consequently avoid fragmentation issues. However, due to the presence of matrix molecules, there is always a high background noise level in the low mass regime. To address this issue, people developed surface-assisted laser desorption/ionization (SALDI) technology, which uses the photothermal effect of micro- and nanostructures to achieve the efficient, background-free desorption and ionization of organic molecules [141].

To date, many different types of nanostructures have been used for SALDI, but surprisingly, plasmonic nanostructures did often not show any advantages over nanostructures made of other materials until very recently [142,143]. This is because most of the plasmonic nanostructures are designed for surface-enhanced spectroscopies and are not optimized for the photothermal effect. Things only started to change in recent years. With the understanding of plasmon photothermal effects becoming deeper, high-performance plasmonic photothermal nanostructures were continuously reported. In particular, porous plasmonic nanostructures have emerged, which exhibit extremely high absorption efficiency, small heat capacity, and low thermal conductivity at the same time [18,19]. Using this structure, the authors successfully demonstrated the ionization of biomolecules. The results showed that its ionization efficiency is several times higher than the case of traditional MALDI substrates without background noise in the low mass charge region (Figure 7) [58].

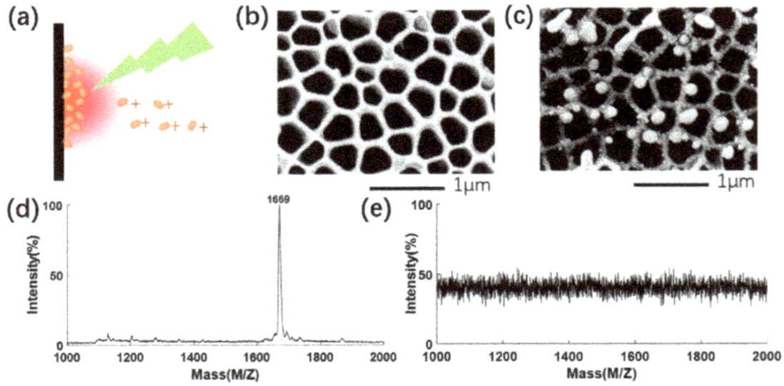

Figure 7. Plasmon-assisted laser desorption/ionization [58]. (**a**) Principle of laser desorption/ionization. (**b**,**c**) Porous plasmonic substrate before and after laser excitation. (**d**) Plasmonic substrate leads to a strong and clean signal. (**e**) No signal can be observed without the substrate.

It is worth noting that the strong photothermal effect can also lead to the complete dissociation of molecules and the formation of plasma. This allows plasmon structures to be applied to atomic spectroscopy (LIPS, laser-induced plasma spectroscopy).

5. Metasurface-Based Chemical Sensing and Analysis

In the previous section, all the applications are based on the properties of the individual nanostructures or uniform structure arrays. In this section, we will discuss the application of nonuniform nanostructure arrays (i.e., metasurfaces). They can map the hidden optical information into directly detectable intensity information, opening up many new possibilities for simplification and miniaturization of conventional analytical techniques.

5.1. Plasmonic Gradient as a Miniaturized Bio-Sensor

With the development of information technologies, particularly the Internet of Things (IoTs), people began to imagine that diagnostic equipment should be more personalized and decentralized, utilizing smart miniaturized point-of-care testing (POCT) equipment [144–146]. In recent years, the emergence of various new high-performance miniaturized chemical and biochemical sensors is making this dream a reality. Especially, plasmonic sensors (e.g., SPR and LSPR sensors) exhibit great potential for building such devices thanks to their small size and high signal strength. Meanwhile, the fast growth of consumer electronics, smartphones, and smartwatches started to integrate communication, computing, and advanced imaging functions together. It makes the integration of SPR/LSPR sensors and mobile phones a very promising direction [146,147].

The major function of SPR and LSPR sensors is to convert the refractive index information into absorption spectra or angular spectra that can be read by optical inspection equipment [148]. However, reading both absorption spectra and angular spectra requires additional equipment. To simplify the measurement, transmission or reflection signals at a single fixed wavelength are often used instead of the full spectra, but it is at the expense of measurement accuracy. To address the above issue, the authors propose an image-based ultrasensitive sensing method using gradient plasmon structures (plasmon metasurfaces), as shown in Figure 8 [149,150]. It is a 2D array of plasmonic nanorods whose resonance wavelength continuously varies from the center to the edge. When illuminated by a monochromic light source, the light will be absorbed at the position where the resonance matches the wavelength of the excitation light, and this will lead to a dark resonant ring. If the surrounding environment changes, the size of the ring will change too due to the resonant wavelength shift of the plasmonic nanorods. Such pattern changes can be accurately recorded and analyzed by mobile phones for ultrasensitive sensing purposes (as shown in Figure 8b,c). The results show that its sensitivity is comparable to large research-level instruments.

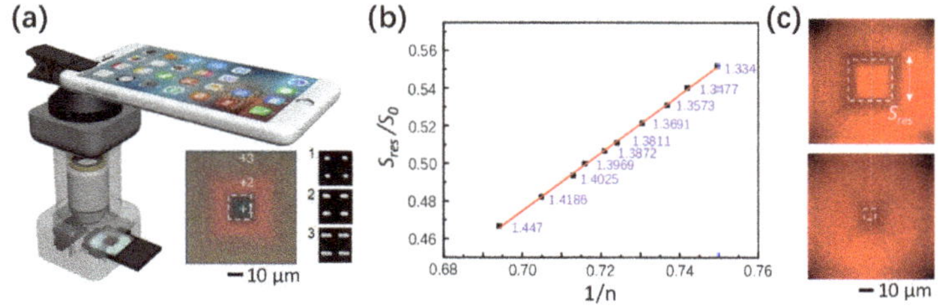

Figure 8. Smartphone-based plasmonic sensors. (**a**) A patterned plasmonic gradient sensor and its reader are capable of quantitatively measuring the environmental refractive index. (**b**) The size of the resonance ring, S_{res}, changes when the refractive index of the surrounding medium, n, changes. (**c**) S_{res} is linearly related to $1/n$ [150].

In addition to refractive index sensors based on local plasmon resonance, people have also tried to combine handheld spectrometers and surface-enhanced fluorescence sensors. In theory, this should lead to higher sensitivity compared to the LSPR-based sensors, but it also requires expensive light sources, filters, and more complex optics.

5.2. Snapshot CD Spectroscopy with a Metasurface

With the emergence of metasurfaces, we are gaining unprecedented light manipulation capability, leading to many new opportunities for the miniaturization of analysis instruments, including optical imaging, polarization measurements, and spectral analysis. Today, there have been a few reviews on optical imaging applications [30,59]; therefore, and we limit our discussion on polarization analysis, more specifically, CD spectroscopy, which is important for biochemical analysis.

CD spectroscopy measures the difference between optical interactions with left- and right-handed circular polarized light. It is the most widely used characterization method for measuring the chirality of molecules, having many important applications in chemistry, life science, and medicine. Because the CD signals are several orders of magnitude weaker than traditional spectroscopic signals, it requires complex optical modulation–demodulation equipment to retrieve them [151]. This makes CD spectroscopy a complex and expensive analytical technique.

Interestingly, with the help of polarization gratings, the measurement for CD spectra can be greatly simplified. The polarization grating is essentially a metasurface, which diffracts left-handed and right-handed circular polarized components of light into different directions with equal efficiencies [152]. The CD measurement can therefore be performed by simply collecting signals with different polarizations at the same wavelength and calculating their differences. In addition, polarization gratings can be coupled to microscopic imaging systems, making it possible to collect signals from a single nanostructure, as shown in Figure 9. Using this method, the authors and collaborators developed the snapshot CD spectroscopy technique and demonstrated the CD spectroscopy of single DNA-assembled 3D nanostructures [35].

It should be noted here that the polarization grating used here is made of liquid crystals instead of metallic nanostructures because liquid crystals are adjustable, transparent, and convenient to process. It is also possible to fabricate polarization gratings with dielectric or metallic nanostructures, but the nanofabrication is much more expensive than the case of liquid crystals.

Finally, it is worth noting that with the emergence of the metasurface technique, the boundaries between imaging, spectrometry, and polarimetry in conventional optical design are becoming blurred. One can map the spectral and polarization information from any spatial point in a sample onto intensity distributions on any given plane, which then can be recorded with imaging devices. This unique capability is particularly important today because high-quality imaging devices with a large format have become available in our daily life. For instance, the pixel number of the CMOS sensor in a smartphone has reached

100 million. One can therefore expect that high-performance and ultracompact spectropolarimetric imaging devices will be integrated into consumer electronic products in the near future. This will dramatically change the landscape of chemical sensing and analysis.

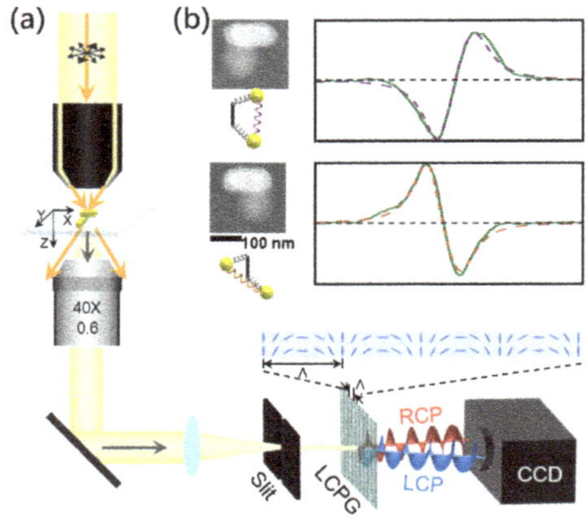

Figure 9. Snapshot CD spectroscopy of single nanostructure using a polarization grating [35]. (**a**) Schematic drawing of the CD spectrometer. (**b**) Scattering CD spectra of right-hand and left-hand enantiomer, respectively. The insets are the SEM images of the nanostructures.

6. Applications in Life Science and Theranostics

As discussed above, plasmons can be used as nanoscale optical sensors, energy converters, and spectral signal amplifiers. These properties offer great potential applications for plasmons in life sciences and diagnostics.

Intracellular plasmonics is a cutting-edge field focusing on the use of plasmonic nanostructures within living cells. By introducing these nanostructures, researchers can enhance imaging contrast, enable targeted therapies, and probe cellular processes at the nanoscale [153–156]. For example, one work reports that by using multifunctional gold nanoparticles, these intracellular sensors can monitor actin rearrangement in live fibroblasts [157]. Another work shows an electrochemical impedance microscope based on surface plasmon resonance that resolves local impedance with submicrometer spatial resolution and monitors the dynamics of cellular processes with millisecond time resolution [158]. Intracellular plasmonics offers novel insights into cell biology and holds great potential for revolutionizing cell-based research and personalized medicine.

In recent years, plasmonic nanoparticles have emerged as key players in advancing theranostics. These nanoparticles enhance imaging contrast in techniques like SERS and photoacoustic imaging while serving as efficient carriers for site-specific drug release [159,160]. Additionally, plasmons enable novel treatments like photothermal therapy, where localized heating selectively destroys cancer cells [20,161,162]. Nowadays, many novel plasmonic nanoparticles have been applied to theranostics. For example, plasmonic nanobubbles lead to rapid heating and vaporization of the surrounding medium, which has been explored for targeted drug delivery and tumor ablation [163]. Another example is gap-enhanced Raman tags (GERTs), which are a class of nanoscale structures that significantly amplify the Raman scattering signal of molecules attached to or within their nanogap regions. These GERTs have been applied as the second near-infrared window (NIR-II) SERS tags and achieved successes from biodetection to theranostics owing to their simultaneous extra-high SERS response, ultra-photostability, and multiplexing capability [164,165]. The development of plasmons in theranostics holds great promise for personalized and effective healthcare.

7. Summary and Outlook

From the above review, it can be seen that optical nanostructures, especially plasmon nanostructures, can significantly enhance the optical effects at the nanoscale, which are otherwise extremely weak. These nanostructures can therefore function as a bridge that connects the nanoscale chemical information in its near field to the far-field propagating waves, allowing researchers to collect and analyze the spectral information from trace amounts of samples or even a single molecule. In the last decade, the field of nano-optics has become mature. With the development of computational electromagnetic technology and nanofabrication technology, the design and fabrication of plasmonic sensors have become standard processes. The focus of research has begun to shift from conventional surface-enhanced spectroscopies to applications that involve complex multi-physical effects, such as LSPR-based sensing and SALDI. The former is associated with the near-field optical trapping phenomenon, and the latter relies on the nanoscale temporal photothermal processes of plasmonics.

Another important trend for nanostructure-based chemical analysis is integration and miniaturization. Today, sensing techniques for physical signals have matured, but there is still a lack of biochemical sensors which are small, integrated, and highly selective. At the same time, information techniques, especially IoT techniques, are growing at a breathtaking speed, and the demand for real-time and reliable monitoring methods of biochemical information for the environment, public safety, and health is becoming increasingly urgent. Optical nanostructures, such as plasmonics and metasurfaces, provide an attractive solution for high-performance, miniaturized, and low-cost sensors. We believe that driven by the market, the development of reliable, highly selectable, and integrable nanostructure-based biochemical sensors will be an important trend in the next decade.

Finally, it is worth mentioning that optical measurement methods have made significant progress in the last decade. Nonlinear optics and quantum optics technologies have begun to make their way from basic research to various application fields, including sensing [166,167]. It has been experimentally demonstrated that nonlinear and quantum effects can improve the performance of optical nanostructure-based sensors, such as the signal-to-noise ratio and accuracy [168–171]. We believe that this will bring many new opportunities in the future.

Author Contributions: Writing—original draft preparation, C.D. and W.Z.; surface-enhanced spectroscopy Y.W.; theoretical part, X.Z.; LSPR sensor, J.B. All authors have read and agreed to the published version of the manuscript.

Funding: This research was funded by National Key Technologies R&D Program of China (No. 2021YFA1400803).

Informed Consent Statement: Not applicable here.

Data Availability Statement: Not applicable.

Conflicts of Interest: The authors declare no conflict of interest.

References

1. McHale, J.L. *Molecular Spectroscopy*, 2nd ed.; CRC Press: Boca Raton, FL, USA, 2017.
2. Koleżyński, A.; Król, M. *Molecular Spectroscopy—Experimental and Theroy*; Springer: Berlin/Heidelberg, Germany, 2019; Volume 26.
3. Lal, S.; Link, S.; Halas, N.J. Nano-optics from sensing to waveguiding. *Nat. Photonics* **2007**, *1*, 641–648. [CrossRef]
4. Willets, K.A.; Van Duyne, R.P. Localized Surface Plasmon Resonance Spectroscopy and Sensing. *Annu. Rev. Phys. Chem.* **2007**, *58*, 267–297. [CrossRef]
5. Homola, J. Surface plasmon resonance sensors for detection of chemical and biological species. *Chem. Rev.* **2008**, *108*, 462–493. [CrossRef] [PubMed]
6. Pohl, D.W.; Kawata, S.; Inouye, Y.; Fischer, U.C.; Dereux, A.; Weeber, J.-C.; Hayashi, S.; Okamoto, T.; Novotny, L.; Sugiura, T.; et al. *Near-Field Optics and Surface Plasmon Polaritons*; Kawata, S., Ed.; Springer: Berlin/Heidelberg, Germany, 2001.
7. Maier, S.A. *Plasmonics: Fundamentals and Applications*; Springer Science & Business Media: Berlin/Heidelberg, Germany, 2007.
8. Novotny, L.; Hecht, B. *Principles of Nano-Optics*; Cambridge University Press: Cambridge, UK, 2012.
9. Bohren, C.F.; Huffman, D.R. *Absorption and Scattering of Light by Small Particles*; John Wiley & Sons, Inc.: Hoboken, NJ, USA, 1983.

10. Raether, H. *Surface Plasmons on Smooth and Rough Surfaces and on Gratings*; Springer: Berlin/Heidelberg, Germany, 1989.
11. Kreibig, U.; Vollmer, M. *Optical Properties of Metal Clusters*; Springer: Berlin/Heidelberg, Germany, 1995; Volume 25.
12. Kneipp, K.; Wang, Y.; Kneipp, H.; Perelman, L.T.; Itzkan, I.; Dasari, R.R.; Feld, M.S. Single Molecule Detection Using Surface-Enhanced Raman Scattering (SERS). *Phys. Rev. Lett.* **1997**, *78*, 1667. [CrossRef]
13. Nie, S.; Emory, S.R. Probing Single Molecules and Single Nanoparticles by Surface-Enhanced Raman Scattering. *Science* **1997**, *275*, 1102–1106. [CrossRef]
14. Englebienne, P. Use of colloidal gold surface plasmon resonance peak shift to infer affinity constants from the interactions between protein antigens and antibodies specific for single or multiple epitopes. *Analyst* **1998**, *123*, 1599–1603. [CrossRef]
15. Mulvaney, P. Surface Plasmon Spectroscopy of Nanosized Metal Particles. *Langmuir* **1996**, *12*, 788–800. [CrossRef]
16. Ament, I.; Prasad, J.; Henkel, A.; Schmachtel, S.; Soennichsen, C. Single Unlabeled Protein Detection on Individual Plasmonic Nanoparticles. *Nano Lett.* **2012**, *12*, 1092–1095. [CrossRef]
17. Pang, Y.; Gordon, R. Optical Trapping of a Single Protein. *Nano Lett.* **2012**, *12*, 402–406. [CrossRef]
18. Hogan, N.J.; Urban, A.S.; Ayala-Orozco, C.; Pimpinelli, A.; Nordlander, P.; Halas, N.J. Nanoparticles Heat through Light Localization. *Nano Lett.* **2014**, *14*, 4640–4645. [CrossRef]
19. Zhou, L.; Tan, Y.L.; Wang, J.Y.; Xu, W.C.; Yuan, Y.; Cai, W.S.; Zhu, S.N.; Zhu, J. 3D self-assembly of aluminium nanoparticles for plasmon-enhanced solar desalination. *Nat. Photonics* **2016**, *10*, 393. [CrossRef]
20. Huang, X.; Jain, P.K.; El-Sayed, I.H.; El-Sayed, M.A. Plasmonic photothermal therapy (PPTT) using gold nanoparticles. *Lasers Med. Sci.* **2008**, *23*, 217–228. [CrossRef]
21. Krioukov, E.; Klunder, D.J.W.; Driessen, A.; Greve, J.; Otto, C. Sensor based on an integrated optical microcavity. *Opt. Lett.* **2002**, *27*, 512–514. [CrossRef]
22. Vollmer, F.; Braun, D.; Libchaber, A.; Khoshsima, M.; Teraoka, I.; Arnold, S. Protein detection by optical shift of a resonant microcavity. *Appl. Phys. Lett.* **2002**, *80*, 4057–4059. [CrossRef]
23. Armani, A.M.; Kulkarni, R.P.; Fraser, S.E.; Flagan, R.C.; Vahala, K.J. Label-free, single-molecule detection with optical microcavities. *Science* **2007**, *317*, 783–787. [CrossRef]
24. Sun, Y.Z.; Fan, X.D. Optical ring resonators for biochemical and chemical sensing. *Anal. Bioanal. Chem.* **2011**, *399*, 205–211. [CrossRef] [PubMed]
25. Rho, D.; Breaux, C.; Kim, S. Label-Free Optical Resonator-Based Biosensors. *Sensors* **2020**, *20*, 5901. [CrossRef] [PubMed]
26. Yu, N.F.; Genevet, P.; Kats, M.A.; Aieta, F.; Tetienne, J.P.; Capasso, F.; Gaburro, Z. Light Propagation with Phase Discontinuities: Generalized Laws of Reflection and Refraction. *Science* **2011**, *334*, 333–337. [CrossRef] [PubMed]
27. Yu, N.F.; Capasso, F. Flat optics with designer metasurfaces. *Nat. Mater.* **2014**, *13*, 139–150. [CrossRef] [PubMed]
28. Khorasaninejad, M.; Chen, W.T.; Devlin, R.C.; Oh, J.; Zhu, A.Y.; Capasso, F. Metalenses at visible wavelengths: Diffraction-limited focusing and subwavelength resolution imaging. *Science* **2016**, *352*, 1190–1194. [CrossRef]
29. Khorasaninejad, M.; Chen, W.T.; Oh, J.; Capasso, F. Super-Dispersive Off-Axis Meta-Lenses for Compact High Resolution Spectroscopy. *Nano Lett.* **2016**, *16*, 3732–3737. [CrossRef]
30. Faraji-Dana, M.; Arbabi, E.; Kwon, H.; Kamali, S.M.; Arbabi, A.; Bartholomew, J.G.; Faraon, A. Hyperspectral Imager with Folded Metasurface Optics. *ACS Photonics* **2019**, *6*, 2161–2167. [CrossRef]
31. Rubin, N.A.; D'Aversa, G.; Chevalier, P.; Shi, Z.J.; Chen, W.T.; Capasso, F. Matrix Fourier optics enables a compact full-Stokes polarization camera. *Science* **2019**, *365*, eaax1839. [CrossRef] [PubMed]
32. Wang, R.X.; Ansari, M.A.; Ahmed, H.; Li, Y.; Cai, W.F.; Liu, Y.J.; Li, S.T.; Liu, J.L.; Li, L.; Chen, X.Z. Compact multi-foci metalens spectrometer. *Light Sci. Appl.* **2023**, *12*, 103. [CrossRef]
33. Solomon, M.L.; Abendroth, J.M.; Poulikakos, L.V.; Hu, J.; Dionne, J.A. Fluorescence-Detected Circular Dichroism of a Chiral Molecular Monolayer with Dielectric Metasurfaces. *J. Am. Chem. Soc.* **2020**, *142*, 18304–18309. [CrossRef]
34. Lin, P.; Chen, W.T.; Yousef, K.M.A.; Marchioni, J.; Zhu, A.; Capasso, F.; Cheng, J.X. Coherent Raman scattering imaging with a near-infrared achromatic metalens. *APL Photonics* **2021**, *6*, 096107. [CrossRef] [PubMed]
35. Zhou, S.; Bian, J.; Chen, P.; Xie, M.; Chao, J.; Hu, W.; Lu, Y.Q.; Zhang, W.H. Polarization-dispersive imaging spectrometer for scattering circular dichroism spectroscopy of single chiral nanostructures. *Light Sci. Appl.* **2022**, *11*, 64. [CrossRef] [PubMed]
36. Graham, D.; Moskovits, M.; Tian, Z.-Q. SERS—Facts, figures and the future. *Chem. Soc. Rev.* **2017**, *46*, 3864–3865. [CrossRef]
37. Zong, C.; Xu, M.; Xu, L.-J.; Wei, T.; Ma, X.; Zheng, X.-S.; Hu, R.; Ren, B. Surface-Enhanced Raman Spectroscopy for Bioanalysis: Reliability and Challenges. *Chem. Rev.* **2018**, *118*, 4946–4980. [CrossRef]
38. Mayer, K.M.; Hafner, J.H. Localized Surface Plasmon Resonance Sensors. *Chem. Rev.* **2011**, *111*, 3828–3857. [CrossRef] [PubMed]
39. Gu, Y.; Chen, L.; Zhang, H.; Gong, Q. Resonance capacity of surface plasmon on subwavelength metallic structures. *EPL* **2008**, *83*, 27004. [CrossRef]
40. Wang, F.; Shen, Y.R. General Properties of Local Plasmons in Metal Nanostructures. *Phys. Rev. Lett.* **2006**, *97*, 206806. [CrossRef] [PubMed]
41. Zhang, W.; Gallinet, B.; Martin, O.J.F. Symmetry and selection rules for localized surface plasmon resonances in nanostructures. *Phys. Rev. B* **2010**, *81*, 233407. [CrossRef]
42. Zhang, W. A general theory for plasmonic nanosensor. In Proceedings of the 2015 International Conference on Optical Instruments and Technology: Optical Sensors and Applications, Beijing, China, 8–10 May 2015; p. 2193396.

43. Muhlschlegel, P.; Eisler, H.J.; Martin, O.J.F.; Hecht, B.; Pohl, D.W. Resonant optical antennas. *Science* **2005**, *308*, 1607–1609. [CrossRef]
44. Novotny, L. Effective wavelength scaling for optical antennas. *Phys. Rev. Lett.* **2007**, *98*, 266802. [CrossRef] [PubMed]
45. Bharadwaj, P.; Deutsch, B.; Novotny, L. Optical Antennas. *Adv. Opt. Photonics* **2009**, *1*, 438–483. [CrossRef]
46. Milligan, T.A. *Modern Antenna Design*, 2nd ed.; John Wiley & Sons, Inc.: Hoboken, NJ, USA, 2005.
47. Le Ru, E.; Etchegoin, P. *Principles of Surface-Enhanced Raman Spectroscopy: And Related Plasmonic Effects*; Elsevier: London, UK, 2009.
48. Buchler, B.C.; Kalkbrenner, T.; Hettich, C.; Sandoghdar, V. Measuring the quantum efficiency of the optical emission of single radiating dipoles using a scanning mirror. *Phys. Rev. Lett.* **2005**, *95*, 063003. [CrossRef] [PubMed]
49. Kuehn, S.; Hakanson, U.; Rogobete, L.; Sandoghdar, V. Enhancement of single-molecule fluorescence using a gold nanoparticle as an optical nanoantenna. *Phys. Rev. Lett.* **2006**, *97*, 017402. [CrossRef]
50. Martin, O.J.F.; Girard, C.; Dereux, A. Generalized Field Propagator for Electromagnetic Scattering and Light Confinement. *Phys. Rev. Lett.* **1995**, *74*, 526–529. [CrossRef]
51. Chance, R.R.; Prock, A.; Silbey, R. Molecular Fluorescence and Energy Transfer Near Interfaces. *Adv. Chem. Phys.* **1978**, *37*, 1–65.
52. Kang, K.A.; Wang, J.; Jasinski, J.B.; Achilefu, S. Fluorescence Manipulation by Gold Nanoparticles: From Complete Quenching to Extensive Enhancement. *J. Nanobiotechnol.* **2011**, *9*, 16. [CrossRef]
53. Brouwer, A.M. Standards for photoluminescence quantum yield measurements in solution (IUPAC Technical Report). *Pure Appl. Chem.* **2011**, *83*, 2213–2228. [CrossRef]
54. Kotnala, A.; Gordon, R. Double nanohole optical tweezers visualize protein p53 suppressing unzipping of single DNA-hairpins. *Biomed. Opt. Express* **2014**, *5*, 1886–1894. [CrossRef]
55. Zhang, W.; Martin, O.J.F. Optical trapping and sensing with plasmonic dipole antennas. In *Plasmonics: Metallic Nanostructures and Their Optical Properties VIII*, Stockman, M.I., Ed.; Proceedings of SPIE; SPIE: Bellingham, WA, USA, 2010; Volume 7757, pp. 85–91.
56. Kotnala, A.; Gordon, R. Quantification of High-Efficiency Trapping of Nanoparticles in a Double Nanohole Optical Tweezer. *Nano Lett.* **2014**, *14*, 853–856. [CrossRef] [PubMed]
57. Zhang, W.H.; Martin, O.J.F. A Universal Law for Plasmon Resonance Shift in Biosensing. *ACS Photonics* **2015**, *2*, 144–150. [CrossRef]
58. Zhu, J.J.; Meng, X.; Zhang, C.; Bian, J.; Lu, Z.D.; Liu, Y.; Zhang, W.H. Tailoring a nanostructured plasmonic absorber for high efficiency surface-assisted laser desorption/ionization. *Phys. Chem. Chem. Phys.* **2018**, *20*, 3424–3429. [CrossRef]
59. Kamali, S.M.; Arbabi, E.; Arbabi, A.; Faraon, A. A review of dielectric optical metasurfaces for wavefront control. *Nanophotonics* **2018**, *7*, 1041–1068. [CrossRef]
60. Xie, X.; Liu, K.P.; Pu, M.B.; Ma, X.L.; Li, X.; Guo, Y.H.; Zhang, F.; Luo, X.G. All-metallic geometric metasurfaces for broadband and high-efficiency wavefront manipulation. *Nanophotonics* **2020**, *9*, 3209–3215. [CrossRef]
61. Jisha, C.P.; Nolte, S.; Alberucci, A. Geometric Phase in Optics: From Wavefront Manipulation to Waveguiding. *Laser Photonics Rev.* **2021**, *15*, 2100003. [CrossRef]
62. Fleischmann, M.; Hendral, P.J.; McQuillan, A.J. Raman spectra of pyridine adsorbed at a silver electrode. *Chem. Phys. Lett.* **1974**, *26*, 163–166. [CrossRef]
63. Jeanmaire, D.L.; Van Duyne, R.P. Surface Raman spectroelectrochemistry: Part I. Heterocyclic, aromatic, and aliphatic amines adsorbed on the anodized silver electrode. *J. Electroanal. Chem.* **1977**, *84*, 1–20. [CrossRef]
64. Moskovits, M. Surface roughness and the enhanced intensity of Raman scattering by molecules adsorbed on metals. *J. Chem. Phys.* **1978**, *69*, 4159–4161. [CrossRef]
65. Panneerselvam, R.; Liu, G.-K.; Wang, Y.-H.; Liu, J.-Y.; Ding, S.-Y.; Li, J.-F.; Wu, D.-Y.; Tian, Z.-Q. Surface-enhanced Raman spectroscopy: Bottlenecks and future directions. *Chem. Commun.* **2017**, *54*, 10–25. [CrossRef]
66. Xu, H.X.; Aizpurua, J.; Kall, M.; Apell, P. Electromagnetic contributions to single-molecule sensitivity in surface-enhanced Raman scattering. *Phys. Rev. E* **2000**, *62*, 4318–4324. [CrossRef]
67. Neubrech, F.; Pucci, A.; Cornelius, T.W.; Karim, S.; Garcia-Etxarri, A.; Aizpurua, J. Resonant Plasmonic and Vibrational Coupling in a Tailored Nanoantenna for Infrared Detection. *Phys. Rev. Lett.* **2008**, *101*, 157403. [CrossRef]
68. Novotny, L.; van Hulst, N. Antennas for light. *Nat. Photonics* **2011**, *5*, 83–90. [CrossRef]
69. Seok, T.J.; Jamshidi, A.; Kim, M.; Dhuey, S.; Lakhani, A.; Choo, H.; Schuck, P.J.; Cabrini, S.; Schwartzberg, A.M.; Bokor, J.; et al. Radiation engineering of optical antennas for maximum field enhancement. *Nano Lett.* **2011**, *11*, 2606–2610. [CrossRef] [PubMed]
70. Zhang, W.; Cui, X.; Yeo, B.-S.; Schmid, T.; Hafner, C.; Zenobi, R. Nanoscale roughness on metal surfaces can increase tip-enhanced Raman scattering by an order of magnitude. *Nano Lett.* **2007**, *7*, 1401–1405. [CrossRef]
71. Zhang, W.; Schmid, T.; Yeo, B.-S.; Zenobi, R. Near-field heating, annealing, and signal loss in tip-enhanced Raman spectroscopy. *J. Phys. Chem. C* **2008**, *112*, 2104–2108. [CrossRef]
72. Zhang, W. *Tip-Enhanced Raman Spectroscopy: Theory, Practice and Applications*; ETH Zurich: Zurich, Switzerland, 2008.
73. Zhang, W.; Ding, F.; Li, W.-D.; Wang, Y.; Hu, J.; Chou, S.Y. Giant and uniform fluorescence enhancement over large areas using plasmonic nanodots in 3D resonant cavity nanoantenna by nanoimprinting. *Nanotechnology* **2012**, *23*, 225301. [CrossRef] [PubMed]
74. Li, W.-D.; Ding, F.; Hu, J.; Chou, S.Y. Three-dimensional cavity nanoantenna coupled plasmonic nanodots for ultrahigh and uniform surface-enhanced Raman scattering over large area. *Opt. Express* **2011**, *19*, 3925–3936. [CrossRef]
75. Zhang, W.; Ding, F.; Chou, S.Y. Large Enhancement of Upconversion Luminescence of NaYF4:Yb3+/Er3+ Nanocrystal by 3D Plasmonic Nano-Antennas. *Adv. Mater.* **2012**, *24*, OP236–OP241. [CrossRef] [PubMed]

76. Moreau, A.; Ciraci, C.; Mock, J.J.; Hill, R.T.; Wang, Q.; Wiley, B.J.; Chilkoti, A.; Smith, D.R. Controlled-reflectance surfaces with film-coupled colloidal nanoantennas. *Nature* **2012**, *492*, 86–89. [CrossRef] [PubMed]
77. Benz, F.; Schmidt, M.K.; Dreismann, A.; Chikkaraddy, R.; Zhang, Y.; Demetriadou, A.; Carnegie, C.; Ohadi, H.; de Nijs, B.; Esteban, R.; et al. Single-molecule optomechanics in "picocavities". *Science* **2016**, *354*, 726–729. [CrossRef]
78. Chen, H.-Y.; Lin, M.-H.; Wang, C.-Y.; Chang, Y.-M.; Gwo, S. Large-Scale Hot Spot Engineering for Quantitative SERS at the Single-Molecule Scale. *J. Am. Chem. Soc.* **2015**, *137*, 13698–13705. [CrossRef]
79. Caridad, J.M.; Winters, S.; McCloskey, D.; Duesberg, G.S.; Donegan, J.F.; Krstić, V. Hot-Volumes as Uniform and Reproducible SERS-Detection Enhancers in Weakly-Coupled Metallic Nanohelices. *Sci. Rep.* **2017**, *7*, 45548. [CrossRef] [PubMed]
80. Li, J.F.; Huang, Y.F.; Ding, Y.; Yang, Z.L.; Li, S.B.; Zhou, X.S.; Fan, F.R.; Zhang, W.; Zhou, Z.Y.; Wu, D.Y.; et al. Shell-isolated nanoparticle-enhanced Raman spectroscopy. *Nature* **2010**, *464*, 392–395. [CrossRef]
81. Lu, H.; Zhu, L.; Zhang, C.; Chen, K.; Cui, Y. Mixing Assisted "Hot Spots" Occupying SERS Strategy for Highly Sensitive In Situ Study. *Anal. Chem.* **2018**, *90*, 4535–4543. [CrossRef]
82. Ji, C.; Lu, J.; Shan, B.; Li, F.; Zhao, X.; Yu, J.; Xu, S.; Man, B.; Zhang, C.; Li, Z. The Origin of Mo2C Films for Surface-Enhanced Raman Scattering Analysis: Electromagnetic or Chemical Enhancement? *J. Phys. Chem. Lett.* **2022**, *13*, 8864–8871. [CrossRef]
83. Zhang, C.; Li, Z.; Qiu, S.; Lu, W.; Shao, M.; Ji, C.; Wang, G.; Zhao, X.; Yu, J.; Li, Z. Highly ordered arrays of hat-shaped hierarchical nanostructures with different curvatures for sensitive SERS and plasmon-driven catalysis. *Nanophotonics* **2022**, *11*, 33–44. [CrossRef]
84. Stone, N.; Kerssens, M.; Lloyd, G.R.; Faulds, K.; Graham, D.; Matousek, P. Surface enhanced spatially offset Raman spectroscopic (SESORS) imaging—The next dimension. *Chem. Sci.* **2011**, *2*, 776–780. [CrossRef]
85. Moody, A.S.; Payne, T.D.; Barth, B.A.; Sharma, B. Surface-enhanced spatially-offset Raman spectroscopy (SESORS) for detection of neurochemicals through the skull at physiologically relevant concentrations. *Analyst* **2020**, *145*, 1885–1893. [CrossRef]
86. Asiala, S.M.; Shand, N.C.; Faulds, K.; Graham, D. Surface-Enhanced, Spatially Offset Raman Spectroscopy (SESORS) in Tissue Analogues. *ACS Appl. Mater. Interfaces* **2017**, *9*, 25488–25494. [CrossRef] [PubMed]
87. Managò, S.; Quero, G.; Zito, G.; Tullii, G.; Galeotti, F.; Pisco, M.; De Luca, A.C.; Cusano, A. Tailoring lab-on-fiber SERS optrodes towards biological targets of different sizes. *Sens. Actuators B* **2021**, *339*, 129321. [CrossRef]
88. Sansone, L.; Campopiano, S.; Pannico, M.; Giordano, M.; Musto, P.; Iadicicco, A. Photonic bandgap influence on the SERS effect in metal-dielectric colloidal crystals optical fiber probe. *Sens. Actuators B* **2021**, *345*, 130149. [CrossRef]
89. Jiang, L.; Yu, Z.; Zhao, W.; Yang, Z.; Peng, Y.; Zhou, Y.; Lin, X.; Jin, S. Self-Assembled MXene-Au Multifunctional Nanomaterials with Various Shapes for Label-free SERS Detection of Pathogenic Bacteria and Photothermal Sterilization. *Anal. Chem.* **2023**, *95*, 1721–1730. [CrossRef]
90. Stuart, D.A.; Yuen, J.M.; Lyandres, N.S.O.; Yonzon, C.R.; Glucksberg, M.R.; Walsh, J.T.; Van Duyne, R.P. In vivo glucose measurement by surface-enhanced Raman spectroscopy. *Anal. Chem.* **2006**, *78*, 7211–7215. [CrossRef]
91. Hsu, P.-H.; Chiang, H.K. Surface-enhanced Raman spectroscopy for quantitative measurement of lactic acid at physiological concentration in human serum. *J. Raman Spectrosc.* **2010**, *41*, 1610–1614. [CrossRef]
92. Ma, K.; Yuen, J.M.; Shah, N.C.; Walsh, J.T., Jr.; Glucksberg, M.R.; Van Duyne, R.P. In Vivo, Transcutaneous Glucose Sensing Using Surface-Enhanced Spatially Offset Raman Spectroscopy: Multiple Rats, Improved Hypoglycemic Accuracy, Low Incident Power, and Continuous Monitoring for Greater than 17 Days. *Anal. Chem.* **2011**, *83*, 9146–9152. [CrossRef]
93. Hu, P.; Zheng, X.-S.; Zong, C.; Li, M.-H.; Zhang, L.-Y.; Li, W.; Ren, B. Drop-coating deposition and surface-enhanced Raman spectroscopies (DCDRS and SERS) provide complementary information of whole human tears. *J. Raman Spectrosc.* **2014**, *45*, 565–573. [CrossRef]
94. Shen, W.; Lin, X.; Jiang, C.; Li, C.; Lin, H.; Huang, J.; Wang, S.; Liu, G.; Yan, X.; Zhong, Q.; et al. Reliable Quantitative SERS Analysis Facilitated by Core-Shell Nanoparticles with Embedded Internal Standards. *Angew. Chem. Int. Ed.* **2015**, *54*, 7308–7312. [CrossRef] [PubMed]
95. Yadav, S.; Senapati, S.; Desai, D.; Gahlaut, S.; Kulkarni, S.; Singh, J.P. Portable and sensitive Ag nanorods based SERS platform for rapid HIV-1 detection and tropism determination. *Colloids Surf. B Biointerfaces* **2021**, *198*, 111477. [CrossRef] [PubMed]
96. Beeram, R.; Vepa, K.R.; Soma, V.R. Recent Trends in SERS-Based Plasmonic Sensors for Disease Diagnostics, Biomolecules Detection, and Machine Learning Techniques. *Biosensors* **2023**, *13*, 328. [CrossRef] [PubMed]
97. Samanta, A.; Maiti, K.K.; Soh, K.S.; Liao, X.J.; Vendrell, M.; Dinish, U.S.; Yun, S.W.; Bhuvaneswari, R.; Kim, H.; Rautela, S.; et al. Ultrasensitive Near-Infrared Raman Reporters for SERS-Based In Vivo Cancer Detection. *Angew. Chem. Int. Ed.* **2011**, *50*, 6089–6092. [CrossRef] [PubMed]
98. Haldavnekar, R.; Venkatakrishnan, K.; Tan, B. Non plasmonic semiconductor quantum SERS probe as a pathway for in vitro cancer detection. *Nat. Commun.* **2018**, *9*, 3065. [CrossRef] [PubMed]
99. Guerrini, L.; Alvarez-Puebla, R.A. Surface-Enhanced Raman Spectroscopy in Cancer Diagnosis, Prognosis and Monitoring. *Cancers* **2019**, *11*, 748. [CrossRef]
100. Pollap, A.; Swit, P. Recent Advances in Sandwich SERS Immunosensors for Cancer Detection. *Int. J. Mol. Sci.* **2022**, *23*, 4740. [CrossRef]
101. Mousavi, S.M.; Hashemi, S.A.; Rahmanian, V.; Kalashgrani, M.Y.; Gholami, A.; Omidifar, N.; Chiang, W.H. Highly Sensitive Flexible SERS-Based Sensing Platform for Detection of COVID-19. *Biosensors* **2022**, *12*, 466. [CrossRef]

102. Lin, Z.S.; He, L.L. Recent advance in SERS techniques for food safety and quality analysis: A brief review. *Curr. Opin. Food Sci.* **2019**, *28*, 82–87. [CrossRef]
103. Zhou, H.L.; Li, X.D.; Wang, L.H.; Liang, Y.F.; Jialading, A.; Wang, Z.S.; Zhang, J.G. Application of SERS quantitative analysis method in food safety detection. *Rev. Anal. Chem* **2021**, *40*, 173–186. [CrossRef]
104. Kumar, S.; Goel, P.; Singh, J.P. Flexible and robust SERS active substrates for conformal rapid detection of pesticide residues from fruits. *Sens. Actuators B Chem.* **2017**, *241*, 577–583. [CrossRef]
105. Okeke, E.S.; Huang, B.; Mao, G.H.; Chen, Y.; Zeng, Z.J.; Qian, X.; Wu, X.Y.; Feng, W.W. Review of the environmental occurrence/analytical techniques, degradation and toxicity of TBBPA and its derivatives. *Environ. Res.* **2022**, *206*, 112594. [CrossRef]
106. Zhang, W.H.; Yeo, B.S.; Schmid, T.; Zenobi, R. Single molecule tip-enhanced Raman spectroscopy with silver tips. *J. Phys. Chem. C* **2007**, *111*, 1733–1738. [CrossRef]
107. Zhang, R.; Zhang, Y.; Dong, Z.C.; Jiang, S.; Zhang, C.; Chen, L.G.; Zhang, L.; Liao, Y.; Aizpurua, J.; Luo, Y.; et al. Chemical mapping of a single molecule by plasmon-enhanced Raman scattering. *Nature* **2013**, *498*, 82–86. [CrossRef] [PubMed]
108. Schmid, T.; Opilik, L.; Blum, C.; Zenobi, R. Nanoscale Chemical Imaging Using Tip-Enhanced Raman Spectroscopy: A Critical Review. *Angew. Chem. Int. Ed.* **2013**, *52*, 5940–5954. [CrossRef] [PubMed]
109. Deckert-Gaudig, T.; Taguchi, A.; Kawata, S.; Deckert, V. Tip-enhanced Raman spectroscopy—From early developments to recent advances. *Chem. Soc. Rev.* **2017**, *46*, 4077–4110. [CrossRef] [PubMed]
110. Shao, F.; Zenobi, R. Tip-enhanced Raman spectroscopy: Principles, practice, and applications to nanospectroscopic imaging of 2D materials. *Anal. Bioanal. Chem.* **2019**, *411*, 37–61. [CrossRef] [PubMed]
111. Kim, S.; Yu, N.; Ma, X.; Zhu, Y.; Liu, Q.; Liu, M.; Yan, R. High external-efficiency nanofocusing for lens-free near-field optical nanoscopy. *Nat. Photonics* **2019**, *13*, 636–643. [CrossRef]
112. Brown, L.V.; Zhao, K.; King, N.; Sobhani, H.; Nordlander, P.; Halas, N.J. Surface-Enhanced Infrared Absorption Using Individual Cross Antennas Tailored to Chemical Moieties. *J. Am. Chem. Soc.* **2013**, *135*, 3688–3695. [CrossRef]
113. Meo, V.D.; Crescitelli, A.; Moccia, M.; Sandomenico, A.; Cusano, A.M.; Portaccio, M.; Lepore, M.; Galdi, V.; Esposito, E. Pixeled metasurface for multiwavelength detection of vitamin D. *Nanophotonics* **2020**, *9*, 3921–3930. [CrossRef]
114. Wang, C.; Zhang, Q.; Song, Y.; Chou, S.Y. Plasmonic Bar-Coupled Dots-on-Pillar Cavity Antenna with Dual Resonances for Infrared Absorption and Sensing: Performance and Nanoimprint Fabrication. *ACS Nano* **2014**, *8*, 2618–2624. [CrossRef]
115. Hu, H.; Yang, X.; Zhai, F.; Hu, D.; Liu, R.; Liu, K.; Sun, Z.; Dai, Q. Far-field nanoscale infrared spectroscopy of vibrational fingerprints of molecules with graphene plasmons. *Nat. Commun.* **2016**, *7*, 12334. [CrossRef] [PubMed]
116. Rodrigo, D.; Limaj, O.; Janner, D.; Etezadi, D.; García de Abajo, F.J.; Pruneri, V.; Altug, H. Mid-infrared plasmonic biosensing with graphene. *Science* **2015**, *349*, 165. [CrossRef] [PubMed]
117. Chou, S.Y.; Li, W.-D. Structures for Enhancement of Local Electric Field, Light Absorption, Light Radiation, Material Detection and Methods for Making and Using of the Same. 9182338, 2011.
118. Zhou, L.; Ding, F.; Chen, H.; Ding, W.; Zhang, W.; Chou, S.Y. Enhancement of Immunoassay's Fluorescence and Detection Sensitivity Using Three-Dimensional Plasmonic Nano-Antenna-Dots Array. *Anal. Chem.* **2012**, *84*, 4489–4495. [CrossRef]
119. Zang, F.; Su, Z.; Zhou, L.; Konduru, K.; Kaplan, G.; Chou, S.Y. Ultrasensitive Ebola Virus Antigen Sensing via 3D Nanoantenna Arrays. *Adv. Mater.* **2019**, *31*, 1902331. [CrossRef]
120. Zijlstra, P.; Paulo, P.M.R.; Orrit, M. Optical detection of single non-absorbing molecules using the surface plasmon resonance of a gold nanorod. *Nat. Nanotechnol.* **2012**, *7*, 379–382. [CrossRef]
121. Al Balushi, A.A.; Gordon, R. A Label-Free Untethered Approach to Single-Molecule Protein Binding Kinetics. *Nano Lett.* **2014**, *14*, 5787–5791. [CrossRef] [PubMed]
122. Al Balushi, A.A.; Gordon, R. Label-Free Free-Solution Single-Molecule Protein Small Molecule Interaction Observed by Double-Nanohole Plasmonic Trapping. *ACS Photonics* **2014**, *1*, 389–393. [CrossRef]
123. Haes, A.J.; Haynes, C.L.; McFarland, A.D.; Schatz, G.C.; Van Duyne, R.R.; Zou, S.L. Plasmonic materials for surface-enhanced sensing and spectroscopy. *MRS Bull.* **2005**, *30*, 368–375. [CrossRef]
124. Dantham, V.R.; Holler, S.; Barbre, C.; Keng, D.; Kolchenko, V.; Arnold, S. Label-Free Detection of Single Protein Using a Nanoplasmonic-Photonic Hybrid Microcavity. *Nano Lett.* **2013**, *13*, 3347–3351. [CrossRef]
125. Bozzola, A.; Perotto, S.; De Angelis, F. Hybrid plasmonic-photonic whispering gallery mode resonators for sensing: A critical review. *Analyst* **2017**, *142*, 883–898. [CrossRef] [PubMed]
126. Liang, F.; Guo, Y.; Hou, S.; Quan, Q. Photonic-plasmonic hybrid single-molecule nanosensor measures the effect of fluorescent labels on DNA-protein dynamics. *Sci. Adv.* **2017**, *3*, 1602991. [CrossRef] [PubMed]
127. Al Balushi, A.A.; Zehtabi-Oskuie, A.; Gordon, R. Observing single protein binding by optical transmission through a double nanohole aperture in a metal film. *Biomed. Opt. Express* **2013**, *4*, 1504–1511. [CrossRef] [PubMed]
128. Al Balushi, A.A.; Kotnala, A.; Wheaton, S.; Gelfand, R.M.; Rajashekara, Y.; Gordon, R. Label-free free-solution nanoaperture optical tweezers for single molecule protein studies. *Analyst* **2015**, *140*, 4760–4778. [CrossRef] [PubMed]
129. Yoo, D.; Gurunatha, K.L.; Choi, H.-K.; Mohr, D.A.; Ertsgaard, C.T.; Gordon, R.; Oh, S.-H. Low-Power Optical Trapping of Nanoparticles and Proteins with Resonant Coaxial Nanoaperture Using 10 nm Gap. *Nano Lett.* **2018**, *18*, 3637–3642. [CrossRef]
130. Ye, W.; Celiksoy, S.; Jakab, A.; Khmelinskaia, A.; Heermann, T.; Raso, A.; Wegner, S.V.; Rivas, G.; Schwille, P.; Ahijado-Guzman, R.; et al. Plasmonic Nanosensors Reveal a Height Dependence of MinDE Protein Oscillations on Membrane Features. *J. Am. Chem. Soc.* **2018**, *140*, 17901–17906. [CrossRef] [PubMed]

131. Rosman, C.; Pierrat, S.; Henkel, A.; Tarantola, M.; Schneider, D.; Sunnick, E.; Janshoff, A.; Soennichsen, C. A New Approach to Assess Gold Nanoparticle Uptake by Mammalian Cells: Combining Optical Dark-Field and Transmission Electron Microscopy. *Small* **2012**, *8*, 3683–3690. [CrossRef]
132. Arcadio, F.; Marzano, C.; Del Prete, D.; Zeni, L.; Cennamo, N. Analysis of Plasmonic Sensors Performance Realized by Exploiting Different UV-Cured Optical Adhesives Combined with Plastic Optical Fibers. *Sensors* **2023**, *23*, 6182. [CrossRef]
133. Principe, M.; Consales, M.; Micco, A.; Crescitelli, A.; Castaldi, G.; Esposito, E.; La Ferrara, V.; Cutolo, A.; Galdi, V.; Cusano, A. Optical fiber meta-tips. *Light Sci. Appl.* **2017**, *6*, e16226. [CrossRef]
134. Martínez-Hernández, M.E.; Rivero, P.J.; Goicoechea, J.; Arregui, F.J. Trends in the Implementation of Advanced Plasmonic Materials in Optical Fiber Sensors (2010–2020). *Chemosensors* **2021**, *9*, 64. [CrossRef]
135. Chiavaioli, F.; Santano Rivero, D.; Del Villar, I.; Socorro-Leránoz, A.B.; Zhang, X.; Li, K.; Santamaría, E.; Fernández-Irigoyen, J.; Baldini, F.; van den Hove, D.L.A.; et al. Ultrahigh Sensitive Detection of Tau Protein as Alzheimer's Biomarker via Microfluidics and Nanofunctionalized Optical Fiber Sensors. *Adv. Photonics Res.* **2022**, *3*, 2200044. [CrossRef]
136. Choudhary, S.; Esposito, F.; Sansone, L.; Giordano, M.; Campopiano, S.; Iadicicco, A. Lossy Mode Resonance Sensors in Uncoated Optical Fiber. *IEEE Sens. J.* **2023**, *23*, 15607–15613. [CrossRef]
137. Del Villar, I.; Arregui, F.J.; Zamarreño, C.R.; Corres, J.M.; Bariain, C.; Goicoechea, J.; Elosua, C.; Hernaez, M.; Rivero, P.J.; Socorro, A.B.; et al. Optical sensors based on lossy-mode resonances. *Sens. Actuators B* **2017**, *240*, 174–185. [CrossRef]
138. Pituła, E.; Janik, M.; Sezemsky, P.; Szymańska, K.P.; Olszewski, M.; Stranak, V.; Koba, M.; Śmietana, M. Smartphone-based dynamic measurements of electro-optically modulated lossy-mode resonance and its biosensing applications. *Measurement* **2023**, *206*, 112349. [CrossRef]
139. Karas, M.; Hillenkamp, F. Laser desorption ionization of proteins with molecular masses exceeding 10,000 daltons. *Anal. Chem.* **1988**, *60*, 2299–2301. [CrossRef] [PubMed]
140. Karas, M.; Kruger, R. Ion formation in MALDI: The cluster ionization mechanism. *Chem. Rev.* **2000**, *103*, 427–439. [CrossRef]
141. Chiang, C.K.; Chen, W.T.; Chang, H.T. Nanoparticle-based mass spectrometry for the analysis of biomolecules. *Chem. Soc. Rev.* **2011**, *40*, 1269–1281. [CrossRef]
142. Sekula, J.; Niziol, J.; Rode, W.; Ruman, T. Gold nanoparticle-enhanced target (AuNPET) as universal solution for laser desorption/ionization mass spectrometry analysis and imaging of low molecular weight compounds. *Anal. Chim. Acta* **2015**, *875*, 61–72. [CrossRef]
143. Hinman, S.S.; Chen, C.Y.; Duan, J.C.; Cheng, Q. Calcinated gold nanoparticle arrays for on-chip, multiplexed and matrix-free mass spectrometric analysis of peptides and small molecules. *Nanoscale* **2016**, *8*, 1665–1675. [CrossRef]
144. Zarei, M. Advances in point-of-care technologies for molecular diagnostics. *Biosens. Bioelectron.* **2017**, *98*, 494–506. [CrossRef]
145. Zarei, M. Portable biosensing devices for point-of-care diagnostics: Recent developments and applications. *TrAC Trends Anal. Chem.* **2017**, *91*, 26–41. [CrossRef]
146. Quesada-Gonzalez, D.; Merkoci, A. Mobile phone-based biosensing: An emerging "diagnostic and communication" technology. *Biosens. Bioelectron.* **2017**, *92*, 549–562. [CrossRef] [PubMed]
147. Gao, X.; Wu, N. Smartphone-Based Sensors. *Electrochem. Soc. Interface* **2016**, *25*, 79–81. [CrossRef]
148. Preechaburana, P.; Gonzalez, M.C.; Suska, A.; Filippini, D. Surface Plasmon Resonance Chemical Sensing on Cell Phones. *Angew. Chem. Int. Ed.* **2012**, *51*, 11585–11588. [CrossRef]
149. Min, S.Y.; Li, S.J.; Zhu, Z.Y.; Liu, Y.; Liang, C.W.; Cai, J.X.; Han, F.; Li, Y.Y.; Cai, W.S.; Cheng, X.; et al. Ultrasensitive Molecular Detection by Imaging of Centimeter-Scale Metasurfaces with a Deterministic Gradient Geometry. *Adv. Mater.* **2021**, *33*, 2100270. [CrossRef] [PubMed]
150. Bian, J.; Xing, X.; Zhou, S.; Man, Z.; Lu, Z.; Zhang, W. Patterned plasmonic gradient for high-precision biosensing using smartphone reader. *Nanoscale* **2019**, *11*, 12471–12476. [CrossRef]
151. Nordén, B.; Rodger, A.; Dafforn, T. *Linear Dichroism and Circular Dichroism: A Textbook on Polarized-Light Spectroscopy*; Royal Society of Chemistry: London, UK, 2010.
152. Chen, P.; Ge, S.J.; Ma, L.L.; Hu, W.; Chigrinov, V.; Lu, Y.Q. Generation of Equal-Energy Orbital Angular Momentum Beams via Photopatterned Liquid Crystals. *Phys. Rev. Appl.* **2016**, *5*, 044009. [CrossRef]
153. Song, J.; Zhou, J.; Duan, H. Self-Assembled Plasmonic Vesicles of SERS-Encoded Amphiphilic Gold Nanoparticles for Cancer Cell Targeting and Traceable Intracellular Drug Delivery. *J. Am. Chem. Soc.* **2012**, *134*, 13458–13469. [CrossRef]
154. Radziuk, D.; Moehwald, H. Prospects for plasmonic hot spots in single molecule SERS towards the chemical imaging of live cells. *Phys. Chem. Chem. Phys.* **2015**, *17*, 21072–21093. [CrossRef]
155. Austin, L.A.; Kang, B.; El-Sayed, M.A. Probing molecular cell event dynamics at the single-cell level with targeted plasmonic gold nanoparticles: A review. *Nano Today* **2015**, *10*, 542–558. [CrossRef]
156. Yuan, H.; Register, J.K.; Wang, H.-N.; Fales, A.M.; Liu, Y.; Vo-Dinh, T. Plasmonic nanoprobes for intracellular sensing and imaging. *Anal. Bioanal. Chem.* **2013**, *405*, 6165–6180. [CrossRef] [PubMed]
157. Kumar, S.; Harrison, N.; Richards-Kortum, R.; Sokolov, K. Plasmonic Nanosensors for Imaging Intracellular Biomarkers in Live Cells. *Nano Lett.* **2007**, *7*, 1338–1343. [CrossRef] [PubMed]
158. Wang, W.; Foley, K.; Shan, X.; Wang, S.; Eaton, S.; Nagaraj, V.J.; Wiktor, P.; Patel, U.; Tao, N. Single cells and intracellular processes studied by a plasmonic-based electrochemical impedance microscopy. *Nat. Chem.* **2011**, *3*, 249–255. [CrossRef]

159. Sharifi, M.; Attar, F.; Saboury, A.A.; Akhtari, K.; Hooshmand, N.; Hasan, A.; El-Sayed, M.A.; Falahati, M. Plasmonic gold nanoparticles: Optical manipulation, imaging, drug delivery and therapy. *J. Control. Release* **2019**, *311–312*, 170–189. [CrossRef]
160. Kim, M.; Lee, J.-H.; Nam, J.-M. Plasmonic Photothermal Nanoparticles for Biomedical Applications. *Adv. Sci.* **2019**, *6*, 1900471. [CrossRef]
161. Singh, S.K.; Mazumder, S.; Vincy, A.; Hiremath, N.; Kumar, R.; Banerjee, I.; Vankayala, R. Review of Photoresponsive Plasmonic Nanoparticles That Produce Reactive Chemical Species for Photodynamic Therapy of Cancer and Bacterial Infections. *ACS Appl. Nano Mater.* **2023**, *6*, 1508–1521. [CrossRef]
162. Ali, M.R.K.; Wu, Y.; El-Sayed, M.A. Gold-Nanoparticle-Assisted Plasmonic Photothermal Therapy Advances Toward Clinical Application. *J. Phys. Chem. C* **2019**, *123*, 15375–15393. [CrossRef]
163. Huang, W.T.; Chan, M.H.; Chen, X.; Hsiao, M.; Liu, R.S. Theranostic nanobubble encapsulating a plasmon-enhanced upconversion hybrid nanosystem for cancer therapy. *Theranostics* **2020**, *10*, 782–796. [CrossRef]
164. Zhang, Y.; Gu, Y.; He, J.; Thackray, B.D.; Ye, J. Ultrabright gap-enhanced Raman tags for high-speed bioimaging. *Nat. Commun.* **2019**, *10*, 3905. [CrossRef]
165. Khlebtsov, N.G.; Lin, L.; Khlebtsov, B.N.; Ye, J. Gap-enhanced Raman tags: Fabrication, optical properties, and theranostic applications. *Theranostics* **2020**, *10*, 2067–2094. [CrossRef]
166. Mesch, M.; Metzger, B.; Hentschel, M.; Giessen, H. Nonlinear Plasmonic Sensing. *Nano Lett.* **2016**, *16*, 3155–3159. [CrossRef]
167. Lee, C.; Lawrie, B.; Pooser, R.; Lee, K.G.; Rockstuhl, C.; Tame, M. Quantum Plasmonic Sensors. *Chem. Rev.* **2021**, *121*, 4743–4804. [CrossRef] [PubMed]
168. Yu, R.W.; Cox, J.D.; de Abajo, F.J.G. Nonlinear Plasmonic Sensing with Nanographene. *Phys. Rev. Lett.* **2016**, *17*, 123904. [CrossRef] [PubMed]
169. Lee, J.S.; Yoon, S.J.; Rah, H.; Tame, M.; Rockstuhl, C.; Song, S.H.; Lee, C.; Lee, K.G. Quantum plasmonic sensing using single photons. *Opt. Express* **2018**, *26*, 29272–29282. [CrossRef] [PubMed]
170. Dowran, M.; Kumar, A.; Lawrie, B.J.; Pooser, R.C.; Marino, A.M. Quantum-enhanced plasmonic sensing. *Optica* **2018**, *5*, 628–633. [CrossRef]
171. Verma, M.S.; Chandra, M. Nonlinear Plasmonic Sensing for Label-Free and Selective Detection of Mercury at Picomolar Level. *ACS Sens.* **2020**, *5*, 645–649. [CrossRef]

chemosensors

Review

Design of DNA-Based Artificial Transmembrane Channels for Biosensing and Biomedical Applications

Wanyu Xu, Hui Chen, Yang Li, Shuangna Liu, Kemin Wang * and Jianbo Liu *

State Key Laboratory of Chemo/Biosensing and Chemometrics, Key Laboratory for Bio-Nanotechnology and Molecular Engineering of Hunan Province, College of Chemistry and Chemical Engineering, Hunan University, Changsha 410082, China; 1349376793@hnu.edu.cn (W.X.); chenhui_0908@163.com (H.C.); ly2353345079@163.com (Y.L.); liushuangna@hnu.edu.cn (S.L.)

* Correspondence: kmwang@hnu.edu.cn (K.W.); liujianbo@hnu.edu.cn (J.L.)

Abstract: Biomolecular channels on the cell membrane are essential for transporting substances across the membrane to maintain cell physiological activity. Artificial transmembrane channels used to mimic biological membrane channels can regulate intra/extracellular ionic and molecular homeostasis, and they elucidate cellular structures and functionalities. Due to their program design, facile preparation, and high biocompatibility, DNA nanostructures have been widely used as scaffolds for the design of artificial transmembrane channels and exploited for ionic and molecular transport and biomedical applications. DNA-based artificial channels can be designed from two structural modules: DNA nanotubes/nanopores as transport modules for mass transportation and hydrophobic segments as anchor modules for membrane immobilization. In this review, various lipophilic modification strategies for the design of DNA channels and membrane insertion are outlined. Several types of DNA transmembrane channels are systematically summarized, including DNA wireframe channels, DNA helix bundle channels, DNA tile channels, DNA origami channels, and so on. We then discuss efforts to exploit them in biosensor and biomedical applications. For example, ligand-gated and environmental stimuli-responsive artificial transmembrane channels have been designed for transmembrane signal transduction. DNA-based artificial channels have been developed for cell mimicry and the regulation of cell behaviors. Finally, we provide some perspectives on the challenges and future developments of artificial transmembrane channel research in biomimetic science and biomedical applications.

Keywords: artificial transmembrane channel; DNA nanostructure; biosensing; biomedical application

Citation: Xu, W.; Chen, H.; Li, Y.; Liu, S.; Wang, K.; Liu, J. Design of DNA-Based Artificial Transmembrane Channels for Biosensing and Biomedical Applications. *Chemosensors* **2023**, *11*, 508. https://doi.org/10.3390/chemosensors11090508

Academic Editor: Barbara Palys

Received: 24 July 2023
Revised: 7 September 2023
Accepted: 15 September 2023
Published: 18 September 2023

1. Introduction

Transmembrane channels are integral membrane proteins with channels or pores that allow particular ions or small molecules to cross a lipid bilayer [1,2]. These channels are critical for regulating ion homeostasis, transporting molecules, maintaining normal cell physiological functions, and performing related life activities [3]. However, their formation requires genetic coding in living cells under strict spatiotemporal control, which is difficult to reproduce in vitro [4]. To date, various materials, such as biological macromolecules, synthetic organic compounds, and inorganic substances, have been successfully used for the design of artificial nanochannels with diverse structures and functions [5]. The design of synthetic biomolecules that mimic the structures and functions of natural transmembrane channels has garnered substantial interest among molecular biology researchers as models for studying fundamental information, creating alternative drugs, and developing advanced biosensors.

Deoxyribonucleic acid (DNA) is an irreplaceable building material for the design of artificial channels [6], since DNA is an easily accessible biomacromolecule with high biocompatibility and programmable self-assembly ability, and it can be safely used in natural biological processes [7]. In addition, each strand used to construct DNA channels can be independently functionalized through the precise design and modification of various biochemical molecules [8]. With excellent shapes, structures, and functions, DNA-based artificial channels have been widely used for the transmembrane transport of ions or molecules, the transduction of intercellular signals, and the regulation of cell physiological activities [9]. Various methods have been reported for the design of DNA-based transmembrane channels [10]. For example, traditional DNA self-assembly technology has been developed as a simple, practical, and rapid route for designing transmembrane channels, which have been used to simulate signal transduction and ion transport [11–13]. Based on DNA origami technology, the pore size, height, and wall thickness values of the designed DNA channels are relatively easy to customize [14]. Hydrophobic modification of DNA nanostructures is a major design difficulty, but it is necessary to make the DNA channels span the phospholipid bilayer [15]. A regulating switch on the DNA channels can modulate transmembrane transport [16–18].

In this review, we summarize the hydrophobic modification strategies for preparing DNA transmembrane channels in terms of the type, number, and location of hydrophobic modification groups. We systematically summarize the construction of different types of DNA artificial transmembrane channels and discuss their applications in biosensors and biomedicine. (Scheme 1) Finally, we describe the challenges and possible remedies of DNA artificial transmembrane channels in biomimetic science and biomedicine applications and propose some prospects for their future development.

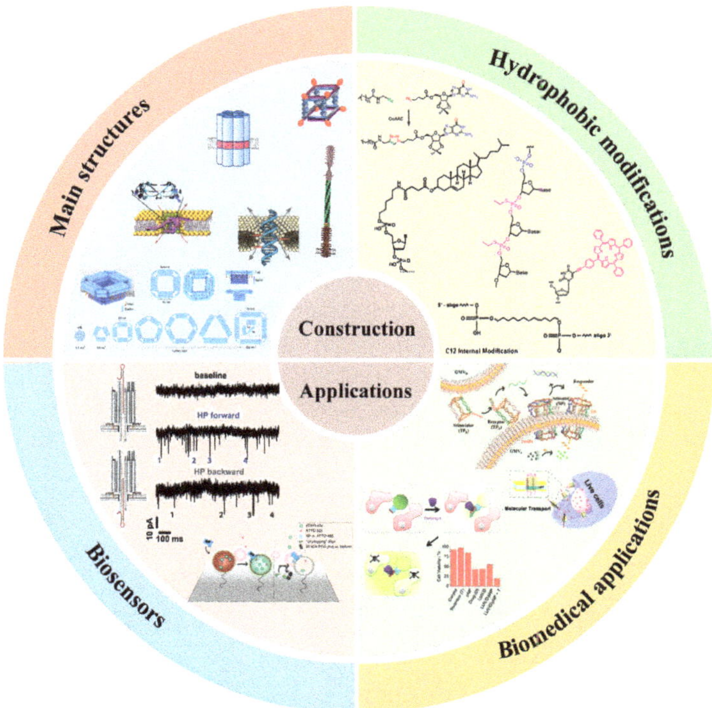

Scheme 1. Diagram of the content of the review. DNA-based artificial channels can be designed from two structural modules: DNA nanotubes/nanopores as main structural modules for mass transportation and hydrophobic modifications as anchor modules for membrane immobilization. We systematically summarize the construction of different types of DNA artificial transmembrane channels and discuss their applications in biosensors and biomedicine.

2. Hydrophobic Modification for Artificial Transmembrane Channels

The purposes of designing DNA-based artificial channels include the transmembrane transport of substances for signal exchange, the adjustment of the concentrations of substances, and the regulation of cell behaviors. The design method, structure (pore size and length), and interaction force between the channel and phospholipid membrane can directly affect the material transport efficiency [19]. DNA-based artificial transmembrane channels transport ions or biomolecules between phospholipid membranes, which mainly rely on the arrangement and assembly of DNA strands to form nanotubes or nanopores with holes [20]. Generally, DNA-based artificial transmembrane channels are composed of two structural modules: DNA nanotubes/nanopores as transport modules for mass transportation; and hydrophobic segments as anchor modules for membrane immobilization (Figure 1A). For example, phospholipids consist of a hydrophilic head and two hydrophobic tails. Hydrophobic fragments can be inserted into the internal hydrophobic site, allowing DNA nanostructures to be well anchored to phospholipid membranes.

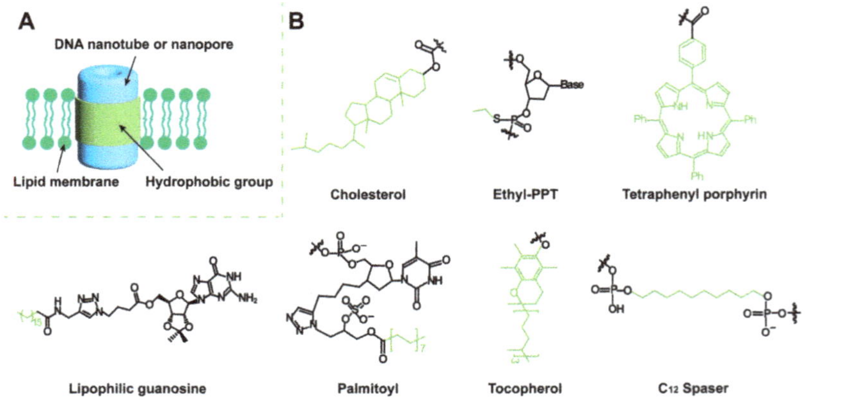

Figure 1. Structure model of artificial transmembrane channels and different hydrophobic modifications. (**A**) Schematic diagram of transmembrane channels formed by hydrophobic-modified DNA nanotubes or nanopores. (**B**) Various hydrophobic modifications: cholesterol, ethyl-PPT, tetraphenyl porphyrin, lipophilic guanosine, palmitoyl, tocopherol, and C_{12} spacer.

A core challenge of DNA nanotubes or nanopores embedded in phospholipid membranes regarding the formation of transmembrane channels is to overcome the adverse interactions between the hydrophilic, negatively charged DNA nanostructures and the hydrophobic membrane environment [6]. Many strategies have been developed to insert DNA nanostructures into the hydrophobic centers of the lipid bilayer. Keyser et al. have proposed several steps for constructing efficient membrane channels, especially for overcoming the high energy barrier of DNA across the bilayer hydrophobic core: (1) The connector between the hydrophobic anchor and the DNA core is shortened to achieve improved control over the anchor position. (2) When large structures are introduced, there should be a large spacing between the hydrophobic anchors to inhibit their simultaneous insertion into the bilayer without inducing crossing [21]. (3) The distance between the end of the structure and the membrane transdomain, as determined by anchor position, is reduced, limiting adverse interactions between the charged material and the hydrophobic core of the membrane [22].

Hydrophobic modification of DNA nanotubes or nanopores is a key step when designing transmembrane channels. Due to the excellent properties resulting from its membrane insertion, cholesterol is the preferred hydrophobic group for DNA modification. Cholesterol is most frequently used because of its strong hydrophobicity and because it can be modified at the DNA end or middle section, which is more conducive to the design of DNA artificial transmembrane channel structures. The amount and location of cholesterol are adjusted according to the designed channel shape and function. The number of cholesterols on artificial channels has been reported to be in the range of 2–60 [23–30]. DNA channels with small pore sizes and simple structures have little resistance to inserting into the membrane, and the corresponding modification method is relatively simple. Small-aperture artificial channels usually require only a few cholesterols at intervals along the sidewalls of the channel. However, as the pore size increases, it becomes increasingly difficult for DNA channels to be inserted into the lipid membrane. Over 10 cholesterols have been reported to be modified in DNA nanochannels with large apertures [25–30]. Inspired by the first DNA nanochannels, cholesterols are typically modified at the brim of the channel, under the flanking structure, or at the bottom of the middle shell [26–28]. It is necessary to add a coating to sequester cholesterol to prevent channel aggregation and ensure that the DNA channels are individually embedded in the membrane [29].

Ethyl-phosphorothioate (ethyl-PPT) [31], tetraphenyl porphyrin (TPP) [32], lipophilic guanosine [33], palmitoyl [28], tocopherol [34], and C_{12} spacer [35] are widely used for DNA hydrophobic modification (Figure 1B). The attachment of the ethyl group to the mercaptan group removes the negative charge of the typical phosphate anion. Ethyl iodide reacts with mercaptan groups through nucleophilic substitution to produce ethyl-protected PPT, which can be used for the hydrophobic modification of DNA channels [31]. To simplify the design of channels and minimize chemical intervention, other chemical labels with improved hydrophobicity have been proposed. It is expected that macroporous channels can be anchored to the membrane with a few chemical labels. TPP meets the requirements of hydrophobicity and can be easily coupled with DNA. Therefore, DNA channels modified with two porphyrin labels are constructed [32]. Acetylene-TPP is rigidly linked to the DNA strand by Sonogashira coupling to deoxyuridine. The T-pore-based artificial channels that were reported in 2016 are hydrophobically modified with tocopherol as a substitute for cholesterol [34]. In addition, biotin-streptavidin-coupled lipid networks have been developed [34]. The biotin-streptavidin connection with the biotinated lipid membrane provides additional channel-to-membrane interactions, enabling the insertion of single DNA channels into the membrane without channel aggregation.

3. Design of DNA Nanostructure for Artificial Transmembrane Channels

Currently, different types of DNA nanostructures have been developed for the design of artificial channels, including DNA wireframes, DNA helix bundles, DNA tiles, DNA origami, and so on.

3.1. *DNA Wireframe-Based Transmembrane Channels*

DNA wireframe nanotubes are promising self-assembled nanostructures that have produced a range of nanotubes or nanopores with different cavity sizes and tube lengths [36–41]. A typical modular assembly process for designing a wireframe structure from short synthetic strands is as follows: First, DNA structures acting as modular building blocks with specific shapes, such as triangles, squares, pentagons, and hexagons, are designed as needed. DNA nanotube rungs are then formed by longitudinal assembly through connecting strands. By controlling the number of DNA strands in each unit and connecting multiple rungs, the tube can be extended, and DNA nanotubes with micrometer lengths and specific pore shapes can be obtained.

DNA wireframe nanotubes can act as artificial transmembrane channels after adding hydrophobic anchors for membrane insertion. Sleiman et al. [23] designed cuboidal DNA channels and found that changing the pattern of the cholesterol unit on the cuboidal DNA significantly alters its interaction pattern with the lipid membrane (Figure 2A). Modification of cholesterol on a single face of the cube results in the peripheral anchoring of the DNA structure, while modification of cholesterol on two opposite faces of the cube enables the DNA pipeline structure to cross the phospholipid membrane. The DNA channels embedded in the membrane function as nanochannels for the transmembrane transport of dyes. Furthermore, researchers have designed a switch for the channel to avoid channel aggregation. This switch is achieved by adjusting the length of the cholesterol-DNA conjugate and the interval between the cubic binding segment of the strand and the cholesterol unit. The cholesterol units are initially hidden inside the cube and then exposed by a conformational switch for membrane insertion. Cholesterol-DNA cubes have become the first open-walled DNA channels that can be used as tools for sensing, drug delivery, and cellular behavior regulation applications. The DNA hexagonal prism constructed by Tan et al. [42] can transport ions across the membrane after modifying cholesterol on the four strands of the sidewall. An additional lock strand is designed on the opening side of the channel, and a key strand is added to open the channel for material transport (Figure 2B). These ligand-gated hexagonal DNA nanochannels have been used to mimic bionic ion channels in cell membranes.

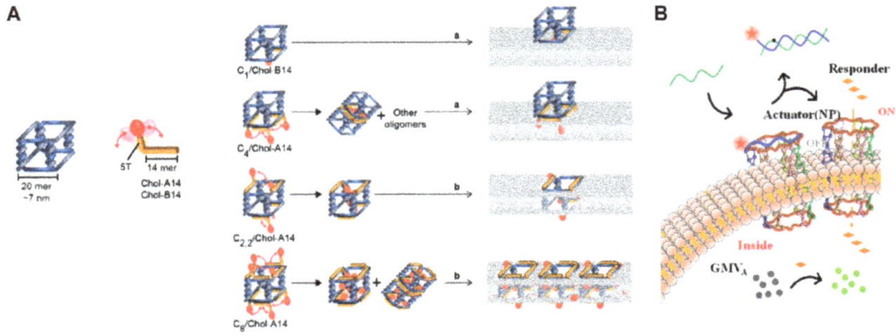

Figure 2. DNA wireframe-based transmembrane channels. (**A**) Schematic representation of DNA cube-based transmembrane channels through wireframe assembly. Different amounts of cholesterol-DNA conjugates (yellow lines) bind to the DNA cube at different positions. The cholesterol unit of the cholesterol-DNA conjugate can be flexibly regulated. Every cholesterol-DNA conjugate has 14 bases in the cuboidal binding segment, and an additional 5-thymidine (T) spacer serves as a chain to link the cholesterol, providing flexibility. From top to bottom are cubes containing 1, 4, 2 + 2, and 8 cholesterol-DNA conjugates. The first two images show cubes with cholesterol modified on a single face (a), eventually leading to extramembrane anchoring, and the last two images show cubes with cholesterol modified on two opposing faces (b), which can puncture the membrane. This figure was adapted with permission from ref. [23], copyright 2019, American Chemical Society. (**B**) Schematic diagram of the ligand-gated transmembrane channel response. This figure was adapted with permission from ref. [42], copyright 2021, American Chemical Society.

3.2. DNA Helix Bundle-Based Transmembrane Channels

Small-pore DNA-based artificial channels can be designed based on helix bundle (HB)-based nanotube assembly through the concatenation of multiple DNA strands [24]. DNA helix bundles are prepared by cross-joining scaffolds and short strands in the middle or at the ends of the materials, and the helix bundles follow the structural layout of polygon arrays [31]. The six DNA helix bundle (6-HB) nanotubes form approximately 2 nm-wide pores at the center (Figure 3A), which are hydrophobically modified to maintain structural stability in the lipid bilayer and to support a constant transmembrane current (Figure 3B). The reported conductivities of these 6-HB DNA transmembrane channels range from approximately 0.3 nS to 1.6 nS [31,32]. Through molecular dynamics (MD) simulations, the following conclusions are found. (1) The chemical modification on the surface of the channel has an extremely large impact on the transport of water and ions across the membrane, and the type, number, and position of the hydrophobic group modification can directly affect the formation of transmembrane channels. (2) DNA nanochannels can be used to transfer charged solute pairs to antistatic gradients through electroosmosis. (3) The porous channel wall allows the transverse leakage of ions and water. (4) The central lumen of the DNA channel is cylindrical and filled with water and ions; the volumes of water at the opening regions of both ends fluctuate in time and exhibit mechanosensitive gating (Figure 3C), creating a force sensor [43–46]. The negatively charged channel lumen has a high degree of control over cargo transport. For example, the lumen can act as an H^+ conductor to control the H^+ current [47] and show effective ATTO655 (dye with −1 charge) flux while blocking calcein (dye with −4 charge) transport [37]. Pore transport is affected by surrounding lipids. Studies on charged neutral lipids have shown that 6-carboxyfluorescein (FAM) dye is not transported, but analysis of negatively charged 1-palmiyl-2-oleoyl phosphatidylglycerol (POPG) shows that FAM dye can be transported by DNA channels [48].

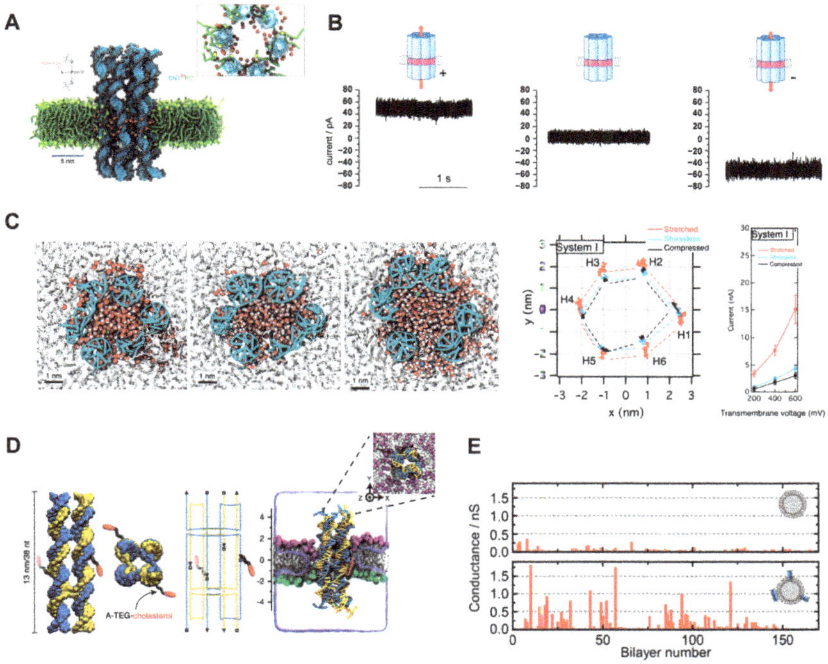

Figure 3. DNA helix bundle-based transmembrane channels. (**A**) Schematic diagram of the molecular simulation of 6-HB-based DNA transmembrane channels modified with ethyl-PPT. Image reproduced with permission from ref. [45], copyright 2017, Springer Nature. (**B**) Transmembrane current test diagram of the 6-HB DNA nanochannel. From left to right, the current traces of the channel under voltages of +100 mV, 0 mV, and −100 mV are shown. Image reproduced with permission from ref. [31], copyright 2013, American Chemical Society. (**C**) Simulation diagram, cross-sectional area, and current of the transmembrane channel under compression, zero tension, and compression. Image reproduced with permission from ref. [43], copyright 2015, American Chemical Society. (**D**) Molecular simulation of the 4-HB transmembrane channel with two cholesterols. Image reproduced with permission from ref. [21], copyright 2018, Springer Nature. (**E**) Conductivity test diagram of the 4-HB DNA channel. Image reproduced with permission from ref. [49], copyright 2015, American Chemical Society.

In addition, Keyser et al. have constructed a DNA nanochannel with a reduced pore size in which four DNA helical bundles (4-HB) are arranged on a square lattice to form a channel [21,49]. Two of these nonadjacent helical bundles contain a DNA strand carrying terminal cholesterol to embed a channel into the lipid membrane (Figure 3D). In this arrangement, the naturally occurring gaps between the helices produce a central channel with a nominal diameter of approximately 0.8 nm. Conductivity tests confirm the ion conduction ability of 4-HB DNA nanochannels embedded on giant unilamellar vesicles (GUVs). The GUVs embedded with DNA nanochannels are significantly more conductive than pure GUVs (Figure 3E).

3.3. DNA Tile-Based Transmembrane Channels

DNA tiles are short intersecting DNA strands that contribute to structural control. They have cohesive ends that can be programmed to self-assemble to form various DNA nanostructures by clinging together [50,51]. DNA tiles have been widely used to prepare DNA nanotubes. The steps of their self-assembly mainly focus on the release of DNA input molecules to trigger the growth of nanostructures, which is a nonautonomous and irreversible reaction [52]. With the maturation of construction technology, several methods have been designed to synthesize DNA tile nanotubes with adjustable, reversible, or controllable termination characteristics [53–55]. By using DNA origami structures as seeds to construct channels, micron-long nanotubes can be obtained through the polymeric growth of DNA tiles. The hydrophobic unit on the seed can directly insert the nanotube into the membrane to form a transmembrane channel (Figure 4A,B) [56]. An additional DNA origami channel cap can be used to terminate the aggregation of tiles. Conductivity measurements reveal that the conductance values of seeds and nanochannels are lower than their uncapped counterparts when the channel caps are attached. The results show that ions move from one end of the channel to the other and that there is partial leakage through the channel wall. However, the observation experiments of fluorescent dyes crossing lipid membranes confirm that molecular transport can occur through DNA nanochannels and that it is mainly end-to-end rather than across the channel wall (Figure 4C) [57].

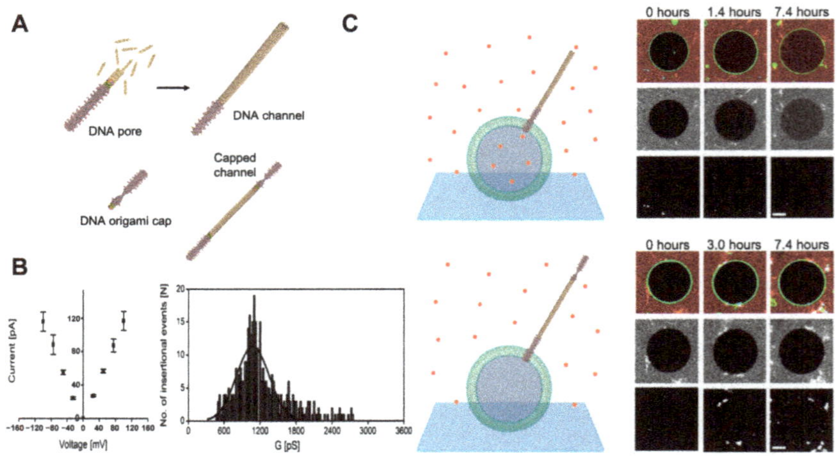

Figure 4. DNA tile-based transmembrane channels. (**A**) Schematic diagram of the channel structure. (**B**) Current-voltage relationship of the DNA tile-based transmembrane channels (left). Histogram of the conductance steps of the DNA tile-based transmembrane channels (right). The images in (**A**,**B**) were reproduced with permission from ref. [56], copyright 2022, bioRxiv. (**C**) Transport of dye molecules is hindered by DNA origami caps. The circles with green fluorescence in the figure refer to GUVs, and the red dots are small molecule dyes. Image reproduced with permission from ref. [57], copyright 2022, American Association for the Advancement of Science.

3.4. DNA Origami-Based Transmembrane Channels

DNA origami technology can be used for designing DNA-based channels; the shapes and sizes of the channels can be adjusted precisely, systematically, and abundantly [19]. DNA origami, proposed by Rothemund in 2006, is a relatively new method for DNA assembly. Based on the principle of complementary base pairing, by utilizing the structural characteristics of DNA molecules, the long DNA strands folded in specific regions are fixed by short strands to construct the expected structure [58]. Due to its simple experimental conditions and high assembly efficiency, DNA origami has become a popular technology for constructing artificial DNA nanochannels. Artificial DNA nanochannels have been comprehensively designed concerning their pore size, length, and morphology characteristics.

Artificial DNA nanochannels can be exploited for size-dependent, selective transmembrane transport. The pore diameter is critical for the selective delivery of substances of different sizes. Transmembrane channels with large pore sizes are required for the transport of macromolecular substances. To this end, long DNA strands that are arranged longitudinally are initially used as basic units to construct channels, and the inner diameters of these channels are <9 nm [25–28]. A method of lateral assembly of DNA strands to form channels has been proposed; the same amount of DNA strands can be used to obtain artificial DNA nanochannels with a relatively large pore size (~35 nm) [29,30]. Inspired by the natural channel protein α-hemolysin, Simmel et al. [25] have constructed a DNA nanochannel based on DNA origami technology. The channel consists of a tube that penetrates the membrane and a barrel cap on the periphery. The tube protrudes from the center of the barrel cap and features six DNA duplexes (Figure 5A), with the inner part of the tube acting as a transmembrane channel. The tube is approximately 2 nm in diameter and 42 nm in length. The brim of the barrel cap is set with 26 cholesterols to mediate adherence of the channel to the membrane. The average ohmic conductance of this channel is 0.87 ± 0.15 nS in a mixed solution containing KCl and $MgCl_2$. DNA nanochannels with pore sizes larger than 2 nm have been designed, and the morphological design of large-pore-size channels is highly complex and diverse. In 2016, the first DNA nanochannel was reported with a large pore size and large conductance embedded in lipid membranes [26]. The channel has been designed as a funnel-shaped structure with three intercovering layers, and the pore size gradually increases from the inside to the outside; the morphology and function are similar to those of the natural protein biofilm channel (Figure 5B). An ionic current recording experiment shows that the conductivity of this channel reaches 30 nS. In addition, Howorka et al. [27] have designed a similar DNA nanochannel with an inner diameter of approximately 7.5 nm and a top opening of 22.5 nm (Figure 5C). This realized the transmembrane transport of enhanced green fluorescent protein (EGFP; 27 kDa) and showed an interception effect on rhodamine B-dextran (70 kDa). These channel walls consist of at most three biphasic layers, and the DNA nanochannels can be stable on the membrane.

The DNA nanochannels synthesized by Kjems et al. [28] have flanks to adjust the cholesterol exposure on demand (Figure 5D). The DNA channels are composed of a double-layer irregular hexagonal cylindrical DNA structure with 46 hydrophobic spots (17 on the walls and 29 on the flanks). Three programmable DNA lobes are on the three nonadjacent sides of the channel and connected approximately 12 nm from the bottom of the channel by a single-stranded DNA hinge. In the closed state, each flank is near the channel wall due to two stable strands that are complementary to the bottom single-strand portion of the channel. The flanking closure can protect the hydrophobic moiety from the aqueous environment and limit hydrophobicity-driven channel aggregation. When a key strand fully complementary to the single-stranded DNA is present around the channel, the flanks are opened and the cholesterol is exposed, thereby driving channel insertion into the membrane. Furthermore, Dekker et al. [29] have designed and assembled a DNA channel with a rigid octagonal ring structure. This channel is formed by folding 7560 base-long scaffold single strands and 240 individual oligonucleotide single strands, arranged in a 4×4 duplex pattern (Figure 5E). A total of 32 cholesterol molecules are uniformly arranged on the surfaces of the channels. In addition, K_{10}-PEG_{5K} molecules are coated to stabilize the octagonal structure of the channels and prevent the aggregation of cholesterol-modified channels. The K_{10}-PEG_{5K} molecule consists of 10 positively charged lysine amino acids (K) linked to a short polyethylene glycol (PEG; 5 kDa). This channel has the largest inner diameter among the reported DNA-based transmembrane channels at 35 nm. Recently, a series of structurally concise DNA nanochannels with different pore sizes have been constructed by using bundled DNA duplex subunits [30]. The channel is designed with an outer membrane cap structure to define the overall pore shape and a barrel structure in the center for puncturing the membrane. Subunits are modularly arranged parallel to the membrane to form pores of specific shapes, such as triangles, squares, pentagons, and hexagons (Figure 5F). Adjusting the number of duplex subunits can regulate the height and shape of the channel, and tuning the number of bases per subunit can regulate the pore size of the channel. The subunits of the cap are linked by short single-stranded (ss) DNA at the innermost double-stranded (ds) position. Furthermore, a short duplex of DNA is added between the double strands of the outermost subunit of the cap. This was conducted to prevent the cap from flipping and deviating from its designed shape.

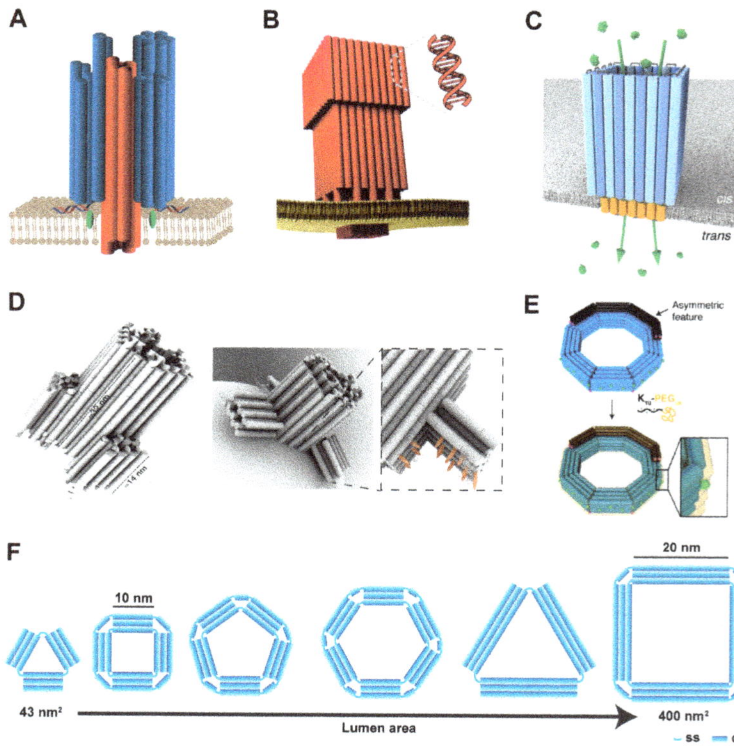

Figure 5. DNA origami-based transmembrane channels. (**A–C**) are schematic structures of various DNA origami transmembrane channels. (**B**) was reproduced with permission from ref. [26], copyright 2016, American Chemical Society. (**C**) was reproduced with permission from ref. [27], copyright 2019, Springer Nature. (**D**) Schematic representation of the DNA origami transmembrane channel with flanks in the closed (**left**) and open (**right**) states. Reprinted with permission from ref. [28], copyright 2019, Springer Nature. (**E**) Schematic diagram of the DNA origami transmembrane channel coated with K_{10}-PEG_{5K} molecules. Reprinted with permission from ref. [29], copyright 2021, American Chemical Society. (**F**) Top view of the DNA nanochannels with different apertures and different shapes formed by the assembly of 10 nm or 20 nm subunits. The pore areas of DNA nanochannels with different polygonal shapes and sizes ranged from 43 nm^2 to 400 nm^2.

3.5. Other DNA-Based Transmembrane Channels

G-quadruplex, a DNA duplex with unique ion transport properties, is utilized to design transmembrane channels. Depending on the hydrogen bond between the nitrogen and oxygen atoms of guanosine and the π-π stacking between the bases, guanine (G)-rich DNA single strands can be deformed or aggregated to form a G-quadruplex structure with a central hole [59]. Moreover, the stability of the G-quadruplex depends heavily on certain cations, such as K^+, Na^+, NH_4^+, and Ca^{2+} [60]. Within the central channel, each ion is completely dehydrated and interacts with the guanine carbonyl O_6 atom around the pore (Figure 6A). This particular feature is reminiscent of the selectivity filters in K^+ ion channel proteins first noted by Feigon et al. Because of this structural similarity, the G-quadruplex is a candidate for the design of artificial ion channels for selective transmembrane transport of K^+ [61–63]. Dash et al. [33] have used telomere DNA to form a G-quadruplex and additionally modified lipophilic guanosine to construct an artificial potassium ion transmembrane transport carrier (Figure 6B). Liu et al. [35] have proposed a G-quadruplex consisting of single-stranded DNA modified with three lipophilic C_{12} spacers and cholesterol as an intelligent transmembrane channel to selectively transport K^+ across the membrane.

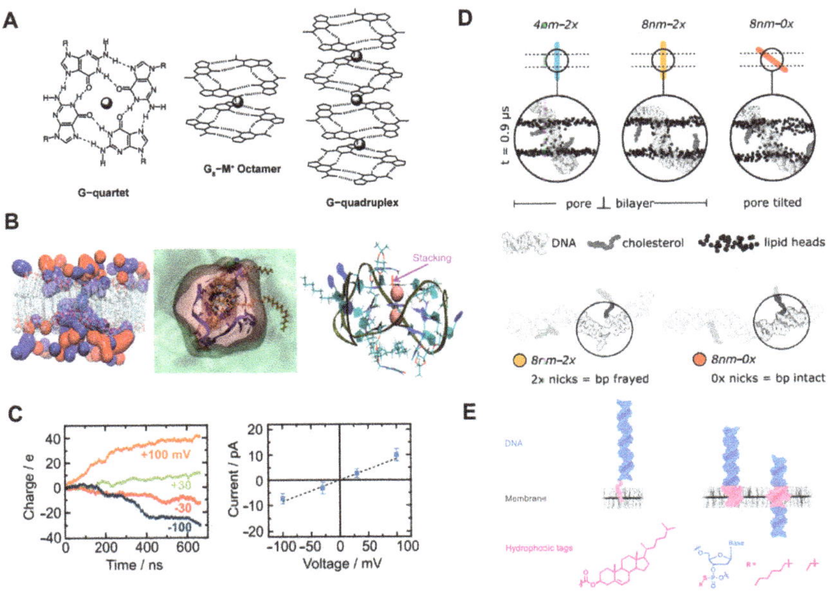

Figure 6. G-quadruplex or single DNA duplex-based artificial transmembrane channels. (**A**) Schematic representation of the structure of the G-quartet and G-quadruplex. Adapted with permission from ref. [59], copyright 2001, American Chemical Society. (**B**) Distribution of K^+ (blue) and Cl^- (red) ions in a channel formed by lipophilic G-quadruplex (left). The middle is the top view of the transmembrane channel, containing the lipid layer (green), water pore (pink), and G-quadruplex (blue) modified by lipophilic guanidine (orange). The structure of the lipophilic G-quadruplex is given on the left. Image was adapted with permission from ref. [33], copyright 2020, Springer Nature. (**C**) Cumulative charge transmitted across the lipid bilayer and the current-voltage characteristics. Images were adapted with permission from ref. [64], copyright 2016, American Chemical Society. (**D**) Simulation of the interaction of DNA duplexes (cholesterol modification distance of 4 nm or 8 nm) with and without nicks with the phospholipid bilayer. Image was adapted with permission from ref. [65], copyright 2021, American Chemical Society. (**E**) Hydrophobic design of lipophilic DNA duplexes and their interactions with lipid bilayers. Image was adapted with permission from ref. [22], copyright 2021, American Chemical Society.

In addition, Keyser et al. [64] have constructed an artificial DNA nanochannel capable of ion conduction using a single long duplex (Figure 6C). MD simulations have shown that cholesterol modification leaves terminal base pairs of DNA structures with gaps, resulting in distortion when embedded in lipid bilayers (Figure 6D). The DNA nanostructure backbone can form stable conductive pores and be inserted into membranes at a higher efficiency than equivalent notched structures. In addition, the notation-free DNA structure can be designed to modulate its tilt orientation within the lipid bilayer. Reducing the degrees of freedom of DNA conformation and adjusting the positions of hydrophobic modifications can control the function of this structure as a synthetic ion channel well (Figure 6E) [22,65].

4. Artificial Transmembrane Channels for Biosensing and Biomedical Applications

DNA-based artificial channels on the cell membrane have been widely used for biosensors and biomedicine. For example, they are used for DNA strand translocation [25,34], ions (K^+, Ca^{2+}, and Pb^{2+}) [31,33,35,42], small molecules (dyes and drugs) [23,66], and large molecules (EGFP, GFP, IgG, and dextran) [27–30,67] transmembrane transport (Table 1). Artificial DNA channels with gates can modulate material transport under specific conditions. These channels provide tools for molecular sensing, artificial cell design, and cell communication, achieving controlled transmembrane transport to cells and regulating cell behavior.

4.1. DNA-Based Transmembrane Channels for Biosensors

4.1.1. Single-Molecule Nanochannel Sensors

Transmembrane DNA channels have been proposed for use as single-molecule nanochannel sensors. In biomimetic sensing experiments, the translocation of analyte molecules leads to current changes in membrane pores and the duration and depth, which are related to the charges and sizes of the analytes. For example, DNA channels designed by Simmel et al. [25] have been used for single-molecule sensing. A stable baseline current is detected on the lipid membrane containing the artificial DNA channels. The additions of the hairpin molecules at the beginning and ~30 min later show transient current blocking, in which the applied voltage can capture, decompress, and translocate the hairpin structure (Figure 7A). Another set of experiments shows that lengthening the tail of the quadruplex increases current blocking. The average current blockades of the quadruplexes with single-stranded tails consisting of 60 and 125 thymidines (Q-T60, Q-T125) are $\Delta I_{Q\text{-}T60} = 5.6 \pm 1.0$ pA and $\Delta I_{Q\text{-}T125} = 15.3 \pm 2.3$ pA, respectively (Figure 7B). Thus, similar to biological pores, artificial DNA nanochannels can be used as sensing devices to distinguish analyte molecules by studying their translocation properties. The T-pore designed by Simmel et al. allows the translocation of double-stranded DNA (dsDNA) molecules (Figure 7C) [34]. As the transmembrane voltage increases, the velocity of the molecule and the frequency of translocation events increase (Figure 7D), while the dwell time within the channel decreases.

DNA channels have been used to identify and detect proteins. Howorka et al. [30] have used DNA nanochannels to identify human SARS-CoV-2 antibodies with a handheld MinION kit (originally intended for portable DNA sequencing). The cognate receptor SARS-CoV-2 spike protein attaches to an adaptor oligonucleotide on the inner wall of the Tri-20 channel through an irreversible metal chelate bridge (Tri-20-spike). In the presence of a human SARS-CoV-2 antibody, the conductivity of Tri-20-spike is significantly reduced, and the average dwell time of the antibody is 1.7 ± 8.6 s.

4.1.2. Ligand-Gated Artificial Transmembrane Channels

The DNA-based artificial transmembrane channels are mostly hollow tubes with openings. Voltage gating is observed in almost all DNA nanochannels, but channels with additional gates are highly flexible and controllable [68]. Therefore, the design of artificial channels with ligand-gated opening or closing properties has become the focus of research. The reported artificial DNA nanochannels with gates can specifically recognize key strands. Howorka et al. [66] built an artificial channel with a gate that can be opened with a key strand. The lock of the channel is tightly bound to the entrance by hybridizing with two docking sites to form a spiral bundle across the channel opening. The docking site is formed by the extension of two duplex support rods in opposite positions. The key can be hybridized with the lock strand to remove it, leaving the channel open. After being modified by cholesterol, such channels can be used to regulate the flow of small organic molecules (including many important drug compounds), which have broad application prospects in biomedicine. A large square DNA origami nanochannel with 0 (open state), 2 (semiclosed state), or 4 (closed state) strips of DNA lock can block the molecular penetration mechanism, improving the selectivity of molecular penetration of artificial DNA nanochannels (Figure 8A) [69].

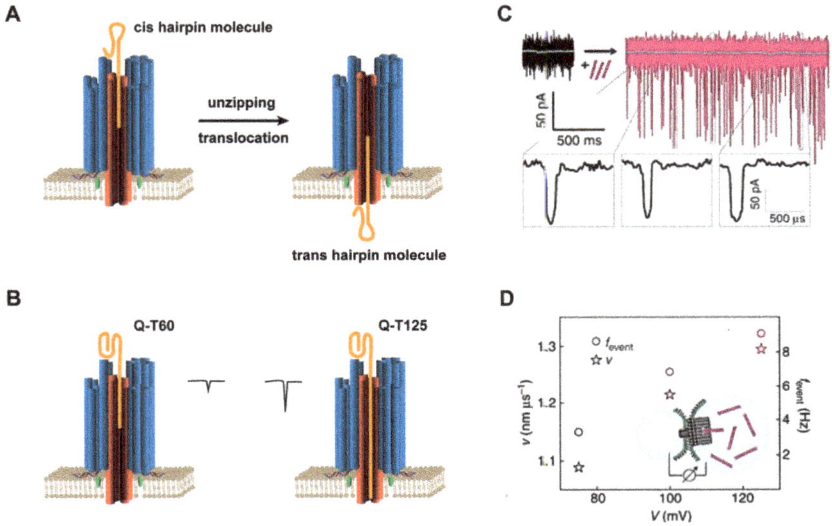

Figure 7. Single-molecule nanochannel sensors. (**A**) Schematic diagram of DNA hairpin strand capture and translocation on transmembrane DNA channels. (**B**) Current blockade of DNA quadruplexes with different tail lengths on transmembrane DNA channels. (**C**) Current on the transmembrane channel in the absence/presence (left/right) of the analyte. The bottom shows three current signals at the time of dsDNA translocation. (**D**) dsDNA translocation velocities and event frequencies detected at 75, 100, and 125 mV. The images in (**C**,**D**) were reproduced with permission from ref. [34], copyright 2016, Springer Nature.

To exploit the potential of DNA-based artificial channels as real-time smart sensing devices, Kjems et al. [28] have designed a bolt on the inside of the channel. PEG is used as a plug that connects the toehold sequence with 8 nucleotides to partially block the gateway (Figure 8B) and allows small molecules (ATTO 655) to pass through (Figure 8C). The unplugged strand is combined with the toehold-mediated strand to remove PEG so that the macromolecular material (dTMR-40k) can pass through the artificial DNA nanochannel. Another reversible gated protein transport membrane channel is constructed based on a horizontal routing DNA origami design strategy with a large pore size of 20.4 nm (Figure 8D) [67]. The passageway opening is designed with a square cover, one side of which is attached to the cap by a flexure hinge. The other side of the cap carries two single strands that can be hybridized with the two single strands on the cap to form a complete double lock. Two key strands are added to the system to open the lock and lid. The lid is switched back to its closed state with a single-strand reverse key pair. This channel allows the precisely timed transport of folded proteins across the membrane. A potential disadvantage of covered channels is that they may leak when closed or when inserted in a double layer. Howorka et al. [70] designed a channel formed by the assembly of two subunits after being triggered by ligands on the membrane. The whole channel consists of two parts, component A and component B, each of which contains a complete duplex and two single strands (Figure 8E). The complementary binding of single strands on the two components allows channel formation. Adding a key unlocks the locks on both components and restores their binding ability. Each component carries one cholesterol, and after the channel forms, the cholesterol is on the opposite side of the channel wall.

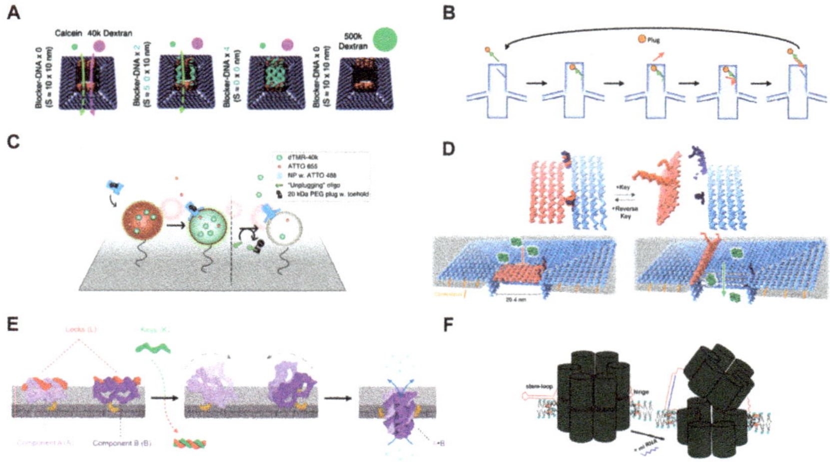

Figure 8. Ligand-gated DNA-based artificial transmembrane channels. (**A**) Schematic representation of the passage/blockage of the dye molecule by the blocker DNA. From left to right are 0, 2, and 4 blocker DNA channels. The channel is designed with a pore size of 10 nm, and it does not allow the passage of 500 k dextran (far right). Image was adapted with permission from ref. [69], copyright 2021, Royal Society of Chemistry. (**B**) Schematic representation of an artificial transmembrane DNA nanochannel with a bolt (blue) recognizing an unplugging oligo (red arrow). In the figure, the orange plug is PEG, and the green arrow is the toehold strand. (**C**) Schematic diagram of an artificial transmembrane DNA channel with a bolt before (left) and after (right) adding an unplugging oligo. (**B**,**C**) were adapted with permission from ref. [28], copyright 2019, Springer Nature. (**D**) A large DNA nanochannel (blue) with a reversible gate (pink). The mechanism of opening and closing by a key and the reverse key is designed. Two sets of locks (red and dark blue) are located on the lid and the cap. Adapted with permission from ref. [67], copyright 2022, Springer Nature. (**E**) Schematic diagram of the formation of a DNA membrane channel in which the key strands trigger assembly. Adapted with permission from ref. [70], copyright 2022, American Chemical Society. (**F**) Schematic representation of DNA transmembrane channels disrupted by tension dependent on the duplex. Adapted with permission from ref. [71], copyright 2022, Wiley-VCH.

Since the DNA duplex is quite stiff and has a persistent length, Elezgaray et al. [72] have designed a DNA channel with two conducting states: closed (low current) and open (high current). The transition of the channel from closed to open is triggered by ssDNA. When the lock at the entrance of the 6-HB DNA channel binds to the complementary strand, the resulting duplex exerts tension on the channel structure, locally widening the pore size. In 2022, Elezgaray et al. [71] designed another channel that relies on the tension of the duplex to switch in the vertical directions, proposing that this design can detect short oligonucleotide sequences (Figure 8F).

4.1.3. Environmental Stimuli-Responsive Artificial Transmembrane Channels

Environmental stimuli-responsive artificial channels have been designed to be sensitive to temperature [73], light [74–76], and ions [77]. The temperature-responsive DNA channel constructed by Howorka et al. [73] has two main parts: a transmembrane barrel-shaped nanotube and a reversibly sealed lid at the top. Biphasic segments 1–4 are designed between the channel and the lid, with a designed melting temperature of approximately 40 °C for segments 2–4 and 62.8 °C for segment 1 (Figure 9A). The lid is hybridized to the two elongated rings of the channel at room temperature to block the mass influx. Temperatures higher than 40 °C selectively separate the lid from loop segments 2–4 to allow the cap to open. By adjusting the temperature, the lid of this channel can achieve reversible on/off functionality. Azobenzene is a reversible cis-trans photoisomerization switchable compound. The conversion of azobenzene from the trans isomer to the cis isomer can be triggered by light irradiation with a wavelength λ < 400 nm, and the reverse effect can be achieved by illumination at λ > 400 nm [74]. The cis-trans isomerization of azobenzene can adjust the on/off state of the channel. Howorka et al. proposed a 6-HB-based DNA channel, and cis-azobenzene corresponded to the closed state (Figure 9B) [75]. Liu et al. designed a DNA transmembrane channel based on a lipophilic G-quadruplex, in which cis-azobenzene causes G-rich DNA strands to assemble into channels for the transmembrane transport of ions [76]. The ion-dependent G-quadruplex channel can be utilized for selective ion transport [77]. Guanine-rich lipophilic ssDNA can form different G-quadruplex isomers with different metal cations [78,79]. Therefore, a biomimetic ion channel is designed based on the G-quadruplex for conformation-dependent selective ion transport to membranes. Specifically, the Pb^{2+}-stable antiparallel conformation G-quadruplex (apG4) preferentially mediates the transmembrane transport of Pb^{2+}, while the K^+-stable parallel conformation G-quadruplex (pG4) promotes the highly selective transport of K^+ (Figure 9C).

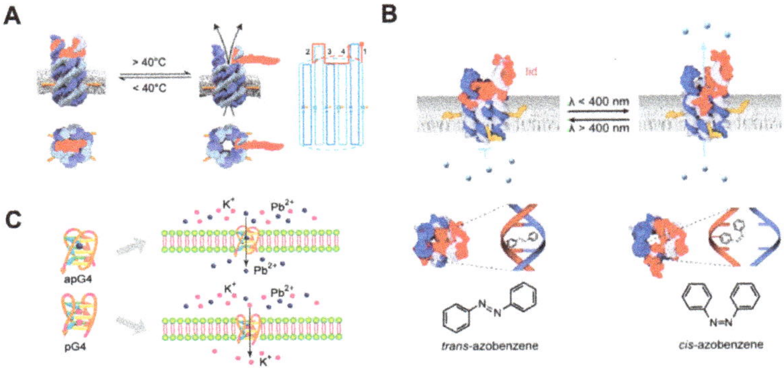

Figure 9. Environmental stimuli-responsive artificial transmembrane channels. (**A–C**) are the channels that depend on temperature, light, and ion recovery transport functions, respectively. (**A**) was adapted with permission from ref. [73], copyright 2019, American Chemical Society. (**B**) was adapted with permission from ref. [75], copyright 2022, John Wiley and Sons. (**C**) was adapted with permission from ref. [77], copyright 2020, American Chemical Society.

4.2. DNA-Based Transmembrane Channels for Biomedical Applications

4.2.1. Cell Mimics for Transmembrane Transport

Artificial DNA nanochannels can serve as synthetic cell membrane components to mimic transmembrane transport. Currently, the transport selectivity of DNA transmembrane channels is largely determined by their pore size. The molecules or ions smaller than their pore size are easily mass-transportable. Meanwhile, the negatively charged DNA ion channel has a poor transport capacity for negatively charged ions. Artificial DNA transmembrane channels with large pores have been proposed for the transmembrane transport of drugs, immune proteins, and so on. In 2016, Howorka et al. [80] were inspired by organelles to create synthetic hybrid nanocontainers composed of polymersomes and DNA nanochannels (Figure 10A). Nanocontainers exhibit size-dependent permeability. These containers enable the transport of the enzyme substrate across the membrane while retaining the relatively large enzyme inside the container. These nanodevices can be used to simulate the site where biocatalytic reactions occur. The 6-HB DNA nanochannels modified with three cholesterol molecules on this container mimic the protein channels in biofilms and enable specific substance transport (Figure 10B). Tan et al. [42] have constructed an artificial signal transduction network using two synthetic cell communities formed by vesicles containing different DNA structures modified with cholesterol (Figure 10C). This system is used to simulate cell-to-cell communication, in which giant membrane vesicles derived from living cells are used as cell models. This signal transduction system consists of two groups of giant membrane vesicles, one as a stimulator and one as a receptor for signal actuation. The stimulation group is modified with a DNA triangular prism stimulator (TP_B). The receptor group contains the DNA triangular prism receptor (TP_A) and transmembrane channel. The transmembrane channel is designed with a lock strand that closes the mass entrance. When a toehold strand displacement reaction is performed with a foreign complementary key strand, the lock strand is released from the transmembrane channel to reopen the channel. In this system, the key is derived from TP_A. When the two groups of artificial cells approach each other, TP_B combines with TP_A. Afterward, the key strand originally on the TP_A breaks away to recombine with the lock strand on the wireframe, thus opening the channel. Then, extramembranous Ca^{2+} flows into the receptor vesicles, allowing calcein, which is encapsulated inside the vesicle, to respond. The DNA nanochannels exhibit high efficiency, accuracy, and programmability as signal receivers and actuators in this system. They make it possible to precisely manipulate signal transduction systems, address the high-order complexity of cell simulations well, regulate the behaviors of natural cells, and control the release of therapeutic drugs at disease sites.

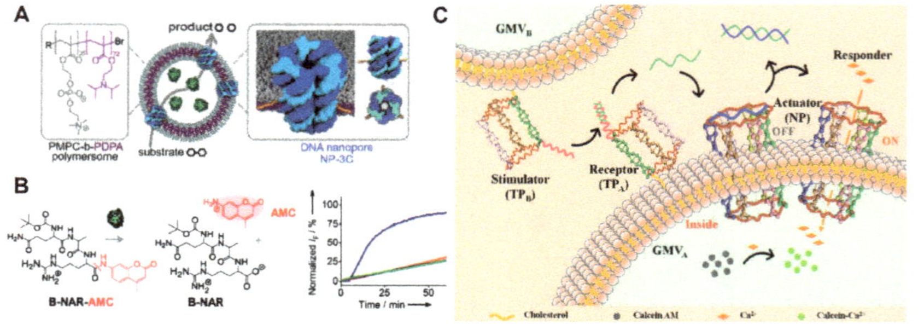

Figure 10. Cell mimics for transmembrane transport. (**A**) Nanocontainers formed by the self-assembly of PMPC-b-PDPA polymersomes coated with enzymes (green). DNA nanochannels (blue) are embedded in the membrane to allow the substrates and products of the enzyme to shuttle across the membrane. (**B**) Schematic diagram of the structure of the trypsin-hydrolyzed substrate peptide and the results of fluorescence detection. The fluorescence kinetics curves of the loaded enzyme hydrolyzed peptide with the addition of three (blue) or zero (red) cholesterol nanochannels or without the addition of a nanochannel (green). (**A**,**B**) were adapted with permission from ref. [80], copyright 2016, John Wiley and Sons. (**C**) Design of the artificial signal transduction system based on DNA triangular prisms, DNA hexagon prism nanochannels, and giant membrane vesicles. Reprinted with permission from ref. [42], copyright 2021, American Chemical Society.

4.2.2. Transmembrane Channels for Cell Death

Artificial transmembrane channels can selectively control ion transport across biological membranes, and artificial channels can disrupt cellular homeostasis of cell death. Howorka et al. [81] designed a DNA channel with a highly hydrophobic 2-nm band composed of ethyl phosphorothioate (EP) at one end, which can penetrate the cell membrane and cause cell cytotoxicity. Tan et al. [82] have found that phosphorothioate (PPT)-modified DNA nanochannels can be spontaneously inserted into the cell membrane (Figure 11A), and they can transport ions and antitumor drugs to neurons and cancer cells, respectively. It has been proposed that their potency can be improved by specifically binding target cancer cells. Loading chemical toxins, such as doxorubicin, with DNA insertion enhances chemical toxicity. In 2019, Zhang et al. [83] proposed a design for the controlled transfer of DNA nanochannels to the plasma membrane. This approach enhances the insertion of DNA nanochannels, inducing membrane depolarization and pyroptosis-like cell death. 3D tumor spheroid experiments show that this process can induce tumor cell death and significantly inhibit tumor growth. A series of experimental results have shown that this device exhibits antitumor ability. It induces a host antitumor immune response by promoting antigen presentation and activating T cells and natural killer cells (Figure 11B).

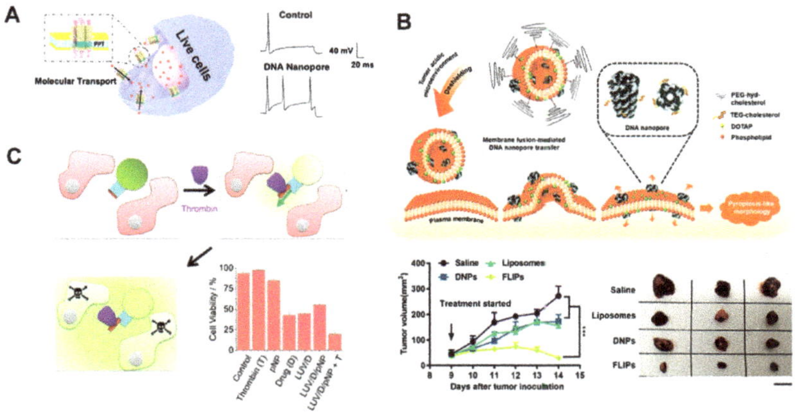

Figure 11. Transmembrane channels for cell killing. (**A**) Schematic illustration of DNA nanochannels embedded in cell membranes to deliver drugs. Dorsal root ganglia (DRG) neurons with channels inserted had more evoked action digits than those without channels. Reproduced with permission from ref. [82], copyright 2020, American Chemical Society. (**B**) Illustration of the transfer of nanochannels from vesicles to the plasma membrane. In the acidic tumor microenvironment, vesicles with DNA nanochannels can be activated to fuse with cell membranes. DNA nanochannels were successfully transferred to the plasma membrane after fusion. Inhibition of tumor growth was detected 9 days after vesicular DNA nanochannel treatment in C57BL/6 mice bearing Hepa1-6 xenografts. Reproduced with permission from ref. [83], copyright 2020, American Chemical Society. (**C**) Schematic representation of protein-triggered DNA nanochannels opening on vesicles to release drugs for cell killing. The bottom right corner shows the survival statistics of HeLa cells treated with different substances for 3 days. Reproduced with permission from ref. [84], copyright 2020, John Wiley and Sons.

Artificial transmembrane channels with large cavities can be exploited for the controlled transport of chemical drugs for cell behavior modulation or cell death. For example, DNA nanochannels with protein-controlled gates have been successfully applied to the off-on transport of topotecans [84]. The DNA-based artificial channels are inserted into GUVs filled with 3 μM topotecan. Topotecan is a clinically used cytotoxic agent with activity against cervical cancer. In the experiment, tri-component vesicles are first added to HeLa cervical cancer cells. The results show that cytotoxic drugs (D) are released to reduce cell viability to 20 ± 2% when both channels and thrombin (T) are present (Figure 11C). To date, artificial nanochannels for tumor cell killing are mostly nontargeted, and their actual application endangers healthy cells to varying degrees [85]. Depending on the diversity of DNA nanostructures, it is feasible and necessary to improve the targeting ability of DNA nanochannels in the application process. If the designed channel has a target recognition function, it can reduce damage to normal cells, improve the binding rate between the channel and target cells, and enhance the cell death effect [86,87].

Table 1. Design and applications of DNA-based artificial transmembrane channels.

Type	Pore Size and Length (nm)	Hydrophobic Modification	Applications	Refs.
Cubical	7; 7	4 or 8 cholesterols on the two opposite faces of the cube	Dye molecular transmembrane transport	[23]
Hexagonal prism	-	4 cholesterols	Transport Ca^{2+} across the membrane as a part of the artificial signal transduction system	[42]
6-helix bundle	2; -	Ethyl-phosphorothioate	1. K^{+} transmembrane transport 2. Selectively cytotoxic to cervical cancer cells 3. Transport ions and antitumor drugs (DOX) to neurons and cancer cells, respectively	[31,44,45]
4-helix bundle	0.8; -	2 cholesterols	Ion conduction	[21,49]
Micron-length	7.3; Micron	12 cholesterols	Leakless end-to-end transport of carboxytetramethyl-rhodamine	[57]
Capped	2; 47	26 cholesterols at the underside of the subunits' caps	DNA hairpin and guanine quadruplex translocation	[25]
Funnel-shaped	6; 54	19 cholesterols	Large ionic conductance	[26]
Protein-conductive	7.5; 46	24 cholesterols	Proteins transmembrane transport	[27]
Large size-selective	9; 32	18 cholesterols and 28 palmitoyls at the terminal of channel and the underside of the flap	1. Size-specific cargo translocation 2. Real-time direct observation of sensing of an oligonucleotide unplugging strand	[28]
Ultrawide	35; 10	32 cholesterols	Transmembrane transport of green fluorescent protein (GFP)	[29]
Highly shape- and size-tunable	8.66~20; -	Underside of the subunits' caps (10 cholesterols on each 10-nm subunit, 15 cholesterols on each 20-nm subunit)	1. IgG sensing 2. Human SARS-CoV-2 anti-bodies detection	[30]
G-quadruplex and liponucleoside	-	Lipophilic guanosine binds to the G-quadruplex with a 3:1 stoichiometry	The transport of K^{+} across CHO and K-562 cell membranes	[33]
Lipophilic G-quadruplex	-	3 C_{12} spacers and 1 cholesterol on each strand	Switchable K^{+} transport	[35]
Single-duplex	-	6 tetraphenyl porphyrins	Ion conduction	[64]
T-like	3.7; 27	57 tocopherols at the bottom of the double-layered plate	1. DNA translocation 2. Molecular transmembrane transport	[34]
Ligand-controlled	-; 9.0 ± 1.5	3 cholesterols	Controlled DNA-triggered and charge-selective release of small-molecule cargo (sulfo-rhodamine B, SRB) from a reservoir	[66]
Large, square	10; 24	47 strands at the bottom of the nanopore, complementary to the anchor DNA on the phospholipid membrane	Selective transmembrane transport of molecules with different sizes (calcein, 40 k or 500 k dextran)	[69]
Reversibly gated	20.4; -	64 cholesterols	1. Dye molecular (Atto633) transmembrane transport 2. Dynamically controlled cargo (Atto633 and GFP) transport across the lipid bilayer	[67]
Trigger-assembled	0.8; 7.5	One cholesterol per component (A or B)	Ion conduction	[70]

Table 1. *Cont.*

Type	Pore Size and Length (nm)	Hydrophobic Modification	Applications	Refs.
Tensegrity driven	-; 4 ± 0.2	4 cholesterols driven channels inserted into the membrane and 2 cholesterols grafted at the ends of two strands facing each other on one side of the barrel	Controlled transmembrane transport of SRB	[72]
Temperature-gated	2; -	4 cholesterols	Permitted temperature-controlled transport of molecular cargo (SRB) across the bilayer	[73]
Light-triggered synthetic	2; 12.5	4 cholesterols	Light-gated small-molecule (SRB) transport	[75]
8-helix bundle	4; 14	4 TEG-cholesterols	1. Depolarized plasma membrane to induce pyroptosis 2. Inhibited tumor growth	[83]
Protein-controlled	Molecular gate approximately 13 nm × 5 nm× 5 nm	4 cholesterols	1. Controlled transport of molecular cargo (SRB) across lipid bilayers 2. Released cytotoxic drug (topotecan) for cell killing	[84]

5. Conclusions and Perspectives

In this review, we describe the design of DNA-based artificial transmembrane channels and their applications in biosensing and biomedicine. Different kinds of artificial transmembrane channels have been developed based on DNA wireframes, DNA helix bundles, DNA tiles, and DNA origami assembly technologies. However, there are still many challenges in the artificial transmembrane DNA nanochannels reported herein for applications. For example, there is some degree of leakage while transmitting material. For example, some positively charged dyes could leak out due to their membrane permeability. Additionally, the channel or gating structure is unstable, and aggregation of channels due to hydrophobic modifications still occurs. Thickening the channel wall or coating it with other polymers can effectively reduce material leakage while stabilizing the channel structure [51]. The controllable assembly of DNA nanochannels prevents the material from leaking in advance; the assembly can be stimulated at a specific time to combine the parts to form channels [72]. The main solution to the problem of channel aggregation is to adjust the type, number, and position of hydrophobic modification groups [25,34]. Designing unique channel structures or aiding surfactants are special approaches that have been proposed [28,29,88].

Previous studies have focused on the biomedical applications of artificial transmembrane channels. However, the mode of action of DNA nanochannels is relatively simple and only depends on the toxicity of the channel to cells or the transmembrane delivery of drugs to kill tumor cells. There is room for improvement in the ability to transport ions, proteins, or drugs across membranes in a form similar to passive transport. Enriching the function of DNA nanochannels, such as enabling the same channel to transport different substances in unique states, may enhance their ability to induce apoptosis [67]. The proposed optimal design is to endow DNA nanochannels with functions similar to ion pumps at the cell membrane or channels that enable active transport, moving the material uphill against a concentration gradient [89,90]. The DNA duplex rotates in one direction at a rate of billions of revolutions per minute under the action of an electric field, and the direction of rotation is determined by the chirality of the duplex [91]. It is challenging to design energy-driven active transport based on the DNA transmembrane channels. Recently, Ma et al. fabricated a bionic micropump for active transmembrane drug delivery by immobilizing urease on the surface of silica-based microtubules [92]. The above results provide evidence for the possibility of the active transport function of DNA nanochannels.

The biomedical applications of transmembrane channels are not limited to cancer therapy [93]. DNA nanochannels with ion transport functions may contribute to the treatment of ion channel diseases [94]. The vast majority of ion channel diseases are associated with genetic mutations that determine the performance of ion channels [95]. The mutations are partly caused by other diseases that damage the membrane channel synthesis pathway or function [96]. In the future, artificial transmembrane channels are expected to be used as biomimetic membrane channels for the controlled transport of substances in the physiological environment to alleviate the symptoms and treatment effects caused by ion channel diseases. We will explore complex physiological operation pathways with artificial transmembrane channels, leading to further research on disease treatment and biomedical applications.

Author Contributions: Conceptualization, W.X., H.C. and J.L.; investigation, Y.L.; resources, H.C. and S.L.; writing—original draft preparation, W.X.; writing—review and editing, K.W. and J.L. All authors have read and agreed to the published version of the manuscript.

Funding: The authors gratefully acknowledge the financial support of the National Natural Science Foundation of China (22377025 and 22177032) and the Natural Science Foundation in Hunan Province (2022RC3047, 2021JJ10013, and 2022JJ40049).

Institutional Review Board Statement: Not applicable.

Informed Consent Statement: Not applicable.

Data Availability Statement: Not applicable.

Conflicts of Interest: The authors declare no conflict of interest.

References

1. Shen, H.; Wang, Y.; Wang, J.; Li, Z.; Yuan, Q. Emerging Biomimetic Applications of DNA Nanotechnology. *ACS Appl. Mater. Interfaces* **2019**, *11*, 13859–13873. [CrossRef] [PubMed]
2. Shen, Q.; Xiong, Q.; Zhou, K.; Feng, Q.; Liu, L.; Tian, T.; Wu, C.; Xiong, Y.; Melia, T.J.; Lusk, C.P.; et al. Functionalized DNA-Origami-Protein Nanopores Generate Large Transmembrane Channels with Programmable Size-Selectivity. *J. Am. Chem. Soc.* **2022**, *145*, 1292–1300. [CrossRef]
3. Jiang, X.; Wang, L.; Liu, S.; Li, F.; Liu, J. Bioinspired artificial nanochannels: Construction and application. *Mater. Chem. Front.* **2021**, *5*, 1610–1631. [CrossRef]
4. Pugh, G.C.; Burns, J.R.; Howorka, S. Comparing proteins and nucleic acids for next-generation biomolecular engineering. *Nat. Rev. Chem.* **2018**, *2*, 113–130. [CrossRef]
5. Luo, Y.; Zhu, C.; Zhang, T.; Yan, T.; Liu, J. Self-assembled Supramolecular Artificial Transmembrane Ion Channels: Recent Progress and Application. *Chem. Res. Chin. Univ.* **2023**, *39*, 3–12. [CrossRef]
6. Langecker, M.; Arnaut, V.; List, J.; Simmel, F.C. DNA nanostructures interacting with lipid bilayer membranes. *Acc. Chem. Res.* **2014**, *47*, 1807–1815. [CrossRef] [PubMed]
7. Seeman, N.C.; Sleiman, H.F. DNA nanotechnology. *Nat. Rev. Mater.* **2017**, *3*, 17068. [CrossRef]
8. Ramezani, H.; Dietz, H. Building machines with DNA molecules. *Nat. Rev. Genet.* **2020**, *21*, 5–26. [CrossRef]
9. Suzuki, Y.; Endo, M.; Sugiyama, H. Mimicking Membrane-Related Biological Events by DNA Origami Nanotechnology. *ACS Nano* **2015**, *9*, 3418–3420. [CrossRef]
10. Wang, D.; Zhang, Y.; Liu, D. DNA nanochannels. *F1000Res* **2017**, *6*, 503. [CrossRef]
11. Wu, N.; Chen, F.; Zhao, Y.; Yu, X.; Wei, J.; Zhao, Y. Functional and Biomimetic DNA Nanostructures on Lipid Membranes. *Langmuir* **2018**, *34*, 14721–14730. [CrossRef] [PubMed]
12. Yin, P.; Hariadi, R.F.; Sahu, S.; Choi, H.M.T.; Park, S.H.; LaBean, T.H.; Reif, J.H. Programming DNA Tube Circumferences. *Science* **2008**, *321*, 824–826. [CrossRef] [PubMed]
13. Frykholm, K.; Muller, V.; KK, S.; Dorfman, K.D.; Westerlund, F. DNA in nanochannels: Theory and applications. *Q. Rev. Biophys.* **2022**, *55*, e12. [CrossRef] [PubMed]
14. Lanphere, C.; Offenbartl-Stiegert, D.; Dorey, A.; Pugh, G.; Georgiou, E.; Xing, Y.; Burns, J.R.; Howorka, S. Design, assembly, and characterization of membrane-spanning DNA nanopores. *Nat. Protoc.* **2021**, *16*, 86–130. [CrossRef]
15. Liu, X.; Zhao, Y.; Liu, P.; Wang, L.; Lin, J.; Fan, C. Biomimetic DNA Nanotubes: Nanoscale Channel Design and Applications. *Angew. Chem. Int. Ed. Engl.* **2019**, *58*, 8996–9011. [CrossRef] [PubMed]
16. Ma, Q.; Chen, L.; Gao, P.; Xia, F. Solid-state nanopore/channels meet DNA nanotechnology. *Matter* **2023**, *6*, 373–396. [CrossRef]
17. Gopfrich, K.; Ohmann, A.; Keyser, U.F. Design and Assembly of Membrane-Spanning DNA Nanopores. *Methods Mol. Biol.* **2021**, *2186*, 33–48.
18. Zhao, N.; Chen, Y.; Chen, G.; Xiao, Z. Artificial Cells Based on DNA Nanotechnology. *ACS Appl. Bio Mater.* **2020**, *3*, 3928–3934. [CrossRef]

19. Niranjan, D.N.; Thiyagarajan, D.; Bhatia, D. DNA Origami in the Quest for Membrane Piercing. *Chem. Asian. J.* **2022**, *17*, e202200591. [CrossRef]
20. Chen, J.; Seeman, N.C. Synthesis from DNA of a molecule with the connectivity of a cube. *Nature* **1991**, *350*, 631–633. [CrossRef]
21. Ohmann, A.; Li, C.Y.; Maffeo, C.; Nahas, K.A.; Baumann, K.N.; Gopfrich, K.; Yoo, J.; Keyser, U.F.; Aksimentiev, A. A synthetic enzyme built from DNA flips 10^7 lipids per second in biological membranes. *Nat. Commun.* **2018**, *9*, 2426. [CrossRef] [PubMed]
22. Jones, S.F.; Joshi, H.; Terry, S.J.; Burns, J.R.; Aksimentiev, A.; Eggert, U.S.; Howorka, S. Hydrophobic Interactions between DNA Duplexes and Synthetic and Biological Membranes. *J. Am. Chem. Soc.* **2021**, *143*, 8305–8313. [CrossRef] [PubMed]
23. Chidchob, P.; Offenbartl-Stiegert, D.; McCarthy, D.; Luo, X.; Li, J.; Howorka, S.; Sleiman, H.F. Spatial Presentation of Cholesterol Units on a DNA Cube as a Determinant of Membrane Protein-Mimicking Functions. *J. Am. Chem. Soc.* **2018**, *141*, 1100–1108. [CrossRef] [PubMed]
24. Burns, J.R.; Howorka, S. Defined Bilayer Interactions of DNA Nanopores Revealed with a Nuclease-Based Nanoprobe Strategy. *ACS Nano* **2018**, *12*, 3263–3271. [CrossRef] [PubMed]
25. Langecker, M.; Arnaut, V.; Martin, T.G.; List, J.; Renner, S.; Mayer, M.; Dietz, H.; Simmel, F.C. Synthetic lipid membrane channels formed by designed DNA nanostructures. *Science* **2012**, *338*, 932–936. [CrossRef]
26. Gopfrich, K.; Li, C.Y.; Ricci, M.; Bhamidimarri, S.P.; Yoo, J.; Gyenes, B.; Ohmann, A.; Winterhalter, M.; Aksimentiev, A.; Keyser, U.F. Large-Conductance Transmembrane Porin Made from DNA Origami. *ACS Nano* **2016**, *10*, 8207–8214. [CrossRef]
27. Diederichs, T.; Pugh, G.; Dorey, A.; Xing, Y.; Burns, J.R.; Nguyen, Q.H.; Tornow, M.; Tampe, R.; Howorka, S. Synthetic protein-conductive membrane nanopores built with DNA. *Nat. Commun.* **2019**, *10*, 5018. [CrossRef]
28. Thomsen, R.P.; Malle, M.G.; Okholm, A.H.; Krishnan, S.; Bohr, S.S.; Sorensen, R.S.; Ries, O.; Vogel, S.; Simmel, F.C.; Hatzakis, N.S.; et al. A large size-selective DNA nanopore with sensing applications. *Nat. Commun.* **2019**, *10*, 5655. [CrossRef]
29. Fragasso, A.; De Franceschi, N.; Stommer, P.; van der Sluis, E.O.; Dietz, H.; Dekker, C. Reconstitution of Ultrawide DNA Origami Pores in Liposomes for Transmembrane Transport of Macromolecules. *ACS Nano* **2021**, *15*, 12768–12779. [CrossRef]
30. Xing, Y.; Dorey, A.; Jayasinghe, L.; Howorka, S. Highly shape- and size-tunable membrane nanopores made with DNA. *Nat. Nanotechnol.* **2022**, *17*, 708–713. [CrossRef] [PubMed]
31. Burns, J.R.; Stulz, E.; Howorka, S. Self-assembled DNA nanopores that span lipid bilayers. *Nano Lett.* **2013**, *13*, 2351–2356. [CrossRef]
32. Burns, J.R.; Gopfrich, K.; Wood, J.W.; Thacker, V.V.; Stulz, E.; Keyser, U.F.; Howorka, S. Lipid-bilayer-spanning DNA nanopores with a bifunctional porphyrin anchor. *Angew. Chem. Int. Ed. Engl.* **2013**, *52*, 12069–12072. [CrossRef]
33. Debnath, M.; Chakraborty, S.; Kumar, Y.P.; Chaudhuri, R.; Jana, B.; Dash, J. Ionophore constructed from non-covalent assembly of a G-quadruplex and liponucleoside transports K+-ion across biological membranes. *Nat. Commun.* **2020**, *11*, 469. [CrossRef]
34. Krishnan, S.; Ziegler, D.; Dietz, H.; Simmel, F.C. Molecular transport through large-diameter DNA nanopores. *Nat. Commun.* **2016**, *7*, 12787. [CrossRef]
35. Li, C.; Chen, H.; Zhou, L.; Shi, H.; He, X.; Yang, X.; Wang, K.; Liu, J. Single-stranded DNA designed lipophilic G-quadruplexes as transmembrane channels for switchable potassium transport. *Chem. Commun.* **2019**, *55*, 12004–12007. [CrossRef]
36. Aldaye, F.A.; Lo, P.K.; Karam, P.; McLaughlin, C.K.; Cosa, G.; Sleiman, H.F. Modular construction of DNA nanotubes of tunable geometry and single- or double-stranded character. *Nat. Nanotechnol.* **2009**, *4*, 349–352. [CrossRef]
37. Lo, P.K.; Karam, P.; Aldaye, F.A.; McLaughlin, C.K.; Hamblin, G.D.; Cosa, G.; Sleiman, H.F. Loading and selective release of cargo in DNA nanotubes with longitudinal variation. *Nat. Chem.* **2010**, *2*, 319–328. [CrossRef]
38. Hamblin, G.D.; Carneiro, K.M.M.; Fakhoury, J.F.; Bujold, K.E.; Sleiman, H.F. Rolling Circle Amplification-Templated DNA Nanotubes Show Increased Stability and Cell Penetration Ability. *J. Am. Chem. Soc.* **2012**, *134*, 2888–2891. [CrossRef]
39. Hariri, A.A.; Hamblin, G.D.; Gidi, Y.; Sleiman, H.F.; Cosa, G. Stepwise growth of surface-grafted DNA nanotubes visualized at the single-molecule level. *Nat. Chem.* **2015**, *7*, 295–300. [CrossRef]
40. Rahbani, J.F.; Vengut-Climent, E.; Chidchob, P.; Gidi, Y.; Trinh, T.; Cosa, G.; Sleiman, H.F. DNA Nanotubes with Hydrophobic Environments: Toward New Platforms for Guest Encapsulation and Cellular Delivery. *Adv. Health Mater.* **2018**, *7*, 1701049. [CrossRef]
41. Saliba, D.; Luo, X.; Rizzuto, F.J.; Sleiman, H.F. Programming rigidity into size-defined wireframe DNA nanotubes. *Nanoscale* **2023**, *15*, 5403–5413. [CrossRef]
42. Yang, Q.; Guo, Z.; Liu, H.; Peng, R.; Xu, L.; Bi, C.; He, Y.; Liu, Q.; Tan, W. A Cascade Signaling Network between Artificial Cells Switching Activity of Synthetic Transmembrane Channels. *J. Am. Chem. Soc.* **2020**, *143*, 232–240. [CrossRef]
43. Yoo, J.; Aksimentiev, A. Molecular Dynamics of Membrane-Spanning DNA Channels: Conductance Mechanism, Electro-Osmotic Transport, and Mechanical Gating. *J. Phys. Chem. Lett.* **2015**, *6*, 4680–4687. [CrossRef]
44. Maingi, V.; Lelimousin, M.; Howorka, S.; Sansom, M.S.P. Gating-like Motions and Wall Porosity in a DNA Nanopore Scaffold Revealed by Molecular Simulations. *ACS Nano* **2015**, *9*, 11209–11217. [CrossRef] [PubMed]
45. Maingi, V.; Burns, J.R.; Uusitalo, J.J.; Howorka, S.; Marrink, S.J.; Sansom, M.S. Stability and dynamics of membrane-spanning DNA nanopores. *Nat. Commun.* **2017**, *8*, 14784. [CrossRef] [PubMed]
46. Birkholz, O.; Burns, J.R.; Richter, C.P.; Psathaki, O.E.; Howorka, S.; Piehler, J. Multi-functional DNA nanostructures that puncture and remodel lipid membranes into hybrid materials. *Nat. Commun.* **2018**, *9*, 1521. [CrossRef] [PubMed]
47. Luo, L.; Manda, S.; Park, Y.; Demir, B.; Sanchez, J.; Anantram, M.P.; Oren, E.E.; Gopinath, A.; Rolandi, M. DNA nanopores as artificial membrane channels for bioprotonics. *Nat. Commun.* **2023**, *14*, 5364. [CrossRef] [PubMed]

48. Diederichs, T.; Ahmad, K.; Burns, J.R.; Nguyen, Q.H.; Siwy, Z.S.; Tornow, M.; Coveney, P.V.; Tampe, R.; Howorka, S. Principles of Small-Molecule Transport through Synthetic Nanopores. *ACS Nano* **2021**, *15*, 16194–16206. [CrossRef]
49. Gopfrich, K.; Zettl, T.; Meijering, A.E.; Hernandez-Ainsa, S.; Kocabey, S.; Liedl, T.; Keyser, U.F. DNA-Tile Structures Induce Ionic Currents through Lipid Membranes. *Nano Lett.* **2015**, *15*, 3134–3138. [CrossRef]
50. Mathieu, F.; Liao, S.; Kopatsch, J.; Wang, T.; Mao, C.; Seeman, N.C. Six-Helix Bundles Designed from DNA. *Nano Lett.* **2005**, *5*, 661–665. [CrossRef]
51. Endo, M.; Seeman, N.C.; Majima, T. DNA Tube Structures Controlled by a Four-Way-Branched DNA Connector. *Angew. Chem. Int. Ed. Engl.* **2005**, *44*, 6074–6077. [CrossRef]
52. Mohammed, A.M.; Šulc, P.; Zenk, J.; Schulman, R. Self-assembling DNA nanotubes to connect molecular landmarks. *Nat. Nanotechnol.* **2016**, *12*, 312–316. [CrossRef] [PubMed]
53. Green, L.N.; Subramanian, H.K.K.; Mardanlou, V.; Kim, J.; Hariadi, R.F.; Franco, E. Autonomous dynamic control of DNA nanostructure self-assembly. *Nat. Chem.* **2019**, *11*, 510–520. [CrossRef] [PubMed]
54. Agarwal, S.; Franco, E. Enzyme-Driven Assembly and Disassembly of Hybrid DNA-RNA Nanotubes. *J. Am. Chem. Soc.* **2019**, *141*, 7831–7841. [CrossRef] [PubMed]
55. Jia, S.; Phua, S.C.; Nihongaki, Y.; Li, Y.; Pacella, M.; Li, Y.; Mohammed, A.M.; Sun, S.; Inoue, T.; Schulman, R. Growth and site-specific organization of micron-scale biomolecular devices on living mammalian cells. *Nat. Commun.* **2021**, *12*, 5729. [CrossRef]
56. Dhanasekar, N.N.; Li, Y.; Schulman, R. The ion permeability of DNA nanotube channels. *BioRxiv* **2022**, 1–33. [CrossRef]
57. Li, Y.; Maffeo, C.; Schulman, R. Leakless end-to-end transport of small molecules through micron-length DNA nanochannels. *Sci. Adv.* **2022**, *8*, eabq4834. [CrossRef]
58. Rothemund, P.W. Folding DNA to create nanoscale shapes and patterns. *Nature* **2006**, *440*, 297–302. [CrossRef]
59. Forman, S.L.; Fettinger, J.C.; Pieraccini, S.; Gottarelli, G.; Davis, J.T. Toward Artificial Ion Channels: A Lipophilic G-Quadruplex. *J. Am. Chem. Soc.* **2000**, *122*, 4060–4067. [CrossRef]
60. Lee, M.P.H.; Parkinson, G.N.; Hazel, P.; Neidle, S. Observation of the Coexistence of Sodium and Calcium Ions in a DNA G-Quadruplex Ion Channel. *J. Am. Chem. Soc.* **2007**, *129*, 10106–10107. [CrossRef]
61. Akhshi, P.; Mosey, N.J.; Wu, G. Free-Energy Landscapes of Ion Movement through a G-Quadruplex DNA Channel. *Angew. Chem.* **2012**, *124*, 2904–2908. [CrossRef]
62. Akhshi, P.; Wu, G. Umbrella sampling molecular dynamics simulations reveal concerted ion movement through G-quadruplex DNA channels. *Phys. Chem. Chem. Phys.* **2017**, *19*, 11017–11025. [CrossRef] [PubMed]
63. Balasubramanian, S.; Senapati, S. Dynamics and Barrier of Movements of Sodium and Potassium Ions Across the Oxytricha nova G-Quadruplex Core. *J. Phys. Chem. B* **2020**, *124*, 11055–11066. [CrossRef]
64. Göpfrich, K.; Li, C.-Y.; Mames, I.; Bhamidimarri, S.P.; Ricci, M.; Yoo, J.; Mames, A.; Ohmann, A.; Winterhalter, M.; Stulz, E.; et al. Ion Channels Made from a Single Membrane-Spanning DNA Duplex. *Nano Lett.* **2016**, *16*, 4665–4669. [CrossRef]
65. Morzy, D.; Joshi, H.; Sandler, S.E.; Aksimentiev, A.; Keyser, U.F. Membrane Activity of a DNA-Based Ion Channel Depends on the Stability of Its Double-Stranded Structure. *Nano Lett.* **2021**, *21*, 9789–9796. [CrossRef] [PubMed]
66. Burns, J.R.; Seifert, A.; Fertig, N.; Howorka, S. A biomimetic DNA-based channel for the ligand-controlled transport of charged molecular cargo across a biological membrane. *Nat. Nanotechnol.* **2016**, *11*, 152–156. [CrossRef]
67. Dey, S.; Dorey, A.; Abraham, L.; Xing, Y.; Zhang, I.; Zhang, F.; Howorka, S.; Yan, H. A reversibly gated protein-transporting membrane channel made of DNA. *Nat. Commun.* **2022**, *13*, 2271. [CrossRef]
68. Seifert, A.; Gopfrich, K.; Keyser, U.F.; Howorka, S. Bilayer-Spanning DNA Nanopores with Voltage-Switching between Open and Closed State. *ACS Nano* **2015**, *9*, 1117–1126. [CrossRef]
69. Iwabuchi, S.; Kawamata, I.; Murata, S.; Nomura, S.M. A large, square-shaped, DNA origami nanopore with sealing function on a giant vesicle membrane. *Chem. Commun.* **2021**, *57*, 2990–2993. [CrossRef]
70. Lanphere, C.; Ciccone, J.; Dorey, A.; Hagleitner-Ertugrul, N.; Knyazev, D.; Haider, S.; Howorka, S. Triggered Assembly of a DNA-Based Membrane Channel. *J. Am. Chem. Soc.* **2022**, *144*, 4333–4344. [CrossRef]
71. Yang, L.; Cullin, C.; Elezgaray, J. Detection of Short DNA Sequences with DNA Nanopores. *Chemphyschem* **2022**, *23*, e202200021. [CrossRef] [PubMed]
72. Mendoza, O.; Calmet, P.; Alves, I.; Lecomte, S.; Raoux, M.; Cullin, C.; Elezgaray, J. A tensegrity driven DNA nanopore. *Nanoscale* **2017**, *9*, 9762–9769. [CrossRef] [PubMed]
73. Arnott, P.M.; Howorka, S. A Temperature-Gated Nanovalve Self-Assembled from DNA to Control Molecular Transport across Membranes. *ACS Nano* **2019**, *13*, 3334–3340. [CrossRef] [PubMed]
74. Li, P.; Xie, G.; Kong, X.Y.; Zhang, Z.; Xiao, K.; Wen, L.; Jiang, L. Light-Controlled Ion Transport through Biomimetic DNA-Based Channels. *Angew. Chem. Int. Ed. Engl.* **2016**, *55*, 15637–15641. [CrossRef]
75. Offenbartl-Stiegert, D.; Rottensteiner, A.; Dorey, A.; Howorka, S. A Light-Triggered Synthetic Nanopore for Controlling Molecular Transport Across Biological Membranes. *Angew. Chem. Int. Ed. Engl.* **2022**, *61*, e202210886. [CrossRef]
76. Li, C.; Chen, H.; Yang, X.; Wang, K.; Liu, J. An ion transport switch based on light-responsive conformation-dependent G-quadruplex transmembrane channels. *Chem. Commun.* **2021**, *57*, 8214–8217. [CrossRef]
77. Li, C.; Chen, H.; Chen, Q.; Shi, H.; Yang, X.; Wang, K.; Liu, J. Lipophilic G-Quadruplex Isomers as Biomimetic Ion Channels for Conformation-Dependent Selective Transmembrane Transport. *Anal. Chem.* **2020**, *92*, 10169–10176. [CrossRef] [PubMed]

78. Lin, C.; Wu, G.; Wang, K.; Onel, B.; Sakai, S.; Shao, Y.; Yang, D. Molecular Recognition of the Hybrid-2 Human Telomeric G-Quadruplex by Epiberberine: Insights into Conversion of Telomeric G-Quadruplex Structures. *Angew. Chem. Int. Ed. Engl.* **2018**, *57*, 10888–10893. [CrossRef]
79. Lu, H.; Li, S.; Chen, J.; Xia, J.; Zhang, J.; Huang, Y.; Liu, X.; Wu, H.; Zhao, Y.; Chaia, Z.; et al. Metal ions modulate the conformation and stability of a G-quadruplex with or without a small-molecule ligand. *Metallomics* **2015**, *7*, 1508–1514. [CrossRef]
80. Messager, L.; Burns, J.R.; Kim, J.; Cecchin, D.; Hindley, J.; Pyne, A.L.; Gaitzsch, J.; Battaglia, G.; Howorka, S. Biomimetic Hybrid Nanocontainers with Selective Permeability. *Angew. Chem. Int. Ed. Engl.* **2016**, *55*, 11106–11109. [CrossRef]
81. Burns, J.R.; Al-Juffali, N.; Janes, S.M.; Howorka, S. Membrane-spanning DNA nanopores with cytotoxic effect. *Angew. Chem. Int. Ed. Engl.* **2014**, *53*, 12466–12470. [CrossRef] [PubMed]
82. Lv, C.; Gu, X.; Li, H.; Zhao, Y.; Yang, D.; Yu, W.; Han, D.; Li, J.; Tan, W. Molecular Transport through a Biomimetic DNA Channel on Live Cell Membranes. *ACS Nano* **2020**, *14*, 14616–14626. [CrossRef] [PubMed]
83. Chen, L.; Liang, S.; Chen, Y.; Wu, M.; Zhang, Y. Destructing the Plasma Membrane with Activatable Vesicular DNA Nanopores. *ACS Appl. Mater. Interfaces* **2020**, *12*, 96–105. [CrossRef] [PubMed]
84. Lanphere, C.; Arnott, P.M.; Jones, S.F.; Korlova, K.; Howorka, S. A Biomimetic DNA-Based Membrane Gate for Protein-Controlled Transport of Cytotoxic Drugs. *Angew. Chem. Int. Ed. Engl.* **2021**, *60*, 1903–1908. [CrossRef] [PubMed]
85. Chen, Q.; Jian, M.; Chen, H.; Zhou, B.; Shi, H.; Yang, X.; Wang, K.; Liu, J. Design of Lipophilic Split Aptamers as Artificial Carriers for Transmembrane Transport of Adenosine Triphosphate. *CCS Chem.* **2021**, *3*, 144–153. [CrossRef]
86. Veetil, A.T.; Chakraborty, K.; Xiao, K.; Minter, M.R.; Sisodia, S.S.; Krishnan, Y. Cell-targetable DNA nanocapsules for spatiotemporal release of caged bioactive small molecules. *Nat. Nanotechnol.* **2017**, *12*, 1183–1189. [CrossRef] [PubMed]
87. Xie, G.; Li, P.; Zhao, Z.; Kong, X.Y.; Zhang, Z.; Xiao, K.; Wang, H.; Wen, L.; Jiang, L. Bacteriorhodopsin-Inspired Light-Driven Artificial Molecule Motors for Transmembrane Mass Transportation. *Angew. Chem. Int. Ed. Engl.* **2018**, *57*, 16708–16712. [CrossRef]
88. List, J.; Weber, M.; Simmel, F.C. Hydrophobic actuation of a DNA origami bilayer structure. *Angew. Chem. Int. Ed. Engl.* **2014**, *53*, 4236–4239. [CrossRef]
89. Bennett, I.M.; Farfano, H.M.V.; Bogani, F.; Primak, A.; Liddell, P.A.; Otero, L.; Sereno, L.; Silber, J.J.; Moore, A.L.; Moore, T.A.; et al. Active transport of Ca^{2+} by an artificial photosynthetic membrane. *Nature* **2002**, *420*, 398–401. [CrossRef]
90. Cui, C.; Chakraborty, K.; Tang, X.A.; Schoenfelt, K.Q.; Hoffman, A.; Blank, A.; McBeth, B.; Pulliam, N.; Reardon, C.A.; Kulkarni, S.A.; et al. A lysosome-targeted DNA nanodevice selectively targets macrophages to attenuate tumours. *Nat. Nanotechnol.* **2021**, *16*, 1394–1402. [CrossRef]
91. Maffeo, C.; Quednau, L.; Wilson, J.; Aksimentiev, A. DNA double helix, a tiny electromotor. *Nat. Nanotechnol.* **2022**, *18*, 238–242. [CrossRef] [PubMed]
92. Wang, L.; Guo, P.; Jin, D.; Peng, Y.; Sun, X.; Chen, Y.; Liu, X.; Chen, W.; Wang, W.; Yan, X.; et al. Enzyme-Powered Tubular Microrobotic Jets as Bioinspired Micropumps for Active Transmembrane Drug Transport. *ACS Nano* **2023**, *17*, 5095–5107. [CrossRef] [PubMed]
93. Yang, J.; Yu, G.; Sessler, J.L.; Shin, I.; Gale, P.A.; Huang, F. Artificial transmembrane ion transporters as potential therapeutics. *Chem* **2021**, *7*, 3256–3291. [CrossRef]
94. Kanner, S.A.; Shuja, Z.; Choudhury, P.; Jain, A.; Colecraft, H.M. Targeted deubiquitination rescues distinct trafficking-deficient ion channelopathies. *Nat. Methods* **2020**, *17*, 1245–1253. [CrossRef] [PubMed]
95. Rastergar, A.; Soleimani, M. Hypokalaemia and hyperkalaemia. *Postgrad. Med. J.* **2001**, *77*, 759–764. [CrossRef]
96. Curran, J.; Mohler, P.J. Alternative paradigms for ion channelopathies: Disorders of ion channel membrane trafficking and posttranslational modification. *Annu. Rev. Physiol.* **2015**, *77*, 505–524. [CrossRef]

chemosensors

MDPI

Review

Research Progress on Bionic Recognition and Biosensors for the Detection of Biomarkers of Diabetic Nephropathy

Ye Tian [1,*,†], Lili Gao [2,†], Abubakar Abdussalam [3,4] and Guobao Xu [3,4,*]

1 The College of Civil Engineering, Shenyang Urban Construction University, Shenyang 110167, China
2 School of Materials Science and Engineering, Shenyang Jianzhu University, Shenyang 110168, China; liligao@sjzu.edu.cn
3 State Key Laboratory of Electroanalytical Chemistry, Changchun Institute of Applied Chemistry, Chinese Academy of Sciences, Changchun 130022, China; aasalam@ciac.ac.cn
4 School of Applied Chemistry and Engineering, University of Science and Technology of China, Hefei 230026, China
* Correspondence: tumugongchengxueyuantian@syucu.edu.cn (Y.T.); guobaoxu@ciac.ac.cn (G.X.)
† These authors contributed equally to this work.

Abstract: Diabetic nephropathy (DN) refers to kidney damage caused by diabetes and is one of the major microvascular complications of diabetes. This disease has a certain degree of concealment in the early stage, with clinical symptoms appearing later and a higher mortality rate. Therefore, the detection of early biomarkers for DN is of great importance in reducing kidney function damage. The common biomarkers for DN mainly include glomerular and tubular lesion markers. At present, clinical diagnosis often uses a combination of multiple indicators and symptoms, and the development of a simple, efficient, and sensitive multi-marker detection platform is particularly important for the early diagnosis of DN. In recent years, with the vigorous development of various biomimetic molecular recognition technologies, biomimetic recognition biosensors (BRBS) have many advantages, such as easy preparation, low cost, high stability, and repeatability under harsh environmental conditions, and have great application potential in the analysis of DN biomarkers. This article reviews the research progress of molecularly imprinted polymers (MIPs) construction technology and aptamers assembly technology developed in the field of biomimetic sensor research in recent years, as well as the detection of DN biomarkers based on BRBS, and prospects for their development.

Keywords: bionic recognition; MIPs; aptamers; diabetic nephropathy; biomarkers

Citation: Tian, Y.; Gao, L.; Abdussalam, A.; Xu, G. Research Progress on Bionic Recognition and Biosensors for the Detection of Biomarkers of Diabetic Nephropathy. *Chemosensors* **2023**, *11*, 510. https://doi.org/10.3390/chemosensors11100510

Academic Editor: Ambra Giannetti

Received: 12 July 2023
Revised: 6 September 2023
Accepted: 17 September 2023
Published: 22 September 2023

1. Introduction

Diabetic nephropathy (DN) refers to kidney damage caused by diabetes, which is one of the main microvascular complications of diabetes. High blood sugar-induced kidney damage leads to a decrease in the glomerular filtration rate (GFR) and the appearance of symptoms such as proteinuria. The disease has a high mortality rate [1]. Currently, DN has seriously threatened human health and the development of socio-economic progress [2]. According to the classification of DN by Mogensen from Denmark, stage I–II lacks obvious clinical manifestations, stage III begins to show trace albuminuria and GFR may be normal, and it is not until stage IV that there is a persistent and progressively worsening amount of albuminuria accompanied by a decline in GFR, which is difficult to reverse, eventually leading to renal failure. Therefore, early DN is reversible, and if diagnosed and treated with early intervention, the development of nephropathy can be halted or reversed. Hence, early screening and diagnosis are particularly important for DN. The screening of DN mainly follows the standards of the American Diabetes Association (ADA), including albuminuria and GFR. However, a large number of clinical observations have found that 20–50% of chronic kidney disease patients have no albuminuria, and the number of such patients is on the rise [3]. This means that these DN patients without albuminuria will be missed, and will not receive timely diagnosis and treatment. Therefore, it is of great practical significance to find highly sensitive and specific early biomarkers for DN and establish accurate and convenient detection methods.

So far, several analytical methods have been developed for the detection of biomarkers (urinary albumin, creatinine, cystatin C, homocysteine, vascular endothelial growth factor, epidermal growth factor receptor, mRNA-21, ceruloplasmin, platelet-derived growth factor, etc.) for DN [4–7], including ultra-high performance liquid chromatography (UHPLC), capillary electrophoresis (CE), liquid chromatography-tandem mass spectrometry (LC-MS), radioimmunoassay (RIA), enzyme-linked immunosorbent assay (ELISA), immunoradiometry (IRMA), and electrochemiluminescent immunoassay (ECLI). Although the results of these methods are reliable and accurate, they have many disadvantages, such as large sample sizes, complex instruments, expensive antibodies, long incubation time, and complicated protocols. These limitations restrict their application in real-time detection [8,9]. Therefore, the introduction of rapid, universal, powerful, and high-throughput sensing technologies for early detection of DN has significant social value.

In recent years, researchers have developed a large number of novel biosensors with high specificity and sensitivity [10,11]. Traditional biosensor recognition elements mainly consist of biological molecules with molecular recognition capabilities, such as antibodies, enzymes, nucleic acids, cells, and semi-synthetic receptors based on them. However, they have inherent drawbacks, such as difficulty in preparation, high cost, and poor stability, which greatly limit their application in the analysis of diabetes kidney disease biomarkers. Compared with traditional biosensors, the biomimetic recognition elements of biomimetic recognition biosensors (BRBS), including MIPs and APT, can be synthesized through in vitro preparation methods. This not only effectively avoids the tedious work of bio-receptor preparation using experimental animals but also reduces the batch-to-batch variability of recognition elements, significantly lowering manufacturing costs. The application of BRBS in the electrochemical sensing field has greatly promoted the development of analytical chemistry and electroanalytical chemistry. Among them, molecularly imprinted electrochemical sensors (MIECS) combine the predetermined recognition and specificity of molecular imprinting technique (MIT) with the high sensitivity, easy operation, low cost, and miniaturization advantages of electrochemical technology [12].

MIT is inspired by the "antigen-antibody" theory and prepares MIPs targeted to specific molecules. Currently, most MIPs are made of vinyl organic compounds or acrylic acid as functional monomers, forming through free radical polymerization under the influence of crosslinking agents, initiators, light or heat, etc. MIPs can selectively recognize template molecules (TM) or their analogs mainly because, after removing the TM in the polymer, highly crosslinked polymers retain "empty holes" with multiple active sites that match the size and conformation of the TM. MIPs demonstrate excellent resistance to harsh environmental interference, but there is a possibility of template leakage when analyzing complex samples, especially for trace analysis, the leakage of residual templates directly affects the accuracy of the analysis results. Moreover, the preparation of MIPs becomes difficult due to the poor solubility of certain specific TM. Using a dummy template molecule that has a similar structure or the same functional groups as the template molecule solves this issue, as the 3D cavity structure and interaction sites left by the dummy molecularly imprinted polymer still have some compatibility with the template molecule. Thus, dummy molecular imprinting technology can solve the problems of template leakage, high costs, and difficulties in preparing MIPs due to the poor solubility of certain specific TM [13]. However, the binding sites, shape, and size of the cavities prepared by the dummy template may not completely match the target molecule, which may reduce sensitivity and detection limits. Therefore, the selection of appropriate functional monomers, crosslinking agents, solvents, polymerization conditions, washing conditions, and preparation methods is crucial for obtaining good specific molecular recognition capabilities during the preparation of molecular imprinting membranes. To achieve breakthrough progress in the field of MIT electrochemical sensing, it is necessary to design and develop better MIP preparation technology strategies.

Research shows that nucleic acid APT are single-stranded RNA or DNA molecules extracted from a library of oligonucleotide ligands through systematic evolution of ligands by exponential enrichment (SELEX) technology. They have advantages such as non-toxicity, ease of preparation, low cost, chemical stability, and high binding efficiency to target molecules [14]. Compared to antibodies, APTs have much lower molecular weight, can recognize subtle structural differences in biomarkers, and are easily dissociable [15]. Furthermore, the shape and length of APT can be controlled and flexibly modified, forming structures with specific physicochemical properties and functions. Research of multifarious adenine-mediated technology (MAMT) suggests that the adsorption capabilities of the four bases of nucleic acid APT to gold follow the order A > C $\geq$ G > T [16]. Among them, adenine has strong adsorption to gold, comparable to Au-S bonds, and Au can even denature the double strands of adenine and thymine. The strong adsorption of adenine to gold electrodes has promising applications in the design of thiol-free APT [17]. Adenine can be used as an anchoring group and density control group to regulate the conformation and grafting density of APT at the sensor interface, reducing the surface density of APT on the electrode with the increase of adenine length. Moreover, adenine preferentially binds to gold atoms, reducing the accessibility of other materials to gold, and significantly hindering the non-specific adsorption of other materials. These aptamer (APT) assembly techniques can be applied to the construction of biosensors.

In recent years, with the development of MIT and APT assembly technology, the application of bionic recognition biosensors in the field of DN biomarker analysis has been effectively expanded. With the discovery of new biomarkers for DN, the development of more efficient, stable, and cost-effective novel biomimetic sensing platforms has significant implications. Not only does it improve our understanding of the pathogenesis of DN and further promote early diagnosis in diabetic nephropathy patients, but it also optimizes the treatment of these patients. Therefore, this article reviews the research progress in the construction of MIPs, APT assembly technology, and the detection of biomarkers for DN based on BRBS in the field of biomimetic sensors in recent years, and forecasts their development.

2. Application of BRBS in the Detection of Diabetic Nephropathy Biomarkers

Early biomarkers of diabetic nephropathy can be divided into glomerular dysfunction markers and tubular dysfunction markers according to the site of lesions. So far, a series of biomarkers have been introduced into the diagnosis of diabetic nephropathy,

2.1. Urinary Albumin

Urinary albumin detection reflects abnormal protein leakage in the kidney and can understand early kidney disease and kidney injury. Pathological increases are seen in diabetic nephropathy, hypertension, and preeclampsia during pregnancy [18]. If the albumin in the urine is in the range of 30–200 mg/L, it belongs to microalbuminuria. Early diabetic nephropathy is a mild increase in urinary albumin excretion (microalbuminuria), which gradually progresses to massive albuminuria and eventually kidney failure, requiring dialysis or kidney transplantation. Up to now, microalbuminuria is still the best non-invasive biomarker for detecting diabetic nephropathy [19].

Zhang [20] successfully constructed a MIP-ECS based on a dual-signal strategy for human serum albumin (HSA) detection in urine (Figure 1). Firstly, the glassy carbon electrode (GE) substrate was modified with AuNPs and polythionine-methylene blue (PTH-MB) sequentially. Then, a dual-functional monomer molecularly imprinted polymer (MIP) was prepared by the electropolymerization method using human serum albumin (HAS) as the TM, HQ, and o-phenylenediamine o-PD as functional monomers. This biosensor showed a dual-signal oxidation of $Fe(CN)_6{}^{4-}$ and PTH-MB at 0.20 V and −0.22 V, respectively. When HSA was captured on MIP, the adsorbed HSA on the electrode surface would hinder electron transfer, reducing the total intensity of dual-signal currents with the increase of HSA concentration. The limit of detection (LOD) of this sensor can reach 3.0×10^{-11} g L^{-1}. Moreover, the selected AuNPs and PTH-MB both have excellent electrocatalytic performance, in which the methylene blue (MB) in PTH-MB molecules acts as an electrocatalyst, which can enhance the electron transfer efficiency of electrochemical reactions on the electrode surface. Furthermore, the synergistic effect between AuNPs and PTH-MB effectively enhances the current signal, thus improving the sensitivity of the biosensor [21]. In practical detection, this biosensor has good selectivity and repeatability, achieving the detection of HSA in urine, and providing a potential universal platform for clinical protein detection.

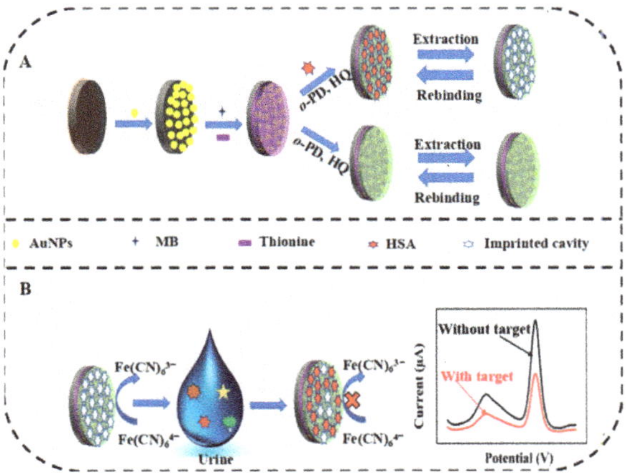

Figure 1. (**A**) Schematic of the proposed procedure for constructing the molecularly imprinted polymer electrochemical biosensor. (**B**) Principle of MIP-ECS detection of HSA [20].

2.2. Creatinine (Crn)

Crn is formed in everyday muscle wear and tear and is an important biomarker of renal qualitative and quantitative function. Crn levels play an important role in calculating muscle and renal function. The normal range of urinary Crn is about 282–2600 ppm. The normal range of serum Crn for women is 0.5–1.1 mg/dL (5–11 ppm) and for men is 0.6–1.2 mg/dL (6–12 ppm). Concentrations above 15 ppm require medical attention, and above 59 ppm indicate serious kidney damage [22]. The kidneys filter plasma through the processes of glomerular filtration and tubular excretion, thereby removing Crn to maintain homeostasis in the body. Crn content is helpful in quantifying the GFR, which is a very important indicator of kidney function and has been proven to be a potential biomarker for DN [23]. Clinically, early detection of renal impairment and the implementation of appropriate treatment measures can slow the progression of diabetic nephropathy by monitoring the levels of the albumin-to-Crn ratio [24].

Prabhu et al. [25] developed an instant medical (Point-of-Care, PoC) sensor device based on LoRaWAN (Long Range Wide Area Network, remote wide area network). In this strategy, Crn was used as the template, methacrylic acid (MAA) as the functional monomer, 2,2-azoisobutronitrile (AIBN) as the reaction initiator, and divinylbenzene (DVB) as the cross-linking agent. MIP prepared by standard chemical methods using precipitation polymerization effectively coated the sensor surface with MIP by acrylic resin, making the sensing interface highly functionalized. The constructed sensor has a high specificity for selective adsorption of Crn molecules, and the electrochemical impedance spectrum (EIS) can accurately detect different levels of Crn content. The detection limit is as low as 0.1 ppm (Table 1). This low-cost, portable device allows patients to self-test at home, and the results obtained from this system can help doctors take necessary measures, which is conducive to early detection and treatment of kidney injury to avoid further deterioration of diabetic nephropathy.

In recent years, the advantages of N-MIT compared with traditional MIP preparation methods have attracted widespread attention. Currently, the common technique for preparing N-MIPs is solid-phase synthesis. In traditional methods for preparing MIPs, the TM are dispersed in the liquid medium, while the solid-phase synthesis method anchors the TM on a specific solid support, and subsequently carries out affinity purification of N-MIPs [26]. Raquel et al. synthesized N-MIPs with advantages such as high affinity, high specificity, excellent stability, and immobilized templates by uniformly distributing imprint nanoparticles in a polyvinyl chloride membrane through precipitation polymerization [27]. In particular, N-MIPs have a large specific surface area, which can expose more binding sites for recognizing small molecules or large biological molecules such as proteins, promoting the development of the imprinting material towards broader application prospects. On the other hand, electrochemiluminescence (ECL) has unique characteristics that differ from other electrochemical methods, such as high signal-to-noise ratio, high sensitivity, wide linear range, high temporal and spatial resolution, and near-zero background light [28]. Developing new chemiluminescence systems is one of the important ways to expand the detection range of ECL [29,30]. Babamiri et al. [31] synthesized nickel nanoclusters (NiNCs) as a new luminescent body and constructed an MIECS basted on ECL for Crn detection. A molecularly imprinted polymer (MIP) @ GO-Fe_3O_4 film was constructed on the surface of an indium tin oxide (ITO) electrode using Crn as the template, tetraethyl orthosilicate (TEOS) as the cross-linking agent, and magnetic graphene oxide (GO-Fe_3O_4) as the modifier and embedded nickel nanoclusters (NiNCs). Due to the excellent conductivity of GO-Fe_3O_4, the MIECS retained the high specificity of the MIP @ GO-Fe_3O_4 template for Crn, while significantly improving its sensitivity. In addition, NiNCs have strong and stable ECL with co-reactant TPA, whereas Crn has a quenching effect on NiNCs' ECL, causing the ECL signal intensity to decrease as Crn concentration increases. The detection limit for Crn in serum and urine samples is 0.5 nmol/L (Table 1).

Traditional imprinting methods prepare MIPs that embed most of the binding sites and nanoscale sensitizing materials, making it difficult for TM to wash off from the highly cross-linked heterogeneous rigid polymer structure, and at the same time, obstructing the electronic transport of nanoscale sensitizing materials, which leads to poor sensitivity of ECS. Qian et al. immobilized bovine serum albumin (BSA) templates on the surface of SiO_2 nanoparticles and synthesized surface-modified MIPs [32]. The results showed that surface MIT can reduce the "embedding" phenomenon, and the MIPs prepared by controlling the template location on the material surface or near the surface have high specificity, high affinity, good reproducibility, as well as advantages such as high imprinting efficiency, high binding capacity, and fast speed, providing a good analytical and detection platform for the imprinting of large molecules such as proteins, cells, and viruses. Li et al. [33] reported a novel graphene nanoplatelet (GNP)/polydopamine (PDA)-MIP-biosensor for ultra-trace detection of Crn in a range of body fluids (Figure 2). Under the catalytic action of ammonium persulfate (AP), dopamine hydrochloride (DA) monomer was polymerized on the surface of GNP forming a thin PDA-MIP layer with high-density Crn recognition sites. This novel surface MIP technology avoids overly thick MIP layers, thereby allowing more TM to anchor on or near the polymer surface during the polymerization process. This ensures that the TM can be more thoroughly removed, leading to a broader detection range (Table 1).

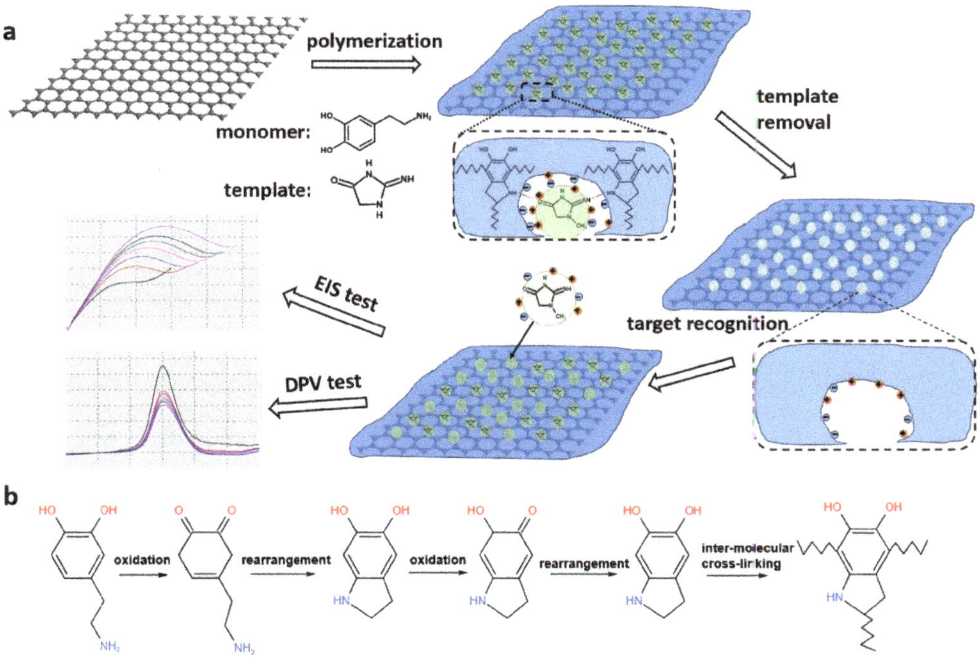

Figure 2. Schematic representation for the preparation of GNP/PDA-MIP. (**a**) The fabrication and application process of GNP/PDA-MIP and the EIS/DPV response of the GNP/PDA-MIP modified electrode towards different concentration of creatinine in electrolyte. (**b**) The formation mechanism of DA spontaneous oxidative polymerization [33].

Table 1. The developed BRBS for the detection of Crn.

Category of Technology	Functional Monomer	Detection Limit	Ref.
Traditional-MIT	methacrylic acid	0.1 ppm	[25]
N-MIT	nickel nanoclusters	0.5 nmol/L	[31]
Surface MIT	dopamine hydrochloride	2×10^{-2} pg/mL	[33]

2.3. *Cystatin C (Cys-C)*

Cys-C is a 13 kDa non-glycosylated polypeptide chain consisting of 120 amino acid residues. It is a small protein that can be freely filtered by the renal glomerulus. The average serum Cys-C level in patients with diabetic nephropathy (1.87 ± 0.51 mg/L) is higher than that in patients with non-diabetic nephropathy (1.025 ± 0.30 mg/L). In addition, serum Cys-C levels are higher in patients with elevated serum Crn levels, suggesting a correlation between Cys-C and Crn. Studies have shown that serum Cys-C detection can help to identify early renal impairment in diabetic patients, with better efficacy than serum Crn [34]. Another study showed that the average serum cystatin C value was 1.73 in type 2 diabetic patients with normal urinary albumin levels, while the average serum cystatin C value was 2.07 in type 2 diabetic patients with microalbuminuria. These results suggest that serum cystatin C, as a renal glomerular filtration biomarker, is an earlier indicator of diabetic nephropathy than urinary microalbumin and serum Crn [35].

Multifunctional monomer imprinting can enhance non-covalent binding between functional monomers and TM via multiple synergistic interaction sites [36]. This is particularly effective in improving the selectivity and adsorption of MIPs for large molecules. Therefore, it is an effective technique to use two or more functional monomers simultaneously, forming multiple complementary interaction sites with TM in different areas. However, the selection, combination, design, synthesis, and full utilization of their synergistic effect to prepare ideal MIPs still require continuous exploration. In addition, polypyrrole (PPy) is a conductive polymer that is well-known for its use in MIP materials [37]. Carbon nanotubes (CNTs) can improve electron transfer and surface area in electrochemical systems, thereby enhancing the electrical properties of electrochemical biosensors. Therefore, after adding CNTs to PPy, a 3D structure with increased electrical conductivity is formed due to the strong π-π bonding between the Ppy-conjugated structure and CNTs [38]. The PPy/CNT nanocomposite exhibits good charge transfer and energy/electron storage performance, making it suitable for supercapacitors [39]. Based on this strategy, Gomes [40] et al. first formed a mixed system of Cys-C, multi-walled carbon nanotubes (MWCNTs), pyrrole (Py), and carboxylated pyrrole (Py-COOH), and electrochemically polymerized it on the surface of a carbon fiber net screen-printed electrode (SPE) to form an MPPy nanocomposite. Subsequently, the Cys-C template was removed by the urea method, constructing a disposable biosensor based on molecular imprinting capable of accurately detecting Cys-C. Because Cys-C has a higher affinity with binding sites when surrounded by bifunctional monomers, the recognition ability of the sensor is enhanced. Additionally, introducing MWCNTs into the PPy matrix resulted in a more porous structure, larger surface area, and better sensitivity. The differential pulse voltammetry (DPV) detection of Cys-C in a pH 6.0 solution containing $Fe(CN)_6{}^{3-/4-}$ using this sensor had a limit of detection (LOD) as low as 0.5 ng/mL, demonstrating its cost-effectiveness and high stability for potential applications in clinical medicine.

2.4. *Homocysteine (Hcy)*

Hcy, cysteine (Cys), and glutathione (GSH) are the three most abundant small-molecule amino acids containing thiol groups, playing important roles in biological processes [41]. The normal level of Cys in human blood plasma is 135.8–266.5 µmol/L, the concentration of Hcy is in the range of 5–15 µmol L^{-1}, and the GSH concentration in most cells is at mmol/L level [42,43]. Among them, Hcy is one of the important biomarkers of diabetic nephropathy. Therefore, it is particularly important to establish sensitive, accurate, and highly selective Hcy sensors.

Currently, electrochemical sensors(ECS) are commonly modified with noble metal nanoparticles on their surfaces. For example, Au nanoparticles (AuNPs) [44] possess good electron density, dielectric performance, and catalytic activity, as well as excellent conductivity, stability, and surface adsorption [45]. In recent years, silver nanoparticles [46] and alloy particles [47,48] have also been widely employed in ECS. In addition, MXene, with a chemical formula generally represented as $M_{n+1}X_nT_x$ [49], is an emerging class of two-dimensional inorganic composite materials. MXene demonstrates immense advantages in areas such as medicine and sensing due to its large surface area, excellent energy storage performance, good bioconductivity, and biocompatibility [50,51]. Moreover, MXene possesses a strong reducing ability, providing new strategies for synthesizing multifunctional nano-hybrids [52,53]. Its exceptional performance offers new possibilities for improving the conductivity, electrocatalytic activity, and sensitivity of the target analyte's signal response in biomimetic recognition bio-ECS interfaces. Liu [54] and others developed an unmarked molecularly imprinted electrochemical sensor (MIECS) for rapid and highly sensitive detection of Hcy. In this strategy, AuNPs were first prepared using chloroauric acid as a raw material in the presence of a reducing agent, and then AuNPs were introduced into the multilayer nanostructure of MXene to form MXene@AuNPs nanocomposites. Finally, the MIP with Hcy-specific recognition was formed by electropolymerization of DA monomer on the MXene@AuNPs-modified GCE surface. The MXene in the sensing interface has strong reducibility and excellent conductivity, and its large surface area accommodates more AuNPs and active sites, significantly enhancing the electrochemical performance of the sensor. The Hcy in the sample was detected by cyclic voltammetry, and the current signal decreased with the increase of Hcy concentration, with an LOD of 11.81 fmol L^{-1}

(Table 2). It is worth noting that due to the inherent two-dimensional structure of MXene with stack ability, the target molecules cannot approach the active region [55], which greatly reduces the sensitivity of detecting target molecules, greatly limiting its electrochemical performance. Therefore, the improvement of MXene function and performance is still the focus of current researchers.

Wen [56] and others prepared an Aptasensor using a graphene sponge (GS) as a material (Figure 3). The GS and AuNPs were separately modified on the glassy carbon electrode (GCE). Then, an Hcy APT with 66 bases of hairpin deoxyribonucleic acid (DNA) was fixed on the GCE by a Au-S covalent bond. Non-linking sites were blocked by bovine serum albumin (BSA) using mechanical filling and adsorption coverage methods. Finally, MB was specifically adsorbed on the APT due to its high affinity with guanine [57,58]. In this study, MB had good biocompatibility and could generate electrochemical signals at a specific voltage in DPV, acting as an electrochemical hybridization indicator. In the presence of Hcy, the conformation of the nucleic acid APT changed after specifically binding with Hcy, causing MB release from the APT or moving away from the GCE surface, and thus changing the relative redox current (ΔI) of MB. The sensor exhibited excellent signal detection, selectivity, and reproducibility, with a good linear relationship between Hcy concentration and ΔI in the linear range of 1~100 μM, a wide linear range, and a detection limit of 1 μM (Table 2).

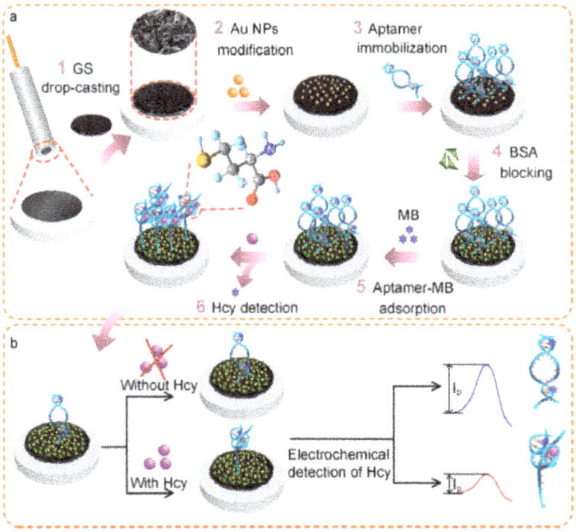

Figure 3. The fabrication procedures and detection mechanism of the aptaseneor. (**a**) The stepwise modifications of the GCE surface. (**b**) The electrochemical response of the aptasensor with/without Hcy. Ip represents the peak current [56].

APT can effectively recognize biomarkers by non-specifically adsorbing on the surface of gold electrodes or AuNPs. Small molecule-assisted technology (SMAT) refers to the formation of an auxiliary interface by small molecules that enable the linear chains of APT to be vertically oriented on the interface. The specificity of APT is enhanced by restricting the non-specific binding sites at the sensor interface through 6-mercaptohexanol, dithiothreitol, and 3-mercaptopropionic acid. Existing research reports that the binary self-assembled monolayer composed of thiol APT and 6-mercaptohexanol can significantly reduce the surface molecular density of APT, confirming that SMAT can reduce the non-specific adsorption of APT [59]. Hcy is a type of redox molecule [60]. Using SMAT Beitollahi et al. [61] developed an unlabeled electrochemical sensor for detecting Hcy (Figure 4). Firstly, the aptamer solution was cast onto the surface of a glassy carbon electrode modified with AuNPs, and stood upright in a humid room to complete the self-assembly of the thiol aptamer. The modified glassy carbon electrode was then immersed in a 6-mercaptohexanol solution and washed with 0.1 M PBS at a pH of 7.4. Small molecule interface technology was used to form a linearly coordinated aptamer chain and limit non-specific binding sites. The Hcy binds through interaction with the aptamer and undergoes an electrochemical reaction at the modified electrode surface, resulting in a significant enhancement of the peak current. This sensor, under optimal conditions, identifies Hcy using the differential pulse voltammetry (DPV) method. The detection limit is 0.01 μM, and the linear range is 0.05–20.0 μM (Table 2).

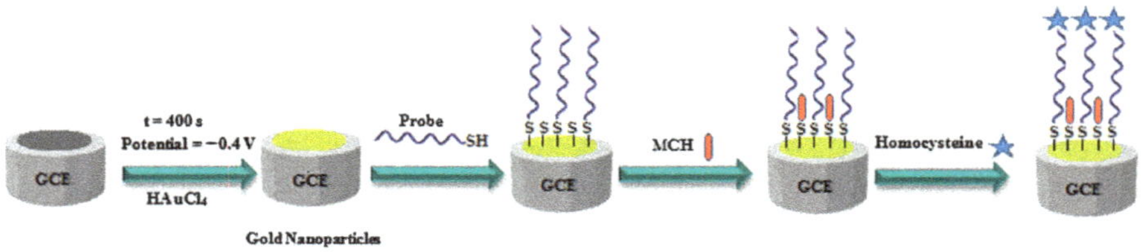

Figure 4. Schematic illustration of stepwise homocysteine aptasensor fabrication process [61].

Table 2. The developed BRBS for the detection of Hcy.

Category of Technology	Functional Monomer/APT	Detection Limit	Ref.
N-MIT	dopamine hydrochloride	11.81 fmol L^{-1}	[54]
MAMT	APT	1 μM	[56]
SMAT	APT	0.01 μM	[61]

2.5. Vascular Endothelial Growth Factor (VEGF)

VEGF is a 45 kDa homodimeric glycoprotein encoded by a single gene. Currently, five human VEGF microRNAs (mRNAs) have been found by splicing of VEGF mRNA, encoding VEGF with 121, 145, 165, 189, and 206 amino acids, respectively [62,63]. It has been found that the glomerular basement membrane has a high affinity for VEGF, which can cause an increase in the permeability of the glomerular basement membrane, thereby promoting the production of NO, endothelin, etc., further changing the renal hemodynamics of diabetic patients, leading to proteinuria and inducing diabetic nephropathy [64]. Therefore, the development of highly sensitive and selective VEGF detection methods is of great significance for early diagnosis and patient recovery assessment.

Palabiyik [65] and others proposed a highly sensitive label-free impedance sensor based on molecularly imprinted polymers (MIP) and graphite screen-printed electrodes (GSPE) as a detection platform for vascular endothelial growth factor (VEGF) (Figure 5). To improve the reproducibility of electrode detection and remove any impurities generated during the electrode manufacturing process, the GSPE was electrochemically pretreated before use. Then, MIP was formed on the pretreated GSPE surface by electropolymerization of ortho-phenylenediamine (o-PD) in the presence of the VGFR template. Under optimized conditions, the sensor used $[Fe(CN)]^{3-/4-}$ as a redox probe and employed a label-free method based on EIS measurement to detect VGFR, with good analytical performance in the range of 20~200 pg·mL^{-1} and a detection limit of 0.08 pg mL^{-1} (Table 3). Due to the disposable nature of the sensor, it is more suitable for clinical needs.

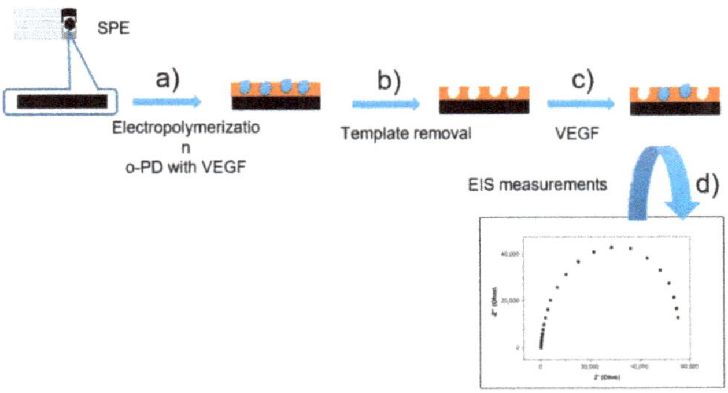

Figure 5. Scheme representation of molecular imprinting and the recognition principle: (**a**) electropolymerization with o-PD and VEGF; (**b**) removal of VEGF template; (**c**) incubation with VEGF sample solution; (**d**) washing and EIS measurements [65].

Cheng [66] and others constructed a convenient and ultra-sensitive ECL APT biosensor for detecting VEGF165. Firstly, the synthesized graphite-like carbon nitride ($g\text{-}C_3N_4$) nanosheets were adhered to CdSe using polydiallyldimethylammonium chloride (PDDA) as an adhesive, followed by assembling the obtained $g\text{-}C_3N_4$/PDDA/CdSe composite material onto the GCE surface. Finally, DNA_1 and AuNP-labeled DNA_2 were sequentially incubated on the modified GCE surface. In this strategy, $AuDNA_2$ showed a significant spectral overlap with the $g\text{-}C_3N_4$/PDDA/CdSe composite, thus allowing for effective energy transfer. In the presence of VEGF165, the sensor impedance decreased significantly after the Au-DNA_2 binds to VEGF165, and the ECL signal increased significantly. The ECL signal of the constructed biosensor increased with the increasing concentration of VEGF165, showing a prominent linear relationship from 2 pg mL^{-1} to 2 ng mL^{-1}, and the detection limit of VEGF165 was as low as 0.68 pg mL^{-1} (Table 3). The excellent sensitivity, stability, repeatability, and selectivity of the ECL biosensor indicate further potential applications in clinical diagnosis.

Table 3. The developed BRBS for the detection of VEGF.

Category of Technology	Functional Monomer/APT	Detection Limit	Ref.
Surface MIT	ortho-phenylenediamine	0.08 pg mL^{-1}	[65]
MAMT	APT	0.68 pg mL^{-1}	[66]

2.6. Epidermal Growth Factor Receptor (EGFR)

In recent years, EGFR has been proven to be one of the potential biomarkers for diabetic nephropathy [67]. Generally, the preparation of MIPs typically involves a single kind of template molecule/ion. However, MIPs based on a single template cannot recognize and remove multiple targets simultaneously. Multi-template MIT uses two or more target substances simultaneously as templates. Li et al. achieved simultaneous detection of dopamine and uric acid using dual-template technology, allowing the formation of MIPs containing multiple specific recognition sites. This provides a feasible strategy for simultaneously enriching, recognizing, detecting, and removing multiple targets [68]. Ahar et al. [69] constructed a MIP biosensor based on gold nanoparticle-modified screen-printed electrodes (Au-SPE) and used antibody-coupled nanoliposome amplification for simultaneous detection of EGFR and vascular VEGF (Figure 6). First, DSP was assembled onto the Au-SPE surface by self-assembly, and then the target proteins (EGFR and VEGF) were attached covalently through amide bonds to the modified SPE using multi-template molecular imprinting. Unbound sites were then blocked with acrylamide (AAM). Under the presence of persulfate, the AAM and *N,N′*-methylenebis(acrylamide) functional monomers were polymerized around the EGFR and VEGF templates, and the polymerization termination time was controlled by mixing the phenol solution with ethanol (w/t, 1%). Finally, by utilizing the ability of oxalic acid (OXA) to destroy peptide bonds, the template molecule is successfully removed from the imprinting layer. This strategy aims to produce reliable electrochemical signals by using nanoliposomes loaded with Cd(II) and Cu(II) ions, and modified with EGFR and VEGF-specific antibodies as the detection targets. In the analysis step, the potentiostatic adsorption analysis electrochemical technique is employed to indirectly determine the trace amounts of EGFR and VEGF in serum based on the content of these ions. Under optimal conditions, the detection limits for EGFR and VEGF analysis are 0.01 and 0.005 pg mL^{-1}, respectively. The sensor has good sensitivity, repeatability, and specificity. In clinical practice, it is often necessary to analyze a series of biomarkers to make a final diagnostic result, and this SPE-based biosensor can be successfully integrated with lab-on-a-chip, microfluidics, or micro-total analytical systems, opening new avenues for the development of multiplexed sensing of biomarkers. However, it is worth noting that due to the dilution of each template's binding sites, the selectivity of multi-template MIPs is lower than MIPs synthesized with a single template.

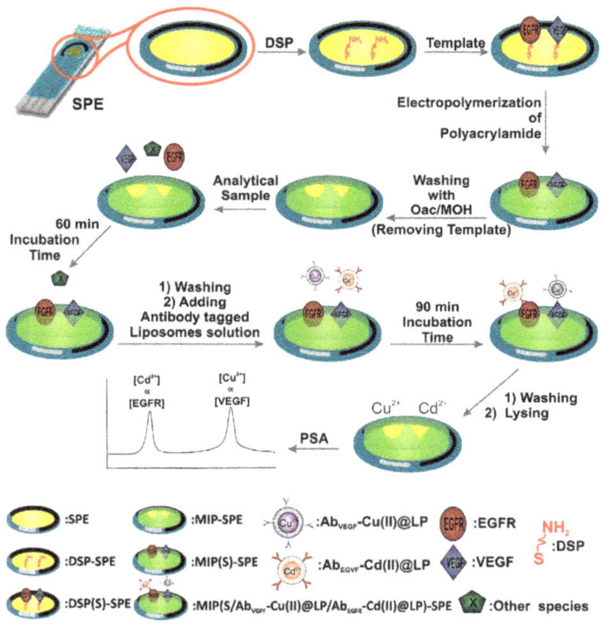

Figure 6. Representation for the engineering processes of EGFR/VEGF biosensors [69].

2.7. mRNA-21

mRNAs are a class of small non-coding RNAs that play important roles in post-transcriptional regulation of gene expression. mRNAs regulate over 60% of protein-coding gene expression. Therefore, changes in their expression are associated with many diseases, including liver diseases, kidney diseases, and diabetes [70–72]. Recently, some of the literature reported that mRNAs, such as mRNA-21, mRNA-223-3p, and mRNA-377, can serve as potential biomarkers for diabetic nephropathy [73,74].

Liu et al. [75] designed a photoelectroactive aptasensor for the detection of mRNA-21. First, chitosan was used to wrap the $CuInS_2$-modified ITO electrode surface, preventing the dissolution of $CulnS_2$ into the solution, and providing amino groups for subsequent electrode modification. Then, the carboxyl-modified DNA_2 reacted with chitosan and was coupled to the electrode via amide bonds. If the hairpin DNA is recognized by mRNA-21, Exonuclease III (Exo III) will cleave the blunt 3′ end of DNA_1, triggering the target-cycled amplification process that releases DNA_3, which can hybridize with DNA_2 on the electrode. The DNA_2–DNA_3 duplex provides a binding site for TATA-binding protein (TBP) attachment. Since TBP induces an 80-degree bend in the TATA sequence by inserting amino acid side chains on the base pairs of the dsDNA sequence [76], this blocks electron transfer, causing a sharp decrease in photocurrent intensity. The decrease in photocurrent due to steric hindrance can be used to quantify microRNA-21. This strategy combines anti-interference photoelectroactive cathode materials, enzyme-assisted target cycle amplification, and TBP-induced signal shutdown, achieving remarkable amplification efficiency. The detection limit of mRNA-21, under optimized conditions, is as low as 0.47 fM, and the linear range is from 1.0×10^{-15} M to 1.0×10^{-9} M. Further research is needed to explore the application of this DNA amplification-based PEC platform in DNA, protein, and small molecule detection.

2.8. *Ceruloplasmin (Cp)*

Cp is a copper-containing α-2 globulin with a molecular weight of 132 kDa, mainly synthesized and secreted by the liver, and plays a crucial role in regulating oxidative stress and iron homeostasis [77]. Some researchers have proposed that the glomerular capillary wall excretion rate of Cp in patients with diabetic nephropathy is almost parallel to the urinary albumin excretion rate. Increased excretion of Cp has also been observed in patients with impaired glucose tolerance and diabetes compared to healthy controls [78]. Increased urinary Cp excretion in diabetic patients with normal urinary protein has also been confirmed [79]. Therefore, detecting urinary Cp has a high value for early diagnosis of diabetic nephropathy.

Haghshenas et al. [80] designed a sensitive electrochemical aptasensor to quantify Cp using Cp-specific recognition APT. First, the diazonium salt of 4-Aminobenzoic acid (ABA) electrochemically reduced was covalently bonded to the MWCNTs/GCE surface [81], and after the carboxyl group was activated by *N*-hydroxysuccinimide(NHS) and *N*-(3-dimethylaminopropyl)-*N*′-ethylcarbodiimide hydrochloride (EDC), the amino APT was immobilized on the electrode surface via amide bond formation. BSA solution was added to prevent non-specific adsorption on the sensor surface. Finally, Cp was detected using DPV and EIS in $[Fe(CN)_6]^{3-/4-}$ solution. Cp-specific binding with APT induces a conformational change in the APT on the electrode surface, and the probe's electron transfer is hindered. The DPV current of the aptasensor decreases with increasing Cp concentration, and the Rct value increases with increasing Cp concentration. Under optimal conditions, Cp concentration shows a linear relationship within the range of 0.02~3.0 ng mL^{-1} and 3.0~80.0 ng mL^{-1}, with a detection limit of 3.7 pg mL^{-1}. This aptasensor has broad application prospects for detecting Cp in human serum.

2.9. *Platelet-Derived Growth Factor (PDGF)*

PDGF is an important cytokine composed of two peptide chains (A and B) connected by disulfide bonds in serum, which can promote the proliferation of vascular endothelial cells, smooth muscle cells, and the biosynthesis of renal tubule matrix, playing a crucial role in regulating cell growth and division [82]. PDGF-AA, PDGF-BB, and PDGF-AB are three subtypes of PDGF [83]. Among them, PDGF-BB is directly involved in cell transformation and tumor growth [84]. Studies have shown that the level of PDGF in the plasma of patients with diabetic nephropathy increases. PDGF can promote the proliferation and matrix generation of renal tubular cells, further accelerating the process of renal fibrosis, which leads to further deterioration of renal function. In addition, PDGF-BB is considered an important marker capable of predicting early deterioration of renal function in patients with diabetic nephropathy. Therefore, the sensitive and selective detection of PDGF-BB in biological samples has significant implications for the diagnosis of diabetic nephropathy.

Zhang et al. [85] introduced an ECL aptasensor based on improved AuNPs to detect PDGF-BB (Figure 7). In this scheme, AuNPs-electrochemically reduced graphene

(AuNPs-EG) nanocomposite material was electrodeposited on the GCE surface, and a large number of PDGF-BB APT (APT_1) were fixed to amplify the detection response, constructing a highly conductive Aptasensor platform that can amplify the Luminol-H_2O_2 system ECL signal. By functionalizing AuNPs with glucose oxidase (GOD), a signal probe for the sandwich sensor of the second aptamer (APT_2) and GOD-modified AuNPs was designed. This ECL sensor has the advantages of high sensitivity, good stability, and good selectivity, and has good analytical performance in the concentration range of 1.0×10^{-13}~5.0×10^{-10} mol L^{-1} PDGF-BB, with a detection limit of 1.7×10^{-14} mol L^{-1} (Table 4). It has the potential for application in biochemical analysis.

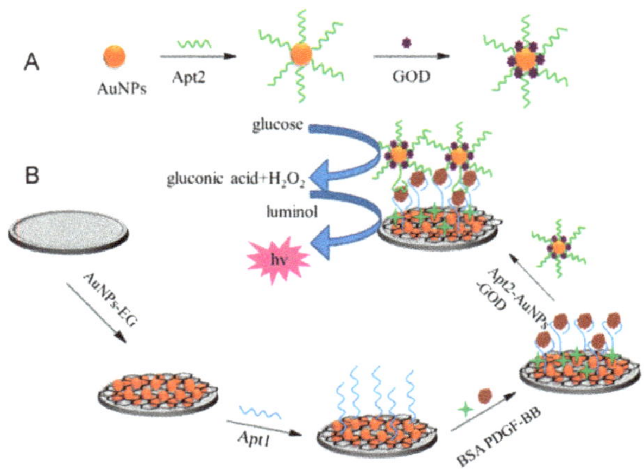

Figure 7. The procedure of the ECL aptasensor preparation. (**A**) Preparation of GOD-Apt2-AuNPs bioconjugates. (**B**) The schematic shows the strategy for constructing the ECL aptasensor [85].

In recent years, two-dimensional layered transition metal dichalcogenide materials have opened a new door in the field of electrochemistry due to their unique electronic and electrochemical properties [86]. VS_2 has significant advantages in large specific surface area, good catalytic adsorption performance, and good biocompatibility. Normally, VS_2 has a stacked-layer structure, and the V atom inserts between two S atoms to form a sandwich structure. Unfortunately, like most transition metal oxides, the weak conductivity of VS_2 severely limits its further application in electrochemical biosensors. In order to improve the electrochemical performance of VS_2, it seems necessary to combine it with good electronic conductive materials. Framework nucleic acids (FNAs) are one-dimensional to three-dimensional (3D) framework structures formed by nucleic acid molecules through self-assembly [87]. The surface performance of electrochemical DNA biosensors modified with FNAs has been significantly improved. Due to the good stability and controllability of FNAs, they can be assembled in an ordered manner on gold surfaces to fabricate more probes [88,89]. The controllable charge transfer characteristics of FNAs can be utilized to optimize the performance of electrochemical biosensors based on FNAs [90]. Compared to traditional electrochemical sensors functionalized with single-stranded DNA, electrochemical biosensors based on FNAs have stronger interference resistance and higher protein resistance. The empty structure allows for the maximization of space utilization, making it highly suitable for interface assembly with various materials [91]. Huang et al. [92] first prepared a novel VS_2-graphene (VS_2-GR) composite material using a simple one-step hydrothermal method and a novel label-free electrochemical aptasensor was designed for the detection of platelet-derived growth factor BB (PDGF-BB) using Exo III-assisted signal amplification technology, VS_2-GR composite material, and tetrahedral structured DNA probe [93,94], which exhibits fast response, high throughput, and current sensitivity. In the absence of PDGE-BB, the APT hybridizes with complementary DNA (cDNA), and the biotin-labeled single-stranded signal DNA at the 5′ end cannot be cleaved by exonuclease III (Exo III). Biotin-labeled signal DNA hybridizes with protruding T-DNA on the AuNP/VS_2-GR electrode, and the specific binding of avidin-biotin absorbs a large amount of avidin-horseradish peroxidase (avidin-HRP) onto the modified electrode, resulting in a strong current response of HRP to the mixture of hydrogen peroxide and hydroquinone (H_2O_2 + HQ). However, when PDGF-BB is present, the APT preferentially binds to PDGF-BB, and duplex DNA is formed between cDNA and signal DNA. The double-stranded DNA is digested by Exo III from the blunt end of the 3′ signal DNA and releases cDNA. Subsequently, the released cDNA undergoes a new cleavage process with the remaining signal DNA in the solution. Finally, the cyclic hybridization-hydrolysis process leads to a reduction in the number of signal molecules HRP on the electrode, resulting in a significant decrease in current response. The proposed detection method takes advantage of the large specific surface area, good conductivity, and excellent biocompatibility of layered VS_2-GR composite materials, providing plenty of binding sites and suitable environments for the immobilization of biomolecules. In addition, the tetrahedral structured DNA probe based on framework nucleic acid assembly technology can minimize non-specific adsorption and obtain upright probes on the electrode surface, improving detection stability and reproducibility. Based on these advantages, the proposed detection method can detect PDGF-BB concentrations as low as 30 fM and has good selectivity for PDGF-BB (Table 4). Due to the inherent characteristics of Exo III, which does not require specific recognition sites and APT, the proposed detection method has a wide range of target molecules and can serve as a universal scheme for detecting DNA, drugs, and even proteins.

Table 4. The developed BRBS for the detection of PDGF.

Category of Technology	Functional Monomer/APT	Detection Limit	Ref.
MAMT	APT	1.7×10^{-14} mol L^{-1}	[85]
FNAs	APT	30 fM	[92]

3. Conclusions and Future Perspectives

Significant progress has been made in the construction technology of biomimetic recognition electrochemical sensors for diabetic nephropathy biomarkers, but there are still many challenges at the current stage. Current challenges include developing new multifunctional recognition elements, simultaneous detection of multiple biomarkers, further studying the binding mechanism between MIPs and target objects, achieving specific recognition of MIPs in the aqueous phase, enhancing the effective recognition rate of recognition elements for large molecules and the anti-interference ability of complex samples, integrating recognition elements and transducers to increase the sensitivity of sensors, making their preparation more convenient and faster, and further promoting their development towards intelligence and portability. With the rapid development of biotechnology, intelligent control technology, and microelectronics technology, especially micro-electro-mechanical systems (MEMS), the future biomimetic sensors will be miniaturized intelligent biomimetic systems supported by continuous system integration technology. Currently, it seems that miniaturized intelligent biomimetic sensors may be one of the main choices for the future construction of the perception layer sensors in the Internet of Things, and could be the only possible choice to replace traditional sensors.

Funding: This project was kindly funded by the National Natural Science Foundation of China (Nos. 22174136, 22004116), and the CAS President's International Fellowship Initiative (PIFI) (No. 2023VBC0020).

Institutional Review Board Statement: Not applicable.

Informed Consent Statement: Not applicable.

Conflicts of Interest: The authors declare no conflict of interest.

References

1. Fawzy, M.S.; Al Beladi, F.I. Association of Circulating Vitamin D, VDBP, and Vitamin D Receptor Expression with Severity of Diabetic Nephropathy in a Group of Saudi Type 2 Diabetes Mellitus Patients. *Clin. Lab.* **2018**, *64*, 1623–1633. [CrossRef] [PubMed]
2. Hayes, A.; Arima, H.; Woodward, M.; Chalmers, J.; Poulter, N.; Hamet, P.; Clarke, P. Changes in Quality of Life Associated with Complications of Diabetes: Results from the ADVANCE Study. *Value Health* **2016**, *19*, 36–41. [CrossRef] [PubMed]
3. Thorn, L.M.; Gordin, D.; Harjutsalo, V.; Hagg, S.; Masar, R.; Saraheimo, M.; Tolonen, N.; Waden, J.; Groop, P.H.; Forsblom, C.M.; et al. The Presence and Consequence of Nonalbuminuric Chronic Kidney Disease in Patients With Type 1 Diabetes. *Diabetes Care* **2015**, *38*, 2128–2133. [CrossRef] [PubMed]
4. Du, Y.; Xu, B.J.; Deng, X.; Wu, X.W.; Li, Y.J.; Wang, S.R.; Wang, Y.N.; Ji, S.; Guo, M.Z.; Yang, D.Z.; et al. Predictive metabolic signatures for the occurrence and development of diabetic nephropathy and the intervention of Ginkgo biloba leaves extract based on gas or liquid chromatography with mass spectrometry. *J. Pharm. Biomed. Anal.* **2019**, *166*, 30–39. [CrossRef]
5. Chen, Y.; Liu, H.; Loh, T.P.; Liu, Q.; Teo, T.L.; Lee, T.K.; Sethi, S.K. Measurement of urine albumin by liquid chromatography-isotope dilution tandem mass spectrometry and its application to value assignment of external quality assessment samples and certification of reference materials. *Clin. Chem. Lab. Med.* **2021**, *59*, 711–720. [CrossRef]
6. Ji, H.; Shen, L.; Shi, X.; Su, J.; Tang, Z.; Wang, H.; Ju, S.; Wang, J. Establishment of an absolute quantitative method for measurement of urinary cystatin C by stable isotope dilution ultra high performance liquid chromatography tandem mass spectrometry. *Anal. Methods* **2018**, *10*, 5236–5241. [CrossRef]
7. Harlan, R.; Clarke, W.; Di Bussolo, J.M.; Kozak, M.; Straseski, J.; Meany, D.L. An automated turbulent flow liquid chromatography-isotope dilution mass spectrometry (LC-IDMS) method for quantitation of serum creatinine. *Clin. Chim. Acta.* **2010**, *411*, 1728–1734. [CrossRef]
8. Wang, H.; Chai, Y.; Li, H.; Yuan, R. Sensitive electrochemiluminescent immunosensor for diabetic nephropathy analysis based on tris(bipyridine) ruthenium(II) derivative with binary intramolecular self-catalyzed property. *Biosens. Bioelectron.* **2018**, *100*, 35–40. [CrossRef]
9. Li, Y.; Wang, Y.; Bai, L.; Lv, H.; Huang, W.; Liu, S.; Ding, S.; Zhao, M. Ultrasensitive electrochemiluminescent immunosensing based on trimetallic Au-Pd-Pt/MoS(2) nanosheet as coreaction accelerator and self-enhanced ABEI-centric complex. *Anal. Chim. Acta.* **2020**, *1125*, 86–93. [CrossRef]
10. Xu, H.; Kou, F.; Ye, H.; Wang, Z.; Huang, S.; Liu, X.; Zhu, X.; Lin, Z.; Chen, G. Highly sensitive antibody-aptamer sensor for vascular endothelial growth factor based on hybridization chain reaction and pH meter/indicator. *Talanta* **2017**, *175*, 177–182. [CrossRef]
11. Wang, H.; Ma, Y.; Guo, C.; Yang, Y.; Peng, Z.; Liu, Z.; Zhang, Z. Templated seed-mediated derived Au nanoarchitectures embedded with nanochitosan: Sensitive electrochemical aptasensor for vascular endothelial growth factor and living MCF-7 cell detection. *Appl. Surf. Sci.* **2019**, *481*, 505–514. [CrossRef]
12. Hassanzadeh, M.; Ghaemy, M.; Amininasab, S.M.; Shami, Z. Molecularly imprinted polymer capped near infrared fluorescent emitting Ag2S-functionalized-COOH quantum dots for detection of creatinine as a nanosensor with high sensitivity and selectivity. *Sens. Actuators A Phys.* **2021**, *331*, 112936. [CrossRef]
13. Zhuang-Fei, J.; Qin, L.; Qing-Yao, L.; Hui-Xian, X.; Jia-Yuan, H.; Chong-Zhi, W.; Lian-Di, Z.; Qi-Hui, Z.; Ling, L.; Chun-Su, Y. Fast exhaustive enrichment and electrochemical quantitative detection of anthocyanins from natural products by using dual responsive and dummy molecularly imprinted polymers. *Microchem. J.* **2022**, *179*, 107545.
14. Taghdisi, S.M.; Danesh, N.M.; Ramezani, M.; Alibolandi, M.; Nameghi, M.A.; Gerayelou, G.; Abnous, K. A novel electrochemical aptasensor for ochratoxin a sensing in spiked food using strand-displacement polymerase reaction. *Talanta* **2021**, *223 Pt 1*, 121705. [CrossRef] [PubMed]
15. Yan, S.R.; Foroughi, M.M.; Safaei, M.; Jahani, S.; Ebrahimpour, N.; Borhani, F.; Rezaei Zade Baravati, N.; Aramesh-Boroujeni, Z.; Foong, L.K. A review: Recent advances in ultrasensitive and highly specific recognition aptasensors with various detection strategies. *Int. J. Biol. Macromol.* **2020**, *155*, 184–207. [CrossRef]
16. Kimura-Suda, H.; Petrovykh, D.Y.; Tarlov, M.J.; Whitman, L.J. Base-dependent competitive adsorption of single-stranded DNA on gold. *J. Am. Chem. Soc.* **2003**, *125*, 9014–9015. [CrossRef] [PubMed]
17. Wang, S.; Cai, X.; Wang, L.; Li, J.; Li, Q.; Zuo, X.; Shi, J.; Huang, Q.; Fan, C. DNA orientation-specific adhesion and patterning of living mammalian cells on self-assembled DNA monolayers. *Chem. Sci.* **2016**, *7*, 2722–2727. [CrossRef]
18. McTaggart, M.P.; Price, C.P.; Pinnock, R.G.; Stevens, P.E.; Newall, R.G.; Lamb, E.J. The Diagnostic Accuracy of a Urine Albumin-Creatinine Ratio Point-of-Care Test for Detection of Albuminuria in Primary Care. *Am. J. Kidney Dis.* **2012**, *60*, 787–794. [CrossRef]
19. MacIsaac, R.J.; Ekinci, E.I.; Jerums, G. Markers of and Risk Factors for the Development and Progression of Diabetic Kidney Disease. *Am. J. Kidney Dis.* **2014**, *63*, S39–S62. [CrossRef]
20. Zhang, G.; Yu, Y.; Guo, M.; Lin, B.; Zhang, L. A sensitive determination of albumin in urine by molecularly imprinted electrochemical biosensor based on dual-signal strategy. *Sens. Actuators B Chem.* **2019**, *288*, 564–570. [CrossRef]
21. Boon, E.M.; Ceres, D.M.; Drummond, T.G.; Hill, M.G.; Barton, J.K. Mutation detection by electrocatalysis at DNA-modified electrodes. *Nat. Biotechnol.* **2000**, *18*, 1096–1102. [CrossRef]
22. Kintzel, P.E. Anticancer drug-induced kidney disorders. *Drug Saf.* **2001**, *24*, 19–38. [CrossRef]
23. Levey, A.S.; Perrone, R.D.; Madias, N.E. Serum Creatinine and Renal Function. *Annu. Rev. Med.* **1988**, *39*, 465–490. [CrossRef]

24. Musa, N.; Ramzy, T.; Hamdy, A.; Arafa, N.; Hassan, M. Assessment of urinary podocalyxin as a marker of glomerular injury in obesity-related kidney disease in children and adolescents with obesity compared to urinary albumin creatinine ratio. *Clin. Obes.* **2021**, *11*, e12452. [CrossRef] [PubMed]
25. Prabhu, S.N.; Mukhopadhyay, S.C.; Gooneratne, C.P.; Davidson, A.S.; Liu, G. Molecularly Imprinted Polymer-based detection of creatinine towards smart sensing. *Med. Devices Sens.* **2020**, *3*, e10133. [CrossRef]
26. Canfarotta, F.; Poma, A.; Guerreiro, A.; Piletsky, S. Solid-phase synthesis of molecularly imprinted nanoparticles. *Nat. Protoc.* **2016**, *11*, 443–455. [CrossRef]
27. Gutiérrez-Climente, R.; Gómez-Caballero, A.; Unceta, N.; Aránzazu Goicolea, M.; Barrio, R.J. A new potentiometric sensor based on chiral imprinted nanoparticles for the discrimination of the enantiomers of the antidepressant citalopram. *Electrochim. Acta* **2016**, *196*, 496–504. [CrossRef]
28. Meng, C.; Knežević, S.; Du, F.; Guan, Y.; Kanoufi, F.; Sojic, N.; Xu, G. Recent advances in electrochemiluminescence imaging analysis. *eScience* **2022**, *2*, 591–605. [CrossRef]
29. Quan, S.; Ji, K.; Liu, F.; Barkae, T.H.; Halawa, M.I.; Hanif, S.; Lou, B.; Li, J.; Xu, G. Chemiluminescence of lucigenin-tetracycline and itsapplication for sensitive determination of procyanidin. *J. Food Drug Anal.* **2022**, *30*, 293–302. [CrossRef]
30. Mostafa, I.M.; Gilani, M.; Chen, Y.; Lou, B.; Li, J.; Xu, G. Lucigenin-pyrogallol chemiluminescence for the multiple detection of pyrogallol, cobalt ion, and tyrosinase. *J. Food Drug Anal.* **2021**, *29*, 510–520. [CrossRef]
31. Babamiri, B.; Salimi, A.; Hallaj, R.; Hasanzadeh, M. Nickel nanoclusters as a novel emitter for molecularly imprinted electrochemiluminescence based sensor toward nanomolar detection of creatinine. *Biosens. Bioelectron.* **2018**, *107*, 272–279. [CrossRef] [PubMed]
32. Chen, L.; Wang, X.; Lu, W.; Wu, X.; Li, J. Molecular imprinting: Perspectives and applications. *Chem. Soc. Rev.* **2016**, *45*, 2137–2211. [CrossRef]
33. Li, Y.; Luo, L.; Nie, M.; Davenport, A.; Li, Y.; Li, B.; Choy, K.L. A graphene nanoplatelet-polydopamine molecularly imprinted biosensor for Ultratrace creatinine detection. *Biosens. Bioelectron.* **2022**, *216*, 114638. [CrossRef]
34. Amer, A.H.; Haridas, N. Early Diagnostic Markers in Diabetic Nephropathy Patients. *J. Clin. Diagn. Res.* **2018**, *12*, BC05–BC09. [CrossRef]
35. Gupta, K.; Nayyar, S.B.; Sachdeva, J.K.; Kumar, P. Cystatin C in the early diagnosis of diabetic nephropathy and its correlation with albuminuria. *Int. J. Adv. Med.* **2017**, *4*, 56–59. [CrossRef]
36. Dingding, D.; Jun, W.; Pengxin, H.; Xin, L.; Luhang, Z.; Shenao, M. Dual-monomer molecularly imprinted electrochemical sensor based on amino-functionalized MOFs and graphene for trace determination of taurine. *Mikrochim. Acta* **2023**, *190*, 126–162.
37. Hoefer, M.; Bandaru, P.R. Determination and enhancement of the capacitance contributions in carbon nanotube based electrode systems. *Appl. Phys. Lett.* **2009**, *95*, 183108. [CrossRef]
38. Ferreira, P.A.B.; Araujo, M.C.M.; Prado, C.M.; de Lima, R.A.; Rodriguez, B.A.G.; Dutra, R.F. An ultrasensitive Cystatin C renal failure immunosensor based on a PPy/CNT electrochemical capacitor grafted on interdigitated electrode. *Colloids Surf. B Biointerfaces* **2020**, *189*, 110834. [CrossRef]
39. Canobre, S.C.; Xavier FF, S.; Fagundes, W.S.; de Freitas, A.C.; Amaral, F.A. Performance of the Chemical and Electrochemical Composites of PPy/CNT as Electrodes in Type I Supercapacitors. *J. Nanomater.* **2015**, *2015*, 560164. [CrossRef]
40. Gomes, R.S.; Gomez-Rodriguez, B.A.; Fernandes, R.; Sales, M.G.F.; Moreira, F.T.C.; Dutra, R.F. Plastic Antibody of Polypyrrole/Multiwall Carbon Nanotubes on Screen-Printed Electrodes for Cystatin C Detection. *Biosensors* **2021**, *11*, 175. [CrossRef] [PubMed]
41. Zhang, B.; Zhang, H.; Zhong, M.; Wang, S.; Xu, Q.; Cho, D.-H.; Qiu, H. A novel off-on fluorescent probe for specific detection and imaging of cysteine in live cells and in vivo. *Chin. Chem. Lett.* **2020**, *31*, 133–135. [CrossRef]
42. Refsum, H.; Ueland, P.M.; Nygard, O.; Vollset, S.E. Homocysteine and Cardiovascular Disease. *Annu. Rev. Med.* **1998**, *49*, 31–62. [CrossRef] [PubMed]
43. Sedgwick, A.C.; Wu, L.; Han, H.H.; Bull, S.D.; He, X.P.; James, T.D.; Sessler, J.L.; Tang, B.Z.; Tian, H.; Yoon, J. Excited-state intramolecular proton-transfer (ESIPT) based fluorescence sensors and imaging agents. *Chem. Soc. Rev.* **2018**, *47*, 8842–8880. [CrossRef] [PubMed]
44. Ben Messaoud, N.; Ghica, M.E.; Dridi, C.; Ben Ali, M.; Brett CM, A. Electrochemical sensor based on multiwalled carbon nanotube and gold nanoparticle modified electrode for the sensitive detection of bisphenol A. *Sens. Actuators B Chem.* **2017**, *253*, 513–522. [CrossRef]
45. Plowman, B.J.; Sidhureddy, B.; Sokolov, S.V.; Young, N.P.; Chen, A.; Compton, R.G. Electrochemical Behavior of Gold-Silver Alloy Nanoparticles. *ChemElectroChem* **2016**, *3*, 1039–1043. [CrossRef]
46. Kumar, N.; Goyal, R.N. Silver nanoparticles decorated graphene nanoribbon modified pyrolytic graphite sensor for determination of histamine. *Sens. Actuators B Chem.* **2018**, *268*, 383–391. [CrossRef]
47. Zhou, Y.-C.; Zhao, M.; Yu, Y.-Q.; Lei, Y.-M.; Chai, Y.-Q.; Yuan, R.; Zhuo, Y. Three-dimensional nano-network composed of Pt nanoparticles functionalized Mn-doped CeO2 and hemin/G-quadruplex as electrocatalysts for cardiovascular biomarker detection. *Sens. Actuators B Chem.* **2017**, *246*, 1–8. [CrossRef]
48. Yang, L.; Xu, B.; Ye, H.; Zhao, F.; Zeng, B. A novel quercetin electrochemical sensor based on molecularly imprinted poly(para-aminobenzoic acid) on 3D Pd nanoparticles-porous graphene-carbon nanotubes composite. *Sens. Actuators B Chem.* **2017**, *251*, 601–608. [CrossRef]

49. Xie, Y.; Gao, F.; Tu, X.; Ma, X.; Xu, Q.; Dai, R.; Huang, X.; Yu, Y.; Lu, L. Facile Synthesis of MXene/Electrochemically Reduced Graphene Oxide Composites and Their Application for Electrochemical Sensing of Carbendazim. *J. Electrochem. Soc.* **2019**, *166*, B1673–B1680. [CrossRef]
50. Wang, H.; Li, H.; Huang, Y.; Xiong, M.; Wang, F.; Li, C. A label-free electrochemical biosensor for highly sensitive detection of gliotoxin based on DNA nanostructure/MXene nanocomplexes. *Biosens. Bioelectron.* **2019**, *142*, 111531. [CrossRef]
51. Nah, J.S.; Barman, S.C.; Zahed, M.A.; Sharifuzzaman, M.; Yoon, H.; Park, C.; Yoon, S.; Zhang, S.; Park, J.Y. A wearable microfluidics-integrated impedimetric immunosensor based on Ti3C2T MXene incorporated laser-burned graphene for noninvasive sweat cortisol detection. *Sens. Actuators B Chem.* **2021**, *329*, 129206. [CrossRef]
52. Ma, X.; Tu, X.; Gao, F.; Xie, Y.; Huang, X.; Fernandez, C.; Qu, F.; Liu, G.; Lu, L.; Yu, Y. Hierarchical porous MXene/amino carbon nanotubes-based molecular imprinting sensor for highly sensitive and selective sensing of fisetin. *Sens. Actuators B Chem.* **2020**, *309*, 127815. [CrossRef]
53. Mohammadniaei, M.; Koyappayil, A.; Sun, Y.; Min, J.; Lee, M.H. Gold nanoparticle/MXene for multiple and sensitive detection of oncomiRs based on synergetic signal amplification. *Biosens. Bioelectron.* **2020**, *159*, 112208. [CrossRef] [PubMed]
54. Liu, M.; Pan, B.; Tang, S.; Wang, W.; Hou, H.; Xie, B.; Liang, A.; Luo, A. A Label-Free Molecularly Imprinted Electrochemical Sensor Based on MXene Nanosheets Modified by Gold Nanoparticles for Sensitive and Selective Detection of Homocysteine. *J. Electrochem. Soc.* **2022**, *169*, 087503. [CrossRef]
55. Li, K.; Liang, M.; Wang, H.; Wang, X.; Huang, Y.; Coelho, J.; Pinilla, S.; Zhang, Y.; Qi, F.; Nicolosi, V.; et al. 3D MXene Architectures for Efficient Energy Storage and Conversion. *Adv. Funct. Mater.* **2020**, *30*, 2000842. [CrossRef]
56. Wen, X.-H.; Zhao, X.-F.; Peng, B.-F.; Yuan, K.-P.; Li, X.-X.; Zhu, L.-Y.; Lu, H.-L. Facile preparation of an electrochemical aptasensor based on Au NPs/graphene sponge for detection of homocysteine. *Appl. Surf. Sci.* **2021**, *556*, 149735. [CrossRef]
57. Idili, A.; Parolo, C.; Ortega, G.; Plaxco, K.W. Calibration-Free Measurement of Phenylalanine Levels in the Blood Using an Electrochemical Aptamer-Based Sensor Suitable for Point-of-Care Applications. *ACS Sens.* **2019**, *4*, 3227–3233. [CrossRef] [PubMed]
58. Sun, Z.; Jin, H.; Sun, Y.; Jiang, X.; Gui, R. Mn-Doping-induced hierarchical petal growth of a flower-like 3D MOF assembled with black phosphorous nanosheets as an electrochemical aptasensor of human stress-induced phosphoprotein 1. *Nanoscale* **2020**, *12*, 14538–14548. [CrossRef] [PubMed]
59. Zhang, L.; Li, Z.; Zhou, X.; Yang, G.; Yang, J.; Wang, H.; Wang, M.; Liang, C.; Wen, Y.; Lu, Y. Hybridization performance of DNA/mercaptohexanol mixed monolayers on electrodeposited nanoAu and rough Au surfaces. *J. Electroanal. Chem.* **2015**, *757*, 203–209. [CrossRef]
60. Lawrence, N.S.; Deo, R.P.; Wang, J. Detection of homocysteine at carbon nanotube paste electrodes. *Talanta* **2004**, *63*, 443–449. [CrossRef]
61. Beitollahi, H.; Zaimbashi, R.; Mahani, M.T.; Tajik, S. A label-free aptasensor for highly sensitive detection of homocysteine based on gold nanoparticles. *Bioelectrochemistry* **2020**, *134*, 107497. [CrossRef]
62. Neufeld, G.; Cohen, T.; Gengrinovitch, S.; Poltorak, Z. Vascular endothelial growth factor (VEGF) and its receptors. *FASEB J. Off. Publ. Fed. Am. Soc. Exp. Biol.* **1999**, *13*, 9–22. [CrossRef]
63. Sullivan, L.A.; Brekken, R.A. The VEGF family in cancer and antibody-based strategies for their inhibition. *MAbs* **2010**, *2*, 165–175. [CrossRef]
64. Aly, M.H.; Arafat, M.A.; Hussein, O.A.; Elsaid, H.H.; Abdel-Hammed, A.R. WITHDRAWN: Study of Angiopoietin-2 and vascular endothelial growth factor as markers of diabetic nephropathy onset in egyptians diabetic patients with non-albuminuric state. *Diabetes Metab. Syndr. Clin. Res. Rev.* **2019**, *13*, 1623–1627. [CrossRef] [PubMed]
65. Bozal-Palabiyik, B.; Lettieri, M.; Uslu, B.; Marrazza, G. Electrochemical Detection of Vascular Endothelial Growth Factor by Molecularly Imprinted Polymer. *Electroanalysis* **2019**, *31*, 1458–1464. [CrossRef]
66. Cheng, J.L.; Liu, X.P.; Chen, J.S.; Mao, C.J.; Jin, B.K. Highly sensitive electrochemiluminescence biosensor for VEGF(165) detection based on a g-C(3)N(4)/PDDA/CdSe nanocomposite. *Anal. Bioanal. Chem.* **2020**, *412*, 3073–3081. [CrossRef] [PubMed]
67. Harris, R.C. The Role of the Epidermal Growth Factor Receptor in Diabetic Kidney Disease. *Cells* **2022**, *11*, 3416. [CrossRef]
68. Li, N.; Nan, C.; Mei, X.; Sun, Y.; Feng, H.; Li, Y. Electrochemical sensor based on dual-template molecularly imprinted polymer and nanoporous gold leaf modified electrode for simultaneous determination of dopamine and uric acid. *Mikrochim. Acta.* **2020**, *187*, 496. [CrossRef]
69. Johari-Ahar, M.; Karami, P.; Ghanei, M.; Afkhami, A.; Bagheri, H. Development of a molecularly imprinted polymer tailored on disposable screen-printed electrodes for dual detection of EGFR and VEGF using nano-liposomal amplification strategy. *Biosens. Bioelectron.* **2018**, *107*, 26–33. [CrossRef]
70. Wang, J.-Y.; Xiao, L.; Wang, J.-Y. Posttranscriptional regulation of intestinal epithelial integrity by noncoding RNAs. *Wiley Interdiscip. Rev. RNA* **2017**, *8*, e1399. [CrossRef]
71. Wang, J.; Chen, J.; Sen, S. MicroRNA as Biomarkers and Diagnostics. *J. Cell Physiol.* **2016**, *231*, 25–30. [CrossRef]
72. Treiber, T.; Treiber, N.; Meister, G. Regulation of microRNA biogenesis and its crosstalk with other cellular pathways. *Nat. Rev. Mol. Cell Biol.* **2019**, *20*, 5–20. [CrossRef] [PubMed]
73. Trionfini, P.; Benigni, A.; Remuzzi, G. MicroRNAs in kidney physiology and disease. *Nat. Rev. Nephrol.* **2015**, *11*, 23–33. [CrossRef] [PubMed]

74. Parrizas, M.; Mundet, X.; Castano, C.; Canivell, S.; Cos, X.; Brugnara, L.; Giraldez-Garcia, C.; Regidor, E.; Mata-Cases, M.; Franch-Nadal, J.; et al. miR-10b and miR-223-3p in serum microvesicles signal progression from prediabetes to type 2 diabetes. *J. Endocrinol. Investig.* **2020**, *43*, 451–459. [CrossRef] [PubMed]
75. Liu, C.; Zhao, L.; Liang, D.; Zhang, X.; Song, W. An $CuInS_2$ photocathode for the sensitive photoelectrochemical determination of microRNA-21 based on DNA-protein interaction and exonuclease III assisted target recycling amplification. *Mikrochim. Acta.* **2019**, *186*, 692. [CrossRef]
76. Ma, Z.Y.; Ruan, Y.F.; Xu, F.; Zhao, W.W.; Xu, J.J.; Chen, H.Y. Protein Binding Bends the Gold Nanoparticle Capped DNA Sequence: Toward Novel Energy-Transfer-Based Photoelectrochemical Protein Detection. *Anal. Chem.* **2016**, *88*, 3864–3871. [CrossRef]
77. Takuma, N.; Mihoko, H.; Masafumi, K.; Seiki, I. Increased urinary excretions of immunoglobulin g, ceruloplasmin, and transferrin predict development of microalbuminuria in patients with type 2 diabetes. *Diabetes Care* **2006**, *29*, 142–144.
78. Narita, T.; Fujita, H.; Koshimura, J.; Meguro, H.; Kitazato, H.; Shimotomai, T.; Kagaya, E.; Suzuki, K.; Murata, M.; Usami, A.; et al. Glycemic control reverses increases in urinary excretions of immunoglobulin G and ceruloplasmin in type 2 diabetic patients with normoalbuminuria. *Horm. Metab. Res. Horm. Und Stoffwechselforschung Horm. Metab.* **2001**, *33*, 370–378. [CrossRef]
79. Yamazaki, M.; Ito, S.; Usami, A.; Tani, N.; Hanyu, O.; Nakagawa, O.; Nakamura, H.; Shibata, A. Urinary excretion rate of ceruloplasmin in non-insulin-dependent diabetic patients with different stages of nephropathy. *Eur. J. Endocrinol.* **1995**, *132*, 681–687. [CrossRef]
80. Haghshenas, E.; Madrakian, T.; Afkhami, A.; Saify Nabiabad, H. An electrochemical ceruloplasmin aptasensor using a glassy carbon electrode modified by diazonium-functionalized multiwalled carbon nanotubes. *J. Iran. Chem. Soc.* **2018**, *16*, 593–602. [CrossRef]
81. Ocana, C.; Hayat, A.; Mishra, R.K.; Vasilescu, A.; Del Valle, M.; Marty, J.L. Label free aptasensor for Lysozyme detection: A comparison of the analytical performance of two aptamers. *Bioelectrochemistry* **2015**, *105*, 72–77. [CrossRef]
82. Nancy, K.; Lipton, A. Platelets as a source of fibroblast growth-promoting activity. *Exp. Cell Res.* **1974**, *87*, 297–301.
83. Hongquan, Z.; Xing-Fang, L.; Chris, L.X. Differentiation and detection of PDGF isomers and their receptors by tunable aptamer capillary electrophoresis. *Anal. Chem.* **2009**, *81*, 7795–7800.
84. Pierce, G.F.; Tarpley, J.E.; Tseng, J.; Bready, J.; Chang, D.; Kenney, W.C.; Rudolph, R.; Robson, M.C.; Berg, J.V.; Reid, P. Detection of platelet-derived growth factor (PDGF)-AA in actively healing human wounds treated with recombinant PDGF-BB and absence of PDGF in chronic nonhealing wounds. *J. Clin. Investig.* **1995**, *96*, 1336–1350. [CrossRef] [PubMed]
85. Zhang, J.J.; Cao, J.T.; Shi, G.F.; Huang, K.J.; Liu, Y.M.; Ren, S.W. A luminol electrochemiluminescence aptasensor based on glucose oxidase modified gold nanoparticles for measurement of platelet-derived growth factor BB. *Talanta* **2015**, *132*, 65–71. [CrossRef] [PubMed]
86. Huang, K.J.; Liu, Y.J.; Zhang, J.Z.; Cao, J.T.; Liu, Y.M. Aptamer/Au nanoparticles/cobalt sulfide nanosheets biosensor for 17beta-estradiol detection using a guanine-rich complementary DNA sequence for signal amplification. *Biosens. Bioelectron.* **2015**, *67*, 184–191. [CrossRef]
87. Ge, Z.; Gu, H.; Li, Q.; Fan, C. Concept and Development of Framework Nucleic Acids. *J. Am. Chem. Soc.* **2018**, *140*, 17808–17819. [CrossRef]
88. Campuzano, S.; Yáñez-Sedeño, P.; Pingarrón, J.M. Tailoring Sensitivity in Electrochemical Nucleic Acid Hybridization Biosensing: Role of Surface Chemistry and Labeling Strategies. *ChemElectroChem* **2019**, *6*, 60–72. [CrossRef]
89. Ge, Z.; Fu, J.; Liu, M.; Jiang, S.; Andreoni, A.; Zuo, X.; Liu, Y.; Yan, H.; Fan, C. Constructing Submonolayer DNA Origami Scaffold on Gold Electrode for Wiring of Redox Enzymatic Cascade Pathways. *ACS Appl. Mater. Interfaces* **2019**, *11*, 13881–13887. [CrossRef]
90. Lu, N.; Pei, H.; Ge, Z.; Simmons, C.R.; Yan, H.; Fan, C. Charge transport within a three-dimensional DNA nanostructure framework. *J. Am. Chem. Soc.* **2012**, *134*, 13148–13151. [CrossRef]
91. Chen, X.; Zhou, G.; Song, P.; Wang, J.; Gao, J.; Lu, J.; Fan, C.; Zuo, X. Ultrasensitive electrochemical detection of prostate-specific antigen by using antibodies anchored on a DNA nanostructural scaffold. *Anal. Chem.* **2014**, *86*, 7337–7342. [CrossRef] [PubMed]
92. Huang, K.J.; Liu, Y.J.; Zhai, Q.F. Ultrasensitive biosensing platform based on layered vanadium disulfide-graphene composites coupling with tetrahedron-structured DNA probes and exonuclease III assisted signal amplification. *J. Mater. Chem. B* **2015**, *3*, 8180–8187. [CrossRef] [PubMed]
93. Ge, Z.; Lin, M.; Wang, P.; Pei, H.; Yan, J.; Shi, J.; Huang, Q.; He, D.; Fan, C.; Zuo, X. Hybridization Chain Reaction Amplification of MicroRNA Detection with a Tetrahedral DNA Nanostructure-Based Electrochemical Biosensor. *Anal. Chem.* **2014**, *86*, 2124–2130. [CrossRef] [PubMed]
94. Abi, A.; Lin, M.; Pei, H.; Fan, C.; Ferapontova, E.E.; Zuo, X. Electrochemical switching with 3D DNA tetrahedral nanostructures self-assembled at gold electrodes. *ACS Appl. Mater. Interfaces* **2014**, *6*, 8928–8931. [CrossRef] [PubMed]

chemosensors

MDPI

Review

Electrochemiluminescence Detection and Imaging of Biomolecules at the Single-Cell Level

Xiaofan He [1], Yufei Deng [1], Dechen Jiang [2] and Danjun Fang [1,*]

[1] School of Pharmacy, Nanjing Medical University, Nanjing 211126, China; hxf@stu.njmu.edu.cn (X.H.); dyf@stu.njmu.edu.cn (Y.D.)

[2] State Key Laboratory of Analytical Chemistry for Life Science, School of Chemistry and Chemical Engineering, Nanjing University, Nanjing 210023, China; dechenjiang@nju.edu.cn

* Correspondence: djf@njmu.edu.cn

Abstract: Electrochemiluminescence (ECL) is an electrochemically induced light produced by the excitation of luminophores in redox reactions. For the past twenty years, ECL analysis has been continuously developed and applied for the sensitive detection of biomolecules at the single-cell level due to its low background interference and the resultant high sensitivity. In recent times, ECL-based microscopy has combined the elements of imaging and has thus emerged as a fast-developed imaging tool to visualize biomolecules in single cells. The surface-confined features of ECL imaging provide detailed information about cell membranes that is not easily obtained using classical fluorescence microscopy. In this review, we summarize the recent works on the detection and imaging of biomolecules at the single-cell level using ECL and discuss the development prospects and challenges in the biological application of this technology in the field of cell analysis.

Keywords: electrochemiluminescence; detection and imaging; single-cell analysis; biomolecules; high sensitivity

Citation: He, X.; Deng, Y.; Jiang, D.; Fang, D. Electrochemiluminescence Detection and Imaging of Biomolecules at the Single-Cell Level. *Chemosensors* **2023**, *11*, 538. https://doi.org/10.3390/chemosensors11100538

Academic Editor: Ilaria Rea

Received: 24 July 2023
Revised: 16 September 2023
Accepted: 26 September 2023
Published: 12 October 2023

1. Introduction

Electrochemiluminescence (ECL) is a kind of light emission triggered by electrochemical stimulation. Under the action of voltage, the luminophore near the electrode surface undergoes an electrochemical oxidation reaction and forms the corresponding reaction intermediates. After a subsequent chemical reaction, the intermediates are transformed into the luminescent excited state, which emits luminescence while returning to the ground state [1–4]. This ECL phenomenon was first observed in Grignard compounds in a high-electric-field experiment conducted by Dufford in 1927 [5]. The next year, Harvey et al. reported the chemiluminescence of luminol under electrolytic conditions [6]. Although much evidence on the ECL phenomenon was revealed in the following years, the first detailed report on ECL was presented by Hercules and Bard in the mid-1960s, which greatly advanced the development of ECL-related studies [7,8]. In the following 50 years, extensive research established ECL-based technology as an effective analytical method, whose applications range from immunoassays, environmental detection, and bioanalysis to commercial clinical diagnosis [9–17].

As a delicate combination of optical and electrochemical methods, ECL inputs the electrical signals into an optical reader, avoiding the interference of charging and discharging currents that occurred in the traditional electrochemical measurement. More importantly, ECL completes the measurement without the need for excitation light, and hence, the influence of background light during the measurement is removed. Therefore, compared with fluorescence and other optical methods, ECL has the advantage of a near-zero background that significantly improves the detection sensitivity [1,18–23]. In addition, the intermediates involved in the ECL high-energy electron transfer reaction are controlled by electrical signals and confined near to the electrode surface. The surface-confined feature allows

for the precise modulation of the ECL duration and location with high spatiotemporal resolution. The integration of high-resolution imaging components enables ECL imaging technology to visualize the spatial distribution and quantitative information of molecules. Thus, ECL-based analysis exhibits significant characteristics, such as high sensitivity, strong versatility, and low detection cost, and becomes a dominant method in clinical tests.

In addition to ECL measurements of molecules in biofluids (e.g., serum and urine) and environmental samples, the application of ECL analysis has been recently expanded to cellular analysis, even at the single-cell level. The cell is the most basic unit of life and is the place where many biochemical reactions and physiological activities take place. Evidence has revealed that under the same physiological conditions, the same type of cells exhibit significant heterogeneity. For example, when cells are exposed to external stimuli or are in a state of disease, distinct differences in the expression of biomolecules within cells are observed, contributing to different cell fates [24–27]. Traditional analytical methods could only provide limited data about the cellular population, which masks these cellular heterogeneities, and thus, much important biological information is lost. Considering the low abundance of biomolecules in one cell, development of an accurate and highly sensitive detection tool to measure biomolecules at the single-cell level is urgently needed. The realization of single-cell analysis is not only a necessary means to prove cell heterogeneity, but it is also an effective way to extensively study the relationship among various molecules and their related signaling pathways in individual cells [28–33].

In view of the shortcomings of the current methods for single-cell analysis, the application of ECL-based analysis with low background, high sensitivity, and strong versatility is particularly promising (Table 1). Over the past 20 years, researchers have successfully established various ECL assays to measure and image many important biomolecules inside single living cells and at the sub-cellular structures. Although many reviews have summarized the studies on single-cell ECL imaging [34–42], most of them only list the advantages of ECL imaging application as a whole, and seldom illustrate the entire procedure, ranging from the measurements to the imaging processes. Also, the most recent breakthrough in single-molecule ECL imaging in single living cells is not included in these reviews, which is now considered as the most important direction in this field. In this review, we comprehensively introduce the development of single-cell ECL measurement and imaging, and emphasize the recent important progress in this field. Finally, the existing challenges and future prospects of ECL imaging in single-cell analysis ae discussed.

Table 1. The advantages and disadvantages of single-cell ECL and fluorescence analysis.

Method	Advantages	Disadvantages
ECL	High sensitivity Low background emission Restricted information on the cell-based membrane	Restricted on the electrode surface Weak luminescence intensity Low spatial–temporal resolution
Fluorescence	Strong fluorescence emission High spatial–temporal resolution Excellent biocompatibility	High background noise Fluorescence quenching

2. ECL Measurement of Biomolecules at the Single-Cell Level

The first single-cell ECL measurement was reported by our group in 2013, which was used to measure active cholesterol in the plasma membrane [43]. The ECL detection strategy relies on the classic luminol-hydrogen peroxide ECL system, including the reaction between active cholesterol in the cellular membrane and cholesterol oxidase in the solution to generate hydrogen peroxide, and the following electrochemical reaction between luminol in the solution and hydrogen peroxide to emit ECL (Figure 1a). To realize single-cell analysis via ECL measurement, one pinhole (i.d. 100 μm) was added between an indium tin oxide (ITO) slide and a photomultiplier tube (PMT). The cells were then cultured on the ITO slide with a low density, and so only one cell was exposed to PMT. The increase in ECL intensity at a single-cell level was observed for the first time upon the addition of cholesterol oxidase, demonstrating the feasibility of detecting active membrane cholesterol in single living cells using ECL. Following the activation of inactive cholesterol at the plasma membrane, the ratio of the active and inactive membrane cholesterol at the single cell was obtained using the ECL measurement.

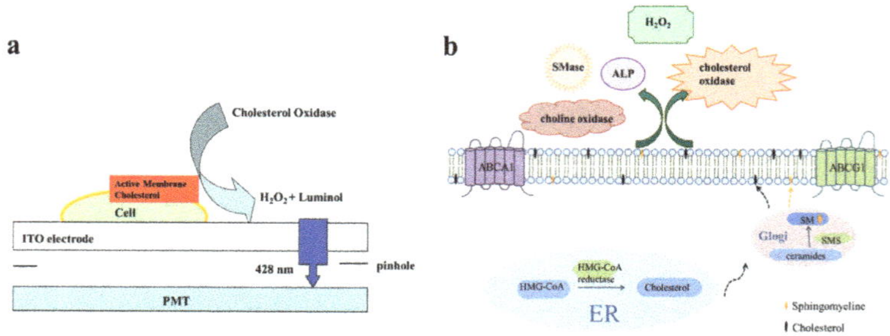

Figure 1. ECL measurement of biomolecules at the single-cell level. (**a**) ECL analysis of active cholesterol at the single-cell level [43]. Copyright 2013 Anal Chem. (**b**) Schematic for co-detection of cholesterol and sphingomyelin in the plasma membrane at the single-cell level [44]. Copyright 2019 Anal Chem.

Considering that the reaction of cholesterol oxidase with membrane cholesterol does not cause the destruction of the cell membrane structure, this strategy can be further applied to analyze other biomolecules in the plasma membrane. For example, co-quantifying membrane cholesterol and lipids at the single cell level is important for understanding lipid interactions. However, due to the difficulty in synthesizing highly fluorescent probes that specifically bind to the desired lipid, this study cannot be easily addressed using fluorescence imaging. In 2019, our group introduced an enzyme cocktail consisting of sphingomyelinase, alkaline phosphatase, and choline oxidase after the ECL measurement of membrane cholesterol. This enzyme cocktail can react with membrane sphingomyelin (SM) to produce hydrogen peroxide for ECL detection (Figure 1b) [44]. The strategy allows for the simultaneous determination of membrane cholesterol and SM in a single cell. This strategy was used to gather fluctuation information about SM and cholesterol during SM-depletion-induced cholesterol efflux. The data obtained from this strategy could help resolve the debate surrounding the ABC transporter's role in cholesterol efflux.

Since the concentration of most biomolecules in a single cell is low, amplifying the ECL signal to increase detection sensitivity is critical for the further application of this method in single-cell analysis. To achieve this, Yang et al. assembled solid zinc co-adsorbed carbon quantum dot (ZnCQDs) nanocomposites using gold nanoparticles and carboxyl functionalized magnetic beads as carriers. The introduction of gold nanoparticles provides a general strategy to enhance the electron-transfer process during the ECL reaction, resulting in a 120-fold increase in ECL compared to that of CQDs [45]. Then, the hyaluronic acid (HA) was functionalized on solid-state probes to enable the specific capture of single cells through the recognition of HA with CD44 on the cell surface. Additionally, the usage of Cu^{2+} ions as a bridge to connect gold nanoparticles and lead-adsorbed carbon quantum dots (PbCQDs) was reported to increase the ECL strength by $37.5 \pm 3.9\%$ [46]. Continuously incorporating new voltage modes and synthesizing highly luminescent ECL probes should be an effective strategy to increase the intensity and stability of ECL. This strategy is being pursued by many research groups [47–59]. These advancements will further advance single-cell ECL measurement in biological studies and potentially even in clinical tests in the near future.

3. ECL Imaging of Biomolecules at the Single-Cell Level

Intensity-based ECL measurement can provide information about single cells with a low detection limit. However, this measurement method does not provide spatial information within a single cell. Therefore, in order to achieve spatial studies of single cells for biological applications, the development of ECL imaging methods specifically designed for single cells is necessary [60,61]. At the end of the last century, researchers began developing ECL imaging techniques, which have been continuously improved over time [62–67]. In 2011, Su et al. made improvements to the ECL imaging setup and designed both negative and positive operation methods to selectively control ECL generation on the electrode surface space [68]. The power of ECL imaging in spatially studying fine features on electrodes was demonstrated by providing high-quality ECL imaging of fingerprints, which was previously unprecedented. In the following years, many research groups have continued to improve the setup of ECL microscopy and the voltage mode to enhance the generation and collection efficiency of ECL emission. Currently, widely used ECL microscopy typically consists of an electrochemical cell, a voltage generator, and a microscope equipped with an electron-multiplying CCD (EMCCD). During the ECL imaging process, the external light source is turned off, and a corresponding voltage is applied at the working electrode to initiate the ECL reaction. The ECL imaging of small molecules, membrane proteins, and intracellular molecules at single cells is summarized and listed in Table 2.

3.1. Imaging of Small Molecules at the Single-Cell Level

Small molecules play important roles in regulating cellular activities and imaging them can provide valuable information about cellular processes [69]. However, imaging small molecules inside cells is challenging due to the lack of specific probes for binding these molecules. A potential strategy to visualize small molecules at the single-cell level is the utilization of a luminol-based ECL assay. This assay relies on the reaction of oxidase with the corresponding small molecule to generate hydrogen peroxide, which then reacts with luminol to emit ECL. In 2015, our team reported the first ECL image of hydrogen peroxide and cholesterol at single living cells. To overcome the information of an inert layer on the electrode during the continuous application of a high positive potential, a voltage switching strategy between 1.0 and -1.0 V was employed. This strategy switching generated more ECL emission and improved the imaging process. In the presence of hydrogen peroxide released or generated from the cellular membrane, it electrochemically reacts with a small amount of L012 (a luminol analog with higher luminescence) beneath the cell to generate ECL. After subtracting the background intensity that was observed at the cell, a higher ECL intensity was observed at the cell, with a spatial resolution of 0.8 μm, which revealed the distribution of hydrogen peroxide efflux and active membrane cholesterol (Figure 2a) [70]. The results demonstrate that ECL can be a new tool for single-cell electrochemical imaging, offering advantages such as high spatiotemporal resolution (compared to scanning electrochemical microscopy), low background, and flexibility in electrode type (compared to surface-plasmon-resonance-based electrochemical imaging techniques).

Despite the advances in ECL imaging, this new strategy still faces several challenges. These challenges include improving the spatial resolution, directly visualizing small molecules, reducing the applied voltage, and achieving stable ECL intensity. Regarding spatial resolution, the initial approach using aqueous luminol to react with hydrogen peroxide leads to the diffusion of these species in the solution, resulting in poor spatial resolution. To improve the spatial resolution, Su's group prepared an ECL nanocage array (ENA) consisting of vertically aligned silica nanoporous films (SNMs). This nanocage array only allowed small-sized coreactants or substances to enter, thereby generating or affecting ECL signals while minimizing lateral diffusion of ECL under the cell. As a result, vertically aligned microtube electrode ensembles were used to achieve highly spatial single-cell ECL imaging, revealing the subcellular heterogeneity of hydrogen peroxide efflux from individual cells [71,72].

Another challenge in the initial ECL imaging strategy is that the ECL emission originates from the reaction between L012 and hydrogen peroxide beneath the cell. The close adherence of the cell to the surface support results in a small amount of L012 under the cell, which generates less ECL intensity compared to the surrounding region with a bare electrode surface. As a result, the cell appears black throughout the imaging process, which does not allow for the direct imaging of small molecules within the cell. To solve this problem, Zhu's group added a porous chitosan layer on the electrode to culture and image the cell [73]. Sufficient ECL reagent is retained under the cell, providing strong ECL emission in the image (Figure 2b). The direct visualization of hydrogen peroxide or some other small molecules facilitates the investigation of their distributions in single living cells.

During cellular analysis, it is important to minimize the stimulation of the cell by applying the lowest possible voltage. However, the voltage required to initiate the ECL process between L012 and hydrogen peroxide is set at 1.0 V, which is too high. To minimize this interruption, single lithium iron phosphate ($LiFePO_4$, LFP) nanoparticles were modified on the electrode. Under a certain voltage, lithium ions from single LFP particles could be de-intercalated and then inserted into ITO to form $LiSnO_2$. This process promotes the electrochemical oxidation of L012, allowing for visualization of the efflux of hydrogen peroxide from single living cells at a voltage as low as 0.5 V [74]. Decreasing the applied voltage in single-cell ECL imaging maximizes cellular activity and provides information that closely resembles the real cellular situation.

Meanwhile, small molecules move quickly within cells to regulate cellular activity. To achieve dynamic observation of small molecules in living cells, it is important to obtain stable ECL emission over a certain time period. To achieve this, Zhu and our groups chose single semiconductor titanium dioxide (TiO_2) nanoparticles (Figure 2c). These nanoparticles have oxygen vacancies on their surface, which have a high affinity for hydrogen peroxides and resist passivation under voltage [75]. Therefore, the continuous electrochemical generation of superoxide and hydroxyl radicals by electrons and surface-trapped holes at the nanoparticles was achieved, resulting in constant ECL under physiological conditions for 90 min. This steady-state luminescence allows for electrochemical visualization of the efflux burst of hydrogen peroxide from single cells, which is not easily achieved with other electrochemical microscopies. The dynamic changes of small molecules are closely related to cellular life activities. Hiramoto's group used the strategy of sequential potential to achieve ECL imaging of cell respiration [76]. The oxygen consumption of cells was dynamically visualized by inhibiting ECL through oxygen consumption in cell respiration, confirming the respiratory activities of cells. The result can distinguish the differences in respiratory activities of different globular cells and enable the visualization of dynamic changes in cell respiratory activities.

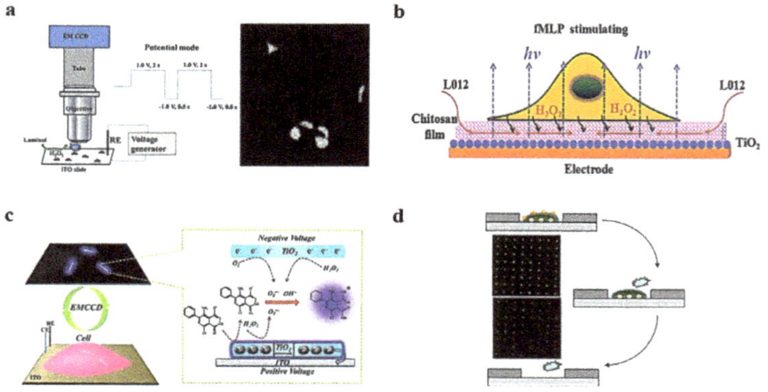

Figure 2. Single-cell small-molecule ECL imaging. (**a**) ECL image of the distribution of hydrogen peroxide efflux and active membrane cholesterol in a single living cell [70]. Copyright 2015 Anal Chem. (**b**) Direct ECL imaging of single cells on a chitosan-modified electrode [73]. Copyright 2018 Anal Chem. (**c**) ECL imaging of local hydrogen peroxide efflux from single cells using TiO_2 nanoparticles (* represents the excited state product) [75]. Copyright 2019 Anal Chem. (**d**) C_3N_4-nanosheet-modified microwell array for single-cell cholesterol ECL imaging analysis [77]. Copyright 2017 Anal Chem.

In addition to spatial imaging of small molecules in single living cells, high-throughput single-cell analysis is also an important direction in this field, as it provides statistical information to uncover cellular processes. Microwell array electrodes, which offer integration, miniaturization, and high analysis throughput, have been developed and applied in this study. On the ITO electrode, 30 μm sized microwells were fabricated to isolate single cells in individual microwells. By employing an oxidase-based assay strategy to convert the corresponding small molecule into hydrogen peroxide, the unique structure of the microwell slows down the diffusion of hydrogen peroxide away from the microwell. Consequently, the L012-hydrogen peroxide ECL reaction was performed to obtain quantitative information about small molecules in single cells. By including multiple microwell-cells in the microscopic field, small molecules from these 64 cells were analyzed using a single ECL image, significantly enhancing the analysis throughput [77] (Figure 2d). To increase the loading and measurement efficiency of single cells into the microwells, Ju et al. combined the advantages of microfluidic technology with ECL imaging to develop a quantitative detection ECL imaging microarray chip. In this method, a single or several cells were captured, and the released secretions were confined to the individual microwell to identify the corresponding ECL probe. As a result, the amount of dopamine release in different microwells was evaluated, providing a general strategy for the quantitative detection of single cell secretions [78].

3.2. *ECL Imaging of Membrane Proteins*

Membrane proteins are important components of the cellular plasma membrane and play a fundamental role in the functioning of biological cells. They are also targets for antibodies in immunoassays and immunohistochemistry [79]. Currently, the imaging of membrane proteins often involves labeling them with fluorescent tags attached on antibodies that specifically recognize the proteins [80–82]. However, this approach can be limited by autofluorescence and high-power laser irradiation, which generate high optical background signals that can affect imaging accuracy. To address this issue, the use of ECL technology, which offers a zero-background feature, has been explored as an alternative method for visualizing membrane proteins at the single-cell level with high sensitivity [83–86]. In this approach, the antibody is linked to an ECL tag, such as $Ru(bpy)_3{}^{2+}$, to enable ECL imaging. The ECL mechanism of the $Ru(bpy)_3{}^{2+}$/TPrA system is illustrated as follows:

$$\mathrm{TPrA} - e^- \rightarrow \mathrm{TPrA}^{\cdot +} \quad (1)$$

$$\mathrm{TPrA}^{\cdot +} \rightleftharpoons \mathrm{TPrA}^{\cdot} + \mathrm{H}^+ \quad (2)$$

$$\mathrm{TPrA}^{\cdot} + [\mathrm{Ru(bpy)_3}]^{2+} \rightarrow \mathrm{P1} + [\mathrm{Ru(bpy)_3}]^{+} \quad (3)$$

$$\mathrm{TPrA}^{\cdot +} + [\mathrm{Ru(bpy)_3}]^{+} \rightarrow \mathrm{TPrA} + [\mathrm{Ru(bpy)_3}]^{2+*} \quad (4)$$

$$[\mathrm{Ru(bpy)_3}]^{2+*} \rightarrow [\mathrm{Ru(bpy)_3}]^{2+} + h\nu \quad (5)$$

In 2017, Sojic's group reported the development of an ECL-based surface-restricted microscope for spatially resolved single-cell ECL imaging [87]. They utilized disposable inkjet-printed carbon nanotube electrodes to culture individual cells and labeled membrane proteins with $Ru(bpy)_3{}^{2+}$ through a biotin-streptavidin linkage. To facilitate sufficient redox reactions for the visual ECL signal, TPrA was introduced into the medium as an oxidative-reductive coreactant. The imaging process is believed to involve a heterogeneous ECL mechanism, where only TPrA undergoes direct oxidation on the electrode surface, followed by deprotonation to form the TPrA radical. This radical species plays a vital role in the generation of excited state $Ru(bpy)_3{}^{2+*}$, ultimately leading to the in situ generation of protons on the membrane protein (Figure 3a). Later, Sojic's group conducted further research on the vertical resolution of single-cell ECL imaging in both reflection and transmission configurations, which expanded the applications of ECL as a surface-confined microscope [88]. The results demonstrated that the ECL process involves short-lived electrogenerated radicals, leading to the emission confined to the immediate proximity (~500 nm) of the electrode surface. As a result, only the basal membrane of the cell exhibits luminescent characteristics in the ECL image. Such fine details observed in the ECL image cannot be resolved using conventional fluorescence microscopy techniques.

Surface-limited ECL microscopy offers the advantage of providing detailed information about cell membranes and the spatial extension of cells, surpassing the capabilities of conventional wide-field fluorescence microscopy. However, the ECL emission from the luminophore is typically weak, and the continuous imaging process can result in a significant loss of ECL intensity. These limitations make it challenging to visualize single biomolecules, particularly in live cells. To address this issue, the development of novel ECL emitters with high ECL intensity is critical. In 2021, Liu, Sojic, and our groups utilized $Ru(bpy)_3^{2+}$-doped silica/gold nanoparticles (RuDSNs/AuNPs) as both the protein tag and ECL nanoemitter [89]. The local surface confinement effect leads to a notable enhancement in ECL at the nanoemitter, facilitating the visualization of individual proteins on the electrode surface and cell membrane (Figure 3b). To gain a deeper understanding of nanoconfinement and mass transport in single-molecule ECL imaging, Liu's group developed a nanoreactor using $Ru(bpy)_3^{2+}$-doped nanoporous zeolite nanoparticles. This allowed for direct visualization of in situ-nanoconfinement-enhanced electrochemical reactions at the single molecule level. This research helped establish the relationship between the design of porous electrochemical materials and their nanoconfinement performance in single cell ECL imaging [90]. Additionally, by leveraging the nanoconfinement effect and high reaction activity within metal–organic frameworks (MOFs), $Ru(bpy)_3^{2+}$-embedded MOF complex (RuMOFs) were designed. These complexes exhibited a bright and consistent ECL emission for up to 1 h, enabling the in operando visualization of protein movements at the cellular membrane [91]. Compared to fluorescence observation, the near-zero ECL background surrounding the target protein with the ECL emitter provided better contrast for dynamic imaging of discrete protein distribution and movement. This advancement has the potential to greatly advance the field of single-cell protein imaging.

In the above work, TPrA acts as the coreactant that is toxic for the cells. To achieve a more biocompatible single-cell ECL co-reactant, Liu's group developed a closed bipolar-electrode-based simple electrochemical luminescence (BPEs-ECL) imaging strategy. They used heterogeneous $Ru(bpy)_3^{2+}@SiO_2$/Au nanoparticles as functional nanoprobes, and successfully achieved the first wireless, sensitive visual immunoassay of single-cell prostate-specific antigen (PSA) through multiple amplification strategies [92]. In recent years, Zhu's group prepared a guanine-rich single-stranded DNA (G-ssDNA)-loaded high-index multi-planar gold nanoflower (Hi-AuNF) as the co-reactant of $Ru(bpy)_3^{2+}$. G-ssDNA, serving as a highly efficient and biocompatible coreactant, could accelerate the electrochemical catalytic pathway from $Ru(bpy)_3^{3+}$ to $Ru(bpy)_3^{2+*}$, and stabilize Hi-AuNF with a relatively sturdy Au-S bond. As a result, this assembly increases the loading capacity of G-ssDNA and promotes the enhancement of the ECL signal of $Ru(bpy)_3^{2+}$ for better ECL visualization of carcinoembryonic antigen (CEA) at cell membranes [93]. To further reduce cytotoxicity and improve ECL imaging intensity, Ju's group designed tertiary amine-conjugated polymer dots (TEA-Pdots) as a novel coreactant-embedded ECL tag at the protein. This design utilizes the superstructure and intramolecular electron transfer inside the probe to emit strong ECL without the coreactant, enabling the in situ imaging of membrane proteins at single cells without additional permeable treatment for transporting the coreactant [94]. This further development of the highly luminescent probe with the biocompatible coreactant or without the coreactant will be an important direction to better fit single-cell analysis in biological studies.

In addition to the labeling strategy mentioned earlier, our group has reported a capacitance imaging method based on ECL to visualize proteins on the cell membrane without the need for tagging the ECL probe to the antibodies [95]. This method relies on the reduction in local capacitance when the antibody binds to the membrane protein, leading to an increase in double-layer potential at the binding site. As the ECL intensity is strongly influenced by the double-layer potential, the binding of the protein–antibody complex enhances the ECL signal, enabling the visualization of the location of the target membrane protein (Figure 3c). This strategy utilizes the ECL reaction from the reagents in the solution and offers the advantages of label-free detection, simple setup, and high sensitivity. Additionally, Sojic's group has developed a label-free shadow ECL (SECL) microscopy technique that confines the ECL emission layer spatially, allowing for the imaging of subcellular organelles and single cells on the electrode surface. By exploiting the local hindrance of ECL reagents by each mitochondrion, SECL enables the detection of mitochondria that are challenging to detect using traditional biomarkers [96]. Moreover, SECL achieves sharp negative contrast imaging at concentrations orders of magnitude lower than conventional methods, offering enhanced sensitivity and the potential for the imaging at the single molecule level. Currently, a bimodal and bicolor approach has been introduced in ECL microscopy, enabling the simultaneous recording of positive ECL (PECL) and SECL images of single cells [97] (Figure 3d). These advancements suggest the feasibility of development high-sensitivity, label-free ECL imaging at the single-cell level, which represents an important direction in this study.

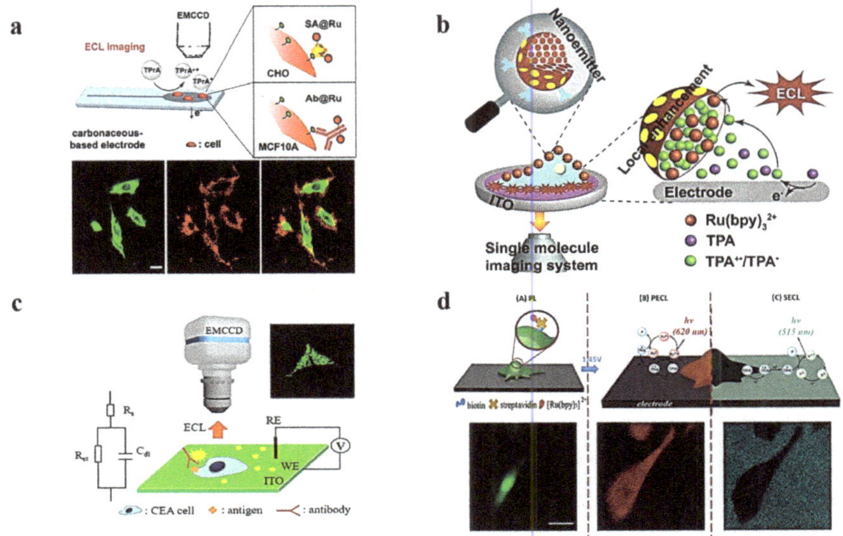

Figure 3. ECL imaging of membrane proteins. (**a**) Surface-confined microscopy based on electrochemical luminescence (ECL), imaging the ECL of the plasma membrane of a single cell at the interface of an electrode [87]. Copyright 2017 J Am Chem Soc. (**b**) Schematic diagram of $Ru(bpy)_3^{2+}$-doped silica/gold nanoparticles (RuDSNs/AuNPs) used as ECL nanoemitters to image individual proteins on the electrode surface and cell membrane [89]. Copyright 2021 J Am Chem Soc. (**c**) Schematic diagram of capacitance microscope based on ECL [95]. Copyright 2019 J Am Chem Soc. (**d**) Multimodal imaging of cells fixed on a glassy carbon electrode (GCE) and imaging of the same CHO-K1 cell by PL, PECL, and SECL [97]. Copyright 2023 Anal Chem.

Cell–matrix adhesions play essential roles in the binding interactions between cells and the extracellular matrix, as well as in various biological processes. Su's group utilized ECL microscopy for label-free imaging of cell–matrix adhesions and further investigated the migration tendencies of cells during collective migration. Their strategy involved using a modified indium tin oxide electrode as the substrate, with a silica nanochannel membrane that enhanced the ECL emission from freely diffusing luminophores. This enhancement provided a distinct visual contrast and non-contacted domains, enabling label-free imaging of cell–matrix adhesions [98]. They further modulated the ECL reaction pathway by adjusting the concentrations of luminophores and/or co-reactants to sequentially image cell–matrix and cell–cell connections at the bottom and near the top surface of the cells. This approach provided an opportunity for developing spatially selective micro-imaging strategies [99]. Recently, they successfully imaged cell–matrix adhesions and collective cell migration by introducing microgrooved electrodes and surface-sensitive ECL microscopy in the system (electrochemiluminescence imaging of cellular contact guidance on microfabricated substrates). Additionally, Xia's group proposed a label-free, rapid method for imaging single-cell electrochemiluminescence using a bipolar nanoelectrode array. By utilizing a closed bipolar electrochemical system with a single-sided platinum-coated gold nanoelectrode array, they employed the dynamic concentration gradient of oxygen as an electroactive probe. This approach achieved label-free, rapid, and positive ECL imaging of single cells, allowing visualization of the local adhesion strength of individual cells [100].

Zhu's group have developed a new dual ECL signaling system for evaluating cell surface receptors and analyzing disease cells at the single-cell level. In this strategy, two ECL probes, Au@L012 and g-C_3N_4, were used to specifically bind to epidermal growth factor (EGFR) and phosphatidylserine (PS) on the cell surface. By scanning the potential, two distinct ECL signals were observed, and the apoptosis rate of normal and cancer cell samples was visualized using ECL microscopy [101]. To enhance the ECL signal future, $Ru(bpy)_3^{2+}$ was used as the luminophore and nitrogen-doped carbon dots (NCDs) were used as the nano-coreactant. This created a catalytic route for the ECL microscopy cell imaging strategy. The catalytic route between $Ru(bpy)_3^{2+}$ and NCDs extended the ECL emitting layer, significantly improving the ECL signal intensity. This enhancement allowed for the vertical resolution imaging of the cell membrane and the imaging of phosphatidylserine (PS) on apoptotic cell membranes. These advancements expand the application of ECL in biological analysis [102].

An interesting phenomenon that must be mentioned in single-cell ECL imaging is the observation of the ECL photobleaching by Sojic's group. Due to the weak luminescence from the ECL probe, it is not easy to observe ECL photobleaching. However, after the photobleaching through sequential stepwise illumination, a linear correlation between ECL reduction and PL loss was recorded in $Ru(bpy)_3^{2+}$ derivatives labeled on the cellular membrane. These losses should be attributed to the photobleaching of the labels, which provide valuable information on the fundamentals of the ECL excited state. In the future, combining ECL microscopy with photobleaching techniques will provide new methods similar to fluorescence recovery after photobleaching (FRAP) or fluorescence loss in photobleaching (FLIP) methods, which can provide kinetic information about protein movement at the cellular membrane [103].

3.3. *ECL Imaging of Intracellular Molecules*

In the past ten years, most of the focus in single-cell ECL imaging has been on the visualization of biomolecules at the cellular membrane, as the ECL reaction can occur at the electrode surface. However, there has been a recent breakthrough in nanoelectrochemistry that allows for intracellular biomolecule analysis, providing more information about the cellular activity. Therefore, many groups are devoted to addressing the challenges in intracellular ECL imaging. In our early work, we fabricated a luminol-gold composite microelectrode that was inserted into the cell to react with intracellular hydrogen peroxide for visualization [104]. Later on, we decorated a Pt deposit on the inside walls of a nanopipette tip to create an open bipolar ECL device. Once positioned inside the cell, the porous structure of the Pt deposit allows for electrochemical loading of intracellular molecules. Coupled with enzymatic reactions and subsequent ECL reaction, we can visualize intracellular concentrations of hydrogen peroxide or glucose, as well as the intracellular sphingomyelinase activity [105].

Table 2. Electrochemiluminescence detection and imaging of biomolecules at single cells.

Single-Cell ECL Method	Category	Luminophore	Cell Type	Target	References
ECL Measurement	small molecules	Luminol	Raw264.7	Cholesterol	[43]
	macromolecule	ZnCQDs	MCF-7, MDA-MB-231	CD_{44}	[45]
		Au@Cu-PbCQD	MCF-7	CD_{44}	[46]
ECL Imaging	small molecules	L-012	Hela	Cholesterol	[70,77]
		$(Ru(bpy)_3{}^{2+}$	PC12	Dopamine	[71,72]
		Pdots	PC12	Dopamine	[78]
		L-012	MCF-7, Hela	H_2O_2	[73–75]
		L-012	HMSC-BM, MCF-7	O_2	[76]
	membrane proteins	$(Ru(bpy)_3{}^{2+}$	CHO-K1	EGFR	[83]
		$(Ru(bpy)_3{}^{2+}$	L-02, MCF-7	CK19	[85]
		$(Ru(bpy)_3{}^{2+}$	MCF-7	EpCAM	[86]
		$(Ru(bpy)_3{}^{2+}$	Hela, Ramos, CCRF-CEM	PTK7	[87]
		$(Ru(bpy)_3{}^{2+}$	MCF-7	CEA	[89]
		L-012	MCF-7	CEA	[91]
		TEA-Pdots	SK-BR-3	HER2	[90]
		L-012	MCF-7	EGFR, PS	[97]
		$(Ru(bpy)_3{}^{2+}$	apoptotic	PS	[98]
	intracellular molecules	Luminol	Hela	H_2O_2	[100]
		L-012	Hela	Glucose, H_2O_2	[101]
		Luminol	Hela	miRNA-21	[102]
		L-012	MCF-7	KDM1/LSD1	[104]
		L-012	Hela	Glucose	[105]

Additionally, Zhang's group established a new ECL imaging technique for the parallel imaging of miRNA-21 in single cancer cells using a Phorbol-12-myristate-13-acetate (PMA)-loaded gold nanocage as a probe. In the cell, the PMA-supported gold nanocages (Au NCs) are sealed by DNA gates and recognized and opened by miRNA-21. PMA is then released, and HeLa cells are further induced to produce reactive oxygen species (ROS), enabling ECL imaging of miRNA-21 within a single HeLa cell [106]. Although this work can visualize the biomolecules inside the cell, the nanoelectrode-based measurement can only provide a visualized image of local information. Imaging the spatial distribution of biomolecules inside the whole cell remains a goal in this field.

In 2021, Zhu's group reported a bio-coreactant-enhanced ECL microscopy to realize the ECL imaging of intracellular structures and dynamic transport [107]. The key design in this strategy is to utilize the electrochemical oxidation of $Ru(bpy)_3{}^{2+}$ into $Ru(bpy)_3{}^{3+}$ at the electrode surface, which enters into the cell to react with intracellular biomolecules with reductive amine moieties, resulting in ECL emission. Consequently, intracellular structures and autophagy involving DNA oxidative damage are imaged using the ECL image for the first time. This strategy effectively employs $Ru(bpy)_3{}^{3+}$ as the molecular antenna, connecting extracellular and intracellular environments and providing a solution for single-cell ECL imaging of intracellular structures and molecules. To further develop intracellular ECL imaging, our group immobilized the cell into agarose hydrogel and loaded antibody-modified single-walled carbon nanotubes into the cells to recognize the corresponding antigen (e.g., KDM1/LSD1 antigen). Typically, an order of tens of kilovolts should be applied to induce the ECL reaction at the nanotube, which is not easily achievable. However, thanks to the confined voltage drop in the micropores of the gel, a low electric field of 1000 V/cm was observed to electrochemically oxidize L012 at one end of the nanotube, resulting in ECL emission. The significant decrease in the applied voltage enables the observation of distinct ECL emission at the nucleus inside a single cell, reflecting the distribution and amount of intracellular KDM1/LSD1 antigen [108] (Figure 4a). As this early work progresses, further advancements in intracellular ECL imaging are expected in the near future.

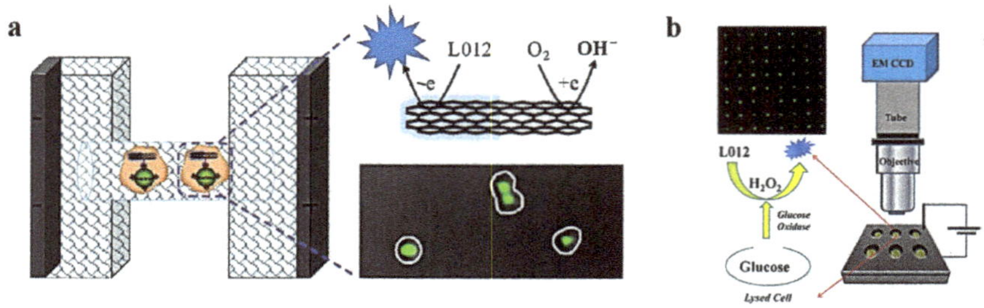

Figure 4. (**a**) Schematic diagram of a BPE-ECL wireless device for the electrochemical visualization of the KDM1/LSD1 antigen in the nucleus [108]. Copyright 2022 CCS Chem. (**b**) Electrochemical luminescence imaging device and detection strategy for rapid analysis of intracellular glucose in single cells [109]. Copyright 2016 Anal Chem.

Besides the acquisition of the spatial distribution of intracellular molecules, high-throughput analysis of these molecules at the single-cell level is also important. Following the previous protocol for the high-throughput ECL analysis of membrane molecules in single cells, the measurement of intracellular molecules only requires the introduction of Triton X-100 to disrupt the cellular membrane inside the microwells. For instance, our group exposed the cells in the microwells to the physiological buffer with glucose oxidase and Triton X-100. Triton X-100 breaks down the cell membrane, releasing glucose, which reacts with glucose oxidase to produce hydrogen peroxide for ECL detection (Figure 4b). By collecting the ECL intensities at the microwells in the image, the concentration of intracellular glucose in single cells was determined, revealing the high heterogeneity of glucose in cells [109]. The same protocol is also applied to measure intracellular cholesterol ester, and the obtained information is crucial for understanding cholesterol trafficking at the single-cell level.

4. Summary and Outlook

Over the past twenty years, single-cell ECL analysis has emerged as a rapidly developing field for determining various molecules at the plasma membrane and inside the cytosol. It has been widely applied in research areas such as cell heterogeneity, disease mechanisms, immunology, and drug response. This technology has brought significant breakthroughs to biomedical research and clinical practice [110–116]. The features of surface confinement, near-zero background, and high sensitivity enable this tool to provide valuable information that is not easily resolved by classic fluorescence microscopy. Despite the tremendous achievement in this field, single-cell ECL measurement and imaging still face some challenges that need to be addressed before their full application in biological and clinic studies. First, the spatial resolution in the image needs to be further improved. Super-resolution fluorescence microscopy has overcome optical diffraction and can provide a spatial resolution down to 20 nm. However, due to the poor luminescence efficiency from the ECL tag, it is difficult to acquire sufficient protons from one ECL molecule. Therefore, the current spatial resolution of a single-cell ECL image is still at the submicron level. The incorporation of super-resolution radial fluctuation (SRRF) algorithms has pushed the resolution down to 100 nm but is still limited in characterizing the electrocatalytic activity at single particles [107,108]. Further development of this super-resolution ECL microscopy is the most urgent direction in single-cell imaging. Second, the temporal resolution of ECL imaging should be improved. As a low-light-emission process, the exposure time for capturing one ECL image of a single cell requires at least a few hundred milliseconds, which may not satisfy the requirements for fast tracking of bioprocesses. Some early work has been conducted to increase the ECL intensity and shorten the exposure time through the usage of new conductive materials [117–121]. More principles and strategies should be established to achieve high-temporal-resolution single-cell ECL imaging. Moreover, obtaining new

ECL reagents with high ECL efficiency, creating a more biocompatible environment to maximize cellular activity, and enabling multivariate assays on more biomolecules within a single cell are also important directions for the development of single-cell ECL imaging.

In summary, single-cell ECL analysis has become a research hotspot in the field of chemistry and bioelectrochemistry and has attracted wide attention from biomedical researchers. Compared to fluorescence methods, ECL single-cell analysis is still in its early stages and faces numerous challenges that need to be addressed. With the continuous application and development of new reagents, mechanisms, equipment, and research strategies, this field should be grown into a broad research prospect [122–124]. We believe that further innovations in ECL measurements will help enable spatiotemporal resolution and high-throughput analysis of single cells, allowing us to gain a deeper understanding of cell behavior and function.

Funding: This research was funded by the National Natural Science Foundation of China, grant number 22274078 and 21974071.

Institutional Review Board Statement: Not applicable.

Informed Consent Statement: Not applicable.

Data Availability Statement: The data are available upon request to the corresponding author.

Conflicts of Interest: The authors declare no conflict of interest.

References

1. Miao, W. Electrogenerated chemiluminescence and its biorelated applications. *Chem. Rev.* **2008**, *108*, 2506–2553. [CrossRef] [PubMed]
2. Hu, L.; Xu, G. Applications and trends in electrochemiluminescence. *Chem. Soc. Rev.* **2010**, *39*, 3275–3304. [CrossRef] [PubMed]
3. Abdussalam, A.; Xu, G. Recent advances in electrochemiluminescence luminophores. *Anal. Bioanal. Chem.* **2022**, *414*, 131–146. [CrossRef] [PubMed]
4. Zanut, A.; Fiorani, A.; Canola, S.; Saito, T.; Ziebart, N.; Rapino, S.; Rebeccani, S.; Barbon, A.; Irie, T.; Josel, H.P.; et al. Insights into the mechanism of coreactant electrochemiluminescence facilitating enhanced bioanalytical performance. *Nat. Commun.* **2020**, *11*, 2668. [CrossRef]
5. Dufford, R.T.; Nightingale, D.; Gaddum, L.W. Luminescence of gridnard compounds in electric and magnetic fields, and related electrical phenomena. *J. Am. Chem. Soc.* **1927**, *49*, 1858–1864. [CrossRef]
6. Harvey, N. Luminescence during Electrolysis. *J. Phys. Chem.* **1929**, *33*, 1456. [CrossRef]
7. Hercules, D.M. Chemiluminescence Resulting from Electrochemically Generated Species. *Science* **1964**, *145*, 808–809. [CrossRef] [PubMed]
8. Santhanam, K.S.V.; Bard, A.J. Chemiluminescence of electrogenerated 9,10-Diphenylanthracene anion radical. *J. Am. Chem. Soc.* **1965**, *87*, 139–140. [CrossRef]
9. Ma, X.; Gao, W.; Du, F.; Yuan, F.; Yu, J.; Guan, Y.; Sojic, N.; Xu, G. Rational Design of Electrochemiluminescent Devices. *Acc. Chem. Res.* **2021**, *54*, 2936–2945. [CrossRef]
10. Zanut, A.; Palomba, F.; Rossi Scota, M.; Rebeccani, S.; Marcaccio, M.; Genovese, D.; Rampazzo, E.; Valenti, G.; Paolucci, F.; Prodi, L. Dye-Doped Silica Nanoparticles for Enhanced ECL-Based Immunoassay Analytical Performance. *Angew. Chem. Int. Ed.* **2020**, *59*, 21858–21863. [CrossRef]
11. Liu, X.; Zhao, S.; Tan, L.; Tan, Y.; Wang, Y.; Ye, Z.; Hou, C.; Xu, Y.; Liu, S.; Wang, G. Frontier and hot topics in electrochemiluminescence sensing technology based on CiteSpace bibliometric analysis. *Biosens. Bioelectron.* **2022**, *201*, 113932. [CrossRef] [PubMed]
12. Chen, M.M.; Xu, C.H.; Zhao, W.; Chen, H.Y.; Xu, J.J. Single Cell Imaging of Electrochemiluminescence-Driven Photodynamic Therapy. *Angew. Chem. Int. Ed.* **2022**, *61*, e202117401. [CrossRef] [PubMed]
13. Yoo, S.M.; Jeon, Y.M.; Heo, S.Y. Electrochemiluminescence Systems for the Detection of Biomarkers: Strategical and Technological Advances. *Biosensors* **2022**, *12*, 738. [CrossRef] [PubMed]
14. Gou, X.; Xing, Z.; Ma, C.; Zhu, J.J. A Close Look at Mechanism, Application, and Opportunities of Electrochemiluminescence Microscopy. *Chem. Biomed. Imaging* **2023**, *1*, 414–433. [CrossRef]
15. Guo, M.; Du, D.; Wang, J.; Ma, Y.; Yang, D.; Haghighatbin, M.A.; Shu, J.; Nie, W.; Zhang, R.; Bian, Z.; et al. Three-Biomarker Joint Strategy for Early and Accurate Diagnosis of Acute Myocardial Infarction via a Multiplex Electrochemiluminescence Immunoarray Coupled with Robust Machine Learning. *Chem. Biomed. Imaging* **2023**, *1*, 179–185. [CrossRef]
16. Trad, F.B.; Delacotte, J.; Guille-Collignon, M.; Lemaître, F.; Arbault, S.; Sojic, N.; Burlina, F.; Labbé, E.; Burie, O. Electrochemiluminescence Imaging of Liposome Permeabilization by an Antimicrobial Peptide: Melittin. *Chem. Biomed. Imaging* **2023**, *1*, 58–65. [CrossRef]

17. Zhu, W.; Dong, J.; Ruan, G.; Zhou, Y.; Feng, J. Quantitative Single-Molecule Electrochemiluminescence Bioassay. *Angew. Chem. Int. Ed.* **2023**, *62*, e202214419. [CrossRef]
18. Hoebe, R.A.; Van Oven, C.H.; Gadella, T.W., Jr.; Dhonukshe, P.B.; Van Noorden, C.J.; Manders, E.M. Controlled light-exposure microscopy reduces photobleaching and phototoxicity in fluorescence live-cell imaging. *Nat. Biotechnol.* **2007**, *25*, 249–253. [CrossRef]
19. Liu, Z.; Qi, W.; Xu, G. Recent advances in electrochemiluminescence. *Chem. Soc. Rev.* **2015**, *44*, 3117–3142. [CrossRef]
20. Forster, R.J.; Bertoncello, P.; Keyes, T.E. Electrogenerated chemiluminescence. *Annu. Rev. Anal. Chem.* **2009**, *2*, 359–385. [CrossRef]
21. Richter, M.M. Electrochemiluminescence (ECL). *Chem. Rev.* **2004**, *104*, 3003–3036. [CrossRef]
22. Mostafa, I.M.; Abdussalam, A.; Zholudov, Y.T.; Snizhko, D.V.; Zhang, W.; Hosseini, M.; Guan, Y.; Xu, G. Recent Applications and Future Perspectives of Chemiluminescent and Bioluminescent Imaging Technologies. *Chem. Biomed. Imaging* **2023**, *1*, 297–314. [CrossRef]
23. Wang, S.; Ren, W.X.; Hou, J.T.; Won, M.; An, J.; Chen, X.; Shu, J.; Kim, J.S. Fluorescence imaging of pathophysiological microenvironments. *Chem. Soc. Rev.* **2021**, *50*, 8887–8902. [CrossRef] [PubMed]
24. Elowitz, M.B.; Levine, A.J.; Siggia, E.D.; Swain, P.S. Stochastic gene expression in a single cell. *Science* **2002**, *297*, 1183–1186. [CrossRef] [PubMed]
25. Keller, L.; Pantel, K. Unravelling tumour heterogeneity by single-cell profiling of circulating tumour cells. *Nat. Rev. Cancer* **2019**, *19*, 553–567. [CrossRef] [PubMed]
26. Pollen, A.A.; Nowakowski, T.J.; Shuga, J.; Wang, X.; Leyrat, A.A.; Lui, J.H.; Li, N.; Szpankowski, L.; Fowler, B.; Chen, P.; et al. Low-coverage single-cell mRNA sequencing reveals cellular heterogeneity and activated signaling pathways in developing cerebral cortex. *Nat. Biotechnol.* **2014**, *32*, 1053–1058. [CrossRef] [PubMed]
27. Chen, H.H.; Chen, T. Probing Oxidant Effects on Superoxide Dismutase 1 Oligomeric States in Live Cells Using Single-Molecule Fluorescence Anisotropy. *Chem. Biomed. Imaging* **2023**, *1*, 49–57. [CrossRef] [PubMed]
28. Stuart, T.; Satija, R. Integrative single-cell analysis. *Nat. Rev. Genet.* **2019**, *20*, 257–272. [CrossRef] [PubMed]
29. Brandman, O.; Ferrell, J.E., Jr.; Li, R.; Meyer, T. Interlinked fast and slow positive feedback loops drive reliable cell decisions. *Science* **2005**, *310*, 496–498. [CrossRef]
30. Piwecka, M.; Rajewsky, N.; Rybak-Wolf, A. Single-cell and spatial transcriptomics: Deciphering brain complexity in health and disease. *Nat. Rev. Neurol.* **2023**, *19*, 346–362. [CrossRef]
31. Labib, M.; Kelley, S.O. Single-cell analysis targeting the proteome. *Nat. Rev. Chem.* **2020**, *4*, 143–158. [CrossRef] [PubMed]
32. Yu, W.; Rush, C.; Tingey, M.; Junod, S.; Yang, W. Application of Super-resolution SPEED Microscopy in the Study of Cellular Dynamics. *Chem. Biomed. Imaging* **2023**, *1*, 356–371. [CrossRef] [PubMed]
33. Ren, X.; Zhang, L.; Zhang, Y.; Li, Z.; Siemers, N.; Zhang, Z. Insights Gained from Single-Cell Analysis of Immune Cells in the Tumor Microenvironment. *Annu. Rev. Immunol.* **2021**, *39*, 583–609. [CrossRef] [PubMed]
34. Knezevic, S.; Bouffier, L.; Liu, B.; Jiang, D.; Sojic, N. Electrochemiluminescence microscopy: From single objects to living cells. *Curr. Opin. Electrochem.* **2022**, *35*, 101096. [CrossRef]
35. Rebeccani, S.; Zanut, A.; Santo, C.I.; Valenti, G.; Paolucci, F. A Guide Inside Electrochemiluminescent Microscopy Mechanisms for Analytical Performance Improvement. *Anal. Chem.* **2022**, *94*, 336–348. [CrossRef] [PubMed]
36. He, Q.-N.; Ma, Z.-Y.; Yang, Y.-X.; Xu, C.-H.; Zhao, W. Recent Advances in Electrochemiluminescence-Based Single-Cell Analysis. *Chemosensors* **2023**, *11*, 281. [CrossRef]
37. Zhang, S.; Liu, Y. Electrochemiluminescence Single-cell Analysis on Nanostructured Interface. *Electroanalysis* **2022**, *34*, 937.
38. Ding, H.; Su, B.; Jiang, D. Recent Advances in Single Cell Analysis by Electrochemiluminescence. *ChemistryOpen* **2023**, *12*, e202200113. [CrossRef]
39. Dong, J.; Feng, J. Electrochemiluminescence from Single Molecule to Imaging. *Anal. Chem.* **2023**, *95*, 374–387. [CrossRef]
40. Yang, Q.; Huang, X.; Gao, B.; Gao, L.; Yu, F.; Wang, F. Advances in electrochemiluminescence for single-cell analysis. *Analyst* **2022**, *148*, 9–25. [CrossRef]
41. Meng, C.; Knežević, S.; Du, F.X.; Guan, Y.R.; Kanoufi, F.; Sojic, N.; Xu, G.B. Recent advances in electrochemiluminescence imaging analysis. *eScience* **2022**, *2*, 591–605. [CrossRef]
42. Zhang, Z.; Ma, C.; Xu, Q.; Zhu, J.J. Recent progress in electrochemiluminescence microscopy analysis of single cells. *Analyst* **2022**, *147*, 2884–2894. [CrossRef] [PubMed]
43. Ma, G.; Zhou, J.; Tian, C.; Jiang, D.; Fang, D.; Chen, H. Luminol electrochemiluminescence for the analysis of active cholesterol at the plasma membrane in single mammalian cells. *Anal. Chem.* **2013**, *85*, 3912–3917. [CrossRef]
44. Huang, S.; Liu, K.; Jiang, D.; Fang, D. Codetermination of Sphingomyelin and Cholesterol in Cellular Plasma Membrane in Sphingomyelin-Depletion-Induced Cholesterol Efflux. *Anal. Chem.* **2019**, *91*, 1501–1506. [CrossRef] [PubMed]
45. Qiu, Y.; Zhou, B.; Yang, X.; Long, D.; Hao, Y.; Yang, P. Novel Single-Cell Analysis Platform Based on a Solid-State Zinc-Coadsorbed Carbon Quantum Dots Electrochemiluminescence Probe for the Evaluation of CD44 Expression on Breast Cancer Cells. *ACS Appl. Mater. Interfaces* **2017**, *9*, 16848–16856. [CrossRef] [PubMed]
46. Long, D.; Chen, C.; Cui, C.; Yao, Z.; Yang, P. A high precision MUA-spaced single-cell sensor for cellular receptor assay based on bifunctional Au@Cu-PbCQD nanoprobes. *Nanoscale* **2018**, *10*, 18597–18605. [CrossRef]
47. Dong, S.; Gao, X.; Fu, L.; Jia, J.; Zou, G. Low-Triggering-Potential Electrochemiluminescence from Surface-Confined $CuInS_2$@ZnS Nanocrystals and their Biosensing Applications. *Anal. Chem.* **2021**, *93*, 12250–12256. [CrossRef] [PubMed]

48. Song, X.; Zhao, L.; Zhang, N.; Liu, L.; Ren, X.; Ma, H.; Kuang, X.; Li, Y.; Luo, C.; Wei, Q. Ultrasensitive Electrochemiluminescence Biosensor with Silver Nanoclusters as a Novel Signal Probe and α-Fe_2O_3-Pt as an Efficient Co-reaction Accelerator for Procalcitonin Immunoassay. *Anal. Chem.* **2023**, *95*, 1582–1588. [CrossRef]
49. Liu, L.; Yang, A.; Luo, W.; Liu, H.; Liu, X.; Zhao, W. Ultrasensitive detection of cyclin D1 by a self-enhanced ECL immunosensor based on Bi_2S_3 quantum dots. *Analyst* **2021**, *146*, 2057–2064. [CrossRef]
50. Gao, X.; Jiang, G.; Gao, C.; Prudnikau, A.; Hübner, R.; Zhan, J.; Zou, G.; Eychmüller, A.; Cai, B. Interparticle Charge-Transport-Enhanced Electrochemiluminescence of Quantum-Dot Aerogels. *Angew. Chem. Int. Ed.* **2023**, *62*, e202214487. [CrossRef]
51. Yang, Y.T.; Liu, J.L.; Sun, M.F.; Yuan, R.; Chai, Y.Q. Highly Efficient Electrochemiluminescence of MnS:CdS@ZnS Core-Shell Quantum Dots for Ultrasensitive Detection of MicroRNA. *Anal. Chem.* **2022**, *94*, 6874–6881. [CrossRef] [PubMed]
52. Chen, C.; Wang, Y.L.; Lin, X.; Ma, S.H.; Cao, J.T.; Liu, Y.M. Cu-MOFs/GOx Bifunctional Probe-Based Synergistic Signal Amplification Strategy: Toward Highly Sensitive Closed Bipolar Electrochemiluminescence Immunoassay. *ACS Appl. Mater. Interfaces* **2023**, *15*, 22959–22966. [CrossRef] [PubMed]
53. Peng, H.; Huang, Z.; Sheng, Y.; Zhang, X.; Deng, H.; Chen, W.; Liu, J. Pre-oxidation of Gold Nanoclusters Results in a 66% Anodic Electrochemiluminescence Yield and Drives Mechanistic Insights. *Angew. Chem. Int. Ed.* **2019**, *58*, 11691–11694. [CrossRef] [PubMed]
54. Peng, H.; Huang, Z.; Deng, H.; Wu, W.; Huang, K.; Li, Z.; Chen, W.; Liu, J. Dual Enhancement of Gold Nanocluster Electrochemiluminescence: Electrocatalytic Excitation and Aggregation-Induced Emission. *Angew. Chem. Int. Ed.* **2020**, *59*, 9982–9985. [CrossRef] [PubMed]
55. Hesari, M.; Ding, Z. Identifying Highly Photoelectrochemical Active Sites of Two Au_{21} Nanocluster Isomers toward Bright Near-Infrared Electrochemiluminescence. *J. Am. Chem. Soc.* **2021**, *143*, 19474–19485. [CrossRef] [PubMed]
56. Wang, Y.; Ding, J.; Zhou, P.; Liu, J.; Qiao, Z.; Yu, K.; Jiang, J.; Su, B. Electrochemiluminescence Distance and Reactivity of Coreactants Determine the Sensitivity of Bead-Based Immunoassays. *Angew. Chem. Int. Ed.* **2023**, *62*, e202216525. [CrossRef] [PubMed]
57. Ouyang, X.; Wu, Y.; Guo, L.; Li, L.; Zhou, M.; Li, X.; Liu, T.; Ding, Y.; Bu, H.; Xie, G.; et al. Self-assembly Induced Enhanced Electrochemiluminescence of Copper Nanoclusters Using DNA Nanoribbon Templates. *Angew. Chem. Int. Ed.* **2023**, *62*, e202300893. [CrossRef]
58. Wu, Z.; Wen, J.; Qin, Y.; Ling, L.; Jiao, L.; Zhang, R.; Luo, Z.; Xi, M.; Hu, L.; Gu, W.; et al. Dual-Site Activation Coupling with a Schottky Junction Boosts the Electrochemiluminescence of Carbon Nitride. *Angew. Chem. Int. Ed.* **2023**, *62*, e202308257. [CrossRef]
59. Han, D.; Goudeau, B.; Jiang, D.; Fang, D.; Sojic, N. Electrochemiluminescence Microscopy of Cells: Essential Role of Surface Regeneration. *Anal. Chem.* **2021**, *93*, 1652–1657. [CrossRef]
60. Chong, Y.T.; Koh, J.L.; Friesen, H.; Duffy, S.K.; Cox, M.J.; Moses, A.; Moffat, J.; Boone, C.; Andrews, B.J. Yeast Proteome Dynamics from Single Cell Imaging and Automated Analysis. *Cell* **2015**, *161*, 1413–1424. [CrossRef]
61. Valm, A.M.; Cohen, S.; Legant, W.R.; Melunis, J.; Hershberg, U.; Wait, E.; Cohen, A.R.; Davidson, M.W.; Betzig, E.; Lippincott-Schwartz, J. Applying systems-level spectral imaging and analysis to reveal the organelle interactome. *Nature* **2017**, *546*, 162–167. [CrossRef]
62. Royce, C.; Kirk, W.; Scott, D. Characterization of electrode heterogeneity with electrogenerated chemiluminescence. *Anal. Chem.* **1987**, *59*, 670–673.
63. Robert, J.; Richard, L.; Christine, M.; Royce, C. Observation of kinetic heterogeneity on highly ordered pyrolytic graphite using electrogenerated chemiluminescence. *Anal. Chem.* **1989**, *61*, 2763–2766.
64. Shultz, L.L.; Stoyanoff, J.S.; Nieman, T.A. Temporal and Spatial Analysis of Electrogenerated $Ru(bpy)_3^{3+}$ Chemiluminescent Reactions in Flowing Streams. *Anal. Chem.* **1996**, *68*, 349–354. [CrossRef]
65. Zu, Y.; Ding, Z.; Zhou, J.; Lee, Y.; Bard, A.J. Scanning optical microscopy with an electrogenerated chemiluminescent light source at a nanometer tip. *Anal. Chem.* **2001**, *73*, 2153–2156. [CrossRef] [PubMed]
66. Szunerits, S.; Tam, J.M.; Thouin, L.; Amatore, C.; Walt, D.R. Spatially resolved electrochemiluminescence on an array of electrode tips. *Anal. Chem.* **2003**, *75*, 4382–4388. [CrossRef] [PubMed]
67. Amatore, C.; Pebay, C.; Servant, L.; Sojic, N.; Szunerits, S.; Thouin, L. Mapping electrochemiluminescence as generated at double-band microelectrodes by confocal microscopy under steady state. *ChemPhysChem* **2006**, *7*, 1322–1327. [CrossRef] [PubMed]
68. Xu, L.; Li, Y.; Wu, S.; Liu, X.; Su, B. Imaging latent fingerprints by electrochemiluminescence. *Angew. Chem. Int. Ed.* **2012**, *51*, 8068–8072. [CrossRef]
69. Rigolot, V.; Biot, C.; Lion, C. To View Your Biomolecule, Click inside the Cell. *Angew. Chem. Int. Ed.* **2021**, *60*, 23084–23105. [CrossRef]
70. Zhou, J.; Ma, G.; Chen, Y.; Fang, D.; Jiang, D.; Chen, H.Y. Electrochemiluminescence imaging for parallel single-cell analysis of active membrane cholesterol. *Anal. Chem.* **2015**, *87*, 8138–8143. [CrossRef]
71. Ding, H.; Guo, W.; Zhou, P.; Su, B. Nanocage-confined electrochemiluminescence for the detection of dopamine released from living cells. *Chem. Commun.* **2020**, *56*, 8249–8252. [CrossRef] [PubMed]
72. Ding, H.; Guo, W.; Ding, L.; Su, B. Confined Electrochemiluminescence at Microtube Electrode Ensembles for Local Sensing of Single Cells. *Chin. J. Chem.* **2021**, *39*, 2911–2916. [CrossRef]
73. Liu, G.; Ma, C.; Jin, B.K.; Chen, Z.; Zhu, J.J. Direct Electrochemiluminescence Imaging of a Single Cell on a Chitosan Film Modified Electrode. *Anal. Chem.* **2018**, *90*, 4801–4806. [CrossRef] [PubMed]

74. Zhang, J.; Jin, R.; Chen, Y.; Fang, D.; Jiang, D. Enhanced electrochemiluminescence at single lithium iron phosphate nanoparticles for the local sensing of hydrogen peroxide efflux from single living cell under a low voltage. *Sens. Actuators B* **2021**, *329*, 129208. [CrossRef]
75. Cui, C.; Chen, Y.; Jiang, D.; Chen, H.Y.; Zhang, J.; Zhu, J.J. Steady-State Electrochemiluminescence at Single Semiconductive Titanium Dioxide Nanoparticles for Local Sensing of Single Cells. *Anal. Chem.* **2019**, *91*, 1121–1125. [CrossRef] [PubMed]
76. Hiramoto, K.; Ino, K.; Komatsu, K.; Nashimoto, Y.; Shiku, H. Electrochemiluminescence imaging of respiratory activity of cellular spheroids using sequential potential steps. *Biosens. Bioelectron.* **2021**, *181*, 113123. [CrossRef] [PubMed]
77. Xu, J.; Qin, Y.; Xia, J.; Jiang, D.; Chen, H.Y. C_3N_4 Nanosheet Modified Microwell Array with Enhanced Electrochemiluminescence for Total Analysis of Cholesterol at Single Cells. *Anal. Chem.* **2017**, *89*, 2216–2220. [CrossRef] [PubMed]
78. Wang, N.; Ao, H.; Xiao, W.; Chen, W.; Li, G.; Wu, J.; Ju, H. Confined electrochemiluminescence imaging microarray for high-throughput biosensing of single cell-released dopamine. *Biosens. Bioelectron.* **2022**, *201*, 113959. [CrossRef]
79. Poudineh, M.; Sargent, E.H.; Pantel, K.; Kelley, S.O. Profiling circulating tumour cells and other biomarkers of invasive cancers. *Nat. Biomed. Eng.* **2018**, *2*, 72–84. [CrossRef]
80. Swiecicki, J.M.; Santana, J.T.; Imperiali, B. A Strategic Approach for Fluorescence Imaging of Membrane Proteins in a Native-like Environment. *Cell Chem. Biol.* **2020**, *27*, 245–251. [CrossRef]
81. Xu, H.; Cai, M.; Gao, J.; Shi, Y.; Chen, J.; Wu, Q.; Zhang, J.; Jiang, J.; Wang, H. Membrane protein density determining membrane fusion revealed by dynamic fluorescence imaging. *Talanta* **2021**, *226*, 122091. [CrossRef] [PubMed]
82. Zhang, J.; Rakhimbekova, A.; Duan, X.; Yin, Q.; Foss, C.A.; Fan, Y.; Xu, Y.; Li, X.; Cai, X.; Kutil, Z.; et al. A prostate-specific membrane antigen activated molecular rotor for real-time fluorescence imaging. *Nat. Commun.* **2021**, *12*, 5460. [CrossRef] [PubMed]
83. Zhou, X.; Zhu, D.; Liao, Y.; Liu, W.; Liu, H.; Ma, Z.; Xing, D. Synthesis, labeling and bioanalytical applications of a tris(2,2′-bipyridyl)ruthenium(II)-based electrochemiluminescence probe. *Nat. Protoc.* **2014**, *9*, 1146–1159. [CrossRef] [PubMed]
84. Miao, W.; Choi, J.P.; Bard, A.J. Electrogenerated chemiluminescence 69: The tris(2,2′-bipyridine)ruthenium(II), ($Ru(bpy)_3^{2+}$)/tri-n-propylamine (TPrA) system revisited-a new route involving TPrA*+ cation radicals. *J. Am. Chem. Soc.* **2002**, *124*, 14478–14485. [CrossRef] [PubMed]
85. Sornambigai, M.; Bouffier, L.; Sojic, N.; Kumar, S.S. Tris(2,2′-bipyridyl)ruthenium (II) complex as a universal reagent for the fabrication of heterogeneous electrochemiluminescence platforms and its recent analytical applications. *Anal. Bioanal. Chem.* **2023**, *415*, 5875–5898. [CrossRef] [PubMed]
86. Yuan, Y.; Han, S.; Hu, L.; Parveen, S.; Xu, G. Coreactants of tris(2,2′-bipyridyl)ruthenium(II) Electrogenerated Chemiluminescence. *Electrochim. Acta* **2012**, *82*, 484–492. [CrossRef]
87. Valenti, G.; Scarabino, S.; Goudeau, B.; Lesch, A.; Jović, M.; Villani, E.; Sentic, M.; Rapino, S.; Arbault, S.; Paolucci, F.; et al. Single Cell Electrochemiluminescence Imaging: From the Proof-of-Concept to Disposable Device-Based Analysis. *J. Am. Chem. Soc.* **2017**, *139*, 16830–16837. [CrossRef] [PubMed]
88. Voci, S.; Goudeau, B.; Valenti, G.; Lesch, A.; Jović, M.; Rapino, S.; Paolucci, F.; Arbault, S.; Sojic, N. Surface-Confined Electrochemiluminescence Microscopy of Cell Membranes. *J. Am. Chem. Soc.* **2018**, *140*, 14753–14760. [CrossRef]
89. Liu, Y.; Zhang, H.; Li, B.; Liu, J.; Jiang, D.; Liu, B.; Sojic, N. Single Biomolecule Imaging by Electrochemiluminescence. *J. Am. Chem. Soc.* **2021**, *143*, 17910–17914. [CrossRef]
90. Lu, Y.; Huang, X.; Wang, S.; Li, B.; Liu, B. Nanoconfinement-Enhanced Electrochemiluminescence for in Situ Imaging of Single Biomolecules. *ACS Nano* **2023**, *17*, 3809–3817. [CrossRef]
91. Li, B.; Huang, X.; Lu, Y.; Fan, Z.; Li, B.; Jiang, D.; Sojic, N. High Electrochemiluminescence from $Ru(bpy)_3^{2+}$ Embedded Metal-Organic Frameworks to Visualize Single Molecule Movement at the Cellular Membrane. *Adv. Sci.* **2022**, *9*, 2204715. [CrossRef] [PubMed]
92. Cao, J.T.; Wang, Y.L.; Zhang, J.J.; Dong, Y.X.; Liu, F.R.; Ren, S.W.; Liu, Y.M. Immuno-Electrochemiluminescent Imaging of a Single Cell Based on Functional Nanoprobes of Heterogeneous $Ru(bpy)_3^{2+}$@SiO_2/Au Nanoparticles. *Anal. Chem.* **2018**, *90*, 10334–10339. [CrossRef] [PubMed]
93. Chen, Y.; Gou, X.; Ma, C.; Jiang, D.; Zhu, J.J. A Synergistic Coreactant for Single-Cell Electrochemiluminescence Imaging: Guanine-Rich ssDNA-Loaded High-Index Faceted Gold Nanoflowers. *Anal. Chem.* **2021**, *93*, 7682–7689. [CrossRef] [PubMed]
94. Wang, N.; Gao, H.; Li, Y.; Li, G.; Chen, W.; Jin, Z.; Lei, J.; Wei, Q.; Ju, H. Dual Intramolecular Electron Transfer for In Situ Coreactant-Embedded Electrochemiluminescence Microimaging of Membrane Protein. *Angew. Chem. Int. Ed.* **2021**, *60*, 197–201. [CrossRef] [PubMed]
95. Zhang, J.; Jin, R.; Jiang, D.; Chen, H.Y. Electrochemiluminescence-Based Capacitance Microscopy for Label-Free Imaging of Antigens on the Cellular Plasma Membrane. *J. Am. Chem. Soc.* **2019**, *141*, 10294–10299. [CrossRef] [PubMed]
96. Ma, Y.; Colin, C.; Descamps, J.; Arbault, S.; Sojic, N. Shadow Electrochemiluminescence Microscopy of Single Mitochondria. *Angew. Chem. Int. Ed.* **2021**, *60*, 18742–18749. [CrossRef] [PubMed]
97. Knežević, S.; Kerr, E.; Goudeau, B.; Valenti, G.; Paolucci, F.; Francis, P.S.; Kanoufi, F.; Sojic, N. Bimodal Electrochemiluminescence Microscopy of Single Cells. *Anal. Chem.* **2023**, *95*, 7372–7378. [CrossRef]
98. Ding, H.; Guo, W.; Su, B. Imaging Cell-Matrix Adhesions and Collective Migration of Living Cells by Electrochemiluminescence Microscopy. *Angew. Chem. Int. Ed.* **2020**, *59*, 449–456. [CrossRef]

99. Ding, H.; Zhou, P.; Fu, W.; Ding, L.; Guo, W.; Su, B. Spatially Selective Imaging of Cell-Matrix and Cell-Cell Junctions by Electrochemiluminescence. *Angew. Chem. Int. Ed.* **2021**, *60*, 11769–11773. [CrossRef]
100. Qin, X.; Jin, H.J.; Li, X.; Li, J.; Pan, J.B.; Wang, K.; Liu, S.; Xu, J.J.; Xia, X.H. Label-Free Electrochemiluminescence Imaging of Single-Cell Adhesions by Using Bipolar Nanoelectrode Array. *Chemistry* **2022**, *28*, e202103964. [CrossRef]
101. Liu, G.; Jin, B.K.; Ma, C.; Chen, Z.; Zhu, J.J. Potential-Resolved Electrochemiluminescence Nanoprobes for Visual Apoptosis Evaluation at Single-Cell Level. *Anal. Chem.* **2019**, *91*, 6363–6370. [CrossRef] [PubMed]
102. Ma, C.; Wang, M.X.; Wei, H.F.; Wu, S.; Zhang, J.R.; Zhu, J.J.; Chen, Z. Catalytic route electrochemiluminescence microscopy of cell membranes with nitrogen-doped carbon dots as nano-coreactants. *Chem. Commun.* **2021**, *57*, 2168–2171. [CrossRef] [PubMed]
103. Han, D.; Goudeau, B.; Manojlovic, D.; Jiang, D.; Fang, D.; Sojic, N. Electrochemiluminescence Loss in Photobleaching. *Angew. Chem. Int. Ed.* **2021**, *60*, 7686–7690. [CrossRef] [PubMed]
104. He, R.; Tang, H.; Jiang, D.; Chen, H.Y. Electrochemical Visualization of Intracellular Hydrogen Peroxide at Single Cells. *Anal. Chem.* **2016**, *88*, 2006–2009. [CrossRef] [PubMed]
105. Wang, Y.; Jin, R.; Sojic, N.; Jiang, D.; Chen, H.Y. Intracellular Wireless Analysis of Single Cells by Bipolar Electrochemiluminescence Confined in a Nanopipette. *Angew. Chem. Int. Ed.* **2020**, *59*, 10416–10420. [CrossRef]
106. Zhang, H.; Gao, W.; Liu, Y.; Sun, Y.; Jiang, Y.; Zhang, S. Electrochemiluminescence-Microscopy for microRNA Imaging in Single Cancer Cell Combined with Chemotherapy-Photothermal Therapy. *Anal. Chem.* **2019**, *91*, 12581–12586. [CrossRef] [PubMed]
107. Ma, C.; Wu, S.; Zhou, Y.; Wei, H.F.; Zhang, J.; Chen, Z.; Zhu, J.J.; Lin, Y.; Zhu, W. Bio-Coreactant-Enhanced Electrochemiluminescence Microscopy of Intracellular Structure and Transport. *Angew. Chem. Int. Ed.* **2021**, *60*, 4907–4914. [CrossRef] [PubMed]
108. Wang, Y.; Jiang, D.; Chen, H.Y. Wireless Electrochemical Visualization of Intracellular Antigens in Single Cells. *CCS Chem.* **2022**, *4*, 2221–2227. [CrossRef]
109. Xu, J.; Huang, P.; Qin, Y.; Jiang, D.; Chen, H.Y. Analysis of Intracellular Glucose at Single Cells Using Electrochemiluminescence Imaging. *Anal. Chem.* **2016**, *88*, 4609–4612. [CrossRef]
110. Li, L.; Chen, Y.; Zhu, J.J. Recent Advances in Electrochemiluminescence Analysis. *Anal. Chem.* **2017**, *89*, 358–371. [CrossRef]
111. Bard, A.J. A life in electrochemistry. *Annu. Rev. Anal. Chem.* **2014**, *7*, 1–21. [CrossRef] [PubMed]
112. Zhang, H.; Liu, Y.; Yao, M.; Han, W.; Zhang, S. Cathodic Electrochemiluminesence Microscopy for Imaging of Single Carbon Nanotube and Nucleolin at Single Tumor Cell. *Anal. Chem.* **2023**, *95*, 570–574. [CrossRef] [PubMed]
113. Valenti, G.; Fiorani, A.; Li, H.; Sojic, N.; Paolucci, F. Essential Role of Electrode Materials in Electrochemiluminescence Applications. *ChemElectroChem* **2016**, *3*, 1990–1997. [CrossRef]
114. Gao, W.; Liu, Y.; Zhang, H.; Wang, Z. Electrochemiluminescence Biosensor for Nucleolin Imaging in a Single Tumor Cell Combined with Synergetic Therapy of Tumor. *ACS Sens.* **2020**, *5*, 1216–1222. [CrossRef] [PubMed]
115. Fu, Y.; Ma, Q. Recent developments in electrochemiluminescence nanosensors for cancer diagnosis applications. *Nanoscale* **2020**, *12*, 13879–13898. [CrossRef] [PubMed]
116. Wu, K.; Zheng, Y.; Chen, R.; Zhou, Z.; Liu, S.; Shen, Y.; Zhang, Y. Advances in electrochemiluminescence luminophores based on small organic molecules for biosensing. *Biosens. Bioelectron.* **2023**, *223*, 115031. [CrossRef] [PubMed]
117. Chen, M.M.; Xu, C.H.; Zhao, W.; Chen, H.Y.; Xu, J.J. Super-Resolution Electrogenerated Chemiluminescence Microscopy for Single-Nanocatalyst Imaging. *J. Am. Chem. Soc.* **2021**, *143*, 18511–18518. [CrossRef] [PubMed]
118. Dong, J.; Lu, Y.; Xu, Y.; Chen, F.; Yang, J.; Chen, Y.; Feng, J. Direct imaging of single-molecule electrochemical reactions in solution. *Nature* **2021**, *596*, 244–249. [CrossRef]
119. Zhao, T.; Zhou, Q.; Lv, Y.; Han, D.; Wu, K.; Zhao, L.; Shen, Y.; Liu, S.; Zhang, Y.J. Ultrafast Condensation of Carbon Nitride on Electrodes with Exceptional Boosted Photocurrent and Electrochemiluminescence. *Angew. Chem. Int. Ed.* **2020**, *59*, 1139–1143. [CrossRef]
120. Gu, W.; Wang, H.; Jiao, L.; Wu, Y.; Chen, Y.; Hu, L.; Gong, J.; Du, D.; Zhu, C. Single-Atom Iron Boosts Electrochemiluminescence. *Angew. Chem. Int. Ed.* **2020**, *59*, 3534–3538. [CrossRef]
121. Chen, M.M.; Zhao, W.; Zhu, M.J.; Li, X.L.; Xu, C.H.; Chen, H.Y.; Xu, J.J. Spatiotemporal imaging of electrocatalytic activity on single 2D gold nanoplates *via* electrogenerated chemiluminescence microscopy. *Chem. Sci.* **2019**, *10*, 4141–4147. [CrossRef]
122. Guo, W.; Ding, H.; Zhou, P.; Wang, Y.; Su, B. Electrochemiluminescence Waveguide in Single Crystalline Molecular Wires. *Angew. Chem. Int. Ed.* **2020**, *59*, 6745–6749. [CrossRef]
123. Gao, N.; Zeng, H.; Wang, X.; Zhang, Y.; Zhang, S.; Cui, R.; Zhang, M.; Mao, L. Graphdiyne: A New Carbon Allotrope for Electrochemiluminescence. *Angew. Chem. Int. Ed.* **2022**, *61*, e202204485. [CrossRef]
124. Chinnadayyala, S.R.; Park, J.; Le, H.T.N.; Santhosh, M.; Kadam, A.N.; Cho, S. Recent advances in microfluidic paper-based electrochemiluminescence analytical devices for point-of-care testing applications. *Biosens. Bioelectron.* **2019**, *126*, 68–81. [CrossRef] [PubMed]

MDPI AG
Grosspeteranlage 5
4052 Basel
Switzerland
Tel.: +41 61 683 77 34

Chemosensors Editorial Office
E-mail: chemosensors@mdpi.com
www.mdpi.com/journal/chemosensors

www.ingramcontent.com/pod-product-compliance
Lightning Source LLC
Chambersburg PA
CBHW041014270926
51850CB00024B/31
9783725814732